SYSTEM IDENTIFICATION
(SYSID'03)

A Proceedings volume from the 13th IFAC Symposium on System Identification,
Rotterdam, The Netherlands, 27 – 29 August 2003

Edited by

P.M.J. Van den HOF
Delft Center for Systems and Control,
Delft University of Technology,
Delft, The Netherlands

B. WAHLBERG
Royal Institute of Technology,
Stockholm, Sweden

S. WEILAND
Department of Electrical Engineering
Eindhoven University of Technology,
Eindhoven, The Netherlands

(In four volumes)

Volume 2

Published for the

INTERNATIONAL FEDERATION OF AUTOMATIC CONTROL

by

ELSEVIER LTD

ELSEVIER Ltd
The Boulevard, Langford Lane
Kidlington, Oxford OX5 1GB, UK

Elsevier Internet Homepage
http://www.elsevier.com

Consult the Elsevier Homepage for full catalogue information on all books, journals and electronic products and services.

IFAC Publications Internet Homepage
http://www.elsevier.com/locate/ifac

Consult the IFAC Publications Homepage for full details on the preparation of IFAC meeting papers, published/forthcoming IFAC books, and information about the IFAC Journals and affiliated journals.

First edition 2004

Library of Congress Cataloging in Publication Data

A catalogue record for this book is available from the Library of Congress

British Library Cataloguing in Publication Data

A catalogue record for this book is available from the British Library

ISBN 0-08-043709 5
ISSN 1474-6670

Printed and bound in the United Kingdom
Transferred to Digital Print 2010

To Contact the Publisher

Elsevier welcomes enquiries concerning publishing proposals: books, journal special issues, conference proceedings, etc. All formats and media can be considered. Should you have a publishing proposal you wish to discuss, please contact, without obligation, the publisher responsible for Elsevier's industrial and control engineering publishing programme:

Christopher Greenwell
Publishing Editor
Elsevier Ltd
The Boulevard, Langford Lane
Kidlington, Oxford
OX5 1GB, UK

Phone: +44 1865 843230
Fax: +44 1865 843920
E.mail: c.greenwell@elsevier.com

General enquiries, including placing orders, should be directed to Elsevier's Regional Sales Offices – please access the Elsevier homepage for full contact details (homepage details at the top of this page).

13th IFAC SYMPOSIUM ON SYSTEM IDENTIFICATION (SYSID 2003)

Sponsored by
International Federation of Automatic Control (IFAC)
IFAC Technical Committees on:
- Modeling, Identification and Signal Processing (MISP)
- Adaptive Control and Tuning (ACT)

Co-sponsored by
IEEE Control Systems Society
Division of Automatic Control (MRBT) of the Royal Institution of Engineers in The Netherlands (KIVI)
The Netherlands Organisation of Scientific Research (NWO)
Royal Netherlands Academy of Arts and Sciences (KNAW)
Dutch Institute of Systems and Control (DISC)
Faculty of Applied Sciences, Delft University of Technology (TUD)
Delft Center for Systems and Control (TUD)
Department of Electrical Engineering, Eindhoven University of Technology, The Netherlands (TU/e)
Stichting Meten en Regelen ER-THE, The Netherlands

Organizing Committee
P.M.J. Van den Hof – Delft University of Technology, Delft, The Netherlands
B. Wahlberg – Royal Institute of Technology, Stockholm, Sweden
S. Weiland – Eindhoven University of Technology, Eindhoven, The Netherlands

IPC Task Force
M. Deistler
M. Gevers
L. Ljung
M. Morari
J. Schoukens
P.M.J. Van den Hof
M. Viberg
B. Wahlberg

International Programme Committee (IPC)
P.M.J. Van den Hof; The Netherlands (Co-Chair)
B. Wahlberg, Sweden (Co-Chair)

P. Albertos; Spain
B. Anderson; Australia
E. Bai; USA
M. Basseville; France
R. Bitmead; USA
S. Bittanti* **; Italy
M. Blanke; Denmark
J. Bokor; Hungary
M. Campi; Italy
H.F. Chen; P.R. China
J. Chen; USA
R. de Callafon; USA
M. Deistler; Austria
B. de Moor; Belgium
J.J. Fuchs; France
K. Godfrey; UK

G. Goodwin; Australia
M. Gevers; Belgium
P. Guillaume; Belgium
L. Guo; P.R. China
H. Hjalmarsson; Sweden
H. Kimura; Japan
R. Kosut; USA
V. Krishnamurty; Australia
K. Kumamaru; Japan
I. Landau; France
J.H. Lee; USA
L. Ljung; Sweden
P. Mäkilä; Finland
T. McKelvey; Sweden
M. Milanese; Italy
M. Morari; Switzerland
B. Ninness; Australia
R. Ortega**; France
G. Picci; Italy
R. Pintelon; Belgium
B. Polyak; Russia
P. Regalia; France
D. Rivera; USA
W. Scherrer; Austria
J. Schoukens; Belgium
R. Schumann; Germany
R. Smith*; USA
T. Söderström*; Sweden
T. Sugie; Japan
R. Tempo; Italy
J. van Schuppen; The Netherlands
M. Verhaegen; The Netherlands
S. Veres; UK
M. Viberg; Sweden
A. Vicino; Italy
E. Walter; France

** Appointed by IFAC Technical Committee MISP*
*** Appointed by IFAC Technical Committee ACT*

National Organizing Committee (NOC)

P.M.J. Van den Hof (Finances, contacts NMO, Public Relations)
S. Weiland (PC Secretariat, Paper handling, Website)
A.C.P.M. Backx (Industrial participation, Sponsors)
M.H.G. Verhaegen (Publications)
Y. Zhu (Exhibitions)
T. Van der Weiden (Local arrangements, Technical and Social Events)

PREFACE

These Proceedings contain all the technical material presented at the 13th IFAC Symposium on System Identification (SYSID 2003), held in the Conference Center "De Doelen", Rotterdam, The Netherlands from 27 – 29 August 2003.

The SYSID symposium is organized every three years and is among the most successful symposia organized by IFAC. This has been the first SYSID symposium in the 3rd millennium and the second SYSID symposium to take place in The Netherlands, following The Hague symposium in 1973.

Being the only worldwide symposium that is fully directed towards system identification, it is the ideal opportunity for researchers and industrial engineers from very many disciplines to present and discuss the developments, the results and the future challenges in all aspects of modelling dynamical systems on the basis of experimental data.

The symposium covered all major aspects of system identification, experimental modelling, signal processing and adaptive control from theoretical and methodological developments to practical applications in a wide range of application areas. For the 13th edition of this symposium, the International Program Committee has taken steps to position SYSID 2003 as a meeting place where scientists and engineers from several research communities can meet to discuss issues related to these areas.

A total of 350 delegates from 40 different countries attended the conference. 100 of the participants were PhD students, showing that system identification is a very vital field of research. Out of a total of 422 papers that were submitted to SYSID 2003, the IPC selected 333 papers and these were incorporated in the final program. The selection was based on two referee reports per paper. The final program of the symposium was composed of 3 plenary papers, 6 semi-plenary papers, 232 papers in oral sessions, 82 posters and 10 software demonstrations. The Preprints of this Symposium appeared on CD-ROM and were distributed among the participants of the symposium. The Proceedings of SYSID-2003 contain 321 papers.

We hope that you, as reader or as researcher in the area of System Identification, will find the contents of these Proceedings useful and informative for your professional work.

We would like to thank all members of the International Program Committee (IPC), members of the IPC Taskforce and members of the National Organizing Committee for their work in the organization of this symposium and in the preparation of these Proceedings. We would also like to thank many friends and colleagues for their help and support in many practical matters related to SYSID 2003.

The editors,

Paul Van den Hof
Bo Wahlberg
Siep Weiland.

CONTENTS

VOLUME 1

PLENARY PAPER
FROM EXPERIMENTS TO CLOSED-LOOP CONTROL

IDENTIFICATION FOR CONTROL

NONLINEAR IDENTIFICATION

IDENTIFICATION OF MIMO COMMUNICATION CHANNELS

IDENTIFICATION OF NONLINEAR SYSTEMS I

MECHANICAL AND AEROSPACE APPLICATIONS

CLOSED-LOOP IDENTIFICATION

INDUSTRIAL APPLICATION OF IDENTIFICATION

PROCESS CONTROL SYSTEMS

CLOSED LOOP AND PERFORMANCE ISSUES

REPRODUCING KERNELS I

VOLUME 2

BLIND ESTIMATION AND EQUALIZATION

CONTINUOUS TIME IDENTIFICATION

INPUT DESIGN

IDENTIFICATION FOR FLIGHT TEST EXPLORATION

IDENTIFIABILITY

PLENARY PAPER
SYSTEM IDENTIFICATION FOR STRUCTURAL DYNAMICS AND VIBROACOUSTICS DESIGN ENGINEERING

SELECTED TOPICS IN IDENTIFICATION

REPRODUCING KERNELS II

IDENTIFICATION OF NONLINEAR BLOCK MODELS

NEW RESULTS IN SUBSPACE IDENTIFICATION

IDENTIFICATION FOR PROCESS CONTROL: INPUT DESIGN

IDENTIFICATION OF MECHANICAL SYSTEMS

SOFTWARE SESSION I

SEMI-PLENARY
DATA-BASED METHODS IN PROCESS CONTROL

IDENTIFICATION METHODS

CONTROLLER TUNING AND IDENTIFICATION

APPLICATIONS OF IDENTIFICATION

BIOENGINEERING SYSTEMS

PARTICLE FILTERS

WIENER HAMMERSTEIN MODELS

IDENTIFICATION USING BASIS FUNCTIONS

SUBSPACE IDENTIFICATION AND APPLICATIONS

IDENTIFICATION IN LARGE SCALE SYSTEMS

INDUSTRIAL APPLICATIONS OF IDENTIFICATION

SOFTWARE SESSION II

PLENARY PAPER
PREDICTION ALGORITHMS: COMPLEXITY, CONCENTRATION AND CONVEXITY

IDENTIFICATION AND PHYSICAL MODELING

IDENTIFICATION OF NONLINEAR SYSTEMS

VOLUME 4

EDUCATION AND TRAINING

RECURSIVE AND SUBSPACE IDENTIFICATION

PROCESS CONTROL: THEORY

APPLICATION OF SYSTEM IDENTIFICATION

OPTIMAL FILTERING

SEMI-PLENARY
IDENTIFICATION OF LINEAR SYSTEMS WITH NONLINEAR DISTORTIONS

SEMI-PLENARY
SOME PROBLEMS IN STATISTICAL INFERENCE FOLLOWING MODEL SELECTION

USER CHOICES IN SUBSPACE IDENTIFICATION

IDENTIFICATION OF STATIC AND DYNAMICAL NONLINEAR SYSTEMS

IDENTIFICATION AND MODEL VALIDATION

MODEL APPROXIMATION

PARAMETER ESTIMATION AND CONVERGENCE

IDENTIFICATION OF HYDROLOGIC SYSTEMS

ERRORS IN VARIABLE IDENTIFICATION

IFAC Publications
www.elsevier.com/locate/ifac

BLIND TURBO EQUALIZATION USING THE CONSTANT MODULUS ALGORITHM

Phillip A. Regalia

Dept. Communications, Image and Information Processing
Institut National des Télécommunications/GET
9, rue Charles Fourier
91011 Evry cedex France
Phillip.Regalia@int-evry.fr

Abstract: A turbo equalizer is modified to allow its operation in a blind manner, i.e., without resorting to training sequences or to channel identification steps. It exploits a recent variant of the constant modulus algorithm, in collaboration with differential encoding, for which the decoder is linked in an iterative scheme with a conventional error correction coder. A characterization of stationary points is obtained, and conditions for proximity to a maximum likelihood decoding rule are identified. *Copyright © 2003 IFAC*

1. INTRODUCTION

Iterative techniques in reception have met with intense interest since the advent of the turbo-decoding algorithm for parallel concatenated codes. More recently, turbo equalization, in which the channel and source coder are interpreted as a serial concatenated coder, have successfully integrated linear equalization and maximum likelihood decoding into iterative algorithms which drop intersymbol interference and channel noise below previously attainable levels.

The technique was first proposed in (Douillard *et al.*, 1995), and is directly related to the iterative decoding of serially concatenated codes (Benedetto and Montorsi, 1996). The "inner" decoder was implemented using a soft-output Viterbi algorithm, for which the computational complexity grows exponentially with the channel length, restricting its practical application to short channels. A modification proposed in (Laot *et al.*, 2001) and subsequently refined in (Tüchler *et al.*, 2002), consists in replacing the Viterbi channel decoder with a linear decision feedback equalizer, whose decision feedback path is driven by the outer decoder's output. This drops the complexity to a function which is linear (or sometimes quadratic) in the channel length, while offering bit-error performance close to the original design from (Douillard *et al.*, 1995).

If the channel is properly identified, then the design equations for the decision feedback equalizer which maximize the signal-to-noise ratio subject to perfect interference cancellation may be obtained with explicit formulas (Laot *et al.*, 2001). As expected, performance may degrade significantly when the channel is not properly identified, requiring more complicated configurations using training sequences in combination with equalizer adaptation strategies [e.g., (Tüchler *et al.*, 2002)]. The use of training sequences, of course, consumes available bandwidth, while blind channel identification schemes can often be ill-conditioned (Delmas *et al.*, 2000), or simply inapplicable when sufficient diversity is lacking.

The technique proposed here uses a completely blind solution, employing the finite-interval variant of the constant modulus algorithm developed in (Regalia, 2002). Since the constant modulus criterion is phase blind, its successful application requires differential encoding/decoding for correct symbol recovery. The technique developed here treats the differential encoder as the inner coder of a serial concatenated code, with the outer coder fournished by a conventional trellis code for error correction

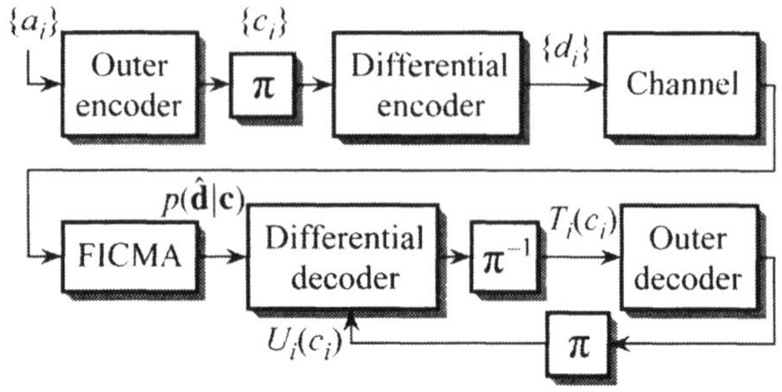

Fig. 1. Using the finite-interval constant modulus algorithm and iterative decoding with the inclusion of the differential coder/decoder pair.

purposes. The decoder stage applied to the output of the constant modulus algorithm implements iterative decoding applied to the serially concatenated pair. Simulations indicate, similar to other iterative decoding techniques, a threshold effect: provided the constant modulus algorithm approximately equalizes the channel, the information exchange between the two decoders iteratively reduces the remaining errors. Although the bit error rate versus signal to noise ratio is perhaps not as impressive as earlier designs exploiting channel knowledge, the proposed scheme works effectively without any channel knowledge.

The next section describes the algorithm, its stationary points, and relations to maximum likelihood decoding.

2. ALGORITHM DESCRIPTION

Figure 1 shows an overview of the proposed scheme, where $\{a_i\}_{i=1}^{K}$ are the information bits, which are fed to an error correction coder (typically a trellis coder) to create the coded bits $\{c_j\}_{j=1}^{N}$ (each being either 0 or 1). We assume that this first coder is systematic, so that $c_j = a_j$ for $j = 1, \ldots, K$. The bits $\{c_j\}$ are coded in turn using a differential encoder, to produce the output bits d_j:

$$d_j = d_{j-1} \oplus c_{\pi(j)}, \quad j = 1, 2, \ldots, N,$$

in which $\{\pi(1), \pi(2), \ldots, \pi(N)\}$ is a permutation of $\{1, 2, \ldots, N\}$, and "$\oplus$" denotes modulo-2 addition; the seed value $d_0 \in \{0, 1\}$ is chosen arbitrarily. The symbols are converted to antipodal signaling $\{-1, +1\}$ and transmitted over a channel with nontrivial delay spread:

$$x_j = (2d_j - 1)$$
$$u_j = \sum_k h_k x_{j-k} + b_j$$

where $\{h_k\}$ denotes the channel impulse response, of unknown length, and b_j is additive noise, assumed white and Gaussian.

The constant modulus algorithm attempts to restore the channel input sequence by adjusting the coefficients of an equalizer; a variant suitable to block processing is given in (Regalia, 2002), and yields an output block $\{\hat{d}_j\}$ which satisfies

$$\hat{d}_j \approx \pm(2d_j - 1)$$

in which the "$\pm$" sign reflects the phase ambiguity which is inherent to the constant modulus algorithm (hence, the need for differential encoding), and the approximation sign reflects how a linear equalizer can at best reduce noise and intersymbol interference, but never eliminate them.

The differential decoder calculates the a posteriori probability ratios

$$\frac{\Pr(c_i = 1|\hat{d}_1, \ldots, \hat{d}_N)}{\Pr(c_i = 0|\hat{d}_1, \ldots, \hat{d}_N)} = \frac{\sum_{\mathbf{c}:c_i=1} \Pr(\mathbf{c}|\hat{\mathbf{d}})}{\sum_{\mathbf{c}:c_i=0} \Pr(\mathbf{c}|\hat{\mathbf{d}})} \qquad i = 1, \ldots, N,$$
$$= \frac{\sum_{\mathbf{c}:c_i=1} p(\hat{\mathbf{d}}|\mathbf{c}) \Pr(\mathbf{c})}{\sum_{\mathbf{c}:c_i=0} p(\hat{\mathbf{d}}|\mathbf{c}) \Pr(\mathbf{c})} \qquad (1)$$

in which the a priori probability function $\Pr(\mathbf{c})$ is assumed to factor into the product of its marginals:

$$\Pr(\mathbf{c}) = \Pr(c_1)\Pr(c_2)\cdots\Pr(c_N).$$

This assumption is, strictly speaking, invalid, since the bits $\{c_j\}$ are interdependent, but the differential decoder cannot exploit these interdependencies without a significant increase in complexity, and so uses the marginals $\Pr(c_i)$ instead.

By virtue of the factorization of $\Pr(\mathbf{c})$, each term of the numerator (resp., denominator) of (1) contains a factor $\Pr(c_i = 1)$ [resp., $\Pr(c_i = 0)$], such that the probability ratio becomes

$$\frac{\Pr(c_i = 1|\hat{\mathbf{d}})}{\Pr(c_i = 0|\hat{\mathbf{d}})} = \underbrace{\frac{\Pr(c_i = 1)}{\Pr(c_i = 0)}}_{\frac{U_i(1)}{U_i(0)}} \underbrace{\frac{\sum_{\mathbf{c}:c_i=1} p(\hat{\mathbf{d}}|\mathbf{c}) \prod_{j\neq i} \Pr(c_j)}{\sum_{\mathbf{c}:c_i=0} p(\hat{\mathbf{d}}|\mathbf{c}) \prod_{j\neq i} \Pr(c_j)}}_{\triangleq \frac{T_i(1)}{T_i(0)}} \qquad (2)$$

The a priori ratio $U_i(1)/U_i(0)$ is initially set to 1, but will be "baised" in subsequent iterations by the extrinsic information to be fed back from the outer decoder below.

The likelihood function $p(\hat{\mathbf{d}}|\mathbf{c})$ is evaluated under the assumption that the constant modulus equalizer has restored the channel input to within additive Gaussian noise (which is, strictly speaking, incorrect here), giving

$$p(\hat{\mathbf{d}}|\mathbf{c}) \sim \exp\left(-\sum_{j=1}^{N} \frac{[\hat{d}_j - (2d_j(\mathbf{c}) - 1)]^2}{2\sigma^2}\right)$$

in which $d_j(\mathbf{c})$ denotes the jth bit of the differential encoder output when the input block is $\mathbf{c}$, and σ^2 is the noise variance. The forward-backward algorithm (Bahl *et al.*, 1974) may be used for the calculations summarized in (2) for $i = 1, 2, \ldots, N$.

The error correction decoder would nominally aim to calculate the a posteriori probability ratios

$$\frac{\Pr(a_i = 1|\hat{c}_1,\ldots,\hat{c}_N)}{\Pr(a_i = 0|\hat{c}_1,\ldots,\hat{c}_N)} = \frac{\sum_{\mathbf{a}:a_i=1} \Pr(\mathbf{a}|\hat{\mathbf{c}})}{\sum_{\mathbf{a}:a_i=0} \Pr(\mathbf{a}|\hat{\mathbf{c}})}$$

$$= \frac{\sum_{\mathbf{a}:a_i=1} p(\hat{\mathbf{c}}|\mathbf{a})\Pr(\mathbf{a})}{\sum_{\mathbf{a}:a_i=0} p(\hat{\mathbf{c}}|\mathbf{a})\Pr(\mathbf{a})} \qquad i = 1,\ldots,K,$$

in which $\Pr(\mathbf{a})$ is taken as a uniform distribution over the 2^K possibilities for $\mathbf{a}$, and so my be omitted. As written, this presupposes a noisy version of the outer encoder's output as $\hat{\mathbf{c}} = (2\mathbf{c} - \mathbf{1}) + \text{noise}$, but this quantity is not immediately available. If it were, then each evaluation of the likelihood function $p(\hat{\mathbf{c}}|\mathbf{a})$ would appear as

$$p(\hat{\mathbf{c}}|\mathbf{a}) \sim \exp\left(-\sum_{j=1}^{N} \frac{[\hat{c}_j - (2c_j(\mathbf{a}) - 1)]^2}{2\sigma^2}\right)$$

in which $c_j(\mathbf{a})$ is either 0 or 1. Therefore, to each hypothetical bit $\hat{c}_j$ we associate two evaluations: $\exp[-(\hat{c}_j \mp 1)/(2\sigma^2)]$, (corresponding to $c_j(\mathbf{a}) = 0$ or 1), which are replaced by the two evaluations of the function T_j calculated in (2) above:

$$\frac{\exp(-(\hat{c}_j - 1)^2/(2\sigma^2))}{\exp(-(\hat{c}_j + 1)^2/(2\sigma^2))} \leftarrow \frac{T_j(1)}{T_j(0)}, \qquad j = 1,\ldots,N.$$

Here "$\leftarrow$" denotes the "usurpation" operator. The forward-backward algorithm may then proceed directly, following this systematic substitution.

To develop an external description of the resulting decoding operation, we note that this substitution is tantamount to replacing the likelihood evaluation by

$$p(\hat{\mathbf{c}}|\mathbf{a}) \leftarrow \prod_{i=1}^{N} T_i(c_i(\mathbf{a})) \qquad (3)$$

in which the right-hand side emphasizes that only those bit combinations $(c_1,\ldots,c_N)$ which lie in the outer coderbook make sense. To this end, let $\phi(\mathbf{c})$ denote the indicator function for the outer code:

$$\phi(\mathbf{c}) = \begin{cases} 1, & \text{if } \mathbf{c} \text{ is an outer codeword;} \\ 0, & \text{if not.} \end{cases}$$

The 2^N configurations of $(c_1, \ldots, c_N)$ generate 2^N evaluations of $\prod_{i=1}^{N} T_i(c_i)$, but only 2^K of these survive in the product $\phi(\mathbf{c})\prod_i T_i(c_i)$. We may then establish a one-to-one correspondence between the 2^K "surviving" evaluations in $\phi(\mathbf{c})\prod_i T_i(c_i)$ and the 2^K "usurped" evaluations of $\mathbf{p}(\hat{\mathbf{c}}|\mathbf{a})$ in which the hypothesis $\mathbf{a}$ varies among 2^K possibilities generated from K bits $(a_1, \ldots, a_K)$, as in (3). The outer decoder then admits an external description as

$$\frac{\sum_{\mathbf{a}:a_i=1} p(\hat{\mathbf{c}}|\mathbf{a})}{\sum_{\mathbf{a}:a_i=0} p(\hat{\mathbf{c}}|\mathbf{a})} \leftarrow \frac{\sum_{\mathbf{c}:c_i=1} \phi(\mathbf{c})\prod_j T_j(c_j)}{\sum_{\mathbf{c}:c_i=0} \phi(\mathbf{c})\prod_j T_j(c_j)}$$

$$= \frac{T_i(1)}{T_i(0)} \underbrace{\frac{\sum_{\mathbf{c}:c_i=1} \phi(\mathbf{c})\prod_{j\neq i} T_j(c_j)}{\sum_{\mathbf{c}:c_i=0} \phi(\mathbf{c})\prod_{j\neq i} T_j(c_j)}}_{\to \frac{U_i(1)}{U_i(0)}} \qquad (4)$$

in which we note that:

- Since the outer coder is systematic, the first k bits $c_1, \ldots, c_K$ coincide with the information bits $a_1, \ldots a_K$. In addition, the formula above may be evaluated as written for the parity-check bits $c_{K+1}, \ldots, c_N$.
- Each term in the numerator (resp., denominator) contains a factor $T_i(1)$ [resp., $T_i(0)$], so that the ratio $T_i(1)/T_i(0)$ naturally factors out. The remaining term (the "extrinsic" information) will usurp the (pseudo-) *a priori* probabilities of the inner decoder for the next iteration:
$$\frac{\Pr(c_i = 1)}{\Pr(c_i = 0)} \leftarrow \frac{U_i(1)}{U_i(0)}$$

If we let a superscript (n) denote an iteration index, then the coupling of the two decoders admits an external description of the form

$$\frac{\sum_{\mathbf{c}:c_i=1} p(\hat{\mathbf{d}}|\mathbf{c})\prod_j U_i^{(n)}(c_i)}{\sum_{\mathbf{c}:c_i=0} p(\hat{\mathbf{d}}|\mathbf{c})\prod_j U_i^{(n)}(c_i)} = \frac{U_i^{(n)}(1)}{U_i^{(n)}(0)} \frac{T_i^{(n)}(1)}{T_i^{(n)}(0)} \qquad (5)$$

$$\frac{\sum_{\mathbf{c}:c_i=1} \phi(\mathbf{c})\prod_j T_i^{(n)}(c_i)}{\sum_{\mathbf{c}:c_i=0} \phi(\mathbf{c})\prod_j T_i^{(n)}(c_i)} = \frac{T_i^{(n)}(1)}{T_i^{(n)}(0)} \frac{U_i^{(n+1)}(1)}{U_i^{(n+1)}(0)} \qquad (6)$$

in which these "pseudo"-posterior probabilities are calculated for $i = 1, \ldots, N$, at each iteration. A stationary point corresponds to $U_i^{(n+1)}(c_i) = U_i^{(n)}(c_i)$ which, by inspection, gives

Property 1. A stationary point occurs if and only if the two decoders reach consensus on the pseudo-posterior probabilities [(5), (6)] for $i = 1, 2, \ldots, N$.

It may not be clear whether the substitution illustrated in (4) above gives an optimal coupling between the two decoders in any sense. To gain greater insight into this question, suppose the likelihood function $p(\hat{\mathbf{d}}|\mathbf{c})$ is scaled so that its outcomes sum to one:

$$\sum_{\mathbf{c}} p(\hat{\mathbf{d}}|\mathbf{c}) = 1.$$

Introduce the N marginals of the likelihood function $p(\hat{\mathbf{d}}|\mathbf{c})$ as

$$p_i(\hat{\mathbf{d}}|c_i = 1) = \sum_{\mathbf{c}:c_i=1} p(\hat{\mathbf{d}}|\mathbf{c}), \quad i = 1, 2, \ldots, N,$$

and similarly for $p_i(\hat{\mathbf{d}}|c_i = 0)$. We may then show:

Property 2. If the likelihood function $p(\hat{\mathbf{d}}|\mathbf{c})$ factors into the product of its marginals, i.e.,

$$p(\hat{\mathbf{d}}|\mathbf{c}) = p_1(\hat{\mathbf{d}}|c_1)p_2(\hat{\mathbf{d}}|c_2)\cdots p_N(\hat{\mathbf{d}}|c_N),$$

then:

(a) The algorithm in (5) and (6) converges in a single iteration;
(b) The resulting pseudo-posteriors coincide with the symbol-by-symbol maximum likelihood estimates for the concatenated code.

To verify, note that if $p(\hat{\mathbf{d}}|\mathbf{c})$ is the product of its marginals, then so is

$$\alpha\, p(\hat{\mathbf{d}}|\mathbf{c}) \prod_{j=1}^{N} U_j^{(n)}(c_j) = \prod_{j=1}^{N} \alpha_j\, p(\hat{\mathbf{d}}|c_j) U_j^{(n)}(c_j)$$

in which the scalars $\{\alpha_j\}$ ensure that the sum over all outcomes equals one. Since the left-hand side of (5) calculates the i-th marginal ratio, we may observe that

$$\frac{\sum_{\mathbf{c}:c_i=1} p(\hat{\mathbf{d}}|\mathbf{c}) \prod_j U_j^{(n)}(c_j)}{\sum_{\mathbf{c}:c_i=0} p(\hat{\mathbf{d}}|\mathbf{c}) \prod_j U_j^{(n)}(c_j)} = \frac{U_i^{(n)}(1)}{U_i^{(n)}(0)} \frac{p(\hat{\mathbf{d}}|c_i = 1)}{p(\hat{\mathbf{d}}|c_i = 0)}$$

whenever $p(\hat{\mathbf{d}}|\mathbf{c})$ factors as the product of its marginals. Upon comparing with (5) we identify

$$\frac{T_i^{(n)}(1)}{T_i^{(n)}(0)} = \frac{p(\hat{\mathbf{d}}|c_i = 1)}{p(\hat{\mathbf{d}}|c_i = 0)}$$

for all iterations n, implying immediately that a stationary point has been attained. We then have

$$\prod_{j=1}^{N} T_j^{(n)}(c_j) = \prod_{j=1}^{N} p(\hat{\mathbf{d}}|c_j) = p(\hat{\mathbf{d}}|\mathbf{c}),$$

and the calculation performed by the outer decoder in (6) becomes, for $i = 1, 2, \ldots, K$,

$$\frac{\sum_{\mathbf{c}:c_i=1} \phi(\mathbf{c}) \prod_j T_j^{(n)}(c_j)}{\sum_{\mathbf{c}:c_i=0} \phi(\mathbf{c}) \prod_j T_j^{(n)}(c_j)} = \frac{\sum_{\mathbf{c}:c_i=1} \phi(\mathbf{c}) p(\hat{\mathbf{d}}|\mathbf{c})}{\sum_{\mathbf{c}:c_i=0} \phi(\mathbf{c}) p(\hat{\mathbf{d}}|\mathbf{c})} = \frac{\sum_{\mathbf{a}:a_i=1} p(\hat{\mathbf{d}}|\mathbf{a})}{\sum_{\mathbf{a}:a_i=0} p(\hat{\mathbf{d}}|\mathbf{a})}$$

in which we note that each nonzero evaluation of $\phi(\mathbf{c})p(\hat{\mathbf{d}}|\mathbf{c})$ may be identified with an evaluation of $p(\hat{\mathbf{d}}|\mathbf{c}(\mathbf{a})) = p(\hat{\mathbf{d}}|\mathbf{a})$, since the indicator function $\phi(\mathbf{c})$ annihilates those evaluations of $p(\hat{\mathbf{d}}|\mathbf{c})$ for which $\mathbf{c}$ is not in the outer codebook. As the outer code is systematic, we have $c_i = a_i$ for $i = 1, \ldots, K$, allowing a direct substitution of the variables of summation. The final ratio which results is recognized as that obtained from a bit-by-bit maximum likelihood metric for the concatenated code. ◇

In practice, the likelihood function $p(\hat{\mathbf{d}}|\mathbf{c})$ need not factor as the product of its marginals, but if it is "close" to such a factorable function, one would expect the algorithm to converge rapidly. This proximity to a factorable likelihood function will in fact hold in extreme conditions:

- *High signal-to noise ratio and good channel diversity.* Let $\mathbf{c}_*$ denote the true input to the differential encoder. If the FICMA algorithm restores a faithful rendition of the encoded sequence $\mathbf{d}(\mathbf{c}_*)$, then $\hat{\mathbf{d}} \approx (2\mathbf{d}(\mathbf{c}_*) - \mathbf{1})$ and

$$\frac{p(\hat{\mathbf{d}}|\mathbf{c} \neq \mathbf{c}_*)}{p(\hat{\mathbf{d}}|\mathbf{c} = \mathbf{c}_*)} = \frac{\exp(-\|\hat{\mathbf{d}} - (2\mathbf{d}(\mathbf{c}) - \mathbf{1})\|^2/2\sigma^2)}{\exp(-\|\hat{\mathbf{d}} - (2\mathbf{d}(\mathbf{c}_*) - \mathbf{1})\|^2/2\sigma^2)} \approx \exp\left(-\frac{m^2}{2\sigma^2}\right)$$

  where $m^2 = \|\hat{\mathbf{d}} - (2\mathbf{d}(\mathbf{c}) - \mathbf{1})\|^2$. As the noise variance σ^2 decreases, this ratio tends to zero, so that

$$p(\hat{\mathbf{d}}|\mathbf{c}) \stackrel{\sigma^2 \to 0}{\longrightarrow} \delta_{\mathbf{c}_*}(\mathbf{c}) = \begin{cases} 1, & \mathbf{c} = \mathbf{c}_*; \\ 0, & \mathbf{c} \neq \mathbf{c}_*. \end{cases}$$

  We note that the Kronecker delta function $\delta_{\mathbf{c}_*}(\mathbf{c})$ can always be written as the product of its marginals (which themselves are Kronecker delta functions of the individual bits). In these favorable conditions, the algorithm converges rapidly, yielding the correct decoding with high probability.
- *Poor signal to noise ratio.* If the noise variance σ^2 is large, the likelihood evaluations $p(\hat{\mathbf{d}}|\mathbf{c})$ become comparable in value, and approach a uniform distribution:

$$p(\hat{\mathbf{d}}|\mathbf{c}) \stackrel{\sigma^2 \to \infty}{\longrightarrow} u(\mathbf{c}) = \frac{1}{2^N} \quad \text{for all } \mathbf{c}$$

  We note that a uniform distribution likewise factors as the product of its marginals (which themselves are uniform distributions). For pessimistic signal-to-noise ratios, then, the algorithm converges rapidly to a solution whose probability of error approaches $\frac{1}{2}$.

For intermediate signal to noise ratios, the convergence properties of iterative decoding are less well understood, owing to the presence of loops in the equivalent belief propagation graph [akin to the situation for parallel concated codes (McEliece *et al.*, 1998) and low-density parity check codes (Fossorier *et al.*, 1999)]. Nonetheless, the existence of stationary points can at least be established, similar to (Richardson, 2000) for parallel concatenated codes.

To this end, we recall that the Brouwer fixed point theorem (Saaty and Bram, 1964) asserts that any continuous map from a closed, bounded and convex set into itself admits a fixed point. Consider the pseudo-prior probabilities $U_i^{(n)}(c_i = 1)$, which lie between zero and one:

$$0 \leq U_i^{(n)}(c_i = 1) \leq 1, \quad i = 1, 2, \ldots, N.$$

This gives a closed, bounded and convex subset of $\mathbb{R}^N$. Since the pseudo-priors $U_i^{(n+1)}(c_i = 1)$ at the next iteration remain within this subset, and since the application which maps $\{U_i^{(n)}(c_i)\}$ to $\{U_i^{(n+1)}(c_i)\}$ is continuous, the Brouwer theorem is satisfied, to show that fixed points of the iteration always exist.

Note that the external description in terms of pseudo-posteriors in (5) and (6) above applies generically to iterative decoding of serially concatenated codes, so that Properties 1 and 2 apply to more general schemes described in (Benedetto and Montorsi, 1996), as does the fixed-point result above.

3. EXAMPLE

We present some results and observations using a rate $1/2$ $(5, 7)$ coder converted to systematic form for the outer coder, a block length of 512 bits for **a**, an all-pole channel of the form $h_k = (0.6)^k$. For a given packet, the limiting pseudo-posteriors $U_i(1)T_i(1)$ from (5) [or (6), cf. Property 1] were observed to display one of two configurations:

- *Case 1*: The pseudo-posteriors tend to a nearly binary $\{0, 1\}$ distribution, as illustrated in Fig. 2.
- *Case 2*: The pseudo-posteriors remain comfortably away from a binary distribution, as illustrated in Fig. 3.

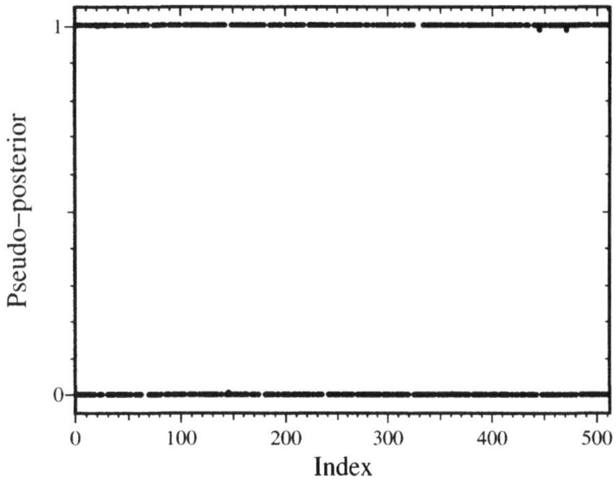

Fig. 2. "Case 1" configuration of the pseudo-posteriors. The decoder decisions are observed, with high reliability, to be correct.

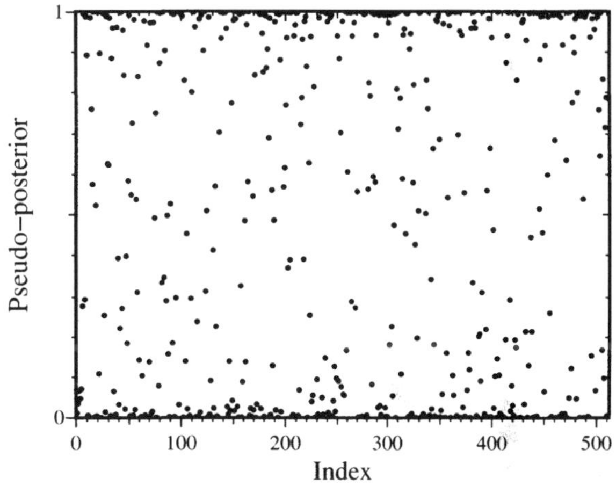

Fig. 3. "Case 2" configuration of the pseudo-posteriors: For such cases, the bit error rate is consistently observed to exceed 40%.

Interestingly, intermediate distributions between figures 2 and 3 were never observed at convergence. For case 2, the bit error rate was consistently found to be in excess of 40%, often approaching a worst-case 50% bit error rate that a "coin-flip" decision device would achieve. For case 1, on the other hand, the decoded symbols are usually correct. Figure 4 shows the percentage of packets for which the algorithm converges to a "Case 1" configuration, versus the (signal plus intersymbol interference)-to-noise ratio [(S+ISI)/N] of the received signal, as determined empirically using 5000 packets. For [(S+ISI)/N] greater than 6 dB, no bit errors were detected for "Case 1" pseudo-posteriors; at 5 dB the bit error rate degrades to about 10^{-3}.

If we average the bit error rate over both "Case 1" and "Case 2" configurations, we obtain values between 10^{-2} and 10^{-1} over the 10 dB to 5 dB range for [(S+ISI)/N], which is hardly impressive. Note that, since the differential encoder has rate 1, the distance spectrum of the concatenated code is basically the same as the distance spectrum for the outer code by itself. Experiments obtained by removing the differential encoder/decoder pair yielded again bit error rates between 10^{-2} and 10^{-1}, provided the phase ambiguity inherent to the constant modulus equalizer is resolved. But the percentage of packets which give the "Case 1" behavior above (i.e., nearly binary pseudo-posteriors) drops by an order of magnitude.

This indicates a potential advantage of the iterative decoding scheme: inspection of the pseudo-posterior distribution gives an indication of the reliability of the result. This can be useful in bidirectional applications: by setting $p_i = \alpha_i T_i(1) U_i(1)$ [where the normalizing constant α_i fulfills $1 - p_i = \alpha_i T_i(0) U_i(0)$], one can examine the average entropy over the packet as

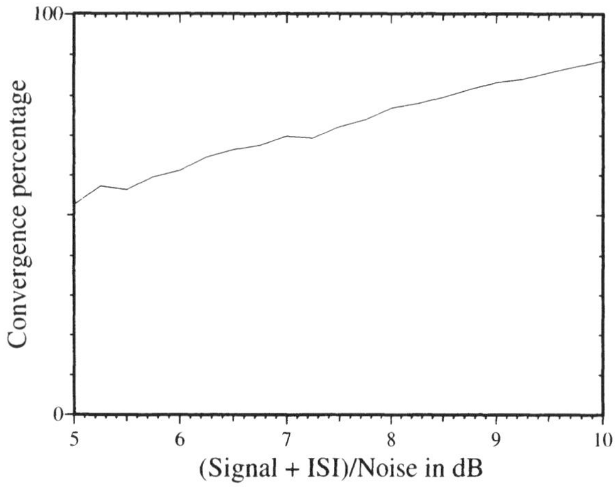

Fig. 4. Percentage of packets which converge to a "Case 1" configuration, versus the (signal + interference)-to-noise ratio of the received signal.

$$H(p) = -\frac{1}{N}\sum_{i=1}^{N} p_i \log p_i$$

When $H(p)$ is less than a given threshold, a "Case 1" configuration is identified and the decoded packet is declared reliable; otherwise a "Case 2" configuration is declared and the receiver requests a packet retransmission. Without the iterative decoding, a significant grey zone persists between Case 1 and Case 2 configurations, and the reliability of a decoded packet is less certain.

4. CONCLUDING REMARKS

A modified turbo equalization scheme has been proposed which uses neither a Viterbi decoder nor a conventional decision feedback equalizer, since the implementation of either in practice requires accurate knowledge of the channel impulse response and/or the use of training sequences. The present scheme, by contrast, works in a completely blind manner, by exploiting a finite-interval constant modulus algorithm (Regalia, 2002). Since the constant modulus criterion is phase blind, differential encoding is included. By interpreting the differential code as a unit rate trellis code, the decoding algorithm can be iteratively linked with the trellis decoder of the outer (or error correction) code, by straightforward adaption of iterative decoding of serial codes (Douillard *et al.*, 1995), (Benedetto and Montorsi, 1996).

The inclusion of a rate-one differential encoder does not improve the distance properties of the code, and therefore should not improve the bit error rate if maximum likelihood decoding were available. But the iterative turbo-decoding algorithm applied here is not, in general, a maximum likelihood decoder, unless the likelihood factorization of Property 2 happens to apply. The converged probabilities furnished by the iterative decoder, however, are observed to approach a binary distribution whenever the decoding appears correct, indicating high reliability, unlike the maximum likelihood decoding rule which need not approach a binary distribution for "correctly decoded" results. This property, however, would appear dependent on the constituent codes used in the concatenation, and is therefore a subject for further study.

REFERENCES

Bahl, L. R., J. Cocke, F. Jelinek, and J. Raviv (1974). Optimal decoding of linear codes for minimizing symbol error rate. *IEEE Trans. Information Theory* **20**(3), 284–287.

Benedetto, S. and G. Montorsi (1996). Iterative decoding of serially concatenated convolutional codes. *Electronics Letters* **32**, 1186–1188.

Delmas, J.-P. H. Gazzah, A. P. Liavas and P. A. Regalia (2000). Statistical analysis of some second-order methods for blind channel identification/equalization with respect to channel undermodeling. *IEEE Trans. Signal Processing* **48**(7), 1984–1998.

Douillard, C., M. Jezequel, C. Berrou, A. Picart, P. Didier and A. Glavieux (1995). Iterative correction of intersymbol interference: Turbo-equalization. *ETT* **6**(5), 507–512.

Fossorier, M. P. C., M. Mihaljević and H. Imai (1999). Reduced complexity iterative decoding of low-density parity check codes based on belief propagation. *IEEE Trans. Communications* **47**(5), 673–680.

Laot, C., A. Glavieux and J. Labat (2001). Turbo equalization: Adaptive equalization and channel decoding jointly optimized. *IEEE J. Selected Areas in Communications* **19**(9), 1744–1752.

McEliece, R. J., D. J. C. MacKay and J.-F. Cheng (1998). Turbo decoding as an instance of Pearl's "belief propagation" algorithm. *IEEE Trans. Selected Areas in Communications* **16**(2), 140–152.

Regalia, P. A. (2002). A finite interval constant modulus algorithm. In: *Proc. ICASSP-02*. Vol. III. pp. 2285–2288. Orlando, FL.

Richardson, T. (2000). The geometry of turbo decoding dynamics. *IEEE Trans. Information Theory* **46**(1), 9–23.

Saaty, T. L. and J. Bram (1964). *Nonlinear Mathematics*. McGraw-Hill. New York.

Tüchler, M., R. Kotter, and A. Singer (2002). Turbo equalization: Principles and new results. *IEEE Trans. Communications* **50**(5), 754–767.

IFAC Publications
www.elsevier.com/locate/ifac

A NEW METHOD FOR CHANNEL ESTIMATION AND DATA DETECTION IN THE CONTEXT OF TURBO EQUALISATION

Sylvie Perreau and Gilles Gorlier *

* *Institute for Telecommunications Research*
University of South Australia
Mawson Lakes SA 5095 Australia
E-Mail: sylvie@spri.levels.unisa.edu.au

Abstract: In this paper, we propose a new turbo equaliser which outputs marginal a-posteriori probabilities of each symbol. These marginal a-posteriori probabilities can be used directly by the decoder in the turbo process. We show how this equaliser is particularly interesting in the context of turbo decoding and also allows to perform blind channel estimation. We will finally show that this new method brings out significant improvements over the classixal Interference Canceller (lower SNR threshold or trigger point) with an acceptable computational cost.

1. INTRODUCTION

In recent years, there has been an growing interest in the so called turbo equalisation. The basic idea is that in systems using channel coding, the sequence produced by the equaliser is not equally likely. Therefore, there is much to be gained during the equalisation process, by using the extra a-priori information provided by the encoder. It is however obvious that an optimal formulation to this problem is not feasible when an interleaver is present. Therefore, instead of considering a one step optimal system which would fully map the allowed sequences at the input of the channel, turbo equalisation proceeds by iteratively passing soft information between the decoder and the equaliser. With each iteration, the Bit Error Rate usually decreases as in turbo decoding. However, it has been shown that there exists an SNR threshold under which iterating between equalisation and decoding in no longer beneficial. Several studies have provided some insight into this trigger point effect. In [1], an analysis tool, based on the extrinsic information transfer (EXIT) chart, was used to compare the performance of several turbo equalisers on a single channel. However, the study was only performed on a single channel realisation and did not allow to predict the performance of turbo equalisation for any given channel. In [2], an interesting study also using the EXIT chart provided an insight into predicting the behavior of a particular equaliser, namely the Interference Canceller (IC), depending on the channel dispersion. This study was very useful since it allowed to predict whether turbo equalisation would be useful and it also provided an estimation of the number fo iterations necessary to reach convergence. The other interesting point raised in this paper was that the performance of the IC canceller over "tough" channels depends on the type of equaliser used for the first iteration: if a DFE is used, turbo equalisation fails, although when a MAP equaliser is used, turbo equalisation is successful. This observation motivates the work presented in this paper. We propose to perform turbo equalisation using the DFE incorporating fixed lag smoothing presented in [3], which can be viewed as a very good trade off (in terms of computational complexity and performance) between an MAP and a DFE equaliser. Simulations show a significant improvment in terms of trigger point

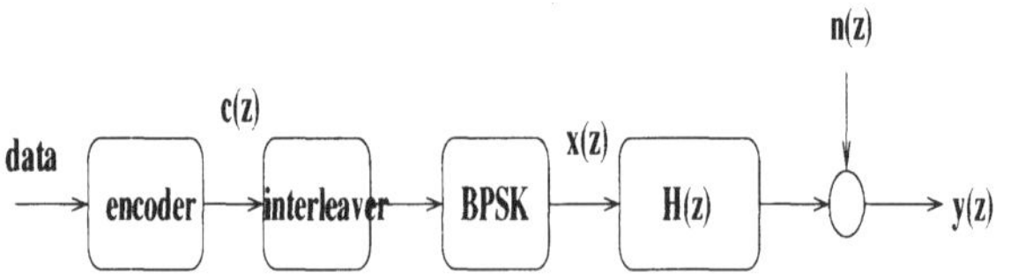

Fig. 1. Transmitter

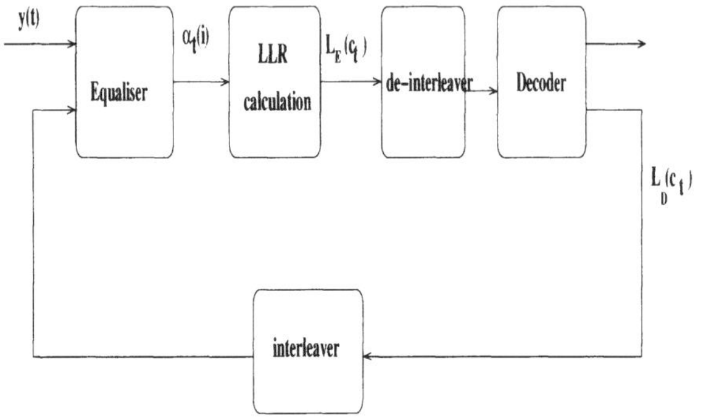

Fig. 2. turbo equalisation

effect as compared to the IC. In this abstract, we present the fixed lag smoother DFE based turbo equaliser.

2. TURBO-EQUALISATION PRINCIPLES

2.1 Problem formulation

We consider the transmitter and channel shown in figure 1. An i.i.d sequence of non coded bits is passed through the channel encoder to produce the sequence c_m which is interleaved and mapped using a BPSK modulation. We assume that the coded and modulated symbols x_t are transmitted through an FIR channel with transfer function $H(z) = \sum_{i=0}^{N-1} h_i\, z^{-i}$ where the h_i are complex-valued coefficients. We assume additive Gaussian white noise with zero mean and variance σ^2. The received signal is thus modeled by

$$y_t = H^T X_t + n_t, \qquad (1)$$

where H and X_t are defined by

$$H^T = [h_0\, h_1\, ...h_{N-1}]$$
$$X_t = [x_t\, x_{t-1}...x_{t-N+1}]^T. \qquad (2)$$

Here, the operator $()^T$ denotes the transposition operation. For sake of simplicity, we also assume throughout this paper a BPSK modulation.

2.2 Decoding and equalisation

Figure 2 depicts the general structure for turbo equalisation. We will see that the equaliser we envisage in this paper produces at time $t+N-1$, the marginal a-posteriori probability for the symbol x_t taking into account observations up to time $t+N-1$. In other words, the equaliser delivers statistical information on symbols when they are "seen" by the channel for the last time.
We denote this a-posteriori probability by

$$\alpha_{t|t+N-1}(i) = Pr(x_t = i|Y_{t+N-1})$$

for $i = 1$ or $i = -1$,
where $Y_{t+N-1} = \{y_1, y_2, \cdots, y_{t+N-1}\}$ denotes the sequence of observations up to time $t+N-1$.

Taking into account the interleaver, each symbol x_t corresponds to a coded bit c_m.

The MAP equalizer takes as prior, soft information on the coded bits (provided by the decoder) in the form of log likelihood ratios(LLRs) and outputs the a posteriori LLR minus the a priori LLR:

$$L(x_t) = \log(\frac{\alpha_{t|t+N-1}(1)}{\alpha_{t|t+N-1}(-1)}) - \log(\frac{P(c_m = 1)}{P(c_m = 0)})(3)$$

Note that the term $\log(\frac{P(c_m=1)}{P(c_m=0)})$ represents prior information the occurence probability of c_m and is provided by the decoder at the previous iteration. Therefore, $L(x_t)$ is independent of this prior information.

Using $L(x_t)$ (i.e $L(c_m)$, with the coded bit c_m corresponding to the modulated symbol x_t) as prior information, the decoder then generates posterior log likelihood on the information bits:

$$L(c_m) = \log(\frac{P(c_m = 1|L(c_1), \cdots L(c_K))}{P(c_m = 0|L(c_1), \cdots L(c_K)))} \qquad (4)$$
$$- \log(\frac{P(c_m = 1)}{P(c_m = 0)}) \qquad (5)$$

Note that at the next iteration, the equaliser recomputes $\alpha_{t|t+N-1}(i)$ using $L(c_m)$ as explained in the following section.

3. THE PROPOSED EQUALISER

In [6], an optimal formulation for maximum *a posteriori* probability estimation of the transmitted symbols was presented. This formulation is based on an HMM formulation. Indeed, the vector X_t can be seen as the state vector of the Markov process described by the following state equation :

$$X_{t+1} = AX_t + x_{t+1} * [1\, 0...0]^T \qquad (6)$$

where A is a shift matrix with $A_{i,j} = 1 \Leftrightarrow i = j + 1$. This Markov process is only observable through the observation equation (1). Suppose that $\hat{H}_t$

the current estimate of the channel, is available at time t. As in [6], define the so-called forward variable, expressing the probability that the state X_t be equal to some realization $[i_0....i_{(N-1)}]^T$ according to the current channel estimate $\hat{H}_t$, and the set of measurements Y_t by:
$\alpha_{t|t}(i_0, i_1, ...i_{N-1}) = Pr(X_t^T = [i_0....i_{(N-1)}]|\hat{H}_t, Y_t)$

The exact computation of this probability involves the so-called forward recursion. We refer the reader to [4] for more details. This recursion requires the calculation of the above probablility for every possible realization of the stochastic process X_t. Such an evaluation obviously requires the computation of M^N probabilities at each step. Thus, it is desirable to seek for a simplified algorithm which permits state revisiting but does not have the exponential complexity in N of the approaches of [5] and [6]. Such an algorithm was presented in [3] This algorithm uses the *marginal* posterior probabilities of the symbols in the channel and has linear computational complexity in the channel duration. We now briefly recall the forward recursion for this algorithm. We assume in the following that the channel is known. (a method for jointly performing turbo equalisation and channel estimation is presented in the next section).
Assume the following quantities be available at time t :

- the approximate filtered probabilities, $\alpha^{(n)}_{t-1|t-1}(i)$ denoting the probability that the $n+1^{th}$ symbol in the channel memory at time $t-1$ (i.e x_{t-n-1}) be equal to i, knowing the observations up to time $t-1$ and the prediction of the other symbols stored in the channel memory at time $t-1$ ($X^{(m)}_{t-1}, \forall m \neq n$, $X^{(m)}_{t-1}$ denotining the $m+1^{th}$ component of vector X_{t-1})

$$\alpha^{(n)}_{t-1|t-1}(i) = P(X^{(n)}_{t-1} = i|Y_{t-1}, X^{(m)}_{t-1} = \hat{X}^{(m)}_{t-1|t-2})$$
$$\forall m \neq n \quad \forall n = 0, \cdots N-1, i = 1 \; or \; -1 \; ;$$

- the current estimate $\hat{X}_{t-1}$ of the vector X_{t-1} as given by the previous recursion. A prediction $\hat{X}_{t|t-1}$ of vector X_t is easily obtained by taking advantage of the shift structure of the process X_t. Clearly we have, for $n = 1...N-1$

$$\hat{X}^{(n)}_{t|t-1} = \hat{X}^{(n-1)}_{t-1|t-1}. \qquad (7)$$

Then by substituting $\hat{X}^{(n)}_{t|t-1}$ for $X^{(n)}_t$ $\forall n = 1 : N-1$ we obtain the approximate filtered probability at time t of the only component of the state vector on which Eq. (7) does not provide information:

$$\alpha^{(0)}_{t|t}(i) = P(c_m = i)P(X^{(0)}_t = i|Y_t, X^{(n)}_t = \hat{X}^{(n)}_{t|t-1}).$$

where $P(c_m = i)$ is evaluated using $L(c_m)$ (i.e the output of the decoder) obtained from the previous iteration of the turbo process.

Substituting from (1) yields

$$\alpha^{(0)}_{t|t}(i) = a^{(0)}_t P(c_m = i)$$
$$N(y_t - \hat{H}^T_t[i, \hat{X}^{(1)}_{t|t-1}, ..., \hat{X}^{(N-1)}_{t|t-1}]^T) (8)$$

where $a^{(0)}_t$ is a normalizing constant and $N(.)$ is a zero mean Gaussian function with variance σ^2. In the forthcoming, $L^{(n)}_t(i)$ denotes the quantity $N(y_t - \hat{H}^T_t[\hat{X}^{(0)}_{t|t}, \hat{X}^{(1)}_{t|t-1}, \ldots, i, \ldots, \hat{X}^{(N-1)}_{t|t-1}]^T)$ where $\hat{X}^{(n)}_{t|t-1}$ has been replaced by i.

The remaining updated probabilities $\alpha^{(n)}_{t|t}(i)$ are also approximated by applying the classical forward recursion of the HMM formulation on conditional instead of joint probabilities. The quantities $\alpha^{(n)}_{t|t}(i)$ recorded as smoothed probabilities are thus obtained as

$$\alpha^{(n)}_{t|t}(i) = a^{(n)}_t \, \alpha^{(n-1)}_{t-1|t-1}(i) \, L^{(n)}_t(i). \qquad (9)$$

The equaliser presented in this section has been shown to provide a much better performance than that of a DFE. Indeed, based on a sub-optimal formulation of HMM theory, it allows to revisit the symbol a-posteriori probabilities as long as the symbol is seen by the channel memory.

Its formulation makes it very easy to use in the context of turbo-equalisation since the prior information provided in the decoder can be used in the calculation of the marginal APPs of a symbol which is seen by the channel for the first time (see equation (8)). This prior information subsequently naturally propagates with the forward recursion (equation (9)). Therefore, when the equaliser delivers the LLR on symbol x_t, it has taken into account information provided by the soft decoder and it has also exploited the time redundancy introduced by the ISI. In a sense, this equaliser takes advantage of the ISI while the IC simply tries to compensate for it.

4. CHANNEL ESTIMATION

Channel and noise variance estimation can be easily implemented using the Expectation-Maximisation (EM) algorithm. In this paper, we only explain how to obtain the channel estimate. However, we believe that future work should concentrate on the noise variance estimate as well since it is involved in every probability computed by the equaliser and therefore, is a crucial parameter for convergence issues.

Since we are operating in the turbo equalisation context, we iteratively process the same block of data. Therefore, channel estimation can be applied off line. In this case, it can be shown that the EM algorithm leads to the following expression for the channel estimate at iteration p:

$$\hat{H}^{(p)} = R^{-1}u \tag{10}$$

Where R is the conditional expectation of the auto-correlation of vector X_t and u is the conditional expectation of the cross correlation between vector X_t and the observation y_t. These quantities are calculated at iteration p, using the marginal a-posteriori Probabilities $\alpha_{t|t}^{(n)}(i)$ to approximate the a-posteriori probability of vector X_t.

More precisely, R and u are computed according to:

$$R = \sum_{t=1}^{T} \hat{X}(t|t)\hat{X}(t|t)^T \tag{11}$$

$$u = \sum_{t=1}^{T} \hat{X}(t|t)y_t \tag{12}$$

with $\hat{X}(t|t) = [\hat{x}(t|t)\ \hat{x}(t-1|t)\cdots\hat{x}(t-N+1|t)]^T$, where the $\hat{(x)}(t-n|t)$ is computed as the conditional expectation of symbol x_{t-n} using $\alpha_{t|t}^{(n)}(i)$.

It has to be pointed out that the EM algorithm may be subject to local minima problems. However, the introduction of a-priori information on the sequence x_t via the decoder output significantly reduces the likelihood of local minima.

5. SIMULATIONS

Simulations have been performed on a type B Proakis channel, i.e, $H = [0.407\ 0.814\ 0.407]$. The main aim of our simulations is to compare our proposed scheme to a turbo equalizer using the IC (DFE is used for the first iteration).

The performance of the proposed turbo equaliser in terms of bit error rate is shown in figures 4 (equaliser output) and 3 (decoder output). The performance of the IC based turbo equaliser can be seen in figure 5 and 6. One can note that the turbo effect is observed for lower SNRs when using our proposed scheme, compared to the IC equaliser. One important observation to be made is that the IC canceller seems to outperform our proposed equaliser seems at the first iteration. However, it is not true for subsequent iterations which seems to indicate that the lower performance of our proposed equaliser at the first iteration is compensated by the fact that the exchange

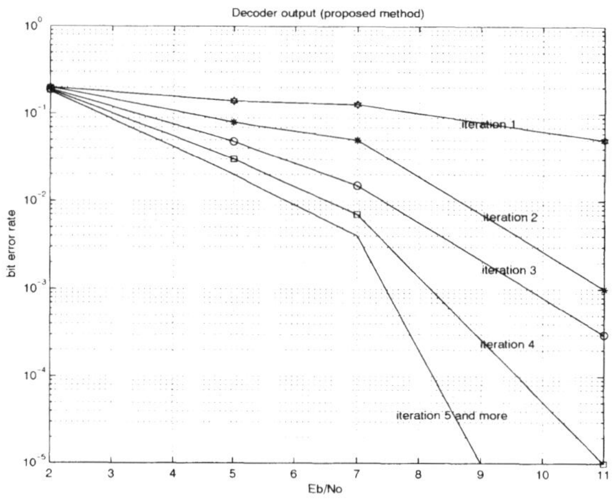

Fig. 3. Decoder output-proposed method

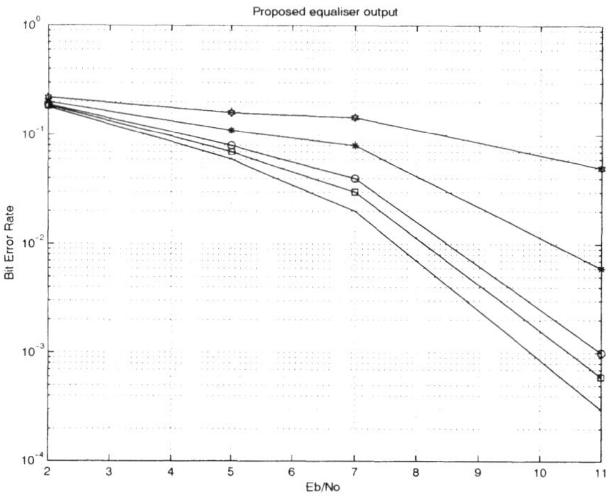

Fig. 4. Output of proposed equaliser

of information between the equaliser and decoder seems much more efficient when using an MAP based equaliser. It would be interested to check this particular point using an EXIT chart.

Another important feature of our proposed method is the channel estimation scheme. Most existing turbo equalisers, such as the IC or DFE assume the knowledge of the channel. Our proposed method allows to relax this requirement as shown in figure 7. The parameters of the channel have been initialised to $\hat{H} = [0.330.330.33]$ and E_b/N_o is set to 9dB. One can see that the channel estimate converges after only 4 iterations approximately to the true values of the channel.

6. CONCLUSION AND FURTHER COMMENTS

In this paper, we have proposed a new turbo equaliser which is based on a reduced complexity MAP equaliser. When compared to the IC based turbo equaliser, a significant improvment has been observed both in terms of robustness

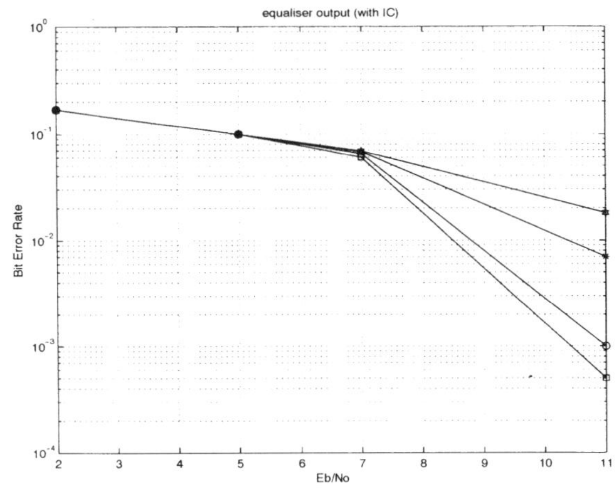

Fig. 5. IC output

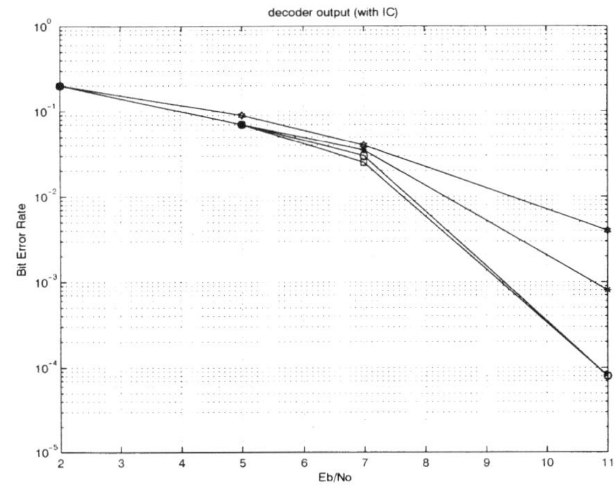

Fig. 6. Decoder output when IC used

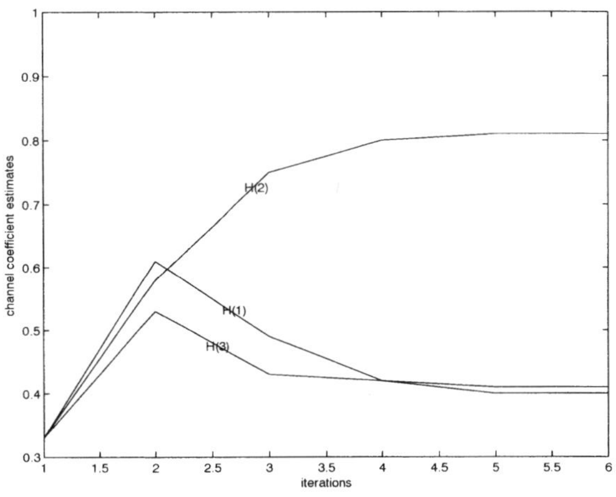

Fig. 7. channel parameter estimates

towards thermal noise effects but also in terms of exchange of information between the equaliser and decoder. This feature should be further explored by studying the EXIT charts related to the turbo equaliser proposed in this paper. In addition, an interesting feature of our proposed scheme is that it allows peforming channel estimation when the channel is unknown. This proposed method offers many directions for future research: indeed, the EM algorithm here used to estimate the channel coefficients could also be used to estimate the noise variance which is an important parameter in the probability computations used in the turbo process. It would be interested to see how using a noise variance estimate instead of the true one affects the turbo process.

REFERENCES

[1] M. Tuchler, R. Koetter and A. Singer *Turbo-Equalisation: Principles and new results* IEEE trans on communications, vol. 50, N0 5, May 2002.

[2] A. Roumy, A. Grant, I. Fijalkow, P. Alexander and D. Pirez *Convergence analysis for turbo equalisation* IEEE int symposium on Inf. Theory (ISIT), June 2001.

[3] S. Perreau, L. White and P. Duhamel *A blind Decision Feedback Equaliser incoporating fixed lag smoothing* IEEE trans on Signal Processing, Vol 48, N0 5, May 2000.

[4] L. Rabiner *A tutorial on hidden markov models and selected applications in speech recognition* Proceedings of the IEEE, v. 77, no. 2, pp.257-285, 1989.

[5] G.K. Kaleh and R. Vallet, "Joint parameter estimation and symbol detection for linear or non-linear unknown dispersive channels," *IEEE Trans. on Comms. January 1994.*

[6] L. White and V. Krishnamurphy, *Adaptive blind equalisation of FIR filters using Hidden Markov Models* IEEE International Conference on Communications, Geneva 1993.

IFAC
Publications
www.elsevier.com/locate/ifac

ON THE APPLICABILITY TO CORRELATED SOURCES OF A BLIND CHANNEL EQUALIZATION METHOD ROBUST TO ORDER OVERESTIMATION

Roberto López-Valcarce [1]

Dept. Teoría Señal y Comunicaciones
Universidad de Vigo, Spain

Abstract: We consider the blind equalization problem in FIR multichannel models from the second-order statistics of the channel output, with a correlated channel input whose statistics are known to the receiver. The few algorithms that handle colored sources require exact knowledge of the channel order, a drawback since order determination is a difficult issue. Recently, Gazzah *et al.* (2002) have presented a channel estimator robust to order overestimation. Although their derivation assumed white sources, we show that it can be suitably modified in order to handle colored inputs. The algorithm is still able to blindly compute an FIR pre-equalizer such that the overall response reduces to an FIR transfer function which is known *a priori* by the receiver, up to a complex phase rotation. Therefore a post-equalizer can be designed in a blind, straightforward manner. The method remains robust to order overmodeling in the correlated source case.

Keywords: Blind equalization, correlated sources, order overestimation.

1. INTRODUCTION

Intersymbol interference is the factor that limits performance in many digital communication systems, and hence adequate processing by an equalizer becomes necessary at the receiver. Blind channel equalization addresses those techniques that estimate the equalizer based only on the channel output waveform and knowledge of the statistics of the transmitted signal. This eliminates the need for training sequences, which decrease system throughput. Second-order statistics (SOS) based methods are particularly attractive, as SOS can be more accurately estimated from finite data records than their higher-order counterparts. The seminal work of Tong *et al.* (1994) showed that under certain conditions, finite impulse response (FIR) single-input multiple-output (SIMO) channels can be equalized by a bank of FIR filters, which can be computed from the channel output SOS. Following (Tong *et al.*, 1994), many SOS-based blind methods have appeared; see (Tong and Perreau, 1998) and the references therein.

We consider systems in which the source statistics *are colored but known to the receiver.* Correlated sources may arise as a result of channel encoding (Mannerkoski and Koivunen, 2000), or from the use of nonlinear modulation formats (Neugebauer, 2002). In either case, knowledge of the transmitter structure will provide the required source statistics to the receiver. Most blind techniques have been designed based on the assumption of a white channel input, failing if the source is correlated. Other methods, such as the one of Moulines *et al.* (1995), require no knowledge of input statistics, which could be an advantage in certain situations. However, this knowledge is often available, and its use should improve performance. Some algorithms exploiting such information are those of Hua *et al.* (1994), Afkhamie and Luo (2000) and López-Valcarce and Dasgupta (2001). One drawback of these methods is that they require precise knowledge of the channel

[1] Supported by the Ramón y Cajal program of the Spanish Ministry of Science and Technology.

order. In practice, order estimation becomes a delicate task (Liavas *et al.*, 1999), so that blind methods robust to order overestimation are of clear interest. Recently, Gazzah *et al.* (2002) have proposed a channel identification method enjoying this robustness property, assuming an uncorrelated source.

Our goal is to show that Gazzah's approach can be suitably modified to allow for colored channel inputs. Blind channel identification seems no longer feasible, but it is shown that the technique is able to blindly determine a 'pre-equalizer' whose convolution with the unknown channel reduces to an FIR filter *which is known a priori up to a complex phase rotation.* Hence, a 'post-equalizer' can be easily (and blindly) designed, and its output provides the desired estimate of the channel input. Even with correlated sources, the method remains robust to channel order overmodeling.

Notation: Bold lowercase letters denote vectors, while bold uppercase or calligraphic letters denote matrices. $(\cdot)^*$, $(\cdot)^T$, $(\cdot)^H$, denote conjugate, transpose and transpose conjugate respectively. $\mathbf{J}$ is the downshift square matrix with ones in the first subdiagonal and zeros elsewhere, $\mathbf{X}$ is the reversal square matrix with ones in the main antidiagonal and zeros elsewhere, and $\mathbf{e}_k$ is the k-th unit vector.

2. CHANNEL MODEL

We consider the FIR single-input p-output model

$$\bar{\mathbf{x}}_n = \sum_{i=0}^{l} \mathbf{h}_i a_{n-i} + \bar{\mathbf{w}}_n, \tag{1}$$

where $\{a_n\}$ is the zero mean, wide sense stationary sequence of transmitted symbols, $\{\bar{\mathbf{x}}_n\}$ is the $p \times 1$ vector of channel outputs, $\{\bar{\mathbf{w}}_n\}$ is a $p \times 1$ zero mean white noise vector with covariance $\sigma_w^2\mathbf{I}$, and the $p \times 1$ vectors $\{\mathbf{h}_i\}$ represent the channel impulse response. This model may arise if multiple sensors are deployed or if the channel output is oversampled. Relation (1) can be recast as

$$\mathbf{x}_n = \mathscr{H}\mathbf{s}_n + \mathbf{w}_n, \tag{2}$$

where, with m the equalizer length,

$$\mathbf{x}_n = [\,\bar{\mathbf{x}}_n^T\ \bar{\mathbf{x}}_{n-1}^T\ \cdots\ \bar{\mathbf{x}}_{n-m+1}^T\,]^T,$$
$$\mathbf{w}_n = [\,\bar{\mathbf{w}}_n^T\ \bar{\mathbf{w}}_{n-1}^T\ \cdots\ \bar{\mathbf{w}}_{n-m+1}^T\,]^T,$$
$$\mathbf{s}_n = [\,a_n\ a_{n-1}\ \cdots\ a_{n-m-l+1}\,]^T,$$

and $\mathscr{H}$ is an $mp \times (m+l)$ generalized Sylvester matrix constructed from the channel impulse response (Tong *et al.*, 1994). A linear equalizer consists of an $mp \times 1$ vector $\mathbf{g}$ whose output is just

$$\mathbf{g}^H\mathbf{x}_n = \mathbf{g}^H\mathscr{H}\mathbf{s}_n + \mathbf{g}^H\mathbf{w}_n.$$

Thus $\mathbf{q} = \mathscr{H}^H\mathbf{g}$ contains the taps of the combined channel-equalizer impulse response. $\mathbf{g}$ constitutes a zero-forcing (ZF) equalizer for the delay δ if $\mathbf{q} = \mathscr{H}^H\mathbf{g}$ has a single nonzero tap at position $\delta + 1$. The following standard assumption is adopted.

Assumption 1. The channel matrix $\mathscr{H}$ is tall and has full column rank. ■

For this, the equalizer length m must satisfy $mp > m + l$ (for which $p \geq 2$ is required). Then, if the transfer functions of the p available subchannels do not present any common root, a celebrated result from Tong *et al.* (1994) states that $\mathscr{H}$ will have full column rank. In that case any combined response $\mathbf{q}$ can be attained by suitably choosing the equalizer vector $\mathbf{g}$. In particular, ZF equalizers of delays 0 through $m + l - 1$ exist and are given by the rows of the pseudoinverse $\mathscr{H}^\#$.

3. PRELIMINARIES

Introduce the lag k autocorrelation matrices

$$\mathscr{R}_x(k) = E[\mathbf{x}_n\mathbf{x}_{n-k}^H], \qquad \mathscr{R}_s(k) = E[\mathbf{s}_n\mathbf{s}_{n-k}^H],$$

of the channel output and the source symbols respectively. Note that

$$\mathscr{R}_x(0) = \mathscr{H}\mathscr{R}_s(0)\mathscr{H}^H + \sigma_w^2\mathbf{I}.$$

Under Assumption 1, the noise variance can be estimated as $\hat{\sigma}_w^2 = \lambda_{\min}[\mathscr{R}_x(0)]$. Then we can construct the *denoised* matrices

$$\tilde{\mathscr{R}}_x(k) = \mathscr{R}_x(k) - \hat{\sigma}_w^2\mathbf{J}^{kp} = \mathscr{H}\mathscr{R}_s(k)\mathscr{H}^H. \tag{3}$$

For convenience, let us define $d = m + l$. The following assumption on the source SOS is made:

Assumption 2.

$$E\left\{\begin{bmatrix}\mathbf{s}_n \\ a_{n-d}\end{bmatrix}\begin{bmatrix}\mathbf{s}_n^H & a_{n-d}^*\end{bmatrix}\right\} > \bigcirc. \tag{4}$$

Let us now introduce the Cholesky factorization

$$\mathscr{R}_s(0) = \mathscr{L}\mathscr{L}^H, \tag{5}$$

with $\mathscr{L}$ lower triangular with positive diagonal elements (note that $\mathscr{R}_s(0) > \bigcirc$ due to Assumption 2) which is known to the receiver. With this, we can introduce the normalized matrices

$$\mathbf{H} = \mathscr{H}\mathscr{L}, \qquad \mathbf{R}_s(1) = \mathscr{L}^{-1}\mathscr{R}_s(1)\mathscr{L}^{-H}. \tag{6}$$

Then from (3) one has

$$\tilde{\mathscr{R}}_x(0) = \mathbf{H}\mathbf{H}^H, \quad \tilde{\mathscr{R}}_x(1) = \mathbf{H}\mathbf{R}_s(1)\mathbf{H}^H. \tag{7}$$

4. REVIEW OF GAZZAH'S APPROACH

Consider vectors $\mathbf{g}_1$, $\mathbf{g}_d$ satisfying

$$\tilde{\mathscr{R}}_x(1)^H \mathbf{g}_1 = \mathbf{0}, \quad \tilde{\mathscr{R}}_x(1)\mathbf{g}_d = \mathbf{0}. \tag{8}$$

Under our assumptions, $\mathbf{H} = \mathscr{H}\mathscr{L}$ has full column rank equal to d. Therefore, from (7), (8) yields

$$\mathbf{R}_s(1)^H \mathbf{H}^H \mathbf{g}_1 = \mathbf{0}, \quad \mathbf{R}_s(1)\mathbf{H}^H \mathbf{g}_d = \mathbf{0}. \tag{9}$$

With $\mathbf{q}_1 = \mathscr{H}^H \mathbf{g}_1$, $\mathbf{q}_d = \mathscr{H}^H \mathbf{g}_d$ the corresponding channel-equalizer combined responses, note that

$$\mathbf{H}^H \mathbf{g}_1 = \mathscr{L}^H \mathbf{q}_1, \qquad \mathbf{H}^H \mathbf{g}_d = \mathscr{L}^H \mathbf{q}_d.$$

If the source is white with variance σ_a^2, then

$$\mathscr{L} = \sigma_a \mathbf{I}, \qquad \mathbf{R}_s(1) = \mathbf{J}. \tag{10}$$

Therefore $\mathbf{q}_1$ and $\mathbf{q}_d$ lie respectively in the left and right null spaces of $\mathbf{J}$. We conclude that $\mathbf{q}_1 = c_1\mathbf{e}_1$ and $\mathbf{q}_d = c_d\mathbf{e}_d$, with c_1, c_d some constants which need not be nonzero. If the zero solution could be avoided, then $\mathbf{g}_1$, $\mathbf{g}_d$ would constitute ZF equalizers with delays 0 (minimal) and $d-1$ (maximal). To do so, Gazzah *et al.* (2002) propose to maximize the SNR at the equalizer output, which is given by

$$\text{SNR} = \frac{\mathbf{g}^H \mathscr{H}\mathscr{R}_s(0)\mathscr{H}^H \mathbf{g}}{\sigma_w^2 \mathbf{g}^H \mathbf{g}}. \tag{11}$$

Thus the problem can be cast as

$$\text{maximize } \frac{\mathbf{g}^H \tilde{\mathscr{R}}_x(0)\mathbf{g}}{\mathbf{g}^H \mathbf{g}} \text{ s.t. } \begin{cases} \tilde{\mathscr{R}}_x(1)^H \mathbf{g} = \mathbf{0} \\ \text{or} \\ \tilde{\mathscr{R}}_x(1)\mathbf{g} = \mathbf{0} \end{cases} \tag{12}$$

Observe that both left and right null spaces of $\tilde{\mathscr{R}}_x(1)$ have dimension $mp-d+1$. Let us focus on the first constraint in (12), as the problem with the second constraint is analogous. Let $\mathscr{U}_1$ be an $mp \times (mp-d+1)$ matrix whose columns form an orthonormal basis of the left null space of $\tilde{\mathscr{R}}_x(1)$. Thus, if $\tilde{\mathscr{R}}_x(1)^H \mathbf{g} = \mathbf{0}$, then $\mathbf{g} = \mathscr{U}_1 \mathbf{f}$ for some vector $\mathbf{f}$. Therefore, the output SNR (11) becomes

$$\text{SNR} = \frac{\mathbf{f}^H \mathscr{U}_1^H \tilde{\mathscr{R}}_x(0)\mathscr{U}_1 \mathbf{f}}{\sigma_w^2 \mathbf{f}^H \mathbf{f}},$$

since $\mathscr{U}_1^H \mathscr{U}_1 = \mathbf{I}$. This is maximized if $\mathbf{f}_1$ is the eigenvector of $\mathscr{U}_1^H \tilde{\mathscr{R}}_x(0)\mathscr{U}_1$ for its largest eigenvalue. Then $\mathbf{g}_1 = \mathscr{U}_1 \mathbf{f}_1$ is a ZF equalizer with zero delay, i.e. the resulting overall response is of the form $\mathbf{q}_1 = c_1\mathbf{e}_1$. c_1 can be found (within a phase ambiguity, inherent to the blind nature of the problem) by noting that $\mathbf{g}_1^H \tilde{\mathscr{R}}_x(0)\mathbf{g}_1 = \sigma_a^2 |c_1|^2$.

5. THE COLORED SOURCE CASE

In the general case, $\mathscr{L} \neq \sigma_a \mathbf{I}$ and $\mathbf{R}_s(1) \neq \mathbf{J}$. However, these matrices present a rich structure which can be exploited (López-Valcarce and Dasgupta, 2001). In particular, they are related by

$$\mathbf{R}_s(1) = \mathscr{L}^{-1}(\mathbf{J} - \mathbf{e}_1 \alpha^H)\mathscr{L}, \tag{13}$$

where $\alpha = [\,\alpha_1 \; \cdots \; \alpha_d\,]^T$ is the coefficient vector of the d-th order forward prediction error filter (FPEF) for the source process $\{a_n\}$. It is given by

$$\alpha = -\mathscr{R}_s(0)^{-1}\mathscr{R}_s(1)^H \mathbf{e}_1.$$

The transfer function of the d-th order FPEF is

$$\alpha(z) = 1 + \sum_{k=1}^{d} \alpha_k^* z^{-k}. \tag{14}$$

Note that $\mathbf{J} - \mathbf{e}_1\alpha^H$ is a companion matrix whose eigenvalues coincide with the zeros of $\alpha(z)$.

If we require now that $\mathbf{g}_1$, $\mathbf{g}_d$ satisfy (8), we see from (9) that $\mathbf{H}^H \mathbf{g}_1$, $\mathbf{H}^H \mathbf{g}_d$ lie in the left and right null spaces of $\mathbf{R}_s(1)$. If $\mathbf{R}_s(1)$ is nonsingular, these null spaces reduce to $\{\mathbf{0}\}$. Then the corresponding equalizer output has no signal component, which is clearly undesirable. To avoid this situation, we shall make an additional assumption.

Assumption 3. $\alpha_d = 0$. ■

This and (13) imply that $\mathbf{R}_s(1)$ is singular. Although this assumption is critical to the following analysis, simulations in Section 7 will show that the method does not break down when $\alpha_d \neq 0$.

Left null space of $\mathbf{R}_s(1)$. When $\alpha_d = 0$, one has

$$\begin{aligned} \mathbf{R}_s(1)^H \mathbf{v} = \mathbf{0} &\Rightarrow \mathscr{L}^H(\mathbf{J}^H - \alpha\mathbf{e}_1^H)\mathscr{L}^{-H}\mathbf{v} = \mathbf{0} \\ &\Rightarrow (\mathbf{J}^H - \alpha\mathbf{e}_1^H)\mathscr{L}^{-H}\mathbf{v} = \mathbf{0} \\ &\Rightarrow \mathscr{L}^{-H}\mathbf{v} = c_1[\,1 \; \alpha_1 \; \cdots \; \alpha_{d-1}\,]^T \end{aligned}$$

where the last line follows from Assumption 3, and c_1 is a constant. Thus if $\tilde{\mathscr{R}}_x(1)^H \mathbf{g}_1 = \mathbf{0}$, then

$$\mathscr{L}^{-H}\mathbf{H}^H \mathbf{g}_1 = \mathscr{H}^H \mathbf{g}_1 = c_1[\,1 \; \alpha_1 \; \cdots \; \alpha_{d-1}\,]^T. \tag{15}$$

Therefore the overall response $\mathbf{q}_1^H = \mathbf{g}_1^H \mathscr{H}$ reduces to a multiple of the FPEF $\alpha(z)$ in (14).

Right null space of $\mathbf{R}_s(1)$. Under Assumption 3,

$$\begin{aligned} \mathbf{R}_s(1)\mathbf{v} = \mathbf{0} &\Rightarrow \mathscr{L}^{-1}(\mathbf{J} - \mathbf{e}_1\alpha^H)\mathscr{L}\mathbf{v} = \mathbf{0} \\ &\Rightarrow (\mathbf{J} - \mathbf{e}_1\alpha^H)\mathscr{L}\mathbf{v} = \mathbf{0} \\ &\Rightarrow \mathscr{L}\mathbf{v} = c_d\mathbf{e}_d \;\Rightarrow\; \mathbf{v} = c_d\beta_0\mathbf{e}_d. \end{aligned}$$

The last line follows from the facts that the right null space of a singular $d \times d$ companion matrix is the span of $\mathbf{e}_d$, and that $\mathscr{L}^{-1}$ is lower triangular; β_0 denotes the (d,d) entry of $\mathscr{L}^{-1}$. (Again, c_d is an unknown constant). Hence, if $\tilde{\mathscr{R}}_x(1)\mathbf{g}_d = \mathbf{0}$, then

$$\mathbf{H}^H \mathbf{g}_d = c_d\beta_0\mathbf{e}_d \;\Rightarrow\; \mathscr{H}^H \mathbf{g}_d = c_d\beta_0\mathscr{L}^{-H}\mathbf{e}_d. \tag{16}$$

In this case the overall response $\mathbf{q}_d^H = \mathbf{g}_d^H \mathscr{H}$ is proportional to the last row of $\mathscr{L}^{-1}$, which contains the coefficients of the FPEF of order $d-1$ for the process $\{a_n\}$. Specifically, if we denote

$$\mathbf{e}_d^H \mathscr{L}^{-1} = [\,\beta_{d-1} \cdots \beta_1\ \beta_0\,] = \beta^T, \quad (17)$$

the $(d-1)$-th order FPEF transfer function is

$$\beta(z) = 1 + \sum_{k=1}^{d-1} \frac{\beta_k^*}{\beta_0} z^{-k} \quad (18)$$

(note that β_0 is positive real). Due to the order-update property of prediction filters (Haykin, 1996), under Assumption 3 it turns out that the FPEFs of orders d and $d-1$ coincide, i.e.

$$\alpha_k = \frac{\beta_k}{\beta_0}, \quad k = 1, \ldots, d-1 \Rightarrow \alpha(z) = \beta(z) \quad (19)$$

Since the β_k coefficients in (17) appear in reverse order, the transfer function of the combined response $\mathbf{q}_d^H = \mathbf{g}_d^H \mathscr{H}$ is proportional to $z^{-(d-1)}\alpha^*(1/z^*)$, that is, the *backward* prediction error filter (BPEF) for the process $\{a_n\}$.

To obtain $|c_d|^2$, note from (16) that

$$\mathbf{g}_d^H \tilde{\mathscr{R}}_x(0)\mathbf{g}_d = \mathbf{g}_d^H \mathbf{H}\mathbf{H}^H \mathbf{g}_d = |c_d|^2 \beta_0^2. \quad (20)$$

To obtain $|c_1|^2$, noting that the vector in the right-hand side of (15) is just $\beta_0^{-1}\mathbf{X}\beta$, one has

$$\mathbf{g}_1^H \tilde{\mathscr{R}}_x(0)\mathbf{g}_1 = \mathbf{g}_1^H \mathbf{H}\mathbf{H}^H \mathbf{g}_1 = \frac{|c_1|^2}{\beta_0^2} \|\mathscr{L}^H \mathbf{X}\beta\|^2 \quad (21)$$

Now the vector $\mathscr{L}^H\mathbf{X}\beta$ has unit norm:

$$\begin{aligned}\beta^H \mathbf{X}^H \mathscr{L}\mathscr{L}^H \mathbf{X}\beta &= \beta^H \mathbf{X}^H \mathscr{R}_s(0)\mathbf{X}\beta \\ &= \beta^H \mathscr{R}_s(0)^* \beta \\ &= \beta^H \mathscr{L}^* \mathscr{L}^T \beta = \mathbf{e}_d^H \mathbf{e}_d = 1,\end{aligned}$$

where the second line follows from the fact that $\mathbf{X}\mathscr{R}_s(0) = \mathscr{R}_s(0)^*\mathbf{X}$ because $\mathscr{R}_s(0)$ is Hermitian Toeplitz, and the third from the definition of β in (17). Therefore, one has $|c_1|^2 = \beta_0^2(\mathbf{g}_1^H \tilde{\mathscr{R}}_x(0)\mathbf{g}_1)$.

In both cases ($\mathbf{g}_1$ or $\mathbf{g}_d$) an output SNR maximization approach can be undertaken exactly as in Section 4, in order to avoid the zero solution.

6. POST-EQUALIZATION

We have shown that the equalizers chosen from left and right null vectors of $\tilde{\mathscr{R}}_x(1)$ result in combined responses that are proportional to the FPEF and BPEF for the source process, which are known to the receiver. That is, the 'pre-equalizers' $\mathbf{g}_1$, $\mathbf{g}_d$ reduce the (unknown) SIMO channel to a known (up to a scaling) SISO one, meaning that a 'post-equalizer' can be designed. Several approaches seem possible. First, by using *both* $\mathbf{g}_1$, $\mathbf{g}_d$, a 1-input 2-output channel would be obtained. These two subchannels do not have common zeros: Assumption 2 guarantees the FPEF and BPEF to be strictly minimum and maximum phase respectively. Hence, a 2-input 1-output ZF post-equalizer could be readily designed.

The problem with this approach is that the residual phase ambiguities in each subchannel are independent, precluding direct design of the post-equalizer. In principle, one could find the phase difference between the two subchannels, since

$$\mathbf{g}_1^H \tilde{\mathscr{R}}_x(0)\mathbf{g}_d = (c_1^* c_d)(\mathbf{e}_1^H \mathscr{R}_s(0)\mathbf{e}_d). \quad (22)$$

However, in most practical cases $|\mathbf{e}_1^H \mathscr{R}_s(0)\mathbf{e}_d|$ will be very small or even zero, making estimation of $c_1^* c_d$ via (22) numerically troublesome.

Another option is to use a single pre-equalizer ($\mathbf{g}_1$ or $\mathbf{g}_d$), and then design a Minimum Mean Squared Error (MMSE) FIR postequalizer. This design can be carried out because the combined response is known in advance, together with the autocorrelation of the source sequence. The autocorrelation function of the noise at the pre-equalizer output is also known: the lag k autocorrelation coefficient is given by $\sigma_w^2(\mathbf{g}^H \mathbf{J}^{kp}\mathbf{g})$. Nevertheless, the fact that the transfer function $\alpha(z)$ of the FPEF is minimum phase (and therefore it can be stably and causally inverted) suggests using $\mathbf{g}_1$ as pre-equalizer and an all-pole filter $1/\alpha(z)$ as post-equalizer. This is computationally much simpler than the MMSE approach, which requires inversion of a covariance matrix. On the other hand, this straightfoward design does not take noise into account, so that noise enhancement problems could appear at low SNRs. With this approach, the algorithm can be summarized as follows:

(1) Estimate $\mathscr{R}_x(0)$, $\mathscr{R}_x(1)$ from the data.
(2) Estimate $\hat{\sigma}_w^2 = \lambda_{\min}[\mathscr{R}_x(0)]$, and compute $\tilde{\mathscr{R}}_x(0)$, $\tilde{\mathscr{R}}_x(1)$ in (3).
(3) Perform an SVD of $\tilde{\mathscr{R}}_x(1)$ and retain in the columns of $\mathscr{U}_1$ the $mp-d+1$ 'smallest' left singular vectors.
(4) Compute $\mathbf{f}_1$ as the unit-norm eigenvector of $\mathscr{U}_1^H \tilde{\mathscr{R}}_x(0)\mathscr{U}_1$ for its largest eigenvalue $\lambda_{\max}$, and let $\mathbf{g}_1 = \mathscr{U}_1\mathbf{f}_1$.
(5) Pre-equalizer output: $y_n = (\mathbf{g}_1^H \mathbf{x}_n)/(\beta_0\sqrt{\lambda_{\max}})$.
(6) Estimate the source data via $\hat{a}_n = [1/\alpha(z)]y_n$.

Regarding order overmodeling, note that the only point in the algorithm at which the (estimated) channel lenght l intervenes is step 3 (since $d = m + l$). If l is overestimated, then fewer singular vectors will be retained in $\mathscr{U}_1$. However, the resulting equalizer will still perform acceptably (unless *all* the singular vectors retained happen to be orthogonal to the signal subspace, an unlikely case), although some degradation is expected.

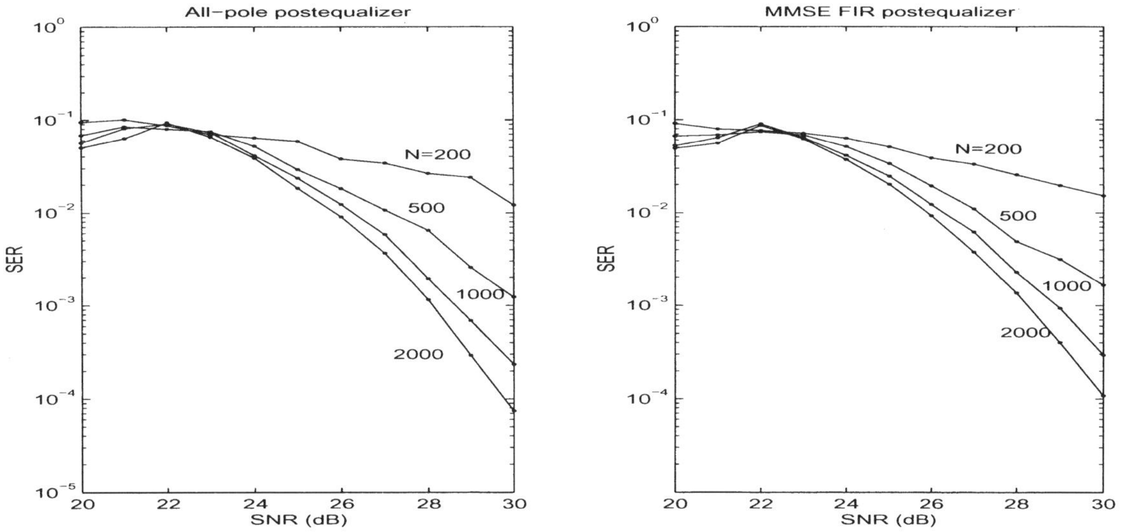

Fig. 1. SER vs. SNR. Pre-equalizer length $m = 10$, assumed channel order $l = 3$.

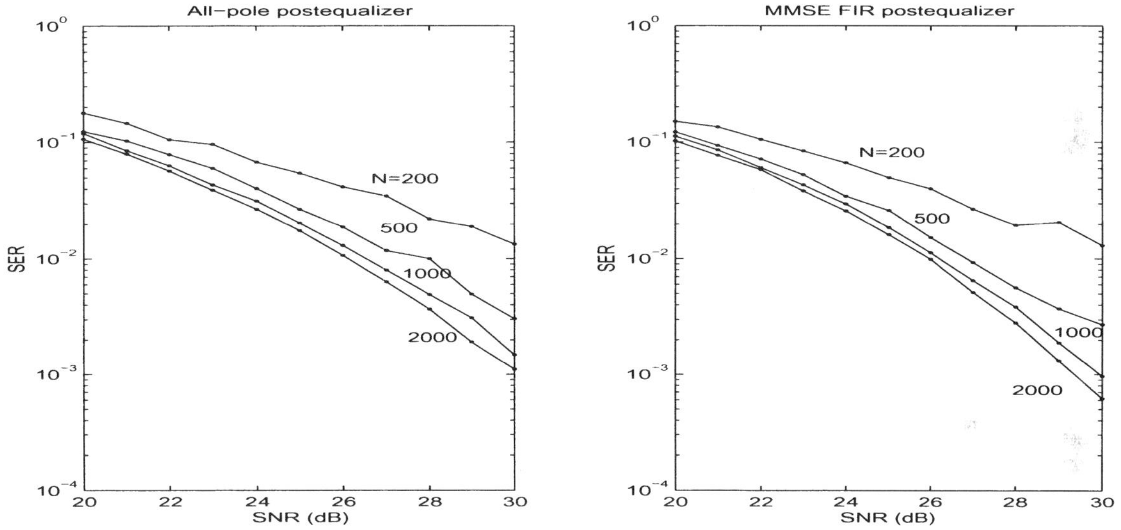

Fig. 2. SER vs. SNR. Pre-equalizer length $m = 5$, assumed channel order $l = 3$.

7. SIMULATION RESULTS

We used a correlated QPSK source $\{a_n\}$ generated as $a_n = b_{n-2} - jb_n$, where $b_n \in \pm 1$ is the input stream of i.i.d. bits. The only nonzero autocorrelation coefficients are

$$E[a_n a_{n-k}^*] = \begin{cases} 2, & k = 0, \\ \pm j, & k = \pm 2. \end{cases}$$

It can be checked that the resulting coefficients α_k of the order d FPEF are, for $d = 2q$ or $d = 2q+1$,

$$\alpha_k = \begin{cases} 0, & k \text{ odd}, \\ \left(1 - \dfrac{k/2}{q+1}\right) j^{k/2}, & k \text{ even}. \end{cases}$$

Note that Assumption 3 is satisfied for odd d, but not for d even. Consider the following $p = 2$ subchannels with $l = 3$ and $[\,\mathbf{h}_0\ \mathbf{h}_1\ \mathbf{h}_2\ \mathbf{h}_3\,] =$

$$\begin{bmatrix} 0.3 & 0.4 - j0.1 & 0.1 + j0.2 & -0.2 + j0.6 \\ -0.2 - j0.1 & 0.5 - j0.2 & -0.5 - j0.1 & 0.4 + j0.2 \end{bmatrix}$$

For performance comparison purposes, the residual phase ambiguity is removed before symbol error rate (SER) evaluation in the simulations (SER was averaged over 100 independent trials).

Fig. 1 shows the SER obtained with the proposed method, using several record lengths ($N = 200$, 500, 1000 and 2000 transmitted symbols) for autocorrelation estimation. The pre-equalizer order was $m = 10$, and $l = 3$ (correct channel order) was assumed. For these values d is odd, so that Assumption 3 holds. The first set of curves in Fig. 1 were computed using a post-equalizer $1/\alpha(z)$, while an MMSE FIR post-equalizer was used for the second set. For a fair comparison, the same number of taps was taken for the FIR and IIR post-equalizers. Since it is known *a priori* that the overall transfer function is minimum phase, an associated delay of zero can be taken in the FIR post-equalizer design without performance loss.

Both approaches perform quite similarly, which makes the 'quick and dirty' all-pole design more attractive

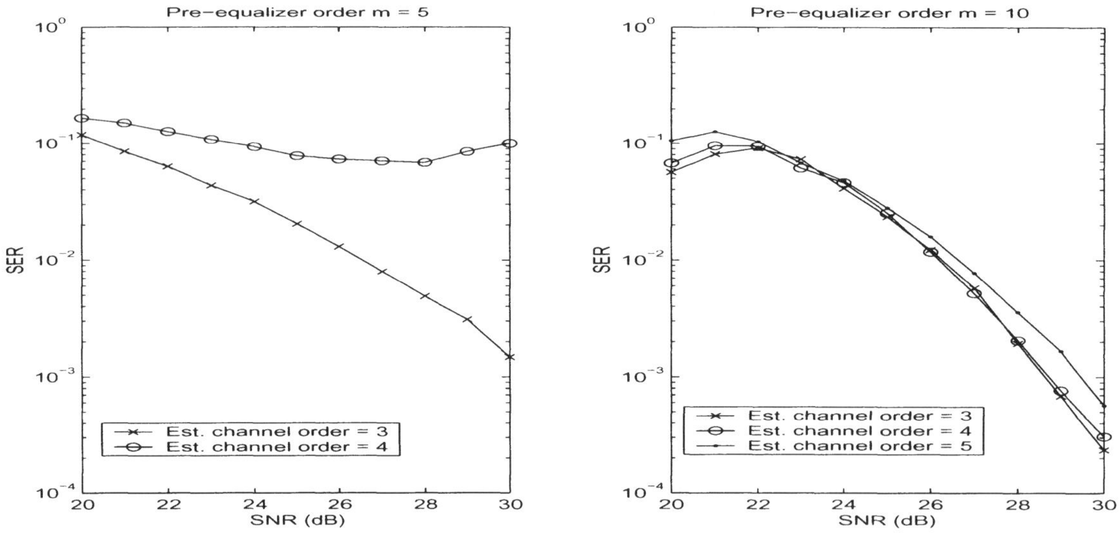

Fig. 3. SER vs. SNR. All-pole post-equalizers, $N = 1000$ symbols used for estimation.

given its smaller computational load. Fig. 2 shows the results obtained under the same conditions, except that now $m = 5$. Then d is even, so that Assumption 3 is violated ($\alpha_d = 0.2$). Significant SER reduction is still observed, so that the algorithm seems to be robust to this effect.

Fig. 3 compares the SER obtained with exact and overestimated orders, for both short and long ($m = 5$ and 10) pre-equalizers with an all-pole post-equalizer. For $m = 5$, performance degrades when the channel order is overestimated, although the scheme does not completely break down. With $m = 10$, channel order overestimation does not translate into performance degradation. This is as expected, since the dimension of the subspace from which the pre-equalizer is extracted is $m(p-1) - l + 1$, which grows linearly with m.

8. CONCLUSIONS

For the colored source case, blind channel identification based on the scheme of Gazzah *et al.* (2002) is not feasible. However, the method is still able to blindly obtain pre-equalizers whose convolution with the channel is known to the receiver (up to a phase rotation). Therefore, design of a post-equalizer is straightforward. The algorithm uses second-order statistics only of the observed signal, and it preserves the appealing property of the original method of being robust to channel order overmodeling. The computational cost of the proposed extension is comparable to that of the original approach as well.

REFERENCES

Afkhamie, K.H. and Z.-Q. Luo (2000). Blind equalization of FIR systems driven by markov-like input signals. *IEEE Trans. Sig. Proc.* **48**, 1726–1736.

Gazzah, H., P.A. Regalia, J.-P. Delmas and K. Abed-Meraim (2002). A blind multichannel identification algorithm robust to order overestimation. *IEEE Trans. Sig. Proc.* **50**, 1449–1458.

Haykin, S. (1996). *Adaptive Filter Theory, 3rd ed.*. Prentice Hall. Upple Saddle River, NJ.

Hua, Y., H. Yang and W. Qiu (1994). Source correlation compensation for blind channel identification based on second-order statistics. *IEEE Sig. Proc. Let.* **1**, 119–120.

Liavas, A., P.A. Regalia and J.-P. Delmas (1999). Blind channel approximation: effective channel order determination. *IEEE Trans. Sig. Proc.* **47**, 3336–3344.

López-Valcarce, R. and S. Dasgupta (2001). Blind channel equalization with colored sources based on second-order statistics: a linear prediction approach. *IEEE Trans. Sig. Proc.* **49**, 2050–2059.

Mannerkoski, J. and V. Koivunen (2000). Autocorrelation properties of channel encoded sequences – applicability to blind equalization. *IEEE Trans. Sig. Proc.* **48**, 3501–3506.

Moulines, E., P. Duhamel, J.-F. Cardoso and S. Mayrargue (1995). Subspace methods for the blind identification of multichannel FIR filters. *IEEE Trans. Sig. Proc.* **43**, 516–525.

Neugebauer, S. (2002). Blind SIMO channel estimation for CPM signals. In: *Proc. 36th Asilomar Conf. on Sig., Sys. and Comp.* Pacific Grove, CA.

Tong, L. and S. Perreau (1998). Multichannel blind identification: from subspace to maximum likelihood methods. *Proc. IEEE* **86**, 1951–1968.

Tong, L., G. Xu and T. Kailath (1994). Blind identification and equalization based on second-order statistics: A time-domain approach. *IEEE Trans. Info. Thry.* **40**, 340–350.

IFAC Publications
www.elsevier.com/locate/ifac

BLIND ESTIMATION WITH SIGNAL SCRAMBLING

Honghui Xu, Xuejie Song, Soura Dasgupta[1]

Department of Electrical & Computer Engineering
The University of Iowa
Iowa City, IA-52242, USA.
hoxu, xsong, dasgupta@engineering.uiowa.edu

ABSTRACT

Traditional blind identification methods deal with the estimation of an unknown system which can be described by a transfer function $H(z)$. In some scenarios, the unknown system is a cascade of two systems with transfer functions $H(z)$ and $G(z)$ and both channels have to be separately estimated. This paper presents a new scrambling aided blind identification method that permits these systems to be unraveled. The identification algorithm is derived. It is shown the identification is unique up to a nonzero scalar and a delay for both channels. *Copyright © 2003 IFAC*

1. INTRODUCTION

The goal of current blind identification methods is to estimate an unknown system $H(z)$ from its outputs [2] [4]. In some scenarios, the unknown system is a cascade of two unknown systems which have to be identified separately. For example, in *ad hoc* networks, signals may transmit in multihop fashion. A relay sensor needs to transmit signals from one device to another device. Due to the disparity of resources residing in different devices, it is desirable for the resource rich device to take the task of channel estimation in two consecutive hops.

Similarly in sensor networks a common problem is that the signal to be reconstructed is processed at a location remote from the sensor. Quite often, neither the communication channel over which the signal is transmitted nor the sensor characteristics is known at the remote processor location. In effect one has the setting of figure 1, with $x(n)$ the signal at the sensor output, $w_i(n)$ the sensor and channel noise, $v(n)$ the sensor output signal and $y(n)$ the signal received at the processor location. For effective processing it is important for the processor to estimate both the channel and sensor characteristics. Should the channel and the sensor be modeled as linear time invariant (LTI) systems, then the commutativity of LTI systems precludes the separate unraveling of the two.

[1]Supported in part by NSF grants ECS-9970105 and CCR-9973133.

One solution to the problem is that the sensor interlace the transmitted signal $v(n)$ with a prearranged training signal. This would permit the receiver to separately identify the channel transfer function and then use blind estimation techniques to estimate the sensor characteristics. The disadvantage of such a scheme is the bandwidth consumed by the training signal whose length must increase with the channel order. Further in a mobile wireless environment the channel changes at a rate proportional to the mobility of the end points, prompting the need to frequently retransmit the training signal.

In this paper, we propose an alternative technique that adopts the arrangement of fig. 2, where rather than sending the sensed signal as is, the sensor is required to transmit a scrambled version of the signal. The scrambling block itself is Linear Time Varying (LTV) and under the right conditions renders the sensor transfer function $H(z)$ and the channel transfer function $G(z)$, noncommutable. Our goal is to estimate the two transfer functions, within this scrambled setting, to within a constant and delay, from the higher order statistics (HOS) of the received signal $y(n)$.

To this end, we build upon the framework of [6], where a similar scrambling techniques was used to unravel two systems, in an *input-output* rather than blind framework. The methodology of [6] employed only second order statistics, (SOS). However, as we argue in this paper, within the blind framework adopted by us, SOS based methods are inadequate. Thus, instead we use HOS based methods.

2. SYSTEM MODEL AND ASSUMPTIONS

Figure 3 depicts the basic arrangement we adopt. The block labelled $\downarrow M$ represents an M-fold decimator discards all but every M-th sample, in particular with $a(k)$ and $b(k)$ its input an output respectively,

$$b(k) = a(Mk).$$

Likewise the block labelled $\uparrow L$ represents an L-fold interpolator that pads $L-1$ zeros between every consecutive

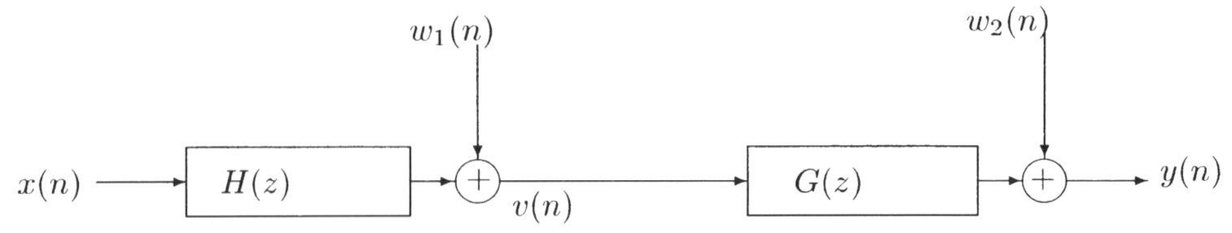

Figure 1: A sensor channel combination

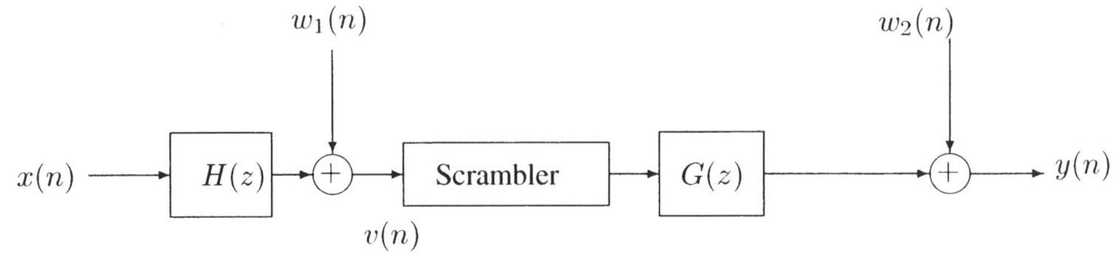

Figure 2: Scrambled transmission

input samples. Thus $a(k)$ and $b(k)$ are its input and output respectively,

$$b(k) = \begin{cases} a(\frac{k}{L}) & \text{if } k \bmod L = 0 \\ 0 & \text{else} \end{cases}.$$

In the sequel we will assume that

$$L > M. \tag{1}$$

Indeed one can choose $L = M + 1$, although the basic theory applies to arbitary positive integers L amd M obeying (1).

Observe that the M-branch delay chain and interpolator arrangement scrambles blocks of M consecutive samples at its input and converts then into L block samples with $L - M$ additional zeros in each block. These additional zeros serve as redundancy. But since L can be $M + 1$, and M can be any positive integer, by choosing M large enough the relative level of redundancy, and the corresponding bandwidth wastage can be kept arbitrarily small. This contrasts to the training based approach where this redundancy is constrained by the channel order and the rapidity with which the channel parameters change.

Observe that this scrambler represents an LTV (in fact periodic) device, and under the right conditions should permit the separate extraction of $G(z)$ and $H(z)$ from the input output relation between $x(n)$ and $y(n)$. However as scaling constants commute with LTV operators such an unraveling can only occur to within a scaling constant. Similarly delays in the two transfer functions can under the right circumstances commute with this operator. Thus at best one can expect separability only to within delays and scaling constants.

In the sequel we assume that $H(z)$ and $G(z)$ are given by $H(z) = \sum_{i=0}^{l_h} h(i)z^{-i}$ and $G(z) = \sum_{i=0}^{l_g} g(i)z^{-i}$ respectively, where l_h and l_g are the orders of $H(z)$ and $G(z)$ respectively. Figure 3 can be transformed into Figure 4 [6], where

$$\mathbf{R}(z) = \begin{pmatrix} R_0(z) & R_1(z) & \cdots & R_{M-1}(z) \\ R_1(z) & R_2(z) & \cdots & R_M(z) \\ \vdots & \vdots & \vdots & \vdots \\ R_{L-1}(z) & z^{-1}R_0(z) & \cdots & z^{-1}R_{M-2}(z) \end{pmatrix} \tag{2}$$

and

$$\mathbf{E}(z) = \begin{pmatrix} E_0(z) & E_1(z) & \cdots & E_{M-1}(z) \\ z^{-1}E_{M-1}(z) & E_0(z) & \cdots & E_{M-2}(z) \\ \vdots & \vdots & \ddots & \vdots \\ z^{-1}E_1(z) & z^{-1}E_2(z) & \cdots & E_0(z) \end{pmatrix}. \tag{3}$$

with $R_i(z)$ and $E_i(z)$ being the polyphase components of $H(z)$ and $G(z)$ such that

$$H(z) = \sum_{i=0}^{M-1} E_i(z^M)z^{-i} \tag{4}$$

$$G(z) = \sum_{i=0}^{L-1} R_i(z^L)z^{-(L-i-1)}. \tag{5}$$

$\mathbf{E}(z)$ is called right pseudocirculant matrix and $\mathbf{R}(z)$ is called left pseudocirculant matrix. In Figure 4, $x_i(n) = x(nM - i + 1)$ for $1 \le i \le M$ and $y_j(n) = y(nL + L - j)$ for $1 \le j \le L$. Then, we have the following input/output relation:

$$\begin{pmatrix} Y_1(z) \\ \vdots \\ Y_L(z) \end{pmatrix} = \mathbf{R}(z)\mathbf{E}(z) \begin{pmatrix} X_1(z) \\ \vdots \\ X_M(z) \end{pmatrix} \tag{6}$$

where $X_i(z)$ and $Y_i(z)$ are the z-transform of $x_i(z)$ and $y_i(z)$ respectively.

For the notation convenience, we denote $\mathbf{C}(z) = \mathbf{R}(z)\mathbf{E}(z)$ and let $\mathbf{C}(z) = \sum_{i=0}^{l_c} C(i)z^{-i}$, where l_c is the order of $\mathbf{C}(z)$. Similarly denote by l_e the order of $\mathbf{E}(z)$ and by l_r the order of $\mathbf{R}(z)$. The orders are related as

$$\lceil l_h + 1 \rceil / M - 1 \leq l_e \leq \lceil l_h + 1 \rceil / M \quad (7)$$

$$\lceil l_g + 1 \rceil / L - 1 \leq l_r \leq \lceil l_g + 1 \rceil / L \quad (8)$$

$$l_c \leq l_e + l_r \quad (9)$$

where $\lceil x \rceil$ denotes the largest number less than or equal to x. If we assume l_g and l_h are known, the upper bounds of l_e, l_r and l_c are known. In the blind identification algorithms presented in later sections, we can use the upper bounds as their orders and this will not affect the effectiveness of the algorithms.

In this paper, we make the following assumptions:

(A1) $L > M$.

(A2) The input signal $x(n)$ is zero mean i.i.d. stationary process with nonzero fourth-order kurtosis.

(A3) The noise sequences $w_i(n)$ are Gaussian and are independent of $x(n)$.

(A4) $g(0)$ and $h(0)$ are nonzero.

(A5) the set of $\{R_i(z)\}$ are coprime.

(A1) ensures enough information is relayed to achieve channel identification. (A2) and (A3) are common in blind identification via higher order statistics. (A4) is crucial to unravel $\mathbf{R}(z)$ and $\mathbf{E}(z)$ [6]. (A4) requires the first element of no delay in $H(z)$ and $G(z)$. *This condition can however, be easily relaxed.*

3. BLIND CHANNEL IDENTIFICATION METHOD

In this section, we focus on the procedure of blind identification of $H(z)$ and $G(z)$. The procedure can identify $H(z)$ and $G(z)$ up to a nonzero constant and is divided into three steps:

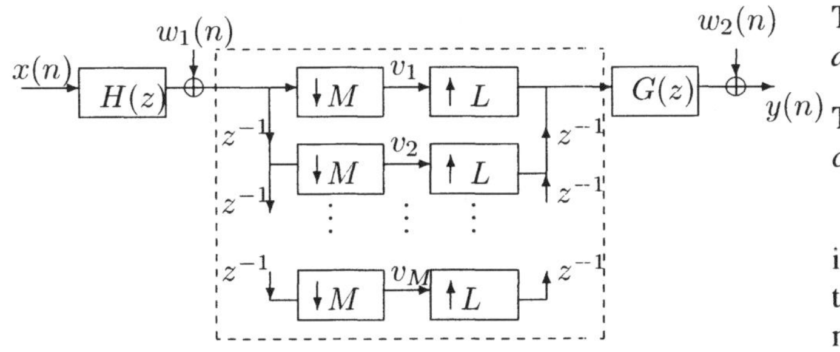

Figure 3: System model.

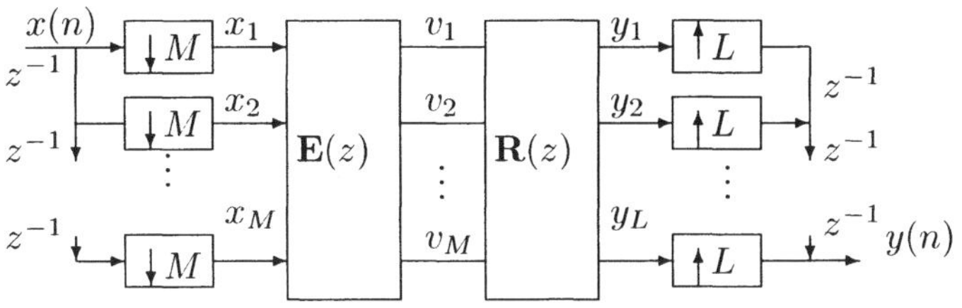

Figure 4: Polyphase Representation.

1. Identification of $\mathbf{C}(z)$.
2. Identification of $G(z)$.
3. Identification of $H(z)$.

In the following three subsections, we will discuss the implementation of the three steps respectively.

3.1. The algorithm of identification of $\mathbf{C}(z)$

Note that $\mathbf{E}(z)$ is an $M \times M$ pseudo-circulant matrix. In general SOS based methods can provide $\mathbf{C}(z)$ to within certain ambiguities only if $\mathbf{C}(z)$ is minimum phase, [4]. However, [1], the zeros of $\mathbf{R}(z)$ and $\mathbf{E}(z)$ cannot be minimum phase if respectively, $G(z)$ and $H(z)$ are nonminimum phase. Thus SOS based methods are not adequate for the estimation task at hand.

Note that in the literature, some methods explore HOS for blind identification of MIMO channels [3] [7] [8] [5]. We are interested to blindly identify $\mathbf{C}(z)$ using the method in [3], which is called LD method in this paper. LD method can identify $\mathbf{C}(z)$ up to a post-multiplication of a monomial matrix[1] based on the following assumptions in addition to assumptions (A1)-(A3):

(A6) $C(0)$ contains no all-zero columns.

(A7) There exists a non-zero z_0 (which may be ∞), such that $\mathbf{C}(z_0)$ has full column rank.

The following theorems show that assumptions (A6) and (A7) can be satisfied.

Theorem 1 *If $g(0)$ and $h(0)$ are nonzero, $C(0)$ contains no all-zero columns.*

Theorem 2 *If $H(z)$ and $G(z)$ are nonzero, $\mathbf{R}(z)$ is of full column rank and $\mathbf{E}(z)$ is of full row rank [6].*

Theorem 2 is presented in [6]. Thus $\mathbf{C}(z) = \mathbf{R}(z)\mathbf{E}(z)$ is full rank and (A7) is satisfied. Therefore, under assumptions (A1)-(A4), LD method identifies $\mathbf{C}(z)$ up to a post-multiplication of a monomial matrix P. That is,

$$\hat{\mathbf{C}}(z) = \mathbf{C}(z)P \quad (10)$$

[1] A monomial matrix is the product of a permutation matrix and a non-singular diagonal matrix.

where $\hat{\mathbf{C}}(z)$ is estimated by LD method.

For general LTI $\mathbf{C}(z)$ this mononmial ambiguity cannot be removed. However, the fact that $\mathbf{C}(z)$ is a product of pseudo-circulant matrices, is shown below to reduce this ambiguity to within a scaling constant.

3.2. The algorithm of identification of $G(z)$

For an $L \times M$ polynomial matrix $\mathbf{C}(z) = \sum_{i=0}^{l_c} C(i)z^{-i}$, define the $mL \times (m + l_c)M$ generalized Sylvester matrix of $\mathbf{C}(z)$ as

$$\mathcal{T}_m(\mathbf{C}) \triangleq \begin{pmatrix} C(0) & \cdots & C(l_c) & & \\ & \ddots & \ddots & \ddots & \\ & & C(0) & \cdots & C(l_c) \end{pmatrix}. \quad (11)$$

Then from $\hat{\mathbf{C}}(z) = \mathbf{C}(z)P = \mathbf{R}(z)\mathbf{E}(z)P$, we have

$$\mathcal{T}_m(\hat{\mathbf{C}}) = \mathcal{T}_m(\mathbf{R})\mathcal{T}_{m+l_r}(\mathbf{E})\mathbf{P} \quad (12)$$

where $\mathbf{P} = \text{diag}\{P, \cdots, P\}$. Since $\mathbf{E}(z)$ and P are full rank matrices, $\mathcal{T}_{m+l_r}(\mathbf{E})\mathbf{P}$ is of full row rank. Hence the left nullspace of $\mathcal{T}_m(\hat{\mathbf{C}})$ is the same as that of $\mathcal{T}_m(\mathbf{R})$. Using the left nullspace of $\mathcal{T}_m(\hat{\mathbf{C}})$, we can estimate $\mathcal{T}_m(\mathbf{R})$ and thus $G(z)$. This method is shown in [6], where assumption (A5) is required to identify $G(z)$ up to a nonzero constant.

3.3. The algorithm of identification of $H(z)$

After the estimation of $\hat{\mathbf{C}}(z)$ and $\mathbf{R}(z)$, $\mathbf{E}(z)$ can be estimated. Note that

$$\hat{\mathbf{C}}^T(z) = P^T\mathbf{E}^T(z)\mathbf{R}^T(z) \quad (13)$$

$$\mathcal{T}_m(\hat{\mathbf{C}}^T) = \mathbf{P}^T\mathcal{T}_m(\mathbf{E}^T)\mathcal{T}_{m+l_e}(\mathbf{R}^T). \quad (14)$$

Since $\mathbf{R}(z)$ is of full column rank, $\mathcal{T}_{m+l_e}(\mathbf{R}^T)$ is of full row rank. It follows that

$$\mathbf{P}^T\mathcal{T}_m(\mathbf{E}^T) = \mathcal{T}_m(\hat{\mathbf{C}}^T)\mathcal{T}_{m+l_e}(\mathbf{R}^T)^\dagger \quad (15)$$

where † stands for pseudo-inverse. From (15), $\mathbf{E}(z)$ can be estimated as

$$\hat{\mathbf{E}}(z) = \mathbf{E}(z)P. \quad (16)$$

In the following, we will show that the monomial matrix P can be reduced to a nonsingular diagonal matrix and hence $H(z)$ can be identified up to a nonzero constant.

Theorem 3 *Suppose $\mathbf{E}(z)$ and $\hat{\mathbf{E}}(z)$ are nonzero right pseudocirculant matrices with the same dimension. If there exists a monomial matrix P such that $\hat{\mathbf{E}}(z) = \mathbf{E}(z)P$, then $P = cI$, where c is a nonzero constant and I is the identity matrix.*

Proof: Since $\mathbf{E}(z)$ is a pseudocirculant matrix, it can be expressed as (3). Suppose $E_i(z) = \sum_{j=0}^{l_e-1} E_{ij}z^{-j}$. Let $\mathbf{E}(z) = \sum_{j=0}^{l_e} D(j)z^{-j}$, where $D(j)$ is the coefficient matrix of $\mathbf{E}(z)$ of order z^{-j}. Then

$$D(j) = \begin{pmatrix} E_{0j} & E_{1j} & \cdots & E_{(M-1)j} \\ E_{(M-1)(j-1)} & E_{0j} & \cdots & E_{(M-2)j} \\ \vdots & \vdots & \ddots & \vdots \\ E_{1(j-1)} & E_{2(j-1)} & \cdots & E_{0j} \end{pmatrix} \quad (17)$$

for $0 \le j \le l_e$, where $E_{ij} = 0$ for $j = -1$ or $j = l_e$. By stacking $D(l_e), \cdots, D(0)$, we have $D = [D(l_e)^T, \cdots, D(0)^T$ We can see D is a toeplitz matrix whose first row is an M-dimensional zero vector and first column is an l_eM-dimensional vector as follows

$$d_1 = [0, e_{l_e-1}, e_{l_e-2}, \cdots, e_0, \underbrace{0, \cdots, 0}_{M-1 \text{ elements}}]^T \quad (18)$$

where $e_j = [E_{(M-1)j}, \cdots, E_{0j}]$.

Similarly, if we stack the coefficient matrices of $\hat{\mathbf{E}}(z)$, we will have a toeplitz matrix $\hat{D}$ with the same structure as that of D. Denote the d_{ij} (resp. $\hat{d}_{ij}$) the elements of D (resp. $\hat{D}$) at i-th row and j-th column. Suppose d_{k1} is the first nonzero element in column 1 of D. Then the first nonzero element in column i of D is $d_{(k+i-1)i}$. Due to the toeplitz structure of D, We have

$$d_{(k+i-1)i} = d_{(k+j-1)j}, 1 \le i, j \le M. \quad (19)$$

Suppose r_i-th element in i-th column of P is a nonzero number p_i. Because P is monomial matrix, $r_i \ne r_j$ for $i \ne j$. From $\hat{\mathbf{E}}(z) = \mathbf{E}(z)P$, we have $\hat{D} = DP$. By computing DP, it can be shown the first nonzero element of column i of $\hat{D}$ is $d_{(k+r_i-1)r_i}p_i$ which is the $(k + r_i - 1)$-th element of column i of $\hat{D}$. Because $\hat{D}$ is a toeplitz matrix, we have

$$d_{(k+r_i-1)r_i}p_i = d_{(k+r_j-1)r_j}p_j \quad (20)$$

$$r_i = r_{i+1} - 1, 1 \le r_i \le M. \quad (21)$$

From (19) and (20), we have $p_i = p_j = c$. From (21), we have $r_i = i$. Therefore, $P = cI$. ∎

The proof of theorem 3 explores the structure of $\mathbf{E}(z)$. By stacking the coefficient matrices of $\mathbf{E}(z)$, we will have a toeplitz matrix D with the first column d_1 as shown in (18). The first column d_1 contains all the coefficients of $\{E_i(z)\}$ and hence those of $H(z)$. The identification algorithm is to estimate d_1 from $\hat{\mathbf{E}}(z)$ and shown as follows.

Let $\hat{\mathbf{E}}(z) = \sum_{i=0}^{l_e} \hat{D}(i)z^{-i}$ and $\hat{D} = [\hat{D}(l_e)^T, \cdots, \hat{D}(0)^T$ Denote by $\hat{d}_{ij}$ the element of $\hat{D}$ at i-th row and j-th column.

First, compute

$$Q_i(k) = \sum_{j=1}^{l_e M} \hat{d}_{1j} \hat{d}^*_{i(j+k)}, \quad (22)$$

for $-(M-1) \leq k \leq (M-1), 1 < i \leq M$

where $\hat{d}_{ij} = 0$ for $j < 1$ or $j > l_e M$ and * indicates conjugate. Find

$$\tau_i = \arg \max_k Q_i(k) \quad (23)$$

$$m = \arg \min_i \tau_i. \quad (24)$$

Then m-th column vector of $\hat{D}$ is d_1 multiplied by an unknown factor. Thus $H(z)$ can be constructed from d_1 as shown in (18) and (4).

4. CONCLUSION

This paper proposes a new model for blind identification of two cascading single input single output systems by using a scrambling technique and higher order statistics. Identification conditions and algorithms are presented. The theory above assumed no delay in either system. In practice the method easily extends to the case where the total input output delay is known. In that case the ambiguity in estimating the $H(z)$ and $G(z)$, extends from a scalar ambiguity to one that includes an unknown delay.

5. REFERENCES

[1] P. P. Vaidyanathan, *Multirate Systems and Filter Banks*, Prentice Hall, 1992.

[2] L. Tong, G. Xu and T. Kailath, "Blind identification and equalization based on second-order statistics: A time-domain approach", *IEEE Trans. Information Theory*, vol. 40 no. 2, pp. 340-350, March 1994.

[3] J. Liang and Z. Ding, "A simple cumulant based aproach for multiuser channel identification", *Proc. IEEE ISCAS 2002*, vol. 3, pp. 659-662, 2002.

[4] Z. Ding and Y. Li, *Bind equalization and identification*, Marcel Dekker, Inc., 2001.

[5] L. Tong, "Identification of multichannel MA parameters using higher-order statistics", *Signal Processing*, vol. 53, pp. 195-209, 1996.

[6] H. Xu, S. Dasgupta, and Z. Ding, "An improved feedback scheme for dual channel identification in wireless communication systems", *Proc. of ICASSP*, 2003.

[7] Y. Inouye, G. B. Giannakis and J. M. Mendel, "Cumulant based parameter estimation of multichannel moving-average processes" *Proc. ICASSP-88*, vol. 2, pp. 1252 -1255, 1988.

[8] Y. Inouye and K. Hirano, "Cumulant-based blind identification of linear multi-input-multi-output systems driven by colored inputs", *IEEE Trans. on Signal Processing*, vol. 45, issue 6, pp. 1543 -1552, Jun 1997

IFAC Publications
www.elsevier.com/locate/ifac

BLIND CHANNEL SHORTENERS

C. R. Johnson, Jr., R. K. Martin, J. M. Walsh, A. G. Klein, C. E. Orlicki, and T. Lin [1]

School of Electrical and Computer Engineering
Cornell University, Ithaca, NY 14853

Abstract: Although blind, adaptive algorithms for equalization are widely studied, hitherto there has been little academic attention given to blind, adaptive algorithms for channel shortening. Channel shortening is needed to preserve subcarrier orthogonality in multicarrier modulation, and it can be used to dramatically reduce the complexity of maximum likelihood sequence estimation and multiuser detection. This paper reviews the channel shortening problem from a tutorial perspective, and shows how it is an extension of traditional equalization. It is shown that traditional methods of devising blind, adaptive equalization algorithms cannot be easily applied to the channel shortening problem. The paper concludes with a discussion of several new property restoral algorithms that enable blind, adaptive channel shortening. *Copyright © 2003 IFAC*

Keywords: Blind, Adaptive, Channel Shortening, Equalization, Inverse Control

1. INTRODUCTION

The ordinary objective of communication channel equalization is to reproduce at the equalizer output a delayed version of the channel input sequence, with the channel output the equalizer input. When using a linear, baud-spaced equalizer, this translates, in the channel-noise-free case, to a channel and equalizer combination with a transfer function of $z^{-\delta}$ with δ a positive integer and z^{-1} a unit sample delay.

In various applications, such as multicarrier communication systems for wired (e.g. DMT for xDSL) and wireless (e.g. OFDM for DVB and WLAN) scenarios, the impulse response of the channel equalizer combination need not be just a single nonzero term. Instead a channel-equalizer combination with an impulse response of suitably limited duration, e.g. $\sum_{i=0}^{P} h_i z^{-(i+\delta)}$, can be adequate for desired system performance. The specific values of the h_i are less important (as long as some are nonzero) than the fact that outside this $P+1$ sample window the impulse response coefficients are all (nearly) zero. Generally, channel-shortening can be achieved with a shorter equalizer and with less noise gain than equalization to a single spike. An extreme example is a finite impulse response (FIR) channel containing a zero on the unit circle. A zero-forcing linear FIR equalizer would have infinite gain at the frequency of the channel null, thereby catastrophically amplifying any channel noise. However, the single zero on the unit circle could be retained in a shortened channel.

While adaptive channel shorteners relying on training have been developed and implemented, prior to last year (2002) no blind method for direct adaptive channel shortening existed. This paper describes the channel-shortening problem formulation as a metamorphosis from a traditional channel equalization problem first into a model-

[1] This work was supported in part by Applied Signal Technology.

following inverse controller, then into a channel-shortener. Trained adaptive algorithms for the linear combiner format that emerges in each case are readily proposed. The inappropriateness of establishing blind channel shortening schemes based on decision-direction or dispersion minimization, which were successful in producing blind channel equalizers, will become apparent.

This paper discusses several blind, adaptive channel shorteners proposed in 2002. These algorithms arose from exploitation of transmitted signal properties found in practical communication systems; in one case whiteness (Balakrishnan *et al.*, 2002) and in the other an inclusion of replicated segments (Martin *et al.*, 2002*a*). Using a classical stochastic gradient descent approach, these algorithms are based on cost functions that penalize deviation from some property of the signal to be recovered, and require only the equalizer input and output to form their parameter updates (and are therefore considered blind).

Other researchers have proposed blind, adaptive algorithms for systems that require channel shortening, although these efforts have focused on equalization, rather than channel shortening. However, since channel shortening is the goal rather than equalization, the use of an equalizer is expected to yield suboptimal performance compared to a channel shortener. In (Jones, 2003), a time domain approach similar to (Martin *et al.*, 2002*a*) was proposed, but the algorithm leads to full equalization rather than just channel shortening. (de Courville *et al.*, 1996) makes use of the common practice with OFDM of transmission of zeros on some carriers as a substitute for training data on those carriers, again leading to a single-spike equalization algorithm. (Romano and Barbarossa, 2003) extended this method through the use of frequency-hopping in the transmitter to allow for blind, adaptive channel shortening.

2. ADAPTIVE PARAMETER ESTIMATION PROBLEM MUTATION

We keep the description basic in drawing a thread from traditional channel equalization to channel shortening via model-following. The intent is to draw attention to the issue of adaptive channel shortening and the need for new approaches to blind solutions.

2.1 Traditional Channel Equalization

In traditional channel equalization the goal is to process the channel output sequence to reproduce the channel input sequence, as depicted in Figure 1. The source s, a delayed version of which is to

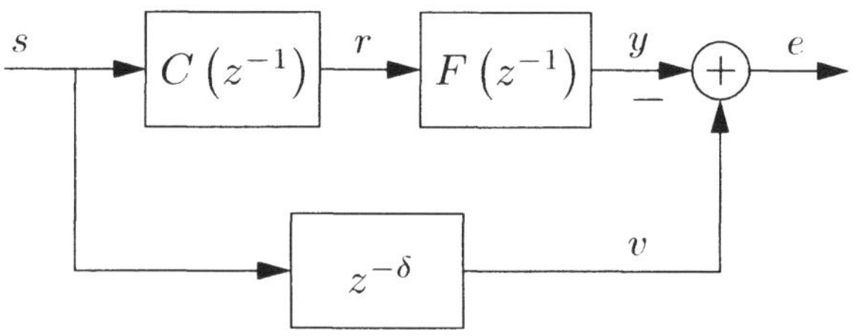

Fig. 1. Traditional Channel Equalization

be recovered so that $\mathrm{E}[e^2]$ is minimized, takes on values from a discrete set, e.g. $\{\pm 1\}$ or $\{\pm 1, \pm 3\}$. The linear channel is presumed to have a causal, FIR model

$$r(k) = \sum_{j=0}^{M-1} c_j s(k-j) \tag{1}$$

with transfer function $C(z^{-1})$. The output of the channel is the received signal in the noise-free idealization of Figure 1. The equalizer also has a causal FIR model

$$y(k) = \sum_{i=0}^{N-1} f_i r(k-i) \tag{2}$$

with transfer function $F(z^{-1})$ that filters the received signal $r(k)$ and produces the output $y(k)$ we would like to have match $s(k-\delta)$. The delayed source recovery error $e(k)$ for delay δ and f_i values at time k is

$$e(k) = s(k-\delta) - y(k) = s(k-\delta) - \sum_{i=0}^{N-1} f_i r(k-i) \tag{3}$$

The recovery error description of (3) fits the format of a linear combiner's prediction error

$$e(k) = d(k) - X^T(k)\theta(k) \tag{4}$$

with its difference between the desired combiner output d and its actual output formed by the inner product of the regressor vector X and the adapted parameter vector θ. The definitions that match (3) to (4) are $d(k) = s(k-\delta)$,

$$X(k) = [r(k) \;\; r(k-1) \;\; ... \;\; r(k-N+1)]^T, \tag{5}$$

and

$$\theta(k) = [f_0(k) \;\; f_1(k) \;\; ... \;\; f_{N-1}(k)]^T. \tag{6}$$

For the linear combiner error of (4), the LMS algorithm (Widrow and Stearns, 1985)

$$\theta(k+1) = \theta(k) + \mu e(k) X(k) \tag{7}$$

is a classic adaptive approach to minimization of $\mathrm{E}[e^2]$.

Implementing (7) requires knowledge of $s(k-\delta)$ at the recevier. This is accomplished by the (periodic) transmission of a prearranged training sequence. We can use the property that the source signal takes on only certain values, e.g. ± 1, to produce blind alternatives that do not require explicit knowledge at the receiver of $s(k-\delta)$.

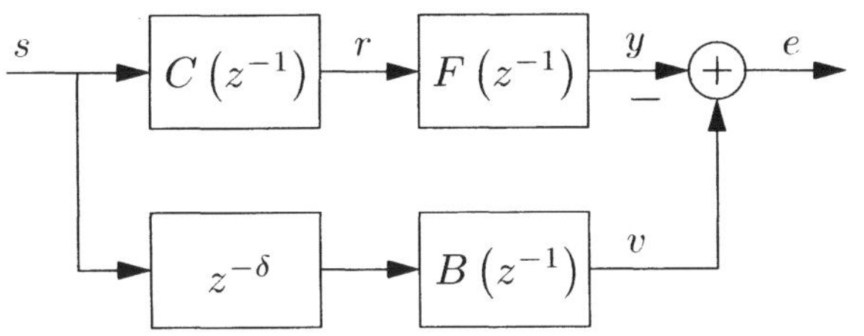

Fig. 2. Model-Following

One strategy presumes that the (initial) equalizer setting, while not perfectly recovering $s(k-\delta)$, generates a signal $y(k)$ that is close to $s(k-\delta)$. In fact, we assume that it is close enough so a quantizer produces error-free delayed source recovery, i.e. $Q[y(k)] = s(k-\delta)$. For $s \in \{\pm 1\}$, a sign operator serves as the quantizer Q, and sign$[y(k)]$ replaces $s(k-\delta)$ in e in (3) which is substituted in (7) to produce the blind decision-directed LMS (DDLMS) algorithm

$$\theta(k+1) = \theta(k) + \mu\left(Q[y(k)] - y(k)\right)X(k). \quad (8)$$

Another blind channel equalization algorithm can be created by differently exploiting the discrete alphabet nature of the source to be recovered. For example, with s either 1 or -1, s^2 is always 1. This suggests the cost function $\mathrm{E}[(1-y^2(k))^2]$. A stochastic gradient descent minimizing this dispersion cost yields the constant modulus algorithm (CMA) (Treichler *et al.*, 2001, Ch. 6)

$$\theta(k+1) = \theta(k) + \mu(1-y^2(k))y(k)X(k) \quad (9)$$

where $y(k) = X^T(k)\theta(k)$. CMA in (9) is termed blind because it does not need knowledge of the specific source sequence $\{s\}$ that $\{y\}$ is to match. Dispersion minimization also works for multilevel discrete alphabet sources.

2.2 Model-Following Inverse Control

Now we will mutate the problem into a version reminiscent of inverse control (Widrow and Stearns, 1985, Ch. 11). Rather than cause the transfer function of the channel-equalizer combination $C(z^{-1})F(z^{-1})$ to equal $z^{-\delta}$, we will choose to have it match a prespecified model transfer function $B(z^{-1})$, as in Figure 2. Writing the error as

$$e(k) = v(k) - \sum_{i=0}^{N-1} f_i r(k-i) \quad (10)$$

reveals a linear combiner format. With this e, but the same definitions of θ and X as in the traditional channel equalization problem, LMS in (7) provides a trained adaptive solution.

The blind algorithms DDLMS and CMA do not survive the mutation because the signal to be matched by y, i.e. v in Figure 2, is not drawn from the same alphabet as the source.

2.3 Channel Shortening

We again mutate the problem. Though the model-following schematic of Figure 2 still applies, we no longer preselect $B(z^{-1})$. Instead the objective is only to have the channel-equalizer combination FIR response have nonzero values only in a window P samples wide. This task only presents a challenge if P is less than the sample width M of the channel. Hence, the labeling of this task as channel shortening.

If we judge our success by minimizing the mean of the square of the model-following error e in Figure 2, we must not admit the trivial solution with the coefficients of both $F(z^{-1})$ and $B(z^{-1})$ set to zero. A simple way to enforce this constraint is to fix $b_\zeta = 1$ for some ζ. Thus,

$$\begin{aligned} e(k) &= \sum_{j=0}^{P-1} b_j(k)s(k-j-\delta) - \sum_{i=0}^{N-1} f_i r(k-i) \\ &= s(k-\zeta-\delta) + \sum_{j=0,j\neq\zeta}^{P-1} b_j(k)s(k-j-\delta) \\ &\quad - \sum_{i=0}^{N-1} f_i r(k-i) \\ &= s(k-\zeta-\delta) - X^T(k)\theta(k) \end{aligned} \quad (11)$$

where

$$\begin{aligned} X(k) = [&-s(k-\delta), \ \ldots, \ -s(k-\delta-\zeta+1), \\ &-s(k-\delta-\zeta-1), \ \ldots, \ -s(k-\delta-P+1), \\ &r(k), \ \ldots, \ r(k-N+1)]^T \end{aligned} \quad (12)$$

and

$$\theta(k) = [b_0, \ldots, b_{\zeta-1}, b_{\zeta+1}, \ldots, b_{P-1}, f_0, \ldots, f_{N-1}]^T \quad (13)$$

Under the assumption that the preselection of δ and ζ leads to an acceptable solution, these definitions of e, X, and θ and (7) provide a trained adaptive channel shortener similar to that in (Falconer and Magee, 1973). But, just as noted in the case of model-following, the desired output y is no longer drawn from the source alphabet and loses the finite-alphabet property exploited by decision direction and dispersion minimization. Can other signal properties can be used to establish a blind channel shortener?

3. THE NEED FOR BLIND CHANNEL SHORTENING

A primary application of channel shortening occurs in multicarrier communication systems (Pollet *et al.*, 2000) present in wired digital subscriber loop (DSL) and wireless local area network (LAN) standards. These standards use a cyclic prefix in the transmitted sequence. Each data block of a prespecified length has its last several samples prepended to the front of the block prior to transmission. Intersymbol (and intercarrier) interference is removed as long as the length of this cyclic prefix segment is greater than the delay spread of the channel-equalizer combination. Adaptation will be needed if the channel is time-varying. Blind adaptation will be desirable if the downtime for (re)training represents a large portion of system operating activity.

Currently, in the DSL application the cyclic prefix set in transmission standards is acknowledged to be much shorter than the channel delay spread encountered in practice. This substantiates the use of a channel shortener. However, the channel is presumed time invariant (aside from glacially-paced, temperature-induced, performance-degrading channel parameter drifts which occur in practice), which obviates the need for periodic retraining and the primary motivation for a blind channel shortener.

In the wireless LAN application, there is general agreement that the channel model is time-varying. However, so far, the belief is that such schemes will not be deployed in circumstances in which the channel delay spread exceeds the standards-mandated cyclic prefix. Thus, while adaptation would be needed in a channel shortener, and a blind scheme would be welcome, the need for a channel shortener is typically denied.

The compelling need for a blind channel shortener remains in the future. Based on a similar experience in the adoption of blind implementation of traditional channel equalization, our contention is that such a need will arise. The previous lack of blind channel shortening schemes and our belief in their eventual utility motivated our search for candidate algorithms for blind channel shortening.

4. TWO PROPERTY RESTORAL ALGORITHMS

Two signal properties that can be exploited include the traditonally assumed whiteness of the transmitted source sequence (especially common with coded and scrambled signals) and the cyclic prefix mentioned that is common with multicarrier system application of a channel shortener. Restoring each of these two properties generates a candidate blind channel shortener.

4.1 The MERRY and FRODO Algorithms

The MERRY algorithm (Multicarrier Equalization by Restoration of Redundancy) makes use of the redundancy of the data introduced by the

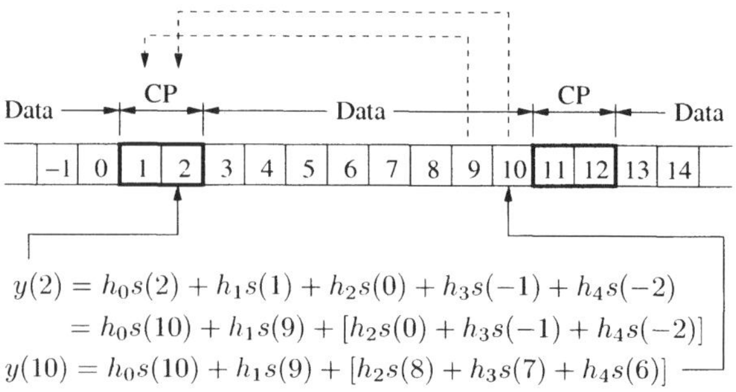

Fig. 3. Illustration of the difference in the ISI at the received CP and at the end of the received symbol.

cyclic prefix (CP). When the CP is added, the last P samples in the length N block are prepended to the start of the block, which makes the transmitted data appear periodic over those $N + P$ samples (called a "symbol"). This periodicity is necessary for maintaining the orthogonality of the subcarriers (Pollet *et al.*, 2000). After the CP is added, the last P samples are identical to the first P samples in the symbol, i.e.

$$s\left(Mk+i\right) = s\left(Mk+i+N\right), \quad i \in \{1, \ldots, P\}, \tag{14}$$

where $M = N + P$ is the total symbol duration and k is the symbol index. Figure 3 shows an example of this, with $N = 8$, $P = 2$, and $M = N + P = 10$. The symbol pictured is for $k = 0$. The received data $\mathbf{r}$ is obtained from $\mathbf{s}$ by

$$r(Mk+i) = \sum_{l=0}^{L_c} c_l \cdot s(Mk+i-l) + n(Mk+i), \tag{15}$$

and the equalized data $\mathbf{y}$ is obtained from $\mathbf{r}$ by

$$y(Mk+i) = \sum_{j=0}^{L_f} f_j \cdot r(Mk+i-j). \tag{16}$$

The effective channel is $\mathbf{h} = \mathbf{c} \star \mathbf{f}$.

The channel destroys the relationship analogous to (14) in the received data, because the ISI that affects the CP is different from the ISI that affects the last P samples in the symbol. Consider the example in Figure 3. The transmitted samples 2 and 10 are identical. However, at the receiver, the interfering samples before sample 2 are not all equal to their counterparts before sample 10. If h_2, h_3, and h_4 were zero, then $y(2) = y(10)$. Trying to force $y(2) = y(10)$ should force $h_2 = h_3 = h_4 = 0$, thus forcing the effective channel to be as short as the CP. The location of the window of P non-zero taps can be varied by comparing $y(3)$ to $y(11)$, or $y(4)$ to $y(12)$, etc.

The MERRY cost function is

$$J = \mathrm{E}\left|y(Mk+P+\delta) - y(Mk+P+N+\delta)\right|^2, \quad \delta \in \{0, \ldots, M-1\}, \tag{17}$$

where δ is the desired delay. A stochastic gradient descent of (17) leads to the blind, adaptive MERRY algorithm:

$$\begin{aligned}
&\text{For symbol } k = 0, 1, 2, \ldots, \\
&\quad \tilde{\mathbf{r}}(k) = \mathbf{r}(Mk + P + \delta) \\
&\qquad\qquad - \mathbf{r}(Mk + P + N + \delta) \\
&\quad e(k) = \mathbf{f}^T(k)\, \tilde{\mathbf{r}}(k) \\
&\quad \hat{\mathbf{f}}(k+1) = \mathbf{f}(k) - \mu\, e(k)\, \tilde{\mathbf{r}}^*(k) \\
&\quad \mathbf{f}(k+1) = \frac{\hat{\mathbf{f}}(k+1)}{\|\hat{\mathbf{f}}(k+1)\|}
\end{aligned} \tag{18}$$

where $\mathbf{r}(i) = [r(i), r(i-1), \ldots, r(i-L_f)]^T$, and * denotes complex conjugation. Note that a constraint (e.g. $\|\mathbf{f}\| = 1$) must be enforced in order to prevent the trivial solution $\mathbf{f} = \mathbf{0}$.

The MERRY algorithm in (18) finds the minimum eigenvector of the matrix $\mathbf{A} = \mathrm{E}\left[\tilde{\mathbf{r}}\tilde{\mathbf{r}}^H\right]$. Let $\mathbf{C}$ be the channel convolution matrix (so that $\mathbf{h} = \mathbf{Cf}$) and let $\mathbf{C}_{wall}$ be obtained from $\mathbf{C}$ by removing rows δ through $\delta + P - 1$. If the input $s(k)$ is white, then $\mathbf{A} = \mathbf{C}_{wall}^T \mathbf{C}_{wall}$, and

$$J_\delta = 2\,\sigma_s^2 \left(\sum_{j=0}^{\delta-1} |h_j|^2 + \sum_{j=\delta+P}^{L_h} |h_j|^2 \right) + 2\,\mathbf{f}^T \mathbf{R}_n \mathbf{f}^*, \tag{19}$$

the energy of the effective channel outside the window plus the noise gain (Martin *et al.*, 2002*b*).

If a blind, non-adaptive channel shortener is required, then the matrix $\mathbf{A} = \mathrm{E}\left[\tilde{\mathbf{r}}\tilde{\mathbf{r}}^H\right]$ can be estimated from the data, and its eigenvector corresponding to its minimum eigenvalue can be computed. This may also provide an initialization technique that avoids slow modes of convergence.

MERRY can be extended to compare multiple samples in the CP to multiple samples at the end of the symbol. As shown in Figure 4, each difference term that is added to the cost function produces a different window with a different delay, and the (somewhat smaller) overall window is the union of the individual windows. This allows the option of using more data, increasing the convergence rate at the expense of over-shortening the channel. This cousin to MERRY is called Forced Redundancy with Optional Data Omission (FRODO). Figure 4 shows an example in which $P = 4$. In this case, comparing three of the four points in the CP to their brethren at the end of the symbol yields a union of three "don't care" windows, in which the impulse reponse doesn't matter so long as it is non-zero. Thus, three times as much data can be used per update, but the shortened channel will be much smaller than necessary. Since the best solution can be found by constraining the filter as little as possible, the "over-shortened" FRODO solution is sub-optimal.

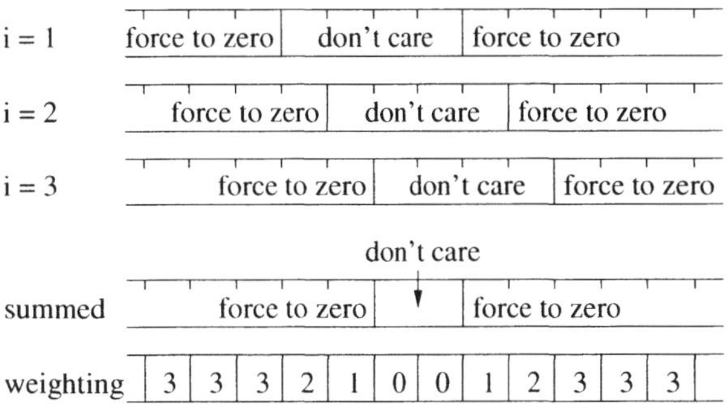

Fig. 4. The relation of the "don't care" windows in the different terms of the FRODO cost function, for $P = 4$. The line "summed" indicates the effect of considering three terms at once, and the line "weighting" indicates how much emphasis the total cost function places on forcing each tap to zero.

4.2 The SAM Algorithm

The SAM algorithm (Sum-squared Auto-correlation Minimization) relies on fourth-order statistics of the received data rather than on properties of multicarrier modulation. The idea is that if the effective channel ($\mathbf{h} = \mathbf{c} \star \mathbf{w}$) is short, its autocorrelation should be short:

$$R_h(l) = \sum_{k=0}^{L_h} h_k h_{k-l} \cong 0, \quad |l| > P. \tag{20}$$

This suggests the cost function

$$\hat{J} = \sum_{l=P+1}^{L_h} |R_h(l)|^2, \tag{21}$$

again with a constraint such as $\|\mathbf{f}\| = 1$ to prevent $\mathbf{f} = \mathbf{0}$. If the source $s(k)$ and channel noise $w(k)$ are white, and if $L_f \leq P$, then

$$R_y(l) = \mathrm{E}[y(n)y(n-l)] = R_h(l), \quad |l| > P, \tag{22}$$

allowing use of the cost function

$$J_{sam} = \sum_{l=P+1}^{L_h} |R_y(l)|^2. \tag{23}$$

If $L_f > P$, (22) is still approximately true so long as the noise is small.

A gradient descent of (23) leads to

$$\begin{aligned}
\mathbf{f}(n+1) = \mathbf{f}(n) - \mu \sum_{l=P+1}^{L_h} & \mathrm{E}\left[y(n)y(n-l)\right] \\
& \cdot \mathrm{E}\left[y(n)\mathbf{r}_{n-l} + y(n-l)\mathbf{r}_n\right]
\end{aligned} \tag{24}$$

For implementation, the expectations can be replaced with instantaneous, moving average, or auto-regressive estimates. The latter yields the fastest convergence rate for the lowest computational cost (Balakrishnan *et al.*, 2002).

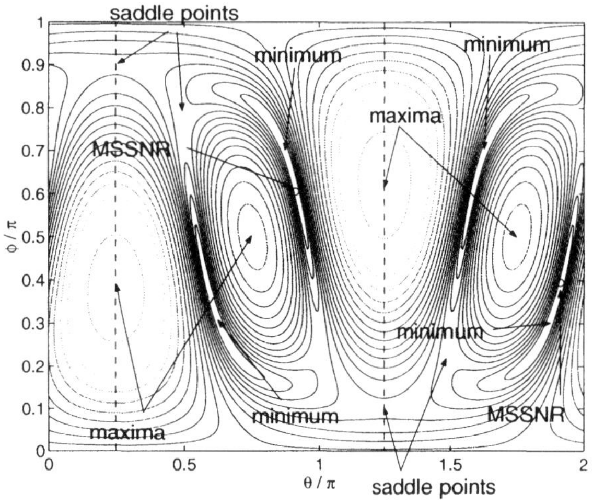

Fig. 5. Contours of the SAM cost function. The two circles are the global maxima of the shortening SNR.

Since (23) involves fourth-order statistics of the data, it is multimodal and difficult to analyze. The auto-correlation is invariant to flipping the filter's zero locations over the unit circle, so there are as many as 2^{L_f} minima that all have the same value of the SAM cost. However, they have very different values of whetever the true cost function is, e.g. MSE, bit rate, or bit error rate. When the filter length is reasonably long[2], the convergence of SAM does not appear to be troubled by the multimodality. One odd effect is that the filter $\mathbf{f}$ is generally symmetric, perhaps because time-reversing $\mathbf{f}$ is equivalent to flipping its zeros over the unit circle (which does not change the cost).

A plot of the SAM cost function is shown in Figure 5. The channel is $\mathbf{c} = [1, 0.3, 0.2]$, $P = 1$ (so a 2-tap channel is desired), there is no noise, and the 3-tap filter $\mathbf{f}$ satisfies $\|\mathbf{f}\| = 1$. $\mathbf{f}$ can be represented in spherical coordinates as $f_0 \stackrel{\triangle}{=} f_x = \cos(\theta)\sin(\phi)$, $f_1 \stackrel{\triangle}{=} f_z = \cos(\phi)$, $f_2 \stackrel{\triangle}{=} f_y = \sin(\theta)\sin(\phi)$. Then time-reversing $\mathbf{f}$ is equivalent to reflecting θ over $\frac{\pi}{4}$ or $\frac{5\pi}{4}$, and $\mathbf{f} \to -\mathbf{f}$ is equivalent to the combination of reflecting ϕ over $\frac{\pi}{2}$ and adding π to θ (mod 2π). The four minima all have equivalent values of the SAM cost, due to the equivalencies of $\mathbf{f} \Leftrightarrow -\mathbf{f}$ and of time-reversing $\mathbf{f}$.

The circles in Figure 5 correspond to the Maximum Shortening SNR (MSSNR) design (Melsa *et al.*, 1996), which maximizes the ratio of the energy inside the window to the energy outside the window. Two of the global minima of the SAM cost nearly match the global maxima of the SSNR. The other two minima can be avoided by reversing the order of taps in the final settings for $\mathbf{f}$.

[2] "Reasonably long" here means possibly shorter than the channel, but comfortably longer than the minimum length needed to achieve a "good" solution.

5. CONCLUSION

Although blind, adaptive equalization algorithms are widely studied, hitherto there has been little academic attention given to blind, adaptive channel shortening algorithms. Traditional approaches to making adaptive equalization algorithms blind cannot be applied to channel shortening, since a channel shortener's output has different signal properties than an equalizer's output. This paper has reviewed some of the salient points of several new blind channel shortening algorithms.

REFERENCES

Balakrishnan, J., R. K. Martin and C. R. Johnson, Jr. (2002). Blind, Adaptive Channel Shortening by Sum-squared Auto-correlation Minimization (SAM). In: *Proc. 36th Asilomar Conf. on Signals, Systems, and Computers.*

de Courville, M., P. Duhamel, P. Madec and J. Palicot (1996). Blind equalization of OFDM systems based on the minimization of a quadratic criterion. In: *Proc. Int. Conf. on Comm.* Dallas, TX. pp. 1318–1321.

Falconer, D. D. and F. R. Magee (1973). Adaptive Channel Memory Truncation for Maximum Likelihood Sequence Estimation. *Bell Sys. Tech. Journal* **52**(9), 1541–1562.

Jones, D. (2003). Property-Restoral Algorithms for Blind Equalization of OFDM. Submitted to *Proc. 37th Asilomar Conf. on Signals, Systems, and Comp.* Pacific Grove, CA.

Martin, R. K., J. Balakrishnan, W. A. Sethares and C. R. Johnson, Jr. (2002*a*). A Blind, Adaptive TEQ for Multicarrier Systems. *IEEE Signal Proc. Letters* **9**(11), 341–343.

Martin, R. K., J. Balakrishnan, W. A. Sethares and C. R. Johnson, Jr. (2002*b*). Blind, Adaptive Channel Shortening for Multicarrier Systems. In: *Proc. 36th Asilomar Conf. on Signals, Systems, and Comp.* Pacific Grove, CA.

Melsa, P. J. W., R. C. Younce and C. E. Rohrs (1996). Impulse Response Shortening for Discrete Multitone Transceivers. *IEEE Trans. on Comm.* **44**(12), 1662–1672.

Pollet, T., M. Peeters, M. Moonen and L. Vandendorpe (2000). Equalization for DMT-Based Broadband Modems. *IEEE Communications Magazine* **38**(5), 106–113.

Romano, F. and S. Barbarossa (2003). Non-data Aided Adaptive Channel Shortening for Efficient Multi-carrier Systems. In: *IEEE International Conf. on Accoustics, Speech, and Signal Processing.* Hong Kong.

Treichler, J. R., C. R. Johnson, Jr. and M. G. Larimore (2001). *Theory and Design of Adaptive Filters.* Prentice Hall. New York.

Widrow, B. and S. D. Stearns (1985). *Adaptive Signal Processing.* Prentice-Hall. Englewood Cliffs, N.J.

IFAC Publications
www.elsevier.com/locate/ifac

MULTIPLE ANTENNA SYSTEM EQUALIZATION USING SEMI-BLIND SUBSPACE IDENTIFICATION METHODS

Chengjin Zhang * **Robert R. Bitmead** **,[1]

* *Department of Electrical and Computer Engineering, University of California, San Diego, La Jolla, CA 92093-0407, USA*
** *Department of Mechanical and Aerospace Engineering, University of California, San Diego, La Jolla, CA 92093-0411, USA*

Abstract: In this paper, we investigate the application of the subspace system identification (SSI) method (e.g. N4SID) to the MIMO frequency-selective fading channel estimation problem. The FIR constraint on the MIMO channel model is suggested to be relieved to draw benefit from possible parsimonious parametrization of the MIMO channel when subchannels become correlated. Also, the criterion for training sequence selection for SSI-based MIMO channel estimation is analyzed. Considering that the formalism of optimal input design is inappropriate for training sequence solution, we suggest still to use the conventional white and spatially uncorrelated sequences for SSI-based (non-FIR) MIMO channel estimation, even if they might be suboptimal. A modification of the SSI methods and a semi-blind approach are proposed to address the issue that only non-contiguous block-wise training sequences are available in practical mobile communication systems.

Keywords: MIMO; Channel estimation; Subspace identification; Training sequences.

1. INTRODUCTION

Digital communication using multiple transmit and receive antennas has been one of the most important technical developments in modern communications. In a rich scattering environment, MIMO systems offer significant capacity gain at no cost of extra spectrum (Foschini and Gans, 1998). So far, most of the proposed MIMO transmission schemes assume channel state information (CSI) is known at the receiver. Therefore, the channel model needs to be identified at the receiver end. The most commonly used model for frequency-selective fading channels is a finite impulse response (FIR) model. FIR models for MIMO frequency-selective fading channels can be very non-parsimonious since the number of parameters (tap gains) to be estimated in a FIR MIMO model increases rapidly with the number of transmit and receive antennas. For a FIR MIMO model with m transmit antennas and p receive antennas, a total number of $m \times p \times L$ parameters have to be estimated, where L is the length of the subchannels assuming all the subchannels have equal length.

The FIR Model for a MIMO channel is not reducible when the subchannels are assumed to be independent, which can be justified in cases for which antennas are separated from each other by some multiple (e.g. 1/4) of the wavelength in both transmitting and receiving ends. However, when a large number of antennas are packed into a limited volume of space, the subchan-

[1] This research was supported by Core Grant No. 02-10109 sponsored by Ericsson and by US National Science Foundation under Grant ECS-0200449.
E-mail addresses: zhangc@ucsd.edu, rbitmead@ucsd.edu.

nels become correlated with each other (Chiurtu *et al.*, 2001; Shiu *et al.*, 2000). Hence the FIR model might be reduced to a more parsimonious state-space model.

Compared to the channel estimation methods based on FIR models of MIMO wireless channels, subspace system identification (SSI) methods (Van Overschee and De Moor, 1996; Verhaegen, 1994; Viberg, 1995), which are based on state-space modelling of the channel, could allow more parsimonious description of the MIMO channel or channel inverse if the subchannels share commonality to some extent. SSI algorithms identify the state-space model in a straightforward way and are numerically robust because they are based on computational tools such as singular value decomposition (SVD) and QR factorization. For MIMO systems with a relatively large number of transmit and receive antennas, the number of parameters to be estimated in the SSI method could be much less than that in methods based on FIR MIMO model.

For time-varying single-input single-output (SISO) wireless channels, training sequence based methods have been widely used to estimate the channel explicitly or implicitly. A pre-selected sequence, known to both the transmitter and the receiver ahead of time, is transmitted through the channel and is captured by the receiver, where it is applied to adjust the adaptive equalizer in accordance with some optimization criterion, e.g. LMS algorithm. Other than training-based approaches, blind channel estimation has recently emerged as a promising technique for channel equalization because no training sequence is needed for this type of approaches (Tong *et al.*, 1991). Instead, the knowledge is used that the transmitted symbols are distributed in a known way over a finite alphabet of fixed characters. However, most blind estimation methods suffer from convergence problem and have not found wide application in mobile communication with rapidly varying channels. This paper focuses on the discussion on the application of subspace system identification methods to the training-based MIMO channel estimation problem. In system identification terms, the presence of a training signal corresponds to knowledge of the input signal to the system (here the channel) being identified.

For broadband FIR MIMO channel with subchannels being independent of each other, the optimal training sequences that achieve the minimum mean square error (MMSE) of channel estimation have an impulse-like auto-correlation sequence and zero cross correlation (Fragouli *et al.*, 2003). However, for a non-FIR MIMO channel with correlated subchannels, it is not clear that white and uncorrelated sequences are "optimal" for channel identification. In fact, the "optimal" choice should depend on the knowledge of the specific channel (Goodwin and Payne, 1977), which implies that the training sequence should adapt to the change of the channel. Considering the high complexity of designing "optimal" training sequence for non-FIR MIMO channel and the fact that training sequences in wireless communication systems are normally selected ahead of time and stay fixed during the transmission, we believe white and uncorrelated training sequences are still the best option.

Another issue with the SSI-based MIMO channel estimation is that the training sequences may not be contiguous in the data stream in practical mobile communication systems. Instead, they appear as the mid-amble of a frame of data. More specifically, in the Groupe Speciale Mobile (GSM) system, a 26-bit long segment in the middle of each 156-bit frame is allocated for the insertion of the training sequence (Steele, 1992). A semi-blind approach can be efficient since it utilizes both the known data (training sequences) and unknown data (information sequences) to estimate the channel. Also, since the traditional SSI methods assume the availability of a contiguous input-output data stream, they need to be modified to suit the situation of MIMO channel estimation. It will be shown that when the length of the training sequence, N_t, is sufficiently large compared to the order or the McMillan degree of the model of the MIMO channel, the modified non-contiguous-data approach retains similar performance to the original contiguous-data approach.

The remainder of the paper is organized as follows. Section 2 overviews subspace system identification with application to MIMO channel estimation. The difference between subspace system identification methods and signal subspace methods that has been used in blind channel estimation is explained. Section 3 discusses the design of training sequences for subspace identification of MIMO frequency-selective fading channels. The formulation of SSI methods for non-contiguous data streams is discussed in Section 4. The conclusion follows in Section 5.

2. SUBSPACE SYSTEM IDENTIFICATION AND MIMO CHANNEL ESTIMATION

2.1 Channel Model

Single-input single-output (SISO) frequency-selective fading channels have been commonly modelled as tapped delay lines to characterize the multipath fading phenomenon. For a SISO FIR channel, the number of channel parameters to be estimated is equal to the length of the impulse response L. This parametrization could be very non-parsimonious for a broadband MIMO channel, which would contain $m \times p \times L$ unknown parameters assuming all the subchannels have equal length L, where m and p are the number of transmit and receive antennas, respectively.

In the case where the subchannels in the MIMO system share commonality to some extent, a state-space model may be able to provide a more parsimonious parametrization of the frequency-selective fading channel than FIR model.

Consider a system that employs m transmit and p receive antennas.

$$\begin{aligned} \mathbf{x}_{k+1} &= A\mathbf{x}_k + B\mathbf{u}_k \\ \mathbf{y}_k &= C\mathbf{x}_k + D\mathbf{u}_k + \mathbf{n}_k \end{aligned} \tag{1}$$

where $\mathbf{u}_k$ is a $m \times 1$ vector that represents the channel input (symbols sent by the m transmit antennas) at time k. $\mathbf{y}_k$ is a $p \times 1$ vector that represents the channel output at time k, i.e. the received symbols by the p receive antennas. $\mathbf{x}_k$ is the q channel state vector where q is the order or the McMillan degree of the MIMO system. Additive white Gaussian noise is assumed and is represented by $\mathbf{n}_k$. A, B, C and D are the system matrices in the state-variable description of the MIMO channel with obvious dimensions.

If the impulse responses of the subchannels are correlated with each other to some extent, the system order q could be much less than the length of the impulse response of a single subchannel. Therefore the state-variable methods could allow a dramatic reduction in the number of parameters to be estimated for the MIMO equalizer compared to the case with a MIMO FIR model. For example, a 4×4 10-tap FIR channel model without other structure would require 160 parameters to be estimated. A state-variable realization with q poles would require no more than 169 parameters for $q = 9$ and 81 parameters for $q = 5$.

2.2 *Subspace System Identification*

Subspace system identification (SSI) refers to a class of recent algorithms, such as N4SID and MOESP, which apply input-output system identification methods to determine directly a state-space realization of system. The key idea of SSI methods is to estimate the extended observability matrix through projection of future input-output data onto past input-output data. Then the system matrices A, B, C and D are computed based on the estimated observability matrix and singular value decomposition (SVD) algorithm. Refer to (Van Overschee and De Moor, 1996; Ljung, 1999) for details about SSI algorithms.

Based on the channel model given in (1), SSI methods require the input to satisfy the following requirements for the channel to be identifiable.

(1) The input $\mathbf{u}_k$ is uncorrelated with the additive Gaussian white noise $\mathbf{n}_k$.
(2) The input $\mathbf{u}_k$ is persistently exciting of order of at least 2 times the maximum order of the channel.
(3) The symbols in the input sequence are contiguous and for consistency the number of input goes to infinity.

The first assumption is usually satisfied for wireless communication systems. The second one requires the training sequence to maintain a certain structure. Also, notice that the third assumption places limitation on the application of SSI methods to channel estimation in wireless communication systems where the training sequences are usually not contiguous in time. Instead, they lie in the mid-amble of a frame and are separated by data symbols, the knowledge of which is not shared between transmitter and receiver. This fact may suggest the use of recursive version of SSI methods. The issue of training sequence design for MIMO channel estimation will be addressed in detail later in Section 3 and 4.

2.3 *Subspace-based MIMO Channel Estimation*

We should point out that despite the similar name, the SSI-based methods differ from "Signal Subspace Methods" for blind MIMO channel estimation which seek to separate the noise and signal subspaces using singular value decomposition on the covariance matrix of the channel output (Moulines *et al.*, 1995).

In (Moulines *et al.*, 1995), channel structure is constrained to be FIR with known input covariance. Moreover, the assumption that the channel matrix is block Toeplitz (FIR) is explored to estimate the channel up to a scale factor through singular value decomposition of the channel output covariance matrix. As for SSI-based methods, since the FIR constraint on the channel is relieved to draw benefit from possible parsimonious parametrization of the channel, the channel estimate cannot be obtained directly by applying SVD on the output covariance matrix. Instead, a more general approach is taken to estimate the extended observability matrix from SVD of the projection of input-output data, and then use the extended observability matrix to compute the channel estimate.

There has been one attempt to use results from both signal subspace methods and SSI methods for blind channel estimation. In (Vandaele and Moonen, 2000), the approach of estimating extended observability matrix is taken under the assumption that the channel is FIR, i.e., matrix A in the state-variable model (1) has a fixed shifting matrix structure.

In wireless communication systems, training sequences are usually placed at the mid-amble of data frames. It is fairly clear that SSI is feasible for a continuous stream of data. But it is less clear that a block-wise sequence of mid-ambles is possible to be used. There are some papers which discuss recursive subspace system identification such as (Lovera *et al.*, 2000).

3. TRAINING SEQUENCE DESIGN FOR SSI-BASED MIMO CHANNEL ESTIMATION

Given that the channel can be estimated with the aid of off-line designed training sequences, the question arises as how to design optimal training sequences so that the error in the channel estimate can be reduced to the minimum.

For an FIR MIMO channel with independent subchannels, it is believed that white and zero spatial cross-correlation training sequences achieve the minimum mean square error (MMSE) of the estimates of channel coefficients (Caire and Mitra, 1998; Fragouli *et al.*, 2003). However, for non-FIR MIMO channels, it is not clear that white and uncorrelated sequences are still the optimal choice. In fact, the optimal choice of the input for identifying a non-FIR channel should depend on the specific channel (Goodwin and Payne, 1977).

Consider a general single-input single-output system

$$y_k = H(z)u_k + n_k \tag{2}$$

where $\{u_k\}$ and $\{y_k\}$ are the input sequence and output sequence, respectively, and $\{n_k\}$ is zero-mean additive white Gaussian noise with variance σ^2. $H(z)$ is the transfer function of the channel which can be non-FIR. If the estimator is assumed to be efficient, so that the parameter covariance matrix achieves the Cramér-Rao lower bound, then a suitable criterion of optimality of the choice of training sequence would be

$$\mathbf{u_o} = \arg\min_{\mathbf{u}}[-\log\det(\mathbf{M})] \tag{3}$$

subject to the input power constraint

$$\frac{1}{N_t}\sum_{k=1}^{N_t} u_k^2 = 1 \tag{4}$$

where N_t is the number of available training symbols, i.e. the length of the training sequence. $\det(\cdot)$ represents the determinant of a matrix. M is Fisher's information matrix given by

$$M = E_{\mathbf{Y}|\theta}\left\{\left(\frac{\partial \log p(\mathbf{Y}|\theta)}{\partial\theta}\right)^T\left(\frac{\partial \log p(\mathbf{Y}|\theta)}{\partial\theta}\right)\right\} \tag{5}$$

where θ is the vector of parameters in $H(z)$ and σ^2. For the system given in (2),

$$M = \frac{1}{\Sigma}\sum_{k=1}^{N_t}\left(\frac{\partial H(z)}{\partial\theta}u_k\right)^T\left(\frac{\partial H(z)}{\partial\theta}u_k\right) + M_c \tag{6}$$

where M_c is a constant matrix which does not depend upon the choice of the input sequence $\mathbf{u}$=$\{u_k\}$. When the channel is FIR and causal, $H(z)$ has the form

$$H(z) = b_0 + b_1 z^{-1} + b_2 z^{-2} + \cdots + b_{L-1} z^{-(L-1)} \tag{7}$$

The cost function (3) is minimized when the training sequence $\{u_k\}$ is white. However, for non-FIR channels (e.g. rational transfer function) the choice of optimal training sequence depends on the structure of the channel transfer function and the complexity of optimal input design is very high (Goodwin and Payne, 1977).

It can be implied from the case of SISO channel that the optimal training sequence design for a general MIMO channel would require certain knowledge of the channel. Furthermore, for time-varying non-FIR channels the optimal training sequences vary in time as well. This suggests the use of iterative and adaptive schemes for the selection of training sequence in broadband MIMO wireless channel, which requires extra communication of channel state information from receiver to transmitter and communication of training sequence selection from the transmitter to the receiver. In current wireless communication systems, training sequences are normally selected ahead of time and stayed fixed during the transmission. It is clear that the complexity of the adaptive schemes is too high to be realistic for practical communication systems. Therefore, the formalism of optimal input design is inappropriate for training sequence selection.

An appropriate choice of training sequences for general MIMO channels seems to be still using white and uncorrelated sequences as in MIMO FIR channel. The advantages of white and uncorrelated training sequences are listed as follows.

(1) *They satisfy the identifiability requirement of subspace system identification.* For SSI methods, such as N4SID, the input is assumed to be persistently exciting (Van Overschee and De Moor, 1996). It is easy to show that white and uncorrelated sequences are persistently exciting of any order.
(2) *They permit the use of simplified SSI algorithm for computing asymptotically unbiased matrices A, B, C and D.* In N4SID method, system matrices are computed based on a certain Kalman filter state sequence. The fact that this Kalman filter sequence cannot be calculated directly from data increases the complexity of the algorithm for computing asymptotically unbiased system matrices. However, if the input sequences are white, it is possible to use another *equivalent* Kalman filter sequence, which can be calculated directly from data, to compute the asymptotically unbiased system matrices, hence simplify the algorithm.
(3) *The performance improvement of optimal training sequence over white & uncorrelated might be small.* As shown in (Goodwin and Payne, 1977), for a typical SISO channel, the improvement in parameter variances achieved by use of the optimal input signal is about 1.49dB compared with the use of the pseudo-random binary signal. This 1.49dB improvement does not seem to be worth the effort made to compute the optimal training sequence iteratively.

Based on the above advantages, white and uncorrelated training sequences are still the best option for the purpose of general frequency-selective fading MIMO channel estimation using SSI methods.

4. SSI FORMULATION FOR NON-CONTIGUOUS DATA STREAMS

In practical mobile communication systems, the known-input data sequences, or training sequences, may not be contiguous in the data stream. In GSM, as stated before, they appear as the 26-bit mid-amble of a 156-bit frame.

One approach to tackling this problem is to treat the equalized data and its received version as "known" input-output data and to use them for SSI channel estimation in addition to the known training sequences. This idea is related to semi-blind adaptation where both training sequences and information data sequences are exploited to estimate the channel.

However, this approach still requires a continuous sequence of frames of data, which is not the case for the TDMA-based GSM system where a singe user is assigned only a part of the 8 TDMA time slots.

In this later circumstance, the state evolution of the received data must be restarted at the frame boundaries. This is at variance with the standard formulation of SSI. We next embark on an introductory foray into the development of a suitable modification of the SSI algorithms.

Consider the evolution of two contiguous N_t-symbol-long blocks of received data, with the first block commencing at time t and the second commencing immediately thereafter at $t+N_t$. Then we may write the blocked state equations as,

$$Y_{t,i,j} = \Gamma X_{t,j} + HU_{t,i,j},$$
$$Y_{t+N_t,i,j} = \Gamma X_{t+N_t,j} + HU_{t+N_t,i,j},$$

where Γ is the extended observability matrix, H is the system block Toeplitz matrix of Markov parameters, and

$$Y_{t,i,j} = \begin{bmatrix} y_t & y_{t+1} & \cdots & y_{t+j-1} \\ y_{t+1} & y_{t+2} & \cdots & y_{t+j} \\ \vdots & \vdots & & \vdots \\ y_{t+i-1} & y_{t+i} & \cdots & y_{t+i+j-2} \end{bmatrix},$$
$$X_{t,j} = \begin{bmatrix} x_t \; x_{t+1} \; \cdots \; x_{t+j-1} \end{bmatrix},$$
$$U_{t,i,j} = \begin{bmatrix} u_t & u_{t+1} & \cdots & u_{t+j-1} \\ u_{t+1} & u_{t+2} & \cdots & u_{t+j} \\ \vdots & \vdots & & \vdots \\ u_{t+i-1} & u_{t+i} & \cdots & u_{t+i+j-2} \end{bmatrix}.$$

In standard SSI approaches, these are combined to form a new matrix equation,

$$Y_{t,i,j+N_t} = \Gamma X_{t,j+N_t} + HU_{t,i,j+N_t}.$$

This absorbs the data vectors into the Hankel structure of the new U and Y matrices. This adds further columns to the equation to be solved for the observability matrix Γ.

Next consider the availability of discontinuous N_t-symbol-long blocks of received data with the first block commencing at time t and the second at some later time $t+M$ with $M > N_t$. Then, we still achieve the relationship,

$$Y_{t,i,j} = \Gamma X_{t,j} + HU_{t,i,j},$$
$$Y_{t+M,i,j} = \Gamma X_{t+M,j} + HU_{t+M,i,j},$$

but now the absorption of the data into individual Hankel matrices is no longer possible, because of the non-contiguity of the received data.

We may, however, write an augmented equation composed from the above set.

$$\begin{bmatrix} Y_{t,i,j} \; Y_{t+M,i,j} \end{bmatrix} = \Gamma \begin{bmatrix} X_{t,j} \; X_{t+M,j} \end{bmatrix} + H \begin{bmatrix} U_{t,i,j} \; U_{t+M,i,j} \end{bmatrix}.$$

This set of equations to be solved for Γ is comparable to the contiguous-data set of equations. It has the same number of rows, i, and has $i-1$ fewer columns. When the length of the training sequence, N_t, is sufficiently large compared to the dimension of the generalized observability matrix, Γ (which depends on the state dimension of the model), then the non-contiguous-data approach is similar in its estimation power to the contiguous-data approach.

5. CONCLUSION

In this paper, we suggest to relieve the FIR constraint on the model of MIMO frequency-selective channel to draw benefit from possible parsimonious parametrization of the channel, and to use subspace system identification (SSI) methods to tackle the channel estimation problem. Also, the selection of training sequences for SSI-based MIMO channel estimation is analyzed. The complexity of using optimal training sequences is found intimidating. Because conventional white and uncorrelated sequences satisfy the persistent excitation requirement of SSI methods and offer comparable performance to the optimal sequences, they are considered still the best choice for general (non-FIR) MIMO channel estimation. Furthermore, a modification of the SSI methods and a semi-blind approach are proposed to address the issue that only non-contiguous block-wise training sequences are available in practical mobile communication systems

REFERENCES

Caire, Giuseppe and Urbashi Mitra (1998). Training sequence design for adaptive equalization of multi-user systems. *Proc. 32nd Asilomar Conf.* **2**, 1479–1483.

Chiurtu, N., B. Rimoldi and E. Telatar (2001). Dense multiple antenna systems. *Proc. IEEE Information Theory Workshop* pp. 108–109.

Foschini, Gerard J. and Michael J. Gans (1998). On limits of wireless communications in a fading environment when using multiple antennas. *Wirelss Personal Communications* **6**(3), 311–335.

Fragouli, Christina, Naofal Al-Dhahir and William Turin (2003). Training-based channel estimation for multile-antenna broadband transmissions. *IEEE Trans. Wireless Communications* **2**(2), 384–391.

Goodwin, Graham and Robert L. Payne (1977). *Dynamic system identification: experiment design and data analysis*. Academic Press. New York, USA.

Ljung, Lennart (1999). *System identification, theory for the user*. second ed.. Prentice Hall PTR. New Jersey, USA.

Lovera, Marco, Tony Gustafsson and Michel Verhaegen (2000). Recursive subspace identification of linear and non-linear wiener state-space models. *Automatica* **36**(11), 1639–1650.

Moulines, Eric, Pierre Duhamel, Jean-Francois Cardoso and Sylvie Mayrargue (1995). Subspace methods for the blind identification of multichannel fir filters. *IEEE Trans. Signal Processing* **43**(2), 516–525.

Shiu, D.-S., G. J. Foschini, M J. Gans and J. M. Kahn (2000). Fading correlation and its effect on the capacity of multielement antenn asystems. *IEEE Trans. Communications* **48**(3), 502–513.

Steele, Raymond (1992). *Mobile Radio Communications*. IEEE Press. Piscataway, USA.

Tong, Lang, Guanghan Xu and Thomas Kailath (1991). A new approach to blind identificaiton and equalization of multipath channels. *Proceedings of the 25th Asilomar Confeerence on Signals, Systems and Computers* **2**, 856–860.

Van Overschee, Peter and Bart De Moor (1996). *Subspace identification for linear systems: theory, implementation, applications*. Kluwer Academic Publishers. Dordrecht.

Vandaele, Piet and Marc Moonen (2000). A stochastic subspace algorithm for blind channel identification in noise fields with unkown spatial covariance. *Signal Processing* **80**(2), 357–364.

Verhaegen, Michel (1994). Identification of the deterministic part of mmo state space models given in innovations form from input-output data. *Automatica* **30**(1), 61–74.

Viberg, Mats (1995). Subspace-based methods for the identification of linear time-invariant systems. *Automatica* **31**(12), 1835–1851.

IFAC
Publications
www.elsevier.com/locate/ifac

THE IDENTIFICATION OF CONTINUOUS-TIME LINEAR AND NONLINEAR MODELS: A TUTORIAL WITH ENVIRONMENTAL APPLICATIONS

Peter C. Young[†*], Hugues Garnier[‡] and Andrew Jarvis[†]

† *Centre for Research on Environmental Systems and Statistics, Lancaster University, Lancaster LA1 4YQ, U.K.*
p.young@lancaster.ac.uk
∗ *Centre for Resource and Environmental Studies, Australian National University, Canberra, Australia.*
‡ *Centre de Recherche en Automatique de Nancy (CRAN - CNRS UMR 7039), Université Henri Poincaré, Nancy, France.*
hugues.garnier@cran.uhp-nancy.fr

Abstract: Initially, the paper will provide a tutorial introduction to the main aspects of existing methods for identifying linear continuous-time models from discrete-time data and show how one of these methods has been applied to the identification and estimation of a model for the transportation and dispersion of a pollutant in a river. It will then go to introduce a widely applicable class of nonlinear, *State-Dependent Parameter* (SDP) models for continuous or discrete-time systems. Finally, the paper will describe how this SDP approach has been used to identify and estimate a nonlinear differential equation model of global carbon cycle dynamics and global warming. *Copyright © 2003 IFAC*

Keywords: continuous-time, linear, instrumental variable, optimal estimation, state dependent parameter, nonlinear environmental.

1. INTRODUCTION

Since the early 1960's, numerous different approaches have been suggested for the identification and estimation [1] of continuous-time, linear *Transfer Function* (TF) models from discrete-time, sampled data (see e.g. the reviews by Young, 1981; Unbehauen and Rao, 1997; and Nielsen *et al.*, 2000). The present paper is in two parts. The first part starts by reviewing the various approaches to linear model identification and estimation, starting with the iterative *Continuous-Time Refined Instrumental Variable* (RIVC) method, that can be interpreted in statistically optimal and quasi-optimal terms. It then proceeds to outline the main aspects of alternative, non-iterative but sub-optimal methods. The utility of this continuous-time methodology, in comparison to the alternative discrete-time approach, is illustrated by the application of the RIVC method to the '*Aggregated Dead-Zone*' (ADZ) modelling of pollutant transport in a river, based on data collected during a tracer experiment.

The second part of the paper considers nonlinear modelling, concentrating on a particular class of *State-Dependent Parameter* (SDP) models which follow logically from the linear methods and can describe a wide variety of nonlinear systems, including chaotic processes. Although applicable to both discrete and continuous-time systems, only the continuous-time case is considered here. The practical utility of this SDP approach is illustrated by its application to a

[1] Here we use the statistical meanings of these words: 'identification' is the definition of the most appropriate model order; and 'estimation' is the estimation of the parameters that characterize this identified model.

well known set of global climate data. This first, non-parametric stage of the SDP analysis identifies the presence and nature of a temperature-dependent, negative feedback effect; and in the second, parametric, stage this nonlinearity is parameterized and the resulting model is estimated from the data.

Note that the paper does not attempt to review the literature on the identification and estimation of models based on Itô stochastic differential equations (see e.g Nielsen *et al.*, 2000, and the references on this topic therein). While this is partly through lack of space, it is also because this alternative approach has not been applied so widely and is theoretically more demanding. By contrast, the methods discussed below have been proven in many practical applications and are available as user-friendly and computationally efficient algorithms in two Matlab™toolboxes.

2. LINEAR CONTINUOUS-TIME MODEL IDENTIFICATION

2.1 *Continuous-Time Models*

The theoretical basis for the statistical identification and estimation of continuous-time models from discrete-time, sampled data can be outlined by considering the following SISO system (MISO and MIMO extensions are straightforward but obviously more complex):

$$\begin{aligned} x(t) &= \frac{B(s)}{A(s)}u(t-\tau) \\ y(t) &= x(t) + e(t) \end{aligned} \tag{1}$$

Here $A(s)$ and $B(s)$ are polynomials in the derivative operator $s = d/dt$ of the form:

$$\begin{aligned} A(s) &= s^n + a_1 s^{n-1} + ... + a_{n-1}s + a_n \\ B(s) &= b_0 s^m + b_1 s^{m-1} + ... + b_{m-1}s + b_m \end{aligned}$$

and τ is any pure time delay in time units. This model structure is denoted by the triad $[n, m, \tau]$. In (1), $u(t)$ is the input signal, $x(t)$ is the 'noise free' output signal and $y(t)$ is the noisy output signal. Initially, the noise $e(t)$ is considered as zero mean, white noise with Gaussian amplitude distribution and variance σ^2, although we will see later that this assumption is not restrictive. Of course, the model (1) can also be written in the following differential equation form, which is often more familiar to physical scientists:

$$\begin{aligned} &\frac{d^n y(t)}{dt^n} + a_1\frac{d^{n-1}y(t)}{dt^{n-1}} + \cdots + a_n y(t) = \\ &b_0\frac{d^m u(t-\tau)}{dt^m} + \cdots + b_m u(t-\tau) + \mu(t) \end{aligned} \tag{2}$$

where $\mu(t)$ is defined as $\mu(t) = A(s)e(t)$.

2.2 *Optimal and Quasi-Optimal Estimation*

The topic of optimal statistical identification and estimation of continuous-time models, such as (1) and (2), from discrete-time series data has received only a small amount of attention (e.g. Young and Jakeman, 1980; Wang and Gawthrop, 2001; Young, 2002). This is rather surprising because optimal statistical methods provide quantification of the uncertainty associated with the parameter estimates (see later examples) that can prove very useful in practice. The more *ad hoc*, sub-optimal methods considered later do not provide such information. The achievement of optimality is made difficult because the input signal is not normally known over the sampling interval and has to be interpolated in some manner. However, in the case where the input is constant over the sampling interval, as in most control applications, it is possible to formulate the optimal estimation solution fairly straightforwardly.

For example, following the usual *Prediction Error Minimization* (PEM) approach (*Maximum Likelihood* (ML) in the present situation because of the Gaussian assumptions), a suitable error function $\varepsilon(t)$ is given by,

$$\begin{aligned} \varepsilon(t) &= y(t) - \frac{B(s)}{A(s)}u(t-\tau) \\ &= \frac{1}{A(s)}\{A(s)y(t) - B(s)u(t-\tau)\} \end{aligned}$$

Minimization of a least squares criterion function in $\varepsilon(t)$, measured at the sampling instants, represents a nonlinear estimation problems and provides the basis for the *response* or *output error* estimation methods. However, since the operators commute in this linear case, the $1/A(s)$ filter can be taken inside the brackets to yield the expression,

$$\varepsilon(t) = A(s)y^*(t) - B(s)u^*(t-\tau) \tag{3}$$

or,

$$\begin{aligned} \varepsilon(t) = s^n y^*(t) + a_1 s^{n-1} y^*(t) + ... + a_n y^*(t) \\ -b_0 s^m u^*(t-\tau) - ... - b_m u^*(t-\tau) \end{aligned}$$

where the * superscript indicates that the associated variable has been '*prefiltered*' by $1/A(s)$. The advantage of this transformation is that (3) is now linear in the unknown parameters $a_i, i = 1, ..., n; b_j, j = 0, ..., m$, so that the associated estimation model can be written in the form:

$$s^n y^*(t) = \mathbf{z}^*(t)^T\mathbf{a} + e(t) \tag{4}$$

where,

$$\mathbf{z}^*(t) = \left[-s^{n-1}y^*(t)... - y^*(t)\; s^m u^*(t-\tau)...u^*(t-\tau)\right]^T$$

$$\mathbf{a} = [a_1 \;...\; a_n \; b_0 \;...\; b_m]^T$$

As a result, all of the prefiltered derivatives appearing as variables in this estimation model are measurable as the inputs of the integrators that appear in the realization of the prefilter $1/A(s)$. Thus, provided we assume that $A(s)$ is known, the estimation model (4) forms a basis for the definition of a likelihood function and ML estimation.

There are two problems with this formulation. The obvious one is, of course, that $A(s)$ is not known *a priori*. The less obvious one is that, in practical applications, we cannot assume that the noise $e(t)$ will have the nice white noise properties assumed above: it is likely that the noise will be a coloured noise process, say $\xi(t)$. Both of these problems can be solved by employing a similar approach to that used in the *Refined Instrumental Variable* (RIV) algorithm for discrete-time (backward shift operator TF) system identification and estimation (see Young, 1984 and the prior references therein). Here, a 'relaxation' optimization procedure is devised that adaptively adjusts an initial estimate $\hat{A}_0(s)$ of $A(s)$ iteratively until it converges on an optimal estimate of $A(s)$. And the coloured noise problem is solved conveniently by exploiting IV estimation within this iterative optimization algorithm. The continuous-time version RIVC of this RIV algorithm is described fully in Young and Jakeman (1980) and outlined in Young (2002). This RIVC algorithm is available in the Lancaster CAPTAIN toolbox for Matlab [2]

Of course, if the noise $e(t)$ is coloured or the input $u(t)$ is not constant between samples, then the above approach to estimation is not optimal in statistical terms, although it is robust and normally yields estimates with reasonable statistical efficiency (i.e. low but not minimum variance). In the former case, it is possible to obtain quasi-optimal estimates by modelling the coloured noise in ARMA or AR terms and expanding the definition of the adaptive prefilters to account for this, as in the optimal RIV method for discrete-time systems. However, since it is well known that there are theoretical and practical problems associated with continuous-time ARMA and AR modelling, it is practically advantageous to use a hybrid approach in which the noise modelling, as well as the noise-derived parts of the prefiltering, are carried out in discrete-time terms (Young and Jakeman, 1980; Johansson, 1994; Pintelon *et al.*, 2000) [3]. As regards the interpolation of the input signal $u(t)$, it is possible to consider optimal interpolation but this is not really worthwhile since experience suggests that simple interpolation normally produces very good estimation results.

2.3 *Alternative Sub-Optimal Approaches*

Initial research on continuous-time model identification and estimation was not formulated in the above optimal manner but was based on the concept of a *State Variable Filter* (SVF) that generated the required prefiltered derivatives. A comprehensive survey of these techniques has been given by (Young, 1981) and then by (Unbehauen and Rao, 1987, 1990, 1998) and (Garnier *et al.*, 2003). A book has also been devoted to these so-called 'direct' methods (Sinha and Rao, 1991). All of the main sub-optimal approaches are available in the CONtinuous-Time System IDentification (CONTSID) toolbox for Matlab (Garnier and Mensler, 2000) [4]. Since the methods have been documented so fully, however, it will suffice here merely to outline the main features of each approach. The advantages of these approaches over indirect methods (where a discrete-time model is first estimated by using standard discrete-time estimation methods available in the SITB toolbox and then converted into a continuous-time model) have been discussed recently (Rao and Garnier 2002, Ljung 2003).

2.3.1. *State-variable Filter (SVF) Methods* These methods originated from the first author's early research in this area (Young 1964, 1965, 1970) and was referred to as the '*Method of Multiple Filters*' (MMF). It involves passing the input and output signals through a chain of (usually identical) first order prefilters with user-specified band-pass, normally selected so that it spans the anticipated bandpass of the system being identified. More recently this MMF approach has been re-named the *Generalized Poisson Moment Functionals* (GPMF) approach (Saha and Rao 1983, Unbehauen and Rao 1987). Recent MMF/GPMF developments have been proposed by Garnier and his co-workers (Garnier *et al.*, 1994, 1995, 1997, 2000; Bastogne *et al.*, 2001).

2.3.2. *Integration-Based Methods* The main idea of these methods is to avoid the differentiation of the data by performing an order n integration. These integral methods can be roughly divided into two groups. The first group, using numerical integration and orthogonal function methods, performs a basic integration of the data and special attention has to be paid to the initial condition issue. The second group includes the *Linear Integral Filter* (LIF: Sagara and Zhao, 1990) and the *Reinitialized Partial Moments* (RPM: Trigeassou, 1987) approaches. Here, advanced integration

[2] http://www.es.lancs.ac.uk/cres/captain/. The CAPTAIN toolbox also contains the equivalent RIV algorithm for discrete-time TF models with white or coloured noise, as well as optimized recursive filtering and fixed interval smoothing algorithms for the estimation of time variable parameters in various models (TF, ARX, linear and harmonic regression models).

[3] Note, however, that continuous-time noise modeling has been considered for models with no input $u(t)$ (Tuan 1977, Fan *et al.*, 1999; Pham, 2000; Söderström and Mossberg, 2000); and some extensions have been made to handle the case of continuous-time ARX models (Söderström et al. 1997).

[4] http://www.cran.uhp-nancy.fr/

methods are used that avoid the initial condition problem either by exploiting a moving integration window (LIF) or a time-shifting window (RPM).

2.3.3. *Modulating Function Methods* This approach was first suggested almost half a century ago by Shinbrot in order to estimate the parameters of linear and nonlinear systems (Shinbrot 1957). Further developments have been based on different modulating functions. These include the Fourier-based functions (Pearson *et al.*, 1994), in either trigonometric or complex exponential form; spline-type functions; Hermite functions and, more recently, Hartley-based functions (Unbehauen and Rao, 1998). A very important advantage of using Fourier- and Hartley-based modulating functions is that the model estimation can be formulated entirely in the frequency domain, making it possible to use efficient DFT/FFT techniques.

2.4 *Linear Example: Pollutant Transport in River Systems*

Both of the practical examples described in this paper are examples of *Data-Based Mechanistic* (DBM) modelling (see Young, 1998, and the prior references therein). This can be contrasted with 'black-box' modelling, since DBM models are only deemed credible if, in addition to explaining the time series data in a statistically efficient, parsimonious manner, they also provide an acceptable physical interpretation of the system under study. They can also be contrasted with 'grey-box' models, because the model structure is inferred inductively from the data, rather than being assumed *a priori* before model identification and estimation.

This first DBM modelling example is based on the analysis data obtained from a tracer experiment in a river system. Tracer experiments[5] are an excellent way of evaluating how a river transports and disperses a dissolved, conservative pollutant (solute). Fig.1 shows a typical set of tracer data from the River Conder, near Lancaster in North West England.

The best known TF model for solute transport and dispersion is the *Aggregated Dead Zone* (ADZ) model introduced by Beer and Young (1983). It has become conventional to identify and estimate this model in discrete-time TF form and then deduce the continuous-time (differential equation) model parameters from the estimated parameters of this discrete-time TF (the 'indirect' method). More recently, however, in related research on imperfect mixing processes (Price *et al.*, 1999), continuous-time models have proven more useful. Moreover, in the present example, discrete-time

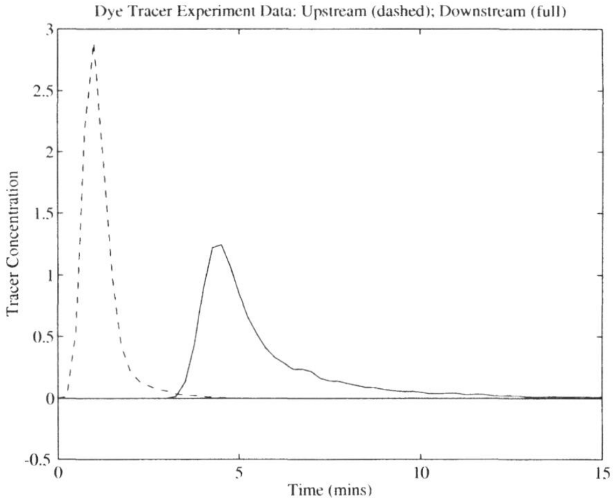

Fig. 1. Dye tracer experiment data for the River Conder, North West England: measured input, upstream concentrations (dashed line); measured output, downstream concentrations (full line).

modelling is not very successful when applied to the data in Fig.1: using a relatively fast sampling interval of 0.25 minutes, both the discrete-time RIV algorithm and the alternative PEM algorithm in Matlab yield second order models which do not explain the data very well. Moreover, while these algorithms produce well fitting third order models, these are clearly over-parameterized and have complex roots, so that the models can be rejected on DBM grounds since they have no obvious physical interpretation.

Continuous-time RIVC modelling of the tracer data is much more successful and also produces models that can be interpreted directly in physically meaningful terms, so satisfying the DBM modelling requirements. This suggests strongly that the dynamic relationship between the measured concentrations at the input (upstream) and at the output (downstream) measurement sites is linear and second order, with the continuous-time TF identified by the RIVC algorithm in the following form,

$$y(t) = \frac{b_o s + b_1}{s^2 + a_1 s + a_2} u(t-3) + \xi(t) \tag{5}$$

or in ordinary differential equation terms,

$$\begin{aligned} \frac{d^2y(t)}{dt^2} + a_1 \frac{dy(t)}{dt} + a_2 y(t) = \\ b_0 \frac{du(t-3)}{dt} + b_1 u(t-3) + \eta(t) \end{aligned} \tag{6}$$

where $\eta(t) = (s^2 + a_1 s + a_2)\xi(t)$. Here, time is measured in minutes and the pure time delay of 3 minutes on the input variable, $u(t-3)$, is the purely advective, 'plug flow' effect. Although there is a little serial correlation and some heteroscedasticity in the estimated residuals $\xi(t)$, the variance is extremely low (0.0001), as reflected in the very high coefficient of determination based on these modelling errors of

[5] the interested reader will find a more complex example of ADZ modelling in Young (2001b), where the same approach used here is applied to data from a tracer experiment conducted in a large Florida wetland area.

$R_T^2 = 0.9984$ (i.e. 99.84% of the $y(t)$ variance is explained by the simulated output of the model) [6].

With such a high R_T^2, the model (5) obviously explains the data very well, as shown in Fig.2 which compares the deterministic (noise free) model output,

$$\hat{x}(t) = \frac{\hat{b}_o s + \hat{b}_1}{s^2 + \hat{a}_1 s + \hat{a}_2} u(t-3) \qquad (7)$$

with the measured tracer concentrations $y(t)$. The estimated parameters are as follows:

$$\hat{a}_1 = 2.051(0.073);\ \hat{a}_2 = 0.603(0.055);$$
$$\hat{b}_0 = 1.194(0.014); \hat{b}_1 = 0.642(0.056)$$

where the figures in parentheses are the estimated standard errors. Introducing the estimated parameter values, the TF model (7) can be decomposed by partial fraction expansion into the following form,

$$\hat{x}(t) = \frac{0.6081}{1+0.5898s} u(t-3) + \frac{0.4575}{1+2.8105s} u(t-3) \qquad (8)$$

which reveals that the model can be considered as a parallel connection of two first order processes which appear to characterize distinctive solute pathways in the system with quite different residence times: one 'quick', with a residence time $T_q = 0.5898$ minutes; and the other 'slow', with a residence time $T_s = 2.8105$ minutes. The associated steady state gains are $G_q = 0.6081$ and $G_s = 0.4575$, respectively. These suggest a parallel partitioning of tracer with a partition percentage of $P_q = 100\ [\ 0.6081/(0.6081 + 0.4575)] = 57.1\%$ for the quick pathway, and $P_q = 100\ [\ 0.4575/(0.6081 + 0.4575)] = 42.9\%$ for the slow pathway.

The decomposition of the TF into the parallel pathway form (8), provides the information required to interpret the model in a physically meaningful manner. The first order model associated with each pathway can be considered as a differential equation describing mass conservation. And if it is assumed that the flow is partitioned in the same way as the dye, then the *Active Mixing Volume* (AMV: see Young and Lees, 1993) of water associated with the dispersion of the solute in each pathway can be evaluated by reference to this equation, the flow rate and the residence times. This yields a quick pathway AMV, $V_q = 26.3m^3$; and a slow pathway AMV, $V_s = 94.1m^3$, respectively. The associated *Dispersive Fraction* (DF), in each case, is calculated as the ratio of the AMV and the total volume of water in the reach, giving $DF_q = 0.12$ and $DF_s = 0.56$: (i.e. the acting mixing volumes are 12% and 56% of the total volume of water in each pathway, respectively). In other words, the slow pathway results in a considerably greater dispersion (and longer-term detention) of the dye than the quick pathway, as one might expect.

Given this quantitative analysis of the model (7), the most obvious physical interpretation of the parallel flow decomposition in (8) is a form of two layer flow, with the slow pathway representing the dye in the water moving adjacent to the cobbled bed and banks of the river, which is being differentially delayed in relation to the quick pathway, which is associated with the more freely moving surface layers of water. The aggregated effect of each pathway is then an advective transportation delay of 3 minutes, associated with non-dispersive 'plug flow'; and an ADZ, defined by the associated AMVs and DFs in each case, which are the main mechanisms for dispersion of the dye (and, therefore, other forms of pollution) in its passage down the river.

This parallel partitioning of the flow and solute also helps to explain the shape of the experimentally measured concentration profile. The individual concentration profiles for the quick and slow pathways, as inferred from the parallel partitioning, are shown as dashed and dash-dot curves, respectively, in Fig.2.

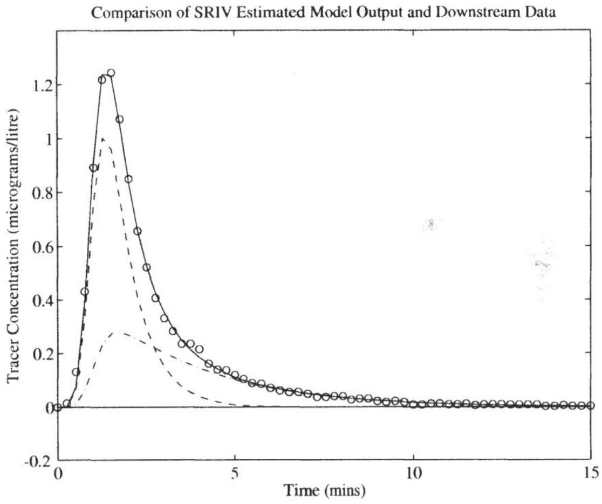

Fig. 2. Comparison of the RIVC identified ADZ model output $\hat{x}(t)$ (full line) and the measured concentration of tracer $y(t)$ (circular points) at the downstream location. Also shown: inferred quick pathway (dashed) and slow pathway (dash-dot) concentration profiles.

3. NONLINEAR CONTINUOUS-TIME MODEL IDENTIFICATION

The identification and estimation of nonlinear continuous-time models is considerably more difficult than linear modelling. First, there is no unified theory for nonlinear systems and so it is necessary to consider a given 'class' of nonlinear model. Secondly, the estimation of time derivatives is more difficult because the

[6] R_T^2 is defined as $R_T^2 = 1 - var\{y(t) - \hat{x}(t)\}/var\{y(t)\}$, where $\hat{x}(t)$ is the deterministic model output from (7).

commutation operation that is so important in defining prefiltered time-derivatives (see section 2.2) is no longer possible in the case of nonlinear systems. Here, we consider the *State Dependent Parameter* (SDP) class of nonlinear models (see Young *et al.*, 2001 and the prior references therein) which can describe a wide variety of nonlinear systems including chaotic processes.

3.1 *State Dependent Parameter Estimation*

As far as the author is aware, the idea of *State Dependent parameter* (SDP) modelling within a stochastic setting was originated by Young (1969) and Mendel (1969). They enhanced recursive estimation performance by assuming that the model parameters could vary over time because of their dependence on the variations in other measured variables. Young (1978) then explored these ideas within a broader SDP setting and Priestley (1988) took them up in a series of papers and a book on the subject. These earlier publications do not, however, exploit the power of recursive fixed interval smoothing (FIS), which provides the main engine for the latest methods of SDP estimation (Young *et al.*, 2001 and the prior references therein).

SDP estimation was originally developed in discrete-time terms (see above references). The simplest SDP continuous-time model is a nonlinear equivalent of the linear TF model (1) and takes the following form:

$$y(t) = \frac{B(s, \mathbf{z}_t)}{A(s, \mathbf{z}_t)} u(t - \tau) + e(t) \qquad \text{(9a)}$$

where $A(s, \mathbf{z}_t)$ and $B(s, \mathbf{z}_t)$ are SDP polynomials in the s operator of the form:

$$A(s, \mathbf{z}_t) = s^n + a_1(z_{1,t})s^{n-1} + \cdots + a_n(z_{n,t})$$
$$B(s, \mathbf{z}_t) = b_0(z_{n+1,t})s^m + \cdots + b_m(z_{n+m+1,t}).$$

while $\mathbf{z}_t$ is a vector of measured variables (states) on which the parameters may be dependent. In estimation equation terms, this model can be written most conveniently as:

$$s^n y(t) = \mathbf{z}_t^T \mathbf{p}_t + e_t \qquad \text{(9b)}$$

where,

$$\mathbf{z}_t^T = \left[-s^{n-1}y(t) \cdots \; -y(t) \; s^m u(t-\tau) \; \cdots \; u(t-\tau)\right]$$
$$\mathbf{p}_t = [p_1(z_{1,t}) \cdots \; p_{n+m+1}(z_{n+m+1,t})]^T,$$

while,

$$p_1(z_{1,t}) = a_1(z_{1,t}); \; p_2(z_{2,t}) = a_2(z_{2,t}), \cdots, p_{n+m+1,t}(z_{n+m+1,t}) = b_m(z_{n+m+1,t})$$

SDP modelling consists of two stages, the full details of which are given in the above references. In the first, non-parametric stage, the recursive SDP estimation algorithm is an extension of the stochastic approach to *Time Variable Parameter* (TVP) estimation, where the time variations in the parameters are assumed to evolve as one of the *Generalized Random Walk* (GRW) class of Gauss-Markov process (e.g. Young, 1999), of which the *Random Walk* (RW) and *Integrated Random Walk* (IRW) processes are the best known. As in the TVP case, SDP estimation exploits the power of recursive FIS estimation to obtain lag-free, smoothed estimates of the parameter variations. However, SDP estimation differs from TVP estimation in two important respects. First, in order to allow for the rapid variation that state dependency can induce in the parameters, the data are sorted into ascending order, so that the rate of change of the parameter variations between samples in this sorted data space is much smaller than in the original observation space. Secondly, in order to allow for the possibility of different state dependency in each parameter, an iterative 'back-fitting' algorithm is used to estimate each SDP separately, based on prior estimates of all the other SDPs in the model. The main estimation engine used in the implementation of this back-fitting algorithm is FIS estimation.

As in the linear situation, the main problem with the above SDP estimation methodology in the continuous-time case is its requirement for measurements of the input and output time derivatives,

$$s^i y(t) = \frac{d^i y(t)}{dt^i}, i = 1, 2, ..., n;$$
$$s^j u(t-\tau) = \frac{d^j u(t-\tau)}{dt^j}, j = 1, 2, ..., m$$

which are not usually available directly. The approach to this time derivative estimation problem used in the later example (see Young, *et al.*, 1993; Young, 1993) is again based on recursive FIS estimation. Each variable in question (here $u(t)$ and $y(t)$) is modelled as a multiple differentiation process in continuous-time and then converted to the discrete-time equivalent of this process. The time derivatives of the variable at the sampling instants are then recovered as the FIS estimated states of this model. In the simplest single derivative case used in the example discussed later, for instance, this takes the form of an IRW process with its parameter defined by the sampling interval, Δt. This process could be formulated more accurately in continuous-time terms but the discrete-time formulation works quite well, as we shall see in the later example (see also the above references). Note that this estimation of time derivatives is only required in the first, nonparametric stage of SDP estimation (see below). The subsequent final parametric estimation stage, discussed below, involves direct integration of the differential equation and PEM optimization, so that no explicit derivative estimation is required.

3.2 *Final Parametric Estimation*

Within DBM modelling, nonparametric SDP modelling normally provides a method for identifying the presence and location of nonlinearities in the SDP models. In this manner, it serves as a prelude to the parameterization of the final nonlinear stochastic model and the more efficient estimation of the (normally constant) parameters that characterize this nonlinear parametric model. Such estimation can be based on various approaches to model optimization, from nonlinear least squares, through maximum likelihood to the latest Monte Carlo-based methods of Bayesian estimation. But the methodology used in any particular application will normally depend on the nature of the system under study, the modelling objectives, and the scientific background of the model builder. In the case of the global carbon cycle example discussed below, optimization is based again on a *Prediction Error Minimization* (PEM) approach (maximum likelihood estimation in the case of Gaussian residuals).

3.3 *Other models*

Although we have outlined the SDP approach to data-based nonlinear model identification and estimation in terms of continuous-time transfer function models, SDP estimation is not restricted to such models. It can be applied to any continuous or discrete-time, nonlinear stochastic model that can be formulated as linear additive sum of nonlinear elements (sometimes termed an 'affine' model). This ranges from static nonlinear regression models (e.g. Young, 2001a) to dynamic state space models. Typical examples of the latter are the well known *Lorenz Strange Attractor*, the model of the *Nicholson Blowfly Data* (Young, 2000), or the simpler global carbon cycle model considered below.

3.4 *Global Carbon Cycle Modelling*

This example provides some of the technical background to a short paper by Jarvis and Young (2002); and a more complete description is given in Young and Jarvis (2002). It investigates the dynamic relationship, over the period 1856 to 2000, between globally averaged annual measures of CO_2 emissions (arising from both the use of carbon fuels and land-use changes); perturbations in atmospheric carbon dioxide partial pressure, pCO_2; and the Northern Hemisphere temperature anomaly. In particular, it reveals the possible presence of a temperature-dependent nonlinearity in the dynamic relationship between CO_2 emissions and atmospheric pCO_2 that has an interesting and potentially important physical interpretation.

Fig. 3 presents the data used in the analysis and shows that there are clear increases in all three variables over the period from 1856 to the end of the last century. The normal statistical procedure would be either to reduce the series to stationarity in some manner (e.g. by differencing). But the climate data have a clear physical meaning and we can be reasonably sure that the increase in the levels of atmospheric pCO_2 are the result of the increases in emissions. In other words, there is an obvious input-output relationship, with the nonstationarity in the input giving rise to nonstationarity in the output. What is much less certain, however, is that the rise in the level of the atmospheric pCO_2 is, in turn, leading to the observed increase in the temperature anomaly.

With these factors in mind, we will analyze the data directly in the form shown in Fig. 3. For this analysis, the input CO_2 emissions, the output atmospheric pCO_2 perturbations about the assumed pre-industrial level (see caption to Fig. 3), and the temperature anomaly will be denoted by $u(t)$, $y(t)$ and $T(t)$, respectively.

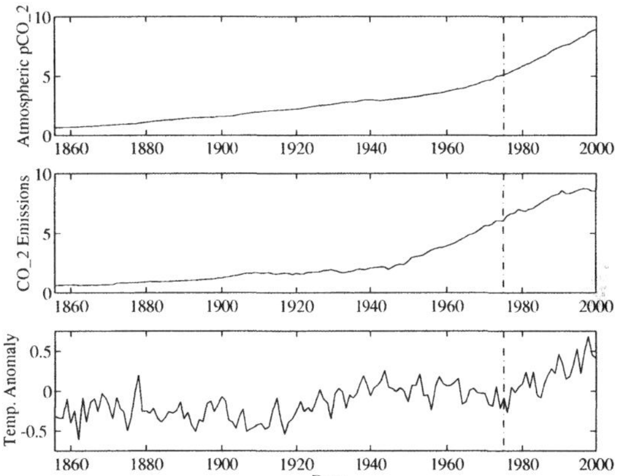

Fig. 3. Annual carbon dioxide and temperature anomaly data 1856-2000. Upper panel, perturbations in atmospheric CO_2, (measured as partial pressure, pCO_2, relative to standard pressure, in pascals, pa) about an assumed pre-industrial level of 28 pa; middle panel, anthropogenic CO_2 emissions arising from fossil fuel usage and land use change ($Gt\ y^{-1}$); lower panel, Northern Hemisphere average temperature anomaly ($°C$). The vertical dash-dot line marks the boundary between model estimation (1856-1975) and predictive validation (1976-2000) data (see text). All series are derived from data available at http://cdiac.esd.ornl.gov/.

3.4.1. *Linear model identification and estimation*

The RIVC algorithm identifies a number of linear, constant parameter, first order models that have good identification and estimation statistics; i.e. coefficients of determination R_T^2 based on the simulated output greater than 0.99 (i.e. the simulated output of the model explains $> 99\%$ of the pCO_2 variance), together with well defined, low standard error parameter estimates and, hence, a satisfactory YIC model order identification criterion (Young, 1989). However, all higher order models are rejected, either because they do not satisfy these statistical criteria or because they are not satisfactory in dynamic terms (e.g. they have unstable or imaginary eigenvalues).

Two important factors emerge from this analysis: first, it is necessary to add a small additional constant input to obtain a good explanation of the $y(t)$ series (see below); second, there is a less well-defined pure time delay of about 5 years. In other words, this initial analysis suggests a model of the form:

$$\frac{dx(t)}{dt} = a_1 x(t) + b_0 u(t-\tau) + c \quad (10a)$$

$$y(t) = x(t) + \xi(t) \quad (10b)$$

where $x(t)$ is the underlying, 'noise-free' pCO_2 perturbation; τ is the pure time delay; $\xi(t)$ is the residual coloured noise at the output of the model, with zero mean value and variance σ_ξ^2; and c is an additional constant input. The addition of c makes sense physically because it will correct for any small inaccuracy in either the assumed pre-industrial level of pCO_2 that has been removed from the pCO_2 data (see caption to Fig. 3), or the CO_2 emissions.

In order to obtain an improved, possibly non-integral, estimate of the time delay, the model (10a) was estimated using the *leastsq* optimization tool in Matlab, with the model simulated in Simulink™, using the *linsim* tool. All integrations of the model used in this optimization were initiated from the measured $y(0)$, since separate optimization of the initial condition had little effect on the estimates. Also, in order to allow for colour in the residuals $\xi(t)$, they were modelled as a *discrete-time* AutoRegressive (AR) process, i.e.,

$$\xi_t = \frac{1}{1 + d_1 z^{-1} + ... + d_n z^{-n}} e_t \quad (10c)$$

where $z^{-r}, r = 1, 2, ...n$ is the backward shift operator; while $\xi_t = \xi(t_i)$ and $e_t = e(t_i)$ are, respectively, $\xi(t)$ and $e(t)$ sampled at the annual sample times $t_i, i = 1, 2, ..., N$, and N is the sample size. The sampled stochastic model residuals e_t are assumed to be a zero mean sequence of serially uncorrelated random variables with variance σ^2 (discrete white noise). This 'hybrid' approach to modelling was used because the discrete-time stochastic model (10c) is easier to handle within the optimization (and in theory). In any case, the noise model estimation makes only a small difference to the final estimated parameter values because the residual noise variance is so small (although it does affect their estimated uncertainty).

Parameter estimation was based on optimization of the following least squares PEM cost function with respect to the unknown parameters:

$$J\{\boldsymbol{\theta}\} = \sum_{t=1}^{t=N} \hat{e}_t^2 \quad \boldsymbol{\theta} = [a_1 \; b_0 \; c \; d_1 \; d_2 \; ... \; d_n \; \tau]^T$$

where $\hat{e}_t$ are the stochastic model residuals (one-year-ahead prediction errors) at the annual sampling interval. The estimated parameters in the most important continuous-time part of the model, based initially on the whole data set ($N = 145$) are as follows:

$$\hat{a}_1 = -0.0167(0.0009); \hat{b}_0 = 0.0371(0.0009);$$
$$\hat{c} = 0.0114(0.0006); \hat{\tau} = 5.0(0.389);$$
$$\sigma_\xi^2 = 0.0041; \quad \sigma^2 = 0.0004$$

The noise model (10c) is identified by the AIC (Akaike, 1974) as either an AR(8) or ARMA(2,2) process, with the final residuals e_t showing no significant auto-correlation and no cross-correlation with the input u_t (the annual samples of the input CO_2 emissions). However, the residuals $\hat{\xi}_t = y_t - \hat{x}_t$ are significantly correlated with the temperature T_t for all lags between 4 and 22 years and a maximum correlation coefficient of 0.4 (0.17) at a lag of 10 years. Moreover, recursive FIS estimation of the local cross correlation is much larger than this at high values of T_t. The model has a coefficient of determination based on the simulated deterministic model residuals $\hat{\xi}_t$ of $R_T^2 = 0.9991$; while the more conventional coefficient of determination based on $\hat{e}_t$ is $R^2 = 0.9999$. This is, of course, important in the present context since it means that the main differential equation which, as we shall see later has physical significance, is explaining the measured output of the dynamic system very well indeed.

3.4.2. *Nonlinear model identification and estimation*

Despite the apparently very good results obtained in linear modelling, the high correlation of the residuals $\hat{\xi}_t$ with the temperature anomaly T_t suggests that there may be a temperature dependent component in the model residuals that is capable of being absorbed within the model and so improving its descriptive ability still further. This is indeed the case: SDP non-parametric estimation suggests strongly that, while the input parameter b_0 is constant, the parameter estimate $\hat{a}_1$ appears to vary as a function of temperature anomaly T_t. This SDP estimate $\hat{a}_1$, plotted in the left hand panel of Fig. 4, suggests that the parameter value is reducing as the temperature increases, although there is a tendency for the variations to 'flatten out' at the lower and upper extremes of the temperature anomaly range. But we must remember that these are non-parametric estimates and tend to be more poorly defined in these regions because of end effects and the paucity of data in these regions.

In order to investigate this problem in a statistically more efficient parametric manner, several different parameterizations of the temperature dependency suggested in Fig. 4 were tried, including polynomial, radial basis function and an exponential decline. However, the best results were obtained with either simple linear (first order polynomial) or sigmoidal laws. In other words, the finally identified nonlinear model takes either of the following forms:

$$\frac{dx(t)}{dt} = \{\alpha + \beta T(t)\}x(t) + b_0 u(t-\tau) + c \qquad \text{(11a)}$$
$$y(t) = x(t) + \xi(t)$$

or,

$$\frac{dx(t)}{dt} = \{\alpha + \frac{1}{1+e^{-\beta T(t)}}(\gamma - \alpha)\}x(t) + b_0 u(t-\tau) + c \qquad \text{(11b)}$$
$$y(t) = x(t) + \xi(t)$$

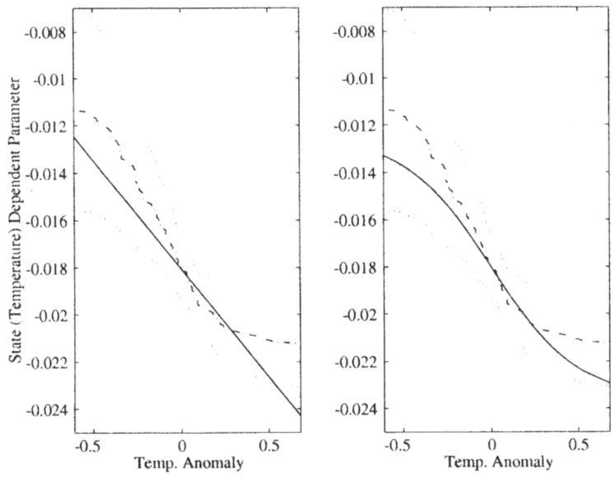

Fig. 4. Initial, non-parametric SDP estimate (dash-dot line) and standard error bounds (dots) compared with the parametric model estimates (full lines) from (11a) and 11(b): (a) linear change with temperature (left panel); (b) sigmoidal change with temperature (right panel) (cf linear model estimate $\hat{a}_1 = -0.0167$).

In the case of (11a), which is the model we shall consider further below, the estimated parameters, again based initially on the whole data set ($N = 145$), are as follows[7]:

$$\hat{\alpha} = -0.0181(0.0016); \hat{\beta} = -0.0092(0.0021);$$
$$\hat{b}_0 = 0.0402(0.0017); c = 0.0069(0.0014);$$
$$\hat{\tau} = 5.0; \sigma_\xi^2 = 0.0025; \sigma^2 = 0.0006$$
$$R_T^2 = 0.9995; R^2 = 0.9999$$

where the AIC once more identifies an AR(8) or ARMA(2,2) model for the noise ξ_t. Also as before, the final model residuals e_t show no significant autocorrelation and no cross-correlation with the input u_t. But now $\hat{\xi}_t$ also shows no significant correlation with T_t (statistically insignificant maximum correlation of $0.11(0.17)$ at a lag of 10 years), as required. Bearing on our earlier discussion, note that the estimate of c is significant but very small, as anticipated.

The linear-in-temperature law obtained in the above estimation is shown as the full line in the left hand panel of Fig. 4 and we see that it coincides well with the non-parametric SDP estimate. The right hand panel in Fig. 4 shows the results obtained with the best competing sigmoidal model (11b). This was investigated to allow for any possible flattening out of the relationship at higher levels of the temperature anomaly (see earlier discussion) which would have most effect on long-term model predictions. However, there is only a minor effect of this type. Also, despite its additional parameter, this model explains the data virtually the same as (11a) and two of its parameters are more poorly defined statistically (particularly β). So, it is sensible on parsimony grounds to proceed with the simpler model (11a).

[7] These results were obtained with a fixed $\tau = 5$: however, this was based on prior estimation with τ allowed to take on non integral values.

3.4.3. *Predictive validation* In order to investigate the predictive capacity of the model (11a), it is re-estimated on the basis of the estimation data set alone (i.e. only the first 120 annual samples up to 1975) and then its performance is evaluated by forecasting the perturbational pCO_2 variations over the last 25 years of the 20th Century, without any re-estimation of the parameters over this period. To make this exercise more demanding, the model integrations are initiated in 1856, with the initial condition for the integrations set to the value of $y(0) = 0.6523$ *pa* on this date and *with no reference to the actual* $y(t)$ *measurements at all after this.* In other words, the forecast is based on a straightforward Monte Carlo Simulation (MCS) of the nonlinear model from this initial condition using only the measured CO_2 emissions and temperature anomaly as inputs to the model. This MCS analysis is based on the estimated covariance matrix of the model parameters and involves 1000 random realizations of the model. The results are shown in Fig. 5, where we see that the $y(t)$ variations are predicted very well indeed ($R_T^2 = 0.9993$, only a little worse than obtained from model estimation based on all 145 samples) and the MCS estimated confidence region is small. For comparison, the sigmoidal model (11b) performed about the same in predictive validation terms; while the linear model (10a,b) was significantly worse (in statistical terms), with $R_T^2 = 0.9985$.

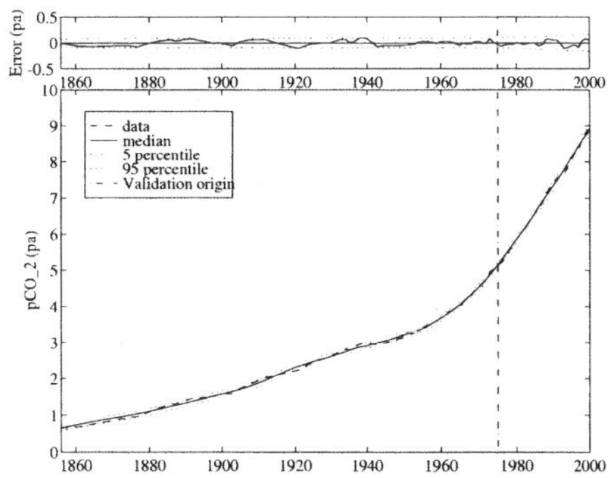

Fig. 5. Monte Carlo Simulation and predictive validation results: perturbations in atmospheric pCO_2 about the assumed pre-industrial level (dashed line); deterministic simulation and MCS median prediction (full line); 5%-95% percentile bounds (dotted lines). The error between the simulated and measured pCO_2 is shown in the top panel at a smaller scale.

4. CONCLUSIONS

This paper provides a tutorial introduction to methods for identifying linear and nonlinear, continuous-time models of stochastic systems from discrete-time sampled data and illustrates the practical utility of these methods in the data-based mechanistic modelling of environmental systems. The main advantage of these methods over the alternative and better known discrete-time methods is that they provide differential equation models whose parameters can be interpreted immediately in physically meaningful terms. As a result, they are of direct use to scientists and engineers who most often derive models in differential equation terms based on natural (e.g. conservation) laws and who are much less familiar with 'black-box' discrete-time models. The continuous-time methods can be adapted easily to handle the case of irregularly sampled data. They are also much superior when applied to rapidly sampled data, where discrete-time methods often perform poorly because the eigenvalues lie close to the unit circle in the complex domain, so that the model parameters are more poorly defined in statistical terms. Finally, all of the methods discussed in the paper are available in the user-friendly Matlab CONTSID and CAPTAIN toolboxes.

5. REFERENCES

Akaike, H. (1974) A new look at statistical model identification, *I.E.E.E. Trans. Auto. Control*, **AC19**, 716-722.

Bastogne, T., H. Garnier and P. Sibille (2001) A PMF-based subspace method for continuous-time model identification. Application to a multivariable winding process. *International Journal of Control*, **74**, 118-132.

Beer, T. and Young, P. C. (1983) Longitudinal dispersion in natural streams, *Jnl. Env. Eng. Div., American Soc. Civ. Eng.*, **102**, 1049-1067.

Fan, H., Söderström, T., Mossberg, M., Carlsson, B. and Zou, Y (1999). Estimation of continuous- time AR process parameters from discrete-time data, *IEEE Trans. on Signal Processing*, **47**, 1232-1244.

Garnier, H., Sibille, P., NGuyen, H.L. and Spott, T. (1994) A bias-compensating least-squares method for continuous time system identification via Poisson moment functionals, *10th IFAC Symposium on System Identification, Copenhagen*. 3675-3680.

Garnier, H., Sibille, P. and Richard, A. (1995) Continuous time canonical state-space model identification via Poisson moment functionals, *34th IEEE Conference on Decision and Control (CDC95), New Orleans, U.S.A.*, 3004-3009.

Garnier, H., Sibille, P. and Bastogne, T. (1997) A bias-free least-squares parameter estimator for continuous-time state-space models, *36th Conference on Decision and Control, San Diego, U.S.A., Vol. 2*, 1860-1865.

Garnier, H. and Mensler, M. (2000) The CONTSID toolbox: a Matlab toolbox for CONtinuous- Time System IDentification, *12th IFAC Symposium on System Identification, Santa Barbara, U.S.A.*.

Garnier, H., Gilson, M. and Zheng, W. X. (2000) A bias-eliminated least-squares method for continuous time model identification of closed-loop systems, *International Journal of Control*, **73**, 38-48.

Garnier, H., Mensler, M and Richard, A. (2003) Continuous-time model identification from sampled data: implementation issues and performance evaluation, *International Journal of Control* (accepted for publication).

Jarvis, A. J. and Young, P. C. (2002) Identification of temperature dependency in the relationship between global anthropogenic CO_2 emissions and atmospheric CO_2, 1856-2000, letter submitted to *Nature*.

Johansson, R. (1994) Identification of continuous-time model, *IEEE Trans. on Signal Processing*, **42**, 887-896.

Ljung, L. (2003) Initialisation aspects for subspace and output error identification methods, *European Control Conference (ECC'2003)*, Cambridge, U.K.

Mendel, J. M. (1969) *A priori* and *a posteriori* identification of time varying parameters, *2nd Hawaii Conference on System Sciences, Hawaii.*

Nielsen, J.N., Madsen, H. and Young, P.C. (2000) Parameter estimation in stochastic differential equations: an overview, *Annual Reviews in Control*, **24**, 83-94.

Pearson, A.E., Y. Shen and V. Klein (1994) Application of Fourier modulating function to parameter estimation of a multivariable linear differential system, in *10th IFAC Symposium on System Identification, Copenhagen*, 49-54.

Pham, D.T. (2000) Estimation of continuous-time autoregressive model from finely sampled data, *IEEE Trans. on Signal Processing*, **48**, 2576-2584.

Pintelon, R., Schoukens, J. and Rolain, Y (2000) Box-Jenkins continuous-time modeling, *Automatica*, **36**, 983-991.

Price, L., P. C. Young, D. Berckmans, K. Janssens and J. Taylor (1999) Data-based mechanistic modelling and control of mass and energy transfer in agricultural buildings, *Annual Reviews in Control*, **23**, 71-82.

Priestley, M. B. (1988) *Nonlinear and Nonstationary Time Series Analysis*, Academic Press: London.

Rao, G. P. and Garnier, H. (2002) Numerical illustrations of the relevance of direct continuous-time model identification, *15th Triennial IFAC World Congress on Automatic Control*, Barcelona, Spain.

Sagara, S. and Z.Y. Zhao (1990) Numerical integration approach to on-line identification of continuous-time systems. *Automatica*, **26**, 63-74.

Saha, D.C. and G.P. Rao (1983) *Identification of Continuous Dynamical Systems - the Poisson Moment Functionals (PMF) approach*, Springer-Verlag: Berlin.

Shinbrot, M. (1957) On the analysis of linear and non linear systems, *Trans. ASME*, **79**, 547-552.

Sinha, N.K. and G.P. Rao (eds.) (1991) *Identification of Continuous-Time Systems: Methodology and Computer Implementation*, Kluwer: Dordrecht.

Söderström, T. and Mossberg, M. (2000). Performance evaluation of methods for identifying continuous-time autoregressive processes, *Automatica*, **36**, 53-59.

Söderström, T., Fan, H., Carlsson, B. and Bigi, S. (1997). Least squares parameter estimation of continuous-time ARX models from discrete-time data. *IEEE Trans. on Automatic Control*, **42**, 659-673.

Trigeassou, J.C. (1987) Contribution à l'extension de la méthode des moments en automatique. Application à l'identification des systèmes linéaires. *Thèse d'Etat, Université de Poitiers, Thèse d'Etat.*

Tuan, P.D. (1977) Estimation of parameters of continuous-time gaussian stationary process with rational spectral density, *Biometrica*, **64**, 385-399.

Unbehauen, H. and G.P. Rao (1987) *Identification of Continuous Systems*, North-Holland: Amsterdam.

Unbehauen, H. and G.P. Rao (1990) Continuous-time approaches to system identification - a survey, *Automatica*, **26**, 23-35.

Unbehauen, H. and G. P. Rao (1997) Identification of continuous-time systems: a tutorial, *11th IFAC Symposium on System Identification, Kitakyushu, Japan*, 1023-1049.

Unbehauen, H. and G.P. Rao (1998) A review of identification in continuous-time systems. *Annual Reviews in Control*, **22**, 145-171.

Wang, L and P. Gawthrop (2001) On the estimation of continuous-time TFs, *International Journal of Control*, **74**, 889-904.

Young, P.C. (1964) In flight dynamic checkout: a discussion, *IEEE Trans. on Aerospace*, **2**, 1106-1111.

Young, P.C. (1965) The determination of the parameters of a dynamic process, *Radio and Electronic Eng.*, **29**, 345-361.

Young, P.C. (1969) Applying parameter estimation to dynamic systems: Part l, Theory, Control Engineering, **16**, 10, 119-125; Part II Applications, **16**, 11, 118-124.

Young, P.C. (1970) An instrumental variable method for real-time identification of a noisy process, *Automatica*, **6**, 271-287.

Young, P.C. (1978) A general theory of modeling for badly defined dynamic systems. In *Modeling, Identification and Control in Environmental Systems*, G.C. Vansteenkiste (Ed.), North Holland, 103-135.

Young, P. C. (1981) Parameter estimation for continuous-time models - a survey, *Automatica*, **17**, 23-39.

Young, P. C. (1984) *Recursive Estimation and Time Series Analysis*, Springer-Verlag: Berlin.

Young, P. C. (1989) Recursive estimation, forecasting and adaptive control, chapter in C.T. Leondes (ed.) *Control and Dynamic Systems: Advances in Theory and Applications*, **30**, Academic Press: San Diego, 119-166.

Young, P.C. (1998) Data-based mechanistic modelling of environmental, ecological, economic and engineering systems, *Environmental Modelling and Software*, **13**, 105-122.

Young, P.C. (1999) Nonstationary time series analysis and forecasting, *Progress in Environmental Science*, **1**, 3-48.

Young, P.C. (2001b) Data-based mechanistic modelling of environmental systems, first plenary session keynote paper, *International Federation on Automatic Control (IFAC) Workshop on Environmental Systems*, Tokyo, Japan.

Young, P. C. (2002) Comments on "On the estimation of continuous-time transfer functions" by L. Wang and P. Gawthrop, *International Journal of Control*, **75**, 693-697.

Young, P. C. and A. J. Jakeman (1980) Refined instrumental variable methods of recursive time-series analysis: part III, extensions, *International Journal of Control*, **31**, 741-764.

Young, P. C. and Jarvis, A. J. (2002) Data-based mechanistic modelling, the global carbon cycle and global warming, *Centre for Research on Environmental Systems and Statistics, CRES Report Number TR/177, Department of Environmental Science, Lancaster University.*

Young, P.C. and Lees, M. J. (1993) The active mixing volume: a new concept in modelling environmental systems. In *Statistics for the Environment*, V. Barnett and K. F. Turkman (Eds.), J. Wiley: Chichester, 3-43.

Young, P.C., Foster, M. and Lees, M. (1993) A direct approach to the identification and estimation of continuous-time systems from discrete-time data based on fixed interval smoothing, *Proc. 12th IFAC World Congress, vol. 10* Pergamon Press: Oxford, 27-30.

Young, P.C., McKenna, P. and Bruun, J. (2001) Identification of nonlinear stochastic systems by state dependent parameter estimation, *International Journal of Control*, **74**, 1837-1857.

IFAC
Publications
www.elsevier.com/locate/ifac

CONTINUOUS-TIME SYSTEM IDENTIFICATION OF A FOOD EXTRUDER: EXPERIMENT DESIGN AND DATA ANALYSIS

Liuping Wang * Peter J Gawthrop ** Charlie Chessari *
Tony Podsiadly *****

** Discipline of Electrical Energy and Control Systems, School of Electrical and Computer Engineering, RMIT University, Melbourne, Victoria 3000, Australia.* ***Liuping.Wang@ems.rmit.edu.au***
*** Centre for Systems and Control and Department of Mechanical Engineering, University of Glasgow, Glasgow. G12 8QQ Scotland.*
P.Gawthrop@eng.gla.ac.uk
**** Food Science Australia, North Ryde, Sydney, Australia*

Abstract: The introduction of product quality self-regulation to food-cooking extrusion is an important aspect of process control within food manufacturing industries. In order to design an automatic control system for product quality, a mathematical model of the food extruder is required. As first-principles models are difficult to obtain in this context, a food extruder is a good candidate for applying system identification tools.
This paper presents the application of continuous time system identification to such a food cooking extruder. More specifically, the reported application features an automated identification experiment apparatus designed using relay feedback control mechanisms and instrumented through existing real time supervisory system for the extruder. Experimental data from the food extruder are obtained and analysed using our identification approach.

Keywords: Continuous time systems, relay feedback control, parameter estimation.

1. INTRODUCTION

Product quality control in the extrusion industry is predominantly based on manual operation. Operators make decisions on process operation conditions based on past experience and a given recipe. As with all food processes, a key objective is to produce a safe, high quality product, while minimising waste arising from process variability; automatic feedback control of product quality provides a means to achieve the the objective. As with any other control applications, a key step is to obtain a valid mathematical model of the process to be controlled. As there are few control engineers in food manufacturing industries, the difficulties associated with obtaining a valid mathematical model in a production plant are formidable. In particular, the design of an identification experiment requires considerable knowledge and experience in the field of system identification (Goodwin and Payne, 1977; Ljung, 1987). Bearing in mind these factors, this paper presents an application of continuous time system identification to food extruder where we consider both automated experiment design and data analysis.

The first step in the project reported here is to design an automated identification experiment apparatus which is instrumented to a Supervisory Control and Data Acquisition System widely used in today's food industry. This apparatus is based on relay feedback control mechanisms (Åström and Hagglund, 1984; Astrom and Hagglund, 1988) with modified periodic oscillations (Wang *et al.*, 1999). With this automated experiment design, the identification experiments are conducted in a closed-loop operation with a nonlinear feedback mechanism, which also provides safe guards for extruder operation. In addition, the relay feedback mechanisms generate a sequence of input signals that target the important frequency region of the plant. The choice of sampling rate is set to be as fast as possible for the data acquisition system in a near continuous fashion, and the input amplitude becomes a choice in relation to the actual plant measurement noise.

The next step is to estimate a mathematical model. Because the extrusion process has mechanical components and chemical reactions, the time constants of the process widely differ. In addition, the capacity of the existing data acquisition equipment allows us to use fast sampling rate in the data collection. At the environment of fast sampling, continuous time models often have the advantages over the discrete models. Thus, in this particular application, we prefer the continuous time model to a discrete time model. Continuous time system identification has been discussed in the literature for the last three decades (Young, 1981; Unbehauen and Rao, 1990). One of the crucial points in continuous time identification is how to convert the discrete time measurement in digital data acquisition system to an equivalent continuous time measurement. The way that this conversion is performed divides the categories of continuous time identification algorithms into direct and indirect approaches. This paper investigates the applications of both types of algorithms in the food extruder identification problem.

In this work, both approaches use the state variable filter (SVF) approach (Young, 1981; Unbehauen and Rao, 1990; Wang and Gawthrop, 2001; Gawthrop and Wang, 2002*b*). In the direct approach, the coefficients of the continuous time transfer function model are estimated directly from the discrete measurement. In contrast, this paper discusses an indirect approach which estimates a non-parametric model (the system step response) followed by fitting of a continuous-time transfer function model using the SVF approach.

2. EXPERIMENT DESIGN FOR EXTRUSION PROCESSES

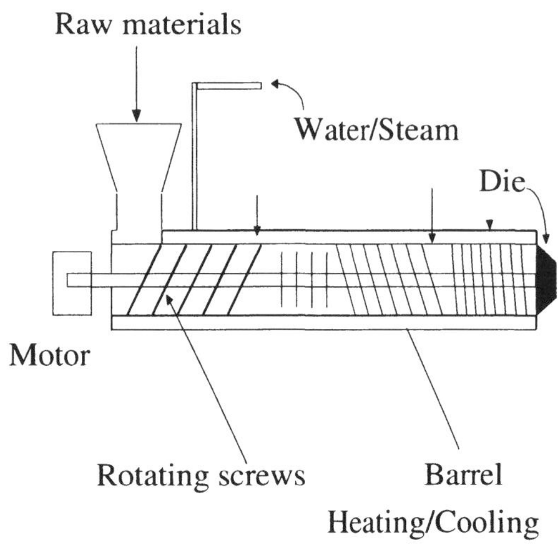

Fig. 1. *Extruder schematic*

Extrusion is a continuous process in which a rotating screw is used to force the food material through the barrel of the machine and out through a narrow die opening. In this process the material is simultaneously transported, mixed, shaped, stretched and sheared under elevated temperature and pressure. Figure 1 shows the schematic diagram of a twin screw food extruder. The extruder in the study is an APV-MPF40 co-rotating twin-screw extruder. The extruder has the following specifications:

- Throughput: 20-75kg/hr
- Screw Diameter: 40mm
- Length to Diameter Ratio: 25:1
- Feeding System: Gravimetric
- Process Monitoring: via Siemens 95U PLC

The extruder was run with the following setup for all experiments:

- Screw Profile No: 12 (High Shear Configuration)
- Die: 3mm diameter, 2mm land length

The factors which can be adjusted on-line for control purpose include the screw speed, the rate at which the raw material enters the extruder (termed feed rate) and the liquid injection rate (measured via liquid pump speed). The process output variables which are important for product quality control (Wang and Gawthrop, 2001) are the Specific Mechanical Energy, Die Pressure, screw motor torque, and Product Temperature at the location which is close to the die. All these variables can be measured on-line by using sensors.

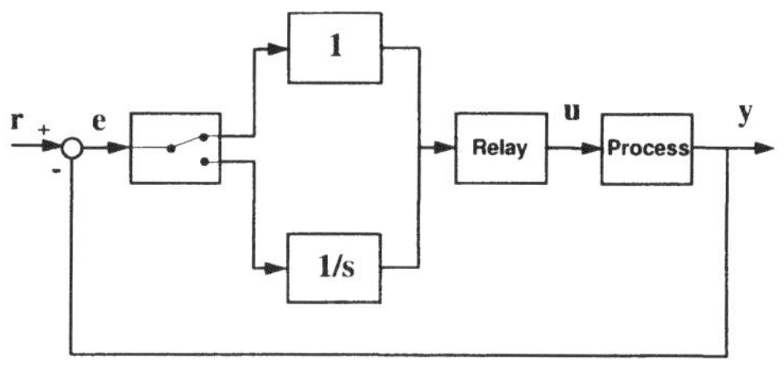

Fig. 2. *Relay feedback block-diagram*

The basic procedure in the identification experiment is to perturb the input signals one at a time and measure the responses in all output signals. The perturbation of the input signal is generated automatically by a nonlinear relay feedback mechanism. In other words, when one relay feedback control system is in operation, the rest of the input signals remains constant. Since the extruder is not a single input and single output system, the first issue in the design of the experiment is the pair of input and output variables for the relay feedback control systems. This issue is addressed by paring the input variable with the output variable that has the largest time constant, also taking consideration of the extruder physics. This approach ensures that sufficient low frequency components exist in the excitation signals for all output variables. More specifically, Screw speed is used to manipulate specific mechanical energy, liquid pump speed is used to manipulate motor torque, feed rate is used to manipulate die pressure, zone 9 temperature setpoint is used to manipulate the temperature at the die.

A simple relay is a nonlinear element that switches between the levels $-a$ and $+a$ based on the error signal e and generates a square wave input signal u to the process. In the extruder case, the process outputs are corrupted with noise, hysteresis is added to the relay to reduce the effect of the noise. Adding hysteresis to the relay produces a dead-zone to prevent the relay signal from switching due to the noise. It is well known that if the width of the hysteresis ϵ equals zero, then the oscillation frequency corresponds to the cross over frequency of the process under the feedback control. An integrator in series to the relay element generates a stable oscillation with the dominant frequency corresponding to -90^0 on the Nyquist plot (Åström and Hagglund, 1984; Astrom and Hagglund, 1988).

A standard relay experiment produces in most cases a limit cycle dominated by a single frequency. However, this information is not sufficient for the estimation of a continuous time transfer function model. The strategy we adopt in the identification experiment design was introduced in (Wang *et al.*, 1999) and applied by (Gawthrop and Wang, 2002*a*) to simulation studies of continuous time system identification, in which we make use of multiple relay experiments to generate frequency response information at several frequencies.

The proposed apparatus combines in parallel a relay element with an integrator in series with a relay element. Figure 2 provides a block diagram of this apparatus. The experiment is performed by alternatively switching the error signal between the relay path and the integrator-relay path. The design of the experiment then reduces to the selection of this switching sequence. The proposed relay experiment on its own provides some interesting ideas about how to design input signals for continuous time identification. One of the main benefits of the apparatus is that the design of an identification experiment suitable for obtaining a mathematical model has now been automated. In addition, choice of sampling rate can be set to near continuous measurement.

This experiment apparatus has been implemented on an extrusion plant through the existing Supervisory Control and Data Acquisition system on the extruder, and has been successfully used in collecting experimental data for identification purposes for all input and output variables under investigation.

3. CONTINUOUS TIME SYSTEM IDENTIFICATION: AN INDIRECT APPROACH

The basic idea of this indirect approach is first to estimate a discrete *non-parametric* model, and then to convert this discrete model into a continuous time *parametric* model. Typical non-parametric models used in the literature include the frequency response and the step response of the process; here we discuss the step response based approach. The procedures for estimating a continuous time model via step response can be summarised as follows.

(1) Estimate a discrete finite impulse response model from the discrete data, yielding a set of impulse response coefficients $h_0, h_1, h_2, \ldots, h_N$.
(2) Convert the set of impulse response coefficients into a set of step response coefficients by using, $i = 0, 1, 2, \ldots, N-1$

$$g_i = \sum_{k=0}^{i-1} h_k \tag{1}$$

(3) Use the State Variable Filter (SVF)(Young, 1981; Unbehauen and Rao, 1990; Wang and Gawthrop, 2001; Gawthrop and Wang, 2002*b*) approach to estimate a continuous-time transfer function model from the step-response data from step 2.

Step 1 is discussed in Section 3.1; step 3 has been discussed previously (Young, 1981; Unbehauen and Rao, 1990; Wang and Gawthrop, 2001; Gawthrop and Wang, 2002*b*).

3.1 *Estimation of Step Response in the Environment of Fast Sampling*

A growing methodology in the field of system identification uses basis functions (Wahlberg (1991), Ninness and Gustafsson (1997), Wang and Cluett (2000), The well known finite impulse response (FIR) model is a special case of the basis function approach. With sufficient model complexity, basis functions can capture complicated model dynamics, i.e. the unmodelled dynamics can be neglected as a small component. Since fast sampling mechanism is used in the data acquisition, an effective model structure is the frequency sampling filtering (FSF) model which maintains the properties of an FIR model and yet uses less coefficients in capturing process dynamics Wang and Cluett (1997). More specifically, a set of frequency sampling filters are pre-chosen so that linear combinations of the form

$$G(z) = \sum_{l=-\frac{n-1}{2}}^{\frac{n-1}{2}} G(e^{j\frac{2\pi l}{N}}) H^l(z) \tag{2}$$

represent a large class of systems, where $H^l(z) = \frac{1}{N}\frac{1-z^{-N}}{1-e^{j\frac{2\pi l}{N}}z^{-1}}$. The goal of discrete system identification is then to estimate the frequency response weights

$G(e^{j\frac{2\pi l}{N}})$ so that $G(z)$ represents an accurate model of a given system. Frequency sampling filter model is totally equivalent to a finite impulse response model when $n = N$. However, its strength lies in the fact that the majority of the coefficients is distributed in the higher frequency region, and hence can be ignored without causing much error in the description of dynamics (Wang and Cluett, 1997), in addition to that the frequency estimates are often severely corrupted by large estimation errors. Particularly in the environment of fast sampling, $n << N$.

We assume that the system response can be described by

$$y(k) = \sum_{l=-\frac{n-1}{2}}^{\frac{n-1}{2}} G(e^{j\frac{2\pi l}{N}})H^l(z)u(k) + \xi(k) \quad (3)$$

for a suitable choice of $G(e^{j\frac{2\pi l}{N}})'s$, where $u(k)$ is the process input, $y(k)$ is the process output and $\xi(k)$ is the disturbance signal. Define

$$F(z) = \begin{bmatrix} H^0(z) \\ H^{-1}(z) \\ H^1(z) \\ H^{-2}(z) \\ \cdots \end{bmatrix}; \theta = \begin{bmatrix} G(e^{j0}) \\ G(e^{j\frac{2\pi}{N}}) \\ G(e^{j\frac{-2\pi}{N}}) \\ G(e^{j\frac{4\pi}{N}}) \\ \cdots \end{bmatrix}$$

Given a set of sampled data $\{y(1), y(2), y(3), \ldots, y(M)\}$ and $\{u(1), u(2), u(3), \ldots, u(M)\}$, we can obtain an estimate of the frequency sampling filter model using Least Squares. Specifically, by using a predictor of the form:

$$\hat{y}(k, \theta) = \theta^T F(z)u(k) \quad (4)$$

we can choose θ to minimise a performance index of the form

$$J = \sum_{k=1}^{M} |y(k) - \hat{y}(k, \theta)|^2 \quad (5)$$

If we define $\phi(k) = F(z)u(k)$ and

$$\Phi_M = \begin{bmatrix} \phi(1) \\ \phi(2) \\ \phi(3) \\ \phi(4) \\ \cdots \end{bmatrix}; Y_M = \begin{bmatrix} y(1) \\ y(2) \\ y(3) \\ y(4) \\ \cdots \end{bmatrix}$$

then the Least Squares estimate is

$$\hat{\theta} = (\Phi_M^* \Phi_M)^{-1} \Phi_M^* Y_M \quad (6)$$

Upon obtaining the estimate of the process frequency response parameters and defining

$$S(m) = [\; S(0,m) \quad S(1,m) \quad S(-1,m) \quad \ldots \; S(\frac{n-1}{2}, m) \; S(-\frac{n-1}{2}, m) \;] \quad (7)$$

the estimated step response of the process at the sample m is (Wang and Cluett, 2000)

$$g_m = S(m)\hat{\theta} \quad (8)$$

where $S(l,m) = \frac{1}{N}\frac{1-e^{j\frac{2\pi l}{N}(m+1)}}{1-e^{j\frac{2\pi l}{N}}}$.

If $\xi(.)$ is white noise with zero mean and variance σ^2, it is easily seen that for a sufficiently large number of frequency $n \leq N$

$$E(\hat{g}(m) - g(m)] = 0 \quad (9)$$

$$E((\hat{g}(m) - g(m))^2) = S(m)(\Phi^*\Phi)^{-1}S(m)^*\sigma^2 \quad (10)$$

However, in the situation of $\xi(.)$ being coloured noise, in order to obtain the estimation of the variances about the step response model, we assume that

$$\xi(k) = \frac{1}{D(z)}\epsilon(k) \quad (11)$$

where $D(z) = 1 + d_1 z^{-1} + d_2 z^{-2} + \ldots + d_q z^{-q}$, and $\epsilon(k)$ is white noise with zero mean. An iterative algorithm in the spirit of relaxed Maximum Likelihood approach Goodwin and Payne (1977) be used for simultaneous estimation of the frequency sampling filter model and the autoregressive noise model. Upon convergence of the algorithm, the variance of the estimated step response at the sample instant m is

$$E((\hat{g}(m) - g(m))^2) = S(m)(\Phi_f^*\Phi_f)^{-1}S(m)^*\sigma_f^2 \quad (12)$$

where Φ_f is the prefiltered data matrix, and σ_f^2 is the variance of the residuals. The estimation algorithm presented here is not only simple and well understood, but also yields unbiased step response estimate with a stochastic description of the error. With the estimated variance for the step response, statistical confidence bounds can be derived.

4. EXPERIMENTAL DATA ANALYSIS

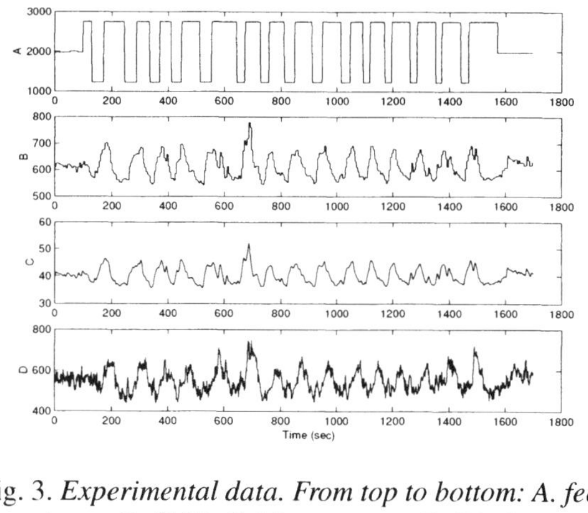

Fig. 3. *Experimental data. From top to bottom: A. feedrate; B. SME; C. Motor toque; D. Die Pressure*

The relay feedback control apparatus has been used in collecting experimental data for identification. This particular set of experimental data analysed here was generated by controlling die pressure with material feed-rate. Figure 3 shows the input signal generated by

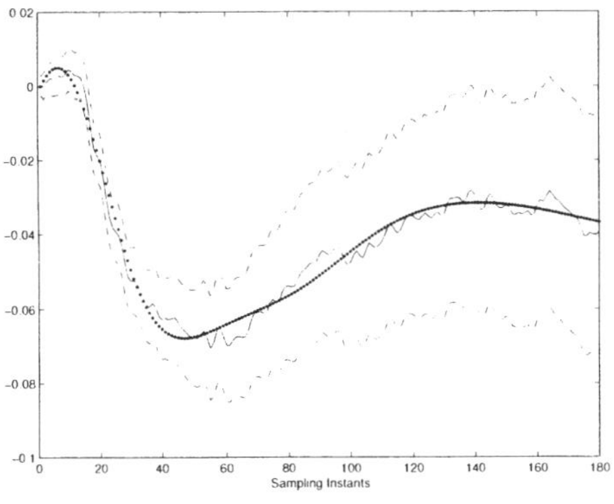

(a) SME

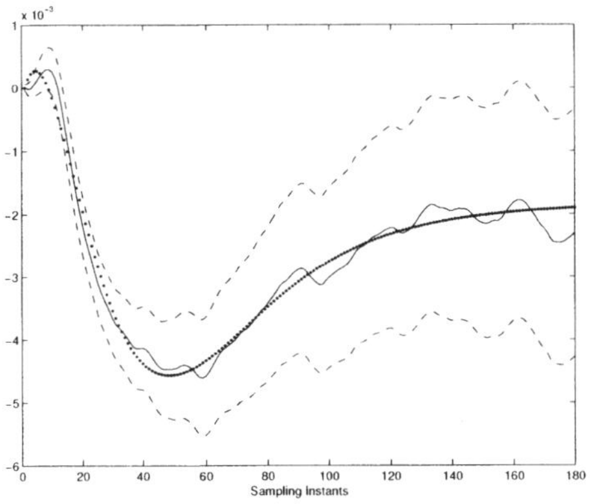

(b) Motor Torque

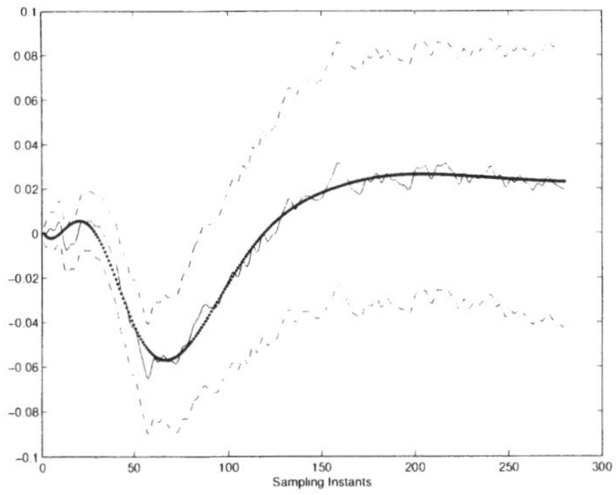

(c) Die Pressure

Fig. 4. *Solid line-estimated step response with* 2σ *confidence bounds; Dark dotted line-step response from the continuous time model*

the relay apparatus and the responses of SME, motor torque and die pressure. Three step response models with confidence bounds were estimated where the relaxed maximum likelihood method Goodwin and Payne (1977) was used to iteratively identify three noise models respectively. The model orders and the converged noise models are given below, respectively, for SME, Motor Torque and Die Pressure. A sufficiently large n was chosen for the three cases to ensure the unmodelled dynamics were negligible.

Model Order	SME	Motor Torque	Die Pressure
$n=1$	0.0562	3.11×10^{-4}	0.3793
$n=2$	0.007	2.52×10^{-5}	
$n=3$	0.0017	7.17×10^{-6}	0.0152
$n=4$	0.0015	6.41×10^{-6}	0.0045
$n=5$	0.0014	6.8×10^{-6}	0.0043

Table 1. Step Resp. Errors (sum squared)

Model Order	SME	Motor Torque	Die Pressure
$n=1$	2.09×10^{6}	8.9×10^{3}	5.25×10^{6}
$n=2$	1.09×10^{6}	4.01×10^{3}	
$n=3$	6.64×10^{5}	2.69×10^{3}	2.39×10^{6}
$n=4$	6.38×10^{5}	2.65×10^{3}	1.77×10^{6}
$n=5$	6.73×10^{5}	2.81×10^{3}	1.75×10^{6}

Table 2. Prediction Errors (sum squared)

- SME: $N = 180$, $n = 119$ and $D(z) = 1 - 0.4757z^{-1} - 0.3543z^{-2}$
- Motor Torque: $N = 180$, $n = 119$ and $D(z) = 1 - 0.7942z^{-1} - 0.0843z^{-2}$
- Die Pressure: $N = 280$, $n = 119$ and $D(z) = 1 - 1.5477z^{-1} + 0.1617z^{-2} + 0.6247z^{-3} - 0.1083z^{-4} - 0.095z^{-5}$

where the noise model orders were selected using the PRESS criterion given in Wang and Cluett (2000). The step response models with confidence bounds are shown in Figures 4(a)– 4(c).

Upon obtaining the continuous time step response data, the identification algorithm with model structure selection given in Wang and Gawthrop (2001) was used in the estimation of a continuous time transfer function model. Both sum of squares of step response errors $E_s = \sum_{k=1}^{N-1}[\hat{g}(t_k) - g(t_k)_m]^2$ and the sum of squares of prediction errors $E_p = \sum_{k=1}^{M} |y(t_k) - \frac{B(p)}{A(p)}u(t_k)|^2$ were used for model structure determination. The relative model order is selected to be one for the three cases; and the denominators of the state variable filters are selected to be in the form of $C(s) = (s+\gamma)^5$, where $\gamma = 0.12, 0.12$ and 0.012 respectively for SME, Motor Torque and Die Pressure. Tables 1–2 summarise the errors.

The step responses of the three continuous time models are compared to their respective responses in Figures 4(a)– 4(c). In addition, the predicted output responses using the continuous time models are compared with experimental data shown in Figures 5(a)-5(c); the actual data is the firm line and the predicted data the dotted line.

5. CONCLUSION

An experimental procedure involving input design has been presented along with an indirect identification procedure. This identification procedure is based on non-parameteric estimation in the discrete time domain and SVF based transfer function estimation in

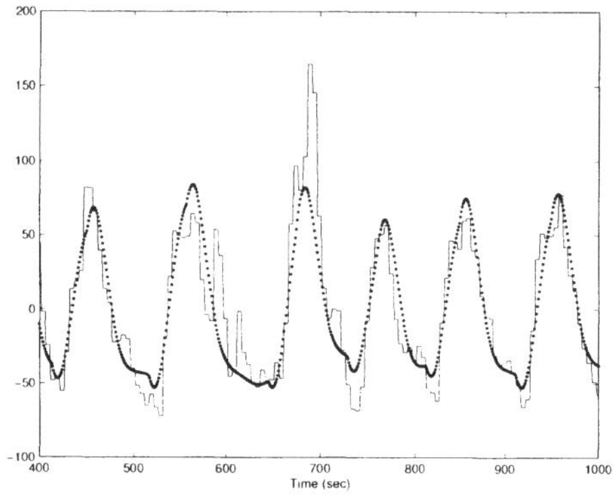

(a) SME

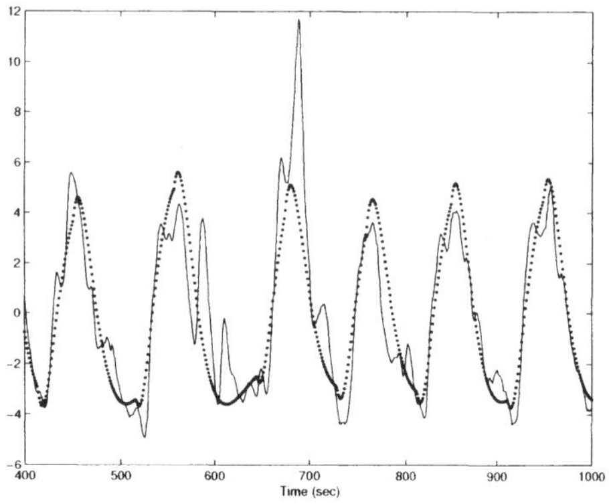

(b) Motor Torque

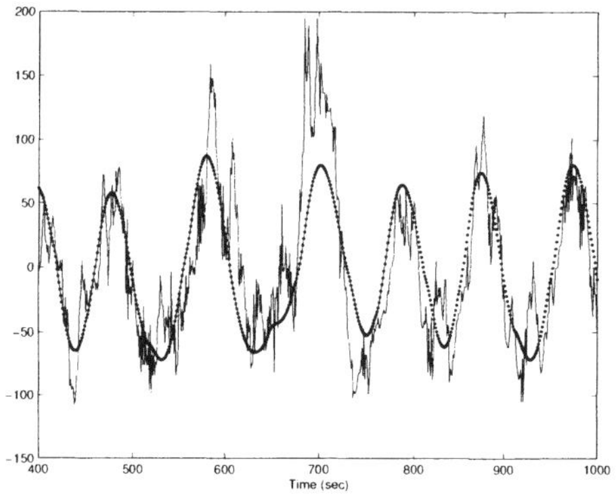

(c) Die Pressure

Fig. 5. *Predicted (dotted) and actual (firm) data*

the continuous domain. It hs been successfully applied to the data from a food cooking extruder.

We believe that this approach is intuitive enough to be used on a regular basis by process engineers not skilled in system identification techniques.

6. REFERENCES

Åström, K. J. and T. Hagglund (1984). Automatic tuning of simple regulators with specifications on gain and phase and amplitude margins. *Automatica* **20**, 645–652.

Astrom, K. J. and T. Hagglund (1988). *Automatic Tuning of PID Controllers*. Instrument Society of America, Research Triangle Park, NC.

Gawthrop, Peter J and Liuping Wang (2002*a*). Infinite-impulse and finite-impulse response filters for continuous-time parameter estimation. In: *Proceedings of the 15th IFAC World Congress*. Barcelona, Spain.

Gawthrop, P.J. and L. Wang (2002*b*). Transfer function and frequency response estimation using resonant filters. *Proceedings of the Institution of Mechanical Engineers Pt. I: Journal of Systems and Control Engineering* **216**(I6), 441–453.

Goodwin, G. C. and R. L. Payne (1977). *Dynamic System Identification*. Academic Press.

Ljung, L. (1987). *System Identification — Theory for the User*. Prentice-Hall. Englewood cliffs, New Jersey, USA.

Ninness, B. and F. Gustafsson (1997). A unifying construction of orthonormal bases for system identification. *IEEEAC* **42**(4), 515–521.

Unbehauen, H. and G. P. Rao (1990). Continuous-time approaches to system identification—a survey. *Automatica* **26**(1), 23–35.

Wahlberg, B. (1991). System identification using Laguerre models. *IEEE Transactions on Automatic Control* **36**, 551–562.

Wang, L. and W. R. Cluett (1997). Frequency-sampling filters: ann improved model structure for step-response identification. *Automatica* **33**, 939–944.

Wang, L and W R Cluett (2000). *From Plant Data to Process Control*. Taylor and Francis. London and New York.

Wang, L., M. Desarmo and W. R. Cluett (1999). Real-time estimation of process frequency response and step response from relay feedback experiments. *Automatica* **Vol. 35**, 1427–1436.

Wang, Liuping and Peter J Gawthrop (2001). On the estimation of continuous time transfer functions. *Int. J. Control* **74**(9), 889–904.

Young, P. C. (1981). Parameter estimation for continuous-time models — a survey. *Automatica* **17**(1), 23–39.

IFAC
Publications
www.elsevier.com/locate/ifac

IDENTIFICATION OF CONTINUOUS TIME MODELS USING DISCRETE TIME DATA

Niels Rode Kristensen *,[1] **Henrik Madsen** **
Sten Bay Jørgensen *

* *Department of Chemical Engineering, Technical University of Denmark, Building 229, DK-2800 Lyngby, Denmark*
** *Informatics and Mathematical Modelling, Technical University of Denmark, Building 321, DK-2800 Lyngby, Denmark*

Abstract: Continuous-discrete stochastic state space models in the form of nonlinear partially observed Itô stochastic differential equations (SDE's) with measurement noise are advocated for modelling dynamic systems in continuous time using discrete time data. Such models provide a decomposition of the noise affecting the system into a process noise term and a measurement noise term, and this *prediction error decomposition* (PED) allows unknown parameters to be estimated from experimental data in a *prediction error* (PE) setting, which gives less biased and more reproducable results in the presence of significant process noise than the more commonly used *output error* (OE) setting. To facilitate the use of continuous-discrete stochastic state space models, a PE estimation scheme that features *maximum likelihood* (ML) and *maximum a posteriori* (MAP) estimation is presented along with a software implementation. To illustrate the superiority of PE estimation over OE estimation a case study is given, which demonstrates the higher sensitivity of OE estimates to process noise. *Copyright © 2003 IFAC*

Keywords: Continuous time systems, stochastic modelling, nonlinear systems, parameter estimation, prediction error methods, maximum likelihood estimators, extended Kalman filters, software tools, parallel computation.

1. INTRODUCTION

Mathematical modelling of dynamic systems based on stochastic differential equations (SDE's) has recieved limited attention in the control and system identification communities since Jazwinski (1970) and Åström (1970). This is evident from a series of review papers on the state of the art of identification of continuous time models (Young, 1981; Unbehauen and Rao, 1990; Unbehauen and Rao, 1998). However, it is the opinion of the authors of the present paper that the topic deserves much more attention, because SDE's, when used in the context of a continuous-discrete stochastic state space model, constitute a very natural framework for modelling real dynamic systems. Continuous-discrete stochastic state space models consist of a set of SDE's describing the dynamics of the system in continuous time and a set of discrete time measurement equations. This model type very correctly reflects the real world of most dynamic systems, where system dynamics are inherently continuous and at the same time

[1] Corresponding author.

subject to random effects, whereas measurements are inherently discrete. Furthermore, many, especially chemical and biological, systems have a large number of state variables, of which usually only a subset can be measured, and this is easily reflected by models of this type as well. The particular structure of continuous-discrete stochastic state space models also implies that they can be easily derived from first principles, that physical insights and other prior information can be incorporated directly, and that the parameters can be easily given a physical interpretation.

A final, and key, advantage of continuous-discrete stochastic state space models is that they allow for a decomposition of the noise affecting the system into a process noise term and a measurement noise term. As a consequence of this *prediction error decomposition* (PED), unknown parameters of such models can be estimated from experimental data in a *prediction error* (PE) setting as opposed to the more commonly used *output error* (OE) setting, which gives more biased and less reproducible results in the presence of significant process noise, because random effects are absorbed into the parameter estimates. Furthermore, PE estimation allows for subsequent application of a number of powerful statistical tools to provide indications for possible improvements to the structure of the model. In particular, estimates of the parameters of the diffusion term can be used to assess the quality of a model and to pinpoint model deficiencies in order to subsequently uncover their structural origin (Kristensen *et al.*, 2002*a*).

The focus of the present paper is on estimation of unknown parameters in continuous time models based on discrete time data. The primary aim of the paper is to present an efficient and flexible scheme for performing the estimation for continuous-discrete stochastic state space models and a software implementation of this scheme. A secondary aim of the paper is to demonstrate the differences between this PE estimation method and standard OE estimation through a case study.

The remainder of the paper is organized as follows: In Section 2 the estimation methodology is outlined, and in Section 3 the software implementation is described. In Section 4 the case study comparing PE and OE estimation is given, and in Section 5 the results are discussed. Finally, in Section 6 the conclusions of the paper are given.

2. ESTIMATION METHODOLOGY

This section contains a condensed outline of a PE estimation scheme for continuous-discrete stochastic state space models. A detailed outline of the scheme is given by Kristensen *et al.* (2002*b*).

2.1 Model structure

In the general case, the continuous-discrete stochastic state space model is a model that consists of a set of nonlinear discretely, partially observed Itô SDE's with measurement noise, i.e.:

$$d\boldsymbol{x}_t = \boldsymbol{f}(\boldsymbol{x}_t, \boldsymbol{u}_t, t, \boldsymbol{\theta})dt + \boldsymbol{\sigma}(\boldsymbol{u}_t, t, \boldsymbol{\theta})d\boldsymbol{\omega}_t \quad (1)$$

$$\boldsymbol{y}_k = \boldsymbol{h}(\boldsymbol{x}_k, \boldsymbol{u}_k, t_k, \boldsymbol{\theta}) + \boldsymbol{e}_k \quad (2)$$

where $t \in \mathbb{R}$ is the time variable; $\boldsymbol{x}_t \in \mathcal{X} \subset \mathbb{R}^n$ is a vector of state variables; $\boldsymbol{u}_t \in \mathcal{U} \subset \mathbb{R}^m$ is a vector of input variables; $\boldsymbol{y}_k \in \mathcal{Y} \subset \mathbb{R}^l$ is a vector of output variables; $\boldsymbol{\theta} \in \Theta \subset \mathbb{R}^p$ is a vector of (possibly unknown) parameters; $\boldsymbol{f}(\cdot) \in \mathbb{R}^n$, $\boldsymbol{\sigma}(\cdot) \in \mathbb{R}^{n \times n}$ and $\boldsymbol{h}(\cdot) \in \mathbb{R}^l$ are nonlinear functions; $\{\boldsymbol{\omega}_t\}$ is an n-dimensional standard Wiener process and $\{\boldsymbol{e}_k\}$ is an l-dimensional white noise process with $\boldsymbol{e}_k \in N(\boldsymbol{0}, \boldsymbol{S}(\boldsymbol{u}_k, t_k, \boldsymbol{\theta}))$.

SDE's may be interpreted both in the sense of Stratonovich and in the sense of Itô, but since the Stratonovich interpretation is less suitable for parameter estimation (Jazwinski, 1970; Åström, 1970), the Itô interpretation is adapted. Furthermore, the diffusion term in (1) is assumed to be independent of the state variables, because this renders parameter estimation more feasible.

2.2 Parameter estimation

The solution to (1) is a Markov process and an estimation scheme based on probabilistic methods can therefore be applied to estimate the unknown parameters of the model in (1)-(2), e.g. *maximum likelihood* (ML) or *maximum a posteriori* (MAP), where the latter can be applied if prior information about the parameters is available. Let:

$$\mathbf{Y} = [\mathcal{Y}^1_{N_1}, \mathcal{Y}^2_{N_2}, \ldots, \mathcal{Y}^i_{N_i}, \ldots, \mathcal{Y}^S_{N_S}] \quad (3)$$

be a set of S stochastically independent sequences of consecutive measurements, where:

$$\mathcal{Y}^i_{N_i} = [\boldsymbol{y}^i_{N_i}, \ldots, \boldsymbol{y}^i_k, \ldots, \boldsymbol{y}^i_1, \boldsymbol{y}^i_0] \quad (4)$$

and let $p(\boldsymbol{\theta})$ be a prior probability density function for the parameters. In the general case, point estimates of the parameters in (1)-(2) can then be found as the parameters $\boldsymbol{\theta}$ that maximize the posterior probability density function:

$$p(\boldsymbol{\theta}|\mathbf{Y}) \propto \left(\prod_{i=1}^{S} p(\mathcal{Y}^i_{N_i}|\boldsymbol{\theta}) \right) p(\boldsymbol{\theta}) \quad (5)$$

or equivalently:

$$p(\boldsymbol{\theta}|\mathbf{Y}) \propto \left(\prod_{i=1}^{S} \left(\prod_{k=1}^{N_i} p(\boldsymbol{y}^i_k|\mathcal{Y}^i_{k-1}, \boldsymbol{\theta}) \right) p(\boldsymbol{y}^i_0|\boldsymbol{\theta}) \right) p(\boldsymbol{\theta}) \quad (6)$$

where the rule $P(A \cap B) = P(A|B)P(B)$ has been applied in a successive manner to form products of conditional probability density functions.

In order to obtain an exact evaluation of the posterior probability density function in (6), the initial probability density functions $p(\boldsymbol{y}_0^i|\boldsymbol{\theta})$, $i = 1, \ldots, S$, must be known and all subsequent conditional probability density functions must be determined by successive prediction based on Kolmogorov's forward equation and updating based on Bayes' rule (Jazwinski, 1970). In practice this approach is computationally infeasible and an alternative is therefore needed. Nielsen *et al.* (2000) have recently given a review of the state of the art with respect to parameter estimation in discretely observed Itô SDE's, which shows that for models of the type in (1)-(2) only methods based on approximate nonlinear filters provide a computationally feasible solution to the problem. However, since the diffusion term in (1) has been assumed to be independent of the state variables, a much simpler alternative can be applied here.

More specifically, since the SDE's in (1) are driven by a Wiener process, and since increments of a Wiener process are Gaussian, it is reasonable to assume, under some regularity conditions, that the conditional probability density functions can be well approximated by Gaussian densities, which means that a method based on the extended Kalman filter (EKF) can be applied. The Gaussian density is completely characterized by its mean and covariance, so by introducing:

$$\hat{\boldsymbol{y}}_{k|k-1}^i = E\{\boldsymbol{y}_k^i|\mathcal{Y}_{k-1}^i, \boldsymbol{\theta}\} \tag{7}$$

$$\boldsymbol{R}_{k|k-1}^i = V\{\boldsymbol{y}_k^i|\mathcal{Y}_{k-1}^i, \boldsymbol{\theta}\} \tag{8}$$

$$\boldsymbol{\epsilon}_k^i = \boldsymbol{y}_k^i - \hat{\boldsymbol{y}}_{k|k-1}^i \tag{9}$$

and by further assuming that the prior probability density function for the parameters is Gaussian as well and therefore also introducing:

$$\boldsymbol{\mu}_{\boldsymbol{\theta}} = E\{\boldsymbol{\theta}\} \tag{10}$$

$$\boldsymbol{\Sigma}_{\boldsymbol{\theta}} = V\{\boldsymbol{\theta}\} \tag{11}$$

$$\boldsymbol{\epsilon}_{\boldsymbol{\theta}} = \boldsymbol{\theta} - \boldsymbol{\mu}_{\boldsymbol{\theta}} \tag{12}$$

the posterior probability density function becomes:

$$p(\boldsymbol{\theta}|\mathbf{Y}) \propto \left(\prod_{i=1}^{S} \left(\prod_{k=1}^{N_i} \frac{\exp\left(-\frac{1}{2}(\boldsymbol{\epsilon}_k^i)^T (\boldsymbol{R}_{k|k-1}^i)^{-1} \boldsymbol{\epsilon}_k^i\right)}{\sqrt{\det\left(\boldsymbol{R}_{k|k-1}^i\right)} \left(\sqrt{2\pi}\right)^l} \right) \times p(\boldsymbol{y}_0^i|\boldsymbol{\theta}) \right) \times \frac{\exp\left(-\frac{1}{2}\boldsymbol{\epsilon}_{\boldsymbol{\theta}}^T \boldsymbol{\Sigma}_{\boldsymbol{\theta}}^{-1} \boldsymbol{\epsilon}_{\boldsymbol{\theta}}\right)}{\sqrt{\det\left(\boldsymbol{\Sigma}_{\boldsymbol{\theta}}\right)} \left(\sqrt{2\pi}\right)^p} \tag{13}$$

The parameter estimates can now be determined by further conditioning on the initial conditions:

$$\mathbf{y_0} = [\boldsymbol{y}_0^1, \boldsymbol{y}_0^2, \ldots, \boldsymbol{y}_0^i, \ldots, \boldsymbol{y}_0^S] \tag{14}$$

and applying nonlinear optimisation to find the minimum of the negative logarithm of the resulting posterior probability density function, i.e.:

$$\hat{\boldsymbol{\theta}} = \arg\min_{\boldsymbol{\theta}\in\Theta} \{-\ln\left(p(\boldsymbol{\theta}|\mathbf{Y}, \mathbf{y_0})\right)\} \tag{15}$$

In the general case the estimation method implied by (15) is MAP, but, if no prior information about the parameters is available, it reduces to ML.

In either case, the innovations $\boldsymbol{\epsilon}_k^i$ and their covariances $\boldsymbol{R}_{k|k-1}^i$ can be computed recursively by means of the EKF for each set of parameters $\boldsymbol{\theta}$ in the optimisation. More specifically, the EKF consists of the output *prediction* equations:

$$\hat{\boldsymbol{y}}_{k|k-1}^i = \boldsymbol{h}(\hat{\boldsymbol{x}}_{k|k-1}^i, \boldsymbol{u}_k^i, t_k^i, \boldsymbol{\theta}) \tag{16}$$

$$\boldsymbol{R}_{k|k-1}^i = \boldsymbol{C}\boldsymbol{P}_{k|k-1}^i\boldsymbol{C}^T + \boldsymbol{S} \tag{17}$$

the *innovation* equation:

$$\boldsymbol{\epsilon}_k^i = \boldsymbol{y}_k^i - \hat{\boldsymbol{y}}_{k|k-1}^i \tag{18}$$

the Kalman *gain* equation:

$$\boldsymbol{K}_k^i = \boldsymbol{P}_{k|k-1}^i\boldsymbol{C}^T(\boldsymbol{R}_{k|k-1}^i)^{-1} \tag{19}$$

the *updating* equations:

$$\hat{\boldsymbol{x}}_{k|k}^i = \hat{\boldsymbol{x}}_{k|k-1}^i + \boldsymbol{K}_k^i\boldsymbol{\epsilon}_k^i \tag{20}$$

$$\boldsymbol{P}_{k|k}^i = \boldsymbol{P}_{k|k-1}^i - \boldsymbol{K}_k^i\boldsymbol{R}_{k|k-1}^i(\boldsymbol{K}_k^i)^T \tag{21}$$

and the state *prediction* equations:

$$\frac{d\hat{\boldsymbol{x}}_{t|k}^i}{dt} = \boldsymbol{f}(\hat{\boldsymbol{x}}_{t|k}^i, \boldsymbol{u}_t^i, t, \boldsymbol{\theta}) \tag{22}$$

$$\frac{d\boldsymbol{P}_{t|k}^i}{dt} = \boldsymbol{A}\boldsymbol{P}_{t|k}^i + \boldsymbol{P}_{t|k}^i\boldsymbol{A}^T + \boldsymbol{\sigma}\boldsymbol{\sigma}^T \tag{23}$$

which are solved for $t \in [t_k^i, t_{k+1}^i[$. In the above equations the following notation has been applied:

$$\boldsymbol{A} = \frac{\partial \boldsymbol{f}}{\partial \boldsymbol{x}_t}\Big|_{\hat{\boldsymbol{x}}_{k|k-1}^i, \boldsymbol{u}_k^i, t_k^i}, \quad \boldsymbol{C} = \frac{\partial \boldsymbol{h}}{\partial \boldsymbol{x}_t}\Big|_{\hat{\boldsymbol{x}}_{k|k-1}^i, \boldsymbol{u}_k^i, t_k^i}$$

$$\boldsymbol{\sigma} = \boldsymbol{\sigma}(\boldsymbol{u}_k^i, t_k^i, \boldsymbol{\theta}), \quad \boldsymbol{S} = \boldsymbol{S}(\boldsymbol{u}_k^i, t_k^i, \boldsymbol{\theta})$$

Initial conditions for the EKF are $\hat{\boldsymbol{x}}_{t|t_0}^i = \boldsymbol{x}_0^i$ and $\boldsymbol{P}_{t|t_0}^i = \boldsymbol{P}_0^i$, which can either be prespecified or estimated as a part of the overall problem.

The EKF is sensitive to nonlinear effects and the approximate solution obtained by solving (22)-(23) may be too crude (Jazwinski, 1970), so in order to obtain a better approximation, the time interval $[t_k^i, t_{k+1}^i[$ is subsampled, i.e. $[t_k^i, \ldots, t_j^i, \ldots, t_{k+1}^i[$, and the equations are linearized at each subsampling instant. This way the numerical solution of (22)-(23) can also be simplified by applying the analytical solutions to the corresponding linearized propagation equations:

$$\frac{d\hat{\boldsymbol{x}}_{t|j}^i}{dt} = \boldsymbol{f}_0 + \boldsymbol{A}(\hat{\boldsymbol{x}}_t^i - \hat{\boldsymbol{x}}_j^i) + \boldsymbol{B}(\boldsymbol{u}_t^i - \boldsymbol{u}_j^i) \tag{24}$$

$$\frac{d\boldsymbol{P}_{t|j}^i}{dt} = \boldsymbol{A}\boldsymbol{P}_{t|j}^i + \boldsymbol{P}_{t|j}^i\boldsymbol{A}^T + \boldsymbol{\sigma}\boldsymbol{\sigma}^T \tag{25}$$

for $t \in [t_j^i, t_{j+1}^i[$, where the notation:

$$\boldsymbol{A} = \frac{\partial \boldsymbol{f}}{\partial \boldsymbol{x}_t}\Big|_{\hat{\boldsymbol{x}}_{j|j-1}^i, \boldsymbol{u}_j^i, t_j^i}, \quad \boldsymbol{B} = \frac{\partial \boldsymbol{f}}{\partial \boldsymbol{u}_t}\Big|_{\hat{\boldsymbol{x}}_{j|j-1}^i, \boldsymbol{u}_j^i, t_j^i}$$

$$\boldsymbol{f}_0 = \boldsymbol{f}(\hat{\boldsymbol{x}}_{j|j-1}^i, \boldsymbol{u}_j^i, t_j^i, \boldsymbol{\theta}), \quad \boldsymbol{\sigma} = \boldsymbol{\sigma}(\boldsymbol{u}_j^i, t_j^i, \boldsymbol{\theta})$$

has been applied. The analytical solutions are:

$$\hat{\boldsymbol{x}}_{j+1|j}^i = \hat{\boldsymbol{x}}_{j|j}^i + \boldsymbol{A}^{-1}(\boldsymbol{\Phi}_s - \boldsymbol{I})\boldsymbol{f}_0 + \left(\boldsymbol{A}^{-1}(\boldsymbol{\Phi}_s - \boldsymbol{I}) - \boldsymbol{I}\tau_s\right)\boldsymbol{A}^{-1}\boldsymbol{B}\boldsymbol{\alpha} \tag{26}$$

$$\boldsymbol{P}^i_{j+1|j} = \boldsymbol{\Phi}_s \boldsymbol{P}^i_{j|j} \boldsymbol{\Phi}_s^T + \int_0^{\tau_s} e^{\boldsymbol{A}s} \boldsymbol{\sigma}\boldsymbol{\sigma}^T e^{\boldsymbol{A}^T s} ds \quad (27)$$

where $\tau_s = t^i_{j+1} - t^i_j$ and $\boldsymbol{\Phi}_s = e^{\boldsymbol{A}\tau_s}$, and where:

$$\boldsymbol{\alpha} = \frac{\boldsymbol{u}^i_{j+1} - \boldsymbol{u}^i_j}{t^i_{j+1} - t^i_j} \quad (28)$$

has been introduced to allow assumption of either *zero order hold* ($\boldsymbol{\alpha} = \mathbf{0}$) or *first order hold* ($\boldsymbol{\alpha} \neq \mathbf{0}$) on the inputs between sampling instants. More details, e.g. about computing the matrix exponential $\boldsymbol{\Phi}_s = e^{\boldsymbol{A}\tau_s}$, are given by Kristensen *et al.* (2002*b*).

The approach described above relies on a number of assumptions and approximations, which may be critical. Methods exist, however, for testing whether or not the identified model consistently describes the estimation data (Holst *et al.*, 1992).

2.3 Uncertainty of parameter estimates

Essential outputs of any sound statistical parameter estimation scheme include an assessment of the uncertainty of the estimates. An estimate of the uncertainty of the parameter estimates can be obtained by using the fact that by the central limit theorem the estimator in (15) is asymptotically Gaussian with mean $\boldsymbol{\theta}$ and covariance matrix:

$$\boldsymbol{\Sigma}_{\hat{\boldsymbol{\theta}}} = \boldsymbol{H}^{-1} \quad (29)$$

where the matrix $\boldsymbol{H}$ is given by the elements:

$$h_{ij} = -E\left\{\frac{\partial^2}{\partial\theta_i\partial\theta_j}\ln\left(p(\boldsymbol{\theta}|\mathbf{Y}, \mathbf{y_0})\right)\right\}$$

for $i, j = 1, \ldots, p$, and where an approximation to $\boldsymbol{H}$ can be obtained from the elements:

$$h_{ij} \approx -\left(\frac{\partial^2}{\partial\theta_i\partial\theta_j}\ln\left(p(\boldsymbol{\theta}|\mathbf{Y}, \mathbf{y_0})\right)\right)\Big|_{\boldsymbol{\theta}=\hat{\boldsymbol{\theta}}}$$

for $i, j = 1, \ldots, p$, which is the Hessian at the minimum of the objective function. A measure of the uncertainty of the individual estimates can be obtained by decomposing the covariance matrix:

$$\boldsymbol{\Sigma}_{\hat{\boldsymbol{\theta}}} = \boldsymbol{\sigma}_{\hat{\boldsymbol{\theta}}} \boldsymbol{R} \boldsymbol{\sigma}_{\hat{\boldsymbol{\theta}}} \quad (30)$$

into $\boldsymbol{\sigma}_{\hat{\boldsymbol{\theta}}}$, which is a diagonal matrix of the standard deviations of the parameter estimates, and $\boldsymbol{R}$, which is the corresponding correlation matrix.

The uncertainty information thus obtained can subsequently be applied to perform tests of various hypotheses, e.g. to determine the significance of the individual parameters through t-tests.

3. SOFTWARE IMPLEMENTATION

The parameter estimation scheme presented in Section 2 has been implemented in a computer program called **CTSM**. This program is available for both Linux, Solaris and Windows platforms (Kristensen *et al.*, 2002*b*), and within its graphical user interface (GUI) unknown parameters of models of the kind in (1)-(2) can be estimated using the methods presented in Section 2. Once a model structure has been set up within the GUI, the program analyzes the model equations to determine the symbolic names of the parameters and displays them to allow the user to specify which parameters to fix, which to estimate, and how each parameter should be estimated (ML or MAP). After specifying which data sets to use, the program then determines the parameter estimates and computes the uncertainty information described in Section 2. The program also provides features to deal with occasional outliers and missing observations, and on Solaris systems the program supports shared memory parallelization to alleviate the extensive computational load often associated with estimation of parameters in continuous-discrete stochastic state space models.

4. CASE STUDY: A COMPARISON OF PE AND OE ESTIMATION

To demonstrate the differences between the PE estimation method presented in Section 2 and standard OE estimation in the sense of nonlinear least squares (NLS) applied to an ordinary differential equation (ODE) model (Bard, 1974), a simulated case study is given in the following.

The problem considered is to estimate 2 parameters and all 3 initial states of a model of a fed-batch bioreactor from data sets with as well as without process noise. The model used for the simulation has the following system equation:

$$d\begin{pmatrix} X \\ S \\ V \end{pmatrix} = \begin{pmatrix} \mu(S)X - \frac{FX}{V} \\ -\frac{\mu(S)X}{Y} + \frac{F(S_F - S)}{V} \\ F \end{pmatrix} dt + \boldsymbol{\sigma} d\boldsymbol{\omega}_t \quad (31)$$

where X and S are the biomass and substrate concentrations, V is the volume, F is the feed flow rate, $Y = 0.5$ is the yield coefficient of biomass and $S_F = 10$ is the feed concentration of substrate. $\mu(S)$ is the biomass growth rate, described by Monod kinetics and substrate inhibition, i.e.:

$$\mu(S) = \mu_{\max}\frac{S}{K_2 S^2 + S + K_1} \quad (32)$$

where $\mu_{\max}$, K_1 and $K_2 = 0.5$ are kinetic parameters. $\boldsymbol{\sigma}$ is a diagonal matrix with elements σ_{11}, σ_{22} and σ_{33}. Using the true parameter and initial state values shown in Tables 1-4, simulated data sets (101 samples each) are generated with stochastic simulation using the Euler scheme by perturbing the feed flow rate along a pre-determined trajectory and subsequently adding measurement noise to the appropriate variables. It is assumed that

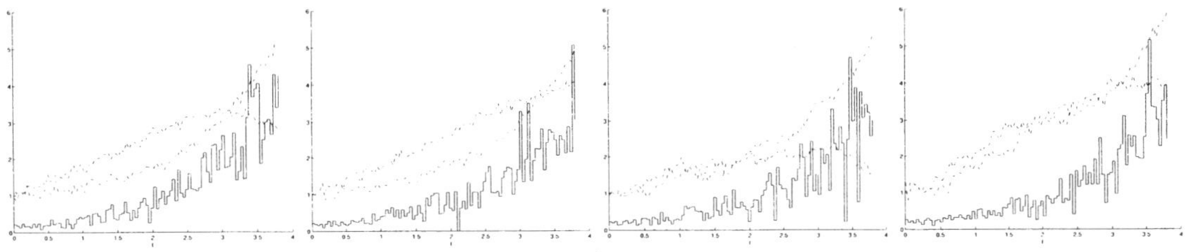

(a) Data set no. 1. (b) Data set no. 2. (c) Data set no. 3. (d) Data set no. 4.

Fig. 1. Simulated data sets for the bioreactor case study. Data sets no. 1-2 are without process noise, and data sets no. 3-4 are with significant process noise. Solid staircase: F, dashed lines: y_1, dotted lines: y_2, dash-dotted lines: y_3. True parameter and initial state values are shown in Tables 1-4.

all state variables can be measured, and the model therefore has the following measurement equation:

$$\begin{pmatrix} y_1 \\ y_2 \\ y_3 \end{pmatrix}_k = \begin{pmatrix} X \\ S \\ V \end{pmatrix}_k + \boldsymbol{e}_k \tag{33}$$

with $\boldsymbol{e}_k \in N(\boldsymbol{0}, \boldsymbol{S})$, where $\boldsymbol{S}$ is a diagonal matrix with elements S_{11}, S_{22} and S_{33}. The generated data sets are all shown in Figure 1. For the PE estimation part of the comparison, the model in (31)-(33) and the methodology presented in Section 2 is applied using the computer program described in Section 3. For the OE estimation part of the comparison, standard NLS is applied, i.e.:

$$\hat{\boldsymbol{\theta}} = \arg\min_{\boldsymbol{\theta} \in \Theta} \sum_{k=0}^{N} (\boldsymbol{y}_k - \hat{\boldsymbol{y}}_{k|0})^2 \tag{34}$$

where $\hat{\boldsymbol{y}}_{k|0}$ is obtained by solving the ODE model:

$$\frac{d}{dt}\begin{pmatrix} X \\ S \\ V \end{pmatrix} = \begin{pmatrix} \mu(S)X - \frac{FX}{V} \\ -\frac{\mu(S)X}{Y} + \frac{F(S_F - S)}{V} \\ F \end{pmatrix} \tag{35}$$

where $\mu(S)$ is given by (32), and by using a measurement equation similar to (33). Estimation results for all 4 data sets are given in Tables 1-4.

Uncertainty information is difficult to obtain with the OE method, so the two methods can only be compared in terms of bias of estimates. The results in Tables 1-2, which are for data sets without process noise, show an almost equal level of performance for the two methods, whereas the results in Tables 3-4, which are for data sets

Table 1. Results for data set no. 1.

Parameter	True value	PE estimate	OE estimate
X_0	1.0000E+00	1.0095E+00	1.0148E+00
S_0	2.4490E-01	2.3835E-01	2.4431E-01
V_0	1.0000E+00	1.0040E+00	1.0092E+00
μ_{max}	1.0000E+00	1.0022E+00	9.9852E-01
K_1	3.0000E-02	3.1629E-02	3.1412E-02
σ_{11}	0.0000E+00	3.6100E-07	-
σ_{22}	0.0000E+00	4.7385E-07	-
σ_{33}	0.0000E+00	7.5881E-14	-
S_{11}	1.0000E-02	7.5248E-03	-
S_{22}	1.0000E-03	1.0636E-03	-
S_{33}	1.0000E-02	1.1388E-02	-

Table 2. Results for data set no. 2.

Parameter	True value	PE estimate	OE estimate
X_0	1.0000E+00	9.8576E-01	9.9595E-01
S_0	2.4490E-01	2.4760E-01	2.3894E-01
V_0	1.0000E+00	1.0137E+00	1.0160E+00
μ_{max}	1.0000E+00	1.0092E+00	1.0184E+00
K_1	3.0000E-02	3.2624E-02	3.6663E-02
σ_{11}	0.0000E+00	8.3976E-06	-
σ_{22}	0.0000E+00	1.9310E-05	-
σ_{33}	0.0000E+00	1.1389E-06	-
S_{11}	1.0000E-02	9.2502E-03	-
S_{22}	1.0000E-03	8.1408E-04	-
S_{33}	1.0000E-02	8.3280E-03	-

Table 3. Results for data set no. 3.

Parameter	True value	PE estimate	OE estimate
X_0	1.0000E+00	9.5255E-01	8.4096E-01
S_0	2.4490E-01	2.3878E-01	4.5647E-02
V_0	1.0000E+00	9.8120E-01	1.2504E+00
μ_{max}	1.0000E+00	9.6795E-01	8.8212E-01
K_1	3.0000E-02	3.1606E-02	1.9189E-02
σ_{11}	3.1623E-01	3.1715E-01	-
σ_{22}	3.1623E-01	2.7524E-01	-
σ_{33}	3.1623E-01	2.5364E-01	-
S_{11}	1.0000E-02	7.9042E-03	-
S_{22}	1.0000E-03	1.2357E-03	-
S_{33}	1.0000E-02	8.4691E-03	-

Table 4. Results for data set no. 4.

Parameter	True value	PE estimate	OE estimate
X_0	1.0000E+00	1.0808E+00	1.3441E+00
S_0	2.4490E-01	2.0078E-01	9.0551E-01
V_0	1.0000E+00	1.1813E+00	1.6106E+00
μ_{max}	1.0000E+00	1.0341E+00	7.9587E-01
K_1	3.0000E-02	4.4851E-02	6.2200E-12
σ_{11}	3.1623E-01	2.7136E-01	-
σ_{22}	3.1623E-01	3.8652E-01	-
σ_{33}	3.1623E-01	3.9257E-01	-
S_{11}	1.0000E-02	1.0219E-02	-
S_{22}	1.0000E-03	1.5330E-04	-
S_{33}	1.0000E-02	9.7136E-03	-

with significant process noise, show that the PE method performs significantly better than the OE method in the sense that more consistent and significantly less biased estimates are obtained.

5. DISCUSSION

The results obtained in Section 4 are in good agreement with theory, because they clearly show that, in the presence of significant process noise, standard OE estimation gives parameter estimates that are severely biased, whereas PE estimation, and, more specifically, the method presented in this paper, performs much better. If, on the other hand, there is no significant process noise, the results show that the two methods perform almost equally well, although, theoretically, OE estimation is better suited in this case.

Process noise due to random variations or model approximations, unmodelled inputs and other model deficiencies is almost unavoidable in practice, so for most practical model building it is the recommendation of the authors of the present paper to use PE estimation and continuous-discrete stochastic state space models. Not only does this allow for the actual process noise to be quantified through estimates of the parameters of the diffusion term, but the properties of the corresponding estimators also allow uncertainty information to be obtained, which facilitates subsequent application of statistical tools to determine if the model can be unfalsified, and, if not, to pinpoint its deficiencies in order to subsequently uncover their structural origin (Kristensen *et al.*, 2002*a*).

A parameter estimation scheme similar to the one presented in this paper has previously been presented by Bohlin and Graebe (1995), who also presented an associated computer program, which has subsequently been developed into an extensive tool for identification of continuous time models called **MoCaVa** (Bohlin, 2001). There are, however, a number of important differences between the estimation scheme presented here and the one implemented in **MoCaVa**, which essentially make the latter equivalent to an OE estimation method. In particular, it has been shown that problems similar to the ones demonstrated for the OE estimation method used in this paper occur with respect to accuracy and consistency in the presence of process noise (Kristensen *et al.*, 2002*b*).

6. CONCLUSION

Continuous-discrete stochastic state space models have been advocated for modelling dynamic systems in continuous time using discrete time data, because such models have several attractive features. In particular, the prediction error decomposition (PED) provided by models of this type facilitates estimation of unknown parameters in a prediction error (PE) setting, which gives less biased and more reproducable results in the presence of significant process noise than estimation in a standard output error (OE) setting. A specific PE estimation scheme and a software implementation of this scheme has also been presented, and to illustrate the superiority of the presented scheme over standard OE estimation, a case study has been given, which demonstrates the higher sensitivity of OE estimates to process noise.

REFERENCES

Bard, Y. (1974). *Nonlinear Parameter Estimation.* Academic Press. New York, USA.

Bohlin, T. (2001). A grey-box process identification tool: Theory and practice. Technical Report IR-S3-REG-0103. Department of Signals, Sensors and Systems, Royal Institute of Technology, Stockholm, Sweden.

Bohlin, T. and S. F. Graebe (1995). Issues in nonlinear stochastic grey-box identification. *International Journal of Adaptive Control and Signal Processing* **9**, 465–490.

Holst, J., U. Holst, H. Madsen and H. Melgaard (1992). Validation of grey box models. In: *Selected Papers from the 4th IFAC Symposium on Adaptive Systems in Control and Signal Processing* (L. Dugard, M. M'Saad and I. D. Landau, Eds.). Pergamon Press. pp. 407–414.

Jazwinski, A. H. (1970). *Stochastic Processes and Filtering Theory.* Academic Press. New York, USA.

Kristensen, N. R., H. Madsen and S. B. Jørgensen (2002*a*). A method for systematic improvement of stochastic grey-box models. Submitted for publication.

Kristensen, N. R., H. Madsen and S. B. Jørgensen (2002*b*). Parameter estimation in stochastic grey-box models. Submitted for publication.

Nielsen, J. N., H. Madsen and P. C. Young (2000). Parameter estimation in stochastic differential equations: An overview. *Annual Reviews in Control* **24**, 83–94.

Unbehauen, H. and G. P. Rao (1990). Continuous-time approaches to system identification - a survey. *Automatica* **26**(1), 23–35.

Unbehauen, H. and G. P. Rao (1998). A review of identification in continuous-time systems. *Annual Reviews in Control* **22**, 145–171.

Young, P. C. (1981). Parameter estimation for continuous-time models - a survey. *Automatica* **17**(1), 23–39.

Åström, K. J. (1970). *Introduction to Stochastic Control Theory.* Academic Press. New York, USA.

IFAC Publications
www.elsevier.com/locate/ifac

ON POSSIBILITIES FOR ESTIMATING CONTINUOUS-TIME ARMA PARAMETERS

Erik K. Larsson * and **Magnus Mossberg** **

* *Systems and Control, Dept. of Information Technology, Uppsala University, P.O. Box 337, SE-751 05 Uppsala, Sweden. E-mail: Erik.K.Larsson@it.uu.se*
** *Dept. of Electrical Engineering, Karlstad University, SE-651 88 Karlstad, Sweden. E-mail: Magnus.Mossberg@kau.se*

Abstract: The problem of estimating the parameters in continuous-time ARMA processes from discrete-time data is considered. Three different approaches, based on the prediction error method, the instrumental variable method and an approximate maximum likelihood method, respectively, are studied. All three techniques provide reliable solutions to the estimation problem. A general discussion of the inherent difficulties of the problem is given together with an extensive numerical study. *Copyright © 2003 IFAC*

Keywords: Continuous-time ARMA process; sampling; estimation.

1. INTRODUCTION

Continuous-time stochastic processes are very important for mathematical modeling purposes. They are most often found in statistics and in econometrics, while their discrete-time counterparts are more common in other disciplines. An obvious reason is that with discrete-time data available, it is natural to work with a discrete-time model. Another reason for the popularity of discrete-time stochastic models is that they are considerably easier to handle mathematically than continuous-time stochastic models. It is important to remember, though, that a discrete-time model only describes the true system at the sampling instants. This means that if the parameters of a continuous-time model can be determined with high enough accuracy, the continuous-time model indeed has something to offer. Further, many physical systems are naturally modeled in continuous-time.

Identification of continuous-time stochastic systems is the topic of (Jones, 1981; Duncan *et al.*, 1999; Pham, 2000; Söderström and Mossberg, 2000; Larsson and Söderström, 2002*b*; Larsson and Söderström, 2002*a*). These papers deal with the problem of estimating the parameters in continuous-time AR (CAR) or continuous-time ARX (CARX) processes. The more difficult problem of estimating the parameters in a continuous-time ARMA (CARMA) process from discrete-time data is studied in this paper. Note that the solution to the problem also provides the solution to the problem of estimating the parameters in a CAR process from sampled data corrupted by discrete-time measurement noise.

2. PROBLEM FORMULATION

Consider a real-valued scalar CARMA process $y(t)$, whose spectrum is given by

$$\Phi(s) = \lambda_c^2 \frac{B(s)B(-s)}{A(s)A(-s)} \tag{1}$$

where λ_c^2 is a positive constant (can be interpreted as the intensity of the process noise) and

$$A(s) = \prod_{i=1}^{n}(s-\tilde{a}_i) = s^n + a_1 s^{n-1} + \ldots + a_n \tag{2}$$

$$B(s) = \prod_{i=1}^{m}(s - \tilde{b}_i) = s^m + b_1 s^{m-1} + \ldots + b_m \quad (3)$$

The model orders, n and m, are assumed to be known. Define $l = n - m$ as the relative degree of the spectral factor $B(s)/A(s)$ and let

$$\theta_c = \left[a_1, \ldots, a_n, b_1, \ldots, b_m, \lambda_c^2\right]^T \quad (4)$$

be the vector of the true and unknown parameters. The objective is to estimate the spectrum (1), or, in particular, the parameter vector (4) from discrete-time measurements $\{y(kh)\}_{k=1}^{N}$, where h denotes the sampling interval.

3. PRELIMINARIES AND SOME SAMPLING RESULTS

To represent the CARMA process as a discrete-time ARMA process, one alternative is to write it on state space form and determine the corresponding discrete-time form and its power spectrum. The representation

$$D(q)y(kh) = C(q)e_d(kh) \quad (5)$$

where $e_d(kh)$ is discrete-time white noise with variance λ_d^2,

$$C(z) = z\prod_{i=1}^{n-1}(z - \tilde{c}_i) = z^n + c_1 z^{n-1} + \ldots + c_{n-1}z, \quad (6)$$

$$D(z) = \prod_{i=1}^{n}(z - \tilde{d}_i) = z^n + d_1 z^{n-1} + \ldots + d_n \quad (7)$$

and where h denotes the sampling interval, is exact at the sampling instants and is given by spectral factorization, see (Åström, 1970; Söderström, 2002). Let

$$\theta_d = \left[d_1, \ldots, d_n, c_1, \ldots, c_{n-1}, \lambda_d^2\right]^T \quad (8)$$

be the parameter–vector of the discrete-time system (5). It should be emphasized that θ_d depends on h.

The mapping between the continuous- and discrete-time poles is given by the well-known relation

$$\tilde{d}_i = e^{\tilde{a}_i h} \quad (9)$$

The corresponding relation between the continuous- and discrete-time zeros can in general not be described by a closed-form expression. However, for small sampling intervals h, some explicit, but approximate, results exist; see (Larsson, 2002; Wahlberg, 1988).

A key ingredient in this work is to find the underlying continuous-time description given a discrete-time model. This type of problem appears in various branches of system science, and both exact and approximate algorithms exist, see (Söderström, 1991) and the references therein. We will primarily use the exact method, referred to as Algorithm E1, presented in (Söderström, 1991). This method requires a modest amount of computations and is recommended as a first choice to use.

Next, a number of approaches for estimating the CARMA parameters (4) from discrete-time data are studied. The methodology we will adopt consists of two steps. First a discrete-time model is estimated. Next, the CARMA parameters are retained by converting the estimated discrete-time model.

4. PREDICTION ERROR METHOD (M1)

The first method we consider is based on the prediction error method (PEM), see (Söderström and Stoica, 1989; Ljung, 1999), which is known to yield consistent and statistically efficient estimates. The method consists of the following steps.

(1) Obtain an estimate of θ_d by using the PEM.
(2) Determine the corresponding continuous-time parameters using Algorithm E1 in (Söderström, 1991).

This approach will be referred to as M1.

5. INSTRUMENTAL VARIABLE METHOD (M2)

The approach in this section, subsequently referred to as M2, is based on the instrumental variable (IV) technique, see (Söderström and Stoica, 1989; Ljung, 1999).

(1) Obtain an estimate of $D(q)$ by using the IV method described in (Stoica *et al.*, 1985).
(2) Let
$$w(kh) = \hat{D}(q)y(kh) \quad (10)$$
and estimate $C(q)$ and λ_d^2 from
$$w(kh) = C(q)e_d(kh) \quad (11)$$
using one of the techniques for estimating MA process parameters described in (Stoica *et al.*, 2000).
(3) For obtaining the continuous-time parameters, apply Step 2 of the approach described in Section 4.

For numerical reasons we have parametrized $D(q)$ using the delta operator $\delta = (q-1)/h$ to get $D(\delta)$. When constructing the instruments in the IV method in Step 1, a filtering operation is involved, and the optimal filter is given by $1/C^2(q)$. Since the optimal filter depends on the unknown $C(q)$, it is not realizable. One idea is to do the estimation iteratively. No filter is used in the first iteration, but in the next iteration, the filter is constructed from the estimate of $C(q)$ from the previous iteration, and so on. Here this idea is used and one iteration is performed.

6. APPROXIMATE MAXIMUM LIKELIHOOD (M3)

The approximate maximum likelihood (ML) method for ARMA spectral estimation described in (Stoica *et al.*, 1987) is considered in this approach. It combines

system	a_1	a_2	poles	b_1	zero
sys1	4	8	$-2\pm 2i$	2	-2
sys2	4	8	$-2\pm 2i$	6	-6
sys3	12	72	$-6\pm 6i$	2	-2
sys4	12	72	$-6\pm 6i$	6	-6

Table 1. Characterization of the systems.

the computational simplicity of the Yule-Walker based methods with the accuracy of the ML techniques.

(1) Determine a large-sample approximate ML estimate of the covariances of the observed ARMA process.
(2) Obtain the approximate ML estimates of the ARMA spectral parameters by using the covariance estimates from Step 1 in a Yule-Walker based procedure; to get $C(q)$ a spectral factorization is required.
(3) For obtaining the continuous-time parameters, apply Step 2 of the approach described in Section 4.

See (Stoica *et al.*, 1987) for a detailed description of Steps 1 and 2. This method will be referred to as M3.

7. NUMERICAL STUDIES

Reconsider the CARMA process $y(t)$, whose spectral density is given by (1). For illustration, four different second-order systems ($n = 2$ and $m = 1$), referred to as sys1–sys4, will be considered; see Table 1 for a description of the process parameters. In all cases λ_c^2 is chosen so that $y(t)$ will have unit variance. The spectrum for the different systems is depicted in Fig. 1. We observe that the effective bandwidth of sys3 and sys4 is considerably larger than for sys1 and sys2. Further, it is clear that sys3 has a relatively large resonance peak. Discrete-time data $\{y(kh)\}_{k=1}^{N}$ are generated by simulating the systems sys1–sys4, respectively. The number of data is fixed $N = 10000$, while the sampling interval h is varying. In the examples to follow, the different methods M1–M3, described in Sections 4–6, will be evaluated for the problem of estimating the process parameters θ_c of sys1–sys4.

Example 1. The first example aims at illustrating the "robustness" of the methods. Here we will consider the estimated discrete-time parameter vector $\theta_d = [d_1, d_2, c_1, \lambda_d^2]$. As a measure of robustness we choose to look at the number of unstable estimates of $D(q)$, as well as the number of complex-valued estimates of c_1 (should be real-valued for the spectrum to be positive) during 250 performed Monte Carlo simulations. The result is shown in Table 2, where a '-' indicates that we were unable to obtain a measure of that specific quantity. Here we also present the evaluation time (the time needed to run the algorithm) for the different methods. It is clear that all methods, except M1, run into problems when the sampling interval is chosen too low or too high. It seems as M3 produces the most unstable estimates (this problem is discussed in (Stoica *et al.*, 1987)), while M2 has the highest tendency to yield complex-valued estimates of $\hat{c}_1$. How to deal with these kind of problems is an issue that deserves research on its own; here we simply reflect an unstable pole with respect to the unit circle, while just taking the real part of a complex-valued estimate of $\hat{c}_1$, see also (Stoica *et al.*, 2000).

Example 2. In the next example we will address issues as bias and variance for the estimate of θ_c. The criteria selected for the performance evaluation are, the normalized mean estimation error

$$NMEE \triangleq \frac{\|\theta_c - \hat{\theta}_c^*\|_2}{\|\theta_c\|_2} \tag{12}$$

and the mean estimation variance

$$MEV \triangleq \frac{1}{m+n}\frac{1}{MC-1}\sum_{k=1}^{MC}\|\hat{\theta}_c^* - \hat{\theta}_c(k)\|_2^2 \tag{13}$$

where

$$\hat{\theta}_c^* = \frac{1}{MC}\sum_{k=1}^{MC}\hat{\theta}_c(k) \tag{14}$$

and $\hat{\theta}_c(k)$ corresponds to the kth estimate of the $MC = 250$ number of Monte-Carlo runs performed.

The criteria $NMEE$ and MEV are shown in Figs. 2–4 and Figs. 5–7, respectively, as functions of the sampling interval. Note that some bars continue above the shown height of the figures. A number of things can be observed from the figures.

- It is clear that the range of possible sampling intervals for an adequate estimation is larger for sys1 and sys2, than for sys3 and sys4, which is natural by comparing effective bandwidths of the systems.
- Compared to sys2, it seems as sys1 is more difficult to estimate for small sampling intervals. This can be understood by notice that $|\tilde{a}_i h - \tilde{b}_1 h|$ is very small in this case, see Section 8.
- We observe that the estimation error variance is worst for M2, in particular for sys3 and sys4. Nevertheless, the result was much improved by using the estimated optimal filter $1/\hat{C}^2$, see Section 5.
- With respect to bias and variance, it is evident that M1 and M3 perform very similar, while M2 in general has the worst performance. In particular, it turns out that M2 has severe problems with estimating λ_c^2, especially for large sampling intervals.

Finally, the spectra obtained by using the averaged estimate $\hat{\theta}_c^*$ (see (14)) are shown in Fig. 8. The sampling interval is $h = 0.15$. The results are very good, except for sys3 estimated with M2. In this case it is λ_c^2 that is estimated poorly.

Some final remarks:

- For estimating $C(q)$ the different methods performed very similar; here a standard spectral factorization was used.

- Step 1 for the methods M2 and M3 are closely related.

8. DISCUSSIONS

A relevant question is why estimation of CARMA parameters is a difficult problem. Some possible answers and explanations will be given in the general discussion that follows next. The discussion also includes a number of issues that are important to take into consideration when solving this estimation problem.

The choice of a proper sampling interval is crucial and in general difficult as it requires some knowledge of the unknown true system. A large effective bandwidth of the system will mean that the acceptable range of sampling intervals will be small. It would be desirable to give guidelines for the choice of sampling interval. An idea is to do a nonparametric identification of the discrete-time spectrum and then choose the sampling interval based on the shape of the spectrum. This is not straightforward since the discrete-time spectrum of course depends on the sampling interval actually used, and is defined only up to the Nyquist frequency. The consequences of sampling too slow in comparison with the effective bandwidth of the true system are:

- The essential dynamics of the system can not be registered.
- Discrete-time poles tend to cluster at $z = 0$, which seems to be the case also for the discrete-time zeros, see Fig. 9. This can give rise to numerical problems.
- Since the discrete-time spectrum is defined for $|\omega|h \leq \pi$, the bandwidth of the discrete-time system will decrease, cf. aliasing.

On the other hand, problems also occur when sampling too fast:

- The process is observed during an interval of length Nh, measured in continuous-time. This means that, for a constant N, the interval will decrease when h becomes smaller. Less information about the system is therefore obtained, in the meaning that the process is observed a shorter time.
- We will inevitably encounter numerical problems, as the discrete-time poles and zeros (see below) will cluster at $z = 1$, see Fig. 9. This will also cause problems with pole-zero cancellations.
- Sensitive transformation from discrete- to continuous-time. As an example, consider the transformation of an estimated discrete-time pole $\hat{q} = q_0 + \Delta q$ into its corresponding continuous-time pole $\hat{p} = p_0 + \Delta p$. It follows that

$$\Delta p = \frac{1}{h}\Big(\frac{\Delta q}{q_0} + O\big((\Delta q/q_0)^2\big)\Big) \qquad (15)$$

which clearly shows that a very small deviation in $\hat{q}$ may be significantly amplified if h is small. This problem can be partly cured using a δ-formalism.

In the above discussion on numerical problems, the locations of the continuous-time poles $\tilde{a}_i$ and zeros $\tilde{b}_i$ must also be taken into consideration. It is a well-known problem that when the product $h\tilde{a}_i$ tends to zero, the discrete-time poles cluster around $z = 1$, see (9). To circumvent this problem to some extent, a δ-formalism can be used instead of the q-formalism. Similarly, it holds that m discrete-time zeros converge to $z = 1$, while the remaining $l - 1$ zeros converge to the zeros of a constant polynomial, see (Larsson, 2002; Wahlberg, 1988). In fact, assuming that $\tilde{b}_i$ is a simple zero of $B(s)$, it can be shown that $\tilde{c}_i = e^{\tilde{b}_i h} + O(h^3)$, indicating that it is the distance $|\tilde{a}_i h - \tilde{b}_i h|$ that is important; a small distance will yield problems with pole-zero cancellations.

Another important aspect is the computational burden for the methods. For the systems studied in Section 7, approaches M2 and M3 have approximately the same, and in general the shortest computational time. For higher order systems and for "difficult" cases, it is expected that approach M1 will need considerable more computational time than the aforementioned methods since it involves a nonlinear multidimensional minimization problem. For fast sampling, the exact transformation from discrete- to continuous-time can be made approximative to reduce the computational complexity. This can be done using either some of the methods described in (Söderström, 1991) or using the results of (Larsson, 2002; Wahlberg, 1988) mentioned earlier.

Let us conclude the paper by the following remarks:

- It is not obvious how to deal with erroneous parameter estimates, e.g. estimates that give unstable systems or estimates that are complex valued (when it is known that it must be real valued). This question was raised in Example 1.
- By using approach M3 together with algorithm E1 of (Söderström, 1991) one can actually circumvent the troublesome step of finding the discrete-time polynomial $C(z)$ when estimating the continuous-time spectrum. This has not been further pursued in this work.
- Increasing the number of samples N will decrease the bias and variance of the estimates significantly, in particular for small sampling intervals. This is natural in view of the discussion above (the observation length will increase), but it also indicates that in order to solve this problem it is important to have a lot of data.
- It is difficult to make a fair comparison of the different methods, as there are several user parameters to choose. Further, implementations can often be made numerically more sound. Also, several other approaches exist; due to lack of space these are not covered in this short paper.

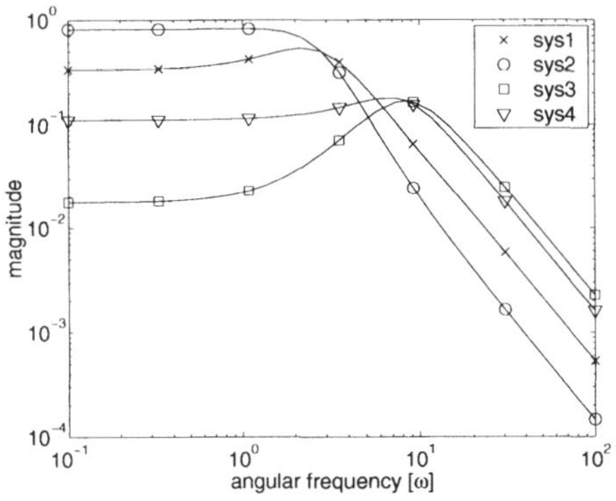

Fig. 1. Spectrum for the different processes.

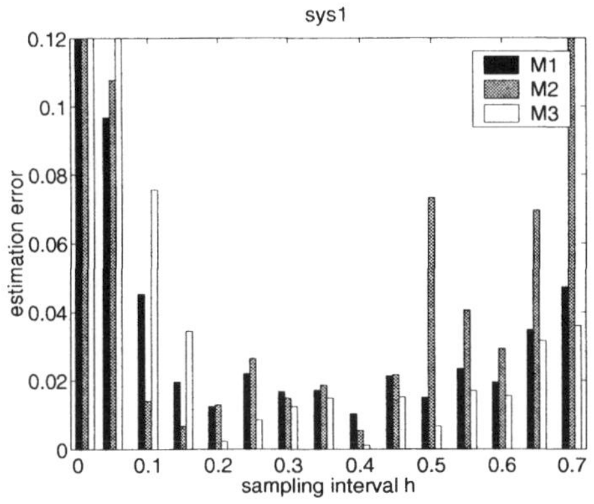

Fig. 2. *NMEE* for M1–M3 applied to sys1.

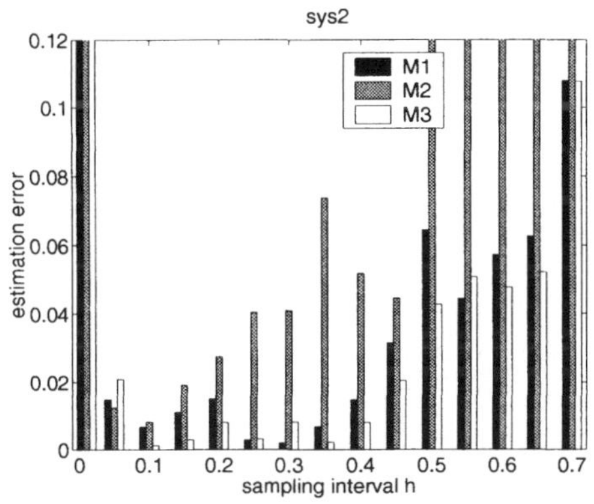

Fig. 3. *NMEE* for M1–M3 applied to sys2.

However, our belief is that they will all suffer from some of the inherent difficulties mentioned above.

ACKNOWLEDGMENT

The authors want to thank Professor Torsten Söderström for valuable discussions.

REFERENCES

Åström, K. J. (1970). *Introduction to Stochastic Control Theory*. Academic Press.

Duncan, T. E., P. Mandl and B. Pasik-Duncan (1999). A note on sampling and parameter estimation in

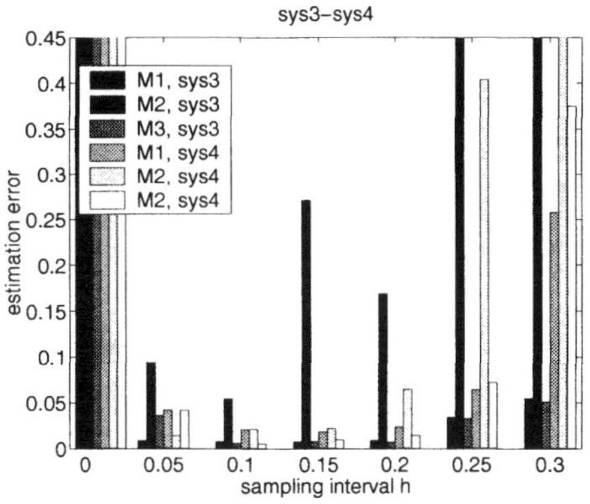

Fig. 4. *NMEE* for M1–M3 applied to sys3 and sys4.

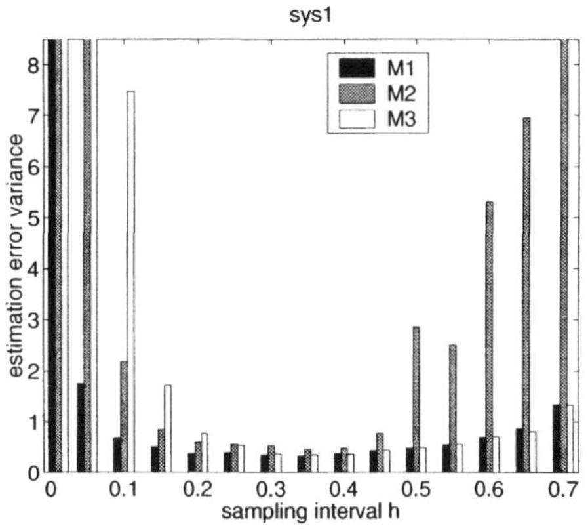

Fig. 5. *MEV* for M1–M3 applied to sys1.

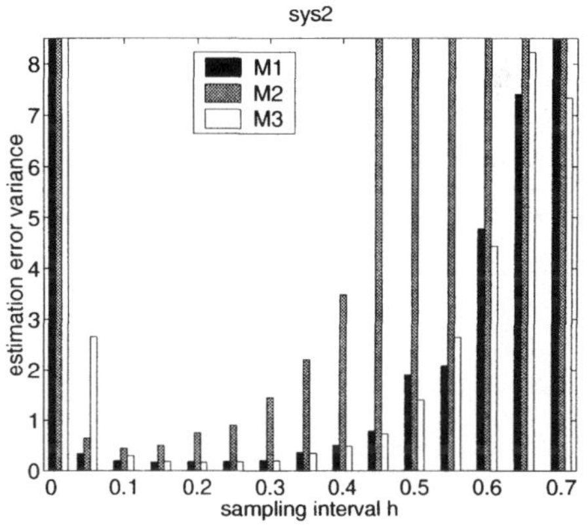

Fig. 6. *MEV* for M1–M3 applied to sys2.

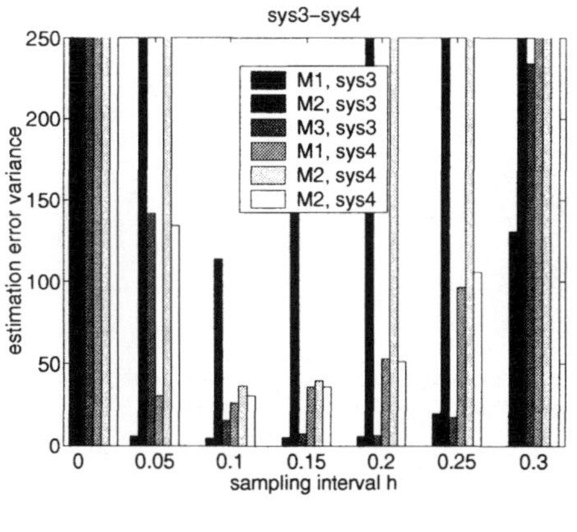

Fig. 7. *MEV* for M1–M3 applied to sys3 and sys4.

system	method	unstable estimations (%)			complex valued $\hat{c}_1$ (%)			evaluation time		
		$h = 0.02$	$h = 0.2$	$h = 0.5$	$h = 0.02$	$h = 0.2$	$h = 0.5$	$h = 0.02$	$h = 0.2$	$h = 0.5$
sys1	M1	0.8	0	0	-	-	-	0.52	0.38	0.35
	M2	24.0	0	0	27.2	0	0	0.09	0.09	0.10
	M3	48.0	0	0	2.0	0	0	0.19	0.19	0.19
sys2	M1	0	0	0	-	-	-	0.48	0.34	0.35
	M2	3.2	0	0	27.6	0	5.2	0.09	0.09	0.10
	M3	31.6	0	0	20.8	0	0	0.19	0.19	0.19
sys3	M1	0	0	2.8	-	-	-	0.21	0.12	0.14
	M2	4.0	1.6	21.2	38.0	15.2	24.8	0.04	0.04	0.04
	M3	26.0	0	48.0	18.0	0	2.0	0.06	0.06	0.06
sys4	M1	0	0	1.6	-	-	-	0.54	0.34	0.40
	M2	6.8	0	19.6	14.4	0	24.0	0.09	0.11	0.09
	M3	19.6	0	51.6	9.6	0	0.4	0.21	0.21	0.19

Table 2. Robustness and evaluation time for the methods.

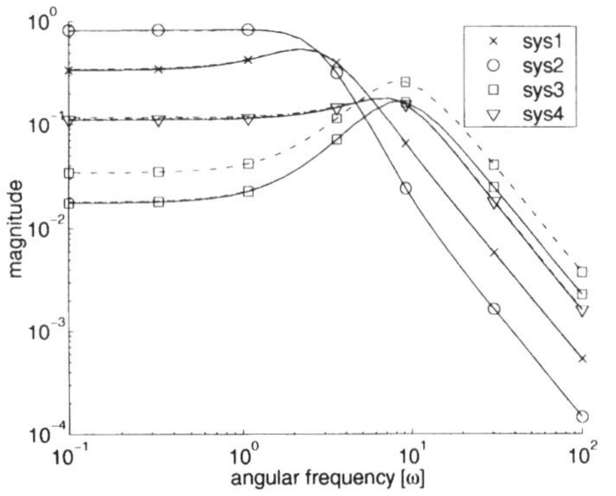

Fig. 8. Estimated spectrum for using M1 (dotted), M2 (dashdotted) and M3 (dashed). True spectrum (solid). Notice that many curves overlap.

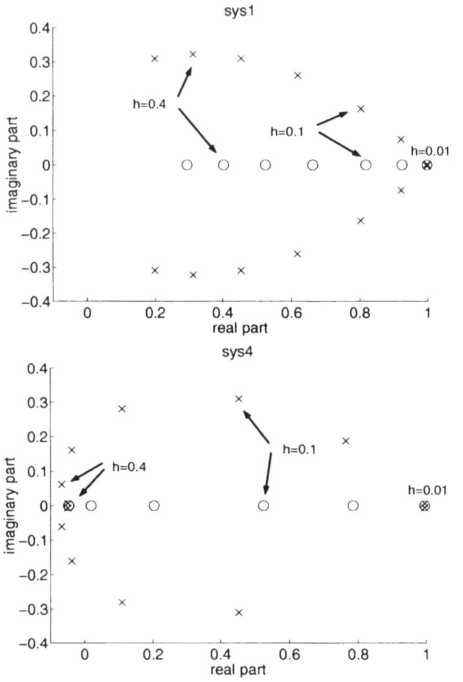

Fig. 9. Poles (×) and zeros (○) as functions of h for sys1 and sys4. Here h takes the values $h \in \{0.01, 0.05, 0.2, 0.3, 0.4, 0.5\}$. Notice that for sys4 the zeros for $h = 0.4$ and $h = 0.5$ almost overlap each other.

linear stochastic systems. *IEEE Trans. on Automatic Control* **44**(11), 2120–2125.

Jones, R. H. (1981). Fitting a continuous time autoregression to discrete data. In: *Applied Time Series Analysis, II*. pp. 651–682. Academic Press.

Larsson, E. K. (2002). On fast sampled stochastic systems. In preparation.

Larsson, E. K. and T. Söderström (2002*a*). Continuous-time AR parameter estimation by using properties of sampled systems. In: *Proc. 15th IFAC World Congress*. Barcelona, Spain. pp. 394–399.

Larsson, E. K. and T. Söderström (2002*b*). Identification of continuous-time AR processes from unevenly sampled data. *Automatica* **38**(4), 709–718.

Ljung, L. (1999). *System Identification*. 2nd ed.. Prentice–Hall. Upper Saddle River, NJ, USA.

Pham, D-T. (2000). Estimation of continuous time autoregressive models from finely sampled data. *IEEE Trans. Signal Processing* **48**(9), 2576–2584.

Söderström, T. (1991). Computing stochastic continuous-time models from ARMA models. *Int. J. Control* **53**(6), 1311–1326.

Söderström, T. (2002). *Discrete-Time Stochastic Systems*. 2nd ed.. Springer-Verlag.

Söderström, T. and M. Mossberg (2000). Performance evaluation of methods for identifying continuous-time autoregressive processes. *Automatica* **36**(1), 53–59.

Söderström, T. and P. Stoica (1989). *System Identification*. Prentice–Hall International. Hemel Hempstead, United Kingdom.

Stoica, P., B. Friedlander and T. Söderström (1987). Approximate maximum-likelihood approach to ARMA spectral estimation. *Int. J. Control* **45**(4), 1281–1310.

Stoica, P., T. McKelvey and J. Mari (2000). MA estimation in polynomial time. *IEEE Trans. on Signal Processing* **48**(7), 1999–2012.

Stoica, P., T. Söderström and B. Friedlander (1985). Optimal instrumental variable estimates of the AR parameters of an ARMA process. *IEEE Trans. on Automatic Control* **30**(11), 1066–1074.

Wahlberg, B. (1988). Limit results for sampled systems. *Int. J. Control* **48**, 1267–1283.

IFAC Publications
www.elsevier.com/locate/ifac

ON THE INTERPRETATION OF A CONTINUOUS–TIME MODEL IDENTIFICATION METHOD IN TERMS OF REGULARIZATION

Saïd Moussaoui, David Brie and Alain Richard

CRAN–CNRS UMR 7039,
Université Henri Poincaré, Nancy 1,
B.P. 239, F-54506 Vandœuvre Cedex, France.
Tel.: +33 3 83 68 44 61, Fax: +33 3 83 68 44 62,
E-mail: {FirstName.LastName}@cran.uhp-nancy.fr

Abstract: This paper presents an interpretation of a continuous–time model identification method in terms of regularization. We show that, in the case of linear filter method, data filtering corresponds to a regularized derivative estimation. We also give the explicit form of the minimized criterion. In addition, we propose a new structure based on the use of a true regularization filter whose performances are compared with the GPMF method. Finally, we propose a new formulation of continuous–time model identification as a joint output signal and model parameter estimation problem. *Copyright © 2003 IFAC*

Keywords: System identification - continuous–time model - differentiation - inverse problem - regularization.

1. INTRODUCTION

Direct identification of systems represented by continuous-time models has been the subject of many works (Young, 1964; Saha and Rao, 1983; Wang and Gawthrop, 2001; Garnier *et al.*, 2002). These studies contributed to the development of several methods allowing to reach a good level of performances and even better than an indirect approach (Rao and Garnier, 2002). The aim of this article is to contribute to the analysis of these methods, by proposing an interpretation in terms of regularization. Based upon this interpretation we present a new structure of filter which can be used in continuous–time model identification.

In continuous–time model identification, it is needed to estimate the derivatives of the system input and output signals. This operation yields two difficulties; first the data are sampled, therefore a numerical differentiation method has to be used. In addition, the output signal is prone to errors, modelled by an additive noise, resulting in a noise amplification in the case of a naive calculation of these derivatives. This difficulty is known to be an ill–posed inverse problem (Tikhonov and Arsénine, 1974). Continuous–time model identification methods use some transformations in the data in order to circumvent this problem, and lead to performances which depend on the transformation used. Nevertheless, a recent contribution gathers most of these methods in a Matlab Toolbox (The CONTSID [1] Toolbox) and gives a comparative analysis of the performances of 17th different methods (Mensler, 1999). This study showed that the linear filter method presents good statistical performances. This is primarily why our analysis is focused on this method.

[1] The CONTSID Toolbox can be downloaded from: http://www.cran.uhp-nancy.fr/francais/savoir_faire/produits/contsid/index.html

This paper is organized as follows: section 2 introduces the concepts of ill-posed inverse problem, regularization and points out the formulation of the regularized derivative estimation. In section 3, we present the principle of continuous-time model identification methods by linear filtering. Then, we provide (section 4) an original interpretation of this method in terms of regularization. Based on this interpretation, we propose a new structure of filters that are used to give some insight into the GPMF phase effect in continuous-time model identification (section 5). Finally, section 6 adresses the problem of optimal filter design by formulating continuous-time model identification as a joint signal-parameter estimation.

2. REGULARIZATION

In order to illustrate this concept, let us consider a dynamic system characterized by its input $x(t)$, output $y(t)$ and disturbances $v(t)$. The model associated to this system expresses a relation between these signals:

$$\mathcal{H}\big(x(t)\big) + v(t) = y(t),$$

where $\mathcal{H}$ is an operator representing the transformation induced by the system to the input signal. In the case of a linear time invariant system, of impulse response $h(t)$ (case that we will consider in this study), $\mathcal{H}$ is a convolution operator; i.e. $\mathcal{H}\big(x(t)\big) = [x\star h](t)$. The inverse problems associated to this formulation are the determination of $\mathcal{H}$ from $x(t)$ and $y(t)$ (identification) and the estimation of $x(t)$ from $\mathcal{H}$ and $y(t)$ (deconvolution). An inverse problem is said to be ill-posed, if the solution does not exist, or is not unique or if a small disturbance on data induces a large variation of the solution. The regularization tries to solve an ill-posed problem by searching a solution meaningful and stable with respect to the variations of the data, (Tikhonov and Arsénine, 1974). In this paper, we consider the regularization techniques which minimize a compound criterion. Instead of minimizing a data fitting criterion, a new term is added in order to make the solution faithful to certain *a priori* specified properties. The Tikhonov regularization is one of the methods suggested from this point of view (Tikhonov and Arsénine, 1974). In the case of deconvolution, this technique consists in minimizing a criterion:

$$\mathcal{J}_\alpha(x(t)) = \big\|y(t) - \mathcal{H}\,(x(t))\,\big\|_2^2 + \alpha \sum_{k=0}^{p} \alpha_k \big\|\mathcal{D}_k\, x(t)\big\|_2^2, \quad (1)$$

where $\alpha, \left\{\alpha_k\right\}_{k=0}^{p}$, are constant regularization parameters and $\mathcal{D}_k = \frac{d^k}{dt^k}$ represents the kth derivative operator. The use of the Tikhonov stabilizers of order p (second part of the criterion (1)) allows to specify that the solution is built in the subspace of functions whose p^{th} derivative is null. The resulting regularized solution is obtained by:

$$\hat{x}_{reg}(t) = \frac{\mathcal{H}^*}{\mathcal{H}^*\mathcal{H} + \alpha \sum\limits_{k=0}^{p} \alpha_k \mathcal{D}_k^*\mathcal{D}_k}\, y(t), \quad (2)$$

where $\mathcal{H}^*$ and $\mathcal{D}_k^*$ are the adjoint operators of $\mathcal{H}$ and $\mathcal{D}_k$ respectively. By applying the Fourier transform to the equation (2), we obtain the regularized solution expressed in the frequency-domain, which corresponds to applying a regularization filter:

$$F(\omega) = \frac{H^*(\omega)H(\omega)}{H^*(\omega)H(\omega) + \alpha \sum\limits_{k=0}^{p} \alpha_k \omega^{2k}}. \quad (3)$$

A particular example of ill-posed inverse problem is the estimation of the nth derivative of a function, where the operator $\mathcal{H}$ is an nth order integrator. According to (3), the expression of the regularization filter is:

$$F(\omega) = \frac{1}{1 + \alpha \sum\limits_{k=n}^{p+n} \alpha_{k-n}\, \omega^{2k}}. \quad (4)$$

This is a low-pass filter, which attenuates the high frequency part where appears noise amplification.

3. CONTINUOUS-TIME MODEL IDENTIFICATION

3.1 Problem Statement

Let us consider a system represented by a continuous-time, linear, stable, causal and time invariant model whose input-output relationship is:

$$\sum_{i=0}^{na} a_i\, y_u^{(i)}(t) = \sum_{i=0}^{nb} b_i\, u^{(i)}(t), \quad (5)$$

where:

- $u(t)$ and $y_u(t)$ are respectively the system input and output signals;
- $x^{(i)}(t)$ is the ith time derivative of $x(t)$ at time t. We suppose that the initial conditions are null;
- $\left\{a_i\right\}_{i=0}^{na}$ and $\left\{b_j\right\}_{j=0}^{nb}$ are the parameters of the model, ($na \geq nb$ and $a_{na} = 1$);

The output and input signals are sampled at a constant frequency $f_s = T_s^{-1}$. Moreover, the measured output signal is prone to disturbances, modelled by an additive noise $v(t)$, assumed to be independent to the input $u(t)$; i.e. $y(t) = y_u(t) + v(t)$.
The available data $\left\{u(kT_s),\ y(kT_s)\right\}_{k=0}^{N-1}$ for the identification are noted $\left\{u(t_k),\ y(t_k)\right\}_{k=0}^{N-1}$, and represent, respectively, the samples of the measured input and output signals. The problem of continuous-time model identification is then stated as follows: knowing the orders na, nb and having the samples $\left\{u(t_k),\ y(t_k)\right\}_{k=0}^{N-1}$, determine the coefficients

$\{a_i\}_{i=0}^{na-1}$ and $\{b_j\}_{j=0}^{nb}$ of the differential equation (5). In order to estimate the model parameters, this differential equation should be evaluated at each time $\{t = t_k\}_{k=0}^{N-1}$ and transformed into a linear regression model. But this operation requires the evaluation of the input and output signal derivatives, from the sampled data. This is the main difficulty due to the noise amplification.

3.2 Linear Filter Methods

One of the solutions used to circumvent derivative estimation consists in applying a linear transformation to the equation (5). This transformation corresponds to a linear filter of impulse response $f(t)$:

$$\sum_{i=0}^{na} a_i\, \mathcal{F}\{y^{(i)}(t)\} = \sum_{i=0}^{nb} b_i\, \mathcal{F}\{u^{(i)}(t)\} + \mathcal{F}\{v(t)\},$$

where $\mathcal{F}\{x^{(i)}(t)\} = \left[\frac{d^i x}{dt^i} \star f\right](t)$. The calculation of the terms $x_F^{(i)}(t_k) = \mathcal{F}\{x^{(i)}(t_k)\}$ is done using an adequate discretization technique (Garnier *et al.*, 2002) and the determination of the parameters of the resulting linear regressor model is carried out by a parametric estimation method.
Many linear filtering methods were developed, only differing on the form of the filter. Within the framework of this paper, we will investigate particularly the *State Variable Filters* method noted SVF, which uses a cascade of first order filters. It originates in the works of (Young, 1964) and was proposed under the name of MMF for *Method of Multiple Filters*. Extensions were introduced leading to the method known as GPMF for *Generalized Poisson Moment Functionals* (Saha and Rao, 1983):

$$F_l(s) = \left(\frac{\beta}{s+\lambda}\right)^{l+1},$$

where s is the Laplace variable, β and λ are design parameters of the filter.
Other improvements were made to this method by using a filter:

$$F(s) = \frac{1}{\hat{A}(s)},$$

where $\hat{A}(s)$ is an estimate of the denominator of the model to identify. This iterative method is known as SRIVC method for *Simplified Refined Instrumental Variable* (Young, 2002).

The analysis of the frequency response of the filters used to estimate the filtered successive derivatives, shows that they behave like derivators, only in the low frequency part and attenuate high frequency band. This filtering has a regularizing effect that we formalize in the next section.

4. INTERPRETATION IN TERMS OF REGULARIZATION

4.1 First Order Approximation

Consider the form of the Poisson filter:

$$F_l(s) = \left(\frac{\beta/\lambda}{1+\frac{1}{\lambda}s}\right)^{l+1}.$$

By posing $\gamma = \frac{1}{\lambda}$, $s = \jmath\omega$ and considering $\beta = \lambda$, the expression of the filter becomes:

$$F_l(\omega) = \left(\frac{1}{1+\gamma^2\omega^2}\right)^{\frac{l+1}{2}} \exp\left[-\jmath(l+1)\arctan(\gamma\omega)\right].$$

A first order approximation of the filter phase around $\omega = 0$ (low-frequencies) gives:

$$\arctan(\gamma\omega) \cong \gamma\omega + \mathcal{O}((\gamma\omega)^3);$$

and a series expansion of the module yields:

$$\left(1+\gamma^2\omega^2\right)^{\frac{l+1}{2}} = 1 + \sum_{k=1}^{M} S_{l+1}^k \gamma^{2k}\omega^{2k},$$

where $M = \begin{cases} \frac{l+1}{2} & \text{if } (l+1) \text{ is even} \\ \infty & \text{if } (l+1) \text{ is odd} \end{cases}$

and $S_{l+1}^k = \prod_{\nu=0}^{k-1} \frac{\left(\frac{l+1}{2}-\nu\right)}{k!}$. Consequently, by setting $\alpha_k(\lambda) = S_{l+1}^k \gamma^{2k}$, the filter is expressed by:

$$F_l(\omega) \cong \frac{1}{1+\sum\limits_{k=1}^{M} \alpha_k(\lambda)\,\omega^{2k}}\, e^{-\jmath(l+1)\gamma\omega}. \quad (6)$$

4.2 Interpretation

While comparing the form (6) with the Tikhonov regularization filter (equation (4)), we note that the expression of the Poisson filter module corresponds to a first derivative regularization filter, which minimizes the criterion:

$$\mathcal{J}_\lambda = \left\|y(t) - \mathcal{H}(x(t))\right\|_2^2 + \sum_{k=0}^{M-1} \alpha_{k+1}(\lambda)\left\|\mathcal{D}_k x(t)\right\|_2^2,$$

where $\mathcal{H}(x(t)) = \int_0^t x(t')\,dt'$ and $\mathcal{D}_k = \frac{d^k}{dt^k}$.

The smoothness constraint imposes that the required solution should be infinitely derivable (in the case where $(l+1)$ is even, the constraint is that the $(M-1)^{th}$ derivative of the solution is null). The regularization parameters $\{\alpha_{k+1}(\lambda)\}_{k=0}^{M-1}$ depend on λ and the filter order. The choice of this constraint is important because it ensures that the estimation of the needed high order derivatives of the signals, using this regularized first derivative, will not yield a noise amplification, if we choose $l \geq na$. Note that this choice corresponds to the use of a minimal order GPMF filter.

On the other hand, one can see that the Poisson filter has a linear phase in the low frequencies. This linear phase can be interpreted as follows. Consider a regularization filter:

$$F(\omega, t_0) = \frac{\|H(\omega)\|^2}{\|H(\omega)\|^2 + \sum_{k=0}^{N} \alpha_k\, \omega^{2k}}\, e^{-\jmath \omega t_0}.$$

We show that this filter corresponds to the minimization of a criterion:

$$\mathcal{J}(x, t_0) = \left\|y(t) - \mathcal{H}\left(x(t+t_0)\right)\right\|_2^2 + \sum_{k=0}^{N} \alpha_k \left\|\mathcal{D}_k x(t+t_0)\right\|_2^2, \qquad (7)$$

where $\mathcal{H}$ is an integrator, and t_0 a time delay.

Proof: The application of the Fourier transform to the equation (7) yields:

$$\begin{aligned} J(X, t_0) &= \|Y(\omega)\|^2 - H^*(\omega)X^*(\omega)Y(\omega)\, e^{-\jmath \omega t_0} \\ &+ \|H(\omega)\, X(\omega)\|^2 - H(\omega)X(\omega)Y^*(\omega)e^{\jmath \omega t_0} \\ &+ \sum_{k=0}^{N} \alpha_k\, \omega^{2k}\, \|X(\omega)\|^2. \end{aligned}$$

By minimizing with respect to $X(\omega)$:

$$\frac{d}{dX} J(X, t_0) \bigg|_{X = \hat{X}_{reg}} = 0$$

$$\Rightarrow \hat{X}_{reg}(\omega) = \frac{H^*(\omega)Y(\omega)}{\|H(\omega)\|^2 + \sum_{k=0}^{N} \alpha_k\, \omega^{2k}}\, e^{-\jmath \omega t_0}.$$

Knowing that $\hat{X}_{reg}(\omega) = F(\omega, t_0)\, \frac{Y(\omega)}{H(\omega)}$, so:

$$F(\omega, t_0) = \frac{\|H(\omega)\|^2}{\|H(\omega)\|^2 + \sum_{k=0}^{N} \alpha_k\, \omega^{2k}}\, e^{-\jmath \omega t_0}.$$

We deduce that a regularization filter which has a linear phase corresponds to the search of a delayed regularized solution. In the case of the Poisson filters, as shown in figure (1), this delay increases when the filter order increases or when the filter cut-off frequency decreases ($t_0 = \frac{l+1}{\lambda}$).

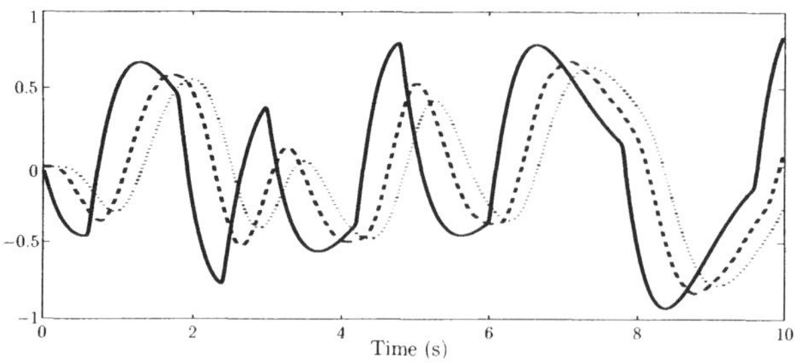

Fig. 1. Exact derivative (full) and filtered derivative by GPMF filter, $l = 2$ (dashed), $l = 4$ (dotted) with $\lambda = \beta = 5\text{rad/s}$ and SNR $= 5\,\text{dB}$.

4.3 Generalization

At the beginning of this section, we concentrated on the GPMF method. The same analysis can be carried out for all linear filters of the form:

$$F(\omega) = \frac{1}{A(\omega)},$$

where $A(\omega)$ is polynomial of order n. This filter can be written as:

$$F(\omega) = \frac{1}{1 + \sum_{k=1}^{\infty} \alpha_k\, \omega^{2k}}\, e^{-\jmath \varphi_A(\omega)}, \qquad (8)$$

where the parameters α_k, and the phase $\varphi_A(\omega)$ depend on $A(\omega)$ coefficients.
The form (8) corresponds to the minimization of a criterion:

$$\mathcal{J}(x) = \left\|y(t) - \mathcal{H}\left(\tilde{x}(t)\right)\right\|_2^2 + \sum_{k=0}^{\infty} \alpha_{k+1} \left\|\mathcal{D}_k \tilde{x}(t)\right\|_2^2,$$

where:

$$\tilde{x}(t) = \left[x \star \mathcal{F}^{-1}\left\{e^{\jmath \varphi_A(\omega)}\right\}\right](t),$$

and $\mathcal{F}^{-1}$ represents the inverse Fourier transform operator.

5. PHASE EFFECT ANALYSIS

In this section, we address the question of the regularizing filter phase effect on continuous-time model identification. The first answer that we can give is that in the case of linear phases (first order development), only a time delay will be introduced on the filtered signals. Consequently, the identification process will not be affected because the same delay is introduced on the system input and output signals. But this is not exactly the case for the Poisson filter, if we take the exact form of the filter:

$$F_l(\omega) = \frac{1}{1 + \sum_{k=1}^{M} \alpha_k(\lambda)\, \omega^{2k}}\, e^{-\jmath \varphi_{F_l}(\omega)},$$

where:

$$\varphi_{F_l}(\omega) = (l+1)\arctan(\gamma\, \omega)$$

In order to get some insights into the filter phase effect, we propose to use a regularization filter which has the same module as a GPMF filter, but a null phase. The synthesis of such a filter can be easily made from the Poisson filter, by considering:

$$\begin{aligned} F_l^{\#}(\omega) &= F_l(\jmath\, \omega)\, F_l(-\jmath\, \omega), \\ &= \left(\frac{\beta^2}{\omega^2 + \lambda^2}\right)^{l+1}. \qquad (9) \end{aligned}$$

Note that the module of this filter is polynomial on ω and of order $2(l+1)$, so the GPMF filter that has the same module is of order $(2\,l+1)$.

Let us pose $\gamma = \frac{1}{\lambda}$, and consider $\beta = \lambda$. By expanding (9), we get:

$$F_l^{\#}(\omega) = \frac{1}{1 + \sum\limits_{k=1}^{l+1} \alpha_k(\lambda)\, \omega^{2k}},$$

where $\alpha_k(\lambda) = C_{l+1}^k \gamma^{2k}$.
This form corresponds exactly to the Tikhonov regularization filter minimizing the criterion:

$$J_\lambda(x(t)) = \left\|y(t) - \mathcal{H}(x(t))\right\|_2^2 + \sum_{k=0}^{l} \alpha_{k+1}(\lambda)\left\|\mathcal{D}_k x(t)\right\|_2^2,$$

where the regularization parameters $\{\alpha_{k+1}(\lambda)\}_{k=0}^{l}$ depend on λ and the filter order. The smoothness constraint imposes that the solution is not delayed (figure (2)) and its lth derivative is null.

5.1 Simulation Example

To asses the effect of the filter phase on continuous–time model identification, we use a simulation example. The example concerns the identification of the following continuous–time transfer function (Wang and Gawthrop, 2001):

$$G(s) = \frac{-2s + 1}{s^3 + 1.6s^2 + 1.6s + 1}.$$

The input signal is a pseudo random binary sequence, which is chosen to excite the system over all its dynamic range. The sampling period is taken equal to 0.02 s and the number of samples is $N = 1260$. The results are obtained for a Monte Carlo simulation of $S = 1000$ trials, with a Signal to Noise Ratio (SNR) equal to 5 dB:

$$\text{SNR} = 10 \log\left(\frac{P_{y_u}}{\sigma_v^2}\right),$$

where P_{y_u} represents the power of the noise free output signal $y_u(t)$ and σ_v is the standard deviation of the additive noise. The parameters are estimated using the instrumental variable method with an auxiliary model which is obtained after an initial estimation by least squares. The empirical mean

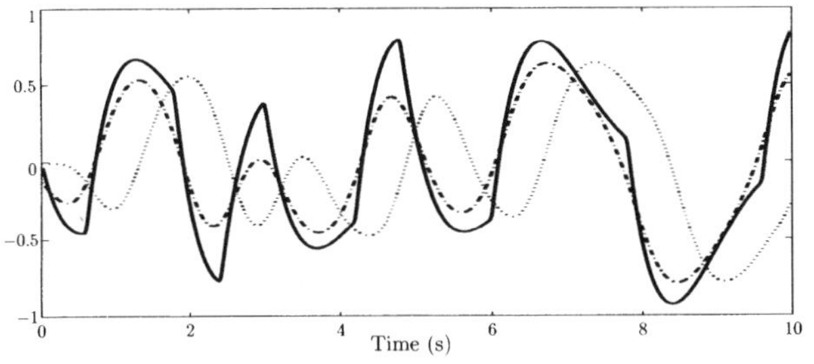

Fig. 2. Exact derivative (full), regularized derivative (dashed), with $l = 2$ and filtered derivative by GPMF (dotted), with $l = 5$, for $\lambda = \beta = 5$ rad/s and SNR = 5 dB

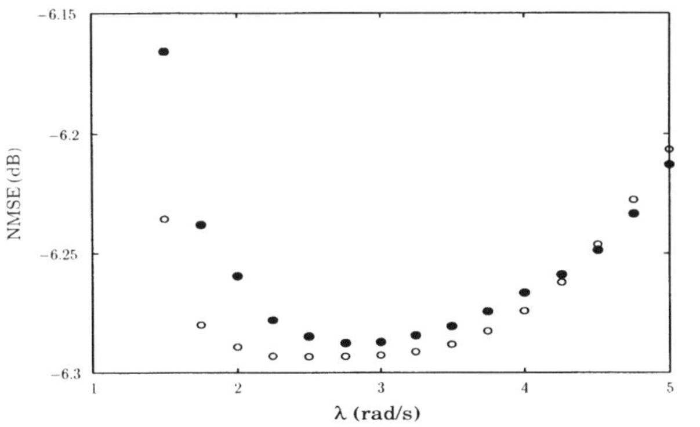

Fig. 3. Normalized Mean Square Error Versus filter bandwidth, (circles) null phase filter and (dots) GPMF filter, for SNR = 5 dB

($\hat{m}_{\hat{\theta}_j}$), standard deviation ($\hat{\sigma}_{\hat{\theta}_j}$) and mean square error ($\widehat{\text{MSE}}_{\hat{\theta}_j}$), estimated for each parameter, are used to discuss the statistical performances.

Figure (3) shows the evolution of the normalized mean square error versus the filter bandwidth. It appears that the use of the null phase filter gives a smaller mean square error than the GPMF filter. We deduce that an improvement is introduced using this filter. The results of estimation for the optimal value of λ, for the two filters, are summarized in table 1. It is noted that the proposed regularization filter can be applied successfully to the identification of a continuous–time model and allows to obtain the parameters of the system. The comparison with the GPMF filter, shows that the two methods are unbiased. In addition the analysis of the standard deviation and the mean square error of the estimated parameters confirms that an improvement is introduced by the use of a true regularization filter.

Table 1. Monte Carlo simulation results

GPMF	λ_{opt} (rad/s)		2.75	
	Criterion	$\hat{m}_{\hat{\theta}_j}$	$\hat{\sigma}_{\hat{\theta}_j}$	$\widehat{\text{MSE}}_{\hat{\theta}_j}$
	$\hat{b}_0$ (1)	1.0010	0.0602	0.0036
	$\hat{b}_1$ (−2)	-2.0058	0.1161	0.0135
	$\hat{a}_0$ (1)	1.0016	0.0713	0.0051
	$\hat{a}_1$ (1.6)	1.6011	0.0448	0.0020
	$\hat{a}_2$ (1.6)	1.6015	0.0941	0.0089
Regularization	λ_{opt} (rad/s)		2.5	
	Criterion	$\hat{m}_{\hat{\theta}_j}$	$\hat{\sigma}_{\hat{\theta}_j}$	$\widehat{\text{MSE}}_{\hat{\theta}_j}$
	$\hat{b}_0$ (1)	0.9980	0.0644	0.0042
	$\hat{b}_1$ (−2)	- 2.0018	0.1105	0.0122
	$\hat{a}_0$ (1)	1.0019	0.0656	0.0043
	$\hat{a}_1$ (1.6)	1.6006	0.0463	0.0021
	$\hat{a}_2$ (1.6)	1.5990	0.0866	0.0075

From this simulation, we deduce that it can be advantageous to use a null phase filter in continuous–time model identification. The second advantage is that this filter correspond to the derivatives estimation by minimizing a criterion which is given explicitly.

6. TOWARDS A JOINT SIGNAL–MODEL PARAMETER ESTIMATION

In this section, we address the problem of the optimal regularization filter design for parameter estimation. Firstly, we note that the GPMF filter as well as the proposed regularization filter separate the signal estimation problem from that of model parameter estimation. However, it is clear that these two problems are strongly coupled. In general, the choice of the filter parameters is made by minimizing the output mean square error. We believe that this decoupling is the main shortcoming of these approaches, which can be overcome by formalizing the problem as a joint system output signal and parameter estimation problem. Note that, the SRIVC method may be interpreted as an attempt to solve that type of problem since the regularization filter depends on the parameters to estimate. However, the relevance of the corresponding criterion can be questioned.

Our approach to that problem consists in finding the values of $(\mathcal{A}, \mathcal{B}, x(t))$ that minimize the following compound criterion:

$$\mathcal{J}_\alpha(x(t)) = \left\|y(t) - x(t)\right\|_2^2 + \alpha\left\|\mathcal{A}(x(t)) - \mathcal{B}(u(t))\right\|_2^2,$$

where $\mathcal{A}$, $\mathcal{B}$ are operators representing, respectively, the numerator and the denominator of the model to identify and $x(t)$ is the filtered output signal. The first part of the criterion is a data fitting measure, while the second part is a model fitting measure that will regularize the solution $x(t)$. This optimisation problem may be solved for example by an iterative approach:

$$(\hat{\mathcal{A}}_{i+1}, \hat{\mathcal{B}}_{i+1}) = \underset{\mathcal{A},\mathcal{B}}{\text{argmin}}\, \mathcal{J}_\alpha(\mathcal{A}, \mathcal{B}, \hat{x}_i(t)) \qquad (10)$$

$$(\hat{x}_{i+1}(t)) = \underset{x}{\text{argmin}}\, \mathcal{J}_\alpha(\hat{\mathcal{A}}_{i+1}, \hat{\mathcal{B}}_{i+1}, x(t)) \qquad (11)$$

where subscript i denotes the estimation obtained in iteration i. This iterative algorithm may be interpreted as a particular form of the SEM algorithm (Celeux and Diebolt, 1985), which ensures that the criterion decreases but does not guarantee the global minimum to be reached. The signal $x(t)$ being fixed to $\hat{x}_i(t)$, the optimisation problem (10) can be solved by an instrumental variable approach. Concerning the problem (11), $\mathcal{A}$ and $\mathcal{B}$ being fixed to $\hat{\mathcal{A}}_{i+1}$ and $\hat{\mathcal{B}}_{i+1}$,we obtain the explicit form of the solution:

$$\hat{x}_{i+1}(t) = \frac{1}{1 + \alpha\hat{\mathcal{A}}^*_{i+1}\hat{\mathcal{A}}_{i+1}} y(t) + \alpha \frac{\hat{\mathcal{A}}^*_{i+1}\hat{\mathcal{B}}_{i+1}}{1 + \alpha\hat{\mathcal{A}}^*_{i+1}\hat{\mathcal{A}}_{i+1}} u(t)$$

where $\hat{\mathcal{A}}^*_{i+1}$ and $\hat{\mathcal{B}}^*_{i+1}$ correspond to the adjoint operators of $\hat{\mathcal{A}}_{i+1}$ and $\hat{\mathcal{B}}_{i+1}$, respectively.

The formulation given above can be viewed as a first attempt to use explicitly regularization techniques in continuous-time model identification. Future works will be directed at investigating more deeply this approach.

7. CONCLUSION

Derivative estimation is an ill–posed inverse problem, which is encountered in continuous–time model identification. This problem is circumvented in an implicit way using some linear transformations on the data. It is interesting to note that, in the case of linear filter methods, these transformations correspond to the regularization of this problem. In addition, the use of a null phase filter shows that the filter phase has an effect on the identification process. A second important consequence, resulting from this analysis, is that we can formulate continuous–time model identification as a joint signal - parameter estimation problem.

ACKNOWLEDGEMENTS

The authors would like to thank gratefully Professor Hugues Garnier for helpful suggestions and discussions.

REFERENCES

Celeux, G. and J. Diebolt (1985). The SEM algorithm: a probabilistic teacher algorithm derived from the EM algorithm for the mixture problem. *Computational statistics* **2**, 73–82.

Garnier, H., M. Mensler and A. Richard (2002). Continuous–time model identification from sampled data. Implementation issues and performance evaluation. Submitted to *Int. J. Control*.

Mensler, M. (1999). Ananlyse et étude comparative de méthodes d'identification des systèmes à représentation continue. Développement d'une boite à outils logicielle. PhD thesis. Université Henri Poincaré, Nancy 1, France.

Rao, G.P. and H. Garnier (2002). Numerical illustration of the relevance of direct continuous–time domain identification. In: *15th Triennial World Congress on Automatic Control.* Barcelona (Spain).

Saha, D.C. and G.P. Rao (1983). *Identification of continuous–time dynamical systems - The Poisson Moment Functionals (PMF) approach.* Springer-Verlag. Berlin.

Tikhonov, A. and V. Arsénine (1974). *Méthodes de résolution de problèmes mal posés.* Editions Mir. Moscou.

Wang, L. and P. Gawthrop (2001). On the estimation of continuous–time transfer functions. *Int. J. Control* **74**(9), 889–904.

Young, P.C. (1964). In flight dynamic checkout. *IEEE Trans. on Aerospace* **AS 2**, 1106–1111.

Young, P.C. (2002). Optimal IV identification and estimation of continuous–time TF models. In: *15th Triennial World Congress on Automat. Control.* Barcelona (Spain).

IFAC Publications
www.elsevier.com/locate/ifac

A SURVEY OF READILY ACCESSIBLE PERTURBATION SIGNALS

K. R. Godfrey[1], A. H. Tan[2] and H. A. Barker[3]

1: *School of Engineering, University of Warwick, Coventry, CV4 7AL, U.K.; Email:* krg@eng.warwick.ac.uk
2: *Faculty of Engineering, Multimedia University, 63100 Cyberjaya, Malaysia; Email:* htai@mmu.edu.my
3: *School of Engineering, University of Wales Swansea, Swansea SA2 8PP, U.K.; Email:* h.a.barker@swansea.ac.uk

Abstract: There has been a rapid increase recently in the number of software packages available to generate different types of perturbation signals for the purpose of system identification. Several can now be downloaded from the World-Wide Web. The objective of this paper is to review what packages are available, and from where, and to outline some of the application areas in which they might be used. *Copyright © 2003 IFAC*

Keywords: Frequency methods; Multi-frequency signals; Multi-level signals; Nonlinear systems; Non-parametric identification; Pseudo-random signals; System identification.

1. INTRODUCTION

Over the last few years, there has been an increasing recognition of the role that suitably designed perturbation signals can play in system identification. With a suitable choice of perturbation signal and method of processing the input-output signals of a system, different aspects of the system can be highlighted. Currently, there are several different designs of perturbation signal that are readily available from easily accessible computer packages. These are classified into computer optimised signals and pseudo-random signals, and they allow certain time or frequency domain properties to be specified by the user, hence providing an advantage over purely random signals. The first part of the paper deals with signals designed for linear system identification. This is followed by examples of nonlinear system identification, for which the signal design criteria are normally quite different.

2. LINEAR SYSTEM IDENTIFICATION

2.1. *Measures of Signal Quality*

For linear system identification, there are two main requirements of a perturbation signal. Firstly, it should be small enough to minimise any nonlinear distortions, and secondly it should be large enough to minimise any noise effects. These requirements are conflicting, and so, for linear system identification, it is desirable to maximise the power in specified harmonics within the amplitude constraints.

Godfrey *et al.* (1999) developed three performance indices which are particularly relevant for linear system identification in the frequency domain. The first of these is the *Performance Index for Perturbation Signals* (PIPS) defined as

$$\text{PIPS} = \frac{200(u_{\text{rms}}^2 - u_{\text{mean}}^2)^{1/2}}{u_{\max} - u_{\min}}\ \% \qquad (1)$$

where u_{rms}, u_{mean}, $u_{\max}$ and $u_{\min}$ are, respectively, the root mean square, mean, maximum, and minimum values of either the continuous signal $u(t)$ or the discrete signal $u(i)$. PIPS is statistically based and independent of both the signal mean and amplitude scale. It is 100% for a signal with the best possible performance, namely a binary signal for which $u_{\max}$ and $u_{\min}$ have equal duration or number of occurrences. PIPS may also be defined as

$$\text{PIPS} = \frac{200}{N(u_{\max} - u_{\min})}\left(\sum_{k=1}^{N-1} |U(k)|^2\right)^{1/2}\ \% \qquad (2)$$

where $U(k)$ is the DFT of the discrete signal, given by

$$U(k) = \sum_{i=0}^{N-1} u(i)\exp(-\text{j}2\pi ki/N) \qquad (3)$$

and N is the period. PIPS is an application-independent measure of perturbation signal performance. In frequency domain identification of continuous systems, not all harmonics are completely effective. The use of a zero-order-hold to generate the continuous signal $u(t)$ from the discrete signal $u(i)$, reduces the effectiveness of all harmonics, and harmonics close to the Nyquist frequency cannot be used because of aliasing. In this case, the power in the kth harmonic in a unilateral spectrum of positive frequencies is given by

$$|C'_u(k)|^2 = \left|\frac{U(k)}{N}\right|^2 \qquad k=0 \qquad (4)$$

$$|C'_u(k)|^2 = 2\left|\left(\frac{\sin(\pi k/N)}{(\pi k/N)}\right)\left[\frac{U(k)}{N}\right]\right|^2 \qquad k>0 \qquad (5)$$

If the first R harmonics are specified for the identification, then the *Effective Performance Index*, PIPSE, is given by

$$\text{PIPSE} = \frac{200}{u_{\max} - u_{\min}}\left(\sum_{k=1}^{R}\left|C'_u(k)^2\right|\right)^{1/2} \% \quad R < N/2 \quad (6)$$

PIPSE is maximised, and the specified harmonics contain the greatest proportion of signal power, when $R = (N-2)/2$ if N is even, and $(N-1)/2$ if N is odd.

The use of PIPSE alone does not account for the fact that the power contained in one of the specified harmonics of a signal might be small, giving rise to low accuracy of estimation at the corresponding frequency. The Frequency Domain System Identification (FDIDENT) Toolbox in MATLAB (Kollár, 1994) uses a quantity $E_{\min}$, which is the minimum ratio between the actual amplitude and the specified amplitude at any of the specified harmonics of a computer optimised signal. The problem with this definition as it stands is that the specified harmonic amplitude is an input to the optimising algorithm, rather than a signal parameter. To overcome this problem, $E_{\min}$ is re-defined here as

$$\text{EMINE} = 100 \underset{k=1,2,\ldots,R}{\text{minimum}} \frac{|C'_u(k)|}{\left(\frac{1}{R}\sum_{k=1}^{R}|C'_u(k)|^2\right)^{1/2}} \% \qquad (7)$$

for $R < N/2$. EMINE is also independent of signal mean and amplitude scale, and its values also range from 0% for the worst possible performance to 100% for the best possible performance.

Note that the power in the specified harmonics appears in the numerator of PIPSE and in the denominator of EMINE, so that signal design is inevitably a compromise. In general, PIPSE is the more important measure of the two, since a low EMINE can result from a lack of power in only one of the specified harmonics.

A fourth performance index considered in this paper is the *Time Factor* $TF(u)$ (Pintelon and Schoukens, 2001; Section 4.2.1). This is designed to give an indication of the time taken to achieve a minimum estimation accuracy of the Frequency Response of a system at any of the specified harmonics in the perturbation signal. In terms of the measures given above, the Time Factor of a signal for which $u_{\min} = -u_{\max}$ is given by

$$Tf(u) = 0.5\left(\frac{100}{\text{PIPSE}}\right)^2\left(\frac{100}{\text{EMINE}}\right)^2 \qquad (8)$$

Note in particular that a large value of $Tf(u)$ can result from a low value of either PIPSE or EMINE.

2.2. *Computer Optimised Signals*

Computer optimised signals are designed to match a specified power spectrum as closely as possible. Three types are currently readily available on the WWW.

2.2.1. *Multisine (Sum of Harmonics) Signals*

A multisine signal can take *any* value within the range between its minimum and maximum values. Such signals are available using the function *msinclip* in the Frequency Domain System Identification Toolbox (Kollár, 1994). This is a commercially available MATLAB Toolbox, details of which may be found at:

http://elecwww.vub.ac.be/fdident/

The user specifies those harmonics in which the signal power is required, together with the desired amplitude pattern. For linear system identification, the objective of the optimisation is to adjust the relative phases of the harmonics in order to minimise the peak-to-peak amplitude of the signal, thus maximising PIPS.

The algorithm used in *msinclip* is based on swapping between the time domain and the frequency domain (Van der Ouderaa *et al.*, 1988). To start the algorithm, the user can specify either Schroeder phases (Schroeder, 1970), which are designed for low peak-to-peak and high PIPS signals, or random phases. Two forms of zero-order-hold sum of harmonics (ZOHSOH) signals are available using *msinclip*, both of which result in a signal with some power in the non-specified harmonics. The first form is a band-limited SOH signal, passed through a zero-order-hold, which has a $(\sin^2 x)/x^2$ power spectrum envelope. The second form has precompensation for the shape of the zero-order-hold spectrum. For this form, the value of EMINE is 100%, regardless of the specification, because the actual and specified harmonic amplitudes are the same at the specified harmonics, but this is at the expense of relatively low PIPS and PIPSE values.

2.2.2. *Discrete-interval Binary and Ternary Signals*

The function *dibs* in the FDIDENT Toolbox can be used to generate discrete-interval binary signals, and the objective of the optimisation in this case is to force as much power as possible into the specified harmonics. The function is based on an algorithm of Van den Bos and Krol (1979). The value of PIPS for such a signal is close to 100%, because the signal is binary and the mean of the signal is either zero or very close to it. However, the value of PIPSE is less than 100% for any harmonic specification. The value of EMINE can also be low for some harmonic specifications, but then the function *dibsimpr* in the FDIDENT Toolbox can be used to try to increase EMINE.

The function *dits*, available in the GUI version of FDIDENT, can be used to generate discrete-interval ternary signals; it is based on the same algorithm as that used in *dibs*.

2.2.3. *Multi-level Multi-harmonic Signals*

The signals obtained from *dibs* and *dibsimpr* are binary, so their peak-to-peak amplitude can be clearly defined, but a significant amount of the total power then appears in non-specified harmonics. In contrast, the signals obtained from *msinclip* generally have more power in the specified harmonics, but have higher peak-to-peak amplitudes for a given harmonic specification. Multi-level multi-harmonic (MLMH) signals are designed to retain the advantages of each type of signal, while reducing the disadvantages. MLMH signals can be designed using the MATLAB function *multilev* (McCormack *et al.*, 1995), which employs the same time-frequency swapping algorithm used in *msinclip*. The user specifies the number of levels M (≥ 2). When $M = 2$, *multilev* can be used as an alternative to *dibs* to obtain discrete-interval binary signals; similarly when $M = 3$, it can be used as an alternative to *dits*. The function *multilev* is not currently available in the FDIDENT Toolbox, but it is freely available and can be accessed at the following location:
http://www.eng.warwick.ac.uk/EED/DSM/multilev.htm

2.2.4. *Inverse-repeat Computer Optimised Signals*

If the user specifies odd harmonics only in any of the types of computer optimised signals described above, and sets N to an even number, the resulting signal will be inverse-repeat, having the second half period the negative of the first half. Such a specification is useful for linear system identification in the presence of nonlinear distortion, as will be discussed later.

2.3. *Pseudo-random (Fixed Spectrum) Signals*

In contrast to the computer optimised signals considered in Section 2.2, pseudo-random signals have fixed spectra. Both binary and multi-level signals are readily available for users, and these will be considered in turn.

2.3.1. *Pseudo-random Binary Signals (PRBS)*

For the signals considered here, generated from pseudo-random binary sequences of period N and levels $\pm V$,

$$\text{PIPS} = 100\left[\left(N^2 - 1\right)/N^2\right]^{1/2}\% \qquad (9)$$

which tends to 100% as N increases. The highest harmonic R to be used in the experiment has the least power, and if $R = (N - 1)/2$, PIPSE is maximised and tends to 88.0% as N increases, while EMINE tends to 72.4% as N increases. To achieve reasonable values of both PIPSE and EMINE in any experiment, it is evident that there needs to be considerable flexibility in the choice of R, and therefore of N.

The function *mlbs* in the FDIDENT Toolbox gives one example for each possible value of N of a PRBS based on a maximum-length binary (MLB) sequence, for which $N = (2^k - 1)$, where k is a positive integer so that $N = 3, 7, 15, 31, 63, 127, 255, 511, 1023, 2047, \ldots$

Although the class of PRBS based on MLB sequences is often thought to be the only kind available, many other classes of PRBS exist. These include quadratic residue binary (QRB) signals, for which $N = (4k - 1)$ and prime ($N = 3, 7, 11, 19, 23, 31, 43, 47, 59, 67, 71, 79, 83, 103, 107, 127, \ldots$); Hall binary (HAB) signals, for which $N = (4k^2 + 27)$ and prime ($N = 31, 43, 127, 223, 283, 811, 1051, 1471, 1627, \ldots$); and Twin Prime binary (TPB) signals, for which $N = k(k + 2)$, with both k and $(k + 2)$ prime ($N = 15, 35, 143, 323, 899, 1763, 3599, 5183, \ldots$).

A MATLAB package *prs* (Tan & Godfrey, 2002b) is freely available on the WWW for generating all the above classes of PRBS with periods up to 50,000. It can be accessed at the following location:

http://www.eng.warwick.ac.uk/EED/DSM/prs.htm

The availability of QRB, HAB and TPB signals greatly increases the number of values of N for which a pseudo-random binary signal is available, compared to if only MLB signals were available. The package is independent of the FDIDENT Toolbox for the generation of MLB, QRB, HAB and TPB signals. However, if *prs* is used in conjunction with the Toolbox, then MLB signals can be generated without the need to know their characteristic polynomials, as the Toolbox provides a default signal for any value of N. There are several different MLB signals for each possible value of $N \geq 7$, and unlike *mlbs*, the *prs* package can generate them all. For the identification of linear systems with negligible nonlinear distortion, the choice of any particular MLB signal of period N is not important, but for nonlinear system identification, some signals may give better results than others (Tan & Godfrey, 2002b) – see Section 3.4.

2.3.2. *Inverse-repeat Pseudo-random Binary Signals*

Inverse-repeat versions of all the signals generated in the *prs* package may be obtained within *prs* by inverting every other member of the sequence to give a new sequence with period $2N$. In the new sequence, the second half of a period is the negative of the first half, and therefore the even harmonics of the signal are suppressed. Signals based on such sequences have a PIPS value of 100%. PIPSE is maximised for $R = (N - 2)/4$, and tends to 88.0% as N increases, while EMINE tends to 72.4% as N increases. These limits are the same as those for the PRBS on which the inverse-repeat signals are based. The suppression of even harmonics in a system input signal ensures that odd-order and even-order nonlinear distortions in the system output may be separated, and the effect of the latter on linear system identification can be completely eliminated.

2.3.3. *Near-binary Pseudo-random Signals*

The *prs* package can also be used to generate quadratic residue ternary (QRT) signals. These are similar to QRB signals, except that the last digit of the period is zero. The signals exist for $N = (4k \pm 1)$ and prime, and some therefore have the same periods as the corresponding QRB signals, while others have the following periods for which no QRB signal exists: N = 5, 13, 17, 29, 37, 41, 53, 61, 73, 89, 97, 101, 109, 113, . . . The PIPS, PIPSE and EMINE values of QRT signals are all close to those of PRBS signals, and so these signals provide a useful addition to the other signals used for linear system identification. Inverse-repeat versions of these signals are also available in *prs*.

2.3.4. *Pseudo-random Multi-level Signals*

Multi-level maximum-length (MLML) signals exist in Galois fields GF(q), where q is a prime or a power of a prime; and their periods are $N = (q^k - 1)$, where k is an integer. Converting the elements of the Galois field into real signal levels then produces a pseudo-random multi-level (PRML) signal. When q is 2, the signals are simply MLB signals. When q is odd and greater than 2, the field elements can be converted into signals, with any odd number of levels from 3 to q, in which even harmonics are suppressed and odd harmonics are uniform (Barker and Zhuang, 1998). The computer software GALOIS (Barker, 2001) is available for generating signals with $q \le 128$. GALOIS has full design facilities for signals with $q \le 31$, and is freely available from the third author of this paper. Although primarily used for nonlinear system identification, the MLML signals in GALOIS may also be used for linear system identification. In particular, ternary signals may be generated for any $q > 2$, but the simplest of these, with q = 3 do not necessarily have the best performance. Godfrey *et al.* (1999) showed that, for ternary signals with q = 3, PIPS tends to 81.6% as N increases, PIPSE is maximised for $R = (N - 2)/4$ when $N/2$ is odd and for $R = N/4$ when $N/2$ is even, tending to 71.8% as N increases, and EMINE tends to 72.4% as N increases. Higher values for PIPS and PIPSE can be obtained for ternary signals with other values of q. For example, a PIPS of 95% and a PIPSE of 84% can be obtained with q = 29.

Harmonic multiples of both 2 and 3 can be suppressed for certain values of q, provided the number of signal levels is greater than two and N is a multiple of six. Within the range of q in the GALOIS software with full design facilities ($2 \le q \le 31$), this can be achieved for q = 7, 13, 19, 25, and 31. For q = 31, it is also possible to suppress harmonic multiples of 5.

2.4. *Comparison of Signals for Different Specifications*

In this Section, the performances of the above signals are compared for three different specifications (A, B, and C), all with 16 equal amplitude harmonics specified. (Such specifications are very useful, for example, in on-line adaptive control schemes, but it should be noted that some identification processes require a larger number of harmonics.) In A, the harmonics were consecutive, 1 to 16; in B, the harmonics were odd consecutive, 1, 3, . . . , 31; and in C, the harmonic multiples of 2 and 3 were suppressed, the specified harmonics being 1, 5, 7, 11, . . . , 47. In all cases, the multisine signals were generated with pre-compensation for the spectrum of the zero-order-hold. The ZOHSOH, DIB, DIT and MLMH signals were generated with 10,000 iterations. The value of R was set close to $0.4N$, which is close to the highest usable harmonic in a spectrum analyser that uses an analog anti-aliasing filter.

The results are given in Table 1. The conclusions drawn were:

- For A and B, the ZOHSOH signal, although having EMINE = 100% by definition, had relatively low PIPS and PIPSE values.
- For A, the *dibs* signal with N = 40 had a mean value of $2/N$, while the *multilev* signal with N = 40 had a mean value of $4/N$. For larger values of N, for which there is greater design flexibility, the mean of these signals is usually small, and, for N even, is often zero.
- None of the EMINE values for the *dibs* or (binary) *multilev* signals were increased using *dibsimpr* in the FDIDENT Toolbox; this is probably due to the large number of iterations (10,000) used in the designs.
- Reasonable PIPSE values for signals from the *prs* package were possible because of the large number of possible values of N in *prs*. For A, a QRB signal was used, while for B the corresponding inverse-repeat signal was used. Had only MLB signals been available, N would need to be increased to 63 and 126 for Specifications A and B respectively, which would lead to low PIPSE values.
- As signals from GALOIS for $q > 2$ are all inverse-repeat, they could not be used in A. For B, there was no point in using a PRML signal, since the Specification could be met with a binary signal – see Section 2.3.4.
- Specification C cannot be met with a binary signal, so that PRML signals should be considered. However, the possible periods of PRML signals with harmonic multiples of 2 and 3 suppressed are relatively sparse. In C, the lowest value of N possible for such a signal was 168 (q = 13, k = 2). While the resulting PIPS value is comparable to those of the other designs, the larger N results in a low PIPSE value but a high EMINE value. Specifications involving greater numbers of harmonics are more suitable for PRML signals, and can usually be framed with sufficient flexibility to allow N to be chosen to meet the specification, rather than *vice-versa*.
- The values of Time Factor for the signals in A and B were similar, except for the *multilev* design in A, for which the value was a little higher. Time Factor values for C were all higher.

Table 1. Comparison of perturbation signals for linear system identification

Spec.	Type	Source	No. of levels	N	PIPS (%)	PIPSE (%)	EMINE (%)	Time Factor
A	ZOHSOH	*msinclip*	multi	40	79.0	72.4	100	0.954
A	DIB	*dibs*	2	40	99.9	89.4	83.4	0.899
A	MLMH	*multilev*	2	40	99.5	90.5	73.8	1.119
A	PRB	*prs*	2	43	100	80.5	85.3	1.060
B	ZOHSOH	*msinclip*	multi	80	83.7	75.9	100	0.868
B	DIB	*dibs*	2	80	100	88.9	79.6	0.998
B	MLMH	*multilev*	2	80	100	89.4	81.0	0.954
B	IRPRB	*prs*	2	86	100	81.1	86.0	1.027
C	ZOHSOH	*msinclip*	multi	120	73.4	66.6	100	1.127
C	DIT	*dits*	3	120	81.6	72.9	80.9	1.438
C	MLMH	*multilev*	3	120	81.7	73.4	81.8	1.387
C	PRML	GALOIS	3	168	78.7	56.9	91.5	1.845

2.5. *Conclusions for Linear System Identification*

For equal amplitude harmonic specifications, which are the most common in practice, the user now has a wide range of readily accessible signals for linear system identification. For many applications, a binary signal is a good choice, because the PIPS value is high, by definition, and through judicious design, the values of PIPSE and EMINE can be made reasonably high. For pseudo-random (fixed spectrum) signals, this involves making R as large as possible; as noted in Section 2.4, in the designs in this paper, it was set close to $0.4N$. This illustrates the value of the MATLAB package described in Section 2.3.1, which yields much more flexibility over possible values of N than if MLB signals alone were available. For binary signals, harmonic suppression is limited to even harmonics, but provided the greater period of the signal is not a problem, this is also a good choice for linear system identification, because distortion from even-order nonlinearities can be removed.

In some applications, the input transducer limits the user to a small number of signal levels, but when there are no such amplitude restrictions, multisine (sum of harmonics) signals (Section 2.2.1) are an attractive choice. Through their definition, they have EMINE = 100%, but their PIPS and PIPSE values tend to be lower than for signals with smaller numbers of levels. However, these signals are the only possible choice if the input spectrum is to be shaped.

3. NONLINEAR SYSTEM IDENTIFICATION

The properties required of a perturbation signal for the identification of a nonlinear system, including identification of the linear dynamics, are often very different from those required for the identification of an inherently linear system. In particular, the user must ensure that the signal is persistently exciting in both frequency and amplitude. In this Section, four examples of nonlinear system identification will be described, with the first three concerned with frequency domain identification and the fourth with time domain identification.

3.1. *Detection of Nonlinear Distortion*

Nonlinear distortion in a system can be detected by perturbing the system with periodic signals containing carefully selected harmonics, and then examining the contributions to the system output at harmonics not contained in the input (Evans *et al.*, 1994; McCormack *et al.*, 1994). Inputs with even harmonics suppressed can be used, because any distortion due to even-order nonlinearities will be detected at even harmonics in the output. If some odd input harmonics are also suppressed, then odd harmonic distortion can also be detected at the non-excited odd harmonics (Pintelon and Schoukens, 2001, Section 3.5.1). An input signal with harmonics 1, 3, 9, 11, 17, 19, . . . is particularly suitable for this because of the combinational properties of the harmonics.

3.2. *Identification of Wiener-Hammerstein Models using NID Multisines*

The nonlinear distortion at the test frequencies can be categorised into Type I and Type II. If the nonlinearity is of odd-order, Type I distortion results from a combination of a test frequency with pairs of equally positive and negative frequencies, with the contributions falling at that particular test frequency. If the nonlinearity is of even-order, this results from a combination of pairs of equally positive and negative frequencies, with the resulting contributions at zero frequency. Type II distortion is caused by all other frequency combinations. No Interharmonic Distortion (NID) multisines (Evans *et al.*, 1996) have no Type II distortion for nonlinearities up to a given order. The resulting spectrum is relatively sparse, and becomes progressively sparser as the order of the nonlinearity increases. Once the required harmonic numbers are known, the signal can be generated using *msinclip*.

Evans *et al.* (1996) used NID multisines for measuring the second-order Volterra kernels of Wiener models, in which linear dynamics precede a static nonlinearity, and Hammerstein models, in which a static nonlinearity precedes linear dynamics. Their examples include a reference nonlinear system and a servo motor system. Tan and Godfrey (2002a) used a method involving linear interpolation in the frequency domain (LIFRED) to identify Wiener-Hammerstein models, which comprise a cascade of a Wiener and a Hammerstein model, from second-order Volterra kernel measurements. This has the advantage that it drastically shortens the experimentation time and reduces the computational burden.

3.3. *Identification of Hammerstein Models with Quadratic Nonlinearity*

The identification of a model comprising a Hammerstein model in parallel with linear dynamics has been considered by Barker *et al.* (2001). A binary signal is not persistently exciting for the Hammerstein model; for example, squaring such a signal could result in only a constant value. In this case, a signal with at least three levels is needed. For the identification of the dynamics in both the linear and the Hammerstein model, it is desirable that the spectrum of both the signal $u(i)$ and its square $u^2(i)$ should be uniform. Signals that possess this property are based on 3-level PRML sequences obtained either from the field GF(3) or higher-order fields GF(3^k), where k is an integer; these can be obtained using the GALOIS software. The importance of this software is clearly seen in this example, where both time and frequency domain specifications are required.

3.4. *Detection of Direction-Dependent Dynamics*

A process with direction-dependent dynamics has dynamics that are different, according to whether the slope of the output is positive or negative. Tan & Godfrey (2001) showed that the departure from linearity can be detected using input-output cross-correlation with an input signal based on an MLB sequence, which can be obtained from the *prs* package described above. In such applications, one MLB signal can be very much better than another of the same length (Tan & Godfrey, 2002b). Hence, it is desirable to have access to MLB signals with different feedback configurations, as in the *prs* package. The presence of direction-dependent behaviour cannot be detected by cross-correlation using other classes of PRBS, and is not evident from the Frequency Response with any class of input signal. To estimate the combined linear dynamics, an inverse-repeat perturbation signal is preferred (Tan & Godfrey, 2001).

3.5. *Conclusions for Nonlinear System Identification*

Although the four examples described have dealt with different aspects of nonlinear system identification, they have shown that the computer packages described in Section 2, although primarily designed for linear system identification, can often provide perturbation signals suitable for nonlinear system identification.

REFERENCES

Barker, H.A. and M. Zhuang (1998). Design of pseudo-random perturbation signals for frequency-domain identification of nonlinear systems, *Proc. 11th IFAC Symposium on System Identification (SYSID 97)*, (Y. Sawaragi and S. Sagara, Eds.), Elsevier, Oxford.

Barker, H.A. (2001). GALOIS - a program for generating pseudo-random perturbation signals, *Proc. 12th IFAC Symposium on System Identification (SYSID 2000)*, (R. Smith, Ed.), Elsevier, Oxford, 505-508.

Barker, H.A., K.R. Godfrey and A.J. Tucker (2001). Nonlinear system identification with multilevel perturbation signals, *Proc. 12th IFAC Symposium on System Identification (SYSID 2000)*, (R. Smith, Ed.) Elsevier, Oxford, 1175-1178.

Evans, C., D. Rees and L. Jones (1994). Nonlinear disturbance errors in system identification using multisine test signals, *IEEE Trans. Instrum. Meas.*, **IM-43**, 238-244.

Evans, C., D. Rees, L. Jones and M. Weiss (1996). Periodic signals for measuring nonlinear Volterra kernels, *IEEE Trans. Instrum. Meas.*, **IM-45**, 362-371.

Godfrey, K.R., H.A. Barker and A.J. Tucker (1999). Comparison of perturbation signals for linear system identification in the frequency domain, *IEE Proc. – Control Theory Appl.*, **146**, 535-548.

Kollár, I. (1994). *Frequency Domain System Identification Toolbox for Use with MATLAB*, The MathWorks, Natick, MA.

McCormack, A.S., K.R. Godfrey and J.O. Flower (1994). The detection of and compensation for nonlinear effects using periodic input signals, *IEE International Conference Control '94 (IEE Publication No. 389)*, Coventry, 21-24 March 1994, 297-302.

McCormack, A.S., K.R. Godfrey and J.O. Flower (1995). Design of multilevel multiharmonic signals for system identification, *IEEE Trans. Instrum. Meas.*, **IM-43**, 232-237.

Pintelon, R. and J. Schoukens (2001). *System Identification: A Frequency Domain Approach*, IEEE Press, New York.

Schroeder, M.R. (1970). Synthesis of low peak factor signals and binary sequences with low autocorrelation, *IEEE Trans. Inform. Theory*, **IT-16**, 85-89.

Tan, A.H. and K.R. Godfrey (2001). Identification of processes with direction-dependent dynamics, *IEE Proc. – Control Theory Appl.*, **148**, 362-369.

Tan, A.H. and K.R. Godfrey (2002a). Identification of Wiener-Hammerstein models using linear interpolation in the frequency domain (LIFRED), *IEEE Trans. Instrum. Meas.*, **IM-51**, 509-521.

Tan, A.H. and K.R. Godfrey (2002b). The generation of binary and near-binary pseudorandom signals: an overview, *IEEE Trans. Instrum. Meas.*, **IM-51**, 583-588.

Van den Bos, A. and R.G. Krol (1979). Synthesis of discrete-interval binary signals with specified Fourier amplitude spectra, *Int. J. Contr.*, **30**, 871-884.

Van der Ouderaa, E., J. Schoukens and J. Renneboog (1988). Peak factor minimization using a time-frequency swapping algorithm, *IEEE Trans. Instrum. Meas.*, **IM-37**, 145-147.

IFAC
Publications
www.elsevier.com/locate/ifac

MULTIPLE INPUT DESIGN FOR REAL-TIME PARAMETER ESTIMATION IN THE FREQUENCY DOMAIN

Eugene A. Morelli

NASA Langley Research Center
Hampton, Virginia USA

Abstract: A method for designing multiple inputs for real-time dynamic system identification in the frequency domain was developed and demonstrated. The designed inputs are mutually orthogonal in both the time and frequency domains, with reduced peak factors to provide good information content for relatively small amplitude excursions. The inputs are designed for selected frequency ranges, and therefore do not require *a priori* models. The experiment design approach was applied to identify linear dynamic models for the F-15 ACTIVE aircraft, which has multiple control effectors.

Keywords: experiment design, parameter estimation, frequency domain, aircraft system identification

1. INTRODUCTION

Frequency domain techniques have been used successfully to identify dynamic models for aircraft, including cases where the aircraft was open-loop unstable (Schkolnik, *et al.*, 1995; Morelli, 2002). Recent work has indicated that real-time parameter estimation for aircraft dynamic models can be done effectively using a recursive chirp-Z transform for a selected frequency band, then employing equation-error parameter estimation in the frequency domain (Morelli, 2000). Advantages of this approach include robustness to measurement biases, noise, and infrequent dropouts in the time-domain data, enhanced signal-to-noise ratio in the frequency domain, accurate parameter estimates and error bounds in real time, with very low computational and memory requirements.

Good experiment design for dynamic modelling in the frequency domain requires that excitation inputs to the dynamic system contain a variety of frequencies. At the same time, the excitation must be such that the amplitudes of the dynamic system responses are not too large, so that the model structure assumption, which is typically linear, is not violated. For the case of real-time parameter estimation on aircraft, inputs are preferably small enough so that the pilot cannot distinguish the aircraft response to excitation inputs from a typical aircraft response to turbulence. Modern aircraft have multiple control effectors, so it would be advantageous if the excitation for frequency domain identification could be applied to multiple control effectors simultaneously, so that the amount of time required to collect data for dynamic modelling could be reduced.

Previous work (Schroeder, 1970) has shown that a phase-shifted sum of sinusoids, called the Schroeder sweep, provides an input with good frequency content and low peak factor. The peak factor is a measure of the ratio of maximum input amplitude to input energy. Inputs with low peak factors are efficient in the sense of providing good frequency content without large amplitudes in the time domain. Comparisons of the Schroeder sweep with conventional linear and logarithmic frequency sweep inputs have indicated that the Schroeder sweep is generally the superior input for frequency domain dynamic model identification (Young and Patton, 1990). The Schroeder sweep has been used successfully in other practical system identification problems (Flower, *et al.*, 1978; Bosworth and Burken, 1997).

This paper describes an extension of the Schroeder sweep input design method to multiple input design with optimised peak factors for real-time parameter estimation. The designed inputs are mutually orthogonal in both the time domain and the frequency domain, and are formulated as perturbation inputs. The only *a priori* information required for the multiple input design is an estimate of the approximate frequency band for the system dynamics and approximate relative control effectiveness for proper relative scaling of the input amplitudes. Investigations were conducted to determine signal-to-noise ratios necessary to achieve good parameter estimation results from the real-time parameter estimator. This information is necessary so that the input amplitude scaling can adapt to different aircraft dynamics throughout the flight envelope.

The multiple input design technique was applied to a lateral / directional linear simulation for the F-15 ACTIVE aircraft, which includes four control effectors. Real-time parameter estimation in the frequency domain (Morelli, 2000) was used to estimate the dynamic model parameters. Parameter estimation results were compiled and analysed.

The next section describes the multiple input design procedure. Following this, real-time parameter estimation results from applying the designed inputs to the F-15 ACTIVE simulation are presented and discussed.

2. MULTIPLE INPUT DESIGN

Each input to the aircraft control surfaces is comprised of a set of summed harmonic sinusoids with individual phase lags. Each input $\boldsymbol{u}_j$, applied to the jth control surface, takes the form

$$\boldsymbol{u}_j = \sum_{k \in \{1,2,\ldots,M\}} A_k \cos\left(\frac{2\pi k \boldsymbol{t}}{T} + \phi_k\right) \qquad (1)$$

where M is the total number of available harmonically-related frequencies, T is the time length of the excitation, and the ϕ_k are phase angles to be chosen for each of the harmonic components to produce a low *peak factor* PF, defined by

$$PF(\boldsymbol{u}_j) = \frac{\left[max(\boldsymbol{u}_j) - min(\boldsymbol{u}_j)\right]/2}{\sqrt{(\boldsymbol{u}_j^T \boldsymbol{u}_j)/N}} \qquad (2)$$

or

$$PF(\boldsymbol{u}_j) = \frac{\left[max(\boldsymbol{u}_j) - min(\boldsymbol{u}_j)\right]}{2\, rms(\boldsymbol{u}_j)} = \frac{\|\boldsymbol{u}_j\|_\infty}{\|\boldsymbol{u}_j\|_2} \qquad (3)$$

where the last equality only holds when $\boldsymbol{u}_j$ oscillates symmetrically about zero. In the literature, the quantity $\|\boldsymbol{u}_j\|_\infty / \|\boldsymbol{u}_j\|_2$ is called the *crest factor*. A single sinusoidal component from the summation in Eq. (1) has $PF = \sqrt{2}$, so the *relative peak factor* *RPF*, defined by

$$RPF(\boldsymbol{u}_j) = \frac{\left[max(\boldsymbol{u}_j) - min(\boldsymbol{u}_j)\right]}{2\sqrt{2}\, rms(\boldsymbol{u}_j)} = \frac{PF(\boldsymbol{u}_j)}{\sqrt{2}} \qquad (4)$$

quantifies the peak factor relative to that of a single sinusoid. For a single sinusoid, *RPF* equals 1. The relative peak factor is a measure of efficiency of a input for parameter estimation purposes, in terms of the amplitude range of the signal divided by a measure of the signal energy. Lower relative peak factors are more desirable for parameter estimation, where the objective is to excite the system without driving it too far away from the nominal operating point, so that model structure assumptions are not violated.

The integers k specifying the frequencies for the jth input $\boldsymbol{u}_j$ are unique to that input, but the integers k used in the summation in Eq. (1) are not necessarily consecutive, as will be explained below. The objective for the experiment design is to excite the aircraft dynamics in a short time period by moving multiple control surfaces simultaneously. Since more than one surface is being moved, it is advantageous for the modelling if the $\boldsymbol{u}_j$ vectors applied to each control surface are mutually orthogonal. This helps the parameter estimation by completely de-correlating the inputs to the aircraft, which improves the accuracy of control effectiveness estimates. It is possible to make the $\boldsymbol{u}_j$ mutually orthogonal in both the time and frequency domains, using inputs designed for low relative peak factor, as will be shown next. This gives the analyst the flexibility to use time domain or frequency domain parameter estimation methods, while retaining the desirable feature of mutually orthogonal inputs in either domain.

In the time domain, a signal composed of a sum of sinusoids is orthogonal to any other sum of sinusoids with harmonically-related frequencies, regardless of the constant phase shift of each sinusoidal component contained in the signals. For example, if two inputs each contain a single, distinct, harmonically-related sinusoidal component,

$$\boldsymbol{u}_1 = cos\left(\frac{2\pi \boldsymbol{t}}{T} + \phi_1\right) \qquad \boldsymbol{u}_2 = cos\left(\frac{2\pi 4\boldsymbol{t}}{T} + \phi_2\right)$$

then the inner product of these inputs, using the discrete-time notation $t_i = i\Delta t$ and $T = (N-1)\Delta t$, is

$$\boldsymbol{u}_1^T \boldsymbol{u}_2 = \sum_{i=0}^{N-1} cos\left(\frac{2\pi t_i}{T} + \phi_1\right) cos\left(\frac{2\pi 4 t_i}{T} + \phi_2\right)$$

$$= \sum_{i=0}^{N-1}\left[cos\left(\frac{2\pi t_i}{T}\right)cos(\phi_1) - sin\left(\frac{2\pi t_i}{T}\right)sin(\phi_1)\right]$$

$$\times\left[cos\left(\frac{2\pi 4 t_i}{T}\right)cos(\phi_2) - sin\left(\frac{2\pi 4 t_i}{T}\right)sin(\phi_2)\right]$$

$$= 0$$

The sine and cosine of the constant phase angles ϕ_1 and ϕ_2 are constants, so the summation equals zero because of the orthogonality of harmonically-related sinusoids. For more than one sinusoidal component in each input, the analysis is similar. So, the inputs assembled as in Eq. (1) are orthogonal in the time domain.

If the frequency indices k that are selected for each input $\boldsymbol{u}_j$ are distinct from those chosen for the other inputs, then the frequency content of each $\boldsymbol{u}_j$ consists of distinct spectral lines in the frequency domain, since the summation in Eq. (1) is a cosine series with harmonically-related frequencies. Therefore, the vectors of Fourier transforms for the inputs as a function of frequency have inner products equal to zero. In this sense, the inputs are also mutually orthogonal in the frequency domain.

The multiple input design procedure is as follows:

1. Select the time period T for the excitation, which determines the smallest harmonic frequency resolution $\Delta f = 1/T$ and the limit on the minimum frequency $f_{min} \geq 2/T$.

2. Select the frequency band of the dynamic system for the excitation frequencies, $[f_{min}, f_{max}]$ Hz. This corresponds to the frequency band where the expected dynamic response of the system will occur. The frequencies are equally spaced by Δf on the interval $[f_{min}, f_{max}]$. The total number of frequencies $M = fix\{(f_{max} - f_{min})/\Delta f\} + 1$, where *fix* indicates rounding to the nearest integer toward zero.

3. Assign approximately an equal number of indices k from the set $\{1,2,\ldots,M\}$ to each input by alternating each consecutive frequency among the multiple inputs. This approach produces lower relative peak factors for the individual inputs, and also ensures that each input has frequency content distributed evenly across the frequency band $[f_{min}, f_{max}]$. Different assignments of the frequency indices could be made for other reasons. For example, lower frequency indices might be assigned to an input that is known to effectively excite a low frequency mode, or particular frequency indices might be omitted to avoid exciting an undesirable structural response. It was found empirically that if the set of selected indices k for a particular input consisted of an integer greater than 1, with 2 or 3 multiples of that integer (e.g., k=2, 4, 6, or k=5, 10, 15, 20), the phase angles could be optimised (in step 5) so that the relative peak factor for that input was very close to 1, and sometimes less than 1. Each frequency index can be assigned to only one of the inputs, to preserve mutual orthogonality of the inputs in both the time and frequency domains.

4. Generate the input $\boldsymbol{u}_j$ for each of the m controls, using Eq. (1) and computing the starting values for the phase angles ϕ_k according to method described by Schroeder (1970), assuming a uniform power spectrum.

5. Use a simplex optimisation algorithm (Press, *et al.*, 1992) to adjust the ϕ_k for each $\boldsymbol{u}_j$ to achieve minimum relative peak factor for that input. The optimisation algorithm does not require gradients.

6. For each input, do a one-dimensional search to find a constant time offset for the components of each input $\boldsymbol{u}_j$, so that the input begins and ends at zero amplitude. This is equivalent to sliding the input along the time axis until a zero crossing is placed at the origin of the time axis. The appropriate phase shift is added to each sinusoidal component phase shift ϕ_k. Note that to implement a constant time shift to all the components, the phase offset for each component will be different, because each component has a different frequency. Since the components of each $\boldsymbol{u}_j$ are harmonics of the base frequency with period T, if all the component phase angles ϕ_k are shifted so that the initial value of the input is zero, then the final value of the input will also be zero. The power spectrum, input orthogonality, and relative peak factor are all unaffected.

7. Return to step 5 until the relative peak factor reaches a pre-defined goal value or until a maximum number of iterations is reached. For this work, the relative peak factor goal was set at 1.01 and the maximum number of iterations was set to 50.

Although there are methods for optimising the frequency spectrum of inputs for parameter estimation, all of them require considerable computation, along with some knowledge of the system dynamics, usually in terms of a nominal model with parameter values. In the current application, there is no use for such methods, because the intent is to identify the aircraft dynamics in real time, for various flight conditions throughout the flight envelope, and for arbitrary failure conditions, to enable control reconfiguration. Under these circumstances, computing an optimised frequency spectrum does not make sense. Instead, the frequency spectra for all inputs were defined to be flat across the selected frequency band, so that the aircraft dynamics would be sufficiently excited, regardless of where the modal frequencies happen to be located within the frequency band.

Step 6 of the input design procedure ensures that each designed input will be a perturbation, so that the designed input can be added to whatever constant value the control may have for another reason, e.g., trim or manoeuvring. The iteration helps to reduce the relative peak factors by adding perturbations to the phase angles ϕ_k (from the phase changes added to make the input start and end at zero), which helps the optimisation algorithm solve this non-convex optimisation problem.

Figure 1 shows the relative peak factor as a function of phase angles for two components in the summation of Eq. (1), using T=15 sec, unit amplitudes, and frequency indices k=2 and k=4. It is clear from the figure that this optimisation problem is not convex, and also that there are several phase angle solutions which are equally good, or nearly so. The data used to make the plot in Figure 1 can be used for a global exhaustive search for the minimum peak factor in this simple case. To generate this data, the relative peak factors were computed for values of the phase angles on a 2-dimensional grid with intervals of 0.0175 rad (1 deg), over the range $[0, \pi]$ rad $([0, 180]$ deg$)$. The minimum relative peak factor of all these computed values was found to be 1.102. Using the technique described above to optimise the phase angles, the relative peak factor achieved was 1.106. This demonstrated that the technique developed here found an input with peak factor very close to the global optimum, in this simple case. The same exercise was then repeated for three harmonic components, with frequency indices k=2, 4, and 6, and the same phase angle intervals, in three dimensions. In this case, the global exhaustive search gave a minimum relative peak factor equal to 1.002, while the optimisation technique described above produced an input with relative peak factor equal to 1.003. These investigations suggest that the procedure described here does an excellent job of designing inputs with very low peak factor.

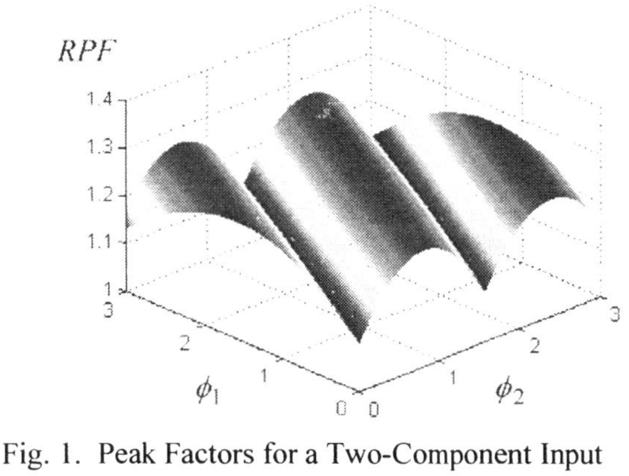

Fig. 1. Peak Factors for a Two-Component Input

3. RESULTS

For small perturbation motions about a reference condition, airplane dynamics can be described by the following linear model equations:

$$\dot{x}(t) = Ax(t) + Bu(t) \quad x(0) = x_o \quad (5)$$

$$y(t) = C\,x(t) + D\,u(t) \quad (6)$$

$$z_i = y_i + v_i \quad i = 0,1,2,\ldots,N-1 \quad (7)$$

Matrices $\boldsymbol{A}$, $\boldsymbol{B}$, $\boldsymbol{C}$, and $\boldsymbol{D}$ in Eqs. (5) and (6) contain stability and control parameters, which are to be estimated from flight data.

For the linearized lateral/directional dynamics of the F-15 ACTIVE aircraft, the state vector **x** and input vector **u** in Eq. (1) are

$$\boldsymbol{x} = [\beta \quad p \quad r \quad \phi]^T \quad \boldsymbol{u} = [\delta_a \quad \delta_r \quad \delta_{dc} \quad \delta_{ds}]^T \quad (8)$$

The output vector is:

$$\boldsymbol{y} = [\beta \quad p \quad r \quad \phi \quad a_y]^T \quad (9)$$

System matrices containing model parameters are:

$$\boldsymbol{A} = \begin{bmatrix} Y_\beta & sin\alpha_o & -cos\alpha_o & g\,cos\theta_o/V_o \\ L_\beta & L_p & L_r & 0 \\ N_\beta & N_p & N_r & 0 \\ 0 & 1 & tan\theta_o & 0 \end{bmatrix} \quad (10)$$

$$\boldsymbol{B} = \begin{bmatrix} 0 & Y_{\delta_r} & Y_{\delta_{dc}} & Y_{\delta_{ds}} \\ L_{\delta_a} & L_{\delta_r} & L_{\delta_{dc}} & L_{\delta_{ds}} \\ 0 & N_{\delta_r} & N_{\delta_{dc}} & N_{\delta_{ds}} \\ 0 & 0 & 0 & 0 \end{bmatrix} \quad (11)$$

$$\boldsymbol{C} = \begin{bmatrix} 1 & 0\;0\;0 \\ 0 & 1\;0\;0 \\ 0 & 0\;1\;0 \\ 0 & 0\;0\;1 \\ \frac{V_o}{g}Y_\beta & 0\;0\;0 \end{bmatrix} \quad \boldsymbol{D} = \frac{V_o}{g}\begin{bmatrix} 0 & 0 & 0 & 0 \\ 0 & 0 & 0 & 0 \\ 0 & 0 & 0 & 0 \\ 0 & 0 & 0 & 0 \\ 0 & Y_{\delta_r} & Y_{\delta_{dc}} & Y_{\delta_{ds}} \end{bmatrix} \quad (12)$$

Figure 2 shows the NASA F-15 ACTIVE aircraft (Doane, *et al.*, 1994). The values of the parameters in the model that were used to generate the simulated data are given in column 2 of Table 1.

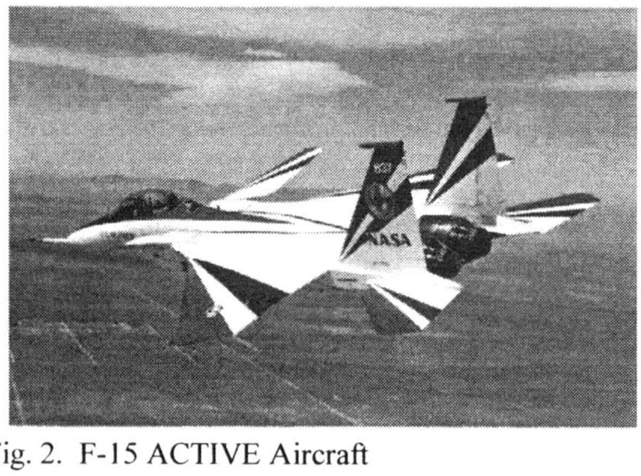

Fig. 2. F-15 ACTIVE Aircraft

The multiple input design for the F-15 ACTIVE aircraft lateral/directional dynamics includes four control effectors: aileron, rudder, differential canard, and differential stabilator. Figure 3 shows the time histories of the multiple input design for an 18 sec manoeuvre, which includes 1 sec for steady trim flight at the start, and 2 sec for free response at the end. The frequency range for the input design was chosen as [0.2, 1.4] Hz. The inputs shown are mutually orthogonal in both the time domain and the frequency domain. These inputs were designed using a flat power spectrum with T=15 sec. Table 2 contains the amplitudes, frequency indices, and the *RPF* achieved for the 15 sec inputs shown in Figure 3. All of the designed inputs achieved an extremely low relative peak factor close to 1.

Figure 4 shows the simulated responses, which include added noise equal to time domain residual sequences from flight test data analysis and modelling at the same flight condition for the real F-15 ACTIVE airplane. This ensures that a realistic noise environment was used in the investigations concerning the effectiveness of the multiple input design. Real-time parameter estimation in the frequency domain (Morelli, 2000) was used to estimate the parameters. The frequencies chosen for the recursive chirp-Z transform were evenly spaced in 0.01 Hz increments on the interval [0.11, 1.5] Hz. This frequency band must include all of the input design frequency band, so that the important control surface information in the frequency domain is included in the parameter estimation. Table 1 contains the parameter estimation results for this case, using only the final values of the real-time parameter estimates and standard errors. The parameter estimation was started with no *a priori* information for the parameter estimates. The results in Table 1 show that the model parameters are estimated very accurately, and that the computed standard errors are representative of the true error in the parameter estimates.

The mean signal-to-noise ratio for the outputs shown in Figure 4 was 30. It is of interest to obtain good parameter estimation results with the lowest possible signal-to-noise ratio on the outputs, so that the aircraft and pilot are disturbed as little as possible by the activity on the control surfaces. Figure 5 shows the effect of output signal-to-noise ratio on the mean of the parameter estimate errors. The data for this plot was generated by reducing the input amplitudes uniformly using the same input forms shown in Figure 3, generating new simulated outputs, then adding the same noise sequences from flight test data used before, so that the output signal-to-noise ratio decreased. For output signal-to-noise ratio equal to 10, the input amplitudes were 0.24 deg, and the mean parameter estimate error was 2.7 percent. This choice of input amplitudes produced excellent parameter estimate accuracy, while keeping the root-mean-square value of the lateral acceleration a_y below 0.02 g, which is less than 0.05 g that is typical for moderate turbulence.

4. CONCLUDING REMARKS

A multiple input design technique for real-time frequency domain parameter estimation was described and demonstrated. The technique can be used to design multiple inputs that are mutually orthogonal in both the time and frequency domains, with very low peak factors. Input energy over a selectable range of frequencies can be efficiently injected into the dynamical system by virtue of low peak factors and the ability to move inputs simultaneously, since the inputs are mutually orthogonal. These features make the inputs attractive for real-time parameter estimation application, where the excitation must have short duration and low output response amplitudes, and the dynamic response and modal frequencies will change because of flight condition changes or failure conditions.

To use this input design technique effectively in flight, the input forms could remain fixed, and a simple feedback could be used to scale the input amplitudes so that the output signal-to-noise ratios are near a chosen acceptable value. The excitation time T should be minimized, and this will be impacted by the output signal-to-noise ratios. In addition, the F-15 ACTIVE airplane has a stability augmentation system, which will distort the input forms designed for parameter estimation. This affects the required input amplitudes, and degrades peak factors and input orthogonality. Further work is required to address these practical considerations.

The multiple input design technique described here can be applied to other problems as well, because there is no dependence on *a priori* models or information, other than the expected frequency range of the dynamic modes. Specific spectral weighting using the component amplitudes can be introduced, if warranted by *a priori* information.

REFERENCES

Bosworth, J.T. and J.J. Burken (1997) "Tailored Excitation for Multivariable Stability-Margin Measurement Applied to the X-31A Nonlinear Simulation," AIAA 94-3361.

Doane, P., R. Bursey, and G. Schkolnik (1994) "F-15 ACTIVE: A Flexible Propulsion Integration Testbed," AIAA 94-3361.

Flower, J.O., G.F. Knott, and S.G. Forge (1978) "Application of Schroeder-phased Harmonic Signals to Practical Identification," *Measurement and Control*, **11**, No. 2, pp. 69-73.

Morelli, E.A. (2000) "Real-Time Parameter Estimation in the Frequency Domain," *Journal of Guidance, Control, and Dynamics*, **23**, No. 5, pp. 812-818.

Morelli, E.A. (2002) "System IDentification Programs for AirCraft (*SIDPAC*)," AIAA-2002-4704.

Press, W.H., S.A. Teukolsky, W.T. Vettering, and B.R. Flannery (1992) *Numerical Recipes in FORTRAN: The Art of Scientific Computing*, 2nd Ed., Cambridge University Press, New York, NY, Chapter 10.

Schroeder, M.R. (1970) "Synthesis of Low-Peak-Factor Signals and Binary Sequences with Low Autocorrelation," *IEEE Transactions on Information Theory*, pp. 85-89.

Schkolnik, G.S., J.S. Orme, and M.A. Hreha (1995) "Flight Test Validation of a Frequency-Based System Identification Method on an F-15 Aircraft," NASA TM 4704.

Young, P. and R.J. Patton (1990) "Comparison of Test Signals for Aircraft Frequency Domain Identification," *Journal of Guidance*, **13**, No. 3, pp. 430-438.

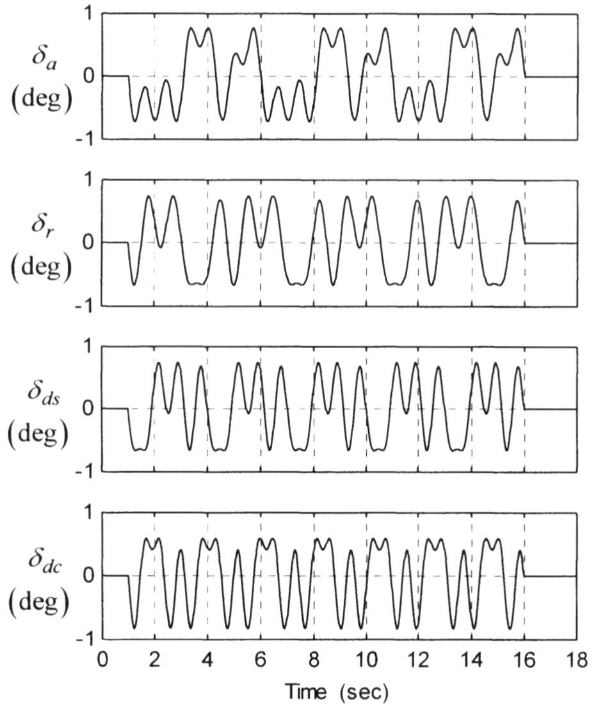

Fig. 3. Multiple Input Design

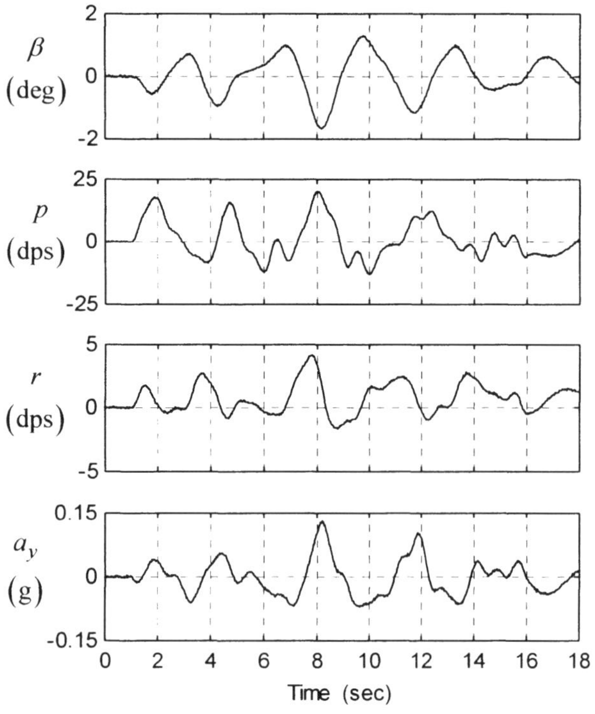

Fig. 4. Simulated F-15 ACTIVE Response

Table 1 F-15 ACTIVE Lateral/Directional Parameter Estimation Results,
$V_o = 793$ fps, $h_o = 20{,}000$ ft, $\alpha_o = \theta_o = 2$ deg

	θ_{true}	$\hat{\theta}$	Std. Error	$\hat{\theta} - \theta_{true}$
Y_β	−0.150	−0.150	0.0002	−0.0001
Y_{δ_r}	0.050	0.050	0.0003	0.0002
$Y_{\delta_{ds}}$	0.035	0.035	0.0003	0.0001
$Y_{\delta_{dc}}$	−0.025	−0.025	0.0003	−0.0002
L_β	−22.5	−22.5	0.08	0.02
L_p	−2.05	−2.04	0.007	0.007
L_r	3.15	3.14	0.029	−0.012
L_{δ_a}	−28.4	−28.2	0.08	0.20
L_{δ_r}	4.20	4.15	0.071	−0.048
$L_{\delta_{ds}}$	−34.2	−34.1	0.07	0.17
$L_{\delta_{dc}}$	5.14	5.09	0.069	−0.112
N_β	4.40	4.42	0.030	0.024
N_p	0.11	0.11	0.002	0.000
N_r	−0.17	−0.17	0.012	0.005
N_{δ_r}	−3.75	−3.72	0.029	0.039
$N_{\delta_{ds}}$	−1.40	−1.41	0.029	−0.015
$N_{\delta_{dc}}$	−2.40	−2.37	0.028	0.044

Table 2 Multiple Input Design

Input	A_k	k	RPF
δ_a	0.707	3, 6, 9, 18	1.055
δ_r	0.707	4, 8, 12, 16	0.995
δ_{ds}	0.707	5, 10, 15, 20	0.995
δ_{dc}	0.707	7, 14, 21	1.003

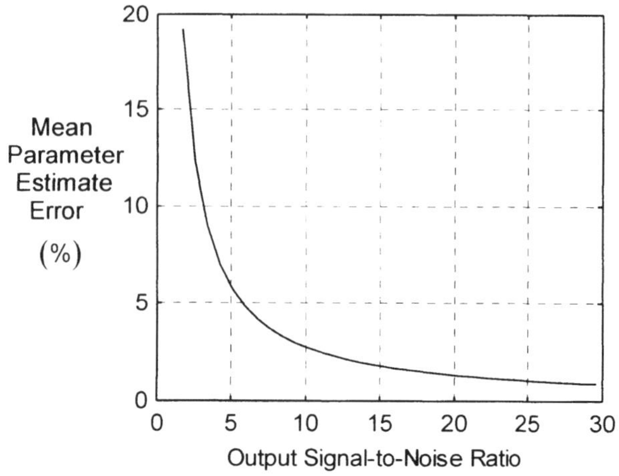

Fig. 5. Parameter Estimation Accuracy Dependence on Output Signal-to-Noise Ratio

IFAC Publications
www.elsevier.com/locate/ifac

MINIMIZING THE WORST-CASE ν-GAP BY OPTIMAL INPUT DESIGN

Roland Hildebrand * **Michel Gevers** **

* *CORE, Université Catholique de Louvain, 34 voie du Roman Pays, 1348 Louvain-la-Neuve, Belgium, Phone +32(0)10-474337,* hildebrand@core.ucl.ac.be
** *CESAME, Université Catholique de Louvain, Bâtiment EULER, 4 av. Georges Lemaitre, 1348 Louvain-la-Neuve, Belgium, Phone +32(0)10-472590,* gevers@csam.ucl.ac.be

Abstract: Parameter identification experiments deliver an identified model together with an ellipsoidal uncertainty region in parameter space. The objective of robust controller design is thus to stabilize all plants in the identified uncertainty region. We design an identification experiment such that the worst-case ν-gap over all plants in the resulting uncertainty region between the identified plant and plants in this region is as small as possible. The experiment design is performed via input power spectrum optimization. Two cost functions are investigated, which represent different levels of trade-off between accuracy and computational complexity. It is shown that the input optimization problem with respect to these cost functions is amenable to standard numerical algorithms used in convex analysis.

Keywords: identification for control, worst-case ν-gap, parametric uncertainty region

1. INTRODUCTION

This contribution continues the line of research that aims at connecting prediction error identification methods with robust control theory ((Bombois *et al.*, 2001), (Gevers *et al.*, 2000)). Subject to investigation are discrete time SISO real-rational stable LTI plants, which are to be identified in open loop within an ARX model structure. We assume the true plant to lie in the model set. Hence the model error is determined only by the covariance of the estimated parameter vector.

Since the aim of the identification experiment is control design, we wish to obtain an uncertainty region with good stability robustness properties. By this is meant that the set of controllers that stabilize all models in the uncertainty set should be as large as possible. A suitable measure of robust stability that connects the "size" of an uncertainty set with a set of robustly stabilizing controllers is the worst-case ν-gap $\delta_{WC}(\hat{G},\mathcal{D})$ introduced in (Gevers *et al.*, 2000). It is the supremum of the Vinnicombe ν-gap (Vinnicombe, 1993) between the identified model $\hat{G}$ and all plants in the uncertainty set $\mathcal{D}$. Specifically, all controllers C that stabilize the model $\hat{G}$ with a stability margin $b_{\hat{G},C} > \delta_{WC}(\hat{G},\mathcal{D})$ stabilize all plants in $\mathcal{D}$.

In previous papers ((Bombois *et al.*, 2001), (Gevers *et al.*, 2000)) a special type of uncertainty sets $\mathcal{D}$ of transfer functions, which emerges from prediction error identification experiments, was described and investigated. It is given by an ellipsoid in parameter space and is determined by the covariance matrix of the parameter vector and the prespecified confidence level, i.e. the probability with which the true plant is lying inside the considered uncertainty set.

[1] The European Commission is herewith acknowledged for its financial support in part to the research reported on in this contribution. The support is provided via the Program Training and Mobility of Researchers (TMR) and Project System Identification (ERB FMRX CT98 0206) to the European Research Network System Identification (ERNSI). This paper presents research results of the Belgian Programme on Interuniversity Poles of Attraction, Phase V, initiated by the Belgian State, Prime Minister's Office for Science, Technology and Culture.

The goal of this contribution is to minimize the worst-case ν-gap of such uncertainty regions $\mathscr{D}$ by choosing a suitable input $u(t)$ for the identification experiment. To restrict the class of admissible inputs we assume the total input energy to be bounded.

The problem setting of experiment design first arose in statistics and was extensively studied (see e.g. (Kiefer, 1974), (Kiefer and Wolfowitz, 1959), (Goodwin *et al.*, 1974), (Zarrop, 1979)). We adopt the most common viewpoint and study input optimization in the frequency domain, i.e. optimize the input power spectrum with respect to a cost function that depends on the average per data sample information matrix $\bar{M}$ of the experiment. This matrix is defined as the limit of the ratio between the information matrix and the number of data as the number of data tends to infinity (see e.g. (Zarrop, 1979)). For typical number of data this leads to a sufficiently good approximation of the optimal input. Thus we will essentially regard the average information matrix instead of the input power spectrum as the quantity that is going to be optimized. Once the optimal average information matrix is found, we proceed by construction of an input power spectrum that produces this information matrix.

For different classes of cost functions iterative procedures were designed to find the optimal input power spectrum up to a prespecified precision. Most of these criteria are analytic in the entries of $\bar{M}$ and Kiefer-Wolfowitz theory (Kiefer, 1974) can effectively be applied to them. All these classical criteria are convex and monotonic with respect to $\bar{M}$ (Zarrop, 1979, p.39).

Our criterion is the worst-case ν-gap of the uncertainty region $\mathscr{D}$. This is a nonstandard cost function, it is nonsmooth and thus more difficult to treat than the classical criteria. We shall also introduce another cost function, which approximates the worst-case ν-gap, but is somewhat simpler. Nevertheless, both cost functions are compound criteria (Kiefer, 1974, section 4G) and application of Kiefer-Wolfowitz theory does not make them more tractable. However, they satisfy the natural condition of monotonicity with respect to $\bar{M}$, as well as the condition of quasiconvexity.

It can be shown (Zarrop, 1979) that under above assumptions the corresponding set of admissible average information matrices, over which the optimization is performed, represents a moment space of a trigonometric Tchebycheff system. The foundations of the theory of moment spaces are classical. It follows from a well-known fact of Tchebycheff system theory (see e.g. (Karlin and Studden, 1966)), restated in Theorem 1 in this contribution, that any admissible average information matrix $\bar{M}$ can be obtained by applying an input with discrete power spectrum, and that there exist admissible $\bar{M}$ which can be realized only by discrete power spectra. In view of this, we propose an algorithm that yields optimal input power spectra which are discrete. There are different ways to choose an input sequence with a desired power spectrum. We can choose the input e.g. as a multisine function. However, in many cases one could use also binary signals (see e.g. (Zarrop, 1979, p.29)) or other functions.

A classical result on moment spaces (Karlin and Studden, 1966, chapter VI, Theorem 4.1) states that the set of possible average information matrices $\bar{M}$ can be represented as the feasible set of a linear matrix inequality (LMI). For a survey on LMI's see e.g. (Boyd *et al.*, 1994). We show that optimization with respect to the worst-case ν-gap and the proposed approximate criterion can be accomplished by application of the apparatus of convex analysis and the theory of LMI's.

Several authors successfully treated input design problems arising in Identification for Control with convex optimization methods. In (Lindqvist and Hjalmarsson, 2001), the input spectrum for an open loop identification experiment was designed to minimize the closed-loop system performance. By a Taylor series truncation, the cost function reduced to a weighted-trace criterion (L-optimality). However, the input spectra were restricted to those which can be realized by white noise filtered through an FIR filter. An LMI description of the corresponding set of information matrices can be derived from the positive-real lemma ((Boyd *et al.*, 1994),(Wu *et al.*, 1996)). Note that in this contribution we optimize over the whole set of nonnegative input power spectra. For recent results in convex optimization see e.g. (Nesterov and Nemirovskii, 1994).

The assumption of an ARX model structure and an input energy constraint are in no way restrictive. The ideas and methods proposed in the present contribution easily carry over to other model structures and to input power or output power/energy constraints.

The remainder is structured as follows. In the next section the considered identification problem as well as the cost functions will be formally defined. In section 3 we show that the set over which the optimization takes place is LMI representable. In section 4 we prove that the optimization problem is quasiconvex. In section 5 we show how to construct cutting planes to the different cost functions. The results obtained in sections 3 to 5 allow the problem to be treated with standard convex analysis methods. Since the optimization takes place in an abstract parameter space, it is necessary to convert values in this space into power spectra and input sequences. This is accomplished in section 6. In section 7 we illustrate the advantages of the proposed procedure by a simulation example. Finally, in section 8 we draw some conclusions.

2. PROBLEM SETTING

Let us consider an ARX model structure with parameters n_a, n_b, n_k: $y(t) + a_1 y(t-1) + \ldots + a_{n_a} y(t - n_a) = b_1 u(t-n_k) + \ldots + b_{n_b} u(t-n_k-n_b+1) + e(t)$, where $u(t)$ is the input signal, $y(t)$ is the output signal, $\theta = (a_1, \ldots, a_{n_a}, b_1, \ldots, b_{n_b})^T$ is the parameter vector, and $e(t)$ is normally distributed white noise with covariance λ_0. Assume that the true system can be described within this structure and corresponds to a parameter value $\theta = \theta_0$, and that it is stable. Denote by z^{-1} the delay operator. Then we can write

$$y = z^{-n_k+1}\frac{B(\theta)}{A(\theta)}u + \frac{1}{A(\theta)}e = G(\theta)u + \frac{1}{A(\theta)}e,$$

where A, B are polynomials in the delay operator with coefficients depending on the parameter vector.

Suppose that an identification experiment with input $(u(1),\ldots,u(N))$ is performed, leading to an observed output $(y(1),\ldots,y(N))$ with N data samples, where $u(t)$ is a realization of a quasistationary stochastic process with power spectrum Φ_u. Suppose a parameter estimate $\hat{\theta}$ is obtained by least squares prediction error minimization. Then it is well-known (Ljung, 1999) that the estimate $\hat{\theta}$ is asymptotically unbiased as $N \to \infty$ and for large N its covariance is proportional to N^{-1}, i.e. $E(\theta_0 - \hat{\theta})(\theta_0 - \hat{\theta})^T \approx \frac{P}{N}$. The matrix P is a function of the input power spectrum and the true values of the coefficients of A and B (Ljung, 1999). The inverse of the parameter covariance matrix is the Fisher information matrix M. Let us denote the asymptotic expression for the average information matrix per data sample $\lim_{N\to\infty}\frac{1}{N}M = P^{-1}$ by $\bar{M}$. Then $\bar{M}$ is given by a convolution of the input power spectrum Φ_u with a rational trigonometric function plus a constant offset stemming from the noise.

Since the parameter estimate $\hat{\theta}$ is asymptotically normally distributed (Ljung, 1999), we can assume, following (Gevers *et al.*, 2000), that the true parameter vector θ_0 lies with a prespecified probability $\alpha \in (0,1)$ in the uncertainty ellipsoid $U = \{\theta | N(\theta - \hat{\theta})^T \bar{M}(\theta - \hat{\theta}) < \chi^2_{n_a+n_b}(\alpha)\}$, where χ^2_l is the χ^2 distribution with l degrees of freedom. The uncertainty ellipsoid U corresponds to an uncertainty set $\mathcal{D} = \{G(z,\theta) = z^{-n_k+1}B(\theta)/A(\theta) | \theta \in U\}$ in the space of transfer functions. The set $\mathcal{D}$ belongs to the class of generic prediction error model uncertainty sets as defined in (Gevers *et al.*, 2000).

The worst-case ν-gap between the identified model $G(\hat{\theta})$ and the uncertainty region $\mathcal{D}$ is defined by $\delta_{WC}(G(\hat{\theta}),\mathcal{D}) = \sup_{\theta\in U}\delta_\nu(G(\hat{\theta}),G(\theta))$, where δ_ν denotes the Vinnicombe ν-gap between two plants (Vinnicombe, 1993). Since $G(\hat{\theta})$ belongs to $\mathcal{D}$, the worst-case ν-gap can be expressed in the following way (Gevers *et al.*, 2000, Lemma 5.1): $\delta_{WC}(G(\hat{\theta}),\mathcal{D}) = \sup_{\omega\in[0,\pi]}\kappa_{WC}(G(e^{j\omega},\hat{\theta}),\mathcal{D})$, where $\kappa_{WC}(G(e^{j\omega},\hat{\theta}),\mathcal{D})$ is called the worst-case chordal distance between $G(\hat{\theta})$ and $\mathcal{D}$ at frequency ω and is defined by $\sup_{\theta\in U}\frac{|G(e^{j\omega},\hat{\theta})-G(e^{j\omega},\theta)|}{\sqrt{(1+|G(e^{j\omega},\hat{\theta})|^2)(1+|G(e^{j\omega},\theta)|^2)}}$.

We have to minimize the quantity $\delta_{WC}(G(\hat{\theta}),\mathcal{D})$ by choosing an input with an appropriate power spectrum. To restrict the class of admissible power spectra we impose an input energy constraint

$$\frac{1}{2\pi}\int_{-\pi}^{\pi}\Phi_u(\omega)d\omega \le c, \tag{1}$$

where $c > 0$ is a prespecified positive constant.

The worst-case ν-gap depends on Φ_u via the average information matrix $\bar{M}$. Via $\bar{M}$ it depends also on the unknown true parameter value θ_0 and noise covariance λ_0. In addition it depends on the identified parameter value $\hat{\theta}$, which is not available before the identification experiment. All these three quantities have to be approximated with values derived from previous knowledge about the system, for instance from a preliminary identification experiment. Since the expectation of $\hat{\theta}$ equals θ_0, these two quantities can be approximated by the same value $\bar{\theta}$.

Problem 1 Find Φ_u satisfying (1) such that $\bar{M}(\Phi_u)$ minimizes the cost function $\mathcal{J}_1 = \delta_{WC}(G(\hat{\theta}),\mathcal{D})$.

Along with the worst-case ν-gap we will consider another cost function, which is easier to compute and is an approximation of $\delta_{WC}(G(\hat{\theta}),\mathcal{D})$. Let us approximate cost function $\mathcal{J}_1 = \mathcal{J}_1(\bar{M})$ by its asymptotic expression for large information matrices, namely

$$\mathcal{J}_2 = \lim_{\varepsilon\to 0}\frac{\mathcal{J}_1(\varepsilon^{-2}\bar{M})}{\varepsilon} \tag{2}$$

$$= \sqrt{\frac{\chi^2_{n_a+n_b}(\alpha)}{N}}\sup_{\omega\in[0,\pi]}\frac{\sqrt{\lambda_{\max}(T(\omega)\bar{M}^{-1}T(\omega)^T)}}{1+|G(e^{j\omega},\hat{\theta})|^2},$$

where $T(\omega)$ is a $2\times(n_a+n_b)$-matrix given by

$$T(\omega) = \begin{pmatrix} Re\dfrac{\partial G(e^{j\omega},\theta)}{\partial\theta}\Big|_{\theta=\hat{\theta}} \\ Im\dfrac{\partial G(e^{j\omega},\theta)}{\partial\theta}\Big|_{\theta=\hat{\theta}} \end{pmatrix}.$$

Problem 2 Find Φ_u satisfying (1) such that $\bar{M}(\Phi_u)$ minimizes cost function $\mathcal{J}_2$ defined by equation (2).

Our goal is to develop numerical algorithms for solving both Problems 1 and 2. There is a two-fold reason for introducing cost function $\mathcal{J}_2$. Beside its much lower computational complexity, it turns out that identification with an input power spectrum minimizing $\mathcal{J}_2$ in many cases gives better results than one with an input power spectrum minimizing $\mathcal{J}_1$. We address this question in detail in the simulation section.

3. LMI DESCRIPTION OF THE SEARCH SPACE

In this section we shall describe the set of possible average information matrices $\bar{M}$, over which the optimization takes place, as the feasible set of an LMI. The following fact is from (Payne and Goodwin, 1974).

Proposition 1. The matrix $\bar{M}$ is contained in a (n_a+n_b)-dimensional affine subspace of the space of symmetric $(n_a+n_b)\times(n_a+n_b)$-matrices.

This subspace can be parameterized by the *trigonometric moments* of the measure $\frac{\Phi_u}{\pi\lambda_0|A|^2}$, i.e. the numbers $x_k = \frac{1}{\pi}\int_0^\pi \frac{\Phi_u}{\lambda_0|A|^2}\cos(k\omega)\,d\omega$, $k = 0,\dots,n$, where $n = n_a + n_b - 1$. Namely, we have $\bar{M} = \sum_{k=0}^n x_k\bar{M}_k + \tilde{M}$, where the matrices $\bar{M}_k$, $\tilde{M}$ are constant and depend only on the coefficients of A and B. While the $\bar{M}_k$ can be obtained immediately from the expression for $\bar{M}$, the matrix $\tilde{M}$ is most easily computed using the method proposed in (Ljung, 1999, p.50).

Let us compose a vector $\bar{x} \in \mathbf{R}^{n+1}$ of the real numbers $x_k, k = 0,\dots,n$. Since $\frac{1}{\pi\lambda_0|A|^2}$ is strictly positive on $\omega \in [0,\pi]$, the set of all $\bar{x}(\Phi_u)$ such that Φ_u is a nonnegative measure on $[0,\pi]$ equals the moment space $\mathscr{M}^{(n+1)}$ of the Tchebycheff system $\{1, \cos\omega, \dots, \cos n\omega\}$ on $[0,\pi]$ (see e.g. (Zarrop, 1979)). Thus the set of feasible information matrices $\bar{M}$ is an affine image of the trigonometric moment cone $\mathscr{M}^{(n+1)}$. It is a classical result that this set is LMI representable (see e.g. (Karlin and Studden, 1966, Chapter VI, Theorem 4.1)). Denote the interior of the feasible set by $\mathscr{M}$.

Definition 1. (see e.g. (Karlin and Studden, 1966)) Let Φ_u be a discrete power spectrum with support $supp\Phi_u \subset [0,\pi]$. The number $\#[supp\Phi_u \cap (0,\pi)] + \frac{1}{2}\#[supp\Phi_u \cap \{0,\pi\}]$, where # denotes the cardinality, is called the *index* of Φ_u.

Theorem 1. (see e.g. (Karlin and Studden, 1966)) Let $\bar{x}$ be a point in $\mathscr{M}^{(n+1)}$. Then $\bar{x} \in Bd(\mathscr{M}^{(n+1)})$ if and only if there exists a discrete nonnegative measure on $[0,\pi]$ with index less than $\frac{n+1}{2}$ that induces $\bar{x}$. This measure is unique. Moreover, $\bar{x} \in Int(\mathscr{M}^{(n+1)})$ if and only if there exists a discrete nonnegative measure on $[0,\pi]$ with index $\frac{n+1}{2}$ that induces $\bar{x}$.

Thus the notion of the index allows us to characterize the interior of the moment space $\mathscr{M}^{(n+1)}$. By the special structure of the matrices $\bar{M}_k$, $\tilde{M}$ we have

Proposition 2. The average information matrix $\bar{M}$ corresponding to a power spectrum Φ_u is singular if and only if Φ_u is discrete with index less than $\frac{n_b}{2}$.

Corollary 1. Any $\bar{M} \in \mathscr{M}$ is strictly positive definite.

This corollary ensures the existence of the inverse $\bar{M}^{-1}$ in the interior of the search space. From the definition of $\mathscr{J}_1$, $\mathscr{J}_2$ we obtain the following monotonicity property.

Proposition 3. Let $\bar{M}_1, \bar{M}_2$ be two positive semidefinite average information matrices, and suppose $\bar{M}_1 \preceq \bar{M}_2$. Then the values of the cost functions $\mathscr{J}_1$, $\mathscr{J}_2$ at $\bar{M}_2$ do not exceed the respective values at $\bar{M}_1$. □

By Proposition 3 the minimum of the considered cost functions under constraint (1) is attained when equality holds, i.e. we can assume in (1) an equality sign. In (Zarrop, 1979) it was shown that this equality reduces the feasible set to an affine section of the trigonometric moment cone. It allows us to express the variable x_0 affinely through $x_1,\dots,x_n$. Thus the feasible set is described by an LMI on the variables $x_1,\dots,x_n$. Denote by $\mathscr{X}$ the set of vectors $x = (x_1,\dots,x_n)^T \in \mathbf{R}^n$ in the interior of the feasible set of this LMI. Any feasible information matrix $\bar{M}$ can hence be represented as $\bar{M} = \bar{M}_0 + \sum_{i=1}^n x_i\bar{M}_i$, where $x = (x_1,\dots,x_n)^T \in \mathscr{X}$ and $\bar{M}_0$, $\bar{M}_i$ are known constant matrices.

Thus we reduced the infinite-dimensional problem of searching the minimum of the cost functions over the set of all admissible input power spectra to the n-dimensional problem of searching the minimum over a convex compact section of the trigonometric moment cone, which can be described by an LMI.

4. QUASICONVEXITY

Proposition 4. On $\mathscr{M}$ cost function $\mathscr{J}_1$ is quasiconvex with respect to $\bar{M}$.

This follows from a general fact about quasiconvexity of functions depending on a quasiconvex constraint. Consider the function $F(y) = \max_{x\in X, g(x,y)\geq 0} f(x)$, where X is an arbitrary set, $f(x)$ is an arbitrary function, and $g(x,y)$ is quasiconvex in y. The following lemma is proven by set-theoretic arguments.

Lemma 1. $F(y)$ is quasiconvex in y.

Since $\mathscr{J}_1(\bar{M})$ is the maximum of a function of θ over the set U, and U is defined by an inequality which is linear in $\bar{M}$, the above lemma applies.

Since there is no restriction imposed on $f(x)$, it is in general impossible to draw computational advantages from the quasiconvexity of the cost function $F(y)$. In order for the problem to be tractable, the function $f(x)$ needs to have some structure. In our case the worst-case chordal distance can be expressed as a solution to a generalized eigenvalue problem (GEVP) (Gevers *et al.*, 2000, Theorem 5.1). We have $\kappa_{WC}(G(e^{j\omega},\hat{\theta}),\mathscr{D}) = \sqrt{\gamma_{opt}}$, where γ_{opt} is the solution of the GEVP

$$minimize\ \gamma\ s.\,t.\ F_0 + \gamma F_1 + \tau R \succeq 0,\ \tau \geq 0. \quad (3)$$

Here F_0, F_1, R are symmetric matrices given by

$$F_0 = V \begin{pmatrix} -1 & 0 & -ImG & ReG \\ 0 & -1 & ReG & ImG \\ -ImG & ReG & -|G|^2 & 0 \\ ReG & ImG & 0 & -|G|^2 \end{pmatrix} V^T, \quad (4)$$

$$F_1 = (1+|G|^2)VV^T,$$

$$R = \begin{pmatrix} I_{n_a+n_b} \\ -\hat{\theta}^T \end{pmatrix} \bar{M} \begin{pmatrix} I_{n_a+n_b} \\ -\hat{\theta}^T \end{pmatrix}^T - \begin{pmatrix} 0 & \cdots & 0 \\ \vdots & \ddots & \vdots \\ 0 & \cdots & \frac{\chi^2_{n_a+n_b}(\alpha)}{N} \end{pmatrix}$$

The function G has to be taken at $z = e^{j\omega}$ and parameter value $\hat{\theta}$. V is a $(n_a+n_b+1)\times 4$-matrix defined by $V = \begin{pmatrix} ReZ_N^T & ImZ_N^T & ImZ_D^T & ReZ_D^T \\ 0 & 0 & 0 & 1 \end{pmatrix}$ with $Z_N = z^{-n_k+1}(0 \cdots 0\, z^{-1} \cdots z^{-n_b})$, $Z_D = (z^{-1} \cdots z^{-n_a}\, 0 \cdots 0)$ being complex row vectors of dimension $n_a + n_b$.

Proposition 5. On $\mathcal{M}$ cost function $\mathcal{J}_2$ is quasiconvex with respect to $\bar{M}$.

The proposition follows from well-known convexity properties of the maximal eigenvalue and the inverse matrix on the positive definite cone.

5. CUTTING PLANES

In this section we provide the necessary tools that allow the user to apply standard convex algorithms to solve Problems 1 and 2 numerically, using the LMI description of the feasible set. Most black-box methods in convex analysis are based on the notion of a cutting plane (see e.g. (Boyd *et al.*, 1994)). If $S \subset \mathbf{R}^m$ is a convex set and $f : S \to \mathbf{R}$ is a quasiconvex function defined on S, then a cutting plane to f at a point $x^{(0)} \in S$ is defined by a nonzero vector $g \in \mathbf{R}^m$ such that $f(x^{(0)}) \leq f(x)$ for any $x \in S$ satisfying the inequality $g^T(x - x^{(0)}) \geq 0$. We will compute cutting planes for cost functions $\mathcal{J}_1, \mathcal{J}_2$ at an arbitrary point $x^{(0)} \in \mathcal{X}$. For a description of different methods see e.g. (Boyd *et al.*, 1994),(Nesterov and Nemirovskii, 1994).

Let $\bar{M}^{(0)}$ be the average information matrix corresponding to $x^{(0)}$. We shall now compute a cutting plane for $\mathcal{J}_1 = \max_{\omega \in [0,\pi]} \kappa_{WC}(G(e^{j\omega}, \hat{\theta}), \mathcal{D})$. Denote by $\omega^{(0)}$ the frequency where the worst-case chordal distance κ_{WC} attains its maximum. The value of $\omega^{(0)}$ can be found e.g. by a grid search. A cutting plane to the function $\kappa_{WC}(G(e^{j\omega^{(0)}}, \hat{\theta}), \mathcal{D})$ or its square will also be a cutting plane to $\mathcal{J}_1$. In the sequel we assume $\omega = \omega^{(0)}$ and omit ω as argument. Thus our goal is to find a cutting plane for the optimum value γ_{opt} of GEVP (3),(4), considered as a function of x. Note that R depends affinely on x, i.e. $R(x) = R_0 + \sum_{i=1}^n x_i R_i$ with known constant matrices R_k. Let $\gamma_{opt}^{(0)}, \tau_{opt}^{(0)}$ be the optimal values for γ, τ in GEVP (3),(4) at $x = x^{(0)}$. Then the matrix $F_0 + \gamma_{opt}^{(0)} F_1 + \tau_{opt}^{(0)} R(x^{(0)})$ is singular. Let V^0 be the nullspace of this matrix.

Proposition 6. If $\tau_{opt}^{(0)} > 0$, then there exists a unit length vector $v \in V^0$ such that $v^T R v = 0$. If $\tau_{opt}^{(0)} = 0$, then there exists a unit length vector $v \in V^0$ such that $v^T R v \leq 0$. In either case the vector $g \in \mathbf{R}^n$ given componentwise by $g_i = -v^T R_i v$, if it is nonzero, defines a cutting plane for the function $\mathcal{J}_1$. If g is zero, then $\mathcal{J}_1$ achieves a minimum at $x^{(0)}$.

The proof is based on inclusion relations between the nullspaces, positive and negative spaces of F_0, F_1, R.

Let us now compute a cutting plane for cost function $\mathcal{J}_2$. Denote by $\omega^{(0)}$ the frequency at which the function $\frac{\lambda_{\max}(T(\omega)\bar{M}^{-1}T(\omega)^T)}{(1+|G(e^{j\omega},\hat{\theta})|^2)^2}$ attains its maximum. Let $v \in \mathbf{R}^2$ be a unit length eigenvector to the maximal eigenvalue of the matrix $T(\omega^{(0)})\bar{M}^{-1}T(\omega^{(0)})^T$.

Proposition 7. Let $g \in \mathbf{R}^n$ be defined componentwise by $g_i = -v^T T(\omega^{(0)})\bar{M}^{-1}\bar{M}_i\bar{M}^{-1}T(\omega^{(0)})^T v$. If $g \neq 0$, then g defines a cutting plane for the cost function $\mathcal{J}_2$ at $x^{(0)}$. If $g = 0$, then $\mathcal{J}_2$ attains a minimum at $x^{(0)}$.

The proof is by computing the gradient of the function $f(x) = \mathrm{tr}(T(\omega^{(0)})^T v v^T T(\omega^{(0)})(\bar{M}(x))^{-1})$.

6. DESIGN OF INPUT SIGNALS

In this section we design an input signal from an obtained solution $x^{(0)} \in \mathcal{X}$. By Theorem 1, any moment point can be realized by a discrete spectrum, and there exist moment points which can be realized only by discrete spectra. Thus we propose a two-step procedure. First a discrete input power spectrum generating the moment point $x^{(0)}$ is computed, and then a multisine input with the desired spectrum is generated.

The point $x^{(0)}$ corresponds to a point $\bar{x} = (x_0, x_1, \ldots, x_n)$ in moment space $\mathcal{M}^{(n+1)}$. By Theorem 1, there exists a discrete realization of $\bar{x}$ with index not greater than $\frac{n+1}{2}$. Its construction can be cast into a standard semidefinite program by exploiting an idea that is used to prove Theorem 1. We omit the details here.

Once we have obtained a discrete realization of $\bar{x}$ with frequencies $\omega_1, \ldots, \omega_m$ and associated weights $\lambda_1, \ldots, \lambda_m$, we can construct the multisine input $u(t) = \sum_{i=1}^m \alpha_i \sin(t\omega_i + \phi_i)$ with $\alpha_i = \sqrt{2c\lambda_i}$, ϕ_i arbitrary, if $\omega_i \neq 0, \pi$; and $\alpha_i = \sqrt{c\lambda_i}$, $\phi_i = \pm\frac{\pi}{2}$, if $\omega_i \in \{0, \pi\}$. It has the input power spectrum defined by the initial realization (see e.g. (Zarrop, 1979)).

7. SIMULATION RESULTS

Consider the true system $y = G_0 u + H_0 e = \frac{B(z)}{A(z)} u + \frac{1}{A(z)} e$ with $G_0 = \frac{B(z)}{A(z)} = \frac{0.1047z^{-1} + 0.0872z^{-2}}{1 - 1.5578z^{-1} + 0.5769z^{-2}}$. Here u is the input, subject to the energy constraint $\bar{E}u^2(t) = 1$, and e is white Gaussian noise with variance 0.1. The system is to be identified within an ARX model structure of order two. The number of collected data is $N = 1000$. The aim is to minimize the worst-case ν-gap of the resulting uncertainty region corresponding to a confidence level of $\alpha = 0.95$.

A Monte-Carlo simulation of 500 runs was performed. Each run consisted of five identification experiments: one preliminary and four mutually independent secondary experiments based on this preliminary experiment. The secondary experiments corresponded to the cost functions $\mathcal{J}_1$, $\mathcal{J}_2$, and, for comparison, the classical criteria D-optimality and E-optimality, respectively. In the preliminary experiment, the input was chosen to be white Gaussian noise with variance

1. The parameter vector and noise variance identified in the preliminary experiment were used as a priori estimates of the true parameter vector and the true noise variance for designing the input power spectrum for the series of second experiments. The actual input sequence was a multisine having the evaluated optimal power spectrum, in each of the four secondary experiments with respect to the corresponding cost function. After each identification experiment the worst-case ν-gap of the identified uncertainty region was recorded.

The mean over 500 runs of the worst-case ν-gap resulting from the preliminary experiments equals 0.1345. The means of the worst-case ν-gap resulting from the experiments with multisine input optimized with respect to the criteria $\mathcal{J}_1, \mathcal{J}_2$ are 0.0937 and 0.0927, respectively. The difference between them is statistically significant (2×1.64 standard deviations). The means of the worst-case ν-gap resulting from the experiments with D- and E-optimal multisine input are equal to 0.1434 and 0.1055, respectively.

It is evident that using inputs optimized with respect to criteria $\mathcal{J}_1, \mathcal{J}_2$ gives better results than using white noise input or input optimized with respect to the classical D- and E-optimality criteria. Note also that the inputs optimized with respect to the cost function $\mathcal{J}_2$, which is a first order approximation of the exact cost function $\mathcal{J}_1$, give better results than $\mathcal{J}_1$, despite the fact that the worst-case ν-gap is in fact $\mathcal{J}_1$. This tendency was observed also in simulations with other systems. The reason is that the optimum of the input power spectrum with respect to $\mathcal{J}_2$ is less dependent on the error in the preliminary estimate $\bar{\theta}$ of the true parameter vector than the optimum with respect to $\mathcal{J}_1$ and that this difference as a rule overweighs the error introduced by approximating cost function $\mathcal{J}_1$ by $\mathcal{J}_2$. Given the lower complexity of $\mathcal{J}_2$ and hence the lower computational effort in comparison with $\mathcal{J}_1$, it is preferable to use primarily the former.

8. CONCLUSIONS

We design an input sequence for an identification experiment that minimizes the worst-case ν-gap between the identified model and the uncertainty region around it. The design is via power spectrum optimization. Two nonstandard cost criteria $\mathcal{J}_1$ and $\mathcal{J}_2$ are defined, which reflect the optimization task with different accuracy. $\mathcal{J}_1$ is the exact worst-case ν-gap, while $\mathcal{J}_2$ is an approximation of $\mathcal{J}_1$. These functions fulfil the natural conditions of monotonicity and quasiconvexity with respect to the power spectrum.

It was shown that optimization of the input power spectrum with respect to these cost criteria can be cast as standard convex optimization problem involving LMI constraints. In Propositions 6 and 7 we demonstrate how to construct cutting planes to the cost functions $\mathcal{J}_1, \mathcal{J}_2$, which is essential for applying standard numerical methods such as the ellipsoid algorithm. We have also briefly touched the problem of designing an input sequence with a prespecified power spectrum.

9. REFERENCES

Bombois, X., M. Gevers, G. Scorletti and B. Anderson (2001). Robustness analysis tools for an uncertainty set obtained by prediction error identification. *Automatica* **37**(10), 1629–1636.

Boyd, S., L. El Ghaoui, E. Feron and V. Balakrishnan (1994). *Linear matrix inequalities in system and control theory*. Vol. 15 of *SIAM Stud. Appl. Math.*. SIAM.

Gevers, M., X. Bombois, B. Codrons, G. Scorletti and B. Anderson (2000). Model validation for control and controller validation: a prediction error identification approach. In: *Proceedings of the 12th IFAC Symposium on System identification (SYSID 2000)*. Pergamon, Elsevier Science. Santa Barbara. pp. 319–324.

Goodwin, G., M. Zarrop and R. Payne (1974). Coupled design of test signal, sampling intervals and filters for system identification. *IEEE Trans. Automat. Control* **AC-19**(6), 748–752.

Karlin, Samuel and William Studden (1966). *Tchebycheff systems: with applications in analysis and statistics*. Vol. XV of *Pure Appl. Math.*. Interscience Publishers.

Kiefer, Jack (1974). General equivalence theory for optimum designs. *Ann. Statist.* **2**(5), 849–879.

Kiefer, Jack and Jacob Wolfowitz (1959). Optimum designs in regression problems. *The Annals of Mathematical Statistics* **30**, 271–294.

Lindqvist, K. and H. Hjalmarsson (2001). Identification for control: Adaptive input design using convex optimization. In: *Proceedings of the 40th CDC*. Orlando, Florida, USA.

Ljung, Lennart (1999). *System identification: theory for the user*. Prentice-Hall Information and System Sciences Series. second ed.. Prentice Hall.

Nesterov, Yurii and Arkadii Nemirovskii (1994). *Interior-point polynomial algorithms in convex programming*. Vol. 13 of *SIAM Stud. Appl. Math.*. SIAM. Philadelphia.

Payne, R. and G. Goodwin (1974). Simplification of frequency domain experiment design for SISO systems. Publication 74/3. Dept. of Computing and Control, Imperial College. London.

Vinnicombe, Glenn (1993). Frequency domain uncertainty and the graph topology. *IEEE Trans. Automat. Control* **AC-38**(9), 1371–1383.

Wu, S.P., Stephen P. Boyd and L. Vandenberghe (1996). FIR filter design via semidefinite programming and spectral factorization. In: *Proceedings of the 35th CDC*. Kobe, Japan.

Zarrop, Martin (1979). *Optimal experiment design for dynamic system identification*. Vol. 21 of *Lecture Notes in Control and Inform. Sci.*. Springer. Berlin, New York.

IFAC Publications
www.elsevier.com/locate/ifac

IDENTIFICATION OF RESONANT SYSTEMS USING PERIODIC MULTIPLICATIVE REFERENCE SIGNALS

Wayne J. Dunstan * **Robert R. Bitmead** *,[1]

* *Department of Mechanical & Aerospace Engineering, University of California San Diego, La Jolla CA 92093-0411, USA.*

Abstract: System identification of the forward path of a linear unity feedback system, which is almost unstable, is considered through the application of a periodic multiplicative reference signal. Motivation for this problem comes from the need to identify the linear acoustics of a combustion process on the verge of instability. The tools of cyclostationary signal analysis provide an entré to develop methods for understanding the problem and for determining requisite properties of the reference signal.

Keywords: Resonant systems, Excitation, Periodic Systems, Cyclostationary.

1. INTRODUCTION

The motivation for this paper derives from a practical problem occurring in combustion chambers of turbomachinery. It is desirable from pollution and economic reasons to run these engines at low fuel-to-air ratios. Operating engines in these ranges can produce strong periodic instabilities in the 100-1000Hz range, believed to stem from the nonlinear interaction between chamber pressure and the heat release rate.

An experimental rig used to observe these instabilities is illustrated in Figure 1. For low fuel-to-air ratio, the resulting spectra of pressure and heat release signals from this experiment demonstrate very strong tonal components consistent with the existence of a limit cycle. As this instability is approached however, we see linear behavior tending to increased resonance (Murray *et al.*, 1998; Dunstan and Bitmead, 2002; Dunstan *et al.*, 2001).

A block diagram of the physical model for this system is shown in Figure 2, where $v(t)$ is a driving noise input, $G(z)$ is the transfer function of interest (believed to consist of linear acoustics plus delay elements), and the memoryless saturation nonlinearity is present in the feedback path. In the linear regime of operation, the saturation is not operative but the closed loop progressively approaches resonance. The ultimate objective is to ameliorate these instabilities using a control design from a low complexity model, which requires identifying $G(z)$. Before the onset of instability the nonlinearity can be replaced by a static gain and the system may be treated using linear tools. In our analysis, the reference $r(t)$ will be periodic and

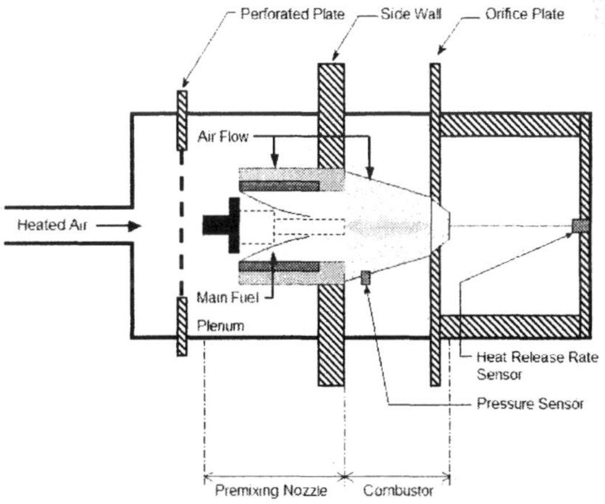

Fig. 1. Single Nozzle Rig experimental setup

[1] Research supported by USA National Science Foundation Grant ECS-0070146 and DARPA Grant N00014-00-1-0799

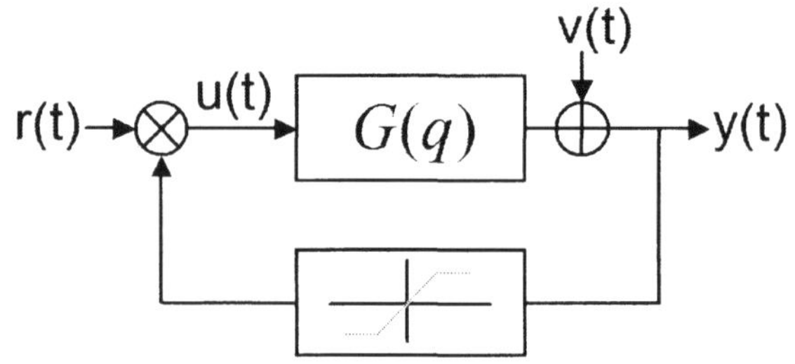

Fig. 2. Combustion instability model.

enter the loop multiplicatively, as shown in Figure 2. Signal notations are as defined by this figure.

The system identification task is then to create an estimate of $G(z)$, denoted $\hat{G}(z)$, with $r(t)$ a known periodic signal and the unity feedback closed loop $[1 - G(z)]^{-1}$ highly resonant. The implications of this are,
(a) the periodic multiplicative reference signal results in a linear time varying system with cyclostationary signals, and,
(b) the highly resonant system is challenging to identify using standard closed loop techniques.

1.1 Limitations in resonant system estimation

Considering the case when $r(t) = 1$, the system reduces to a stationary noise-driven system with transfer function description,

$$y(t) = [1 - G(z)]^{-1} v(t). \tag{1}$$

An Auto-Regressive (AR) model,

$$y(t) = \frac{1}{A(z)} v(t)$$

of arbitrary order may be fitted to output data, $\{y(t)\}$, from this system. The resulting estimate, $A(z)$ and therefore $\hat{G}(z) = 1 - A(z)$, will be biased due to the inability of the AR model to capture adequately the noise dynamics, $v(t)$, added to the signal leaving $G(z)$ (Quinn and Fernandes, 1991). Further, as the resonance of the closed-loop becomes stronger, $y(t)$ becomes more strongly tonal and the bias worsens. For the computed model $\hat{G}(z)$ the lowest bias will be where $G(e^{j\omega})$ is close to 1, that is, where the signal spectrum is most persistently exciting.

Equation (1) implies the following the power spectral density (PSD) relationship.

$$\Phi_y(\omega) = [I - G(e^{j\omega})]^{-1} \; \Phi_v(\omega) \; [I - G(e^{-j\omega})]^{-T}$$

where $\Phi_y(\omega)$ and $\Phi_v(\omega)$ are the PSD's of $y(t)$, $v(t)$ respectively.

Example 1. Consider the system in (1) with,

$$G(z) = \frac{0.1}{z^2 + 2\alpha cos(\beta) z + \alpha^2}$$

where $\alpha = 0.9$ and $\beta = \pi/7$.

Fitting a 20th order AR model to the $\{y(t)\}$ data results in the comparative fit shown in Figure 3. Notice

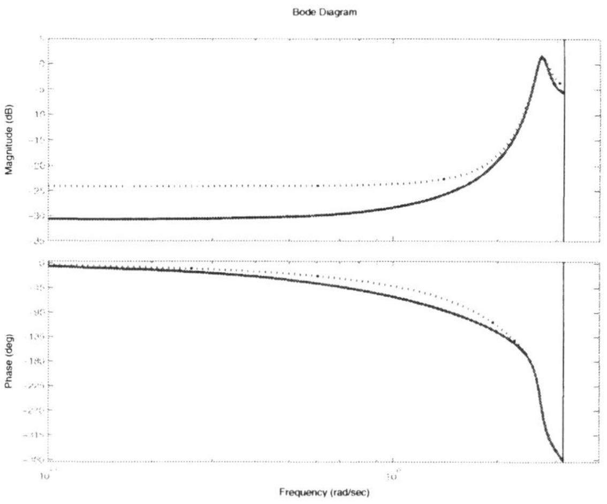

Fig. 3. Discrete Bode comparison of $G(z)$ (solid) and $\hat{G}(z)$ (dotted) for Example 1

the low bias in regions where $G(e^{j\omega})$ is close to 1, and the large bias elsewhere. This illustrates the problems which occur when $\Phi_y(\omega)$ contains strong tonal components.

Our aim is to quantify the improvements to the transfer function estimation problem introduced by a multiplicative reference signal.

The paper proceeds as follows:
• Establish that the signals are cyclostationary when a periodic multiplicative reference signal is used.
• Introduce tools of cyclostationary signal analysis based upon the lifting to a stationary vector process.
• Consider Least Squares parameter estimation of the stationary vector process.
• Determine properties of the reference signal to guarantee identifiability of $G(z)$.

2. CYCLOSTATIONARY SYSTEM ANALYSIS

Suppose that the multiplicative reference $r(t)$ is a deterministic periodic signal of period N and denote by z the z-transform variable and by q the forward-shift operator. Then the closed-loop system in Figure 2 is described by

$$y(t) = G(q)u(t) + v(t), \tag{2}$$
$$u(t) = r(t)y(t). \tag{3}$$

Suppose further that this stable. Then $y(t)$ is the output of a stable, linear, periodically time-varying, (LPTV) system driven by noise and thus is asymptotically a cyclostationary process (Gardner, 1994; Hale, 1969).

2.1 *Lifting*

The scalar LPTV system in (2) and (3) is transformed to a LTI vector system, of dimension N, using an isomorphism called *lifting* (Khargonekar *et al.*, 1985). Denote the unit delay operator q^{-1} and the N-step delay operator $\zeta = q^{-N}$, and the equivalent transform variable

$$\sigma = z^{-N}.$$

We identify the time variable as

$$t = kN + \ell.$$

Variable t is in the original single-sample domain and variable k is in the lifted N-sample domain. Define scalar signals $y_\ell(k) = y(kN+\ell)$ for $\ell = 0, \ldots, N-1$ and the *lifted* N-vector signal

$$y_k = \begin{bmatrix} y(kN+0) \\ y(kN+1) \\ \vdots \\ y(kN+N-1) \end{bmatrix}. \quad (4)$$

Similarly define the σ-transforms

$$Y_\ell(\sigma) = \sum_{j=0}^{\infty} y_\ell(j)\sigma^j$$
$$= \sum_{j=0}^{\infty} y(jN+\ell)\sigma^j, \quad \text{for } \ell = 0, \ldots, N-1.$$

Then define

$$Y(\sigma) = \sum_{k=0}^{\infty} y_k \sigma^k = \begin{bmatrix} Y_0(\sigma) \\ Y_1(\sigma) \\ \vdots \\ Y_{N-1}(\sigma) \end{bmatrix}.$$

This is the σ-transform of the lifted process.

The N-vector lifted process y_k is a stationary stochastic process in time-variable k. This is the lifted form of the cyclostationary scalar process $y(t)$. Clearly lifting is an isomorphism, which may be applied conformably to $u(t)$ and $v(t)$ to yield lifted N-vector sequences, u_k and v_k. Using these N-vector sequences and the identity $\zeta = q^{-N}$, the relations of (2) and (3) may be rewritten

$$y_k = \mathcal{G}(\zeta)u_k + v_k, \quad (5)$$
$$u_k = Ry_k. \quad (6)$$

To determine $\mathcal{G}$ and R, write $G(\zeta) = \sum_{i=1}^{\infty} g_i\zeta^{-i}$ so that g_i are the impulse response parameters of G then matrix $\mathcal{G}(\sigma)$ is Toeplitz and square.

$$\mathcal{G}(\zeta) = \begin{bmatrix} \tilde{g}_0(\zeta) & \zeta\tilde{g}_{N-1}(\zeta) & \cdots & \zeta\tilde{g}_1(\zeta) \\ \tilde{g}_1(\zeta) & \tilde{g}_0(\zeta) & \cdots & \zeta\tilde{g}_2(\zeta) \\ \vdots & \vdots & \ddots & \vdots \\ \tilde{g}_{N-1}(\zeta) & \tilde{g}_{N-2}(\zeta) & \cdots & \tilde{g}_0(\zeta) \end{bmatrix},$$

where

$$\tilde{g}_i(\zeta) = (g_i + \zeta^1 g_{N+i} + \zeta^2 g_{2N+i} + \ldots)$$

and,

$$R = \begin{bmatrix} r_0 & 0 & 0 & \cdots & 0 \\ 0 & r_1 & 0 & \cdots & 0 \\ 0 & 0 & r_2 & \cdots & 0 \\ \vdots & \vdots & \vdots & \ddots & \vdots \\ 0 & 0 & 0 & \cdots & r_{N-1} \end{bmatrix}.$$

Equation (5) and (6) imply the following the PSD relationship.

$$\Phi_y(\Omega) = [I - \mathcal{G}(e^{j\Omega})R]^{-1}\Phi_v(\Omega)[I - \mathcal{G}(e^{-j\Omega})R]^{-T} \quad (7)$$

where $\Phi_y(\Omega)$ and $\Phi_v(\Omega)$ are the PSD's of y_k, v_k respectively in the lifted frequency domain Ω. The mapping in (7),

$$[\Phi_v(\Omega), G(z), r(t)] \Rightarrow \Phi_y(\Omega)$$

alludes to the influence of the noise spectrum and reference signal $r(t)$ on the spectrum of y_k.

This section establishes an isomorphism under which the scalar cyclostationary system is replaced by an N-vector stationary lifted system with a matrix transfer function $\mathcal{G}(\sigma)$ and a relationship between the spectrum of this vector signal and the reference and noise signals.

3. LEAST SQUARES PARAMETRIZATION AND ESTIMATION

We develop a Least Squares (LS) estimator for vector systems, appropriate for application to the estimation of the parameters of this N-vector stationary y_k process. Write the first row of (5),

$$y(kN) = \tilde{g}_0(\zeta)\,u(kN) + \zeta\tilde{g}_{N-1}(\zeta)\,u(kN+1) + \cdots$$
$$+ \zeta\tilde{g}_1(\zeta)\,u(kN+N-1) + v(kN) \quad (8)$$

Observation 1. All $\tilde{g}_i(\zeta)$, $i=0..N-1$ have the same scalar denominator polynomial $A(\zeta)$ of degree m.

Write (8) in polynomial form (note the reordering of the $u(kN+1)\ldots u(kN+N-1)$ terms),

$$A(\zeta)y(kN) = B_0(\zeta)\,u(kN) + B_1(\zeta)\,\zeta u(kN+N-1)(\zeta) + \cdots$$
$$+ B_{N-1}(\zeta)\,\zeta u(kN+1) + \varepsilon(kN)$$

where,

$$\varepsilon(kN) = A(\zeta)v(kN),$$
$$A(\zeta) = 1 + a_1\zeta + \ldots + a_m\zeta^m,$$
$$B_i(\zeta) = b_{i,0} + b_{i,1}\zeta + \ldots + b_{i,m}\zeta^m \quad i=0..N-1,$$

and it is assumed that all $B_i(\zeta)$ polynomials also have degree m.

This is a scalar AutoRegressive with an eXogenous input (ARX) model, for which a parametrization can be written,

$$\begin{aligned}y(kN) = &-a_1 y((k-1)N) - \cdots - a_m y((k-m)N)\\ &+b_{0,0}u(kN) + \cdots + b_{0,m}\, u((k-m)N)\\ &+b_{1,0}u((kN-1) + \cdots + b_{1,m}\, u((k-m)N-1)\\ &+b_{2,0}u((kN-2)) + \cdots + b_{2,m}\, u((k-m)N-2)\\ &\vdots\\ &+b_{N-1,0}u((kN-(N-1)) + \cdots + b_{N-1,m}u((k-m)N-(N-1))\\ &+\varepsilon(kN)\end{aligned}$$

Writing the other rows in a similar way yields,

$$y_k = \Psi_k^T\,\theta + \varepsilon_k, \tag{9}$$

where the $N \times [(N+1)m - 1]$ regressor and $N \times N$ reference matrices,

$$\Psi_k^T = \left[\,\bar{\mathcal{Y}}_0 \mid R_0\mathcal{Y}_0 \mid R_1\mathcal{Y}_1 \mid \cdots \mid R_{N-1}\mathcal{Y}_{N-1}\right] \tag{10}$$

and,

$$\mathcal{Y}_i = \begin{bmatrix} y((k-0)N+0-i) & \cdots & y((k-m)N+0-i)\\ y((k-0)N+1-i) & & y((k-m)N+1-i)\\ \vdots & & \vdots\\ y((k-0)N+N-1-i) & \cdots & y((k-m)N+N-1-i)\end{bmatrix}$$

$$R_i = \begin{bmatrix} r_{0-i} & 0 & 0 & \cdots & 0\\ 0 & r_{1-i} & 0 & \cdots & 0\\ 0 & 0 & r_{2-i} & \cdots & 0\\ \vdots & \vdots & \vdots & \ddots & \vdots\\ 0 & 0 & 0 & \cdots & r_{N-1-i}\end{bmatrix}$$

where for $x < 0$ $r_x = r_{N-x}$, and,

$\bar{\mathcal{Y}}_0 = -\mathcal{Y}_0$ with first column removed.

$$\theta = \left[a_1 \cdots a_m \mid b_{0,0} \cdots b_{0,m} \mid b_{1,0} \cdots b_{N-1,m}\right]^T$$

$$\varepsilon_k = \begin{bmatrix}\varepsilon(kN)\\ \varepsilon(kN+1)\\ \vdots\\ \varepsilon(kN+N-1)\end{bmatrix}.$$

The structure of $\mathcal{Y}_i$ is summarized:

- The elements going down the rows represent increasing samples along each period of $r(t)$.
- Stepping across the columns represents an N-step time shift in the signals.

Defining a scalar cost function for $M = D \times N$ data,

$$V_M = \sum_{k=0}^{D-1}(y_k - \Psi_k^T\theta)^T(y_k - \Psi_k^T\theta).$$

Note that this lifted form differs from the regular least squares formulation through the appearance of an N-vector measurement y_k and an $N \times [(N+1)m-1]$ regressor matrix Φ_k. The minimum is given by the standard LS solution structure with increased dimension elements,

$$\hat{\theta}_M = \left[\frac{1}{D}\sum_{k=0}^{D-1}\Psi_k\Psi_k^T\right]^{-1}\frac{1}{D}\sum_{k=0}^{D-1}\Psi_k y_k. \tag{11}$$

Equation (11) forms the link,

$$\boxed{\mathbf{E}\left[\Psi\Psi^T\right] \Rightarrow \hat{\theta}.}$$

We note this parametrization of $\mathcal{G}(\sigma)$ is isomorphic to a parametrization of $G(z)$. Identifiability of $\mathcal{G}$ relies on properties of the stationary N-vector signal y_k which are inherited (or descended) from the properties of cyclostationary $y(t)$, whose analysis lacks tools.

4. IDENTIFIABILITY

Identifiability imposes conditions on the input and output signals to ensure an estimate $\hat{G}(z)$ exists and is unique. In this context, we consider the identifiability of $\mathcal{G}(z)$ from the lifted N-vector LS problem of the previous section. The lifting admits appeal to tools from stationary signals analysis, which are inaccessible for the original cyclostationary processes. Further, (7) establishes the spectral/correlation properties of the N-vector stationary signals.

Theorem 1. Parameter θ in (9) is identifiable if

$$\mathcal{I} = \frac{1}{D}\sum_{k=0}^{D-1}\Psi_k\Psi_k^T \tag{12}$$

has full rank.

The identifiability matrix, $\mathcal{I}$, for (10) is,

$$\mathcal{I} = \frac{1}{D}\times \begin{bmatrix} \bar{\mathcal{Y}}_0^T\bar{\mathcal{Y}}_0 & \bar{\mathcal{Y}}_0^T R_0\mathcal{Y}_0 & \cdots & \bar{\mathcal{Y}}_0^T R_{D-1}\mathcal{Y}_{N-1}\\ \mathcal{Y}_0^T R_0\bar{\mathcal{Y}}_0 & \mathcal{Y}_0^T R_0^T R_0\mathcal{Y}_0 & \cdots & \mathcal{Y}_0^T R_0^T R_{N-1}\mathcal{Y}_{N-1}\\ \mathcal{Y}_1^T R_1\bar{\mathcal{Y}}_0 & \mathcal{Y}_1^T R_1^T R_0\mathcal{Y}_0 & \cdots & \mathcal{Y}_1^T R_1^T R_{N-1}\mathcal{Y}_{N-1}\\ \vdots & \vdots & & \vdots\\ \mathcal{Y}_{N-1}^T R_{N-1}\bar{\mathcal{Y}}_0 & \mathcal{Y}_{N-1}^T R_{N-1}^T R_0\mathcal{Y}_0 & \cdots & \mathcal{Y}_{N-1}^T R_{N-1}^T R_{N-1}\mathcal{Y}_{N-1}\end{bmatrix}$$

This Toeplitz, symmetric, nonnegative definite matrix contains elements which depend on the correlation properties of y_k and $r(t)$. This can be exploited to establish an alternative identifiability condition. This, in turn, provides conditions necessary for determining how $r(t)$ affects the identifiability of θ and therefore $G(z)$.

Define the auto-correlation function of $y(t)$

$$\mathcal{R}_y(t,\tau) = E(y(t)y(t-\tau)).$$

This is both t- and τ-dependent and is N-periodic in t (Gardner, 1994). We now consider the relationship between the elements of $\mathcal{I}$, $\mathcal{R}_y(t,\tau)$ and $r(t)$ by computing representative expansions. The sub-blocks of $\mathcal{I}$ are Toeplitz matrices comprised of combinations of $\mathcal{R}_y(t,\tau)$ and r_i.

Elements of the average of $\bar{\mathcal{Y}}_0^T\bar{\mathcal{Y}}_0$ have composition,

$$(1,1) = \mathcal{R}_y(1,0)+\mathcal{R}_y(2,0)+\cdots+\mathcal{R}_y(N-1,0)$$
$$(1,2) = \mathcal{R}_y(1,N)+\mathcal{R}_y(2,N)+\cdots+\mathcal{R}_y(N-1,N)$$
$$(1,N) = \mathcal{R}_y(1,mN)+\mathcal{R}_y(2,mN)+\cdots+\mathcal{R}_y(N-1,mN)$$

Elements of the average of $\mathcal{Y}_0^T R_0^T R_0 \mathcal{Y}_0$ have composition:

$$(1,1) = r_0^2\mathcal{R}_y(0,0)+r_1^2\mathcal{R}_y(1,0)+\cdots+r_{N-1}^2\mathcal{R}_y(N-1,0)$$
$$(1,2) = r_0^2\mathcal{R}_y(0,N)+r_1^2\mathcal{R}_y(1,N)+\cdots+r_{N-1}^2\mathcal{R}_y(N-1,N)$$
$$(1,N) = r_0^2\mathcal{R}_y(0,mN)+r_1^2\mathcal{R}_y(1,mN)+\cdots+r_{N-1}^2\mathcal{R}_y(N-1,mN)$$

Elements of the average of $\bar{\mathcal{Y}}_1^T R_1^T R_2 \bar{\mathcal{Y}}_2$ have composition,

$$(1,1) = r_{N-1}r_{N-2}\mathcal{R}_y(N-1,1)$$
$$+r_0 r_{N-1}\mathcal{R}_y(0,1)+\cdots+r_1 r_2\mathcal{R}_y(1,1)$$
$$(1,2) = r_{N-1}r_{N-2}\mathcal{R}_y(N-1,N-1)$$
$$+r_0 r_{N-1}\mathcal{R}_y(0,N-1)+\cdots+r_1 r_2\mathcal{R}_y(1,N-1)$$
$$(1,N) = r_{N-1}r_{N-2}\mathcal{R}_y(N-1,mN-1)$$
$$+r_0 r_{N-1}\mathcal{R}_y(0,mN-1)+\cdots+r_1 r_2\mathcal{R}_y(1,mN-1)$$

Since the matrices are Toeplitz only the first row and column need to be calculated.

We have shown by exhausting example that the identifiability conditions for the forward-path plant $G(z)$ depend explicitly on the autocorrelation of $y(t)$ and the reference $r(t)$, which in turn are connected through (7). That is,

$$\Phi_y(\Omega) \Rightarrow \mathcal{R}_y(t,\tau), r(t) \Rightarrow \mathcal{I}.$$

5. ROAD MAP

This paper has investigated the system identification of a linear unity feedback system, which is almost unstable, through the application of a periodic multiplicative reference signal. Tools have been developed and applied to move toward a solution in which one can determine the identifiability properties of the forward path transfer function $G(z)$ through the appropriate selection of periodic reference signal $r(t)$.

Tools of cyclostationary signal analysis have been used to create,

PSD relationship

$$[\Phi_v(\Omega), G(z), r(t)] \Rightarrow \Phi_y(\Omega)$$

Vector Least Squares Parametrization

$$\mathbf{E}\left[\Psi\Psi^T\right] \Leftrightarrow \hat{\theta}$$

and to explore,

Identifiability Condition

$$\Phi_y(\Omega) \Rightarrow \mathcal{R}_y(t,\tau), r(t) \Rightarrow \mathcal{I}$$

for which analysis continues on the link between the Identifiability Matrix and the Correlation Matrices,

$$\mathcal{I} \Leftrightarrow \mathbf{E}\left[\Psi\Psi^T\right]$$

Some representative expansions of $\mathcal{I}$ begin to unveil the how the structure of $r(t)$ affects the rank properties of the $\mathcal{Y}_i^T\mathcal{Y}_j$ matrices. These matrices are composed of correlation matrices, $\mathcal{R}_y(t,\tau)$, which are indicative of the shape of $\Phi_y(\Omega)$. This path allows us to study the influence of $r(t)$ on resonant system identification. It is expected that the multiplicative effect of $r(t)$ will introduce sidebands into $\Phi_y(\Omega)$, which are observable in $\mathcal{R}_y(t,\tau)$.

Further work will link properties of the correlation matrices, $\mathcal{R}_y(t,\tau)$ which ultimately will show how to design $r(t)$ to improve the identification of resonant systems. How does this work?

Resonance and Sidebands

Resonance of the unexcited ($r(t) = 1$) loop, which for the practical combustion instability problem represents the onset of limit cycling, is associated with there existing a single base-rate frequency ω_0 such that

$$G(e^{j\omega_0}) \approx 1. \qquad (13)$$

In signal terms, this corresponds to the appearance of strongly dominant single frequency in the closed-loop scalar signal spectrum $\Phi_y(\omega)$.

In the unexcited lifted framework (13) translates to the property that

$$\det\left[I - \mathcal{G}(e^{-jN\omega_0})\right] \approx 0, \qquad (14)$$

or $\lambda_i\left(\mathcal{G}(e^{-jN\omega_0})\right) \approx 1$. Analysis of (13) and the structure of $\mathcal{G}$ indicates that a left eigenvector associated with this eigenvalue of one is

$$v_0 = \left[1 \ e^{j\omega_0} \ \ldots \ e^{j(N-1)\omega_0}.\right] \qquad (15)$$

With the introduction of non-constant N-periodic $r(t)$ composed of a constant plus small sinusoidal variation

$$r(t) = 1 + \beta\cos(\nu t), \ \ \beta \ll 1, \ \ \nu = \frac{\alpha}{N}, \ \alpha \in \mathbb{N}$$

this eigenvector is perturbed by diagonal matrix R to two others with eigenvalues close to one and within a β-neighborhood of v_0. The signal frequencies associated with these perturbed eigenvalues are $\omega_0 + \nu$ and $\omega_0 - \nu$. This is the appearance of sidebands in the closed-loop signals. Note that the original resonant signal at frequency ω_0 remains.

Improved Identifiability
The splitting of the dominant closed-loop signal to yield persistent frequency content at new adjacent frequencies through sideband formation has an immediate effect on the identifiability of $G(z)$. As was demonstrated in the introduction via example, the estimation of G is accurate and unbiased at frequencies where the signal-to-noise ratio is high in closed loop. Sidebands extend the region of identifiability of G from just its frequency response at ω_0 (which is fully captured by just two real parameters) to include also the frequency response at $\omega_0 \pm \nu$, thereby extending the information content to six real parameters. This permits the improved estimation of the forward path in the resonant system.

This conceptual analysis needs to be extended to a quantified treatment of the properties of the identifiability matrix $\mathcal{I}$. This remains to do. Further work will focus on the specific design of the signal $r(t)$.

Combustion Instability Modeling for Control
In the practical example of combustion instability, the multiplicative reference is achieved through fuel modulation - a small proportion of the fuel flow is modulated at frequency ν. This can be a difficult actuation problem except at relatively low frequencies $\nu \ll \omega_0$. However, this is precisely where the excitation can have the desired effect of exposing the frequency response of G in the neighborhood of the resonance ω_0.

A control solution to the combustion instability would most likely rely on the correct phasing of a small fuel flow at the resonant frequency to assist in quenching the instability (Banaszuk *et al.*, 2003; Zhang *et al.*, 1998). To achieve this diminution of the instability with a model-based controller, accurate phase information is needed in the neighborhood of ω_0. Thus, low-frequency reference excitation is an approach well-suited to this practical problem.

REFERENCES

Banaszuk, A., K. B. Ariyur, M. Krstic and C. Jacobson (2003). An adaptive algorithm for control of combustion instability. *Automatica*.

Dunstan, W.J. and R.R. Bitmead (2002). Model confidence fo nonlinear systems. *IFAC World Congress on Automatic Control, Barcelona, Spain*.

Dunstan, W.J., R.R. Bitmead and S.M. Savaresi (2001). Fitting nonlinear low-order models for combustion instability control. *Control Engineering Practice* **9**, 1301–1317.

Gardner, W. (1994). *Cyclostationarity in Communications and Signal Processing*. IEE Press, 345 East 47th Street, New York, NY 10017-2394.

Hale, J. (1969). *Ordinary Differential Equations*. Wiley Interscience, New York, NY.

Khargonekar, P.P., K. Poolla and A. Tannenbaum (1985). Robust control of linear time-invariant plants using periodic control. *IEEE Transactions on Automatic Control* **AC-30**(11), 1088–1096.

Murray, R.M., C.A. Jacobson, R. Casas, A.I. Khibnik, C.R. Johnson Jr, R.R. Bitmead, A.A. Peracchio and W.M. Proscia (1998). System identification for limit cycling systems: a case study for combustion instabilities. *American Control Conference, Philadelphia, USA* pp. 2004–2008.

Quinn, B.G. and J.M. Fernandes (1991). A fast efficient technique for the estimation of frequency. *Biometrika* **78**(3), 489–497.

Zhang, Y., P.G. Mehta, R.R. Bitmead and Jr C.R. Johnon (1998). An adaptive algorithm for control of combustion instability. *American Control Conference 1998, Philadelphia PA* pp. 1480–1482.

IFAC Publications
www.elsevier.com/locate/ifac

AIRCRAFT PARAMETER ESTIMATION BY USING THE OPTIMAL INPUT DESIGN AND LINEAR MATRIX INEQUALITIES

Carine Jauberthie, * **Lilianne Denis-Vidal,** ** **and Ghislaine Joly-Blanchard** ***

* *University of Technology Compiègne, Dpt GI*
BP 20 529, 60205 Compiègne, France
Carine.Jauberthie@utc.fr, fax (33)344 23 44 77
** *Univ. Sciences and Tech. Lille, UFR Math(M2)*
59655 Villeneuve d'Ascq, France
denvid@attglobal.net, fax (33)320 43 69 93
*** *University of Technology Compiègne, Dpt GI*
BP 20 529, 60205 Compiègne, France
Ghislaine.Joly-Blanchard@utc.fr, fax (33)344 23 44 77

Abstract: System identification based on physical laws often involves parameter estimation. Even if parameters are theoretically identifiable, they may be poorly estimable for a given experiment. Thus a significant increase in accuracy of the parameter estimation may be obtained by a suitable choice of experimental conditions. The original idea of this paper is the combination of a dynamical programming method with a gradient algorithm in the solution of the optimal design. After getting an optimal input, the parameter estimation is performed by minimizing a weighted least square criterion. Weights are either based on the known measurement noise or given by the solution of a linear matrix inequality problem. *Copyright © 2003 IFAC*

Keywords: Non-linear systems, Control systems, Input design, Parameter estimation, Identification algorithm.

1. INTRODUCTION

In this paper, controlled non-linear dynamical systems and more precisely aircraft dynamics are considered. They are described as:

$$\begin{cases} f(\dot{x}(t,p), x(t,p), p, u(t)) = 0, \; x(0) = x_0, \\ y(t,p) = h(x(t,p), p). \end{cases} \quad (1)$$

The measures are given by:

$$z(t_i) = y(t_i, p) + \nu(t_i), \quad i = 1, ..., N. \quad (2)$$

In these equations, $x(t,p) \in \mathbb{R}^n$, and $y(t,p) \in \mathbb{R}^m$ denote the state variables and the measured outputs. The function f (resp. h) is analytic in x, p, u (resp. in x, p). The vector $p = (p_1, ..., p_l)$ represents the parameters to be estimated and (p belongs to an admissible set). The given integer N is the total number of sample times. The vector z represents the data. The input vector $u(t) \in \mathbb{R}^q$ is such as $u \in \mathcal{U}[0, \infty[$, the set of bounded and measurable functions on $[0, \infty[$. The measurement noise $\nu(t_i)$ is assumed to be white gaussian with zero mean and covariance matrix $E[\nu(t_i)\nu^\top(t_j)] = R\delta_{ij}$ where R is a known diagonal matrix and $i, j = 1, ..., N$.

The constraints on inputs and outputs are given by:

$$|u_j(t) - u_{j_0}| \leq \mu_j, \quad \forall t, \; j = 1, 2, ..., q, \quad (3)$$

$$|y_k(t) - y_{k_0}| \leq \eta_k, \quad \forall t, \; k = 1, 2, ..., m, \quad (4)$$

where μ_j et η_k are positive constants, u_{j_0} and y_{k_0} are the trim values of u_j and y_k. The test time is assumed to be known and it will be denoted by T.

This paper is organized as follows. In Section 2, an approach based on two successive steps is given to calculate optimal input functions. An analytic work for similar problems (Chen 1975) indicates that square wave type inputs are better than sinusoidal type inputs for parameter estimation. The initial step uses dynamic programming, which gives a first result but requires a restrictive admissible input set. Then the square wave inputs are approximated by regular differential inputs which involve some new parameters. That leads to the second step based on a gradient algorithm whose initial conditions are given by the input obtained by the first procedure. It allows all the admissible inputs to be taken into account and it leads to a local result. This approach is applied to the input design for the identification of the aircraft model presented in Section 4. In Section 3, two techniques of weighting least-square criterion are presented. The first one uses the measurement noise covariance matrix inverse and the second one computes the weights of the least square criterion by the linear matrix inequality approach (LMI). To estimate parameters, these criteria are considered with two different experimental conditions (optimal input and an other admissible input). In the last section, these techniques are applied to an example of aircraft longitudinal motion and results are compared.

2. EXPERIMENT DESIGN

2.1 *Introduction*

Dynamical systems are often identified by exciting them with suitable input signals and by observing the resulting response of the system. Experiment design is important for identifying high-fidelity mathematical models of real process from test data. A good experiment design must account for practical constraints during the test. The overall goal is to design an experiment that produces data from wich model parameters can be estimated accurately.

In order to obtain the optimal input and consequently obtain the most accurate estimates of model parameters, the information content in the system response during the test must be maximized. The information contained in the response is embodied in the information matrix, whose elements are combinations of partial derivatives of the system response variables with respect to the model parameters. The information matrix elements, the sensitivities, are obtained by solving the so-called sensitivity equations.

Input designs for parameter estimation experiments are evaluated by examining the Cramer-Rao bounds on the parameter standard errors, which are a function of the information content in the response to a given input. In the following we consider optimal input design for aircraft parameter estimation.

2.2 *Theoretical development*

By using an asymptotically unbiased and efficient estimator, we can find the minimum achievable parameter standard errors which are given by the square root of the diagonal elements of the dispersion matrix $S(p,\Xi)$ ((Morelli 1993), (Morelli 1999), (Morelli and Klein 1990)). The vector $\Xi \in \mathcal{E}$ corresponds to the experimental conditions to design. In our case Ξ represents the input and $\mathcal{E}$ is an admissible experimental input set. The matrix $S(p,\Xi)$ is the inverse of the information matrix $F(p,\Xi)$, where $F(p,\Xi)$ is given by

$$F(p,\Xi) = \sum_{i=1}^{N} (\frac{\partial y(t_i,p)}{\partial p})^T R^{-1} \left(\frac{\partial y(t_i,p)}{\partial p}\right) \quad (5)$$

where the sensitivities:

$$\frac{\partial y(t_i,p)}{\partial p} \quad (6)$$

are solutions of:

$$\frac{d}{dt}(\frac{\partial x}{\partial p_j}) = \sum_{i=1}^{n} \left[\frac{\partial f}{\partial x_i}\frac{\partial x_i}{\partial p_j}\right] + \frac{\partial f}{\partial p_j}, \quad (7)$$

$$\frac{\partial y}{\partial p_j} = \sum_{i=1}^{n} \left[\frac{\partial h}{\partial x_i}\frac{\partial x_i}{\partial p_j}\right] + \frac{\partial h}{\partial p_j}, \; j = 1,...,l. \quad (8)$$

We know that the values of parameter accuracies represent goals for Cramer-Rao bounds of every model parameter. The cost function to be optimized is a function:

$$j(\Xi) = \phi(F(p,\Xi)),$$

where the vector $\Xi \in \mathcal{E}$ and ϕ is often chosen convex monotonous. In this paper A-optimality criterion (see (Walter and Pronzato 1994)) is considered and the cost function is the trace of $S(p,\Xi)$.

2.3 *First step: dynamic programming*

This approach is based on dynamic programming principles. Dynamic programming allows the realistic practical constraints on the input and output variables to be included. More it is a very efficient method to do a global exhaustive search. But the drawback is the computational requirement which expands rapidly by increasing the dimensions of control, state, output and parameter vectors. Thus, the application of the method has brought us to restrict the admissible input form to full amplitude square waves only (Chen 1975).

The dynamic programming consists in a test splitting into stages. For every stage, the information matrix inverse is computed by a sequential calculation (Chen 1975). An initial parameter value p_0 is assumed to be known. In the case of flight mechanical models, p_0 can be obtained by wind tunnel measures. The associated criterion and the outputs are also computed (for more details see (Morelli 1993), (Morelli 1999), (Morelli and Klein 1990)).
This technique discards any input among square wave sequence whose output trajectory exceeds the constraint limits. And the obtained result is an optimal square wave input obtained in a single-pass solution. Inputs tested by this procedure are described as:

$$u(t) = u_0 + \sum_{i=0}^{r} (a_i \varepsilon_i - a_{i-1}\varepsilon_{i-1}) h(t - t_i) \quad (9)$$

where h is the Heaviside function, u_0 is an input trim value. The variables a_i represent the square wave amplitudes and ε_i belongs to $\{-1, 0, +1\}$, $i = 0, ..., r$ and $\varepsilon_{-1} = 0$. In this section, the admissible experimental input set is given by $\mathcal{E} = \{\varepsilon_i |\ i = 0, ...r\}$. The variables t_i are the switching times for the input with $t_0 < t_1 < ... < t_r$. In this part, they are assumed to be known constants.

2.4 *Second step: use of regular differentiable inputs*

The non-differentiable input, obtained by the dynamic programming method, is approximated by a differential function. Therefore, it can be used as a starting point of a gradient method for solving problems with inequality contraints. The input (9) can be approximated by:

$$\tilde{u}(t) = u_0 + \sum_{i=0}^{r} \frac{a_i \varepsilon_i - a_{i-1}\varepsilon_{i-1}}{1 + e^{k(t_i - t)}}, \quad (10)$$

as k tends to infinity.
Since these functions are regular, one can implement a gradient algorithm for optimizing the cost function given in Section 2.3 with starting values given by the first step. In this section, the admissible experimental input set is given by $\mathcal{E} = \{a_i, t_i,\ i = 0, ...r\}$.

3. ESTIMATION PROCEDURES

The aim of this section is to estimate parameters by using a classical least-square objective functions based on the residuals at times t_i. The first procedure uses the measurement noise covariance matrix inverse. In the second procedure, the least-square objective function is modified in order to improve the condition number of the estimation problem.

3.1 *Parameter estimation procedure using the covariance matrix inverse*

Let us consider a weighted least square method where the cost function $J(p)$ is given by:

$$\sum_{i=1}^{N} (z(t_i) - y(t_i, p))^T R^{-1} (z(t_i) - y(t_i, p)) \quad (11)$$

and p_0 as starting point.
The cost function is minimised with respect to the unknown parameters p. This problem is solved by a Quasi-Newton method which is implemented in the toolbox optimization of MATLAB 5.3.

3.2 *Parameter estimation procedure using linear matrix inequalities*

The least-square objective function is modified by means of a weighting operator in order to improve the condition number of the minimization problem. Then the cost function $J_W(p)$ is given by:

$$\sum_{i=1}^{N} (z(t_i) - y(t_i, p))^T W(t_i) (z(t_i) - y(t_i, p)) \quad (12)$$

where $W = diag(W_1, \ldots, W_N)$ is a symmetric, definite positive diagonal matrix.
M. Ouarit et al. propose in (Ouarit *et al.* 2001) to formulate this problem in terms of LMI. The problem consists in looking for the weight matrix W and the following criterion is defined:

$$\begin{cases} \min\limits_{W_1, \ldots, W_N, t} \quad t \\ \text{submitted to} \\ \begin{bmatrix} tI & G_W - I \\ (G_W - I)^\top & tI \end{bmatrix} > 0 \\ W > 0, \end{cases} \quad (13)$$

where G_W is the Gauss-Newton approximation of the hessian matrix. Thus the following algorithm is introduced:

initialization Choose p_0.
Step k
1 Calculation of the weighting matrix W by solving (13) for ε given,
2 Minimization of the criterion (12).

The problem (13) is solved by using the function **mincx** of the toolbox LMI of MATLAB 5.3.

4. EXAMPLE

This example is concerned with the aircraft longitudinal motion. The projection of the general equations of motion (Wanner 1983) onto the aerodynamic reference frame of the aircraft gives the following system :

$$\left\{\begin{array}{l} m\dot{V} = -mg\sin(\theta-\alpha) - \frac{1}{2}\rho SV^2(C_x^0 \\ +C_{x\alpha}(\alpha-\alpha_0) + C_{x\delta_m}(\delta_m - \delta_{m_0})), \\ \\ mV(\dot{\alpha}-\dot{\theta}) = mg\cos(\theta-\alpha) \\ -\frac{1}{2}\rho SV^2(C_{zD} + C_{z\dot{\alpha}}\frac{\dot{\alpha}l}{V}), \\ \\ B\dot{q} = \frac{1}{2}\rho SlV^2(C_{m\alpha}^{25}(\alpha-\alpha_0) \\ +\frac{x_{25}-x_g}{l}C_{zD} \\ +\frac{ql}{V}(C_{mq}^{25} - \frac{x_{25}-x_g}{l}C_{m\alpha}^{25}) \\ +C_{m\dot{\alpha}}^{25}\frac{\dot{\alpha}l}{V} + C_{m\delta_m}^{25}(\delta_m - \delta_{m_0}) \\ +\frac{x_{25}-x_g}{l}C_{z\dot{\alpha}}\frac{\dot{\alpha}l}{V}), \\ \dot{\theta} = q, \end{array}\right. \tag{14}$$

with :

$$C_{zD} = C_z^0 + C_{z\alpha}(\alpha-\alpha_0) + C_{z\delta_m}(\delta_m - \delta_{m_0}) + \frac{ql}{V}(C_{zq} - \frac{x_{25}-x_g}{l}C_{z\alpha}).$$

In these equations, $x = (V, \alpha, q, \theta)^T$ is the state variable vector, $y = (V, \alpha, q, \theta)^T$ is the observation, $u = \delta_m$ is the input and $(C_{z\dot{\alpha}}, C_{zq}, C_{m\dot{\alpha}}^{25}, C_{mq}^{25})^T$ are the parameters to be identified. V denotes the speed of aircraft, α the angle of attack, θ the pitch angle, q the pitch rate, δ_m the deflection angle, ρ the air density, g the acceleration due to gravity and l a reference length.

The final aim of this work is the parameter estimation. Thus the following steps have to be done: identifiability analysis, search of a good starting point p_0, input optimization and finally estimation problem solution. The first two steps have been completely solved in (Denis-Vidal *et al.* 2001).

4.1 *Optimal input*

The dynamic programming consists in a flight test splitting into stages. For every stage, the matrix (5) is computed with the matrix R given by:

$$R = \begin{bmatrix} 25.10^{-4} & 0 & 0 & 0 \\ 0 & 0.04 & 0 & 0 \\ 0 & 0 & 0.04 & 0 \\ 0 & 0 & 0 & 0.04 \end{bmatrix}.$$

The initial parameter value is computed by wind tunnel experiments or by an integration method (Denis-Vidal *et al.* 2001). Here, p_0 is given by $(1.8,\ 5,\ -5,\ -22)^T$. For the system (14) the flight test is split into respectively four and six stages. This method leads to the following inputs (correponding respectively to four and six stages):

$$u_4(t) = \delta_{m0} + a_4 - 2a_4h(t - t_{2_4}) + 2a_4h(t - t_{3_4}), \tag{15}$$

$$u_6(t) = \delta_{m0} + a_6h(t - t_{1_6}) - 2a_6h(t - t_{3_6}) + 2a_6h(t - t_{4_6}) - a_6h(t - t_{5_6}), \tag{16}$$

Only two switching times (resp. four) are obtained with four (resp. six) stages instead of the three (resp. five) expected times.
In these equations, δ_{m0} is equal to -2.6 degrees, and:

$$\left\{\begin{array}{l} a_4 = 1.6^o, \\ t_{1_4} = 0.25s, \\ t_{2_4} = 0.5s, \\ t_{3_4} = 0.75s, \end{array}\right. \quad \left\{\begin{array}{l} a_6 = 1.6^o, \\ t_{1_6} = 0.1667s, \\ t_{2_6} = 0.3334s, \\ t_{3_6} = 0.5001s, \\ t_{4_6} = 0.6668s, \\ t_{5_6} = 0.8335s. \end{array}\right. \tag{17}$$

The output trajectories corresponding to (15) and (16) are given on the left part of Fig.1 and Fig.2.

The non-differentiable optimal input, obtained by the dynamic programming method, is approximated by a differential function (10). Therefore, it can be used as a starting point of a gradient method for solving problems with inequality constraints. After approximation of the non-optimal inputs (15) and (16), the software second step leads to:

$$\tilde{u}_4(t) = \delta_{m0} + \tilde{a}_{40} + \frac{\tilde{a}_{41} - \tilde{a}_{40}}{1 + e^{k(\tilde{t}_{1_4} - t)}} + \frac{\tilde{a}_{42} - \tilde{a}_{41}}{1 + e^{k(\tilde{t}_{2_4} - t)}} + \frac{\tilde{a}_{43} - \tilde{a}_{42}}{1 + e^{k(\tilde{t}_{3_4} - t)}}, \tag{18}$$

$$\tilde{u}_6(t) = \delta_{m0} + \frac{\tilde{a}_{61}}{1 + e^{k(\tilde{t}_{1_6} - t)}} + \frac{\tilde{a}_{62} - \tilde{a}_{61}}{1 + e^{k(\tilde{t}_{2_6} - t)}} + \frac{\tilde{a}_{63} - \tilde{a}_{62}}{1 + e^{k(\tilde{t}_{3_6} - t)}} + \frac{\tilde{a}_{64} - \tilde{a}_{63}}{1 + e^{k(\tilde{t}_{4_6} - t)}} + \frac{\tilde{a}_{65} - \tilde{a}_{64}}{1 + e^{k(\tilde{t}_{5_6} - t)}}, \tag{19}$$

where k is given and with:

$$\left\{\begin{array}{l} \tilde{a}_{4_0} = -0.5558^o, \\ \tilde{a}_{4_1} = 1.5298^o, \\ \tilde{a}_{4_2} = -0.8594^o, \\ \tilde{a}_{4_3} = 1.5699^o, \\ \tilde{t}_{1_4} = 0.2607s, \\ \tilde{t}_{2_4} = 0.4794s, \\ \tilde{t}_{3_4} = 0.8472s, \end{array}\right. \quad \left\{\begin{array}{l} \tilde{a}_{6_1} = 1.0829^o, \\ \tilde{a}_{6_2} = 0.2521^o, \\ \tilde{a}_{6_3} = -1.5928^o, \\ \tilde{a}_{6_4} = 1.3006^o, \\ \tilde{a}_{6_5} = -0.55^o, \\ \tilde{t}_{1_6} = 0.1500s, \\ \tilde{t}_{2_6} = 0.3390s, \\ \tilde{t}_{3_6} = 0.4656s, \\ \tilde{t}_{4_6} = 0.6812, \\ \tilde{t}_{5_6} = 0.8220s. \end{array}\right. \tag{20}$$

The corresponding output trajectories to (18) and (19) are given on the right part of Fig. 1 and Fig. 2.

Remark 4.1. If the starting point of the gradient method is any point in $\mathcal{E}$, the results are much less good than the previous ones.

The optimal input design produced by our software seems to be really suitable for aircraft parameter estimation as shown in the following table:

Input	trace(S)
Figure 1 (left)	0.0684
Figure 1 (right)	0.0452
Figure 2 (left)	0.0485
Figure 2 (right)	0.0286

Table 1. Input design results.

4.2 *Parameter estimation*

The following tables give the relative errors for each estimated parameter obtained by the procedures presented in Section 3:

Parameter	Procedure 1	Procedure 2
$C_{z\dot{\alpha}}$	0.6334	0.5179
C_{zq}	0.4401	0.3675
$C^{25}_{m\dot{\alpha}}$	0.1217	0.0385
C^{25}_{mq}	0.0230	0.0328

Table 2. Relative errors in parameter estimation by using an nonoptimal admissible input.

Parameter	Procedure 1	Procedure 2
$C_{z\dot{\alpha}}$	0.1072	0.0149
C_{zq}	0.0788	0.0605
$C^{25}_{m\dot{\alpha}}$	0.0648	0.0452
C^{25}_{mq}	0.0188	0.0083

Table 3. Relative errors in parameter estimation by using the optimal input.

The second column gives the relative errors when the system is excited respectively by an nonoptimal admissible input (Table 2) or an optimal input (Table 3) and the criterion (11) minimized by using the covariance matrix inverse. The third column denotes the relative errors when the system is excited by an nonoptimal admissible input (Table 2) or an optimal input (Table 3) and the identification method based on linear matrix inequalities.

The results presented in Fig. 3, Fig. 4, Fig. 5 and Fig. 6 show a good trajectory reconstruction when the inputs are specified and the weighting is calculated via the linear matrix inequalities. They pointed out the efficiency of the method. In these reconstructions, we note $\hat{p}$ the obtained estimated parameters by the second procedure and by using an optimal input.

5. CONCLUSION

In this contribution, a complete procedure for estimating aircraft parameters is pointed out. Indeed, successive steps leading to optimal inputs are described. Thanks to this procedure, most accurate estimates of parameters are obtained which implies good trajectory reconstructions.

A least-square criterion improved by introducing the linear matrix inequalities seems to be especially interesting in the case of specified inputs.

Only the aircraft longitudinal motion has been considered, but this work is a part of a more important analysis which takes account of aircraft lateral motions. Thus the successive steps, described above, could be applied to the complete aircraft model.

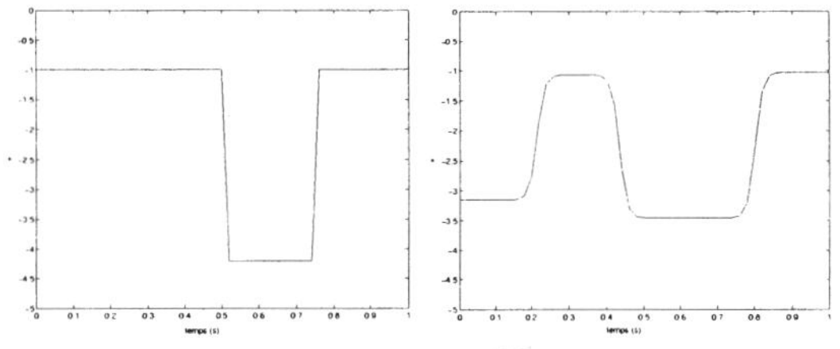

Fig. 1. Step 1 obtained input (left) / Step 2 optimal input (right) (four stages).

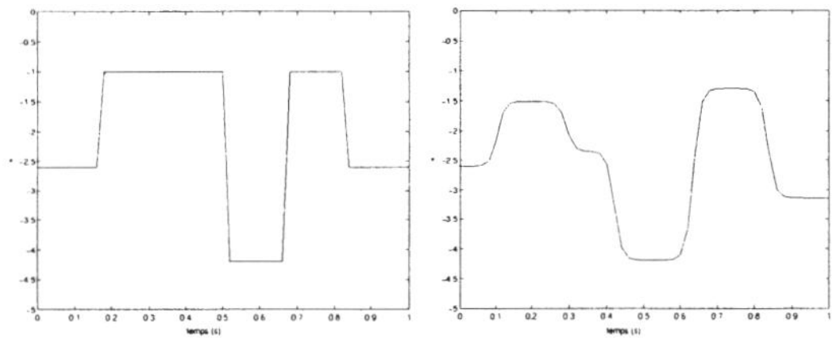

Fig. 2. Step 1 obtained input (left) / Step 2 regular differentiable optimal input (right) (six stages).

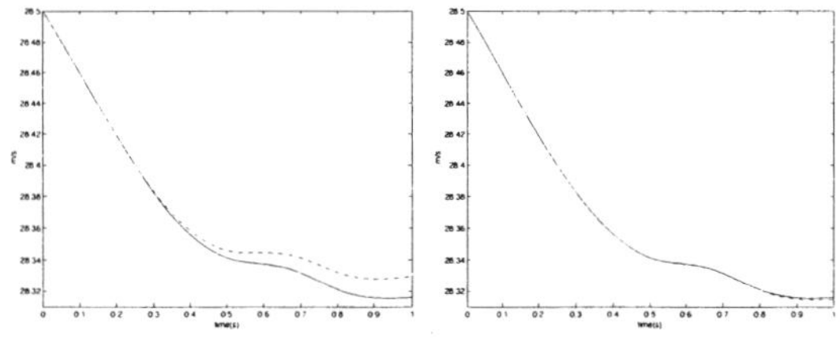

Fig. 3. Speed reconstruction with p_0 (left) / with $\hat{p}$ (right).

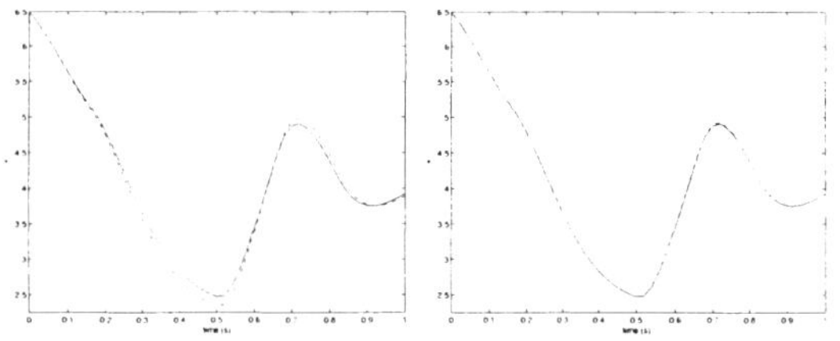

Fig. 4. Angle of attack reconstruction with p_0 (left) / with $\hat{p}$ (right).

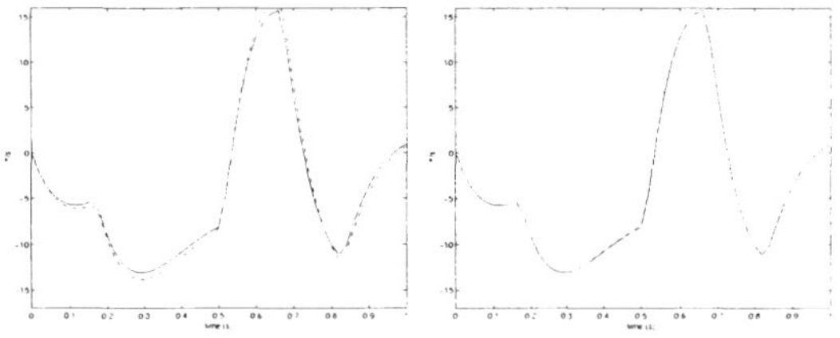

Fig. 5. Pitch rate reconstruction with p_0 (left) / with $\hat{p}$ (right).

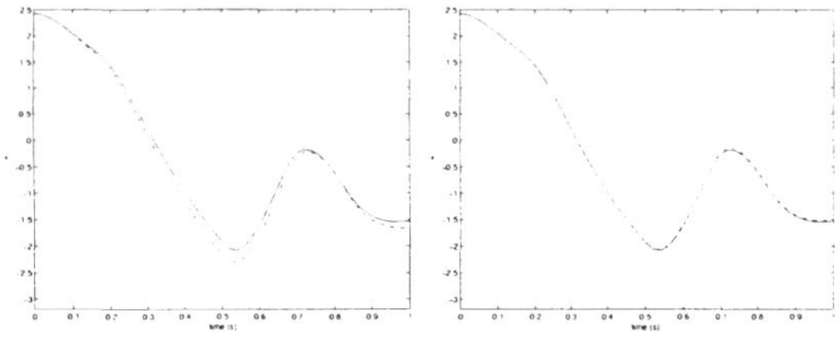

Fig. 6. Pitch angle reconstruction with p_0 (left) / with $\hat{p}$ (right).

6. REFERENCES

Chen, R.T (1975). Input design for aircraft parameter identification : using time optimal control formulation. *Advisory Group for Aerospace Research and Development.*

Denis-Vidal, L., C. Jauberthie, G. Joly-Blanchard and P. Coton (2001). Aircraft parameter estimation : successive steps. In: *Proc.5th IFAC NOLCOS.* St Petersburg.

Morelli, E.A. (1993). Practical input optimization for aircraft parameter estimation experiments. *Technical report CR 191462.*

Morelli, E.A. (1999). Flight test of optimal inputs and comparison with conventional inputs. *Journal of Aircraft.*

Morelli, E.A. and V. Klein (1990). Optimal input design for aircraft parameter estimation using dynamic programming principle. *Technical report 90 2081 – CP, AIAA.*

Ouarit, M, J.P. Yvon and J. Henry (2001). Optimal weighting design for distributed parameter system estimation. *Optimal Control Application and Methods* **22**, 37–49.

Walter, E. and L. Pronzato (1994). *Identification de modèles paramétriques à partir de données expérimentales.* MASSON.

Wanner, J.C. (1983). Dynamique du vol et pilotage des avions. *Office National d'Etudes et de Recherches Aérospatiales.*

IFAC Publications
www.elsevier.com/locate/ifac

THE PERFORMANCE OF MULTILEVEL PERTURBATION SIGNALS FOR NONLINEAR SYSTEM IDENTIFICATION

H. A. Barker[1], A. H. Tan[2] and K. R. Godfrey[3]

1: *School of Engineering, University of Wales Swansea, Swansea, SA2 8PP, U.K.*
2: *Faculty of Engineering, Multimedia University, 63100 Cyberjaya, Malaysia.*
3: *School of Engineering, University of Warwick, Coventry, CV4 7AL, U.K.*

Abstract: A method for determining the optimal levels of multilevel perturbation signals for nonlinear system identification is described. A performance index for the optimized signals, directly related to the identification accuracy, is based on the condition number of a Vandermonde submatrix of the perturbation signal levels vector. The signal levels are optimized under the constraint that the signal has appropriate harmonic properties for system identification. The performance of all pseudo-random signals with 3, 5 or 7 levels, generated in Galois fields from GF(3) to GF(13), when used for identifying nonlinearities of order up to 6, is determined and compared.

Keywords: Hammerstein models; Maximum-length sequences; Multilevel signals; Nonlinear systems; Perturbation signals; Pseudo-random signals; System identification.

1. INTRODUCTION

For the identification of a nonlinear system, the type of perturbation signal required depends on the system structure. When the system is modelled by a block-oriented structure comprising interconnected dynamic linear and static nonlinear blocks, the optimal type of perturbation signal may be defined. If the signal is applied directly to a dynamic linear block, as in the case of a Wiener model where linear dynamics precede a nonlinearity (Billings and Fakhouri, 1977), then the optimal perturbation signal is binary (Barker *et al.*, 2001). If, however, the perturbation signal is applied directly to a static nonlinear block, as in the case of a Hammerstein model where a static nonlinearity precedes linear dynamics (Narendra and Gallman, 1966), then the optimal signal is multilevel (Barker *et al.*, 2001).

In a recent paper (Barker *et al.*, 2003a), it was shown that the optimal levels of a multilevel perturbation signal applied directly to a static nonlinear block could be determined by minimising the condition number (Hill, 1988) of a Vandermonde submatrix of the perturbation signal levels vector. The results obtained were virtually identical to those obtained by a more complex method, described by Nowak and Van Veen (1994), for determining the optimal levels of a perturbation signal applied to a Volterra series model of a nonlinear system. This demonstrates both the accuracy and generality of the new method.

Unfortunately, the optimal perturbation signal obtained by this method does not necessarily have correlation or harmonic properties suitable for nonlinear system identification. It is therefore necessary to adapt the method to obtain the optimal perturbation signal levels vector for which the perturbation signal has such properties, and to do this appropriate signal levels vectors must first be determined. A method for determining these vectors, for pseudo-random perturbation signals that are generated from maximum-length sequences in Galois fields by converting each field element into a signal level (Barker, 1993; Barker, 2001), is described in a recent paper (Barker *et al.*, 2003b). This paper is therefore concerned with optimising these signal levels vectors, for which the corresponding signals have uniform spectra in which harmonics that are multiples of either 2, or of 2 and 3, are suppressed. Such properties are highly desirable for nonlinear system identification, and have been exploited directly in applications ranging from a wafer reactor (Braun *et al.*, 1999) to a neutralization process (Lara and Milani, 2003) and indirectly by embedding in general purpose identification software (Braun *et al.*, 2001). The new design method not only ensures that a system can be uniformly and persistently excited over a specified bandwidth, with interaction between nonlinearities either eliminated or reduced, but also that the levels of the perturbation signal used can be optimised to give the most accurate identification possible. Results are given for signals with 3, 5 and 7 levels generated in Galois fields GF(3), GF(5), GF(7), GF(9), GF(11) and GF(13).

2. SIGNAL LEVEL OPTIMISATION

In a multilevel perturbation signal u, the signal takes the set of q levels $u_1, u_2, \ldots, u_q$. As shown previously (Barker *et al.*, 2003a), the optimal levels may be obtained from the case in which each signal level occurs only once during a measurement period. When such a signal is the input of a static nonlinearity of order r the output y is

$$y = a_0 + a_1 u + a_2 u^2 + \ldots + a_r u^r \qquad (1)$$

and the outputs corresponding to each input level are given by the set of equations

$$\begin{aligned} y_1 &= a_0 + a_1 u_1 + a_2 u_1^2 + \ldots + a_r u_1^r \\ y_2 &= a_0 + a_1 u_2 + a_2 u_2^2 + \ldots + a_r u_2^r \\ y_3 &= a_0 + a_1 u_3 + a_2 u_3^2 + \ldots + a_r u_3^r \\ &\vdots \\ y_q &= a_0 + a_1 u_q + a_2 u_q^2 + \ldots + a_r u_q^r \end{aligned} \qquad (2)$$

If $r \geq q$ then this set of equations is underdetermined and cannot be solved, but if $r < q$ then solutions may be obtained. Eq. 2 may be expressed in matrix form by defining an input signal levels vector **u**, an output measurements vector **y** and a nonlinear coefficients vector **a** as

$$\begin{aligned} \mathbf{u} &= [\, u_1 \;\; u_2 \;\; u_3 \;\; .\,. \;\; u_q \,] \\ \mathbf{y} &= [\, y_1 \;\; y_2 \;\; y_3 \;\; .\,. \;\; y_q \,] \\ \mathbf{a} &= [\, a_0 \;\; a_1 \;\; a_2 \;\; .\,. \;\; a_r \,] \end{aligned} \qquad (3)$$

Eq. 2 may then be written as

$$\mathbf{y} = \mathbf{a}\mathbf{V}_{\text{sub}} \qquad (4)$$

where $\mathbf{V}_{\text{sub}}$ is a submatrix of the Vandermonde matrix **V** of the input signal levels vector **u**, obtained by deleting the last $q - r - 1$ rows of **V** to give

$$\mathbf{V}_{\text{sub}} = \begin{bmatrix} 1 & 1 & 1 & .\,. & 1 \\ u_1 & u_2 & u_3 & .\,. & u_q \\ u_1^2 & u_2^2 & u_3^2 & .\,. & u_q^2 \\ : & : & : & :\,: & : \\ u_1^r & u_2^r & u_3^r & .\,. & u_q^r \end{bmatrix} \qquad (5)$$

If $r < q - 1$, then $\mathbf{V}_{\text{sub}}$ is not square and Eq. 5 is solved by using the pseudo-inverse pinv($\mathbf{V}_{\text{sub}}$), while if $r = q - 1$, then $\mathbf{V}_{\text{sub}}$ is the Vandermonde matrix **V** and Eq. 5 is solved by using the inverse inv(**V**). It is convenient to take the second case as a special case of the first, and to use $\mathbf{V}_{\text{sub}}$ throughout, even when $\mathbf{V}_{\text{sub}} = \mathbf{V}$. Eq. 5 may then be solved if at least $r + 1$ of the signal levels in **u** are all different, that is if not more than $q - r - 1$ of the levels occur more than once. If Eq. 5 is to be solved accurately, however, it is also necessary that $\mathbf{V}_{\text{sub}}$ is well conditioned. If it is not, then small errors in either the input signal levels **u** or the output measurements **y** may result in relatively large errors in the nonlinear coefficients **a** (Hill, 1988). In this application, input errors commonly result from actuator errors, roundoff or noise, while output errors commonly result from transducer errors, roundoff or noise.

A measure of the conditioning of $\mathbf{V}_{\text{sub}}$ is its norm-2 condition number cond($\mathbf{V}_{\text{sub}}$) (Hill, 1988). The minimum value of cond($\mathbf{V}_{\text{sub}}$) is 1, and up to $\log_{10}$(cond($\mathbf{V}_{\text{sub}}$)) digits of accuracy may be lost when its value is greater than this. It is therefore convenient to define a performance index for this application using a logarithmic scale. A suitable index, with an upper value of 100% for the best performance and a lower value of 0%, is the Levels Index for Perturbation Signals (LIPS), defined as

$$\begin{aligned} &\text{LIPS} = (100 - 20\log_{10}(\text{cond}(\mathbf{V}_{\text{sub}})))\% \\ &\qquad\qquad\qquad\qquad \text{if } \text{cond}(\mathbf{V}_{\text{sub}}) \leq 100000 \\ &\text{LIPS} = 0\% \qquad\quad\; \text{if } \text{cond}(\mathbf{V}_{\text{sub}}) > 100000 \end{aligned} \qquad (6)$$

LIPS should be as close as possible to 100%, and ideally not less than 80%, so that at most one digit of accuracy may be lost. There can be little confidence in any identification results obtained when LIPS is less than 60%, as more than two digits of accuracy may be lost. As cond($\mathbf{V}_{\text{sub}}$), and therefore LIPS, depend only on the input signal levels **u**, the primary problem is therefore to determine the optimal signal levels vector **u** that minimises cond($\mathbf{V}_{\text{sub}}$), and therefore maximises LIPS.

In Barker *et al.*, (2003a), it was shown that this could be accomplished by using the MATLAB function *fminsearch* to optimise each signal level in **u** to obtain the optimal signal levels $\mathbf{u}_{\text{opt}}$ and the corresponding optimal condition number cond($\mathbf{V}_{\text{sub}}$)$_{\text{opt}}$. The optimal signal levels $\mathbf{u}_{\text{opt}}$ obtained by this free optimisation of the signal levels were always antisymmetric, and the values obtained for

$\text{cond}(\mathbf{V}_{sub})_{opt}$ showed that little confidence could be placed on identification results obtained with signals of more than 7 levels. Both $\mathbf{u}_{opt}$ and $\text{cond}(\mathbf{V}_{sub})_{opt}$ were tabulated for signals with 3, 5 or 7 levels, generated in Galois fields GF(3), GF(5) and GF(9), when used to identify nonlinearities with orders at least one less than the number of levels. These tabulations define the global minimum of $\text{cond}(\mathbf{V}_{sub})_{opt}$, and hence the global maximum of LIPS, that can be obtained for signals generated in these fields by free optimisation of the signal levels. The optimal signal levels $\mathbf{u}_{opt}$ do not, however, necessarily define signals with desirable properties for system identification.

3. SIGNAL LEVELS FOR REQUIRED HARMONIC PROPERTIES

In order to ensure that only input signal levels **u** with desirable properties for nonlinear system identification are involved in the maximisation of LIPS, it is necessary to constrain the values of the signal levels during optimisation. This can be easily accomplished if each signal level is a known function of a few variables, because then it is only necessary to optimise those variables, rather than the individual signal levels.

A method for obtaining the necessary functions in the appropriate symbolic form has only recently been developed (Barker *et al.*, 2003b). It uses the fact that the harmonic and correlation properties of a pseudo-random signal, generated from a maximum-length sequence in the Galois field GF(q) by converting the q field elements into the signal levels **u**, depend only on corresponding properties of **u** (Barker, 1993). If each signal level $\text{u}(i)$ in **u** is converted from a field element i, for $i = 0, 1, \ldots, q-1$, then **u** is an ordered vector of signal levels. The relevant properties involve only the signal levels converted from the nonzero field elements. They may usefully be developed when q is odd and the signal level u(0), converted from the zero field element 0, is itself zero. Removing u(0) from **u** leaves an ordered subvector $\mathbf{u}_{sub}$ of the $q-1$ signal levels converted from the nonzero field elements, and it is necessary to change the order of the signal levels in $\mathbf{u}_{sub}$ to obtain an ordered signal levels vector **m** with the required properties, using

$$\text{m}(i) = \text{u}(g^{i-1}) \qquad \text{for } i = 1, 2, \ldots, q-1 \qquad (7)$$

where g is a primitive element of the field (Barker, 1993). As the powers of a primitive element generate all the nonzero field elements, all the signal levels in $\mathbf{u}_{sub}$ appear in **m**, but in a different order, so Eq. 7 allows **m** to be obtained from $\mathbf{u}_{sub}$, and *vice-versa*.

The conditions which ensure that the pseudo-random signal generated in GF(q) has the desirable properties described in Section 1 were given in Barker *et al.*, (2003b). Harmonics that are multiples of 2 are suppressed if **m** is antisymmetric, that is

$$\text{m}(i) + \text{m}(i + (q-1)/2) = 0 \qquad \text{for } i = 1, 2, \ldots, (q-1)/2 \qquad (8)$$

This condition can always satisfied by $\mathbf{u}_{opt}$, in which the signal levels are antisymmetric. The remaining harmonics are then uniform if the periodic autocorrelation function of **m** and its periodic extension are primitive, that is

$$\phi_{\text{mm}}(j) = \frac{1}{q-1}\sum_{i=1}^{q-1} \text{m}(i)\text{m}(i+j) = 0 \qquad \text{for } j \neq r(q-1)/2 \qquad (9)$$

This condition is not necessarily satisfied by $\mathbf{u}_{opt}$.

Harmonics that are multiples of 2 and 3 are suppressed if Eq. 8 is satisfied and the equation

$$\text{m}(i) + \text{m}(i + (q-1)/3) + \text{m}(i + 2(q-1)/3) = 0 \qquad \text{for } i = 1, 2, \ldots, (q-1)/3 \qquad (10)$$

is also satisfied. In this case $q - 1$ must be a multiple of both 2 and 3, so the lower Galois fields involved are GF(7) and GF(13). The condition that the remaining harmonics are uniform then becomes

$$\begin{aligned} \phi_{\text{mm}}(j) &= 0 \qquad \text{if } j \neq r(q-1)/6 \\ \phi_{\text{mm}}((q-1)/6) &= \phi_{\text{mm}}(0)/2 \end{aligned} \qquad (11)$$

Symbolic tabulations of signal levels **m** that satisfy the conditions in Eqs. 8 and 9 are given in Barker *et al.*, (2003b) for pseudo-random signals with 3, 5 or 7 levels generated in Galois fields from GF(3) to GF(13). Similar tabulations of signal levels **m** that satisfy the conditions in Eqs. 8, 10 and 11 were also given for signals generated in GF(7) and GF(13).

4. SIGNAL LEVELS VECTOR OPTIMISATION

To optimise the levels of pseudo-random signals with the properties in Section 3, the signal levels vectors tabulated in Barker *et al.*, (2003b) were used. From Eq. 8, each signal levels vector **m** with $q-1$ signal levels may be expressed as

$$\mathbf{m} = [\mathbf{m}_{sub} \quad -\mathbf{m}_{sub}] \qquad (12)$$

where $\mathbf{m}_{sub}$ is the subvector of $(q-1)/2$ signal levels m(1), m(2), . . . , m($(q-1)/2$). The greatest number of independent signal levels involved in the optimisation is therefore $(q-1)/2$. This number is reduced by at least 1 by the constraints in Eqs. 9, 10 and 11. It may be reduced further by the constraint that the number of signal levels is 3, 5 or 7, for which the numbers of independent signal levels are at most 1, 2 or 3 respectively. All remaining signal levels in $\mathbf{m}_{sub}$ are functions of the independent signal levels. For example, when $q = 13$, two typical subvectors $\mathbf{m}_{sub}$, with one and two independent signal levels respectively, are

$$[m(1)\ \ 3m(1)\ \ 5m(1)\ \ -3m(1)\ \ m(1)\ \ \ \ 0\ \ \ ]$$

and

$$[m(1)\ \ m(1)\ \ m(3)\ \ -m(1)\ \ m(1)\ \ m(3)-m(1)].$$

A signal levels vector $\mathbf{u}$ is therefore formed as

$$\mathbf{u} = [0 \quad \mathbf{m}] = [0 \quad \mathbf{m}_{sub} \quad -\mathbf{m}_{sub}] \qquad (13)$$

and the Vandermonde submatrix $\mathbf{V}_{sub}$ of the signal levels vector $\mathbf{u}$ is obtained as described in Section 2. The MATLAB function *fminsearch* is then used to optimise the signal levels in $\mathbf{m}_{sub}$, and to obtain the corresponding optimal condition number cond($\mathbf{V}_{sub}$)$_{opt}$ and LIPS. All the signal level vectors tabulated in Barker *et al.*, (2003b) were optimised by this method. The optimal subvectors $\mathbf{m}_{sub}$ with greatest values of LIPS, for signals with a uniform spectrum in which harmonics that are multiples of 2 are suppressed, are shown in Table 1. Those for signals with a uniform spectrum in which harmonics that are multiples of 2 and 3 are suppressed are shown in Table 2. For linear system identification, the greatest values of LIPS are always 100%, so these results are not shown.

For generating the pseudo-random signals defined by the tabulations, the conversions of all the nonzero elements of the Galois fields are required. These are defined through Eq. 13, and obtained through Eq. 7 as shown in Section 3. In GF(7), for example, using the primitive element 3 gives the ordered vector $\mathbf{u}_{opt}$ as

$$\mathbf{u}_{opt} = [0\ \ m(1)\ \ m(3)\ \ m(2)\ \ m(5)\ \ m(6)\ \ m(4)] \quad (14)$$

In Barker *et al.*, (2003a), the individual levels of pseudo-random signals were optimised to obtain the global maxima of LIPS for signals with 3, 5 or 7 levels generated in GF(3), GF(5) and GF(7). Applying signal levels vector optimisation to signals generated in the same fields, the LIPS values obtained are only a few percent less than the global maxima, while for signals generated in GF(9), GF(11) and GF(13), the LIPS values may be greater.

A detailed examination of the results in Tables 1 and 2 shows that LIPS values are not particularly dependent on whether the suppressed harmonics are multiples of 2, or multiples of 2 and 3, or on the number of signal levels, or on the Galois field in which the pseudo-random signals are generated. They are, however, dependent on the order of the nonlinearity to be identified. Fig. 1 shows the average of the LIPS values for each order of nonlinearity from 1 to 6, and the negative correlation shows that there can be little confidence in any identification results obtained for nonlinearities with orders greater than 5.

5. CONCLUSIONS

The method described in Sections 2, 3 and 4 allows the optimal levels of multilevel perturbation signals with desirable properties for system identification to be determined, and the performance of these signals in nonlinear system identification to be compared. The Levels Index for Perturbation Signals (LIPS), based on the condition number of a Vandermonde submatrix of the perturbation signal levels vector, is ideal for these comparisons. The signal levels vector can be optimised under constraints which ensure that the resulting signal has a uniform spectrum in which harmonics that are multiples of either 2, or of 2 and 3, are suppressed. Using this method, the performance of every pseudo-random signal with 3, 5 or 7 levels generated in Galois fields GF(3), GF(5), GF(7), GF(9), GF(11) and GF(13) has been determined and compared, and the levels of those signals with the best performance have been tabulated.

The results obtained show that good performance can be obtained by appropriate choice of the signal levels. Constraints on the signal levels vector result in little loss of performance compared with free optimisation of the levels, and any reductions in performance can be recovered by generating the signals in higher-order fields. The factor that limits performance is the order of the nonlinearity to be identified, rather than the number of signal levels or the Galois field in which the signal is generated. The LIPS values obtained show that little confidence can be placed in results obtained in the identification of nonlinearities with orders greater than 5.

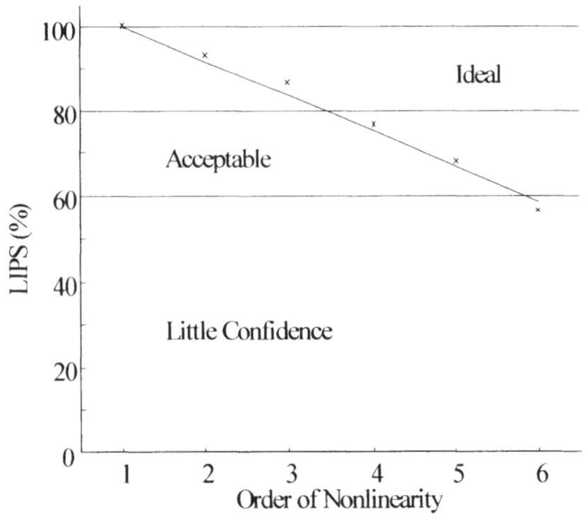

Fig. 1. Dependence of LIPS on the order of the nonlinearity to be identified.

Table 1. Optimal LIPS and signal levels $\mathbf{m}_{sub}$ for pseudo-random signals with a uniform spectrum in which harmonics that are multiples of 2 are suppressed.

Number of Signal Levels	Order of Non-linearity	Field GF( )	LIPS %	Optimal Signal Levels					
				m(1)	m(2)	m(3)	m(4)	m(5)	m(6)
3	2	3	90.0	1.107					
		5	93.5	1.257	0				
		7	94.8	1.368	0	0			
		9	95.1	1.456	0	0	0		
		11	94.8	1.531	0	0	0	0	
		13	94.6	1.343	0	0	1.343	0	0
5	2	5		3-Level Solution Only					
		7	89.5	1.141	-1.141	-0.571			
		9	95.2	1.455	0	-0.364	0		
		11	87.8	0.883	-0.883	0.883	1.324	0.883	
		13	95.2	1.551	0	0	0.919	0	0
5	3	5	84.6	1.147	-0.599				
		7	82.0	1.085	-1.085	-0.543			
		9	84.2	1.252	0	-0.662	0		
		11	85.9	0.840	-0.840	0.840	1.260	0.840	
		13	89.1	0.671	0.671	-0.671	1.342	0.671	0.671
5	4	5	76.8	1.183	-0.857				
		7	70.9	1.156	-1.156	-0.578			
		9	78.7	1.252	0	-0.856	0		
		11	76.5	0.825	-0.825	0.825	1.238	0.825	
		13	77.7	0.752	0.752	0	0	1.216	-1.216
7	2	7		3-Level Solution Only					
		9		5-Level Solution Only					
		11	93.7	0.548	-0.548	0	0.748	1.496	
		13	95.2	0.422	0.422	-0.422	0.117	0.422	-1.589
7	3	7	86.4	1.210	-0.658	-0.426			
		9		5-Level Solution Only					
		11	87.4	0.477	-0.477	0	0.651	1.302	
		13	89.1	0.625	0.625	-0.718	-0.625	0.625	-1.344
7	4	7	76.6	1.214	-0.873	-0.508			
		9	75.5	1.187	1.187	-0.183	0.871		
		11	79.1	0.784	0.784	0.485	0	1.269	
		13	80.1	0.434	0	0.868	1.293	-0.868	0
7	5	7	69.2	1.118	-0.896	-0.497			
		9	64.2	1.091	1.091	-0.215	0.732		
		11	66.7	0.716	0.716	0.442	0	1.158	
		13	69.2	1.173	0.599	0.918	-0.599	0	0.599
7	6	7	61.9	1.152	-1.011	-0.538			
		9	48.7	1.167	1.167	-0.343	0.637		
		11	57.6	0.744	0.744	0.460	0	1.203	
		13	56.1	0.762	0.762	-0.457	-0.762	0.762	-1.220

Table 2. Optimal LIPS and signal levels $\mathbf{m}_{sub}$ for pseudo-random signals with a uniform spectrum in which harmonics that are multiples of 2 and 3 are suppressed.

Number of Signal Levels	Order of Non-linearity	Field GF( )	LIPS %	Optimal Signal Levels					
				m(1)	m(2)	m(3)	m(4)	m(5)	m(6)
3	2	7	91.4	1.150	1.150	0			
		13	94.6	1.343	0	0	0	-1.343	0
5	2	7	91.4	0.664	1.328	0.664			
		13	94.6	0.775	0	-0.775	0	-1.550	0
5	3	7	86.8	0.606	1.213	0.606			
		13	86.3	0.666	0	-0.666	0	-1.332	0
5	4	7	74.3	0.606	1.212	0.606			
		13	78.7	0.644	0	-0.644	0	-1.289	0
7	2	7	91.4	1.306	0.443	-0.863			
		13	94.6	1.096	0	-0.402	0	-1.498	0
7	3	7		5-Level Solution Only					
		13	88.6	0.670	0.670	-0.670	0.335	-1.340	-0.335
7	4	7	77.2	1.223	0.321	-0.902			
		13	78.9	0.839	0	-0.451	0	-1.290	0
7	5	7	69.2	1.106	0.289	-0.818			
		13	69.2	1.107	0.289	0.818	1.107	-0.289	0.818
7	6	7	56.5	1.173	0.424	-0.749			
		13	58.5	0.793	0	-0.406	0	-1.199	0

REFERENCES

Barker, H. A. (1993). Design of multi-level pseudo-random signals for system identification, Chapter 11 of *Perturbation Signals for System Identification*, (K. R. Godfrey, Ed.), Prentice-Hall, Hemel Hempstead.

Barker, H. A. (2001). GALOIS - a program for generating pseudo-random perturbation signals, *Proc. 12th IFAC Symposium on System Identification* (*SYSID 2000*), (R. Smith, Ed.), Elsevier Science, Oxford, 505-508.

Barker, H. A., K. R. Godfrey and A. J. Tucker (2001). Nonlinear system identification with multilevel perturbation signals, *Proc. 12th IFAC Symposium on System Identification* (*SYSID 2000*), (R. Smith, Ed.), Elsevier Science, Oxford, 1175-1178.

Barker, H. A., A. H. Tan and K. R. Godfrey (2003a). Criteria for determining the optimal levels of multilevel perturbation signals for nonlinear system identification, *American Control Conference 2003*, Denver, Colorado, USA, *Paper ACC03-IEEE0105.*

Barker, H. A., A. H. Tan and K. R. Godfrey (2003b). The design of multilevel perturbation signals with harmonic properties suitable for nonlinear system identification, *Paper submitted to Proceedings IEE.*

Billings, S. A. and S. Y. Fakhouri (1977). Identification of nonlinear systems using the Wiener model, *Electron. Lett.*, **13**, 502-504.

Braun, M. W., D. E. Rivera, A. Stenman, W. Foslien, and C. Hrenya (1999). Multi-level pseudo-random signal design and 'model-on-demand' estimation applied to nonlinear identification of a RTP wafer reactor, *Proc. American Control Conference 1999*, San Diego, California, USA, 1573-1577.

Braun, M. W., D. E. Rivera and A. Stenman (2001). A 'model-on-demand' identification methodology for non-linear process systems, *Int. J. Control*, **74**, (18), 1708-1717.

Hill, D. R. (1988). *Experiments In Computational Matrix Algebra*, Random House, New York.

Lara, J. M. V. and B. E. A. Milani (2003). Identification of a neutralisation process using multi-level pseudo-random signals, *American Control Conference 2003*, Denver, Colorado, USA, *Paper ACC03-IEEE0416.*

Narendra, K. S. and P. G. Gallman (1966). An iterative method for the identification of nonlinear systems using the Hammerstein model, *IEEE Trans. Automat. Contr.*, **11**, 546-550.

Nowak, R. D. and B. D. Van Veen (1994). Random and pseudorandom inputs for Volterra filter identification, *IEEE Trans. Signal Processing*, **42**, 2124-2135.

IFAC Publications
www.elsevier.com/locate/ifac

APPLYING SYSTEM IDENTIFICATION TO ASSESS THE VIBRO-ACOUSTIC BEHAVIOUR OF AIRPLANES

Bart Peeters[(1)], Romualdo Ruotolo[(2)], Antonio Vecchio[(1)], Herman Van der Auweraer[(1)]

(1) LMS International, Interleuvenlaan 68, B-3001 Heverlee, Belgium.
(2) Dept. of Aerospace Engineering, Politecnico di Torino, Corso Duca degli Abruzzi 24, I-10129 Torino, Italy.

Effective aircraft interior noise reduction measures require an in-depth understanding of the operational noise and vibration fields as well as the intrinsic system characteristics. The former requires detailed mapping of the in-flight sound and vibration responses, whereas the latter requires the proper modelling of the vibro-acoustic system behaviour. The latter is discussed in this paper where frequency-domain system identification is applied to acoustical transfer functions, measured in a fully-trimmed aircraft cavity.

Keywords: aircraft, transfer function matrices, smoothing, modal, parameter estimation, identification.

1. INTRODUCTION

When performing a study of aircraft interior acoustics, it is important to relate the observed in-flight vibration and acoustics response to the intrinsic system behaviour of fuselage and cabin cavity. The classical approach to experimentally model the vibro-acoustic system behaviour of a mechanical structure, consists of the identification of the modal system model parameters. The system behaviour is divided into a set of individual resonance phenomena, each characterised by a resonance frequency, damping ratio, and mode shape. The experimental data set to derive this model from consists of a set of transfer functions between a limited set of reference (i.e. input) degrees of freedom and all response (i.e. output) degrees of freedom. For structural (vibration) responses, this technique is widely spread. Its application to aircraft dynamics is however mainly limited to the low frequency range of global wing/fuselage/tail modes, which are of importance for validating and improving analytical models, flutter studies and structural integrity.

For acoustic response, the situation is more complex. It is not straightforward that the same model formulation also holds for acoustic variables. Also, for the case of systems like trimmed aircraft cabins, the damping of the modes is high, resulting in highly overlapping modes with complex mode shapes. The high damping is explained by the friction at the interface between frames and trim panels, including the thermal insulation. Furthermore, measured frequency response functions usually show even at resonance, a propagating acoustic field instead of a standing wave pattern. Again, the trim and the resulting non-uniform damping properties of the cavity walls are the probable cause to these phenomena.

The conclusion from all this is that an experimental modal analysis of a trimmed aircraft in the acoustically relevant frequency regions is far from trivial. In section 2, the technique of experimental acoustic modal analysis is introduced. In section 3, a frequency-domain system identification technique that is able to cope with high modal damping is briefly explained. In section 4, the discussed principles will be illustrated by means of test results obtained on an ATR42, a twin propeller aircraft.

2. EXPERIMENTAL ACOUSTIC MODAL ANALYSIS

An important consideration in the application of experimental modal analysis techniques for acoustic problems is the validity of the modal model formulation and the selection of proper input/output variables. In addition, equipment considerations need to be made to perform the required frequency response function measurements correctly.

2.1. Basic formulation

Consider a three-dimensional closed acoustic system with rigid or finite impedance but non-vibrating boundaries. The governing equation of this system, excited by a point monopole of volume velocity at $\vec{r}_0$, can be written in the form (Fahy, 1985):

$$\nabla^2 p(\vec{r},t) - \frac{1}{c^2}\ddot{p}(\vec{r},t) = -\rho\,\dot{q}\,\delta(\vec{r}-\vec{r}_0) \qquad (1)$$

where p is the sound pressure, which is a function of space $\vec{r}$ and time t; c is the speed of sound; ρ is the density of the medium; and $\dot{q}$ is the volume velocity.

Assuming now that a number of point monopoles of known volume velocity are placed in the cavity and the sound pressure across the volume is sampled at an appropriate number of points, it can be shown that the continuous wave equation can then be substituted by its discrete equivalent:

$$A\,\ddot{p} + B\,\dot{p} + C\,p = \dot{q} \qquad (2)$$

No direct physical meaning can be attributed to the matrices A, B, C, but the discrete governing equation (2) is equivalent to the discrete mechanical equations of motions, with A, B, C in the role of mass, damping, and stiffness matrices; p in the role of displacement; and $\dot{q}$ in the role of force. Taking the Laplace-transform and assuming zero initial conditions we get:

$$\left(s^2 A + s\,B + C\right) p(s) = s\,q \qquad (3)$$

As usual in structural dynamics, the inverse of the matrix term can be substituted by the frequency response matrix $H(s)$:

$$p(s) = H(s)\,s\,q \qquad (4)$$

The frequency response function matrix can in its turn be expressed as a partial fraction expansion of modal parameters:

$$H(s) = \sum_{i=1}^{n} \frac{v_i\,l_i}{s-\lambda_i} + \frac{v_i^*\,l_i^*}{s-\lambda_i^*} \qquad (5)$$

where n is the number of modes (occurring in complex conjugated pairs); v_i is a mode shape; l_i is a modal participation factor; λ_i is a complex pole.

Now it becomes obvious that the modal parameters of the system can be gained from frequency response measurements where the sound pressures across the volume are referenced to the volume velocities of the sources. Further considerations regarding the calculation of forced acoustic fields and the equivalence between structural and acoustic values can be found in Auguztinovicz (1993).

2.2. Equipment requirements

In principle, no correct experimental acoustic modal analysis can be conducted without using a well-controlled volume velocity source. A few experimental systems have been reported on in literature, out of which the converted acoustic driver method seems and actually has been found to be the most practicable.

Imagine an electrodynamic loudspeaker, which is provided with a closed, sealed housing behind the diaphragm. The most obvious realisation could be to use a horn driver. Unfortunately, these loudspeakers are generally designed for high frequency sound reproduction and sometimes cannot radiate sufficient acoustic power in the frequency range relevant for acoustic modal analysis applications. A good quality medium-range loudspeaker with closed housing or, in case of even lower frequencies, a closed box loudspeaker unit may be helpful.

An important element in this approach is the measurement of an appropriate reference signal. This is extensively discussed in Auguztinovicz (1993). If the analyst is interested in the eigenfrequencies and the mode shapes of the system only, and a correct modal model is of no importance, the volume velocity source can be substituted by a simple loudspeaker. Then the reference signal can be taken directly from the input clamps of the speaker. It should be remarked that the reference signal cannot be derived using a microphone in the close vicinity of the source. The sound pressure measured in any point of the volume is a response rather than an excitation signal.

3. GLOBAL SMOOTHING TECHNIQUE

The method was originally proposed by Dat and Meurzec (1972) to deal with single-input single-output systems and showed to be a very interesting way of analysing data from structures, which exhibit close modes, and/or a high level of damping. Recently this method was extended so that it could deal simultaneously with multiple inputs and multiple outputs (Ruotolo and Storer, 2001; Ruotolo, 2001). Also some attention was paid to the improvement of the numerical conditioning of the parameter estimation problem. The term "smoothing technique" seems to stem from aeronautical engineering. It expresses that the measured non-parametric frequency response functions are smoothed by fitting a parametric model to these.

There exist two variants of the method. In the first one, a common-denominator model is fitted to measured frequency response functions (Ruotolo and Storer, 2001). In a first step, the poles λ_i are estimated. The residues, i.e. $v_i\, l_i$ in (5), are estimated in a second, linear least squares step. From these residues, the mode shapes v_i and participation factors l_i are obtained as the rank-one approximations using a singular value decomposition. In the second variant, a right matrix fraction model is fitted to measured frequency response functions (Ruotolo, 2001). This formulation yields in a first step both the poles λ_i and the participation factors l_i. In a second step, the mode shapes v_i are obtained by solving (5) in a least squares sense.

Referring to the classification of frequency-domain parameter estimation methods as presented in Guillaume, *et al.* (1996) and Verboven (2002), the global smoothing technique has the following characteristics. The method operates in the Laplace domain (as opposed to the *z*-domain). The advantage is that the frequency axis has not to be mapped on the unit circle, which would result in discontinuity errors at the lowest and highest frequencies. The disadvantage is the numerical conditioning problem when solving the equations. This conditioning problem can be solved by using orthogonal polynomials, scaling the frequency axis, scaling the parameters, and/or dividing the frequency axis in different sub-bands.

The terms in the cost function are weighted by the inverse of the magnitude of the denominator polynomial. This both solves the problem of over-emphasizing the high frequency measurements in the cost function and allows minimizing the true fitting error between the mathematical model and experimental data, but results in an iterative procedure.

A final characteristic of the smoothing technique is that it can be considered as a deterministic method. The variances of the measurements are not used in the parameter estimation process as would be the case in stochastic methods (Guillaume, *et al.*, 1996).

4. APPLICATION TO THE ATR42 TWIN PROPELLER AIRCRAFT

Specific ground tests have been executed on the ATR42 aircraft (Fig. 1), to derive the intrinsic system information, which should render it possible to explain observed in-flight behaviour. In this case, simultaneous excitation was applied at four loudspeakers, two longitudinal and two lateral ones. Microphones at 20 positions captured the responses simultaneously. A total of 12 sections of the plane cavity were measured, resulting in 240 response locations. These are shown in Fig. 2.

In Van der Auweraer, *et al.* (1993), it was stated that experimental modelling in the acoustic frequency ranges is not straightforward due to the high modal density and the relatively high damping of many of the system modes. In Fig. 3, a stabilisation diagram is showed that is obtained by applying the classical LSCE modal parameter estimation method to the ATR42 acoustical data. The LSCE method and the stabilisation diagram concept are explained in another paper of these proceedings (Van der Auweraer, 2003). Fig. 3 shows that it is indeed difficult to extract modal parameters using a classical parameter estimation method.

Fig. 1. ATR42 twin propeller aircraft.

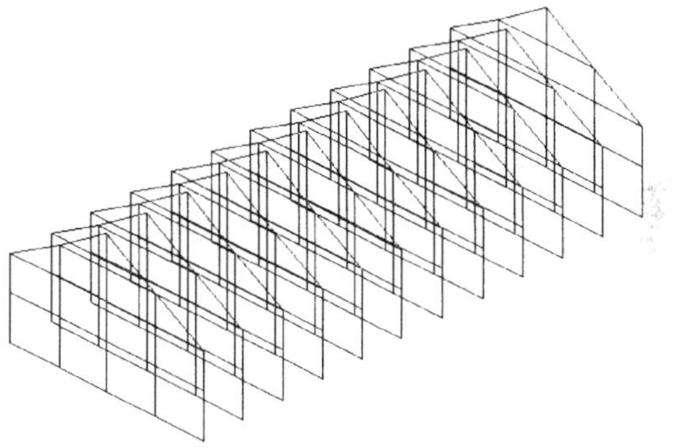

Fig. 2. Microphone locations inside the ATR42.

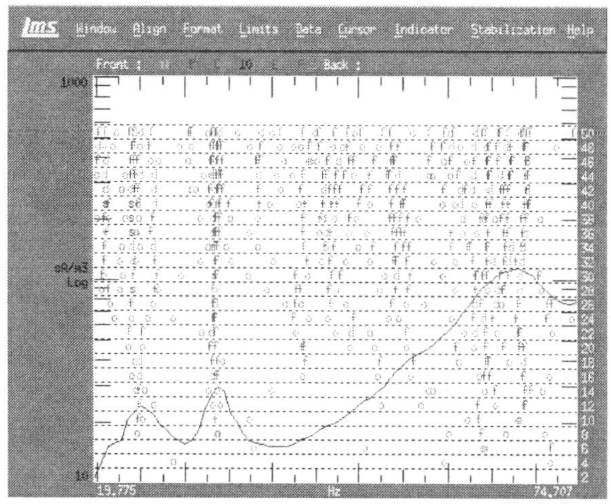

Fig. 3. LSCE Stabilisation diagram.

In this paper, the global smoothing technique, outlined in previous section and which is expected to be more suited for the type of system considered here, is applied. The common-denominator version was applied to the 960 measured frequency response functions simultaneously, aiming to estimate in a first step the poles of the acoustic cavity. The analysis was performed by subdividing the whole frequency range of interest, from 20 to 200 Hz, in sub-bands 40 to 50 Hz wide and applying the algorithm in each sub-band. Subsequently, the residues, i.e. $v_i l_i$ in (5), are estimated in a second, linear least squares step. From these residues, the mode shapes v_i and participation factors l_i are obtained as the rank-one approximations using a singular value decomposition.

Figs. 4–7 are showing four measured and the corresponding, from the identified poles synthesized frequency response functions relating the four inputs to a selected output. The correspondence highlights the good quality of the fitted mathematical model.

Typical acoustical mode shapes, i.e. v_i in (5), which were identified from the measurements are shown in Figs. 8–13. Each figure shows two snapshots of the mode. The snapshots differ by 180°. The corresponding resonance frequencies and damping ratios are represented in Table 1. As can be observed, the damping ratios are rather high (4–12 %).

Table 1. Identified acoustical poles, expressed as eigenfrequencies and damping ratios, of the ATR42.

Mode	Eigenfrequency [Hz]	Damping ratio [%]
1	32.9	8.5
2	61.5	8.8
3	67.3	4.4
4	78.3	7.0
5	89.0	5.3
6	107	7.0
7	149	9.3
8	167	12.3
9	198	9.6

5. CONCLUSIONS

This paper showed how system identification can assist in determining the aircraft interior acoustics. Experimental acoustic modal analysis was introduced as well as the smoothing technique which yields the modal parameters from measured frequency response functions and is specifically suited for highly damped systems. This technique was successfully applied to a ground acoustical test on the ATR42 aircraft. These results confirm earlier good experience with the smoothing technique applied to a trimmed car body, which is another highly damped system (Ruotolo and Storer, 2001).

ACKNOWLEDGEMENT

This research is conducted in the context of the EUREKA project 2419, FLITE. The financial support of the Flemish Institute for the Improvement of the Scientific and Technological Research in Industry (IWT) is gratefully acknowledged.

REFERENCES

Auguztinovicz, F. (1993). Acoustic modal analysis. In: *Proceedings of the 4th International Seminar on Applied Acoustics*. Department of Mechanical Engineering, Leuven, Belgium.

Dat, R. and J.L. Meurzec (1972). Exploitation par lissage mathematique des mesures d'admittance d'un systeme lineaire. *La Recherche Aerospatiale*, **4**, 209-215.

Fahy, F. (1985). *Sound and structural vibration. Radiation, transmission and response*. Academic Press, London, UK.

Guillaume, P., R. Pintelon, and J. Schoukens (1996). Parametric identification of multivariable systems in the frequency domain – a survey. In: *Proceedings of ISMA 21, the International Conference on Noise and Vibration Engineering*. Department of Mechanical Engineering, Leuven, Belgium.

Ruotolo, R. and D.M. Storer (2001). A global smoothing technique for FRF data fitting. *Journal of Sound and Vibration*, **239**, 41-56.

Ruotolo, R. (2001). A multiple-input multiple-output smoothing technique: theory and application to aircraft data. *Journal of Sound and Vibration*, **247**, 453-469.

Van der Auweraer, H., D. Otte, and F. Augusztinovicz (1993). Vibroacoustic analysis of trimmed aircraft through modal and principal field modelling. In: *Proceedings of the 15th AIAA Aeroacoustics Conference*. Long Beach, CA, USA.

Van der Auweraer, H. (2003). System identification for structural dynamics and vibroacoustics design engineering. In: *Proceedings of SYSID 2003, the 13th IFAC Symposium on System Identification*. Rotterdam, The Netherlands.

Verboven, P. (2002) *Frequency domain system identification for modal analysis*. PhD Thesis, Vrije Universiteit Brussel, Belgium.

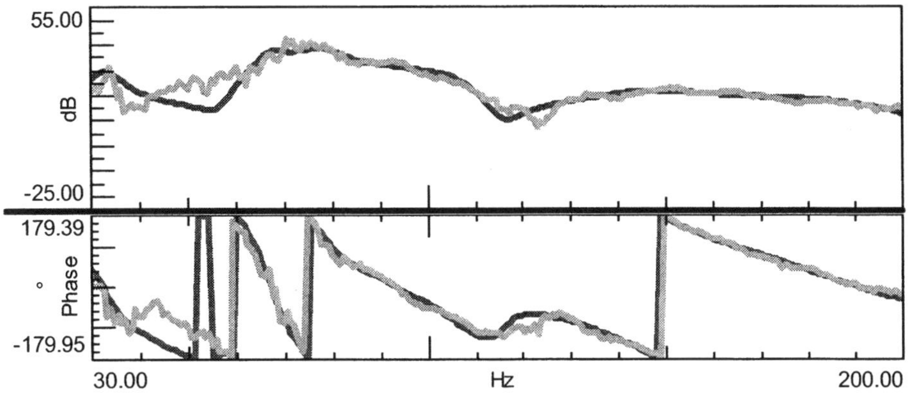

Fig. 4. Measured (grey) and fitted (black) FRF of first longitudinal loudspeaker to location 01:0116:+X.

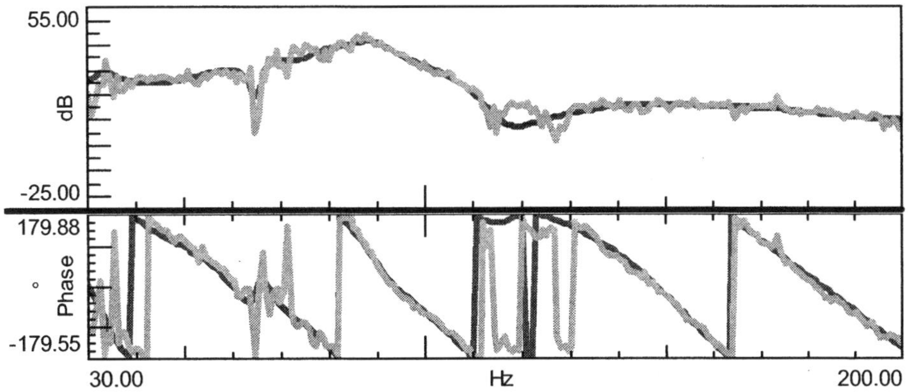

Fig. 5. Measured (grey) and fitted (black) FRF of second longitudinal loudspeaker to location 01:0116:+X.

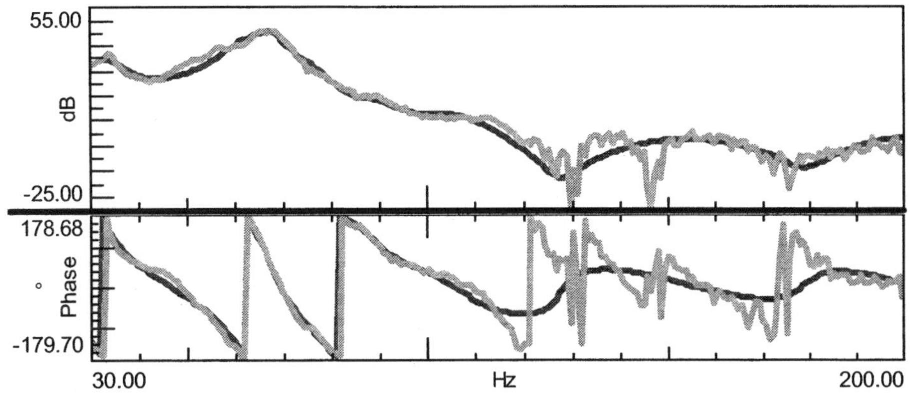

Fig. 6. Measured (grey) and fitted (black) FRF of first transversal loudspeaker to location 01:0116:+X.

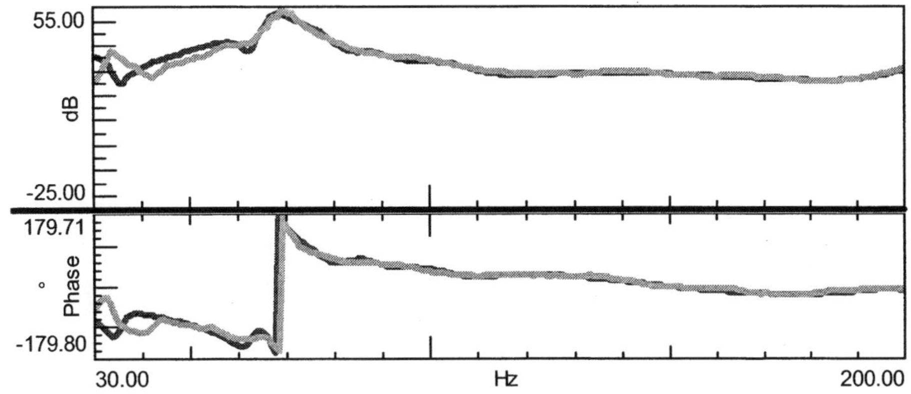

Fig. 7. Measured (grey) and fitted (black) FRF of second transversal loudspeaker to location 01:0116:+X.

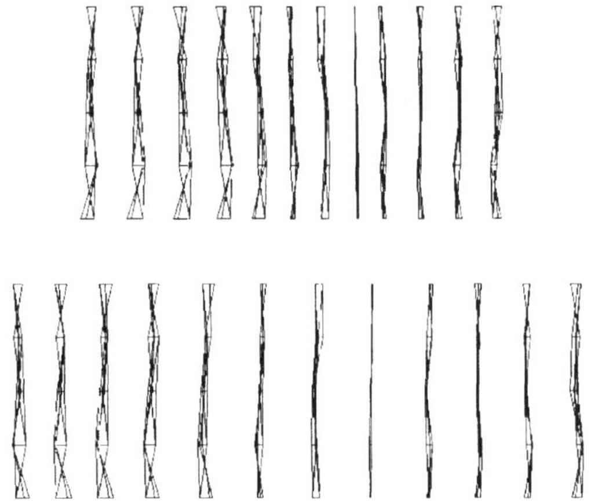

Fig. 8. Mode shape at 32.9 Hz.

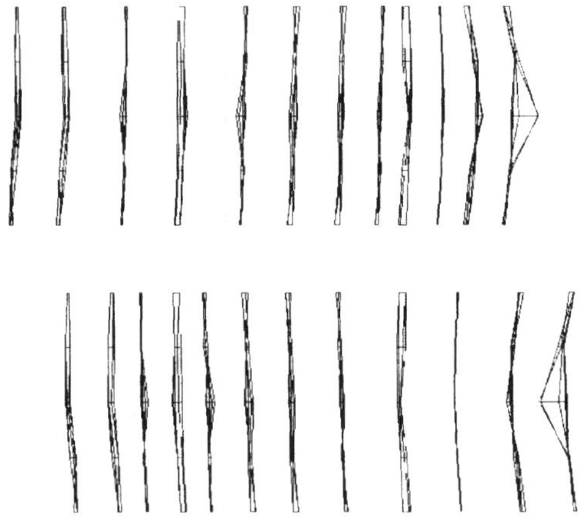

Fig. 9. Mode shape at 61.5 Hz.

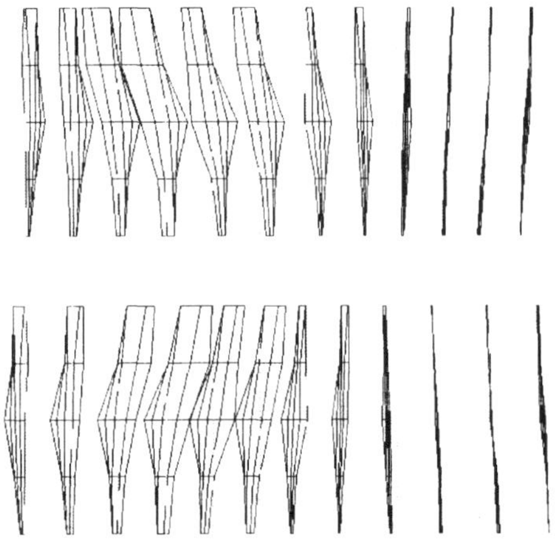

Fig. 10. Mode shape at 78.3 Hz.

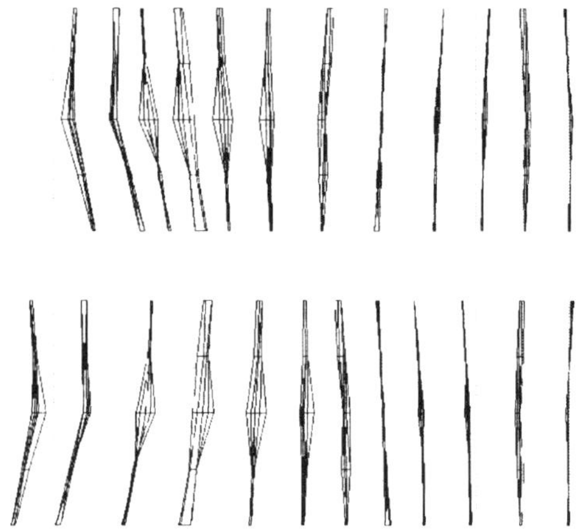

Fig. 11. Mode shape at 88.9 Hz.

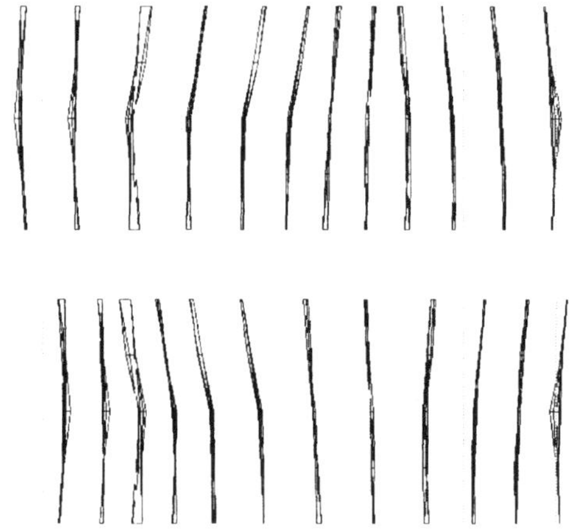

Fig. 12. Mode shape at 107 Hz..

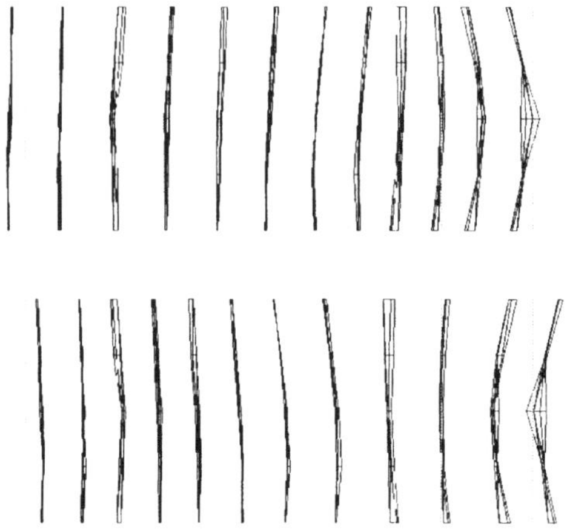

Fig. 13. Mode shape at 149 Hz.

IFAC
Publications
www.elsevier.com/locate/ifac

SUBSPACE IDENTIFICATION COMBINED WITH NEW MODE SELECTION TECHNIQUES FOR MODAL ANALYSIS OF AN AIRPLANE

Goethals Ivan*[1] and Bart De Moor*[2]

* *ESAT-SCD, KULeuven, Kasteelpark Arenberg 10, B3001 Heverlee-Leuven, Belgium*
{ivan.igoethals, bart.demoor}@esat.kuleuven.ac.be
http://www.esat.kuleuven.ac.be/sista-cosic-docarch

Abstract: Linear system identification is an important tool in experimental modal analysis. It allows for the extraction of resonance frequencies, damping ratios and mode shapes of a vibrating structure. In general, the model order is chosen quite high so as to catch all the important characteristics of the structure, even in the presence of large amounts of measurement noise. This often results in the appearance of non-physical, or so-called spurious modes. In this paper we will present a set of heuristic techniques to remove spurious modes from a previously identified model. The advantage of the techniques that will be presented is that they do not rely on statistical information, making them ideally suited for use in combination with subspace identification. The quality of the techniques will be assessed using simulated data and observations from in flight flutter tests. *Copyright © 2003 IFAC*

Keywords: Parameter estimation, Subspace methods.

1. INTRODUCTION

System identification is a standard tool for the analysis of forcefully or ambient excited vibrating structures (Van der Auweraer, 2001). A linear model for the structure is built from available observations, based on which modal characteristics as resonance frequencies and modal shapes can be estimated. Typically, the vibrating structure is equipped with tens to hundreds of sensors and the average identification order needed to obtain a suitable model is reasonably high. Since measurements on the structure are often disturbed by large amounts of measurement noise, and unknown inputs do not always satisfy the white noise assumption, eg. during in flight tests with an aircraft subjected to turbulence, choosing a model order is often a difficult task, certainly when the amount of available measurements is limited. Not seldomly do models in which the order was chosen according to some order selection technique, therefore prove inadequate to describe all relevant characteristics of the structure. In modal analysis, one therefore typically uses an identification order that is guaranteed to be larger than necessary. Unfortunately however, as the order of the model is increased, so will the amount of identified modes. This will in many cases inevitably result in the appearance of so-called spurious modes which bear no immediate physical relevance. A common technique to remove spurious modes from a model is the stabilization diagram (Van der Auweraer, 2001), where models of increasing order are compared, and modes that are repeatedly found in these models with about the same characteristics are considered to be physical. A problem however is that the comparisson is highly user interactive, making the stabilization diagram unsuited for use in an online envionment. In this paper we will describe several automatic techniques to detect spurious modes and remove them from the model. Although the techniques are in general quite heuristic in nature, as is the stabilization diagram, we will show by means of a simulation and an example from the avionics industry that in many practical cases a quick discrimination between spurious and physical modes can effectively be made, without reverting to an analysis of the stabilization diagram. An advantage of the techniques that will be presented is that they are ideally suited to be applied to models obtained using subspace identification. Subspace identification is a popular concept that

[1] I. Goethals is a Research Assistant with the Fund for Scientific Research-Flanders (FWO-Vlaanderen)
[2] B. De Moor is a full professor with the KULeuven

allows for a fast and robust identification of MIMO systems by using projections of subspaces spanned by the rows and columns of Hankel matrices containing input- and output measuremens (Verhaegen 1996; Van Overschee and De Moor, 1996; Bauer and Ljung, 2002), but does unfortunately not return stochastic information as confidence bounds around poles and zeros, which are used by some recently proposed mode selection techniques (Verboven *et al.*, 2002).

In section 2 we will briefly refresh the basic concepts of subspace identification, state space models and the theory of balanced model reduction from a practical point of view. Several mode selection techniques will then be presented in section 3. In section 4, we will assess the quality of the proposed techniques by means of a simulated example, and an analysis of data obtained from a test flight of an airplane. We will show that the methods presented in section 3 can effectively be applied to detect and remove spurious modes from a linear model, even in the presence of large amounts of measurement noise. Finally, in Section 5, some conclusions will be drawn.

Some common notations that will be used throughout this text are the following: $E\{\cdot\}$ will be used to denote the expected value of an expression. $A(i:j,k:l)$ denotes a submatrix of A, bounded by the ith and j[th] row and k[th] and l[th] column. If a colon (:) is used on its own (eg. A(:,k:l)) all available rows and/or columns are included in the submatrix.

2. SUBSPACE IDENTIFICATION AND THE STATE SPACE FORMULATION

The aim of subspace identification is to identify models of the form:

$$\begin{aligned} x_{k+1} &= Ax_k + Bu_k + w_k, \\ y_k &= Cx_k + Du_k + v_k, \end{aligned} \tag{1}$$

with

$$E\left\{\begin{bmatrix} w_p \\ v_p \end{bmatrix}\begin{bmatrix} w_q^T & v_q^T \end{bmatrix}\right\} = \begin{bmatrix} Q & S \\ S^T & R \end{bmatrix}\delta_{pq} \geq 0, \tag{2}$$

where $E\{\cdot\}$ denotes the expected value operator and δ_{pq} the Kronecker delta. It is assumed that:

$$E\left\{\begin{bmatrix} w_p \\ v_p \end{bmatrix} x_k^T\right\} = 0, \forall p \geq k. \tag{3}$$

The elements of the vectors $y_k \in \mathbb{R}^l$ and $u_k \in \mathbb{R}^m$ are given observations of the outputs and inputs of the system at the discrete time index *k*. The vector $x_k \in \mathbb{R}^n$ is the unknown state vector at time *k*. The unobserved process and measurement noise $w_k \in \mathbb{R}^n$ and $v_k \in \mathbb{R}^l$ are assumed to be white, zero mean, gaussian with covariance matrices as given in (2). The system matrices *A,B,C,D* and the covariance matrices *Q, S*, and *R* have appropriate dimensions.

In subspace identification, the model (1) is obtained using projections of rows and columns of so-called block Hankel matrices containing the inputs and outputs of the system. These projections can typically be calculated using basic tools as QR- and SVD-decompositions, eliminating the need for a costly optimization of a non convex cost function as in many predictor-error methods, and making the method inherently robust.

The obtained state-space representation (1) is not unique. Applying a basis transformation $x \rightarrow Tx$ and a corresponding transformation of the state space matrices, $(A,B,C,D) \rightarrow (TAT^{-1}, TB, CT^{-1}, D)$, the model (1) can be written in a multitude of forms, which all describe the same input-output behavior. A common representation in modal analysis is the so-called modal representation.

$$\begin{aligned} x_{k+1}^m &= \Lambda x_k^m + B^m u_k + w_k, \\ y_k &= C^m x_k^m + Du_k + v_k, \end{aligned} \tag{4}$$

where the system matrix Λ is diagonal and mainly consists of pairs of complex conjugated eigenvalues $\lambda, \bar{\lambda}$ being the poles of the system. For this to be possible the original system matrix *A* needs to be diagonalizable which is in practical applications usually the case. The modal characteristics of the structure under study can then easily be obtained from (6) as follows:

$$\begin{aligned} f_i &= \arg\left(\lambda_i \tfrac{T_s}{2\pi}\right), \\ d_i &= \frac{\ln(|\lambda_i|)}{\sqrt{\ln(|\lambda_i|)^2 + \arg(\lambda_i)^2}}, \\ v_i &= C^m(:,i), \end{aligned} \tag{5}$$

with f_i, d_i and v_i the resonance frequency, damping and mode shapes corresponding to the ith pole $\Lambda(i,i) = \lambda_i$. T_s is the sampling rate.

Another commonly used representation is the so-called Balanced representation (Obinata and Anderson, 2001). The idea of the Balanced representation is to decompose the controllabity and observability grammians of the model into principal components in order to evaluate the contributions of each mode to the overall input/output behavior of the model. The controllability Grammian *P* and observability Grammian *Q* can easily be obtained as solutions to the following Lyapunov equations:

$$\begin{aligned} APA^T + P - BB^T &= 0 \\ A^TQA + Q - C^TC &= 0 \end{aligned} \tag{6}$$

The key property of a balanced realization is that a state transformation $x^b = Tx$ and a corresponding similarity transformation $(A^b, B^b, C^b, D) = (TAT^{-1}, TB, CT^{-1}, D)$ is selected such that the controllability and observability grammians are both equal to a diagonal matrix Σ.

$$A^b \Sigma A^{b^T} + \Sigma - B^b B^{b^T} = 0 \qquad (7)$$
$$A^{b^T} \Sigma A^b + \Sigma - C^{b^T} C^b = 0$$

The larger a diagonal entry of the grammians, the bigger the contribution of the corresponding entry of the state vector to the overall input/output behavior of the model. The diagonal entries are therefore usually sorted on the diagonal in descending order. The so called concept of Balanced model reduction is then nothing else than the removal of the last entries of the state. More concretely, if the balanced system matrices are partitioned as follows:

$$A^b = \begin{bmatrix} A_{11}^b & A_{12}^b \\ A_{21}^b & A_{22}^b \end{bmatrix} \qquad (8)$$

$$B^b = \begin{bmatrix} B_1^b \\ B_2^b \end{bmatrix} \qquad (9)$$
$$C^b = \begin{bmatrix} C_1^b & C_2^b \end{bmatrix}$$

The reduced model would be $(A_{11}^b, B_1^b, C_1^b, D)$.

3. METHODS FOR MODE DISCRIMINATION

3.1. Introduction

In this section we will describe some techniques to remove spurious modes from a model of the form (1). We will thereby make extensive use of the modal and balanced representations of a system, as introduced in section 2.

3.2. H_2 and H_∞ modal truncation

A first naive approach would be to write the model in its modal form, as given in (4), remove a certain mode, and assess the "damage" done to the model in H_2 and H_∞ norm. Hence, if the full order model is called H_{full}, and H_{reduced} is the lower order model formed by removing the complex conjugated poles λ_i and $\overline{\lambda}_i$ from (4), the following expressions are evaluated:

$$\|H_{\text{full}} - H_{\text{reduced}}\|_2, \quad \|H_{\text{full}} - H_{\text{reduced}}\|_\infty, \qquad (10)$$

where

$$H_{\text{full}}(z) - H_{\text{reduced}}(z) = H_i(z) =$$
$$\begin{bmatrix} C(:,i) & \overline{C(:,i)} \end{bmatrix} \left(zI_2 - \begin{bmatrix} \lambda_i & 0 \\ 0 & \overline{\lambda}_i \end{bmatrix} \right)^{-1} \begin{bmatrix} B(i,:) \\ \overline{B(i,:)} \end{bmatrix}. \qquad (11)$$

The expressions in (10) can be calculated as:

$$\|H_{\text{full}} - H_{\text{reduced}}\|_2 = Tr\left(C(:,i) P_i C(:,i)^T\right), \qquad (12)$$
$$\|H_{\text{full}} - H_{\text{reduced}}\|_\infty = \max_\omega \overline{\sigma}\left(H_i(e^{j\omega})\right),$$

where P_i is the controllability matrix of the second order model (11), and can be obtained by solving a Lyapunov equation as in (6). Numerical procedures are widely available for the calculation of the infinity norm (Boyd *et al.*, 10). After repeatedly calculating (12), once for each mode, the distance measures obtained are divided by their maximal value, this is, the maximal distance that can be obtained by removing 1 mode. The result is a number between 0 and 1 for each mode, and each criterium (H_2 and H_∞) which will be used as a significance parameter describing the importance of the mode in section 4.

In general it seems reasonable to assume that the more important a mode is, the bigger will be the influence of its removal from the model, and hence its significance parameter. Furthermore it is shown in (Jonckheere, *1984*) that for nearly undamped structues, the grammians of the modal form are almost diagonal, meaning that the modal and the balanced form are "close" to each other in some sense. Hence, for such structures, modal truncation makes perfect sense since it leads to similar results as balanced model reduction. In many practical cases, however, the structure under study is not nearly undamped, and modal truncation may lead to an inadequate rejection of spurious modes due to phenomena as mode coupling, which make it very hard to assess the importance of a mode by examining a single second order subsystem. This will also be shown in the examples in section 4. In order to draw better conclusions for complicated structures, we will therefore look into the connection between the balanced and the modal representation of the identified model.

3.3. Connection between the balanced and the modal form

Since the balanced and the modal form are both representations of the same model, there is always a similarity transformation T linking one form to the other:

$$\Lambda = TA^bT^{-1},$$
$$B^m = TB^b, \qquad (13)$$
$$C^m = C^bT^{-1},$$
$$D^m = D^b.$$

From $\Lambda = TA^bT^{-1}$ it follows that the diagonal elements of Λ are a linear combination of the entries of A^b, where we know that the entries of A^b that are most relevant for the input/output behavior of the model in the sense of Moore are situated in its upper left part. A formal way to exploit this fact in a mode selection context is to replace A^b with a significance matrix S of the same dimensions, where the elements of S give a measure of the importance of the corresponding entries in A^b, eg.

$$S = \begin{bmatrix} n & n-1 & n-2 & \cdots \\ n-1 & n-1 & n-2 & \cdots \\ n-2 & n-2 & n-2 & \cdots \\ \vdots & \vdots & \vdots & \ddots \end{bmatrix}, \quad (14)$$

and inspect the diagonal elements of $|T| \cdot S \cdot |T^{-1}|$, with $|T|$ the elementwise absolute value of T, to obtain a measure for the significance of the corresponding pole in Λ. Again, the significance parameters are rescaled so as to lie between 0 and 1.

3.4. *Continuous extension to balanced truncation*

Closely related to the former technique is the concept of a continuous extension to balanced truncation. Instead of truncating the model and completely removing the last entries of the state, one might opt to change the balanced system matrix,

$$A^b = \begin{bmatrix} a_{11}^b & a_{12}^b & \cdots & a_{1n}^b \\ a_{21}^b & a_{22}^b & \cdots & a_{2n}^b \\ \vdots & \vdots & & \vdots \\ a_{n1}^b & a_{n2}^b & \cdots & a_{nn}^b \end{bmatrix} \quad (15)$$

and introduce a parameter ε to continuously remove the last entries of the state, eg. as follows:

$$\tilde{A}^b(\varepsilon) = \begin{bmatrix} a_{11}^b & \varepsilon a_{12}^b & \cdots & \varepsilon^{n-1} a_{1n}^b \\ \varepsilon a_{21}^b & \varepsilon a_{22}^b & \cdots & \varepsilon^{n-1} a_{2n}^b \\ \vdots & \vdots & & \vdots \\ \varepsilon^{n-1} a_{n1}^b & \varepsilon^{n-1} a_{n2}^b & \cdots & \varepsilon^{n-1} a_{nn}^b \end{bmatrix} \quad (16)$$

While ε is continuously decreased, starting from one, the influence on the system poles can be assessed, eg. by using the euclidean distance measure in the complex plane. Again it is assumed that modes that are mainly related to the least important elements of A^b will be influenced more and can hence be classified as spurious.

3.5. *Pole/zero cancellations*

Pole/zero cancellations, a zero of a rational entry in the transfer function matrix that is almost or completely equal to a system pole, rendering a mode nearly uncontrollable or unobservable with respects to some or all of the inputs and outputs, are not uncommon in models identified from vibrating structures, especially when a high modeling order is used, and are often an indication that the cancelled pole is spurious. It is however seldom a good idea to revert to measures as the distance between a pole and some nearby transfer function zeros as the basis of mode selection techniques, as lowly damped, weakly excited modes may well be accompanied by a nearby zero, even if the mode is quite important for the physical characteristics of the structure as a whole. In (Verboven *et al.* 2002), it was therefore proposed to take extra statistical information as the variance of a pole into account when examining pole/zero cancellations. One of the techniques described in this work is to construct a confidence region around every pole and count the number of transfer function zeros within this region. Such confidence regions, however, are not available if the model is obtained using subspace identification. It is however reasonably acceptable that if we started moving a pole in our model, the influence on the model as a whole would be inversely proportional to the unknown variance on the pole position. As a mode selection rule, similar to the one presented in (Verboven *et al.* 2002) we therefore propose to move each pole to its $m \times l$ closest transfer function zeros and assess the influence on the model as a whole in each of the $m \times l$ cases, eg. by calculating the sum of the 2- or infinity-norms of the differences between the adjusted and the original models. As usual, a significance parameter is obtained by dividing the distances so obtained by their maximal value resulting in a value between 0 and 1.

3.6. *Cross correlations with SISO models*

An obvious critertium, proposed in (Verboven et al. 2002), is to check to what extent poles of the full MIMO model (1) can be retrieved from smaller models obtained from individual or groups of input and output sequences. For our examples in section 4, we constructed l MISO models from the available observations, one for each output, and compared the poles so obtained with the ones from the full MIMO model using the standard euclidean distance measure in the complex plane. To speed up the procedure, the individual models can for instance be obtained using a fast ARX modeling procedure.

4. EVALUATION

In order to evaluate the techniques outlined in section 3, we applied them to two datasets. The first was generated by filtering white noise through a known twelfth order model and adding 40% of measurement noise to the simulated output so obtained. A second example involves observation data from an in flight flutter test of an aircraft. This dataset was also analyzed by the aircraft's manufacturer to allow for a comparisson with our selection methods.

4.1. *A simulation*

Six dominant modes, obtained from a ground vibration test of an airplane were used to create a twelfth order model with 38 outputs, one known input and three unknown noise sources. White, zero mean, stationary, gaussian noise with unit variance was applied to the known, as well as the three unknown inputs in order to create 64 seconds of output data, sampled at 256 Hz. 40% of measurement noise was added and the output data together with the known input where thereafter used for identification using a robust N4SID subspace algorithm, described in (Van Overschee *et al.*, 1996), where the modelling order was set to 30. Frequencies, dampings, and significance parameters as returned by the different heuristic mode selection techniques are given in table 1, where the six true modes are located on top. The heuristic mode selection techniques are ordered from left to right in the same order as described in this paper, and the last column (sum) is nothing else than the sum of all the significance parameters for a certain mode. It is important to note here that a significance parameter smaller than 0.5 does not necessarily mean that the mode is unimportant. The significance parameters are mostly rescaled distances which means that their absolute value has little meaning. In a modal analysis context a proper way to remove spurious modes would be to sort the modes in descending order of significance and look for a sudden decrease in significance or, if not available, set a treshhold for the maximal number of modes you are willing to consider in your further analysis. From table 1 it is clear that the six true modes are correctly identified as the six most important ones by all heuristic techniques except for H_2, which can be explained by the effect of mode coupling of the modes around 5 Hz, where individual modes reinforce each-other so as to create a clear resonance, although the individual modal subsystems have a low H_2 norm.

4.2. *In Flight flutter testing of an aircraft*

67 seconds of measurement data, sampled at 256 Hz, obtained during in flight flutter tests of a fly-by-wire airplane equipped with 12 accelerometers and excited by white noise where identified using the same subspace algorithm as in the previous example, with the modelling order set to 40. As in the previous example, frequencies, dampings, and results for the different heuristic techniques are given in table 2. The most important modes, as given by the airplane's manufacterer are printed in bold, and a stabilization diagram is added for comparison in figure 1. Note that all modes are classified correctly, except for the mode at 3.897 Hz, printed in italic, which is stabilized in figure 1, but was not accepted as such by our algorithms. Further analysis revealed that the mode in question was extremely poorly excited during the test-flight, which clarifies its classification as insignificant. For a correct classification of such weak modes, a combination of the proposed techniques with an automatic analysis of the stabilization diagram, as presented in (Vechhio *et al.*, 2003) might be useful

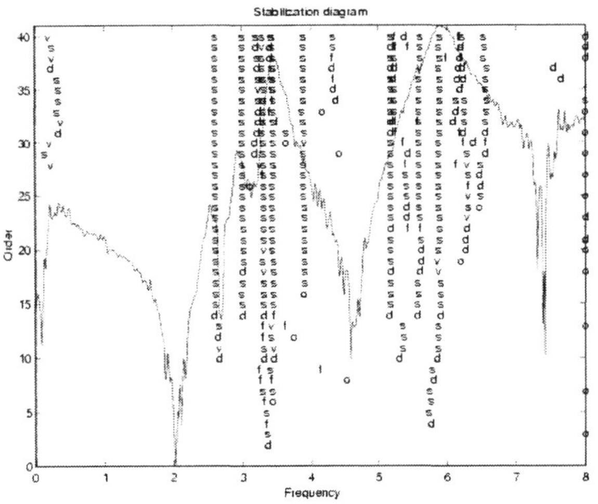

Fig. 1. Stabilization diagram from in-flight flutter test measurements. 'f' is used for stable frequencies (1%), 'd' for stable damping ratios (5%) and 'v' for stable vectors (2%). If all conditions are satisfied, the pols is labeled as stable 's'.

5. CONCLUSIONS

In this paper, several techniques were proposed for the removal of spurious mode from a previously identified model. A special property of the presented techniques is that they do not rely on statistical information, making them ideally suited for use with models obtained using subspace identification. Although the techniques presented are quite heuristic in nature, it was shown by a simulation and an example from the avionics inductry that in many cases, a distinction between true and spurious modes can effectively be made, even in the presence of large amounts of measurement noise. Problems may however occur if modes are very poorly excited.

ACKNOWLEDGEMENTS

Dr. Bart De Moor is a full professor at the Katholieke Universiteit Leuven, Belgium. Our research is supported by grants from several funding agencies and sources:

Research Council KUL: Concerted Research Action GOA-Mefisto 666 (Mathematical Engineering), IDO (IOTA Oncology, Genetic networks), several PhD/postdoc & fellow grants;

Flemish Government: Fund for Scientific Research Flanders (several PhD/postdoc grants, projects G.0256.97 (subspace), G.0115.01 (bio-i and microarrays), G.0240.99 (multilinear algebra), G.0197.02 (power islands), G.0407.02 (support vector machines), research communities ICCoS, ANMMM), AWI (Bil. Int. Collaboration Hungary/ Poland), IWT (Soft4s (softsensors), STWW-Genprom (gene promotor prediction), GBOU-McKnow (Knowledge management algorithms), Eureka-Impact (MPC-control), Eureka-FLiTE (flutter modeling), several PhD grants);

Belgian Federal Government: DWTC (IUAP IV-02 (1996-2001) and IUAP V-10-29 (2002-2006): Dynamical Systems and Control: Computation, Identification & Modelling), Program Sustainable Development PODO-II (CP-TR-18: Sustainibility effects of Traffic Management Systems);

Direct contract research: Verhaert, Electrabel, Elia, Data4s, IPCOS;

REFERENCES

Boyd, S., K. Balakrishnan and P. Kabamba (1989). A bisection method for computing the H_∞ norm of a transfer matrix and related problems, *Mathematics of Control, Signals and Systems,* **2(3)**, pp. 207-219.

Bauer, D. and L. Ljung (2002). Some facts about the choice of the weighting matrices in Larimore type of subspace algorithms, *Automatica*, **38**, pp. 763-773

Jonckheere, E. A. (1984). Principal Component Analysis of Flexible Systems – Open loop case, *IEEE, Trans. Automat. Contr.*, **vol. AC-29, No 12**, december, pp. 382-38.

Obinata G., and B.D.O. Anderson (2001). *Model Reduction for Control System Design*. Springer-Verlag, London.

Van der Auweraer, H. (2001). Structural Dynamics Modeling using Modal Analysis: Applications, Trends and Challenges, in H. Van der Auweraer, editor, *Proceedings of the 2001 IEEE Instrumentation and Measurement Technology Conference*, Budapest, Hungary.

Van Overschee P. and B. De Moor (1996). *Subspace Identification for Linear Systems: Theory – Implementation – Applications*. Kluwer Academic Publishers, Boston/London/Dordrecht.

Verboven, P., E. Parloo, P. Guillaume and M. Van Overmeire (2002). Autonomous Structural Health monitoring – part 1: Modal Parameter Estimation and Tracking. *Mechanical Systems and Signal Pocessing* (In Press).

Lancelots, J., I. Goethals, A. Vecchio, H. Van der Auweraer and B. De Moor (2003). Advanced tools to improve detection of structural changes from in-flight flutter data. Submitted for publication to: *ISVR 2003, conference on Recent Advances in Structural Dynamics*.

Verhaegen, M.H.G. (1996). A subspace model identification solution to the identification of mixed causal, anti-causal LTI systems. S*iam J. Matrix Anal. Appl.*, **17, 2**, pp. 332-347.

Table 1 Frequencies, dampings, and significance parameters for a simulation using a 12th order model. Modes are ordered according the the sum of their significance paramaters. The true modes are printed in bold.

Frequency	Damp	H2	Hinf	Balanced	Continuous	Pole/zero	Cross-corr.	Sum
3.169Hz	**0.69%**	**0.3565**	**0.1303**	**1.0000**	**1.0000**	**1.0000**	**1.0000**	**4.4867**
3.743Hz	**0.82%**	**0.2871**	**0.3605**	**0.9288**	**0.9288**	**0.7273**	**0.6563**	**3.8887**
4.627Hz	**1.30%**	**0.1401**	**0.1043**	**0.7485**	**0.7485**	**0.0853**	**0.1802**	**2.0069**
5.099Hz	**1.95%**	**0.2760**	**0.1951**	**0.7557**	**0.7557**	**0.1395**	**0.0778**	**2.1998**
5.885Hz	**1.27%**	**1.0000**	**1.0000**	**0.8474**	**0.8474**	**0.2756**	**0.2607**	**4.2311**
8.392Hz	**1.54%**	**0.1237**	**0.1107**	**0.6425**	**0.6425**	**0.0515**	**0.0673**	**1.6381**
6.161Hz	6.23%	0.2324	0.0322	0.5486	0.5486	0.0132	0.0076	1.3826
5.610Hz	7.48%	0.1752	0.0247	0.4871	0.4871	0.0059	0.0084	1.1885
0.541Hz	55.73%	0.0224	0.0042	0.3780	0.3779	0.0016	0.0048	0.7888
1.923Hz	13.29%	0.0139	0.0025	0.3368	0.3368	0.0006	0.0056	0.6961
9.030Hz	2.38%	0.0070	0.0019	0.2630	0.2630	0.0002	0.0061	0.5412
3.175Hz	9.30%	0.0195	0.0037	0.2277	0.2277	0.0001	0.0072	0.4859
3.490Hz	25.20%	0.0602	0.0099	0.1845	0.1844	0.0004	0.0026	0.4419
7.976Hz	11.05%	0.0121	0.0021	0.1516	0.1516	0.0002	0.0069	0.3245
3.414Hz	8.56%	0.0119	0.0035	0.1318	0.1318	0.0000	0.0071	0.2862

Table 2 Frequencies, dampings, and significance parameters for data obtained from an in-flight flutter test. Modes are ordered according the the sum of their significance paramaters. The true modes are printed in bold. The status of the mode at 3.897 Hz, printed in italic, is unsure. It is well stabilized in the stabilization diagram, but not recognized as such by the selection procedures. Further analysis revealed that the mode in question is very poorly excited.

Frequency	Damping	H2	Hinf	Balanced	Continuous	Pole/zero	Cross-corr.	Sum
3.296Hz	**4.92%**	**1.000**	**1.000**	**0.948**	**0.244**	**1.000**	**0.473**	**4.665**
3.439Hz	**2.75%**	**0.692**	**0.338**	**1.000**	**0.590**	**0.678**	**1.000**	**4.298**
5.901Hz	**4.09%**	**0.748**	**0.891**	**0.914**	**1.000**	**0.499**	**0.241**	**4.293**
2.609Hz	**2.53%**	**0.017**	**0.023**	**0.859**	**0.187**	**0.163**	**0.847**	**2.096**
5.590Hz	**2.77%**	**0.183**	**0.104**	**0.836**	**0.059**	**0.144**	**0.289**	**1.615**
5.179Hz	**2.73%**	**0.112**	**0.072**	**0.790**	**0.096**	**0.066**	**0.165**	**1.302**
3.006Hz	**3.05%**	**0.032**	**0.020**	**0.715**	**0.124**	**0.072**	**0.212**	**1.175**
6.597Hz	5.44%	0.035	0.018	0.548	0.157	0.043	0.076	0.877
3.418Hz	4.22%	0.114	0.066	0.436	0.060	0.042	0.078	0.796
5.259Hz	4.98%	0.141	0.064	0.429	0.072	0.045	0.036	0.786
6.008Hz	2.99%	0.059	0.055	0.479	0.074	0.018	0.090	0.775
3.220Hz	3.93%	0.067	0.035	0.489	0.078	0.015	0.013	0.697
6.416Hz	1.24%	0.008	0.023	0.437	0.075	0.003	0.134	0.679
6.118Hz	6.27%	0.093	0.055	0.189	0.215	0.029	0.039	0.619
4.813Hz	4.60%	0.013	0.006	0.368	0.069	0.028	0.121	0.605
3.897Hz	***4.11%***	***0.005***	***0.013***	***0.124***	***0.052***	***0.000***	***0.226***	***0.421***
4.514Hz	1.72%	0.002	0.001	0.245	0.057	0.001	0.082	0.388

IFAC Publications
www.elsevier.com/locate/ifac

FLIGHT FLUTTER ANALYSIS USING FREQUENCY-DOMAIN SYSTEM IDENTIFICATION TECHNIQUES

Patrick Guillaume, Peter Verboven, Bart Cauberghe

Vrije Universiteit Brussel (VUB)
Department of Mechanical Engineering
Acoustics & Vibration Research Group
Peinlaan 2, B-1050 Brussel, BELGIUM

Abstract: In this paper frequency-domain estimators will be presented for application in the field of flight flutter analysis. Flight flutter tests are expensive and not without risks even when approached with caution. The complete session of flight tests must be as fast as possible but this often results in low quality data. Dedicated frequency-domain estimators will be presented to estimate and track the modal parameters as a function of the flight conditions.

Keywords: Aerospace systems, modal analysis, parameter estimation, vibration, monitoring, multivariable.

1. INTRODUCTION

In the framework of the EUREKA-FLITE project new methods are developed for the automated analysis, validation and interpretation of structural dynamics data. The focus of the project is on in-flight aircraft flutter testing. Flutter can be defined as the dynamic instability of an elastic body in an air stream. It is most commonly encountered on bodies subjected to large lateral aerodynamic loads of the lift type, such as aircraft wings, tails, and control surfaces. Modern aircraft are subject to many kinds of flutter phenomena, the most common involving the coupling of two or more degrees of freedom. The speed at which such instability takes place for a given aircraft is called the flutter speed and constitutes one of the most important design parameter for an aircraft wing. Before an airplane is released, flight flutter tests have to be performed to detect possible onset of flutter. The classical flight flutter testing approach is to expand the flight envelope of a airplane by performing a vibration test at constant flight conditions, curve-fit the data to estimate the resonance frequencies and damping ratios, and then to plot these frequencies and damping estimates against flight speed or Mach number. The damping values are then extrapolated in order to determine whether it is save to proceed to the next flight test point. Flutter will occur when one of the damping values tends to become negative. Before starting the flight tests, ground vibration tests as well as numerical simulations and wind tunnel tests are used to get some prior insight into the problem. The costs of test-flights are enormous and hence as many flight conditions as possible should be verified in one single flight. Moreover flight flutter tests can be highly dangerous even when approached with caution. An aircraft represent a huge investment in terms of time and money and a flutter occurrence can be spectacularly destructive. Several fatal cases are reported in literature.

An important goal of the project is to reduce the required test time and to improve the accuracy of the damping monitoring and flutter prediction by means of improved system identification techniques. This will be the topic of the present paper. Dedicated frequency-domain estimators for flutter testing will be proposed.

2. MULTIVARIABLE TRANSFER FUNCTION MODELLING

Many multivariable transfer-function models are available (Kalaith, 1980). In this paper, a common-denominator transfer function model will be considered. The relationship between output o ($o=1,\ldots,N_o$) and input i ($i=1,\ldots,N_i$) can be modelled in the frequency domain as

$$\hat{H}_k(\omega)=\frac{N_k(\omega)}{d(\omega)} \quad (1)$$

for $k = 1,\ldots,N_oN_i$ (where $k = (o-1)N_i + i$). The numerator polynomial between output o and input i equals

$$N_k(\omega) = \sum_{j=0}^{n} \Omega_j(\omega) B_{kj} \tag{2}$$

while

$$d(\omega) = \sum_{j=0}^{n} \Omega_j(\omega) A_j \tag{3}$$

is the common-denominator polynomial. The real-valued coefficients A_j and B_{kj} are the parameters to be estimated.

All these coefficients are grouped together in one column vector $\boldsymbol{\theta} = [\boldsymbol{\beta}_1^T, \ldots, \boldsymbol{\beta}_{N_oN_i}^T, \boldsymbol{\alpha}^T]^T$ with

$$\boldsymbol{\beta}_k = \begin{Bmatrix} B_{k0} \\ B_{k1} \\ \vdots \\ B_{kn} \end{Bmatrix}, \quad \boldsymbol{\alpha} = \begin{Bmatrix} A_0 \\ A_1 \\ \vdots \\ A_n \end{Bmatrix} \tag{4}$$

Several choices are possible for the polynomial basis functions $\Omega_j(\omega)$. For a discrete-time domain model, the functions $\Omega_j(\omega)$ are usually given by $\Omega_j(\omega) = \exp(-i\omega T_s \cdot j)$ (with T_s the sampling period) while for a continuous-time domain model $\Omega_j(\omega) = (i\omega)^j$.

The bad numerical conditioning of the continuous-time domain approach can be improved by using for instance orthogonal Forsythe polynomials, but, this will increase the computation time. As time is an important aspect in flutter testing, discrete-time domain models will be used in this paper.

3. FREQUENCY-DOMAIN MODAL PARAMETER ESTIMATION USING FRF DATA

In modal analysis, measurements of Frequency Response Functions (FRFs) are commonly used. Starting from these measured FRFs, $H_k(\omega_f)$ (with $k = 1,\ldots,N_oN_i$ and $f = 1,\ldots,N_f$), estimates of the transfer-function coefficients can be obtained by minimizing the following nonlinear least-squares (NLS) cost function with respect to the parameter vector $\boldsymbol{\theta}$

$$\ell_{\text{NLS}}(\boldsymbol{\theta}) = \sum_{k=1}^{N_oN_i} \sum_{f=1}^{N_f} \left| \varepsilon_k^{\text{NLS}}(\omega_f, \boldsymbol{\theta}) \right|^2 \tag{5}$$

where the (weighted) NLS equation error, $\varepsilon_k^{\text{NLS}}(\omega_f, \boldsymbol{\theta})$, is defined as

$$\varepsilon_k^{\text{NLS}}(\omega_f, \boldsymbol{\theta}) = W_k(\omega_f) \left(\frac{N_k(\omega_f, \boldsymbol{\beta}_k)}{d(\omega_f, \boldsymbol{\alpha})} - H_k(\omega_f) \right) \tag{6}$$

with $W_k(\omega_f)$ an arbitrary weighting functions. The quality of the estimate can often be further improved by using an adequate weighting function such as

$$W_k(\omega_f) = \frac{1}{\sqrt{\text{var}\{H_k(\omega_f)\}}} \tag{7}$$

By doing so, the quality of the measured FRFs can be into account: FRF measurements with a small variance, $\text{var}\{H_k(\omega_f)\}$, have an important contribution to the cost function while noisy FRF measurements are penalized.

To solve this NLS problem, good starting values are required. Starting values can be obtained via (sub-optimal) linear least-squares estimators

3.1 The Least-Squares Complex Frequency-domain (LSCF) Approach for FRF Data

Note that the nonlinear least-squares problem can be approximated by a (sub-optimal) linear least-squares one. Indeed, by multiplying $\varepsilon_k^{\text{NLS}}(\omega_f, \boldsymbol{\theta})$ with $d(\omega, \boldsymbol{\alpha})$, one obtains an equation error that is linear in the parameters (Pintelon, *et al.*, 1994)

$$\begin{aligned} \varepsilon_k^{\text{LS}}(\omega_f, \boldsymbol{\theta}) &= d(\omega_f, \boldsymbol{\alpha}) \cdot \varepsilon_k^{\text{NLS}}(\omega_f, \boldsymbol{\theta}) \\ &= W_k(\omega_f) \left(N_k(\omega_f, \boldsymbol{\beta}_k) - d(\omega_f, \boldsymbol{\alpha}) H_k(\omega) \right) \\ &= W_k(\omega_f) \sum_{j=0}^{n} \left(\Omega_j(\omega_f) B_{kj} - \Omega_j(\omega_f) A_j H_k(\omega_f) \right) \end{aligned} \tag{8}$$

Because the equations (8), for $f = 1,\ldots,N_f$, are "linear-in-the-parameters", they can be reformulated in matrix notations as

$$\boldsymbol{\varepsilon}_k^{\text{LS}}(\boldsymbol{\theta}) = \begin{Bmatrix} \varepsilon_k^{\text{LS}}(\omega_1, \boldsymbol{\theta}) \\ \vdots \\ \varepsilon_k^{\text{LS}}(\omega_{N_f}, \boldsymbol{\theta}) \end{Bmatrix} = [\mathbf{X}_k \quad \mathbf{Y}_k] \cdot \begin{Bmatrix} \boldsymbol{\beta}_k \\ \boldsymbol{\alpha} \end{Bmatrix} = \mathbf{J}_k \cdot \begin{Bmatrix} \boldsymbol{\beta}_k \\ \boldsymbol{\alpha} \end{Bmatrix} \tag{9}$$

with

$$\mathbf{X}_k = \begin{bmatrix} W_k(\omega_1)[\Omega_0(\omega_1), \Omega_1(\omega_1), \ldots, \Omega_n(\omega_1)] \\ \vdots \\ W_k(\omega_{N_f})[\Omega_0(\omega_{N_f}), \Omega_1(\omega_{N_f}), \ldots, \Omega_n(\omega_{N_f})] \end{bmatrix} \tag{10}$$

$$\mathbf{Y}_k = \begin{bmatrix} -W_k(\omega_1) H_k(\omega_1)[\Omega_0(\omega_1), \Omega_1(\omega_1), \ldots, \Omega_n(\omega_1)] \\ \vdots \\ -W_k(\omega_{N_f}) H_k(\omega_{N_f})[\Omega_0(\omega_{N_f}), \Omega_1(\omega_{N_f}), \ldots, \Omega_n(\omega_{N_f})] \end{bmatrix} \tag{11}$$

The (weighted) linear least-squares problem is found by minimizing

$$\ell_{\text{LS}}(\boldsymbol{\theta}) = \sum_{k=1}^{N_oN_i} \sum_{f=1}^{N_f} \left| \varepsilon_k^{\text{LS}}(\omega_f, \boldsymbol{\theta}) \right|^2$$

$$
\begin{aligned}
&= \sum_{k=1}^{N_oN_i} \mathrm{Re}\left(\left(\boldsymbol{\varepsilon}_k^{\mathrm{LS}}(\boldsymbol{\theta})\right)^H \cdot \boldsymbol{\varepsilon}_k^{\mathrm{LS}}(\boldsymbol{\theta})\right) \\
&= \sum_{k=1}^{N_oN_i} \left(\begin{bmatrix} \boldsymbol{\beta}_k^T & \boldsymbol{\alpha}^T \end{bmatrix} \cdot \begin{bmatrix} \mathbf{R}_k & \mathbf{S}_k \\ \mathbf{S}_k^T & \mathbf{T}_k \end{bmatrix} \cdot \begin{Bmatrix} \boldsymbol{\beta}_k \\ \boldsymbol{\alpha} \end{Bmatrix} \right)
\end{aligned} \tag{12}
$$

with $\mathbf{R}_k = \mathrm{Re}(\mathbf{X}_k^H \mathbf{X}_k)$, $\mathbf{S}_k = \mathrm{Re}(\mathbf{X}_k^H \mathbf{Y}_k)$, and $\mathbf{T}_k = \mathrm{Re}(\mathbf{Y}_k^H \mathbf{Y}_k)$.

In the minimum of the cost function the derivatives of (12) with respect to the unknown coefficients $\boldsymbol{\beta}_k$ and $\boldsymbol{\alpha}$ have to be zero

$$
\frac{\partial}{\partial \boldsymbol{\beta}_k} \ell_{\mathrm{LS}}(\boldsymbol{\theta}) = 2\left(\mathbf{R}_k \boldsymbol{\beta}_k + \mathbf{S}_k \boldsymbol{\alpha}\right) = \mathbf{0}, \quad k = 1, \ldots, N_o N_i \tag{13}
$$

$$
\frac{\partial}{\partial \boldsymbol{\alpha}} \ell_{\mathrm{LS}}(\boldsymbol{\theta}) = 2\left[\sum_{k=1}^{N_oN_i} \left(\mathbf{S}_k^T \boldsymbol{\beta}_k + \mathbf{T}_k \boldsymbol{\alpha}\right)\right] = \mathbf{0} \tag{14}
$$

Substitution of (13), $\boldsymbol{\beta}_k = -\mathbf{R}_k^{-1}\mathbf{S}_k \cdot \boldsymbol{\alpha}$, in (14) yields

$$
\left[2\sum_{k=1}^{N_oN_i} \left(\mathbf{T}_k - \mathbf{S}_k^T \mathbf{R}_k^{-1} \mathbf{S}_k\right)\right] \cdot \boldsymbol{\alpha} = \mathbf{M} \cdot \boldsymbol{\alpha} = \mathbf{0} \tag{15}
$$

with $\mathbf{M} = 2\sum_{k=1}^{N_oN_i} \left(\mathbf{T}_k - \mathbf{S}_k^T \mathbf{R}_k^{-1} \mathbf{S}_k\right)$.

Equations (13) and (14) are the so-called normal equations, which are usually formulated as

$$
2\begin{bmatrix} \mathbf{R}_1 & 0 & \cdots & \mathbf{S}_1 \\ 0 & \mathbf{R}_2 & & \mathbf{S}_2 \\ \vdots & & \ddots & \vdots \\ \mathbf{S}_1^T & \mathbf{S}_2^T & \cdots & \sum_{k=1}^{N_oN_i} \mathbf{T}_k \end{bmatrix} \cdot \begin{Bmatrix} \boldsymbol{\beta}_1 \\ \boldsymbol{\beta}_2 \\ \vdots \\ \boldsymbol{\beta}_{N_oN_i} \\ \boldsymbol{\alpha} \end{Bmatrix} = 2\,\mathrm{Re}(\mathbf{J}^H \mathbf{J}) \cdot \boldsymbol{\theta} = 0 \tag{16}
$$

This structure is due to the use of a common-denominator transfer-function model. The size of the square matrix $\mathbf{M}$ in the "reduced" normal equations (15) is $n+1$, and thus much smaller than the size of $\mathrm{Re}(\mathbf{J}^H \mathbf{J})$ in (16).

If a discrete time-domain model is used, i.e. $\Omega_j(\omega_f) = \exp(-i\omega_f T_s \cdot j)$, and if the frequencies are uniformly distributed (i.e. $\omega_f = f \cdot \Delta\omega$, $f = 1, \ldots, N_f$, with $\Delta\omega = 2\pi / N T_s$), then, the entries of $\mathbf{R}_k$, $\mathbf{S}_k$, and $\mathbf{T}_k$ can be written as

$$
\begin{aligned}
R_k(r,s) &= \mathrm{Re}\left[\sum_{f=1}^{N_f} \left(\left|W_k(\omega_f)\right|^2 \cdot e^{i2\pi(r-s)f/N}\right)\right] \\
S_k(r,s) &= -\mathrm{Re}\left[\sum_{f=1}^{N_f} \left(\left|W_k(\omega_f)\right|^2 H_k(\omega_f) \cdot e^{i2\pi(r-s)f/N}\right)\right] \\
T_k(r,s) &= \mathrm{Re}\left[\sum_{f=1}^{N_f} \left(\left|W_k(\omega_f) H_k(\omega_f)\right|^2 \cdot e^{i2\pi(r-s)f/N}\right)\right]
\end{aligned} \tag{17}
$$

One can readily verify that $\mathbf{R}_k$, $\mathbf{S}_k$, and $\mathbf{T}_k$ have a Toeplitz structure. Moreover, their entries can be computed in a time-efficient way by means of the Fast Fourier Transform (FFT) algorithm (Van der Auweraer, *et al.*, 2001).

3.2 The Maximum Likelihood (ML) Approach for FRF Data

When the data is noisy and accuracy is important, which is the case for flight flutter testing, more sophisticated estimators have to be used. The multivariable implementation of the frequency-domain maximum likelihood estimator is an example of such estimators (Guillaume, *et al.*, 1998; Verboven, 2002).

Assuming the FRFs to be uncorrelated, the (negative) log-likelihood function reduces to

$$
\ell_{\mathrm{ML}}(\boldsymbol{\theta}) = \sum_{k=1}^{N_oN_i} \sum_{f=1}^{N_f} \frac{\left|\dfrac{N_k(\omega_f, \boldsymbol{\beta}_k)}{d(\omega_f, \boldsymbol{\alpha})} - H_k(\omega_f)\right|^2}{\mathrm{var}\{H_k(\omega_f)\}} \tag{18}
$$

Note that the maximum likelihood (ML) problem is equivalent to a weighted nonlinear least-squares one with equation error (see (6))

$$
\varepsilon_k^{\mathrm{ML}}(\omega_f, \boldsymbol{\theta}) = \frac{1}{\sqrt{\mathrm{var}\{H_k(\omega_f)\}}} \left(\frac{N_k(\omega_f, \boldsymbol{\beta}_k)}{d(\omega_f, \boldsymbol{\alpha})} - H_k(\omega_f)\right) \tag{19}
$$

The maximum likelihood estimator takes the quality of the measured FRFs into account: FRF measurements with a small variance, $\mathrm{var}\{H_k(\omega_f)\}$, have an important contribution to the cost function (18) while noisy FRF measurements are penalized.

The maximum likelihood estimate $\hat{\boldsymbol{\theta}}_{\mathrm{ML}}$ is obtained by minimizing (18). This can be done by means of a Gauss-Newton optimization algorithm, which takes advantage of the quadratic form of the cost function (18). The Gauss-Newton iterations are given by

$$
\begin{aligned}
&\text{(a) solve } \mathrm{Re}(\mathbf{J}_p^H \mathbf{J}_p)\boldsymbol{\delta}_p = -\mathrm{Re}(\mathbf{J}_p^H \mathbf{r}_p) \text{ for } \boldsymbol{\delta}_p \\
&\text{(b) set } \boldsymbol{\theta}_{p+1} = \boldsymbol{\theta}_p + \boldsymbol{\delta}_p
\end{aligned} \tag{20}
$$

with $\mathbf{r}_p = \mathbf{r}(\boldsymbol{\theta}_p)$, $\mathbf{J}_p = \partial \mathbf{r}(\boldsymbol{\theta}) / \partial \boldsymbol{\theta}\big|_{\boldsymbol{\theta}_p}$ and

$$
\mathbf{r}(\boldsymbol{\theta}) = \begin{Bmatrix} \varepsilon_1^{\mathrm{ML}}(\omega_1, \boldsymbol{\theta}) \\ \vdots \\ \varepsilon_1^{\mathrm{ML}}(\omega_{N_f}, \boldsymbol{\theta}) \\ \varepsilon_2^{\mathrm{ML}}(\omega_1, \boldsymbol{\theta}) \\ \vdots \\ \varepsilon_{N_oN_i}^{\mathrm{ML}}(\omega_{N_f}, \boldsymbol{\theta}) \end{Bmatrix} \tag{21}
$$

Also here it is possible to form the normal equations (i.e. $\mathrm{Re}(\mathbf{J}_p^H \mathbf{J}_p)$ and $\mathrm{Re}(\mathbf{J}_p^H \mathbf{r}_p)$) in a time-efficient way.

4. FREQUENCY-DOMAIN MODAL PARAMETER ESTIMATION USING INPUT/OUTPUT DATA

All above results for FRF data can be generalized to input/output data. The relationship between the spectrum of output o and the input force vector $\mathbf{F}(\omega_f)$ is given by

$$d(\omega_f)X_o(\omega_f) = \mathbf{N}_o\mathbf{F}(\omega_f) - T_o(\omega_f) \tag{22}$$

for $o = 1,\ldots,N_o$.

4.1 The Least-Squares Complex Frequency-domain (LSCF) Approach for Input/Output Data

Starting from (22), one obtains an equation error that is linear in the parameters

$$\varepsilon_o^{\text{LS-IO}}(\omega_f,\boldsymbol{\theta}) = W_o(\omega_f)\left(\mathbf{N}_o(\omega_f,\boldsymbol{\beta}_o)\mathbf{F}(\omega_f) - d(\omega_f,\boldsymbol{\alpha})X_o(\omega_f)\right) \tag{23}$$

The (weighted) linear least-squares problem is found by minimizing

$$\ell_{\text{LS-IO}}(\boldsymbol{\theta}) = \sum_{o=1}^{N_o}\sum_{f=1}^{N_f}\left|\varepsilon_o^{\text{LS-IO}}(\omega_f,\boldsymbol{\theta})\right|^2 \tag{24}$$

4.2 Leakage and Transient effects

Since one of the research aims is to obtain accurate estimates from short data sequences the influence of leakage and transient phenomena is an important issue. Until recently, frequency-domain system identification assumed that input and output signals are periodic or time-limited within the observation window. Usually, a Hanning window is applied to reduce the influence of leakage. Nevertheless the estimates will be biased. In Pintelon (1997) it has been shown that leakage behaves like transient effects that can be modelled by means of an additional polynomial. This can be understood by considering the Laplace transform of the n-th derivative of a time signal $x(t)$

$$L\left\{\frac{d^n x(t)}{dt^n}\right\} = s^n X(s) - \sum_{r=0}^{n-1} s^r \left.\frac{d^{n-r-1}x(t)}{dt^{n-r-1}}\right|_{t=0} \tag{25}$$

where the sum is a polynomial in the Laplace variable s of order n-1 taking the initial conditions of the system into account. Hence, it is readily verified that the Laplace transform of the input-output differential equation of order n can be written for output o as

$$d(s)X_o(s) = \mathbf{N}_o\mathbf{F}(s) - T_o(s) \tag{26}$$

with $o = 1,\ldots,N_o$ and

$$T_o(\omega) = \sum_{j=0}^{n-1}\Omega_j(\omega)C_{oj} \tag{27}$$

4.3 The Least-Squares Complex Frequency-domain (LSCF) Approach for Input-Output Data with Transient Effects

Based on (26), the following equation error can now be build

$$\begin{aligned}\varepsilon_o^{\text{LS-IO-T}}(\omega_f,\boldsymbol{\theta}) = W_o(\omega_f)\big(&\mathbf{N}_o(\omega_f,\boldsymbol{\beta}_o)\mathbf{F}(\omega_f) + \\ &T_o(\omega_f,\boldsymbol{\gamma}_o) - d(\omega_f,\boldsymbol{\alpha})X_o(\omega_f)\big)\end{aligned} \tag{28}$$

with

$$\boldsymbol{\theta} = [\boldsymbol{\beta}_1^T,\ldots,\boldsymbol{\beta}_{N_o}^T,\boldsymbol{\gamma}_1^T,\ldots,\boldsymbol{\gamma}_{N_o}^T,\boldsymbol{\alpha}^T]^T \tag{29}$$

the parameter vector containing all parameters to be estimated and

$$\boldsymbol{\gamma}_o = \begin{Bmatrix} C_{o0} \\ C_{o1} \\ \vdots \\ C_{o(n-1)} \end{Bmatrix} \tag{30}$$

the coefficients of the transient polynomials. The (weighted) linear least-squares problem is found by minimizing

$$\ell_{\text{LS-IO-T}}(\boldsymbol{\theta}) = \sum_{o=1}^{N_o}\sum_{f=1}^{N_f}\left|\varepsilon_o^{\text{LS-IO-T}}(\omega_f,\boldsymbol{\theta})\right|^2 \tag{31}$$

4.4 The Maximum Likelihood (ML) Approach for Input-Output Data with Transient Effects

In a similar way as Section 3.2, one can show that the ML estimates of $\boldsymbol{\theta} = [\boldsymbol{\beta}_1^T,\ldots,\boldsymbol{\beta}_{N_o}^T,\boldsymbol{\gamma}_1^T,\ldots,\boldsymbol{\gamma}_{N_o}^T,\boldsymbol{\alpha}^T]^T$ are obtained by minimizing

$$\ell_{\text{ML}}(\boldsymbol{\theta}) = \sum_{o=1}^{N_o}\sum_{f=1}^{N_f}\left|\varepsilon_o^{\text{ML-IO-T}}(\omega_f,\boldsymbol{\theta})\right|^2 \tag{32}$$

with

$$\begin{aligned}\varepsilon_o^{\text{ML-IO-T}}(\omega_f,\boldsymbol{\theta}) = W_o^{\text{ML-IO-T}}(\omega_f,\boldsymbol{\theta})\big(&\mathbf{N}_o(\omega_f,\boldsymbol{\beta}_o)\mathbf{F}(\omega_f) + \\ &T_o(\omega_f,\boldsymbol{\gamma}_o) - d(\omega_f,\boldsymbol{\alpha})X_o(\omega_f)\big)\end{aligned} \tag{33}$$

and

$$W_o^{\text{ML-IO-T}}(\omega_f,\boldsymbol{\theta}) = \frac{1}{\sqrt{\mathbf{N}_o(\omega_f,\boldsymbol{\beta}_o)\mathbf{C}_{\mathbf{F}}(\omega_f)\mathbf{N}_o^H(\omega_f,\boldsymbol{\beta}_o) + \left|d(\omega_f,\boldsymbol{\alpha})\right|^2 \operatorname{var}\{X_o(\omega_f)\}}} \tag{34}$$

4. MONITORING OF DAMPING RATIOS

Based on real Ground Vibration Test performed on an airplane, simulated Flight Flutter data was generated for continuously varying flight conditions.

This data have been used to compare the performances of different parameter estimation techniques as well as the developed mode tracking tools.

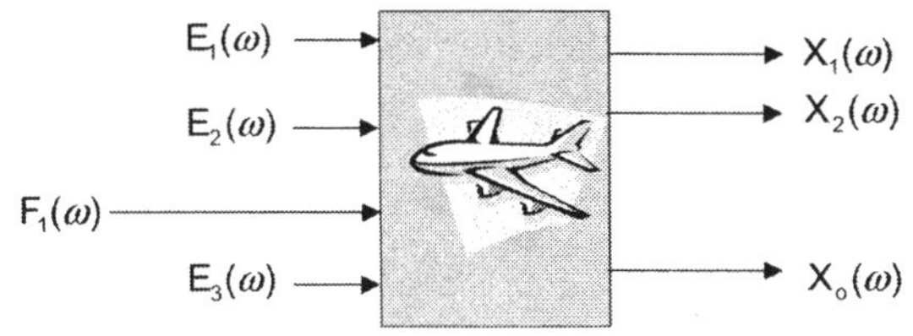

Fig. 1. Simulated model.

The airplane model contains 6 modes in the frequency band of 0 to 10 Hz. The system is excited with one know force, F_1, located on the left wing of the airplane and a set of unknown random (turbulence) forces, E_i, located all over the surface of the plane. The evolution of the (true) poles as a function of time is given in Figure 2.

In Figure 3 the ML damping estimates (using input/output data without transient polynomials) are given as a function of time using a sliding window of 64s. In Figure 4 the same results are given for a sliding window of only 16s. These results are clearly not accurate enough and cannot be used to detect flutter. In Figure 5 and 6 the ML damping estimates using input/output data together with a transient polynomials are given as a function of time. Contrary to the previous results, these estimates can be used to detect flutter. Even for small record sizes, acceptable results are obtained. Small record sizes are needed in order to obtain a fast detection of flutter and to reduce the bias error on the damping estimates (time-varying system).

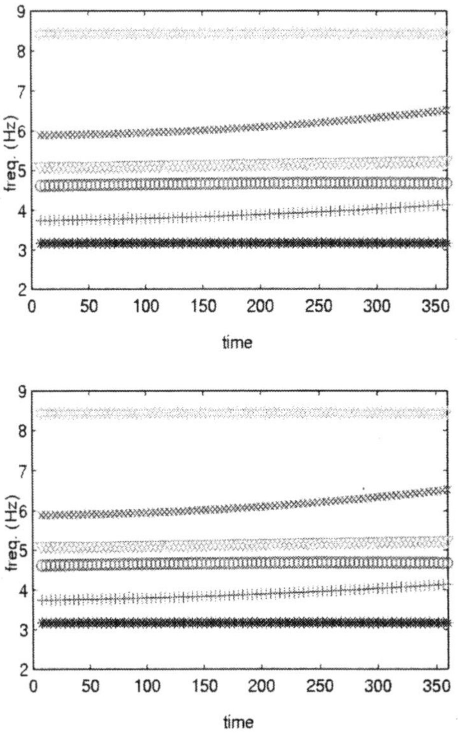

Fig. 2. True modal frequencies and damping ratios as a function of time.

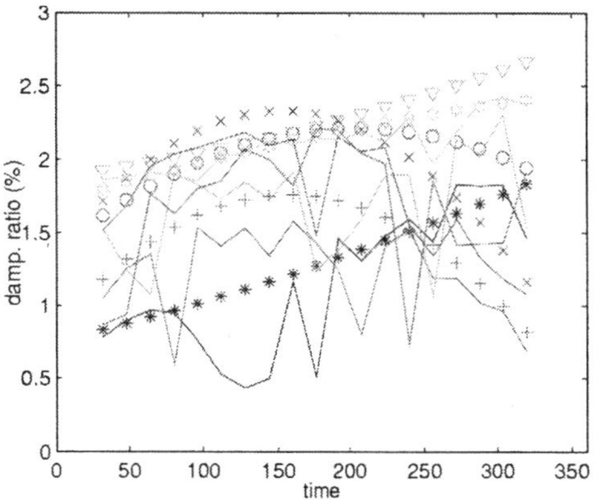

Fig. 3. Monitoring of the damping ratios using input/output sequences of 64s (ML estimation without transient polynomials). Dots: true damping ratios; Solid line: estimated values.

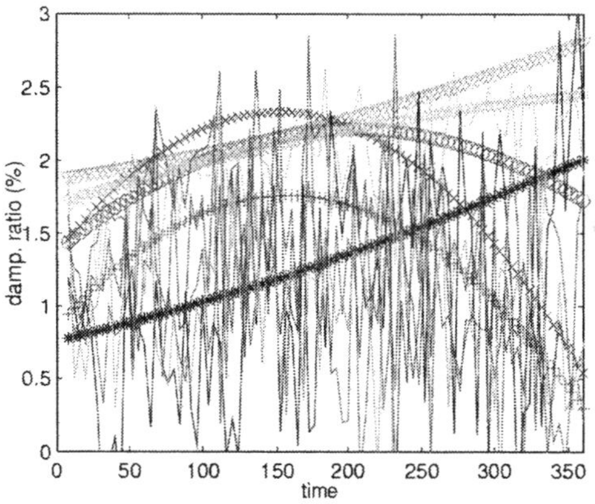

Fig. 4. Monitoring of the damping ratios using input/output sequences of 16s (ML estimation without transient polynomials). Dots: true damping ratios; Solid line: estimated values.

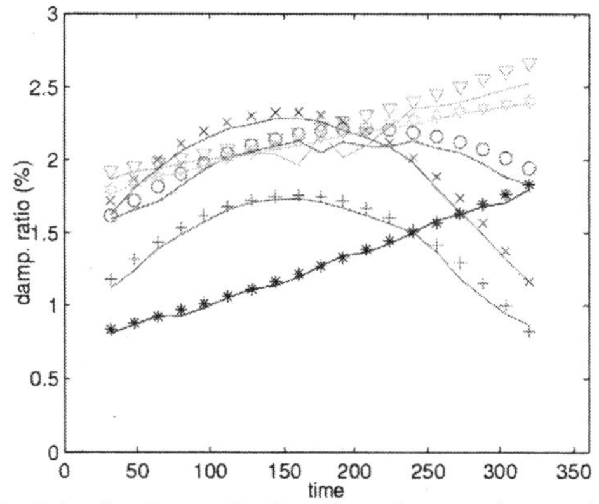

Fig. 5. Monitoring of the damping ratios using input/output sequences of 64s (ML estimation with transient polynomials). Dots: true damping ratios; Solid line: estimated values.

In Figure 7 the damping estimates are given when neglecting the ML frequency weighting (i.e., when minimizing the ML cost function with $W_o^{\text{ML-IO-T}}(\omega_f,\boldsymbol{\theta})=1$). This results in an increase of the variability of the estimates. Hence, a proper frequency weighting is important to reduce the variability of the damping estimates.

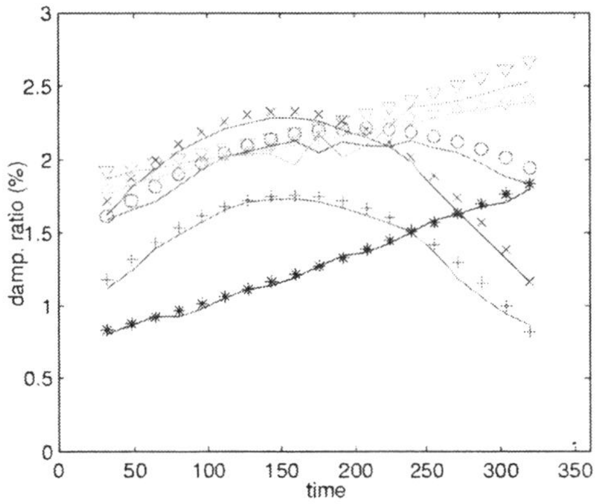

Fig. 6. Monitoring of the damping ratios using input/output sequences of 16s (ML estimation with transient polynomials). Dots: true damping ratios; Solid line: estimated values.

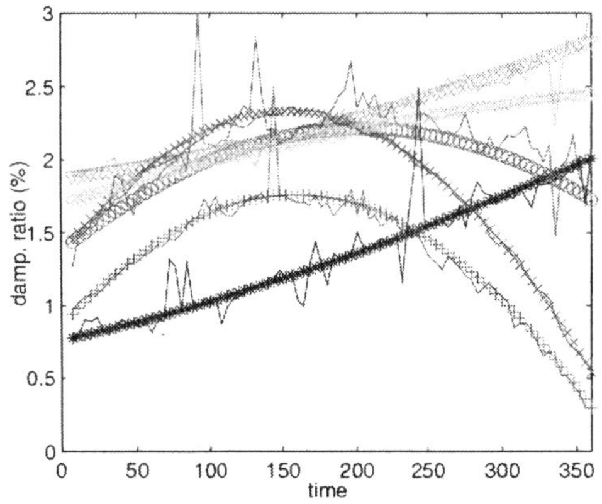

Fig. 7. Monitoring of the damping ratios using input/output sequences of 16s (ML estimator with transient polynomials and $W_o^{\text{ML-IO-T}}(\omega_f, \boldsymbol{\theta}) = 1$). Dots: true damping ratios; Solid line: estimated values.

5. CONCLUSIONS

An important goal of the EUREKA-FLITE project is to reduce the required test time and to improve the accuracy of the damping monitoring and flutter prediction by means of improved system identification techniques. It has been shown that this goal can be obtain by means of frequency-domain estimators by introducing transient polynomials. By doing so, small records can be processed without suffering from leakage. An advantage of frequency-domain estimators is the fact that the quality of the data can be taken into account via a frequency weighting function. A proper frequency weighting turns out to be important to reduce the variability of the damping estimates. All proposed estimators are based on a common-denominator discrete-time transfer-function model. These models are well conditioned and optimized algorithms have been design to reduce the computation time.

REFERENCES

Cauberghe B., P. Guillaume, P. Verboven, and E. Parloo. Identification of modal parameters including unmeasured forces and transient effects. *Journal of Sound and Vibration*, accepted for publication.

Guillaume P., P. Verboven and S. Vanlanduit (1998), Frequency-Domain Maximum Likelihood Identification of Modal Paramaters with Confidence Intervals. *Proceedings of the International Conference on Noise and Vibration Engineering (ISMA-23)*, Leuven (Belgium), September 16-18, pp. 359-366.

Kailath, T. (1980). *Linear Systems*. Prentice Hall.

Pintelon R., P. Guillaume, Y. Rolain, J. Schoukens and H. Van hamme (1994). Parametric Identification of Transfer Functions in the Frequency Domain - a Survey. *IEEE Trans. Autom. Control*, vol. 39, no. 11, pp. 2245-2260.

Pintelon R., J. Schoukens, and G. Vandersteen (1997). Frequency domain system identification using arbitrary signals. *IEEE Trans. Autom. Control*, vol. 42, no. 12, 1717-1720.

Pintelon R. and J. Schoukens (2001). *System Identification: A Frequency Domain Approach*. IEEE Press and John Wiley & Sons.

Van der Auweraer H., P. Guillaume, P. Verboven and S. Vanlanduit (2001). Application of a Fast-Stabilizing Frequency Domain Parameter Estimation Method. *ASME Journal of Dynamic Systems, Measurement and Control*, vol. 123, no. 4, pp. 651-658.

Verboven P. (2002). *Frequency Domain System Identification for Modal Analysis*, PhD Thesis, Vrije Universiteit Brussel (VUB).

IFAC
Publications
www.elsevier.com/locate/ifac

REAL-TIME MODAL ANALYSIS AND ITS APPLICATION FOR FLUTTER TESTING

Tadeusz Uhl, Mariusz Bogacz

University of Mining and Metallurgy, 30 Mickiewicza Avenue, Kraków, 30-059, Poland
e-mail: tuhl@rob.wibro.agh.edu.pl mbogacz@agh.edu.pl

Abstract: This paper presents two methods for flutter testing during an aircraft flight. Proposed procedures are based on time signal windowing and recursive identification method. There are discussed the simulation results of proposed procedures for two degrees of freedom system with varying damping. The methods have been applied to identify the flutter region for trainer jet directly from flight vibration measurements. The flutter was detected on the left and right aft stabilator. *Copyright © 2003 IFAC*

Keywords: Recursive Least Square, on-line identification, ARMA model, aircraft.

1. INTRODUCTION

Flutter is a complex phenomenon, where in the classical approach, two or more structural modes are coupled and excited trough time dependent on aerodynamic loads. The flutter phenomenon is related to self-excited vibration phenomena present at certain forward speed of flow. For the aircraft, flutter can occur if critical velocity of airplane is exceeded. When flight takes place below the critical flutter speed any vibrations will be damped. When flying above that velocity, the structural vibrations will grow. If those dynamic vibrations are not be limited by nonlinear aerodynamic or structural behavior, they will cause structural failure. It is mandatory to design an airplane structure to operate in the operational flight envelope outside critical flutter speed. Safety margin of operational velocity envelope is usually at least 15% of critical flutter speed.

The modal parameter which is responsible for increasing of vibrations amplitude for given flight speed is damping. Basically, the damping ratio of the critical mode has been employed as the index of flutter stability margin. An additional information regarding estimation of the modal damping can be found in literature (Torii, and Matsuzaki, 2001).

If damping is less then zero, amplitude of vibrations increasing immediately (explosive flutter). In order to predict the flutter speed for the system, the variation of the damping ratios from airspeed shall to be monitored. Flutter occurs, when at least one of the damping ratios becomes zero. The mathematical model, which is commonly used for flutter investigation of the airplane, is a modal model. Ground Vibration Test (GVT) provides a preliminary check of a mathematical model according to flutter. The modes with a very small damping ratio should be carefully investigated according to flutter phenomenon.

Nowadays, there is a possibility to monitor these modes on-line. There are techniques for modal analysis of the structures during a flight as well. The methods of modal analysis with ambient excitation during experiments are widely used in many area for diagnostic and design purposes (Lisowski, and Uhl, 2001). There are several methods available in time as well as frequency domains. Frequency domain techniques that were applied for output-only vibration measurement, consist of application procedures employing for standard modal model identification methods. These procedures use auto-, cross-spectra or transmissibility functions instead of FRF.

Application of the methods based on Fourier transform requires stationary conditions of the measured signals. This demand is actually hardly ever satisfied during ambient excitation test, for operational excitation (rotating machinery) or during vehicle/aircraft motion/flight. Therefore, in order to obtain more accurate results, time domain identification methods were applied. Some of these methods uses introductory step consists of correlation function determination. This approach enables to reduce the database size and improves data quality with respect to raw data. The LSCE method is an example of correlation driven method using decomposition of correlation functions into decaying sine components.

The second approach uses Stochastic Subspace Identification methods, applicable to row data as well as correlation functions. A great variety of algorithms might be found in (James, et al., 1995).

The Operational Modal Analysis (OMA) method was successfully applied to civil engineering structures and aviation (Lisowski, and Uhl, 2001). In presented system, time domain OMA algorithms are implemented. Time domain identification methods using in-operation measurements can be divided into three basic categories:

- methods using auto-correlation and cross-correlation of signals,
- methods using auto-regression function for the response signals,
- methods realized stochastic subspace.

There are no modal analysis methods which can yield results in real time for airplanes in order to identify flutter.

Damping parameters of mechanical structures for particular mode can be estimated based on several methods. The most common are: half-power method (Brown, et al., 1979), circle fitting method, resonance amplitude method (Uhl, 1998) and method based on wavelet transform. The methods formulated for state space model (like DSRA, Extended Kalman filtering) or discrete parametric model of a structure are applied as well. Some of these methods can be used on-line to help detect early changes of damping for monitoring modes.

In this case the method base on data segment from measurement system. The damping is identified for each segment of the data independently. The methods based on on-line parameters updating for ARMA/AR model can be also applied to damping estimation. The main problem is to calculate the roots of polynomial on-line. The calculation of all the roots is not easy task and it is usually too time consuming. However, the stability margin can be investigated without solution of characteristic equation (Torii, and Matsuzaki, 2001).

Both methods, based on signal segmentation as well as parameters updating for varying damping ratios were tested and applied for real aircraft structures. The results are presented in the paper.

2. FLUTTER ON-LINE IDENTIFICATION METHOD

The following methods are applied for on-line flutter testing:

- a method based on measured signal segmentation,
- a method based on recursive parameters updating.

Both methods presented in the paper are based on difference equation of a model, which was form for MDOF system. The difference equation describes system dynamics as follows:

$$y(i) = -a_1 y(i-1) - a_2 y(i-2), ..., -a_{nA} y(i-nA) + ... \\ ... + e(i) + c_1 e(i-1) + c_2 e(i-2), ..., + c_{nC} e(i-nC). \quad (1)$$

where:
y is the system output,
e is a white noise,
a_i (AR part) and c_i (MA part) for i=1..n are model parameters.

This model is commonly used for modeling of time series and dynamic system when only output is measured. The AR part describes system properties and MA disturbances which corrupted measured signal. Stationary model, which is assumed to be used in time series segmentation technique, can be transformed to form of discrete transfer function by using z-transform (Ljung, 1987; Ljung, and Söderström, 1986):

$$y(i) = H(z^{-1})e(i) = \frac{C(z^{-1})}{A(z^{-1})} e(i) \quad (2)$$

where:
$A(z^{-1}) = 1 + a_1 z^{-1} + a_2 z^{-2} +, ..., + a_{na}(i) z^{-na}$ is AR part of the model in a form of polynomial with order na,
$C(z^{-1}) = 1 + c_1 z^{-1} + c_2 z^{-2} +, ..., + c_{nc} z^{-nc}$ is MA part of the model in a form of polynomial with order nc.

In order to find damping ratio for particular modes, poles of the system should be found. The poles can be calculated by finding roots of polynomial $A(z^{-1})$. The order of polynomial depends on number of modes which should be monitored. For non-stationary case, which is considered, if parameter updating method is used the model has the following form (Söderström, and Stoica, 1988; Ljung, and Söderström, 1986):

$$y(i) = H(z^{-1},i)e(i) = \frac{C(z^{-1},i)}{A(z^{-1},i)}e(i) \qquad (3)$$

where:

$A(z^{-1},i) = 1 + a_1(i)z^{-1} + a_2(i)z^{-2} +, \ldots, + a_{na}(i)z^{-na}$

$C(z^{-1},i) = 1 + c_1(i)z^{-1} + c_2(i)z^{-2} +, \ldots, + c_{nc}(i)z^{-nc}$

As it can be easily noticed the model parameters depend on time. To find model parameters, particularly coefficient of $A(z^{-1})$ polynomial, algorithms specific to every case can be applied. Such algorithms are presented in section 2.1 and 2.2. of this paper.

The next step of damping identification procedure is determination of system poles positions. To get numerical value of the poles the characteristic equation should be solved on-line. In the paper approximation of complex roots for discrete system using Bairstow's method has been applied. As a result real and imaginary part of roots have been obtained. In order to avoid influence of the modes outside the interesting frequency band, the measured signal was filtered. The limitation of a frequency band enables to select model order easier as well. The filtering operation was done before approximation of the model parameters and roots extraction. In this particular case, only two modes will be considered, therefore model order is assumed as fourth. To find modal damping ratio, the system roots for continuous time are needed. The relation between continuous and discrete roots is as follows:

$$z_i = \exp(s_i T) \qquad (4)$$

where:

z_i is a root of characteristic polynomial of discrete time,

s_i is a root of characteristic polynomial of continuous time,

T is a sampling period.

The damping ratio for continuous case can be found by using following formula:

$$s_r = a_r + jb_r \qquad (5)$$

$$\omega_r = \sqrt{a_r^2 + b_r^2} \qquad (6)$$

$$\xi_r = \frac{-a_r}{\omega_r} \qquad (7)$$

where:

ω_r and ξ_r are natural frequency and damping for r-th mode respectively.

For discrete system equivalent formula can be used:

$$\xi_i = -(\ln|z_i| / |\ln z_i|) \qquad (8)$$

Using formula (8), the damping ratio is calculated in proposed algorithm.

2.1 Method based on signal segmentation technique

The time-frequency method is based on assumption that for short time interval damping is not changed. The idea of signal segmentation and analysis of period overlapping is presented in figure 1. The data from short signal segment are used to estimate ARMA model's parameters by using Least Square procedure or more sophisticate algorithm such as Instrumental Variable or Sub-space method (Söderström, and Stoica, 1988). Based on ARMA model's parameters, damping ratio for particular modes can be calculated by system poles extraction. The procedure is repeated for each segment independently. The algorithm of the method is depicted in figure 2.

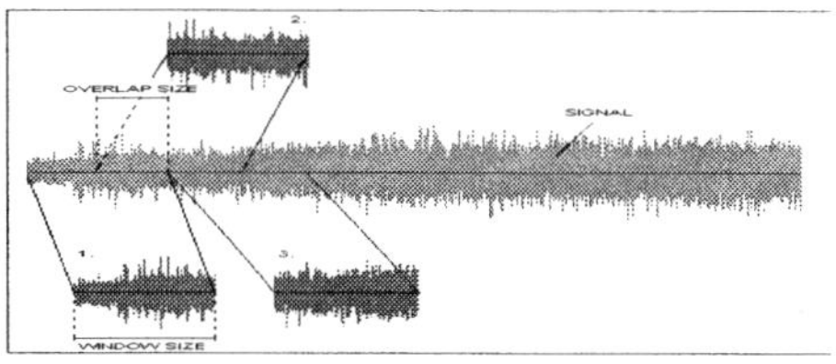

Fig. 1. Idea of signal segmentation for flutter testing.

The ARMA model's parameters for one signal segment can be found by using Least Square estimator for N samples in the form:

$$\theta(N) = (\Phi^T(N)\Phi(N))^{-1}\Phi^T(N)y(N) \qquad (9)$$

where:

$\{\varphi(k-1)\}^T = \{-y(k-1), -y(k-2), \cdots, -y(k-N), e(k-1), e(k-2), \cdots, e(k-N)\}$

is the data vector,

$\{\theta\} = \{a_1, a_2, \cdots a_{na}, c_1, c_2, \cdots, c_{nc}\}^T$

is parameters vector.

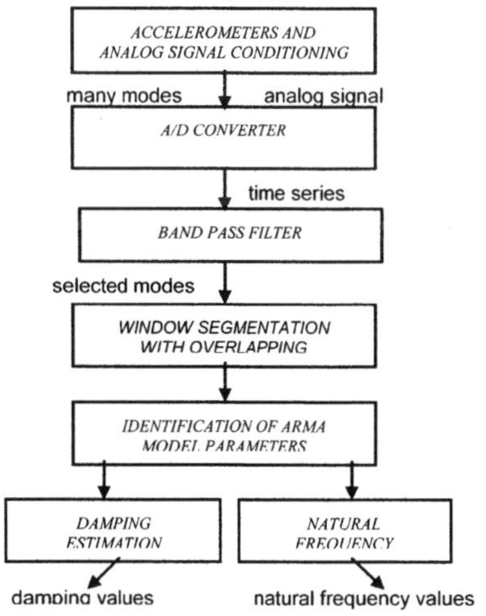

Fig. 2. Flowchart of damping identification method based on signal segmentation.

To apply the method for any case, the choice of using larger window for better accuracy or smaller one for better parameters tracking should be decided.

The method was tested on the system with two degrees of freedom and applied to the data from trainer jet flight.

2.2 Model based on recursive parameters updating

The time domain method is formulated for linear system with time-varying parameters. The new model parameters are estimated for each step of iteration based on values from previous calculation. The recursive identification methods are widely used in adaptive control system (Söderström, and Stoica, 1988). The flowchart of the recursive procedure for damping estimation is depicted in figure 3.
Parameters estimator is computed at each sampling interval according to minimum of objective function which is defined as follows:

$$V\left[\hat{\theta}(i)\right] = \sum_{j=1}^{i} \alpha^{i-j}\left[y(j) - \varphi^{T}(j-1)\theta(i)\right]^{2} \qquad (10)$$

where:
α is forgetting factor, which describes influence of j-th sample on i-th estimator, $0<\alpha\leq 1$.

If model parameters are not changing α should be set as 1. The proper choice of α value gives better parameters tracking properties of the algorithm. Speed of adaptation of the algorithm is measured by T_α given by formula:

$$T_{\alpha} = \frac{1}{f_{s}(1-\alpha)} \qquad (11)$$

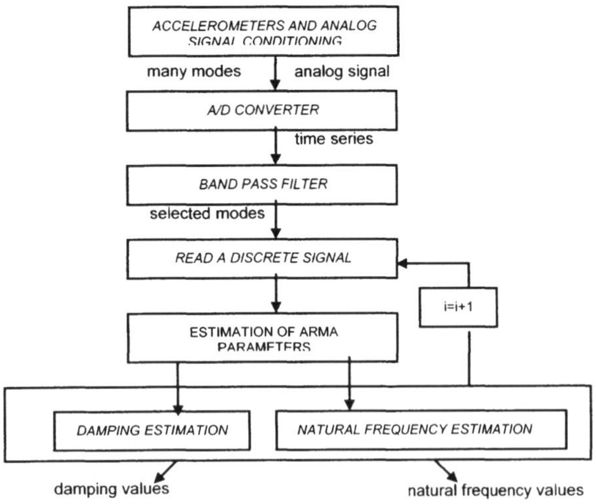

Fig. 3. Flowchart of recursive method for damping estimation.

The basic assumption for the algorithm is that the matrix P is positive definite at each iteration step:

$$\mathbf{P}^{-1}(i) \overset{def}{=} \sum_{j=1}^{i} \alpha^{i-j}\varphi(j-1)\varphi^{T}(j-1) \qquad (12)$$

The matrix P^{-1} can be singular if:

- system is not enough good excited,
- direct influence of input to output is observed,
- overfitting of polynomial A and C,
- sampling frequency is to big,
- not proper value of α was chosen.

The ARMA model's parameters in presented method are estimated from iterative formula:

$$\begin{aligned} &\theta(i+1) = \theta(i) + K(i)(y(i+1) - \phi(i+1)^{T}\theta(i)) \\ &K(i) = P(i+1)\phi(i+1) \\ &P(i+1) = (I - K(i)\phi(i+1)^{T})P(i) \end{aligned} \qquad (13)$$

The procedure was tested based on simulation of two degrees of freedom model with varying damping parameters. The same test was carried out with time-varying forgetting factor.

For tracking of the time-varying parameters of the system the proper forgetting factor λ should be used. If $\lambda < 1$, the algorithm is more sensitive to changes and estimate parameter changes more quickly. From practical point of view the forgetting factor can be changed as follows:

- to let λ tend to exponentially to 1, or

$$\lambda(t) = 1 - \lambda_{t}^{0}(1-\lambda_{0}) \qquad (14)$$

This equation can be implemented in more efficient way as a recursive routine:

$$\lambda(t) = \lambda_{0}\lambda(t-1) + (1-\lambda_{0}) \qquad (15)$$

The typical value λ_0 ranges from 0.95 to 0.995.

- λ is the function of the estimation error given in the following form

$$\lambda(nT) = \begin{cases} \lambda((n-1)T)\lambda_{d} + (1-\lambda_{d}) & if(\hat{e}(nT)^{2} < e_{d}) \\ \lambda_{r} & if(\hat{e}(nT)^{2} \geq e_{d}) \end{cases} \qquad (16)$$

The values of λ_d, λ_r range from 0.95 to 0.99. The value e_d should be chosen during experiment.

3 VERIFICATION OF PROPSED DAMPING ESTIMATION ALGORITHMS

To verify proposed algorithms simulation study of two degrees of freedom mechanical system has been done. The scheme of the system is shown in figure 4. White noise was applied as an excitation. Damping coefficients for simulated system have been changed at sample equals 6000. The damping profile change is shown in figure 5. The simulation results in the form of displacement and velocity for all degrees of freedom are depicted in figure 6. To identify damping for each mode two procedures presented above are employed. The results of damping estimation are shown in figure 7 for recursive algorithm and in figure 8 for window segmentation method.

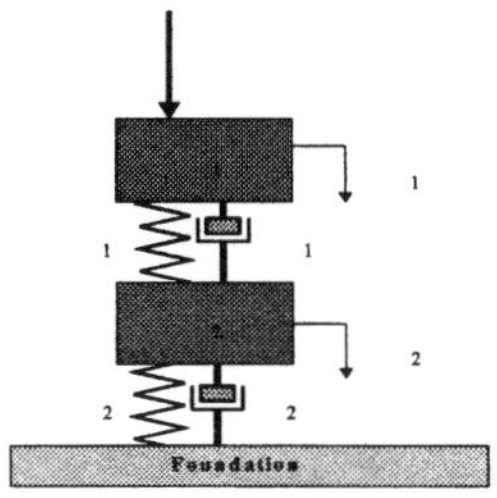

Fig. 4. Scheme of simulated system.

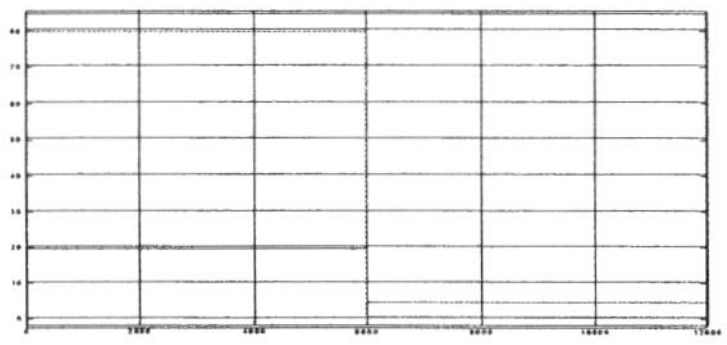

Fig. 5. Damping coefficient time profile.

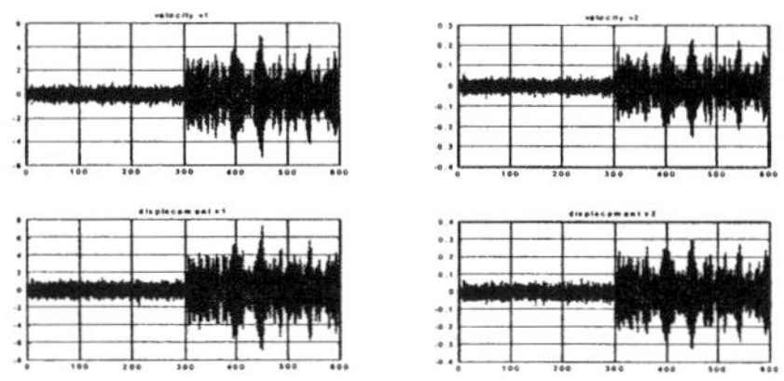

Fig. 6. Simulation results in a form of displacement and velocity of particular mass.

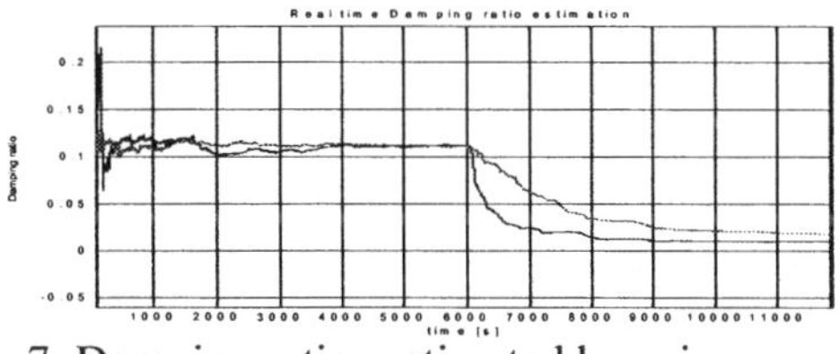

Fig. 7. Damping ratios estimated by using recursive method.

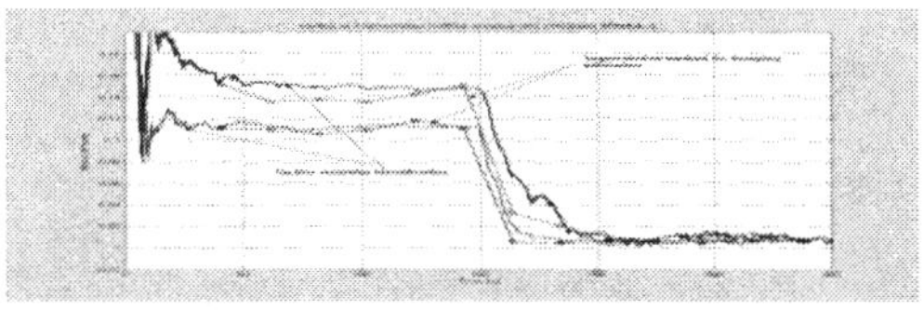

Fig. 8 Damping ratios for one selected mode estimated employing windows segmentation method as well as RLS algorithm with time-varying forgetting factor.

APPLICATION OF PRESENTED METHODES FOR FLUTTER TESTING OF TREINER JET

Flight tests were performed on the airplane, which is: a single-engine, two-seat trainer jet, used for preliminary and advanced fighter pilot training.

Diving was used as a structural excitation. Dive maneuvers are required to obtain airspeeds greater than those which can be achieved in a straight and level flight. Such flight conditions may result in structural instability and flutter induced vibration. The biggest advantage of this type of excitation is the fact that no special on-board exciter hardware is required. Unfortunately, there are also several disadvantages. The observed turbulence is often not intense enough to produce sufficient excitation compared with that obtained from on-board exciters. Turbulence usually excites only the lower frequency modes for the most airplanes.

Ten accelerometers were installed to measure dynamic response of the airplane during the flight flutter tests. The accelerometers were located in vertical and horizontal tails and wings of the aircraft.

To check flutter at speeds higher than 730km/h it was necessary for the pilot to dive the airplane. The procedure was to fly the plane to a high altitude. Then the airplane was put in a dive using maximum power so that the desired speed was reached (ranging from 730 km/h to 870 km/h). During the dive the accelerometers output signals were continuously recorded. The signal obtained from the measurement (DAS) is a time series with sampling frequency equals 160Hz. The recorded signal at point 7 is shown in figure 10. The data were processed using a band pass digital filter in frequency band between 20Hz and 55 Hz. to eliminate noise and higher modes influence. The time history of the signal indicates two vibration modes which are excited well. These two modes have been analyzed to detect damping and natural frequency variations.

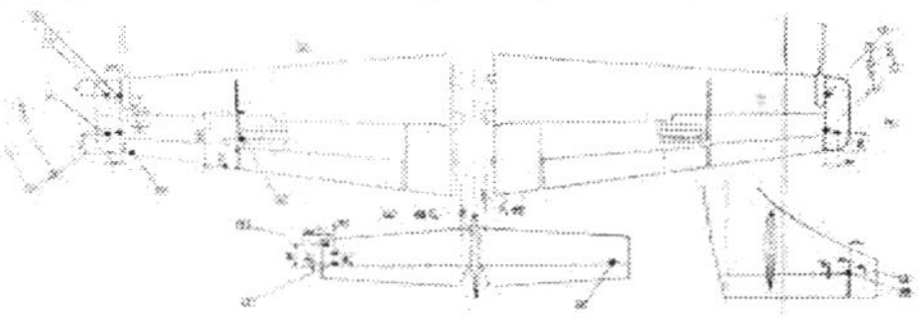

Fig. 9. Locations of accelerometers during a flight flutter test.

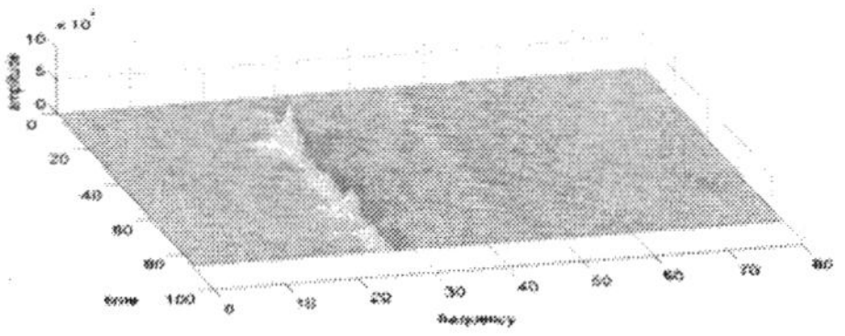

Fig. 10. Time frequency plot for signal measured during flight at measurement point #7.

The results obtained from damping estimation procedure based on time window segmentation are shown in figure 11. The structure response #7, measured on left aft stabilator tip presents variation of damping for first mode and some instabilities region, which indicate occurrence of the flutter at this point. Similar results have been obtained for point #8. But for rest of the point no instabilities have been observed.

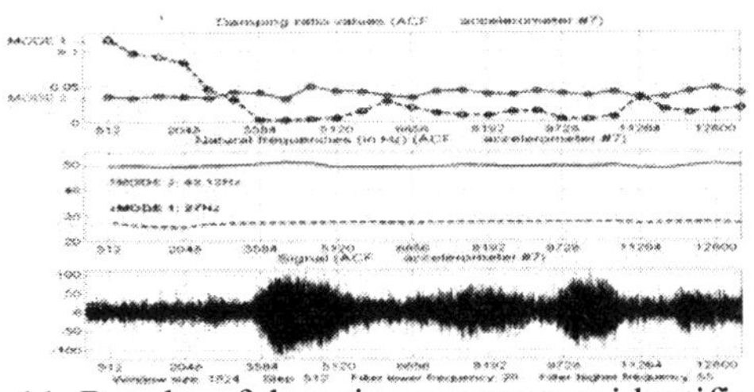

Fig. 11. Results of damping parameters identification from flight data using signal segmentation method.

The above estimations indicate that flutter was detected only on left and right aft stabilator. It occurred when modal damping of mode 1 of these signals became close-to-zero values, which resulted in structural instability. All damping estimates of mode 2 are above the flutter region, therefore it can be considered as "free" from flutter. The discovered instabilities were analyzed with bigger resolution. For this purpose the window size and overlapping were set to 256 and 200 samples respectively (step – 56 samples) and only a portion of the signal was selected for analysis. The results of the estimation are shown in figure 12.

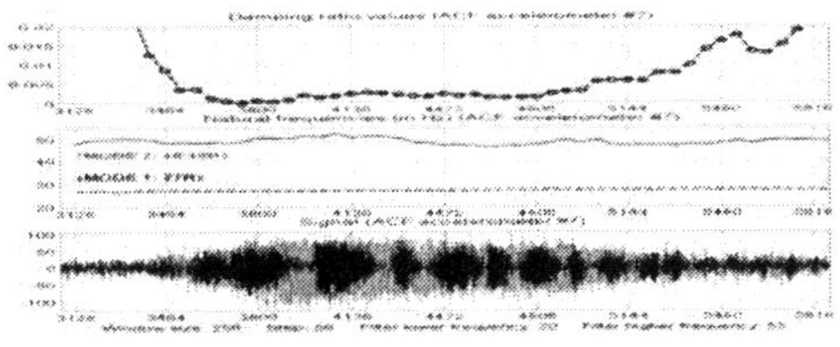

Fig. 12. Damping variations for first mode in instability region for point #7.

The result obtained from recursive identification algorithm is depicted in figure 13.

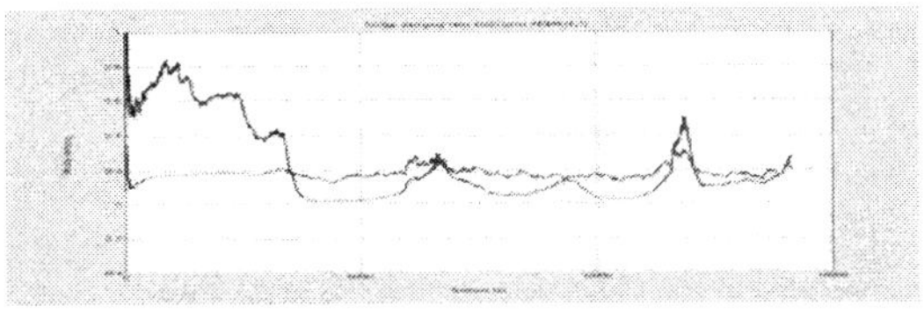

Fig. 13. Damping identification results obtained using recursive identification method.

The results are very similar to obtain from windows segmentation method but there is no observed smoothing effect of time windowing.

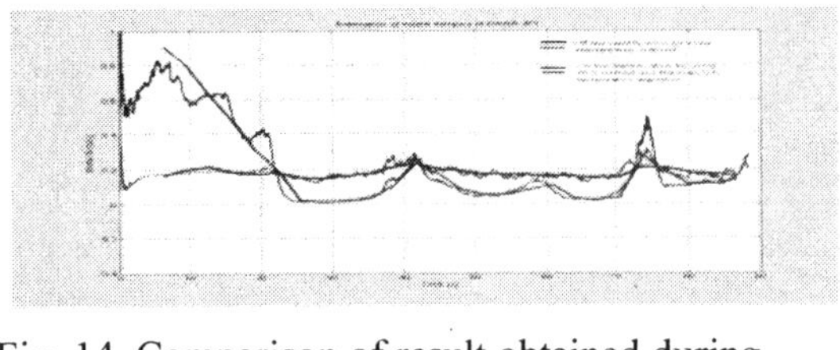

Fig. 14. Comparison of result obtained during damping identification by using recursive identification and segmentation method.

4. CONCLUSION AND FINAL REMARKS

The authors proposed methods for flutter testing based on in-flight measurements. Two methods for real time damping estimation have been tested successfully. The signal segmentation algorithm gives smoother results but delayed on time depends on time window and overlapping size. This method can be applied to on-line identification in non-time critical system. The estimation based on recursive identification algorithm gives damping ratio quicker but results strongly depends on algorithm parameters, mainly forgetting factor. The recursive identification method can be useful for practical flutter estimation after very careful algorithm parameters adjustment.

ACKNOWLEDGEMENTS

This research was supported by FLITE EUREKA project EU 1246.

REFERENCES

Brown, D., R. Allemang, R. Zimmerman, M. (1979), *Parameter Estimation Techniques for Modal Analysis*, SAE Paper 7902221, pp. 19.

Haylen, W., S., Lammens, P., Sas (1997), *Moda Analysis Theory and Practice*, KU Leuven.

James, G.H.III, T.G., Carne, J.P., Laufer (1995), *The Natural Excitation Technique (NexT) for Modal Parameter Extraction from Operating Structures*, Int. Journal of Analytical and Experimental Modal Analysis, V.10, No. 4, pp. 260-277.

Lisowski, W., T., Uhl (2001), *Application of modal analysis to diagnostics of complex mechanical systems, Exploitation Problems of Machines*, Polish Academy of Science, Vol. 36, No. 4.

Lisowski, W., T., Uhl (2001), *In-operation modal analysis and it's application for diagnostics*, AGH, Kraków.

Ljung, L. (1987), *System identification theory for the user*, Prentice Hall, Englewood Cliffs, NJ.

Ljung, L., T., Söderström (1986), *Theory and practical of recursive identification*, The MIT Press, London.

Pinkelman, J. K., S. M., Batill, M. W., Kehoe (1996), *Total Least Squares Criteria in Parameter Identification for Flight Flutter Testing*, Journal of Aircraft, Vol. 33, No. 4, pp. 784-792.

Söderström, T., P., Stoica (1988), *System identification*, Prentice-Hall International, Hemel Hempstead, U.K.

Torii, H., Y., Matsuzaki (1997), *Flutter boundary prediction based on nonstationary data measurement*, Journal of Aircraft, vol.34, No. 3.

Torii, H., Y., Matsuzaki (2001), *Flutter margin evaluation for discrete-time systems*, Journal of Aircraft, vol.38, No. 1, pp. 42-47.

IFAC Publications
www.elsevier.com/locate/ifac

STATISTICAL APPROACH TO FLUTTER MONITORING [1]

Laurent Mevel [*,2] **Michèle Basseville** [*,3] **Albert Benveniste** [*,2]

[*] IRISA, *Campus de Beaulieu, 35042 Rennes Cedex, F.-*
`Firstname.Name@irisa.fr`

Abstract: We investigate the flutter monitoring problem, stated as a statistical hypotheses testing problem regarding a specified damping coefficient. We previously advocated for a modal monitoring algorithm based on a residual associated with subspace-based covariance-driven identification and on the statistical local approach to the design of detection algorithms. In this paper, we describe two types of flutter monitoring tests, based on such a subspace-based residual designed for a given damping coefficient. The first type of tests comes up from variations on the local approach and the GLR test adapted to several null and alternative hypotheses on the damping. The second test builds on a different approximation for the residual and on the CUSUM test.

Keywords: Subspace-based identification algorithms, change detection, statistical tests, local approach, flutter monitoring, damping coefficient.

1. INTRODUCTION

The development of new aircrafts requires a careful exploration of the dynamical behaviour of the structure subject to vibration and aeroelastic forces. This is achieved via a combination of ground and in-flight tests. Investigations aiming at improving the exploitation of flight test data have been undertaken, in two directions. First, the use of output-only system identification methods is investigated for exploiting tests under natural excitation conditions (e.g., turbulent), without resorting to artificial control surface excitation. Second, algorithms achieving the on-line in-flight exploitation of flight test data are expected to allow a more direct exploration of the flight domain, with improved confidence and reduced costs. One important issue to be addressed on-line is the flight flutter monitoring problem.

While frequencies and mode shapes are usually the most important parameters in structural analysis, the most critical ones in flutter analysis are the damping factors, for some critical modes. It is known, e.g. from Cramer-Rao bounds (Gersch, 1974) that damping factors are difficult to estimate accurately. For improving the estimation of damping factors, and moreover for achieving this in real-time during flight tests, one possible although unexpected route is to resort to monitoring algorithms able to detect whether some damping factor decreases below some critical value. The rationale is that monitoring algorithms usually have a much shorter response time than identification algorithms.

The purpose of this paper is to describe and analyze two types of statistical tests for monitoring a damping factor. Both tests types are based on a statistics associated with subspace-based output-only covariance driven identification. The first type of tests handles *local* statistical hypotheses, and comes up from variations on the local approach and the GLR test adapted to several null and alternative hypotheses on the damping. The second test handles *non local* statistical hypotheses and builds on a different approximation for the residual and on the CUSUM test.

The paper is organized as follows. Section 2 summarizes the main lines of the subspace-based monitoring statistics we proposed and investigated in earlier

[1] This work has been carried out within the framework of Eureka project no 2419 FLITE (**Flight Test Easy**), coordinated by Sopemea, Velizy-Villacoublay, France.
[2] Also with INRIA.
[3] Also with CNRS.

papers. In section 3, we introduce the first type of tests for flutter monitoring, based on local hypotheses. The second test for flutter monitoring, based on non local hypotheses, is described in section 4. Section 5 is devoted to implementation issues and experimental results. Some conclusions are drawn in section 6.

2. SUBSPACE-BASED MODAL IDENTIFICATION AND MONITORING

We first recall the main equations and parameters of the models we use, and we recall the main lines of the subspace-based covariance-driven identification and monitoring algorithms we advocate for.

Modeling and parameterizations. The structure's behavior is assumed to be described by a stationary linear system:

$$M\ddot{\mathcal{Z}}(t) + C\dot{\mathcal{Z}}(t) + K\mathcal{Z}(t) = \nu(t), Y(t) = L\mathcal{Z}(t) \quad (1)$$

where t denotes continuous time, M, C, K are the mass, damping and stiffness matrices respectively, (high dimensional) vector $\mathcal{Z}$ collects the displacements or accelerations of the degrees of freedom of the structure; the external (non measured) force ν is modeled as a non-stationary white noise with time-varying covariance matrix $Q_\nu(t)$, measurements are collected in the (often, low dimensional) vector Y, and matrix L indicates which components of the state vector are actually measured (where the sensors are located). The modes or eigen-frequencies denoted generically by μ, and the modeshapes or eigenvectors denoted generically by $L\Psi_\mu$, are solutions of:

$$\begin{cases} \det\,(\mu^2\, M + \mu\, C + K) &= 0 \\ (\mu^2\, M + \mu\, C + K)\, \Psi_\mu &= 0 \end{cases} \quad (2)$$

Sampling model (1) at rate $1/\tau$ yields the discrete time model in state space form:

$$\begin{cases} X_{k+1} &= F\, X_k + V_{k+1} \\ Y_k &= H\, X_k \end{cases} \quad (3)$$

where $X_k = \begin{pmatrix} \mathcal{Z}(k\tau) \\ \dot{\mathcal{Z}}(k\tau) \end{pmatrix}, Y_k = Y(k\tau), H = (L\; 0), F = e^{\mathcal{L}\tau}$, $\mathcal{L} = \begin{pmatrix} 0 & I \\ -M^{-1}K & -M^{-1}C \end{pmatrix}$, and state noise V_{k+1} is *unmeasured*, Gaussian, zero-mean, white, with covariance matrix: $Q_{k+1} \stackrel{\text{def}}{=} \mathbf{E}(V_{k+1}\; V_{k+1}^T) = \int_{k\tau}^{(k+1)\tau} e^{\mathcal{L}s}\, \widetilde{Q}(s)\, e^{\mathcal{L}^T s} ds$, where $\mathbf{E}(.)$ denotes the expectation operator and $\widetilde{Q}(s) = \begin{pmatrix} 0 & 0 \\ 0 & M^{-1}Q_\nu(s)M^{-T} \end{pmatrix}$. State X and observed output Y have dimensions $2m$ and r respectively, with r (often much) smaller than $2m$ in practice. The modal parameters defined in (2) are equivalently found from the eigenstructure $(\lambda, \varphi_\lambda)$ of the state transition matrix F:

$$e^{\tau\mu} = \lambda, \quad L\,\Psi_\mu = \phi_\lambda \stackrel{\text{def}}{=} H\,\varphi_\lambda$$

The frequency f and damping coefficient ρ are recovered from a given eigenvalue λ through:

$$f = a/2\pi\tau, \quad \rho = 100\,|b|/\sqrt{a^2 + b^2}$$
$$\text{where } a = |\arctan \Im(\lambda)/\Re(\lambda)|\,,\; b = \ln|\lambda| \quad (4)$$

Eigenvectors are real if proportional damping is assumed, that is $C = \alpha M + \beta K$. The λ's and ϕ_λ's are pairwise complex conjugate. We assume that the system has no multiple eigenvalues. In addition, 0 is *not* an eigenvalue of state transition matrix F. The collection of modes (λ, ϕ_λ) form a canonical parameterization[4] of the pole part of the system in (3), referred to as the system eigenstructure. From now on, the collection of modes is also considered as the system parameter θ:

$$\theta \stackrel{\text{def}}{=} \begin{pmatrix} \Lambda \\ \text{vec}\Phi \end{pmatrix} \quad (5)$$

In (5), Λ is the vector whose elements are the eigenvalues λ, Φ is the matrix whose columns are the ϕ_λ's, and vec is the column stacking operator. Parameter θ has size $(r+1)m$.

Residual associated with subspace identification. The system eigenstructure is related to the SVD of the output-only covariance-driven Hankel matrix $\mathcal{H}_{p+1,q}$

$$\mathcal{H}_{p+1,q} \stackrel{\text{def}}{=} \text{Hank}\,(R_i)\,,\; R_i \stackrel{\text{def}}{=} \mathbf{E}Y_k Y_{k-i}^T \quad (6)$$

as follows. The left factor of the SVD of $\mathcal{H}_{p+1,q}$ yields the observability matrix $\mathcal{O}$. Then, (H, F) are recovered from the shift invariance property of $\mathcal{O}$ and (λ, ϕ_λ) from eigenstructure analysis of (H, F).

Choosing the eigenvectors of matrix F as a basis for the state space of model (3) yields the following *modal* representation of the observability matrix:

$$\mathcal{O}_{p+1}(\theta) = \begin{pmatrix} \Phi \\ \Phi\Delta \\ \vdots \\ \Phi\Delta^p \end{pmatrix} \quad (7)$$

where diagonal matrix Δ is defined as $\Delta = \text{diag}(\Lambda)$, and Λ and Φ are as in (5). Whether a nominal parameter θ_0 is in agreement with a given output covariance sequence $(R_j)_j$ is characterized by (Basseville *et al.*, 1987; Viberg, Wahlberg and Ottersten, 1997) $\mathcal{O}_{p+1}(\theta_0)$ *and* $\mathcal{H}_{p+1,q}$ *having the same left kernel space.* The latter property can be checked as follows. From the nominal θ_0, compute $\mathcal{O}_{p+1}(\theta_0)$ using (7), and perform e.g. a SVD of $W_1\mathcal{O}_{p+1}(\theta_0)$ for extracting a matrix U such that $U^T\; U = I_s$ and $U^T\; W_1\; \mathcal{O}_{p+1}(\theta_0) = 0$. Matrix U is not unique (two such matrices are related through a post-multiplication

[4] A parameterization invariant w.r.t. changes in the state basis.

with an orthonormal matrix), but can be regarded as a function of θ_0. Then the characterization writes:

$$U(\theta_0)^T \, W_1 \, \mathcal{H}_{p+1,q} \, W_2^T = 0 \tag{8}$$

where W_1 and W_2 are arbitrary invertible weighting matrices.

Assume now that *a reference* θ_0 *and a new sample* $Y_1, \ldots, Y_n$ *are available.* For checking whether the data agree with θ_0, the idea is to compute $\widehat{\mathcal{H}}_{p+1,q}$:

$$\widehat{\mathcal{H}}_{p+1,q} \stackrel{\text{def}}{=} \text{Hank}\left(\widehat{R}_i\right), \; \widehat{R}_i \stackrel{\text{def}}{=} 1/n \sum_{k=1}^{n} Y_k Y_{k-i}^T \tag{9}$$

and to define the residual:

$$\zeta_n(\theta_0) \stackrel{\text{def}}{=} \sqrt{n}\, \text{vec}(U(\theta_0)^T W_1 \widehat{\mathcal{H}}_{p+1,q} W_2^T) \tag{10}$$

Let θ be the actual parameter value for the system which generated the new data sample, and $\mathbf{E}_\theta$ be the expectation when the actual parameter is θ. From (8), we know that: $\mathbf{E}_\theta\,(\zeta_n(\theta_0)) = 0 \;\text{ iff }\; \theta = \theta_0$. Thus, $\zeta_n(\theta_0)$ *has zero mean when no change occurs in* θ*, and nonzero mean if a change occurs,* and thus it plays the role of a residual.

On-board χ^2*-test for modal monitoring.* Testing if $\theta = \theta_0$ holds true requires the knowledge of the distribution of $\zeta_n(\theta_0)$, usually unknown. One manner to circumvent this is to assume, for large n, *small* deviations in θ:

$$\theta = \theta_0 + \delta\theta/\sqrt{n} \tag{11}$$

where vector $\delta\theta$ is unknown, but fixed. This is the statistical local approach (Benveniste, Basseville and Moustakides, 1987; Zhang, Basseville and Benveniste, 1994; Basseville, 1998). Let $\mathbf{E}_\theta$ be the expectation when the actual system parameter is θ, and define the mean deviation:

$$\mathcal{J}(\theta_0) \stackrel{\text{def}}{=} -1/\sqrt{n}\; \partial/\partial\theta\, \mathbf{E}_{\theta_0}\, \zeta_n(\theta)\, |_{\theta=\theta_0} \tag{12}$$

and covariance $\Sigma(\theta_0) \stackrel{\text{def}}{=} \lim_{n\to\infty} \; \mathbf{E}_{\theta_0}(\zeta_n \;\, \zeta_n^T)$. Then, provided that $\Sigma(\theta_0)$ is positive definite, and for all $\delta\theta$, the residual ζ_n in (10) is asymptotically Gaussian distributed (Basseville, Abdelghani and Benveniste, 2000) when assuming θ as in (11):

$$\zeta_n(\theta_0) \xrightarrow[n \to \infty]{} \mathcal{N}\left(\mathcal{J}(\theta_0)\,\delta\theta, \Sigma(\theta_0)\right) \tag{13}$$

Thus a deviation $\delta\theta > 0$ in the system parameter θ is reflected into a change in the mean value of the residual ζ_n. Note that matrices $\mathcal{J}(\theta_0)$ and $\Sigma(\theta_0)$ depend on neither n nor the deviation $\delta\theta$. They can be estimated prior to testing, using data on the safe system, just as the reference θ_0. Consistent estimates of $\mathcal{J}$, based on a data sample, are given in (Basseville, Abdelghani and Benveniste, 2000), and do *not* depend on the particular normalization of the eigenvectors φ_λ. The estimation of Σ is somewhat tricky (Zhang, Basseville and Benveniste, 1994).

Let $\widehat{\mathcal{J}}$ and $\widehat{\Sigma}$ be consistent estimates of $\mathcal{J}(\theta_0)$ and $\Sigma(\theta_0)$, and assume that $\mathcal{J}(\theta_0)$ is f.c.r. Then, deciding whether residual ζ_n is *significantly* different from zero, stated as testing the presence of a small deviation (11), can be achieved with the following χ^2-test:

$$\chi_n^2 \stackrel{\text{def}}{=} \zeta_n^T\, \widehat{\Sigma}^{-1}\, \widehat{\mathcal{J}} \left(\widehat{\mathcal{J}}^T\, \widehat{\Sigma}^{-1}\, \widehat{\mathcal{J}}\right)^{-1} \widehat{\mathcal{J}}^T\, \widehat{\Sigma}^{-1}\, \zeta_n \tag{14}$$

which should be compared to a threshold. The only term in (14) which is computed after data collection is ζ_n. Test χ_n^2 is asymptotically a χ^2-variable, with $\text{rank}(\mathcal{J})$ degrees of freedom and non-centrality parameter: $\delta\theta^T\, \mathcal{J}^T \Sigma^{-1} \mathcal{J}\; \delta\theta$.

Modal diagnosis. The modal diagnosis problem is to decide which components of θ have changed. Determining which eigenfrequencies and associated modeshapes have been affected by the damage is often addressed as an estimation problem, based on modal identification in the pre- and post-damage stages (Farrar, Doebling and Nix, 2001). We address modal diagnosis as a detection problem instead, using the estimated Jacobian matrix $\widehat{\mathcal{J}}_i$ corresponding to mode and modeshape i. The directional test focussed on this mode i writes:

$$\chi_n^{(i)2} \stackrel{\text{def}}{=} \zeta_n^T \widehat{\Sigma}^{-1} \widehat{\mathcal{J}}_i \left(\widehat{\mathcal{J}}_i^T \widehat{\Sigma}^{-1} \widehat{\mathcal{J}}_i\right)^{-1} \widehat{\mathcal{J}}_i^T \widehat{\Sigma}^{-1} \zeta_n \tag{15}$$

Such a directional test restricted to the damping coefficient ρ of a given mode could be designed as an answer to the flutter monitoring problem. However, the relevant hypotheses for that problem do not have the same form as in (11). This calls for a further step in the design of tests for flutter monitoring. In what follows, the $\widehat{}$ above $\mathcal{J}$ and Σ are dropped for simplicity.

3. FLUTTER MONITORING - FIRST TESTS

As explained in the introduction, the flutter monitoring problem is basically *to decide whether some damping coefficient* ρ *defined in (4) decreases below some specified critical value* ρ_c. In this section, we describe a first class of statistical tests for solving this problem. These tests come up from variations on the local approach and the GLR test adapted to several null and alternative hypotheses on the damping.

The simpler case $\rho_c = \rho_0$*.* We first assume that $\rho_c = \rho_0$, where ρ_0 is the actual value of the monitored damping coefficient, in the reference parameter θ_0. The observed data is the residual $\zeta \stackrel{\text{def}}{=} \zeta(\theta_0)$ defined in (10). Let $\mathcal{J} \stackrel{\text{def}}{=} \mathcal{J}(\rho_0)$ be the Jacobian matrix corresponding to the damping coefficient[5] ρ_0, and $\Sigma \stackrel{\text{def}}{=} \Sigma(\theta_0)$. We introduce:

$$\overline{\zeta} \stackrel{\text{def}}{=} \mathcal{J}^T \Sigma^{-1} \zeta, \quad \overline{\Sigma} \stackrel{\text{def}}{=} \mathcal{J}^T \Sigma^{-1} \mathcal{J} \tag{16}$$

which are scalar numbers. It results from (13) that, for all $\delta\rho$, the residual ζ_n in (10) has the following asymptotic behavior when assuming $\rho = \rho_0 + \delta\rho/\sqrt{n}$:

$$\overline{\zeta} \xrightarrow[n \to \infty]{} \mathcal{N}(\overline{\Sigma}\, \delta\rho, \overline{\Sigma}) \tag{17}$$

Consider first the following *local* hypotheses:

$$\mathbf{H}_0 : \delta\rho \geq 0 \quad \text{and} \quad \mathbf{H}_1 : \delta\rho < 0 \tag{18}$$

which express that the damping coefficient decreases below ρ_0 while remaining in close neighborhoods. The GLR (generalized likelihood ratio) test for the testing problem (18), namely:

$$\begin{aligned} l(\theta_0) \stackrel{\text{def}}{=} & \sup_{\delta\rho<0} \left(-(\overline{\zeta} - \overline{\Sigma}\, \delta\rho)^T \overline{\Sigma}^{-1} (\overline{\zeta} - \overline{\Sigma}\, \delta\rho) \right) \\ & - \sup_{\delta\rho \geq 0} \left(-(\overline{\zeta} - \overline{\Sigma}\, \delta\rho)^T \overline{\Sigma}^{-1} (\overline{\zeta} - \overline{\Sigma}\, \delta\rho) \right) \end{aligned} \tag{19}$$

can be shown (Lehmann, 1986) to boil down to:

$$l(\theta_0) = -\operatorname{sign}(\overline{\zeta}) \cdot \overline{\chi}^2, \quad \overline{\chi}^2 \stackrel{\text{def}}{=} \overline{\zeta}^T \overline{\Sigma}^{-1} \overline{\zeta} \tag{20}$$

The test statistics $l(\theta_0)$ enjoys the following property:

$$\mathbf{E}_{\mathbf{H}_0}\, l(\theta_0) \leq 0, \quad \mathbf{E}_{\mathbf{H}_1}\, l(\theta_0) > 0 \tag{21}$$

and thus it is a good candidate for solving the hypotheses testing problem in (18).

The realistic case $\rho_c < \rho_0$. Of course, the practical situation of interest is when the critical value ρ_c for the damping is such that: $0 < \rho_c < \rho_0$. Thus the hypotheses (18) must be replaced by:

$$\mathbf{H}_0 : \rho \geq \rho_c \quad \text{and} \quad \mathbf{H}_1 : \rho < \rho_c \tag{22}$$

Unfortunately, these hypotheses are *non local* for the following reason. The nominal value for the damping is ρ_0, thus local hypotheses must involve small deviations *from* ρ_0, namely in the form $\rho_0 + \delta\rho/\sqrt{n}$. But $\rho_c - \rho_0$ is fixed, and independent of the sample size n. Hence a deviation below ρ_c cannot be captured in the form of some $\rho_0 + \delta\rho/\sqrt{n}$! Thus we cannot address our problem *via* a local approach, by sticking to ρ_o as a nominal damping coefficient.

[5] This matrix is obtained by selecting the corresponding column of the matrix which results from the multiplication of $\mathcal{J}(\theta_0)$ with the Jacobian of the transformation (4) from the $\lambda's$ to the $\rho's$ (Basseville, Mevel and Goursat, 2003).

The proposed test. From the above discussion, we see that we need to cheat in some way or another. Let:

$$\widetilde{\theta}_0 \stackrel{\text{def}}{=} \theta_0 \text{ except that } \rho_0 \leftarrow \rho_c\,, \tag{23}$$

namely we substitute the critical value ρ_c for the monitored damping coefficient ρ_0. Now we build our test statistics *as if our nominal model was* $\widetilde{\theta}_0$. Note that this is coherent with the sensitivity approach to monitoring taken in (15) (Basseville, 1998).

Let $\zeta(\widetilde{\theta}_0)$ and $l(\widetilde{\theta}_0)$ be respectively the residual (10) and test statistics (20)-(16) computed with this $\widetilde{\theta}_0$. The first proposed test exploits property (21) for $l(\widetilde{\theta}_0)$. It should be noticed that in practice this residual is biased, and the bias cannot be estimated (Basseville, Mevel and Goursat, 2003). Consequently, the decision function $l(\widetilde{\theta}_0)$, which is basically a quadratic form - see (20), is biased too.

4. FLUTTER MONITORING - SECOND TEST

In this section, we introduce another decision function, which overcomes this bias issue. We assume again that the reference parameter is $\widetilde{\theta}_0$ (23). The second test builds on a different approximation for the *same* residual $\zeta(\widetilde{\theta}_0)$, and on the CUSUM test statistics. Formula (10) for $\zeta(\widetilde{\theta}_0)$ rewrites as:

$$\zeta(\widetilde{\theta}_0) = \sum_{k=1}^{n} Z_k(\widetilde{\theta}_0)/\sqrt{n}, \; Z_k(\widetilde{\theta}_0) \stackrel{\text{def}}{=} \check{\mathcal{Z}}_k(\widetilde{\theta}_0)\, \check{\mathcal{Y}}_k^T \tag{24}$$

where:

$$\check{\mathcal{Y}}_k \stackrel{\text{def}}{=} W_2\, \mathcal{Y}_{k,q}^-, \; \check{\mathcal{Z}}_k(\widetilde{\theta}_0) \stackrel{\text{def}}{=} U(\widetilde{\theta}_0)^T W_1\, \mathcal{Y}_{k,p+1}^+ \tag{25}$$

and

$$\mathcal{Y}_{k,p+1}^+ \stackrel{\text{def}}{=} \begin{pmatrix} Y_k \\ \vdots \\ Y_{k+p} \end{pmatrix}, \quad \mathcal{Y}_{k,q}^- \stackrel{\text{def}}{=} \begin{pmatrix} Y_k \\ \vdots \\ Y_{k-q+1} \end{pmatrix} \tag{26}$$

From (13), we know that ζ, *and thus* $\sum_{k=1}^{n} Z_k/\sqrt{n}$, *is asymptotically Gaussian distributed, with zero mean under* $\rho = \rho_c$. Following the arguments in (Benveniste, Métivier and Priouret, 1990)[5.4.1], it is legitimate to deduce from this fact the following approximation: for k large enough, Z_k *is asymptotically Gaussian distributed, with zero mean under* $\rho = \rho_c$, *and the* Z_k*'s are independent.*

Consider now the following *non local* hypotheses:

$$\widetilde{\mathbf{H}}_0 : \rho = \rho_c + \epsilon \quad \text{and} \quad \widetilde{\mathbf{H}}_1 : \rho = \rho_c - \epsilon \tag{27}$$

with $\epsilon > 0$. Testing between these hypotheses can be addressed with the aid of the CUSUM test statistics g_n (Basseville and Nikiforov, 1993):

$$S_n(\widetilde{\theta}_0) \stackrel{\text{def}}{=} \sum_{k=1}^{n} Z_k(\widetilde{\theta}_0), \quad T_n(\widetilde{\theta}_0) \stackrel{\text{def}}{=} \max_{1\leq k\leq n} S_k(\widetilde{\theta}_0)$$

$$g_n(\widetilde{\theta}_0) \stackrel{\text{def}}{=} T_n(\widetilde{\theta}_0) - S_n(\widetilde{\theta}_0) \geq 0 \qquad (28)$$

Now, from (24) and (25), we have:

$$\begin{aligned} S_n(\widetilde{\theta}_0) &= \sqrt{n}\, \zeta(\widetilde{\theta}_0) \qquad (29) \\ &= \sum_{k=1}^{n} \text{vec}\left(U(\widetilde{\theta}_0)^T W_1 \mathcal{Y}_{k,p+1}^{+} \mathcal{Y}_{k,q}^{-\,T} W_2^T\right) \end{aligned}$$

From (29), we deduce that test statistics g_n in (28) actually handles the *time-unnormalized* residual $\zeta(\widetilde{\theta}_0)$. Furthermore, since g_n monitors *on-line* the deviations of S_n, and thus $\zeta(\widetilde{\theta}_0)$ w.r.t. its current maximum value, we deduce that g_n is *not* affected by a bias on $\zeta(\widetilde{\theta}_0)$.

For numerical efficiency, it is preferable to normalize $S_n(\widetilde{\theta}_0)$ in order to handle identity covariance matrices, see (17). This is achieved by using the *scale-normalized* residual $\overline{\zeta}(\widetilde{\theta}_0)$ as in (16), and computing:

$$\begin{aligned} \overline{S}_n(\widetilde{\theta}_0) &\stackrel{\text{def}}{=} \sqrt{n}\ \Sigma(\widetilde{\theta}_0)^{-1/2}\ \zeta_n(\widetilde{\theta}_0) \\ \overline{g}_n(\widetilde{\theta}_0) &\stackrel{\text{def}}{=} \overline{T}_n(\widetilde{\theta}_0) - \overline{S}_n(\widetilde{\theta}_0) \qquad (30) \end{aligned}$$

Test statistics $\overline{g}_n(\widetilde{\theta}_0)$ enjoys the following properties:

$$\text{under } \widetilde{\mathbf{H}}_0 : \overline{g}_n(\widetilde{\theta}_0) \approx 0 \qquad (31)$$

$$\text{under } \widetilde{\mathbf{H}}_1 : \overline{g}_n(\widetilde{\theta}_0) > 0 \qquad (32)$$

$$\text{if } \rho \text{ decreases, } \overline{g}_n(\widetilde{\theta}_0) \text{ increases}$$

$$\text{if } \rho \text{ increases, } \overline{g}_n(\widetilde{\theta}_0) \text{ decreases}$$

5. EXPERIMENTAL ISSUES AND RESULTS

The above tests statistics $l(\widetilde{\theta}_0)$ and $\overline{g}_n(\widetilde{\theta}_0)$ are currently experimented on different data sets available within the Eureka project FLITE. In this paper, we discuss results obtained on a nonstationary data set recorded on a flight structure, provided by one of the industrial partners. Automatic modal analysis and flutter monitoring (monitoring one damping coefficient) have been performed using the Scilab Modal Analysis toolbox (Mevel *et al.*, 2003). For obvious reasons, the actual values of the damping coefficient and the frequency are not displayed.

The results of the automatic modal analysis for the considered mode (frequency and damping coefficient) are displayed on Fig. 1. It appears clearly that, whereas the frequency is slowly increasing, the damping coefficient exhibit large fluctuations. Notice that the change in the frequency is linear except that in two places, where unexpected jumps happen in the estimation, which implies overestimation of the corresponding damping values (see the narrow peaks in the plot).

The first test statistics $l(\widetilde{\theta}_0)$, not displayed here, has shown up a very poor behavior, mainly due to the bias problem. The behavior of the second test statistics $\overline{g}_n(\widetilde{\theta}_0)$ is displayed for two different values of the critical damping: $\rho_c = \rho_1$ and $\rho_c = \rho_2$ on Fig.s 2 and 3, respectively, with $\rho_2 << \rho_1$.

Behavior of test (30) for $\rho_c = \rho_1$. The first value ρ_1 is between the first and second graduation of the damping plot in Fig. 1. The general trend in the damping behavior in that figure is roughly flat first, then under some threshold, and then over the threshold. This overall trend is reflected in the time behavior of test statistics $\overline{g}_n(\widetilde{\theta}_0)$ in Fig. 2. The test is flat, near zero, during a first stage corresponding to a damping near to the critical value ρ_c. Then the damping decreases and the test increases during the entire period with damping lower than the critical value. Then the damping increases, and, at the same time, the test stops rising up and starts decreasing towards zero reasonably fast. finally, the test remains close to zero, since the damping remains over the critical value until the end of the dataset. Note that the damping enters a very short dip near the end of the dataset, before increasing, and the test seems to react to this by slightly increasing for a while. Of course, improved evidence of the actual correlation between both events is mandatory.

Behavior of test (30) for $\rho_c = \rho_2$. The second value ρ_2 is close to zero, and is a reasonable candidate for a flutter monitoring critical value. In this case, from the plot of the damping estimates in Fig. 1, it is obvious that the test should exhibit a much lower sensitivity: since the critical value is much lower, the damping coefficient is most of the time over the critical value. Note the different scales in the ordinates of Fig.s 2 and 3. Also, because we plug the critical value of the damping into the reference parameter $\widetilde{\theta}_0$, we are likely to generate instabilites in the computations somehow. This might be an explanation for the noisy behavior of the plot in Fig. 3, especially on the right handside albeit the value of the test is mostly around zero. A smoothed version might be of interest to get rid of that undesirable feature.

However, a remarkable feature in Fig. 3 is that test $\overline{g}_n(\widetilde{\theta}_0)$ exhibits two visually distinct positive patterns during the two periods where the damping is close to zero, and thus to the critical value ρ_2.

6. DISCUSSION

Motivated by the flutter monitoring problem, we have described the rationale for the design of two tests statistics for monitoring a damping coefficient. On a data set from a *real flight structure*, the second test gives promising results. It should be stressed that the proposed flutter monitoring algorithm can be computed in real time, whereas automatic modal identification is both computationally expensive and subject

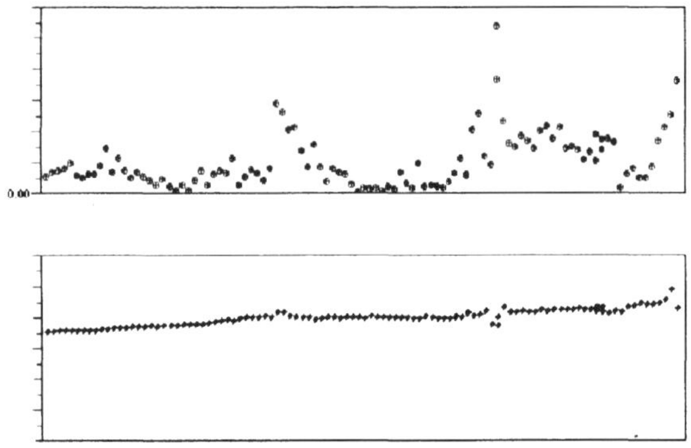

Figure 1. Automatic identification of one frequency (bottom) and damping coefficient (top).

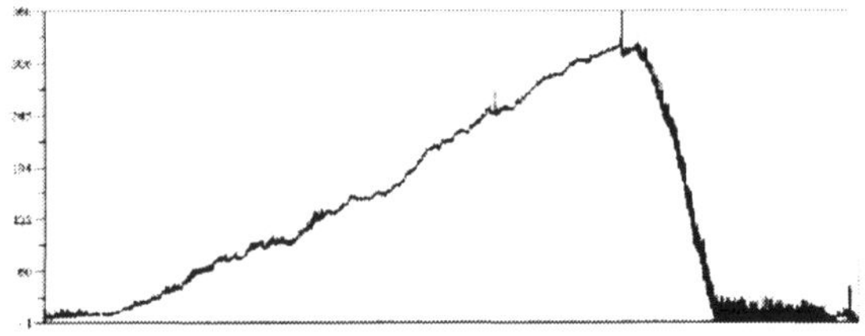

Figure 2. The corresponding test $\overline{g}_n$ for $\rho_c = \rho_1$.

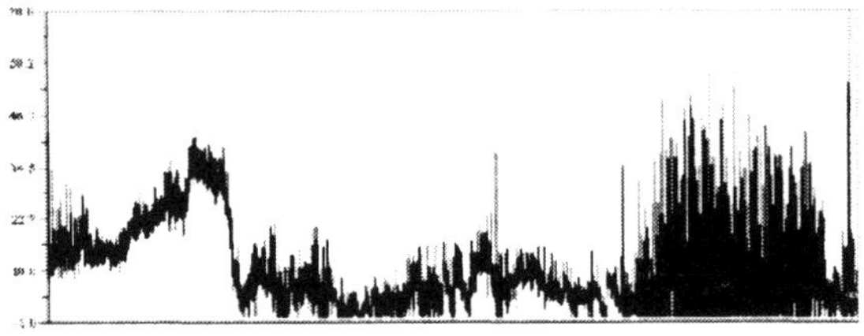

Figure 3. The corresponding test $\overline{g}_n$ for $\rho_c = \rho_2$.

to estimation errors especially in the damping factors. Actually, the proposed damping test does *not* require re-identification of the damping coefficient at every step. Thus is much faster, and free from estimation error propagation. For tuning the proposed algorithm, the user has to choose a threshold for the test, where it is assumed something is happening.

Future investigations include extensive tests of the methods on different applications, multiple damping monitoring, and careful evaluation of the sensitivity of the test w.r.t. changes in the critical value ρ_c, and also w.r.t. the choice of θ_0 and of the data sets used for the estimation of the reference $\theta_0, \mathcal{J}, \Sigma$.

7. REFERENCES

M. Basseville (1998). On-board component fault detection and isolation using the statistical local approach. *Automatica*, **34**(11), 1391–1416.

M. Basseville and I.V. Nikiforov (1993). *Detection of Abrupt Changes - Theory and Application*, Prentice-Hall. http://www.irisa.fr/sigma2/kniga/.

M. Basseville, M. Abdelghani, A. Benveniste (2000). Subspace-based fault detection algorithms for vibration monitoring. *Automatica*, **36**(1), 101–109.

M. Basseville, L. Mevel, M. Goursat (2003). Statistical model-based damage detection and localization: subspace-based residuals and damage-to-noise sensitivity ratios. *Jal Sound and Vibration*, to appear.

M. Basseville, A. Benveniste, G. Moustakides, A. Rougée (1987). Detection and diagnosis of changes in the eigenstructure of non-stationary multivariable systems. *Automatica*, **23**, 479–489.

M. Basseville, A. Benveniste, M. Goursat, L. Hermans, L. Mevel, H. Van der Auweraer (2001). Output-only subspace-based structural identification: from theory to industrial testing practice. *ASME Journal of Dynamic Systems Measurement and Control*, Special Issue on Identification of Mechanical Systems, **123**(4), 668–676.

A. Benveniste, M. Basseville, G. Moustakides (1987). The asymptotic local approach to change detection and model validation. *IEEE Trans. Automatic Control*, **AC-32**(7), 583–592.

A. Benveniste, M. Métivier and P. Priouret (1990). *Adaptive Algorithms and Stochastic Approximations*. Springer, New York.

C.R. Farrar, S.W. Doebling and D.A. Nix (2001). Vibration-based structural damage identification. *The Royal Society, Philosophical Transactions: Mathematical, Physical and Engineering Sciences*, **359**(1778), 131–150.

W. Gersch (1974). On the achievable accuracy of structural parameter estimates. *Jal Sound and Vibration*, **34**(1), 63–79.

E.L. Lehmann (1986). *Testing Statistical Hypotheses*. Wiley. Series in Probability and Mathematical Statistics, 2nd Ed.

L. Mevel, L. Hermans, H. Van der Auweraer (1999). Application of a subspace-based fault detection method to industrial structures. *Mechanical Systems and Signal Processing*, **13**(6), 823–838.

L. Mevel, M. Goursat, M. Basseville and A. Benveniste (2003). Subspace-based modal identification and monitoring of large structures, a Scilab Toolbox. These Proceedings.

B. Peeters, G. De Roeck (1999). Reference-based stochastic subspace identification for output-only modal analysis. *Mechanical Systems and Signal Processing*, **13**(6), 855–878.

P. Van Overschee, B. De Moor (1996). *Subspace Identification for Linear Systems: Theory – Implementation – Methods*. Kluwer.

M. Viberg, B. Wahlberg, B. Ottersten (1997). Analysis of state space system identification methods based on instrumental variables and subspace fitting. *Automatica*, **33**(9), 1603–1616.

Q. Zhang, M. Basseville, A. Benveniste (1994). Early warning of slight changes in systems and plants with application to condition based maintenance. *Automatica*, **30**(1), 95–114.

IFAC
Publications
www.elsevier.com/locate/ifac

RELIABLE SYSTEM IDENTIFICATION FOR LARGE FLEXIBLE SPACE STRUCTURES

Vít Babuška*[1], Seth L. Lacy[2], R. Scott Erwin[2], Alexander M. Melin[3]

[1]Veridian Systems, 14700 Lee Road, Chantilly, VA 20151, USA
[2]Air Force Research Laboratory, Space Vehicles Directorate, 3550 Aberdeen Ave SE, Kirtland AFB, NM 87117, USA
[3]Graduate Student, Dept. of Mechanical & Aerospace Engineering, Univ. of Missouri, Columbia, MO, 65211, USA

Abstract: This paper describes considerations for identifying models of large, lightly damped, flexible space structures to be used for high-performance control system design and analysis. The specific aspects of these problems are in turn used to derive a candidate system identification process for use on such systems. Particular attention is focused on two parts of this process, namely model order selection and model tuning. Three methods of model order selection are evaluated utilizing on-orbit data from the Satellite Ultraquiet Isolation Technologies Experiment (SUITE), which also provides a sample application for illustration of the proposed process. *Copyright © 2003 IFAC*

Keywords: Control Oriented Models; Frequency Response Methods; Least-squares Identification; Structure Systems; Subspace Methods

1. INTRODUCTION

Satellites and their instruments are subject to vibration throughout their lifetimes. Although the space vibration environment is much less severe than that encountered during launch, on-orbit vibration levels can still cause significant optical performance degradation. The performance of payloads on large remote sensing satellites is particularly sensitive to vibration since a small amplitude vibration at the payload can significantly degrade image quality.

The trend in remote sensing and sensitive astronomical platforms is toward ever-larger structures. While payload size is increasing, the electro-magnetic spectrum of interest remains the same and perhaps is becoming more ambitious. This implies that payloads will require even greater relative precision. Furthermore, for these large payloads, it may be impractical to perform fully deployed system tests prior to launch.

These large precision spacecraft present a challenging control problem. They are modally dense (more than 10 modes/Hz may not be uncommon), have a very large frequency bandwidth (typically four decades or more) over which disturbances can affect payload performance, and are lightly damped. The vibration management system will be distributed among disturbance isolation, vibration suppression in the structural load path and payload isolation, as shown in Figure 1.

The performance of passive vibration isolation systems will be limited by the constraints of physics and mechanical components, particularly at low frequencies. This motivates the need for an active isolation system. To take advantage of modern control methods, high fidelity system models will be required. This is a challenge for experiments, algorithms, and computational resources.

This paper describes the system identification process shown in Figure 2, and discusses its application to the identification of large precision space structures for control system

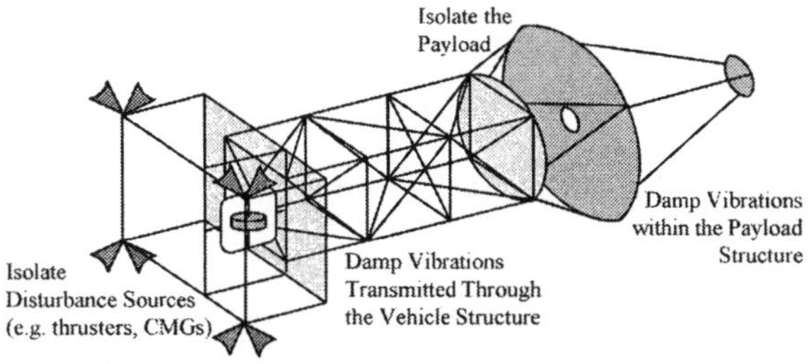

Fig. 1 Layered Vibration Management Approach

applications. The process is not merely algorithmic. It requires user experience as well as efficient computational elements. The main part of the paper addresses the model order selection and model tuning steps. The process was applied to the SUITE flight experiment (Anderson, et.al., 2000), and data from it are used to illustrate the results.

2. SYSTEM IDENTIFICATION PROCESS

The first step of any system identification process is data acquisition. Much work has been performed on selection and design of input signals for system identification. An overview of the subject from the practical point of view can be found in Pintelon (2001). The primary objective of input design is to provide a large signal to noise ratio (i.e., amount of response due to experimental input, in the face of exogenous inputs and measurement noise) while being subject to several significant practical constraints (e.g., actuator saturation, limitations on total applied power, etc).

For space applications, input signal design is driven by several considerations. First, due to tight modal spacing of structural dynamics, very tight frequency resolution may be required, which in turn yields limits on the sweep rate for chirp signals and dictates the number of component sinusoids for sum-of-sine inputs. Next, lightly damped space structures can exhibit extremely large dynamic ranges in the system response, with frequency response amplitude variations of 100 dB or more not uncommon for precision applications. This in turn provides saturation issues for random inputs and generates complex frequency-weighting issues for periodic inputs. Flight experiments, often used to demonstrate new technologies prior to their inclusion in new systems, impose additional constraints. These include limited computational resources - both data storage (which affects "pre-generated" signals such as sum-of-sines) and processor power (which affects the ability to generate truly nonperiodic, random signals) – and limited data download rates.

The second step in the system identification process is data reduction and manipulation, the most commonly used example of which is averaging. Space applications again present several interesting issues in the area of data reduction and processing. The most important of these are limited on-board data storage, limited processing capability, and constraints on test schedules. Limited data storage forces a trade between the amount of data collected for any particular input-output pair, and the number of input-output pairs that can be collected simultaneously in an experiment. Spacecraft operations affect the scheduling of dynamic testing and restrict test durations. This can mean that significant time may elapse between experiments, which can lead to significant environmental differences that affect the experimental data. Limited processing capability inhibits real-time high-resolution FFT processing. Periodic inputs permit a large amount of data to be time-domain averaged on-board. Alternately, less data can be downloaded and processed in the frequency domain with corresponding impacts on the amount of averaging that can be done.

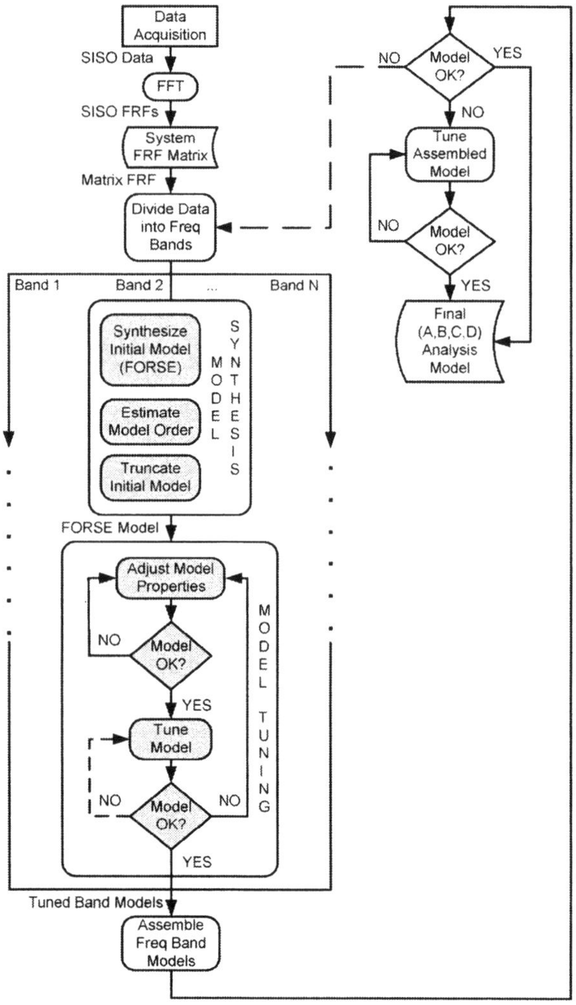

Fig. 2: Proposed System Identification Process for Precision Structural Dynamics Space Systems.

Finally, space-based structural dynamic systems have several characteristics that drive the choice of suitable algorithms or techniques. The lightly-damped resonances of space-structures yield systems that require an extensive amount of data in order to characterize them through impulse responses, which has implications in terms of on-board resource requirements for methods such as ERA (Juang, 1994).

On the other hand, precision space systems are designed and built carefully with an emphasis on predictability and linearity, employ high quality sensors and actuators, and have significant terrestrial resources available. This means that system responses can be measured with very small variances in the entire frequency band, and ground processing can make use of powerful computer systems.

For the SUITE flight experiment, the above conditions led to the following experimental design.

Data were collected using frequency-weighted chirp signals, with three distinct signals (experiments) being needed to provide enough frequency resolution and address signal-to-noise problems adequately over the 1000 Hz bandwidth of interest. The decision to utilize periodic time-domain signals was driven by the limited storage capacity, limited processing power, and use of data averaging to acquire good signal-to-noise ratio characteristics. Approximately 12 time-domain averages of the system response to these signals were performed on-orbit, a number that was selected based on the total amount of time the experiment could remain on. Finally, the data were downloaded via telemetry for further processing.

Once on the ground, the data from the three frequency bands was processed using FFT techniques to provide estimates of the system frequency-response function (FRF) over each bandwidth. Care was taken to have the appropriate signal content and processing parameters to minimize or eliminate the degrading effects of leakage on the resulting FRF estimates.

The data from these multiple frequency bands was stitched together (using a linear interpolation in the overlap regions) and assembled into a complex frequency response function (FRF) matrix spanning the entire frequency bandwidth. This composite FRF matrix contained the data from which models were identified. The system identification process in Figure 2 is an explicit two-step process in which a model is synthesized from the FRF data in the first step and then tuned in the second step. The Frequency domain Observability Range Space Extraction (FORSE) method of Liu, et.al. (1996) was used for model synthesis. This method, based on frequency domain subspace identification techniques, was adopted due to the unique considerations that space-based structural dynamics pose, as outlined above.

Some of the data sets used on the SUITE experiment were very large (up to 16000 points per SISO FRF) so the data were partitioned into frequency bands in order to reduce the computing capabilities required for the system identification task. Models were generated for each frequency band (see Section 3), and these sub-models were concatenated into a single model of the plant over the entire frequency range, which was then ready for final tuning (see Section 4).

3. MODEL ORDER SELECTION METHODS

In most practical applications the process of deciding an order for the system model is a tradeoff between model accuracy, computational resources, model complexity, and the modeling of non-system dynamic characteristics, such as noise effects. If the SNR of the data used is adequate, the order of a finite-dimensional system can be determined from the break-point in a system singular value plot (e.g., van Overschee, 1996). However, for complex lightly damped structures the boundary between noise effects and weakly observable modes is not clear; there is no obvious "cliff" and model order is indiscernible in this way. An example of this phenomenon, based on data from the SUITE experiment, is shown in Section 5, Figure 6. The figure illustrates the difficulty in discerning the order of a complex system using the textbook approach.

Three model order determination algorithms were assessed using the SUITE data. The first is a MIMO extension of the Akaike Information Criterion (AIC), (Akiake, 1974; Kashyap 1980), the second is a MIMO extension of the Minimum Description Length (MDL) criterion, (Rissanen, 1978, Barron 1998; Zarowski 1998), and the third is a balanced realization error method similar to error chart methods used in experimental modal analysis (Allemang, 1999).

Cost functions for the AIC and MDL methods are given by:

$$V_{\mathrm{AIC}} = V + n(m+p) + mp \tag{1a}$$

$$V_{\mathrm{MDL}} = V + (n(m+p) + mp)\frac{\log(2mpF)}{2} \tag{1b}$$

respectively, where

$$V = \sum_{k=1}^{F}\sum_{i=1}^{p}\sum_{j=1}^{m}\frac{\left|G_k(i,j) - \hat{G}_k(i,j)\right|^2}{V^2_{\hat{G}_k(i,j)}}, \tag{2}$$

$G_k(i,j) \in \mathrm{C}$ is the frequency response of the model transfer function from input j to output i at frequency k, $\hat{G}_k(i,j) \in \mathrm{C}$ is the measured frequency response, F is the corresponding number of frequencies, $V^2_{\hat{G}k(i,j)}$ is the variance of the measured frequency response, p is the number of outputs, and m is the number of inputs. It has been reported that the AIC generally selects higher order models than the MDL criterion (Rissanen, 1978; Barron 1998; Zarowski 1998).

With the balanced realization error method, the objective is to find the lowest order model whose weighted RMS error with respect to the original data is below some threshold. The specific metric used is a weighted operator 2-norm:

$$J_1 = \sqrt{\frac{1}{F}\sum_{k=1}^{F}\frac{\left[G_k - \hat{G}_k\right]_F^2}{\underline{\sigma}^2\left(\hat{G}_k\right)}} \tag{3}$$

where $G_k \in \mathrm{C}^{p\times m}$ is the FRF matrix of the model at frequency k, $\hat{G}_k \in \mathrm{C}^{p\times m}$ is the measured FRF matrix at frequency k, and $\underline{\sigma}\left(\hat{G}_k\right)$ is the smallest singular value of the measured FRF matrix at frequency k. Use of $\underline{\sigma}\left(\hat{G}_k\right)$ as a weighting term implies that the SNR of the measured data is high and $\underline{\sigma}\left(\hat{G}_k\right)$ represents system dynamics.

This metric is consistent with the least-squares cost functions used in subspace model synthesis methods, is straightforward to compute, and does not use variance information. Compatibility with the synthesis method is important since models generated with respect to one metric have been observed to be poor when evaluated with another metric.

The balanced realization error method consists of the following steps: synthesize an overparameterized (i.e. very high order) model; transform the model into a balanced realization; compute the weighted RMS error between truncated models and the measured data; select the model order from a plot of model order vs. weighted RMS error, which usually has an obvious flat zone (based on empirical observations).

This model order selection method was based on experience from experimental modal analysis and observations from the model synthesis step. The error between a truncated model of order M and one of order N, $N>M$, is bounded by the sum of the residual singular values. When these are very small, they do not substantially improve the fit between the model and the FRF data.

4. MODEL TUNING

Once a model of reasonable order has been synthesized, the next step in the identification process is to tune the model. This is also a two-step that involves manual adjustment followed by optimization with respect to an error metric that cannot be implemented effectively in the model synthesis step. Often a synthesis algorithm will identify (or omit) poles that do not agree with the known physics of the system. Here the engineer must intervene to adjust the model based on his knowledge of the physics of the structure.

For models that are to be used for control system development, metrics based on minimizing the size of an uncertainty set associated with the model are attractive tuning metrics because the residual error is directly applicable for control system analysis and possibly synthesis. Identification using H_∞-norm minimization techniques is an example of this type of approach.

Model tuning itself is performed as a two-step process. First, the synthesized model(s) can be tuned using the logarithmic cost function described in Sidman (1991) and Jacques (1996). The logarithmic cost function improves the fit where the response magnitude is low such as the transmission zeros. This can be done efficiently and usually yields a model that matches the data well enough to initialize the final tuning step.

The purpose of the final tuning step is to minimize a cost function that is related to modern control synthesis methods. It is desirable to make the cost function flat as a function of frequency to simplify the uncertainty weighting functions that augment the plant model in modern controller synthesis methods. A straightforward approach is to find a model that minimizes the largest singular value of the additive error, namely:

$$\min_{(A,B,C,D)}\left(\max_{\omega_k}\left[\bar{\sigma}\left(G_k-\hat{G}_k\right)\right]\right) \tag{4}$$

In simulations, resulting models fit the FRF data very well near the poles (large magnitude response) while sacrificing the FRF fit at the zeros (low magnitude response). Instead, the weighted cost

$$J_F=\max_{\omega_k}\left(\frac{\left\|G_k-\hat{G}_k\right\|_F}{\underline{\sigma}\left(\hat{G}_k\right)}\right) \tag{5}$$

is used. This cost function amplifies the contributions of the zeros, and is traceable to the unstructured multiplicative uncertainty used in modern robust control design methods,

$$\begin{aligned}
&\hat{G}(I+\Delta)=G\\
&\hat{G}\Delta=G-\hat{G}\\
&\Delta=\hat{G}^{-1}(G-\hat{G})\\
&\bar{\sigma}(\Delta)=\bar{\sigma}(\hat{G}^{-1}(G-\hat{G}))\le\bar{\sigma}(\hat{G}^{-1})\bar{\sigma}(G-\hat{G})\\
&\bar{\sigma}(\hat{G}^{-1})\bar{\sigma}(G-\hat{G})=\frac{\bar{\sigma}(G-\hat{G})}{\underline{\sigma}(\hat{G})}\\
&\le\frac{\sum_{i=1}^{\min p,m}\sigma_i(G-\hat{G})}{\underline{\sigma}(\hat{G})}=J_F.
\end{aligned} \tag{6}$$

Two computational approaches for the final tuning step minimizing J_F were investigated. The first was the iterative re-weighted non-linear least squares approach suggested by Bohn and Unbehauven (1998) for model synthesis. This method becomes a tuning method after the first iteration. The second was the sequential quadratic programming method implemented in the MATLAB function `fminimax`.

Two approaches for managing the computational load in the final tuning step are restricting the tunable parameters those in the neighborhood of the maximum error, and evaluating the cost function only in the neighborhood of the maximum error during the optimization. The former is straightforward when the model is in modal canonical form. With these approaches, the final tuning becomes an iterative process as the error is "squeezed down" in different frequency bands. While there is no guarantee that this process will converge, it does provide the engineer with the flexibility to apply his judgment about the importance of the error in different frequency bands.

5. IDENTIFICATION OF SUITE MODELS

The SUITE isolation platform is shown in Figure 3. A flight unit flew as a payload on the PICOSat satellite, and concluded operations in November 2002. A nominally identical ground unit resides at the AFRL/VS facility. The SUITE hardware is composed of two major subsystems, a hexapod assembly (HXA) and an electronic data control system (DCS). A block diagram of the system is shown in Figure 4.

The hexapod unit is an electromechanical assembly composed of six struts, with a piezo-stack actuator (u) and a geophone sensor (y) in each strut. The struts contain active and passive isolation stages in series. The assembly includes an additional six sensors - one tri-axial geophone set on the base (identified by subscript w) and another on the platform (identified by subscript z) - for measuring vibration. The piezostack strut actuators have a stroke of 30 μm peak-to-peak. Two proof mass actuators (PMAs) - one on the base and one on the platform - for generating disturbances are also part of the HXA. The DCS is a separate custom circuit board that performs signal processing.

Figure 6 shows the model errors defined in Eqs. 1-3, as a function of model order for P_{zu} model synthesis. The AIC and MDL model order selection cost functions both indicate a model order of 379 states. The penalty term was very small compared to the ML term, V. The normalized RMS model error was computed using synthesized overparameterized models of various orders from 250 to 400 states. The error curve is flat above approximately 180 states regardless of the initial model order. A 184-state model order was selected. The original synthesized models failed to include the three payload geophone suspension modes so they were added manually. The 190 state model was tuned using the log-least squares

Fig. 3 SUITE Isolation Platform

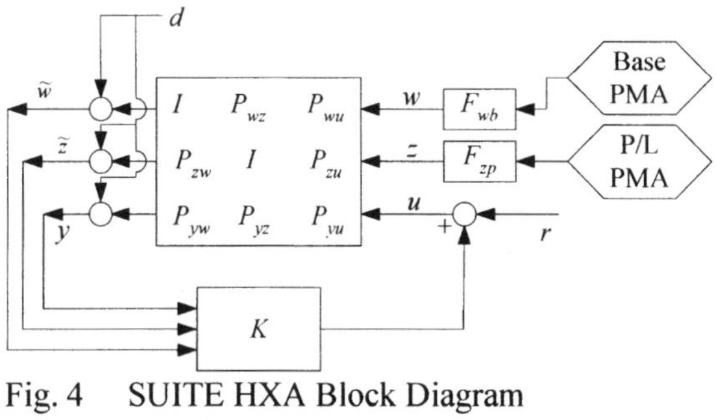

Fig. 4 SUITE HXA Block Diagram

cost. Figure 7 shows the error between the measured transfer function and the transfer function from the synthesized 190 state model, prior to final tuning. The weighted error is largest at low frequencies where the response magnitude is small. Fitting the small magnitude response at low frequencies was difficult.

The results of final model tuning are illustrated in Figure 8, which shows the magnitude of the measured and modeled SISO P_{yu} transfer functions of Strut 0 of the ground unit before and after tuning using the Bohn algorithm and the `fminimax` approach. All parameters and the entire frequency range were used in the iterative Bohn result. With the `fminimax` approach, the final tuning focused on the 10 – 100 Hz frequency band. Both algorithms

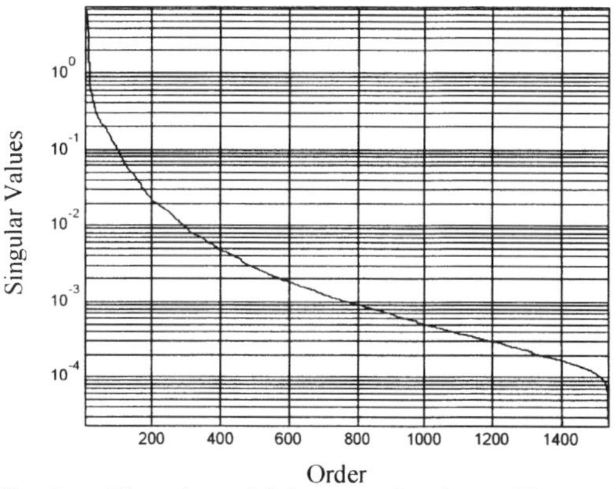

Fig. 5 Singular Values of the Observability Projection Matrix for the SUITE P_{zu} Transfer Function

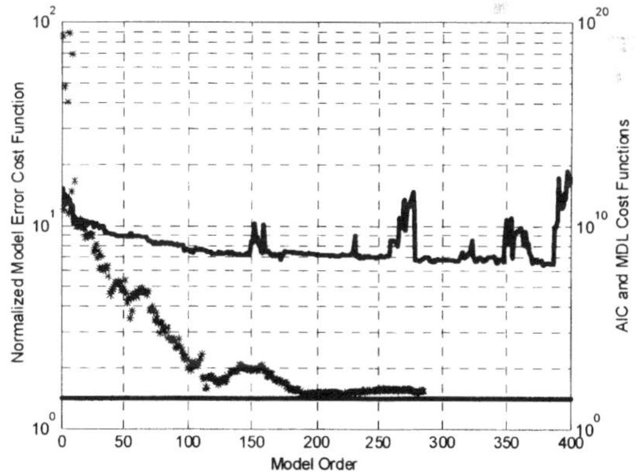

Fig. 6 Normalized RMS Model Error of SUITE P_{zu} Plant

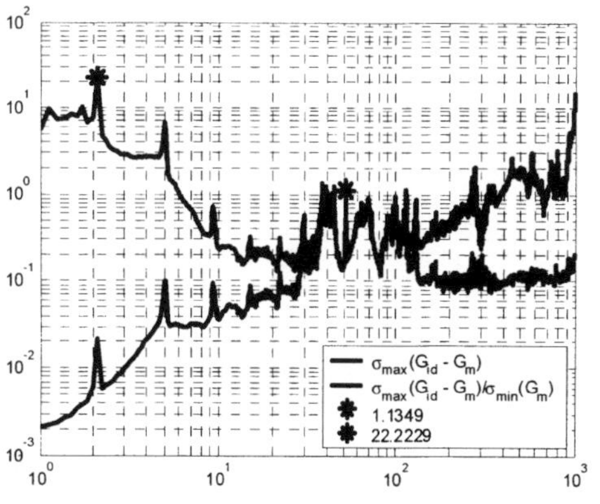

Fig. 7 SUITE P_{zu} Model Error Principal Gain

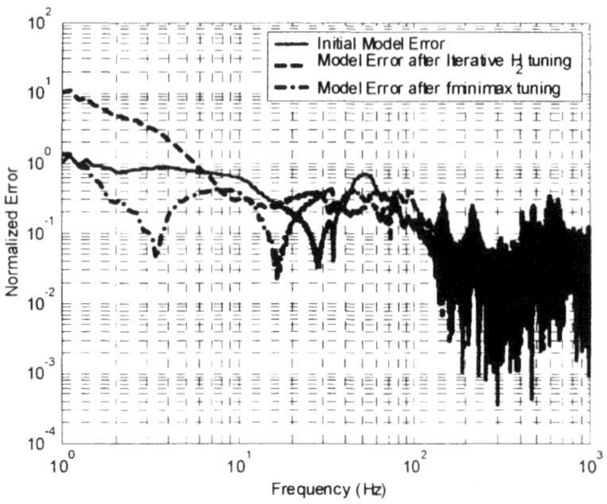

Fig. 8 Strut 0 P_{yu} Model Error after Final Tuning

were successful, however both approaches increased the error below 10 Hz. The static gain was adjusted manually to improve the low frequency fit. Both methods were sensitive to the selection of tuning parameters.

6. SUMMARY

Some practical issues associated with system identification of large, lightly damped, flexible structures were discussed. These issues include large bandwidth, large dynamic range, high modal density, low damping, and large model orders. A two-step system identification approach was introduced. The process was demonstrated on the SUITE flight experiment.

Model order selection using the singular values of the observability projection matrix was not feasible on the SUITE FRF data. Three alternate model order selection techniques were examined based on the issues involved in space-based structural dynamics problems, and a comparison of the results was presented. The balanced realization error method selected a model of reasonable order.

Two methods for model tuning were examined: the iterative re-weighted non-linear least squares approach suggested by Bohn and Unbehauven (1998) for model synthesis; and the sequential quadratic programming method implemented using the `fminimax` function in MATLAB's optimization toolbox. The methods can be applied sequentially to limited numbers of parameters, thereby reducing the computational costs. The methods were applied for generating models for various transfer functions of the SUITE hexapod. Both approaches performed comparably.

ACKNOWLEDGEMENTS

The authors thank Ms. Leslie A. Sullivan of Boeing-SVS, who led mission operations for the SUITE flight experiment.

REFERENCES

Akaike, H. (1974) A New Look at the Statistical Model Identification. *IEEE Transactions on Automatic Control* **6** 716-723.

Allemang, R.J., (1999) *Vibrations: Experimental Modal Analysis,* University of Cincinnati Report UC-SDRL-CN-20-263-663/664.

Anderson, E.H, Fumo, J.P. and R.S. Erwin, (2000) Satellite Ultraquiet Isolation Technology Experiment (SUITE), *Proceedings of the 2000 IEEE Aerospace Conference.*

Barron, A., Rissanen, J., and B. Yu, (1998) The Minimum Description Length Principle in Coding and Modeling. *IEEE Transactions on Information Theory* **6** 2743-2760.

Bohn, C. and H. Unbehauen, (1998) Minmax and Least Squares Multivariable Transfer Function Curve Fitting: Error Criteria, Algorithms and Comparisons, *Proceedings of the 1998 American Automatic Control Conference.*

Jacques, R.N., K. Liu, and D.W. Miller, (1996) Identification of Highly Accurate Low Order State Space Models in the Frequency Domain, *Signal Processing*, **52**, 195-207.

Juang, J-N., (1994) *Applied System Identification,* Prentice-Hall PTR.

Kashyap R. L. (1980) Inconsistency of the AIC Rule for Estimating the Order of Autoregressive Models. *IEEE Transactions on Automatic Control* AC-**25**(5) 996-998.

Liu, K, R.N. Jacques, and D.W. Miller, (1996) Frequency Domain Structural System Identification by Observabiity Range Space Extraction, *ASME Journal of Dynamic Systems, Measurement, and Control,* **118**, 211-220.

Pintelon, R., and J. Schoukens, (2001) *System Identification, a Frequency Domain Approach* IEEE Press.

Rissanen, J. (1978) Modeling by Shortest Data Description. *Automatica* **5** 465-471.

Sidman, M.D., DeAngelis, F. E., and G.C. Verghese, (1991) Parametric System Identification on Logarithmic Frequency Response Data, *IEEE Transactions on Automatic Control,* **39**(9), 1065-1070.

van Overschee, P., and B. de Moor, (1996) *Subspace Identification for Linear Systems: Theory, Implementation, Applications,* Kluwer Academic Publishers.

Zarowski, C. (1998) The MDL Criterion for Rank Determination Via Effective Singular Values *IEEE Transactions on Signal Processing* **46**(6) 1741-1744.

IFAC Publications
www.elsevier.com/locate/ifac

IDENTIFIABILITY ANALYSIS OF A CLASS OF SYSTEMS DESCRIBED BY CONVOLUTION EQUATIONS

Lotfi Belkoura

LAIL, Laboratory of control and Industrial informatic of Lille, Villeneuve d'Ascq, France

Abstract: Parameter identifiability is studied for a class of finite and infinite dimensional systems described by convolution equations. For linear differential delay equations of neutral type and with distributed delays, it is shown how the identifiability property can be formulated in terms of controllability conditions, namely approximate controllability. The notion of sufficiently rich input which enforces identifiability is also adressed, and the results are obtained assuming knowledge of the solution on a bounded time interval. *Copyright © 2003 IFAC*

Keywords: Identifiability, differential delay equations, convolution equations, approximate controllability.

1. INTRODUCTION

This paper proposes sufficient conditions for the identifiability of a class of linear systems described by an input/output convolution equation of the form

$$P * y = Q * u, \tag{1}$$

where P and Q are respectively $n \times n$ and $n \times m$ matrices with entries in the space $\mathcal{E}'$ of distributions with compact support. In this equation, and also to avoid computational complexity, we implicitly assume zero initial state. We shall also use the equivalent representation

$$R * w = 0, \tag{2}$$

$$R := [P, -Q]\,, \quad w := (y^t, u^t)^t. \tag{3}$$

In the behavioral approach of systems described by convolution equations (see (Vettori and Zampieri, 2000)), $R(s)$, the Laplace transform of R (see notations), provides a kernel representation of the behavior $\mathcal{B}$ which consists of the set $\tilde{w}$ of all admissible trajectories in $(\mathcal{E})^{n+m}$. Here, $\mathcal{E}$ denotes the space of $C^\infty(\mathbb{R}, \mathbb{R})$ functions, and $\tilde{w} \in \mathcal{B} = \ker_{\mathcal{E}} R(s)$.

However, although the convolutional algebra of distributions with compact support is isomorphic, via Laplace transform, to the multiplicative algebra of the so called Paley-Wiener functions (i.e entire functions that satisfy a certain growth conditions), we adopt here a distributional formulation, as one can find in (Kamen, 1975) and (Yamamoto, 1989) for delay differential systems. This choice is motivated by the fact that the theorem of supports of Titcmarsh-Lions (Hormander, 1990*a*, Chp. IV), presented in section 3 and extended to the matrix-valued case, generalizes the results one can obtain via the exponential type calculus for entire functions (O.Diekmann and Walther, 1995). Moreover, a nonsmoothness property will be used in the design of a sufficiently rich input. In the behavioral approach, the use of trajectories $\tilde{w} \notin C^\infty(\mathbb{R}, \mathbb{R}^m)$ would require, for instance, notions such as weak solutions as defined in (Polderman and Willems, 1998), giving rise to computational complexity problems.

Identifiability analysis imposes more restrictive conditions than those related to the equivalent kernel representation in the behavioral framework (see e.g. (Polderman and Willems, 1998) for differential equations and (Habets, 1999) for more general systems over rings). In this paper, we shall consider the identifiability problem for an input/output convolution equation of the form (1) in which P satisfies the following assumptions:

H1 P^{-1} exists in $\mathcal{D}'_+$,
H2 $ord(P^{-1}) = -ord(P)$,
H3 $P(0) = I$,

where P^{-1} denotes the inverse of P with respect to convolution, $ord(P)$ is the order of P (see section 2.3), $\mathcal{D}'_+$ is the space of distributions with support bounded on the left, and $P(0)$ denotes the Laplace transform of P at $s = 0$ (see notations). Following the definition of (Yamamoto, 1989), distributions satisfying *H1* and *H2* will be said to be of *normal type*. Note that these assumptions are not very severe restrictions since, in addition to ordinary differential equations, they include finite and infinite dimensional cases such as (a) pure delay equations, and (b) differential delay equations of neutral type and with distributed delays, for which P is of the form

$$P = \delta I + \sum_{i=1}^{N} P_i\, \delta_{h_i}, \qquad (4)$$

$$P = \delta' I + \sum_{i=0}^{N} P_i\, \delta_{h_i} + \sum_{i=1}^{N} P_{-i}\, \delta'_{h_i} + P_c(\theta) \quad (5)$$

for some reals $0 \leq h_0 < \ldots < h_N$, and $n \times n$ matrices P_i, P_{-i} and P_c with entries respectively in $\mathbb{R}$, $\mathbb{R}$, and $L^2(0, h_N)$. Although it requires $(\det P)(0) \neq 0$, assumption *H3* is only introduced for the sake of normalization, and particularly, every stable system in the sense that $(\det P)(s)$ is zero-free in the domain $\Re(s) \geq 0$ can be assumed to be in a normalized form.

The reference model which is used for the identifiablilty of the plant is given by

$$\hat{P} * \hat{y} = \hat{Q} * u \qquad (6)$$

where $\hat{P}$ and $\hat{Q}$ are $n \times n$ and $n \times m$ matrices with entries in $\mathcal{E}'$, and $\hat{P}$ is invertible and normalized in the sense of *H1* and *H3*. We do not impose explicit formulation for these distributions. Instead, denoting conformably $\hat{R} = \left[\hat{P}, -\hat{Q}\right]$, we shall consider a reference model for which:

C1 $ord(\hat{P}) = ord(P)$,
C2 $conv(\hat{R}) \subseteq conv(R)$,

where $conv(R)$ is the smallest closed interval that contains the support of R (see section 3).

Definition 1. The linear system described by (1) is called identifiable by (6) if one can find a input u and a bounded time interval I such that the equality $\hat{R} = R$ results from $\hat{y} = y$ on I.

In Section 4.1 we will show how to explicitly construct a appropriate input signal $u(t)$ which enforces identifiability of the system, and give more information about the required time interval I.

2. PRELIMINARIES

2.1 Notations

We recall some language from distribution theory. Let Ω be an open subset of $\mathbb{R}$. The space of $\mathcal{C}^\infty$ functions with support in Ω is denoted by $C_0^\infty(\Omega)$, and $\mathcal{D}'(\Omega)$ is the space of distributions on Ω.

The complement of the largest open subset of Ω in which T vanishes is called the support of T and is denoted by *supp* T. When T is a matrix valued distribution, *supp* T is the union of the support of its entries.

$\mathcal{D}'_+$ (resp. $\mathcal{E}'$, $\mathcal{E}'(\mathbb{R}_-)$) is the space of distributions with support bounded on the left (resp. compact support, compact support contained in $(-\infty, 0]$). It is an algebra with respect to convolution with identity δ, the Dirac distribution. When concentrated at a point $\{\tau\}$, the latter distribution is denoted δ_τ.

With no danger of confusion, if $T \in (\mathcal{E}')^{n\times m}$, we denote $T(s)$, $s \in \mathbb{C}$, the $n \times m$ matrix of entire functions obtained by Laplace transform of T.

2.2 Invertible distributions

The development of this section can be found in (Ehrenpreis, 1960) where the general problem of division in various spaces is studied.

Definition 2. A distribution $u \in \mathcal{E}'$ is called invertible for $\mathcal{D}'$ (in the sense of Ehrenpreis) if the map $\mathcal{D}' \ni v \to u * v \in \mathcal{D}'$ is onto.

Let $\mathcal{O}$ denote the space of all entire functions and $E' \subset \mathcal{O}$ the Laplace transform of $\mathcal{E}'$. (E' is thus the space of all entire functions (of s) of exponential type which are of polynomial increase in $\Re(s)$). Denote also (in this paragraph) $\tilde{u} \in E'$ the Laplace transform of $u \in \mathcal{E}'$.

Theorem 3. (Ehrenpreis, 1960) The following statements are equivalent:

(1) $u \in \mathcal{E}'$ is invertible.
(2) For any $g \in \mathcal{O}$, if $\tilde{u}\, g \in E'$, then $g \in E'$.

Such distributions are also referred to as completely invertible distributions and, in particular, if u admits an elementary solution in $\mathcal{D}'$ (i.e.

$w = \delta$), then it is invertible in the sense of this definition. Examples of non invertible distributions are the elements of C_0^∞ (See also (Hormander, 1990*b*)).

2.3 Order of distributions

We use here an extended notion of the order of a distribution as one can find in (Zemanian, 1965, p162) and (Yamamoto, 1984). A distribution is said to be of order r (> 0) if it acts continuously on C^r-functions but not on C^{r-1}-functions. Measures which are not functions are of order 0. A function ψ is of order $-r$ if r is the largest integer such that $D^r\psi$ is a measure.

We know that $ord(\alpha * \beta) \leq ord\,\alpha + ord\,\beta$ is always valid for scalar valued α, β, and, letting $ord\,P := \max ord\,p_{ij}$, it is clear that this inequality still holds for the matrix-valued case. Now if P is of normal type and Q has its entries in $\mathcal{D}'_+$, one easily gets the converse inequality under assumption *H2* by considering $Q = P^{-1} * (P * Q)$. Therefore for such P, Q,

$$ord(P * Q) = ord\,P + ord\,Q. \tag{7}$$

3. THE THEOREM OF SUPPORTS

Let us first recall a well known result on scalar valued distributions. For $u, v \in \mathcal{E}'$, one has

$$conv\,u * v = conv\,u + conv\,v, \tag{8}$$

where $conv\,w$ is the smallest closed interval that contains the support of w, and the sum $T_1 + T_2$ in the right hand side is defined by

$$\{x + y\,;\, x \in T_1,\ y \in T_2\}. \tag{9}$$

The result in (8) is known as the Theorem of Titcmarsh-Lions, also referred as the Theorem of Supports, and the aim of this section is the study of its analog in the matrix-valued case.

Definition 4. For $A \in (\mathcal{E}')^{n\times m}$, we let $E(A) := sup\,\{t \in supp\,A\}$, $I(A) := \inf\{t \in supp\,A\}$, and $conv(A) := [I(A), E(A)]$.

Using (8) and the definition, if A and B are matrix-valued distributions (of compatible size) with entries in $\mathcal{E}'$, it is not difficult to obtain the inclusion

$$conv\,A * B \subset conv\,A + conv\,B. \tag{10}$$

As we shall see, the properties obtained for the supremum of the support easily applies to the infimum by considering the distribution $\check{A}(-t) := A(t)$. Hence we first consider the case of the supremum and recall some algebraic notions introduced in (Yamamoto, 1989).

Let $\mathcal{F}$ be the quotient field of the quotient ring $\mathcal{A} := \mathcal{E}'(\mathbb{R}_-)/J$, where

$$J = \{\varphi \in \mathcal{E}'(\mathbb{R}_-); E(\varphi) < 0\}, \tag{11}$$

and denote $\theta : \mathcal{E}'(\mathbb{R}_-) \to \mathcal{F}$ the composition of the canonical projection $\mathcal{E}'(\mathbb{R}_-) \to \mathcal{A}$ with the inclusion $\mathcal{A} \to \mathcal{F}$. In this setting, an element $w \in (\mathcal{E}'(\mathbb{R}_-))^n$ is nonzero when considered over $\mathcal{F}$ if and only if $E(w) = 0$. The convention we shall use here slightly differs from (Yamamoto, 1989) since, in what follows, when we speak of a rank of a matrix $W \in (\mathcal{E}')^{n\times m}$ over $\mathcal{F}$, we shall always mean the rank of the matrix $\theta(\delta_{-E(W)} * W)$ (and not $\theta(W)$, $W \in (\mathcal{E}'(\mathbb{R}_-))^{n\times m}$) considered over $\mathcal{F}$.

Proposition 5. For $A \in (\mathcal{E}')^{n\times n}$ non singular, the following statements are equivalent:

(1) $E(\det A) = nE(A)$
(2) $E(A * B) = E(A) + E(B), \quad \forall B \in (\mathcal{E}')^{n\times p}$
(3) $rank_{\mathcal{F}}\,A = n$

Proof. See Appendix. ■

Remark 6. The regularity assumption ($\det A \neq 0$) only avoids the case of null distribution for which $E(0)$ is not defined.

With the convention mentioned above, the properties introduced in this proposition are now translational invariant, since $E(I\delta_\tau * A) = \tau + E(A)$ and $E\det(I\delta_\tau * A) = n\tau + E(\det A)$. Note also that they only depend on the behavior of A near $E(A)$, and hence take a very simple form in case of distributions with a punctual support. Hence, if for example, $P := \sum_{i=0}^N P_{-i}\delta'_{hi} + P_i\delta_{hi}$, one only requires $\det(\delta' P_{-N} + \delta P_N) \neq 0$, or equivalently, by Laplace transform, $rank(s\,P_{-N} + P_N) = n$ for *some* $s \in \mathbb{C}$.

For a non square element $A \in (\mathcal{E}')^{n\times m}$, if $rank_{\mathcal{F}}(A) = n$, this equivalently means that $rank_{\mathcal{F}}(M) = n$ with $E(M) = E(A)$ for some $n\times n$ block M of A. It is then clear that property (2) of the proposition still holds.

Now letting $\check{A}(-t) := A(t)$, one clearly has $supp\,\check{A} = -supp\,A$, but also (see e.g. (Hirsh and Lacombes, 1999, p334))

$$(A * B)\check{} = \check{A} * \check{B}. \tag{12}$$

The above proposition can then be straightforwardly applied to $I(A) = E(\check{A})$. In particular, if $\check{A}$ has rank n over $\mathcal{F}$, then, equivalently, for any B (of compatible size) with compact support, $I(A * B) = I(A) + I(B)$. Using the definition and the above considerations, we can now extend the theorem of supports to the matrix-valued case as follows:

Theorem 7. For $A \in (\mathcal{E}')^{n\times m}$ and $B \in (\mathcal{E}')^{m\times p}$, the following statements are equivalent:

(1) $rank_{\mathcal{F}}(A) = rank_{\mathcal{F}}(\check{A}) = n$
(2) $conv\, A * B = conv\, A + conv\, B \quad \forall B.$

For general delay differential systems with P defined as in (5), Proposition 5 provides a necessary and sufficient condition for the spectral completeness of the plant (i.e. the space spanned by the eigenfunctions is dense in the state space (see (Yamamoto, 1989) or (O.Diekmann and Walther, 1995, Chp V)). On the other hand, the restriction of P to any sufficiently small interval $[I(P), I(P)+\varepsilon]$ being of the form $P_{[0,\varepsilon]} = \delta' + W$ with W a measure, we clearly have $I(\det P) = nI(P) = 0$, and hence P satisfies the rank conditions of the theorem in case of spectral completeness.

4. IDENTIFIABILITY

4.1 Time interval of observation and Sufficiently rich input

In order to restrict the identifiability study to a bounded time interval, we shall assume that the input has also a compact support,

$$supp\ u \subset [t_1, t_2]\,. \tag{13}$$

The "richness" of u will be defined within this time interval. We also let

$$h := mes\ conv\, R, \tag{14}$$

where $mes\ [a,b] = b - a$ is simply the Lebesgue measure of $[a,b]$. One can interpret h as the memory length of the plant. The next proposition is a simple application of relation (10).

Proposition 8. If $I := [r_1, r_2]$ is such that

$$r_1 + (n+1)h < t_1 < t_2 < r_2 - (n+1)h \tag{15}$$

then $y = \hat{y}$ on I implies $y = \hat{y}$ on $\mathbb{R}$.

Proof. See Appendix ∎

Now we turn to the design of an input u such that the equality of the outputs results in that of the impulse responses (or transfer functions). Define u as a piece-wise $\mathbb{R}^m$-valued polynomial function and let

$$\Lambda_u = \{s_0, s_1, ..., s_L\} \tag{16}$$

denote its singular support (i.e.the set of points in $\mathbb{R}$ having no open neighborhood to which the restriction of u is a C^∞ function). Denote also

$$\sigma_u^k(s_l) = u^{(k)}(s_l + 0) - u^{(k)}(s_l - 0) \tag{17}$$

the jump of $u^{(k)}(t)$ at $t = s_l$. One of the fundamental properties of distributions is that by differentiation, we do not miss something essential as discontinuity. In our case, if all the polynomials defining u are of order $\leq N$ for some $N \in \mathbb{N}$, differentiation at the order $N+1$ results in the singular distribution

$$u^{(N+1)} = \sum_{l=0}^{L} U_l(D)\,[\delta] * \delta_{s_l}, \tag{18}$$

where $D = d/dt$ and

$$U_l(D) = \sum_{i=0}^{N} \sigma_u^{N-i}(s_l)\, D^i. \tag{19}$$

Now form from (17) and (19) , the $m \times L$ polynomial matrix

$$U(D) = [U_0(D), ..., U_L(D)]\,. \tag{20}$$

Using a Smith form factorization and invertibility of differential operators in $\mathcal{D}'_+$, we have the

Proposition 9. Let u defined as above be such that

(1) $rank\, U(D) = m$,
(2) $s_l - s_{l-1} > (n+1)h, \quad l = 1, .., L.$

Then $y = \hat{y}$ implies $P^{-1} * Q = \hat{P}^{-1} * \hat{Q}$.

Proof. See Appendix ∎

Note that with *supp* u and a time of observation I defined as in proposition 8, we have obtained an input u for which

$$y = \hat{y} \quad on\ I \Rightarrow P^{-1} * Q = \hat{P}^{-1} * \hat{Q}. \tag{21}$$

The simplest example consists of a piece-wise constant $\mathbb{R}^m$-valued function with appropriate discontinuities. More generally, we can design an input u of class $\mathcal{C}^r$ for an arbitrary finite integer r, but the main property required in this approach remains the non smoothness of u. Before proceeding, we note that condition (2) in the previous proposition depends on the system under study, since it required an a priori knowledge of the memory length h (or at least an upper bound), and one could prefer the use of the term "peristently exciting input" instead. However, this approach extends to systems with memory and multi-inputs a class of "sufficiently rich functions" introduced in (Miller and Michel, 1990).

4.2 Identifiability of the plant

Equality of the impulse response can be written equivalently as

$$\hat{R} = \alpha * R, \tag{22}$$

$$\alpha = \hat{P} * P^{-1}. \tag{23}$$

First observe that in (22), and for any invertible α in $(\mathcal{D}'_+)^{n\times n}$, $\hat{P} = \alpha * P$ and $\hat{Q} = \alpha * Q$ yields

$\hat{P}^{-1} * \hat{Q} = P^{-1} * Q$. This shows in particular that possibly unbounded supported distributions may result in the same impulse response (or transfer function) as that of the initial plant. However, it is clear from (10) and (22) that $\hat{R}$ is compact supported if α is, but the converse is not true in general.

Proposition 10. If $rank\, R(s) = n$, $s \in \mathbb{C}$, then $\alpha \in (\mathcal{E}')^{n\times n}$.

Proof. Since $\hat{R}$ and R are compact supported distributions, $\hat{R}(s)$ and $R(s)$ are with entries in $E' \subset \mathcal{O}$. Hence, taking the Laplace transform of (22), we deduce that, if $rank\ R(s) = n$, $s \in \mathbb{C}$, then $\alpha(s)$ must be in $\mathcal{O}^{n\times n}$. It remains to show that $\alpha(s) \in E'^{n\times n}$. Since $(\det P)(s).\alpha(s) = \hat{P}(s).Adj(P)(s) \in E'^{n\times n}$, and $\det P$ is invertible, theorem 3 applies componentwise and thus $\alpha(s) \in E'^{n\times n}$. ■

The property $rank\, R(s) = n$, $s \in \mathbb{C}$, is usually referred as spectral controllability. Clearly, this condition is not sufficient to guarantee uniqueness of the representation of the impulse response in terms of the kernel R. We can now state our main result.

Theorem 11. Let $R \in (\mathcal{E}')^{n\times(n+m)}$ be such that

(1) $rank\, R(s) = n, \quad s \in \mathbb{C}$,
(2) $rank_{\mathcal{F}}(R) = rank_{\mathcal{F}}(\check{R}) = n$,

then the system (1) is identifiable by (6).

Proof. From (1) and proposition 10, α has a compact support. Using (2), the theorem of supports applied to (22) yields $conv\, \hat{R} = conv\, \alpha + conv\, R$. This contradicts the statement *C2* unless $conv\, \alpha = \{0\}$, which means that α is a linear combination of Dirac distributions and its derivatives. From *C1* and (7), $ord\, \alpha = 0$, hence $\alpha = \alpha_0\delta$ for some invertible $n \times n$ real matrix α_0. The normalization *H3* of P and $\hat{P}$ yields $\alpha_0 = I$. ■

Note that conditions (2) are automatically satisfied if $n = 1$, but also for ordinary differential equations, since $conv\, R = \{0\}$, $E(\det P) = nE(P) = 0$, and $\check{P} = P$. Identifiability in these cases reduces to spectral controllability. For distributions with punctual support, this condition take also a very simple form if one recalls that $rank_{\mathcal{F}}(R)$ and $rank_{\mathcal{F}}(\check{R})$ uniquely depend on the behavior of R in the vicinity of $I(R)$ and $E(R)$.

The following example shows the link between identifiability and controllability results for delay differential equations. Let P as in (5), $Q = Q_0\delta$, and consider the system described by

$$P * x = Q * u, \quad y(t) = x(t - h_N). \tag{24}$$

The kernel is given here by $R = [\delta_{-h_N} * P, -Q]$. From section 3 and translational invariance, it is clear that $rank_{\mathcal{F}}(\check{R}) = n$ is always satisfied and the remaining conditions for identifiablity are

$$rank\, R(s) = n, \quad s \in \mathbb{C}, \tag{25}$$
$$rank_{\mathcal{F}}(R) = n. \tag{26}$$

These are nothing but the necessary and sufficient conditions for the system (24) to be approximately controllable in the sense that the reachable space is dense in the state space (Yamamoto, 1989).

Remark 12. There is another direct consequence of Theorem 11 if one considers the problem of realization. Indeed, keeping only the assumption of invertibility of P (in the sense of definition 2), one can easily show that, if there exists such R, it corresponds to the (non unique) realization with the *smallest memory length* h (as defined in (14)), since among all other $\hat{R}$ for which $\hat{P}^{-1} * \hat{Q} = P^{-1} * Q$, we necessarily have $\hat{h} \geq h$.

5. CONCLUSION

We have presented a sufficient condition for the identifiability of a general class of linear systems described by an input/output convolution equation. One of the key results is the theorem of supports extended to the matrix-valued case. However, we have only considered distributions of one variable (i.e the time axis), and the possible extension to partial differential equations is under investigation. Another open problem concerns the robustness of these results since, throughout this paper, we have assumed perfect data.

6. APPENDIX

Proof of proposition 5: For $u, v \in \mathcal{E}'$, $E(u * v) = E(u) + E(v)$ is a first consequence of (8). The following inequalities are also easily derived from (8) and (10),

$$E(\det A) \leq nE(A), \tag{27}$$
$$E(adj(A)) \leq (n-1)E(A), \tag{28}$$
$$E(A * B) \leq E(A) + E(B). \tag{29}$$

We first start showing that (1) is equivalent to the statement:

$$(1') : E(A) + E(adj(A)) = E(\det A).$$

$(1' \Rightarrow 1)$ Applying (27) to $adj(A)$ yields

$$E \det(adj(A)) \leq nE(adj(A)).$$

Since $\det adj(A) = (\det A)^{*n-1}$, we obtain using (1'), $(n-1)E\det(A) \leq n(E\det(A) - E(A))$, hence

$E\det(A) \geq nE(A)$ and using (27), $E\det A = nE(A)$.

$(1 \Rightarrow 1')$ If (1) holds, then using $A * adj(A) = \det(A)$ and (29) yields $E\det(A) = nE(A) \leq E(A) + E(adj(A))$ so $E(adj(A)) \geq (n-1)E(A)$ which gives with (28), $E(adj(A)) = (n-1)E(A)$ and (1') follows.

$(2 \Rightarrow 1')$ Take $B = adj(A)$.

$(1' \Rightarrow 2)$ Let $C = A * B$. A left convolution with $adj(A)$ yields $adj(A) * C = (\det A) * B$ so $E\det(A) + E(B) \leq E(adj(A)) + E(C)$. If (1') holds, then $E(C) \geq E(A) + E(B)$ and (2) follows using (29)

$(3 \Leftrightarrow 2)$ We know from (Yamamoto, 1989, Thm 3.9) that (3) is equivalent to the statement: for A with entries in $\mathcal{E}'(\mathbb{R}_-)$, there exists no $\psi \in (\mathcal{E}')^n$ such that $E(\psi) > 0$ and $E(\psi^T * A) \leq 0$. This is clearly equivalent to (2).

Proof of proposition 8: Denote $\bar{y} := y - \hat{y}$, $\Delta := \det P$, and consider $\Delta * \hat{P} * \bar{y}$. We get after some simple manipulations

$$S(Q) * u = \Delta * \hat{P} * \bar{y}, \qquad (30)$$

where we have denoted formally

$$S(.) = \Delta * \hat{P} * (P^{-1} * (.) - \hat{P}^{-1} * \widehat{(.)}) \qquad (31)$$
$$= \hat{P} * Adj(P) * (.) - \Delta * \widehat{(.)}. \qquad (32)$$

Write $\bar{y} = \bar{y}_1 + \bar{y}_2$ with $supp\,\bar{y}_1 \subset]-\infty, r_1]$ and $supp\,\bar{y}_2 \subset [r_2, \infty[$ so (30) is rewritten

$$S(Q) * u = \Delta * \hat{P} * \bar{y}_2 + \Delta * \hat{P} * \bar{y}_1. \qquad (33)$$

Using (14), *C2*, and (10), $S(Q)$ and $\Delta * \hat{P}$ have their support included in some interval $[\nu_1, \nu_2]$ with $\nu_2 - \nu_1 \leq (n+1)h$. Hence, the respective supports in (33) are included in $[t_1 + \nu_1, t_2 + \nu_2]$, $]-\infty, r_1 + \nu_2]$, and $[r_2 + \nu_1, \infty[$. If (15) holds, these supports are disjoint, so each corresponding distribution must identically vanish. Invertibility of $\Delta * \hat{P}$ results in $\bar{y}_1 = \bar{y}_2 = 0$.

Proof of proposition 9: In (31), $\Delta * \hat{P}$ is invertible so it suffices to show that $S(Q) * u = 0 \Rightarrow S(Q) = 0$. Using (18), differentiation of $S(Q) * u = 0$ yields

$$\sum_{l=0}^{L} S(Q) * U_l(D)\,[\delta] * \delta_{s_l} = 0, \qquad (34)$$

with $supp\ S(Q) \subset [\nu_1, \nu_2]$, $\nu_2 - \nu_1 \leq (n+1)h$, while $supp\ U_l(D)\,[\delta] = \{0\}$. If (2) holds, all the terms of the sum in (34) have disjoint supports and then must identically vanish. Hence,

$$S(Q) * [U_0(D), ..., U_L(D)]\,[\delta] = 0. \qquad (35)$$

Now if $rank\ U(D) = m$, $U(D)$ admits a Smith form factorization (Ayres, 1973), $U(D) = W(D)\,[\Lambda(D), 0]\,V(D)$ where $W(D), V(D)$ are unimodular matrices (i.e. they admit an inverse which is also polynomial), and $\Lambda(D) = diag(\lambda_i(D))$, with no identically zero polynomial on its diagonal. Since every non zero polynomial $\lambda(D)\,[\delta]$, $D = \partial/\partial t$ is invertible in $\mathcal{D}'_+$ ((Yger, 2001, Thm 3.3)), the conclusion follows.

REFERENCES

Ayres, F. Jr. (1973). *Theory and problems of matrices.* McGraw-Hill. New York.

Ehrenpreis, L. (1960). Solutions of some problems of division, part iv, invertible and elliptic operators. *Amer. Journal of Mathematics* **82**, 522–588.

Habets, L.C.G.J.M. (1999). System equivalence for ar-systems over rings, with an application to delay-differential systems. *Journal of Mathematics of Control, Signals, and Systems* **12**, 219–244.

Hirsh, F. and G. Lacombes (1999). *Elements of functional analysis.* Springer-Verlag. New York.

Hormander, L. (1990*a*). *The Analysis of Linear Partial Differential Operators I.* Springer-Verlag. Berlin.

Hormander, L. (1990*b*). *The Analysis of Linear Partial Differential Operators II.* Springer-Verlag. Berlin.

Kamen, E. W; (1975). On an algebraic theory of systems defined by convolution operators. *Journal of Mathematical systems theory* **9, No 1**, 57–74.

Miller, R.K. and A.N. Michel (1990). An invariance theorem with application to adaptive control. *IEEE Trans. on Automatic Control* **35**, 744–748.

O.Diekmann, S.A. van Gils, S.M.V. Lunel and H.O. Walther (1995). *Delay equation, Functional-, Complex-, and Nonlinear Analysis.* Springer-Verlag. New York.

Polderman, J. W. and J.C. Willems (1998). *Introduction to Mathematical Systems Theory. A Behavioral Approach.* Springer-Verlag. New York.

Vettori, P. and S. Zampieri (2000). Controllability of systems described by convolutionnal or delay-differential equations. *SIAM Journal of Control and Optimization* **39, No 3**, 728–756.

Yamamoto, Y. (1984). A note on linear input/output maps of bounded-type. *IEEE Trans. on Automatic Control* **AC-29**, 733–734.

Yamamoto, Y. (1989). Reachability of a class of infinite-dimensional linear systems: An external approach with application to general neutral systems. *SIAM J. of Control and Optimization* **27**, 217–234.

Yger, A. (2001). *Analyse complexe et distributions.* Ellipses. Paris.

Zemanian, A.H. (1965). *Distribution theory and transform analysis.* McGraw-Hill. New York.

IFAC Publications
www.elsevier.com/locate/ifac

IDENTIFICATION OF FULLY PARAMETERIZED LINEAR AND NONLINEAR STATE-SPACE SYSTEMS BY PROJECTED GRADIENT SEARCH

Vincent Verdult * Niek Bergboer ** Michel Verhaegen *

* *Delft University of Technology, Faculty of Information Technology and Systems, Control Systems Engineering, P. O. Box 5031, NL-2600 GA Delft, The Netherlands*
** *Maastricht University, Department of Computer Science, St Jacobstraat 6, Maastricht, The Netherlands*

Abstract: A nonlinear optimization-based identification procedure for fully parameterized multivariable state-space models is presented. The method can be used to identify linear time-invariant, linear parameter-varying, composite local linear, bilinear, Hammerstein and Wiener systems. The nonuniqueness of the full parameterization is dealt with by a projected gradient search to solve the nonlinear optimization problem. Both white and nonwhite measurement noise at the output can be dealt with in a maximum likelihood setting. It is proposed to use subspace identification methods to initialize the nonlinear optimization problem. A computationally efficient and numerically reliable implementation of the procedure is discussed in detail. *Copyright © 2003 IFAC*

Keywords: System identification, state-space methods, nonlinear models, nonlinear programming, numerical algorithms, maximum likelihood, bilinear systems.

1. INTRODUCTION

Efficient and reliable identification algorithms for multivariable linear and nonlinear systems are of considerable interest to both academic research groups and industry to solve large-scale practical problems. Although a multitude of identification methods for multivariable linear time-invariant (LTI) systems exist, the choices are limited when dealing with multivariable nonlinear systems.

To deal with multivariable systems, state-space models offer considerable advantages over input-output descriptions. State-space models for LTI systems can be identified using the classical maximum-likelihood methods or using subspace methods (Ljung, 1999). In the maximum-likelihood methods the system is parameterized and the optimal values of the parameters are obtained by minimizing a cost function. In general, the cost function is a nonconvex nonlinear function that has to be minimized by an iterative procedure. Hence, maximum likelihood methods require a reasonable initial guess of the parameters. Subspace identification methods (Van Overschee and De Moor, 1996; Verhaegen, 1994) can be used to provide such an initial guess. The subspace methods are noniterative, but not optimal in a maximum likelihood sense. The drawbacks of maximum likelihood and subspace methods can be overcome by combining the two. For LTI systems the combination of subspace and optimization based identification methods has been proposed by several authors (for example Ljung, 1999).

A delicate issue in maximum-likelihood methods is the choice of system parameterization. Using canonical forms like the observer canonical form can lead to bad numerical properties. To avoid the choice of parameterization McKelvey and Helmersson (1997) proposed to use a fully parameterized state-space model and deal with the nonuniqueness of this model using a special gradient projection method for minimizing the maximum likelihood cost function. This projection can be regarded as choosing a local parameterization of the system at each iteration step of the numerical minimization of the the maximum-likelihood cost function. Therefore, this method is also referred to as the local parameterization method or the data driven local coordinates (DDLC) method. Currently, the topological and geometrical properties of this approach are investigated by Ribarits and Deistler (2002) (see also Deistler and Ribarits, 2001). A similar method has been proposed by Lee and Poolla (1999) for the identification of linear parameter-varying (LPV) systems.

In this paper we show that the gradient projection method of McKelvey and Helmersson (1997) and Lee and Poolla (1999) can also be used for the identification of the following multivariable nonlinear state-space systems: bilinear systems, composite local linear systems, Hammerstein systems and Wiener systems. For most of these systems subspace methods are available to generate the initial starting point, and thus, a practical identification method is available by combining a maximum-likelihood and a subspace method. The paper also discusses computationally efficient and numerically reliable implementation of the identification procedures.

The paper is organized as follows: Section 2 describes the types of linear and nonlinear models that we consider. Section 3 describes the output error identification problem; the output is only disturbed by white measurement noise. A projected gradient algorithm is presented to minimize the output error cost function. As mentioned above a reasonable initial starting point for the projected gradient algorithm can be obtained using subspace identification; this is discussed in Section 4. Two possibilities to deal with nonwhite output disturbances are discussed in Section 5. A computationally efficient and numerically reliable implementation of the proposed identification procedure is discussed in Section 6. Due to lack of space, we do not present any examples of the proposed approach. The reader can find bilinear, LPV and composite local linear examples in: Verdult (2002), Verdult *et al.* (2001), Verdult *et al.* (2002).

2. IDENTIFICATION OF LINEAR AND NONLINEAR STATE-SPACE SYSTEMS

The identification framework that is presented is suitable for a number of different model structures. In all models $x(k) \in \mathbb{R}^n$ is the state sequence, $u(k) \in \mathbb{R}^m$ the input, and $y(k) \in \mathbb{R}^\ell$ the output.

The first model structure to be considered is the linear parameter-varying system

$$x(k+1) = A\begin{bmatrix} x(k) \\ p(k) \otimes x(k) \end{bmatrix} + B\begin{bmatrix} u(k) \\ p(k) \otimes u(k) \end{bmatrix},$$
$$y(k) = C\begin{bmatrix} x(k) \\ p(k) \otimes x(k) \end{bmatrix} + D\begin{bmatrix} u(k) \\ p(k) \otimes u(k) \end{bmatrix},$$

where the signal $p(k) \in \mathbb{R}^s$ is the time-varying parameter, which is assumed to be known, $\otimes$ is the Kronecker matrix product, and $A = [A_0, A_1, \dots, A_s]$ with $A_i \in \mathbb{R}^{n \times n}$, $i = 1, 2, \dots, s$. The matrices B, C and D can be partitioned similarly with $B_i \in \mathbb{R}^{n \times m}$, $C_i \in \mathbb{R}^{\ell \times n}$, and $D_i \in \mathbb{R}^{\ell \times m}$.

This LPV model structure also accommodates other model structures as follows:

- For linear time-invariant systems, we have $s = 0$, and thus the dependency on $p(k)$ disappears.
- For bilinear systems, we have $s = m$, $p(k) = u(k)$, and $B_i = 0$, $C_i = 0$ and $D_i = 0$ for $i = 1, 2, \dots, m$.

Another model structure that can be dealt with is a composite local linear system (Murray-Smith and Johansen, 1997) in which the state equation consists of a weighted combination of LTI models

$$x(k+1) = \sum_{i=1}^{s} f_i\Big(\phi(k)\Big)\Big(A_i x(k) + B_i u(k) + O_i\Big),$$
$$y(k) = C_0 x(k) + D_0 u(k),$$

where O_i represent offsets on the state. The weights $f_i(\phi(k))$ are taken as radial basis functions and $\phi(k)$ is the scheduling vector which is assumed to be a known function of the input $u(k)$ and the output $y(k)$ The scheduling vector represents the operating point of the system. The weights determine which combination of local models is active based on the operating point of the system. Note that by taking the parameters $p(k)$ in the LPV structure equal to $f_i(\phi(k))$, the composite local model structure can be regarded as a special case of the LPV system.

In addition to these model structures, the identification framework presented in this paper can handle nonlinear Hammerstein and Wiener systems. In Hammerstein systems, the input first passes through a smooth nonlinear function before entering an LTI system. We assume the following special structure for the Hammerstein system

$$x(k+1) = A_0 x(k) + Bf\Big(u(k)\Big),$$
$$y(k) = C_0 x(k) + Df\Big(u(k)\Big),$$

with $f : \mathbb{R}^m \to \mathbb{R}^{m(s+1)}$ a fixed and smooth nonlinear function. For example, the elements of f can be polynomial functions of $u(k)$ and then B and D, contain the weights to combine these functions.

In Wiener systems, the output of an LTI system passes through a smooth nonlinear function. We consider Wiener systems with the following structure:

$$x(k+1) = A_0 x(k) + B_0 u(k),$$
$$z(k) = C_0 x(k) + D_0 u(k),$$
$$y(k) = Wf\Big(z(k)\Big),$$

with $f : \mathbb{R}^\ell \to \mathbb{R}^\ell$, and W the corresponding weights.

Other structures of the nonlinear parts in the Hammerstein and Wiener systems can be used, but are not discussed here. It is also possible to deal with a Hammerstein-Wiener combination, where the LTI part is between two static nonlinearities. Although we will only discuss static nonlinearities in combination with LTI systems, we can also handle static nonlinearities in combination with LPV systems.

3. OUTPUT ERROR IDENTIFICATION

This section introduces the output error identification problem. It is assumed that the output of the system is disturbed by white measurement noise. More general disturbances are dealt with in Section 5.

3.1 *System Parameterization*

In this paper we adopt a full parameterization of the system matrices which is given by

$$\theta = P\text{vec}\left(\begin{bmatrix} A & B \\ C & D \end{bmatrix}\right), \tag{1}$$

where P is a selection matrix that discards the entries that are zero by definition (as in the bilinear model for example). Note that the LTI parts of the Hammerstein and Wiener systems can also be described in this way.

In the composite local linear system, the radial basis functions f_i need to be determined as well. These functions are parameterized by their means and variances, collected in the parameter vector ρ.

For the Wiener systems the weights W also need to be determined, these weights stored in the parameter vector $\rho = \text{vec}(W)$.

3.2 *The Cost Function*

To identify one of the models of Section 2 we search for a set of parameters θ and ρ such that the output of the model $\widehat{y}(k;\theta,\rho)$ approximates the output $y(k)$ of the real system sufficiently accurately. To achieve this goal, the output error is minimized with respect to the parameters θ and ρ. The output-error cost function is given by

$$\begin{aligned} V_N(\theta,\rho) &:= \sum_{k=1}^{N} ||y(k) - \widehat{y}(k;\theta,\rho)||_2^2 \\ &= E_N^T(\theta,\rho)E_N(\theta,\rho), \end{aligned}$$

where $E_N(\theta,\rho) = \begin{bmatrix} e(1)^T & e(2)^T & e(3)^T & \cdots & e(N)^T \end{bmatrix}^T$, $e(k) = y(k) - \widehat{y}(k;\theta,\rho)$, and N is the total number of measurements available.

Minimization of (2) is a nonlinear, nonconvex optimization problem because of the nonlinear dependence of $\widehat{y}(k;\theta,\rho)$ on θ and ρ. In the next section we present an algorithm to numerically search for a solution to this optimization problem.

The cost function for the Wiener system can be modified and the weights W can be eliminated from the problem using the principle of separable least squares (Golub and Pereyra, 1973). This was pointed out by Bruls *et al.* (1999).

3.3 *Projected Gradient Search*

The input-output behaviors of the state-space systems of Section 2 do not change under a nonsingular linear similarity transformation $T \in \mathbb{R}^{n\times n}$ of the state: $x_T(k) = T^{-1}x(k)$. Since we use a full system parameterization, the minimization of $V_N(\theta,\rho)$ does not have a unique solution: there exist different parameter values θ that yield the same input-output behavior of the model and hence the same value of the cost function. Below, we present an iterative projected gradient search method that deals with this nonuniqueness by restricting the parameter update at each iteration to directions in which the cost function changes. Such a method has been previously described by Lee and Poolla (1999) for LPV models, by McKelvey and Helmersson (1997) for LTI models and by Verdult *et al.* (2002) for bilinear models.

The nonuniqueness due to the similarity transformation can be characterized by the *similarity map* which is defined as:

$$\begin{aligned} S(\theta,T) := &\begin{bmatrix} T^{-1} & 0 \\ 0 & I_\ell \end{bmatrix} \begin{bmatrix} A(\theta) & B(\theta) \\ C(\theta) & D(\theta) \end{bmatrix} \\ &\times \begin{bmatrix} I_{s+1}\otimes T & 0 \\ 0 & I_{m(s+1)} \end{bmatrix}. \end{aligned}$$

Lee and Poolla (1999) have shown that by linearizing the similarity map, the directions in which the cost function change are given by the left null space of the matrix

$$\begin{aligned} M(\theta) := &\sum_{i=1}^{s+1} \begin{bmatrix} \Pi_i^T \\ 0_{m(s+1)\times n} \end{bmatrix} \otimes \begin{bmatrix} A(\theta)\Pi_i^T \\ C(\theta)\Pi_i^T \end{bmatrix} \\ &- \begin{bmatrix} A(\theta)^T \\ B(\theta)^T \end{bmatrix} \otimes \begin{bmatrix} I_n \\ 0_{\ell\times n} \end{bmatrix}, \end{aligned}$$

where $\Pi_i := \begin{bmatrix} 0_{n\times(i-1)n} & I_n & 0_{n\times(s+1-i)n} \end{bmatrix}$. The matrix $M(\theta)$ has full column rank if the similarity map $S(\theta,T)$ for a fixed θ is locally one-to-one around I_n. For LTI, Wiener and Hammerstein systems, the similarity map is one-to-one if the pair (A,B) is controllable or if the pair (A,C) is observable; conditions for LPV systems have been described by Lee and Poolla (1999), conditions for bilinear systems and composite local linear models by Verdult (2002).

The matrix $M(\theta)$ does not take into account that certain blocks of the matrices A, B, C, and D can be zero by definition (as is the case for the LTI, bilinear, composite local linear models, Hammerstein and Wiener models). Since these zero blocks do not change the degrees of freedom of the similarity transformation, these zero blocks can be taken into account by simply discarding the corresponding rows of the matrix $M(\theta)$. In other words the left null space of the matrix $PM(\theta)$ will be determined, where P is the selection matrix of equation (1).

Other constraints on the parameters θ, of the form $\Gamma\theta = \theta_0$, with Γ and θ_0 given, can be taken into account by incorporating them in the derivation of the matrix $M(\theta)$ as discussed by Avdeenko (2002).

The left null-space of a matrix is usually obtained using a singular value decomposition. However, since $PM(\theta)$ is of full rank we can use a QR factorization, which is computationally faster:

$$PM(\theta) = \begin{bmatrix} \mathcal{Q}_1(\theta) & \mathcal{Q}_2(\theta) \end{bmatrix} \begin{bmatrix} \mathcal{R}_1(\theta) \\ 0 \end{bmatrix}. \tag{2}$$

The columns of the matrix $Q_2(\theta)$ form a orthonormal basis for the left null-space of $M(\theta)$. A gradient search will be performed in this resulting local parameter subspace. We propose to use a dedicated trust-region implementation of the Levenberg-Marquardt algorithm (Moré, 1978). For this algorithm the Jacobian of the error vector $E_N(\theta, \rho)$ is needed; it is given by:

$$\begin{aligned}\Psi_N(\theta, \rho) &:= \left[\Psi_{N,\theta}(\theta, \rho) \;\; \Psi_{N,\rho}(\theta, \rho)\right] \\ &= \left[\frac{\partial E_N(\theta, \rho)}{\partial \theta^T} \;\; \frac{\partial E_N(\theta, \rho)}{\partial \rho^T}\right].\end{aligned}$$

After integrating the local gradient search into this algorithm, subsequent iterations update the parameters as

$$\begin{bmatrix}\theta^{(i+1)} \\ \rho^{(i+1)}\end{bmatrix} = \begin{bmatrix}\theta^{(i)} \\ \rho^{(i)}\end{bmatrix} + d(\theta^{(i)}, \rho^{(i)}, \lambda^{(i)}),$$

in which $d(\theta^{(i)}, \rho^{(i)}, \lambda^{(i)})$ is given by

$$\begin{aligned}d(\theta, \rho, \lambda) = &-\overline{Q}_2(\theta)\Phi(\theta, \rho, \lambda)^{-1}\overline{Q}_2(\theta)^T \\ &\times \Psi_N(\theta, \rho)^T E_N(\theta, \rho), \qquad (3)\end{aligned}$$

with

$$\Phi(\theta, \rho, \lambda) := \overline{Q}_2(\theta)^T \Psi_N(\theta, \rho)^T \Psi_N(\theta, \rho)\overline{Q}_2(\theta) + \lambda I,$$

$$\overline{Q}_2 := \begin{bmatrix} Q_2(\theta) & 0 \\ 0 & I_q\end{bmatrix}^T,$$

and q the number of parameters in ρ. The Levenberg-Marquardt regularization parameter $\lambda^{(i)}$ is determined in each iteration and depends on the linearity of the cost function in the vicinity of the point $(\theta^{(i)}, \rho^{(i)})$. The regularization in the Levenberg-Marquardt takes care of nonuniqueness in the nonlinear system representations that is not due to the similarity transformation.

Besides $Q_2(\theta)$, the vector $E_N(\theta)$ and its Jacobian $\Psi_N(\theta, \rho)$ are needed to compute (3). According to (2), $E_N(\theta)$ follows from the output of the model $\hat{y}(k; \theta^{(i)})$ which can be obtained by simulating the state-space system. The Jacobian can also be obtained by simulating a dynamic state-space model. This will be discussed in detail in Section 6.2.

4. INITIALIZATION

As mentioned before, the initial starting point of the projected gradient search has a big influence on the final result. A good initial starting point for most of the model structures of Section 2 can be obtained using subspace identification. Subspace identification methods have been described for LTI (Van Overschee and De Moor, 1996; Verhaegen, 1994), LPV (Verdult and Verhaegen, 2002; Verdult, 2002), bilinear (Favoreel *et al.*, 1999; Favoreel, 1999; Verdult and Verhaegen, 2001; Chen and Maciejowski, 2000; Verdult, 2002), Wiener (Westwick and Verhaegen, 1996), and Hammerstein systems (Verhaegen and Westwick, 1996) The parameters ρ (W) in the Wiener system can be initialized from an estimate of the signal $z(k)$ provided by subspace identification.

For the local linear model structure no subspace identification methods are available. A natural way to initialize the model is to estimate a global LTI state-space model using subspace identification and to take all the local models equal to this linear model. The initial weighting functions are distributed uniformly over the operating range (Verdult *et al.*, 2001). Using linear models for initialization has been proposed and motivated by Sjöberg (1997). More sophisticated ways of initializing these local linear model structures, including subspace identification are a topic for further research.

5. DEALING WITH COLORED NOISE

In Section 3 the output measurement noise was assumed to be a white-noise sequence. If the output measurement noise is nonwhite the optimization of the cost function (2) leads to estimates of the parameters θ, and ρ that are not of minimum variance in a maximum likelihood sense. This problem can be overcome by modifying the cost function.

Minimum variance estimates can be obtained by minimizing a weighted cost function

$$V_N(\theta, \rho) = E_N^T \Sigma_v^{-1} E_N = (\Sigma_v^{-1/2} E_N)^T (\Sigma_v^{-1/2} E_N),$$

in which the weighting matrix Σ_v is based upon a model of the measurement noise v_k. If the weighting matrix Σ_v is taken equal to the covariance matrix of the residual vector E_N that results from the measurement noise, maximum-likelihood estimates of the system matrices can be obtained.

The computation of the huge dimensional inverse covariance matrix Σ_v^{-1} can be done in a computationally efficient way by modeling the noise v_k by an multivariable AR model and exploiting the Gohberg-Heinig explicit inverse of a Toeplitz matrix. This was recently pointed out by David and Bastin (2001). Bergboer *et al.* (2002) described a computationally efficient implementation of this method.

Another way to obtain minimum variance estimates is by minimizing the prediction error instead of the output error. For LTI systems this is well known and the Kalman filter can be used for deriving the predictor. For the other systems, especially the nonlinear ones, deriving the predictor is far from trivial. The LPV and bilinear systems would require a time-varying Kalman gain (Fnaiech and Ljung, 1987).

6. EFFICIENT IMPLEMENTATION

It is well-known that computing the search-direction using (3) is inefficient and inaccurate. The QR factorization can be used to implement the parameter update rule (3) in a numerically reliable way, in which only the R factor needs to be computed, and which is such that only one factorization is needed for several values of the regularization parameter λ. Furthermore, for huge data lengths, it is possible with this QR factorization to process the data in batches to compute the residuals $E_N(\theta, \rho)$ and the gradients $\Psi_N(\theta, \rho)$. In this way the identification method can deal with huge data lengths, because only a small batch of the data needs to be stored in memory. Details of these steps have been presented by Bergboer *et al.* (2002).

6.1 *Obtaining the Local Parameter Subspace*

The determination of the left null space of the matrix $M(\theta)$ accounts for much of the computation time of the method. It can be computed efficiently using a QR factorization (2) based on Householder rotations. Given the Householder vectors, the matrix Q_2 can be calculated directly, without calculating Q_1, by applying all Householder rotations to $[0 \quad I]^T$. Efficient numerical implementations exist for this (Golub and Van Loan, 1996, p. 211).

Recently, McKelvey (2002) proposed an alternative way to obtain the local parameter subspace for LTI systems which results in considerable computational savings. It is based on using the impulse response of the LTI system, and results in performing a QR factorization on a matrix of which the number of rows grows linearly with n instead of quadratically as in $M(\theta)$. This method can also be used for Hammerstein and Wiener systems. Based on the results of Isidori (1973) an extension to bilinear systems is possible, but not useful, because it results in a matrix of which the number of rows grows exponentially with n.

6.2 *Computing the Projected Gradient*

Memory usage and computation time can be reduced by computing the product $\Psi_{N,\theta} Q_2$ directly, rather than first calculating $\Psi_{N,\theta}$ and then multiplying by Q_2. Since the number of columns in $M(\theta)$ is n^2, the number of columns in $\Psi_{N,\theta} Q_2$ is n^2 less than the number of columns in $\Psi_{N,\theta}$. For systems having a large order this difference can be substantial, especially for LTI systems since the number of columns in $\Psi_{N,\theta} Q_2$ will be proportional to n whereas the number of columns in $\Psi_{N,\theta}$ is proportional to n^2.

We start with the LPV model structure. The columns of $\Psi_{N,\theta} Q_2$ are the directional derivatives of E_N in the directions specified by the columns of Q_2,

$$\Omega_N := \Psi_{N,\theta} Q_2 = \frac{\partial E_N}{\partial \theta^T} Q_2.$$

The kth element of the jth column of Ω_N can be written as

$$\Omega_N(k, j) = \sum_{i=1}^{p} \frac{\partial \hat{y}(k;\theta)}{\partial \theta_i} Q_2(i, j), \tag{4}$$

where p is the number of parameters in θ. As differentiation is a linear operation, this sum can be obtained from

$$\Omega_N(k, j) = C(\theta) \begin{bmatrix} \mathcal{X}_j(k;\theta) \\ p(k) \otimes \mathcal{X}_j(k;\theta) \end{bmatrix} + \left(\sum_{i=1}^{p} \frac{\partial C(\theta)}{\partial \theta_i} Q_2(i, j) \right) \begin{bmatrix} \hat{x}(k;\theta) \\ p(k) \otimes \hat{x}(k;\theta) \end{bmatrix} + \left(\sum_{i=1}^{p} \frac{\partial D(\theta)}{\partial \theta_i} Q_2(i, j) \right) \begin{bmatrix} u(k) \\ p(k) \otimes u(k) \end{bmatrix},$$

where the state sequence

$$\mathcal{X}_j(k;\theta) := \sum_{i=1}^{p} \frac{\partial \hat{x}(k;\theta)}{\partial \theta_i} Q_2(i, j)$$

follows by simulating the dynamic equation

$$\mathcal{X}_j(k+1;\theta) = A(\theta) \begin{bmatrix} \mathcal{X}_j(k;\theta) \\ p(k) \otimes \mathcal{X}_j(k;\theta) \end{bmatrix} + \left(\sum_{i=1}^{p} \frac{\partial A(\theta)}{\partial \theta_i} Q_2(i, j) \right) \begin{bmatrix} \hat{x}(k;\theta) \\ p(k) \otimes \hat{x}(k;\theta) \end{bmatrix} + \left(\sum_{i=1}^{p} \frac{\partial B(\theta)}{\partial \theta_i} Q_2(i, j) \right) \begin{bmatrix} u(k) \\ p(k) \otimes u(k) \end{bmatrix}.$$

The weighted sums of the derivatives of the various system matrices can be obtained from a column of Q_2 as follows:

$$\begin{aligned} &\text{vec}\left(\begin{bmatrix} \sum_{i=1}^{p} \frac{\partial A}{\partial \theta_i} Q_2(i, j) & \sum_{i=1}^{p} \frac{\partial B}{\partial \theta_i} Q_2(i, j) \\ \sum_{i=1}^{p} \frac{\partial C}{\partial \theta_i} Q_2(i, j) & \sum_{i=1}^{p} \frac{\partial D}{\partial \theta_i} Q_2(i, j) \end{bmatrix} \right) \\ &= \sum_{i=1}^{p} \frac{\partial}{\partial \theta_i} \text{vec}\left(\begin{bmatrix} A & B \\ C & D \end{bmatrix} \right) Q_2(i, j) \\ &= \sum_{i=1}^{p} \frac{\partial \theta}{\partial \theta_i} Q_2(i, j) = \sum_{i=1}^{p} \mathbf{e}_i Q_2(i, j) = Q_2(:\,, j), \end{aligned}$$

in which $\mathbf{e}_i$ denotes a vector which contains zeros, except for the ith component, which equals one.

For LTI, bilinear models, and local linear models some of the entries in the system matrices are zero by definition, and the weighted sum is obtained as

$$P \sum_{i=1}^{p} \frac{\partial}{\partial \theta_i} \text{vec}\left(\begin{bmatrix} A & B \\ C & D \end{bmatrix} \right) Q_2(i, j) = Q_2(:\,, j).$$

For Hammerstein systems, the input signal in the procedure outlined above should be replaced by $f(u(k))$. For Wiener systems, the chain rule for differentiation must be applied to the output equation.

For composite local linear and Wiener systems also $\Psi_{N,\rho}$, the part of the Jacobian related to the parameters ρ must be computed. It is not difficult to show that for composite local linear models $\Psi_{N,\rho}$ can also be obtained by simulating a dynamic equation (Verdult *et al.*, 2001). For Wiener systems, the system's states do not depend on ρ and thus $\Psi_{N,\rho}$ can be obtained from a static relation.

Successful identification requires that the Jacobian $\Psi_N(\theta, \rho)$ is bounded. Thus, the dynamic equations for the Jacobian computation must be stable. It is easy to verify for LTI, Wiener, Hammerstein, and bilinear systems that the dynamics governing the Jacobian computations are stable if the model corresponding to θ is stable. Since the optimization method aims at minimizing the output error, it is very unlikely that the parameters θ describing the model are modified towards instability of the model.

For LPV and local linear models the dynamics governing the Jacobian computations are stable if the model corresponding to θ is stable and in addition the parameter $p(k)$ or the scheduling vector $\psi(k)$ does not depend on the output or state of the model (Verdult *et al.*, 2001; Verdult, 2002).

7. CONCLUSIONS

Multivariable linear and nonlinear state-space systems can be identified by numerically solving a nonlinear optimization problem in which the system is fully parameterized. The nonuniqueness of the state-space representation is taken into account by solving the optimization problem using a projected gradient search that restricts the update of the parameters at each iteration to directions that change the input-output behavior. This paper shows that such a method can be used for LTI, LPV, bilinear, composite local linear, Hammerstein and Wiener systems. Colored output noise can be taken into account in a maximum likelihood procedure by estimating the inverse covariance matrix of the residuals and using the inverse of this matrix as a weighting in the cost function, or alternatively by minimizing the prediction error if a predictor can be derived. It is pointed out that the optimization-based method can be initialized by a model obtained from subspace identification.

REFERENCES

Avdeenko, T. (2002). On structural identifi ability of system parameters of linear models. In: *Preprints of the 15th IFAC World Congress*. Barcelona, Spain.

Bergboer, N., V. Verdult and M. Verhaegen (2002). An effi cient implementation of maximum likelihood identifi cation of LTI state-space models by local gradient search. In: *Proceedings of the 41st IEEE Conference on Decision and Control*. Las Vegas, Nevada.

Bruls, J., C. T. Chou, B. Haverkamp and M. Verhaegen (1999). Linear and non-linear system identifi cation using separable least-squares. *European Journal of Control* **5**(1), 116–128.

Chen, H. and J. Maciejowski (2000). An improved subspace identifi cation method for bilinear systems. In: *Proceedings of the 39th IEEE Conference on Decision and Control*. Sydney, Australia.

David, B. and G. Bastin (2001). An estimator of the inverse covariance matrix and its application to ML parameter estimation in dynamical systems. *Automatica* **37**(1), 99–106.

Deistler, M. and T. Ribarits (2001). Parametrizations of linear systems by data driven local coordinates. In: *Proceedings of the 40th IEEE Conference on Decision and Control*. Orlando, Florida. pp. 4754–4759.

Favoreel, W. (1999). Subspace Methods for Identifi cation and Control of Linear and Bilinear Systems. PhD thesis. Faculty of Engineering, K. U. Leuven. Leuven, Belgium.

Favoreel, W., B. De Moor and P. Van Overschee (1999). Subspace identifi cation of bilinear systems subject to white inputs. *IEEE Transactions on Automatic Control* **44**(6), 1157–1165.

Fnaiech, F. and L. Ljung (1987). Recursive identifi cation of bilinear systems. *International Journal of Control* **45**(2), 453–470.

Golub, G. H. and C. F. Van Loan (1996). *Matrix Computations*. third ed.. The Johns Hopkins University Press. Baltimore, Maryland.

Golub, G. H. and V. Pereyra (1973). The differentiation of pseudo-inverses and nonlinear least squares problems whose variables separate. *SIAM Journal of Numerical Analysis* **10**(2), 413–432.

Isidori, A. (1973). Direct construction of minimal bilinear realizations from nonlinear input-output maps. *IEEE Transactions on Automatic Control* **18**(6), 626–631.

Lee, L. H. and K. Poolla (1999). Identifi cation of linear parameter-varying systems using nonlinear programming. *Journal of Dynamic Systems, Measurement and Control* **121**(1), 71–78.

Ljung, L. (1999). *System Identification: Theory for the User*. second ed.. Prentice-Hall. Upper Saddle River, New Jersey.

McKelvey, T. (2002). A new minimal local parametrization for multivariable linear systems. In: *Preprints of the 15th IFAC World Congress*. Barcelona, Spain.

McKelvey, T. and A. Helmersson (1997). System identifi cation using an over-parametrized model class: Improving the optimization algorithm. In: *Proceedings of the 36th IEEE Conference on Decision and Control*. San Diego, California. pp. 2984–2989.

Moré, J. J. (1978). The Levenberg-Marquardt algorithm: Implementation and theory. In: *Numerical Analysis* (G. A. Watson, Ed.). Vol. 630 of *Lecture Notes in Mathematics*. pp. 106–116. Springer Verlag. Berlin.

Murray-Smith, R. and T. A. Johansen (1997). *Multiple Model Approaches to Modelling and Control*. Taylor and Francis. London.

Ribarits, T. and M. Deistler (2002). Data driven local coordinates: Some new topological and geometrical results. In: *Preprints of the 15th IFAC World Congress*. Barcelona, Spain.

Sjöberg, J. (1997). On estimation of nonlinear black-box models: How to obtain a good initialization. In: *Proceedings of the 1997 IEEE Workshop Neural Networks for Signal Processing VII*. Amelia Island Plantation, Florida. pp. 72–81.

Van Overschee, P. and B. De Moor (1996). *Subspace Identification for Linear Systems; Theory, Implementation, Applications*. Kluwer Academic Publishers. Dordrecht, The Netherlands.

Verdult, V. (2002). Nonlinear System Identifi cation: A State-Space Approach. PhD thesis. University of Twente, Faculty of Applied Physics. Enschede, The Netherlands.

Verdult, V. and M. Verhaegen (2001). Identifi cation of multivariable bilinear state space systems based on subspace techniques and separable least squares optimization. *International Journal of Control* **74**(18), 1824–1836.

Verdult, V. and M. Verhaegen (2002). Subspace identifi cation of multivariable linear parameter-varying systems. *Automatica* **38**(5), 805–814.

Verdult, V., L. Ljung and M. Verhaegen (2001). Identifi cation of composite local linear state-space models using a projected gradient search. Technical report. Accepted for publication in *International Journal of Control*.

Verdult, V., N. Bergboer and M. Verhaegen (2002). Maximum likelihood identifi cation of multivariable bilinear state-space systems by projected gradient search. In: *Proceedings of the 41st IEEE Conference on Decision and Control*. Las Vegas, Nevada.

Verhaegen, M. (1994). Identifi cation of the deterministic part of MIMO state space models given in innovations form from input-output data. *Automatica* **30**(1), 61–74.

Verhaegen, M. and D. Westwick (1996). Identifying MIMO Hammerstein systems in the context of subspace model identifi cation methods. *International Journal of Control* **63**(2), 331–349.

Westwick, D. and M. Verhaegen (1996). Identifying MIMO Wiener systems using subspace model identifi cation methods. *Signal Processing* **52**(2), 235–258.

IFAC Publications
www.elsevier.com/locate/ifac

A DIFFERENTIAL GEOMETRIC VIEWPOINT ON LOCAL IDENTIFIABILITY AND IDENTIFICATION PART I: THEORY

Bernhard Eitzinger * **Kurt Schlacher** **

* *WFT Research, Fabrikstrasse 20, 4050 Traun, Austria*
** *Johannes-Kepler University, Institute for Automatic Control and Electrical Drives and CD-Laboratory for Automatic Control of Mechatronic Systems in Steel Industries, Altenbergerstrasse 69, 4040 Linz, Austria*

Abstract: The questions of local identifiability and identification of nonlinear systems are treated from a differential geometric point of view. It is shown how identifiability can be interpreted in the framework of Lie groups, which provides convenient tools for a unifying concept of identifiability. Three different notions of identifiability are interpreted in this framework. Furthermore parameter identification itself is also treated. It is shown how a parameter estimator can be formulated in a coordinate-free setting. In this setting it can be clearly determined to what extent the underlying geometric structure and the coordinate-specific parameterization of the system influence the properties of the estimator. *Copyright © 2003 IFAC*

Keywords: Differential geometric methods; Identification algorithms; Identifiability; Nonlinear systems; Parameter Estimation.

1. INTRODUCTION

Differential geometry has earned great merits in the analysis and synthesis of nonlinear control systems, see for example (Isidori, 1995) and (Nijmeier and van der Schaft, 1990) and many other publications. While some areas like controller and observer design have greatly profited from differential geometric methods, other areas such as feedback linearization would be virtually impossible without it. It seems, however, that nonlinear system identification slightly lags behind this progress. This is insofar surprising as differential geometry has also entered the area of statistics and parameter estimation as shown in (Amari and Nagaoka, 2000), (Amari, 1985) and (Murray and Rice, 1993). Therefore the present investigation aims at improving this situation by investigating two main issues in nonlinear system identification from a differential geometric point of view. The first issue is identifiability, i.e. the question whether a locally unique solution of the parameter identification problem can be found. This will be investigated by the theory of Lie groups which allows to put various notions of identifiability in a unifying concept. This question was also investigated, for example in (Kozlowski, 1998), (Chappell *et al.*, 1990) and (Walter, 1982), and an important global result for real analytic systems can be found in (Ljung and Glad, 1994). An interesting contribution concerning identfiability and estimability of econometric models can be found in (Deistler and Seifert, 1978). Algorithms to check for local identifiability are, for example, presented in (Ollivier and Sedoglavic, 2002) and (Sedoglavic, 2002).

The second issue is parameter identification itself. It will be shown how the underlying geometric concept of parameter identification can be extracted by removing all references to a specific coordinate system. This will allow to determine how far typical properties of estimators like unbiasedness and robustness can be formulated without resorting to coordinates and it therefore provides some opportunities for additional insight.

2. PROBLEM STATEMENT

A large class of nonlinear dynamical systems can be described by an explicit system of ordinary differential equations together with a set of output functions. Usually the model of the system contains a few parameters which could not be determined during the modelling phase or which undergo changes during operation of the system. These parameters may be assumed to be an element $\boldsymbol{\theta} \in \Theta$ of some finite dimensional smooth manifold Θ. The system S is therefore given by

$$\begin{aligned} \dot{\mathbf{x}}(t,\boldsymbol{\theta}) &= \mathbf{f}(\mathbf{x}(t),\mathbf{u}(t),\boldsymbol{\theta},\mathbf{d}(t)) \\ \mathbf{y}(t,\boldsymbol{\theta}) &= \mathbf{g}(\mathbf{x}(t),\mathbf{u}(t),\boldsymbol{\theta},\mathbf{n}(t)). \end{aligned} \tag{1}$$

Apart from the state vector $\mathbf{x} \in X$ and the input signal $\mathbf{u} \in U^0$ the output $\mathbf{y} \in \mathbb{R}^k$ may also be affected by disturbance $\mathbf{d} \in D$ and noise $\mathbf{n} \in N$. In order to be able to determine the unknown parameters $\boldsymbol{\theta}$ two questions need to be answered. Firstly, whether it is at all possible to uniquely determine the parameters and, secondly, how the parameters can be determined from the measured output. The former question is known as identifiability and the latter as identification. In the following both questions will be investigated from a differential geometric viewpoint.

3. IDENTIFIABILITY

Here identifiability is considered as a question 'in principle', it describes a system-specific property which does not depend on noise and disturbances. Therefore noise and disturbances may be set to zero. Let Z be a set of experimental conditions to which the system has been exposed. These experimental conditions can be an arbitrary collection of states and inputs or a trajectory and they will be fixed in more detail later on. As shown in Figure 2, the system S then maps $\Theta \times Z$ to some output space Y, which can also be considered a manifold. A system will be called locally identifiable at θ^* according to the following definition.

Definition 1. Let $S : \Theta \times Z \to Y$ be a system. The system is called locally identifiable at $\theta^* \in \Theta$ if there exists an open set $U \subseteq \Theta$ containing θ^* such that for every $\theta \in U$, $\theta \neq \theta^*$, there exists a $z \in Z$ such that

$$S(\theta, z) \neq S(\theta^*, z).$$

This definition means that, if a parameter θ sufficiently close to θ^* is given, there exist experimental conditions z such that by exposing the system to the experimental conditions z and measuring the output $S(\vartheta, z)$ it is possible to decide whether $\vartheta = \theta$ or $\vartheta = \theta^*$. As will soon become obvious, Lie groups are the appropriate mathematical tool to investigate such a situation. Therefore the notion of a Lie group will be introduced as presented in (Olver, 2000).

3.1 *Lie Groups*

A Lie group G is a manifold M together with a group operation $() \cdot () : M \to M$ and an inversion $()^{-1} : M \to M$ such that $\cdot$ respects the group axioms and both $\cdot$ and $^{-1}$ are smooth maps between manifolds. The identity element of G will be denoted by e. Lie groups are the basis for defining a local group of transformations.

Definition 2. Let Θ be a smooth manifold. A local group of transformations acting on Θ is given by a Lie group G, an open set U with

$$\{e\} \times \Theta \subset U \subset G \times \Theta$$

and a C^∞-map $\Phi : U \to \Theta$ such that the following properties hold.

(1) If $(h, \theta) \in U$ and $(g, \Phi(h, \theta)) \in U$ and also $(g \cdot h, \theta) \in U$ then $\Phi(g, \Phi(h, \theta)) = \Phi(g \cdot h, \theta)$.
(2) For all $\theta \in \Theta$, $\Phi(e, \theta) = \theta$ holds.
(3) If $(g, \theta) \in U$ then $(g^{-1}, \Phi(g, \theta)) \in U$ and $\Phi(g^{-1}, \Phi(g, \theta)) = \theta$.

The trivial group $G = \{e\}$ induces the identity mapping $\Phi(e, \theta) = \theta$ as a group of transformations. Based on this a definition for a G-invariant mapping can be given.

Definition 3. Let G be a local group of transformations acting on Θ. Let S be a C^∞-function mapping Θ to a manifold Y. Then S is called G-invariant if for all $\theta \in \Theta$ and all $g \in G$ such that $\Phi(g, \theta)$ is defined

$$S(\Phi(g, \theta)) = S(\theta).$$

Note that locally the flow $\exp(t\vartheta)\theta$ of a vector field $\vartheta \in T\Theta$ passing through $\theta \in \Theta$ constitutes a local group of transformations. Indeed every local Lie group can be infinitesimally generated by a vector field. Therefore to investigate group-invariance of functions only infinitesimal conditions need to be tested. This allows easy application of the results. With the Lie derivative of S along ϑ denoted by $L_\vartheta S$ the following theorem, proved in (Olver, 2000), sums this up.

Theorem 4. Let G be a local group of transformations acting on Θ. Let $S : \Theta \to \mathbb{R}$ be a C^∞-function then S is G-invariant if and only if $L_\vartheta S = 0$ for every infinitesimal generator ϑ of G.

In the theorem S maps Θ to $\mathbb{R}$, but it is obvious that if S maps Θ to some $\mathbb{R}^n$ then S is G-invariant if and only if each component S^i, $i = 1, 2, ..., n$, is G-invariant. In case S is vector-valued the Lie derivative of S will also be denoted by $L_\vartheta S$, meaning that the Lie derivative is applied to each component of S. The following theorem then answers the question of local identifiability in general.

Theorem 5. Let $S : \Theta \times Z \to Y$ be a system. The system is locally identifiable at $\theta^* \in \Theta$ if and only if the only local group of transformations acting on Θ which leaves S invariant for all $z \in Z$ is the trivial group.

Proof. Let G be a non-trivial local group of transformations infinitesimally generated by ϑ, which leaves S invariant for all $z \in Z$. Then for the corresponding flow $\exp(t\vartheta)\theta^*$ the condition $S(\exp(t\vartheta)\theta^*, z) = S(\theta^*, z)$ holds for all $z \in Z$ and by the continuity of flows every open set of Θ containing θ^* also contains another point θ such that $S(\theta, z) = S(\theta^*, z)$ for all $z \in Z$. Therefore existence of non-trivial transformation groups which leave S invariant implies that the system S is not locally identifiable. Conversely, let S be locally identifiable then there is an open set $U \in \Theta$ containing θ^* such that $S(\theta, z) \neq S(\theta^*, z)$ for all $\theta \in U$ and some $z \in Z$ which may depend on θ. Therefore by the continuity of flows no non-zero vector field ϑ can exist, whose flow $\exp(t\vartheta)\theta^*$ leaves S invariant. As every group is infinitesimally generated by the flow of a vector field the only group leaving S invariant is the trivial group, which proves the theorem. ■

For a practical implementation of a test for local identifiability the following theorem may be used.

Theorem 6. Let $S : \Theta \times Z \to Y = \mathbb{R}^n$ be a system. The system is locally identifiable at $\theta^* \in \Theta$ if and only if there exists a subset $V \subseteq Z$ such that

$$\int_V \left(\frac{\partial S(\theta^*, z)}{\partial \theta}\right)^T \left(\frac{\partial S(\theta^*, z)}{\partial \theta}\right) \mathrm{d}z \qquad (2)$$

is positive definite.

Proof. Note that for mappings between real spaces $L_\vartheta S = (\partial S / \partial \theta)\, \vartheta$ and that $L_\vartheta S = 0$ holds for all $z \in Z$ only if S is G-invariant and therefore not locally identifiable. The proof is then rather trivial. ■

Because of the assumed continuity of all expressions one may conclude from local identifiability at θ^* that there exists an open subset $U \subseteq \Theta$ containing θ^* such that the system is locally identifiable at all $\theta \in U$. The Theorem 6 contains also the special case, where Z is a trajectory of (1). Now various notions of identifiability can easily be constructed by selecting the set Z appropriately and three examples will be presented below, but before, the notion of a jet space will be introduced.

3.2 *Jet Space*

Consider the output equation $y = g(x, u, \theta, n) = g^0(x, u, \theta, n)$ of system (1) and remember that $d(t) = n(t) = 0$ and that $\dot{\theta} = 0$. Differentiating the output equation once with respect to time yields

$$\dot{y} = \frac{\partial g^0}{\partial x}\dot{x} + \frac{\partial g^0}{\partial u}\dot{u}$$

and f may be substituted for $\dot{x}$ such that

$$\dot{y} = \frac{\partial g^0}{\partial x} f + \frac{\partial g^0}{\partial u}\dot{u} = g^1(x, u, \dot{u}, \theta, n)$$

is obtained. This process is also called prolongation. Now, while g^0 is a function $X \times U^0 \times \Theta \times N \to \mathbb{R}^k$, the first derivative g^1 has an extended domain of definition $X \times U^0 \times U^1 \times \Theta \times N \to \mathbb{R}^k$, where $X \times U^0 \times U^1$ is known as the 1-jet space. It is clear that the time derivation can be repeated several times, e.g up to r times, and the set product $X \times U^0 \times U^1 \times \cdots \times U^r$ is called the r-jet space. In general a Lie group or its generating vector field needs to be prolonged to the r-jet space. In the case at hand, however, this prolongation of the group action is rather trivial.

3.3 *General Identifiability*

General identifiability asks whether there exist at all experimental conditions to which the system can be exposed such that the parameters can be uniquely identified. This means that

$$Z = X \times U^0 \times U^1 \times \cdots \times U^r, \qquad (3)$$

that is, all possible states, inputs and their time derivatives are taken into consideration. The system S is then a mapping

$$S : \Theta \times Z \to Y = \mathbb{R}^{kr}, \qquad (4)$$

of the parameter space Θ and the experimental conditions Z to the k real-valued outputs of the system and their time derivatives up to order r. General identifiability immediately follows if the conditions of theorem 6 are satisfied. In all technically relevant applications the r-jet space Z will be bounded and the integral can be evaluated. Identifiability can then be judged by calculating the condition number of the matrix in equation

(2). Note, however, that this matrix will be rather ill conditioned even if the system is identifiable. Furthermore this test does not deliver conditions when the system is not identifiable, for no bound on r is given.

3.4 *Offline Identifiability*

For offline identifiability it is assumed that the initial state is known and the input $u(t) : [0,T] \to U^0$ is given on the interval $[0,T]$. This is equivalent to asking whether the input $u(t)$ is sufficiently exciting. In this case the system can be written as $S : \Theta \times [0,T] \to Y = \mathbb{R}^k$ and obviously $Z = [0,T]$. Local identifiability can again be judged by resorting to theorem 6. Note that in this case the output $S(\theta,t)$ is viewed as a function of the parameters θ and time t alone. Therefore the Lie derivative L_ϑ of a vector field ϑ on $T\Theta$ and the time derivative $d()/dt$, which is just the Lie derivative along a special vector field of the tangential space of $[0,T]$, commute because $\dot{\theta} = 0$. Hence if $L_\vartheta S = 0$ for all $t \in [0,T]$ then also

$$L_\vartheta(\frac{d}{dt}S) = \frac{d}{dt}(L_\vartheta S) = 0$$

and no additional information about the parameters can be obtained by differentiating the output with respect to time. This is a beneficial result, for in practical applications the repeated numerical differentiation of the output in the presence of noise and disturbances usually poses some problems. As a consequence theorem 6 works in both directions and can be used to prove identifiability as well as non-identifiability.

3.5 *Online Identifiability*

For online identification it is desirable to be able to uniquely identify the parameters at every state $x \in X$. The question is therefore whether it is possible to choose an input at state x such that the parameters can be uniquely identified. Therefore

$$Z = U^0 \times U^1 \times \cdots \times U^r$$

and the system is given by $S : \Theta \times \{x\} \times Z \to Y = \mathbb{R}^{kr}$, and yet again identifiability can be proved by invoking theorem 6. Note that more frequently the state itself is also unknown and needs to be reconstructed. This can also be accounted for by extending the parameter space $\Theta' = \Theta \times X$. The system is then given by $S : \Theta' \times Z \to Y = \mathbb{R}^{kr}$ and the theory applies as presented previously. Note that, even if the system is locally identifiable for a known state $x \in X$ and it is also locally observable for known parameters $\theta \in \Theta$, the combined identification/observation problem need not be uniquely solvable at $\theta' = (\theta, x)$. Furthermore note that, even if there does not exist a single state at which the system is locally identifiable, there may still exist sufficiently exciting inputs, that is, the notions of online and offline identifiability are independent. Examples showing applications of the presented theory can be found in (Eitzinger, 2002).

4. IDENTIFICATION

Once it has been proved that the system is at least locally identifiable the next obvious question is how to identify the parameters. In contrast to identifiability the assumption of zero noise and disturbances will be dropped and consequently the system in general loses its injectivity. Without going into technical details it is assumed that noise and disturbances are sufficiently well behaved such that, e.g., time integrals of the output quantities can be evaluated. The system then takes parameters θ, experimental conditions (e.g. states, inputs and their time derivatives) z, noise n and disturbance d to some technically measurable output y. That is,

$$S : \Theta \times Z \times N \times D \to Y.$$

If all the arguments entering S were known, identification would be trivial and could formally be stated as the natural projection $\pi_{trivial} : \Theta \times Z \times N \times D \to \Theta$ onto the parameter space. But this is usually not possible because only the output y is available for measurement and due to the presence of noise and disturbances the inverse system S^{-1} cannot be constructed. The goal of identification is therefore to find an estimator $F : Y \to \Theta$ which optimally approximates the mapping $\pi_{trivial} \circ S^{-1}$. This situation is shown in Figure 2. In order to be able to define the optimality of the approximation statistical assumptions on noise and disturbance are made such that qualities like unbiasedness, minimum-variance or robustness can be ascribed to the estimator. One may ask how far these qualities can be formulated in a coordinate-free manner. The first step is therefore to give a coordinate-free formulation of parameter estimation, which is the topic of the following sections.

4.1 *A Geometric Interpretation*

Assume a parameter estimator $F : Y \to \Theta$ is given, then, for a known value θ^*, one can calculate the set $U(\theta^*) = \{y \in Y, F(y) = \theta^*\}$ of all possible outputs in Y which are mapped to the same parameter value θ^*. In some cases such a set can have a rather pathological structure, for example it need not even be a Hausdorff space. In many practical situations, however, the sets $U(\theta^*)$

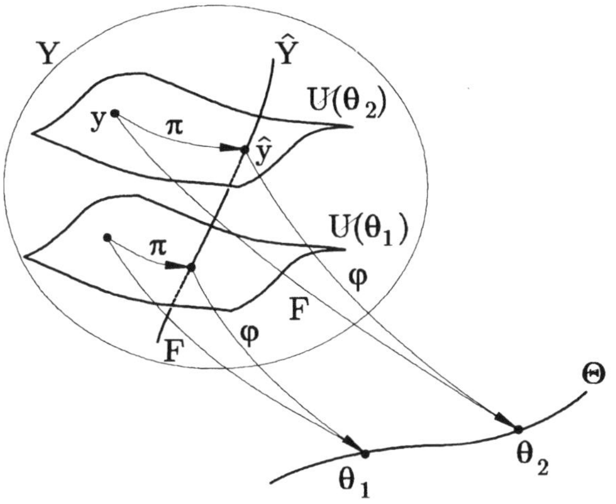

Fig. 1. A foliation of the output manifold Y can be constructed from the equivalence classes of outputs being mapped to the same parameter values.

will carry the structure of a manifold. Clearly the sets are disjoint, i.e. $U(\theta_1) \cap U(\theta_2) = \{\}$, whenever $\theta_1 \neq \theta_2$ and every $y \in Y$ belongs to exactly one set $U(F(y))$. Therefore the sets $U(\theta)$ are a foliation of the manifold Y.

The set $U(\theta^*)$ can also be understood as an equivalence class of all outputs being mapped to the same parameter, with the equivalence relation provided by the mapping F. Now, a representative output $\hat{y}$ of each equivalence class $U(\theta)$ can be chosen and the set of all representatives, if chosen appropriately, can be combined in a manifold of representative outputs denoted by $\hat{Y}$. It is obvious that under mild technical restrictions $\hat{Y}$ carries the structure of a manifold because there is a one-to-one mapping to the parameter space. Therefore $\hat{Y}$ inherits the topology from the parameter space. While this construction is always possible the topology may significantly deviate from the standard topology. Usually if the system equations are sufficiently often continuously differentiable with respect to the parameters the topology remains reasonable. The manifold $\hat{Y}$ is also a subset of Y but, even more than that, it can be mapped to Y by the natural inclusion and forms an embedded submanifold of Y. Or, put the other way, $\hat{Y}$ is a quotient manifold of Y. This construction is shown in Figure 1.

The three manifolds Y, $\hat{Y}$ and Θ are connected by two mappings. First, every element $\hat{y} \in \hat{Y}$ is uniquely linked to a parameter value θ^* by $F(\hat{y}) = \theta^*$. This means that there is a one-to-one mapping φ,

$$\varphi : \hat{Y} \rightarrow \Theta,$$

relating the representative output to its corresponding parameter value. Secondly, the manifold Y needs to be mapped to the submanifold of representative outputs by some projection π,

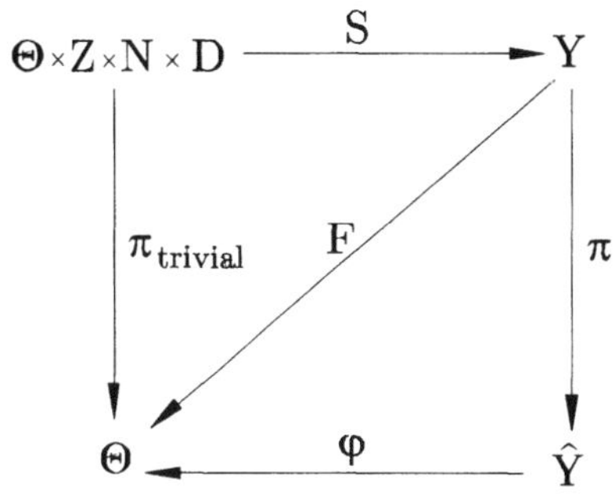

Fig. 2. The estimator F can be interpreted as a projection π onto a submanifold $\hat{Y}$ and the evaluation of a chart φ.

$$\pi : Y \rightarrow \hat{Y}.$$

It can be required that $\pi \circ \pi = \pi$ or, equivalently, that the restriction of π to $\hat{Y}$ is the identity mapping. The parameter estimator is then the composition of φ and π, i.e.

$$F = \varphi \circ \pi.$$

This completes the diagram shown in Figure 2. Note that, if $\Theta = \mathbb{R}^d$, as is most often the case, then φ can be understood as a global chart for the manifold $\hat{Y}$. Thus the design of an estimator requires two choices, which are coordinate-free and completely independent of the specific parameterization of the system: The manifold $\hat{Y}$ and the projection π onto this manifold. In general these choices will be made such that the equality

$$\pi_{trivial} = \varphi \circ \pi \circ S \tag{5}$$

is fulfilled as far as possible. How the difference between the right and left hand side of equation (5) is measured, depends on statistical assumptions on noise and disturbance or on other practical considerations and will not be investigated here.

4.2 *Example*

A simple example will illustrate the construction proposed above. Consider the weighted least squares problem for a linear regression

$$\min_{\boldsymbol{\theta} \in \mathbb{R}^d} \frac{1}{2} (\mathbf{A}\boldsymbol{\theta} - \mathbf{y})^T \mathbf{H} (\mathbf{A}\boldsymbol{\theta} - \mathbf{y}),$$

with $\mathbf{H} \in \mathbb{R}^{n \times n}$, $\mathbf{H}$ positive definite, $\mathbf{y} \in \mathbb{R}^n$ and an $\mathbf{A} \in \mathbb{R}^{n \times d}$ of rank d and $n \geq d$. The solution is found by equating the gradient to zero, such that the estimator is given by

$$\boldsymbol{\theta} = \mathbf{F}(\mathbf{y}) = (\mathbf{A}^T \mathbf{H} \mathbf{A})^{-1} \mathbf{A}^T \mathbf{H} \mathbf{y}.$$

This estimator can be decomposed into φ and π. The mapping $\varphi : \hat{Y} = \text{img}(\mathbf{A}) \rightarrow \Theta$ is given by

$$\varphi = (\mathbf{C}^T \mathbf{A})^{-1} \mathbf{C}^T,$$

where $\mathbf{C} \in \mathbb{R}^{n \times d}$ is arbitrary as long as $\mathbf{C}^T\mathbf{A}$ is regular and its inverse is given by $\varphi^{-1} = \mathbf{A}$. For the projection $\pi : Y \to \hat{Y}$ one obtains

$$\pi = \mathbf{A}(\mathbf{A}^T\mathbf{H}\mathbf{A})^{-1}\mathbf{A}^T\mathbf{H}.$$

Note that $\pi \circ \pi = \pi$. The sets $U(\boldsymbol{\theta})$, with $\boldsymbol{\theta} = \varphi(\hat{\mathbf{y}})$, as discussed above, are given by

$$U(\boldsymbol{\theta}) = \{\mathbf{y} \in \mathbb{R}^n, \mathbf{A}^T\mathbf{H}(\hat{\mathbf{y}} - \mathbf{y}) = \mathbf{0}\}$$

and the manifold $\hat{Y}$ is simply given by

$$\hat{Y} = \{\mathbf{y} \in \mathbb{R}^n, \mathbf{y} = \mathbf{A}\boldsymbol{\theta}\}.$$

5. CONCLUSIONS

A differential geometric viewpoint of local identifiability and identification was presented. It was shown how the issue of local identifiability can be answered in the framework of Lie groups and group invariant mappings. It was proved that if the only transformation group acting on the parameter space which leaves the system invariant is the trivial transformation group then the system is locally identifiable. Various notions of identifiability which occur in practical problems were put under this unifying concept.

It is noteworthy that the entire investigation of identifiability as well as identification does not need continuous-time nonlinear dynamical systems as a background. It works also for identification of discrete-time systems or curve-fitting applications. Indeed the assumption of a finite dimensional output manifold was never made and therefore this investigation includes applications where the output manifold is, for example, an appropriately chosen function space. Likewise only minor modifications are necessary to also include the case when the parameter space itself is not of finite dimension. While the differentiability properties of the system were not investigated it is obvious that rather weak conditions suffice. For example the system needs to be once continuously differentiable with respect to the parameters and it may be only integrable with respect to the experimental conditions. This further broadens the applicability of the presented theory.

The problem of parameter identification of nonlinear dynamical systems was also investigated. By interpreting the parameter estimator as a projection onto an embedded submanifold of the output space all references to a specific coordinate system could be removed. The actual parameterization of the system was interpreted as a chart for this submanifold. Therefore without referring to specific coordinates two things can be chosen for a parameter estimator: the embedded submanifold and the projection onto this submanifold. In many applications, especially those, where the parameters have a technical meaning, the chart is already given by technical considerations and not left as a choice. In this case all desired properties of the estimator have to be established by an appopriate choice of the submanifold and the projection. It is of interest to investigate how these two choices affect the estimator. In a subsequent paper it will be shown, for example, how the submanifold of the output manifold needs to be chosen such that the estimator is asymptotically unbiased, it will also be shown how robustness can be secured by an appropriate choice of the projection and a technically relevant application will demonstrate the effectiveness of the theory presented here.

6. REFERENCES

Amari, S. (1985). *Differential-Geometrical Methods in Statistics.* Springer. Heidelberg.

Amari, S. and H. Nagaoka (2000). *Methods of Information Geometry.* Oxford University Press.

Chappell, M.J., K.R. Godfrey and S. Vajda (1990). Global identifiability of the parameters of nonlinear systems with specified inputs: A comparison of methods. *Mathematical Biosciences* **102**, 41–73.

Deistler, M. and H.G. Seifert (1978). Identifiability and consistent estimability in econometric models. *Econometrica* **46**, 969–980.

Eitzinger, B. (2002). *Parameter Identification for Nonlinear Dynamical Systems.* Dissertation. Johannes-Kepler University, Institute for Automatic Control and Electrical Drives.

Isidori, A. (1995). *Nonlinear Control Systems.* Springer. London.

Kozlowski, K. (1998). *Modelling and Identification in Robotics.* Springer. London.

Ljung, L. and T. Glad (1994). On global identifiability for arbitrary model parametrizations. *Automatica* **30**, 265–276.

Murray, M.K. and J.W. Rice (1993). *Differential Geometry and Statistics.* Chapman Hall. New York.

Nijmeier, H. and A.J. van der Schaft (1990). *Nonlinear Dynamical Control Systems.* Springer. London.

Ollivier, F. and A. Sedoglavic (2002). Algorithmes efficaces pour tester l'identifiabilite locale. *Proceedings CIFA'2002* pp. 811–816.

Olver, P.J. (2000). *Applications of Lie Groups to Differential Equations.* Springer. New York.

Sedoglavic, A. (2002). A probabilistic algorithm to test local algebraic observability in polynomial time. *Journal of Symbolic Computation* **33**, 735–755.

Walter, E. (1982). *Identifiability of State-Space Models.* Springer. Berlin.

IFAC
Publications
www.elsevier.com/locate/ifac

A DIFFERENTIAL GEOMETRIC VIEWPOINT ON LOCAL IDENTIFIABILITY AND IDENTIFICATION PART II: APPLICATION

Bernhard Eitzinger * **Kurt Schlacher** **

* *WFT Research, Fabrikstrasse 20, 4050 Traun, Austria*
** *Johannes-Kepler University, Institute for Automatic Control and Electrical Drives and CD-Laboratory for Automatic Control of Mechatronic Systems in Steel Industries, Altenbergerstrasse 69, 4040 Linz, Austria*

Abstract: The robustness of parameter identification for nonlinear dynamical systems is investigated by interpreting the parameter estimation problem from a differential geometric point of view. Results on robust identification are proved in a deterministic setting and an algorithm is developed to derive a robust estimator for the parameters of nonlinear dynamical systems. The proposed algorithm is tested on the offline identification of the rotor resistance of an electric machine for stochastic and deterministic disturbances. It could be shown that the robustness is significantly improved compared to conventional algorithms. *Copyright © 2003 IFAC*

Keywords: Robust estimation; Differential geometric methods; Identification algorithms; Electric machines; Condition numbers.

1. INTRODUCTION

A standard approach to solve parameter estimation problems for dynamical systems is to make some statistical assumptions on the output error and to resort to the theory of point estimation. These methods are well documented in the literature, e.g. in (Ljung, 2000) and (Nelles, 2001) or in a more general setting in (Lehmann and Casella, 1998). For nonlinear systems the statistical assumptions on the output error deserve a closer look, because even if the disturbance acting on the system is normally distributed, the final probability distribution of the output error will be rather arbitrary after the disturbance has been propagated through the nonlinear system. Indeed only in simple cases can the probability distribution of the output error actually be calculated. Therefore a least squares estimator need not be statistically optimal or deliver an unbiased estimate. Nevertheless due to their simplicity least squares algorithms are attractive and some advances have aimed at improving their performance for nonlinear systems, e.g. (Billings and Voon, 1984), (Stortelder, 1996). Once, however, it is accepted that the probability distribution of the output error is unknown, the robustness of the identification algorithm becomes immediately relevant. The approach to robustness taken here is mapping-based, i.e. the system is viewed as a map between the parameter space and an output space, e.g. as in (Lyshevski, 1998). No statistical assumptions on noise and disturbances will be made, but it is assumed that noise and disturbances are bounded, e.g. as in (Boutayeb, 2000).

The novelty of the presented approach lies in the rigorous application of differential geometry to the parameter estimation problem in a deterministic setting. A more statistical approach using differential geometry was developed by Amari in (Amari and Nagaoka, 2000), although not for nonlinear dynamical systems. An approach similar to Amari's can also be found in (Murray and

Rice, 1993). These ideas will be used as a basis to derive a robust identification algorithm in a deterministic setting.

2. PROBLEM STATEMENT

It is assumed that a model of the system under investigation, based on physical, mechanical or other considerations, is available. In many cases such a model can be formulated as a system of ordinary nonlinear differential state-space equations and a set of nonlinear output functions. Both, the static and the dynamic part of the model may be parameterized by a parameter vector $\boldsymbol{\theta} \in \Theta$, element of some finite dimensional smooth Riemannian manifold. Most frequently $\Theta = \mathbb{R}^d$, $d < \infty$. The model structure is then as follows

$$\dot{\mathbf{x}}(t, \boldsymbol{\theta}) = \mathbf{f}(\mathbf{x}(t), \mathbf{u}(t), \boldsymbol{\theta}, \mathbf{d}(t)),$$
$$\mathbf{y}(t, \boldsymbol{\theta}) = \mathbf{g}(\mathbf{x}(t), \mathbf{u}(t), \boldsymbol{\theta}, \mathbf{n}(t)),$$

where $\mathbf{x}(t) = [x_1(t), x_2(t), ..., x_n(t)]^T$ is the state vector, $\mathbf{u}(t) = [u_1(t), u_2(t), ..., u_m(t)]^T$ is the input vector and $\mathbf{y}(t) = [y_1(t), y_2(t), ..., y_k(t)]$ is the output vector and $\mathbf{d}(t) = [d_1(t), d_2(t), ..., d_n(t)]$ and $\mathbf{n}(t) = [n_1(t), n_2(t), ..., n_k(t)]$ denote disturbance and noise, respectively. No statistical assumptions on noise and disturbance will be made, but it will be assumed that noise and disturbance are not necessary for the identification task, i.e. output equations like $\mathbf{y}(t, \boldsymbol{\theta}) = \mathbf{n}^T(t)\mathbf{g}(\mathbf{x}(t), \boldsymbol{\theta})$ will be excluded.

The parameter estimation problem is to estimate the parameters $\boldsymbol{\theta}$ for a known initial state $\mathbf{x}(0)$, a known input signal and a measured output signal up to some final time T. If everything is sufficiently well behaved the measured output may be assumed to be an element of a signal space Y which can be made into a Riemannian manifold by introducing a metric, which in many cases will be simply the metric corresponding to the 2-norm, but considerations are not limited to this case. A parameter estimator is then a mapping

$$F : Y \to \Theta,$$

and a diffeomorphism φ^{-1} can be introduced

$$\varphi^{-1} : \Theta \to \hat{Y} \subset Y$$

mapping the parameter space to an embedded submanifold $\hat{Y}$ of Y. The submanifold $\hat{Y}$ will be called the manifold of representative outputs. Provided the system is identifiable for the given input, such a mapping can, e.g., be constructed by mapping the parameters to the corresponding output with zero noise and disturbance or alternatively to the expected output. Once this manifold is fixed, all that remains to be chosen is a projection π of the manifold Y onto the embedded submanifold $\hat{Y}$

$$\pi : Y \to \hat{Y}.$$

The estimator F is then the composition of φ and π,

$$F = \varphi \circ \pi. \tag{1}$$

How φ, π and the manifold $\hat{Y}$ can be chosen to achieve a robust estimator is the subject of the following section.

3. ROBUST IDENTIFICATION

First the following rather obvious theorem provides information on how to choose $\hat{Y}$.

Theorem 1. The estimator $F = \varphi \circ \pi : Y \to \Theta$ is asymptotically unbiased if $\varphi^{-1} : \Theta \to \hat{Y}$ maps Θ onto the manifold of expected outputs.

Proof. Let $y^i \in Y$ denote the output on the interval $[0, T]$ in the i-th experiment, $i = 1, 2, ..., N$, let E() denote the expected value and let $\hat{Y}$ be the manifold of expected outputs. With probability one

$$\lim_{N \to \infty} \frac{1}{N} \sum_{i=1}^{N} y^i = \mathrm{E}(y) \in \hat{Y}$$

and hence, by noting that $\pi \circ \pi = \pi$,

$$\lim_{N \to \infty} F(\frac{1}{N} \sum_{i=1}^{N} y^i) = \varphi \circ \pi(\mathrm{E}(y))$$
$$= \varphi(\mathrm{E}(y))$$

and the correct parameters are obtained. ∎

Let $\Gamma(\hat{y})$ denote the set of all smooth paths $\gamma : [0, 1] \to Y$ with $\gamma(0) = \hat{y} \in \hat{Y}$. As the manifold Y is assumed to be Riemannian there exists the notion of length of the path γ from $\gamma(0) = \hat{y}$ to $\gamma(\tau)$. This length will be denoted by $\ell_Y(\gamma(\tau), \hat{y})$. Likewise, the length of a smooth path $\vartheta(\tau)$ on Θ will be denoted by $\ell_\Theta(\vartheta(\tau), \hat{\theta})$ where $\vartheta(0) = \hat{\theta}$. With this notation a robust estimator can be defined.

Definition 2. The estimator $F = \varphi \circ \pi : Y \to \Theta$ is called robust if F is such that the minimum of

$$\max_{\gamma \in \Gamma(\hat{y})} \lim_{\tau \to 0} \frac{\ell_\Theta(F(\gamma(\tau)), F(\hat{y}))}{\ell_Y(\gamma(\tau), \hat{y})}$$

is achieved for all $\hat{y} \in \hat{Y}$.

In other words the estimator is robust if the ratio of the length of a path γ in Y and the length of its corresponding path $F(\gamma)$ in Θ is minimal, or in yet other words, if a deviation from $\hat{y}$ has a minimal effect on the parameter estimate. In order to make the problem of deriving such an estimator theoretically tractable the limit $\tau \to 0$ was taken in the above definition.

Let F be a mapping between two manifolds Y and Θ, then its tangential mapping will be denoted by $F_* : TY \to T\Theta$. The tangential spaces TY_y and $T\Theta_\theta$ attached to the points y and θ, respectively, are equipped with norms induced by the respective Riemannian metrics on Y and Θ, and these norms will be denoted by $\|\cdot\|_{Y,y}$ and $\|\cdot\|_{\Theta,\theta}$. Then by $\dot{\gamma}(0) = v \in TY_{\hat{y}}$ it is obvious that the above definition is equivalent to minimizing

$$\max_{v \in TY_y} \frac{\|F_* v\|_{\Theta,F(\hat{y})}}{\|v\|_{Y,\hat{y}}} \tag{2}$$

over all $\hat{\mathbf{y}} \in \hat{Y}$. Therefore an estimator is robust according to the above definition if the operator norm of its tangential mapping is minimal over the entire submanifold $\hat{Y}$ of representative outputs. To shorten notation the operator norm will be written as

$$\max_{v \in TY_y} \frac{\|F_* v\|_{\Theta,F(\hat{y})}}{\|v\|_{Y,\hat{y}}} = \|F_*\|_{Y,\Theta} .$$

Similarly to the Cramér-Rao bound the operator norm is bounded from below for unbiased estimators.

Theorem 3. Let $F = \varphi \circ \pi : Y \to \Theta$ be an unbiased estimator then

$$\|F_*\|_{Y,\Theta} \geq \|\varphi_*\|_{\hat{Y},\Theta}$$

for all $\hat{y} \in \hat{Y}$.

Proof. Consider the restriction $F|_{\hat{Y}}$ of F to the manifold $\hat{Y}$ of expected outputs. As the estimator is unbiased $F|_{\hat{Y}} = \varphi$. Consequently,

$$\left\|F_*|_{\hat{Y}}\right\|_{\hat{Y},\Theta} = \|\varphi_*\|_{\hat{Y},\Theta} .$$

As the maximum in expression (2) can only become larger by extending F from $\hat{Y}$ to Y,

$$\|F_*\|_{Y,\Theta} \geq \left\|F_*|_{\hat{Y}}\right\|_{\hat{Y},\Theta} = \|\varphi_*\|_{\hat{Y},\Theta} ,$$

which proves the theorem. ∎

Furthermore it can be seen that the operator norm of F_* can be bounded from above

$$\begin{aligned}\|F_*\|_{Y,\Theta} &= \|\varphi_* \circ \pi_*\|_{Y,\Theta} \\ &\leq \|\pi_*\|_{Y,\hat{Y}} \, \|\varphi_*\|_{\hat{Y},\Theta} ,\end{aligned} \tag{3}$$

and robustness can thus be achieved if $\|\pi_*\|_{Y,\hat{Y}} = 1$. The following theorem states when this is the case.

Theorem 4. Let $F = \varphi \circ \pi : Y \to \Theta$ be an unbiased estimator. The estimator is robust if the projection π is orthogonal onto the manifold of expected outputs.

Proof. Consider a tangential vector $v \in TY_{\hat{y}}$ for some arbitrary $\hat{y} \in \hat{Y}$. The tangential vector can be decomposed by $v = \pi_* v + u$.

$$\begin{aligned}\|v\|^2_{Y,\hat{y}} &= \langle \pi_* v + u, \pi_* v + u\rangle_{Y,\hat{y}} \\ &= \langle \pi_* v, \pi_* v\rangle_{Y,\hat{y}} + 2\,\langle \pi_* v, u\rangle_{Y,\hat{y}} + \langle u, u\rangle_{Y,\hat{y}}\end{aligned}$$

As $\hat{Y}$ inherits the metric from Y it is possible to write $\|\pi_* v\|^2_{\hat{Y},\hat{y}} = \|\pi_* v\|^2_{Y,\hat{y}}$ for all $v \in TY_{\hat{y}}$. Now, if the projection is orthogonal then $\langle \pi_* v, u\rangle_Y = 0$ and

$$\begin{aligned}\|v\|^2_{Y,\hat{y}} &= \|\pi_* v\|^2_{\hat{Y},\hat{y}} + \|u\|^2_{Y,\hat{y}} \\ \|v\|^2_{Y,\hat{y}} &\geq \|\pi_* v\|^2_{\hat{Y},\hat{y}} .\end{aligned} \tag{4}$$

The operator norm of π_* can be bounded from above by inequality (4)

$$\|\pi_*\|_{Y,\hat{Y}} = \max_{v \in TY_{\hat{y}}} \frac{\|\pi_* v\|_{\hat{Y},\hat{y}}}{\|v\|_{Y,\hat{y}}} \leq 1.$$

This bound is actually attained because from the unbiasedness $\pi_* v = v$ for all $v \in T\hat{Y}_{\hat{y}}$. Invoking Theorem 3 and inequality (3) proves the theorem. ∎

The conclusion of Theorem 4 is that for a robust identification the operator norm of the projection has to be minimized. In the next sections an algorithm is presented to solve this problem approximately and its effectiveness is demonstrated on a technically relevant example. For this algorithm the following lemma is needed.

Lemma 5. Let $\mathbf{A} \in \mathbb{R}^{N \times d}$ and $\mathbf{H} \in \mathbb{R}^{N \times N}$ with $\mathbf{A}^T\mathbf{A} > 0$ and $\mathbf{H} > 0$. If $\mathbf{A}^T\mathbf{H}\mathbf{A} = \mathbf{I}$ then

$$\|\mathbf{A}(\mathbf{A}^T\mathbf{H}\mathbf{A})^{-1}\mathbf{A}^T\mathbf{H}\| = 1. \tag{5}$$

Proof. Set $\Pi = \mathbf{A}(\mathbf{A}^T\mathbf{H}\mathbf{A})^{-1}\mathbf{A}^T\mathbf{H}$. Every vector $\mathbf{v} \in \mathbb{R}^N$ may be decomposed by $\mathbf{v} = \mathbf{A}\mathbf{x} + \mathbf{z}$ with $\mathbf{A}^T\mathbf{z} = \mathbf{0}$. Note that $\Pi\mathbf{A}\mathbf{x} = \mathbf{A}\mathbf{x}$. Set $\mathbf{H} = \mathbf{A}(\mathbf{A}^T\mathbf{A})^{-2}\mathbf{A}^T$ then $\mathbf{A}^T\mathbf{H}\mathbf{A} = \mathbf{I}$ and, by substituting into Π, it follows that $\Pi\mathbf{z} = \mathbf{0}$ if $\mathbf{A}^T\mathbf{z} = \mathbf{0}$, which proves that

$$\|\Pi\| = \max_{\mathbf{v} \in \mathbb{R}^N} \frac{\|\Pi\mathbf{v}\|_2}{\|\mathbf{v}\|_2} = \max_{\mathbf{v} \in \mathbb{R}^N} \frac{\|\Pi\mathbf{A}\mathbf{x}\|_2}{\|\mathbf{A}\mathbf{x} + \mathbf{z}\|_2} = 1.$$

∎

4. AN ALGORITHM

The algorithm is based on the least squares estimator, i.e. a cost function

$$J(\boldsymbol{\vartheta}, \boldsymbol{\theta}) = \frac{1}{2}\sum_{i=1}^{N}(\tilde{y}_i - y_i(\boldsymbol{\vartheta}))^2 h_i(\boldsymbol{\theta}) \tag{6}$$

is minimized, where $\tilde{y}_i$ denotes the measured output sampled at time $i \cdot T_{sample}$ and $y_i(\boldsymbol{\vartheta})$ denotes the simulated output for parameters $\boldsymbol{\vartheta}$.

The measured and simulated output will be combined in a vector $\tilde{\mathbf{y}} = [\tilde{y}_1, \tilde{y}_2, ..., \tilde{y}_N]^T$ and $\mathbf{y} = [y_1, y_2, ..., y_N]^T$, respectively. In contrast to a standard least squares algorithm a parameter dependent weighting $h_i(\boldsymbol{\theta}) > 0$ is introduced. The weights will be combined in a matrix $\mathbf{H}$ =diag(h_i). In the notation of the previous section, minimizing equation (6) corresponds to minimizing the distance between the measured output and the manifold of representative outputs $\hat{Y}$. Roughly, the algorithm then proceeds as follows.

1. $k = 1$; `Select initial value` $\boldsymbol{\theta}_1$;
2. `Determine weights` $h_i(\boldsymbol{\theta}_k)$, $i = 1, 2, ..., N$;
3. `Solve` $\boldsymbol{\theta}_{k+1}$ =`argmin` $J(\boldsymbol{\vartheta}, \boldsymbol{\theta}_k)$ `w.r.t.` $\boldsymbol{\vartheta}$;
4. $k = k + 1$;
5. `Go to step` 2 `until convergence.`

The solution of the minimization problem in step 3 can be found by any suitable optimization algorithm. In the application described below a quasi-Newton method with BFGS-update was used. Details are in (Dennis and Schnabel, 1996).

To determine the weights in step 2, the following observation is useful. At the minimizing value $\boldsymbol{\theta}_{\min}$ of equation (6) the cost function $J(\boldsymbol{\vartheta}, \boldsymbol{\theta}_{\min})$ can be expanded in a Taylor series with respect to $\boldsymbol{\vartheta}$ at $\boldsymbol{\vartheta} = \boldsymbol{\theta}_{\min}$ to evaluate the effect of some small changes $\Delta\tilde{\mathbf{y}}$ in the measured output $\tilde{\mathbf{y}}$ on the parameter estimate.

$$J(\boldsymbol{\theta}_{\min} + \Delta\boldsymbol{\vartheta}, \boldsymbol{\theta}_{\min}) \approx J(\boldsymbol{\theta}_{\min}, \boldsymbol{\theta}_{\min}) - \Delta\boldsymbol{\vartheta}^T \left(\frac{\partial \mathbf{y}}{\partial \boldsymbol{\vartheta}}\right)^T \mathbf{H}\Delta\tilde{\mathbf{y}} + \frac{1}{2}\Delta\boldsymbol{\vartheta}^T \left(\frac{\partial \mathbf{y}}{\partial \boldsymbol{\vartheta}}\right)^T \mathbf{H} \left(\frac{\partial \mathbf{y}}{\partial \boldsymbol{\vartheta}}\right) \Delta\boldsymbol{\vartheta}$$

Let $\mathbf{A} = \partial\mathbf{y}/\partial\boldsymbol{\vartheta}$ then equating the gradient with respect to $\Delta\boldsymbol{\vartheta}$ to zero yields, in the notation of equation (1),

$$\Delta\boldsymbol{\vartheta} = F_*\Delta\tilde{\mathbf{y}} = (\mathbf{A}^T\mathbf{H}\mathbf{A})^{-1}\mathbf{A}^T\mathbf{H}\Delta\tilde{\mathbf{y}}.$$

The diffeomorphism φ_* and the projection operator π_* are therefore given by

$$\varphi_* = (\mathbf{C}^T\mathbf{A})^{-1}\mathbf{C}^T$$
$$\pi_* = \mathbf{A}(\mathbf{A}^T\mathbf{H}\mathbf{A})^{-1}\mathbf{A}^T\mathbf{H},$$

where $\mathbf{C}$ is arbitrary as long as $\mathbf{C}^T\mathbf{A}$ is regular. Minimizing the operator norm of π_*, a - usually large - $N \times N$ matrix, is not feasible but Lemma 5 motivates instead the minimization of the condition number

$$\text{cond}(\mathbf{A}^T\mathbf{H}\mathbf{A}) = \left\|(\mathbf{A}^T\mathbf{H}\mathbf{A})^{-1}\right\| \left\|\mathbf{A}^T\mathbf{H}\mathbf{A}\right\|,$$

for cond$(\cdot) \geq 1$ and cond$(\mathbf{I}) = 1$. Indeed an approximate solution can be found by the following algorithm.

1. $j = 1$; $h_1(\boldsymbol{\theta}_k) = 1$; $\alpha = 1.1$; $\mathbf{M}_0 = \mathbf{0}$;
2. $\mathbf{Z} = \frac{1}{2}(\frac{\partial y_j}{\partial \boldsymbol{\vartheta}}|_{\theta_k})^T h_j(\boldsymbol{\theta}_k)(\frac{\partial y_j}{\partial \boldsymbol{\vartheta}}|_{\theta_k})$;
3. $\mathbf{M}_j^+ = \mathbf{M}_{j-1} + \mathbf{Z}\cdot\alpha$; c^+ =`cond`$(\mathbf{M}_j^+)$;
4. $\mathbf{M}_j^- = \mathbf{M}_{j-1} + \mathbf{Z}/\alpha$; c^- =`cond`$(\mathbf{M}_j^-)$
5. $\mathbf{M}_j = \mathbf{M}_{j-1} + \mathbf{Z}$; c =`cond`$(\mathbf{M}_j)$;
6. `if` $c^+ < c$ `then`
   $\mathbf{M}_j = \mathbf{M}_j^+$; $h_j(\boldsymbol{\theta}_k) = h_j(\boldsymbol{\theta}_k)\cdot\alpha$; `else`
7. `if` $c^- < c$ `then`
   $\mathbf{M}_j = \mathbf{M}_j^-$; $h_j(\boldsymbol{\theta}_k) = h_j(\boldsymbol{\theta}_k)/\alpha$;
8. $h_{j+1}(\boldsymbol{\theta}_k) = h_j(\boldsymbol{\theta}_k)$; $j = j + 1$;
9. `Go to step` 2 `until` $j = N$.

The algorithm proceeds from $j = 1$ to $j = N$ and tries to reduce the condition number by increasing or decreasing the weight $h_j(\boldsymbol{\theta}_k)$. The constant α can be chosen arbitrarily, but values close to 1 are useful. Note that this algorithm works only for a single output quantity, but a generalization to more outputs is easily possible and the details can be found in (Eitzinger, 2002). Note also that, if the output $\mathbf{y}(\boldsymbol{\vartheta})$ is calculated with noise and disturbance set to zero, then the proposed algorithm does not produce an unbiased or asymptotically unbiased estimator because the projection is not onto the submanifold of expected outputs. Furthermore the operator norm of the projection is minimized only approximately. Nevertheless a significant improvement in robustness can be achieved in the following application.

5. APPLICATION

The rotor resistance of a squirrel-cage induction motor is to be identified. Knowledge of the rotor resistance can be used to improve the performance and the energy consumption of such motors. Additionally, the rotor resistance changes its value during operation and is not available for direct measurement. Hence it is a quantity which can only be identified. An identification by measurement of the stator currents alone is especially important, because these currents are known in the drive system and no additional sensors are needed at the motor. It can be shown, however, that the rotor resistance is not identifiable under constant load from the stator currents and hence the identification problem is usually rather ill conditioned.

A squirrel-cage induction motor can be described by the following fifth-order system of differential equations.

$$\begin{bmatrix} \dot{I}_a \\ \dot{I}_b \\ \dot{\Psi}_a \\ \dot{\Psi}_b \\ \dot{\omega} \end{bmatrix} = \begin{bmatrix} Z_3\omega\Psi_b - Z_1 I_a + Z_2\Psi_a + \frac{1}{\sigma L_s} U_a \\ -Z_3\omega\Psi_a - Z_1 I_b + Z_2\Psi_b + \frac{1}{\sigma L_s} U_b \\ -n_p\omega\Psi_b + \frac{L_{sr}R_r}{L_r} I_a - \frac{R_r}{L_r}\Psi_a \\ n_p\omega\Psi_a + \frac{L_{sr}R_r}{L_r} I_b - \frac{R_r}{L_r}\Psi_b \\ \frac{1}{D_m}((I_b\Psi_a - I_a\Psi_b) - (M_L + d)) \end{bmatrix}$$

$$y = \frac{1}{2}(I_a^2 + I_b^2) + n,$$

$$Z_1 = \frac{1}{\sigma L_s}\left(R_s + \frac{L_{sr}^2 R_r}{L_r^2}\right) \quad Z_2 = \frac{L_{sr}R_r}{\sigma L_s L_r^2}$$

$$Z_3 = n_p \frac{L_{sr}}{\sigma L_s L_r} \quad Z_4 = \frac{L_{sr}}{L_r} n_p$$

where a disturbance denoted by d has been included in the load M_L and noise n has been added to the output. The origin of this model will not be detailed, it can be found in (Schroeder, 2000). For the system parameters the following values are used

$$\begin{aligned} L_{sr} &= 154 \text{ mH} & L_r &= 164.7 \text{ mH} \\ L_s &= 164.3 \text{ mH} & n_p &= 2 \\ R_s &= 4.5\ \Omega & R_r &= 3.0\ \Omega \\ D_m &= 0.0016 \text{ kgm}^2 & \sigma &\approx 0.12358. \end{aligned}$$

The input voltage is sinusoidal with an amplitude of 120 V and the torque load is a step from 0 Nm to 2 Nm at $t = 0$. Simulation time is 0.5 seconds, i.e. $T = 0.5$ and the output was sampled with $T_{sample} = 0.5$ ms. The torque and the rotor resistance are assumed to be unknown, i.e. $\boldsymbol{\theta} = [\, R_r \ M_L \,]^T$. With the above notation $\Theta = \mathbb{R}^+ \times \mathbb{R}$, the tangential spaces TY and $T\Theta$ will be equipped with the trivial norm $\|\cdot\|_2$ and the spaces Θ and Y are thus Euclidean spaces.

5.1 *Results*

The weighted least squares estimator (WLSE), proposed above, was compared with a standard, i.e. $h_i(\boldsymbol{\theta}) = 1$, least squares estimator (LSE) in two different settings.

5.1.1. *Stochastic Setting* Noise n and disturbance d are assumed to be normally distributed zero mean independent random variables with standard deviation σ_n and σ_d, respectively. The standard deviations of noise and disturbance were simultaneously increased in ten levels from 0.1 to 1. At each level 20 simulations were made and the parameters were estimated for each simulation giving a total of 200 estimates. The identification of the unknown torque load M_L is possible with sufficient precision for all practical purposes

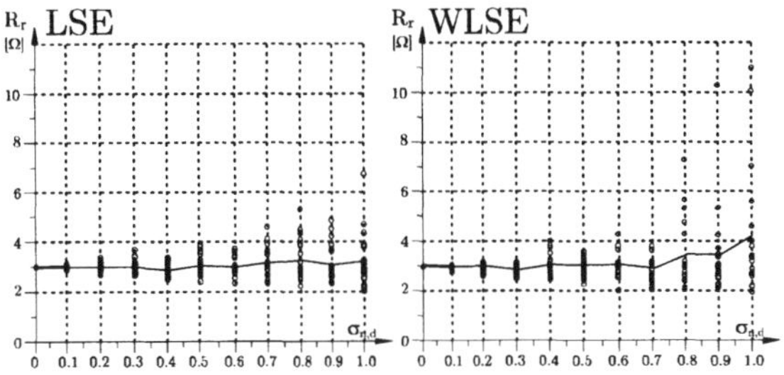

Fig. 1. Estimated values for the rotor resistance for the least squares and the weighted least squares estimator with increasing levels of stochastic noise and disturbance.

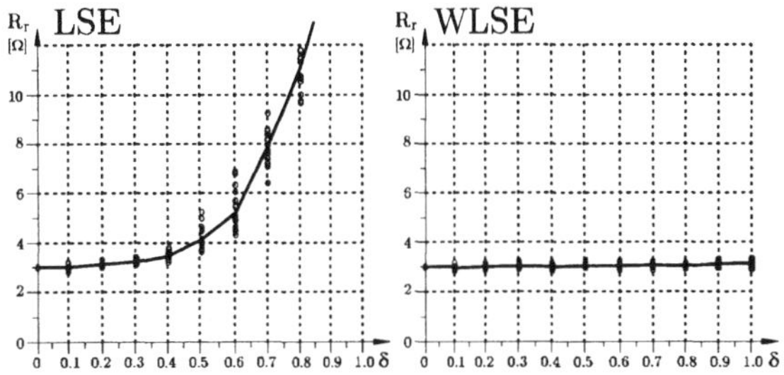

Fig. 2. Estimated values for the rotor resistance for the least squares and the weighted least squares estimator with increasing slope of a ramp on the torque and stochastic noise.

therefore only the results for the rotor resistance are shown in Figure 1. As can be seen both algorithms perform fairly equal, with possibly a slight advantage in favor of the standard least squares algorithm. In this setting a standard least squares algorithm is close to a maximum likelihood estimator and therefore it performs statistically rather well. But it can be seen that by introducing the weighting not much is lost with respect to the performance for stochastic noise and disturbances.

5.1.2. *Deterministic Setting* The noise is again assumed to be a normally distributed zero mean independent random variable with standard deviation 0.1, but the disturbance is now a slight ramp rising with slope δ, i.e. $d(t) = \delta t/T$. The slope was chosen in ten levels from 0.1 to 1, at each level 20 simulations were made and the parameters were estimated by both algorithms. The results for the rotor resistance are shown in Figure 2. A drastic difference can be seen between the least squares and the weighted least squares algorithm. With increasing slope the estimated rotor resistance increases quickly and reaches mean values - not shown in the figure - of 27.1 Ω and 38.4 Ω for $\delta = 0.9$ and $\delta = 1.0$, respectively. The weighted least squares algorithm, however, delivers mean values between 3.00 Ω and 3.15 Ω at low variance even for the highest slopes. Taking into consideration that the rotor resistance in such a motor changes by roughly $\pm 50\%$ only the estimates with the weighted least squares algorithm will deliver reasonable results in the presence of unmodelled non-stochastic disturbances.

6. DISCUSSION

The proposed algorithm shows a significant improvement in robustness compared to a least squares algorithm, therefore it may be useful for parameter estimation in situations where the stochastic properties of the output error are unknown or non-stochastic disturbances are to be expected. Unlike statistically based estimators a hard bound on the parameter error can be calculated for sufficiently small errors. The algorithm, however, requires norms or Riemannian metrics on the parameter and output space. In many cases a loss function for parameter errors, e.g. in terms of energy, money or time, can be determined with properties similar to a metric on the parameter space. On the output space statistical assumptions or experimentally derived information on the output error may serve as a guideline to find a metric, e.g. the inverse covariance matrix of the output error could be used.

It has to be noted that the algorithm is not specific to continuous-time nonlinear dynamical systems. It can be equally well applied to discrete-time systems or curve fitting applications and similar improvements in robustness are to be expected since the theorems and their proofs do not make use of the structure of a dynamical system.

The algorithm needs slightly more time for computation than a standard least squares algorithm, but if a good guess for the true parameters is available, the difference is small.

In situations where sufficient and reliable statistical information on noise and disturbances is available an estimator based on statistical considerations may perform better.

7. CONCLUSIONS

A differential geometric viewpoint of parameter estimation for nonlinear dynamical systems was presented. The estimator was decomposed into a projection onto a submanifold of the output space and a diffeomorphism between this submanifold and the parameter space. A robust estimator was defined and bounds on the minimal error sensitivity of an unbiased estimator were derived. An algorithm was proposed to approximately achieve this bound. As an application the rotor resistance of a squirrel-cage induction motor was identified and it could be demonstrated that for some deterministic disturbances an estimator according to this algorithm is significantly more robust than a standard least squares algorithm.

Future research could be directed, firstly, at improving the approximation in determining the weights and, secondly, at converting this offline algorithm to an online algorithm. In light of the presented application such an online algorithm would clearly be of some practical relevance.

8. REFERENCES

Amari, S. and H. Nagaoka (2000). *Methods of Information Geometry*. Oxford University Press.

Billings, S.A. and W.S.F. Voon (1984). Least squares parameter estimation algorithms for non-linear systems. *Int. J. Systems Sci.* **15**, 601–615.

Boutayeb, M. (2000). Identification of nonlinear systems in the presence of unknown but bounded disturbances. *IEEE Transactions on Automatic Control* **45**, 1503–1507.

Dennis, J.E. and R.B. Schnabel (1996). *Numerical Methods for Unconstrained Optimization and Nonlinear Equations*. SIAM.

Eitzinger, B. (2002). *Parameter Identification for Nonlinear Dynamical Systems*. Dissertation. Johannes-Kepler University, Institute for Automatic Control and Electrical Drives.

Lehmann, E.L. and G. Casella (1998). *Theory of Point Estimation*. Springer. Berlin.

Ljung, L. (2000). *System Identification - Theory for the User*. Prentice Hall. Upper Saddle River.

Lyshevski, S.E. (1998). Nonlinear identification of control systems. *Proceedings of the American Control Conference* pp. 2366–2370.

Murray, M.K. and J.W. Rice (1993). *Differential Geometry and Statistics*. Chapman Hall. New York.

Nelles, O. (2001). *Nonlinear System Identification*. Springer. Berlin.

Schroeder, D. (2000). *Elektrische Antriebe - Grundlagen*. Springer. Berlin.

Stortelder, W.J.H. (1996). Parameter estimation in nonlinear models by using total least squares. In: *Parameter Identification and Inverse Problems in Hydrology, Geology and Ecology* (J. Gottlieb and P. DuChateau, Eds.). 1st ed.. pp. 249–259. Kluwer Academic Publishers. Netherlands.

IFAC Publications
www.elsevier.com/locate/ifac

IDENTIFIABILITY OF NONLINEAR HOMOGENEOUS POLYNOMIAL SYSTEMS

Ralf Peeters * **Bernard Hanzon** **

* *Mathematics Department, Universiteit Maastricht, P.O. Box 616, 6200 MD Maastricht, The Netherlands, E-mail:* ralf.peeters@math.unimaas.nl
** *CWI, Kruislaan 413, P.O. Box 94079, 1090 GB Amsterdam, The Netherlands, E-mail:* hanzon@cwi.nl

Abstract: New results are presented concerning the state isomorphism approach to global identifiability analysis of parameterized classes of nonlinear homogeneous systems with specified initial states. For such systems, the local state isomorphism for a pair of indistinguishable parameter vectors is homogeneous of degree one. Under certain conditions, which may only be satisfied for homogeneous polynomial systems, the local state isomorphism is linear. Here, the key issue is whether or not the observability rank condition holds at the origin. The controllability rank condition is shown to play a truly secondary role. The results are generalized to the multivariable case and a worked example demonstrates how identifiability analysis may be simplified along these lines. *Copyright © 2003 IFAC*

Keywords: Identifiability, Nonlinear systems, Observability, Controllability.

1. INTRODUCTION

Consider a parameterized class of nonlinear dynamical state-space systems with initial conditions, described by:

$$\dot{x}(t,p) = f(x(t,p),p) + g(x(t,p),p)u(t), \quad (1)$$

$$y(t,p) = h(x(t,p),p), \quad (2)$$

$$x(0,p) = x_0(p). \quad (3)$$

The vector fields $f(.,p) : M_p \to \mathbb{R}^n$ and $g(.,p) : M_p \to \mathbb{R}^n$ are assumed to be real analytic on some state space domain M_p, which is an open connected subset of $\mathbb{R}^n$ containing the initial state $x_0(p)$. The scalar output function $h(.,p) : M_p \to \mathbb{R}$ is also assumed to be real analytic and the parameter vector p takes its values in some parameter set $\Omega \subseteq \mathbb{R}^q$. The scalar input function $u(\cdot) : [0,\tau] \to \mathbb{R}$ is chosen from some set $U[0,\tau]$ of bounded and measurable controls. Here $\tau > 0$ is a fixed constant, and it is assumed that the system of equations (1)-(3) has a unique solution on M_p over the entire time interval $[0,\tau]$. (Some subtleties of a technical nature are involved in this set-up, but we will not go into this; see also (Sussmann, 1977).)

For this class of systems, parameter indistinguishability and (global) identifiability can be studied by various methods. One particular approach of interest is based on the *local state isomorphism theorem*, see also (Sussmann, 1977; Hermann and Krener, 1977; Tunali and Tarn, 1987). This theorem can be viewed as the nonlinear counterpart of a well-known result from realization theory, which states that if two minimal linear state-space systems are equivalent from an input-output point of view then there exists a nonsingular linear state-space transformation which takes one system into the other.

To make these notions precise, let Σ_p denote the input-output mapping which maps the input space $U[0,\tau]$ to a suitable space $Y[0,\tau]$ of output functions, according to the rule $u(.) \mapsto y(.,p)$ as generated by the equations

(1)-(3). (The dependence of the input-output mapping on τ will be suppressed in the notation for the sake of readability.) Then two parameter vectors p and $\widetilde{p}$ are said to be *indistinguishable* if $\Sigma_p = \Sigma_{\widetilde{p}}$. Indistinguishability is an equivalence relation. The system (1)-(3) is called *globally identifiable* if the mapping $p \mapsto \Sigma_p$ is injective on Ω, i.e., no pair of distinct indistinguishable parameter vectors exists.

Reformulated for the purpose of identifiability analysis, the following form of the local state isomorphism theorem can essentially be found in (Vajda *et al.*, 1989).

Theorem 1. Let p and $\widetilde{p} \in \Omega$ be two parameter vectors for each of which the system (1)-(3) is locally reduced at the initial state, i.e., both the controllability rank condition (CRC) and the observability rank condition (ORC) are satisfied. Then p and $\widetilde{p}$ are indistinguishable if and only if there exists an open neighborhood $\widetilde{V}$ of $x_0(\widetilde{p})$, an open neighborhood V of $x_0(p)$ and a real analytic diffeomorphism $\lambda : \widetilde{V} \to V$ with the following properties:
(i) rank $(\nabla\lambda(\widetilde{x})) = n$, for all $\widetilde{x} \in \widetilde{V}$,
(ii) $\lambda(x_0(\widetilde{p})) = x_0(p)$,
(iii) $f(\lambda(\widetilde{x}), p) = \nabla\lambda(\widetilde{x}) \cdot f(\widetilde{x}, \widetilde{p})$, for all $\widetilde{x} \in \widetilde{V}$,
(iv) $g(\lambda(\widetilde{x}), p) = \nabla\lambda(\widetilde{x}) \cdot g(\widetilde{x}, \widetilde{p})$, for all $\widetilde{x} \in \widetilde{V}$,
(v) $h(\lambda(\widetilde{x}), p) = h(\widetilde{x}, \widetilde{p})$, for all $\widetilde{x} \in \widetilde{V}$,
where $\nabla\lambda(\widetilde{x})$ denotes the Jacobian matrix of the mapping λ evaluated at the point $\widetilde{x}$, and the symbol $\cdot$ denotes matrix-vector multiplication. For a given choice of $\widetilde{V}$, this real analytic diffeomorphism λ is unique.

See (Isidori, 1989; Hermann and Krener, 1977; Sussmann, 1977) for more details on the CRC and ORC.

The state isomorphism approach to identifiability analysis is centered around the following procedure. Assume that p and $\widetilde{p} \in \Omega$ are indistinguishable and let the systems in the model class be locally reduced at the initial states. Then Thm. 1 guarantees the existence of a unique local state isomorphism λ. From the literature, an explicit representation of λ can often be computed, see for instance (Isidori, 1989), which may be useful. The five conditions on λ stated in Thm. 1 give rise to a set of equations which involve p and $\widetilde{p}$, and which hold on an open neighborhood $\widetilde{V}$ of the initial state $x_0(\widetilde{p})$. If these equations imply that λ is the identity mapping and $p = \widetilde{p}$ yields the only feasible solution, global identifiability holds. Otherwise, unidentifiability follows, and the construction provides the various conditions for indistinguishable pairs of parameters and their associated state isomorphisms.

Now, if λ is known in advance to have a certain structure, such as being linear or affine, one may exploit this to substantially reduce the complexity of the computations in the various stages of the procedure sketched above. Thus, it becomes of interest to study particular subclasses of systems for which such properties of λ can be established. One such class is that of *homogeneous systems*. For a definition and some properties of homogeneous functions used in this paper, see App. A. A worked example is included in Sect. 6. Because of space limitations, all the proofs are given elsewhere; see (Peeters and Hanzon, 2002).

2. PRELIMINARY RESULTS FOR HOMOGENEOUS SYSTEMS

In a recent paper, see (Hanzon and Peeters, 2001), it has been established for homogeneous vector fields $f(.,p)$ and $g(.,p)$ and linear $h(.,p)$, that if the systems are all locally reduced at the initial states $x_0(p)$, then any local state isomorphism λ for a pair of indistinguishable parameters is homogeneous of degree 1.

Now, if a real analytic function is homogeneous of degree 1 on an open domain containing the origin, then it has to be linear. In (Hanzon and Peeters, 2001) this fact is exploited, yielding that λ is linear under the additional condition that the ORC always holds at the origin. Note that to facilitate the ORC to be properly evaluated at the origin, smoothness of $f(.,p)$, $g(.,p)$ and $h(.,p)$ at the origin is required. Together with homogeneity and real analyticity this implies that $f(.,p)$, $g(.,p)$ and $h(.,p)$ are *homogeneous polynomials* and that their degrees of homogeneity k, ℓ and m are nonnegative integers.

The following two theorems summarize and slightly extend these results by also allowing the output function $h(.,p)$ to be homogeneous on M_p, which notably may or may not contain the origin.

Theorem 2. Consider the parameterized class of systems given by (1)-(3). For all $p \in \Omega$, let the CRC and the ORC be satisfied at $x_0(p)$ and let $f(.,p)$, $g(.,p)$ and $h(.,p)$ be real analytic and homogeneous on M_p, of degrees k, ℓ and $m \in \mathbb{R}$, respectively. Let p and $\widetilde{p} \in \Omega$ be indistinguishable and let λ be the associated real analytic local state isomorphism as in Thm. 1. Then λ is homogeneous of degree 1.

Theorem 3. Consider the parameterized class of systems given by (1)-(3). For all $p \in \Omega$, let $f(.,p)$, $g(.,p)$ and $h(.,p)$ be homogeneous polynomials on M_p containing the origin, of nonnegative integer degrees k, ℓ and m, respectively. Let the CRC and the ORC be satisfied at $x_0(p)$ and let the ORC be satisfied at the origin too. Let p and $\widetilde{p} \in \Omega$ be indistinguishable and let λ be the associated real analytic local state isomorphism as in Thm. 1. Then λ is linear.

At this point it is not required that the homogeneous polynomial systems also satisfy the CRC at the origin. In fact, if the degrees k and ℓ are both strictly positive, then the CRC cannot hold at the origin. In the next section it will be made clear to what extent the roles played by CRC and the ORC at the origin are different, and how they are interrelated.

Remark 4. In (Joly-Blanchard and Denis-Vidal, 1998) and (Hanzon and Peeters, 2001) it has recently been demonstrated that a number of results reported in the literature on the local isomorphism approach to identifiability analysis are invalid. Identifiability of a class of polynomial systems was previously studied in (Chappell *et al.*, 1990) but Thm. 2 in that paper is invalid: it is based on an invalid result of (Vajda and Rabitz, 1989) and a counter example is provided by Example 1 in (Peeters and Hanzon, 2002).

3. ON THE ROLES PLAYED BY THE CRC AND THE ORC AT THE ORIGIN

As mentioned above, if $f(.,p)$, $g(.,p)$ and $h(.,p)$ are homogeneous and real analytic and if the CRC and the ORC can be properly evaluated at the origin, then $f(.,p)$, $g(.,p)$ and $h(.,p)$ are actually *polynomial.*

The following theorem indicates that for a homogeneous polynomial system which is locally reduced at the initial state, requiring the CRC to hold at the origin is a stronger condition than requiring the ORC to hold at the origin.

Theorem 5. For a homogeneous polynomial system, if the ORC holds at some state x_0 and the CRC holds at the origin, then the CRC and the ORC hold everywhere.

Thm. 3 and Thm. 5 jointly lead to the following result.

Corollary 6. Consider the parameterized class of systems given by (1)-(3). For all $p \in \Omega$, let $f(.,p)$, $g(.,p)$ and $h(.,p)$ be homogeneous polynomials on M_p containing the origin, of nonnegative integer degrees k, ℓ and m, respectively. Let the ORC and the CRC be satisfied at $x_0(p)$ and let the CRC be satisfied at the origin too. Let p and $\widetilde{p} \in \Omega$ be indistinguishable and let λ be the associated real analytic local state isomorphism as in Thm. 1. Then λ is linear.

This shows that, for homogeneous polynomial systems, evaluation of the CRC at the origin does not bring any additional prior information on the possible linearity of λ compared to the information already obtainable by evaluation of the ORC at the origin. However, if it is easier to verify the CRC at the origin than the ORC (and if the CRC happens to hold) Cor. 6 may still prove to be of practical value.

Conversely, situations may occur where the ORC holds at the origin and Thm. 3 applies, while the CRC does not hold at the origin. An obvious class of systems for which this happens is that of the *bilinear systems*, where f, g and h are all linear. In that case the CRC cannot hold at the origin, but it is well possible for the ORC to hold at the origin (and also for the CRC to hold outside the origin). Another example is constituted by the following SISO system, given by

$$\dot{x}_1 = x_1 x_3,$$
$$\dot{x}_2 = x_3^2,$$
$$\dot{x}_3 = x_2^2 + u,$$
$$y = x_1 + x_2,$$

with the non-zero initial state $x_0(p) = (1,0,0)^T$. For this system the ORC holds everywhere and the CRC holds at all states with $x_1 \neq 0$. To see this, first consider the functions:

$$\omega_1(x) := h(x) = x_1 + x_2,$$
$$\omega_2(x) := L_g L_f h(x) = x_1 + 2x_3,$$
$$\omega_3(x) := L_g L_f L_g L_f h(x) = x_1.$$

Together, their gradients constitute a constant invertible 3×3 matrix, which makes clear that the ORC holds everywhere.
With respect to the CRC note that the first component of the vector field f has a factor x_1, while the first component of g is zero. Now suppose that v is a homogeneous polynomial vector field of which the first component has a factor x_1, so that it can be written as $v(x) = (x_1\widetilde{v}_1(x), v_2(x), v_3(x))^T$, with $\widetilde{v}_1$, v_2 and v_3 polynomial. Then it is straightforward to show that the first components of the Lie brackets $[f,v]$ and $[g,v]$ both also contain a factor x_1. As a consequence, the first component of any repeated Lie bracket of f and g is always zero when evaluated at a point for which $x_1 = 0$ and the CRC does not hold. On the other hand the CRC does hold if $x_1 \neq 0$ since $[f,g](x) = (-x_1, -2x_3, 0)^T$ and $[[f,g],g](x) = (0,2,0)^T$.

If $f(.,p)$ and $g(.,p)$ are polynomial vector fields, one may also investigate the situation where the CRC holds at the origin directly, without explicit reference to the output function $h(.,p)$ or the ORC at the origin. For such systems, the local state isomorphism λ is merely affine.

Theorem 7. Consider the parameterized class of systems given by (1)-(3). For all $p \in \Omega$, let $f(.,p)$ and $g(.,p)$ be homogeneous polynomials on M_p containing the origin, of nonnegative integer degrees k and ℓ, respectively. Let the ORC and the CRC be satisfied at $x_0(p)$ and let the CRC be satisfied at the origin too. Let p and $\widetilde{p} \in \Omega$ be indistinguishable and let λ be the associated real analytic local state isomorphism as in Thm. 1. Then λ is affine.

Note that nothing special other than real analyticity is required from the output function $h(.,p)$. In case $h(.,p)$ is also homogeneous on M_p (and therefore polynomial, since M_p contains the origin) it follows from Thm. 2 that λ is also homogeneous of degree 1. Hence, λ is actually linear. This constitutes an alternative proof of Cor. 6.

4. ON THE SCOPE OF THE RESULTS

Starting from the assumptions of real analyticity and homogeneity of the system, for the CRC and the ORC to be properly defined at the origin the functions f, g and h are required to be homogeneous polynomials and their degrees of homogeneity k, ℓ and m to be nonnegative integers.

Then the repeated Lie brackets of f and g, as well as the repeated Lie derivatives of h along f and g, are all homogeneous polynomials too. Their exact degrees of homogeneity depend on how many times f and g occur in the repeated operations. Note that the roles of f and g are entirely interchangeable, both with respect to the CRC and the ORC. Each time a Lie bracket is formed with f (resp. g) and each time a Lie derivative is taken along f (resp. g) the degree of homogeneity increases by $k-1$ (resp. $\ell-1$). Clearly, for $k=0$ (or $\ell=0$) the degree decreases by 1. For $k=1$ (or $\ell=1$) the degree remains the same, and for $k \geq 2$ (or $\ell \geq 2$) the degree (strictly) increases.

At the origin, a homogeneous polynomial vector field evaluates to zero unless its degree of homogeneity is zero, i.e., unless it is constant. Therefore, for the CRC to hold at the origin a necessary and sufficient condition is that n independent constant vector fields are contained in the Control Lie Algebra generated by f and g. With respect to the ORC one needs to consider the observability co-distribution $d\mathcal{O}$ spanned by the gradient fields associated with the repeated Lie derivatives of h along f and g. At the origin these gradients evaluate to zero, except for the repeated Lie derivatives which are are linear. Thus, for the ORC to hold at the origin, a necessary and sufficient condition is that n independent linear functions are contained in the Observability Space. With respect to the degrees k, ℓ and m the following observations can now be made.

(1) The CRC may hold at the origin only if $k=0$ or $\ell=0$. If $k=\ell=0$, then all the repeated Lie brackets of f and g are identically zero, so one should also have that $n \leq 2$.

(2) The ORC may hold at the origin in each of the following three situations:
(a) $k=0$ or $\ell=0$ (and m arbitrary);
(b) $k=m=1$ and $\ell \geq 1$, or $k \geq 1$ and $\ell=m=1$;
(c) $k \geq 2$ and $\ell \geq 2$ and $n=m=1$.
Case (a) contains the linear systems (for $k=1$, $\ell=0$ and $m=1$), which are well-known to involve linear state space transformations. Case (b) includes the bilinear systems (for $k=\ell=m=1$) which are also well-known to involve linear state space transformations (see (D'Alessandro *et al.*, 1974)). Case (c) is not of much interest with respect to identifiability analysis, since $n=m=1$. One then deals with a scalar first-order differential equation in x, which is directly observable from the output y being a linearly scaled version of x.

5. GENERALIZATION TO THE MULTIVARIABLE CASE

To generalize the results of the previous section to the multi-input multi-output case is straightforward. The nonlinear systems under consideration are required to have the form

$$\dot{x}(t) = f(x(t)) + \sum_{i=1}^{r} g_i(x(t))u_i(t), \tag{4}$$

$$y_j(t) = h_j(x(t)), \quad j = 1, \ldots, s, \tag{5}$$

$$x(0) = x_0. \tag{6}$$

Here one is dealing with r input signals u_i, $i = 1, \ldots, r$, entering the system via the dynamical equation in a linear way, each having its own associated input vector field g_i. Also, one now has s output signals y_j, $j = 1, \ldots, s$, each obtained from the state x by means of an associated output function h_j. This constitutes a common class of nonlinear systems often studied in the literature. Note that the concepts of the CRC and the ORC are defined in the literature to apply to these multivariable systems too (see, e.g., (Isidori, 1989)).

For the theorems of the previous sections to hold, the input vector fields $g_1, \ldots, g_r$ are required to be real analytic and homogeneous of degree $\ell_1, \ldots, \ell_r$, respectively. The output functions $h_1, \ldots, h_s$ are required to be real analytic and homogeneous of degree $m_1, \ldots, m_s$, respectively. The Control Lie Algebra $\mathcal{C}$ and the Observation Space $\mathcal{O}$ have to be constructed in the usual way, by taking repeated Lie brackets of $f, g_1, \ldots, g_r$ and by taking repeated Lie derivatives of all the output functions $h_1, \ldots, h_s$ with respect to f, $g_1, \ldots, g_r$. Obviously, more complicated sequences of Lie brackets and Lie derivatives can now be formed, but their total number remains countable and systematic procedures are easily developed to enumerate them. All the theorems of the previous sections are then seen to allow for immediate generalization to this multivariable set-up.

With respect to the issue of the limitations imposed on the degrees of homogeneity $k, \ell_1, \ldots, \ell_r, m_1, \ldots, m_s$ in case the CRC or the ORC holds at the origin, the following holds.

(1) For the CRC to hold at the origin, at least one of the degrees $k, \ell_1, \ldots, \ell_r$ must be equal to zero. If all of them are equal to zero, then n should not exceed $r+1$.

(2) For the ORC to hold at the origin, one of the following three situations must hold:
(a) at least one of the degrees $k, \ell_1, \ldots, \ell_r$ is equal to 0;
(b) the degrees $k, \ell_1, \ldots, \ell_r$ are all ≥ 1, at least one of them is equal to 1, and moreover at least one of the degrees $m_1, \ldots, m_s$ is equal to 1;
(c) the degrees $k, \ell_1, \ldots, \ell_r$ are all ≥ 2, and at least n of the degrees $m_1, \ldots, m_s$ are equal to 1.

6. A WORKED EXAMPLE

Consider the following class of homogeneous polynomial systems, given by

$$\begin{aligned}
\dot{x}_1 &= -px_1^2x_3 + x_2x_3^2,\\
\dot{x}_2 &= -2p^2x_1^3 + 2px_1x_2x_3 + x_1x_3^2 + x_3^3,\\
\dot{x}_3 &= u,\\
y_1 &= x_1,\\
y_2 &= x_3,
\end{aligned}$$

with the initial state $x_0(p) = (0, 0, 1)^T$.

To verify the ORC, the following functions ω_1^p, ω_2^p and ω_3^p are determined, involving Lie derivatives of the components of h along the vector fields f and g:

$$\begin{aligned}
\omega_1^p(x) &:= h_1(x) = x_1,\\
\omega_2^p(x) &:= L_fh_1(x) = -px_1^2x_3 + x_2x_3^2,\\
\omega_3^p(x) &:= h_2(x) = x_3.
\end{aligned}$$

The corresponding matrix of gradient vectors takes the form

$$\begin{pmatrix} 1 & 0 & 0 \\ -2px_1x_3 & x_3^2 & -px_1^2 + 2x_2x_3 \\ 0 & 0 & 1 \end{pmatrix},$$

having full rank 3 whenever $x_3 \neq 0$, which holds for instance for the initial state vector $x_0(p)$.

To verify the CRC, one may compute $[[[f, g], g], g](x) = \begin{pmatrix} 0 \\ -6 \\ 0 \end{pmatrix}$ and $[[[[[[f, g], g], g], f], g], g](x) = \begin{pmatrix} 12 \\ 0 \\ 0 \end{pmatrix}$.

Together with $g = \begin{pmatrix} 0 \\ 0 \\ 1 \end{pmatrix}$ this shows that the CRC is satisfied everywhere.

An explicit representation of λ can be computed from ω_1^p, ω_2^p and ω_3^p, since according to Thm. 1 it holds on an open neighborhood of the initial state that $\omega_i^p \circ \lambda = \omega_i^{\widetilde{p}}$ for all i. One has:

$$\begin{aligned}
\widetilde{x}_1 &= x_1,\\
-\widetilde{p}\widetilde{x}_1^2\widetilde{x}_3 + \widetilde{x}_2\widetilde{x}_3^2 &= -px_1^2x_3 + x_2x_3^2,\\
\widetilde{x}_3 &= x_3.
\end{aligned}$$

This gives the mapping $x = \lambda(\widetilde{x})$ described by

$$\begin{aligned}
x_1 &= \widetilde{x}_1,\\
x_2 &= \widetilde{x}_2 + (p - \widetilde{p})\frac{\widetilde{x}_1^2}{\widetilde{x}_3},\\
x_3 &= \widetilde{x}_3,
\end{aligned}$$

which is indeed homogeneous of degree 1 but nonlinear for $p \neq \widetilde{p}$. But since the CRC holds everywhere, and in particular at the origin, it follows from Cor. 6 that λ is linear. Hence one can conclude that $p = \widetilde{p}$, whence identifiability holds.

It is illustrative to note that without the prior knowledge of linearity of λ provided by Cor. 6, one would normally proceed by imposing the properties (i)–(v) of λ as given by Thm. 1. It then happens that the conditions (i) and (ii) and (v) are obviously satisfied.

However, condition (iv) imposes the restriction

$$-(p - \widetilde{p})\frac{\widetilde{x}_1^2}{\widetilde{x}_3^2} = 0.$$

This again leads to the conclusion $p = \widetilde{p}$, since this identity should hold on an open neighborhood of the initial state. As before, only the trivial solution with λ equal to the identity mapping is obtained and once again identifiability is concluded to hold.

From Thm. 5 it actually follows that the ORC apparently holds everywhere too, despite the partial analysis above which so far has only shown the ORC to hold at points for which $x_3 \neq 0$. To verify this, one can compute the function ω_4^p defined by

$$\omega_4^p(x) := L_gL_gL_fh_1(x) = 2x_2.$$

Then the matrix of gradient vectors of ω_1^p, ω_3^p and ω_4^p becomes equal to the constant nonsingular matrix

$$\begin{pmatrix} 1 & 0 & 0 \\ 0 & 0 & 1 \\ 0 & 2 & 0 \end{pmatrix}.$$

Indeed, the ORC is satisfied everywhere, and in particular at the origin so that Thm. 3 applies. If one now chooses the functions ω_1^p, ω_3^p and ω_4^p to establish again an explicit representation of λ (instead of the functions ω_1^p, ω_2^p and ω_3^p that were used above) then it is found directly that the identity mapping is the unique candidate for the mapping λ in this example.

As the example demonstrates, some care should be taken in computing a candidate function λ. Note how the interplay between the various results of this paper can sometimes avoid certain computations, such as the verification of the conditions of Thm. 1 on λ, or the computation of the function ω_4^p (for which there are no a priori guidelines which sequence of Lie derivatives of which output components to consider).

7. CONCLUSIONS

In this paper several new theorems are presented concerning the class of homogeneous systems. The first result shows that the state isomorphism for a pair of indistinguishable systems is, under certain conditions, homogeneous of degree one. The second result shows that, under certain conditions and for homogeneous polynomial systems, this state isomorphism is linear in case the ORC holds at the origin. They generalize preliminary results reported in (Hanzon and Peeters, 2001) to the multivariable case, and with homogeneous rather than linear output functions.

The third result shows that, again for homogeneous polynomial systems, linearity of the state isomorphism also holds in case the CRC holds at the origin, but this is due to the fact that the ORC then also holds at the origin. Fourth, if the CRC holds at the origin and the vector fields making up the dynamical equation of the system are homogeneous polynomials, then the state isomorphism is affine, for any real analytic output function.

A worked example is presented, demonstrating the construction of the state isomorphism λ and illustrating how the results may be used in system identifiability analysis to reduce the complexity of the computations. In the given examples calculations can be done by hand, but for more complex cases the calculations can be done (at least in principle) by constructive algebra methods and exact and symbolic computation.

REFERENCES

Chappell, M.J., K.R. Godfrey and S. Vajda (1990). Global identifiability of the parameters of nonlinear systems with specified inputs: A comparison of methods. *Mathematical Biosciences* **102**, 41–73.

D'Alessandro, P., A. Isidori and A. Ruberti (1974). Realization and structure theory of bilinear dynamical systems. *SIAM Journal of Control* **12**, 517–535.

Hanzon, B. and R.L.M. Peeters (2001). On the state isomorphism approach to identifiability of homogeneous systems. *Proceedings of the 40th IEEE Conference on Decision and Control, December 4-7, 2001, Orlando (Fla), USA* pp. 3104–3105.

Hermann, R. and A.J. Krener (1977). Nonlinear controllability and observability. *IEEE Transactions on Automatic Control* **AC-22**, 728–740.

Isidori, A. (1989). *Nonlinear Control Systems, 2nd ed.*. Springer Verlag. Berlin.

Joly-Blanchard, G. and L. Denis-Vidal (1998). Some remarks about an identifiability result of nonlinear systems. *Automatica* **34**(9), 1151–1152.

Peeters, R.L.M. and B. Hanzon (2002). Identifiability of homogeneous systems using the state isomorphism approach. Technical Report M02-10. Dept. Mathematics, Universiteit Maastricht.

Sussmann, H.J. (1977). Existence and uniqueness of minimal realizations of nonlinear systems. *Mathematical Systems Theory* **10**, 263–284.

Tunali, E.T. and T.J. Tarn (1987). New results for identifiability of nonlinear systems. *IEEE Transactions on Automatic Control* **AC-32**, 146–154.

Vajda, S. and H. Rabitz (1989). State isomorphism approach to global identifiability of nonlinear systems. *IEEE Transactions on Automatic Control* **AC-34**, 220–223.

Vajda, S., K.R. Godfrey and H. Rabitz (1989). Similarity transformation approach to identifiability analysis of nonlinear compartmental models. *Mathematical Biosciences* **93**, 217–248.

Appendix A. HOMOGENEOUS FUNCTIONS

The following definition of a homogeneous function of n real variables is employed in this paper.

Definition 8. Let $f : D \rightarrow \mathbb{R}$ be a function on some open non-empty domain $D \subseteq \mathbb{R}^n$. Let $k \in \mathbb{R}$. Then f is called homogeneous of degree k on the domain D if for all $x \in D$ there exists an open neighborhood I_x of 1 in the interval $[0, \infty)$, on which it holds that $\phi x \in D$ for all $\phi \in I_x$ and for which

$$f(\phi x) = \phi^k f(x), \qquad \forall \phi \in I_x. \qquad \text{(A.1)}$$

If f is homogeneous and $x \in D$, let $I_x^{\max}$ be the largest subinterval of $[0, \infty)$ containing 1 on which $\phi x \in D$ for all $\phi \in I_x^{\max}$. Then the validity of property (A.1) extends to all $\phi \in I_x^{\max}$ and not just to $\phi \in I_x$.

The set of all *differentiable* homogeneous functions f of degree k on some open domain D can be characterized as the solution set of a first order partial differential equation. This is the content of Euler's Theorem.

Theorem 9. (Euler's Theorem.) Let $f : D \rightarrow \mathbb{R}$ be differentiable on some open non-empty domain $D \subseteq \mathbb{R}^n$. Let $k \in \mathbb{R}$. Then f is homogeneous of degree k if and only if

$$\nabla f(x) \cdot x = k f(x), \qquad \forall x \in D. \qquad \text{(A.2)}$$

In the literature one sometimes finds that the degree of homogeneity k is restricted to be a nonnegative integer. While the nonnegative integers are the only relevant values in case of real analyticity of f at the origin, it may be useful to allow k to attain other real values too for functions which are defined on a domain outside the origin.

Proposition 10. Let f be homogeneous of degree $k \in \mathbb{R}$ on an open connected domain $D \subseteq \mathbb{R}^n$ containing the origin. Moreover, let f be real analytic on the punctured domain $D - \{0\}$ and at least $[k] + 1$ times continuously differentiable at the origin, where $[k]$ denotes the largest integer smaller than or equal to k. Then f is a homogeneous polynomial on D. In particular, f is real analytic at the origin too, and if f is not identically zero, then k is a nonnegative integer which is equal to the total degree of f.

If f is real analytic on an open connected domain D which includes the origin, then it is homogeneous of degree 0 if and only if it is constant, and it is homogeneous of degree 1 if and only if it is linear.

IFAC
Publications
www.elsevier.com/locate/ifac

SYSTEM IDENTIFICATION FOR STRUCTURAL DYNAMICS AND VIBROACOUSTICS DESIGN ENGINEERING

Herman Van der Auweraer
herman.vanderauweraer@lms.be

LMS International
Interleuvenlaan 68, B – 3001 Leuven, Belgium
www.lmsintl.com

Abstract: System identification plays a crucial role in structural dynamics and vibro-acoustic system optimization. A number of industrially as well as socially relevant applications will be discussed. The most popular modeling approach is based on the "Modal Analysis" concept, leading to an interpretation in terms of visualized Eigenmodes. Using the modal models, design improvements can be predicted and the structure optimized. The main modal testing procedures and modal parameter identification methods are reviewed, including both input/output and output-only approaches. The current critical elements in system identification for these applications are outlined and discussed in the broader context of the changing role of testing in the product engineering process. New trends in modal analysis that specifically address these problems are reviewed and illustrated with case studies. This includes the issues of instrumentation, test definition, measurement principles and parameter estimation. *Copyright © 2003 IFAC*

Keywords: Noise, vibration, modal, eigenvectors, system identification, stochastic realisation, structural parameters, subspace, maximum likelihood

1. SYSTEM IDENTIFICATION AND MODAL ANALYSIS OF STRUCTURES

The vibration and acoustical behavior of a mechanical structure is determined by its dynamic characteristics. This dynamic behavior is typically described using a linear system modeling approach. The inputs to the system in general are forces ("loads"), the outputs the resulting displacement or acceleration responses. Using these variables, classical system analysis can be applied.

Specific to the mechanical problem is the physical interpretation that can be given to the system's Eigenvalues and Eigenvectors. System poles in structural dynamics usually occur in complex conjugate pairs, each corresponding to a structural vibration "mode". The pole's imaginary part relates to the resonance frequency and the real part to the damping. Structural damping is typically very low (a few % of the critical damping). As a consequence, resonance effects are very outspoken, easily observed, and directly linked to many structural dynamics problems. The system's Eigenvectors, expressed in the basis of the physical coordinates on the structure, then correspond to characteristic structural vibration patterns, referred to as the system's vibration "mode-shapes".

Each vibration mode can be considered as an independent single-degree-of-freedom system. The dynamic system response $\{x(t)\}$ of a mechanical system to an arbitrary load $\{f(t)\}$ can hence be written as a linear superposition of the independent single-degree-of-freedom "modal" responses $\{q_i(t)\}$. The derivation and use of such system model based on resonance frequencies, damping ratios and mode-shapes is the essence of the "modal analysis" approach. An example of a bending mode of a rectangular plate is shown in Figure 1.

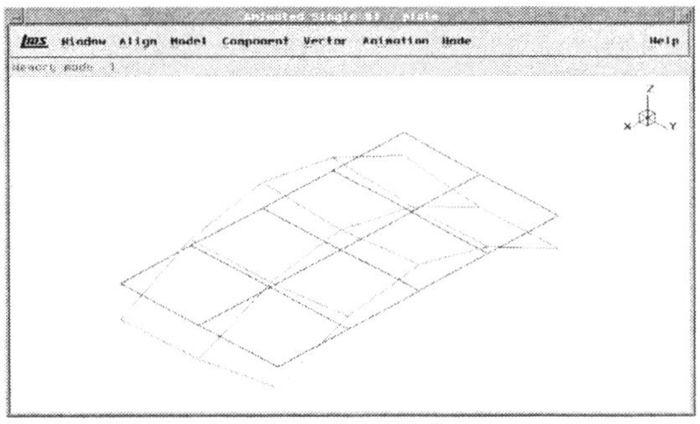

Figure 1: Mode shape of a plate

The modal representation of a mechanical structure can be determined analytically for a lumped mass-spring system. In the general case of a continuous structure, a numerical approximation by means of a Finite Element Model (FEM) is made, discretizing the structure in a finite number of physical coordinates. The equations of motion describing this system in the time and Laplace domain are given by:

$$[M]\{\ddot{x}(t)\}+[C]\{\dot{x}(t)\}+[K]\{x(t)\}=\{f(t)\} \quad (1)$$

$$\left[s^2[M]+s[C]+[K]\right]\{X(s)\}=\{F(s)\} \quad (2)$$

with $[M]$, $[C]$, $[K]$ resp. mass, damping and stiffness matrices. These equations leads to an Eigenvalue problem solved in terms of the modal parameters (Craig, 1981).

The limitations of the FEM approach lie in the huge model size required to properly describe complex structures (models with over 1 million degrees of freedom are commonly used). This leads to high model construction and calculation times, while inherent modeling accuracy limitations remain, related to the modeling of structural junctions, non-homogeneous elements, complex materials etc.

To address these limitations, an experimental approach to modal analysis was developed yielding results which can be used either as a model by itself, or to validate and improve the FE models. The resulting Experimental Modal Analysis (EMA) approach has become a standard element of the mechanical product design and engineering process.

Using the EMA models or the updated FE models, structural design problems related to vibrations, fatigue and noise can be addressed, a diagnosis on the causes made and appropriate design modifications designed and their effect predicted.

2. EXPERIMENTAL MODAL ANALYSIS

Experimental modal analysis can be regarded as a "Black Box" approach, extracting the modal model from measurements. Hereto, the structure under consideration is discretized by a grid of test locations (which is much coarser than a FE discretization).

Typically, artificial excitation forces are applied at a subset of locations and the corresponding excitation force signals (the "inputs") as well as the vibration responses at all locations (the "outputs") are measured. Mechanical reciprocity makes that it is not necessary to excite all inputs as long as all outputs have been measured. Broadband (Gaussian noise, impulse or multisine) as well as swept or frequency-stepped sinusoidal signals are applied to excite the structure over a frequency range of interest. Electro-magnetic or hydraulic exciters, or even impact hammers (for impulsive excitation) are used hereto.

Several particular constraints however make the process of system identification for structural dynamics largely different from this in electrical engineering or process control.

First of all, a continuous structure will have an infinite number of modes. In practice, the analyst will be interested only in a limited number of these modes, up to a certain frequency or only in a certain frequency band. Still, model orders of 50 to 100 are no exception. While some of the modes are rather separated in resonance frequency, others may be very close leading to highly overlapping responses. A typical Frequency Response Function (FRF) on the engine block of a car is shown in Figure 2.

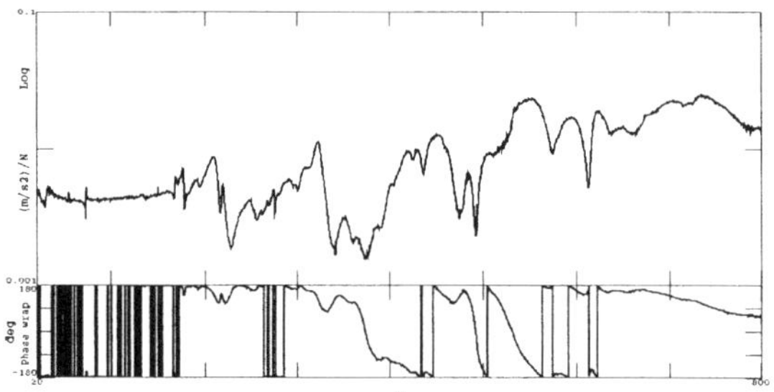

Figure 2: Typical FRF on a car engine block

A second issue is the large number of spatial locations, which typically have to be measured. While the system poles are global parameters, the Eigenvectors –and hence the mode shapes- require a sufficiently dense grid of responses to be measured. Figure 3 shows a number of FRFs measured on an aircraft. The global nature of several resonance frequencies can be clearly observed.

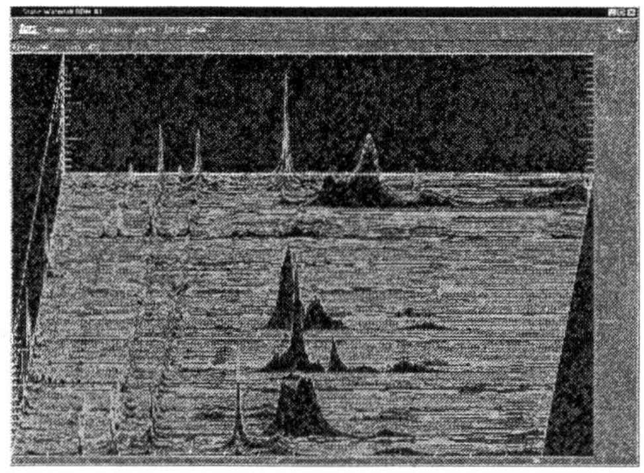

Figure 3: Waterfall display of FRFs on an aircraft

A car body may require over 1000 measurement locations, at each of which accelerations in 3 directions are measured. In most cases, such large number of measurements cannot be made for all responses simultaneously due to equipment (number of sensors, number of data-acquisition channels) or setup (mass-loading) limitations. Hence a full modal analysis will often be performed in several "patches", over periods ranging from one day to several weeks. The combined processing of all "patches" (often with inconsistencies in between them) is hence a must.

Finally, limitations to the level of applicable excitation forces combined with the large size and distributed damping of many structures may lead to low response levels contaminated by measurement noise. The reduction of the variance on the result requires very long data observations to be made.

The consequence of these constraints is that classical system identification approaches, extracting the parameters of a discrete-time state-space model or of an ARMA model directly from the sampled input-output data, are often not practical or not feasible. In general system theory terms we are speaking of systems with 4 to 8 inputs, over 1000 outputs, on which long time histories (often several thousands of samples) are measured non-simultaneously.

3. MODAL PARAMETER ESTIMATION

3.1 Parameter estimation methods

Because of the complexity of the structural dynamics identification problem, a large variety of dedicated modal parameter estimation methods were proposed over the years. In most methods, the raw input-output data are first processed into a non-parametric system description matrix, consisting of FRFs (Frequency Response Functions) or IRs (Impulse Responses), resulting in a significant data reduction and allowing the use and combination of non-simultaneously measured data. This also fits the test reality where the modal testing process is often separated in time (and location) from the actual analysis process.

The matrix of FRFs or IRs can then be used as input data to establish the parameters of a system model such as a frequency or time domain state-space model (referred to as Eigensystem Realization (ERA) (Juang and Pappa, 1985) or Subspace Identification (Van Overschee and De Moor, 1996), due to the inherent SVD model reduction), or to directly identify the structural parameters of a model described by the constitutive equations (Direct Parameter Estimation (Leuridan, 1984). Both approaches have been implemented in the time as well as in the frequency domain in many different algorithms.

Alternatively, model formulations can be used which directly exploit specific properties of the FRFs (or IRs). For example, the FRF and IR matrices can be expressed directly in terms of the modal parameters:

$$[H(j\omega)]=\sum_{r=1}^{N}\left(\frac{Q_r\{\psi\}_r\{\psi\}_r^t}{(j\omega-\mu_r)}+\frac{Q_r^*\{\psi\}_r^{*t}}{(j\omega-\mu_r^*)}\right) \quad (3)$$

$$[h(t)]=\sum_{r=1}^{N}\left(Q_r\{\psi\}_r\{\psi\}_r^t e^{\mu_r t}+Q_r^*\{\psi\}_r^*\{\psi\}_r^{*t} e^{\mu_r^* t}\right) \quad (4)$$

where

Q_r : modal scaling factor

$\{\psi\}_r$: modal vector r

$\mu\lambda_r$: system pole: $\sigma_r+j\omega_r$

σ_r : damping factor

ω_r : damped natural frequency

$[V]:=\left[\{\psi\}_1\cdots\{\psi\}_N\ \{\psi\}_1^*\cdots\{\psi\}_N^*\right]$: modal vector matrix

Directly solving Eq. (3) leads to the non-linear frequency domain method (Busturia and Gimenez, 1985), which, when used in its original form, requires a large non-linear set of equations to be solved in an iterative way. This proves to be unpractical for most modal data sets hence alternative methods have been developed. Below, the most popular ones are presented.

3.2. Least Squares Complex Exponential method.

The still most widely used approach is the Polyreference Least Squares Complex Exponential method which dates from the early 80-ties (Vold, *et al*, 1982). The method starts from the Impulse Responses (IR) between the measured inputs and outputs, Eq. (4) and yields global estimates of poles and the modal participation factors.

Define the impulse response function matrix $[R_k]$ at time lag k between N_{resp} responses and N_{ref} inputs. Mathematically, the Polyreference LSCE will decompose the correlation functions as a sum of decaying sinusoids. So,

$$[R_k]=\sum_{r=1}^{N_m}\{\psi\}_r e^{\mu_r k\Delta t}\{L\}_r^T+\{\psi\}_r^* e^{\mu_r^* k\Delta t}\{L\}_r^{T*} \quad or$$

$$[R_k]=\sum_{r=1}^{N_m}\{\psi\}_r\lambda_r^k\{L\}_r^T+\{\psi\}_r^*\lambda_r^{k*}\{L\}_r^{T*} \quad (5)$$

where $\lambda_r=e^{\mu_r\Delta t}$ and $\{L\}_r$ is a column vector of N_{ref} modal participation factors, which are multipliers which are constant for all response stations for the r-th mode. The combinations of complex exponential and constant multipliers, $\lambda_r\{L\}_r^T\ or\ \lambda_r^*\{L\}_r^{T*}$ are a solution of the following matrix finite difference equation of order t:

$$\lambda_r^k\{L\}_r^T[I]+\lambda_r^{k-1}\{L\}_r^T[W]_1+\ldots+\lambda_r^{k-t}\{L\}_r^T[W]_t=\{0\}^T \quad (6)$$

where $[W]_1\ldots[W]_t$ are coefficient matrices with dimension N_{ref} x N_{ref}. In case the system has N_m physical modes, the order t in equation (6) has to be greater than or equal to $2N_m/N_{ref}$ in order to find $2N_m$ characteristic poles.

Since the correlation functions are a linear combination of the characteristic solutions of equation (6), $\lambda_r\{L\}_r^T\ or\ \lambda_r^*\{L\}_r^{T*}$, they are also a solution of that equation. Hence,

$$[R_k][I]+[R_{k-1}][W]_1+\ldots+[R_{k-t}][W]_t=[0] \quad (7)$$

Eq. (7) uses all response stations simultaneously enabling a global least squares estimate of the coefficient matrices $[W]_1\ldots[W]_t$. Overdetermination is achieved by considering all available or selected time intervals. Once the coefficient matrices are known, equation (6) can be reformulated into a generalized Eigenvalue problem resulting into $N_{ref}t$ Eigenvalues

λ_r, as estimates for the system poles μ_r and the corresponding left eigenvectors $\{L\}_r^T$.
As the Polyreference LSCE does not yield the mode shapes, a second step is needed to extract these using the identified modal frequencies, damping ratios and participation factors as known parameters in Eq. (3), which now becomes a simple linear set of equations.

3.3. Least Squares Frequency Domain Methods

As the LSCE method requires the transformation of FRFs into IRs, methods directly working in the frequency domain have been researched. Most such methods express the FRF in a rational matrix polynomial or a common denominator formulation, using least squares or maximum likelihood estimators for the polynomial (matrix) coefficients. In a second step, poles and Eigenvectors are derived. For example, in a common-denominator model, a scalar matrix-fraction description is used for the measured set of FRFs. The FRF between output o and input i is modeled as:

$$\hat{H}_{oi}(\omega_f) = \frac{N_{oi}(\omega_f)}{D(\omega_f)} \quad (8)$$

for $i = 1,\ldots,N_i$ and $o = 1,\ldots,N_o$ with

$$N_{oi}(\omega_f) = \sum_{j=0}^{n} \Omega_j(\omega_f) B_{oij}$$

the numerator polynomial between output o/ input i

and $$D(\omega_f) = \sum_{j=0}^{n} \Omega_j(\omega_f) A_j$$

the common-denominator polynomial. Several choices are possible for the polynomial basis functions $\Omega_j(\omega_f)$. A discrete-time or a continuous time formulation can be used and special polynomials such as Orthogonal Forsythe Polynomials can be applied (Van der Auweraer and Leuridan, 1987). But for every formulation, the parameters to be estimated are the real-valued coefficients A_j and B_{oij}. The most promising method today appears to be the discrete-time model formulation proposed in (Guillaume, *et al*, 1998), as this leads to a very stable equation solution.

Replacing the model $\hat{H}_{oi}(\omega_f)$ in Eq. (8) by the measured FRF $H_{oi}(\omega_f)$ gives, after multiplication with the denominator polynomial,

$$\sum_{j=0}^{n} \Omega_j(\omega_f) B_{oij} - \sum_{j=0}^{n} \Omega_j(\omega_f) H_{oi}(\omega_f) A_j \approx 0 \quad (9)$$

for $i = 1,\ldots,N_i$, $o = 1,\ldots,N_o$ and $f = 1,\ldots,N_f$.

Note that every equation in Eq. (9) can be weighted. The equations (9) can be reformulated (Guillaume, *et al.*, 1998) as

$$\begin{bmatrix} X_1 & 0 & \cdots & 0 & Y_1 \\ 0 & X_2 & & 0 & Y_2 \\ \vdots & & \ddots & & \vdots \\ 0 & 0 & & X_{N_oN_i} & Y_{N_oN_i} \end{bmatrix} \begin{bmatrix} B_1 \\ B_2 \\ \vdots \\ B_{N_oN_i} \\ A \end{bmatrix} \approx 0 \quad (10)$$

or

$$[J] \begin{bmatrix} B_1 \\ B_2 \\ \vdots \\ B_{N_oN_i} \\ A \end{bmatrix} \approx 0 \quad (11)$$

with $$B_k = \begin{bmatrix} B_{oi0} \\ B_{oi1} \\ \vdots \\ B_{oin} \end{bmatrix}, \quad A = \begin{bmatrix} A_0 \\ A_1 \\ \vdots \\ A_n \end{bmatrix}$$

$$X_k = \begin{bmatrix} X_k(\omega_1) \\ \vdots \\ X_k(\omega_{N_f}) \end{bmatrix}, \quad Y_k = \begin{bmatrix} Y_k(\omega_1) \\ \vdots \\ Y_k(\omega_{N_f}) \end{bmatrix}$$

$$X_k(\omega_f) = W_{oi}(\omega_f)[\Omega_0(\omega_f), \Omega_1(\omega_f), \ldots, \Omega_n(\omega_f)]$$

$$Y_k(\omega_f) = -X_k(\omega_f) \cdot H_{oi}(\omega_f)$$

and $k = (o-1)N_i + i = 1,\ldots,N_oN_i$.

The Least Squares estimate can be computed efficiently via a "structured" QR decomposition of the Jacobian matrix [*J*] of Eq. (10). However, most estimators used in modal analysis form the normal equations and compute $\mathrm{Re}(J^H J)$ as this results in a faster implementation. As the discrete time-domain model generally leads to a well-conditioned Jacobian matrix *J*, this explicit calculation is justified. The resulting method is referred to as the Least Squares Complex Frequency Domain (LSCF) technique.

3.4 Maximum Likelihood Method

Starting from the same scalar matrix-fraction model formulation as in Eq. (8), (Guillaume, *et al.*, 1995) proposes a different parameter estimation approach based on a maximum likelihood approach. Assuming the different FRFs to be uncorrelated, the (negative) log-likelihood function reduces to:

$$\ell_{\mathrm{ML}}(\theta) = \sum_{o=1}^{N_o} \sum_{i=1}^{N_i} \sum_{f=1}^{N_f} \frac{\left|\hat{H}_{oi}(\theta,\omega_f) - H_{oi}(\omega_f)\right|^2}{\mathrm{var}\{H_{oi}(\omega_f)\}} \quad (12)$$

The ML estimate of $\theta = [B_1^T, \ldots, B_{N_oN_i}^T, A^T]^T$ is given by minimizing (12). This can be done using a Gauss-Newton optimization algorithm, which takes advantage of the quadratic form of the cost function Eq. (12). The Jacobian matrix J_m has the same structure as the one of Eq. (10). Also here the normal equations (i.e., $\mathrm{Re}(J_m^H J_m)$ and $\mathrm{Re}(J_m^H r_m)$) can be formed in a time efficient way.

Obviously, the optimization approach leads to an iterative procedure, which increases the duration of the estimation process. A proper choice of the starting values (preferably by a first step linear parameter estimation process) is important to ensure rapid convergence (Verboven, *et al.*, 1998). But the high-quality results, the capabilities to let the method perform a semi-automated analysis and the capability to complement the estimation results with uncertainties (Schoukens and Pintelon, 1991), give the ML approach unique features.

3.5. The stabilization diagram

All developed modal parameter estimation methods remain highly interactive. The key problems are the selection of the correct model order and the fact that not all resulting Eigenvalues have a physical meaning. Many are due to mathematical effects or to noise and result from the forced fulfillment of the specified model order rather than from dynamic system properties. To overcome this, the concept of the "stabilization diagram" is used. Hereto, a repeated analysis of the same data set is performed, each time for a different model order. The pole values from each analysis are combined in one single diagram, with as horizontal axis the pole frequency and as vertical axis the solution order. The pole is indicated by a symbol in this diagram. Poles corresponding to the physical system appear at nearly identical locations for every analysis, which is readily visible in the diagram. To point out that the frequency (resp. damping and mode shape vector) of a pole falls within certain bounds of the result obtained at a lower system order, this is additionally indicated (for example by a "f", "d" or "v"). This is illustrated in Figure 4 for the case of a vehicle body. Two methods, the LSCE and LSCF are compared.

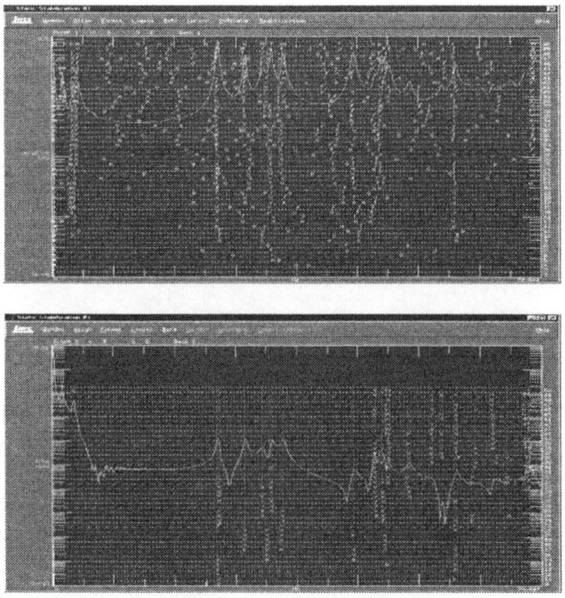

Figure 4: LSCE (upper) and LSCF (lower) stabilization diagram

The drastically improved stabilization characteristics of the LSCF method are clearly visible.

4. APPLICATIONS OF MODAL ANALYSIS

Modal analysis has become a standard procedure in today's structural dynamics optimization studies. A number of typical examples are discussed.

4.1. Vehicle engineering

Modal analysis is routinely performed on prototypes of every new design, variant or design alternative of a car. This includes modal tests of the "naked" car body (so-called "body-in-white"), on various critical car components (engine, suspension, exhaust, brake,...), and on the fully equipped car.

The modal analysis of the body-in-white is used to assess the global vibration modes, which are very important in view of the dynamic stiffness of the vehicle, as well as local modes related to e.g. car panels. Over the last decade, the resonance frequency of the first bending and torsion modes of a typical car body has increased by no less than 50%, indicating the greatly improved dynamic stiffness (which on its turn are essential for the vehicle handling and noise behavior).

Figure 5 shows an example of a car body modal test using a shaker (upper) and an impact hammer (lower).

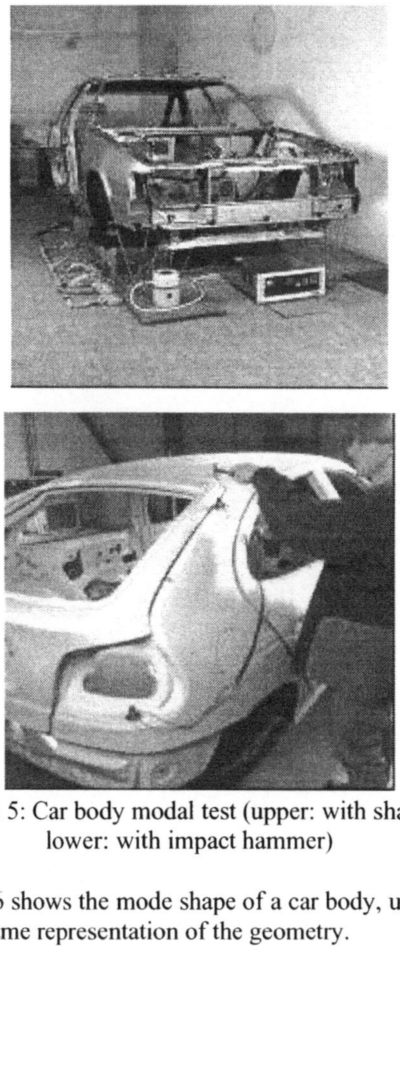

Figure 5: Car body modal test (upper: with shaker; lower: with impact hammer)

Figure 6 shows the mode shape of a car body, using a wire-frame representation of the geometry.

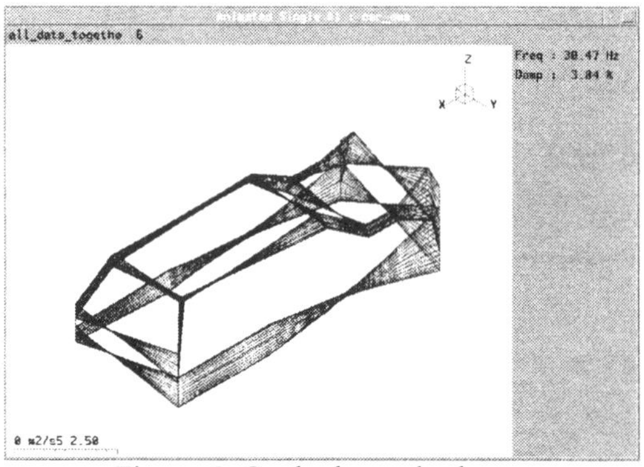

Figure 6: Car body mode shape

Another important modal test is this of the vehicle on its suspension. The low-frequency modes (pitch, roll, heave) are determined by the suspension properties, while the car body itself behaves rigid. These modes are very important in view of the driving comfort and the vehicle handling. In such case, the excitation may be applied through wheel platforms. An example of a test of a Formula-1 car is shown in Figure 7, and some basic mode in Figure 8.

Figure 7: Formula 1 suspension modal test

Figure 8: Formula-1 suspension pitch mode

Also components are subjected to modal tests. This may be done for the optimization of the intrinsic component behavior, including the modal displacement at connection points to the main structure. Furthermore, a "virtual" coupling of the modal models of various substructures can be performed to predict the behavior of an assembly.

An example of an FRF and a mode shape of an exhaust system are shown in Figure 9.

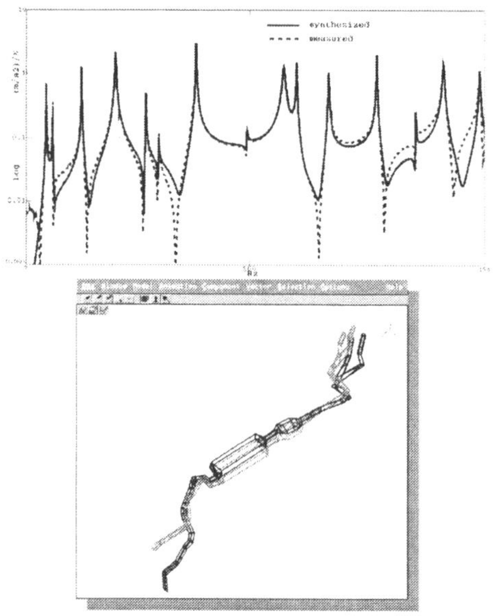

Figure 9: Exhaust system FRF and mode shape

The resulting modal models can be used to identify parts with excessive vibrations, to identify critical frequencies, but also as a basis to design structural modifications. Typical resulting actions are increasing the stiffness of one of the structure's parts to shift a resonance frequency out of a critical region (away from the frequencies excited by the engine harmonics, or away from a strongly coupled resonance of another component), or increasing the damping to reduce the peak response amplitude.

A last example of vehicle system identification is the simulation of road tests on multi-axis test-rigs. Next to removing any dependence on weather or traffic, shortened load sequences can be derived preserving the same damage potential as the originally measured signals and customer-specific load profiles can be synthesized. Essential hereto is the capability to reproduce on the test-rig exactly the same vibrations as on the road. This requires an accurate on-line system identification of the shaker-vehicle system. An example of such test-rig is shown in Fig. 10.

Figure 10: Multi-axis test rig for road simulation

4.2. Aerospace applications

Modal analysis is performed on many parts of the aircraft, from critical components such as landing gear, control surfaces, engine components (turbine blades, nacelle...) to the complete fuselage.

A full aircraft modal test is one of the essential tasks to be performed on any new aircraft before it is cleared for test flights. Such tests are referred to as GVT (Ground Vibration Tests) and serve to validate (and possibly improve) the numerical (FE) models, which have guided the aircraft design. Such models are for example used to assess the aero-elasticity stability (flutter) of the aircraft within its flight envelope and hence these are safety-critical. An example of a modal test on an aircraft structure is shown in Figure 11. Some typical aircraft mode shapes are shown in Figure 12.

Figure 11: Twin-propeller aircraft modal test

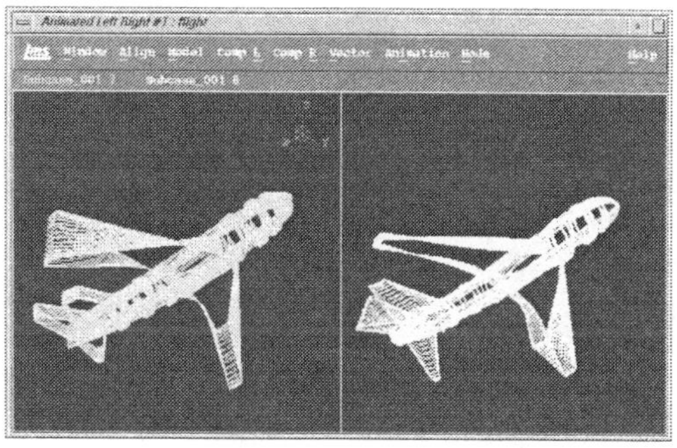

Figure 12: Jet-engine mode shapes

Also spacecraft structures (launchers, payloads, ...) and components (payload fixtures, equipment racks, antennas, ...) are subjected to modal tests to assess their structural behavior in view of the very high loads occurring during launch and certain flight phases (re-entry, stage separation). Full satellites are usually tested on large shaker tables, which requires specific analysis procedures using transmissibilities in stead of force/acceleration FRFs (Peeters, *et al.*, 2002a). Figure 13 shows an example of a satellite modal test and a corresponding mode shape.

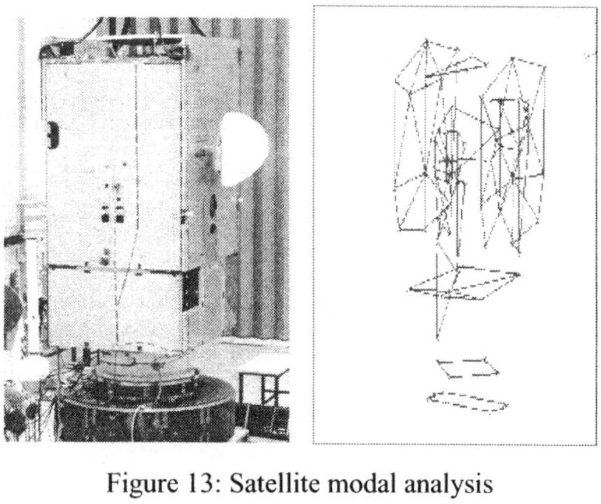

Figure 13: Satellite modal analysis

4.3. Process industry

Also in the process industry, system identification is widely used for structural dynamics and modal analysis. Applications include the vibration analysis of pumps, compressors, piping systems, shafts and bearings, turbine blades, machinery foundations, etc. Vibration behavior is for example critical for the accuracy in high precision equipment or for the noise generated by manufacturing machinery. Excessive vibrations may also be the cause of structural failure. Figures 14 and 15 show a gantry on which a modal test was performed and one of its mode shapes.

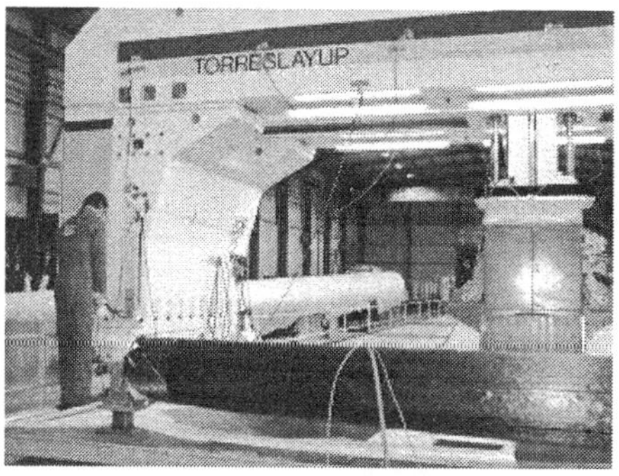

Figure 14: Gantry machine setup

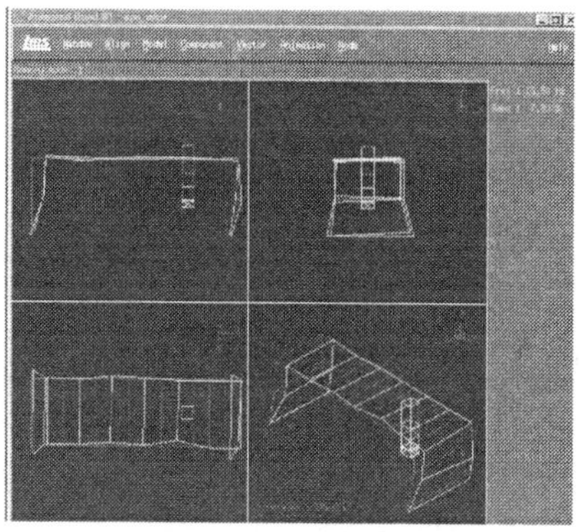

Figure 15: Gantry mode shape

4.4. Civil construction

Structural dynamics characteristics are also of great importance in civil engineering, for example for bridges, dams, high-rise buildings, off-shore platforms etc. Of particular importance is the damping ratio of the modes which may be excited by wind or swell, which may lead to unstable vibrations similar to flutter. Resonance frequencies have to be separated clearly from excitation phenomena. Typical examples of such unstable vibrations are the destroyed TACOMA bridge (US, 1940), or more recently the Millennium bridge in London.

Of course, performing tests on such structures requires specific equipment in terms of high-force excitation systems (hydraulic shakers, drop-force impact systems or using unbalance shakers) and robust all-weather measurement equipment. An example of some modeshapes obtained on a highway bridge are shown in Figure 16. Results from tests on e.g. the Vasco Da Gama and the Øresund bridges can be found in (Peeters, *et al.*, 2002b; Peeters, *et al.*, 2003).

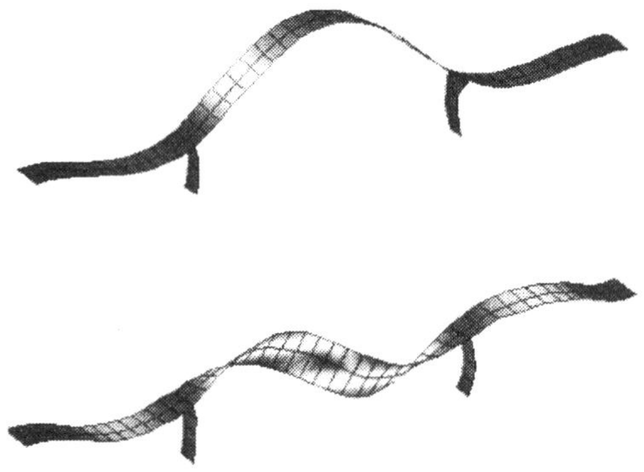

Figure 16: Bending and torsion mode shape of a highway bridge

4.5. Other applications

While above discussed cases comprise some major modal analysis application areas, the technology is used in a even wider field, from household systems such as washing machines or refrigerators to audio systems such as loudspeakers, CD-drives, or even computer rack, printed circuit boards etc. which leads to the conclusion that modal analysis has become one of the most important tools in the design and structural optimization of mechanical products.

In all these applications, it is the purpose to obtain adequate system models to perform troubleshooting as well as system optimization in terms of mission-critical functional performance parameters such as safety, stability, fatigue life, vibration comfort, interior and exterior noise etc.

The modal results may be used as such, to get physical insight in the structural dynamics problems by evaluating the modal parameters and especially the mode shapes. But more and more are the resulting modal models used as a basis for further system improvement using modal modification techniques, FE model validation or even for model building, using modal substructuring methods.

The technology has become a de facto standard in product development and is part of the curricula of most mechanical engineering education programs.

5. RECENT EVOLUTIONS

The critical role of modal analysis also makes that the method is under constant pressure for improvement in terms of efficiency and effectiveness and that several new challenges have been identified and are addressed in the scientific modal community. In order to illustrate this, the critical steps and main trends in the application and technology of modal analysis in industry will be reviewed below.

5.1 Test Definition

A modern modal analysis test can easily comprise more than 1000 degrees of freedom. This is due to the complexity of many built-up products, where many components can exhibit a local behavior. In particular, when modal analysis is applied to validate and update a FE model, a sufficiently dense grid of test points is needed. The FE model grid is anyway much finer than the test one, requiring a substantial reduction. Optimally selecting a minimal set of test points is required in optimizing the test efficiency.

In case a FE model is available, it can be used to this purpose: to select the optimal location of (a minimal set of) measurement sensors and excitations for the optimal control and observation of a number of selected target modes (Heylen, *et al.*, 1995). This "pre-test analysis" maximizes the test efficiency and limits the unavailability of the prototype under test. Figure 17 shows an example where the optimal sensors for a modal test on a prototype for the space-station return vehicle are derived from a FE model.

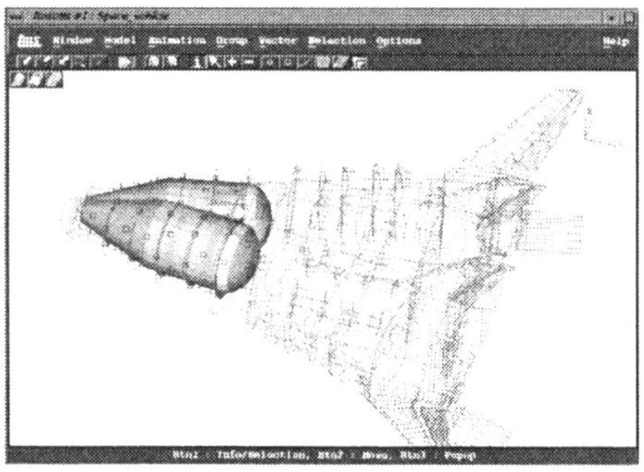

Figure 17: Sensor selection for a space structure

5.2 Test Instrumentation

The increasing test complexity –and size- make that instrumentation set-up takes up a major part of the total test duration. State-of-the-art equipment consists of multi-channel data-acquisition front-ends with local ADC and signal processing, connected to powerful computers. Channel counts range form 16 up to as much as 1000 and bandwidths from 100 Hz to 20 kHz, allowing to use audio DSP components.

As cabling and sensor identification errors become more difficult to be noticed, automated procedures using "Smart Transducers" with embedded position and calibration information become more widespread. The concept of a "Transducer Electronic Data Sheet" or TEDS, embedded in the transducer is being taken up in the IEEE 1451 standard. Current research focuses on the use of MEMS-based transducers, including on-chip AD conversion and wireless transmission capability. Signal bandwidth and power supply however remain bottlenecks.

But even with advanced measurement equipment, a major problem remains with ensuring the adequate validity ("plausibility") of the measured data before any attempt to extract the modal parameters is made.

Errors in the set-up or during the test must be detected and identified before the test is closed and the set-up dismantled. Often there is no second chance to redo the tests and in other cases, this would be too expensive and block the critical path. "Test-Right-First-Time" is the paradigm. The detection of test data problems has to be done on several levels:

- Transducers: are they properly connected to the structure and to the test system?
- Are the test conditions in accordance to the specifications (excitation, boundary conditions)?
- Is the quality of the measurement signals OK (level, spectrum, linearity, noise, DSP errors..)?
- Data errors (drop-out, bias, drift, spikes)?
- Does the structure fulfil the modelling assumptions (Linearity: Schoukens *et al*, 1998)?
- Is the consistency between the signals adequate in view of the analysis (coherences, FRFs etc)?

In order to yield data with a maximal usability, indicators regarding various test plausibility factors have to be defined, calculated and monitored. Deviations have to be immediately recognised and decisions on whether to redo part of the test, correct data afterwards, or just flag data as being invalid or with decreased plausibility have to be made. Figure 18 shows the example of the geometric visualisation of the average coherence from an aircraft GVT, indicating a sensor problem near the tail.

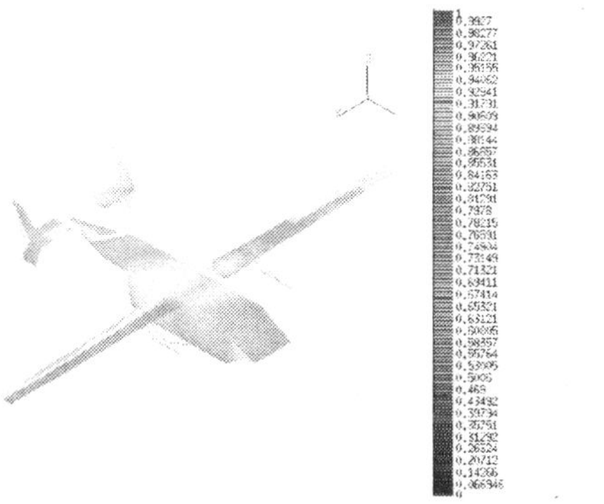

Figure 18: Averaged coherence form a GVT test

5.3 Optical Measurements

One of the instrumentation constraints with modal analysis is the effect of the transducer mass on the structure. With increasing frequency, this influence increases, while the number of transducers has to increase (higher spatial complexity of the modes). Various non-contact optical measurement techniques have been developed hereto.

One is based on Laser Doppler Vibrometers (LDV), which scan the vibrating surface in a sequential way. This approach supports broadband as well as sinusoidal testing and is frequently used. The second approach is based on full field Electronic Speckle Pattern Interferometry (ESPI) (Freymann, *et al.*, 1996). This approach uses strobed continuous or pulsed lasers and requires sinusoidal excitation. Special CCD cameras are used and specific image processing transforms the qualitative interferograms to quantitative FRF values at the excitation frequency. The complete FRF is then built up frequency by frequency. Special data reduction methods allow to reduce the pixel-density vibration response fields to the spatial resolution needed for a proper modal analysis. Fig. 19 shows an ESPI-obtained vibration shape from the floor plate of a car. The high spatial resolution can be clearly seen.

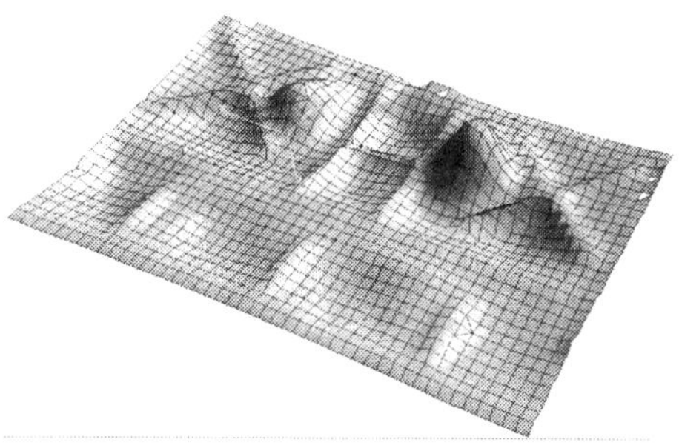

Fig. 19 ESPI mode shape of a car plate

5.4 Vibro-Acoustic Modal Analysis

In many interior noise problems, not only the structure, but also the cavity acoustics show resonance behavior. An example is the "booming" noise in a car at certain speeds, where a cavity resonance is excited by engine induced vibrations. The concept of modal analysis can also be applied to acoustical or mixed structural-acoustical systems using as acoustical input variable a source volume velocity and as acoustical output the pressure. The system equations, expressed again in a finite element formulation, are (Wyckaert, *et al.*, 1996):

$$\begin{bmatrix} K^s & -K^c \\ 0 & K^f \end{bmatrix}\begin{Bmatrix} x \\ p \end{Bmatrix} - j\omega\begin{bmatrix} C^s & 0 \\ 0 & C^f \end{bmatrix}\begin{Bmatrix} x \\ p \end{Bmatrix} - \omega^2\begin{bmatrix} M^s & 0 \\ M^c & M_f \end{bmatrix}\begin{Bmatrix} x \\ p \end{Bmatrix} = \begin{Bmatrix} f \\ p\dot{q} \end{Bmatrix} \quad (13)$$

The Eigenvalues are due to the coupled behavior of structure and acoustics. The Eigenvectors each have a specific acoustic and structural part. Special scaling relations due to the vibro-acoustic reciprocity principle have to be taken into account. An acoustic mode of a car interior is shown in Figure 20.

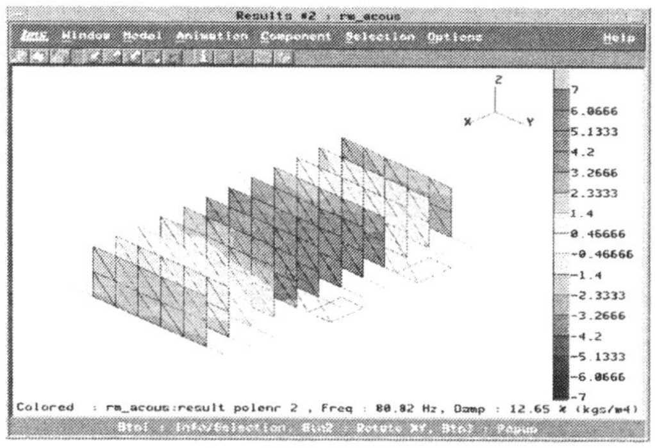

Figure 20: Acoustic mode shape of a car cabin

The acoustic pressure in the car sections is represented by color levels, showing the modal behavior. In this particular case, the acoustic cavity mode coincided with a structural suspension mode, resulting in a strong transmission of wheel vibrations as noise into the car interior (Wyckaert, *et al.*, 1994).

5.5 In-Operation Modal Analysis

The classical modal parameter estimation approach starts from FRFs or IRs measured in laboratory conditions. However, in many applications, the real operating conditions may differ significantly from those applied during the modal test. As all real-world systems are to a certain extent non-linear, the models obtained in laboratory conditions will be linearized for different working points than applicable under real loading, requiring an in-operation identification. This would also properly take into account environmental influences on the system behavior (pre-stress of suspensions, load-induced stiffening, aero-elastic interaction, ...). In many cases (off-shore platforms, civil constructions) forced excitation tests may even be very difficult to conduct, and operating data are often the only ones available.

Hence, a considerable interest exists in extracting valid models directly from operating data. In most cases, only response data are measurable while the actual loading conditions are unknown. As a result, several techniques have been studied over the last years for estimating modal parameter from output-only data (Cooper and Deforges, 1997; Hermans and Van der Auweraer, 1999). They include ARMA (Auto-Regressive Moving Averaging) models, or EMA derivatives, replacing FRFs or IRs by Cross-power or Correlation functions. Examples are the Natural Excitation Technique (NexT) (James III, *et al.* 1995) and the Stochastic Realization methods (Desai, *et al.*, 1985; Basseville, *et al.*, 2001).

The time domain Stochastic Subspace method is presently one of the main accepted techniques, in particular in its covariance driven form, allowing a data reduction using covariances with respect to a set of reference sensors. It was shown that a limited set of well-selected references is sufficient, avoiding the huge system matrices of the data-driven methods.

The following stochastic discrete time state space model is considered:

$$\begin{aligned}\{x_{k+1}\} &= [A]\{x_k\} + \{w_k\} \\ \{y_k\} &= [C]\{x_k\} + \{v_k\}\end{aligned} \qquad (14)$$

where $\{x_k\}$ represents the state vector of dimension n and $\{w_k\}$, $\{v_k\}$ are zero-mean, white vector sequences, respectively representing the process noise and measurement noise. The matrices $[A]$ and $[C]$ are respectively the state space matrix and the output matrix. Along with this model, an observability matrix $\lfloor O_p \rfloor$ and a controllability matrix $\lfloor C_p \rfloor$ of order p are defined:

$$[O_p] = \begin{bmatrix} [C] \\ [C][A] \\ \vdots \\ [C][A]^{p-1} \end{bmatrix}; [C_p] = \begin{bmatrix} [G] & [A][G] & \cdots & [A]^{p-1}[G] \end{bmatrix} \qquad (15)$$

where $[G] = E\left[\{x_{k+1}\}\{y_k\}^T\right]$. $E[.]$ denotes the expected value operator. The matrices $[O_F]$ and $[C_F]$ are assumed to be of rank n. The problem considered here is the estimation of the matrices $[A]$ and $[C]$ in equation (14), up to a similarity transformation, using only the output measurements $\{y_k\}$. This identification problem is also known as the stochastic realization problem (Van Overschee and De Moor, 1996; Desai, *et al.*, 1985).

Key element in the approach is the establishment and decomposition of a block Hankel matrix of the measured correlations. Defining the empirical correlation matrix of the measured output vector

$$[R_k] = \frac{1}{M}\sum_{t=0}^{M}\{y_{t+k}\}\{y_t\}^T \qquad (16)$$

where M is the number of observations, the following block Hankel matrix can be defined and decomposed into its singular values;

$$[H_{p,p}] = \begin{bmatrix} [R_1] & [R_2] & \cdots & [R_p] \\ [R_2] & [R_3] & \cdots & [R_{p+1}] \\ \vdots & \vdots & \ddots & \vdots \\ [R_p] & [R_{p+1}] & \cdots & [R_{2p-1}] \end{bmatrix} = \qquad (17)$$

$$\begin{bmatrix} [U_1] & [U_2] \end{bmatrix} \begin{bmatrix} [S_1] & 0 \\ [0] & [S_2] \end{bmatrix} \begin{bmatrix} [V_1^T] \\ [V_2^T] \end{bmatrix}$$

where

$$S_1 = diag(\sigma_1 \cdots \sigma_n), \sigma_1 \geq \sigma_2 \cdots \sigma_n \geq 0$$

$$S_2 = diag\left(\sigma_{n+1} \cdots \sigma_{pN_{resp}}\right) \quad with \quad \sigma_{n+1} \rangle\rangle \sigma_n$$

p is a user defined parameter such that $p > 2N_m$. where N_m represents the number of physical modes. $[S_1]$ and $[U_1]$ contain respectively the n first singular values and the corresponding left singular vectors. From the stochastic realization theory, one can factor out $[H_{p,p}] = [\hat{O}_p][\hat{C}_p]$, yielding as an estimate of the observability matrix $[\hat{O}_p] = [U_1][S_1]^{1/2}$. The system matrices $[A]$ and $[C]$ are then estimated up to similarity transformation, using the shift structure of $[\hat{O}_p]$ (Basseville, *et al.*, 2001).

The dynamics of the system are completely characterized by the eigenvalues and the observed parts of the eigenvectors of the $[A]$ matrix. Note that the extracted mode shapes cannot be mass-normalized as this requires knowing the input force.

Formulations using the LSCF and Maximum Likelihood approach to output-only data have also been proposed (Hermans et al., 1998b). A full review of stochastic system identification methods for modal analysis is presented in Peeters and De Roeck (2001).

Figure 21 shows vibration data measured on Ariane flight 501, during and shortly after launch. The launcher and internal components such as fuel tanks were instrumented with over 100 accelerometers. All data were sent by telemetry to a ground station.

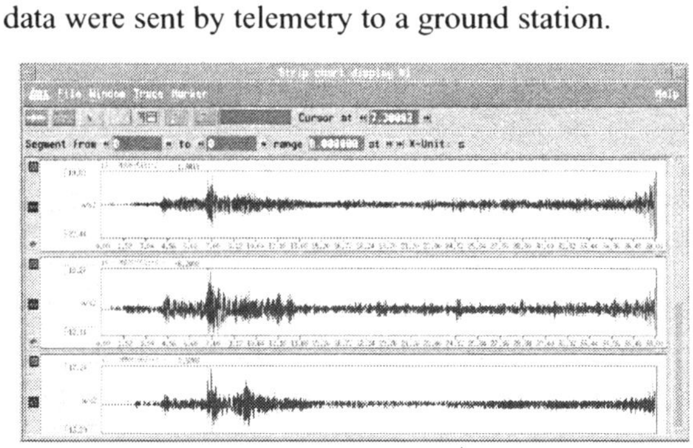

Figure 21: Ariane Flight 501 vibration data.

Modal parameters were estimated in different flight segments, allowing to assess the evolution of resonance frequencies as a function of the fuel amount, Fig. 22 (Hermans, *et al.*, 1998a).

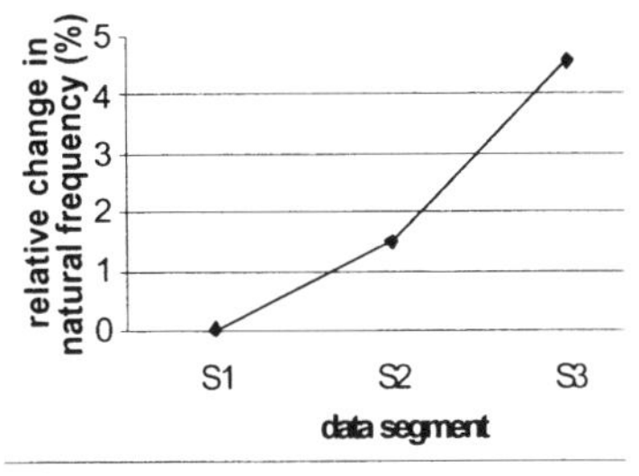

Figure 22: Resonance evolution over 3 segments

Figure 23 compares helicopter mode-shapes, resulting from ground vibration tests as well as from in-flight tests under different conditions (cruising speeds, climbing, descent, hover). Though the flight data are of somewhat lower quality, identical mode-shapes are identified (Hermans, *et al.*, 1999).

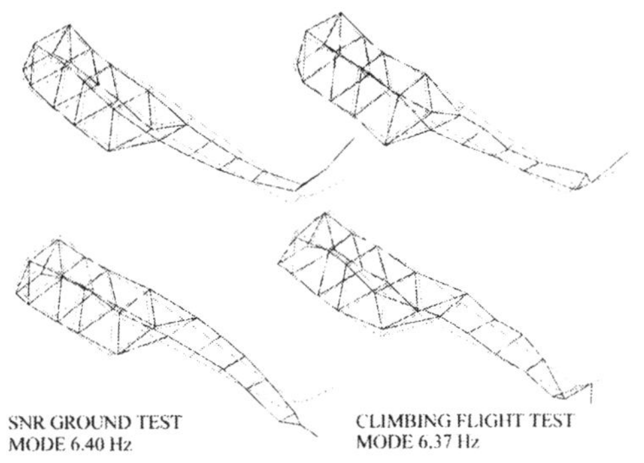

Figure 23: Helicopter mode shapes (ground and flight tests)

5.6 Automatic Modal Analysis

Due to the high system orders, the high degree of coupling between the modes and the large variation in modal responses per output, modal parameter estimation for structural dynamics remains a very interactive and heuristic process. Each major car or aircraft manufacturer has its own group of "modal experts". With the increasing use of modal analysis as a standard tool by many, also less-experienced, users, the strong need is expressed to automate the process. Researched solutions include estimation methods that are much more robust with respect to the appearance of spurious poles, discrimination methods to distinguish physical from mathematical poles and fully automated, self-tuning, algorithms such as the maximum likelihood estimator. Research is performed on capturing the user knowledge by Neural Networks and on implementing the heuristic pole selection rules in automated procedures. An example is the automated modal identification done on each Space Shuttle after landing (Pappa, *et al.*, 1997). This topic will remain a major research area in the coming years (Verboven et al., 2001).

In many applications, structural identification is closely related to the problem of detecting changes in the system dynamics. An example is the flight qualification of aircraft, requiring the repeated in-flight modal analysis at different airspeeds. At each air-speed, resonance frequencies and damping ratios of critical modes are checked to verify the absence of aero-elastic instability (flutter). As during the change from one flight condition to the next one, the dynamics may change due to imminent flutter, the damping must be monitored continuously. Another example is the use of changes in the modal system model to detect structural damage (e.g. in the Shuttle survey's, or in civil structures) or in the assessment of the integrity of a structure after the forced loading during a qualification test. Automatic modal analysis as well as statistical-test methods are pursued (Basseville, 2000).

CONCLUSIONS

Experimental Modal Analysis has evolved into a standard tool for structural dynamics analysis and optimization. It is based on a "black-box" system identification approach, but several specific elements (large number of outputs, high system orders...) distinguish this problem from the classical one encountered in electrical engineering. The research area is very dynamic, with a focus on performance improvement, test cost reduction and the changing role of testing in the modern, virtual prototype based, design process (Van der Auweraer, 2002). Many new system identification techniques are studied, but the main interest of the modal analysis community is to critically evaluate the added value of each method for addressing their specific problems. Automation of the process is the first research priority.

ACKNOWLEDGMENT

This research is conducted in the context of the EUREKA project 2419, FLITE. The financial support of the Flemish Institute for the Improvement of the Scientific and Technological Research in Industry (IWT) is gratefully acknowledged.

REFERENCES

Baseville, M., Abdelghani, M. and A. Benveniste (2000). Subspace-based fault detection algorithms for vibration monitoring. *Automatica*, **36**(1), 101-109.

Basseville, M., Benveniste, A., Goursat, M., Hermans, L., Mevel, L. and H. Van der Auweraer (2001). Output-Only Subspace-Based Structural Identification: From Theory to Industrial Testing Practice. *ASME Journal of Dynamic Systems, Measurement, and Control*, **123**(4), 668-676.

Busturia, J. and J. Gimenez (1985). Multiexcitation multiresponse non-linear least squares algorithm. *Proceedings of 10th ISMA*, Leuven, Belgium.

Cooper, J. and M. Deforges (1997). Modal parameter identification from ambient response data using advanced identification methods. *Proceedings of DAMAS '97*, Sheffield (UK), 309-318.

Craig, R. (1981). *Structural Dynamics*, John Wiley & Sons.

De Cuyper, J., Coppens, D., Liefooghe, C., Swevers, J., Verhaegen, M. (1998). Advanced Drive File Development Methods for Improved Service Load Simulation on Multi Axial Durability Test Rigs, *Proceedings Int. Acoustics and Vibration Asia Conf.*, Singapore, 339-354.

Desai, U., Debajyoti, P. and R. Kirkpatrick (1985). A realization approach to stochastic model reduction. *International Journal of Control*, **42**(4), 821-838.

Freymann, R., Honsberg, W., Winter, F. and H. Steinbichler (1996). Holographic modal analysis. In: *Laser in Research and Engineering*, Springer Verlag Berlin, 530-542.

Guillaume, P., Pintelon R., and J. Schoukens (1995). Robust parametric transfer function estimation using complex logarithmic frequency response data. *IEEE Transactions on Automatic Control*, **40**(7), 1180-1190.

Guillaume, P., Verboven, P. and S. Vanlanduit (1998). Frequency domain maximum likelihood identification of modal parameters with confidence intervals. *Proceedings of ISMA23*, Leuven, Belgium, 359-366.

Hermans, L., Van der Auweraer, H., Benveniste, A., Goursat, M., Haerens, D., Mourey, P., (1998a). Estimation of in-flight structural dynamics models of a spacecraft launcher. *Proceedings of ISMA23*, Leuven, Belgium, 427-433.

Hermans, L., Van der Auweraer, H., and Guillaume, P., (1998b). A frequency-domain maximum likelihood approach for the extraction of modal parameters from output-only data, *Proceedings of ISMA23*, Leuven, Belgium, 367-376.

Hermans, L. and H. Van der Auweraer (1999). Modal testing and analysis of structures under operational conditions: industrial applications. *Mechanical Systems and Signal Processing*, **13**(2), 193-216.

Hermans, L., Van der Auweraer, H., Hatami, A., Cooper, J., Uhl, T., Lisowski, W., Wasilak, A., (1999). In-flight modal testing and analysis of a helicopter. *Proceedings of IMAC17*, 80-86.

Heylen, W., Lammens, S. and P. Sas (1995). *Modal Analysis Theory and Testing*, Society of Experimental Mechanics.

James III, G., Carne, T. and J. Laufer (1995). The natural excitation technique (NExT) for modal parameter extraction from operating structures. *Int. J. of Analytical and Experimental Modal Analysis*, **10**(4), 260-277.

Juang, J.-N. and R. Pappa (1985). An eigensystem realization algorithm for modal parameter identification and reduction. *Journal of Guidance, Control and Dynamics*, **8**(5), 620-627.

Leuridan, J. (1984). *Some Direct Parameter Model Identification Methods Applicable for Multiple Input Modal Analysis*, PhD dissertation, Univ. of Cincinnati, USA.

Pappa, R., James III, G. and D. Zimmerman (1997). Autonomous model identification of the space shuttle tail rudder. *Proceedings of ASME Design Engineering Technical Conf.*, DETC97/VIB-4250.

Peeters, B. and G. De Roeck (2001). Stochastic system identification for operational modal analysis: a review. *ASME Journal of Dynamic Systems, Measurement, and Control*, **123**(4), 2001, 659-667.

Peeters, B., Van der Auweraer H. and P. Guillaume (2002a). The integration of operational modal analysis in vibration qualification testing. *Proceedings of IMAC 20*, Los Angeles, CA, USA, 977-983.

Peeters, B., De Roeck, G., Caetano, E. and A. Cunha (2002b). Dynamic study of the Vasco da Gama Bridge. *Proceedings of ISMA 2002*, Leuven, Belgium, 545-554.

Peeters, B., Couvreur, G., Razinkov, O., Kündig, C., Van der Auweraer, H. and G. De Roeck (2003). Continuous monitoring of the Øresund Bridge: system and data analysis. *Proceedings of IMAC 21*, Kissimmee, FL, USA.

Schoukens, J. and R. Pintelon (1991). *Identification of linear systems: a practical guideline to accurate modeling*. Pergamon Press.

Schoukens, J., Dobrowiecki, T., and R. Pintelon, R. (1998). Identification of linear systems in the presence of non-linear distortions. A frequency domain approach. *IEEE Trans. on Automatic Control*, **43**(2), 176-190.

Van der Auweraer, H. and J. Leuridan (1987). Multiple input orthogonal polynominal parameter estimation. *Mechanical Systems and Signal Processing*, **1**(3), 259-272.

Van der Auweraer, H. (2002). Requirements and opportunities for structural testing in view of hybrid and virtual modelling, *Proceedings of ISMA25*, Leuven, Belgium.

Van Overschee, P. and B. De Moor (1996). *Subspace identification for linear systems: theory, implementation, applications*. Kluwer Academic Publishers, Dordrecht.

Verboven, P., Guillaume, P., and M. Van Overmeire (1998). Modal parameter identification: estimation of starting values for MLE-like algorithms. *Proceedings of ISMA23*, Leuven, Belgium, 409-418.

Verboven, P., Parloo, E., Guillaume, P., Van Overmeire, M. (2001). Autonomous Modal Parameter Identification based on a Statistical Frequency-Domain Maximum Likelihood Approach, *Proc. of IMAC19*, 1511-1517.

Vold, H., Kundrat, J., Rocklin, T. and R. Russel (1982). A multi-input modal parameter estimation algorithm for mini-computers. SAE paper 820194, *Trans. SAE*, **91**(1), 815-821.

Wyckaert, K., Van der Auweraer, H. and W. Hendricx (1994). Correlation of acoustical modal analysis with operating data for road noise problems. *Proceedings 3rd Int.l Congress on Air- and Structure-Borne Sound and Vibration*, Montreal (CND), 931-940.

Wyckaert, K., Augusztinowicz, F. and P. Sas (1996). Vibro-acoustical modal analysis: Reciprocity, model symmetry and model validity. *J. Ac. Soc. Am.*, 3172-3181.

IFAC Publications
www.elsevier.com/locate/ifac

A PERSONAL VIEW ON THE DEVELOPMENT OF SYSTEM IDENTIFICATION [1]

Michel Gevers *

* *Center for Systems Engineering and Applied Mechanics (CESAME)*
Université Catholique de Louvain
Bâtiment Euler, 4 Av. Georges Lemaître,
B-1348 Louvain-la-Neuve, Belgium
Phone: +32-10-472590
Email: gevers@csam.ucl.ac.be

Abstract: This paper presents the author's personal view on the development of identification theory in the control community, starting from the year 1965. We show how two landmark papers, (Ho and Kalman, 1965) and (Åström and Bohlin, 1965), gave birth to two main streams of research that have dominated the development of system identification over the last fourty years. The Ho-Kalman paper, which gave a first solution to state-space realization theory, led to stochastic realization, and much later to subspace identification. The Åström-Bohlin paper laid the foundations for Maximum Likelihood methods based on parametric input-output models, which later became known as the highly successful Prediction Error Identification framework. The present paper also shows how the thinking in the identification community moved from a search for the "true system" to the formulation of identification as an approximation problem. This led to the view of identification as a design problem, in which the ultimate use of the model plays a paramount role in the formulation of the experiment design and in the choice of the identification criterion. *Copyright © 2003 IFAC*

Keywords: Identification, history

1. INTRODUCTION

The development of identification theory in the control literature followed on the heels of the development of model-based control design around 1960. Up until the late 1950's, much of control design relied on Bode, Nyquist or Ziegler-Nichols charts, or on step response analyses. These techniques were limited to control design for single-input single-output (SISO) systems. Around 1960, Kalman introduced the state-space representation and laid the foundations for state-space based optimal filtering and optimal control theory, with Linear Quadratic (LQ) optimal control as a cornerstone for model-based control design.

It is the availability of these model-based control design techniques that put pressure on the scientific community to extend the fields of application of "modern" control design beyond the realm of mechanical, electrical and aerospace applications, for which reliable models were easily available. Thus the need arose to develop data-based techniques that would allow one to develop dynamical models for such diverse fields as process control, environmental systems, biological and biomedical systems, transportation systems, etc.

Much of the early work on identification was developed by the statistics, econometrics and time series

[1] This paper presents research results of the Belgian Programme on Interuniversity Poles of Attraction, initiated by the Belgian State, Prime Minister's Office for Science, Technology and Culture. The European Commission is herewith also acknowledged for its financial support via the Program Training and Mobility of Researchers (TMR) and Project System Identification (ERB FMRX CT98 0206) to the European Research Network System Identification (ERNSI). The scientific responsibility rests with its author.

community. Even though the statistical theory of parameter estimation has its roots in the work of Gauss (1809) and Fisher (1912), most of the theory of stationary stochastic processes was developed during the period 1920 to 1970. We shall not describe this work here, because we want to focus on the engineering views and developments of system identification. An excellent review of the history of system identification and time series analysis in the statistics community can be found in (Deistler, 2002).

Even though a lot of results had already been established in the statistics and econometrics literature, one can view 1965 as the birthyear for identification theory in the control community, with the publication of two seminal papers, (Ho and Kalman, 1965) and (Åström and Bohlin, 1965). These two papers paved the way for the development of the two mainstream identification techniques that still dominate the field today: subspace identification and prediction error identification. The former is based on projection techniques in Euclidean space, the latter is based on the minimization of a parameter dependent criterion.

The Ho-Kalman paper provided the first solution to the determination of a minimal state-space representation from impulse response data. The solution of this deterministic realization problem was later extended by Akaike (Akaike, 1974) and others to stochastic realization, where a Markovian model is obtained for a purely random process on the basis of covariance data. This technology, based on canonical correlation analysis, was extended in the early nineties to processes that also contain a measurable (control) input, and became known as subspace state-space identification.

The Åström-Bohlin paper introduced into the control community the Maximum Likelihood framework that had been developed by time series analysts for the estimation of the parameters of difference equation models. These were known in the statistical literature by such esoteric names as ARMA (AutoRegressive Moving Average) or ARMAX model (AutoRegressive Moving Average with eXogeneous inputs). These models, and the Maximum Likelihood framework, were there to stay, since they gave rise to the immensely successful Prediction Error Identification framework.

In 1970, Box and Jenkins published their book "Time series analysis, forecasting and control" (Box and Jenkins, 1970), which gave a major impetus to applications of identification. Indeed, the book gave a rather complete recipe for identification, all the way from initial data analysis to the estimation of a model. In the spirit of the time series analysis methods of the time, it relied a lot on correlation analysis for the determination of model structure. For about 15 years, it remained the major high quality reference book on system identification. Other important references of the time were the survey paper (Åström and Eykhoff, 1971) and the special issue on system identification and time series analysis published by the IEEE Transactions on Automatic Control in December 1974. The Åström and Eykhoff survey was to be used by many young researchers of the time as a stepping stone for future work. It explained the state of the art as much as it displayed some of the important open questions of the time. One of these was the identification of closed-loop systems, for which the Hankel-based projection methods (based on cross-correlation information) had been shown to fail.

From about the middle of the seventies, the prediction error framework completely dominated identification theory and, perhaps more importantly, identification applications. Much of the research activity focused on identifiability problems for both multivariable systems and closed-loop systems. Just about all of the activity at that time was focused on the search for the "true system", i.e. it dealt with questions of identifiability, convergence to the "true parameters", and asymptotic normality of the estimated parameters.

Around 1978 the first attempts were made to view system identification as an approximation theory, in which one searches for the best possible approximation of the "true system" within some model class (Anderson *et al.*, 1978; Ljung and Caines, 1979). The prevailing view in the identification community changed consequently from a search for the "true system" to a search for and characterization of the "best approximation". Hence, the characterization of the model errors (bias error and variance error) became the focal point of research. For control engineers, the object of primary interest is the model, in particular the transfer function model, rather than the parameters which are just a vehicle for the description of this model. As it turns out, the research on bias and variance error moved remarkably swiftly from the characterization of parameter errors to that of transfer function errors, thanks to some remarkable analysis of Ljung based on the idea of letting the model order go to infinity (Ljung, 1985; Wahlberg and Ljung, 1986).

The work on bias and variance analysis of identified models of the eighties then led, almost naturally, to a new perspective in which identification became viewed as a "design problem". With an understanding of the effect of the experimental conditions, the choice of model structure, the choice of criterion, etc on the quality of the identified model, one can tune these design variables towards the objective for which the model is being identified (Gevers and Ljung, 1986). The book "System identification: Theory for the user" (Ljung, 1987) has had a profound impact on the engineering community of system identifiers. It squarely put forward the view of system identification as a design problem, in which the model use plays a central role. This viewpoint clearly distinguishes the field from the statistical literature on system identification and time series analysis, where the prevailing view is that the model must "explain" the data as best as possible.

The observation that the model quality could be influenced, through the choice of appropriate design variables, towards the eventual objective for which the model was being built opened the way to a flood of new activity that took place in the nineties and continues up to this day. The major application of this new paradigm went to the situation where a model is being built with the view of designing a model-based controller. Thus, identification for control has blossomed, since its early beginnings around 1990. Because that topic embraces many aspects of identification and robust control theory, it has also opened or reopened new research interest in areas such as experiment design, closed-loop identification, frequency domain identification, uncertainty estimation, and data-based robust control analysis and design.

The present paper attempts to exhibit both the continuity and the motivation for the developments that took place in system identification in the last fourty years, and also the significant new departures and insights that came as the result of some important breakthroughs.

2. THE MILESTONE PAPERS

2.1 *Deterministic realization theory*

In 1965, (Ho and Kalman, 1965) provided a first solution to a challenging system theoretical problem that became known as the state-space realization problem. It can be stated as follows:

How to construct a minimal state-space realization

$$\begin{cases} x_{t+1} &= Ax_t + Bu_t \\ y_t &= Cx_t \end{cases}$$

for an input-output model described by its impulse response matrices (also called Markov parameters) H_k

$$y_t = \sum_{k=1}^{\infty} H_k y_{t-k}.$$

The problem is to replace the infinite description

$$H(z) = \sum_{k=1}^{\infty} H_k z^{-k}$$

with $H_k \in \mathbb{R}^{p\times m}$, by a finite description A, B, C, with $A \in \mathbb{R}^{n\times n}, B \in \mathbb{R}^{n\times m}, C \in \mathbb{R}^{p\times n}$, in such a way that

$$H(z) = C(zI - A)^{-1}B$$

with $\dim(A)$ minimal. This problem can be split up in two parts: (i) find the McMillan degree of $H(z)$, which is then the minimal dimension of A; (ii) compute the matrices A, B, C. The key tool for the solution of this problem is the Hankel matrix, and its factorization into the product of an infinite observability matrix times an infinite controllability matrix:

$$\mathcal{H} = \begin{bmatrix} H_1 & H_2 & H_3 & H_4 & \dots \\ H_2 & H_3 & H_4 & H_5 & \dots \\ H_3 & H_4 & H_5 & H_6 & \dots \\ \vdots & \vdots & \vdots & \vdots & \end{bmatrix} = \begin{bmatrix} C \\ CA \\ CA^2 \\ \vdots \end{bmatrix} \begin{bmatrix} B & AB & A^2B & \dots \end{bmatrix}$$

If the McMillan degree of $H(z)$ is n, then

(1) rank $\mathcal{H} = n$
(2) $\exists\ A, B, C$ such that $H_k = CA^{k-1}B$ with $A \in \mathbb{R}^{n\times n}, B \in \mathbb{R}^{n\times m}, C \in \mathbb{R}^{p\times n}$.

It took years of research to go from the theoretical results described in (Ho and Kalman, 1965) to a numerically reliable realization algorithm. However, all the key insights were present in the 1965 paper, and they were to have a profound impact on linear system theory, and on realization and identification theory.

2.2 *The Maximum Likelihood framework*

In complete contrast to the state-space formulation proposed by Ho and Kalman, the landmark 1965 paper of Åström and Bohlin (Åström and Bohlin, 1965) introduced the Maximum Likelihood method for the estimation of the parameters of input-output models in ARMAX form:

$$A(z^{-1})y_t = B(z^{-1})u_t + \lambda C(z^{-1})e_t$$

where $\{e_t\}$ is a sequence of independent identically distributed Normal $(0, 1)$ random variables. The Maximum Likelihood (ML) method was of course well known and had been widely studied in mathematical statistics, including in its application to a range of time series models. However, what is remarkable about the Åström-Bohlin paper is that the authors not only gave a complete algorithmic derivation of the ML identification method for ARMAX models, but also presented all analysis results that were available at that time, such as the consistency, asymptotic efficiency and asymptotic normality of the parameter estimates, the persistence of excitation conditions on the input signal in connection with the order of the model, the model order validation on the basis of the whiteness of the residuals, etc.

The concepts and notations introduced by Åström and Bohlin in 1965 have been with us for almost 40 years now. Indeed, the following household notations of the identification community can all be found in this milestone paper:

- the residuals $C(z^{-1})\varepsilon_t = A(z^{-1})y_t - B(z^{-1})u_t$

- the cost criterion $V(\theta) = \frac{1}{2}\sum_{t=1}^{N} \varepsilon_t^2$
- the parameter estimate $\hat{\theta} = \arg\min V(\theta)$
- the white noise variance estimate $\hat{\lambda}^2 = \frac{2}{N}V(\hat{\theta})$.

The publication of (Åström and Bohlin, 1965) gave rise to a flurry of activity in parametric identification. It also established the basis for the adoption of the Prediction Error framework. The step from Maximum Likelihood to Prediction Error essentially consists in observing that, under an assumption of white Gaussian noise in the ARMAX model, the maximization of the likelihood function of the observations is equivalent to the minimization of the sum of the prediction errors. The Prediction Error framework then consists in adopting the minimization of a norm of the prediction errors as a reasonable criterion for parameter estimation, even in the absence of any known probability distribution for the observations. Such suggestion had already been made by Mr. Gauss himself (Gauss, 1809), as observed in the fascinating paper (Åström, 1980).

3. FROM DETERMINISTIC TO STOCHASTIC REALIZATION THEORY

The combination of the deterministic realization theory based on the factorization of the Hankel matrix, and of the theory of Markovian and innovations representations gave rise to the stochastic theory of minimal realizations. The stochastic realization problem can be stated as follows:

Given a covariance sequence $R_k \triangleq E\{y_t y_{t-k}^T\}$, $k = 1, 2, \ldots$ *of a zero-mean stochastic process* $\{y_t\}$, *find a minimal Markovian representation for the process* $\{y_t\}$:

$$\begin{cases} x_{t+1} &= Ax_t + Gw_t \\ y_t &= Cx_t + v_t \end{cases} \tag{1}$$

where $\begin{pmatrix} w_t \\ v_t \end{pmatrix}$ *is a zero-mean white noise sequence with covariance matrix*

$$W = E\left\{\begin{pmatrix} w_t \\ v_t \end{pmatrix}\begin{pmatrix} w_t \\ v_t \end{pmatrix}^T\right\} = \begin{pmatrix} Q & S \\ S^T & R \end{pmatrix},$$

i.e. find the state-space matrices $\{A, G, C\}$ *with* $n = dim(A)$ *minimal, and the elements* Q, S, R *of the covariance matrix W such that the covariance of the output of (1) is exactly* R_k.

Observe that the covariance of the output of the Markovian representation (1) is given by $R_k = CA^{k-1}N$ with $N = A\Pi C^T + GS$ for $k > 0$, and $R_0 = C\Pi C^T + R$, where Π is the state covariance: $\Pi \triangleq E\{x_t x_t^T\}$.

The stochastic realization problem was studied very intensively during the early seventies in connection with innovations theory and spectral factorization theory: (Akaike, 1974; Gevers and Kailath, 1973; Faurre, 1976). The first step of the solution consists in observing that the Hankel matrix made up of the covariance sequence can be factored as

$$\mathcal{H} = \begin{bmatrix} R_1 & R_2 & R_3 & \ldots \\ R_2 & R_3 & R_4 & \ldots \\ R_3 & R_4 & R_5 & \ldots \\ \vdots & \vdots & \vdots & \end{bmatrix}$$

$$= \begin{bmatrix} C \\ CA \\ CA^2 \\ \vdots \end{bmatrix} \begin{bmatrix} N & AN & A^2N & A^3N \ldots \end{bmatrix}$$

where $R_k = CA^{k-1}N$ with N defined as above. The Ho-Kalman algorithm mentioned above allows one to determine the minimal dimension n as the rank of $\mathcal{H}$, and to compute the matrices C, A and N from the factorization of $\mathcal{H}$.

There are various ways of performing the second step, which consists in computing the missing elements of the Markovian representation (1) from C, A, N and the output variance R_0. One way is to use a particular version of the Markovian representation called an *innovations model*:

$$\begin{cases} \xi_{t+1} &= A\xi_t + K\varepsilon_t \\ y_t &= C\xi_t + \varepsilon_t \end{cases} \tag{2}$$

where $\{\varepsilon_t\}$ is a white noise sequence with covariance matrix $\Sigma = E\{\varepsilon_t\varepsilon_t^T\}$. Denoting $\hat{\Pi} = E\{\xi_t\xi_t^T\}$ the covariance of the state of this model, and imposing that the covariance sequence of the output of this innovations model (2) is $\{R_k\}$, one gets the following three constraints on the unknown quantities $\hat{\Pi}$, K and Σ:

$$\begin{cases} \hat{\Pi} &= A\hat{\Pi}A^T + K\Sigma K^T \\ N &= A\hat{\Pi}C^T + K\Sigma \\ R_0 &= C\hat{\Pi}C^T + \Sigma \end{cases}$$

Observe that the first (Lyapunov) equation follows directly from the state equation of the Markovian model, while the other two constraints are imposed by the matching of the output covariance R_k: see above. The solution of this second step is obtained via an associated Riccati equation. A complete and tutorial presentation of the solution, including the stochastic realization of minimal ARMA models from covariance sequences, can be found in (Gevers and Wouters, 1978).

An interesting aspect of the solution of the stochastic realization problem given by Akaike was the definition of the state of the innovations model as the set of canonical correlations obtained by projecting the space of future outputs onto the space of past outputs (Akaike, 1975). This insight formed the basis for the later work on *subspace identification*; it also gave rise to extensive studies of the interface between past and future observation spaces (Katayama and Picci, 1999).

Another important outcome of the stochastic realization and innovations theories of the seventies was the covariance equivalence established between the Markovian realization (1) and its innovations realization (2). In particular, this means that, even if a Markovian model (1) has been constructed as a *physical model* for y_t, driven by two independent white noise sources w_t and v_t (i.e. $S = 0$, say), it can be rewritten as the Markovian innovations model (2) driven by a unique white noise source ε_t with the same output covariance $\{R_k\}$. The same equivalence applies to models with deterministic inputs. This means that any (physical) state-space model

$$\begin{cases} x_{t+1} &= Ax_t + Bu_t + Gw_t \\ y_t &= Cx_t + v_t \end{cases}$$

can be rewritten as an *innovations model* for y_t driven by a unique noise source ε_t:

$$\begin{cases} \xi_{t+1} &= A\xi_t + Bu_t + K\varepsilon_t \\ y_t &= C\xi_t + \varepsilon_t \end{cases}$$

The input-output equation of this state space innovations model is:

$$\begin{aligned} y_t &= C(zI - A)^{-1}Bu_t + [C(zI - A)^{-1}K + I]\varepsilon_t \\ &= G(z)u_t + H(z)\varepsilon_t \end{aligned}$$

where

$$G(z) = C(zI-A)^{-1}B,\ H(z) = I+C(zI-A)^{-1}K.$$

Observe that the transfer functions $G(z)$ and $H(z)$ have the same poles, and that $H(z)$ is monic. Thus, this model is equivalent to an ARMAX model:

$$A(z^{-1})y_k = B(z^{-1})u_k + C(z^{-1})\varepsilon_k.$$

This theory established a nice link between Markovian models obtained from 'first principles modeling', their corresponding state-space innovations models, and the ARMAX input-output models used in Maximum Likelihood (ML) or Prediction Error (PE) identification. It also gave a solid theoretical justification to the use of ARMAX models for the representation of stationary linear Markov processes, whether these processes are driven by one or several noise sources.

4. THE GOLDEN YEARS: 1975-1985

4.1 *The frantic years and the big cleanup*

The years 1975 to 1985 saw a frantic activity in system identification in the engineering community. The parametric input/ouput methods, for which asymptotic efficiency results existed, completely took over the field, in large part because increased computer speed and the development of special purpose identification software made it more and more feasible to perform the iterative minimization of a cost criterion over a range of possible model structures. This is the period where many authors were trying to put their name on a new combination of model structure and method, with the inevitable claim about the supremacy of their new combination over existing methods. New "methods" were pouring out constantly in the scientific journals. Other authors were making comparisons among the flurry of such methods through simulations, with titles that were typically reading like 'Comparison of six online identification methods'.

Some solid cleaning work was required, and it was one of L. Ljung's major contributions to perform this cleaning work. It consisted of clearly separating two independent concepts: the choice of a parametric model structure, which was seen as just a vehicle for computing predictions and hence parameter dependent prediction errors, and the choice of an identification criterion, which was to be chosen as some positive function of these prediction errors and hence of the parameter vector: see e.g. (Ljung, 1978). All existing parametric identification methods could then be seen as particular cases of this prediction error framework.

Thus Ljung introduced the generic input-output model structure

$$y_t = G(z,\theta)u_t + H(z,\theta)e_t \tag{3}$$

with $G(z,\theta)$ and $H(z,\theta)$ parametrized as rational transfer functions, and with e_t white noise. All commonly used model structures were special cases of the structure (3). From such model structure, one can derive the parameter-dependent one-step ahead prediction $\hat{y}_{t|t-1}(\theta)$, and hence the one-step ahead prediction error $\varepsilon_t(\theta) = y_t - \hat{y}_{t|t-1}(\theta)$. From a set of N data, Z^N, and hence of N prediction errors, one can then define a criterion

$$V_N(\theta, Z^N) = \frac{1}{N}\sum_{t=1}^{N} l(\varepsilon_t(\theta)), \tag{4}$$

where $l(.)$ is a positive scalar-valued function. The minimization of $V_N(\theta, Z^N)$ with respect to θ in some domain D_θ then yields the parameter estimate:

$$\hat{\theta}_N = \arg\min_{\theta\in\mathcal{D}_\theta} V_N(\theta, Z^N). \tag{5}$$

The methodological work of Ljung culminated with the publication of his book in 1987 (Ljung, 1987), which has become the standard reference book in system identification. His work was complemented by that of Stoica and Söderström, who adopted the same clear distinction between choice of model structure and choice of criterion; their book focused less on design issues, but more on analysis and on alternative criteria, in particular criteria based on correlation methods and instrumental variables (Söderström and Stoica, 1989).

4.2 *Breakthroughs for MIMO and closed-loop systems*

During that same period, some important theoretical breakthroughs were made in two directions. The first was the elucidation of the manifold structure of multi-input multi-output systems. The second consisted in a range of identifiability results for linear systems identified under closed-loop conditions.

Many authors contributed to the solution of both problems; too many to quote in this short paper. Let us just mention that the manifold structure of MIMO systems, together with the key role of the Kronecker (or structure) indices, was elucidated in (Hazewinkel and Kalman, 1976; Clark, 1976). Subsequently, many authors worked on methods for the estimation of these structure indices; others studied the relationship between the canonical (or pseudo-canonical) forms in state-space and in ARMA form (Deistler and Gevers, 1981; Van Overbeek and Ljung, 1982; Picci, 1982; Rissanen, 1982; Wertz *et al.*, 1982).

As for the identifiability of closed-loop systems, one of the earliest solutions was provided by the famous Swedish trio made up of Gustavsson, Ljung and Söderström[2], all PhD students of K.J. Åström at the time (Söderström *et al.*, 1976). They showed, among other things, that in many situations of practical interest, the direct application of a PE identification method to input-output data allows one to identify the open loop plant, despite the presence of a feedback controller. Other closed-loop identifiability results covered indirect methods (in which the closed-loop transfer function is identified first, and the plant model is then derived from it using knowledge of the controller) and the so-called 'joint input-output method' (Ng *et al.*, 1977; Gustavsson *et al.*, 1977; Sin and Goodwin, 1980; Anderson and Gevers, 1982)

4.3 *System identification viewed as approximation*

For most of the sixties and seventies, the prevailing assumption was that the system was in the model set: $\mathcal{S} \in \mathcal{M}$. Thus, the focus of research was on questions of convergence to the *true system* and of statistical efficiency of the parameter estimates. In the late seventies, the first attempts were made to view system identification in the context of *approximation* (Anderson *et al.*, 1978; Ljung and Caines, 1979). This marked the beginning of an entirely new era, in which the elusive search for a linear time-invariant 'true system' was progressively abandoned to give way to the search for a 'best approximate model' within some a priori chosen model set $\mathcal{M}$. With the idea of model approximation came of course the idea of model error, and hence the desire to characterize this model error.

4.4 *The birth of* θ^*

In statistics, the natural way to analyse estimation errors is through the concepts of bias and variance errors. However, in the context of model sets that do not contain the true system, the concept of parameter error becomes meaningless, since there are no 'true parameters'. The object of interest is the transfer function, not the parameters that are used to represent it. By observing that, under reasonable conditions, the parameter estimate $\hat{\theta}_N$ converges to a well-defined θ^*,

$$\lim_{N\to\infty} \hat{\theta}_N = \theta^* = \arg\min_{\theta\in\mathcal{D}_\theta} \lim_{N\to\infty} EV_N(\theta),$$

Ljung introduced the following decomposition of the total transfer function error, at some frequency ω:

$$G_0(e^{j\omega}) - G(e^{j\omega}, \hat{\theta}_N) = \underbrace{G_0(e^{j\omega}) - G(e^{j\omega}, \theta^*)}_{bias\ error} + \underbrace{G(e^{j\omega}, \theta^*) - G(e^{j\omega}, \hat{\theta}_N)}_{variance\ error}$$

Within this framework, he and his collaborators derived approximate asymptotic expressions for the transfer function variance, and integral expressions for the transfer function bias that were to become the cornerstone of the next major phase of the development of system identification: the design phase (Ljung, 1985; Wahlberg and Ljung, 1986).

5. IDENTIFICATION AS A DESIGN PROBLEM

If identification is viewed as approximation, then one knows and accepts that the model is erroneous. If the model is to be used for a specific purpose (as is most often the case), then perhaps one can construct a model in such a way that the model errors do not penalize too much the goal for which the model is being built. This is the whole idea behind goal-oriented identification. More precisely, if one can understand the connection between

- identification design (experiment design, choice of model structure, choice of criterion) and model quality on the one hand,
- the effect of model quality on the intended model application on the other hand,

then one can formulate the identification problem as a *goal-oriented design problem*.

The first few steps of this new paradigm were laid in (Gevers and Ljung, 1986; Wahlberg and Ljung, 1986; Ljung, 1987). This new (engineering) way of looking at the identification problem opened up a vast new window of opportunities for research. In particular, this viewpoint was instrumental in the development, from around 1990, of a field that has seen an enormous activity ever since, both on the theoretical front and in practical applications: *identification for control*.

6. SYSTEM IDENTIFICATION IN THE NINETIES

At the triennial IFAC Symposium on System Identification held in Budapest in 1991, there was a feeling

[2] Over a 4-year period they jointly published no less than 6 important papers on various aspects of system identification.

among many participants that most of the important problems in system identification had been solved and that the golden age of identification was over. This prediction proved to be wrong. The research on system identification was pulled all through the nineties essentially by two catalysts, whose first feeble signs emerged around 1990: subspace-based identification, and identification for control. In addition, important new research activity took place in frequency domain identification, in closed-loop identification, and in the development of a range of new methods for the quantification of model uncertainty.

6.1 *Subspace-based identification*

The reasons for the emergence of subspace identification are almost certainly to be found in the state of the art of identification of multivariable systems in the eighties. Even though the manifold structure of MIMO (multi-input multi-output) systems had been completely characterized in the late seventies, the practical problem of identifying MIMO systems remained wide open. Indeed, the estimation of the structure indices that characterize the parametrizations of multivariable systems remained very tricky, and led to ill-conditioned numerical procedures. Thus, there was a great incentive for the development of simple, albeit suboptimal procedures, based on the numerically robust Singular Value Decomposition and Least Squares techniques, that completely bypass the need for the estimation of structure indices. The development of subspace-based identification methods filled a much needed gap, because in that framework the handling of MIMO systems caused no additional difficulty.

A major difficulty was that the projection methods developed by Akaike, which were based on canonical correlation analysis, were not easily extendable to output data that contained, besides the stochastic components, also a contribution due to a measured input. In the early nineties, several research teams managed to crack this nut; they provided several closely related solutions to the problem (Larimore, 1990; Van Overschee and De Moor, 1994; Verhaegen, 1994; Viberg, 1995). These first few solutions opened the way for a lot of research on the properties of subspace-based identification, their connection with stochastic realization theory, and on improvements of the numerical procedures: see e.g. (Chui and Maciejowski, 1996; Katayama and Picci, 1999; Chiuso and Picci, 2002).

6.2 *Identification for control*

Identification for control has been the major outlet for the new paradigm of system identification as a design problem. The reasons for this are many: (i) in the systems and control community of system identification, control is very often the main motivation for model building; (ii) it has been observed that high performance control can often be achieved with very simple models, provided some basic dynamical features of the system are accurately reflected; (iii) a powerful robust control theory, based on models and uncertainty sets, had been developed all through the eighties but the models and uncertainty sets used were not data-based for lack of a proper theory; (iv) the identification for control research delivered iterative model and controller tuning tools that were intuitive, practical and easy to implement by the process engineers.

Whereas the building blocks for goal-oriented identification were laid around 1986, the first specific contributions in which identification and control design were looked upon as a combined design problem appeared only around 1990. The plenary (Gevers, 1991) at the 1991 IFAC Symposium on System Identification addressed many of the key issues about the interplay between the identification of a reduced order model and the design of a controller from such model; however, it was more an agenda for research than a presentation of solutions. Indeed, in 1990 there was very little understanding about the interplay between system identification and robust control. The two theories had been developed by two separate communities that had had very little contact with one another.

The robust control community had developed a robust analysis and design theory that was based on uncertainty descriptions which were not based on data but on prior assumptions. The identification community had delivered bias and variance error descriptions that were not explicit, and that were certainly not transferable at the time to the toolboxes used in robust control analysis and design. More importantly, neither community had given much attention to the interaction between model building and control design: what are the qualities that a model must possess (or, conversely, what are the plant-model errors that are acceptable) if the model is to be used for the design of a controller that must achieve a given level of performance on the plant?

In July 1992, the IEEE Transactions on Automatic Control devoted a special issue to 'system identification for robust control design'. In retrospect, and in keeping with the observation just made, such issue was perhaps premature, given the paucity of results that were available at that time. About half of the papers in that special issue did not really deal with identification for control, but with the estimation of uncertainty sets, without any account taken of control-oriented design issues. A few papers in that special issue did address the joint 'identification and control design' paradigm (Bayard *et al.*, 1992; Schrama, 1992), and they produced one of the first key results in identification for control: the necessity of an iterative scheme for the design of a control-oriented nominal model; this observation had in fact been first made in (Liu and Skelton, 1990)

The first half of the nineties produced a string of results on the design of control-oriented nominal models. These results were produced by different teams who used their favorite combinations of identification criteria and control design criteria (Schrama and Bosgra, 1993; Lee *et al.*, 1993; Åström and Nilsson, 1994; Zang *et al.*, 1995; de Callafon and Van den Hof, 1997). They confirmed the necessity of using an iterative scheme of model updates and controller updates, and they produced significant evidence about the advantages of performing the identification in closed loop, rather than in open loop, when the model is to be used for the design of a new controller. Of course, the closed-loop experimental conditions that produce a specific input signal spectrum can always be mimicked by an open loop experiment with the same input spectrum, but the advantage of the closed-loop experiment is that the appropriate input spectrum is automatically generated by the unknown system itself.

The iterative schemes of identification and control design had a remarkably fast transfer into the world of applications. There were two main reasons for this:

- whereas the industrial world was still living with the belief that one should 'open the loop' to perform a valid identification experiment, here was a new theory that showed the benefits of closed-loop identification; this came as welcome news to process control engineers who had never really liked the idea of opening the loop;
- in the process industry, thousands of measurements are flowing into the computer; here was a theory that showed how these data could be used for the design of a better controller.

The early work on identification for control focused on control-oriented identification criteria, i.e. on obtaining a nominal model whose bias error distribution was tuned for control design. This means that the nominal control performance obtained with the optimal controller computed from the nominal model is close to the actual control performance obtained with the same controller on the actual plant.

More recently, attention has shifted to the distribution of the variance error of the identified models, i.e. to the estimation of control-oriented uncertainty sets (Kosut and Anderson, 1994; Mäkilä *et al.*, 1995; de Vries and Van den Hof, 1995; Bombois *et al.*, 2000). The idea is that, since one can manipulate the shape of the model uncertainty set by the choice of the experimental conditions under which the new model is identified, one should attempt to obtain a model uncertainty set that is tuned for control design. Even though many new insights have been gained on the interplay between uncertainty sets estimated from data and corresponding sets of stabilizing controllers, there is at this point no clear view as to the most operational definition of a 'control-oriented uncertainty set'. One view is that the corresponding set of controllers achieving stability and the required performance with all models of that set should be as large as possible (Gevers *et al.*, 2003). Another view is that the worst-case performance achieved by some optimal robust controller with all models in this uncertainty set should be as close as possible to the performance achieved with the central (nominal) model.

The work on identification for control had two beneficial side effects:

- it triggered an enormous new research activity on the estimation of data-based uncertainty sets for identified models;
- it also reopened the debate and the activity on closed-loop identification.

6.3 *Quantification of model uncertainty*

The demands of robust control theory for adequate uncertainty sets triggered a lot of interest for the estimation of uncertainty sets from data. It is fair to say that most of the robust control theory developed in the eighties had been based on a priori assumed uncertainty bounds on model errors and on the noise. In the context of system identification, the estimation of model errors from data is of course of interest in its own right; indeed a reputable engineer should never deliver a product (a model, in this case) without a statement about its error margins. In this authors's opinion, this activity on estimation of uncertainty sets was erroneously put under the umbrella of 'identification for control'; indeed, in most of the work on data-based estimation of uncertainty sets of the last decade, the control objective is not taken into account in the identification design.

A wide range of new techniques were developed to provide error bounds on estimated models, in all kinds of shapes and frameworks, using time domain, frequency domain, H_∞, l_1, probabilistic, worst case, set membership and other methods (Helmicki *et al.*, 1991; Goodwin *et al.*, 1992; Poolla *et al.*, 1994; Smith and Dahleh, 1994; de Vries and Van den Hof, 1995; Kosut, 1995; Hakvoort and Van den Hof, 1997; Giarré *et al.*, 1997). Even though it is now a few years old, probably one of the best presentations of these different methods is the survey paper (Ninness and Goodwin, 1995).

6.4 *Closed-loop identification revisited*

For some reason, the work of the seventies on closed-loop identification had stopped at the question of identifiability, i.e. under what conditions do the parameter estimates converge to the 'true' parameters in the case where the system is in the model set ($\mathcal{S} \in \mathcal{M}$). The influence of the experimental conditions on bias error distribution in the case of restricted complexity models, as well as on asymptotic variance, had not been resolved.

One of the important lessons that emerged from the study of the interplay between identification and control is the benefit of closed-loop identification when the model is to be used for control design. Until the late eighties, it was commonly accepted that closed-loop identification was preferably to be avoided. In identification for control with reduced order models, the required connection between the control performance criterion (obviously a closed-loop criterion) and the identification criterion, established the need for closed-loop identification, as described above. In the ideal context of optimal experiment design with full order models, optimality of closed-loop identification was actually established when the model is to be used for control design with a noise rejection objective (Gevers and Ljung, 1986; Hjalmarsson *et al.*, 1996; Forssell and Ljung, 2000).

This observation triggered an important new activity in the design of special purpose closed-loop identification methods, the main goal pursued by these new methods being to obtain a better handle on the bias error in closed-loop identification (Hansen *et al.*, 1989; Van den Hof and Schrama, 1993; Van den Hof *et al.*, 1995; Forssell and Ljung, 1999).

6.5 *Frequency domain identification*

Another area of important activity in the nineties has been frequency domain identification. In the identification community, there has historically been no more than a polite attention paid to frequency domain identification. Every now and then, some of the experts in the area published a paper whose main aim was to reassure the community with a message that sounded essentially like 'yes, indeed, we can also handle frequency domain data', or 'whatever properties we have obtained with time domain data do also apply to models obtained from frequency domain data' (Ljung and Glover, 1981).

Things changed drastically at the end of the eighties, with a convergence of efforts arising from two completely different horizons.

- During the eighties, the robust control community had developed most of its analysis and design tools in the frequency domain; thus, there was a great demand for tools that would enable one to obtain frequency domain models and, even more importantly, frequency domain uncertainty descriptions from data. This led, at the end of the eighties, to the development of a number of new interpolation techniques that were using noisy pointwise frequency domain transfer function measurements as their data (Parker and Bitmead, 1987; Partington, 1991; Gu and Khargonekar, 1992);
- At about the same time, Pintelon and Schoukens had independently developed frequency domain identification techniques, essentially based on the Maximum Likelihood principle. With their instrumentation and measurement background, they were interested in methods that would deliver reliable models for devices under test, through the application of short input data sequences (Pintelon and Schoukens, 1990).

Aware of the interest for frequency domain identification emanating from the robust control community, Schoukens and Pintelon continued their work on frequency domain identification all through the nineties, with specific and important contributions on the use of periodic excitation and of maximally informative input signals. Their recent book covers almost everything one always wanted to know about frequency domain identification (Pintelon and Schoukens, 2001).

6.6 *Some other areas of activity in the last decade*

The areas mentioned above have been areas of major activity in the nineties. With the exception of subspace-based identification, most of these areas have had their origin in the new engineering view of identification as a design problem. This has created a lot of interest in all aspects of *identification for control*, which turned out to require better methods for the quantification of uncertainty, a better understanding of the bias issues in closed-loop identification and the development of frequency domain identification methods that were as much as possible compatible with frequency domain based robust control analysis tools. But these were by no means the only areas of activity in system identification.

• Interesting progress has been made on the use of alternative basis functions (other than the shift operator) for the representation of input-output models, such as Laguerre, Kautz, and other generalized orthonormal basis functions. Such alternative bases can not only lead to more compact descriptions when some prior knowledge is available about the system, but they have also led to improved formulas for the estimation of the variance of black box transfer function models (Heuberger *et al.*, 1995; Ninness *et al.*, 1999*a*; Ninness *et al.*, 1999*b*).

• One other important byproduct of the approach to system identification as a design problem has been a renewed interest in experiment design issues. This subject had been very active in the early years of research in system identification: see e.g. (Goodwin and Payne, 1977). The work of the seventies, which led to the formulation of a number of optimal design criteria and results for the estimation of the parameters of dynamical systems, ground to a halt for about fifteen years. The new paradigm on identification as a design problem gave it a new lease of life, and the recent work on the connection between model uncertainty sets obtained by identification and corresponding sets of robust controllers have clearly put this subject in the limelight again (Gevers *et al.*, 1999; Forssell and

Ljung, 2000; Hildebrand and Gevers, 2003). Experiment design is expected to be a major subject of activity for the next few years.

• Finally, this overview of recent activity would not be complete without a mention of the topic of identification of nonlinear systems. As is very nicely stated in (Deistler, 2002), *"identification of nonlinear systems" is like a statement about "non-elephant zoology"*. Indeed, the area is vast, and the approaches are many, ranging from semi-parametric identification using, e.g, neural network methods, to the development of ad hoc techniques for specific classes of nonlinear systems such as, e.g. compartmental models. The paper (Sjöberg *et al.*, 1995), even though no longer recent, is still a very valuable overview and introduction to the subject.

7. CONCLUDING REMARKS

In this paper I have presented my personal views on the development of system identification in the control community, as I have observed it over a period of thirty years, both as a student of the subject eager to learn and understand the work of my colleagues, and as an active participant in these developments. As a result, this paper should not be read as a survey of the subject, but rather as a story told with the eyes and the prejudices of one of the actors in the field. I tend to believe that the way a particular field of science develops depends on a combination of two forces: the socio-technical environment created by the evolution of the neighbouring fields of science and by the demands of the applications world, and the creative role played by a few individuals who suddenly make it possible to venture into a totally new direction or to establish a useful link with another field of science that sheds a totally new light on the subject. During these thirty years, it has helped me a great deal to see the evolution of system identification in this socio-historical perspective. The main reason for writing this 'story' is to share my experience with others, in the hope that it might also help them, particularly the newcomers to the field. As for the senior members of the community who do not fully recognize their own contributions in this 'personal view', let me reassure them about the respect I have for their work; they will almost surely be part of somebody else's story.

8. REFERENCES

Akaike, H. (1974). Stochastic theory of minimal realization. *IEEE Trans. Automatic Control* **26**, 667–673.

Akaike, H. (1975). Markovian representation of stochastic processes by canonical variables. *SIAM Journal on Control and Optimization* **13**, 162–173.

Anderson, B.D.O. and M. Gevers (1982). Identifiability of linear stochastic systems operating under linear feedback. *Automatica, 18* **2**, 195–213.

Anderson, B.D.O., J.B. Moore and R.M. Hawkes (1978). Model approximation via prediction error identification. *Automatica* **14**, 615–622.

Åström, K.J. (1980). Maximum likelihood and prediction error methods. *Automatica* **16**, 551–574.

Åström, K.J. and J. Nilsson (1994). Analysis of a scheme for iterated identification and control. In: *Proc. IFAC Symp. on Identification*. Copenhagen, Denmark. pp. 171–176.

Åström, K.J. and P. Eykhoff (1971). System identification – A survey. *Automatica* **7**, 123–162.

Åström, K.J. and T. Bohlin (1965). Numerical identification of linear dynamic systems from normal operating records. In: *Proc. IFAC Symposium on Self-Adaptive Systems*. Teddington, UK.

Bayard, D.S., Y. Yam and E. Mettler (1992). A criterion for joint optimization of identification and robust control. *IEEE Trans. Automatic Control* **37**, 986–991.

Bombois, X., M. Gevers and G. Scorletti (2000). A measure of robust stability for an identified set of parametrized transfer functions. *IEEE Transactions on Automatic Control* **45**(11), 2141–2145.

Box, G.E.P and G.M. Jenkins (1970). *Time Series Analysis, forecasting and control*. Holden-Day. Oakland, California.

Chiuso, A. and G. Picci (2002). Geometry of oblique splitting subspaces, minimality and hankel operators. In: *'Directions in Mathematical Systems Theory and Optimization', A. Rantzer and C. Byrnes Eds*. Springer-Verlag. New York.

Chui, N.L.C. and J.M. Maciejowski (1996). Realization of stable models with subspace methods. *Automatica* **32**(11), 1587–1595.

Clark, J.M.C. (1976). The consistent selection of parametrizations in systems identification. In: *Proc. Joint Automatic Control Conference*. Purdue University.

de Callafon, R.A. and P.M.J. Van den Hof (1997). Suboptimal feedback control by a scheme of iterative identification and control design. *Mathematical Modelling of Systems* **3**(1), 77–101.

de Vries, D.K. and P.M.J. Van den Hof (1995). Quantification of uncertainty in transfer function estimation: a mixed probabilistic - worst-case approach. *Automatica* **31**, 543–558.

Deistler, M. (2002). System identification and time series analysis: Past, present, and future. In: *'Stochastic Theory and Control', Festschrift for Tyrone Duncan*. Springer. Kansas, USA. pp. 97–108.

Deistler, M. and M. Gevers (1981). Some properties of the parametrization of arma systems with unknown order. *Multivariate Analysis* **11**, 474–484.

Faurre, P. (1976). Stochastic realization algorithms. In: *'System identification: Advances and Case*

Studies' R.K. Mehra and D.G. Lainiotis Eds.. Academic Press. New York.

Forssell, U. and L. Ljung (1999). Closed-loop identification revisited. *Automatica* **35**, 1215–1241.

Forssell, U. and L. Ljung (2000). Some results on optimal experiment design. *Automatica* **36**, 749–756.

Gauss, C.F. (1809). *Teoria Motus Corporum Coelestium in Sectionibus Conicis Solem Ambientium*. Reprinted translation: 'Theory of Motion of the Heavenly Bodies Moving about the Sun in Conic Sections'. Dover. New York.

Gevers, M. (1991). Connecting identification and robust control: A new challenge. In: *Proc. IFAC/IFORS Symposium on Identification and System Parameter Estimation*. Budapest, Hungary. pp. 1–10.

Gevers, M. and L. Ljung (1986). Optimal experiment designs with respect to the intended model application. *Automatica* **22**, 543–554.

Gevers, M. and T. Kailath (1973). An innovations approach to least-squares estimation, part vi: Discrete-time innovations representations and recursive estimation. *IEEE Transactions Automatic Control* **18**, 588–600.

Gevers, M. and W.R. Wouters (1978). An innovations approach to the discrete-time stochastic realization problem. *Journal A* **19**(2), 90–110.

Gevers, M., X. Bombois, B. Codrons, F. De Bruyne and G. Scorletti (1999). The role of experimental conditions in model validation for control. In: *Robustness in Identification and Control - Proc. of Siena Workshop, July 1998* (A. Garulli, A. Tesi and A. Vicino, Eds.). Vol. 245 of *Lecture Notes in Control and Information Sciences*. Springer-Verlag. pp. 72–86.

Gevers, M., X. Bombois, B. Codrons, G. Scorletti and B.D.O. Anderson (2003). Model validation for control and controller validation in a prediction error identification framework - Part I: theory. *Automatica* **39**(3), 403–415.

Giarré, L., M. Milanese and M. Taragna (1997). H_∞ identification and model quality evaluation. *IEEE Trans. Automatic Control* **42**(2), 188–199.

Goodwin, G. and R.L. Payne (1977). *Dynamic System Identification: Experiment Design and Data Analysis*. Academic Press. New York.

Goodwin, G.C., M. Gevers and B. Ninness (1992). Quantifying the error in estimated transfer functions with application to model order selection. *IEEE Trans. Automatic Control* **37**, 913–928.

Gu, G. and P.P. Khargonekar (1992). A class of algorithms for identification in H_∞. *Automatica* **28**, 299–312.

Gustavsson, I., L. Ljung and T. Söderström (1977). Identification of processes in closed loop - identifiability and accuracy aspects. *Automatica* **13**, 59–75.

Hakvoort, R.G. and P.M.J. Van den Hof (1997). Identification of probabilistic system uncertainty regions by explicit evaluation of bias and variance errors. *IEEE Trans. Automatic Control* **42**(11), 1516–1528.

Hansen, F., G. Franklin and R. Kosut (1989). Closed-loop identification via the fractional representation: Experiment design. *Proc. American Control Conference* pp. 1422–1427.

Hazewinkel, M. and R.E. Kalman (1976). On invariants, canonical forms and moduli for linear constant, finite-dimensional dynamical systems. In: *Lecture Notes in Economics and Mathematical Systems*. Vol. 131. Springer-Verlag. Berlin. pp. 48–60.

Helmicki, A.J., C.A. Jacobson and C.N. Nett (1991). Control oriented system identification: A worst-case/deterministic approach in H_∞. *IEEE Trans. Automatic Control* **AC-36**, 1163–1176.

Heuberger, P.S.C., O.H. Bosgra and P.M.J. Van den Hof (1995). A generalized orthonormal basis for linear dynamical systems. *IEEE Trans. Automatic Control* **AC-40**(3), 451–465.

Hildebrand, R. and M. Gevers (2003). Identification for control: optimal input design with respect to a worst-case ν-gap cost function. *SIAM Journal on Control and Optimization* **41**(5), 1586–1608.

Hjalmarsson, H., M. Gevers and F. De Bruyne (1996). For model-based control design, closed-loop identification gives better performance. *Automatica* **32**, 1659–1673.

Ho, B.L. and R.E. Kalman (1965). Effective construction of linear state-variable models from input-output functions. *Regelungstechnik* **12**, 545–548.

Katayama, T. and G. Picci (1999). Realization of stochastic systems with exogenous inputs and subspace identification methods. *Automatica* **35**, 1635–1652.

Kosut, R.L. (1995). Uncertainty model unfalsification : a sytem identification paradigm compatible with robust control design. In: *Proc. 34th IEEE Conference on Decision and Control*. New Orleans,LA.

Kosut, R.L. and B.D.O. Anderson (1994). Least-squares parameter set estimation for robust control design. In: *Proc. American Control Conference*. Baltimore,Maryland. pp. 3002–3006.

Larimore, W.E. (1990). Canonical variate analysis in identification, filtering, and adaptive control. In: *Proc. 29th IEEE Conf. on Decision and Control*. Honolulu, Hawaii. pp. 596–604.

Lee, W.S., B.D.O. Anderson, R.L. Kosut and I.M.Y Mareels (1993). A new approach to adaptive robust control. *Int. Journal of Adaptive Control and Signal Processing* **7**, 183–211.

Liu, K. and R.E. Skelton (1990). Closed loop identification and iterative controller design. *29th IEEE Conf on Decision and Control* pp. 482–487.

Ljung, L. (1978). Convergence analysis of parametric identification methods. *IEEE Trans. Automatic Control* **AC-23**, 770–783.

Ljung, L. (1985). Asymptotic variance expressions for identified black-box transfer function models. *IEEE Trans. Automatic Control* **AC-30**, 834–844.

Ljung, L. (1987). *System Identification: Theory for the User*. Prentice-Hall. Englewood Cliffs, NJ.

Ljung, L. and K. Glover (1981). Frequency domain versus time domain methods in system identification. *Automatica* **17**(1), 71–86.

Ljung, L. and P.E. Caines (1979). Asymptotic normality of prediction error estimators for approximative system models. *Stochastics* **3**, 29–46.

Mäkilä, P.M, J.R. Partington and T.K. Gustafsson (1995). Worst-case control-relevant identification. *Automatica* **31**(12), 1799–1819.

Ng, T.S., G.C. Goodwin and B.D.O. Anderson (1977). Identifiability of mimo linear dynamic systems operating in closed loop. *Automatica* **13**, 477–485.

Ninness, B. and G.C. Goodwin (1995). Estimation of model quality. *Automatica* **31**(12), 32–74.

Ninness, B., H. Hjalmarsson and F. Gustafsson (1999*a*). Asymptotic variance expressions for output error model structures. In: *14th IFAC World Congress*. Beijing, P.R. China.

Ninness, B., H. Hjalmarsson and F. Gustafsson (1999*b*). On the fundamental role of orthonormal bases in system identification. *IEEE Transactions on Automatic Control* **44**(7), 1384–1407.

Parker, P.J. and R.R. Bitmead (1987). Adaptive frequency response identification. *Proc. CDC*.

Partington, J.R. (1991). Robust identification and interpolation in H_∞. *International Journal of Control* **54**, 1281–1290.

Picci, G. (1982). Some numerical aspects of multivariable systems identification. *Math. Programming Studies* **18**, 76–101.

Pintelon, R. and J. Schoukens (1990). Robust identification of transfer functions in the s- and z-domains. *IEEE Trans. on Instrumentation and Measurement* **39**(4), 565–573.

Pintelon, R. and J. Schoukens (2001). *System Identification - A Frequency Domain Approach*. IEEE Press. New York.

Poolla, K., P.P. Khargonekar, A. Tikku, J. Krause and K. Nagpal (1994). A time-domain approach to model validation. *IEEE Trans. Automatic Control* **39**, 951–959.

Rissanen, J. (1982). Estimation of structure by minimal description length. *Circuits, Systems, and Signal Processing* **1**, 395–406.

Schrama, R.J.P. (1992). Accurate identification for control: The necessity of an iterative scheme. *IEEE Trans. on Automatic Control* **37**, 991–994.

Schrama, R.J.P. and O.H. Bosgra (1993). Adaptive performance enhancement by iterative identification and control design. *Int. Journal of Adaptive Control and Signal Processing* **7**(5), 475–487.

Sin, K.S. and G.C. Goodwin (1980). Checkable conditions for identifiability of linear systems operating in closed loop. *IEEE Trans. Automatic Control* **AC-25**, 722.

Sjöberg, J., Q. Zhang, L. Ljung, A. Benveniste, B. Deylon, P-Y. Glorennec, H. Hjalmarsson and A. Juditsky (1995). Nonlinear black-box modeling in system identification: a unified overview. *Automatica* **31**(12), 1691–1724.

Smith, R. S. and M. Dahleh (1994). *The Modeling of Uncertainty in Control Systems*. Lecture Notes in Control and Information Sciences, vol. 192, Springer-Verlag.

Söderström, T. and P. Stoica (1989). *System Identification*. Prentice-Hall International. Hemel Hempstead, Hertfordshire.

Söderström, T., L. Ljung and I. Gustavsson (1976). Identifiability conditions for linear multivariable systems operating under feedback. *IEEE Trans. Automatic Control* **AC-21**, 837–840.

Van den Hof, P.M.J. and R.J.P. Schrama (1993). An indirect method for transfer function estimation from closed loop data. *Automatica* **29**, 1523–1527.

Van den Hof, P.M.J., R.J.P. Schrama, R.A. de Callafon and O.H. Bosgra (1995). Identification of normalized coprime plant factors from closed-loop experimental data. *European Journal of Control* **1**, 62–74.

Van Overbeek, A.J.M. and L. Ljung (1982). On-line structure selection for multivariable state-space models. *Automatica* **18**, 529–544.

Van Overschee, P. and B. De Moor (1994). N4SID: Subspace algorithms for the identification of combined deterministic-stochastic systems. *Automatica* **30**(1), 75–93.

Verhaegen, M. (1994). Identification of the deterministic part of mimo state space models given in innovations form from input-output data. *Automatica* **30**(1), 61–74.

Viberg, M. (1995). Subspace-based methods for the identification of linear time-invariant systems. *Automatica* **31**(12), 1835–1851.

Wahlberg, B. and L. Ljung (1986). Design variables for bias distribution in transfer function estimation. *IEEE Trans. Automatic Control* **AC-31**, 134–144.

Wertz, V., M. Gevers and E.J. Hannan (1982). The determination of optimum structures for the state-space representation of multivariable stochastic processes. *IEEE Transactions on Automatic Control* **27**(6), 1200–1211.

Zang, Z., R.R. Bitmead and M. Gevers (1995). Iterative weighted least-squares identification and weighted LQG control design. *Automatica* **31**(11), 1577–1594.

IFAC Publications
www.elsevier.com/locate/ifac

SYSTEM IDENTIFICATION VIA A COMPUTATIONAL BAYESIAN APPROACH

Brett Ninness [*,1] **Soren Henriksen** [*]

[*] *Dept. of Elec. & Comp. Eng, Uni. Newcastle, Australia.*
*email:*brett@ee.newcastle.edu.au.
eesjh@ee.newcastle.edu.au *FAX: +61 49 21 69 93*

Abstract: This paper takes a Bayesian approach to the problem of dynamic system estimation, and illustrates how posterior densities for system parameters, or more abstract and rather arbitrary system properties (such a frequency response, phase margin etc.) may be numerically computed. In achieving this, the key idea of constructing an ergodic Markov chain with invariant distribution equal to the desired posterior is fundamental, and it is inspired by recent developments in the mathematical statistics literature. An essential point of the work here is that via the associated posterior computation from the Markov chain, error bounds on estimates are provided that do not rely on asymptotic in data length arguments, and hence they apply with arbitrary accuracy for arbitrarily short data records.

Keywords: Parameter Estimation, Bayesian Estimation, Markov Chain Monte Carlo, Metropolis Hastings Method.

1. INTRODUCTION

A dominant force in the practise and understanding of modern methods for dynamic system identification has been work within a Maximum Likelihood framework and associated approximations, such as prediction error approaches (Ljung, 1999; Hannan and Deistler, 1988).

A key aspect of this strategy is that any quantification of the accuracy of the associated system estimates relies on employing asymptotic in data length expressions as if they applied for finite data lengths. For example, the Gaussian distribution commonly achieved by estimates in the infinite limit, is usually assumed to hold for whatever finite data length is available.

While these techniques enjoy widespread acceptance, there has recently arisen a body of work under terms such as "bounded error estimation", "robust estimation", and "estimation for control" (amongst others) that has sought to derive methods and supporting theory which apply for arbitrarily short data records.

This paper is directed at the same issue of dynamic system identification in a finite data record setting, but takes a different approach to the problem. In particular, the perspective here is that, especially for very short data lengths, it is sensible to take a Bayesian approach to quantifying the manner in which prior knowledge and data-based information are combined to yield posterior information about system properties.

While there are strong scientific and philosophical arguments for this strategy, it has historically foundered on the difficulty of actually computing the posterior distribution. However, the introduction of so-called Markov Chain Monte–Carlo (MCMC) methods in the mathematical statistics literature has recently caused something of a minor revolution in that field by offering a means for numerically computing posterior distributions for very complex modelling scenarios (Tierney, 1994; Gilks *et al.*, 1996). The essential idea used there (which incidentally has a half century history in the physics literature (Metropolis *et al.*, 1953)) is the invention of a method for constructing a Markov Chain which converges to an invariant density equal to the desired posterior. Sampling from this simulated chain then provides a means for computing posteriors with respect to this density via sample averages from the chain, hence the 'Monte–Carlo' epithet.

The contribution of this paper is to show how these ideas may be applied to the problem of Bayesian estimation of dynamic systems, and in doing so provide estimation error quantifications that apply

[1] This work was supported by the Australian Research Council. It was originally accepted for the IEEE Conference on Decision 2002, and Control, but withdrawn due to an inability of either author attending the conference.

for arbitrarily short data records. As well, we show how posterior density functions, and hence finite data error quantifications, may be computed for rather arbitrary functions of any underlying estimated system. Examples of these posterior functions include system frequency response, phase margin, pole position, robustness margin etc.

The ideas behind this paper have been considered in a context of analysing dynamic systems before, mostly for the purposes of state estimation; see, for example (Liu and Chen, 1998; Doucet *et al.*, 2001; Doucet *et al.*, 2000; Handschin and Mayne, 1969) However, there seems to be no extant literature examining the use of the so-called 'Metropolis–Hastings' based MCMC methods employed in this paper for the applications of parameter estimation, subsequent controller design and estimation-error quantification for finite data sets that are considered here.

2. BACKGROUND

Consider the following linear and time invariant relationship between an observed input data record $\{u_t\}$ and output data record $\{y_t\}$ expressed in an innovations form as

$$x_{t+1} = Ax_t + Bu_t + Ky_t, \tag{1}$$

$$y_t = Cx_t + Du_t + e_t. \tag{2}$$

This system is possibly multivariable in that $u_t \in \mathbf{R}^m$ and $y_t \in \mathbf{R}^p$ are not necessarily scalar. Of interest then is the estimation of the matrices $A \in \mathbf{R}^{n\times n}$, $B \in \mathbf{R}^{n\times m}$, $C \in \mathbf{R}^{p\times n}$ and $D \in \mathbf{R}^{p\times m}$ that parameterise the model (1), (2); this estimation to be performed on the basis of observing an input-output data record $\{u_t\}$, $\{y_t\}$ of length N samples.

For this purpose, it is usual to parameterise A, B, C, D via a vector θ according to some mapping

$$\theta \longmapsto \{A_\theta, B_\theta, C_\theta, D_\theta\} \tag{3}$$

which implies a corresponding innovations form model

$$\widehat{x}_{t+1} = A_\theta \widehat{x}_t + B_\theta u_t + K_\theta y_t, \tag{4}$$

$$y_t = C_\theta \widehat{x}_t + D_\theta u_t + e_t. \tag{5}$$

and hence an estimate (based on N data points) of $\{A, B, C, D\}$ in (1), (2) via an estimate $\widehat{\theta}_N$ of θ.

With this in mind, the efforts of many researchers over the last few decades has produced a very substantial body of work centred on the theme of finding the estimate $\widehat{\theta}_N$ of θ via the aforementioned maximum-likelihood technique; see the monographs (Ljung, 1999; Hannan and Deistler, 1988) for detailed treatments. This latter approach relies on the specification of a probability density function $p_e(\cdot)$ for the stochastic component e_t so that the 'likelihood' of the observed N point data record may be computed as the probability

$$p(Y_N \mid \theta) = p(y_0 \mid \theta) \prod_{t=1}^{N} p(y_t \mid Y_{t-1}, \theta) \tag{6}$$

where the notation $Y_t \triangleq \{y_0, y_1, \cdots, y_t\}$ has been introduced and Bayes' rule

$$p(A \mid B)\,p(B) = p(B \mid A)\,p(A)$$

has been used. Furthermore, according to (4), (5)

$$p(y_t \mid Y_{t-1}, \theta) = p_e(y_t - \widehat{y}_{t|t-1}(\theta)) \tag{7}$$

where

$$\widehat{y}_{t|t-1}(\theta) \triangleq C_\theta \widehat{x}_t + D_\theta u_t \tag{8}$$

which can be used as a means for evaluating the likelihood (6), and hence the maximum-likelihood estimate $\widehat{\theta}_N$ as

$$\widehat{\theta}_N \triangleq \arg\max_\theta p(Y_N \mid \theta). \tag{9}$$

A particularly appealing feature of this approach to parameter estimation is that it enjoys some desirable properties (Ljung, 1999; Hannan and Deistler, 1988). Firstly, it is convergent to a quantity $\theta_\circ$ that is characterisable as follows w.p.1

$$\lim_{N\to\infty} \widehat{\theta}_N = \theta_\circ \triangleq \arg\max_\theta \lim_{N\to\infty} \frac{1}{N}\mathbf{E}\left\{\log p(Y_N \mid \theta)\right\} \tag{10}$$

Secondly, the asymptotic distributional properties of $\widehat{\theta}_N$ also obey a normal law

$$\sqrt{N}(\widehat{\theta}_N - \theta_\circ) \xrightarrow{\mathcal{D}} \mathcal{N}(0, P) \quad \text{as } N \to \infty. \tag{11}$$

where, provided the data was in fact generated by a system obeying (1), (2) and with respect to the mapping (3) for some $\theta = \theta_\circ$, then

$$P^{-1} = \frac{1}{\sigma^2} \lim_{N\to\infty} \frac{\mathrm{d}^2}{\mathrm{d}\theta \mathrm{d}\theta^T} \frac{1}{N}\mathbf{E}\left\{p(Y_N \mid \theta)\right\}\Big|_{\theta=\theta_\circ} \tag{12}$$

where $\sigma^2 = \mathbf{E}\{e_t^2\}$. In particular, note that these results can still hold (modulo multiplication by a scalar in (12)) even if the function $p_e(\cdot)$ is not equal to any underlying true one, but instead merely satisfies some mild regularity conditions (Ljung, 1999).

Nevertheless, despite these recommendations, there are some limitations to this approach that, over the last decade, have been of increasing importance to workers in the systems and control community. Firstly, even though the results (10), (11), (12) are strictly asymptotic in data length N results, in practise it so happens that the only means available for quantification of possible estimation error involves assuming that they hold approximately for the finite data available.

Secondly, when the estimated quantities of interest are not the parameters themselves, but one or more functions of the parameters, such as frequency response, phase margin, etc., then the ubiquitous strategy is to form a first order Taylor expansion in this function of interest about the assumed true parameter, and then couple this with (11) in order to derive estimates of the function of interest together with error bounds. The success of this approach depends on the Taylor expansion being accurate by virtue of $\|\widehat{\theta}_N - \theta_\circ\|$ being small, and again this depends on the data length N being large.

As mentioned in the introduction, concern about these issues has led to many new estimation methods. As a possible solution to resolve these difficulties, the paper here proposes a new Bayesian-computational approach that employs recently developed techniques from the statistics literature known as 'Markov–Chain Monte–Carlo' (MCMC) methods.

3. A COMPUTATIONAL BAYESIAN APPROACH

The key idea of this paper is to provide estimates together with error bounds that apply for arbitrarily short finite data lengths N by taking a Bayesian perspective and considering the posterior density $p(\theta \mid Y_N)$ as a means of quantifying all information that can be extracted from a combination of the observed data and prior information. For the model structures (4), (5) considered here, this essential quantity may be computed using Bayes' rule applied to (6) which leads to

$$p(\theta \mid Y_N) = \frac{p(y_0 \mid \theta)p(\theta)}{p(Y_N)} \prod_{t=1}^{N} p_e(y_t - \widehat{y}_{t|t-1}(\theta)) \quad (13)$$

where $p(\theta)$ is the a-priori distribution of θ and $p(Y_N)$ is a normalising constant given as

$$p(Y_N) = \int_{\mathbf{R}^n} p(Y_N \mid \theta)p(\theta)\, \mathrm{d}\theta. \quad (14)$$

Now, the left hand side of (13) can, in principle, be computed for any desired θ, and hence the posterior density $p(\theta \mid Y_N)$ may be directly evaluated.

If, as would commonly be the case, the marginal density of only a particular i'th parameter element θ^i is required, then this too can be evaluated via numerical computation of the integral

$$p(\theta^i \mid Y_N) = \int_{\mathbf{R}^{n-1}} p(\theta \mid Y_N) \mathrm{d}\theta^1..\mathrm{d}\theta^{i-1}\mathrm{d}\theta^{i+1}..\mathrm{d}\theta^n \quad (15)$$

For a large dimensional model, this could involve a rather heavy computational load. For example, since typically around thirty points on a histogram are needed to represent it accurately, then (15) implies the numerical evaluation of around thirty multidimensional integrals, and this latter dimension could be relatively large: a fifth order model would imply a nine dimensional integral.

Furthermore, since (13) involves a product term over N terms, then for large N this product can be very large when p_e is greater than one, and very small when it is less than one. This large dynamic range implies great numerical difficulties in its evaluation. One strategy to circumvent this is to instead work with $\log p(\theta \mid Y_N)$, but then the marginal density (15) cannot be computed since the logarithm and integral operator don't commute.

Nevertheless, if possible, the evaluation of (13) and (15) provide a means for providing parameter estimation error quantifications applicable to arbitrarily short data records. Indeed, from a Bayesian perspective, the calculation of the posterior (13) is an optimal strategy in that it completely characterises the information available from the data and prior knowledge about the parameter θ.

However, suppose interest is centred not on the parameters themselves, but on a function of them such as (for example) system phase margin $\phi_m(K)$ for a given closed loop controller $K(q)$. Then it is not at all clear how one might tractably compute the posterior density

$$p(\phi_m(K) \mid Y_N). \quad (16)$$

As a solution to these difficulties, this paper proposes an approach of numerically computing posterior densities, or quantities that depend on these densities (such as expectations) by a strategy of first generating a random realisation $\{\theta_t\} = \{\theta_0, \theta_1, \cdots, \theta_M\}$ with limiting distribution equal to the desired posterior density; viz.

$$\lim_{t\to\infty} p(\theta_t = \theta \mid \theta_0) = p(\theta \mid Y_N) \qquad \forall \theta_0. \quad (17)$$

This simulated realisation $\{\theta_t\}$ is then used as if it were a random sample from $p(\theta \mid Y_N)$. Provided (as will be shown presently) that the required distributional convergence holds, then via a law of large numbers argument, this will lead to consistent estimates of various quantities. For example, it allows the numerical computation and consistent estimation of rather arbitrary posterior densities via the sample histogram:

$$p(g(\theta) \in A \mid Y_N) \approx \frac{1}{M} \sum_{t=k+1}^{k+M} \chi_{g^{-1}(A)}(\theta_t). \quad (18)$$

Here, g is an arbitrary measurable function, χ is the indicator function, and A is a g-measurable set.

While, this may seem like a reasonable approach, it may initially appear to be impossible to implement due to the difficulty of sampling from the rather arbitrary multivariable posterior density $p(\theta \mid Y_N)$. Perhaps surprisingly then, it so happens that a Markov chain can be constructed with limiting density equal to the required posterior $p(\theta \mid Y_N)$, and in a manner that is far more straightforward than might prima facie be thought possible.

More specifically, with the notation θ^i being used to denote an i'th block of $\theta \in \mathbf{R}^n$, together with θ^{-i} denoting all of θ *except* for θ^i, then as will be presently established, the following algorithm implements a Markov chain with the properties mentioned above.

Algorithm 3.1.

(1) Initialise θ_0 at some value;
(2) At iteration k, consider a candidate value ξ_k^i for the i'th block θ_k^i of θ_k which is drawn from a somewhat arbitrary *proposal* density $\gamma_i(\xi_k^i \mid \theta_{k-1})$. That is, find a possible realisation for θ_k^i as

$$\xi_k^i \sim \gamma_i(\cdot \mid \theta_{k-1}); \qquad (19)$$

(3) Set $\phi_k = \xi_k^i \cup \theta_{k-1}^{-i}$, $\xi_k^{-i} = \theta_{k-1}^{-i}$ and compute the acceptance probability

$$\alpha(\xi_k^i \mid \theta_{k-1}) = \min\left\{1, \frac{p(y_0 \mid \phi_k)p(\phi_k)}{p(y_0 \mid \theta_k)p(\theta_k)} \times \right.$$

$$\left. \frac{\prod_{t=1}^N p_e(y_t - \widehat{y}_{t|t-1}(\phi_k))}{\prod_{t=1}^N p_e(y_t - \widehat{y}_{t|t-1}(\theta_k))} \cdot \frac{\gamma_i(\theta_{k-1}^i \mid \xi_k)}{\gamma_i(\xi_k^i \mid \theta_{k-1})}\right\}; \qquad (20)$$

(4) Accept the proposed ξ_k^i and set $\theta_k^i = \xi_k^i$ with probability $\alpha(\xi_k^i \mid \theta_{k-1})$;
(5) Move to another i'th block of θ_k and return to step 2. If steps 2-4 have already been performed for all blocks θ^i of θ for the current value of k, then increment k and return to step 2.

This algorithm is in fact an instance of the so-called 'Metropolis–Hastings algorithm', which appears to have initially been proposed within the physics community by M. Metropolis in a paper co-authored with Edward Teller as part of their efforts to develop the hydrogen bomb (Metropolis *et al.*, 1953). The method was later generalised by Hastings (Hastings, 1970), and was then overlooked for almost another twenty years before work such as (Tierney, 1994) re-invigorated interest in the method in the statistics community. Since then it has attracted great interest, with several monographs dealing with it and related topics now being available (Liu and Chen, 1998; Gilks *et al.*, 1996).

Note the simplifying aspect of the acceptance probability $\alpha(\xi_k^i \mid \theta_{k-1})$ in (20) depending only on the *ratio* of probabilities, and not their actual values. Not only does this imply that normalising constants $p(Y_N)$ such as occur in (13) need not be computed, it also permits previously mentioned numerical problems associated with the large dynamic ranges of the ratio components to be avoided, since they need only be computed in a ratio-combined fashion.

4. CONVERGENCE ANALYSIS

With the definition of Algorithm 3.1 now given, the purpose of this section is to establish that it possesses the desired property of generating samples from the posterior density $p(\theta \mid Y_N)$. For this purpose, it is crucial to observe that the mechanism of generating a new sample θ_k^i is in fact a Markov Chain, with time-homogeneous transition density $K(\theta_k \mid \theta_{k-1})$ determined by the transition density $K^i(\theta_k^i \mid \theta_{k-1})$ for the i'th component as

$$K^i(\theta_k^i = \xi_k^i \mid \theta_{k-1}) = \alpha(\xi_k^i \mid \theta_{k-1})\gamma_i(\xi_k^i \mid \theta_{k-1}) + \delta(\xi_k^i - \theta_{k-1}^i)r(\theta_{k-1}^i) \qquad (21)$$

where

$$r(\theta_{k-1}^i) = \left[1 - \int_{-\infty}^{\infty} \alpha(\xi_k^i \mid \theta_{k-1})\gamma_i(\xi_k^i \mid \theta_{k-1})\, d\xi_k^i\right] \qquad (22)$$

is the probability of no change in the value of θ_k^i, and in (21) the delta function is of the Dirac type. Note that since (by construction) the proposal random variables ξ_k^i are jointly independent (w.r.t. i), and since $\xi_k^{-i} = \theta_{k-1}^{-i}$ in the computation of the acceptance probability $\alpha(\xi_k^i \mid \theta_{k-1})$, then

$$K(\theta_k \mid \theta_{k-1}) = \prod_{i=1}^n K^i(\theta_k^i \mid \theta_{k-1}). \qquad (23)$$

Now, suppose that θ_{k-1} is drawn randomly according to a probability density function $\pi_{k-1}(\theta)$. Then clearly, the probability density function $\pi_k(\theta)$ for an ensuing element θ_k in the Markov chain given by Algorithm 3.1 is then

$$\pi_k(\theta_k) = \int_{\mathbf{R}^n} K(\theta_k \mid \theta_{k-1})\pi_{k-1}(\theta_{k-1})\, d\theta_{k-1}. \qquad (24)$$

Therefore, if the realisations $\{\theta_k\}$ generated by Algorithm 3.1 are to converge in a distributional sense to realisations having some constant density $\pi(\theta)$, then that density must satisfy

$$\pi(\theta) = \int_{\mathbf{R}^n} K(\theta \mid \xi)\pi(\xi)\, d\xi \qquad (25)$$

in which case $\pi(\theta)$ is termed an *invariant* (or stationary) density with respect to the transition kernel $K(\theta_k \mid \theta_{k-1})$.

With this in mind, we now establish that Algorithm 3.1 is specifically constructed so that it yields a Markov chain with invariant density equal to the posterior $p(\theta \mid Y_N)$ of interest.

Lemma 4.1. Define the support Θ of $p(\theta \mid Y_N)$ as

$$\Theta = \{\theta \in \mathbf{R}^n : p(\theta \mid Y_N) \in [0, 1]\} \qquad (26)$$

and suppose that this support is a subset of that of the proposal density

$$\gamma(\xi \mid \theta) = [\gamma_1(\xi^i \mid \theta), \cdots, \gamma_n(\xi^n \mid \theta)]^T$$

in that

$$\Theta \subset \bigcup_{\xi \in \Theta} \{\theta \in \mathbf{R}^n : \gamma(\xi \mid \theta) \in [0, 1]\}. \qquad (27)$$

Then $p(\theta \mid Y_N)$ defined by (13) is an invariant density of the Markov chain realised by Algorithm 3.1 for all $\theta \in \Theta$.

Proof: See (Ninness and Henriksen, 2002). ■

Therefore, the desired posterior density $p(\theta \mid Y_N)$ is a candidate for any density that realisations of Algorithm 3.1 might converge to. Note that in establishing this, the condition (27) which states that the support of $p(\theta \mid Y_N)$ must be contained within that of the proposal density $\gamma(\xi \mid \theta)$ is essential, since otherwise the algorithm would be unable to generate samples within certain regions of the support of $p(\theta \mid Y_N)$.

Using this lemma, it is then possible to establish that sample path averages from this algorithm converge to expectations with respect to $p(\theta \mid Y_N)$.

Theorem 4.1. Suppose that the proposal density $\gamma_i(\xi^i \mid \theta)$ of Algorithm 3.1 satisfies condition (27) of Lemma 4.1. Suppose further that $f : \mathbf{R}^n \to \mathbf{R}$ satisfies

$$\int_{\mathbf{R}^n} |f(\theta)|\, p(\theta \mid Y_N)\, d\theta < \infty. \qquad (28)$$

Then for the sequence $\{\theta_k\}$ generated by Algorithm 3.1

$$\lim_{N\to\infty} \sum_{k=1}^{N} f(\theta_k) = \int_{\mathbf{R}^n} f(\theta) p(\theta \mid Y_N)\, d\theta \qquad (29)$$

except possibly on a subset $\mathcal{G}$ of the underlying probability space $\{\Omega, \mathcal{F}, \mathbf{P}\}$ on which $\{\theta_k\}$ is defined for which

$$\int_{\theta \in \mathcal{G}} p(\theta \mid Y_N)\, d\theta = 0. \qquad (30)$$

Proof: See (Ninness and Henriksen, 2002). ■

Note that with the choice of f as the indicator function in (18), then this theorem asserts that the sample histogram converges almost surely to the underlying density that it seeks to estimate.

5. SIMULATION EXAMPLE

To illustrate the application of these ideas, we begin by addressing the very simple illustrational example of

$$y_t = \left(\frac{b}{q-a}\right) u_t + e_t \qquad (31)$$

where $b = 0.2$, $a = 0.8$ and $\{e_t\}$ is a zero mean i.i.d. process with $\sigma^2 = \mathbf{E}\{e_t^2\} = 0.01$. Suppose further that the available data consists of $N = 20$ samples of $\{y_t\}$ and $\{u_t\}$ when u_t is a piecewise constant signal, transiting $1 \mapsto 0$ at $t = 10$. This is illustrated in figure 1, where the solid line is the noise free response, and the samples around this line are the noise corrupted data assumed to be available.

In the case where the density $p_e(\cdot)$ governing e_t is uniform, and with prior distribution on $\theta = [b, a]$ being one that assigns zero weight to $b < 0$ and $|a| > 1$, then the posterior distributions for these parameters given the data realisation shown in figure 1 are illustrated in figure 2.

There, the solid line shows the marginal posterior density for b and a computed via numerical computation of the integral (15) via the use (13) to obtain the posterior joint density. The bar graph, is the sample histogram of 10^5 realisations from the Markov chain, constructed via the material presented in the previous sections so as to have invariant density $p(\theta \mid Y_N)$. Note the close agreement between these two numerically computed estimates of the posterior.

If an Output Error model structure is fitted to this data via the methods outlined earlier, then the asymptotic results (11)–(12) are, as is commonly done, used in a finite data setting in order to provide error quantification, then the results of this strategy are as shown by the dash-dot Gaussian-curve lines in figure 2. While these quantifications are not strictly comparable to the posterior distributions, since they evaluate different quantities, it would still seem interesting to compare the two in terms of their utility for informing a user of what system information can be extracted from the available data.

Finally, suppose that a closed loop PI controller $K(q)$ is to be designed for this plant, so as to achieve at least $\phi_m(G, K) = 105°$ of phase margin, and suppose that

$$K(q) = 1 + \frac{0.1}{q-1} \qquad (32)$$

is a candidate for this task. Then to assess the suitability of this proposal, one could seek to know the posterior density

$$p(\phi_m(G, K) \mid Y_N) \qquad (33)$$

in order to assess what information the data contains in relation to the question of whether (32) will attain the design objective. While this would seem a daunting task from an analytical point of view, the same Markov chain realisation $\{\theta_k\}$ used to form the histogram estimates of posterior densities in figure 2 can be simply employed to estimate (33) via its sample histogram, which is shown in the top diagram of figure 3. Clearly, there seems to be good evidence from the data that the controller (32) will achieve the required 105° phase margin. Finally, for the sake of completeness, the computed posterior density for the gain margin $g_m(G, K)$ for this same scenario is shown in the bottom diagram of figure 3, and indicates that the data strongly supports a conclusion that the controller (32) achieves a gain margin of greater than 7.5.

6. CONCLUSION

This paper presents a preliminary investigation of the employment, in a systems and control setting, of new ideas and techniques developed in the mathematical statistics literature. While the early results presented here appear to have promise, there are many questions and issues to be investigated, such as convergence rate, choice of proposal density, estimation of parameters quantifying p_e, model order estimation and so on. The authors are currently engaged on these issues.

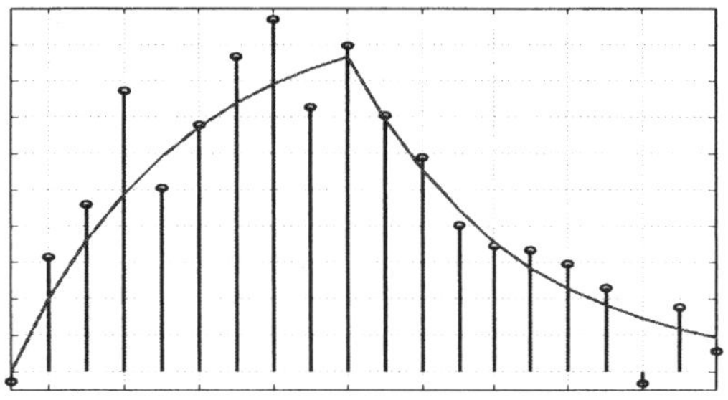

Fig. 1. *System response: Solid line is noise free, sampled dots are the noise corrupted measurements actually available.*

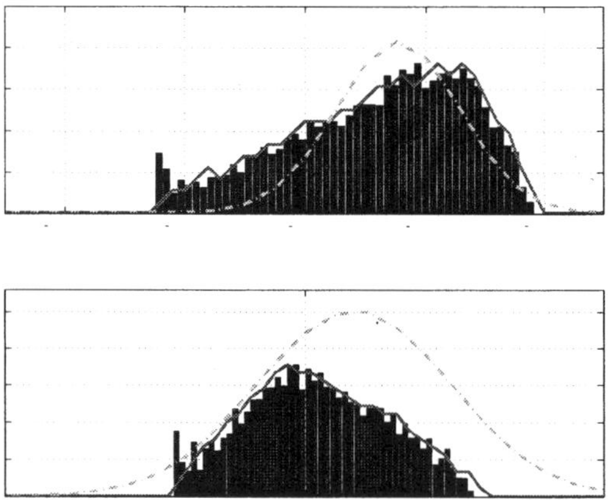

Fig. 2. *Sample histogram estimation of marginal posteriors together with (solid) numerical integration of computation, and (dash-dot line), the parameter information inferred via a prediction error approach.*

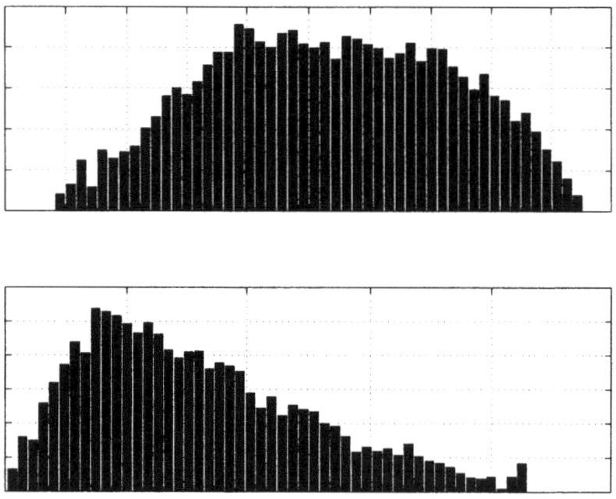

Fig. 3. *Posterior distributions for phase margin $\phi_m(G, K)$ and gain margin $g_m(G, K)$ for a given PI controller.*

7. REFERENCES

Doucet, A., A. Logothetis and V. Krishnamurthy (2000). 'Stoch. sampling alg's for state est. of jump Markov linear systems'. *IEEE Trans. Auto. Cont* **45**(2), 188–202.

Doucet, A., N.J. Gordon and V. Krishnamurthy (2001). 'Particle filters for state est. of jump Markov linear systems'. *IEEE Trans. Sig. Proc* **49**(3), 613–524.

Gilks, W., S. Richardson and D.J. Spiegelhalter (1996). *Markov Chain Monte Carlo in practice.* Chapman and Hall.

Handschin, J. E. and D. Q. Mayne (1969). 'Monte Carlo techniques to estimate cond. exp'n in multi-stage non-lin. filtering'. *Int. J. Control (1)* **9**, 547–559.

Hannan, E. and M. Deistler (1988). *The Statistical Theory of Linear Systems.* John Wiley and Sons. New York.

Hastings, W. (1970). 'Monte Carlo sampling methods using Markov chains and their applications'. *Biometrika* **57**, 97–109.

Liu, J. S. and Rong Chen (1998). 'Sequential Monte Carlo methods for dynamic systems'. *J. Amer. Statist. Assoc.* **93**(443), 1032–1044.

Ljung, L. (1999). *Syas. Id: Theory for the User,.* Prentice-Hall, Inc.. New Jersey.

Metropolis, M., A.W. Rosenbluth, M.N. Rosenbluth, A.H. Teller and E. Teller (1953). 'Equations of state calculations by fast computing machines'. *Journal of Chemical Physics* **21**, 1087–1091.

Ninness, B. and S. Henriksen (2002). Sys. Id. via a comp. Bayesian approach. Tech. rep. School of Elec. Eng, Uni. Newc.

Tierney, L. (1994). 'Mark. chains for exploring post. dist's'. *Ann. Statist.* **22**(4), 1701–1762.

IFAC Publications
www.elsevier.com/locate/ifac

A NEW INFORMATION THEORETIC APPROACH TO ORDER ESTIMATION PROBLEM

Soosan Beheshti * **Munther A. Dahleh** *

* *Massachusetts Institute of Technology, Cambridge, MA 02139, U.S.A.*

Abstract: We introduce a new method of model order selection: minimum description complexity (MDC). The approach is motivated by the Kullback-Leibler information distance. The method suggests to choose the model set for which the "model set relative entropy" is minimum. The proposed method is comparable with the existing order estimation methods such as AIC and MDL. We elaborate on the advantages of MDC over the available information theoretic approaches. *Copyright © 2003 IFAC*

Keywords: Kullback-Leibler distance, AIC, MDL, Order Estimation

1. INTRODUCTION

Classical problem of model selection among parametric model sets is considered. The goal is to choose a model set which best represents an observed data. The critical task is the choice of a criterion for model set comparison. Pioneer information theoretic based approaches to this problem are Akaike information criterion (AIC) and different forms of minimum description length (MDL) (Akaike, 1974), (Barron *et al.*, 1998). The prior assumption for calculation of these in criteria is that the unknown true model is a member of all the competing sets.

The new approach, minimum description complexity(MDC), is based on a new distance measure defined for the elements of the model sets. The distance of the true model and each model set is the minimum Kullback-Leibler distance of the true model and the elements of the model set. We provide a probabilistic method of MDC estimation for a class of parametric model sets. In this calculation the key factor is our prior assumption: unlike the existing methods no necessary assumption of the true parameter being a member of the competing model sets is needed. The main strength of the MDC calculation method is in its method of information extraction from the observed data.

Because of MDL's consistency, it has been widely used in practical problems. However, lack of a proper prior assumption in calculation of the criterion causes some defects such as high sensitivity to large signal to noise ratio when the true model does not belong to any of the model sets (A.P. Liavas and J.Delmas, 1999). Here we compare MDC with MDL and AIC in application. MDC is able to answer the challenging question of quality evaluation in identification of stable LTI systems under a fair prior assumption on the unmodeled dynamics. It also provides a new solution to a class of signal denoising problems (Beheshti and Dahleh, 2002).

2. IMPORTANT INFORMATION THEORETIC CRITERIA

Consider the following problem: Given a finite set of observed data y^N of length N, which is an element of set Y^N, a family of models which are parameterized by elements of a compact set S_M with order M, and a family of probability density functions $f(Y^N;S_M)$, select the model that best fits the data (Wax and Kailath, 1985).

In the following order estimation methods the estimate of the true parameter θ^* is calculated in S_m, a subset of S_M of order m. The estimate, $\hat{\theta}_{S_m}(y^N)$, is the maximum likelihood (ML) estimate of θ^* in S_m.

Akaike information criterion(AIC) is the estimate of the Kullback-Leibler distance of the true density

$f(Y^N;\theta)$, and the estimated density $f(Y^N;\hat{\theta}_{S_m}(y^N))$ in S_m (Akaike, 1974).

The AIC estimate is given by

$$\mathrm{AIC}_{S_m}(y^N) = -\frac{1}{N}\log f(y^N;\hat{\theta}_{S_m}(y^N)) + \frac{m}{N} \quad (1)$$

This estimate is calculated with the assumption that N is large enough and that the parameter estimate, $\hat{\theta}_{S_m}(y^N)$, approaches the true parameter θ^* in subset S_m. The method suggests to select the model set S_m which minimizes the AIC.

Any model defined by a parameter in set S_M can be used to encode the observed data by using the Shannon coding method. The two-stage minimum description length(MDL) method is defined based on this coding. In the two-stage MDL approach the description length of the data in each subset is defined as (Rissanen, 1984)

$$\mathrm{DL}_{S_m}(y^N) = -\frac{1}{N}\log f(y^N;\hat{\theta}_{S_m}(y^N)) + m\frac{\log N}{2N}. \quad (2)$$

Similar to AIC the main assumption in calculation of MDL is that the ML estimate $\hat{\theta}_{S_m}(y^N)$ approaches θ^* as the length of data grows.

Bayesian information criterion(BIC) is another order estimation method which was proposed in (Schwarz, 1978). In this method a prior probability for the competing model sets is assumed. It is suggested to select the model that yields the maximum posterior probability. Note that the criterion in this approach is similar to MDL criterion in (2).

The two important prior assumptions for calculation of AIC and MDL in (1) and (2), for subset S_m, are

$$1)\ \theta^* \in S_m, \quad 2)\ \hat{\theta}_{S_m}(y^N) \rightarrow \theta^* \quad (3)$$

The second condition in most cases implies that $M << N$.

Note that in practical problems we do not know whether or not the unknown θ^* is an element of a given S_m. However, in application of MDL and AIC the calculated criteria in (1) and (2) are used for all the subsets regardless of validity of the two prior assumptions in (3).

3. MINIMUM DESCRIPTION COMPLEXITY

We introduce a new method of subset selection by using the observed data of length N. Unlike the existing approaches none of the conditions in (3) are needed as our prior assumption. The set Y^N need not to be stationary. However, for each parametric probability distribution function (pdf), the expected value of Y^N and its covariance are finite. Also the pdfs $f(y^N;\theta)$ are continuous functions of Y^N.

Before providing a method of order estimation using the observed data we define a notion of distance for the pdfs. Note that in the following discussion the length of data is assumed to be fixed N.

For a given compact set S_M, use a positive cost function $V(\theta, y^N)$ for which $E_{\theta_1}\frac{1}{N}V(\theta_2, Y^N)$ is a finite nonnegative number and

$$E_{\theta_1}\frac{1}{N}V(\theta_2, Y^N) \geq E_{\theta_1}\frac{1}{N}V(\theta_1, Y^N) \quad (4)$$

for any θ_1 and θ_2 in S_M. The equality holds only for $\theta_1 = \theta_2$. Choose the cost function such that it is a continuous function of both θ and y^N.

Definition 1 The description complexity of Y^N using pdf $f(Y^N;\theta_1)$, when the data is generated by θ, is defined by

$$\mathrm{DC}_N(\theta,\theta_1) \triangleq E_\theta\frac{1}{N}V(\theta_1, Y^N) \quad (5)$$

For any element of S_M, define $\bar{\theta}_{S_m}$ in set S_m as

$$\bar{\theta}_{S_m}(N) = \arg\min_{\theta_1\in S_m} E_\theta\frac{1}{N}V(\theta_1, Y^N). \quad (6)$$

The description complexity of the data set using subset S_m, when the data is generated by θ, is then defined as

$$\mathrm{DC}_N(\theta,S_m) \triangleq \min_{\theta_{S_m}\in S_m}\mathrm{DC}_N(\theta,\theta_{S_m}) \quad (7)$$

$$= \mathrm{DC}_N\left(\theta,\bar{\theta}_{S_m}(N)\right) \quad (8)$$

Definition 2 The minimum description complexity (MDC) of Y^N, when the data is generated by θ, is provided by subset S_m^*

$$S_m^* = \arg\min_{S_m}\mathrm{DC}_N\left(\theta,\bar{\theta}_{S_m}(N)\right). \quad (9)$$

In general the set of all possible cost functions depends on the structure of the parametric model set. One example of such cost function for any parametric pdf is

$$V(\theta, y^N) = -\log f(y^N;\theta) \quad (10)$$

This function is well defined for all Y^N with prior assumption that $f(y^N;\theta) \neq 0$ for any y^N. For this cost function

$$\mathrm{DC}_N(\theta,\theta) = E_\theta\frac{1}{N}V(\theta, Y^N) = \frac{1}{N}H_\theta(Y^N)\log 2 \quad (11)$$

where $H_\theta(Y^N)$ is the differential entropy of Y^N when it is generated by θ

$$H_\theta(Y^N) = -E_\theta\log_2 f(y^N;\theta) \quad (12)$$

If we want the description complexity function to be more like a distance measure we add the extra condition $\mathrm{DC}(\theta,\theta) = 0$. For example the new description complexity using the defined DC in (5) can be defined as

$$I_N(f(:;\theta), f(:,\bar{\theta}_{S_m}(N)) = \mathrm{DC}_N(\theta,\bar{\theta}_{S_m}(N)) - \mathrm{DC}_N(\theta,\theta) \quad (13)$$

where the cost function is defined in (10). In this case $I_N(\cdot)$ is the Kullback-Leibler distance of θ and $\bar{\theta}_{S_m}(N)$.

3.1 *MDC and Data Observation*

Based on the defined description complexity, consider a family of parameter estimators for which

$$E_\theta(\hat{\theta}_{S_m}(Y^N)) = \bar{\theta}_{S_m}(N) \quad (14)$$

where $\bar{\theta}_{S_m}(N)$ is defined in (6). Note that for this set of estimators we have

$$E_\theta(\hat{\theta}_{S_M}(Y^N)) = \theta \quad (15)$$

and therefore the estimator is unbiased in S_M. The observed data y^N is generated by the unknown parameter θ^* and in each subset $\hat{\theta}_{S_m}(y^N)$ and $V(\hat{\theta}_{S_m}(y^N), y^N)$ are available. For order estimation the goal is to first use this information to find an estimate for $\mathrm{DC}(\theta^*,\hat{\theta}_{S_m}(y^N))$ for each subset and then choose the subset for which this error is minimum.

The first step is to validate θ's given the available estimate $\theta_{S_m}(y^N)$. The random variable $\mathrm{DC}(\theta,\hat{\theta}_{S_m}(Y^N))$ for each θ has a mean and a variance which are functions of θ, S_m and N. If the data is generated with θ then with probability p, the bound $\varepsilon_p(\theta, S_m, N)$ is available such that for a set of $x^N \in Y^N$

$$\Pr\{|(\mathrm{DC}(\theta,\hat{\theta}_{S_m}(x^N)) - \mathrm{E}_\theta \mathrm{DC}(\theta,\hat{\theta}_{S_m}(Y^N))| \le \varepsilon_p(\theta,S_m,N)\} = p \quad (16)$$

and subset $T_p(\theta, S_m, N)$ in Y^N is defined as

$$T_p(\theta,S_m,N) = \{x^N \in Y^N : |(\mathrm{DC}(\theta,\hat{\theta}_{S_m}(x^N)) - \mathrm{E}_\theta \mathrm{DC}(\theta,\hat{\theta}_{S_m}(Y^N))| \le \varepsilon_p(\theta,S_m,N)\}.$$

The validation with probability p, and based on the observed data, provides the following set of parameters in S_M

$$\Theta(y^N,S_m,p) = \{\theta \in \Theta | y^N \in T_p(\theta,S_m,N)\}. \quad (17)$$

Therefore, with validation probability p in each subset S_m, the desired DC, $\mathrm{DC}_N(\theta^*,\hat{\theta}_{S_m}(y^N))$, is bounded by

$$\min_{\theta\in\Theta(y^N,S_m,p)} \mathrm{DC}_N(\theta,\hat{\theta}_{S_m}(y^N)) \le \quad (18)$$
$$\mathrm{DC}_N(\theta^*,\hat{\theta}_{S_m}(y^N)) \le \max_{\theta\in\Theta(y^N,S_m,p)} \mathrm{DC}_N(\theta,\hat{\theta}_{S_m}(y^N))$$

Note that if the observed y^N can be produced by all elements of S_M, then for $p = 1$,

$$T_1(\theta,N,S_m) = S_M \quad (19)$$

and we have

$$0 \le \mathrm{DC}_N(\theta^*,\hat{\theta}_{S_m}(y^N)) \le \max_{\theta\in S_M} \mathrm{DC}_N(\theta,\hat{\theta}_{S_m}(y^N)) \quad (20)$$

However, if $p \ne 1$ the value of $\varepsilon_p(\theta, S_m, N)$ and therefore the set $T_p(\theta, N, S_m)$ depends on the *variance* of random variable $\mathrm{DC}(\theta,\hat{\theta}_{S_m}(Y^N))$. In most cases the variance of this error is a function of dimension (order) of S_m. With a fixed data of length N as the dimension m grows, the variance of error also grows. However the estimate bias is a decreasing function of order. Therefore for a given finite length data there is a tradeoff between the error variance and bias.

The MDC order estimation method suggests to choose the following subset

$$S_m^* = \arg\min_{S_m} \max_{\theta\in\Theta(y^N,S_m,p)} \mathrm{DC}(\theta,\hat{\theta}_{S_m}(y^N)) \quad (21)$$

which provides the MDC with validation probability p.

3.2 *Impulse Response Identification of an LTI system*

Finite length input and corrupted output of an LTI system, which is at rest, is available. The system output is corrupted by an additive white Gaussian noise (AWGN) which is zero mean and has variance σ_w^2. The goal is to find the best estimate of the impulse response of the system.

Note that for a system which is at rest the input and output of length N are related to each other only by h^*, the first N elements of the impulse response. Therefore the unknown h^* is an element of a set of order N, S_N ($M = N$). By implementing the MDC we want to choose an estimate of h^* of proper length $m^* \le N$ which minimizes the description complexity of the true system.

Subset S_m of R^N represents one of the spaces of impulse responses of length m. The input-output relationship of the system is

$$\begin{aligned} y &= \bar{y} + w = h^* * u + w \quad (22)\\ &= h^*_{S_m} * u + \Delta_{S_m} * u + w \\ &= A_{S_m} h^*_{S_m} + B_{S_m}\Delta_m + w \end{aligned}$$

where u is the input, $\bar{y}$ is the noiseless output and y is the noisy output. Also A_{S_m} and B_{S_m} are functions of input u. $h^*_{S_m}$ is the projection of h^* in S_m. It is an element of S_m which is a vector of length N with only m nonzero elements. In each subset S_m, $\hat{h}_{S_m}(y^N)$ is the ML estimate of h.

Here we use the cost function of form (10) where the logarithm is a natural logarithm. The description complexity of random variable Y^N in (5), when the data is generated by h_1,

$$\mathrm{DC}_N(h_1,h_2) = \log\sqrt{2\pi\sigma_w^2} + E_{h_1}\left(\frac{||Y^N - \bar{y}_{h_2}||^2}{2N\sigma_w^2}\right) \tag{23}$$

where $\bar{y}_{h_2} = u * h_2$. Note that in this scenario we have

$$\mathrm{DC}_N(h,h) = H_\theta(Y^N)\log 2 = \log\sqrt{2\pi\sigma_w^2} + \frac{1}{2} \tag{24}$$

which is the same for all elements of S_M. Therefore, the comparison of the DC and comparison of Kullback-Leibler distance in (13) are the same.

In each subset $\bar{h}_{S_m}(N)$, defined in (6), is

$$\bar{h}_{S_m}(N) = h_{S_m} + \left(\frac{1}{N}A_{S_m}^T(N)A_{S_m}(N)\right)^{-1}\frac{1}{N}A_{S_m}^T(N)B_{S_m}(N)\Delta_{S_m}. \tag{25}$$

where h_{S_m} is the projection of h in subset S_m. The minimum description complexity of h in S_m is

$$\mathrm{DC}_N(h,\bar{h}_{S_m}(N)) = \log\sqrt{2\pi\sigma_w^2} + \frac{1}{2}(1 + \frac{1}{N\sigma_w^2}||G_{S_m}(N)B_{S_m}(N)\Delta_{S_m}||^2) \tag{26}$$

where

$$G_{S_m}(N) = I - \frac{1}{N}A_{S_m}(N)\left(\frac{1}{N}A_{S_m}^T(N)A_{S_m}(N)\right)^{-1}A_{S_m}^T(N) \tag{27}$$

is a projection matrix.

3.3 *MDC and order estimation*

The ML estimate in this example is an efficient estimator which satisfies the necessary condition in (14)

$$\hat{h}_{S_m}(y^N) = \arg\min_{g\in S_m} ||y^N - y_g||^2 \tag{28}$$

where $y_g = u * g$. The observed data is generated by h^*, the unknown elements of S_M.

The goal is to find probabilistic bounds on $\mathrm{DC}_N(h^*,\hat{h}_{S_m}(y^N))$ based on the observed data in each subset S_m. The first step is the validation step in which $\mathrm{DC}_N(h^*,\bar{h}_{S_m}(N))$ is validated. This calculation is based on the observed error

$$V(\hat{h}_{S_m}(y^N),y^N) = \log\sqrt{2\pi\sigma_w^2} + \frac{1}{2}\left(1 + \frac{||y - \hat{y}_{S_m}||^2}{N\sigma_w^2}\right) \tag{29}$$

where $\hat{y}_{S_m} = u * \hat{h}_{S_m}(y^N)$. This is a sample of a Chi-square random variable with the following expected value and variance

$$E_h(V(\hat{h}_{S_m}(Y^N),Y^N)) = \mathrm{DC}_N\left(h,\bar{h}_{S_m}(N)\right) + \frac{1}{2}\frac{M-m}{N} \tag{30}$$

$$\mathrm{var}\left(V(\hat{h}_{S_m}(Y^N),Y^N)\right) = \frac{1}{2N}\frac{M-m}{N} + \frac{1}{N\sigma_w^2}\left(\mathrm{DC}_N\left(h,\bar{h}_{S_m}(N)\right) - \mathrm{DC}_N(h,h)\right) \tag{31}$$

Therefore, for a chosen p_1, the set $\mathrm{DC}_N(h^*,\bar{h}_{S_m}(N))$ is validated by using the Chi-square distribution table. This set is then used to find bounds on the DC criterion $\mathrm{DC}_N(h^*,\hat{h}_{S_m}(y^N))$ in (21) for each subset. MDC chooses the subset which minimizes the obtained upper bound on the description complexity.

3.4 *Estimation of MDC*

Here we use the properties of the second order statistics of the random variable $V(\hat{h}_{S_m}(Y^N),Y^N))$. The expected value and variance of this random variable is such that the validation step in calculation of $\Theta(y^N,S_m,p)$ can provide bounds on $\mathrm{DC}_N\left(h^*,\bar{h}_{S_m}^*(N)\right)$

$$L_{S_m}(y^N,p_1) \le \mathrm{DC}_N\left(h^*,\bar{h}_{S_m}^*(N)\right) \le U_{S_m}(y^N,p_1) \tag{32}$$

On the other hand $\mathrm{DC}_N(h,\hat{h}_{S_m}(y^N))$ itself is a random variable

$$\mathrm{DC}_N(h,\hat{h}_{S_m}(y^N)) = \log\sqrt{2\pi\sigma_w^2} + E_h\left(\frac{||Y^N - \hat{y}_{S_m}||^2}{2N\sigma_w^2}\right) \tag{33}$$

which is a Chi-square random variable with the following expected value and variance

$$E_h\mathrm{DC}_N(h,\hat{h}_{S_m}(Y^N)) = \mathrm{DC}_N\left(h,\bar{h}_{S_m}(N)\right) + \frac{m}{2N} \tag{34}$$

$$\mathrm{var}_h\mathrm{DC}_N(h,\hat{h}_{S_m}(Y^N)) = \frac{m}{2N^2} + \frac{1}{N\sigma_w^2}\left(\mathrm{DC}_N(h,\bar{h}_{S_m}(N)) - \mathrm{DC}_N(h,h)\right) \tag{35}$$

The second order statistics of this random variable depends only on $\mathrm{DC}_N(h,\bar{h}_{S_m}(N))$, m, N, and $\mathrm{DC}_N(h,h)$, which is fixed for all h. Therefore, by using $\mathrm{DC}_N(h,\bar{h}_{S_m}(N))$ we can provide probabilistic bounds on $\mathrm{DC}_N(h,\hat{h}_{S_m}(y^N))$

$$|\mathrm{DC}_N(h,\hat{h}_{S_m}(y^N)) - E_h\mathrm{DC}_N(h,\hat{h}_{S_m}(Y^N))| \le \varepsilon_p(h,S_m,N) \tag{36}$$

The probability p is the probability that this DC is at most in $\varepsilon_p(h,S_m,N)$ distance of its expected value.

Hence, with probability p_1 bounds on $\mathrm{DC}_N(h^*, \bar{h}^*_{S_m}(N))$ can be *validated* and without calculation of the set $\Theta(y^N, S_m, p)$ *probabilistic* bounds, with probability p, on $\mathrm{DC}_N(h^*, \hat{h}_{S_m}(Y^N))$ can be calculated. The provided bounds are

$$d_L(y^N, S_m, p, p_1) \leq \mathrm{DC}_N(h^*, \hat{h}_{S_m}(y^N)) \leq d_U(y^N, S_m, p, p_1) \quad (37)$$

The optimum subset, using this estimate of MDC, is

$$S^*_m(y^N) = \arg\min_{S_m} d_U(y^N, S_m, p, p_1). \quad (38)$$

When m and $N - m$ are large enough, the Chi-square distributions of $V(\hat{h}_{S_m}(y^N), y^N)$ and $\mathrm{DC}_N(h^*, \hat{h}_{S_m}(y^N))$ can be well estimated with Gaussian distributions. In this case the validation probability p_1 and the confidence probability p can be defined in term of $Q(\cdot)$ function [1]. The following theorem provides bounds on the desired DC, $\mathrm{DC}_N(h^*, \hat{h}_{S_m}(y^N))$, for this scenario. The calculation of the bounds is similar to the quality evaluation of LTI system estimates in (Beheshti and Dahleh, 2000)

Theorem When m, the order of S_m, and $M - m$ are large enough the Chi-square distributions in of $V(\hat{h}_{S_m}(y^N), y^N)$ and $\mathrm{DC}_N(h^*, \hat{h}_{S_m}(y^N))$ can be estimated with Gaussian distributions. Consider $p_1 = Q(\alpha)$ and $p = Q(\beta)$. Then for the LTI system in (22), the upper and lower bounds $d_L(y^N, S_m, p, p_1)$ and $d_U(y^N, S_m, p, p_1)$ are

$$d_U(y^N, S_m, Q(\alpha), Q(\beta)) = \frac{m}{2N} + \frac{1}{2} + \log\sqrt{2\pi\sigma_w^2} + 2\sigma_w^2 U_{S_m} + \frac{\beta}{\sqrt{N}}\sqrt{\frac{m}{2N} + 2U_{S_m}} \quad (39)$$

and

$$d_L(y^N, S_m, Q(\alpha), Q(\beta)) = \max\{0, \frac{m}{2N} + \frac{1}{2} + \log\sqrt{2\pi\sigma_w^2} + 2\sigma_w^2 L_{S_m} - \frac{\beta}{\sqrt{N}}\sqrt{\frac{m}{2N} + 2U_{S_m}}\} \quad (40)$$

where L_{S_m} and U_{S_m} are defined as follows

$$U_{S_m} = x_{S_m} - m_w + \frac{2\alpha^2\sigma_w^2}{N} + K_{S_m}(\alpha). \quad (41)$$

where $m_w = (1 - \frac{m}{N})\sigma_w^2$, and

$$x_{S_m} = \frac{1}{N}||y^N - \hat{y}^N_{S_m}||^2 \quad (42)$$

and

$$K_{S_m}(\alpha) = 2\alpha\frac{\sigma_w}{\sqrt{N}}\sqrt{\frac{\alpha^2\sigma_w^2}{N} + x_{S_m} - \frac{1}{2}m_w}. \quad (43)$$

[1] $Q(x) = \frac{1}{\sqrt{2\pi}}\int_{-x}^{x} e^{-t^2/2}dt$

If $(m_w - \alpha\sqrt{v_m}) \leq x_{S_m} \leq (m_w + \alpha\sqrt{v_m})$, where $v_m = \frac{2}{N}(1 - \frac{m}{N})\sigma_w^4$, the lower bound L_{S_m} is zero and if $(m_w + \alpha\sqrt{v_m}) \leq x_{S_m}$ then

$$L_{S_m} = x_{S_m} - m_w + \frac{2\alpha^2\sigma_w^2}{N} - K_{S_m}(\alpha) \quad (44)$$

Consider the following conditions on α and β

$$\alpha_N \geq \sqrt{\frac{N}{2}}\left(1 - \frac{x_m}{(1 - \frac{m}{N})\sigma_w^2}\right), \quad (45)$$

$$\lim_{N\to\infty} \alpha_N = \infty, \lim_{N\to\infty} \beta_N = \infty, \quad (46)$$

$$\lim \frac{\alpha_N}{\sqrt{N}} = 0, \lim_{N\to\infty} \frac{\beta_N}{\sqrt{N}} = 0. \quad (47)$$

These are the sufficient conditions for the bounds on the DCs to approach each other for when $m << N$. Also the conditions guarantee that the validation and confidence probabilities $p_1 = Q(\alpha)$ and $p = Q(\beta)$ approach one as length of data, N, grows.

3.5 *Comparison of Order Estimation Methods*

AIC in (1) and description length in two-stage MDL (2) for the LTI system in (22) are

$$\mathrm{AIC}_{S_m}(y^N) = -\log\left(\frac{1}{\sqrt{2\pi}\sigma_w}e^{-\frac{||y - \hat{y}_{S_m}||^2}{2N\sigma_w^2}}\right) + \frac{m}{N} \quad (48)$$

$$\mathrm{DL}_{S_m}(y^N) = -\log\left(\frac{1}{\sqrt{2\pi}\sigma_w}e^{-\frac{||y - \hat{y}_{S_m}||^2}{2N\sigma_w^2}}\right) + m\frac{\log N}{2N} \quad (49)$$

Similar to the MDC criterion, these criteria are functions of the output error (42), the variance of the additive noise, length of the data and order of the subset. However, unlike MDC calculation, to calculate these criteria, it is assumed that the true impulse response is an element of the subset S_m!.

In practical applications one important method of order estimation evaluation is to check if the method is consistent. A consistent method is able to point to the subset with smallest order which includes the true model set as the length of the data grows. It is known that MDL is a consistent order estimation method and AIC is not a consistent method. For MDC, the consistency of the method is guaranteed by the proper choice of α, and β. As the length of the data grows these parameters have to be chosen such that the validation and estimation probabilities approach one. Therefore, an improper choice of $\alpha = \beta = 0$ leads to a criterion which is not consistent. It is important to note that for subset S_m which includes the true model set, the MDC criterion in (39) with $\alpha = \beta = 0$ is the AIC in (48). Also, it should be mentioned that the calculated MDC for LTI systems in this paper is the same as the new MDL criterion for linear models which is introduced in (Beheshti and Dahleh, 2003) and is comparable with the two-stage MDL.

When the signal to noise ratio is considerably large and the true system has an infinite length impulse response, the behavior of a consistent method might not be desirable. In this case a practical method should be able to suggest a threshold on the criterion, otherwise the consistent method chooses the model set with the highest order. For this scenario while MDC is able to provide a thresholding method, MDL thresholding is not possible. More detailed discussion on these practical issues is in (Beheshti, 2002) and (Beheshti and Dahleh, 2002).

4. CONCLUSION

In this paper we presented MDC, a new method of order estimation. We elaborated on the advantages of this consistent method over the available information theoretic solutions. It was shown that AIC is a special case of MDC criterion. In this paper the proposed method calculated the description complexity of noisy data for a family of Gaussian distributions. The approach can be extended for calculation of the description complexity for more general classes of linear models with additive noises and also for when the variance of the additive noise is unknown.

5. REFERENCES

Akaike, H. (1974). A new look at the statistical model identification. *IEEE Trans. on Automatic Control* **19**, 716–723.

A.P. Liavas, P.A. Regalia and J.Delmas (1999). Blind channel approximation: effective channel order estimation. *IEEE Trans. on Signal Processing* **47**, 3336–3344.

Barron, Y., J. Rissanen and B. Yu (1998). he minimum description length principle in coding and modeling. *IEEE Trans. on Information Theory* **44**, 2743–2760.

Beheshti, S. (2002). *Minimum Description Complexity, Ph.D. thesis*. MIT.

Beheshti, S. and M.A. Dahleh (2000). On model quality evaluation of stable lti systems. *Proceeding of the 39th IEEE Conference on Decision and Control.*

Beheshti, S. and M.A. Dahleh (2002). On denoising and signal representation. *Proceedings of the 10th Mediterranean Conference on Control and Automation.*

Beheshti, S. and M.A. Dahleh (2003). A new minimum description length. *Proceeding of the IEEE Conference on American Control Conference.*

Rissanen, J. (1984). Universal coding, information, prediction, and estimation. *IEEE Trans. on Information Theory* **30**, 629–636.

Schwarz, G. (1978). Estimating the dimension of a model. *The Annals of Statistics* **6**, 461–464.

Wax, M. and T. Kailath (1985). Detection of signals by information theoretic criteria. *IEEE Trans. Acoust., Speech, Signal Processing* **33**, 387–392.

IFAC Publications
www.elsevier.com/locate/ifac

CONDITIONS FOR LOCAL CONVERGENCE OF MAXIMUM LIKELIHOOD ESTIMATION FOR ARMAX MODELS

Graham C. Goodwin * **Juan Carlos Agüero** *
Robert E. Skelton **

* *The University of Newcastle, Australia*
** *University of California, San Diego*

Abstract: This paper analyzes the conditions for local convergence of Maximum Likelihood estimation for ARMAX models. We do this by examining the region in which the steepest descent direction leads to a reduction in the Euclidean norm of the parameter error. Inter-alia this gives new insights into the question of existence of local maxima of the likelihood function for various commonly used model structures. *Copyright © 2003 IFAC*

Keywords: Parameter estimation, Maximum likelihood principle, Convergence analysis.

1. INTRODUCTION

Many of the methods used in the field of system identification are based on the principle of Maximum Likelihood (**ML**) (Kendall and Stuart, 1967; Goodwin and Payne, 1977; Ljung, 1999). This procedure transforms the estimation problem into an optimization problem. The basic idea of the maximum likelihood algorithm can be described as follows: Say that we have a parametrized family of probability density functions $p(Y|\theta)$ which describe the probability of the random variable $Y = [y_1, y_2, \cdots, y_N]$ given the parameter vector θ, having "true value", θ_0. Then, given a set of data $Y = \boldsymbol{y}$ the likelihood function is simply $\mathcal{L}(\theta) = p(Y = \boldsymbol{y}|\theta)$. The maximum likelihood estimate is obtained by maximizing $p(Y = \boldsymbol{y}|\theta)$ with respect to θ or equivalently by minimizing the negative log-likelihood function, $l(\theta)$, i.e.

$$\theta_{ML} = \underset{\theta}{arg\,min}\; l(\theta)$$
$$l(\theta) = -ln\left[p(Y = \boldsymbol{y}|\theta)\right] \qquad (1)$$

Well known advantages of this procedure are that, subject to mild conditions,

(i) it is consistent, i.e. θ_{ML} converges to the true value θ_0 (in a suitable probabilistic sense), and
(ii) it is asymptotically efficient, in the sense that for large sample sizes it achieves the Cramér Rao inequality which gives a lower bound on the covariance of the parameter estimates for all unbiased estimators.

However, the **ML** technique often leads to a non convex optimization problem. This is typically solved by using an iterative algorithm, e.g. Gradient-based method or a more sophisticated method like Expectation-Maximization (**EM**) (Dempster *et al.*, 1977).

Here our emphasis will be on linear single-input single-output stochastic systems.

1.1 *Model Structure*

Model structure is one of the first decisions to take when we are trying to identify a process. A general family of model structure is (Ljung, 1999):

$$A_o(q^{-1})y_t = \frac{B_o(q^{-1})}{F_o(q^{-1})}u_t + \frac{C_o(q^{-1})}{D_o(q^{-1})}w_t \qquad (2)$$

where

$$\begin{aligned} A_o(q^{-1}) &= 1 + a_1^o q^{-1} + \cdots + a_{na}^o q^{-na} \\ B_o(q^{-1}) &= b_1^o q^{-1} + \cdots + b_{nb}^o q^{-nb} \\ C_o(q^{-1}) &= 1 + c_1^o q^{-1} + \cdots + c_{nc}^o q^{-nc} \\ D_o(q^{-1}) &= 1 + d_1^o q^{-1} + \cdots + d_{nd}^o q^{-nd} \\ F_o(q^{-1}) &= 1 + f_1^o q^{-1} + \cdots + f_{nf}^o q^{-nf} \end{aligned} \tag{3}$$

where the subindex "o" denotes true value. In this paper we assume that the input u_t is known, and that $w_t \sim N(0, \sigma^2)$. In the sequel, we will omit the subindex "o" when we refer to the current estimate.

The model structure described above has been given different names in the literature depending on which of the five polynomial are used. For example when $C = D = F = 1$ the model structure is called ARX (See Ljung (1999) for more details).

1.2 *Local Minima*

Another issue in **ML** estimation concerns the question of convergence to local minima. It is known that, depending on the model structure and the signal-to-noise ratio the likelihood function can have local minima. Some specific cases have been analyzed in the literature; e.g.:

- For ARMA models ($B = 0, D = F = 1$) there are no local minima (Åström and Söderström, 1974).
- For ARX models ($C = D = F = 1$) there are no local minima.
- For ARARX ($C = F = 1$) there are no false local minima if the signal-to-noise ratio is large enough (Söderström, 1975).
- If $A = 1$, there are no false local minima if $nf = 1$.
- If $A = C = D = 1$, there are no false local minima provided the input is white noise (Söderström, 1975).
- For ARMAX models ($\mathbf{F} = \mathbf{D} = 1$), there are no false local minima if $Re\left\{\frac{\mathbf{C}_o(e^{-jw})}{\mathbf{C}(e^{-jw})}\right\} > 0$ (Ljung, 1999).

In this paper we analyze the ARMAX ($F = D = 1$) model. We will present what we believe to be a new result, which gives a necessary and sufficient condition for the region of parameter space in which a gradient algorithm will move the estimated parameters closer (in a Euclidean sense) to the true parameter value. Using the definition of this region we will investigate conditions to obtain, or not, local minima. We show, via the tools developed here, that for ARX ($C = D = F = 1$) models there are no local minima. This re-establish a well known result.

2. MAXIMUM LIKELIHOOD ESTIMATION FOR ARMAX MODELS

Consider the ARMAX version of (2) where we take $n_a = n_b = n_c = n$, and $D = F = 1$.

The likelihood function is:

$$\begin{aligned} \mathcal{L}(\theta) &= p(y_0, y_1, \cdots, y_n|\theta) \\ &= \prod_{k=1}^{N} p(y_k|y_0, \cdots, y_{k-1}, \theta) \end{aligned} \tag{4}$$

where θ is the vector that contains all the parameters of the polynomials A, B, and C.

Based on a Gaussian assumption for w_t, the negative-log-likelihood function for the system is readily seen to be

$$l(\theta) = constant + N log[\sigma] + \sum_{k=1}^{N} \frac{e_k(\theta)^2}{2\sigma^2} \tag{5}$$

where

$$e_k(\theta) = \frac{A_\theta(q^{-1})y_k - B_\theta(q^{-1})u_k}{C_\theta(q^{-1})} \tag{6}$$

and where the subscript θ refers to the parameter vector, $\theta = (a_1, \ldots, a_n, b_1, \ldots, b_n, c_1 \ldots, c_n)^T$.

Once θ_{ML} has been found, then the estimate for σ is readily found. Indeed, by differentiating (5) we find

$$\frac{\partial l(\theta)}{\partial \sigma} = \frac{N}{\sigma} - \sum_{k=1}^{N} \frac{e_k(\theta)^2}{\sigma^3} \tag{7}$$

Hence, setting $\frac{\partial l}{\partial \sigma}$ to zero, we find

$$\sigma_{ML}^2 = \frac{1}{N} \sum_{k=1}^{N} e_k(\theta_{ML})^2 \tag{8}$$

Also, the maximum likelihood estimate for θ can be obtained by minimising

$$J(\theta) = \frac{1}{2N} \sum_{k=1}^{N} e_k(\theta)^2 \tag{9}$$

since σ is simply a scaling factor.

We can see from (6) and (9) that $l(\theta)$ is, in general, a non-quadratic function of θ and thus we might well anticipate that multiple local maxima points may exist. Hence, iterative algorithms (such as a gradient based algorithm) can only be expected to converge to a local maximum of the likelihood function. To study this problem, we next examine the gradient of the likelihood function:

Lemma 1. The gradient of $J(\theta)$ with respect to θ is:

$$\frac{\partial J}{\partial \theta} = \frac{1}{N} \sum_{k=1}^{N} e_k \frac{\partial e_k}{\partial \theta} \tag{10}$$

$$\frac{\partial e_k}{\partial \theta} = -\frac{1}{C(q^{-1})} \varphi_{k-1} \tag{11}$$

where

$$\varphi_{k-1} = \begin{bmatrix} -y_{k-1} \\ \vdots \\ -v_{k-n} \\ u_{k-1} \\ \vdots \\ u_{k-n} \\ e_{k-1} \\ \vdots \\ e_{k-n} \end{bmatrix} \tag{12}$$

PROOF.

Equation (10) is immediate from (9). Also, from (6), we have

$$\frac{\partial e_k}{\partial \theta} = \begin{bmatrix} \frac{y_{k-1}}{C(q^{-1})} \\ \vdots \\ \frac{y_{k-n}}{C(q^{-1})} \\ \frac{u_{k-1}}{C(q^{-1})} \\ \vdots \\ \frac{u_{k-n}}{C(q^{-1})} \\ -\frac{A(q^{-1})y_{k-1} - B(q^{-1})u_{k-1}}{C(q^{-1})^2} \\ \vdots \\ -\frac{A(q^{-1})y_{k-n} - B(q^{-1})u_{k-n}}{C(q^{-1})^2} \end{bmatrix} \tag{13}$$

$$= -\frac{1}{C(q^{-1})} \begin{bmatrix} -y_{k-1} \\ \vdots \\ -y_{k-n} \\ u_{k-1} \\ \vdots \\ u_{k-n} \\ e_{k-1} \\ \vdots \\ e_{k-n} \end{bmatrix} \tag{14}$$

□

3. THE REGION OF INTEREST

The region of interest is defined as follows:

Definition 2. The region Γ of Decreasing Euclidean Parameter Error Norm (DEPEN) in parameter space, is defined to be those values of θ such that an infinitesimal step in the negative gradient direction of $J(\theta)$ takes θ closer (in a Euclidean sense) to the true value θ_o.

Lemma 3. $\theta \in \Gamma$ if and only if the inner product between the vector $\tilde{\theta} \triangleq (\theta - \theta_o)$ and the gradient direction at θ, i.e. $\frac{\partial J}{\partial \theta}$, is positive.

PROOF.

If θ_i is the current estimate parameter, the gradient algorithm can be expressed as:

$$\theta_{i+1} = \theta_i + \beta \left[-\frac{\partial J^i}{\partial \theta} \right] \tag{15}$$

Subtracting θ_o from both sides yields

$$\tilde{\theta}_{i+1} = \tilde{\theta}_i - \beta \left[\frac{\partial J^i}{\partial \theta} \right] \tag{16}$$

Squaring both sides gives

$$\tilde{\theta}_{i+1}^T \tilde{\theta}_i = \tilde{\theta}_i^T \tilde{\theta}_i - 2\beta \tilde{\theta}_i^T \left[\frac{\partial J^i}{\partial \theta} \right] + \beta^2 \left[\frac{\partial J^i}{\partial \theta} \right]^T \left[\frac{\partial J^i}{\partial \theta} \right] \tag{17}$$

For β sufficiently small, the last term is negligible and the result follows. □

Remark 4. The definition of DEPEN region is closely related to the region where the likelihood function is *quasi-convex*. A cost function is said to be *quasi-convex* in a convex region $\mathcal{D}_f$ if, for all $x, y \in \mathcal{D}_f$, and $0 \leq \lambda \leq 1$, $J(\lambda x + (1-\lambda)y) \leq max(J(x), J(y))$ (Roberts and Varberg, 1973; Rockafellar, 1972; Boyd and Vandenberghe, 2002). Thus, for a convex region $\mathcal{D}_f$ ($\theta_o \in \mathcal{D}_f$) that is not DEPEN, the likelihood function is not *quasi-convex* on $\mathcal{D}_f$. ▽▽▽

Lemma 5. If the DEPEN region Γ is the complete parameter space, then there are no local minima.

PROOF. If there are local minima, in a ball around a local minimum the gradient must take one closer to the local minimum and away from the global minimum. □

We recall the following result from the analysis of recursive maximum likelihood algorithms (Solo, 1979; Ljung, 1999).

Lemma 6. The vector φ_{k-1} in equation (12) satisfies the following dynamical relationship:

$$C_o(q^{-1})\{e_k - w_k\} = -(\varphi_{k-1})^T \tilde{\theta} \tag{18}$$

PROOF.

$$\begin{aligned} e_k &= \frac{A(q^{-1})y_k - B(q^{-1})u_k}{C(q^{-1})} \\ &= y_k - (\varphi_{k-1})^T \theta \\ &= (\varphi_{k-1}^*)^T \theta_o + w_k - (\varphi_{k-1})^T \theta \end{aligned} \tag{19}$$

where φ_{k-1}^* is the following vector

$$\varphi_{k-1}^* = \begin{bmatrix} -y_{k-1} \\ \vdots \\ -y_{k-n} \\ u_{k-1} \\ \vdots \\ u_{k-n} \\ w_{k-1} \\ \vdots \\ w_{k-n} \end{bmatrix} \quad (20)$$

Adding and subtracting $(\varphi_{k-1})^T\theta_o$ to the right hand side of (19) we obtain

$$\begin{aligned} e_k =& (\varphi_{k-1})^T\theta_o - (\varphi_{k-1})^T\theta + (\varphi_{k-1}^*)^T\theta_o \\ & - (\varphi_{k-1})^T\theta_o + w_k \\ =& -(\varphi_{k-1})^T\tilde{\theta} \\ & + [0, \cdots 0, w_{k-1} - e_{k-1}, \cdots w_{k-n} - e_{k-n}]\theta_o \\ & + w_k \\ =& -(\varphi_{k-1})^T\tilde{\theta} + [C_o(q^{-1}) - 1][w_k - e_k] + w_k \end{aligned}$$

Rearranging this equation gives (18). □

We can then state our key result:

Theorem 7. (a) In the case of open loop identification and for large N, a necessary and sufficient condition for θ to belong to Γ (Definition 2) is that the following frequency domain inequality holds for the various polynomials

$$\int_0^\pi Re\{G_1(e^{-jw})\} \{|G_2(e^{-jw})|^2\phi_{uu}(w) + |G_3(e^{-jw})|^2\sigma^2\}dw > 0 \quad (21)$$

where

$$G_1 = \frac{C_o(q^{-1})}{C(q^{-1})} \quad (22)$$

$$G_2 = \frac{A(q^{-1})B_o(q^{-1}) - A_o(q^{-1})B(q^{-1})}{A_o(q^{-1})C(q^{-1})} \quad (23)$$

$$G_3 = \frac{A(q^{-1})C_o(q^{-1}) - A_o(q^{-1})C(q^{-1})}{A_o(q^{-1})C(q^{-1})} \quad (24)$$

(b) A sufficient condition for (21) to hold for all θ is that the transfer function $\left[\frac{C_o(e^{-jw})}{C(e^{-jw})}\right]$ be strictly positive real.

PROOF.

(a) From Lemma 3, we require

$$\tilde{\theta}^T\left[\frac{\partial J}{\partial \theta}\right] > 0 \quad (25)$$

Now using Lemma 1 and Lemma 6 we have

$$\begin{aligned} \tilde{\theta}^T\left[\frac{\partial J}{\partial \theta}\right] &= \frac{1}{N}\sum_{k=1}^N e_k\left[-\frac{1}{C(q^{-1})}(\varphi_{k-1})^T\tilde{\theta}\right] \\ &= \frac{1}{N}\sum_{k=1}^N e_k\left[\frac{C_o(q^{-1})}{C(q^{-1})}(e_k - w_k)\right] \quad (26) \\ &= \frac{1}{N}\sum_{k=1}^N [e_k - w_k + w_k]\left[\frac{C_o(q^{-1})}{C(q^{-1})}(e_k - w_k)\right] \end{aligned}$$

For large N, the term $\frac{1}{N}\sum \omega_k\left[\frac{C_o(q^{-1})}{C(q^{-1})}(e_k - \omega_k)\right]$ goes to zero since ω_k is a white noise sequence and the second term depends on $\omega_{k-1}, \omega_{k-2}\cdots$

Hence for large N

$$\tilde{\theta}^T\left[\frac{\partial J}{\partial \theta}\right] = \lim_{N\to\infty}\frac{1}{N}\sum_{k=1}^N \alpha_k\left[\frac{C_o(q^{-1})}{C(q^{-1})}\alpha_k\right] \quad (27)$$

where

$$\alpha_k = e_k - w_k \quad (28)$$

Using the definition of e_k we have

$$e_k = \frac{A(q^{-1})y_k - B(q^{-1})u_k}{C(q^{-1})} \quad (29)$$

where

$$y_k = \frac{B_o(q^{-1})u_k + C_o(q^{-1})w_k}{A_o(q^{-1})} \quad (30)$$

Hence

$$\begin{aligned} e_k - w_k =& \frac{A(q^{-1})B_o(q^{-1}) - A_o(q^{-1})B(q^{-1})}{A_o(q^{-1})C(q^{-1})}u_k \\ & + \frac{A(q^{-1})C_o(q^{-1}) - A_o(q^{-1})C(q^{-1})}{A_o(q^{-1})C(q^{-1})}w_k \end{aligned} \quad (31)$$

Substituting (31) into (27) and using Parseval's theorem and the orthogonality of $\{u_k\}$ and $\{w_k\}$ (a consequence of open loop operation) gives

$$\tilde{\theta}^T\left[\frac{\partial J}{\partial \theta}\right] = \frac{1}{\pi}\int_0^\pi Re\{G_1(e^{-jw})\} \quad (32)$$

$$\{|G_2(e^{-jw})|^2\phi_{uu}(w) + |G_3(e^{-jw})|^2\sigma^2\}dw \quad (33)$$

where G_1, G_2 and G_3 are as in (22), (23) and (24).

(b) Follows from the properties of positive real functions. □

Remark 8. (1) Notice that $\tilde{\theta}^T\frac{\partial J}{\partial \theta}$ is zero when $A = A_o$, $B = B_o$ and $C = C_o$.
(2) G_2 and G_3 reflect the frequency domain location of the modelling errors. The real part of G_1 is required to be positive where these model errors are significant.

(3) Similar results can be obtained for the case of closed loop identification when u_k and ω_k are correlated (via the feedback loop). ▽▽▽

For an ARX ($C = D = F = 1$) model there are no local minima since the cost function $l(\theta)$ is quadratic. In this case $G_1(q^{-1}) = 1$, and hence the DEPEN region is the complete parameter space, and we can conclude that there are no local minima from Lemma 5. We also have the following result for first order ARMAX systems:

Corollary 9. For a first order ($n = 1$) ARMAX model ($F = D = 1$) there are no local minima.

PROOF. Consider $G_1(q^{-1}) = \frac{1+c_1^o q^{-1}}{1+c_1 q^{-1}}$, where $-1 < c_1^o < 1$, and $-1 < c_1 < 1$. We need to calculate the real part of $G_1(e^{-jw})$. To do that we convert the problem into one in the continuous time domain using the transformation $q = \frac{1+s}{1-s}$.

$$G_1^c(s) = \frac{s(1-c_1^o)+1+c_1^o}{s(1-c_1)+1+c_1} \tag{34}$$

$$G_1^c(jw) = \frac{1+c_1^o+jw(1-c_1^o)}{1+c_1+jw(1-c_1)} \tag{35}$$

and then the real part of $G_1^c(jw)$ is given by

$$Re\{G_1^c(jw)\} = \frac{(1+c_1^o)(1+c_1)+w^2(1-c_1^o)(1-c_1)}{(1+c_1)^2+(1-c_1)^2} \tag{36}$$

which is always greater than zero. Hence, applying Lemma 5 we obtain the result. □

Using the insight of the DEPEN region, we next show by means of an example how the form of the likelihood function for ARMAX models can be affected due to a bad choice of the input signal.

4. EXAMPLE

Consider the following "true" model

$$A_o(q^{-1}) = 1 + 0.2q^{-1} + 0.4q^{-2} + 0.1q^{-3} \tag{37}$$

$$B_o(q^{-1}) = 0.5q^{-1} + 0.3q^{-2} + 0.1q^{-3} \tag{38}$$

$$C_o(q^{-1}) = 1 + 1q^{-1} + 0.8q^{-2} + 0.2q^{-3} \tag{39}$$

We will focus on a specific point in the parameter space where the different polynomials are given by $C(q^{-1}) = 1 + 0.5q^{-1} + 0.9q^{-2} + 0.5q^{-3}$, $A(q^{-1}) = 1 - 0.0404q^{-1} + 0.4302q^{-2} + 0.0041q^{-3}$, and $B(q^{-1}) = 0.4983q^{-1} + 0.1830q^{-2} + 0.0648q^{-3}$. From the Nyquist plot of $\frac{C_o(q^{-1})}{C(q^{-1})}$ in figure 1, we can see that the real part of $\frac{C_o(q^{-1})}{C(q^{-1})}$ is negative for $w \in [1.59, 1.91]$.

Since the system $\frac{B_o(q^{-1})}{A_o(q^{-1})}$ is of third order, we choose the input signal as the sum of 3 sinusoidal

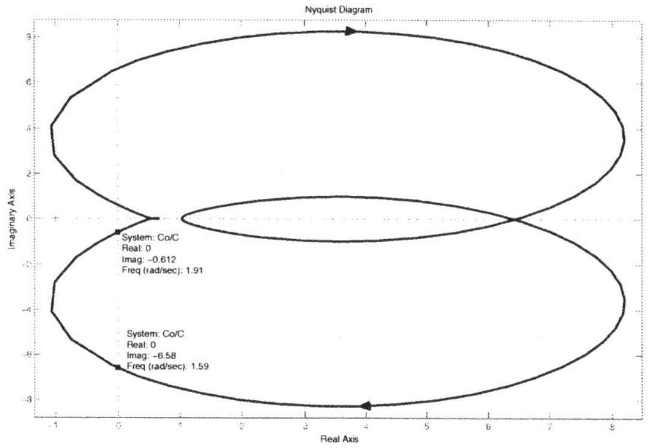

Fig. 1. Nyquist diagram of $\frac{C_o(q^{-1})}{1+0.5q^{-1}+0.9q^{-2}+0.5q^{-3}}$

signals with frequencies chosen where the Nyquist diagram (figure 1) has a negative real part:

$$u_k = 2\left[sin(1.7k) + sin(1.75k) + sin(1.83k)\right] \tag{40}$$

We assume that the white noise w_k is normally distributed with zero mean and variance σ^2. We will analyze two cases, one when the variance of the noise is $\sigma^2 = 2.25$, and another when $\sigma^2 = 10^{-4}$. The respective plots are shown in Figure 2. We can see that for $\sigma^2 = 2.25$ the likelihood function is a decreasing function on the segment joining θ and θ_o. On the other hand, when $\sigma^2 = 10^{-4}$, we can see that the likelihood function is an increasing function on a section of the segment joining θ and θ_o. This positive increase is clear from (21) for σ^2 small.

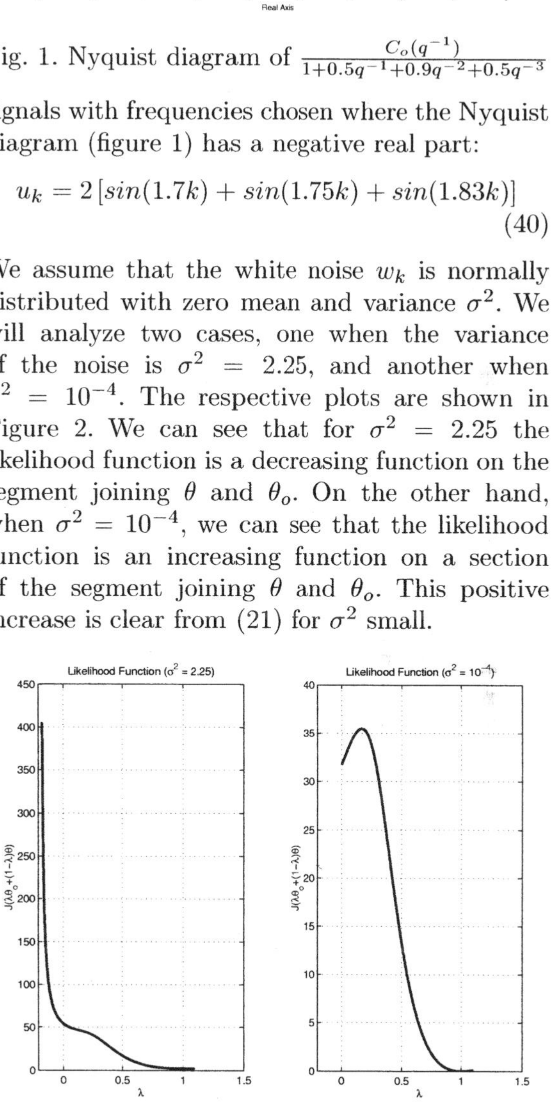

Fig. 2. Likelihood Function $J(\lambda\theta_o + (1-\lambda)\theta)$ for different values of σ^2.

5. CONCLUSIONS

In this paper we have defined the DEPEN region Γ as those values of θ such that an infinitesimal step in the negative gradient direction of the negative-log-likelihood takes θ closer (in a Euclidean sense) to the true value θ_o. We have used this result to establish necessary and sufficient condition for θ to belong to Γ. We show that the DEPEN region can be a useful tool to investigate the existence of local minima. Finally, using the ideas proposed in this paper, we show an example where the form of the Likelihood Function for ARMAX models change depending on the input spectrum. Extensions to the MIMO case are currently under investigation. There are also interesting links, not yet fully explored to the well known sufficient condition for global convergence of recursive maximum likelihood algorithms.

6. REFERENCES

Åström, K. and T. Söderström (1974). Uniqueness of the maximum likelihood estimates of the parameters of an ARMA model. *IEEE Transactions on Automatic Control* **AC-19**(6), 769–773.

Boyd, S. and L. Vandenberghe (2002). *Convex Optimization.* www.stanford.edu/~boyd/cvxbook.html.

Dempster, A. P., N. M. Laird and D. B. Rubin (1977). Maximum likelihood from imcomplete data via the EM algorithm. *Journal of the Royal Statistical Society* **39, Issue 1, Series B**, 1–38.

Goodwin, G. C. and R. Payne (1977). *Dynamic System Identification: Experiment design and data analysis.* Academic Press.

Kendall, M. and A. Stuart (1967). *The advanced theory of statistics.* Vol. 2. C. Griffin. London.

Ljung, L. (1999). *System Identification: Theory for the user.* 2nd. ed.. Prentice Hall.

Roberts, A. W. and D. E. Varberg (1973). *Convex Functions.* Academic Press. New York, USA.

Rockafellar, R. T. (1972). *Convex Analysis.* Princeton University Press. New Jersey, USA.

Söderström, T. (1975). On the uniqueness of maximum likelihood identification. *Automatica* **11**, 193–197.

Solo, V. (1979). The convergence of AML.. *IEEE Transactions on Automatic Control* **AC-24**, 958–963.

IFAC Publications
www.elsevier.com/locate/ifac

A NONPARAMETRIC APPROACH TO MODEL SELECTION

Maiza Bekara *,[1] **Abd-Krim Segouane** * **Fleury Gilles** *

* *École Supérieure d'Électricité - Service des Mesures,
Plateau de Moulon, 3 rue Joliot Curie,
91192 Gif sur Yvette Cedex, FRANCE
Tel: +33 [0]1 69 85 14 13; fax: +33 [0]1 69 85 14 29
Email: {firstname.lastname@supelec.fr}*

Abstract: We consider the problem of learning a regression function from samples based on a sequence of candidate models from which an optimal one is to be selected. In the absence of any reliable a priori information about the data generating process, we adopt a nonparametric approach to functionally characterize the data statistics. This nonparametric description is then used to derive a *nonparametric reference* model whose complexity is automatically determined by data-driven procedure. The *nonparametric complexity* can be used as a benchmark to select a suitable *parametric complexity* from the class of candidates. The proposed method is highly effective against overfitting and largely outperforms previous approaches in experimental study of polynomial curve fitting. The only requirement is to have access to a collection of *unlabeled* data. *Copyright © 2003 IFAC*

Keywords: Model selection, nonparametric smoothing, polynomial regression

1. INTRODUCTION

Many data modelling problems are characterized by the absence of an *a priori* model of the data generating process, such as models frequently available in physics (Utans and Moody, 1991). To overcome this difficulty, a class of candidate models $\mathscr{M}$ is constructed and it remains to select the model that "best explain" the data in a well defined sense. This is a major issue in machine learning in general and regression in particular, where the problem here can be viewed as belonging to the second stage of statistical inference, the first one corresponds to estimating parameters of various models in classical statistical approach.

It would be advantageous to make $\mathscr{M}$ as expressive as possible, since it would offer the greatest chance of representing the true model. However by making $\mathscr{M}$ too expressive we run the risk of *overfitting*. On the other hand restricting the complexity of $\mathscr{M}$ can introduce the opposite problem of *underfitting*. Thus, there is a trade off between the ability to represent a good model and the ability to identify a good model. This is the fundamental dilemma of machine learning (German *et al.*, 1992).

Model selection strategies are based on choosing the model that gives the smallest loss . Often, the loss is taken to be the *generalization error*, known also as the *prediction risk*. For some applications, the numerical value of the prediction risk is important in its own (Utans and Moody, 1991). Yet evaluating the generalization error requires the knowledge of the true generating model. To solve this problem, several schemes have been proposed in the literature leading to different model selection criteria (Linhart and Zucchini, 1986). These can be broadly classified into three major categories.

First, criteria based on minimizing an estimate of the generalization error. This estimate can be obtained by some data *resampling* techniques like cross validation (Stone, 1974) or the bootstrap (Efron and Tibshirani, 1995). It can also be obtained by *asymptotic expansion*, which usually relay on the fact that under certain regularity conditions, asymptotically, estimated parameters have a well known distribution,

[1] Corresponding author

which then permits to have an *asymptotically unbiased* estimate of the generalization error (Akaike, 1970).
Second, *structured risk minimization*, SRM, where an upper bound of the generalization error is obtained using some form of the uniform law of large numbers (Vapnik and Chernovenkins, 1971) and the criterion is based on minimizing this upper bound. SRM was successfully applied in may situations (Chapelle *et al.*, 2001), (Cherbassky *et al.*, 1997).
Finally, *metric based* approaches , like ADJ and TRI (Schuurmans, 1997) that exploit the metric structures of the class of candidate models by using a set of *unlabeled* data [2]. The idea behind this approaches is to readjust the estimate of the generalization error by multiplying the empirical loss estimate with an index that penalizes models that behave differently on and off training data.
In this paper, the problem of model selection for regression purpose is treated, in a general setting where the generating model does not necessarily belong to the class of candidate $\mathcal{M}$. In such a case, the model is said to be *unfaithful* (Murata *et al.*, 1994).
The proposed approach is based on using a *nonparametric* estimator of the data's probability density function to compute the generalization error. For regression, this leads to the construction of a *nonparametric reference* model, whose *complexity* is determined via data driven procedure. This nonparametric complexity is a trustful measure of model's complexity when we lack a reliable a priori information about the true model and can be used as a benchmark to select an appropriate parametric complexity. One can exploit additional information about the learning task when we dispose of an unlabeled data set to prevent catastrophic overfitting in a less heuristic way than the metric based approach.
The remainder of this paper is organized as follows. In section 2, a brief problem formulation is presented. Section 3, presents the nonparametric approach to model selection. Section 4 presents simulation results for polynomial regression. We end up by concluding remarks.

2. PROBLEM FORMULATION

Consider a training data set $D_n = \{(x_i,y_i)\}_{i=1}^n$ where x_i's are generated according to a distribution $q(x)$ and y_i's are assumed to be generated by unknown function f and a noise process,

$$y_i = f_0(x_i) + \varepsilon_i \qquad i = 1,\dots,n,$$

were ε_i are i.i.d random variable with a distribution $\pi(\varepsilon)$ that satisfies

$$\int \varepsilon\pi(\varepsilon) = 0 \quad \text{and} \quad \int \varepsilon^2\pi(\varepsilon) = \sigma_\varepsilon^2 < \infty.$$

In this context, data are represented by n independent and identically distributed pairs drawn according to a fixed but unknown joint probability density function $q(x,y) = \pi(y - f_0(x))\, q(x)$.
In the same way, a candidate family $\mathbf{M}_k$ of complexity index k is constructed as

$$\mathbf{M}_k = \left\{f(x;\theta_k), \theta_k \in \Theta_k \subset \mathbf{R}^{dim(\theta_k)}\right\}.$$

The index k, is an integer that is assigned to the family $\mathbf{M}_k$ as a label of its complexity and it increases with complexity increasing [3]. For simplicity, we assume $k = 1,2,\dots,k_{max}$ and the class of candidate models is $\mathcal{M} = \bigcup_{k=1}^{k_{max}} \mathbf{M}_k$.
We further dispose of a set of estimate $\hat{\theta}_k$ for the best fitted model $\hat{m}_k = \{f(x;\hat{\theta}_k)\} \in M_k$ using the training set D_n . Here, there is no assumption about the type of estimator used to obtain $\hat{\theta}_k$. The loss as defined by the generalization error

$$L(\hat{\theta}_k) = E_{q(x,y)}\left[(y - f(x;\hat{\theta}_k))^2\right], \tag{1}$$

where the expectation is with respect to the generation model's joint distribution. $L(\hat{\theta}_k)$ would provide a suitable measure of how good the fitted model $\hat{m}_k$ on the average generalizes the learning to unseen points. Yet evaluating $L(\hat{\theta}_k)$ is not possible, since doing so requires the knowledge of the generating model and further the knowledge of the noise distribution also. Using the empirical distribution

$$q^*(x,y) = \frac{1}{n}\sum_{i=1}^n \delta(x - x_i, y - y_i),$$

an estimate of the loss function called, the *empirical loss*, is obtained

$$L_{emp}(\hat{\theta}_k) = \frac{1}{n}\sum_{i=1}^n (y_i - f(x_i;\hat{\theta}_k))^2. \tag{2}$$

The empirical loss decreases when model's complexity increases and therefore leads to a biased model selection strategy (Cadez and Smyth, 2000).

3. NONPARAMETRIC APPROACH

3.1 *Derivation*

We assume that we dispose of a nonparametric estimate of $q(x,y)$, which we denote $\hat{q}(x,y)$. Nonparametric estimate is a suitable solution in such cases, where we lack reliable a priori information about the generating model.
First of all, let's introduce the regressor $f_{np}(x)$

$$f_{np}(x) = \int y\hat{q}(y|x)dy = \int \frac{y\hat{q}(y,x)}{\hat{q}(x)}dy.$$

We define the *smoothed* loss estimate by

---

[2] We consider data as input-output variables. Unlabeled data are those for which only the input is available

[3] Complexity is a measure of the degree of freedom of a given family. For $k = dim(\theta_k)$ it is the number of parameters in the model

$$L_s(\hat{\theta}_k) = \iint (y - f(x;\hat{\theta}_k))^2 \hat{q}(x,y)dxdy$$
$$= \iint (y - f_{np}(x) + f_{np}(x) - f(x;\hat{\theta}_k))^2 \hat{q}(x,y)dxdy$$
$$= \int (f_{np}(x) - f(x;\hat{\theta}_k))^2 \hat{q}(x)dx$$
$$+ \iint (y - f_{np}(x))^2 \hat{q}(x,y)dxdy.$$

The *smoothed* loss estimate in its final picture is given by

$$L_s(\hat{\theta}_k) = \int (f(x;\hat{\theta}_k) - f_{np}(x))^2 \hat{q}(x)dx + \hat{\sigma}^2_{np}.$$

The first part of $L_s(\hat{\theta}_k)$ measures the approximation error between the candidate model and the regressor $f_{np}(x)$. The second part is a nonparametric estimate of the noise variance and is irrelevant for ranking candidate models.

Finally the proposed model selection criterion is

$$k_{opt} = \arg\min_{k \in I} \int (f(x;\hat{\theta}_k) - f_{np}(x))^2 \hat{q}(x)dx$$
$$I = \{1, 2, \ldots, k_{max}\}. \quad (3)$$

Two objects are required to use this criterion: the regressor $f_{np}(x)$ and the input p.d.f $\hat{q}(x)$. Both terms depend on the type of the nonparametric estimator $\hat{q}(x,y)$ [4].

We propose to use a *bivariate kernel* density estimate with a diagonal smoothing matrix

$$\hat{q}(x,y) = \frac{1}{nh_x h_y} \sum_{i=1}^{n} K\left(\frac{x - x_i}{h_x}\right) K\left(\frac{y - y_i}{h_y}\right). \quad (4)$$

where h_x and h_y, called *global bandwidth* and control the level of smoothing along the x- and the y direction respectively (Wand and Jones, 1993). $K(z)$ is a standard Gaussian kernel

$$\int K(z)dz = 1, \int zK(z) = 0 \quad and \int z^2 K(z)dz = 1.$$

Now the two required objects are

- a *kernel* estimate of the input probability density function (Rosenblat, 1956),(Parzen, 1962).

$$\hat{q}(x) = \frac{1}{nh_x} \sum_{i=1}^{n} K\left(\frac{x - x_i}{h_x}\right), \quad (5)$$

- the regressor

$$f_{np}(x) = \frac{\sum_{i=1}^{n} y_i K\left(\frac{x - x_i}{h_x}\right)}{\sum_{i=1}^{n} K\left(\frac{x - x_i}{h_x}\right)}, \quad (6)$$

which is the famous Nadayara-Watson regressor (Nadayara, 1964),(Watson, 1964).

[4] Even though $\hat{q}(x)$ as presented here is derived from $\hat{q}(x,y)$ by marginalization, however it can be estimated separately since it requires only input data. *Unlabeled* data can be used also to improve the estimation. In fact many practical learning applications like image, speech or text databases have a huge collection of unlabeled training set.

The computation of $f_{np}(x)$ needs to determine the optimal amount of smoothing h_x. There exist different methods that propose a reliable data-driven global bandwidth selector. An effective plug-in bandwidth selection strategy for local least squares regression was proposed by Ruppert that seems to perform adequately for a wide range of situations (Ruppert *et al.*, 1995). In this paper, we adopt his method to compute h_x.

3.2 *Large sample penalty*

Let us replace $\hat{q}(x,y)$ by its expression in (4) to compute the smoothed loss estimate $L_s(\hat{\theta}_k)$

$$L_s(\hat{\theta}_k) = \frac{1}{nh_x h_y} \sum_{i=1}^{n} \iint (y - f(x;\hat{\theta}_k))^2 K\left(\frac{x - x_i}{h_x}\right) \ldots$$
$$K\left(\frac{y - y_i}{h_y}\right) dxdy$$
$$= \frac{1}{nh_x} \sum_{i=1}^{n} \int (y_i - f(x;\hat{\theta}_k))^2 K\left(\frac{x - x_i}{h_x}\right) dx$$
$$+ \frac{1}{n} \sum_{i=1}^{n} y_i^2.$$

When the number of data n gets large, the optimal bandwidth h_x tends to zero and the smoothing became more local around the data points x_i's. This justifies a first order Taylor expansion of $f(x;\hat{\theta}_k)$ around x_i (Silverman, 1986)

$$f(x;\hat{\theta}_k) = f(x_i;\hat{\theta}_k) + (x - x_i) f'(x_i;\hat{\theta}_k).$$

Making a change of variable $z = \frac{x - x_i}{h_x}$, and dropping the constant term from the above equation,

$$L_s(\hat{\theta}_k) = \frac{1}{n} \sum_{i=1}^{n} \int \left[(y_i - f(x_i;\hat{\theta}_k)) + zh_x f'(x_i;\hat{\theta}_k) \right]^2 K(z)dz$$
$$= \frac{1}{n} \sum_{i=1}^{n} (y_i - f(x_i;\hat{\theta}_k))^2 \int K(z)dz$$
$$+ \frac{1}{n} \sum_{i=1}^{n} 2f'(x_i;\hat{\theta}_k)(y_i - f(x_i;\hat{\theta}_k)) h_x \int zK(z)dz$$
$$+ \frac{1}{n} \sum_{i=1}^{n} f'(x_i;\hat{\theta}_k)^2 h_x^2 \int z^2 K(z)dz$$
$$= L_{emp}(\hat{\theta}_k) + \frac{h_x^2}{n} \sum_{i=1}^{n} f'(x_i;\hat{\theta}_k)^2 \quad (7)$$

The smoothed loss $L_s(\hat{\theta}_k)$ consists of two parts. The first is the *empirical loss* $L_{emp}(\hat{\theta}_k)$ which penalizes the lack of fit. The second part acts as a penalty term, which penalizes *non smooth* or *complex* models. Here, the model's complexity is measured by a term that looks like the energy of the derivative function of the candidate model.

It is worth to mention that this approach permits to compare models from different parametric classes

(polynomials with Fourier series), since it uses a non-parametric reference model.
One referee mentioned the similarity of (7) with regularization criteria. The same approach can be adopted to treat the problem of learning with regularized function, but with a slight different modification. Since the primary objective of this paper is model selection, the problem of regularization will not be considered here.

4. SIMULATION RESULTS

To investigate the effectiveness of the proposed model selection procedure, we consider the classical problem of fitting a polynomial to a set of points. The motivation for studying this example is that polynomials create a difficult model selection problem that has a strong tendency to produce catastrophic overfitting effects (Figure 1). An other benefit is that polynomials are an interesting class of models, for which there are efficient techniques for computing the best fit. A referee suggests to use orthogonal polynomials to avoid overfitting. We argue that the two problems are equivalent in the sense that $(1, x, x^2, \ldots, x^p)$ form a base of $\mathbb{R}^p$, but it is not an orthogonal one. By using orthogonal base, we only perform a linear nonsingular transformation of the original problem leading to a diagonal design matrix. This will not improve the problem's conditioning (major cause of overfitting), just it simplifies the formulation.
To investigate the efficiency of the proposed method, we compare its performance to a number of standard model selection strategies like FPE (Akaike, 1970), MDL (Rissanen, 1978), structured risk minimization SRM (Cherbassky *et al.*, 1997), the leave-one-out cross validation CV and finally the metric based methods TRI and ADJ.
We conduct a series of experiments by fixing a *uniform* domain distribution for $q(x)$ on the unit interval $[0, 1]$ and then fixing a various target functions. To determine the effectiveness of the proposed method, which we call M1, compared with the strategies listed above, we measure the *ratio* of the true loss of the polynomial they selected to the best true loss among the least squares fitted polynomials in the sequence $\hat{m}_1, \ldots, \hat{m}_{k_{max}}$. The rational for doing this is that we wish to measure the model selection strategy's ability to approximate the true generalization error (Schuurmans, 1997). This approach is also adopted in (Chapelle *et al.*, 2001). (Schuurmans, 1997).

4.1 *Example 1*

We reproduce the same example as in (Schuurmans, 1997) by fitting a step function $f(x) = step(x \geq 0.5)$ corrupted by a Gaussian noise. We obtain the results by repeatedly generating a training sample of fixed size for a 1000 trails and recording the ratio achieved by each strategy. Figure 2 displays using boxplots the

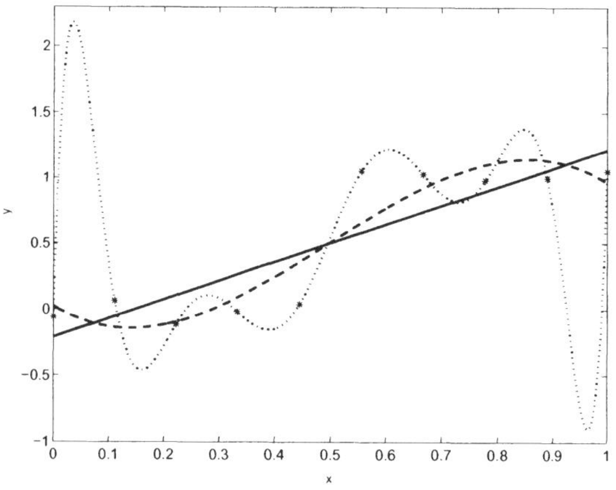

Fig. 1. An example of minimum square error polynomial regression of degree 1,4 and 9 for a set of 10 training points. Clearly the high order polynomial demonstrates erratic behavior on and off training set

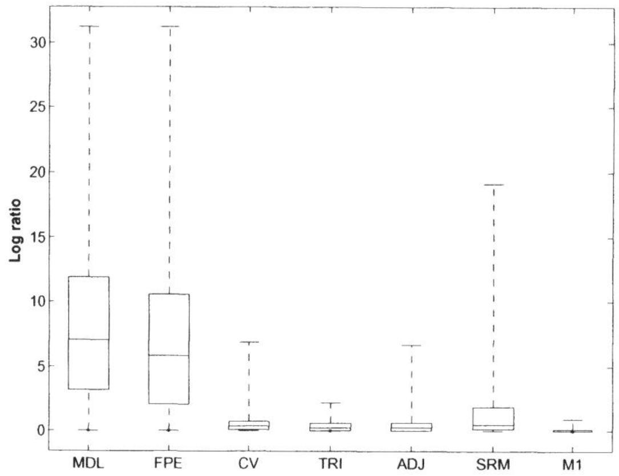

(a)

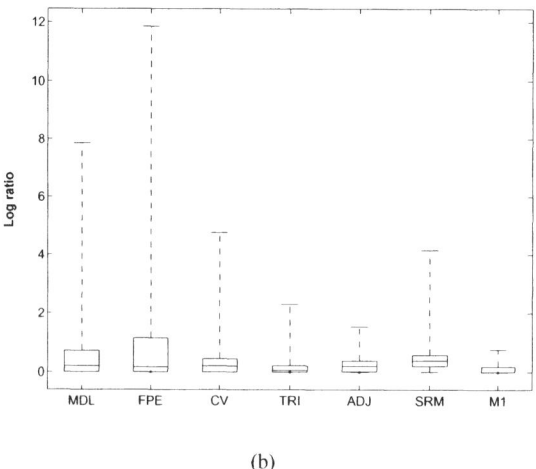

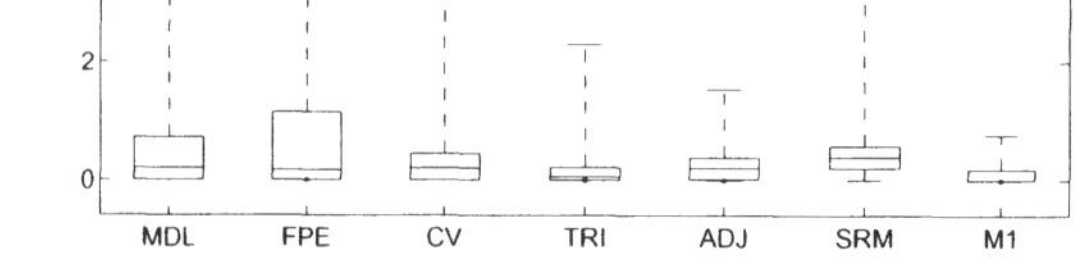

(b)

Fig. 2. Approximation ratio for fitting a step function achieved for $\sigma_\varepsilon = 0.05$ and data size, (a) $n = 10$. (b) $n = 30$.

distribution of the recorded ratio for training samples of size $n = 10$ and 30 respectively and noise level $\sigma_\varepsilon = 0.05$.
The results are quite interesting, M1 achieves a *median* log ratio of 0 for training samples $n = 10$ and 30 compared with the second best records of 0.22 and

0.05 respectively, achieved by TRI. The strength of M1 is in its *robustness* against overfitting. Although CV and SRM performed reasonably good, they often made catastrophic overfitting as recorded by the *maximum* value of log ratio. These basic results are unchanged when noise level increases as shown on Figure 3.

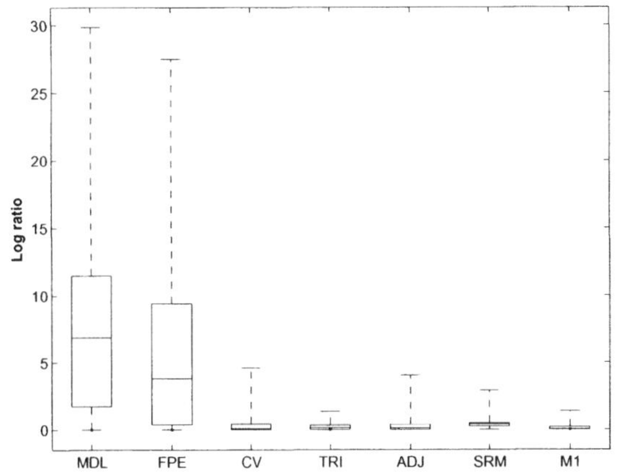

(a)

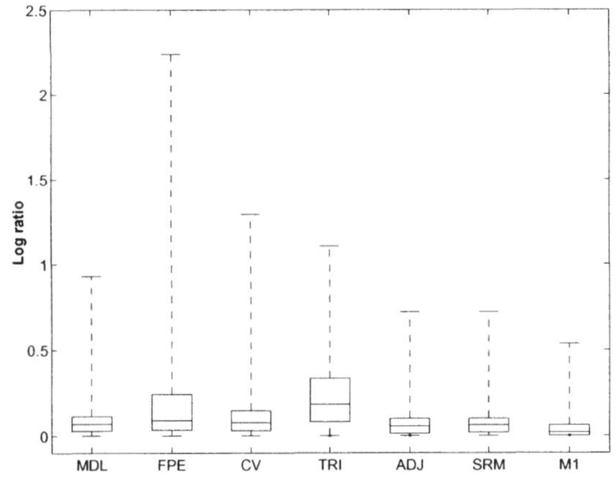

(b)

Fig. 3. Approximation ratio for fitting a step function achieved for $\sigma_\varepsilon = 0.5$ and data size, (a) $n = 10$. (b) $n = 30$.

4.2 *Example 2*

One may argue that the step function is rather a pathological target to fit with polynomials and therefore it is important to consider a more "natural target". Others may add that a step function is almost everywhere a highly *smooth* target which can be easily reproduced with nonparametric tools. By repeating the previous experiment with the target function $f(x) = \sin(\frac{\pi}{x+0.3})$ and a sample size of $n = 40$, a slight different result is obtained. Even though M1 achieves the smallest median of log ratio as shown in Figure 4, it achieves also a relatively large 75 percentile. This variability will surely reduce its effectiveness. The reason of

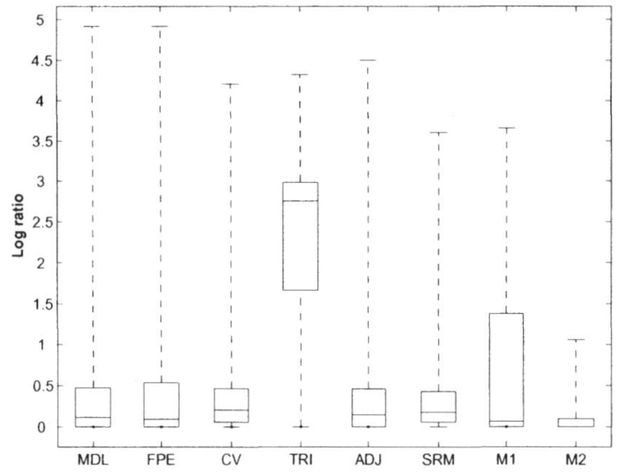

Fig. 4. Approximation ratio for fitting $f(x) = \sin(\frac{\pi}{x+0.3})$ achieved for training data of size $n = 40$ and noise level $\sigma_\varepsilon = 0.05$

this bad performance is the nature of the Nadayara-Watson regressor which suffers from bias increase in the boundary region and in the regions where the derivative of the true function or the input p.d.f $q(x)$ is large (Fan, 1992). The Nadayara-Watson can be seen as a *local constant* approximation of the true regression function (Tsybakov, 2002). More generally we can introduce *local polynomials approximation* which are shown to overcome the bias problem of Nadayara-Watson estimator. They adapt automatically to various types of design points, highly clustered and nearly uniform and there is an absence of boundary effect (Fan and Gijbels, 1996). All these interesting features of local polynomials lead to consider the use of an other *nonparametric reference* function.

Now we face the problem of selecting the order p of the local polynomial. Locally fitting polynomials of higher orders leads to a possible reduction in bias, but also to an increase of variability caused by introducing more local parameters. It should be noted that there is no reduction in bias when moving from even to odd orders due to the symmetric nature of the kernel which leads to a zero odd moments. For that reason it was recommended in many applications to take $p = 1$ (Fan and Gijbels, 1996).

Replacing the Nadayara-Watson regressor by a local linear fit in (3) leads to a modified model selection criterion which we denote M2. It is clear from Figure 4 that M2 achieves the best results among all the methods and was highly robust against overfitting.

5. CONCLUSION

In this paper, it was suggested that in the absence of a reliable a priori information about the data generating process, using a nonparametric approach can prevent catastrophic overfitting in a simple regression problem. One can argue that since we can get a reference model f_{np}, why should we try to find an optimal parametric model $\hat{m}_{opt}$ from a set of candidates?. The answer is that, parametric learning methods, PL, as compared with nonparametric learning methods, NPL, offer a data compression ability, which is of practical importance in may applications. Therefore, we are often in the need to use a parametric modeling of our data. Moreover PL methods provide a faster convergence rate or *learning speed* to the true model with respect to data size.

PL methods make assumptions about the global structure of the model to be learned. In contrast NPL methods tries to avoid that by focusing only on the local structure around a point of interest. However, PL methods fail dramatically when the global assumptions are not fulfilled (case of unreliable a priori information about the data generating process). To still take advantage of their strength, NPL methods are pooled with PL methods to construct a Hybrid Learning rules that are highly robust against overfitting as it was shown in a simulation study.

For further work, we hope to investigate in more details three important issues. First, the use of other loss functions, which are adapted for a given application. For example, in pattern recognition a 1 (for correct classification) and 0 (for miss classification) loss is adopted instead of the L_2 loss. Second, to investigate the effect of using different types of nonparametric estimators on the final result of model selection. Finally to apply the proposed approach for learning functions from high dimensional data space.

6. REFERENCES

Akaike, H. (1970). Statistical predictor identification. *Annals of the Institute of Statistical Mathematics* **22**, 202–217.

Cadez, I. V. and P. Smyth (2000). Model complexity, goodness of fit and diminishing returns. In: *In Proceedings of NIPS 2000*. pp. 388–394.

Chapelle, O., V. Vapnik and Y. Bengio (2001). Model selection for small sample regression. *Machine Learning* pp. 9–23.

Cherbassky, V., F. Mulier and V. A. Vapnik (1997). Comparaison of vc-methods with classical methods for model selection. In: *Proc. World Congress on Neural Networks*. pp. 957–962.

Efron, B. and R. Tibshirani (1995). *The Bootstrap*. Chapman & Hall.

Fan, J. (1992). Design-adaptive nonparametric regression. *Journal of the American Statistical Association* **87**, 998–1004.

Fan, J. and I. Gijbels (1996). *Local Polynomial and its Applications*. Chapman Hall. London.

German, S., E. Bienenstock and R. Doursat (1992). Neural networks and the bias/variance dilemma. *Neural Computation.* **41**, 1–58.

Linhart, H. and W. Zucchini (1986). *Model Selection*. Wiley series in Probability and Mathematical Statistics. John Wiley & Sons. NY.

Murata, N., S. Yoshizawa and A. Amari (1994). Network information criterion-determining the number of hidden units for an artificial neural network model. *IEEE Transactions on Neural Networks* **5**(6), 865–872.

Nadayara, E. A. (1964). On estimating regression. *Theory of Probability and Application* **10**, 186–190.

Parzen, E. (1962). On estimation of a probability density and mode. *Annals of Mathematical Statistics* **33**, 1065–1076.

Rissanen, J. (1978). Modeling by the sortest data description. *Automatica* **14**, 455–471.

Rosenblat, M. (1956). Remarks on some nonparametric estimates of density function. *Annals of Mathematical Statistics*.

Ruppert, D., S. J. Sheather and M. P. Wand (1995). An effective bandwidth selector for local least squares regression. *Journal of the American Statistical Association* **90**, 1257–1270.

Schuurmans, D. (1997). A new metric based approach to model selection. In: *In Proceedings of the 14th National Conference on Artificial Intelligence Providence RI*.

Silverman, B. W. (1986). *Density Estimation for Statistics and Data Analysis*. Chapman & Hall. London.

Stone, M. (1974). Cross validation choice and assessement of statistical predictions. *Journal of the Royal Statistical Society.* **10**(4), 1040–1053.

Tsybakov, A. (2002). Estimation non-paramétrique. Manuel de cours de DEA probabilitès et Applications, Université de Paris XI.

Utans, J. and J. Moody (1991). Selecting neural networks architecture via the prediction risk: Application to corporate bond rating prediction. In: *Proceeding of the first International Conference on Artificial Intelligence Application on Wall Street*. IEEE Computer Society Press Los Alamitos CA.

Vapnik, V. N. and A. Y. Chernovenkins (1971). On the uniform convergence of relative frequencies of events to thier probabilities. *Theory of Probability and Application* **35**(3), 486–499.

Wand, M. P. and M. C. Jones (1993). Comparaison of smoothing paramatrization in bivariate density estimation. *Journal of the American Statistical Association* **88**(422), 520–528.

Watson, G. S. (1964). Smooth regression analysis. *Sanklya Series* **A26**, 359–372.

IFAC Publications
www.elsevier.com/locate/ifac

AN INTRODUCTION TO LEARNING WITH REPRODUCING KERNEL HILBERT SPACES

Massimiliano Pontil

* *Dept of Computer Science, University College London, Gower Street, London WC1E, U.K.*

Abstract: After a brief introduction to learning theory, we review the elements of reproducing kernel Hilbert spaces and discuss learning algorithms which work thereby.

Keywords: Learning theory, Regularization, Nonlinear regression.

1. INTRODUCTION

Over the past ten years learning theory has undergone a significant progress in the development of learning algorithms and in their theoretical foundations. The theory builds on mathematical concepts which combine ideas from probability and statistics, and functional analysis. The formers are the natural tools to study the performance of a learning algorithm. This has been formalized by the work of Vapnik and Chervonenkis which we briefly touch in Section 2. The latter provides us with families of function spaces where a learning algorithm comes to play. In this paper we focus on a general class of function spaces, called reproducing kernel Hilbert spaces (RKHS) (Aronszajn, 1950). Among recent algorithms, an increasing number makes use of RKHS as the key computational ingredient. This has also renewed the interests of mathematicians on this older subject.

Section 3 provides an short introduction to RKHS. Section 4 reviews learning algorithms which work in RKHS. Special focus is devoted to regularization-based algorithms.

2. THE LEARNING PROBLEM

The central theme of learning theory is to compute - *learn* - a function on the base of a finite sample. The typical case studied is learning a real valued function (the related binary classification problem is treated as a special case). There is a large literature on the subject: useful reviews are (Wahba, 1990; Devroye *et al.*, 1996; Vapnik, 1998; Evgeniou *et al.*, 2000; Cristianini and Shawe-Taylor, 2000; Cucker and Smale, 2001; Hastie *et al.*, 2001; Scholkpof and Solla, 2002), and references therein. In the following we briefly explain the problem.

We consider two sets of random variables $x \in X$ and $y \in Y \subseteq \mathbb{R}$ which are related by a probabilistic relationship. The relationship is probabilistic because generally an element of X does not determine uniquely an element of Y, but rather a probability measure on Y. This can be formalized by assuming that an unknown probability measure $\rho(x,y)$ is defined over the set $X \times Y$.

We are provided with *examples* of this probabilistic relationship, that is with a training set D_m of m pairs (x_i, y_i) sampled in $X \times Y$ according to $\rho(x,y)$. The goal of learning is to provide an *estimator*, that is a function $f : X \to Y$ able to predict a value y from any value of $x \in X$.

The standard way to solve this problem consists in defining an error functional, which measures the average amount of error associated with a function, and then looking for the function with the lowest error. Let $V(y, f(x))$ be a *loss function* measuring the error we make when we predict y by $f(x)$. The *expected error*, is defined by

$$E[f] \equiv \int_{X,Y} V(y, f(x))\rho(x,y)\ dxdy.$$

Our desired function is the minimizer of the expected error. We denote this function with f_ρ to emphasize that it depends on the measure ρ. If for example $V(y,f) = (f-y)^2$ it is easy to see that f_ρ is the regression function, $f_\rho(x) = \int_Y y\rho(y|x)dy$.

Unfortunately E can not be computed because the measure ρ is unknown. We are only provided with the training set D_m. A natural approach is to replace the expected error with the empirical error

$$E_m(f) = \frac{1}{m}\sum_{i=1}^{m} V(f(x_i),y_i).$$

We then minimize E_m in a space $\mathcal{H}$, also called the *hypothesis space*. This space reflect our guess about where a good solution could be found. Let f_m be a minimizer of E_m. A main issue in the theory is to study how well f_m "imitates" the true function, f_ρ, i.e. to estimate the generalization error: $E(f_m) - E(f_\rho)$. This quantity depends on two competing factors: the number of examples and the size or "capacity" of the hypothesis space. This issue is addressed in the recent work by (Cucker and Smale, 2001) where a general answer is provided. Let $f_{\mathcal{H}}$ be the minimizer of E within the hypothesis space $\mathcal{H}$. The generalization error can be decomposed in two parts: the *sample error* $E(f_m) - E(f_{\mathcal{H}})$ and the *approximation error* $E(f_{\mathcal{H}}) - E(f_\rho)$.

The latter depends only on $\mathcal{H}$ and ρ but not on the sampled examples. It can be studied with tools from approximation theory. Recent results for RKHS were derived by (Cucker and Smale, 2001) in the case that V is the square loss. However there is need for more development in this direction.

The former is well developed. Its study is rooted in empirical processes theory and go back to the work of Vapnik and Chervonenkis (see, e.g., (Vapnik, 1982)) - see also (Devroye and Lugosi, 2001) for a nice summary of recent developments in this direction. The general statement of these results is that the bound

$$E(f_m) - E(f_{\mathcal{H}}) \le \varepsilon(m, 1/h, 1/delta)$$

holds true with a probability at least $1-\delta$, with $\delta \in (0,1)$, and ε is a non-decreasing function. The symbol h denotes a collection of parameters which measure the size of $\mathcal{H}$. Appropriate capacity quantities are defined in the theory, the most popular one being the VC-dimension or scale sensitive versions of it (see (Alon *et al.*, 1997)).

Intuitively, if the capacity of the function space in which we minimize the empirical error is very large and the number of examples is small, then the sample error can be large and *overfitting* is very likely to occur. The approximation error, at the contrary, decreases with the size of the hypothesis space. Then, in order to achieve good generalization, it is important to find a good trade-off between approximation error and sample error. In Section 4 we discuss regularization-based approaches which provide a general solution to this problem.

3. REPRODUCING KERNEL HILBERT SPACES

A reproducing kernel Hilbert space (RKHS) is a function space associated to a Mercer kernel.

Definition 1. A function $K: X \times X \to \mathbb{R}$ is called a Mercer kernel if: (a) K is symmetric, $K(x,y) = K(y,x)$ for all $x,y \in X$, (b) K is positive definite, meaning that for all $x_1,\ldots,x_\ell \in X$, and $\ell \ge 1$, the matrix with entries $K(x_i,x_j)$ is non-negative definite.

For $x \in X$, we define $K_x : X \to \mathbb{R}$ as $K_x(t) = K(x,t)$. Let H_0 be the space formed by all finite linear combinations of functions $K_x, x \in X$ (i.e., H_0 is the span of functions K_x). If $f,g \in H_0$, $f(x) = \sum_{i=1}^m \alpha_i K_{x_i}$ and $g(x) = \sum_{i=1}^{\ell} \beta_i K_{t_i}$, we define the scalar product:

$$(f,g)_K \equiv \sum_{i=1}^{m}\sum_{j=1}^{\ell} \alpha_i\beta_j K(x_i,t_j).$$

The name reproducing kernel is due to the following property, which follows immediately from the definition of the scalar product:

$$(f,K_x)_K = f(x) \text{ for every } f \in H_0,\ x \in X.$$

$(\cdot,\cdot)_K$ is well defined. In fact, it is easy to check that $(f,g)_K = (g,f_K)$ and $(af+bh,g)_K = a(f,g)_K + b(h,g)_K$. Since K is positive definite it also follows that $(f,f)_K \ge 0$. It remains to verify that $(f,f)_K = 0$ implies $f = 0$. Using the reproducing property we have

$$(f+aK_x, f+aK_x)_K = (f,f)_K + 2a(f,K_x)_K + a^2(K_x,K_x)_K = 2af(x) + a^2K(x,x) \ge 0.$$

The choice $a > 0$ gives $f(x) \ge -\frac{a}{2}K(x,x)$, while $a < 0$ gives $f(x) \le \frac{|a|}{2}K(x,x)$. It follows that, since a can be any real number, $f(x)$ must be zero. This argument is true for every $x \in X$. We conclude that $f = 0$.

Definition 2. The RKHS is the closure of space H_0 with respect to the norm $\|\cdot\|_K = \sqrt{(\cdot,\cdot)_K}$ induced by the scalar product.

Remark: It can be also shown (Aronszajn, 1950) that if a Hilbert space H admits a kernel function K : $X \times X \to \mathbb{R}$, such that: $K_x \in H$ for every $x \in X$ and $(f,K_x) = f(x)$, for every $f \in \mathcal{H}$, $x \in X$, then K is a Mercer kernel.

3.1 *Properties of the RKHS*

Besides the reproducing property 3, the RKHS enjoys few more key properties.

Proposition 3. Let K be a Mercer kernel and $\mathcal{H}$ the associated RKHS. Then, for every $x,y \in X$

(a) $K(x,x) \ge 0$.

(b) $|K(x,y)| \leq \sqrt{K(x,x)}\sqrt{K(y,y)}$.
(c) $|f(x)| \leq ||f||_K \sqrt{K(x,x)}$ for every $f \in \mathscr{H}$.

Proof: (a): Note that $||K_x||_K^2 = K(x,x)$. (b): We have $K(x,y) = (K_x, K_y)_K$. The result follows by the Cauchy-Schwartz inequality. (c): We first note that $||K_x||_K^2 = (K_x,K_x)_K = K(x,x)$. Let $f(x) = \sum_{i=1}^m \alpha_i K_{x_i}$. By the reproducing property (3), $f(x) = (f, K_x)_K$. The result follows by applying the Cauchy-Schwartz inequality:

$$|f(x)| \leq ||f||_K ||K_x||_K = ||f||_K \sqrt{K(x,x)}.$$

3.2 *Feature space and Mercer Theorem*

Let $\varphi_n : X \to \mathbb{R}, n = 1, \ldots, N$ be a set of functions. For every $x \in X$ let $\Phi : X \to \mathbb{R}^N$ be given by

$$\Phi(x) = (\varphi_1(x), \ldots, \varphi_N(x)).$$

A general type of Mercer kernel is

$$K(x,t) = \Phi(x) \cdot \Phi(t) \equiv \sum_{n=1}^{N} \varphi_n(x)\varphi_n(t) \qquad (1)$$

In fact, by definition K is symmetric and it is easy to verity that

$$\sum_{i,j=1}^{\ell} \alpha_i \alpha_j K(x_i,x_j) = \left(\sum_{i=1}^{\ell} \alpha_i \Phi(x_i) \right)^2$$

showing that K is also positive definite.

The map Φ is called the *feature map* and the space $\mathscr{X} = \{\Phi(x) : x \in X, \ ||\Phi(x)|| \leq \infty\}$ the *feature space*.

Example 4. (Homogeneous polynomial kernels). Let $X \subset \mathbb{R}^{n+1}$, $x = (x_0, x_1, \ldots, x_n)$,

$$K(x,y) = (x \cdot y)^d$$

with d a positive integer and $\cdot$ the scalar product in $\mathbb{R}^{n+1}$. K is of the form in Eq. (1) with $\Phi(x) = \{x^q \sqrt{C_q^d}\}_{|q|=d}$ where we used the notation $q = (q_0, q_1, \ldots, q_n)$, $|q| = \sum_{i=0}^n q_i$, $C_q^d = \frac{d!}{q_0! q_1! \cdots q_n!}$. The feature space is made of all the monomials in $\mathbb{R}^{n+1}$ of degree d. There are $\frac{(n+d)!}{n!d!}$ such monomials.

Example 5. (Non-homogeneous polynomial kernel). Let $X \subset \mathbb{R}^n$, $a > 0$, $d \in N, d \geq 1$:

$$K(x,y) = (a + x \cdot y)^d, \ a > 0.$$

This is the same as the above kernel if we define $x' = (\sqrt{a}, x) \in \mathbb{R}^{n+1}$ and $K'(x',y') = K(x,y)$. The feature space consists of all monomials in $\mathbb{R}^n$ of degree *at most d*. For instance, if $n = d = 2$ we have

$$\Phi(x) = (\sqrt{a}, \sqrt{2a}x_1, \sqrt{2a}x_2, x_1^2, x_2^2, \sqrt{2}x_1x_2)$$

If we set $a = 0$ we obtain the feature map corresponding to the homogeneous polynomial kernel.

In general the number of features, N, may be infinite (see below) provided that the series in r.h.s. of Eq. (1) converges for every $x, y \in X$. In this case $\mathscr{X}$ is a subset of ℓ^2, the Hilbert space of real sequences. In fact, under some general conditions on the space X any Mercer kernel can be equivalently written in the form in Equation (1), with $N \in \mathbb{N} \bigcup \{\infty\}$. We now discuss this fact.

3.2.1. *Integral operators* Let X be a compact metric space and $\mathscr{L}_\nu^2(X)$ the Hilbert space of square integrable functions on X (wrt. a positive measure ν, e.g. the Lebesgue measure). Let $C(X)$ be the space of continuous functions on X wrt. the norm $||f||_\infty = \sup_{x \in X} |f(x)|$.

Definition 6. If K is a continuous Mercer kernel, we define the operator $L_K : \mathscr{L}_\nu^2(X) \to \mathscr{L}_\nu^2(X)$ such that $(L_K f)(x) = \int K(x,t) f(t) dt$.

Theorem 7. (Mercer Theorem). L_K admits a system $\{(\lambda_n, u_n)\}_{n=1}^\infty$ of eigen-values/functions: $L_K u_n = \lambda_n u_n$, $n \geq 1$, with $\lambda_n \geq \lambda_{n+1} \geq 0$. In addition for all $x, y \in X$, $K(x,y) = \sum_{n=1}^\infty \lambda_n u_n(x) u_n(y)$, where the convergence is absolute and uniform on $X \times X$.

Thus, a continuous Mercer kernel is of the form in Equation (1) with $\varphi_n = \sqrt{\lambda_n} u_n$. Note that the decomposition depends on the measure ν used in $\mathscr{L}_\nu^2(X)$ and that the basis functions u_n do not need to be neither orthogonal (e.g, in Examples 4–5 below, they are not) nor linearly independent.

The map Φ is continuous too. In fact it easily follows that

$$||\Phi(x) - \Phi(y)||_{\ell^2}^2 = K(x,x) + K(y,y) - 2K(x,y)$$

and, since K is continuous, the l.h.s. tends to zero when x tends to y.

If $f \in H_0$, $f = \sum_{i=1}^{\ell} \alpha_i K_{x_i}$, it is immediate to verify that f can be equivalently written as $f(x) = \sum_{n=1}^N a_n u_n(x)$, with $a_n = \sqrt{\lambda_n} \sum_{i=1}^{\ell} \alpha_i u_n(x_i)$, and $||f||_K^2 = \sum_{n=1}^\infty \frac{a_n^2}{\lambda_n}$. The theorem below makes this connection precise.

Theorem 8. If $f, g \in \mathscr{L}_\nu^2$, with $f = \sum_{n=1}^\infty a_n u_n$, and $g = \sum_{n=1}^\infty b_n u_n \in \mathscr{H}$, we define $\langle f, g \rangle = \sum_{n=1}^\infty \frac{a_n b_n}{\lambda_n}$. Then, the space

$$H_K = \{f = \sum_{n=1}^\infty a_n u_n \in \mathscr{L}_\nu^2 \mid \sum_{n=1}^\infty \frac{a_n^2}{\lambda_n} < \infty\}.$$

coincides with the RKHS $\mathscr{H}$.

This different representation of $\mathscr{H}$ helps understanding the meaning of RKHS. The case of periodic kernels is particularly instructive.

3.3 *Translation invariant and periodic kernels*

Take $X = [0,\pi]$ and $K(x,y) = h(x-y)$, where h is defined on $[-\pi,\pi]$, it is continuous and periodic. Since K is symmetric, h is even ($h(x) = h(-x)$). It follows that the Fourier expansion of h involves only cosines:

$$h(x) = a_0 + \sum_{n=1}^{\infty} a_n \cos nx$$

where $a_n = 1/\pi \int_{-\pi}^{\pi} h(x)\cos nx$, $n \geq 1$, and $a_0 = \frac{1}{2\pi}\int_{-\pi}^{\pi} h(x)dx$. Using the property $cos(x-y) = \sin x \sin y + \cos x \cos y$, we have

$$K(x,y) = a_0 + \sum_{n=1}^{\infty} a_n \cos nx \cos ny + \sum_{n=1}^{\infty} a_n \sin nx \sin ny.$$

Assuming $a_n \geq 0$, K is of the form in Eq. (1)$\langle\Phi(x),\Phi(y)\rangle$ with

$$\Phi(x) = (\sqrt{a_0}, \sqrt{a_1}\sin x, \sqrt{a_1}\cos x, \ldots$$
$$\ldots, \sqrt{a_n}\sin nx, \sqrt{a_n}\cos nx, \ldots)$$

We have then proved:

Theorem 9. Let $K(x,y) = h(x-y)$. Then K is a Mercer kernel iff h is even and its Fourier coefficients are non-negative.

What is the associate RKHS? Denote the Fourier coefficients of a function f by

$$f_n^c = \frac{2}{\pi}\int_0^{\pi} f(x)\cos nx dx, \quad f_n^s = \frac{2}{\pi}\int_0^{\pi} f(x)\sin nx dx$$

According to Theorem 8, the scalar product in the RKHS is

$$\langle f,g\rangle_{\mathcal{H}} \equiv \sum_{n=0}^{\infty} \frac{f_n^c g_n^c + f_n^s g_n^s}{a_n}$$

Periodic kernels provides an intuition about the meaning of norm $||f||_K^2$ as a measure of the smoothness of function f: since $\lambda_n \to 0$ when n goes to infinity, components with higher frequencies are more penalized, and, thus, functions in $\mathcal{H}$ cannot oscillate too much. See (Wahba, 1990) for more details.

The analysis can be extended to $X = [a,b]^k$. In this case we have

$$\Phi(x) \equiv \{\sqrt{a_n}\sin(n\cdot x), \sqrt{a_n}\cos(n\cdot x)\}_{n\in\mathbb{N}^k}$$

where $n = (n_1,\ldots,n_k)$, and $n_i \geq 0$ for $i = 1,\ldots,k$.

Periodic kernels are a special case of translation invariant kernels. The latter are of the type $K(x,t) = K(x-t)$ but are not necessarily periodic. The next example is well known but clarifies this important difference.

Example 10. (Gaussian Kernel). Let $X \subset \mathbb{R}^n$, $K(x,t) = h(x-t)$, with $h(x) = e^{-\beta||x||^2}$. We will show below that K is a Mercer kernel. Choose $X = [0,\pi]$, clearly $h(0) \neq h(\pi)$, showing that K is not periodic.

3.4 *Form of the Kernels*

If we are given a feature map $\Phi(x)$, we can immediately build a kernel by setting $K(x,y) = \langle\Phi(x),\Phi(y)\rangle$. However in many case this feature map is unknown or may not exist properly. We then need to verify directly whether a given K is a Mercer kernel. Here we discuss a general result which characterize families of positive definite functions.

Suppose $K_1,\ldots K_n$ are some Mercer kernels. Let $F : \mathbb{R}^n \to \mathbb{R}$. We ask the following question: which properties of F guarantee that $F(K_1,\ldots,K_n)$ is also a Mercer kernel?

The next theorem by (Fitzgerald *et al.*, 1995) provides a complete answer to this question. We first introduce some new notations. Let $\mathcal{P}^n$ be the set of functions $F : \mathbb{R}^n \to \mathbb{R}$ such that for every $\ell \in \mathbb{N}$ the following property is true: if $A_1,\ldots,A_n$ are arbitrary $\ell \times \ell$ positive definite matrices, then also $F(A_1,\ldots,A_n)$ is positive definite. If $z = (z_1,\ldots,z_n) \in \mathbb{R}^n$, and $\beta = (\beta_1,\ldots,\beta_n) \in \mathbb{N}^n$, we set $z^\beta = z_1^{\beta_1}\cdots z_n^{\beta_n}$. Finally we denote by $\mathcal{M}_X$ the set of Mercer kernels on $X \times X$.

Theorem 11. The function $F : \mathbb{R}^n \to \mathbb{R}$ belongs to $\mathcal{P}^n$ iff F is real entire of the form

$$F(z) = \sum_{\beta\in\mathbb{N}^n} c_\beta z^\beta$$

where $c_\beta \geq 0$ for all $\beta \in \mathbb{N}^n$.

We discuss few examples which show the value of this result.

Example 12. Let $F(z) = (a+z)^d$, $a \geq 0$, $d \in \mathbb{N}$. It is easy to verify that $F \in \mathcal{P}^1$. Then, if we choose $X \subset \mathbb{R}^n$ $(a + x\cdot y)^d$ is a Mercer kernel. If $a > 0$ we have the dishomogeneous polynomial kernel of degree d in Example 5. Setting $a = 0$ gives the homogeneous polynomial kernel in Example 4.

Example 13. Let $F(z) = e^{\lambda z}$, $\lambda > 0$. $F \in \mathcal{P}^1$ since:

$$e^z = \sum_{n=0}^{\infty} \frac{z^n}{n!}$$

Then, if we choose $X \subset \mathbb{R}^n$, the function $\exp\{\lambda x\cdot y\}$ is a Mercer kernel. This analysis also shows that the feature consists of all monomials with a scaling factor $1/n!$, being n the degree of the monomial.

The next example shows that neural networks do not implement Mercer kernels.

Example 14. $F(z) = \tanh\{\lambda z\}$ does not belong to $\mathcal{P}^1$ for every choice of $\lambda \in \mathbb{R}$.

Note that Theorem 11 implies that $\mathcal{M}_X$ is closed under addition and multiplication by a positive constant. It is also closed by pointwise limits, meaning that if

$K_n \in \mathscr{C}$, and $\lim_n K_n(x,y) = K(x,y)$, then K is in $\mathscr{C}$. We conclude that $\mathscr{C}$ is a closed convex cone.

4. LEARNING ALGORITHMS IN RKHS

The discussion at the end of Section 2 suggests that in order to achieve good generalization it is important to find the best trade-off between sample error and approximation error. This observation leads to the method of *structural risk minimization (SRM)* and ultimately to regularization.

The idea of SRM (Vapnik, 1998) is to define a nested sequence of hypothesis spaces $H_1 \subset H_2 \subset \cdots \subset H_p$, where each hypothesis space H_i has finite capacity. Here we choose H_i to be a subset of a RKHS. A natural choice with RKHS is $H_i = \{f \in \mathscr{H} \mid \|f\|_K \leq A_i\}$ and $A_1 < A_2 < \cdots < A_p$. Let $f_{m,i}$ be the minimizer of the empirical error in $\mathscr{H}_i$, $i = 1,\ldots,p$. Using such a nested sequence of more and more complex hypothesis spaces, SRM consists in choosing the minimizer of the empirical error in the space $H_{\beta*}$ for which the bound on the generalization error

$$E(f_{m,i}) - E(f_\rho) = \text{approx. error}(\mathscr{H}_i) + \varepsilon(m,h_i,\delta)$$

is minimized. Further information about the statistical properties of SRM can be found in (Devroye *et al.*, 1996). Unfortunately, the implementation of the SRM method is not practical because it requires to look for the solution of a large number of constrained optimization problems. An alternative approach is, instead of looking for the solution of many optimization problems, to search for the minimum of

$$H(f) = \frac{1}{m}\sum_{i=1}^{m} V(y_i, f(x_i)) + \lambda\|f\|_K^2. \tag{2}$$

The functional $H(f)$ contains both the empirical error and the norm (complexity or smoothness) of f in the RKHS, similarly to functionals considered in Regularization Theory (Tikhonov and Arsenin, 1977). The *regularization parameter* λ can be seem as a penalty for functions with high capacity: the larger λ, the smaller the RKHS norm of the solution will be.

The key issue in SRM is the choice of the hypothesis space, i.e. the element i^* of the structure where the structural error is minimized. In the case of the functional of equation (2), the key issue becomes the choice of the regularization parameter λ. These two problems, as discussed in (Evgeniou *et al.*, 2000), are related, and the SRM method can in principle be used to choose λ (Vapnik, 1998). In practice, however, more practical statistical methods are used such as cross-validation, Generalized Cross Validation, Finite Prediction Error are used - see (Hastie *et al.*, 2001) for a review.

4.1 *The representer theorem*

An important feature of the above regularization functional is that, independently of the loss function V, any minimizer has the same general form

$$f(x) = \sum_{i=1}^{m} \alpha_i K(x,x_i). \tag{3}$$

There are different proofs of this fact. The most general we found is based on the idea of reducing the minimization of 2 to a *minimal interpolation problem* (see, e.g., (Micchelli and Pontil, 2002)).

Lemma 15. The solution to the problem:

$$\min_f\{\|f\| \text{ such that} : f(x_i) = y_i,\ i = 1,\ldots,m\}$$

is unique and has the form $f = \sum_{i=1}^m \alpha_i K(x_i,x)$.

Now, let f_m be a minimizer of 3. Consider the minimum interpolation problem:

$$\min_f\{\|f\| \text{ such that} : f(x_i) = f_m(x_i),\ i = 1,\ldots,m\}$$

The lemma tells us that the solution is unique, call it f and has the form $f = \sum_{i=1}^m \alpha_i K(x_i,x)$. Now set $g = f_m - f$. By definition $g(x_i) = 0$ for $i = 1,\ldots,m$ and, thus, $V(f_m(x_i),y_i) = V(f(x_i),y_i)$. Note that $\|f_m\|^2 = \|f\|^2 + 2(f,g)_K + \|g\|^2$. But: $(f,g)_K = \sum_{i=1}^m \alpha_i (K_{x_i},g)_K = \sum_{i=1}^m \alpha g(x_i) = 0$. We conclude that $H(f_m) = H(f) + \lambda\|g\|^2$, and, so, $g = 0$.

Equation (3) establishes a representation of the function f as a linear combination of kernels centered in each data point. This compact representation is of great advantage for learning. It allows to avoid working with the representation $f = \sum_{n=1}^\infty a_n u_n$, which requires estimating an infinite number of parameters.

4.2 *Learning techniques*

We now turn to the discuss of a few learning techniques based on the minimization of functionals of the form (2) by specifying the loss function V. In particular, we will consider Regularization Networks ((Poggio and Girosi, 1990)) and Support Vector Machines (SVM) (see (Vapnik, 1998) and references therein). We now briefly discuss each of these two techniques.

4.2.1. *Regularization Networks* The approximation scheme that arises from the minimization of the quadratic functional

$$\frac{1}{m}\sum_{i=1}^{m}(y_i - f(x_i))^2 + \lambda\|f\|_K^2 \tag{4}$$

for a fixed λ is a special form of regularization. It can be easily verified (see for example (Evgeniou *et al.*, 2000)) that the coefficients α_i of the minimizer of

(4) in equation (3) satisfy the following linear system of equations:

$$(G+\lambda I)c = y,$$

where I is the identity matrix, and we have defined

$$(y)_i = y_i, \ (c)_i = \alpha_i, \ (G)_{ij} = K(x_i, x_j).$$

4.2.2. *Support Vector Machines* We distinguish between real output (regression) and binary output (classification, $y \in \{-1, 1\}$) problems. SVM classification corresponds to the following loss function

$$V(y_i, f(x_i)) = (1 - y_i f(x_i))_+, \tag{5}$$

where $|x|_+ = x$ if $x > 0$ and zero otherwise. SVM regression uses the loss

$$V(y_i, f(x_i)) = (|y - f(x_i)| - \varepsilon)_+$$

These two loss functions are closely related. First, they are both "scale sensitive", meaning that the loss $V(y, f)$ is zero below a certain resolution. For SVM regression the mechanism is clear: if $|y - f(x)|$ is below "scale" ε the loss is zero. In the classification case the loss is zero if the distance of point x to the hyperplane $f(x) = 0$ is larger than the scale "$1/\|f\|$". The interpretation of SVM classification is less evident by looking at functional 2. In fact SVM were originally introduced as quadratic programming problems (Vapnik, 1998). Second, it can be shown (see, e.g., (Evgeniou *et al.*, 2000)) that by solving a classification problem by means of the SVM regression loss gives the same solution provided that ε is close to 1.

A remarkable property of SVM loss functions is that they lead to *sparse* solutions. This means that, unlike the case of regularization networks, typically only a small fraction of the coefficients α_i in equation (3) are nonzero. The data points x_i associated with the nonzero α_i are called *support vectors*. Those are the points which have a positive loss or are at the "edge" of zero and positive loss.

5. DISCUSSION

In this last section we discuss some open problems on the use of RKHS in learning theory. There are three main research directions on this topic which we would like to discuss.

We already mentioned in Section 2 the need for more studies on the approximation error in learning theory. Assume that the target function f_ρ belongs to a large space $\mathscr{F}$ (in this paper we assumed this is the space of continuous functions). Then, it would be interesting to study families of RKHS which are dense in $\mathscr{F}$.

A second important open problem is to develop Mercer kernels for specific data domains X. In particular an important area which is waiting for better solutions is learning in structured domains. Examples are spaces of graphs, e.g. trees or grids. Major progress needs to be made in this direction.

Finally, it is important to develop theory and methods for problems beyond the standard classification and regression ones. There are, however, other learning problems which have received much less attentions: learning order relations, multi-label classification, multiple output regression. A better understanding and theoretical development of these tasks will open the way to new applications areas.

6. REFERENCES

Alon, N., S. Ben-David, N. Cesa-Bianchi and D. Haussler (1997). Scale-sensitive dimensions, uniform convergnce, and learnability. *J. of the ACM* **44**(4), 615–631.

Aronszajn, N. (1950). Theory of reproducing kernels. *Trans. Amer. Math. Soc.* **686**, 337–404.

Cristianini, N. and J. Shawe-Taylor (2000). *An Introduction to Support Vector Machines.* Cambridge University Press.

Cucker, F. and S. Smale (2001). On the matematical foundations of learning. Preprint.

Devroye, L. and G. Lugosi (2001). *Combinatorial methods in density estimation.* Springer-Verlag.

Devroye, L., L. Györfi and G. Lugosi (1996). *A Probabilistic Theory of Pattern Recognition.* number 31 In: *Applications of mathematics.* Springer. New York.

Evgeniou, T., M. Pontil and T. Poggio (2000). Regularization networks and support vector machines. *Advances in Computational Mathematics* **13**, 1–50.

Fitzgerald, C.H, C.A. Micchelli and A. Pinkus (1995). Functions that preserve families of positive definite functions. *Linear algebra and its applications* **221**, 83–102.

Hastie, T., R. Tibshirani and J. Friedman (2001). *The elements of statistical learning: datamining, inference, and prediction.* Springer-Verlag.

Micchelli, C.A and M. Pontil (2002). On learning vector valued functions. Preprint.

Poggio, T. and F. Girosi (1990). Regularization algorithms for learning that are equivalent to multilayer networks. *Science* **247**, 978–982.

Scholkpof, B. and A. Solla (2002). *Learning with kernels.* MIT Press.

Tikhonov, A. N. and V. Y. Arsenin (1977). *Solutions of Ill-posed Problems.* W. H. Winston. Washington, D.C.

Vapnik, V. N. (1982). *Estimation of Dependences Based on Empirical Data.* Springer-Verlag. Berlin.

Vapnik, V. N. (1998). *Statistical Learning Theory.* Wiley. New York.

Wahba, G. (1990). *Splines Models for Observational Data.* Series in Applied Mathematics, Vol. 59, SIAM. Philadelphia.

IFAC Publications
www.elsevier.com/locate/ifac

SPARSE GAUSSIAN PROCESSES: INFERENCE, SUBSPACE IDENTIFICATION AND MODEL SELECTION

Lehel Csató, Manfred Opper

Neural Computing Research Group, Dept of Information Engineering,
Aston University, Birmingham, U.K.
{csatol,opperm}@aston.ac.uk

Abstract:
Gaussian Process (GP) inference is a probabilistic kernel method where the GP is treated as a latent function. The inference is carried out using the Bayesian online learning and its extension to the more general iterative approach which we call TAP/EP learning (short for TAP (Opper and Winther, 2001) and "expectation-propagation" (EP) (Minka, 2000)).
Sparsity is introduced in this context to make the TAP/EP method applicable to large datasets. We address the prohibitive scaling of the number of parameters by defining a subset of the training data that is used as the support the GP, thus the number of required parameters is independent of the training set, similar to the case of "Support–" or "Relevance–Vectors".
An advantage of the full probabilistic treatment is that allows the computation of the marginal data likelihood or evidence, leading to hyper-parameter estimation within the GP inference.
An EM algorithm to choose the hyper-parameters is proposed. The TAP/EP learning is the E-step and the M-step then updates the hyper-parameters. Due to the sparse E-step the resulting algorithm does not involve manipulation of large matrices. The presented algorithm is applicable to a wide variety of likelihood functions. We present results of applying the algorithm on classification and nonstandard regression problems for artificial and real datasets.

Keywords: kernel methods, Bayesian inference, sparse representation

1. INFERENCE WITH GAUSSIAN PROCESSES

Gaussian Processes (GPs) are probabilistic kernel methods which combine the flexibility provided by the generic kernel framework with the advantage of a full probabilistic treatment of the problem, e.g. besides the *most probable latent function* it also allows us to assess the *uncertainties* associated to the function values at each location.

To have probabilistic treatment, we encode the data $\mathscr{D} = \{(\boldsymbol{x}_n, y_n)\}_{n=1}^N$ using a likelihood function. For independent data we have a factorising likelihood as

$$P(\mathscr{D}|\boldsymbol{f}) = \prod_{n=1}^{N} P(y_n|\boldsymbol{x}_n, \boldsymbol{f}) = \prod_{n=1}^{N} P(y_n|f_{\boldsymbol{x}_n}) \qquad (1)$$

where we defined $\boldsymbol{f} = [f_{\boldsymbol{x}_1}, \ldots, f_{\boldsymbol{x}_N}]^T$. The likelihood is conditioned on a *latent function* $\boldsymbol{f}$ which we model as a GP, ie. a random function characterised by a joint Gaussian distribution of the function values for any finite collection of inputs. Notice that the general conditioning of a single likelihood on the *whole* random function $\boldsymbol{f}$ is simplified to dependence on a single random variable: the value of the random function at location $\boldsymbol{x}_n$ (we use $f_n \doteq f_{\boldsymbol{x}_n}$).

To perform Bayesian inference we need a *GP prior* $p_0(\boldsymbol{f})$ for the function $\boldsymbol{f}$. A GP is fully specified by its prior mean function $\langle f_{\boldsymbol{x}} \rangle_0$ and the prior covariance kernel $K_0(\boldsymbol{x}, \boldsymbol{x}')$. In the following we assume zero prior mean function, thus the choice of the covariance kernel fully encodes our class of functions. The *posterior process* is derived from Bayes' rule and is written as

$$p_{post}(\boldsymbol{f}) = \frac{1}{Z} p(\mathscr{D}|\boldsymbol{f}) p_o(\boldsymbol{f}) \qquad (2)$$

where $Z = P(\mathscr{D}) = \int d\boldsymbol{f}\; p(\mathscr{D}|\boldsymbol{f}) p_o(\boldsymbol{f})$ is the normalising constant or the free energy.

The simple expression from eq. (2) describing the posterior process can seldom be applied in practise: two fundamental problems need to be addressed. The first problem appears with non-Gaussian likelihoods which leads to non-Gaussian posterior processes. This implies that we need approximations resulting in a tractable posterior, a possible approximation is presented in this section. The second problem is the super-linear increase of the parameters with the size of the dataset, addressed in Section 2.

To have computational tractability we approximate the non-Gaussian posterior process with a Gaussian one by retaining only the posterior mean and covariance kernel functions of the non-tractable posterior. The posterior mean $\langle f_{\boldsymbol{x}} \rangle_{post}$ and covariance kernel $K_{post}(\boldsymbol{x},\boldsymbol{x}')$ functions are given by the following expressions (Csató and Opper, 2002):

$$\begin{aligned} \langle f_{\boldsymbol{x}} \rangle_{post} &= \langle f_{\boldsymbol{x}} \rangle_0 + \textstyle\sum_{n=1}^{N} \alpha_n K_0(\boldsymbol{x},\boldsymbol{x}_n) \\ K_{post}(\boldsymbol{x},\boldsymbol{x}') &= K_0(\boldsymbol{x},\boldsymbol{x}') \\ &\quad + \textstyle\sum_{m,n=1}^{N} K_0(\boldsymbol{x},\boldsymbol{x}_m) C_{mn} K_0(\boldsymbol{x}_n,\boldsymbol{x}') \end{aligned} \qquad (3)$$

where the vector $\boldsymbol{\alpha} = [\alpha_1,\ldots,\alpha_N]^T$ and matrix $\boldsymbol{C} = \{C_{mn}\}_{m,n=1}^{N}$ are the scalar parameters given as:

$$\begin{aligned} \alpha_n &= \frac{\partial}{\partial \langle f_n \rangle_0} \ln Z \\ C_{mn} &= \frac{\partial^2}{\partial \langle f_m \rangle_0 \partial \langle f_n \rangle_0} \ln Z \end{aligned} \qquad (4)$$

where Z is the complete data likelihood or the normalising constant in eq. (2) and the derivatives are with respect to the prior mean $\langle f_n \rangle_0$ of the GP at the training points $\boldsymbol{x}_n$.

It is important that the *functional form* of the posterior approximation does not depend on the particular likelihood – they are expressed as weighted single and double sums of the prior kernel function taken at the location of the data points. The result for the posterior mean was also obtained in the Kimeldorf-Wahba representer theorem (Kimeldorf and Wahba, 1971). Additionally to that, we have a representation for a *full Gaussian* process by providing the covariance kernel of the posterior approximation. The estimation of both moments of the posterior allows the probabilistic treatment, leading to eg. Bayesian error bars for prediction.

The expression of the posterior moments from eq. (3) merely states the *existence* of the parameters $(\boldsymbol{\alpha},\boldsymbol{C})$ since the free energy Z cannot be computed nor the derivatives can be taken. Therefore we have to consider approximations to find them. Exact result exists only for Gaussian likelihood, however efficient approximations are feasible for various other likelihood functions (Csató and Opper, 2002; Minka, 2000; Csató, 2002).

We are focusing on the *joint* approximation of the coefficients for both the mean and the covariance kernel functions and mention that the approximation of the parameters of the mean $\boldsymbol{\alpha}$ alone for different likelihoods is an active research area in the kernel community, see eg. in (Schölkopf *et al.*, 1999).

The approximation we use is the extension of the Bayesian online learning, called the TAP/EP algorithm (Minka, 2000; Opper and Winther, 2001). The "expectation-propagation" (EP) algorithm proposed a sequential optimisation procedure which updates the GP coefficients $(\boldsymbol{\alpha},\boldsymbol{C})$ by considering a single likelihood term $P(\boldsymbol{y}_k|f_k)$ from eq. (1) at each iteration – this iterative update is the online update from (Opper, 1998). At each step the approximated posterior with moments $(\langle f_{\boldsymbol{x}} \rangle_p, K_p(\boldsymbol{x},\boldsymbol{x}'))$ obtained from a previous iteration, was viewed as prior in the current step. The parameter update is performed using $\ln Z_k$ computed from the single likelihood and the current prior. If we denote the first and second derivative of $\ln Z_k$ with q_k and r_k respectively, then we immediately have from eq. (3) the following updates for the GP mean and covariance functions:

$$\begin{aligned} \langle f_{\boldsymbol{x}} \rangle_{new} &= \langle f_{\boldsymbol{x}} \rangle_p + q_k\, K_p(\boldsymbol{x},\boldsymbol{x}_k) \\ K_{new}(\boldsymbol{x},\boldsymbol{x}') &= K_p(\boldsymbol{x},\boldsymbol{x}') + K_p(\boldsymbol{x},\boldsymbol{x}_k)\, r_k\, K_p(\boldsymbol{x}_k,\boldsymbol{x}'). \end{aligned} \qquad (5)$$

The online approximation of the full posterior is possible for a variety of likelihood functions, being an important extension over the Kalman- (Kalman and Bucy, 1961) and the extended Kalman filtering techniques, values for the coefficients q_k and r_k are provided in Appendix A.

In developing the EP algorithm Minka observed that at each individual step the ratio of the prior and the approximated posterior *defines* a Gaussian approximation to each likelihood term $P(\boldsymbol{y}_k|f_k)$, these approximations are however *dependent* on each other, the dependence is coming from setting the current prior to the previous posterios, ie. the approximation using other likelihood terms. These quadratic approximations allowed to go beyond the conventional online learning and perform iterative updates. In online learning each term in the likelihood can only be used once. In the iterative extension first a "subtraction" of the previous Gaussian approximation to the likelihood is performed, followed by the online learning of the current input (see (Minka, 2000; Opper and Winther, 2001) for details). An other important benefit of the EP algorithm is the approximation to the marginal data likelihood, ie. model evidence and the possibility to perform model selection, presented in Section 3.

An important limiting factor of the framework presented so far is that it cannot be applied to large datasets, the *quadratic* scaling of the parameters with the size of the data is far too prohibitive. Therefore a second approximation is performed, presented next.

2. OBTAINING SPARSE SOLUTION

The kernels we use might not be infinite-dimensional, this implies that often the *representation* of the approximating posterior is *redundant*. Several techniques to exploit the redundancy of the MAP solution, ie. the mean function from eq. (3) have been proposed (Wahba, 1990; Smola and Bartlett, 2001; Williams and Seeger, 2001). These approaches did not take into account the posterior covariance. Our aim in this section, similar in philosophy to the earlier work, is to describe the posterior GP using a *constant and small* number of parameters and to have this number independent from the size of the dataset.

We want to write the posterior mean and covariance considering a *small subset* of the training inputs, called *"basis vector set"* or $\mathcal{BV}$ set (similar to Support Vectors (Vapnik, 1995) or Relevance Vectors (Tipping, 2000)). Thus the posterior mean is:

$$\langle f_{\boldsymbol{x}} \rangle^*_{post} = \sum\nolimits_{n \in \mathcal{BV}} \alpha^*_n K_0(\boldsymbol{x}, \boldsymbol{x}_n) \qquad (6)$$

and similarly to this the posterior covariance kernel is replaced with one that involves a reduced matrix $\boldsymbol{C}^*$, where we assume that $(\boldsymbol{\alpha}^*, \boldsymbol{C}^*)$ are chosen such that the new process is "as close as possible" to the Bayesian posterior whilst $|\mathcal{BV}| = d$ is constant. Since both the original GP, given by $(\boldsymbol{\alpha}, \boldsymbol{C})$ from eq. (3), and its approximation which uses only the elements from the $\mathcal{BV}$ set are Gaussians, their KL-distance (Cover and Thomas, 1991) is computable. We perform a minimisation to find $(\boldsymbol{\alpha}^*, \boldsymbol{C}^*)$ from the original $(\boldsymbol{\alpha}, \boldsymbol{C})$ and also evaluate the KL-distance which gives an efficient criterion to select the elements of the $\mathcal{BV}$ set (Csató, 2002; Csató and Opper, 2002). The evaluation of the KL-distance is important since it provides means to estimate the usefulness of each element in the $\mathcal{BV}$ set. It is important to mention that the computation of the KL-distance *does not* require the computation of $(\boldsymbol{\alpha}^*, \boldsymbol{C}^*)$, additionally to the "old" parameters it only requires the inverse of the kernel matrix. In implementation iteratively update the inverse kernel matrix, thus further reducing the computational cost and providing computationally cheap criterion to assess the "usefulness" of each $\mathcal{BV}$.

An efficient algorithm is obtained by *combining* the TAP/EP iterations with the KL-optimal projections. In the TAP/EP step, when processing the selected input, the $\mathcal{BV}$ set is *increased*. After the TAP/EP update, in the *pruning phase*, we consider the following optimisation problem: find the parameters of a new GP containing *one less* $\mathcal{BV}$ and it is the closest possible in the KL-sense to the GP resulting from the TAP/EP update step. The resulting algorithm is efficient and avoids the inversion of large kernel matrices, a frequent problem when using kernel methods. The algorithm is a greedy one in choosing the elements of the $\mathcal{BV}$ set, i.e. at each time it only looks at the possibility of exchanging a single value.

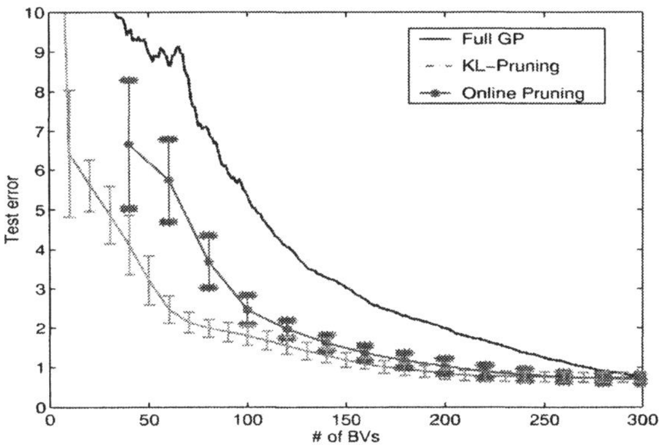

Fig. 1. Results for the Friedman dataset #1 using 300 training and 500 test data. The lines show the average test errors with error bars showing the empirical variance across different training sets.

The approximated posterior is used for predicting the probability of the unseen output $\boldsymbol{y}$ at an unseen point $\boldsymbol{x}$. Applying again Bayes' rule, we have the *predictive distribution* of the outputs:

$$p(\boldsymbol{y}|\boldsymbol{x}) = \int df_{\boldsymbol{x}}\, P(\boldsymbol{y}|f_{\boldsymbol{x}})\, p_{post}(f_{\boldsymbol{x}}) \qquad (7)$$

where $P(\boldsymbol{y}|f_{\boldsymbol{x}})$ is conditioned on the function at the input $\boldsymbol{x}$. The latent process is a GP, meaning that the integral in eq. (7) is one-dimensional and it is with respect to a Gaussian, often doable exactly or approximately. A second observation is that a Gaussian approximation to the latent variables does not mean a Gaussian predictive distribution, this depends on the choice of the likelihood function.

We show the performance of the sparse algorithm for normal regression using the Friedman dataset in Fig. 1. The continuous line on the top shows the result of the batch GP regression when the inputs were only *partially* used, up to the size of the $\mathcal{BV}$ set. The bottom, dash-dotted line shows the performance when sparsification was applied *to the full GP solution* ie. all inputs were added to the $\mathcal{BV}$ set and then only the specific number of basis vectors has been retained. The middle, continuous line shows the results of combining the online learning with sparsity. We obtain a stable performance for $\mathcal{BV}$ set sizes exceeding 120. The continuous line, labelled "Full GP" shows us that the error for the sparse GP is never worse than the performance of a GP where we stop at the $\mathcal{BV}$ set size. The fact that the two bottom lines overlap means that the for $\mathcal{BV}$ sizes matching the "effective" dimension of the data the sparse GP is optimal.

We compared the sparse GP with the SVM and RVM methods. The SVM used 116 support vectors that lead to a test error of 2.92 and similar test error performance of 2.80 was obtained using 59 relevance vectors (Tipping, 2001), stating that our algorithm compares well to these other ones.

A few differences between these methods need to be mentioned. A first one is that both the SVM and the

RVM start with a full solution and then obtain the sparse result without having control over the size of the result. On the contrary, in the sparse GP method we start with an empty $\mathcal{BV}$ set add training inputs up to the capacity of the machine used for the experiment or up to the dimension of the kernel. A second observation is that due to the storage of the covariance, the number of parameter is quadratic with respect to $|\mathcal{BV}|$ and this number is linear for the other two methods.

In this experiment the hyper-parameters were not selected automatically. The next section we introduce a method for hyper-parameter selection.

3. MODEL SELECTION

For model selection we use the expectation-maximisation (EM) algorithm (Bishop, 1995) which aims at maximising the complete data likelihood $P(\mathcal{D})$, the normalisation constant in eq. (2). We see that the set of model parameters can be divided in two groups: a first group related to the likelihood function $p(y|f_{\mathbf{x}})$ and a second set that specifies the kernel function $K_0(\mathbf{x},\mathbf{x}')$.

To optimise **the likelihood parameters** we perform the optimisation of a lower bound on the model evidence using the EM-algorithm with the GP posterior having the "old" likelihood parameters and we want to find the new ones contained in $P(\mathcal{D}|\boldsymbol{f})$:

$$\ln p(\mathcal{D}) \geq \int d\boldsymbol{f}\, p_{post}(\boldsymbol{f}) \ln P(\mathcal{D}|\boldsymbol{f}) \qquad (8)$$

The GP inference is the *E-step* which gives the GP with a fixed set of hyper-parameters.

In the *M-step*, assuming that the posterior GP $p_{post}(\boldsymbol{f})$ is fixed, we optimise eq. (8) with respect to the likelihood parameters.The first term *does not* depend on the likelihood parameters, we only need to optimise the second expression.

Since the likelihood is factorising, the complete log-likelihood is rewritten as a sum. Each component of the sum depends on a single value of the latent GP, meaning that the integrals involved are one-dimensional, usually computable. After the update of the likelihood parameters we re-run the TAP/EP algorithm to find the new posterior $p_{post}(\boldsymbol{f})$ and alternate the steps until convergence.

To test the EM-algorithm, we applied the GP inference for non-symmetric and non-Gaussian additive noise, with the single data likelihood given by:

$$P(y|f_x,\lambda) = \begin{cases} \lambda \exp[-\lambda(y-f_x)] & \text{if } y > f_x \\ 0 & \text{otherwise} \end{cases} \qquad (9)$$

We can integrate the likelihood for a single data, we can apply thus the TAP/EP learning. This example shows the benefits of the online learning, i.e. that approximation can be performed even in cases where the MAP solution cannot be obtained. Fig. 2 shows the results for this toy problem (left-hand sub-figure)

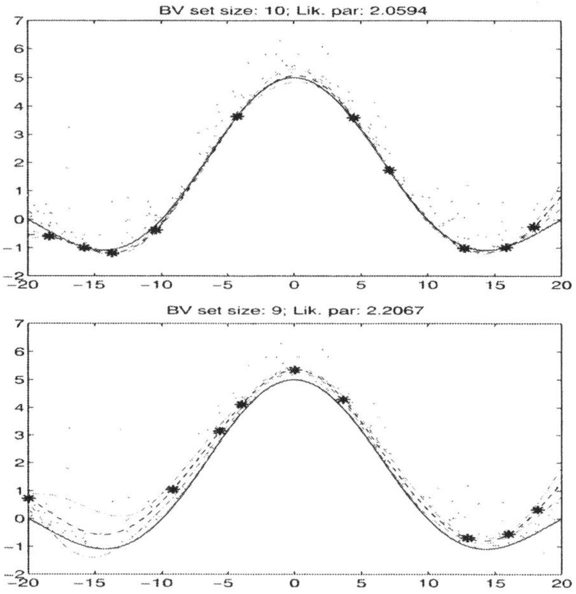

Fig. 2. Finding likelihood parameters of a robust and one-sided regression problem. The left sub-figure shows the result if the correct noise model is assumed and on the right we see the approximation for symmetrical noise. The noise parameter for generating the data was $\lambda = 2$ for both cases.

and an other GP regression where we used a symmetric Laplace noise. The non-symmetric likelihood provides a very good fit and infers the correct likelihood parameter, this is in contrast with symmetric noise assumption (we used the same asymptotic behaviour) where there is a constant bias and larger Bayesian error bars which are shown with thin dash-dotted lines.

Next we address the optimisation of **the kernel parameters**. The *E-step* is the same as before. In the *M-step* however, instead of upper-bounding the marginal data likelihood, we compute the gradient of the log-evidence with respect to a generic kernel parameter θ. Using matrix algebra and the parameters of the posterior GP ($\boldsymbol{\alpha}$,$\boldsymbol{C}$), we have the following relation:

$$\begin{aligned} \frac{\partial \log \mathrm{P}(\mathcal{D})}{\partial \theta} &= \mathrm{tr}\left[\frac{\partial \log \mathrm{P}(\mathcal{D})}{\partial \boldsymbol{K}} \cdot \frac{\partial \boldsymbol{K}}{\partial \theta}\right] \\ \frac{\partial \log \mathrm{P}(\mathcal{D})}{\partial \boldsymbol{K}} &= -\frac{1}{2}\left(\boldsymbol{\alpha}\boldsymbol{\alpha}^T + \boldsymbol{C}\right) \end{aligned} \qquad (10)$$

where θ is a parameter of the kernel and $\partial \boldsymbol{K}$ is the matrix derivative. An exact update is not possible, we used conjugate gradient algorithm (SCG) to find the optimal kernel parameters. Here needs to be emphasised the sparse usefulness of the approximation: the size of the matrices involved is small, only the size of the $\mathcal{BV}$ set, making the proposed EM algorithm feasible.

In the experiments we used the RBF kernels defined as:

$$K(\mathbf{x},\mathbf{x}') = A \exp\left[-\frac{\sum_i \lambda_i (\mathbf{x}_i - \mathbf{x}'_i)^2}{2}\right] \qquad (11)$$

where the A is a scaling constant and λ_i are are the relevance parameters (ARD) specifying the impor-

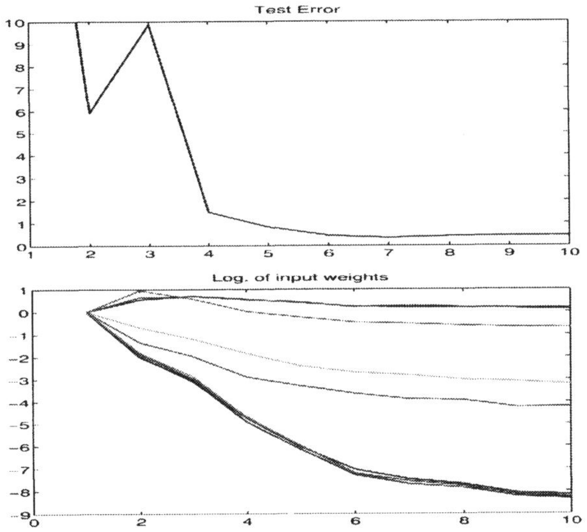

Fig. 3. Learning kernel parameters for noiseless Friedman data. On the top sub-figures the evolution of the average test error is shown. The bottom sub-figure shows the evolution of the $\ln \lambda_i$. The two rows show the results for no noise (top) and additive noise with $\sigma^2 = 1$ (bottom) respectively.

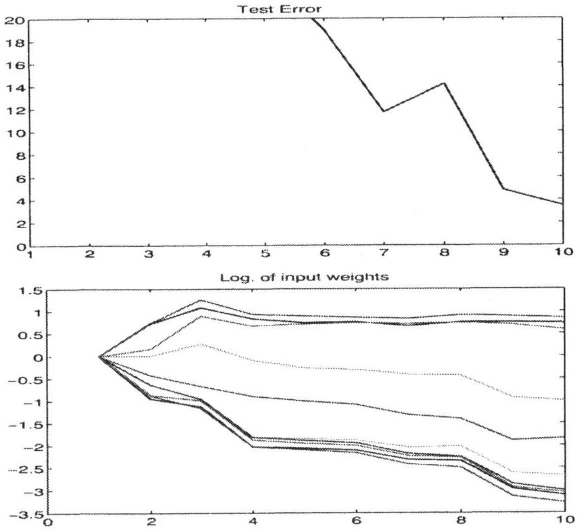

Fig. 4. Learning kernel parameters for noisy Friedman data. For explanation see Fig. 3.

tance of the i-th input dimension in predicting the output. We tested the performance of the method first for regression and the Friedman #1 dataset. This is a benchmark for the relevance parameters since only 5 out of the 10-dimensional inputs were used to generate the output. Additionally to this source of uncertainty, the outputs are also corrupted with Gaussian noise. In Fig. 3 and 4 we see the test error and the resulting ARD kernel parameters for no noise and $\sigma^2 = 1$ respectively. As it can be seen, the separation of the irrelevant inputs from the relevant ones is clearer for the noiseless case.

Learning kernel parameters can also be done for classification tasks. We applied the EM algorithm for learning Crab data with 6-dimensional inputs (Csató

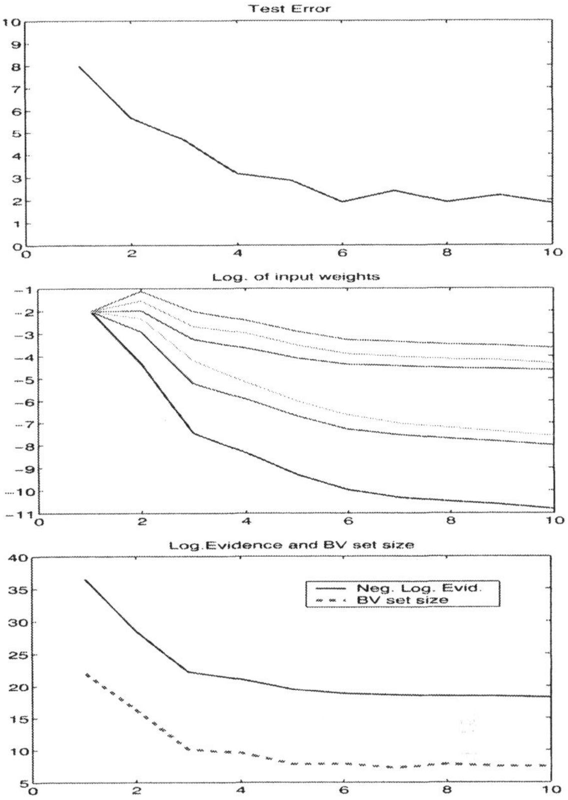

Fig. 5. Results for the Crab data. All figures show the averages over 20 random permutation of the training set. The figures show the test errors (top), the logarithm of the input weights (middle) and on the bottom, in the same plot, the negative log-evidence of the model and $\mathcal{BV}$ set size respectively. On the X-axis the number of the EM-steps is counted.

and Opper, 2002). This dataset has been widely used to assess various machine learning methods and most of the results confirmed that optimal performance is achieved using three of the six components. This is seen in the middle plot of Fig. 5 where three components are close together (approx -4 on a log-scale) with the rest several orders of magnitude smaller, practically removing the corresponding input dimension from the kernel function. It is important that simultaneously to this the the test error (upper plot of Fig. 5) attains a minimum value that outperforms other cited methods (see (Csató, 2002) for details).

It is interesting that, starting with a $\mathcal{BV}$ set size of 25, at the end of the iterations the size of the $\mathcal{BV}$ set is less than 8 on average and the log-evidence is also significantly smaller. We believe that this result encourages further investigation.

4. CONCLUSIONS

A method for probabilistic and sparse GP inference is presented. The inference is based on a representation that is independent of the likelihood function. Joined with the general representation, we present a method to derive a sparse solution for the problem.

The presented method of selecting the reduced representation is constructive and flexible. It is applicable to problems with non-differentiable likelihood function as well as to classification and Gaussian regression.

Further research is aimed to test the method for other likelihoods and to try to establish conditions when we can expect good performance. Open question is still the performance of the sparse GP inference if *all* hyper-parameters are simultaneously optimised.

5. REFERENCES

Bishop, Christopher M. (1995). *Neural Networks for Pattern Recognition*. Oxford University Press. New York, N.Y.

Cover, Thomas M. and Joy A. Thomas (1991). *Elements of Information Theory*. John Wiley & Sons.

Csató, Lehel (2002). Gaussian Processes – Iterative Sparse Approximation. PhD thesis. Neural Computing Research Group. www.ncrg.-aston.ac.uk/Papers.

Csató, Lehel and Manfred Opper (2002). Sparse online Gaussian Processes. *Neural Computation* **14**(3), 641–669.

Kalman, R.E. and R.S. Bucy (1961). New results in linear filtering and prediction theory. *Trans ASME, Journal of Basic Engineering, Ser. D* **83**, 95–108.

Kimeldorf, George S. and Grace Wahba (1971). Some results on Tchebycheffian spline functions. *J. Math. Anal. Applic.* **33**, 82–95.

Minka, Thomas P. (2000). Expectation Propagation for Approximate Bayesian Inference. PhD thesis. Dep. of El. Eng. & Comp. Sci.; MIT. vismod.www.media.mit.edu/~tpminka.

Opper, Manfred (1998). A Bayesian approach to online learning. In: *On-Line Learning in Neural Networks*. pp. 363–378. Cambridge Univ. Press.

Opper, Manfred and Ole Winther (2001). Adaptive and self-averaging TAP mean field theory for probabilistic modeling. *Physical Review E* **64**(056131), 1–14.

Schölkopf, Bernhard, Burges, Christopher J.C. and Smola, Alexander J., Eds.) (1999). *Advances in kernel methods (Support Vector Learning)*. The MIT Press.

Smola, Alexander J. and Peter Bartlett (2001). Sparse greedy Gaussian pocess regression. In: *NIPS* (Todd K. Leen, Thomas G. Diettrich and Volker Tresp, Eds.). Vol. 13. The MIT Press. pp. 619–625.

Tipping, Michael (2000). The Relevance Vector Machine. In: *NIPS* (Sara A. Solla, Todd K. Leen and Klaus-Robert Müller, Eds.). Vol. 12. The MIT Press. pp. 652–658.

Tipping, Michael E. (2001). Sparse bayesian learning and the relevance vector machine. *Journal of Machine Learning Research* **1**, 211–244.

Vapnik, Vladimir N. (1995). *The Nature of Statistical Learning Theory*. Springer-Verlag. New York, NY.

Wahba, G. (1990). *Splines Models for Observational Data*. Series in Applied Mathematics, Vol. 59, SIAM. Philadelphia.

Williams, Christopher K. I. and Matthias Seeger (2001). Using the Nyström method to speed up kernel machines. In: *NIPS* (Todd K. Leen, Thomas G. Diettrich and Volker Tresp, Eds.). Vol. 13.

Appendix A. UPDATES FOR DIFFERENT LIKELIHOODS

The table below summarises the likelihood functions and their log-averages used in experiments. We assume that the likelihood depends only on the marginal of a GP, a Gaussian characterised by mean and variance $(\mu_{\mathbf{x}}, \sigma_{\mathbf{x}}^2)$. The first and second order derivatives with respect to $\boldsymbol{\mu}$ are the update coefficients q_k and r_k for the online learning step.

Reg. Gauss	$P(y \mid f_{\mathbf{x}}) = \frac{1}{(2\pi\sigma_0^2)^{1/2}} \exp\left[-\frac{(y-f_{\mathbf{x}})^2}{2\sigma_0^2}\right]$
$\log\langle P(y \mid f_{\mathbf{x}})\rangle = \frac{1}{2}\left[\log(2\pi(\sigma_0^2+\sigma_{\mathbf{x}}^2)) + \frac{(y-\mu_{\mathbf{x}})^2}{\sigma_0^2+\sigma_{\mathbf{x}}^2}\right]$	
Reg. Laplace	$P(y \mid f_{\mathbf{x}}) = \frac{\lambda}{2}\exp[-\lambda \lvert y-f_{\mathbf{x}}\rvert]$
$\log\langle P(y \mid f_{\mathbf{x}})\rangle = \log\frac{\lambda}{2}\,[g(s,u)+g(s,-u)]$ $g(s,u) = \Phi(-u-s)\cdot\exp(\frac{s^2}{2}+su)$ where $s = \lambda\sigma_{\mathbf{x}}$ and $u = (y-\mu_{\mathbf{x}})/\sigma_{\mathbf{x}}$	
Reg. Pos. Exp.	$P(y \mid f_{\mathbf{x}}) = \begin{cases}\lambda\exp[-\lambda(y-f_{\mathbf{x}})] & \text{if } y > f_{\mathbf{x}} \\ 0 & \text{otherwise}\end{cases}$
$\log[\lambda g(s,u)]$ where $g(z)$, u, and s taken from **Reg. Lapl**	
Classification	$P(y \mid f_{\mathbf{x}}) = \Phi\left[\frac{y f_{\mathbf{x}}}{\sigma_0}\right]$
$\Phi\left[\frac{y\,\mu_{\mathbf{x}}}{\sqrt{\sigma_0^2+\sigma_{\mathbf{x}}^2}}\right]$	

where $\Phi(z)$ is the cumulative Gaussian distribution defined as

$$\Phi(z) = \frac{1}{\sqrt{2\pi}}\int_{-\infty}^{z}\exp\left[-\frac{t^2}{2}\right]dt$$

and we find the first and second derivatives using the exponential function and efficient implementations of the cumulative Gaussian distribution function $\Phi(z)$.

IFAC Publications
www.elsevier.com/locate/ifac

SPARSE KERNEL METHODS

Steve R. Gunn

*ISIS Group, Dept. of Electronics and Computer Science,
University of Southampton, U.K.*

Abstract: A disadvantage of many statistical modelling techniques is that the resulting model is extremely difficult to interpret. A number of new concepts and algorithms have been introduced by researchers to address this problem. They focus primarily on determining which inputs are relevant in predicting the output. This work describes a transparent, non-linear modelling approach that enables the constructed predictive models to be visualised, allowing model validation and assisting in interpretation. The technique combines the representational advantage of a sparse ANOVA decomposition, with the good generalisation ability of a kernel machine. It achieves this by employing two forms of regularisation: a 1-norm based structural regulariser to enforce transparency, and a 2-norm based regulariser to control smoothness. The resulting model structure can be visualised showing the overall effects of different inputs, their interactions, and the strength of the interactions. *Copyright © 2003 IFAC*

Keywords: Kernel, Sparse, SVMs.

1. INTRODUCTION

In empirical data modelling a process of induction is used to build up a model of a system from examples. Ultimately the quantity and quality of the observations will govern the performance of this model. However, the choice of modelling approach will also influence the performance of the model. By its observational nature, data obtained is finite and sampled; typically this sampling is non-uniform and due to the high dimensional nature of the problem, the data will form only a sparse distribution in the input space. Consequently, the problem is nearly always ill-posed (Poggio *et al.*, 1985) in the sense of Hadamard (1923). To address the ill-posed nature of the problem it is necessary to convert the problem to one that is well-posed. For a problem to be well-posed, a unique solution must exist that varies continuously with the data. Conversion to a well-posed problem is typically achieved with some form of capacity control, which aims to balance the fitting of the data with constraints on the model flexibility, producing a robust model that generalises successfully. Previous approaches to restoring the well posedness have included regularisation methods (Tikhonov and Arsenin, 1977). In this paper, the method chosen is based around kernel methods due to their rigorous formulation and good generalisation ability for small sample sizes. Girosi (1997) and Smola (1998) have shown that kernel methods can be placed in a regularisation framework, which guarantees their well posedness; Support Vector Machines (SVMs) (Vapnik, 1995) and Gaussian Processes (GPs) (Rasmussen, 1996) are examples. For tutorial introductions to SVMs see Burges (1998), Cristianini and Shawe-Taylor (2000) or Gunn (1998). Given a dataset, $\mathcal{D} = \{(x^1, y^1), \ldots, (x^l, y^l) \mid x^i \in \mathbb{R}^d, y^i \in \mathbb{R}\}$, the model is a weighted linear summation of kernels,

$$f(x) = \sum_{i=1}^{l} \alpha_i K(x^i, x), \tag{1}$$

where these kernels are 'centred' on the data points. Consequently, the solution is opaque due to the large number of terms that will typically exist in this expansion. Furthermore, the multivariate basis functions can be difficult to interpret. The number of terms in the expansion can be reduced in some circumstances by enabling a proportion of the kernel multipliers to become zero. This can be achieved using a loss function that has a 'dead-zone', such as an ϵ insenstive loss function (Vapnik, 1995).

Whilst a predictive model may be the ultimate goal of modelling, it is often desirable and sometimes even essential to be able to interpret the final model structure. This is especially true in medical domains, where *black-box* models, such as traditional neural networks, bagged descision trees as well as kernel methods, are viewed with great suspicion (Wyatt, 1995; Plate, 1999). In situations where model interpretation is important, many researchers revert to using simpler, but more interpretable modelling methods, for example logistic regression. As Plate observes (Plate, 1999) there is a danger in using such simple models, since they typically suffer from the problem of model mismatch, and hence they may fail to discover an important relationship in the data because they lack the flexibility to model it. In this work we introduce interpretability, or *transparency*, by producing a parsimonious model, which has a sparse structural representation, but is flexible enough to avoid problems of model mismatch. The transparency is beneficial in that it enables the model to be validated and interpreted. Features that aid transparency are input selection and ways of decomposing the model into smaller more interpretable pieces that can be easily visualised. To address this issue we introduce a modified kernel model of the form, where the kernel is formed from a weighted sum of simpler kernels, which are aligned to the inputs. Transparency can then be introduced by a careful choice of the additive kernels, K_j and by making their weighting coefficients, c_j, sparse. In this paper we focus on the integration of an ANOVA (ANalysis Of Variance) representation to provide a transparent approach to modelling. ANOVA kernels (Stitson *et al.*, 1999) have previously been used with SVMs, with promising performance. However, the difference here is to develop a technique that will select a sparse ANOVA kernel producing strong transparency. The ANOVA representation is motivated by the decomposition of a function into additive components, with the goal of representing the function by a subset of the terms from this expansion. A function may be decomposed into

$$f(x) = f_0 + \sum_{i}^{d} f_i(x_i) + \sum_{i<j}^{d} f_{i\otimes j}(x_i, x_j) + \cdots + f_{1\otimes 2\otimes \ldots \otimes d}(x), \qquad (2)$$

where d is the number of inputs, f_0 represents the bias and the other terms represent the univariate, bivariate, etc., components. The notation x_i denotes the scalar value of input i. The basis functions are semi-local and are similar to the approaches used by Friedman (1991) in the Multivariate Adaptive Regression Splines (MARS) technique and in the Adaptive Spline Modelling of Observational Data (ASMOD) technique (Kavli and Weyer, 1995). The decomposition of a function into this form is unique since the univariate terms are constrained to pass through the origin, the bivariate terms are constrained to be zero along the axes, and similarly for the higher order terms. The additive representation is advantageous when the higher order terms can be ignored, so that the resulting model is represented by a small subset of the ANOVA terms, which may be easily visualised. This produces a transparent model, in contrast to the majority of neural network models, providing the modeller with structural knowledge that can be used for both validation and model interpretation. Due to the curse of dimensionality (Bellman, 1961), an exhaustive search of the possible model structures is demanding. Even in the highly restrictive scenario, that the solution is a weighted linear combination of *fixed* basis functions, the parameter space has size 2^d. Extension to flexible basis functions, which is required for typical modelling, will only compound this dimension. Accordingly, greedy methods are typically used. ASMOD employs an evolutionary strategy to search the model space using a forward selection/backward elimination algorithm to select suitable refinements to a model. The MARS algorithm employs a recursive partitioning procedure to search the model space for an appropriate model. The drawback with both approaches is that they can become entrapped by local minima, due to the greedy nature of their search algorithms. A problem with deploying additive models in flexible non-linear modelling methods is that they cannot provide a transparent model if the phenomenon being modelled contains high dimensional interactions. One possibility is to enforce transparency by constraining the order of possible interactions (e.g. restriction to univariate and bivariate terms only), providing a coarse, but interpretable structure, at the expense of structural integrity.

The aim of this work is to produce transparent models that generalise well, using a global approach to the modelling problem. This paper details the SUpport vector Parsimonious ANOVA (SUPANOVA) technique to realise this goal. It will be shown that the technique is attractive since it can employ a wide range of loss functions (Smola *et al.*, 1998), can produce interpretable models, and it is solved by breaking the problem down into simple convex optimisation problems, which can be implemented using readily available mathematical programming optimisers (Mészáros, 1998).

2. SPARSE KERNEL METHODS

An additive sparse kernel model extends a standard kernel model by replacing the kernel with a weighted linear sum of kernels,

$$f(x) = \sum_{i=1}^{l} \alpha_i \sum_{j=1}^{m} c_j K_j(x^i, x), \quad c_j \geq 0, \qquad (3)$$

where K_j are positive definite functions and where the positivity constraints on the kernel coefficients, c_j, ensure that the complete kernel function is positive definite. Here, the term sparse refers to sparseness in the kernel coefficients c_j rather than the usual sparseness in the multipliers, α_i; sparseness in these multipliers can still be obtained by employing an appropriate loss function. A conventional kernel model regulariser will not enforce sparsity in the kernel coefficients and hence a more complex regulariser is required. The goal in selecting a sparse representation is to minimise the number of non-zero coefficients, c_i. This can be achieved with a p-norm on the kernel coefficients. As p increases the solution becomes less sparse and the computational complexity of the resulting optimisation problem is relaxed. Ideally a value of $p = 0$, which counts the number of terms in the expansion is attractive. This case is employed in the atomic decomposition of (Chen, 1995), but it results in a computationally hard combinatorial optimisation problem. Alternatively choosing a value of $p = 2$ produces a straightforward optimisation problem. This case is referred to as the method of frames or ridge regression, but crucially the sparseness within the expansion is now lost. A good compromise occurs when $p = 1$ producing a sparse solution, with a practical implementation. This penalty function has successfully been used in basis-pursuit de-noising (Chen, 1995). To enforce sparsity in the kernel expansion we consider a regularised cost functional of the form

$$\Phi(\alpha, c) = L(y, K(c)\alpha) + \lambda_\alpha \|\alpha\|^2_{K(c)} + \lambda_c \|c\|_1, \quad c_i \geq 0, \lambda_\alpha, \lambda_c > 0 \tag{4}$$

where L is the loss function, and λ_α, λ_c are regularisation parameters controlling the smoothness and sparsity of the kernel expansion respectively. The direct solution of this problem is non-trivial, so an iterative method is introduced, whereby we solve two separate sub-problems: $\min_\alpha \Phi$ with c fixed; $\min_c \Phi$ with α fixed. The solution for a quadratic loss, $L(y, \hat{y}) = (y - \hat{y})^T(y - \hat{y})$, is given by Equation 5,

$$\Phi(\alpha, c) = \left\| y - \sum_i c_i K_i \alpha \right\|_2^2 + \lambda_\alpha \sum_i c_i \alpha^T K_i \alpha + \lambda_c \sum_i c_i, \quad \forall_p \, c_p \geq 0. \tag{5}$$

$$\alpha^* = \arg\min_\alpha \; \alpha^T \Big(\sum_i \sum_j c_i c_j K_i K_j + \lambda_\alpha \sum_k c_k K_k\Big)\alpha - \Big(2y^T \sum_l c_l K_l\Big)\alpha \tag{6}$$

$$c^* = \arg\min_c \; \sum_i \sum_j c_i c_j (\alpha^T K_i K_j \alpha) + \sum_k c_k (\lambda_\alpha \alpha^T K_k \alpha + \lambda_c - 2y^T K_k \alpha), \quad \forall_p \, c_p \geq 0, \tag{7}$$

where y and $\hat{y}$ are vectors of target and predicted values respectively. The solution for an ϵ-Insensitive Loss, $L(y, \hat{y}) = \sum_i \max(0, |y_i - \hat{y}_i| - \epsilon)$ is given by Equation 8,

$$\Phi(\alpha, c) = \left\| y - \sum_i c_i K_i \alpha \right\|_{1,\epsilon} + \lambda_\alpha \sum_i c_i \alpha^T K_i \alpha + \lambda_c \sum_i c_i, \quad \forall_p \, c_p \geq 0. \tag{8}$$

$$\begin{aligned} \alpha^* = \arg \min_{\alpha = \alpha^+ - \alpha^-} \; & (\alpha^+ - \alpha^-)^T \Big(\lambda_\alpha \sum_k c_k K_k\Big)(\alpha^+ - \alpha^-) \\ & - \sum_i (\alpha^+ - \alpha^-) y_i + \sum_i (\alpha^+ + \alpha^-)\epsilon, \\ & \forall_i \; 0 \leq \alpha_i^+, \alpha_i^- \leq \frac{1}{2\lambda_\alpha}. \end{aligned} \tag{9}$$

$$\begin{aligned} c^* = \arg \min_{c, \zeta^+, \zeta^-} \; & \sum_i (\zeta_i^+ + \zeta_i^-) + \sum_j c_j (\lambda_\alpha \alpha^T K_j \alpha + \lambda_c), \\ & \forall_{i,j} \; c_j \geq 0, \zeta_i^+, \zeta_i^- \geq 0, -\zeta^- - \epsilon \leq \sum_k c_k K_k \alpha \leq \zeta^+ + \epsilon. \end{aligned} \tag{10}$$

where ζ^+ and ζ^- are slack variables. An attraction of this iterative technique is that it decomposes the problem into two simple convex optimisation problems. In the quadratic loss case the solution for α^* is given by simple matrix inversion, and for c^* by a bound constrained quadratic program. In the ϵ-insensitive case the solution for α^* is given by a box constrained quadratic program, and for c^* by a bound constrained linear program with linear constraints. Consequently, they can all be solved readily using a standard quadratic programming optimiser (Mészáros, 1998). A similarity can be drawn between this approach and Bayesian methods (MacKay, 1995) that employ a two stage iterative procedure, a parameter update and 'hyperparameter' update. However, unlike most Bayesian methods, the update stages consist of convex optimisation problems.

If λ_α and λ_c are known the solution can be obtained by,

$$\text{Initialise}: \alpha_0^* = \arg\min_\alpha \Phi(\alpha, c_0^*), \quad c_0^* = \mathbf{1}$$

$$\text{Iteration}: \begin{array}{ll} \text{(a)} & c_{i+1}^* = \arg\min_c \Phi(\alpha_i^*, c) \\ \text{(b)} & \alpha_{i+1}^* = \arg\min_\alpha \Phi(\alpha, c_{i+1}^*). \end{array}$$

In the quadratic case the second order partial derivatives with respect to α and c are always positive ensuring that every slice is convex. This fact combined with the knowledge that the solution is finite in α and c should ensure convergence to the global minimum. A similar result should be obtainable for the ϵ-insensitive loss function. The convergence properties of this algorithm will be studied in future work. In practice the situation is more complicated since λ_α and λ_c will not be known but will need to be estimated. Intuitively, both λ_c and λ_α should initially be set large and reduced gradually; reducing λ_α too quickly will over smooth the space making the sparse selection harder; reducing λ_c too quickly will tend to produce an over-sparse model. To provide a workable solution the method used in this paper uses an initialisation step and one iteration. In the initilisation step and part (b) of the iteration, λ_c does not enter the optimisation and as such does not need to be determined; λ_α can be determined using cross-validation. The difficult part is determining the parameters in part (a) of the iteration. A possible method could fix λ_α at the value used in the initialisation step and select λ_c to obtain a comparable loss to that of the initialisation step. However, the method chosen, which was based on the best empirical performance, was to set $\lambda_\alpha = 0$ and to select λ_c such that the loss was equal to that of the validation error in the initialisation step. Alternative methods for determining these parameters will be investigated in future work. In the next section a particular class of sparse additive kernel model is introduced with some attractive transparency properties.

3. SUPANOVA

The SUPANOVA technique is designed to select a parsimonious model representation by selecting a small set of terms from the complete ANOVA representation. The technique is an additive kernel model, Equation 3 with a particular choice of ANOVA kernels can be expressed as and hence we can employ the sparse kernel method described in the previous section to obtain its solution. This section considers some possibilities for ANOVA kernel models. The following theory is based upon Reproducing Kernel Hilbert Spaces (RKHS) (Aronszajn, 1950; Wahba, 1990). If K is a symmetric positive definite function, which satisfies Mercer's Conditions, then the kernel represents a legitimate inner product in feature space and it may be deployed within Equation 3. The following two theorems (Aronszajn, 1950) are required in proving that ANOVA kernels satisfy Mercer's Conditions.

Theorem 1: If k_1 and k_2 are both positive definite functions then so is $k_1 + k_2$

Theorem 2: If k_1 and k_2 are both positive definite functions then so is $k_1 \otimes k_2$

It follows from theorem 2 that multidimensional kernels can be obtained by forming tensor products of univariate kernels. A multivariate ANOVA kernel is given by the tensor product of a univariate kernel plus a bias term,

$$\begin{aligned} K_{\text{ANOVA}}(u, v) &= \prod_{i=1}^{d} (1 + k(u_i, v_i)) \\ &= 1 + \sum_{i}^{d} k(u_i, v_i) + \sum_{i<j}^{d} k(u_i, v_i)k(u_j, v_j) + \cdots + \prod_{i=1}^{d} k(u_i, v_i). \end{aligned} \quad (11)$$

It follows from theorems 1 and 2 that if k is a valid kernel then so is K_{ANOVA}. Considering Equation 11 it is evident that the tensor product produces the ANOVA terms, producing a flexible model. Another consequence of theorems 1 and 2 is that each of the additive terms in the expansion Equation 11 is also positive definite, and hence a valid kernel in its own right. This enables partial forms of Equation 11 to be used as valid kernels, and this is the method employed within the SUPANOVA technique to produce parsimonious kernels. The choice of univariate kernel, k, will control the form of the final model. For simplicity, we shall restrict ourselves to the case where the same kernel is used for each dimension, although different univariate kernels could be deployed.

An attractive property of the kernel-based approach is that many functions commonly employed within modelling have kernels that satisfy Mercer's Conditions. Gaussian Radial Basis Function kernels have been successfully deployed in kernel methods. However, whilst they have some attractive properties from a regularisation perspective they are poor at modelling functions with different degrees of smoothness, and require the determination of an additional smoothing parameter. Multi-Layer Perceptron (MLP) kernels, using a set of sigmoidal functions, have also been used. However, the MLP kernel is only positive definite for particular values of its two controlling parameters, making deployment more difficult. Polynomial kernels have often been used and are cheap to compute. Their disadvantage is that in an ANOVA framework a high order polynomial will be required to model arbitrary functions. Splines are an attractive choice for modelling (Wahba, 1990) due to their ability to approximate arbitrary functions. Many types of splines have kernel representations, such as odd order B-splines and infinite splines. B-splines have been used in other modelling approaches and are favourable when a rule-base interpretation is desired (Brown and Harris, 1994). However, whilst they can have some computational advantages, the regularisation operator corresponding to a B-spline kernel representation has some weaknesses (Smola, 1998). This has been observed experimentally by the production of models with a tendency to oscillation (Gunn, 1998). An infinite spline incorporates the flexibility of a spline approach without the oscillation problem associated with B-splines, and this motivates it use within an ANOVA framework. Another advantage of the infinite spline kernel is that is has no scale, and therefore no associated scale parameter to determine. This is of great advantage in the SUPANOVA technique, since the ANOVA decomposition would introduce a multitude of such parameters which would need to be determined. The first order infinite spline kernel, which passes through the origin, is defined on the interval $[0, \infty)$ by,

$$k_{\text{spline}}(u, v) = \int_0^\infty (u - \tau)_+ (v - \tau)_+ d\tau, \tag{12}$$

where $(x)_+$ is equal to the positive part of x. The solution has the form of a piece-wise cubic polynomial,

[t]

$$k_{\text{spline}}(u, v) = uv + \frac{1}{2}(u + v)\min(u, v) - \frac{1}{6}(\min(u, v))^3, \tag{13}$$

and therefore the form of the SVM solution is a piecewise cubic with knots located at a subset of the data points. Multivariate spline kernels obtained from Equation 13 will produce a lattice of piecewise multicubic functions.

Using a complete ANOVA kernel Equation 11 has drawbacks when it comes to interpretation of the model, due to the large number of terms within the expansion. To introduce enhanced transparency we employ a parsimonious ANOVA kernel. Considering the expansion of Equation 11 an additional set of positive coefficients, c_i, are introduced,

$$K_{\text{ANOVA}}(u, v) = c_0 + \sum_i^d c_i k(u_i, u_i) + \sum_{i<j}^d c_{i,j} k(u_i, u_i) k(u_j, u_j) + \cdots + c_{1,2,\ldots,d} \prod_{i=1}^d k(u_i, u_i). \tag{14}$$

Consequently the resulting kernel is a weighted linear sum of kernels, and a parsimonious model solution can be obtained by using the method of the previous section.

Since the univariate ANOVA term is constrained to pass through the origin, bivariate and higher order terms will be constrained to be zero along their axes. Consequently the parsimonious model will not simply consist of the single highest order ANOVA term, but will favour low order terms in preference to high order terms. The ANOVA terms in the parsimonious model can be recovered from the final SVM expansion. For example, the univariate terms are given by,

$$f_g(x) = c_g \sum_{i=1}^l \alpha_i k(x_g^i, x_g), \tag{15}$$

and the bivariate terms are given by,

$$f_{g\otimes h}(x) = c_{g,h} \sum_{i=1}^l \alpha_i k(x_g^i, x_g) k(x_h^i, x_h), \tag{16}$$

where α_i are the Lagrange multipliers obtained from the complete ANOVA kernel solution. However, the computation required to solve the optimisation problem is extremely demanding due to the combinatorial nature of the problem and the curse of dimensionality (Bellman, 1961) associated with the full ANOVA expansion. To overcome this problem the ANOVA expansion can be truncated to simplify the problem, since if transparency is to be obtained the selected terms should be of low order. This technique contrasts with other parsimonious techniques, such as MARS and ASMOD, in that it aims to find a full model and sub-select the significant terms. The drawback with the MARS and ASMOD approaches is that they are local, and can suffer from entrapment in local minima within the construction process. Additionally, they may not be strictly well-posed. A further attraction of the SUPANOVA technique is that it decomposes the problem into two simple convex optimisation problems. An important issue is the form of solution produced when highly correlated inputs exist. The combination of the regularisers, Equation 4 will produce a model that is distributed for two or more identical inputs; if a $\|c\|_0$ regulariser was used the model would not be distributed. In the case when the inputs are only highly correlated, the technique will produce a sparse model, and therefore a simple correlation test could be employed to identify the limiting case.

4. ACKNOWLEDGEMENT

Parts of this paper have been published in the journal of Machine Learning, published by Kluwer.

5. REFERENCES

Aronszajn, N. (1950). Theory of reproducing kernels. *Trans. Amer. Math. Soc.* **686**, 337–404.

Bellman, R.E. (1961). *Adaptive Control Processes*. Princeton University Press. Princeton, NJ.

Brown, M. and C. J. Harris (1994). *Neurofuzzy Adaptive Modelling and Control.*. Prentice Hall. Hemel Hempstead.

Burges, C. J. C. (1998). A tutorial on support vector machines for pattern recognition. *Data Mining and Knowledge Discovery*.

Chen, S. (1995). Basis Pursuit. PhD thesis. Department of Statistics. Stanford University.

Cristianini, N. and J. Shawe-Taylor (2000). *An Introduction to Support Vector Machines*. Cambridge University Press.

Friedman, J.H. (1991). Multivariate adaptive regression splines. *The Annals of Statistics* **19**, 1–141.

Girosi, F. (1997). An equivalence between sparse approximation and Support Vector Machines. A.I. Memo 1606. MIT Artificial Intelligence Laboratory.

Gunn, S.R. (1998). Support vector machines for classification and regression. Technical Report, Dept. of Electronics and Computer Science. University of Southampton. Southampton, U.K.

Hadamard, J. (1923). *Lectures on the Cauchy Problem in Linear Partial Differential Equations*. Yale University Press.

Kavli, T. and E. Weyer (1995). On ASMOD - an algorithm for building multivariable spline models. In: *Advances in Neural Networks for Control Systems* (G.R. Irwin K.J. Hunt and K. Warwick, Eds.). pp. 83–104. Springer series on Advances in Industrial Control. Springer Verlag.

MacKay, D. J. C. (1995). Ensemble learning and evidence maximisation. Techical report. University of Cambridge.

Mészáros, C. (1998). The bpmpd interior point solver for convex quadratic problems. Technical Report WP 98-8. Computer and Automation Research Institute, Hungarian Academy of Sciences. Budapest.

Plate, T. A. (1999). Accuracy versus interpretability in flexible modelling: implementing a tradeoff using gaussian process models. *Behaviourmetrika special issue on Interpreting Neural Network Models* (26), 29–50.

Poggio, T., V. Torre and C. Koch (1985). Computational vision and regularization theory. *Nature* **317**, 314–319.

Rasmussen, C. E. (1996). Evaluation of Gaussian Processes and other Methods for Non-Linear Regression. PhD thesis. University of Toronto. Available from http://bayes.imm.dtu.dk/pub/.

Smola, A., B. Schölkopf and K.-R. Müller (1998). General cost functions for support vector regression. In: *Proc. of the Ninth Australian Conf. on Neural Networks* (T. Downs, M. Frean and M. Gallagher, Eds.). University of Queensland. Brisbane, Australia. pp. 79 – 83.

Smola, A. J. (1998). Learning with Kernels. PhD thesis. Technische Universität Berlin.

Stitson, M. A. Gammerman, V. Vapnik, V. Vovk, C. Watkins and J. Weston (1999). Support vector regression with ANOVA decomposition kernels. In: *Advances in Kernel Methods — Support Vector Learning* (B. Schölkopf, C.J.C. Burges and A.J. Smola, Eds.). MIT Press. Cambridge, MA. pp. 285–292.

Tikhonov, A. N. and V. Y. Arsenin (1977). *Solutions of Ill-posed Problems*. W. H. Winston. Washington, D.C.

Vapnik, V. (1995). *The Nature of Statistical Learning Theory*. Springer. N.Y.

Wahba, G. (1990). *Spline Models for Observational Data*. Series in Applied Mathematics, Vol. 59, SIAM. Philadelphia.

Wyatt, J. (1995). Nervous about artificial neural networks? (commentary). *The Lancet* (346), 1175–1177.

IFAC Publications
www.elsevier.com/locate/ifac

A GENERALISED LS–SVM

József Valyon and Gábor Horváth

Budapest University of Technology and Economics
Department of Measurement and Information Systems
(Budapest, Hungary, H-1521, P. O. Box 91. valyon@mit.bme.hu, horvath@mit.bme.hu)

Abstract: The Least–Squares Support Vector Machine (LS–SVM) solution can be obtained by solving a linear equation set. This allows us to address the problems in a linear fashion. In this paper we present a generalised view of the Least–Squares Support Vector Regression (SVR), which enables us to develop new formulations and algorithms to this regression technique. The modifications are based on manipulating the linear equation set, which embodies all information about the regression in the learning process. These modifications simplify the formulations, speed up the calculations and/or provide better results.[1]

Keywords: Function estimation, Least–Squares Support Vector Machines, Regression, Support Vector Machines, System Modelling.

1. INTRODUCTION

The primary advantage of SVM methods is that the resulting network structure, which is "optimal" in respect to the generalization error, is automatically derived. Another benefit of these methods is that the network guaranties an upper bound on the generalization error.

Among the SVM methods, the Least–Squares Support Vector Machine (LS–SVM) is attracting increasing attention, mostly because it is computationally more effective than the standard SVM methods. In this case training means solving a set of linear equations instead of the long and computationally hard quadratic programming problem involved by the standard SVM (Vapnik, 1995). It must also be emphasized, that LS-SVM is closely related to Gaussian processes and regularisation networks in that the obtained linear systems are equivalent (Suykens, *et al.*, 2002).

The method effectively reduces the algorithmic complexity, however for really large problems, comprising a very large number of training samples, even this least-squares solution can become highly memory and time consuming.

According to some experiments sometimes the LS-SVM can produce better results, which is probably due to the fact, that it builds all available information into the resulting net.

The price paid for this is the network size, which may be much larger than it would be with a traditional SVM, which incorporates a support vector selection method. While the least–squares version incorporates all training data in the network to produce the result, the traditional SVM selects some of them (the support vectors) that are important in the regression. The use of only a subset of all vectors is a desirable property of SVM, because it provides additional information regarding the training data, and concludes in a more effective solution formulating a smaller network. The sparseness of traditional SVM can also be reached with LS-SVM by applying a pruning method (Suykens, *et al.*, 2000b). Pruning techniques are also well known in the context of traditional neural networks. Their purpose is to reduce the complexity of the networks by eliminating as many hidden neurons as possible. Unfortunately if the traditional LS-SVM pruning method is applied, the performance declines proportionally to the eliminated training samples, since the information (input–output relation) they described is lost. Another problem is that this iterative method multiplies the algorithmic complexity.

The LS–SVM method should also be able to handle outliers (e.g. resulting from non–Gaussian noise). Another modification of the method, called weighted LS–SVM, is aimed at reducing the effects of this type of noise. The biggest problem is that pruning and weighting –although their goals do not rule out each other– cannot be used at the same time, because they work in opposition. The generalised

[1]This work was sponsored by the Hungarian Fund for Scientific Research (OTKA) under contract T033058

approach, presented in this paper, enables us to accomplish both goals by allowing a more universal construction and formulation of the kernel matrix or more precisely, the LS–SVM equation set. Our objectives include a simpler formulation, noise reduction, sparseness, and further reduction of algorithmic complexity, while maintaining the quality of the results.

The LS–SVM method is capable of solving both classification and regression problems. The classification approach is easier to understand and more historic. However, the present study concerns regression, therefore only this is introduced in the sequel, along with the most common extensions (pruning, weighting, and a fixed size version). Before going into the details the main and distinguishing features of the basic procedures are summarized. Only a brief outline of the method is presented, for a detailed description see (Suykens, *et al.*, 2000a; Suykens, *et al.*, 2000b; Suykens, *et al.*, 2002; Suykens, 2001).

2. A BRIEF OVERVIEW OF THE LS–SVM METHOD

Given the $\{\mathbf{x}_i, d_i\}_{i=1}^{N}$ training data set, where $\mathbf{x}_i \in \Re^p$ represents a p–dimensional input vector and $d_i \in \Re$ is the scalar target output, our goal is to construct an $y = f(\mathbf{x})$ function, which represents the dependence of the output d_i on the input $\mathbf{x}_i$. Let's define the form of this function as formulated below:

$$y = \sum_{j=0}^{h} w_j \varphi_j(\mathbf{x}) = \mathbf{w}^T \boldsymbol{\varphi}(\mathbf{x}), \qquad \mathbf{w} = [w_0, w_1, ..., w_h]^T, \quad \boldsymbol{\varphi} = [\varphi_0, \varphi_1, ..., \varphi_h]^T. \tag{1}$$

The $\varphi(.): \Re^p \to \Re^h$ is a mostly non-linear function, which maps the data into a higher h-dimensional (possibly infinite) feature space. The $\varphi_0(\mathbf{x})$ basis function is assumed to be 1, therefore w_0 represents the bias b. The main difference from the standard SVM is in the constraints. The optimisation problem and the inequality constraints are replaced by the following equations ($k = 1,...,N$):

$$\min_{\mathbf{w},b,e} J_p(\mathbf{w}, e) = \frac{1}{2}\mathbf{w}^T\mathbf{w} + C\frac{1}{2}\sum_{k=0}^{N} e_k^2 \tag{2}$$

with constraints: $d_k = \mathbf{w}^T \varphi(\mathbf{x}_k) + b + e_k$.

The first term forces a smoother solution, while the second one minimizes the training errors ($C \in \Re$ is the trade–off parameter between the two terms).From this, the Lagrangian:

$$L(w,b,e;\alpha) = J_p(w,e) - \sum_{k=0}^{N} \alpha_k \{\mathbf{w}^T \varphi(\mathbf{x}_k) + b + e_k - d_k\} \tag{3}$$

can be formed, where the α_k parameters are the Lagrange multipliers. The solution concludes in a constrained optimization, where the conditions for optimality are the followings:

$$\begin{aligned}
\frac{\partial L}{\partial w} &= 0 \quad \to \quad w = \sum_{k=0}^{N} \alpha_k \varphi(x_k) \\
\frac{\partial L}{\partial b} &= 0 \quad \to \quad \sum_{k=0}^{N} \alpha_k = 0 \\
\frac{\partial L}{\partial e_k} &= 0 \quad \to \quad \alpha_k = C e_k \qquad k = 1,...,N \\
\frac{\partial L}{\partial \alpha_k} &= 0 \quad \to \quad \mathbf{w}^T \varphi(\mathbf{x}_k) + b + e_k - d_k = 0 \quad k = 1,...,N
\end{aligned} \tag{4}$$

The solution concludes in a constrained optimisation, which leads to the following overall solution:

$$\begin{bmatrix} 0 & \vec{\mathbf{1}}^{\mathrm{T}} \\ \vec{\mathbf{1}} & \boldsymbol{\Omega} + C^{-1}\mathbf{I} \end{bmatrix} \begin{bmatrix} b \\ \boldsymbol{\alpha} \end{bmatrix} = \begin{bmatrix} 0 \\ \mathbf{d} \end{bmatrix}, \quad d = [d_0, d_1, ..., d_N]., \ \boldsymbol{\alpha} = [\alpha_0, \alpha_1, .., \alpha_N], \quad \vec{\mathbf{1}} = [1,...,1], \ \Omega_{i,j} = K(\mathbf{x}_i, \mathbf{x}_j) = \boldsymbol{\varphi}^T(\mathbf{x}_i)\boldsymbol{\varphi}(\mathbf{x}_j), \tag{5}$$

where $K(\mathbf{x}_i, \mathbf{x}_j)$ is the kernel function, and $\boldsymbol{\Omega}$ is the kernel matrix. Throughout this paper the RBF like

$$K(\mathbf{x}, \mathbf{x}_j) = \exp\left(-\frac{1}{2\sigma^2}\|\mathbf{x} - \mathbf{x}_i\|^2\right), \tag{6}$$

kernel function is assumed. For other kernels see (Suykens, *et al.*, 2002).

The result is:

$$y = \sum_{i=0}^{N} \alpha_i K(\mathbf{x}, \mathbf{x}_i) + b \tag{7}$$

It is easy to see, that the training of a support vector machine is a series of mathematical calculations, but the equation used for determining the answer (eq. **7.**) represents exactly the same calculations as a one hidden layer neural network. Although in practice SVMs are rarely formulated as actual networks, this neural interpretation is very important, because it provides an easier discussion framework than the purely mathematical point of view. This paper uses the neural interpretation throughout the discussions, because the points and statements of this work can be more easily understood this way.

It is important to emphasize, that according to eq. 4. the α_k weights are proportional to the e_k errors in the training points: $\alpha_k = Ce_k$. The following iterative methods are based on this property of the LS–SVM.

LS–SVM pruning (Suykens, *et al.*, 2000a, Suykens, *et al.*, 2000b): One of the main drawbacks of the least–squares solution is that the solution is not sparse in the sense, that it incorporates all training vectors in the resulting network. Practically this means that the net consists of–in its hidden layer– as many neurons, as the numbers of training vectors. In most real life situations the resulting networks are unnecessarily large. This drawback can be eliminated by applying a pruning method, which eliminates some training samples based on the sorted support vector spectrum (Suykens, *et al.*, 2000b). The α_k weighting of the Least–Squares SVM reflects the importance of the training samples, therefore by eliminating some vectors, represented by the smallest values from this $|\alpha_k|$ spectrum, the number of neurons can be reduced. The irrelevant points are left out by iteratively leaving out the least significant vectors. These are the ones corresponding to the smallest $|\alpha_k|$ values. The algorithm is the following (Suykens, *et al.*, 2002):

1. Train the LS–SVM based on N points. (N is the number of all available training vectors.)
2. Remove a small amount of points (e.g., 5% of the set) with the smallest values in the sorted $|\alpha_k|$ spectrum.
3. Re-train the LS–SVM based on the reduced training set.
4. Go to 2, unless the user–defined performance index degrades. If the performance becomes worse, it should be checked whether an additional modification of C, or the kernel parameters might improve the performance.

In SVM sparseness is achieved by the use of such loss functions, that errors smaller than ε are ignored (e.g., ε–insensitive loss function). This method reduces the difference between the SVM and LS–SVM, because the omission of some data points implicitly corresponds to creating an ε–insensitive zone (Suykens, 2001).

The described method leads to a sparse model, but some questions arise: How many neurons are needed in the final model? How many iterations it should take to reach the final model? Another problem is that a usually large linear equation set must be solved in all iterations. The pruning is especially important if the number of training vectors is large. In this case however, the iterative method is not very effective. The proposed Generalised LS-SVM method, leads to a sparse solution, automatically answers the questions and solves the problem described above.

Fixed LS–SVM (Suykens, *et al.*, 2002)**:** Another solution to this problem is called fixed LS–SVM , which uses an entropy based iterative method to select a predefined (fixed) size subset of the training samples as support vectors. After determining this optimal –in the sense of the defined entropy measure– subset a final LS–SVM network is trained.

Weighted LS–SVM (Suykens, *et al.*, 2002)**:** This method addresses the problem of noisy data –like outliers in a dataset– by using a weighting factor in the calculation, based on the error variables determined from a previous –first unweighted– solution. The method uses a bottom–up approach by starting from a standard solution, and calculating one or more weighted LS–SVM based on the previous result. The weighting is designed such, that the results improve in view of robust statistics. Large e_i–s mean a small weight and *vice versa.*

A common property of the described methods, is that they are all iterative, where every step is based on the result of an LS-SVM learning. This means, that the entire large problem must be solved at least once, and a relatively large one in every further iteration step. Another drawback is that pruning and weighting can not be easily combined, because the methods favour contradictory types of points. While pruning drops the training points belonging to small α_i-s, the weighted LS–SVM increases the effects of these points.

3. THE PROPOSED METHODS

This section proposes some modifications and extensions to the standard LS–SVM. Their purpose is to gain control over network size, to reduce complexity, and to improve the quality of the results.

3.1 Using an overdetermined equation set

If the training set consists of N samples, then our original linear equation set will have $(N+1)$ unknowns, the α_i-s, $(N+1)$ equations and $(N+1)^2$ multiplication coefficients. These factors are mainly the values of the $K(\mathbf{x}_i,\mathbf{x}_j)$ kernel function calculated for every combination of the training inputs. The cardinality of the training set, therefore determines the size of the coefficient matrix, which plays a major part in the solution, as algorithmic complexity; network size, etc., depends on this. It is easy to see that, in order to reduce complexity, this matrix has to be manipulated. Let's take a closer look at the linear equation set of eq. 5..

$$\left[\begin{array}{c|c} 0 & \vec{\mathbf{1}}^{\mathrm{T}} \\ \hline \vec{\mathbf{1}} & \mathbf{\Omega}+C^{-1}\mathbf{I} \end{array}\right]\begin{bmatrix} b \\ \boldsymbol{\alpha} \end{bmatrix}=\begin{bmatrix} 0 \\ \mathbf{d} \end{bmatrix}. \qquad \textbf{(8)}$$

The first row means:

$$\sum_{i=0}^{N}\alpha_i=0\,, \qquad \textbf{(9)}$$

and the j-th row stands for the:

$$\begin{aligned} &b+\alpha_0K(\mathbf{x}_j,\mathbf{x}_0)+.+\alpha_k\left[K(\mathbf{x}_j,\mathbf{x}_j)+C^{-1}\right]+\ldots \\ &+\alpha_NK(\mathbf{x}_j,\mathbf{x}_N)=y_j\,, \end{aligned} \qquad \textbf{(10)}$$

condition.

To reduce the equation set, columns and/or rows may be omitted.

1. If the k–th **column** is left out, then the corresponding α_k is also deleted, therefore the resulting network will be smaller. The condition of eq. 9. automatically adapts, since the remaining α–s will still add up to zero.
2. If the j–th **row** is deleted, then the condition defined by the $(\mathbf{x}_j,d_j)$ training point is lost, because the j–th equation is removed. This was the only one that comprised d_j and therefore the information of this training example.

The most important component of the main matrix $\mathbf{\Omega}$ kernel matrix; its elements are the results of the kernel function for two training inputs:

$$\Omega_{i,j}=K(\mathbf{x}_i,\mathbf{x}_j). \qquad \textbf{(11)}$$

To reduce the size of $\mathbf{\Omega}$ usually some training samples should be omitted. Each column of the kernel matrix represents a neuron of the final solution, with a kernel function centred on the corresponding $\mathbf{x}_i$ input. The rows, however, represent the input–output relations, described by the training points. The solution ($\boldsymbol{\alpha}$–weighting) is determined to satisfy these. It can be seen, that the network size is determined by the number of columns, which -in order to reach sparseness- must

be reduced. The following reduction techniques can be used on the kernel matrix (the names of these techniques are introduced here for easier discussion):
<u>*Full reduction*</u> – a training sample $(\mathbf{x}_k, d_k)$ is fully omitted, therefore both the column and the row corresponding to this sample are eliminated.
<u>*Partial reduction*</u> – a training sample $(\mathbf{x}_k, d_k)$ is only partially omitted, by only eliminating the corresponding k–th column, but keeping the k-th row, which defines an input-output relation. It means, that the weighted sum of that row should still meet the d_k goal (as closely as possible).
If <u>*full reduction*</u> is applied –which means that only the remaining vectors will play part in the solution– than these vectors must be the ones representing the function as accurately as possible. The least noisy vectors seem to be the best choice. In this case however, reduction also means that the knowledge represented by the numerous other samples are lost. The next figure demonstrates how the equation changes by fully omitting some training points. The deleted elements are coloured grey.

$$\left[\begin{array}{c|ccccc} 0 & & & \vec{1} & & \\ \hline & \Omega_{00}+\frac{1}{C} & \Omega_{01} & \cdots & \Omega_{0N} \\ & \Omega_{10} & \Omega_{11}+\frac{1}{C} & \cdots & \Omega_{1N} \\ \vec{1}^T & \vdots & \vdots & \vdots & \\ & \Omega_{(N-1)0} & \Omega_{(N-1)1} & \cdots & \Omega_{(N-1)N} \\ & \Omega_{N0} & \Omega_{N1} & \cdots & \Omega_{NN}+\frac{1}{C} \end{array}\right] \begin{bmatrix} b \\ \alpha_0 \\ \alpha_1 \\ \vdots \\ \alpha_N \end{bmatrix} = \begin{bmatrix} 0 \\ d_0 \\ d_1 \\ \vdots \\ d_N \end{bmatrix} \qquad \textbf{(12)}$$

This is exactly the case in traditional LS–SVM pruning, because pruning iteratively omits some training points. The information embodied in these points is entirely lost.
To avoid this information loss, one may use the technique referred here as partial reduction. . This proposition resembles to the basis of the Reduced Support Vector Machines (RSVM) introduced for standard SVM classification (Lee and Mangasarian, 2001). In partial reduction, the omission of a training sample means that only the corresponding column is eliminated, while the row is kept. By selecting some (e.g. M , $M < N$) vectors as "*support vectors*", the number of α_i variables are reduced, resulting in more equations than unknowns. The effect of partial reduction is shown on the next figure, where the removed elements are coloured grey.

$$\left[\begin{array}{c|ccccc} 0 & & & \vec{1} & & \\ \hline & \Omega_{00}+\frac{1}{C} & \Omega_{01} & \cdots & \Omega_{0N} \\ \vec{1}^T & \Omega_{10} & \Omega_{11}+\frac{1}{C} & \cdots & \Omega_{1N} \\ & \vdots & \vdots & \vdots & \vdots \\ & \Omega_{(N-1)0} & \Omega_{(N-1)1} & \cdots & \Omega_{(N-1)N} \\ & \Omega_{N0} & \Omega_{N1} & \cdots & \Omega_{NN}+\frac{1}{C} \end{array}\right] \begin{bmatrix} b \\ \alpha_0 \\ \alpha_1 \\ \vdots \\ \alpha_N \end{bmatrix} = \begin{bmatrix} 0 \\ d_0 \\ d_1 \\ \vdots \\ d_N \end{bmatrix} \qquad \textbf{(13)}$$

By applying this *partial reduction* our equation set becomes overdetermined, which can be solved as a *linear least–squares problem*, consisting of only (M+1)x(N+1) coefficients. The solution is calculated from an equation of the form

$$\mathbf{A}^T\mathbf{A}\mathbf{u} = \mathbf{A}^T\mathbf{v} \qquad \textbf{(14)}$$

equation, where **A** and **u** are the reduced $\left[\begin{array}{c|c} 0 & \vec{1}^{\mathrm{T}} \\ \hline \vec{1} & \boldsymbol{\Omega}+C^{-1}\mathbf{I} \end{array}\right]$ and $\begin{bmatrix} b \\ \boldsymbol{\alpha} \end{bmatrix}$ matrices respectively and $\mathbf{v} = \begin{bmatrix} 0 \\ \mathbf{d} \end{bmatrix}$.

The modified matrix **A** has (N+1) rows and (M+1) columns. After the matrix multiplications the results are obtained from a reduced equation set, incorporating $\mathbf{A}^T\mathbf{A}$, which is only of size (M+1)x(M+1). The omission of columns with keeping the rows, means that the network size is reduced; still all the known constraints are taken into consideration. This is the key concept of keeping the quality, while the equation set is simplified.

3.2 Complexity issues

This section deals with the algorithmic issues of the described solutions. LS–SVM training requires the solution of a linear equation set. In case of N training vectors this may be solved using the LU decomposition in $1/3\,N^3 + N^2$ steps, each with one multiplication and one addition. If the training set comprises N points, than the actual size of the equation set is $N+1$, but to keep the formulas simple, the effect of the one additional row is neglected. The reduced row echelon form of a matrix can be reached in about N^2 steps. The proposed "support vector" selection is made based on the result of this transformation. Let's assume that the reduction method leads to M selected vectors, and partial reduction is used. In this case, the calculation of $\mathbf{A}^T\mathbf{A}$ (defined in eq. 14.) requires M^2N^2 steps. Solving this new equation set costs $1/3\,M^3 + M^2$ steps. So the total cost of the proposed algorithm adds up to: $N^2 + M^2N^2 + 1/3\,M^3 + M^2$. If $M << N$ this means a smaller complexity compared to that of traditional LS–SVM. It is important to mention, that even if there is no algorithmic gain or it is rather small, this calculation provides a sparse solution with a good performance. If –in order to reach sparseness– the iterative pruning algorithm is applied to the traditional LS–SVM, than an equation set –slowly decreasing in size– must be solved in every step, which multiplies the complexity, whilst the errors may grow.

3.3 The use of customized kernels

The above described technique shows that the columns and rows may be handled independently, therefore the equation set may be generalized further. Each column (j) stands for a neuron, with a kernel centered on the corresponding input ($\mathbf{x}_j$). However in a generalized case:

- The kernels may be centered around any point (not just input samples), so the columns may be represented by any chosen $\mathbf{c}_j$ vector. For example the most simple construction of a fixed LS–SVM is to define the centers (e.g., M uniformly positioned vectors in the input space), and solve the equation set formulated accordingly (see eq. 16).

- The kernel functions may be different form column to column.

The formulation of $\mathbf{\Omega}$ and the equation set changes as follows:

$$\Omega_{i,j} = K_j(\mathbf{x}_i, \mathbf{c}_j) \tag{15}$$

$$\left[\begin{array}{c|ccc} 0 & & \vec{\mathbf{1}}^{\mathbf{T}} & \\ \hline & K_1(\mathbf{x}_1,\mathbf{c}_1)+C^{-1} & \cdots & K_M(\mathbf{x}_1,\mathbf{c}_M) \\ & \vdots & \ddots & \vdots \\ \vec{\mathbf{1}} & K_1(\mathbf{x}_M,\mathbf{c}_1) & \cdots & K_M(\mathbf{x}_M,\mathbf{c}_M)+C^{-1} \\ & \vdots & \ddots & \vdots \\ & K_1(\mathbf{x}_N,\mathbf{c}_1) & \cdots & K_M(\mathbf{x}_N,\mathbf{c}_M) \end{array}\right] \left[\begin{array}{c} b \\ \hline \alpha_0 \\ \vdots \\ \alpha_M \end{array}\right] = \left[\begin{array}{c} 0 \\ \hline d_0 \\ \vdots \\ d_M \\ \vdots \\ d_N \end{array}\right] \tag{16}$$

and the result will be calculated from:

$$y = \sum_{i=0}^{M} \alpha_i K_i(\mathbf{x}, \mathbf{c}_i) + b\,, \tag{17}$$

where M is the number of kernels used and K_i is the i-th kernel function.

This equation set is solved according to eq. 14. .

There is a slight problem with the regularisation parameter C, since it can only be inserted in the first M rows. This does not exactly reflect the same theoretical meaning as in the original equation (eq. 5.), but it is enough to ensure M linearly independent rows, so the equation set can be solved.

The selection of the $\mathbf{c}_i$ kernel centers is a complex problem, which has been extensively studied, mostly in respect of RBF networks. Briefly, it can be stated, that the $\mathbf{c}_i$ centers

- may be distributed uniformly for the simplest solution (e.g. for Fixed LS–SVM),
- may be selected from the training sample set (just as in the original SVM),
- may be selected by utilizing a clustering method,

etc. . In the followings a selection method is proposed, as a possible way to determine a subset of the training vectors for $\mathbf{c}_i$ centers.

3.4 Selecting support vectors

Standard SVM automatically selects the support vectors. To achieve sparseness by partial reduction, the linear equation set has to be reduced in such a way, that the solution of this reduced (overdetermined) problem is the closest to what the original solution would be.

As the matrix is formed from columns we can select a linearly independent subset of column vectors and omit all others, which can be formed as linear combinations of the selected ones. This can be done by finding a "basis" (the quote indicates, that this basis is only true under certain conditions defined later) of the coefficient matrix, because the basis is by definition the smallest set of vectors that can solve the problem.

The basic idea of doing a feature selection in the kernel space is not new. The nonlinear principal component analysis technique, the Kernel PCA uses the same idea (Schölkopf, et al., 1999). A basis selection from the kernel matrix has been shown in (Baudat and Anouar 2001).

This reduced input set (the support vectors) is (are) selected automatically by finding a "basis" of the $\mathbf{\Omega}$ (or the $\mathbf{\Omega} + C^{-1}\mathbf{I}$) matrix. A slight modification of the common mathematical method, used for bringing the matrix to the reduced row echelon form, can be utilized to find a set of vectors, that are linearly independent. The linear dependence discussed here, does not mean exact linear dependence, because the method uses an adjustable tolerance value when determining the "resemblance" (parallelism) of the column vectors. The use of this tolerance value is essential, because none of the columns of the coefficient matrix will likely be exactly dependent (parallel). This tolerance (ε') can be related to the ε parameter of the standard SVM, because it has similar effects. The larger the tolerance, the fewer vectors the algorithm will select. If the tolerance is chosen too small, than a lot of vectors will seem to be independent, resulting in a larger network. For a more detailed discussion of this method see (Valyon and Horváth, 200x)

3.5 Noise reduction

This section discusses a noise reduction technique for the case when the training samples contain some outliers (due to some non–Gaussian noise). This solution is based on the fact that once our equation set is overdetermined some (*N–M*) rows (constraints) may be removed and the equation set can still be solved. But which rows should be removed? Let's define some (*M*) vectors as support vectors. Every training sample, input–output pair, defines a constraint, which is represented by an equation. Provided that enough training points are available to learn the function, than the addition of new samples –equations– should not change the solution. This means, that this equation linearly dependent from the others. Our goal is remove some equations, such that the error of the least–squares solution is minimised. This can be achieved by using a "linearly dependent" subset of equations and leaving out the most linearly independent ones. Just like in the SV selection method, the linear dependence discussed here, does not mean exact linear dependence, only in a sense of a "resemblance" (parallelism) measure. The removed equations are the ones that are the least resembling to the others. The selection can be illustrated as follows. Fit an *N*+1–dimensional linear hyperplane on the points defined by the rows of the matrix $[\mathbf{\Omega} \;\; \mathbf{d}]$. Calculate the distance of every point from this plane. Leave out the points with the largest distance.

If the number of the kernel functions is M, and the training set has N samples, than at most *N–M* equations may be removed, otherwise the equation set will become underdetermined. In most cases it is unnecessary to leave out that many points, since the more constraints are considered, the better results can be expected.

4. EXPERIMENTS

The next figures show the results for a *sinc*(x) regression. The training set contains 55 data points burdened with Gaussian noise.

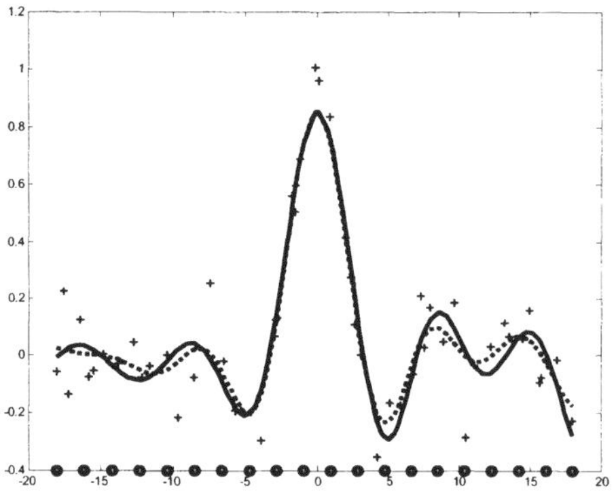

Fig. 1. The continuous black line plots the result for a generalised LS-SVM using 20 evenly distributed $\mathbf{c}_i$ kernel centers. The dashed line is the original LS-SVM.

The next figure demonstrates, that by using partial reduction along with the described support vector selection, our method can lead to almost the same quality result, but to much less neurons. If the number of training samples is very high for the problem complexity, than the gain in the network size can be very large.

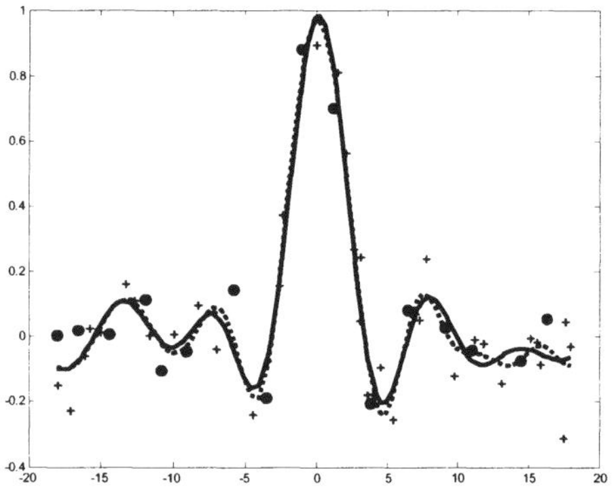

Fig. 2. The continuous black line plots the result for a generalised LS-SVM using the SV selection method described in *3.4*. The dashed line is the original LS-SVM. The 16 selected training samples (SVs) are also marked.

In the next experiment 5 input points are changed to outliers. These outliers are detected by the use of the described noise reduction method. The support vectors are selected from the training samples as described in section *3.4* . The input points are plotted, where the detected outliers are marked with large dots. It is important to mention that the results are based on a much smaller network (11 neurons) , since the number of columns was also reduced.

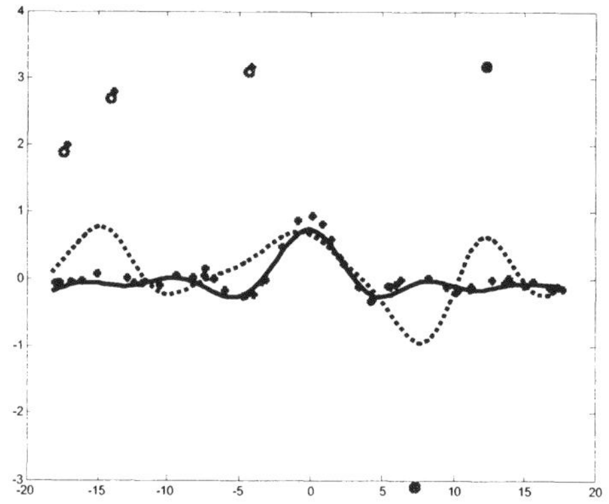

Fig. 3. The dashed line is the original LS-SVM result, while the continuous black line plots the result of the proposed method.

5. CONCLUSION

In this paper a generalised view of the least–squares support vector machine was presented. The basic idea is that the vectors chosen to be centers of kernels must not coincide with those, which the constraints are based on. In fact, even their numbers may be different, hence this equation set may be overdetermined. By eliminating variables pruning can be achieved, and by omitting equations, the effects of noise can be reduced.

REFERENCES

Baudat G. and Anouar F. (2001) "Kernel-based methods and function approximation". In *International Joint Conference on Neural Networks*, pages 1244–1249, Washington, DC July 15–19.

Lee Yuh–Jye and Mangasarian Olvi L. (2001) "RSVM: Reduced Support Vector Machines", *Proceedings of the First SIAM International Conference on Data Mining*, Chicago, April 5–7.

Suykens J. A. K., Lukas L., and Vandewalle J. (2000a) "Sparse approximation using least squares support vector machines" In: *IEEE International Symposium on Circuits and Systems ISCAS'2000.*

Suykens J. A. K., Lukas L., and Vandewalle J. (2000b) "Sparse least squares support vector machine classifiers" In: *ESANN'2000 European Symposium on Artificial Neural Networks*, pp. 37–42.

Suykens J. A. K., (2001) "Nonlinear Modeling and Support Vector Machines", *IEEE Instrumentation and Measurement Technology Conference*, Budapest, Hungary.

Suykens J. A. K., Gestel V. T., De Brabanter J., De Moor B., Vandewalle J. (2002) "Least Squares Support Vector Machines", *World Scientific.*

Valyon, J., and Horvath G. (200x) Reducing the complexity and network size of LS–SVM solutions. Submitted to: *IEEE Transactions on Neural Networks.*

Vapnik V. (1995) "The Nature of Statistical Learning Theory", New–York: Springer–Verlag.

IFAC
Publications
www.elsevier.com/locate/ifac

ADAPTIVE KERNEL METHODS

Anthony Kuh

Department of Electrical Engineering
University of Hawaii
Honolulu, HI 96822
email: kuh@spectra.eng.hawaii.edu

Abstract: This paper discusses the Least Squares Support Vector Machine and implementing adaptive on-line algorithms based on recursive least squares algorithms. The algorithms are of moderate complexity and can implement nonlinear decision regions which make it suitable for many applications in communication and signal processing. *Copyright © 2003 IFAC*

Keywords: Kernel methods, Least Squares Support Vector Machine, recursive least squares algorithms.

1. INTRODUCTION

This paper discusses kernel methods and implementation using adaptive on-line algorithms. We focus on the Least Squares - Support Vector Machine (LS)-SVM developed by (Suykens *et al.*, 1999*a*). The algorithm has many potential signal processing and communication applications where we are interested in adaptive nonlinear algorithms of moderate complexity that can perform better than adaptive linear algorithms. In many applications from channel equalization to multiuser detection there can be a significant gap in performance between optimal algorithms and optimal linear algorithms. Adaptive LS-SVM algorithms can approach the performance of optimal nonlinear algorithms.

In the last several years Support Vector Machines (SVM) and kernel methods (Boser *et al.*, 1992; Cortes and Vapnik, 1995; Cristianini and Shawe-Taylor, 2000; Scholkopf *et al.*, 1999; Vapnik, 1998) have received much attention and have been applied to a number of practical problems ranging from image processing to optical character recognition to analyzing DNA data (Cristianini and Shawe-Taylor, 2000; Scholkopf *et al.*, 1999; Muller *et al.*, 2001). SVM are based on principles of structural risk minimization and statistical learning theory (Vapnik, 1998). A nice feature about using SVM is that the computational capabilities of the machine are controlled by using kernel functions. The solution is found by working in the dual space where the computational complexity of the problem depends on the number of observation examples and not the dimension of the input feature space, (Scholkopf *et al.*, 1999).

The LS-SVM has many nice properties as

(1) they are based on a quadratic error criterion and are solved by finding solutions to a set of linear equations.
(2) they can easily implement nonlinear decision regions or functions by using nonlinear kernel functions.
(3) the complexity is moderate even for machines that use nonlinear kernels as solutions can be found in the dual space.
(4) the algorithm is amenable to analysis as it is based on linear minimum mean squared estimation.
(5) on-line adaptive implementation is relatively simple using gradient algorithms or recursive

least squares algorithms. These algorithms can track time varying-changes.

We examine detection problems involving two hypotheses. The LS-SVM algorithm is based on linear analog mean-squared estimation methods, but is easily applied to the detection problem. Although criteria such as the minimum error probability criterion and the minimum mean squared error technique are different they can yield similar solutions depending on the distribution of the data. Linear minimum mean squared error solutions have been applied to practical detection problems from multi-user detection (Verdu, 1998) to digital channel equalization (Haykin, 1996) because of many of the reasons listed above: good performance, a variety of adaptive implementations, and these algorithms can often be analyzed. The LS-SVM has these advantages and can also implement nonlinear machines with moderate complexity.

The paper is organized as follows. Section 2 discusses the two hypotheses detection problem and using detectors based on linear minimum mean squared error methods. Section 3 formulates both the LS-SVM solution in the primal space and the dual observation space. Section 4 discusses the performance of the LS-SVM. Section 5 discusses adaptive on-line implementation of the LS-SVM algorithm and issues associated with windowing, numerical stability, and complexity. Finally, Section 6 discusses communication and signal processing applications of on-line LS-SVM algorithms.

2. BAYESIAN DETECTION PROBLEM

The Bayesian detection problem (Van Trees, 1968) can be modeled with two random variables: $X \in \mathcal{R}^n$, describes the input and $Y \in \{-1, 1\}$, describes the output. For the detection problem we see a sample (or several samples) drawn from X and then make a decision $\hat{Y}$ on what Y is. We are typically given the likelihood conditional density function $f_{X|Y}(x|y)$ and the prior probability distribution, $P(Y = 1) = p$ and the $P(Y = -1) = 1 - p$.

There are also costs associated with making decisions where $C_{i,j}$ is the cost for deciding that the output is i given that it is j. The goal of Bayesian detection is to minimize the expected cost averaged over the input distribution X. The decision that minimizes the expected cost is called the Bayesian optimal solution. For the minimum error probability problem we have that

$$C_{1,1} = C_{-1,-1} = 0, \quad C_{1,-1} = C_{-1,1} = 1.$$

The optimal decision depends on maximizing the posterior probability, $f_{Y|X}(y|x)$ or by considering the likelihood function defined as

$$L(x) = f_{X|Y}(x|1)/f_{X|Y}(x|-1).$$

If $L(x) \geq (1-p)/p$ decide that $\hat{Y} = 1$ otherwise decide that $\hat{Y} = -1$. The decision region can be a simple linear threshold function or a more complex decision function depending on the two conditional likelihood densities.

If we do not have the conditional density functions or the posterior probabilities we must estimate the appropriate sufficient statistic. A sufficient statistics for the binary Bayesian detection problem is the log-likelihood function, $l(x) = \log(L(x))$ and the posterior probability, $P(x) = f_{Y|X}(1|x)$. Here we make our estimates based on observing labeled training data. We would like to estimate either $l(x), P(x)$ or the optimal Bayesian decision rule, $\mathrm{sgn}(P(x) - .5)$. In this paper we compare the optimal Bayesian decision rule with the LS-SVM solution.

3. LEAST SQUARES SUPPORT VECTOR MACHINES

3.1 LS-SVM Solution

The standard SVM are solved using quadratic programming methods, however these methods are often time consuming and are difficult to implement adaptively (Scholkopf *et al.*, 1999; Cristianini and Shawe-Taylor, 2000). Research has been undertaken to use a quadratic error criterion instead of the $\mathcal{L}_1$ norm used for the SVM. Ridge regression methods using a quadratic error criterion were developed for classification problems (Cristianini and Shawe-Taylor, 2000) and recently (Mangasarian and Musicant, 2001) use a quadratic error criterion to find an iterative solution to their Lagrangian SVM networks. These methods still have inequality constraints, however (Suykens *et al.*, 1999*a*) formulated a modified least squares SVM (LS-SVM) based on using a quadratic error criterion with equality constraints. The LS-SVM is given by

$$\min J(w,b) = \min_{w,b} \frac{1}{2}||w||^2 + \frac{1}{2}C\sum_{i=1}^{l} e_i^2 \quad (1)$$

subject to $y_i(w \cdot \phi(x_i) + b) = 1 - e_i, \quad i = 1, \ldots, l$. This problem is easily solved by setting the partial derivatives $\partial J(w,b)/\partial(w) = 0$ and $\partial J(w,b)/\partial(b) = 0$ and solving for w and b. We get that

$$(R(l) + I/(lC))w = P(l) \quad (2)$$

and

$$m_X(l)^T w + b = m_Y(l) \tag{3}$$

where $m_X(l) = (1/l)\sum_{i=1}^{l} \phi(x_i)$ and $m_Y(l) = (1/l)\sum_{i=1}^{l} y_i$ are the sample first order statistics. The second order statistics are given by

$$R(l) = (1/l)\sum_{i=1}^{l} \phi(x_i)\phi(x_i)^T - m_X(l)m_X(l)^T$$

and

$$P(l) = (1/l)\sum_{i=1}^{l} y_i\phi(x_i) - m_Y(l)m_X(l)$$

If $(R(l)+I/(lC))$ is nonsingular we then have that

$$w = (R(l) + I/(lC))^{-1}P(l)$$

and $b = m_Y(l) - m_X(l)^T(R(l) + I/(lC))^{-1}P(l)$. The LS-SVM solution depends on the first and second order statistics of $\phi(x(i))$ and $y(i)$. As the number of samples, l grows large we can state the following result.

Proposition Under the assumptions of Section 2 where samples are independently drawn from the same distribution let ϕ be a Mercer kernel with $\phi(X)$ and Y being second order random variables. Let $R = \mathbf{E}(\phi(X)\phi(X)^T)$ be positive definite, $P = \mathbf{E}(Y\phi(X))$, $m_x = \mathbf{E}(\phi(X))$, and $m_Y = \mathbf{E}(Y)$. Then the LS-SVM solution converges almost surely to the kernel minimum mean squared error solution defined by weight vector, $w = R^{-1}P$ and threshold value, $b = m_Y - {m_X}^T R^{-1}P$.

Proof: A random variable $\phi(X)$ is a second order random variable if $\mathbf{E}(|\phi(X)|^2)$ is finite (Grimmett and Stirzaker, 2001). From the assumptions and the Law of Large Numbers (Grimmett and Stirzaker, 2001), we therefore have that the time averages all converge almost surely to ensemble averages. We then use the fact that if two sequences of random variables converge almost surely then their products will also converge almost surely (Grimmett and Stirzaker, 2001). Then from equations (2-3) and that R is positive definite we have that the time averages will converge almost surely to ensemble averages with $w = R^{-1}P$ and $b = m_Y - {m_X}^T R^{-1}P$.

Remark 1: For the binary hypothesis detection problem let $m_i = \mathbf{E}(\phi(X)|Y = i)$ and

$$R_i = \mathbf{E}(\phi(X)\phi(X)^T|Y = i), \quad i = -1, 1.$$

Also let $m = pm_1 + (1-p)m_{-1}$ and $R = pR_1 + (1-p)R_{-1} - mm^T$. Then under assumptions of Section 2 and the Proposition above, for l large we will have that

$$w = (R)^{-1}(pm_1 - (1-p)m_{-1}) , \tag{4}$$

and

$$b = 2p - 1 - m^T R^{-1}(pm_1 - (1-p)m_{-1}). \tag{5}$$

We then get that $\hat{Y} = \text{sgn}(w^T\phi(x) + b$ with the mean squared error given by $4p(1-p) - (pm_1 - (1-p)m_{-1})^T R^{-1}(pm_1 - (1-p)m_{-1})$. In general, this solution will be different from the minimum probability of error solution which depends on the likelihood ratio $L(x)$.

Remark 2: Computation of the weight vector w and threshold value b depends on the dimensionality of the vector, $\phi(x) \in \mathcal{R}^d$. For polynomial kernels this can be quite high and for Gaussian kernels the dimensionality is infinite. This is a reason that we use the kernel trick to work in the dual observation space when $l < d$. The formulation in the dual observation space is given below.

3.2 LS-SVM Solution in Dual Observation Space

The dual problem can be implemented (Suykens *et al.*, 1999*a*) by considering equation (6) and augmenting it with the equality constraints and lagrange multipliers, α_i to get

$$\min J(w, b, \alpha, e) = \min_{w,b} \frac{1}{2}||w||^2 + \frac{1}{2}C\sum_{i=1}^{l} {e_i}^2$$
$$- \sum_{i=1}^{l} \alpha_i(y_i(w^T\phi(x_i) + b) - 1 + e_i)$$

Using the KKT conditions we set the partial derivatives equal to zero (Suykens *et al.*, 1999*a*) to get

$$\frac{\partial J}{\partial w} = 0 \;\rightarrow\; w = \sum_{i=1}^{l} \alpha_i y_i \phi(x_i)$$

$$\frac{\partial J}{\partial b} = 0 \;\rightarrow\; \sum_{i=1}^{l} \alpha_i y_i = 0$$

$$\frac{\partial J}{\partial e_i} = 0 \;\rightarrow\; \alpha_i = Ce_i$$

$$\frac{\partial J}{\partial \alpha_i} = 0 \;\rightarrow\; y_i(w^T\phi(x_i) + b) - 1 + e_i = 0.$$

We can express these equations in more compact form to get

$$\alpha^T y = 0 \tag{6}$$

where $y = (y_1 \ldots y_l)^T$, $\alpha = (\alpha_1 \ldots \alpha_l)^T$, and

$$yb + (\Omega + C^{-1}I)\alpha = \mathbf{1}_l \tag{7}$$

where $\mathbf{1}_l$ is a one vector with l components and

$$\Omega(i,j) = y_i y_j K(x_i, x_j), \ \ 1 \leq i,j \leq l \quad (8)$$

The formulation and solution is presented in (Suykens *et al.*, 1999*a*) and because the constraints are all equality constraints the solution involves solving a set of linear equations. Here C is a regularization factor. If C is large, then the solution approaches the standard minimum mean squared error estimate. If C is made small, then the regularization term, $||w||^2$ becomes more important and outlier points are deemphasized.

3.3 Comments about LS-SVM solution

Remark 3: The algorithm complexity depends on d and l. Assuming R and K are invertible, the algorithm involves inverting either R or K. If linear machines are used or the number of training examples is low or moderate, the LS-SVM algorithm complexity will be relatively low and have advantages over standard SVM. For linear machines solution is usually found in primal space. When nonlinear kernels are used we often have d greater than l and LS-SVM solution is found in dual observation space.

Remark 4: The LS-SVM algorithm is very similar to many other kernel algorithms including Kernel Fisher Discriminant Analysis (KFDA) discussed in (Baudat and Anouar, 2000; Mika *et al.*, 1999; Muller *et al.*, 2001). The LS-SVM algorithm is easily transformed into KFDA. In KFDA the goal in the kernel space is to choose a projection of the data to maximize the separation of the means of the two different data sets while minimizing the variances.

4. LS-SVM PERFORMANCE

From the previous section the LS-SVM will converge to the linear minimum mean squared error solution in kernel space as the number of training examples gets large. This is a different solution than the Bayesian detector where optimal solutions depend on sufficient statistics such as the likelihood ratio.

However, in many situations where the reproducing kernel Hilbert Space (RKHS) is rich enough and the conditional densities are of certain forms the LS-SVM detector can closely approximate the Bayesian detector. In (Lin, 2002) they show conditions where the LS-SVM and the SVM using a quadratic cost function converges to the optimal Bayesian solution.

The linear minimum mean squared error solution minimizes the variance of the error given by

$$e = Y - w^T \phi(X) - b.$$

Recall that $\hat{Y} = \text{sgn}(w^T \phi(X) + b)$. For the binary Bayesian detection problem there are two sources of errors that occur: when a positively labeled example is miclassified by $\hat{Y} = -1$ or when a negatively labeled example is misclassified by $\hat{Y} = 1$. Probability of error can be caluclated by integrating the tails of the conditional errors. The probability of error is given by $P_e = pP(e < -1|Y = 1) + (1-p)P(e > 1|Y = -1)$. If the conditional error has certain distributions, then the linear minimum mean squared error solution is the same as the minimum probability of error solution. Therefore the LS-SVM will converge to the minimum probability of error solution. As a simple example, consider a detection problem where under each hypothesis the observations are Gaussian with same second order statistics, but different means. Then the optimal detector is a linear threshold function and the error will also have a Gaussian distribution. Minimizing the variance of the error will also minimize the probability of error.

The performance of the LS-SVM depends on the distribution of the conditional error. For many practical situations, the LS-SVM solution with the right kernels and certain error distributions can approximate the optimal Bayesian solution.

5. ADAPTIVE LS-SVM

As discussed in Section 3, the LS-SVM can be solved in either the primal or dual space by inverting R in the primal space or solving equations (6) and (7) in the dual observation space. Both methods involve solving a set of linear equations making it amenable to adaptive on-line implementation. These algorithms can work quite well for moderate sized problems (one hundred to ten thousand observation examples). For problems with larger number of observation examples there has been recent success developing decomposition algorithms for the standard Support Vector Machine to solve the quadratic programming problem in the dual space such as the sequential minimization optimization algorithm (Platt, 1998) and methods developed in (Chang *et al.*, 2000).

A number of adaptive algorithms are possible including using gradient methods discussed in (Suykens *et al.*, 1999*b*) where the conjugate gradient algorithm is implemented. Here we focus on recursive least squares (RLS) algorithms. RLS algorithms have the advantage of having relatively quick convergence. In (Kuh, 2001) we studied a windowed recursive least squares algorithm that was solved in the dual observation space. A window size of length N was used. The training data are described by the inputs $x(k) = [x_k|\ldots|x_{k+N-1}]$ and targets $y(k) = (y_k \ldots y_{k+N-1})^T$. Let $U(k) = \Omega_k + C^{-1}I$ and

$$\Omega_k(i,j) = y_{i+k}y_{j+k}K(x_{i+k}, x_{j+k}) \tag{9}$$

where $0 \leq i,j \leq N-1$. The parameters of the LS SVM algorithm at time k are described by the threshold value $b(k)$ and the Lagrangian multipliers $\alpha(k) = (\alpha_k \ldots \alpha_{k+N-1})^T$. We then get the matrix equation described by

$$\begin{bmatrix} 0 & y(k)^T \\ y(k) & U(k) \end{bmatrix} \begin{bmatrix} b(k) \\ \alpha(k) \end{bmatrix} = \begin{bmatrix} 0 \\ \mathbf{1}_N \end{bmatrix} \tag{10}$$

Using simple matrix identities and assuming $U(k)$ is nonsingular we get that

$$lb(k) = g(k) = \frac{y(k)^T P(k)\mathbf{1}_N}{y(k)^T P(k) y(k)} \tag{11}$$

where $P(k) = U(k)^{-1}$ and

$$\alpha(k) = P(k)(\mathbf{1}_N - g(k)y(k)). \tag{12}$$

A key to this algorithm is the computation of $P(k) = U(k)^{-1}$. In (Kuh, 2001) we partitioned $U(k)$ into old data, $D(k)$ and new data to eventually get the following algorithm that find $\alpha(k)$ and $b(k)$ recursively:

(1) Initialization: $k = 1$,
(2) Get data $(x(k), y(k))$ and compute $P(k)$ (from $Q(k-1) = D(k-1)^{-1}$ or initialization).
(3) Compute $b(k)$ from equation (11) and $\alpha(k)$.
(4) Compute $Q(k)$ from $P(k)$.
(5) $k \leftarrow k+1$ go to 2.

Each update of algorithm runs in $\mathcal{O}(N^2)$ time where N is the size of the window. There are a number of issues with implementing a recursive LS-SVM.

- When working in the dual space the windowing of data becomes crucial. We cannot have exponential weighting with an increasing number of training points as the size of the matrix dimension depend directly on the number of training points. This is why we window the data to a fixed length N. In primal space the dimension of R is d regardless of window size and in the dual space the dimension of K is $N+1$. It is a simple modification to have exponential weighting with this fixed window. If the error criterion is exponentially weighted and given by

$$\min_{w,b} \frac{1}{2}||w||^2 + \frac{1}{2}C \sum_{i=k+1}^{k+N} \lambda^{k+N-i} e_i^2 \tag{13}$$

where $0 < \lambda < 1$ then we have that $U(k) = \Omega_k + \text{diag}(1/(C\lambda^{N-1}), 1/(C\lambda^{N-2}, \ldots, 1/C)$.
- Numerical stability is always a concern when working with RLS algorithms. Methods to improve stability (Haykin, 1996), include using square root algorithms such as the QR-RLS algorithm. These algorithms can be implemented for solving the dual LS-SVM. Algorithms such as the QR-RLS have better numerical stability problems than standard RLS algorithms.
- Complexity of algorithm has been addressed extensively for RLS algorithms and variants (Haykin, 1996). Fast algorithms with low complexity have been developed for the RLS algorithm that run in $\mathcal{O}(N)$ time such as the Fast Transversal Filter (FTF) (Cioffi and Kailath, 1984). The algorithm exploits structure in the filter structure to create four transversal filters that have linear complexity. It is an ongoing research project to explore whether fast convergence algorithms can be constructed for the LS-SVM to solve problems in the dual observation space. Another issue of concern is the stability of these class of low complexity fast algorithms (Slock and Kailath, 1991).

6. APPLICATIONS

In the introduction we mentioned that many problems in communications and signal processing involve nonlinear decision region boundaries. Examples include channel equalization and multiuser detection. Here we give a brief discussion of how we can implement algorithms developed in the previous section for a multiuser detection problem.

CDMA systems have received an increasing amount of attention because of performance benefits when compared to alternate TDMA and FDMA options and the possibilities of using sophisticated signal processing solutions for interference suppression. Current CDMA systems (e.g. IS-95) use limited signal processing capabilities, but many third-generation wireless networks will use CDMA and will require more complex signal processing in order to meet the increasing user requirements for bandwidth and mobility. There has been substantial research in using signal processing to improve the performance of CDMA systems by reducing the effects of multiple access interference (MAI). When all signature sequences are available optimal receivers can be designed to minimize the effects of MAI and additive noise, but these receivers often have high complexity (Verdu, 1998). Tradeoffs can be obtained by reducing the complexity of receivers by using linear receivers such as the decorrelator receiver or the linear minimum mean squared error receiver. These receivers are superior to matched filters in reducing MAI and have lower complexity than optimal nonlinear receivers (Verdu, 1998).

In (Kuh, 2001) we implemented a nonlinear SVM receiver to detect a CDMA signal in additive noise with high MAI. Nonlinear SVM using Gaussian kernels have superior performance to linear receivers and approach the performance of optimal maximum likelihood receivers. We have extended the work to include adaptive on-line LS-SVM receivers. Their performance is similar to SVM receivers with the advantage that these receivers can be implemented adaptively on-line. These receivers can also be implemented at the mobile station as they do not require signature sequences of other users.

REFERENCES

Baudat, G. and F. Anouar (2000). Generalized discriminant analysis using a kernel approach. *Neural Computation* **12**, 2385–2404.

Boser, B., I. Guyon and V. Vapnik (1992). A training algorithm for optimal margin classifiers. In: *Proceedings of the 5th Annual ACM Workshop on Computational Learning Theory.* ACM. pp. 144–152.

Chang, C.-C., C.-W. Hsu and C.-J. Lin (2000). The analysis of decomposition methods for support vector machines. *IEEE Trans. on Neural Networks* **11**, 1003–1008.

Cioffi, J. and T. Kailath (1984). Fast, recursive-least squares transversal filters for adaptive filtering. *IEEE Trans. Acoust. Speech Signal Process.* **32**, 304–337.

Cortes, C. and V. Vapnik (1995). Support vector networks. *Machine Learning* **20**, 273–297.

Cristianini, N. and J. Shawe-Taylor (2000). *An Introduction to Support Vector Machines.* Cambridge University Press. Cambridge, United Kingdom.

Grimmett, G. and D. Stirzaker (2001). *Probability and Random Processes.* 3rd ed. ed.. Oxford Science. New York City, NY.

Haykin, S. (1996). *Adaptive filter theory, 3rd Ed..* Prentice Hall.

Kuh, A. (2001). Adaptive kernel methods for cdma systems. In: *International Joint Conference on Neural Networks, 2001.* pp. 1404–1409.

Lin, Y. (2002). Support vector machines and the bayes rule in classification. *Data Mining and Knowledge Discovery* **6**, 259–275.

Mangasarian, O. and D. Musicant (2001). Lagrangian support vector machines. *Journal of Machine Learning Research* **1**, 161–177.

Mika, S., G. Ratsch, J. Weston, B. Scholkopf and K. Muller (1999). Fisher discriminant analysis with kernels. In: *Neural Networks for Signal Processing* (Y.-H. Hu, J. Larsen, E. Wilson and S. Douglas, Eds.). pp. 41–48.

Muller, K., S. Mika, G. Ratsch, K. Tsuda and B. Scholkopf (2001). An introduction to kernel-based learning algorithms. *IEEE Trans. on Neural Networks* **12**, 181–202.

Platt, J. (1998). Sequential minimal optimization: a fast algorithm for training support vector machines. Technical Report MSR-TR-98-14. Microsoft Research.

Scholkopf, B., C. Burges and A. Smola (1999). *Advances in kernel methods support vector learning.* MIT Press. Cambridge, MA Kingdom.

Slock, D. and T. Kailath (1991). Numerically stable fast transversal filters for recursive least squares adaptive filtering. *IEEE Trans. Signal Process.* **39**, 92–114.

Suykens, J., C. Burges and A. Smola (1999*a*). Least squares support vector machine classifiers. *Neural Processing Letters* **9**, 293–300.

Suykens, J., L. Lukas, P. Van Dooren, B. De Moor and J. Vandewalle (1999*b*). Least squares support vector machine classifiers : a large scale algorithm. In: *Proc. of ECCTD'99.* pp. 839–842.

Van Trees, H. (1968). *Detection, Estimation, and Modulation Theory Part I.* John Wiley. New York City, NY.

Vapnik, V. (1998). *Statistical learning theory.* John Wiley. New York City, NY.

Verdu, S. (1998). *Multiuser Detection.* Cambridge University Press. Cambridge, United Kingdom.

IFAC
Publications
www.elsevier.com/locate/ifac

SUBSPACE REGRESSION IN REPRODUCING KERNEL HILBERT SPACE

L. Hoegaerts * J.A.K. Suykens * J. Vandewalle * B. De Moor *

** KU Leuven, Dept. of Electrical Engineering ESAT-SCD-SISTA,*
Kasteelpark Arenberg 10, B-3001 Heverlee, Belgium
Email: {luc.hoegaerts, johan.suykens}@esat.kuleuven.ac.be

Abstract:
We focus on three methods for finding a suitable subspace for regression in a reproducing kernel Hilbert space: kernel principal component analysis, kernel partial least squares and kernel canonical correlation analysis and we demonstrate how this fits within a more general context of subspace regression. For the kernel partial least squares case a least squares support vector machine style derivation is given with a primal-dual optimization problem formulation. The methods are illustrated and compared on a number of examples. *Copyright © 2003 IFAC*

Keywords: RKHS, kernels, KPCA, KPLS, KCCA, primal-dual LS-SVM formulations

1. INTRODUCTION

Over the last years one can see certain learning algorithms being transferred to a kernel representation (Schölkopf and Smola, 2002; Suykens *et al.*, 2002*a*). The benefit lies in the fact that nonlinearity can be allowed, while avoiding to solve a nonlinear optimization problem. In this paper we focus on least squares regression models in the kernel context. By means of a nonlinear map into a Reproducing Kernel Hilbert Space (RKHS) (Wahba, 1990) the data are projected to a high-dimensional space.

Estimation of regression coefficients in the RKHS will then be performed in a new basis constructed by optimization of (co)variance criteria. By reducing the new basis and projecting data to a subspace, the number of regression parameters is controlled. This scheme can reduce multicollinearity between the new variables by reducing variance on the least squares estimators. A better generalization can be obtained in this way for regression models.

This paper is organized as follows. In section 2 we introduce some minimal background of kernel methods in relation to reproducing kernel Hilbert spaces. In section 3 we introduce regression in a feature subspace. In section 4 we consider three closely related methods for finding an optimal subspace. In section 5 the kernel version of PLS is derived in two manners. In section 6 we demonstrate the method with the three criteria by some results on an artificial and a real-world dataset.

2. REPRODUCING KERNEL HILBERT SPACE

The central idea in kernel algorithms in learning theory context is to change the representation of a data point into a higher-dimensional mapping in a reproducing kernel Hilbert space (RKHS) by means of a kernel function. When appropiately chosen, the kernel function with arguments in the original space corresponds to a dot product with arguments in the RKHS. This allows to circumvent working explicitly with the new representation as long as one can express the computation in the RKHS as inner products.

Assume data $\{\{(\mathbf{x_i}, y_i)\}_{i=1}^{n} \in \mathbb{R}^p \times \mathbb{R}\}$ have been given. A kernel k provides a similarity measure between pairs of data points

$$k : X \times X \rightarrow \mathbb{R} : (\mathbf{x_1}, \mathbf{x_2}) \mapsto k(\mathbf{x_1}, \mathbf{x_2}). \quad (1)$$

Once a kernel is chosen, one can associate to each $x \in X$ a mapping $\varphi : X \rightarrow H_k : \mathbf{x} \mapsto k(\mathbf{x}, \cdot)$, which can be evaluated at $\mathbf{x}'$ to give $\varphi(\mathbf{x})(\mathbf{x}') = k(\mathbf{x},\mathbf{x}')$. The set of these mappings can be extended by including all possible finite linear combinations $\sum_{i=1}^{l} a_i k(\mathbf{x},\cdot)$, adjoining the limits and constructing an inner product

$$\varphi(\mathbf{x})(\mathbf{x}') = \langle k(\mathbf{x},\cdot), k(\mathbf{x}',\cdot)\rangle = k(\mathbf{x},\mathbf{x}') \quad (2)$$

based on the chosen kernel. One obtains a RKHS on X under the condition that the kernel is positive definite (Aronszajn, 1950). Remark that the kernel function coincides with an inner product function and that it reproduces itself. It is also called a representer of f at $\mathbf{x}$ because $f(\mathbf{x}) = \langle f, k(\mathbf{x},\cdot)\rangle$.

The resulting RKHS has then the property that every evaluation operator and norm of any element in H_k is bounded (Wahba, 1990). This makes the elements of a RKHS well-suited to interpolate pointwise known functions that must be smooth. In the context of regularization one tries to do this by minimizing a pointwise cost function c over data and monotonic smoothness function g:

$$\min_{f\in H_k} c((\mathbf{x_i}, y_i, f(\mathbf{x_i}))_{i=1}^{n}) + g(||f||). \quad (3)$$

The representer theorem (Kimeldorf and Wahba, 1971) says that the solution is constrained to the subspace spanned by the mapped data points:

$$f(\mathbf{x}) = \sum_{i=1}^{n} w_i \varphi(\mathbf{x_i})(\mathbf{x}) = \sum_{i=1}^{n} w_i k(\mathbf{x_i}, \mathbf{x}). \quad (4)$$

Thus the solution can always be expanded as a linear function of dot products. The w_i are typically found by solving the optimization problem that involves the specific regularization functional.

The Mercer-Hilbert-Schmidt theorem reveals more about the nature of φ by stating that for each positive definite kernel there exists an orthonormal set $\{\phi_i\}_{i=1}^{d}$ with non-negative λ_i such that we have following spectral decomposition:

$$k(\mathbf{x},\mathbf{x}') = \sum_{i=1}^{d} \lambda_i \phi_i(\mathbf{x})\phi_i(\mathbf{x}') = \langle \varphi(\mathbf{x}), \varphi(\mathbf{x}')\rangle \quad (5)$$

where $d \leq \infty$, and λ_i and ϕ_i are the eigenvalues and eigenvectors of the kernel, respectively, and solutions to the integral equation

$$\int k(\mathbf{x},\mathbf{x}')\phi_i(\mathbf{x})p(\mathbf{x})\mathrm{d}\mathbf{x} = \lambda_i \phi_i(\mathbf{x}'), \quad (6)$$

on $L_2(\mathbb{R}^p)$. This allows to formulate the inner product in terms of expansion coefficients:

$$\langle f, g\rangle = \langle \sum_{i=1}^{d} a_i \phi_i, \sum_{j=1}^{d} a_j \phi_j \rangle = \sum_{i=1}^{d} \frac{a_i b_i}{\lambda_i}, \quad (7)$$

where $\langle \phi_i, \phi_j \rangle = \delta_{ij}/\lambda_i$. A proper scaling of the basis vectors ϕ_i with factor $\sqrt{\lambda_i}$ will transform the inner product to its most simple canonical form of an Euclidean dot product so that $k(\mathbf{x},\mathbf{x}') = \varphi(\mathbf{x})^T\varphi(\mathbf{x}')$. Furthermore one can express $\varphi(\mathbf{x})$ in this basis, with components $\langle \varphi(x), \sqrt{\lambda_i}\phi_i\rangle_H$, so that the φ mapping can be identified with a $d \times 1$ *feature* vector $\varphi(\mathbf{x}) = [\sqrt{\lambda_1}\phi_1(\mathbf{x}) \quad \sqrt{\lambda_2}\phi_2(\mathbf{x}) \quad \dots \quad \sqrt{\lambda_d}\phi_d(\mathbf{x})]^T$.

For regression purposes, just like one builds up a $n \times p$ data matrix X from the $\mathbf{x_i}$ inputs, one constructs with the mapped data points an $n \times d$ *feature matrix* :

$$\Phi = \begin{pmatrix} \sqrt{\lambda_1}\phi_1(\mathbf{x_1}) & \sqrt{\lambda_2}\phi_2(\mathbf{x_1}) & \dots & \sqrt{\lambda_n}\phi_d(\mathbf{x_1}) \\ \sqrt{\lambda_1}\phi_1(\mathbf{x_2}) & \sqrt{\lambda_2}\phi_2(\mathbf{x_2}) & \dots & \sqrt{\lambda_n}\phi_d(\mathbf{x_2}) \\ \vdots & \vdots & \ddots & \vdots \\ \sqrt{\lambda_1}\phi_1(\mathbf{x_n}) & \sqrt{\lambda_2}\phi_2(\mathbf{x_n}) & \dots & \sqrt{\lambda_n}\phi_d(\mathbf{x_n}) \end{pmatrix}. \quad (8)$$

3. KERNEL SUBSPACE REGRESSION

The goal is to obtain a regression estimate of the underlying function, given the data $\{\{(\mathbf{x_i}, \mathbf{y_i})\}_{i=1}^{n} \in \mathbb{R}^p \times \mathbb{R}^q\}$. Here, we consider the standard linear multivariate regression model in feature space

$$Y = \Phi W + E, \quad (9)$$

where Y represents a $n \times q$ matrix of observations of the dependent variable, Φ is a $n \times d$ matrix of regressors $\{\varphi(\mathbf{x_i})\}_{i=1}^{n}$, W is the unknown $d\times q$ matrix of regression coefficients and E is a $d \times q$ matrix of errors with zero-mean Gaussian i.i.d. values of equal variance σ^2 (unknown). We will assume all mapped data variables have been mean-centered.

A first difficulty in this setup, is that d may be very large compared to the number of data points n, even $d = +\infty$ potentially. In order to restrict the infinite number of regression coefficients, one can introduce a projection of the data into a subspace of finite dimension $m \ll d$, so as to obtain an overdetermined system. Hence the name subspace regression. If we gather the basis vectors $\{\mathbf{v_i}\}_{i=1}^{m}$ of the subspace in the columns of a $d \times m$ transformation matrix V, we can express the $\varphi(\mathbf{x_i})$ in the new coordinates $z_i = \varphi(\mathbf{x_i})^T\mathbf{v}$, so that in matrix notation $Z = \Phi V$, where Z is a $n\times m$ matrix of transformed regressors (scores). If $m = d$ then a mere basis change is performed, but when $m < d$ we project into a real subspace and the confined regression model becomes $Y = ZW + E$, where W is now a $m \times q$ matrix of regression coefficients.

The particular choice of the basis vectors, which will affect the quality of the model, is here assumed to be given and its determination on the basis of the data will be discussed in section 4. To estimate the unknown true model parameters, the elements of W, we choose here to minimize the error E in least squares sense:

$$\min_{W} ||Y - ZW||_2^2. \quad (10)$$

A second difficulty is that in general the elements of matrix Φ are unknown because the explicit expression for φ is not available. An elegant and optimal approach is to make use of the so-called kernel trick (Schölkopf and Smola, 2002). The need for a direct expression is typically avoided by formal manipulation towards scalar products so that kernels come into play. Matrix Φ may be unknown, but the $n \times n$ elements of $\Phi\Phi^T = K$ are easily computable. In the above setup, this can be achieved by introducing a decomposition $V = \Phi^T A$. In section 5 we derive the expression of V in terms of Φ in two ways. This allows to write $Z = \Phi V = \Phi\Phi^T A = KA$ and our model equation becomes:

$$Y = (KA)W + E. \tag{11}$$

Here the $n \times n$ kernel matrix K can be easily calculated and the computation of the $n \times m$ matrix A will be discussed in detail in section 4. So in order to obtain an estimate of our $n \times q$ matrix Y, we have to find the $m \times q$ matrix W of regression coefficients. In the univariate case ($q = 1$) we can write $Y = \mathbf{y}$ as

$$f(\mathbf{x}) = \sum_{i=1}^{n}(\sum_{j=1}^{m} a_{ij}w_j)k(\mathbf{x_i}, \mathbf{x}) = \sum_{i=1}^{n} \alpha_i \, k(\mathbf{x_i}, \mathbf{x}). \tag{12}$$

Since we assumed that the data are centered, we need to adjust the expression $\varphi(\mathbf{x})$ with $\varphi(\mathbf{x}) - \frac{1}{n}\sum_{i=1}^{n}\varphi(\mathbf{x_i})$ everywhere. For the kernel matrices this has the consequence of replacing K by (Schölkopf and Müller, 1998)

$$K := (I_n - \frac{1}{n}1_n 1_n^T)K(I_n - \frac{1}{n}1_n 1_n^T) \tag{13}$$

where I_n is the identity matrix and 1_n a vector of ones.

4. SUBSPACE CONSTRUCTION

Basically, we need to find directions in the variable space that form a new basis, upon which we can project. Several criteria can be chosen to arrive at a subspace to confine the regression. We take an overview of two important criteria and an intermediate one. Then we motivate a natural transfer to a kernel context in the next section.

4.1 *Minimization of within-space correlation*

A more common name for within-space correlation is multicollinearity, the degree of covariance between the data vectors in x space. It occurs often when using multiple regression on data that one has collected that one has no full control over the design of the experiment. A high degree of multicollinearity produces unacceptable uncertainty (large variance) in regression coefficient estimates.

The commonly used tool for the purpose of multicollinearity reduction is Principal Component Analysis (PCA) (Jolliffe, 1986). It is mostly stated as a problem in which one maximizes the variance of the new variables $\mathbf{s} = \mathbf{v^T x}$:

$$\max_{\mathbf{v}} \text{var}\,(\mathbf{v^T x}) = \mathbf{v^T} C_{xx}\mathbf{v} \tag{14}$$

subject to $||\mathbf{v}|| = 1$ and $V^T V = I_p$, with $C_{xx} = \frac{1}{n-1}X^T X$ the $p \times p$ sample covariance matrix. This involves a diagonalization procedure which requires solving an eigenvalue problem

$$C_{xx}\mathbf{v} = \lambda\mathbf{v}. \tag{15}$$

From the viewpoint of dimension reduction one can also describe PCA as the search for a best fitting subspace in a least squares sense. So PCA is equivalent to successive minimization of

$$J_{\text{PCA}}(\mathbf{v}) = \sum_{i=1}^{n} ||\mathbf{x_i} - \mathbf{v}\mathbf{v^T}\mathbf{x_i}||^2, \tag{16}$$

subject to the constaints.

4.2 *Maximization of between-space correlation*

On the other hand, the goal is maximization of between-space correlation. For the purpose of prediction one wishes to select subspace input vectors that have strong correlation with the target vectors.

The specific method that implements this criterion is Canonical Correlation Analysis (CCA) (Gittins, 1985). Here, one considers the $p \times q$ sample covariance matrix $C_{xy} = \frac{1}{n-1}X^T Y$ between two spaces. To minimize the cross-covariances, again diagonal elements are maximized

$$\max_{\mathbf{v},\mathbf{w}} \text{corr}(\mathbf{v^T x}, \mathbf{w^T y}) = \frac{\mathbf{v^T C_{xy} w}}{\sqrt{\mathbf{v^T C_{xx} v}}\sqrt{\mathbf{w^T C_{yy} w}}} \tag{17}$$

subject to $||\mathbf{v^T x}|| = \text{var}(\mathbf{v^T x}) = \mathbf{v^T C_{xx} v} = 1$ and $||\mathbf{w^T y}|| = \text{var}(\mathbf{w^T y}) = \mathbf{w^T C_{yy} w} = 1$. Essentially this requires the solution of the system

$$\mathbf{C_{xy} w} = \lambda\,\mathbf{C_{xx} v} \tag{18}$$

$$\mathbf{C_{yx} v} = \lambda\,\mathbf{C_{yy} w}. \tag{19}$$

The new basises in both spaces are chosen such that the vector components (projections) of all data maximally coincide. In CCA one successively minimizes

$$J_{\text{CCA}}(\mathbf{v}, \mathbf{w}) = \sum_{i=1}^{n} ||\mathbf{v^T x_i} - \mathbf{w^T y_i}||^2 \tag{20}$$

subject to the constraints. Thus the difference between the cosine of angles of lines in both spaces is minimized.

4.3 *An intermediate criterion*

The two above criteria can be taken as two extremes, and an optimal subspace choice may involve a tradeoff. The Partial Least Squares (PLS) (Wold, 1966) method can be positioned in between these two. PLS is a multivariate technique that delivers an optimal basis in x-space for y onto x regression. Reduction to a certain subset of the basis introduces a bias, but reduces the variance.

In general, PLS is based on a maximization of the covariance between successive linear combinations in x and y space, $\langle \mathbf{v}, \mathbf{x} \rangle$ and $\langle \mathbf{w}, \mathbf{y} \rangle$, where coefficient vectors are normed to unity and constrained to be orthogonal in x space:

$$\max_{\mathbf{v},\mathbf{w}} \text{cov}\,(\mathbf{v}^{\mathbf{T}}\mathbf{x}, \mathbf{w}^{\mathbf{T}}\mathbf{y}) = \mathbf{v}^{\mathbf{T}} C_{xy} \mathbf{w} \quad (21)$$

subject to $||\mathbf{v}|| = 1 = ||\mathbf{w}||$ and $V^T V = I_p$. Solutions can be obtained by using Lagrange multipliers, which leads to solving the following system

$$C_{xy}\mathbf{w} = \lambda \mathbf{v} \quad (22)$$

$$C_{yx}\mathbf{v} = \lambda \mathbf{w}. \quad (23)$$

As a least squares cost function, PLS turns out to be a sum of the least squares formulation of each of the above methods:

$$J_{\text{PLS}}(\mathbf{v}, \mathbf{w}) = J_{\text{PCA}}(\mathbf{v}) + J_{\text{CCA}}(\mathbf{v}, \mathbf{w}) + J_{\text{PCA}}(\mathbf{w}) \quad (24)$$

subject to the constraints. Indeed, by simplifying this expression, we obtain the covariance term only: the two variance minimizations of the CCA criterion are compensated by the PCA variance maximizations.

5. INTRODUCING THE KERNEL FUNCTION

In this section we will first follow the typical approach used for obtaining a kernel version. Next, we will give an alternatively motivated approach, along the same line of (Suykens *et al.*, 2002*b*), where PCA was originally reported in a support vector machine formulation and (Suykens *et al.*, 2002*a*) for the kernel CCA case which will be related here to PLS.

5.1 *Feature space with kernel trick*

Starting from criterion (21), we can proceed likewise in feature space, but now with $\mathbf{v}$ and $\mathbf{w}$ as $d \times 1$ feature space vectors. To arrive at calculation with kernels instead of feature vectors one typically expands the new basis vectors as follows

$$\mathbf{v} = \sum_{i=1}^{n} \alpha_i \varphi(\mathbf{x_i}) = \Phi^T A \quad (25)$$

$$\mathbf{w} = \sum_{i=1}^{n} \beta_i \varphi(\mathbf{y_i}) = \Phi^T B. \quad (26)$$

By substitution in (21) and use of some algebra one arrives at the following criterion

$$\max_{\alpha,\beta} \frac{\alpha^T K_{xx} K_{yy} \beta}{\sqrt{\alpha^T K_{xx} \alpha} \sqrt{\beta^T K_{yy} \beta}} \quad (27)$$

where $[K_{xx}]_{ij} = \varphi(\mathbf{x_i})^{\mathbf{T}} \varphi(\mathbf{x_j})$ and $[K_{yy}]_{ij} = \varphi(\mathbf{y_i})^{\mathbf{T}} \varphi(\mathbf{y_j})$ are the kernel Gram matrices and $\mathbf{v}$ and $\mathbf{w}$ were divided by their norm. Taking then derivatives with respect to α and β and equating to zero leads to the following coupled system of equations

$$\begin{aligned} K_{yy}\beta &= \lambda \alpha \\ K_{xx}\alpha &= \lambda \beta. \end{aligned} \quad (28)$$

In the same approach KPCA (Schölkopf and Müller, 1998) was derived as the eigenproblem

$$K_{xx}\, \alpha = \lambda\, \alpha, \quad (29)$$

and also a kernel version of PLS1 (Rosipal and Trejo, 2001).

5.2 *Primal-dual optimization*

A more principled approach can be taken by starting from the PLS least squares formulation (24). This primal cost criterion aims at optimizing the coefficient vectors $\mathbf{v}$ and $\mathbf{w}$, searching simultaneously for the maximal projection of a data point in x space, maximal covariation with the corresponding projection of the point in y space, and maximal projection of a data point in y space. Since the coefficients could become arbitrarily large, these are typically constrained or at least regularized. By simplifying expression (24) we obtain $\langle \mathbf{v}^{\mathbf{T}}\mathbf{x}, \mathbf{w}^{\mathbf{T}}\mathbf{y} \rangle$ and by adding (soft) regularization, we arrive at the following primal form problem:

$$\max_{\mathbf{v},\mathbf{w}} J_{PLS}(\mathbf{v}, \mathbf{w}, e, r) = \gamma \sum_{i=1}^{n} e_i r_i - \frac{1}{2}\mathbf{v}^{\mathbf{T}}\mathbf{v} - \frac{1}{2}\mathbf{w}^{\mathbf{T}}\mathbf{w} \quad (30)$$

such that $e_i = \mathbf{v}^{\mathbf{T}}\varphi(\mathbf{x_i})$ and $r_i = \mathbf{w}^{\mathbf{T}}\varphi(\mathbf{y_i})$ for $i = 1, \ldots, n$ with hyperparameter $\gamma \in \mathbb{R}_0^+$. Introducing α_i, β_i as Lagrange multiplier parameters, the Lagrangian is written as

$$L(\mathbf{v}, \mathbf{w}, e, r; \alpha, \beta) = \gamma \sum_{i=1}^{n} e_i r_i - \frac{1}{2}\mathbf{v}^{\mathbf{T}}\mathbf{v} - \frac{1}{2}\mathbf{w}^{\mathbf{T}}\mathbf{w}$$

$$- \sum_{i=1}^{n} \alpha_i (e_i - \mathbf{v}^{\mathbf{T}}\varphi(\mathbf{x_i})) - \sum_{i=1}^{n} \beta_i (r_i - \mathbf{w}^{\mathbf{T}}\varphi(\mathbf{y_i})). \quad (31)$$

A given optimization problem has a corresponding dual formulation. The number of constraints in the original problem becomes the number of variables in the dual problem. One optimizes the Lagrangian subject to the following optimality conditions:

$$\frac{\partial L}{\partial \mathbf{v}} = 0 \to \mathbf{v} = \sum_{i=1}^{n} \alpha_i \varphi(\mathbf{x_i}) = \Phi^T A \quad (32)$$

$$\frac{\partial L}{\partial \mathbf{w}} = 0 \to \mathbf{w} = \sum_{i=1}^{n} \beta_i \varphi(\mathbf{y_i}) = \Phi^T B \quad (33)$$

$$\frac{\partial L}{\partial \mathbf{e_i}} = 0 \to \gamma r_i = \alpha_i \quad i = 1, \ldots, n \quad (34)$$

$$\frac{\partial L}{\partial \mathbf{r_i}} = 0 \to \gamma e_i = \beta_i \quad i = 1, \ldots, n \quad (35)$$

$$\frac{\partial L}{\partial \alpha_\mathbf{i}} = 0 \to e_i = \mathbf{v}^{\mathbf{T}}\varphi(\mathbf{x_i}) \quad i = 1, \ldots, n \quad (36)$$

$$\frac{\partial L}{\partial \beta_\mathbf{i}} = 0 \to r_i = \mathbf{w}^{\mathbf{T}}\varphi(\mathbf{y_i}) \quad i = 1, \ldots, n. \quad (37)$$

By elimination of the variables $\mathbf{e}, \mathbf{r}, \mathbf{v}, \mathbf{w}$, we can simplify the dual problem further, which results into the very same system (28).

We point out that this KPLS solution can be considered in the optimization context as a special case of the 'regularized' KCCA variant which was proposed within the context of primal-dual LS-SVM formulations in (Suykens *et al.*, 2002*a*)

$$\begin{aligned} K_{yy}\,\beta &= \lambda\,(\nu_1 K_{xx} + I)\,\alpha \\ K_{xx}\,\alpha &= \lambda\,(\nu_2 K_{yy} + I)\,\beta. \end{aligned} \tag{38}$$

If one chooses the parameters $\nu_1 = \nu_2 = 0$. The 'regular' KCCA was originally reported by (Lai and Fyfe, 2000) and (Bach and Jordan, 2002), in an independent component analysis (ICA) context.

These three closely related methods, by partial solution of the (generalized) eigenproblems (29), (28) and (38), deliver us useful subspaces spanned by eigenvectors $\{\alpha_i\}_{i=1}^{m}$ (ordered corresponding to nondecreasing values of the eigenvalues λ_i). Together with expression (32), it allows to introduce the kernel matrix K in the regression model equation (11).

6. EXPERIMENTS

6.1 *Artificial data example*

To demonstrate some characteristics of these subspace regression methods, we apply it first to a sinc function. We started from a noiseless dataset. We considered a domain dataset of 100 equally-spaced points in the interval [-10,10]. The corresponding output values were centralized. We used a Gaussian kernel $\exp(-\|\mathbf{x_i} - \mathbf{x_j}\|_2^2/h^2)$ with width parameter $h = 1$ (and for kCCA always $\nu_1 = \nu_2 = 1$). In Fig. 1 the first three components of KCCA qualitatively show a good correlation with the targets. Non-equally-spaced sampling causes the components to be more irregular and oscillatory, while prediction will be less performant in undersampled regions and near boundaries. In Fig.2 we show an example of KCCA subspace regression on the same data, but with added Gaussian noise with standard deviation $\sigma = 0.2$. The other two methods give likewise component profiles and prediction results.

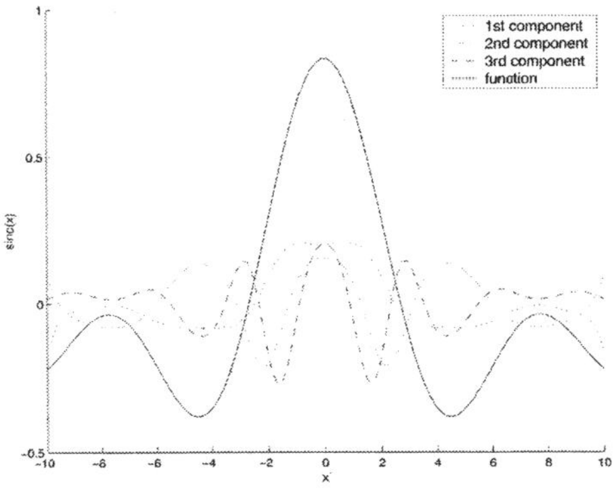

Fig. 1. Visualisation of the first three components extracted by KCCA for a noiseless sinc function.

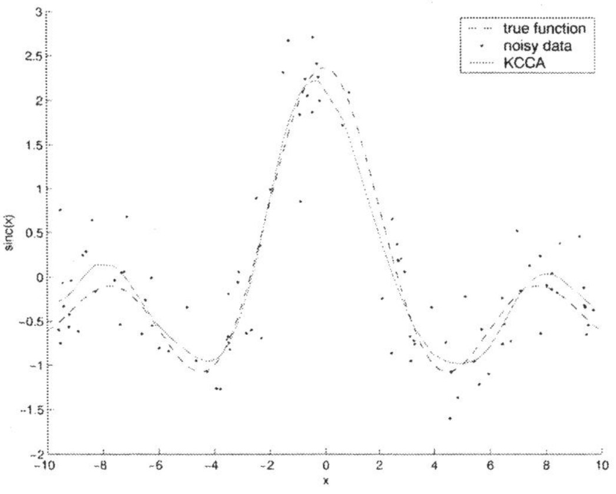

Fig. 2. Example of prediction obtained by KCCA subspace regression on a noisy sinc ($\sigma = 0.2, m = 4, h = 1$).

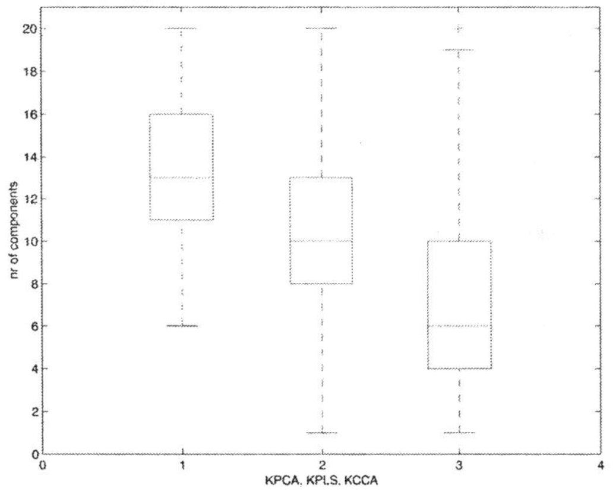

Fig. 3. Boxplots of number of components for KPCA, PLS, KCCA respectively, on a noisy sinc ($\sigma = 0.2$ and fixed $h = 1$).

We compared for 100 trials on sinc data sets (noise added with $\sigma = 0.2$) the number of components m for the three methods. Parameters $m \in \{1, 2, .., 20\}$ and $h \in [0.01, 5]$ were determined by 10-fold cross-validation (CV). On average the three methods perform equally well in terms of mean square error (MSE) on an independent test data set, with comparable variance. When lowering the number of components further, the best bias-variance tradeoff is for KCCA and then KPLS. From the boxplots in Fig. 3 we see that KCCA needs on average less components. This is also reflected in the non-increasing ordered eigenvalue spectrum that has large first eigenvalues, most quickly decaying for KCCA. As for the parameters ν_1 and ν_2 we may conclude that the regression result is fairly insensitive to ν_2, but that large values of ν_1 cause overfitting. The use of other kernels, like the polynomial or the sigmoidal kernel, did not produce such good results as the Gaussian kernel.

6.2 *Real-world data*

The Boston Housing data set (Harrison and Rubinfeld, 1978) consists of 506 cases having $p = 13$ input variables. The aim is to predict the housing prices. We standardized the data to zero mean and unit variance.

We picked at random a training set of size $n = 400$ and a test set of size $n_t = 106$. From the literature (Schölkopf *et al.*, 1997) we adopted the commonly used value of $h^2 = 0.15p = 3.9$ as fixed Gaussian kernel width for comparison reasons. We performed 10-fold CV to estimate the number of components $m \in \{1, 2, ..., 300\}$. After CV the best model was evaluated on an independent test set. The randomisation trials were repeated 10 times. See table 1 for details. The results are similar and not significantly different from LS-SVMs for regression (kernel ridge regression or regularization network with additional bias term) using the LS-SVMlab software http://www.esat.kuleuven.ac.be/sista/lssvmlab/.

method	nr of components	Mean Square Error
KPCA	220 ± 35	0.2292 ± 0.0525
KPLS	272 ± 14	0.2327 ± 0.0479
KCCA	245 ± 30	0.2230 ± 0.0445

Table 1. Boston Housing data set. Obtained parameters and independent test set performance of the three subspace regression methods using a Gaussian kernel with $h = \sqrt{3.9}$.

7. CONCLUSIONS

We presented three related methods -PCA, PLS and CCA- in a unified view from the viewpoint of RKHS regression, where they deliver a suitable subspace in order to reduce the potentially high number of regression parameters in high-dimensional feature space. Starting from the least squares criterion we showed that also PLS can be motivated in a primal-dual LS-SVM optimization fashion, in addition to KPCA and KCCA. Future issues that can be investigated in this context are different PLS variants with other constraints and making the methods adaptive.

Acknowledgements

This research work was carried out at the ESAT laboratory of the Katholieke Universiteit Leuven. It is supported by grants from several funding agencies and sources: Research Council KU Leuven: Concerted Research Action GOA-Mefisto 666 (Mathematical Engineering), IDO (IOTA Oncology, Genetic networks), several PhD/postdoc & fellow grants; Flemish Government: Fund for Scientific Research Flanders (several PhD/postdoc grants, projects G.0407.02 (support vector machines), G.0256.97 (subspace), G.0115.01 (bio-i and microarrays), G.0240.99 (multilinear algebra), G.0197.02 (power islands), research communities ICCoS, ANMMM), AWI (Bil. Int. Collaboration Hungary/ Poland), IWT (Soft4s (softsensors), STWW-Genprom (gene promotor prediction), GBOU-McKnow (Knowledge management algorithms), Eureka-Impact (MPC-control), Eureka-FLiTE (flutter modeling), several PhD grants); Belgian Federal Government: DWTC (IUAP IV-02 (1996-2001) and IUAP V-10-29 (2002-2006) (2002-2006): Dynamical Systems and Control: Computation, Identification & Modelling), Program Sustainable Development PODO-II (CP/40: Sustainibility effects of Traffic Management Systems); Direct contract research: Verhaert, Electrabel, Elia, Data4s, IPCOS. Luc Hoegaerts is a PhD student supported by the Flemish Institute for the Promotion of Scientific and Technological Research in the Industry (IWT). Johan Suykens is a postdoctoral researcher with the National Fund for Scientific Research FWO - Flanders and a professor with KU Leuven. Bart De Moor and Joos Vandewalle are full professors at KU Leuven, Belgium.

8. REFERENCES

Aronszajn, N. (1950). Theory of reproducing kernels. *Transactions of the American Mathematical Society* **686**, 337–404.

Bach, F.R. and M.I. Jordan (2002). Kernel independent component analysis. *Journal of Machine Learning Research* **3**, 1–48.

Gittins, R. (1985). *Canonical analysis*. Vol. 12 of *biomathematics*. Springer-Verlag.

Harrison, D. and D.L. Rubinfeld (1978). Hedonic prices and the demand for clean air. *J. Environ. Economics & Management* **5**, 81–102.

Jolliffe, I.T. (1986). *Principal Component Analysis*. Springer Verlag.

Kimeldorf, G.S. and G. Wahba (1971). Tchebycheffian spline functions. *J. Math. Ana. Applic.* (33), 82–95.

Lai, P.L. and C. Fyfe (2000). Kernel and nonlinear canonical correlation analysis. *International Journal of Neural Systems, submitted.*

Rosipal, R. and L.J. Trejo (2001). Kernel partial least squares regression in reproducing kernel hilbert space. *Journal of Machine Learning Research* **2**, 97–123.

Schölkopf, B. and A. Smola (2002). *Learning with kernels*. MIT Press, Cambridge MA.

Schölkopf, B., K. Sung, C. Burges, F. Girosi, P. Niyogi, T. Poggio and V. Vapnik (1997). Comparing support vector machines with gaussian kernels to radial basis function classiers. *IEEE Trans. on Signal Processing* **45**(11), 2758–2765.

Schölkopf, B., Smola A.J. and K.R. Müller (1998). Nonlinear component analysis as a kernel eigenvalue problem. *Neural Computation* **10(5)**, 1299–1319.

Suykens, J.A.K., T. Van Gestel, J. De Brabanter, B. De Moor and J. Vandewalle (2002*a*). *Least Squares Support Vector Machines*. World Scientific, Singapore.

Suykens, J.A.K., T. Van Gestel, J. Vandewalle and B. De Moor (2002*b*). A support vector machine formulation to pca analysis and its kernel version. *IEEE Transactions on Neural Networks*, in press.

Wahba, G. (1990). *Spline Models for Observational Data*. Vol. 59 of *CBMS-NSF Regional Conference Series in Applied Mathematics*. SIAM. Philadelphia, PA.

Wold, H. (1966). Estimation of principal components and related models by iterative least squares. In: *Multivariate Analysis* (P.R. Krishnaiah, Ed.). pp. 391–420. Academic Press. New York.

IFAC
Publications
www.elsevier.com/locate/ifac

Frequency Domain Identification of Wiener Models

Er-Wei Bai
Dept. of Electrical and Computer Engineering
University of Iowa
er-wei-bai@uiowa.edu

Abstract: This paper discusses Wiener model identifications in frequency domain using the sampled data. By exploring the fundamental frequency and harmonics generated by the unknown nonlinearity, we propose a frequency domain approach and show its convergence for both the linear and nonlinear subsystems in the presence of noise. No a priori knowledge of the structure of the nonlinearity is required and the linear part can be nonparametric. *Copyright © 2003 IFAC*

1. Introduction

The Wiener model is a nonlinear dynamic system which has applications in many engineering problems and thus identification of the Wiener model has been an active research area for many years.

In this paper, we extend our work in [2] for frequency domain identification of Hammerstain models to Wiener models. In the literature, the idea of frequency domain identification for the Wiener and Hammerstein models has been discussed [3, 11]. Though there were several approaches, they are more or less the same ideas as in [3, 6, 7]. In [7], the nonlinearity is assumed to be a polynomial with known order. The reason is that once the order is known, the highest harmonic is known and thus the relationship between the input and the highest harmonic presented in the output may be derived [7]. The problem is that many nonlinearities are not polynomials, though they may be approximated by polynomials. To have a reasonably good approximation, the order has to be high which results in small coefficients with high order terms. This has a significant impact on the signal to noise ratio in identification and makes the method sensitive. The frequency approach of [3] also assumes that the nonlinearity is a polynomial with known order and the linear part is a rational transfer function with known order. With known structures and orders, the output is a known nonlinear function of input magnitude and frequency with some unknown parameters. By repeatedly applying different magnitudes and frequencies, these unknown parameters can be uniquely solved. Without the exact structure or order information, uniqueness is however lost. The other frequency approach [6] also needs a parameterization of the unknown nonlinearity. In this paper, we propose a different frequency approach. The key is the use of a non-standard basis for the Fourier series representation. Then, we show the relationship between the phase of the linear part and the output at the fundamental frequency. Moreover, because of the non-standard Fourier series representation, the phase information of the unknown linear part can be easily extracted based on discrete Fourier transforms of the output measurements. The distinctive feature of the method proposed in this paper is that *no a priori information on the structure of the nonlinearity is assumed and the linear part can be nonparametric.* The results are particularly useful when a priori knowledge about the system before identification is so poor that the family of all possible characteristics can not be parameterized.

2. Problem statement and preliminaries

2.1. The Wiener model

Consider the Wiener model,

$$X(s) = G(s)U(s), \ \ y(t) = f(x(t)) + v(t)$$

where $u(t)$, $v(t)$ $y(t)$ and $y_f(t)$ are the system input, noise, output and filtered output respectively. $x(t)$ denotes the unavailable internal signal. $y(mT_s)$ and $y_f(mT_s)$ denote the sampled output and filtered output signals respectively with the sampling interval T_s that will be specified later. The filter is a lowpass filter at designer's disposal. The transfer function $G(s)$ represents the unknown stable linear system and $f(\cdot)$ is the unknown nonlinearity.

The goal of the frequency domain identification is to apply inputs of the form

$$u(t) = ACos(\omega_k t), \ \ t \in [0, T_k], \ \ k = 1, 2, ..., q \qquad (2.1)$$

for some A, $T_k > 0$, and then to determine a pair of the estimates $\hat{f}(\cdot)$ and $\widehat{G}(j\omega_k)$ based on the known inputs of (2.1) and the sampled outputs $y(mT_s)$ and $y_f(mT_s)$ so that

$$\hat{f}(\cdot) \to f(\cdot), \ \ \widehat{G}(j\omega_k) \to G(j\omega_k)$$

Here, the linear part $G(\cdot)$ is not necessarily a finite order rational transfer function. Also, no a priori information on the structure of $f(\cdot)$ is assumed. Only condition imposed on the unknown nonlinearity is the continuity and piecewise smoothness.

Assumption 2.1 *Throughout the paper, we assume*

1. *The linear part $G(s)$ is a stable continuous time system so that $G(j\omega)$ is continuous in ω and* $\sup|G(j\omega)| < \infty$. *$G(s)$ can be non-parametric.*

2. *The nonlinearity $f(x)$ is a static function which is continuous and piecewise smooth in $x \in$* $[-A\sup|G(j\omega)|, A\sup|G(j\omega)|]$. *No structural information is assumed.*

2.2. Fourier series representations

Now, for each input $u(t) = ACos(\omega_k t)$, $k = 1, 2, ...q$, the internal variable $x(t)$ is given by

$$x(t) = A|G(j\omega_k)|Cos(\omega_k t + \angle G(j\omega_k))$$

which is periodic with period $2\pi/\omega_k$. This implies that $f(x)$ is also periodic with the same period. Moreover, the unknown nonlinearity $f(x)$ is assumed to be continuous and piecewise smooth and thus, it permits a Fourier series representation. Usually, the basis for the Fourier series representation for a function with period $2\pi/\omega_k$ is

$$1,\ Cos(i\omega_k t),\ Sin(i\omega_k t),\ i = 1, 2...$$

We however adopt the following non-standard basis

$$1,\ Cos(i[\omega_k t + \angle G(j\omega_k)]),\ Sin(i[\omega_k t + \angle G(j\omega_k)]),\ i = 1, 2...$$

It is easily verified that this basis is also orthogonal and complete in L_2 sense over $t \in [-\pi/\omega_k, \pi/\omega_k]$. We now write $f(x)$ as

$$f(x(t)) = f(A|G(j\omega_k)|Cos(\omega_k t + \angle G(j\omega_k)))$$

$$= r_0 + \sum_{i=1}^{\infty}\{a_i Cos(i[\omega_k t + \angle G(j\omega_k)]) + b_i Sin(i[\omega_k t + \angle G(j\omega_k)])\}$$

where r_0, a_i's and b_i's are Fourier coefficients. Consider,

$$b_i = \frac{\omega_k}{\pi}\int_{-\frac{\angle G(j\omega_k)}{\omega_k}}^{\frac{2\pi}{\omega_k} - \frac{\angle G(j\omega_k)}{\omega_k}} f(A|G(j\omega_k)|Cos(\omega_k t + \angle G(j\omega_k)))$$

$$Sin(i[\omega_k t + \angle G(j\omega_k)])dt,\ i = 1, 2, ...$$

Let $\tau = t - \frac{\pi - \angle G(j\omega_k)}{\omega_k}$ and it follows that

$$b_i = \pm\frac{\omega_k}{\pi}\int_{-\pi/\omega_k}^{\pi/\omega_k} f(-A|G(j\omega_k)|Cos(\omega_k \tau))Sin(i\omega_k \tau)d\tau,$$

where $\pm$ depends on whether i is even or odd. Note that $Cos(\omega_k \tau)$ is an even function and so is $f(-A|G(j\omega_k)|Cos(\omega_k \tau))$. $Sin(i\omega_k \tau)$ is odd and hence $f(-A|G(j\omega_k)|Cos(\omega_k \tau))Sin(i\omega_k \tau)$ is odd which implies $b_i = 0,\ i = 1, 2, ...$

By summarizing the above discussion, we have

$$y(t) == r_0 + \sum_{i=1}^{\infty} r_i Cos(i\omega_k t)Cos(i\angle G(j\omega_k)) - \quad (2.2)$$

$$\sum_{i=1}^{\infty} r_i Sin(i\omega_k t)Sin(i\angle G(j\omega_k)) + v(t)$$

for some unknown r_i's. This equation relates the phase information $\angle G(j\omega_k)$ to the output and becomes the center piece in identification. Moreover, because continuity and piecewise smoothness assumptions on $f(\cdot)$, the Fourier series converges.

2.3. Discrete Fourier transforms

For most applications, only sampled input-output data are available. We will show that $G(j\omega)$ and $f(\cdot)$ can be calculated based on the sampled input-output measurement data. To this end, we need three preliminary steps: the choice of lowpass filter, the determination of the sampling interval T_s and the calculation of discrete Fourier transforms DFTs on the sampled data..

Filter choice:

To prevent alias due to sampling, let the cutoff frequency $\bar{\omega}$ of the lowpass filter be

$$0 < \bar{i}\omega_k < \bar{\omega} < (\bar{i} + 1)\omega_k \quad (2.3)$$

for some integer $\bar{i} \geq 1$. Then, the output $y_f(t)$ of the lowpass filter is in the form of

$$y_f(t) = r_0 + \sum_{i=1}^{\bar{i}} r_i Cos(i\omega_k t)Cos(i\angle G(j\omega_k)) - \quad (2.4)$$

$$\sum_{i=1}^{\bar{i}} r_i Sin(i\omega_k t)Sin(i\angle G(j\omega_k)) + v_f(t)$$

where $v_f(t)$ is the filtered noise plus some residue due to imperfect filtering.

Determination of the sampling interval T_s.

Since the highest frequency remaining in $y_f(t)$ due to the input is $\bar{i}\omega_k$, we define the sampling interval by

$$T_s = \frac{2\pi}{\bar{i}\omega_k}\frac{1}{M}, \quad M > 2 \quad (2.5)$$

The choice of the integer $M > 2$ is to make sure that the sampling frequency is always higher than the Nyquist frequency $\bar{i}\omega_k/\pi$. Obviously, we have

$$T_k = L\frac{2\pi}{\omega_k} = L\bar{i}MT_s,\ T_s/T_k = \frac{1}{L\bar{i}M},\ \omega_k T_s = \frac{2\pi}{\bar{i}M}$$

DFT implementation:

With the sampled filtered output data $y_f(mT_s)$, $m = 0, 1, ..., L\bar{i}M - 1$, we now define the DFT of $y_f(mT_s)$.

$$Y_{f,DFT}(p\omega_k) = \frac{1}{L\bar{i}M}\sum_{m=0}^{L\bar{i}M-1} y_f(mT_s)e^{-jp\omega_k mT_s} \quad (2.6)$$

$$V_{f,DFT}(p\omega_k) = \frac{1}{L\bar{i}M}\sum_{m=0}^{L\bar{i}M-1} v_f(mT_s)e^{-jp\omega_k mT_s} \quad (2.7)$$

These DFTs can be efficiently calculated by many commercially available softwares. The following lemma is a well known result in the literature [9].

Lemma 2.1 *Suppose $v_f(t)$ is a continuous time random signal that is the output of an unknown stable and proper linear system driven by a white noise of zero mean and finite variance. Let $v_f(mT_s)$ and $V_{f,DFT}(pj\omega_k)$ denote its samples and the DFT. Then, uniformly in p,*

$$\mathbf{E}V_{f,DFT}(p\omega_k) = 0, \quad \mathbf{E}|V_{f,DFT}(p\omega_k)|^2 = O(\frac{1}{T_k})$$

3. Frequency domain identification

Consider the DFT of $Y_{f,DFT}(p\omega_k)$ at $p = 1$,

$$Y_{f,DFT}(\omega_k) = \frac{T_s}{T_k}\sum_{m=0}^{L\bar{i}M-1}\{r_0 +$$

$$\sum_{i=1}^{\bar{i}} r_i Cos(i\omega_k mT_s)Cos(i\angle G(j\omega_k)) -$$

$$\sum_{i=1}^{\bar{i}} r_i Sin(i\omega_k mT_s)Sin(i\angle G(j\omega_k))e^{-j\omega_k mT_s}\} + V_{f,DFT}(\omega_k)$$

It is easily verified that

$$\frac{T_s}{T_k}\sum_{m=0}^{L\bar{i}M-1} Cos(i\omega_k mT_s)e^{-j\omega_k mT_s} = \begin{cases} 1/2 & i=1 \\ 0 & i \neq 1 \end{cases}$$

$$\frac{T_s}{T_k}\sum_{m=0}^{L\bar{i}M-1} Sin(i\omega_k mT_s)e^{-j\omega_k mT_s} = \begin{cases} -j/2 & i=1 \\ 0 & i \neq 1 \end{cases}$$

Therefore,

$$Y_{f,DFT}(\omega_k) = \frac{r_1}{2}Cos(\angle G(j\omega_k)) + j\frac{r_1}{2}Sin(\angle G(j\omega_k)) + V_{f,DFT}(\omega_k)$$

Since $V_{f,DFT}(\omega_k) \to 0$ in probability, it follows that

$$Y_{f,DFT}(\omega_k) \to \frac{r_1}{2}Cos(\angle G(j\omega_k)) + j\frac{r_1}{2}Sin(\angle G(j\omega_k))$$

3.1. Estimation of the phase

Let $\angle\widehat{G}(j\omega_k)$ denote the estimate of $\angle G(j\omega_k)$

$$\angle\widehat{G}(j\omega_k) = tan^{-1}\{\frac{Imag(Y_{f,DFT}(\omega_k))}{Real(Y_{f,DFT}(\omega_k))}\}, \quad k = 1,2,...,q \quad (3.1)$$

Since

$$Imag(Y_{f,DFT}(\omega_k)) \to \frac{r_1}{2}Sin(\angle G(j\omega_k))$$

$$Real(Y_{f,DFT}(\omega_k)) \to \frac{r_1}{2}Cos(\angle G(j\omega_k))$$

the following theorem is obvious.

Theorem 3.1 *Consider the estimate (3.1). Then, under the assumption of Lemma 2.1, we have*

$$\angle\widehat{G}(j\omega_k) \to \angle G(j\omega_k), \quad k = 1,2,...,q$$

in probability as $L \to \infty$.

Note that in the definition of the estimate $\angle\widehat{G}(j\omega_k)$, only the information provided by the fundamental frequency ω_k is used. In fact, $\angle\widehat{G}(j\omega_k)$ may also be defined as

$$\angle\widehat{G}(j\omega_k) = \frac{1}{i}tan^{-1}\{\frac{Imag(Y_{f,DFT}(i\omega_k))}{Real(Y_{f,DFT}(i\omega_k))}\}, \quad k = 1,2,...,q$$

based on the ith harmonic for $i \leq \bar{i}$. Therefore, an average $\angle\widehat{G}(j\omega_k)$ using the fundamental frequency as well as harmonics may be used to average out the effect of noise.

For each input $u(t) = ACos(\omega_k)$, equation (3.1) provides a reliable estimate of the phase $\angle\widehat{G}(j\omega_k)$ for large L. Note that in the estimation, no information on the structure of the unknown nonlinearity $f(\cdot)$ is assumed.

3.2. Estimation of the nonlinearity $f(\cdot)$

It is important to point out that the characterization of the Wiener model shown in Figure 1 is in fact not well defined for identification purposes. The gains of $f(\cdot)$ and $G(\cdot)$ can not be uniquely determined. Any pair $(\alpha f(\cdot), G(\cdot)/\alpha)$, $\alpha \neq 0$, would produce identical input and output measurements. Therefore, one of the gains must be normalized for identification purpose. There are several ways to make the representation unique, e.g., either the gain of $f(\cdot)$ or $G(\cdot)$ can be fixed to be unit. In this paper, we take a different approach and assume

Assumption 3.1 *For some $\bar{k} \in [1,2,...,q]$,*

$$A|G(j\omega_{\bar{k}})| = 1$$

The choice of $\bar{k}$ is arbitrary. If indeed $A|G(j\omega_{\bar{k}})| = 1$, the estimates provided would be $(G(\cdot), f(\cdot))$ and if the actual $A|G(j\omega_{\bar{k}})| = \alpha \neq 1$, the estimates provided would be $(G(\cdot)/\alpha, \alpha f(\cdot))$. This scaling constant is indistinguishable by the inherent property of the Wiener model.

Although the structure of the unknown nonlinearity $f(\cdot)$ is unknown, the nonlinearity is static and its structure can be determined from the graph of $y(t)$ versus $x(t)$. The output $y(t)$ is available and the difficulty is the unavailability of the internal signal $x(t)$. Thus, the key is to recover the internal signal $x(t)$. To this end, note that the phase estimates $\angle\widehat{G}(j\omega_k)$'s are available from (3.1). Also from Assumption 3.1, $A|G(j\omega_{\bar{k}})| = 1$. Hence, the estimate $\hat{x}(t)$ of $x(t)$ for the input $u(t) = ACos(\omega_{\bar{k}}t)$ can be constructed as

$$\hat{x}(t) = A|G(j\omega_{\bar{k}})|Cos(\omega_{\bar{k}}t + \angle\widehat{G}(j\omega_{\bar{k}})) = Cos(\omega_{\bar{k}}t + \angle\widehat{G}(j\omega_{\bar{k}}))$$

Note that $\angle\widehat{G}(j\omega_{\bar{k}}) \to \angle G(j\omega_{\bar{k}})$ and the true

$$x(t) = A|G(j\omega_{\bar{k}})|Cos(\omega_{\bar{k}}t + \angle G(j\omega_{\bar{k}})) = Cos(\omega_{\bar{k}}t + \angle G(j\omega_{\bar{k}}))$$

As a consequence $\hat{x} \to x(t)$ in probability.

Once $\hat{x}(t)$ is constructed, the identification of $f(\cdot)$ can be carried out. We discuss two cases separately. In the first case, the structure of the unknown $f(\cdot)$ can be determined from the graph of $y(mT_s)$ versus $\hat{x}(mT_s)$. For the second case, no structural information can be extracted from the graph.

The structure of $f(\cdot)$ can be determined from the graph

In this scenario, the unknown $f(\cdot)$ can be determined to have a certain structure $f(d, \cdot)$ from the graph parameterized by the unknown coefficient vector d. Accordingly, the estimate $\hat{f}$ is defined to have the same structure $\hat{f} = f(\hat{d}, \cdot)$ and then the estimation of the nonlinearity becomes the parameter estimation of $\hat{d}$. For instance, if the graph of $y(mT_s)$ versus $\hat{x}(mT_s)$ shows that the unknown nonlinearity $f(\cdot)$ is a deadzone

$$f(d, x) = \frac{1 - sgn(d - sgn(x))}{2}(x - d \cdot sgn(x))$$

for some unknown threshold d. The estimate $\hat{f}$ can be consequently defined as

$$\hat{f} = f(\hat{d}, x) = \frac{1 - sgn(\hat{d} - sgn(x))}{2}(x - \hat{d} \cdot sgn(x))$$

Next, the coefficients $\hat{d}$ can be determined by minimizing the least squares errors

$$\hat{d} = arg \min_{\hat{d}} J_1(\hat{d}) = arg \min \sum_{m=n}^{L\bar{i}M-1} (y(mT_s) - f(\hat{d}, \hat{x}(mT_s)))^2 \quad (3.2)$$

where $\hat{x}(mT_s) = Cos(\omega_{\bar{k}} mT_s + \angle\widehat{G}(j\omega_{\bar{k}}))$ is the estimate of $x(mT_s)$ and $n > 0$ is used to remove the effect of transients.

Two questions need to be answered concerning the estimate $\hat{d}$ of (3.2): (1) how to find the global minimum of the nonlinear optimization problem (3.2)? (2) if the global minimum is achieved, what is its convergence and consistency? In general, how to find the global minimum for a nonlinear optimization problem is not easy. For several frequently encountered nonlinearities, e.g., polynomial nonlinearities, the optimization of (3.2) is convex in the unknown variable $\hat{d}$ and the global minimum can be easily obtained. For some non-smooth nonlinearities, e.g., the saturation and deadzone nonlinearities, the optimization is no longer convex. However, the dimension of d is low and in fact is one for many non-smooth nonlinearities including the saturation and deadzone nonlinearities, either the saturation level or the threshold. In such a case, the optimization is one-dimensional and there are some efficient algorithms. Alternatively, one may simply plot the graph of $J_1(\hat{d})$ versus $\hat{d}$. From the graph, an accurate information on the value of the global minimum can be visually found. More discussions will be provided later in the simulation section.

In term of the consistency, it is well known from the literature [9] that the strong consistency results hold for the estimate $\hat{d}$ in the sense that with probability one

$$\lim_{L \to \infty} \hat{d} = d_0 \quad (3.3)$$

$$d_0 = arg \min_{\hat{d}} \lim_{L \to \infty} (y(mT_s) - f(\hat{d}, \hat{x}(mT_s)))^2$$

provided that the noise $v_f(mT_s)$ is i.i.d., zero mean and $\mathbf{E}|v_f(mT_s)|^2$, $\mathbf{E}|v_f(mT_s)|^{4+\rho} < \infty$ for some $\rho > 0$. Thus, the estimate $\hat{d}$ is well behaved in the presence of noise. In particular, if $\angle\widehat{G}(j\omega_{\bar{k}}) = \angle G(j\omega_{\bar{k}})$ and the global minimum of (3.2) is unique, we have $\hat{d} = d$ for large L.

No structural information can be extracted from the graph

If no structural information can be extracted from the graph, we approximate $f(\cdot)$ by a finite order polynomial

$$\hat{f}(x) = \sum_{i=0}^{l} \hat{\alpha}_i x^i \quad (3.4)$$

The coefficients $\hat{\alpha}_i$'s can be determined again from the least squares method

$$(\hat{\alpha}_1, ..., \hat{\alpha}_l) = arg \min_{\hat{\alpha}_i} J_2(\hat{\alpha}_i) \quad (3.5)$$

$$= arg \min \sum_{m=n}^{L\bar{i}M-1} (y(mT_s) - \sum_{i=0}^{l} \hat{\alpha}_i \hat{x}^i(mT_s))^2$$

This is a convex optimization problem in $\hat{\alpha}_i$ and the solution is unique. We comment that because the structure of $f(\cdot)$ is unknown and $\hat{f}(x) = \sum_{i=0}^{l} \hat{\alpha}_i x^i$ is an approximation, the choice of the order is essential. In general, with a higher order approximation, a smaller $J_2(\hat{\alpha}_i)$ is expected. The important thing is to investigate whether or not the improvement is significant. If the order l is already high enough to adequately describe the unknown nonlinearity $f(\cdot)$, any increment in l only produces a small reduction in J_2. Therefore, if the increment from l to $l+1$ only gives rise a marginal improvement in J_2, the lower order l can be chosen. Otherwise, a higher order is preferred. This observation can also be translated into the size of the estimates $\hat{\alpha}_l$. In general, if the last coefficient $\hat{\alpha}_l$ is small, any increment in the order l is likely to produce a marginal improvement. On the other hand, if $\hat{\alpha}_l$ is large, the contribution of the term with the order l is significant and likely, increment in the order l would probably give rise a sizable improvement in J_2. In terms of consistency, the same strong consistency result of (3.3) holds for $\hat{\alpha}_i$.

3.3. Estimation of $|G(j\omega_k)|$

Once the estimates $\angle\widehat{G}(j\omega_k)$ and $\hat{f}(\cdot)$ are obtained, the magnitude of $G(j\omega_k)$ can be estimated. For each input $u(t) = ACos(\omega_k t)$, the output $y(t)$ is available. We define the estimate $|\widehat{G}(j\omega_k)|$ as

$$|\widehat{G}(j\omega_k)| = arg\min_{\eta\geq 0} J_3(\eta) = arg\min \sum_{m=n}^{L\bar{i}M-1} (y(mT_s) \tag{3.6}$$
$$-\hat{f}(A\eta Cos(\omega_k mT_s + \angle\widehat{G}(j\omega_k))))^2$$

In the absence of noise and assuming that $\angle\widehat{G}(j\omega_k) = \angle G(j\omega_k)$ and $\hat{f} = f$, $J_3(|G(j\omega_k)|)$ clearly equals to zero. Again, we comment that the optimization is nonlinear but one-dimensional and can be solved easily. For instance, the graph of $J_3(\eta)$ versus η provides the information on where the global minimum is. Similarly, the strong consistency result of (3.3) holds for the estimate $|\widehat{G}(j\omega_k)|$.

Now, we are in a position to summarize the identification algorithm for the Wiener model.

Frequency domain identification algorithm

Consider the Wiener model in Figure 1. For each input $u(t) = ACos(\omega_k t)$, choose the filter, the sampling interval T_s as described before and collect the data $y(mT_s)$ and $y_f(mT_s)$.

Step 1: Calculate the DFT $Y_{f,DFT}(\omega_k)$.

Step 2: Determine the phase estimate $\angle\widehat{G}(j\omega_k)$ using (3.1).

Step 3: When $k = \bar{k}$, i.e., $A|G(j\omega_{\bar{k}})| = 1$, plot the graph of $y(mT_s)$ versus $\hat{x}(mT_s) = Cos(\omega_{\bar{k}} mT_s + \angle\widehat{G}(j\omega_{\bar{k}}))$. Determine the structure of $f(\cdot)$ from the graph.

Step 4: Depending on the outcome of Step 3, find the nonlinearity estimate $\hat{f}$ using either (3.2) or (3.4) and (3.5).

Step 5: Computer the estimate $|\widehat{G}(j\omega_k)|$ from (3.6).

The algorithm provides the nonlinearity estimate $\hat{f}$ and the linear part estimates $\widehat{G}(j\omega_k) = |\widehat{G}(j\omega_k)|e^{j\angle\widehat{G}(j\omega_k)}$, $k = 1, 2, ..., q$.

4. Simulation

The nonlinearity is a deadzone nonlinearity with threshold $d = 0.3$ and the unknown linear part $G(s)$ is

$$G(s) = \frac{2s+1}{s^2+2s+4},$$

The input frequencies are $\omega_k = [0.1, 0.2, 0.35, 0.5, 1, 2, 3, 5, 10]$. For simulation, $A = 1.6125$ so that $A|G(j)| = 1$, i.e., $\bar{k} = 5$, and $L = 200$, $n = 150$ and $L\bar{i}M = 15,000$. A uniformly distributed i.i.d. measurement noise with magnitude 0.2 is added. Note that the noise is significant, e.g., at $\omega_5 = 1$, $max|y(t)| < 1$ and $max|v(t)| = 0.2$.

First, the phase estimates $\angle\widehat{G}(j\omega_k)$'s are obtained. Then, the graph of $y(mT_s)$ versus $\hat{x}(mT_s) = Cos(\omega_5 mT_s + \angle\widehat{G}(j\omega_5))$ is plotted shown in Figure 1. From the figure, it is clear that the unknown nonlinearity is a deadzone with the threshold between 0.2 and 0.4. The dispersion around the unknown but true deadzone is caused by the added noise. Thus, the nonlinearity estimate is defined as

$$\hat{f}(x) = \frac{1 - sgn(\hat{d} - sgn(x))}{2}(x - \hat{d}\cdot sgn(x))$$

To find $\hat{d}$, we simply plot the graph of $J_1(\hat{d})$ versus $\hat{d}$ shown in Figure 2. Clearly, the global minimum $\hat{d} = 0.3$ is found which happens to be the exact value of $d = 0.3$. Finally, to solve optimization of (3.6) in order to find $|\widehat{G}(j\omega_k)|$, we plot $J_3(\eta)$ versus η at each input frequency ω_k.

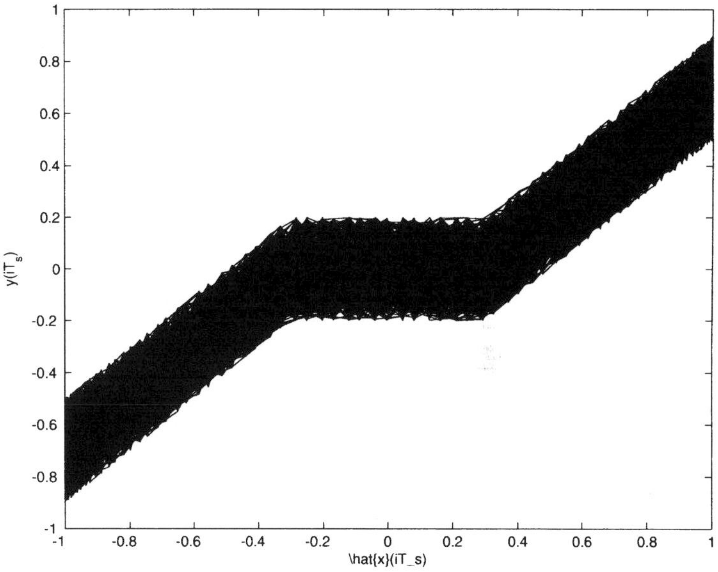

Figure 1: The graph of $y(mT_s)$ versus $\hat{x}(mT_s)$.

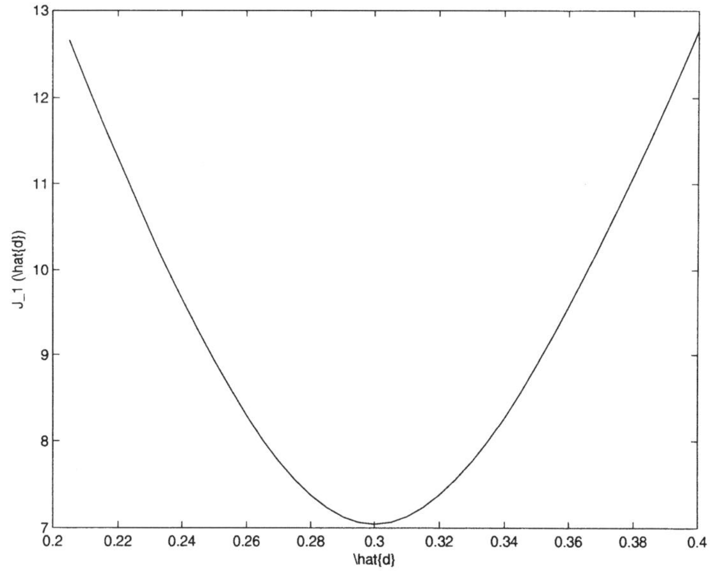

Figure 2: The graph of $J_1(\hat{d})$ versus $\hat{d}$.

5. Discussions

Use of lowpass filters

In the simulation example, no filter is used, i.e., $y(t) = y_f(t)$. We comment that a large number of simulation seems to suggest that in many cases identification results are similar with or without the lowpass filter. One explanation is that r_i's are small for $i \geq \bar{i}M - 1$, the use of the lowpass does not make too much difference.

Characterization of $G(s)$

Because of page limit, only point estimation of $G(j\omega_k)$, $k = 1, 2, ..., q$, was discussed and there was no attempt made to identify $G(s)$. If identification of $G(s)$ is of interests, some well known techniques available in the literature can be applied. Suppose the unknown $G(s)$ is an *nth* order transfer function, the technique of [8] may be used. In the case that $G(s)$ is non-parametric, either the spectral analysis method [9] or the interpolation method [5] can be applied. The spectral analysis aims to determine the transfer function based on spectral estimation and smoothing while the interpolation method is to make the worst case error bounded. For details, see [9] and [5].

Polynomial nonlinearities

No a priori structural information on the unknown $f(\cdot)$ is assumed in identification. However, if such prior information becomes available, it could be used in identification. For instance, suppose the unknown nonlinearity is a polynomial with known order l, i.e., $f(x) = \sum_{i=0}^{l} \beta_i x^i$. Then, for input $u(t) = ACos(\omega_{\bar{k}} t)$ and $x(t) = A|G(j\omega_{\bar{k}})|Cos(\omega_{\bar{k}} t) = Cos(\omega_{\bar{k}} t)$, it can be easily verified that

$$y(t) = \sum_{i=0}^{l} r_i Cos(i\omega_{\bar{k}} t)$$

$$r_0 = \beta_0 + \sum_{m=1}^{l_1} \beta_{2m} \frac{1}{2^{2m}} c_{2m}^{m}, \quad r_{2i+1} = \sum_{m=i}^{l_2} \beta_{2m+1} \frac{1}{2^{2m}} c_{2m+1}^{m-i},$$

$$r_{2i} = \sum_{m=i}^{l_1} \beta_{2m} \frac{1}{2^{2m-1}} c_{2m}^{m-i}$$

$$c_m^k = \frac{m(m-1)...(m-k+1)}{1 \cdot 2 ... \cdot ... \cdot k}.$$

$$l_1 = \lfloor \frac{l}{2} \rfloor, \quad l_2 = \lfloor \frac{l}{2} \rfloor - rem(l+1, 2)$$

where $\lfloor \frac{l}{2} \rfloor$ rounds $l/2$ to the nearest integer towards zero and $rem(l+1, 2)$ is the remainder after division $(l+1)/2$.

Since there is one-to-one map between β_i's and r_k's, the estimates $\hat{\beta}_i$'s and $\hat{f} = \sum_{i=0}^{l} \hat{\beta}_i x^i$ can be obtained based on the estimates of $\hat{r}_i$'s which can be easily calculated based on the equation (2.4) and $Y_{f,DFT}$ in a similar way as (3.1).

Fourier series representation

Note that the estimation starts with the equation (2.2) that is a Fourier series representation using a non-standard basis. This non-standard basis provides a clean relationship between the phase of $G(j\omega_k)$ and the output. In fact, other basis may also be used and this brings the question of what basis could provide clearer information about the system and have a faster convergence rate for the nonlinearity concerned. This study will be a very interesting topic in the context of identification for Wiener models.

6. Concluding remarks

In this paper, we have proposed a frequency domain identification approach for Wiener models. By exploring the frequencies contained in the output, the phase of the unknown transfer function can be calculated based on DFTs of the sampled output. Then, the nonlinearity as well as the magnitude of the linear part can be estimated. No information on the form of the nonlinearity is assumed and the linear part could be non-parametric. The method is simple and efficient.

References

[1] Bai, E.W. (2002), "A blind approach to the Hammerstein-Wiener model identification" *Automatica*, **38**, pp.967-979

[2] Bai, E.W. (2002), "Frequency domain identification of Hammerstein models", *Proc of 41st IEEE Conf. on Control and Decision*, Las Vegas, pp.1011-1016, also to appear in *IEEE Trans. on Auto. Contr.*

[3] Baumgartner, S. and W. Rugh (1975), "Complete identification of a class of nonlinear systems from steady state frequency response", *IEEE Trans. on Circuits and Systems*, **22**, pp.753-759

[4] Brigham, E.O. (1974) *The fast Fourier transform*, Prentice-Hall, Englewood Cliffs

[5] Chen, J. and G. Gu (2000), *Control Oriented System Identification: An H_∞ Approach*, John Wiley & Sons, New York

[6] Crama, Ph. And J. Schoukens (2001), "First estimates of Wiener and Hammerstein systems using multisine excitation", *Proc. of IEEE Instrumentation and Measurement Conf*, Budapest, Hungary, pp.1365-1369

[7] Gardiner, A. (1973), "Frequency domain identification of nonlinear systems", *3rd IFAC Symp. on Identification and System Parameter Estimation*, Hague, Netherlands, pp.831-834

[8] Levi, E.C. (1959), "Complex curve fitting", *IEEE Trans. On Auto. Contr.*, **4**, pp.37-43

[9] Ljung, L. (1999) *System Identification: Theory for the users*, 2nd Ed. Prentice-Hall, Upper Saddle River

[10] Pintelon, R. and J. Schoukens (2001), *System Identification: A Frequency Domain Approach*, IEEE Press, Piscataway, NJ

[11] Zadeh, L. (1956), "On the identification problem", *IRE Trans on Circuit Theory*, **3**, pp.277-281

IFAC Publications
www.elsevier.com/locate/ifac

NON-PARAMETRIC IDENTIFICATION OF NON-LINEARITY IN HAMMERSTEIN SYSTEMS

Włodzimierz Greblicki*, **Przemysław Śliwiński**†

*† *Institute of Engineering Cybernetics, Wrocław University of Technology*
Wybrzeże Wyspiańskiego 27, 50-370 Wrocław, Poland
* *wgre@ict.pwr.wroc.pl*, † *slk@ict.pwr.wroc.pl*

Abstract. In the paper non-parametric algorithms identifying the non-linearity in Hammerstein system are presented. About the non-linearity it is only assumed that it is a piecewise Lipschitz function. It is shown that the algorithms converge to the non-linearity with growing number of measurements n. For Lipschitz and smoother non-linearities the algorithms attain the convergence rate $O\left(n^{-2p/(2p+1)}\right)$ – the best possible for non-parametric algorithms. Both, convergence and its rate are independent of the regularity of the input probability density function.
Copyright © 2003 IFAC

Key Words. non-linear systems, system identification, non-parametric identification, regression estimates, accuracy

1. INTRODUCTION

In statistical inference we deal with two kind of information:

- A priori information – possessed before experiment.
- Empirical information – obtained from measurements.

If the former is reach enough to build a reliable parametric model – the parametric approach is applied. However, it is often the case that this information is not available and we can mostly rely on the empirical data — then a non-parametric approach appears to be more suitable.

In the paper we present non-parametric algorithms identifying a non-linearity in a Hammerstein system. About the non-linearity we assume that it is piecewise Lipschitz (i.e. it may have discontinuities). From practical point of view – we admit in algorithms virtually all non-linearities. It is clear that there is no parametric model able to handle all possible non-linearities from such a class.

We show that in spite of the poor a priori information our algorithms converge to the non-linearity. We also examine the rates of algorithms convergence and show that they are not much slower than those achieved by the parametric approach, thus the non-parametric algorithms accuracy is not much worse than their parametric counterparts.

The problem of identification of Hammerstein system has been widely investigated; see e.g. Narendra and Gallman (1966), Chang and Luus (1971), Haist *et al.* (1973), Thathachar and Ramaswamy (1973) for parametric algorithms and e.g. Greblicki and Pawlak (1986), Krzyżak (1989), Pawlak (1991), Hasiewicz (1999), Hasiewicz and Śliwiński (2002) or Śliwiński (2000) for non-parametric ones.

Hammerstein systems have been also applied in various fields: e.g. control, Zi-Qiang (1993), chemistry, Eskinat *et al.* (1991) or biology Emerson *et al.* (1992), Hunter and Korenberg (1986), Korenberg and Hunter (1986).

2. IDENTIFICATION PROBLEM

Consider the Hammerstein system, i.e. the cascade system with non-linear static element followed by dynamic linear one (Fig. 1). Denote by m the non-linear characteristic of the former and by $\{k_i\}_{i=0}^{\infty}$ the discrete impulse response of the latter.

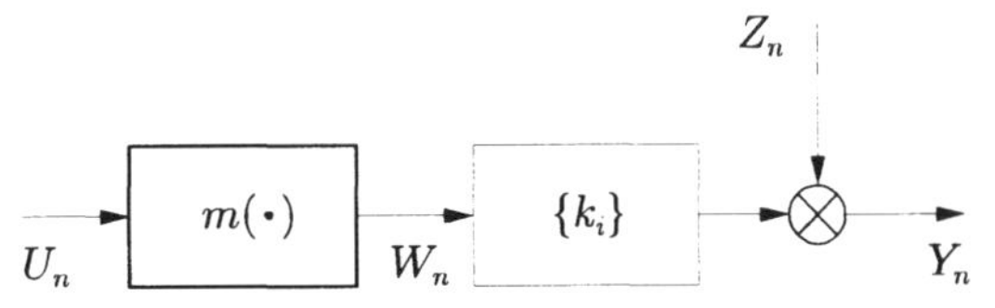

Fig. 1. Identified Hammerstein system

The goal is to recover the non-linearity from the random input-output measurements $(U_1, Y_1), (U_2, Y_2), \cdots, (U_n, Y_n)$ of the whole system (the internal signal W_n is not available for measurements) under the following assumptions:

1. The input signal $\{U_n; n = \cdots - 1, 0, 1, 2, \cdots\}$ is a stationary white random process with unknown probability density function f. We assume that $-1 \leq U_n \leq 1$ and that f satisfies the restriction

$$|f(u) - f(v)| \leq c|u - v| \quad (1)$$

some $c > 0$, for each u where $f(u) = 0$.

2. The non-linearity m is piecewise Lipschitz function (see Appendix **A**).

3. The dynamic subsystem is asymptotically stable, i.e. $\sum_{i=0}^{\infty} |k_i| < \infty$.

4. The external noise $\{Z_n; n = \cdots - 1, 0, 1, 2, \cdots\}$ is a stationary white process with zero mean and finite variance.

2.1. Comments upon assumptions

By assumption 1, the class of admitted input signals is confined to those which have smooth distribution function (i.e. for which exists the probability density function). For identification algorithms converging for any distribution of U (i.e. distribution-free or universally consistent algorithms) we refer to the paper by Greblicki and Pawlak (1989); see also Györfi *et al.* (2002) for the up-to-date results in statistics. This assumptions also limits the range of input signals to white noise processes — for algorithms applicable for correlated input we refer in turn to the paper by Greblicki and Pawlak (1994*b*). Finally, the restriction in (1) admits densities which are not bounded from zero (e.g. triangle and parabolic densities) (cf. Greblicki (1996), Greblicki and Pawlak (1994*a*)).

The pivotal for our approach is assumption 2, which makes the problem *non-parametric*, and applicable in these practical situations, in which the a priori knowledge about the non-linearity is poor.

Due to assumption 3, the structure of the dynamic part can be arbitrary ARMA system with unknown order, and therefore, it can remain unknown during identification routine. Similarly, assumption 4 imposes no restrictions on the distribution of the external noise.

3. IDENTIFICATION PROBLEM

The input-output equation of the system can be written as

$$Y_n = \mu(U_n) + \xi_n + Z_n$$

where μ is scaled and shifted version of the genuine characteristic m, i.e. $\mu(u) = k_0 m(u) + d$ with $d = \mathsf{E}m(U_1) \sum_{i=1}^{\infty} k_i$. The signal ξ_n is defined as

$$\xi_n = \sum_{i=1}^{\infty} k_i [m(U_{n-i}) - \mathsf{E}m(U_1)]$$

and is an additional distortion acting, together with noise Z_n, on the output of μ.

Construction of identification algorithms starts from the observation that for Hammerstein systems the following holds:

$$\mathsf{E}\{Y_n | U_n = u\} = \mu(u) \quad (2)$$

i.e. that the regression of Y_n on U_n is equal to the scaled and shifted non-linear characteristic m. Hence, estimating the regression function we recover the non-linearity μ. Below we present two non-parametric algorithms identifying the non-linearity μ. The kernel algorithm and the orthogonal one.

Remark 1 *The fact that the non-linearity can be recovered up to some constants is a consequence of the assembled structure of the system and of the inaccessibility of the interconnecting signal W_n — it is independent of the identification algorithms. Furthermore, the constants cannot be estimated without additional a priori information (e.g. known values of m in some points).*

3.1. Preprocessing

Both algorithms operate on the measurement data rearranged into a new sequence $(U_{(1)}, Y_{[1]})$, $(U_{(2)}, Y_{[2]}), \cdots, (U_{(n)}, Y_{[n]})$ for which $U_{(1)} < U_{(2)} < \cdots < U_{(n)}$, that is, on the measurement pairs sorted increasingly according to the values of U_i's.

3.2. Algorithms

The first algorithm is the *kernel algorithm* of the following form (see Greblicki (1996)):

$$\hat{\mu}(u) = \frac{1}{h(n)} \sum_{j=1}^{n} Y_{[j]} \int_{U_{(j-1)}}^{U_{(j)}} K\left(\frac{u-v}{h(n)}\right) dv \quad (3)$$

where K is a kernel function (see Appendix A) and $\{h(n)\}$ is a positive number sequence. The second is the *orthogonal algorithm* of the form (see Greblicki and Pawlak (1994*a*)):

$$\tilde{\mu}(u) = \sum_{k=0}^{q(n)} \tilde{c}_k \varphi_k(u) \quad (4)$$

with

$$\tilde{c}_k = \sum_{j=1}^{n} Y_{[j]} \int_{U_{(j-1)}}^{U_{(j)}} \varphi_k(v)\, dv \quad (5)$$

where $\{\varphi_k; k = 0, 1, 2, \cdots\}$ is a set of complete orthogonal functions on $[-1, 1]$ (e.g. trigonometric or Legendre set) and $\{q(n)\}$ is a positive number sequence.

Remark 2 *If a system has a l-step delay, i.e.,* $k_r = 0, r = 0, \ldots, l-1$, *then it holds that (cf. (2)):*

$$\mathsf{E}\{Y_{n+l} \,|U_n = u\} = \mu(u)$$

and the kernel algorithm turns into the form

$$\hat{\mu}(u) = \frac{1}{h(n)} \sum_{j=1}^{n} Y_{[j+l]} \int_{U_{(j-1)}}^{U_{(j)}} K\left(\frac{u-v}{h(n)}\right) dv$$

while coefficients in orthogonal algorithm are calculated according to the formula

$$\tilde{c}_k = \sum_{j=1}^{n} Y_{[j+l]} \int_{U_{(j-1)}}^{U_{(j)}} \varphi_k(v)\, dv.$$

3.3. Convergence

The following theorems characterize the conditions of the global convergence of the algorithms for the assumptions **1-4**:

Theorem 1 *Let*

$$h(n) \to 0 \quad \textit{and} \quad nh(n) \to \infty \quad \textit{as} \quad n \to \infty \quad (6)$$

then

$$\mathsf{E}\int_{-1}^{1} [\mu(u) - \hat{\mu}(u)]^2 du \to 0 \quad \textit{as} \quad n \to \infty$$

Theorem 2 *Let*

$$q(n) \to \infty \quad \textit{and} \quad \frac{q(n)}{n} \to 0 \quad \textit{as} \quad n \to \infty \quad (7)$$

then

$$\mathsf{E}\int_{-1}^{1} [\mu(u) - \tilde{\mu}(u)]^2 du \to 0 \quad \textit{as} \quad n \to \infty$$

Since (6) and (7) hold for $h(n) = n^{-\alpha}$ and $q(n) = n^{\alpha}$, respectively, where α is arbitrarily chosen from the interval $(0, 1)$, then the theorems say that under very mild conditions imposed on the sequences $\{h(n)\}$ and $\{q(n)\}$, the respective algorithms converge globally, in the mean square sense, to the non-linearity μ with growing measurement number n.

3.4. Convergence rates

It is clear that the rate the algorithms converge with, corresponds to the their accuracy: the faster rate the better accuracy can be expected from the same measurement number. The theorems below unveil the convergence rates for exemplary algorithms:

Theorem 3 *Let m have p derivatives and let pth derivative be bounded in* $[-1, 1]$. *Let K in (3) be compactly supported and have p vanishing moments (see Appendix **B**). If*

$$h(n) \sim n^{-\frac{1}{2p+1}}$$

Then, for any $0 < \epsilon < 1$,

$$\mathsf{E}\int_{-(1-\epsilon)}^{1-\epsilon} [\mu(u) - \hat{\mu}(u)]^2 du = O\left(n^{-\frac{2p}{2p+1}}\right) \quad (8)$$

Theorem 4 *Let m have p derivatives and let pth derivative be bounded in* $[-1, 1]$. *Let* $\{\varphi_k\}$ *in (4)-(5) be a trigonometric series (i.e.* $\{\varphi_k = e^{ik\pi}; k = 0, \pm 1, \pm 2, \ldots\}$*). If*

$$q(n) \sim n^{\frac{1}{2p+1}}$$

Then, for any $0 < \epsilon < 1$,

$$\mathsf{E}\int_{-(1-\epsilon)}^{1-\epsilon} [\mu(u) - \tilde{\mu}(u)]^2 du = O\left(n^{-\frac{2p}{2p+1}}\right) \quad (9)$$

From the theorems one can infer that:

- Convergence rate is related to the smoothness of the non-linear characteristic: the smoother m (and hence μ) the faster algorithm's convergence rate and the higher the algorithm accuracy for the same number of measurement. For instance, for $p = 1$ the error is of order $O\left(n^{-2/3}\right)$ and decreases to $O\left(n^{-6/7}\right)$ for $p = 3$.
- For regular m with large p the rates approach the order $O\left(n^{-1}\right)$ which is typical for parametric inference (i.e. for situation where the a priori information is very rich).

- The convergence rate is independent of the regularity of the input probability density f. This property is the advantage of the presented algorithms over other known in the literature (see e.g. Greblicki and Pawlak (1986, 1987)) which convergence rates are worsened by irregular input probability density functions.
- The convergence rate $O\left(n^{-2p/(2p+1)}\right)$ is *asymptotically optimal*, i.e. the best possible for algorithms with non-parametric a priori knowledge (see Stone (1982) or Härdle (1990)). It should be emphasized that for our algorithms this rate holds true in the presence of the correlated noise $\{\xi_n\}$.

Remark 3 *Convergence rates hold in the narrower interval* $(-1+\varepsilon, 1-\varepsilon)$ *rather than the entire one due to the well known (for non-parametric inference) boundary effects (see e.g. Härdle (1990)).*

Remark 4 *The conclusions reported above hold for large measurements sets. To the best of authors knowledge there are no formal results concerning their behaviour for small sample sets.*

4. MODIFICATIONS OF THE ALGORITHMS

To calculate algorithms (3) and (4)-(5) one should compute definite integrals which can be time-consuming. Below we present their computationally simpler counterparts.

The kernel algorithm has now the following form (see Greblicki (1996)):

$$\check{\mu}(u) = \frac{1}{h(n)} \times \sum_{j=1}^{n} Y_{[j]} K\left(\frac{u-U_{(j)}}{h(n)}\right)\left(U_{(j)}-U_{(j-1)}\right) \tag{10}$$

while the orthogonal algorithm takes the form (see Greblicki and Pawlak (1994*a*)):

$$\bar{\mu}(u) = \sum_{k=0}^{q(n)} \bar{c}_k \varphi_k(u) \tag{11}$$

with

$$\bar{c}_k = \sum_{j=1}^{n} Y_{[j]} \varphi_k\left(U_{(j)}\right)\left(U_{(j)}-U_{(j-1)}\right) \tag{12}$$

4.1. Convergence

For these algorithms the following theorems hold:

Theorem 5 *Let*

$$h(n) \to 0 \quad and \quad n^2 h^3(n) \to \infty \quad as \quad n \to \infty \tag{13}$$

then

$$\mathsf{E}\int_{-1}^{1} \left[\mu(u) - \check{\mu}(u)\right]^2 du \to 0 \quad as \quad n \to \infty$$

Theorem 6 *Let*

$$q(n) \to \infty \quad and \quad \frac{q^3(n)}{n^2} \to 0 \quad as \quad n \to \infty \tag{14}$$

then

$$\mathsf{E}\int_{-1}^{1} \left[\mu(u) - \bar{\mu}(u)\right]^2 du \to 0 \quad as \quad n \to \infty$$

Observe that the conditions imposed in (13) and (14) are more restrictive then they counterparts in (6) and (7), respectively. The algorithms are now consistent for $h(n) = n^{-\alpha}$ and $q(n) = n^{\alpha}$ where α is from the interval $(0, 2/3)$ only.

4.2. Convergence rates

As it can be expected (since these new algorithms are 'more rough'), the obtained convergence rate are slower than their origins:

Theorem 7 *Let m have p derivatives and let pth derivative be bounded in $[-1,1]$. Let K in (3) be compactly supported and have p vanishing moments. If*

$$h(n) \sim n^{-\frac{1}{2p+3}}$$

Then, for any $0 < \epsilon < 1$,

$$\mathsf{E}\int_{-(1-\epsilon)}^{1-\epsilon} \left[\mu(u) - \check{\mu}(u)\right]^2 du = O\left(n^{-\frac{2p}{2p+3}}\right) \tag{15}$$

Theorem 8 *Let m have p derivatives and let pth derivative be bounded in $[-1,1]$. Let $\{\varphi_k\}$ in (11)-(12) be a trigonometric series. If*

$$q(n) \sim n^{\frac{1}{2p+3}}$$

Then, for any $0 < \epsilon < 1$,

$$\mathsf{E}\int_{-(1-\epsilon)}^{1-\epsilon} \left[\mu(u) - \bar{\mu}(u)\right]^2 du = O\left(n^{-\frac{2p}{2p+3}}\right) \tag{16}$$

To compare the convergence rates of the algorithms (10) and (11)-(12) with (3) and (4)-(5) assume that m have one bounded derivative. We get $O\left(n^{-2/5}\right)$ versus the previously obtained optimal $O\left(n^{-2/3}\right)$. However, for more regular m's, for instance for m with $p = 3$, we get $O\left(n^{-2/3}\right) \sim O\left(n^{-0.67}\right)$ and $O\left(n^{-6/7}\right) \sim O\left(n^{-0.86}\right)$, respectively. In general, the difference between accuracy of these algorithms diminishes with increasing regularity of m.

5. CONCLUSIONS — ACCURACY OF ALGORITHMS

Taking into consideration presented properties, the following conclusions about the accuracy of algorithms can be drawn from the convergence conditions:

- The algorithms accuracy grows with increasing number of measurements for virtually all nonlinearities which can be met in practice.
- Accuracy can be arbitrarily high (the algorithm error can be made arbitrarily small) with growing n.
- It is not affected either by the regularity of the input density function f or the presence of the unknown dynamics (neither f nor $\{k_i\}$ has to be known to recover μ).

From the properties of the convergence rates we conclude additionally that:

- Accuracy grows with growing regularity of the non-linearity m — the higher p the smaller algorithms' errors for the same size of measurement set.
- For large n, the accuracy of the algorithms (3) and (4)-(5) is the best possible amongst all algorithms with non-parametric a priori information (accuracy of (10) and (11)-(12) is only slightly worse).
- In spite of the poor a priori knowledge convergence rates are not so far from $O\left(n^{-1}\right)$, which is typical for parametric inference. (The rates of non-parametric algorithms are slower than those achieved by the parametric one due to fact, that the poorer a priori knowledge has to be compensated by the richer empirical knowledge, that is, the greater number of measurements.)

It should be finally emphasized that the proposed algorithms (especially the versions (10) and (11)-(12)) are computationally simple (they involve only elementary arithmetic operations). The latter, together with the properties examined earlier in the paper, make them a valuable proposition from the theory in all situations in which the a priori knowledge does not allow to propose a reliable model.

APPENDIX A. A function g is piecewise Lipschitz if for some r exists a partition $\{A_i; i \in 1, \ldots, r\}$ such that $A_1 \cup A_2 \cup \cdots \cup A_r = \mathbf{R}$ in which g is Lipschitz; i.e. $|g(u) - g(v)| \leq \alpha_i |u - v|, u, v \in A_i$ some α_i's.

APPENDIX B. The following kernel functions with compact supports can be used in the algorithms

$$\left.\begin{array}{lll} \bullet & \text{rectangle:} & 1/2 \\ \bullet & \text{triangle:} & 1-|u| \\ \bullet & \text{parabolic:} & 3\left(1-u^2\right)/4 \end{array}\right\} \times I_{[|u|\leq 1]}(u)\cdot$$

I is the indicator function. Observe that the latter has $p = 1$ vanishing moments. To get $p = 3$ one should use e.g. $\left[(-63/32)\, u^4 - (45/16)\, u^2 + 27/32\right] \times I_{[|u|\leq 1]}(u)$.

REFERENCES

Chang, F. H. I. and R. Luus (1971). A noniterative method for identification using Hammerstein model. *IEEE Transactions on Automatic Control* **16**, 464–468.

Emerson, R. C., M. J. Korenberg and M. C. Citron (1992). Identification of complex-cell intensive nonlinearities in a cascade model of cat visual cortex. *Biological Cybernetics* **66**, 291–300.

Eskinat, E., S. H. Johnson and W. L. Luyben (1991). Use of hammerstein models in identification of nonlinear systems. *American Institute of Chemical Engineers Journal* **37**, 255–268.

Greblicki, W. (1996). Nonlinearity estimation in Hammerstein systems based on ordered statistics. *IEEE Transactions on Signal Processing* **44**, 1224–1233.

Greblicki, W. and M. Pawlak (1986). Identification of discrete Hammerstein system using kernel regression estimates. *IEEE Transactions on Automatic Control* **31**, 74–77.

Greblicki, W. and M. Pawlak (1987). Hammerstein system identification by non-parametric regression estimation. *International Journal of Control* **45**, 343–354.

Greblicki, W. and M. Pawlak (1989). Nonparametric identification of Hammerstein systems. *IEEE Transactions on Information Theory* **35**, 409–418.

Greblicki, W. and M. Pawlak (1994*a*). Dynamic system identification with order statistics. *IEEE Transactions on Information Theory*.

Greblicki, W. and M. Pawlak (1994*b*). Nonlinear system identification with nonparametric deconvolution. In: *Proceedings of IEEE Internationa Symposium on Information Theory*. Trondheim. p. 124.

Györfi, L., M. Kohler, A. Krzyżak and H. Walk (2002). *A Distribution-Free Theory of Nonparametric Regression*. Springer-Verlag. New York.

Haist, N. D., F. H. L. Chang and R. Luus (1973). Nonlinear identification in the presence of the correlated noise using hammerstein model. *IEEE Transactions on Automatic Control* **18**, 552–555.

Härdle, W. (1990). *Applied Nonparametric Regression*. Cambridge University Press. Cambridge.

Hasiewicz, Z. (1999). Hammerstein system identification by the Haar multiresolution approximation. *International Journal of Adaptive Control and Signal Processing* **13**, 697–717.

Hasiewicz, Z. and P. Śliwiński (2002). Identification of non-linear characteristics of a class of block-oriented non-linear systems via Daubechies wavelet-based models. *International Journal of Systems Science* **14**, 1121–1144.

Hunter, I. W. and M. J. Korenberg (1986). The identification of non-linear biological systems: Wiener and hammerstein cascade models. *Biological Cybernetics* **55**, 135–144.

Korenberg, M. J. and I. W. Hunter (1986). The identification of nonlinear biological system: Lnl cascade models. *Biological Cybernetics* **55**, 125–134.

Krzyżak, A. (1989). Identification of discrete Hammerstein systems by the Fourier series regression estimate. *International Journal of System Science* **20**, 1729–1744.

Narendra, K. S. and P. G. Gallman (1966). An iterative method for the identification of nonlinear systems using the hammerstein model. *IEEE Transactions on Automatic Control* **11**, 546–550.

Pawlak, M. (1991). On the series expansion approach to the identification of Hammerstein systems. *IEEE Transactions on Automatic Control* **36**, 763–767.

Śliwiński, P. (2000). Non-linear system identification by wavelets. PhD dissertation. Wrocław University of Technology. Institute of Engineering Cybernetics.

Stone, C. J. (1982). Optimal global rates of convergence for nonparametric regression. *Annals of Statistics* **10**, 1040–1053.

Thathachar, M. A. L. and S. Ramaswamy (1973). Identification of a class of nonlinear systems. *International Journal of Control* **18**, 741–752.

Zi-Qiang, L. (1993). Controller design oriented model identification method for Hammerstein system. *Automatica* **29**, 767–771.

IFAC Publications
www.elsevier.com/locate/ifac

GENERATION OF ENHANCED INITIAL ESTIMATES FOR WIENER SYSTEMS AND HAMMERSTEIN SYSTEMS ⋆

Philippe Crama *,[1] **Johan Schoukens** * **Rik Pintelon** *

* *Vrije Universiteit Brussel (Fakulteit TW—ELEC); Pleinlaan, 2; B–1050 Brussel; Belgium; Philippe.Crama@vub.ac.be*

Abstract: Wiener systems and Hammerstein systems are nonlinear models that are used in many domains for their simplicity and physical meaning. However, when these systems are identified, a cost function which is highly non-quadratic in the system parameters needs to be minimized. This paper presents a simple iterative method for generating good starting values which can be used to initialize the numerical nonlinear optimization of the cost function. *Copyright © 2003 IFAC*

Keywords: Wiener System, Hammerstein System, Nonlinear System, System Identification, initial estimate

1. INTRODUCTION

Nonlinear models can provide an accurate description and prediction of physical systems that have a nonlinear behavior. Modeling nonlinear systems has become an important issue with many practical applications. However, finding the model parameters for given measurements is still an open question that has been studied in many papers (Bai, 2002; Bai and Fu, 2002; Celka *et al.*, 2001; Crama and Schoukens, 2001; Giri *et al.*, 2001; Greblicki, 2000; Lang, 1997; Pawlak *et al.*, 2000; Verhaegen and Westwick, 1996; Vörös, 1997; Westwick and Verhaegen, 1996). The identification problem isn't trivial because the cost function that needs to be optimized is strongly non-quadratic in the parameters. The iterative algorithms used to fit the model to the measurements require a proper initialization for rapid and reliable convergence.

The main contribution of this paper is the enhancement of the initial guesses of (Crama and Schoukens, 2001) by an iterative procedure. In each iteration, the linear part of the system is estimated while the parameters of the static nonlinearity are kept constant. This is followed by estimating the static nonlinearity using the knowledge of the new estimate of the linear part of the system.

Section 2 presents some definitions and notations that will be used throughout the paper. The iterative process is presented in Section 3.2 for Wiener systems. Section 6 contains a short comparison of existing methods with this paper. Hammerstein systems can be treated similarly and the modifications with respect to the procedure for Wiener system are explained in Section 4. Due to the iterative nature of the presented method, the question of convergence arises : (Crama *et al.*, 2002) proves that the proposed methods converge in the noise-less case when there are no model errors.

2. SOME DEFINITIONS

Figures 1 and 2 define the systems that will be considered in this paper. The systems studied are discrete

⋆ This work was supported by FWO-Vlaanderen, by the Flemish community under concerted action ILiNoS and by the Belgian government under IUAP-V/22

[1] Philippe Crama is Aspirant of the Fund for Scientific Research Flanders (FWO, Fonds voor Wetenschappelijk Onderzoek Vlaanderen)

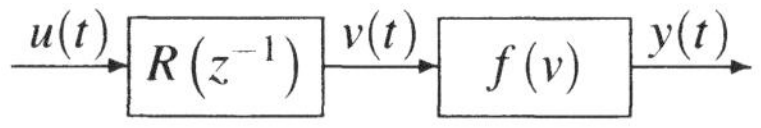

Fig. 1. A Wiener system

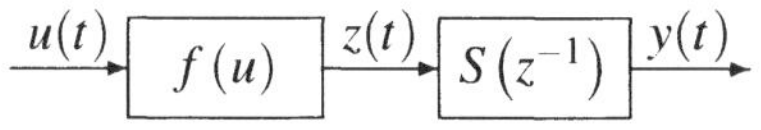

Fig. 2. A Hammerstein system

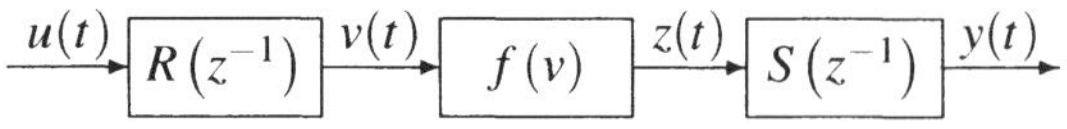

Fig. 3. A Wiener-Hammerstein system

time single input, single output (SISO), so all signals $u(t)$, $v(t)$, $z(t)$ and $y(t)$ are scalars. Both consist of a linear dynamic system ($R(z^{-1})$ or $S(z^{-1})$) and a static nonlinearity $f(\cdot)$. The difference between both is their relative position. In the Wiener system, the input is filtered first and next distorted by the static nonlinearity.

Identification of linear systems in the frequency domain in the presence of nonlinear distortions is studied in (Schoukens *et al.*, 1998) and (Pintelon and Schoukens, 2001). There, the influence of the nonlinearity on the estimate of the best linear approximation of nonlinear systems is studied. Since these results are crucial for the understanding of the proposed method, they will be summarized here, and applied to a Wiener-Hammerstein system (note that the systems shown in Figures 1 and 2 are special cases of the Wiener-Hammerstein system in Figure 3).

Definition 1. A signal $x(t)$ is a random phase multisine if

$$x(t) = \sum_{k=-N}^{N} \frac{X_k}{\sqrt{N}} e^{j2\pi k \frac{f_{\max}}{N} t} \tag{1}$$

where $X_k = X_{-k}^* = X(k)e^{j\varphi_k}$ (* denotes the complex conjugate), $f_{\max}$ is the maximum frequency of the excitation signal, $N \in \mathbb{N}$ is the number of frequency components, $X(k) \in \mathbb{R}^+$, and the phases φ_k are the realization of an independent uniformly distributed random process on $[0, 2\pi)$. $x(t)$ has a constant power for every N (∞ included).

Definition 2. The nonparametric Frequency Response Function (FRF) of a nonlinear system excited by a random phase multisine is defined as

$$G(l) = \frac{Y(l)}{U(l)} \tag{2}$$

where $Y(l)$ and $U(l)$ are the values of the l-th line of the discrete Fourier transform of the output $y(t)$ and the input $u(t)$ of the system.

In (Schoukens *et al.*, 1998), the FRF defined by (2) is split into two parts :

$$G(l) = G_R(l) + G_S(l) \tag{3}$$

G_R contains the contributions of the nonlinear system that are independent of the particular realization of the phases of the input signal used to measure $G(l)$. G_S collects all other nonlinear contributions. This term looks like measurement or process noise (i.e. it has zero mean and is uncorrelated frequency per frequency with the input signal), but cannot be reduced by measuring more periods of the same input signal. Instead, because it is a function of the input signal, it can be reduced by averaging the measured $G(l)$ over different realizations of the excitation (see (Schoukens *et al.*, 1998) for a proof).

Instead of using a time consuming averaging process to retrieve G_R, a parametric model can be fitted through the nonparametric FRF. Because G_S behaves as noise for the parameter identification step, the identified FRF will converge strongly to G_R (Schoukens *et al.*, 1998) as $N \to \infty$.

2.1 *A special case : the Wiener and the Hammerstein system*

For Wiener-Hammerstein systems, (Schoukens *et al.*, 1998; Pintelon and Schoukens, 2001) show that G_R (3) ($N \to \infty$ with N the number of components in the excitation signal) converges to the underlying linear system within a real frequency independent scale factor that depends on the power spectrum of the excitation signal and the linear system R. This property holds for signals like Gaussian band-limited noise, random phase multisines and periodic noise.

$$G_R(l) = R(l)S(l)C(U \cdot R) + O(N^{-1}) \tag{4}$$

In (4), $C(U \cdot R)$ depends on the power spectrum of the input of the static nonlinearity. This result can be applied to the special cases of the Wiener system and the Hammerstein system :

$$G_{R,\mathrm{W}}(l) = R(l)C_{\mathrm{W}}(U \cdot R) \tag{5}$$

$$G_{R,\mathrm{H}}(l) = C_{\mathrm{H}}(U)S(l) \tag{6}$$

where C_{W} and C_{H} are both independent of the frequency. Both (5) and (6) show that the linear part of a Wiener system or a Hammerstein system can be estimated very easily by identifying the best linear approximation G_R of the system. This is the starting point for the proposed methods.

3. WIENER SYSTEM

The full procedure to get an enhanced initial estimate of a Wiener system will be described in Section 3.2 and briefly compared to existing methods in Section 6. Section 3.3 illustrates the method on a real test system.

3.1 Assumptions and parameterization of the model

Assumption 1.1 *G_R is not identically 0.*

Assumption 1.2 *$f(\cdot)$ must be invertible for Wiener systems.*

There is no special requirement on the parameterization of the static nonlinearity : the user may choose any method to fit a static nonlinearity to the cloud of (v, y) couples produced at each iteration by the method (see Section 3.2.1). However, since the inverse of the static nonlinearity is needed in the iteration, it is easier to model the inverse of the static nonlinearity linearly in the parameters ψ as in :

$$g(y, \psi) = \sum_{k=1}^{q} g_k(y)\psi_k \tag{7}$$

with $g_k(y)$ some well-chosen basic function : e.g. polynomial or piecewise linear.

The linear model R is given by a rational form in z :

$$R\left(z^{-1}, \theta\right) = \frac{\sum_{k=0}^{n_b} b_k z^{-k}}{\sum_{k=0}^{n_a} a_k z^{-k}}$$

3.2 The method

3.2.1. Starting the iterative process The goal of this step is to estimate the best linear approximation G_R of the Wiener system to recover the unknown linear dynamic part R of the system with (5). This estimation step can be done using any technique for linear system identification (Ljung, 1999; Pintelon and Schoukens, 2001) in the time or in the frequency domain or using subspace identification techniques (Verhaegen and Westwick, 1996; Viberg, 1995).

$$\hat{\theta}^{[1]} = \arg\min_{\theta} \sum_{l=1}^{N} \left|\hat{G}(l) - R\left(e^{-j2\pi\frac{l}{N}}, \theta\right)\right|^2 \tag{8}$$

$$\hat{R}^{[1]}(l) = R\left(e^{-j2\pi\frac{l}{N}}, \hat{\theta}^{[1]}\right) = \hat{G}_R(l) \tag{9}$$

The very first estimate $\hat{R}^{[1]}$ of the linear part of the Wiener system is given by (9) by virtue of (5). The unknown scaling factor C_W isn't important because the gain distribution between the linear system R and the static nonlinearity $f(\cdot)$ isn't unique : the systems $(\alpha R, f(\frac{\cdot}{\alpha}))$ and $(R, f(\cdot))$ produce the same output given the same input. Hence, both estimated subsystems ($R(z^{-1})$ and $f(\cdot)$) are rescaled such that R has a gain of 1 at a frequency specified by the user.

Suppose that n samples were measured, then

$$\hat{v}^{[1]}(t) = \hat{R}^{[1]}[u(t)] \tag{10}$$

$$\hat{\psi}^{[1]} = \arg\min_{\psi} \sum_{t=1}^{n} \left(\hat{v}^{[1]}(t) - g(y(t), \psi)\right)^2 \tag{11}$$

Once R is known, the static nonlinearity can be estimated very easily : the parametric model $\hat{R}^{[1]}$ allows to filter $u(t)$ to estimate the input $v(t)$ of the static nonlinearity (10). The couples $(\hat{v}^{[1]}(t), y(t))$ are a nonparametric representation $\hat{f}_{NP}^{[1]}$ of the static nonlinearity. At this stage, a parameterization may be chosen to fit a function $g(\cdot, \hat{\psi}^{[1]})$ through all the data points e.g. by minimizing a cost function as in (11). If the set of nonlinear characteristics defined by the parameterization includes the true system, the algorithm converges to the true values in the noise-less case (Crama *et al.*, 2002).

3.2.2. Enhance the initial guess The stochastic nonlinear contributions G_S (3) disturbed the estimate of $\hat{R}^{[1]}$. Hence, the goal is to identify the linear system using the data sets $u(t)$ and $v(t)$ instead of $u(t)$ and $y(t)$. This should result in a better estimate of the linear part (which will become $\hat{R}^{[2]}$). It is assumed that the static nonlinearity is invertible, so the algorithm estimates $v(t)$ through the estimate of $g(\cdot, \hat{\psi}^{[1]})$:

$$\hat{v}^{[2,a]}(t) = g\left(y(t), \hat{\psi}^{[1]}\right) \tag{12}$$

$$\hat{V}^{[2,a]}(l) = \text{DFT}\left(\hat{v}^{[2,a]}(t)\right) \tag{13}$$

$$\hat{R}_{NP}^{[2]}(l) = \frac{\hat{V}^{[2,a]}(l)}{U(l)} \tag{14}$$

where NP stands for nonparametric estimate.

The errors in the estimate of the static nonlinearity ($\hat{\psi}^{[1]} \neq \psi_0$) introduce stochastic nonlinearities in $\hat{R}_{NP}^{[2]}$ (14). Their impact is reduced by fitting a parametric model to $\hat{R}_{NP}^{[2]}$.

$$\hat{\theta}^{[2]} = \arg\min_{\theta} \sum_{l=1}^{N} \left|\hat{R}_{NP}^{[2]}(l) - R(e^{-j2\pi\frac{l}{N}}, \theta)\right|^2 \tag{15}$$

The result is an improved estimate of the linear part of the Wiener system that can be used to compute the new estimate $g(\cdot, \hat{\psi}^{[2]})$ (similar to Section 3.2.1) :

$$\hat{v}^{[2,b]}(t) = \hat{R}^{[2]}[u(t)] \tag{16}$$

$$\hat{\psi}^{[2]} = \arg\min_{\psi} \sum_{k=1}^{n} \left(\hat{v}^{[2,b]}(k) - g(y(k), \psi)\right)^2 \tag{17}$$

It is trivial to repeat the steps of this section until some convergence criterion is met (e.g. variation of the prediction error small enough, variation of the parameter vectors small enough, ...).

Algorithm 1. Iterative procedure to produce initial estimates of Wiener systems

(1) Compute G_R and then compute an estimate of the static nonlinearity (see Section 3.2.1).
(2) Invert $\hat{f}^{[k]}$ to estimate the signal $\hat{v}^{[k+1,a]}(t)$.
(3) Build a parametric model $\hat{R}^{[k+1]}$ of the linear system that approaches $\hat{v}^{[k+1,a]}(t)$ best with $u(t)$ as input.

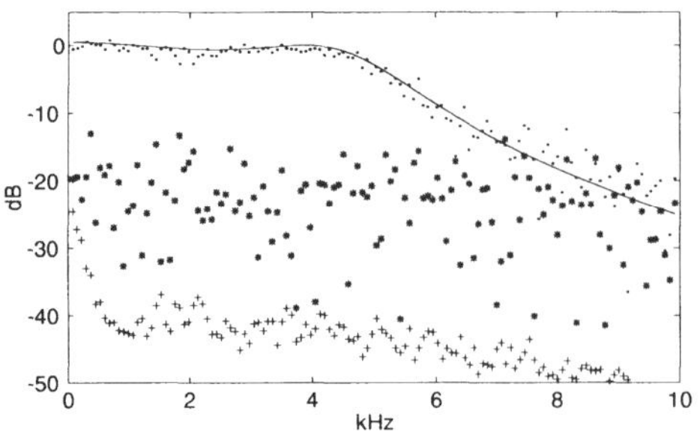

Fig. 4. Estimate of G_R ($\cdot$: measured nonparametric FRF $\hat{G}$, $-$: transfer function $\hat{R}^{[1]}$ fitted to the measurements, $*$: estimate of $\hat{G}_S = \hat{G} - \hat{R}^{[1]}$, $+$: standard deviation of measurement noise)

(4) Use $\hat{R}^{[k+1]}$ of step 3 to estimate $\hat{v}^{[k+1,b]}(t)$ by filtering of $u(t)$.

(5) Consider the couples $(\hat{v}^{[k+1,b]}(t), y(t))$ and fit a parametric model through the obtained static nonlinearity.

(6) If the convergence criterion is not met, jump to step $2 : k = k+1$.

3.3 The method applied to measurements

The first steps of the presented method will be applied to data that was measured for (Crama and Schoukens, 2001).

3.3.1. Parameterization of the model in the example In this example, the linear system R has been estimated with a discrete-time rational transfer function of order 3 for both numerator and denominator. The linear identification step was done in the frequency domain using FDIDENT, a frequency domain identification toolbox for MATLAB (Kollár *et al.*, 1991). The static nonlinearity was estimated using a piecewise linear approximation with 150 breakpoints.

3.3.2. Results Figure 4 compares G, $\hat{R}^{[1]}$, the estimate of G_S and the standard deviation of the noise of the measurement of G. The standard deviation of the measurement noise was estimated using 13 consecutive periods of the periodic input and output signal. In this example, the nonlinear contributions dominate the process and measurement noise. The contribution of G_S to G can be clearly seen at the lower frequencies : G looks very noisy, though the SNR is 40dB in nearly all of the pass-band.

Figure 5 illustrates on the same measurement example how the static nonlinearity is estimated at the first iteration.

The estimate of $f(\cdot, \hat{\psi}^{[1]})$ is used to invert the nonlinearity : Figure 6 compares the nonparametric FRF from (14) with $\hat{R}^{[2]}$. Note that the nonparametric representation of the FRF is already much smoother in Figure 6 than in Figure 4. This indicates that the inversion of the static nonlinearity was nearly perfect

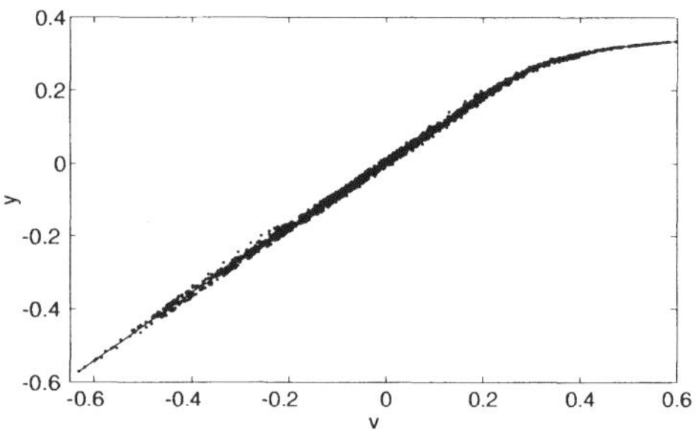

Fig. 5. Static nonlinearity ($\cdot$: $\hat{f}_{\mathrm{NP}}^{[1]}$, $-$: $f(\cdot, \hat{\psi}^{[1]})$)

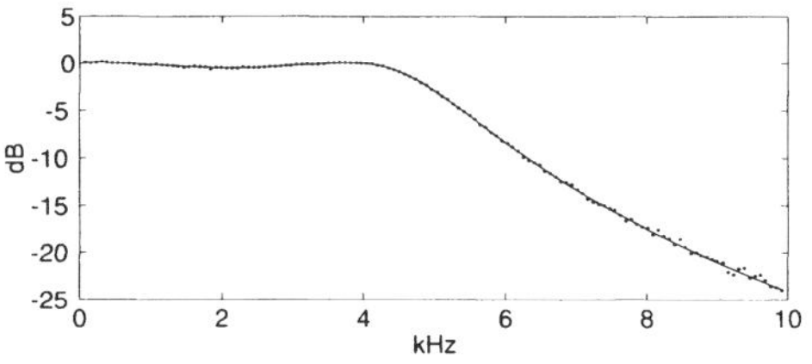

Fig. 6. Estimate of R at iteration 2 ($\cdot$: $\hat{R}_{\mathrm{NP}}^{[2]}$, $-$: $\hat{R}^{[2]}$)

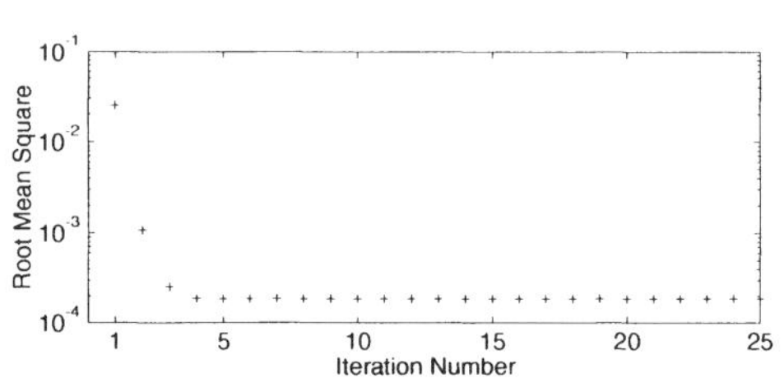

Fig. 7. Root mean square of the output simulation error of the identified models on a validation set

since the nonlinear stochastic contributions are much smaller. The new estimate of R is then used as in Section 3.2.1 to form a new estimate $f(\cdot, \hat{\psi}^{[2]})$ of the static nonlinearity with (16) and (17).

Figure 7 shows the root mean square error of the model output (simulation error) after each iteration step on validation data. If the measurement and process noise and the model errors had been 0, then the assumptions of the convergence proof would have been met and the simulation error would have gone to 0.

4. HAMMERSTEIN SYSTEM

The basic ideas explained in Section 3 can be reused for the case of the Hammerstein system : the algorithm is stated in Section 4.2 and Section 4.3 shows some results.

4.1 Assumptions and parameterization

Assumption 2.1 *G_R is not identically 0.*

The static nonlinearity has to be modelled linear in the parameters ψ :

$$f(u, \psi) = \sum_{k=1}^{q} \psi_k f_k(u)$$

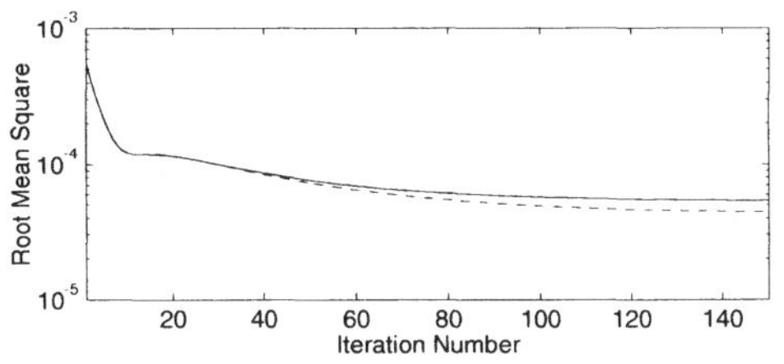

Fig. 8. Root mean square of the output simulation error of the identified models (– : on a validation set, -- : on the identification data)

where $f_k(u)$ are some basis functions (which have to be chosen a priori).

The linear part is a rational transfer function :

$$S(z^{-1}, \theta) = \frac{\sum_{k=0}^{n_b} b_k z^{-k}}{\sum_{k=0}^{n_a} a_k z^{-k}}$$

The following notation will be used for the parametric estimate of the linear part :

$$\hat{S}^{[k]}(l) = S\left(e^{-j2\pi\frac{l}{N}}, \hat{\theta}^{[k]}\right)$$

4.2 The method

4.2.1. Starting the iterative process Estimate the linear system :

$$\hat{\theta}^{[1]} = \arg\min_{\theta} \sum_{l=1}^{N} \left| \frac{Y(l)}{U(l)} - S(e^{-j2\pi\frac{l}{N}}, \theta) \right|^2 \quad (18)$$

$$Z_k(l) = \mathrm{DFT}(z_k(t)) = \mathrm{DFT}(f_k(u(t)))$$

Estimate the nonlinear system :

$$\hat{\psi}^{[1]} = \arg\min_{\psi} \sum_{l=1}^{N} \left| Y(l) - \sum_{k=1}^{q} \psi_k Z_k(l) \hat{S}^{[1]}(l) \right|^2 \quad (19)$$

4.2.2. Enhancing the initial guess Estimate the intermediary signal $z(t)$:

$$\hat{Z}^{[1]}(l) = \mathrm{DFT}\left(\sum_{k=1}^{q} \hat{\psi}_k^{[1]} f_k(u(t)) \right) = \sum_{k=1}^{q} \hat{\psi}_k^{[1]} Z_k(l) \quad (20)$$

$$\hat{S}_{\mathrm{NP}}^{[2]}(l) = \frac{Y(l)}{\hat{Z}^{[1]}(l)} \quad (21)$$

Estimate the linear system :

$$\hat{\theta}^{[2]} = \arg\min_{\theta} \sum_{l=1}^{N} \left| \hat{S}_{\mathrm{NP}}^{[2]}(l) - S\left(e^{-j2\pi\frac{l}{N}}, \theta\right) \right|^2 \quad (22)$$

Estimate the nonlinear system :

$$\hat{\psi}^{[2]} = \arg\min_{\psi} \sum_{l=1}^{N} \left| Y(l) - \sum_{k=1}^{q} \psi_k Z_k(l) \hat{S}^{[2]}(l) \right|^2 \quad (23)$$

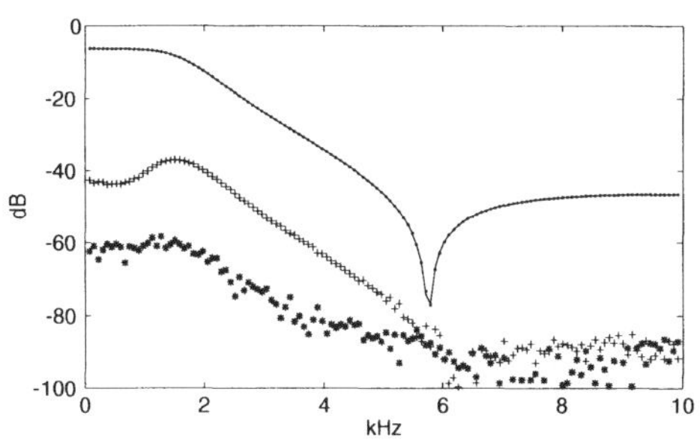

Fig. 9. Comparison of the true value of S and its estimate (· : true value, – : best estimate, + : $S - \hat{S}^{[1]}$, $*$: $S - \hat{S}_{\mathrm{best}}$)

Algorithm 2. Iterative procedure to produce initial estimates of Hammerstein systems

(1) Compute G_R and then compute an estimate of the static nonlinearity (Section 4.2.1).
(2) Estimate the values of the signal $z(t)$ using the most recent estimate of $f(\cdot)$.
(3) Build a parametric model $\hat{S}^{[k+1]}$ of the linear system that approaches $y(t)$ best with $\hat{z}^{[k]}(t)$ as input.
(4) Use $\hat{S}^{[k+1]}$ to obtain a parametric model of the static nonlinearity $\hat{f}^{[k+1]}$ (23).
(5) If the convergence criterion is not met, jump to step 2 : $k = k+1$.

4.3 Measurement Results

The results of this algorithm are shown in Figure 8. The algorithm used 1000 iterations (of which only 150 are shown in Figure 8) to identify a rational transfer function model of order 3 for the linear part and a piecewise linear approximation of the static nonlinearity with 60 breakpoints. This takes about 91 minutes on a PIII clocked at 500MHz (this is an average of 5s per iteration). Figures 8 and 9 prove that the iterative scheme improves the estimate.

5. CONVERGENCE OF THE ITERATIVE METHOD

Due to the iterative nature of the presented method, the question of convergence arises : (Crama *et al.*, 2002) proves that the proposed methods converge in the noise-less case when there are no model errors.

6. COMPARING THE PROPOSED METHOD WITH EXISTING METHODS

None of the already existing methods for generating initial estimates, except (Vörös, 1997) for Hammerstein systems, iterate to improve their initial estimates. In these methods, three different types are encountered : nonparametric (Greblicki, 2000; Lang, 1997), semi-parametric (Celka *et al.*, 2001; Pawlak *et al.*,

2000) and parametric (Bai, 2002; Bai and Fu, 2002; Crama and Schoukens, 2001; Giri *et al.*, 2001; Verhaegen and Westwick, 1996; Vörös, 1997; Westwick and Verhaegen, 1996). The semi-parametric methods differ from the parametric methods by their representation of the linear part of the system : these methods describe the linear part by a finite impulse response, which implies that many parameters are estimated, and hence that long data records are needed.

It has been shown (Crama *et al.*, 2002) that if the linear part is estimated nonparametrically, the method converges (in one step) to a biased solution.

In the class of the parametric methods, the advantage of subspace methods (Verhaegen and Westwick, 1996; Westwick and Verhaegen, 1996) is that they also address the MIMO case. However, no iterative improvement has been presented.

The main point of this paper is the iteration process to generate enhanced estimates. This iterative process converges to the true system (for SISO systems) if the estimation step of the linear dynamic part is consistent. Because the subspace method is a consistent estimator, the frequency domain identification method can also be replaced by the subspace method (or any other consistent estimator) in Algorithms 1 and 2, still leading to the same results.

The method presented in (Bai, 2002) is simpler for Hammerstein systems when the static nonlinearity is expressed with few parameters (1 or 2) but has special requirements on the input amplitude distribution when more parameters are needed to describe the static nonlinearity. The method presented in (Bai and Fu, 2002) handles blind identification but is limited to systems driven by a zero order hold excitation.

7. CONCLUSION

A fast and effective method has been presented to compute initial estimates of Wiener systems or Hammerstein systems. They are obtained by optimizing first the parameters of the linear system while keeping the parameters of the static nonlinearity constant and next by estimating the parameters of the static nonlinear part while the linear system is kept constant. This process can be repeated iteratively. The resulting estimates can be used as starting values for optimization techniques varying all parameters at once.

REFERENCES

Bai, E. W. (2002). Identification of linear systems with hard input nonlinearities of known structure. *Automatica* **38**(5), 853–860.

Bai, E. W. and M. Fu (2002). A blind approach to Hammerstein model identification. *IEEE Transactions on Signal Processing* **50**(7), 1610–1619.

Celka, P., N. J. Bershad and J-M Vesin (2001). Stochastic gradient identification of polynomial Wiener systems : Analysis and application. *IEEE Transactions on Signal Processing* **49**(2), 301–313.

Crama, P. and J. Schoukens (2001). First estimates of Wiener and Hammerstein systems using multisine excitation. *IEEE Transactions on Instrumentation and Measurement* **50**(6), 1791–1795.

Crama, P., J. Schoukens and R. Pintelon (2002). Generation of enhanced initial estimates for Wiener systems and Hammerstein systems. Technical report. Vrije Universiteit Brussel. ftp://elecftp.vub.ac.be/Papers/PhCRAMA/CramaTR1.pdf.

Giri, F., F. Z. Chaoui and Y. Rochidi (2001). Parameter identification of a class of Hammerstein plants. *Automatica* **37**(5), 749–755.

Greblicki, W. (2000). Continuous-time Hammerstein system identification. *IEEE Transactions on Automatic Control* **45**(6), 1232–1236.

Kollár, I., R. Pintelon and J. Schoukens (1991). Frequency domain system identification toolbox. In: *Proceedings of the 9th IFAC/IFORS Symposium on Identi˘cation and System Parameter Estimation, Budapest (Hungary)*. pp. 1243–1246.

Lang, Z. Q. (1997). A nonparametric polynomial identification algorithm for the Hammerstein system. *IEEE Transactions on Automatic Control* **42**(10), 1435–1441.

Ljung, L. (1999). *System Identi˘cation — Theory for the User*. Prentice-Hall (Upper Saddle River).

Pawlak, M., R. K. Pearson, B. A. Ogunnaike and F. J., III Doyle (2000). Nonparametric identification of generalized Hammerstein models. In: *Proceedings of the IFAC System Identi˘cation, Santa Barbara, California, USA*.

Pintelon, R. and J. Schoukens (2001). *System Identi˘cation — A Frequency Domain Approach*. IEEE Press (New York).

Schoukens, J., T. Dobrowiecki and R. Pintelon (1998). Parametric and nonparametric identification of linear systems in the presence of nonlinear distortions — a frequency domain approach. *IEEE Transactions on Automatic Control* **43**(2), 176–190.

Verhaegen, M. and D. Westwick (1996). Identifying MIMO Hammerstein systems in the context of subspace model identification methods. *International Journal of Control* **63**(2), 331–349.

Viberg, M. (1995). Subspace-based methods for the identification of linear time-invariant systems. *Automatica* **31**(12), 1835–1851.

Vörös, J. (1997). Parameter identification of discontinuous Hammerstein systems. *Automatica* **33**(6), 1141–1146.

Westwick, D. and M. Verhaegen (1996). Identifying MIMO Wiener systems using subspace model identification methods. *Signal Processing* **52**(2), 235–258.

IFAC Publications
www.elsevier.com/locate/ifac

USER CHOICES AND MODEL VALIDATION IN SYSTEM IDENTIFICATION USING NONLINEAR WIENER MODELS

Torbjörn Wigren

Systems and Control, Department of Information Technology,
Uppsala University, PO Box 337, SE-751 05 Uppsala, SWEDEN.
torbjorn.wigren@it.uu.se. http://www.uu.se/Graduates/tw/tw.html

Abstract: The issue of user choices in system identification is of paramount importance. This paper therefore attempts to systematically discuss user choices for algorithms based on a specific class of nonlinear models, namely the Wiener model. In particular, the paper addresses model selection, user choices in algorithms, sampling, input signal selection as well as disturbance handling and modelling errors. Validation methods applicable to Wiener type systems are also discussed. A new method based on mean residual analysis is presented. Parts of the discussion of the paper applies also to general nonlinear system identification. *Copyright © 2003 IFAC*

Keywords: Nonlinear systems, System identification, Validation.

1. INTRODUCTION

How should a real world system identification problem be attacked? Occasionally, the answer to the question may require the design of new algorithms. However, more often, the practitioner faces the task of selecting not only a suitable model and a corresponding algorithm, but also signals and algorithmic tuning parameters. In order to succeed, the user first needs to have knowledge of model structures, related algorithms and the effect of the relevant user choices. Secondly, the system at hand must be understood "well enough", where "well enough" refers to the capability to make an educated selection of inputs, outputs, model, identification method and related user choices. Thirdly, the user needs to be able to verify the identified models by applying appropriate tools for model validation.

The first and third of the general conditions above are the ones related strictly to system identification. They have been extensively discussed in the case of linear systems, see e.g. (Ljung and Söderström, 1977; Söderström and Stoica, 1989), chapter 5 and chapter 11 respectively. However, the results in the nonlinear field appear to be more scattered. The present paper attempts to address this, and so doing it assumes that the model is the Wiener model of Fig. 1. The main contribution of the paper is hence the discussion of user choices and model validation tools (one of them novel), related to the use of the Wiener model in system identification.

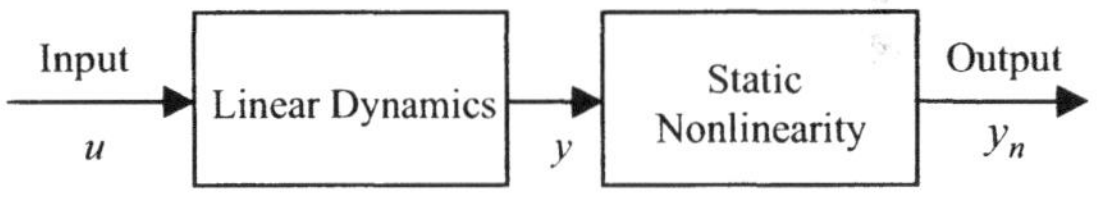

Fig. 1: The nonlinear Wiener model. The signal y is not available for measurement.

Block oriented models like the Wiener model restrict the model structure of nonlinear systems, thereby also restricting applicability. The benefit is the relaxed algorithmic complexity and the fact that some very basic issues like stability can be handled by linear methods. Furthermore, many cases of systems with Wiener model structures have been reported. Examples include nonlinear sensor handling, e.g. sensors insensitive to sign and compensation for sensor saturation, blind equalisation, quantization in adaptive filtering, control valve modelling, pH-control systems, extremum control, (biological) vision, associative memories and direction dependent systems. See e.g. the references in (Westwick and Verhagen, 1996; Wigren, 1998). Because of these facts, Wiener model based system identification has been an active field of research for many years. Early contributors include e.g. Billings and Fakhouri, (1978), who developed a correlation based algorithm. Non-parametric methods have also been studied extensively by Greblicki, see e.g. (Greblicki, 1992). In (Pajunen, 1984, 1992) a recursive and parametric algorithm was developed using the principle to model the inverse system by a Hammerstein model. Recursive and parametric identification algorithms based on the Wiener model appeared in (Wigren, 1990), together with an analysis of convergence properties. The subspace based identification methods that are able to attack also MIMO problems in a systematic way, were extended to the Wiener model case in (Westwick and Verhagen, 1996). This latter contribution is important since many subsystems in e.g. the chemical process industry are nonlinear at the same time as they require MIMO modelling.

The paper is organised with a definition of reference algorithms and data sets in section 2. User choices are discussed in section 3, followed by model validation methods in section 4. Conclusions appear in section 5. Whenever the applicability of the discussion extends beyond the Wiener model, this is remarked.

2. REFERENCE MATERIAL

2.1 Reference algorithms

The reference algorithms used in the present paper appear in (Wigren, 1990, 1993, 1994). The reference algorithms are parametric, recursive prediction error methods (RPEMs) and the search direction is of Gauss-Newton type. A known *or* a piecewise linearly parameterized model of the nonlinearity is utilized.

2.2 Reference model and reference system

The reference model of the paper is taken from (Wigren, 1990, 1994). It describes a valve for fluid flow control. Briefly, the model is obtained from a continuous time one using zero order hold sampling with a sampling period of 0.075 s, resulting in

$$y(t)=\frac{0.0616q^{-1}+0.0543q^{-2}}{1-1.5714q^{-1}+0.6873q^{-2}}u(t) \tag{1}$$

$$y_n(t)=\frac{y(t)}{\sqrt{0.1+(1-0.1)y^2(t)}}+w(t) \tag{2}$$

The signal definitions can be found in Fig. 1. $w(t)$ denotes additive coloured noise and q^{-1} denotes the backward shift operator.

The reference system that generated the live data was the CE8 coupled electric drives (Wellstead, 1979). This educational control laboratory system consists of two electric motors that drive a pulley using a flexible belt. The pulley is held by a spring and it can be moved up and down by controlling the motors to apply different tensions to the belt. The angular speed of the pulley is measured with a pulse counter and this sensor is *insensitive to the sign of the velocity.*

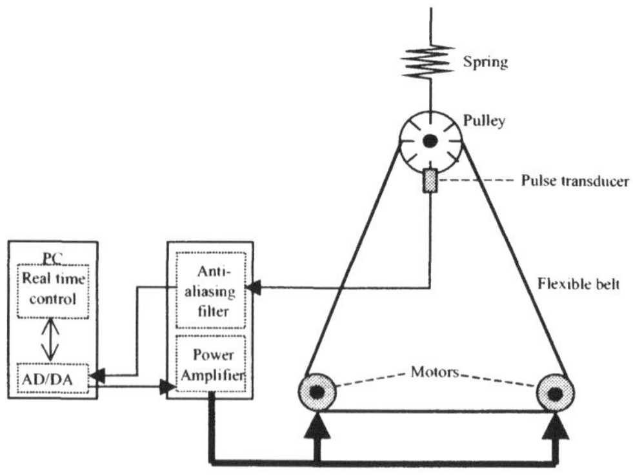

Fig. 2: The CE8 coupled electric drives.

By considering the sum of the voltages applied to the motors as the input, physical modelling results in a lightly damped linear third order system, as measured from voltages to pulley velocity. Tentative descriptions of this system is

$$y(t)=\frac{b_1q^{-1}+\ldots+b_nq^{-n}}{1+f_1q^{-1}+\ldots+f_nq^{-n}}u(t) \tag{3}$$

$$y_n(t)=|y(t)|+w(t) \tag{4}$$

where $\{b_i\}_1^n, \{f_i\}_1^n$ are parameters and n is the order.

2.3 Reference data

Reference data were collected in experiments performed with the equipment of Fig. 2. The sampling period was 20 ms and the bandwidth of the anti-aliasing filter was 12 Hz.

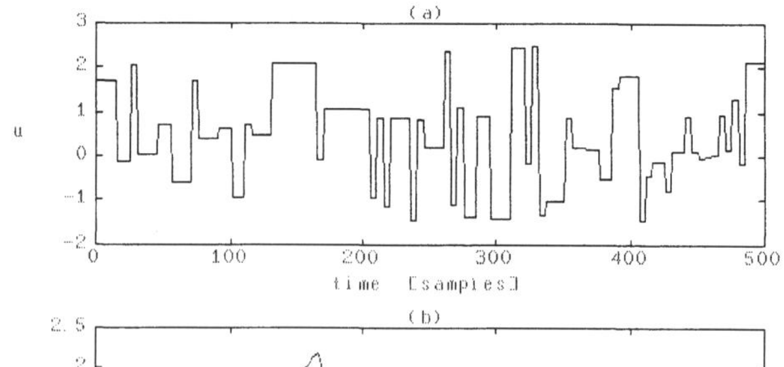

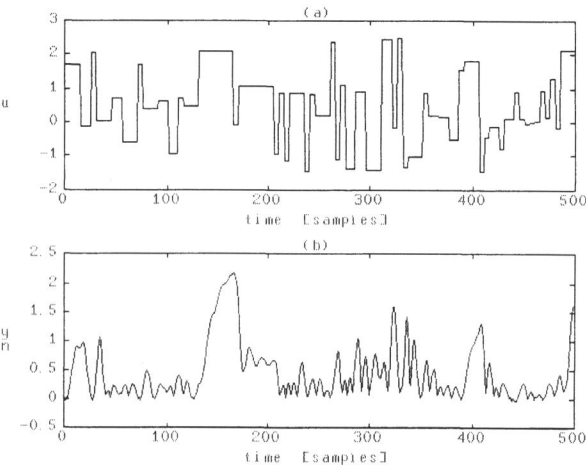

Fig. 3: Uniformly distributed (input) data set.

A PRBS with a basic clock period equal to 5 times the sampling period was the starting point. Each interval of constant signal level and of length equal to integer multiples of 5 times the sampling period was then multiplied by a random factor, uniformly distributed. This creates an input signal uniformly distributed in amplitude (Wigren, 1990). The first set of data used a factor that gave inputs between -1.5 V and +2.5 V. The second set of data gave inputs between -1.0 V and + 3.0 V. The data sets are short, the system poorly damped and the nonlinear effect is noninvertible and large. This makes identification challenging.

3. USER CHOICES

3.1 General aspects

Due to the fact that the Wiener model has linear dynamics, a number of things are less troublesome than in general nonlinear system identification The stability issue is e.g. a linear one, a simplification that reduces the stability monitoring effort to a standard problem. Furthermore, ideas for parameterisation of the linear block can be reused. Other block oriented models have similar advantages. On the other hand much becomes different than in the field of linear systems. First, the superposition principle does not apply. Hence standard techniques like subtraction of biases, trends, filtering and frequency shaping (Wahlberg and Ljung, 1986) of input and output data are less useful. Aspects like signal biases, operating points, the amplitude of signals and the treatment of disturbances immediately become more critical than in linear system identification. Note also that the output nonlinearity may not be invertible. The Wiener model is therefore in a sense more nonlinear than e.g. the Hammerstein model which can be transformed to a linear MISO model.

3.2 Wiener models

A classification is often done into parametric and non-parametric Wiener models. The non-parametric models are often thought of as linear, continuous time pulse response functions followed by a static nonlinear function. In the non-parametric framework these two functions are estimated rather than a discrete set of parameters, see e.g. (Greblicki, 1992). Parametric models on the other hand, are parameterised in terms of a usually small set of parameters that define models of the blocks of the Wiener model. Discrete time transfer function operators have been used for the linear block, a fact that has mostly limited the applicability to SISO problems. Polynomial or piecewise polynomial functions have been used for the static nonlinear block, e.g. (Pajunen, 1992). More recently, a systematic approach to MIMO Wiener modelling has been introduced by applying subspace based algorithms (Westwick and Verhagen, 1996). These methods exploit a parameterization of a state space model for the linear and nonlinear blocks.

All modelling of Wiener systems must handle one fundamental complication. This complication is due to the cascade structure of Fig. 1, from which it is clear that a static gain parameter can be arbitrarily distributed between the two blocks without affecting the input-output relationship. When parametric modelling is used, the total number of degrees of freedom may therefore have to be reduced by one. This is true for the SISO transfer function operator based models as well as for the MIMO state space models used for subspace identification (Pajunen 1984, 1992; Wigren, 1990, 1993; Gomez and Baeyens, 2002). Non-parametric methods circumvent this issue by exploiting certain correlation results, see e.g. (Billings and Fakhouri, 1978). The same problem is in place for more advanced block oriented models and also between the state space and output equations in the state space setting of the general nonlinear system identification problem.

Note that many algorithms assume that the static nonlinearity is invertible to allow for inverse modelling. Such an assumption limits applicability. However, Wigren (1990, 1993, 1994) as well as Westwick and Verhagen (1996) present algorithms that do not rely on this assumption. However, the subspace based algorithms of (Westwick and Verhagen, 1996) impose the further constraint of an odd nonlinearity, in their basic setting. Ways to circumvent this and to handle even nonlinearities are presented, but the performance is then reduced.

Regarding accuracy, the literature has in the past put much effort on algorithmic properties. However, in practice the accuracy in nonlinear system identification can be expected to be more affected by the *ability of the model to describe the data well*, i.e. by undermodelling, see e.g. (Nordsjö and Wigren, 2002). This argument may be in favour of the flexible non-parametric models. On the other hand it is true that an unnecessarily high number of parameters is not beneficial for the accuracy and should be avoided (the parsimony principle). This trade-off is believed to be valid for a wide class of nonlinear identification situations.

3.3 Algorithms

Aspects that usually govern the choice of algorithm include convergence time, convergence properties, bandwidth and misadjustment under tracking, safety net requirements as well as complexity.

The non-parametric methods are in general based on an analysis of correlations between measured signals, exploiting the Bussgang theorem. No parametric underlying model needs to be exploited. As a result, the algorithms cannot become unstable and do not suffer from false local convergence points, meaning that the need for safety nets is minimal. The drawback is sometimes found on the performance side, with relatively long convergence times, a need for relatively large data sets and large (fine discretised) models.

Often, parametric methods exploit criterion minimisation. Optimality then contributes to good performance in terms of quick convergence and high performance also for small data sets. On the other hand problems with convergence to false local minima do occur, in particular for output error methods. Stability monitoring is required when the Wiener model exploits a pole/zero linear transfer function operator. The best performance (in case of convergence to the correct minimum point of the criterion) is in general offered by PEM and RPEM algorithms exploiting Newton type minimisation methods, at least for SISO pulse transfer function operator based models. One way to reduce the risk for convergence to local suboptimal minimum points of the criterion is here to exploit a finite impulse (FIR) filter model for the linear block, cf. (Wigren, 1998). The basic subspace algorithms for Wiener model identification can often be cast in a form that allows the conventional algorithms for linear systems, based on the singular value decomposition (SVD), to be reused. No false local minima can then result. Related output error methods are subject to problems with false local minima though (Westwick and Verhagen, 1996).

As in linear system identification, computational complexity is a highly application dependent issue, affected by the size of the model, the algorithm, the sampling period and/or the number of data points.

3.4 Initialisation

When the algorithm is not guaranteed to converge, initialisation becomes more important. In case of insufficient prior information initial values must be obtained from measurements on the system. One way is then to perform an initial identification run, e.g. with a non-parametric method. Alternatively, initial values may be obtained with methods from linear system identification, neglecting nonlinear effects.

Example 1. The model (1)-(2) was used to generate data in this example. The input was generated as described in section 2.3. The operating point and amplitude were selected to cover most of the support of the nonlinearity between 0 and 1. Identification was then performed with the RPEM algorithm of (Wigren, 1990, 1994), assuming a known static nonlinearity. The nonlinearity was selected to be *linear*, thereby neglecting nonlinear effects. The algorithmic setting was the same as in section 4.5 of (Wigren 1990). The result appears in Fig. 4. The estimated linear block of the Wiener model is quite close to the true one. The identified parameters are highly likely to be useful for initiation of a Wiener model based identification algorithm.

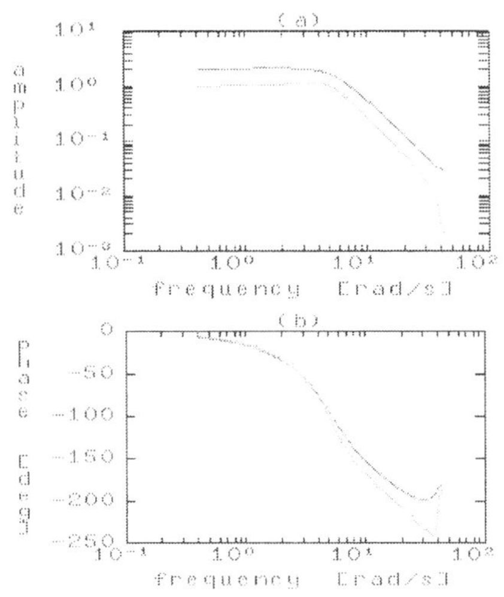

Fig. 4: Bode plot of system (solid) and the estimated model (dashed)

3.5 Sampling

Sampling for nonlinear systems has been studied in (Billings and Aguirre, 1995). In the general nonlinear sampling problem, numerical errors seem to be unavoidable. However, due to the use of a linear dynamic block, approximations can be avoided when sampling a Wiener model. Using conventional zero order hold sampling of the input signal, a discrete time system can be computed exactly by a solution of the vector differential equation describing the underlying continuous time system. Since the output of the Wiener model is obtained by a static nonlinear transformation of the continuous time system *at the time of sampling*, it follows that the discrete time linear model, followed by the static nonlinear transformation, describe the discrete time Wiener model.

Regarding the selection of the sampling period, some caution needs to be exercised. This fact is tightly related to the applied anti-aliasing filtering. This filtering results in a linear block being connected after the nonlinear output equation. It is important to select the bandwidth of the anti-aliasing filter high enough, so that the significant overtone frequency contents in the output, generated by the nonlinearity, is not filtered out. In general, it may therefore be wise to select higher sampling frequencies when identifying Wiener systems than when identifying linear systems with comparable (linear) time constants.

3.6 Input Signal

A consequence of the application of a nonlinear model is that a small variation around the operating point is no necessity. Hence the input signal amplitudes should be increased to suppress the effect of random disturbances. In case the Wiener model is a valid assumption, this will be possible until other nonlinear effects become dominant.

In cases where the input signal to the system cannot be selected, many of the non-parametric algorithms suffer from the drawback that their construction is based on a white Gaussian noise input. This fact reduces applicability e.g. in embedded on-line applications. When the input signal restriction is not close to fulfilment, conventional parametric methods may offer a more widely applicable alternative. However, also in the parametric case a number of theoretical properties need to be fulfilled in order to secure a well-behaved algorithmic performance. Such properties have been determined in e.g. (Wigren, 1990, 1993). The input signal requirement that is related to the linear block is one of persistency of excitation of high enough order. A similar requirement is in effect for subspace based identification algorithms. Regarding the nonlinear block, the analysis of (Wigren, 1990, 1993) shows that the input signal needs to be such that the output from the linear block of the Wiener model has signal energy in all amplitude intervals of (the support of) the piecewise linear static nonlinearity. It can be conjectured that similar requirements are valid for other models of the nonlinear block. Extrapolating the results and insights of (Wahlberg and Ljung, 1986) the following guideline results: The input signal should concentrate the energy to the frequency band and amplitude interval where accuracy is important.

3.7 Disturbance handling and modelling errors

Non-parametric methods often rely on the assumption that also the disturbance is white Gaussian noise. The situation is more scattered regarding parametric algorithms. In general, least squares type methods do rely on white noise assumptions, while so called output error based methods can handle arbitrarily coloured zero mean disturbances (Wigren, 1990).

In the linear case, parametric pulse transfer operator based output error type algorithms have been shown to be more robust with respect to undermodelling than least squares type algorithms (Wahlberg and Ljung, 1986). The results regarding the estimated pole variation of (Westwick and Verhagen, 1996) indicate that output error methods may also have an advantage as compared to the SVD based subspace methods. Therefore, parametric output error methods for Wiener system identification may be expected to be more robust to undermodelling than least squares algorithms and SVD based subspace methods.

In cases where the static nonlinearity is known (or estimated in previous steps) and where the disturbance enters between the two blocks of the Wiener model, it may be advantageous to invert the nonlinearity and to apply linear system identification tools. However, as shown in (Wigren, 1994), this approach can introduce a severe noise amplification in situations where the disturbance enters after the static nonlinear block.

4. MODEL VALIDATION

4.1 General aspects

As indicated above, undermodelling effects are likely to be a main source of error in nonlinear system identification, in particular if signal amplitudes are increased to exploit the nonlinear model structure. Therefore the statistical tools developed for validation of linear dynamic systems, based on residual correlation analysis are believed to be of less value in the nonlinear case. Needless to say, all model validation should be performed on a different set of data than the set used for identification. The parameters obtained at the end of the run should be used for recursive algorithms.

4.2 Loss function

The value of a loss function, e.g. a sum of squared prediction errors, captures all types of modelling errors. When plotted against model orders, it is often an effective tool in assessment of the relative merit of estimated models of all kinds, including the nonlinear Winer model, cf. (Wigren, 1990).

4.3 Simulation

Another useful tool that is applicable to all types of models is simulation of the estimated model. Plots of the measured output signal together with the prediction obtained from the simulated model has the advantage to illustrate the achieved quality for different signal amplitudes (Wigren, 1990).

4.4 Mean residual analysis

One difference in Wiener model based identification as compared to linear system identification is the increased importance of the amplitude contents of the signals. Hence it would be advantageous to validate the quality of the identified model as a function of the amplitude of the output signal of the Wiener system. This subsection describes a novel method for this.

In order to describe the method, the output signal range of the system is first divided into a number of subintervals I_l, $l = 1,\dots,L$. The residual,

$$\varepsilon(t,\theta) = y_n(t) - y_n(t,\theta) \tag{5}$$

corresponding to the output signal $y_n(t)$ of the system is then sorted in L classes depending on in which subinterval $y_n(t)$ is located. The estimated model signal is denoted by $y_n(t,\theta)$, where θ is the estimated parameter vector. The average residual of each class is then computed as

$$\bar{\varepsilon}_l = \frac{1}{N_l} \sum_{\{t|y_n(t)\in I_l\}} \varepsilon(t,\theta), \quad l = 1,\ldots,L. \qquad (6)$$

N_l denotes the number of residuals that belong to the class l. The $\bar{\varepsilon}_l$ are then plotted as a function of the amplitude of the corresponding sorting interval. The number of residuals in each subinterval is plotted similarly. In cases where the random disturbances on the output signal are small, it follows that the amplitude characteristics of the model is close to the system in amplitude intervals where the $\bar{\varepsilon}_l$ are small. Note that the input signal cannot be used as independent parameter since a specific value of the input signal $u(t)$ can correspond to any output signal value $y_n(t)$ because of dynamic effects.

Example 3. The live data described in section 2.3 was used in this example. The algorithm of (Wigren, 1990, 1993) was used to estimate the system, using a setting similar to that of section 5.6.6 of (Wigren, 1990). The results were then validated by the mean residual analysis method. The results are shown in Fig. 5. As can be seen from the plot, the accuracy is best where the signal energy is concentrated, cf. section 3.6.

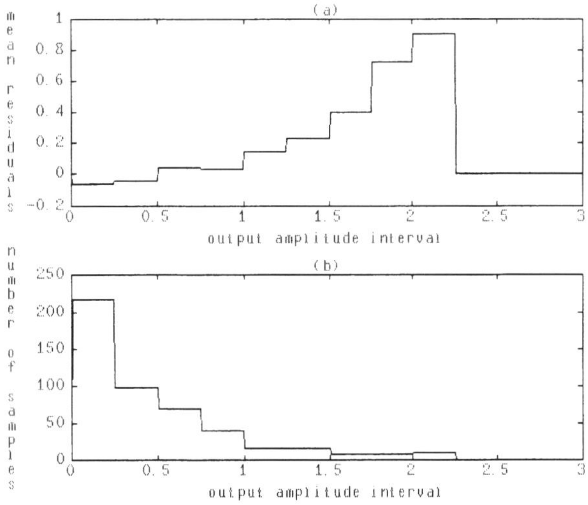

Fig. 5: Mean residuals (a) and the number of residuals (b) of each subinterval.

6. CONCLUSIONS

The paper has discussed user choices in Wiener model based system identification, with a focus on effects tied to the nonlinearity of this model. In general, all discussed methods seem to have limitations and the choice must be governed by the properties required for the application at hand. Model validation for the Wiener model case was also discussed. A novel method, measuring the accuracy as a function of the amplitude was presented. To make further progress in this area it is believed that more experimental results are needed. The live data used in the paper is available from the author for this purpose.

REFERENCES

Billings, S. A. and S. Y. Fakhouri (1978). Identification of nonlinear systems using the Wiener model. *Electronics Letters,* **13**, 502-504.

Billings, S. A. and L. A. Aguirre (1995). Effects of the sampling time on the dynamics and identification of nonlinear models. *Int. J. Bifurcation and Chaos,* **5**, 1541-1556.

Greblicki, W. (1992). Nonparametric identification of Wiener systems. *IEEE Trans. Inform. Theory,* **IT-38**, 1487-1493.

Gomez, J. C. and E. Baeyens (2002). Subspace identification of multivariable Hammerstein and Wiener models. *15:th IFAC World Congress*, Barcelona, Spain.

Ljung, L. and T. Söderström (1983). *Theory and Practice of Recursive Identification.* MIT Press, Cambridge, MA.

Nordsjö, A. E. and T. Wigren (2002). On estimation of model errors caused by nonlinear undermodeling in system identification. *Int. J. Contr.,* **75**, 1100-1113.

Pajunen, G. A. (1984). Application of a model reference adaptive technique to the identification and control of Wiener type nonlinear processes. Doctoral Dissertation, Department of Electrical Engineering, Helsinki University of Technology, Helsinki, Finland.

Pajunen, G. A. (1992). Adaptive control of Wiener type nonlinear systems. *Automatica,* **28,** 781-785.

Söderström, T and P. Stoica (1989). *System Identification.* Prentice Hall, Hemel Hempstead, UK.

Wahlberg, B. and L. Ljung (1986). Design variables for bias distribution in transfer function estimation. *IEEE Trans. Automat. Contr.,* **AC-31**, 134-144.

Wellstead, P. E. (1979). *Introduction to Physical System Modelling.* Academic Press, London, UK.

Westwick, D. and M. Verhagen (1996). Identifying MIMO Wiener systems using subspace model identification methods. *Signal Processing,* **52**, 235-258.

Wigren, T. (1990). Recursive identification based on the nonlinear Wiener model. Doctoral Dissertation, Department of Technology, Uppsala University, Uppsala, Sweden.

Wigren, T. (1993). Recursive prediction error identification using the nonlinear Wiener model. *Automatica*, **29**, 1011-1025.

Wigren, T. (1994). Convergence analysis of recursive identification algorithms based on the nonlinear Wiener model. *IEEE Trans. Automat. Contr.* **39**, 2191-2206.

Wigren, T. (1998). Adaptive filtering using quantized output measurements. *IEEE Trans.Signal Processing.,* **46,** 3426.

IFAC Publications
www.elsevier.com/locate/ifac

APPROXIMATION OF FEASIBLE PARAMETER SET IN WORST CASE IDENTIFICATION OF BLOCK-ORIENTED NONLINEAR MODELS

L. Giarré * and G. Zappa **,[1]

* *Dipartimento di Ingegneria dell'Automazione e dei Sistemi - Università di Palermo, Palermo, Italy.*
giarre@unipa.it *39-(091)-481019-17*
** *Dipartimento di Sistemi e Informatica,- Università di Firenze, via di Santa Marta, 50139 Firenze, Italy.*
zappa@dsi.unifi.it *39-055-4796574*

Abstract: The estimation of the Feasible Parameter Set for block-oriented nonlinear models in a worst case setting is considered. A bounding procedure is determined both for polytopic and ellipsoidic sets, consisting in the projection of the FPS $\subset \mathbb{R}^{MN}$ of the extended parameter vector onto suitable M or N-dimensional subspaces and in the solution of convex optimization problems which provide the extreme points of the Parameter Uncertainties Intervals of the model parameteres. Bounds obtained are tighter then in the previous approaches. *Copyright © 2003 IFAC*

Keywords: Block-oriented nonlinear systems, identification for nonlinear systems, Hammerstein-Wiener models, set membership identification

1. INTRODUCTION: BLOCK-ORIENTED MODELS

In this paper we consider modeling and identification of discrete-time nonlinear systems whose input (u)-output (y) relationship can be approximated with Volterra series (Volterra, 1959). The usefulness of Volterra series is based on their ability to models a very wide class of nonlinear operators; (Sandberg, 1983) demonstrated that systems whose nonlinearities are *analytic* can be modeled with Volterra series i/o operators, moreover (Boyd and Chua, 1985) showed that any *fading* memory operator can be approximated by a (finite) Volterra series operator which can be realized as a finite dimensional linear dynamical system.

Such models have been proven to be successful as simple nonlinear models for a wide number of applications (nonlinear filtering, actuator saturations, audio-visual processing, signal analysis, biologic systems, chemical processes). There exists a large body of work on identification of these models, exploring different approaches and frameworks. Because the evaluation of high-order kernels is cumbersome and time-consuming, *block structured network models* have been largely used (see (Chen, 1985) for a complete and detailed characterization) where interconnections of memoryless nonlinear blocks $\mathcal{N}$ and linear dynamics blocks $\mathcal{L}$ are considered.

In particular, as shown in details in (Doyle *et al.*, 2001), the block-oriented models can be seen as highly structured special cases of the Volterra-like model class. The most common block models are: the series of $\mathcal{N}$ and $\mathcal{L}$ known as Hammerstein (H) model; the series of $\mathcal{L}$ and $\mathcal{N}$ ('dual' Wiener model W); the Lur'e model where $\mathcal{L}$ is interconnected with $\mathcal{N}$ in the feedback loop. Moreover, in (Chen, 1985) the parallel $\mathcal{P}$ of r pathways of

[1] Corresponding author: G. Zappa, Research supported by MURST Grant ex-40%

H models ($\mathcal{N}^1\mathcal{L}^1 \dots \mathcal{N}^r\mathcal{L}^r$, indicated as $\mathcal{PNL}_r$) and the parallel of r pathways of W models (indicated as $\mathcal{PLN}_r$) have been considered. The $\mathcal{PNL}_r$ models are also known as Uryson models (see (Gallmann, 1975)). The $\mathcal{PLN}_r$ models are also known as Projection Pursuit models (see (Doyle *et al.*, 2001)). In (Chen, 1985) all the Volterra kernel relations of the above SISO block-oriented models are given.

In order to identify block oriented models, the following parametrization has been adopted.

$$\mathcal{L}(z) = \sum_{i=1}^{M} h_i B_i(z^{-1}), \quad h \stackrel{\triangle}{=} [h_1 \dots h_M]^T \quad (1)$$

where $B_i(z^{-1})$ depends on the chosen basis function as Laguerre, Kautz, othonormal, etc. For the sake of simplicity we consider FIR (finite impulse response) parametrization, i.e. $B_i(z^{-1}) = z^{-i}$.

$$\mathcal{N}(u) = \sum_{i=1}^{N} p_i g_i(u), \quad p \stackrel{\triangle}{=} [p_1 \dots p_N]^T \quad (2)$$

where $g_i(\cdot) : \mathbb{R} \to \mathbb{R}$ are a set of specified basis functions (for instance polynomial or trigonometric functions). Assuming that the nonlinearity is invertible (i.e. $\mathcal{N}$ monotonic function), for the W models we consider:

$$\mathcal{N}^{-1}(u) = \sum_{i=1}^{N} p_i g_i(u) \quad (3)$$

Block-oriented models usually provide i/o relations in which the components of p, h enter in a multilinear way. However they can be embedded into larger classes with i/o relations linearly parameterized. For istance $\mathcal{PNL}_r$ models can be embedded in the class of diagonal Volterra models D (Doyle *et al.*, 2001) (also known as Generalized Hammerstein models (Garulli *et al.*, 2002)):

$$\begin{aligned} y_k &= \sum_{i=1}^{N} p_i(\mathcal{L}^i(g_i(u_k))) \\ &= \sum_{i=1}^{N}\sum_{j=1}^{M} p_i h_j^i g_i(u_{k-j}) \end{aligned} \quad (4)$$

which can be put in standard regressor form $y_k = \phi_k^T \theta$ by introducing the parameter vector

$$\theta = [\theta_1^T \; \theta_2^T \dots \theta_N^T]^T \in \mathbb{R}^{MN} \quad (5)$$

where $\theta_i = p_i h^i$ and the regressor vector:

$$\phi_k \doteq [g_1(u_{k-1}) \dots g_1(u_{k-M}), \dots g_N(u_{k-M})]^T \quad (6)$$

In fact, defining the parameter matrix

$$\Theta \stackrel{\triangle}{=} \begin{bmatrix} \theta_1^T \\ \theta_2^T \\ \vdots \\ \theta_N^T \end{bmatrix} \in \mathbb{R}^{N \times M} \quad (7)$$

it turns out that the $\mathcal{P}NL_r$ models are diagonal models in which

$$\Theta = \sum_{\ell=1}^{r} p^\ell (h^\ell)^T \quad (8)$$

where p^ℓ and h^ℓ, $\ell = 1, \dots, r$ parametrize the nonlinear $\mathcal{N}^\ell$ and linear $\mathcal{L}^\ell$ block of each parallel Hammerstein model. Therefore the $\mathcal{PNL}_r$ models satisfy the following constraint:

$$rank(\Theta) = r \quad (9)$$

θ will be called the *extended* parameter vector.

2. INTRODUCTION: PARAMETRIC IDENTIFICATION

The literature on parametric identification of block oriented systems, based on i/o noisy measurements, can be essentially divided in three categories:

- *Minimum Prediction Error (MPE)* approach, giving raise to iterative procedures.
  For $\mathcal{PNL}_1$ models, for example, the most popular approach is the Narendra-Galmann algorithm (Narendra and Gallman, 1966). an iterative least square procedure, recently revisited in (Rangaan *et al.*, 1995).
- *Gray Box* procedures, where a decoupling of the estimation of the nonlinear and the linear part is obtained through a particular choice of the input signal.
  For $\mathcal{PNL}_1$, $\mathcal{PLN}_1$ and Lur'e models in (Pearson and Pottmann, 2000) the linear and nonlinear parts are decoupled assuming the static-gain of the linear part to be unitary, and the steady-state characteristic a-priori known.
- *Two stage* procedures, where the Θ matrix is estimated ignoring rank constraints and subsequently the estimate is projected onto the manifold of rank r matrices. See (Bai, 1998) for $\mathcal{PNL}_1$ systems.

Almost all these contributions assume a statistical description of the noise. Conversely, worst-case identification of block-oriented nonlinear plants in presence of unknown but bounded (UBB) noise has not been addressed with only few exceptions. In (Garulli *et al.*, 2002) the identification of a $\mathcal{PNL}_1$ model for a class of nonlinear systems of D type was considered. In particular the estimate with minimum worst-case error was computed; it is provided by the Chebichev center of the Feasible Parameter Set (FPS) *conditioned* to the *non-linear* manifold of rank-one matrices.

Conversely, upper and lower bounds on the Parametric Uncertainty Intervals (PUI) of $\mathcal{PNL}_1$ models can be found in (Belforte and Gay, 2001). Here the noise represents the equation error (ARX structure). Finally, in (Cerone and Regruto, 2003), a *gray box* procedure is exploited for $\mathcal{P}NL_1$ models with OE structure: the nonlinear part is first identified using steady-state measurements; subsequently the FPS of the linear part is evaluated.

In the present work upper and lower bounds of the FPS are computed for $\mathcal{PNL}_1$ models affected by UBB noise in ℓ_2 and ℓ_∞ norm. The bounds are much less conservative than in (Belforte and Gay, 2001). The proposed method can be also easily extended for Lur'e and $\mathcal{PLN}_1$ models under the assumptions on the noise adopted in (Pearson and Pottmann, 2000).

3. PROBLEM FORMULATION AND RESULTS

Given m i/o measurements $\{y_k, u_k; k = 1, \ldots, m\}$ on the nonlinear system, suppose that they can be recasted in the multilinear regression form (with ARX structure)

$$y_k = \phi_k^T \theta + e_k, \tag{10}$$

This multilinear regression (10) is very general since $\mathcal{PNL}_r$, $\mathcal{PLN}_r$ and Lur'e models admit such representation. In the previous paragraph we have derived it specifically for $\mathcal{PNL}_r$ models, where θ is given by (5) and ϕ_k by (6).
Assuming that the noise $\{e_k\}$ is unknown but bounded (UBB), we can define the FPS for the extended vector θ, i.e. the set Ω_θ of the parameters θ not falsified by the measurements, ignoring the constraints among the components of θ provided by its multilinear structure. In particular, we consider the following two cases:
A.1 The noise is bounded in the ℓ_2 norm, i.e.

$$\frac{1}{m}\sum_{k=1}^{m} e_k^2 \leq \epsilon \, . \tag{11}$$

Then the FPS

$$\Omega_\theta = \{\theta : \frac{1}{m}\sum_{k=1}^{m}(y_k - \phi_k^T\theta)^2 \leq \epsilon\} \tag{12}$$

is an ellipsoid in the parameter space $\mathbb{R}^{NM}$.

$$\Omega_\theta = \{\theta : (\theta - \hat{\theta})^T Q (\theta - \hat{\theta}) \leq 1\} \tag{13}$$

where $\hat{\theta}$ is the least square estimate of θ and

$$\begin{aligned} Q &= \Phi^T\Phi\delta^{-1} \\ \hat{\theta} &= (\Phi^T\Phi)^{-1}\Phi^T Y \\ \delta &= m\epsilon - Y^T Y + Y^T\Phi(\Phi^T\Phi)^{-1}\Phi^T Y \end{aligned} \tag{14}$$

$Y = [y_1 \ldots y_m]^T$ and $\Phi = [\phi_1 \quad \phi_2 \ldots \phi_m]^T$ □

A.2 The noise is bounded in the ℓ_∞ norm, i.e.

$$|e_k| \leq \epsilon. \tag{15}$$

Then the FPS

$$\Omega_\theta = \{\theta : \; |y_k - \phi_k^T\theta| \leq \epsilon \;\; , \;\; k \; = \; 1 \ldots m\} \tag{16}$$

is a convex polytope in $\mathbb{R}^{NM}$ □

The block oriented models considered here are specified by the constraint $rank(\Theta) = r$. Therefore, partitioning Θ as $\begin{bmatrix} \Theta_r \\ \Theta_{N-r} \end{bmatrix}$, $\Theta_r \in \mathbb{R}^{r \times M}$, a possible minimal parametrization is provided by Θ_r (assumed full rank) and by the coefficients of the expansion of the rows of Θ_{N-r} in the basis provided by Θ_r, i.e. by the matrix $P \in \mathbb{R}^{r\times(N-r)}$ such that:

$$P\Theta_{N-r} = \Theta_r \tag{17}$$

Therefore we are interested in determining the FPS for X, i.e. the set

$$\Omega_P \overset{\triangle}{=} \{P : [-I_r \quad P]\Theta = 0 \;\; \text{for} \;\; \Theta \in \Omega_\theta\} \tag{18}$$

The exact determination of Ω_P is a difficult task since, in general, Ω_P is non convex. Hence outerbounds should be investigated. Hereafter outer bounds are obtained by exploiting the following assumptions:

- $r = 1$; hence P reduces to the row vector $[p_2, p_3, \ldots, p_N] \in \mathbb{R}^{N-1}$ and Θ_r to the M-dimensional row vector θ_1^T.
- Instead of Ω_p, the FPS Ω_{p_i} of each scalar component p_i is evaluated.

Notice that the computation of Ω_{p_i} requires, in particular, the solution of the following problems

$$\min_{p,h} p_i \quad , \quad \max_{p,h} p_i \quad , \quad i = 2, \ldots, N$$
such that:
$$p = [p_1, p_2, \ldots, p_N]^T, \quad p_1 = 1, \quad h \in \mathbb{R}^M \tag{19}$$
$$ph^T \in \Omega_\theta$$

Unfortunately these problems are nonconvex. We shall prove that convexification can be obtained by replacing the set Ω_θ with the outer approximation $\Omega_{\theta_1} \times \Omega_{\theta_2} \times \ldots \times \Omega_{\theta_N}$ where $\times$ denotes Cartesian product and Ω_{θ_i} the projection of Ω_θ along the coordinates of θ_i. Notice that

(1) If Ω_θ is a polytope, then Ω_{θ_i} is a polytope as well and can be represented by:

$$\Omega_{\theta_i} = \{\theta : M_i\theta \leq m_i\} \qquad (20)$$

where the matrix M_i and the vector m_i can be computed, for instance, by exploiting the Fourier-Motzkin elimination algorithm (Keerthi and Gilbert, 1987);

(2) If Ω_θ is an ellipsoid then Ω_{θ_i} is an ellipsoid as well

$$\Omega_{\theta_i} = \{\theta_i : (\theta_i - \hat{\theta}_i)^T Q_i(\theta_i - \hat{\theta}_i) \leq 1\}$$

where the matrix $Q_i \in \mathbb{R}^{M\times M}$ is the Schur complement of a suitable submatrix of Q and therefore is computed analytically.

The following Lemma holds (proofs are omitted due to space limitations):
Lemma 1 - Ω_{p_i}, if non empty, is an interval.

Consider now the computation of the extreme points p_i^-, p_i^+ of the uncertainty interval Ω_{p_i}, $i = 2,\ldots,N$.
For *polytopic uncertainties* it can be easily shown that p_i^- p_i^+ are provided by the solution of two LP problems:

$$\alpha^- = \min_{\alpha,h} \alpha_i \quad , \quad \alpha^+ = \max_{\alpha,h} \alpha_i$$
such that:
$$\alpha = [\alpha_1, \alpha_2, \ldots, \alpha_N], \quad \alpha_1 = 1, \quad h \in \mathbb{R}^M \qquad (21)$$
$$M_j h \leq \alpha_j m_j, \quad j = 1,\ldots,N$$

It turns out that $p_i^+ = (\alpha^-)^{-1}$ and $p_i^- = (\alpha^+)^{-1}$.

Remark - Clearly one can remove from (21) those constraints for which $m_j \geq 0$, corresponding to politopes Ω_{θ_j} containing the origin.

For *ellipsoidic uncertainties*, the extreme points p_i^-, p_i^+ of the uncertainty interval Ω_{p_i} are the solutions of two mathematical programming problems:

$$p_i^- = \min_{p,h} p_i \quad , \quad p_i^+ = \max_{p,h} p_i$$
such that:
$$p = [p_1, p_2, \ldots, p_N]^T, \quad p_1 = 1, \quad h \in \mathbb{R}^M \qquad (22)$$
$$(p_j h - \hat{\theta}_j)^T Q_j(p_j h - \hat{\theta}_j) \leq 1, \quad j = 1,\ldots,N$$

We shall prove that this problem is convex. Consider the $j - th$ constraint in (22), $j \neq 1, i$. Clearly h is feasible for such a constraint if and only if there exist scalars $\bar{p}_j$ such that $(\bar{p}_j h - \hat{\theta}_j)^T Q_j(\bar{p}_j h - \hat{\theta}_j) = 1$. In turn, this holds if and only if

$$h^T M_j h \geq 0, \quad \text{and} \quad h^T Q_j\hat{\theta}_j \geq 0 \qquad (23)$$

where

$$M_j \doteq Q_j\hat{\theta}_j\hat{\theta}_j^T Q_j - Q_j(1 - \hat{\theta}_j^T Q_j\hat{\theta}_j) \qquad (24)$$

Notice that M_j is a symmetric but non definite matrix. Should M_j be positive definite, any vector h become admissible for the $j - th$ constraint. In turn, this implies that the ellipsoid Ω_{θ_j} contains the origin. Consider now the set:

$$K_j^+ \doteq \{h; \ h^T M_j h \geq 0, \ \ h^T Q_j\hat{\theta}_j \geq 0\}, \qquad (25)$$

Lemma 2 - K_j^+ is a convex cone.

Remark - A feasible vector h for (22) represents a line in $\mathbb{R}^M$, passing through the origin, which intersects all the projected ellipsoids Ω_{θ_i}. By Lemma 2 the set of all such lines is still a convex cone.

Hence (22) amounts to the following two convex programming problems:

$$p_i^- = \min_{p,h} p_i \quad , \quad p_i^+ = \max_{p,h} p_i$$
such that:
$$p = [p_1, p_2, \ldots, p_N]^T, \quad p_1 = 1, \quad h \in \mathbb{R}^M \qquad (26)$$
$$(p_i h - \hat{\theta}_i)^T Q_i(p_i h - \hat{\theta}_i) \leq 1,$$
$$h \in \Omega_{\theta_1} \bigcap \{ \bigcap_{j\neq 1,i} K_j^+ \} \doteq C(i)$$

These problems have the following geometric interpretation: found those values of p_i for which the scaled ellipsoid $p_i\Omega_{\theta_i}$ is tangent to the convex set $C(i)$. This can be easily solved by computing the distance $d(p_i)$ between $p_i\hat{\theta}_i$, center of the scaled ellipsoid $p_i\Omega_{\theta_i}$, and $C(i)$ (this is a simple convex problem) and looking for those values of p_i for which $d(p_i) = 1$. Since the set Ω_{p_i} is an interval there exist only two solutions p_i^+ and p_i^- to the equation $d(p_i) = 1$, which can be found by a bisection procedure.

This technique improves previous results (Belforte and Gay, 2001) in two respects:

1) Ellipsoidal uncertainties are considered as well.
2) Less conservative bounds are obtained. In fact correlations among the components of θ_i are explicitly taken into account here, while are ignored in (Belforte and Gay, 2001).

Remark - Notice that the above technique computes, for rank-one matrices $\Theta = ph \in \Omega_\theta$, an overbound for the uncertainty set of the vector

h, namely Ω_{θ_1}, and the PUI's of the components of p. Actually, interchanging the rows and columns of Θ, one can also compute an overbound on the uncertainty set of the vector p and the PUI's of the components of h. Merging these two descriptions, tighter bounds can be obtained. □

Remark - The degree of conservativeness of the proposed technique depends on the "shape" of Ω_θ, i.e. on the choice of input signal. In particular, if the input signal is a random steps signal (i.e. a sequence of random amplitude steps of sufficiently long duration such that steady-state behaviour is achieved), then the projection of Ω_θ in the directions spanned by the coefficients of p is a tight description for Ω_p and no conservatism is introduced at this stage. Notice that this input selection has been often adopted in the parameter identification of block oriented models since it allows the decoupling in the estimate of the linear and nonlinear part. □

4. NUMERICAL EXAMPLE

The effectiveness of the bounding procedure is illustrated in a very simple example. Let us consider a plant with $h = [1 \quad 2]^T$ and $p = [1 \quad 2 \quad 3]^T$

corresponding to the extended parametr matrix

$$\Theta_{\text{true}} = \begin{bmatrix} 1 & 2 \\ 2 & 4 \\ 3 & 6 \end{bmatrix}$$

The nonlinearity is of polynomial type: $\mathcal{N}(u_k) = [u_k \quad u_k^2 \quad u_k^3]^T$, i.e. $g_i(u) = u^i$. An experiment is carried out with $m = 100$ measurements. The noise is bounded in the ℓ_2 norm, $m\epsilon^2 = 0.5$. The input is a square wave of amplitude ± 1 with noise super imposed with variance 0.001. It turns out that the FPS is the ellipsoid $\Omega_\theta = (\theta - \hat{\theta})^T Q (\theta - \hat{\theta}) \le 1$ with

$$Q = \begin{bmatrix} 4.08 & 3.54 & -0.02 & 3.55 & 4.01 & 3.48 \\ 3.54 & 7.16 & 3.36 & 3.07 & 7.04 & 7.05 \\ -0.02 & 3.36 & 11.11 & 3.18 & 3.03 & 6.84 \\ 3.55 & 3.07 & 3.18 & 14.21 & 6.67 & 6.07 \\ 4.01 & 7.04 & 3.03 & 6.67 & 18.16 & 10.11 \\ 3.48 & 7.05 & 6.84 & 6.07 & 10.11 & 21.15 \end{bmatrix}$$

and

$$\hat{\theta} = \begin{bmatrix} 0.9901 \\ 1.9995 \\ 2.0014 \\ 2.9985 \\ 4.0011 \\ 6.0027 \end{bmatrix}$$

.

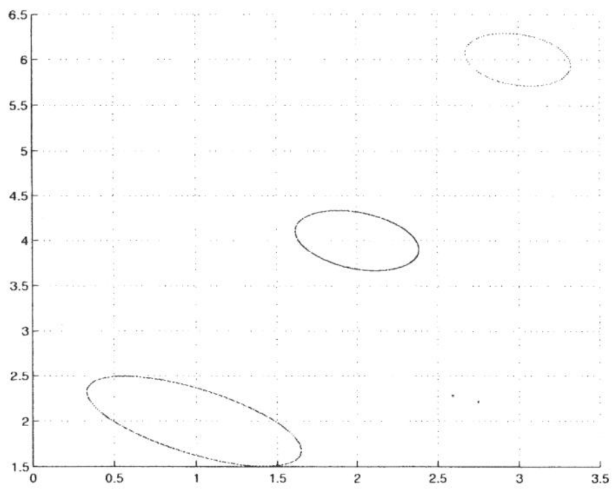

Fig. 1. Projected Ellipsoids: Ω_{θ_1}, Ω_{θ_2} and Ω_{θ_3}

The projected ellipsoids $\Omega_{\theta_i}, i = 1, 2, 3$ are represented in figure 1.

According to (26), the intervals $\Omega_{p_i} = [p_i^+ \quad p_i^-]$ have been computed for $i = 2, 3$, giving raise to the following interval: $\Omega_{p_2} = [1.63 \quad 2.33]$ and $\Omega_{p_3} = [2.52 \quad 3.45]$.

For the same example, the bounding algorithm, introduced in (Belforte and Gay, 2001), was adapted to the ellipsoidal case and the uncertainty intervals computed. Let call this algorithm *BG* and *GZ* our algorithm.

Using interval mathematics notations, the *BG* algorithm can be summarized by the expression:

$$\Omega_{p_i} = \bigcap_{j=1}^{N} \frac{[\ \theta_{1j}\ ]}{[\ \theta_{ij}\ ]} \tag{27}$$

where $[\ \theta_{ij}\ ]$ denotes the uncertainty interval of each component of Θ, obtained by projecting Ω_θ along the $i, j-th$ direction. Notice the ratio of two intervals coincides with $\mathbb{R}$ if the denominator contains the origin.

BG-algorithm gives the following intervals: $\Omega_{p2} = [1.02 \quad 2.88]$ and $\Omega_{p_3} = [0.6153 \quad 4.1836]$. In the following table we report the length of the two intervals (Ω_{p_i} for $i = 2, 3$), for the two algorithms:

i	GZ	BG
2	0.7	1.86
3	0.93	3.57

It is clear that *BG*-algorithm computes intervals that are much more conservative.

5. CONCLUSIONS

A bounding procedure has been proposed for the estimation of the Feasible Parametr Set for Hammerstein models in a worst case setting. Under suitable assuption it can be applied also to Wiener and Lure models. It consists in the projection of the FPS $\subset \mathbb{R}^{MN}$ of the extended parameter vector onto suitable M or N-dimensional subspaces and in the solution of convex optimization problems which provide the extreme points of the Parameter Uncertainties Intervals of the model parameteres. In particular for polytopic FPS the optimization problem reduces to a linear problem; however the computation of the projection is hard. Conversely for ellipsoidic FPS the computation of the projection is straightforward. Bounds obtained are tighter then in the previous approaches. Extension to parallel block-oriented models is currently under investigation.

6. REFERENCES

Bai, E.-W. (1998). An optimal two-stage identification algorithm for Hammerstein-Wiener nonlinear systems. *Automatica* **34**(3), 333–338.

Belforte, G. and P. Gay (2001). Discrete time Hammerstein model identification with Unknown but bounded errors. *IEE Proc. of Control theory and applications* pp. 523–529.

Boyd, S. and L. O. Chua (1985). Fading memory and the problem of approximating nonlinear operators with Volterra series. *IEEE Trans. on Cyrcuits and Systems* **32**, 1150–1161.

Cerone, V. and D. Regruto (2003). A two stage parameter bounding procedure for hammertein models with bounded output errors. *To appear in Proceedings ACC03.*

Chen, H.W. (1985). Modeling and identification of parallel nonlinear systems: structural classification and parameter estimation models. *Proc. of IEEE* **83**, 39–66.

Doyle, F. J., R. K. Pearson and B. A. Ogunnaike (2001). *Identification and Control Using Volterra Models.* Springer Verlag.

Gallmann, P. G. (1975). An iterative method for the identification of nonlinear systems using a Uryson model. *IEEE Trans. on Automatic Control* **20**, 771–775.

Garulli, A., L. Giarrè and G. Zappa (2002). Identification of approximated Hammerstein models in a worstcase setting. *IEEE Trans. on Automatic Control.*

Keerthi, R.K. and E. G. Gilbert (1987). Computation of minimum-time feedback control laws for discrete-time systems with state-control constraints. IEEE Trans. on Automatic Control (32), 432–435.

Narendra, K. S. and P. G. Gallman (1966). An iterative method for the identification of nonlinear systems using a Hammerstein model. *IEEE Trans. on Automatic Control* **11**, 546–550.

Pearson, R.K. and M. Pottmann (2000). Graybox identification of block oriented nonlinear models. *Journal of Process Control* pp. 301–315.

Rangaan, S., G. Wolodkin and K. Poolla (1995). New results for Hammerstein system identification. *Proc. of* 34th *Conf. on Dec. and Control.* pp. 697–702.

Sandberg, I. W. (1983). Series expansions for nonlinear systems. *Circuits, Systems and Signal Processing* pp. 77–87.

Volterra, V. (1959). *Theory of functionals and of integral and integro differential equations.* New-York:Dover.

IFAC
Publications
www.elsevier.com/locate/ifac

PARAMETERS SET EVALUATION OF WIENER MODELS FROM DATA WITH BOUNDED OUTPUT ERRORS

V. Cerone M. Milanese D. Regruto

Dipartimento di Automatica e Informatica, Politecnico di Torino, corso Duca degli Abruzzi 24, 10129 Torino, Italy — E-mail: vito.cerone@polito.it, mario.milanese@polito.it, diego.regruto@polito.it, Phone: + 39 11 5647064, Fax: + 39 11 5647099

Abstract: In this paper a procedure is presented for deriving parameters bounds in SISO Wiener models when the nonlinear block can be modeled by a polynomial and the output measurement errors are bounded. First, using steady-state input-output data, parameters of the nonlinear block are tightly bounded. Next in order to estimate the parameters of the linear block, the evaluation of the inner unmeasurable signal is considered. No invertibility assumption of the nonlinearity is required. Then, through a suitable design of the identification experiment, bounds on the unmeasurable inner signal are evaluated. Finally, such bounds together with the input sequence are used for bounding the parameters of the linear model. *Copyright © 2003 IFAC*

Keywords: Wiener model, bounded uncertainty, output errors, parameter bounding, linear programming.

1. INTRODUCTION

A wide class of simple nonlinear systems, also called block-oriented systems, can be modeled by interconnected memoryless nonlinear gains and linear subsystems. The configuration considered in this paper, commonly referred to as Wiener model, is shown in Figure 1; it consists of a linear dynamic system followed by static nonlinear block $\mathcal{N}$. The identification of such a model is carried out on the basis of the sequences u_t and y_t, while the inner signal x_t, i.e. the output of the linear block, is not assumed to be available. The Wiener model has been successfully used in a large variety of fields. The identification of Wiener structure has attracted the attention of many authors, as can be seen in Billings (1980), Billings and Fakhouri (1977), Haber and Unbehauen (1990), Bai (2002), Wigren (1998). The main difficulty in the identification of nonlinear block-oriented system is that the internal signals are not available for measurement. As far as Wiener systems are concerned most of contributions assume invertibility of the nonlinearity. As a matter of fact under such an assumption the inner signals can be recovered from the output measurements through inversion of the previously estimated nonlinearity. However, many output nonlinearity encountered in real world problem are non-invertible (see, e.g., Wigren, 1998), thus the invertibility assumption appear to be quite restrictive. Removal of such an hypothesis made the consistent evaluation of the inner signal sequence a difficult task even in the case of exactly known nonlinearity. In all the papers mentioned above, the authors assume that the measurement error η_t is statistically described. A worthwhile alternative to the stochastic description of measurement errors is the bounded-errors characterization, where uncertainties are assumed to belong to a given set. The interested reader can find further details on this approach in a number of survey papers (see, e.g., (Milanese and Vicino, 1991; Walter and Piet-Lahanier, 1990)), in the book edited by Milanese *et al.* (Milanese *et al.*, 1996) and the special issues edited by Norton (Norton(Ed.), 1994; Norton(Ed.), 1995). To the author's best knowledge, no contribution can be found which address the identification of Wiener models when the measurement error η_t is supposed to be bounded. In this paper the identification of single-input single-output (SISO) Wiener models is considered, when the nonlinear block can be modeled by a polynomial, with finite and known order, and when the output measurement errors are bounded. First, using steady-state input-output data, parameters of the nonlinear block are tightly bounded. Next in order to estimate the parameters of the linear block, the evaluation of the inner unmeasurable signal x_t is considered. No invertibility assumption of the nonlinearity is required, thus the inner signal x_t providing a given measured output y_t is not unique even in the uncertainty free case, i.e., the estimated inner sequence $\{x_t\}$ to be used in the evaluation of the feasible parameter set of the linear block, might not be the output of a unique linear systems. In the paper it is shown how such a fundamental problem can be addressed through a suitable design of the identification experiment. Finally, through the designed dynamic experiment, for all y_t belonging to a given output transient sequence $\{y_t\}$, upper and lower bounds on the inner signal x_t are computed. Such bounds, together with the input sequence $\{u_t\}$ are used for bounding the parameters of the linear block.

2. PROBLEM FORMULATION

Consider the SISO discrete-time Wiener model shown in Figure 1, where the linear dynamic block, which is modeled by a discrete-time system, maps the input signal u_t into the unmeasurable inner variable x_t according to

$$x_t = \frac{B(q^{-1})}{A(q^{-1})} u_t = G(q^{-1}) u_t, \qquad (1)$$

or, equivalently, in terms of a linear difference equation

$$A(q^{-1}) x_t = B(q^{-1}) u_t, \qquad (2)$$

where $A(\cdot)$ and $B(\cdot)$ are polynomials in the backward shift operator q^{-1}, $(q^{-1} w_t = w_{t-1})$,

$$A(q^{-1}) = 1 + a_1 q^{-1} + \ldots + a_{na} q^{-na}, \qquad (3)$$
$$B(q^{-1}) = b_0 + b_1 q^{-1} + \ldots + b_{nb} q^{-nb}. \qquad (4)$$

The nonlinear block transforms x_t into the noise-free output w_t through the following polynomial function

$$w_t = \sum_{k=1}^{n} \gamma_k x_t^k, \quad t = 1, \ldots, N; \qquad (5)$$

whose order n is taken to be finite and a-priori known; N is the length of the input sequence. In line with the work done by a number of authors, it is assumed that (i) the linear system is asymptotically stable (see, e.g. Stoica and Söderström, 1982; Krzyżak, 1993; Lang, 1993; Sun *et al.*, 1999); (ii) $\sum_{j=0}^{nb} b_j \neq 0$, that is, the steady-state gain is not zero (see, e.g. Lang, 1993; Sun *et al.*, 1999); (iii) the only *a priori* information needed is an estimate of the process settling-time (see, e.g. Kalafatis *et al.*, 1997). Let y_t be the noise-corrupted measurements of w_t

$$y_t = w_t + \eta_t. \qquad (6)$$

Measurements uncertainty is known to range within given bounds $\Delta\eta_t$, i.e.,

$$|\eta_t| \leq \Delta\eta_t. \qquad (7)$$

Unknown parameter vectors $\gamma \in R^n$ and $\theta \in R^p$ are defined, respectively, as

$$\gamma^T = [\gamma_1 \quad \gamma_2 \quad \ldots \gamma_n], \qquad (8)$$

$$\theta^T = [a_1 \quad \ldots \quad a_{na} \quad b_0 \quad b_1 \quad \ldots b_{nb}], \qquad (9)$$

where $n_a + n_b + 1 = p$. It is easy to show that the parameterization of the structure of Figure 1 is not unique. As a matter of fact, any parameters set $\tilde{b}_j = \alpha^{-1} b_j, j = 1, 2, \ldots, nb$, and $\tilde{\gamma}_k = \alpha\gamma_k, k = 1, 2, \ldots, n$, for some nonzero and finite constant α, provides the same input-output behaviour. Thus, any identification procedure cannot perceive the difference between parameters $\{b_j, \gamma_k\}$ and $\{\alpha^{-1} b_j, \alpha\gamma_k\}$. In this work, it is assumed, without loss of generality, that the steady-state gain of the linear part be one, that is

$$g = \frac{\sum_{j=0}^{nb} b_j}{1 + \sum_{i=1}^{na} a_i} = 1 \qquad (10)$$

In this paper the problem of deriving bounds on parameters γ and θ consistently with given measurements, error bounds and the assumed model structure is addressed.

3. ASSESSMENT OF TIGHT BOUNDS ON THE NONLINEAR STATIC BLOCK PARAMETERS

In this work we exploit steady-state operating conditions to bound the parameters of the nonlinear static block. The known input and noise corrupted output sequences are collected from the steady-state response of the system to a set of step inputs with different amplitude. It is only assumed that a rough information on the settling time of the system under consideration is available, in order to know when steady-state conditions are reached, so that steady-state data can be collected. Indeed, combining equations (5), (2), (6) and (10) in steady-state operating conditions the following input-output description is obtained.

$$\bar{y}_s = \sum_{k=1}^{n} \gamma_k \bar{u}_s^k + \eta_s, \quad s = 1, \ldots, M \qquad (11)$$

where $\bar{u}_s$ and $\bar{y}_s$ are steady-state values of the known input signal and output observation respectively, while η_s is the measurement error; $M \geq n$ is the length of the steady-state sequences. A block diagram description of equation (11) is depicted in Figure 2. Thus, the feasible parameter region of the static nonlinear block is defined as

$$\mathcal{D}_\gamma = \{\gamma \in R^n : \bar{y}_s = \sum_{k=1}^{n} \gamma_k \bar{u}_s^k + \eta_s, \quad |\eta_s| \leq \Delta\eta_s; \quad s = 1, \ldots, M\}. \qquad (12)$$

where $\{\Delta\eta_s\}$ is the sequence of bounds on measurements uncertainty. From the definition of $\mathcal{D}_\gamma$ one gets

$$\bar{\varphi}_s^{\mathrm{T}} \gamma \leq \bar{y}_s + \Delta\eta_s \qquad (13)$$
$$\bar{\varphi}_s^{\mathrm{T}} \gamma \geq \bar{y}_s - \Delta\eta_s \qquad (14)$$

where

$$\bar{\varphi}_s = \left[\bar{u}_s \quad \bar{u}_s^2 \quad \bar{u}_s^3 \ldots \bar{u}_s^n\right]^{\mathrm{T}} \tag{15}$$

for $s = 1, 2, \ldots, M$. The above exact description of $\mathcal{D}_\gamma$ will be used in the next section when deriving tight bounds on the unmeasurable inner signal x_t. Since $\mathcal{D}_\gamma$ is a convex polytope, whose shape may result quite complex for increasing n and M, an outer bound to it such as an ellipsoid or a box is often computed. In this paper we consider an orthotope-outer bounding set $\mathcal{B}_\gamma$ containing $\mathcal{D}_\gamma$

$$\begin{aligned}\mathcal{B}_\gamma = \{\gamma \in R^n : \gamma_j = \gamma_j^c + \delta\gamma_j,\\ \mid \delta\gamma_j \mid\leq \Delta\gamma_j/2, j = 1, \ldots, n\},\end{aligned} \tag{16}$$

where

$$\gamma_j{}^c = \frac{\gamma_j^{min} + \gamma_j^{max}}{2}, \tag{17}$$

$$\Delta\gamma_j = \mid \gamma_j^{max} - \gamma_j^{min} \mid, \tag{18}$$

and

$$\gamma_j^{min} = \min_{\gamma\in\mathcal{D}_\gamma} \gamma_j, \quad \gamma_j^{max} = \max_{\gamma\in\mathcal{D}_\gamma} \gamma_j. \tag{19}$$

The evaluation of $\mathcal{B}_\gamma$ requires the solution of $2n$ linear programming problems with n variables and $2M$ constraints.

4. EVALUATION OF FEASIBLE INTERVALS FOR OUTPUT MEASUREMENTS AND INNER SIGNALS

In the previous section it has been shown how the nonlinear block can be characterized using steady-state data. In order to estimate the parameters of the linear model, one should first evaluate the inner signal x_t from the output records y_t via the polynomial nonlinearity. Unfortunately, one must consider the fact that nonlinearity (5) is, in general, noninvertible, which means that, given the measured output y_t, the inner signal x_t cannot be evaluated uniquely even in the case of exactly known polynomial and noise free measurements. As a matter of fact, considering (5) at a single time t one gets

$$p_t(x_t, w_t) = 0 \tag{20}$$

where

$$p_t(x_t, w_t) = w_t - \sum_{k=1}^{n} \gamma_k x_t^k. \tag{21}$$

belongs to the following family of polynomials

$$\mathcal{P}_t = \{p_t(x_t, w_t) : w_t \in R\} \tag{22}$$

For given w_t and exactly known γ_k, $k = 1, 2, \ldots, n$, equation (20) shows, in general, n different solutions in the unknown x_t. Nonuniqueness, unfortunately, is responsible of nonconsistent inner signal estimates, i.e., estimated $\{x_t\}$ might not be the output of a unique linear system. The main idea for overcoming this problem is to design the input sequence $\{u_t\}$ which will force an output sequence $\{w_t\}$ so that polynomial (21) will show either only one real root when n is odd or two real roots when n is even in the case of exactly known parameters of the nonlinear block and noiseless measurements. In the case of uncertain polynomial parameters the following family of polynomials can be defined

$$\Pi_t = \{p_t(x_t, w_t, \gamma) : w_t \in R, \ \gamma \in \mathcal{D}_\gamma\} \tag{23}$$

where

$$p_t(x_t, w_t, \gamma) = w_t - \sum_{k=1}^{n} \gamma_k x_t^k. \tag{24}$$

It is assumed that all polynomials in Π_t have degree equal to n. In this case, in order to evaluate the inner signal x_t one has to find the real roots of the uncertain polynomial (24). The *real spectral set* of the family of polynomial Π_t, is defined as

$$\begin{aligned}\mathcal{S}_{\mathrm{R}}(\Pi_t) = \{x_t \in R : p_t(x_t, w_t, \gamma) = 0,\\ \text{for some } \gamma \in \mathcal{D}_\gamma, w_t \in R\}\end{aligned} \tag{25}$$

4.1 Uncertain polynomial nonlinearity — Here it is assumed that $\gamma \in \mathcal{D}_\gamma$ and the problem of evaluating the inner signal x_t in terms of roots of the uncertain polynomial (24) is addressed. One shall look for conditions under which each polynomial $p_t(x_t, w_t, \gamma) \in \Pi_t$ shows either only one real root when n is odd or two real roots when n is even.

Proposition 1. Each polynomial $p_t(x_t, w_t, \gamma) \in \Pi_t$, shows either only one real root when n is odd or two real roots when n is even if and only if

$$w_t > \bar{w} \ \text{ or } \ w_t < \underline{w} \ \text{ when } n \text{ is odd} \tag{26}$$

or

$$\operatorname{sign}(\gamma_n)w_t > \frac{1 + \operatorname{sign}(\gamma_n)}{2}\bar{w} - \frac{1 - \operatorname{sign}(\gamma_n)}{2}\underline{w} \quad \text{when } n \text{ is even} \tag{27}$$

where

$$\bar{w} = \max_{x_t\in\Upsilon_t} \max_{\gamma\in\mathcal{D}_\gamma} \sum_{k=1}^{n} \gamma_k x_t^k \tag{28}$$

$$\underline{w} = \min_{x_t\in\Upsilon_t} \min_{\gamma\in\mathcal{D}_\gamma} \sum_{k=1}^{n} \gamma_k x_t^k \tag{29}$$

$$\Upsilon_t = \left\{x_t \in R : \frac{d}{dx_t}\sum_{k=1}^{n} \gamma_k x_t^k = 0, \text{for some } \gamma \in \mathcal{D}_\gamma\right\} \tag{30}$$

Moreover, under conditions (26) and (27) the real roots of polynomial (24) satisfy

$$x_t \in]\bar{x}, +\infty] \tag{31}$$

$$x_t \in [-\infty, \underline{x}[\tag{32}$$

$$\begin{aligned}\bar{x} = \max\{x_t \in R : \frac{1 + \operatorname{sign}(\gamma_n)}{2}\bar{w} +\\ \frac{1 - \operatorname{sign}(\gamma_n)}{2}\underline{w} - \sum_{k=1}^{n} \gamma_k x_t^k = 0,\\ \text{for some } \gamma \in \mathcal{D}_\gamma\}\end{aligned} \tag{33}$$

$$\begin{aligned}\underline{x} = \min\{x_t \in R : \frac{1 + (-1)^n\operatorname{sign}(\gamma_n)}{2}\bar{w} +\\ \frac{1 - (-1)^n\operatorname{sign}(\gamma_n)}{2}\underline{w} - \sum_{k=1}^{n} \gamma_k x_t^k = 0,\\ \text{for some } \gamma \in \mathcal{D}_\gamma\}.\end{aligned} \tag{34}$$

The proof of Proposition 1 can be found in (Cerone *et al.*, 2002).

The computational aspects of quantities and sets involved in Proposition 1 are now discussed.

Computation of Υ_t — First consider the set defined by equation (30), i.e. the set of real valued x_t for which the uncertain polynomial shows stationary points (relative maxima, relative minima or points of inflexion). The first derivative of the uncertain polynomial is still an uncertain polynomial, namely

$$p_t{}'(x_t,\gamma)=-\frac{d}{dx_t}\sum_{k=1}^{n}\gamma_k x_t^k=-\sum_{k=1}^{n}k\gamma_k x_t^{k-1} \quad (35)$$

which, clearly, shows nonlinear relations in the unknown x_t and the uncertain γ. A one-dimensional gridding procedure for finding the roots of (35) is proposed. It is noticed that for a given $x_t \in R$, in order to find the real spectral set of polynomial (35) one must solve a set of $2M$ linear inequalities (i.e., $\gamma \in \mathcal{D}_\gamma$) and one linear equality (i.e., $\sum_{k=1}^{n} k\gamma_k x_t^{k-1} = 0$) in the unknown $\gamma \in R^n$.

Computation of $\bar{w}$ *and* $\underline{w}$ — Next equations (28) and (29) which define two nonlinear programming problems are considered. However, when x_t is given, they simplify to linear programs. Thus, to compute $\bar{w}$ and $\underline{w}$, for each value of $x_t \in \Upsilon_t$ the solution of two linear programming problems with n variables and $2M$ constraints is required. A one-dimensional gridding procedure is used in order to carry out the optimization over a finite number of $x_t \in \Upsilon_t$.

Computation of $\bar{x}$ *and* $\underline{x}$ — Finally, equation (33) and equation (34) are considered. In order to simplify the discussion, odd order polynomial with $\gamma_n > 0$ are first considered. In this case one gets

$$\bar{x} = \max\{x_t \in R : \bar{w} - \sum_{k=1}^{n}\gamma_k x_t^k = 0, \text{ for some } \gamma \in \mathcal{D}_\gamma\} \quad (36)$$

$$\underline{x} = \min\{x_t \in R : \underline{w} - \sum_{k=1}^{n}\gamma_k x_t^k = 0, \text{ for some } \gamma \in \mathcal{D}_\gamma\} \quad (37)$$

Equations (36) and (37) show nonlinear relations in the unknown x_t and the uncertain γ. In this case a search procedure is proposed for finding the roots of uncertain polynomials involved in the above equations. It is noticed that for a given $x_t \in R$, in order to solve each one of the above equations one must solve a set of $2M$ linear inequalities (i.e., $\gamma \in \mathcal{D}_\gamma$) and one linear equality (i.e., $\sum_{k=1}^{n}\gamma_k x_t^k = 0$) in the unknown $\gamma \in R^n$. It is also seen that the search can be started from the unique real root of one nominal polynomial obtained, e.g., setting $\gamma = \gamma^c$; next, only right side of the nominal root of equation (36) and only left side of the nominal root of equation (37) are explored in order to find a suitable approximation of $\bar{x}$ and $\underline{x}$ respectively. Analogous considerations can be made in all other cases ($\gamma_n > 0$, $\gamma_n < 0$, n odd, n even).

4.2 Input sequence design — One is left with the problem of the choice of the input sequence $\{u_t\}$. In order to drive the inner signal $\{x_t\}$ into the desired interval, the input signal $\{u_t\}$ should contain a DC component u_{DC} (offset) and a dynamic exciting signal $\{u_{td}\}$ whose amplitudes should be chosen in such a way that $x_t = x_{DC} + x_{td}$ satisfies either (31) $\forall t$ or (32) $\forall t$. Since the steady-state gain of the linear subsystem is constrained to be one, the amplitudes of the DC components in $u_t = u_{DC} + u_{td}$ and x_t are the same, i.e., $u_{DC} = x_{DC}$. Guidelines for the design of the dynamic exciting signal $\{u_{td}\}$ are provided by the following two propositions.

Proposition 2. For given $u_{DC} \geq \bar{x}$, the sequence $\{x_t\}$ satisfies (31) if and only if:

$$\|\{u_{td}\}\|_\infty \leq \frac{|u_{DC} - \bar{x}|}{\|g\|_1} \quad (38)$$

where g and $\|g\|_1$ are, respectively, the impulse response and the ℓ_1 norm of the linear block; $\|\cdot\|_\infty$ is the ℓ_∞ norm of a sequence.

Proposition 3. For given $u_{DC} \leq \underline{x}$, the sequence $\{x_t\}$ satisfies (32) if and only if:

$$\|\{u_{td}\}\|_\infty \leq \frac{|u_{DC} - \underline{x}|}{\|g\|_1} \quad (39)$$

Propositions 2 and 3 are straightly derived from the definition of ℓ_∞-norm/ℓ_∞-norm system gain which equals the ℓ_1-norm of g.

Remark 2 — Note that Proposition 2 and Proposition 3 give necessary and sufficient conditions for $\{w_t\}$ to satisfy either inequality (26) or (27).

Remark 3 — Note that even if only a rough upper bound g_{up} of $\|g\|_1$ is known, inequality (38) is satisfied by choosing an input dynamic exciting sequence $\{u_{td}\}$ such that

$$\|\{u_{td}\}\|_\infty \leq \frac{|u_{DC} - \bar{x}|}{g_{up}} \quad (40)$$

while, (39) is satisfied by choosing an input dynamic exciting sequence $\{u_{td}\}$ such that

$$\|\{u_{td}\}\|_\infty \leq \frac{|u_{DC} - \underline{x}|}{g_{up}} \quad (41)$$

Otherwise, when no a priori information on the ℓ_1-norm of the linear systems is available, inequalities (38) and (39) can be indirectly satisfied varying the amplitude of the sequence $\{u_{td}\}$ by trial and error until inequalities (26) and (27) are met by the output sequence $\{w_t\}$, i.e., until the measured output sequence $\{y_t\}$ satisfies the following inequalities $\forall t$:

$$(y_t - \Delta\eta_t) > \bar{w} \text{ or } (y_t + \Delta\eta_t) < \underline{w}, \text{ when } n \text{ is odd} \quad (42)$$

or

$$\mathrm{sign}(\gamma_n)(y_t - \mathrm{sign}(\gamma_n)\Delta\eta_t) > \frac{1+\mathrm{sign}(\gamma_n)}{2}\bar{w} + - \frac{1-\mathrm{sign}(\gamma_n)}{2}\underline{w}, \text{ when } n \text{ is even} \quad (43)$$

5. EVALUATION OF BOUNDS ON THE UNMEASURABLE INNER SIGNAL.

Given the polynomial nonlinearity as characterized in Section 3 and a feasible sequence of measured outputs $\{y_t\}$ as characterized in Section 4, in this section it is shown how upper and lower bounds on the unmeasurable inner signal x_t are evaluated. Such bounds, together with the input sequence u_t will be used to bound the parameters of the linear dynamic block in Section 6. Combining equations (5), (6) and (7) one obtains

$$\left| y_t - \sum_{k=1}^{n} \gamma_k x_t^k \right| \leq \Delta\eta_t, \quad t = 1, 2, \ldots, N \tag{44}$$

Given the output measurement y_t, its uncertainty bounds $\Delta\eta_t$ and the feasible parameter set $\mathcal{D}_\gamma$, upper and lower bounds on the unmeasurable inner signal x_t are defined as

$$x_t^{max} = \max \left\{ x_t \in R : y_t - \sum_{k=1}^{n} \gamma_k x_t^k \leq \Delta\eta_t, \right. \\ \left. y_t - \sum_{k=1}^{n} \gamma_k x_t^k \geq -\Delta\eta_t, \text{ for some } \gamma \in \mathcal{D}_\gamma \right\} \tag{45}$$

$$x_t^{min} = \min \left\{ x_t \in R : y_t - \sum_{k=1}^{n} \gamma_k x_t^k \leq \Delta\eta_t, \right. \\ \left. y_t - \sum_{k=1}^{n} \gamma_k x_t^k \geq -\Delta\eta_t, \text{ for some } \gamma \in \mathcal{D}_\gamma \right\} \tag{46}$$

Equations (45) and (46) show nonlinear relations in the unknown x_t and the uncertain γ. The computation of x_t^{max} and x_t^{min} can be carried out through a one dimensional searching procedure. It is noticed that for a given $x_t \in R$, in order to solve each one of the above equation one must solve a set of $2M+2$ linear inequalities. It is also seen that the search can be started from the unique real root of one nominal polynomial obtained, e.g., setting $\gamma = \gamma^c$ and $\Delta\eta = 0$; next, right side and left side of that root are explored to find x_t^{max} and x_t^{min} respectively. If the following quantities are defined

$$x_t^c = \frac{x_t^{min} + x_t^{max}}{2} \tag{47}$$

$$\Delta x_t = \frac{x_t^{max} - x_t^{min}}{2} \tag{48}$$

a compact description of x_t in terms of x_t^c and δx_t is as follows

$$x_t = x_t^c + \delta x_t \tag{49}$$

$$|\, \delta x_t \,| \leq \Delta x_t. \tag{50}$$

6. BOUNDING THE PARAMETERS OF THE LINEAR DYNAMIC MODEL

In this section bounds on the parameters of the linear dynamic block are evaluated. The identification of the linear block can be formulated in the frame of output error models, i.e., in terms of the known input sequence $\{u_t\}$ and the uncertain inner sequence $\{x_t\}$ as shown in Figure 3. Combining equation (2), (3), (4) and (49) one gets

$$x_t^c = -\sum_{i=1}^{na} (x_{t-i}^c - \delta x_{t-i}) a_i + \sum_{j=0}^{nb} u_{t-j} b_j + \delta x_t. \tag{51}$$

The feasible parameter region for the linear system is defined as

$$\mathcal{D}_\theta = \{\theta \in R^p : A(q^{-1})[x_t^c - \delta x_t] = B(q^{-1}) u_t; \\ g = 1; |\, \delta x_t \,| \leq \Delta x_t; t = 1, \ldots, N\}. \tag{52}$$

Due to serial dependence between x_t samples at different time, exact parameter bounds for model (51) are no longer linear (Veres and Norton, 1991). Thus, in this paper, a polytopic outer approximation $\mathcal{D}'_\theta$ of the exact *FPR* $\mathcal{D}_\theta$, i.e. $\mathcal{D}'_\theta \supset \mathcal{D}_\theta$, will be presented, together with an orthotope-outer bounding set $\mathcal{B}_\theta$ of $\mathcal{D}'_\theta$, which provides parameters uncertainties intervals.

The dynamic model (51) with bounds on the inner signal uncertainty δx_t (50) fits in the framework of the bounded output error model outlined by Clement and Gentil (1988) and in the more general framework of the bounded-errors-in-variables model (see, e.g. Cerone, 1993; Veres and Norton, 1991). Indeed, the feasible parameter region $\mathcal{D}'_\theta$ can be described by

$$(\phi_t - \Delta\phi_t)^{\mathrm{T}} \theta \leq y_t + \Delta\eta_t \tag{53}$$

$$(\phi_t + \Delta\phi_t)^{\mathrm{T}} \theta \geq y_t - \Delta\eta_t \tag{54}$$

where

$$\phi_t^{\mathrm{T}} = \left[-x_{t-1}^c \ldots - x_{t-na}^c \; u_t \; u_{t-1} \ldots u_{t-nb} \right] \tag{55}$$

$$\Delta\phi_t^{\mathrm{T}} = [\Delta x_{t-1} \mathrm{sgn}(a_1) \; \ldots \; \Delta x_{t-na} \mathrm{sgn}(a_{na}) \\ 0 \; 0 \; \ldots \; 0] \tag{56}$$

A further significant reduction of $\mathcal{D}'_\theta$ is obtained adding the constraint about the steady-state gain. As a matter of fact, equation (10), which can be written in the following form

$$[1 \; \ldots \; 1 \; -1 \; -1 \; \ldots \; -1]\, \theta = -1 \tag{57}$$

forces the feasible parameter region to belong to the hyperplane described by (57).

The orthotope-outer bounding set $\mathcal{B}_\theta$ is defined as

$$\mathcal{B}_\theta = \{\theta \in R^p : \theta_j = \theta_j^c + \delta\theta_j, \\ |\, \delta\theta_j \,| \leq \Delta\theta_j / 2, j = 1, \ldots, p\}, \tag{58}$$

where

$$\theta_j^c = \frac{\theta_j^{min} + \theta_j^{max}}{2}, \tag{59}$$

$$\Delta\theta_j = |\, \theta_j^{max} - \theta_j^{min} \,|, \tag{60}$$

and

$$\theta_j^{min} = \min_{\theta \in \mathcal{D}'_\theta} \theta_j, \quad \theta_j^{max} = \max_{\theta \in \mathcal{D}'_\theta} \theta_j. \tag{61}$$

Parameter vectors γ^c and θ^c are Chebishev centers in the ℓ_∞ norm of $\mathcal{D}_\gamma$ and $\mathcal{D}'_\theta$ respectively and are commonly referred to as central estimates.

7. CONCLUSIONS

In this paper the identification of SISO Wiener models has been considered when the nonlinear block can be modeled by a polynomial, with finite and known order, and when the output measurements are corrupted by unknown but bounded noise. First, using steady-state input-output data, parameters of the nonlinear block have been tightly bounded. Next in order to estimate the parameter of the linear block, the evaluation of the inner unmeasurable signal has been considered under the hypothesis of non-invertibility of the nonlinear block. Conditions under which the inner signal can be consistently estimated by input-output data have been established and, on the basis of such conditions, the design of a suitable input sequence has been outlined. Then, through a dynamic experiment, upper and lower bounds on the inner signal have been computed. Finally, such bounds, together with the input sequence, have been used for bounding the parameters of the linear block.

8. ACKNOWLEDGMENTS

This research was partly supported by the italian Ministry of Universities and Research in Science and Technology (MURST), under the plan "Robustness techniques for control of uncertain systems".

9. REFERENCES

Bai, E.W. (2002). A blind approach to the Hammerstein - Wiener model identification. *Automatica* **38**, 967–979.

Billings, S.A. (1980). Identification of nonlinear systems — a survey. *IEE Proc. Part D* **127**(6), 272–285.

Billings, S.A. and S.Y. Fakhouri (1977). Identification of nonlinear systems using Wiener model. *Electron. Lett.* **17**, 502–504.

Cerone, V. (1993). Feasible parameter set for linear models with bounded errors in all variable. *Automatica* **29**(6), 1551–1555.

Cerone, V. M. Milanese and D. Regruto (2002). Parameters bounds for Wiener models. *Internal Report DAI0215.*

Clement, T. and S. Gentil (1988). Recursive membership set estimation for armax models: an output-error approach. In: *Proc. 12th IMACS Congress on Scientific Computation.* pp. 484–486.

Haber, R. and H. Unbehauen (1990). Structure identification of nonlinear dynamic systems - a survey on input/uotput approaches. *Automatica* **26**(4), 651–677.

Kalafatis, A.D. L. Wang and W.R. Cluett (1997). Identification of Wiener-type nonlinear systems in a noisy enviroment. *Int. J. Control* **66**(6), 923–941.

Krzyżak, A. (1993). Identification of nonlinear block-oriented systems by the recursive kernel estimate. *Int. J. Franklin Inst.* **330**(3), 605–627.

Lang, Z.Q. (1993). Controller design oriented model identification method for Hammerstein system. *Automatica* **29**(3), 767–771.

Milanese, M. and A. Vicino (1991). Optimal estimation theory for dynamic sistems with set membership uncertainty: an overview. *Automatica* **27(6)**, 997–1009.

Milanese, M. Norton, J. Piet-Lahanier, H. and Walter, E. Eds.) (1996). *Bounding approaches to system identification.* Plenum Press. New York.

Norton(Ed.), J.P. (1994). Special issue on bounded-error estimation. *Int. J. of Adaptive Control & Signal Processing.*

Norton(Ed.), J.P. (1995). Special issue on bounded-error estimation. *Int. J. of Adaptive Control & Signal Processing.*

Stoica, P. and T. Söderström (1982). Instrumental-variable methods for identification of hammerstein systems. *Int. J. Control* **35**(3), 459–476.

Sun, L. W. Liu and A. Sano (1999). Identification of a dynamical system with input nonlinearity. *IEE Proc. Part D* **146**(1), 41–51.

Veres, S.M. and J.P. Norton (1991). Parameter-bounding algorithms for linear errors in variables models. In: *Proc. of IFAC/IFORS Symposium on Identification and System Parameter Estimation.* pp. 1038–1043.

Walter, E. and H. Piet-Lahanier (1990). Estimation of parameter bounds from bounded-error data: a survey. *Mathematics and Computers in simulation* **32**, 449–468.

Wigren, T. (1998). Output error convergence of adaptive filters with compensation for output nonlinearities. *IEEE Trans. Automatic. Control* **43**(7), 975–978.

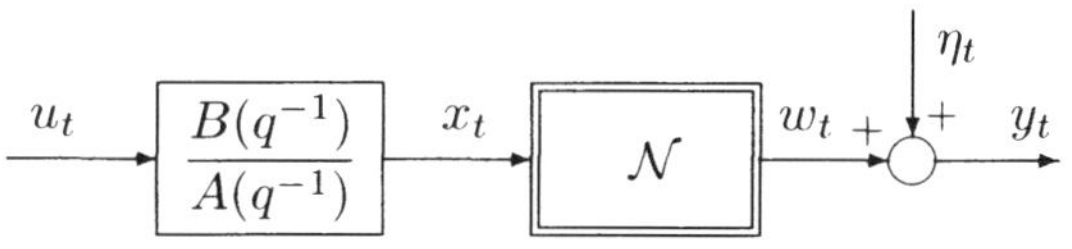

Fig. 1. Single-input single-output Wiener model.

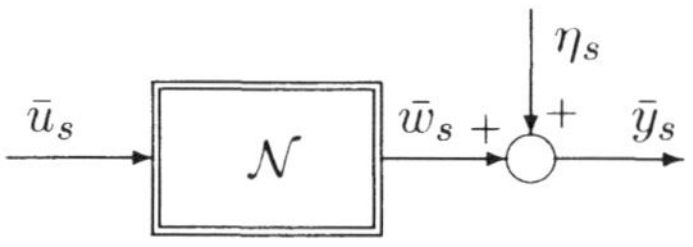

Fig. 2. Steady-state behaviour of the Wiener model when $g = 1$.

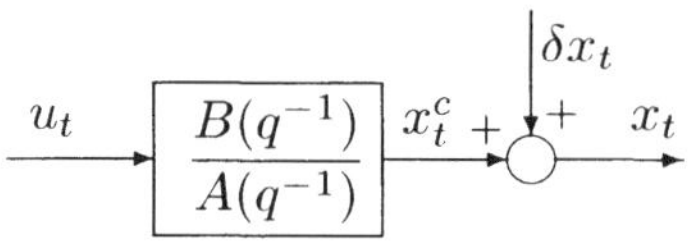

Fig. 3. Output error set-up for bounding the parameters of the linear system.

IFAC Publications
www.elsevier.com/locate/ifac

CONSTRUCTING THE STATE OF RANDOM PROCESSES WITH FEEDBACK

Alessandro Chiuso and Giorgio Picci *

* *Department of Information Engineering, University of Padova, Padova, Italy; Email:* {chiuso,picci}@dei.unipd.it

Abstract: Subspace identification, is based on **state space construction**, i.e. stochastic realization theory. However state space construction in the presence of feedback is still an open problem. In this paper we provide a geometric construction based on a *oblique predictor space* which yields a bona-fide (stationary) state space in the presence of feedback, if the open loop transfer functions of the system satisfy certain stability conditions. *Copyright © 2003 IFAC*

Keywords: Stochastic realization theory with feedback, Subspace identification, Identification with feedback

1. INTRODUCTION

Stochastic realization theory lies at the grounds of subspace identification methods, which have shown superior performance especially for multivariable state-space model-building, and have been intensively investigated in the past fifteen years (Larimore, 1983; Overschee and Moor, 1993; Overschee and Moor, 1994; Lindquist and Picci, 1996; Verhaegen, 1994). The basic steps of subspace algorithms are geometric operations on certain vector spaces generated by observed input-output time series. These operations can be interpreted as "sample versions" of certain abstract geometric operations of stochastic realization theory (Lindquist and Picci, 1996; Picci, 1996; Picci, 1997). In fact, stochastic realization theory provides a natural theoretical background for subspace identification of time-series (no inputs) and also in the presence of inputs but assuming absence of feedback. The main step of any subspace algorithm is the geometric construction of the state space of the system, say by projection of the future of the output process onto its past (in the time series case). In spite of a rather massive effort to apply subspace methods to closed-loop systems (A. C. Van der Klauw and den Bosch, 1997; Verhaegen, 1993; Overschee and Moor, 1997; Chou and Verhaegen, 1999), how to accomplish this basic construction in the presence of feedback, is still an open problem. Some basic conceptual issues underlying state space construction remain unclear.

In this paper we shall discuss procedures for constructing the state space and state space models for a stationary process $\mathbf{y}$, assuming it is described by a finite-dimensional stationary model of the form

$$\begin{cases} \mathbf{x}(t+1) = A\mathbf{x}(t) + B\mathbf{u}(t) + G\mathbf{w}(t) \\ \mathbf{y}(t) = C\mathbf{x}(t) + D\mathbf{u}(t) + J\mathbf{w}(t) \end{cases} \quad (1)$$

"driven" by an exogenous observable stationary input signal $\mathbf{u}$. Here $\mathbf{w}$ is a normalized white noise process. We shall allow feedback from $\mathbf{y}$ to $\mathbf{u}$.

[1] This work has been supported in part through the national project *Identification and adaptive control of industrial systems* funded by the Italian ministry for higher education (MIUR). Part of this work has been also supported by the TMR-European Research Network *System Identification* (ERB FMRX CT98 0206).

We will especially be interested in coordinate-free (i.e. "geometric") procedures by which one could abstractly construct state space models of the form (1), starting from certain Hilbert spaces of random variables generated by the "data" of the problem, namely the processes $\mathbf{y}$ and $\mathbf{u}$. We shall also characterize some structural properties of state space models of this kind, like minimality, absence of feedback etc. Due to space limitations the proofs will be skipped. Some can be found in the paper (Chiuso and Picci, 2002) and others will appear elsewhere.

2. STOCHASTIC LINEAR MODELS WITH FEEDBACK

The state space system (1) can equivalently be described in the (symbolical) I/O form:

$$\mathbf{y}(t) = F(z)\mathbf{u}(t) + G(z)\mathbf{w}(t) \qquad (2)$$

where $F(z)$ and $G(z)$ are rational transfer matrices. For instance one could define $F(z)\mathbf{u}(t)$ to be the (causal) Wiener filter

$$\mathrm{E}\,[\mathbf{y}(t)|\mathbf{u}(s),\ s \leq t]$$

and $G(z)$ to be a minimal outer spectral factor of the difference (or error) process

$$\mathbf{d}(t) = \mathbf{y}(t) - F(z)\mathbf{u}(t).$$

From this construction one sees that input-output models are only half of the story as we can always write also an I/O model for $\mathbf{u}$ of the form

$$\mathbf{u}(t) = H(z)\mathbf{y}(t) + K(z)\mathbf{v}(t) \qquad (3)$$

where $\mathbf{v}$ is another normalized white noise. In fact, it is known that there is a whole equivalence class of joint I/O models (2), (3), which represent the joint process $(\mathbf{y},\,\mathbf{u})$ up to second order statistics (Gevers and Anderson, 1981). One can restrict this class of models by imposing that $\mathbf{v}$ and $\mathbf{w}$ are *uncorrelated.* In this case we shall call the model (2), (3), a *feedback representation* of the joint process $(\mathbf{y},\,\mathbf{u})$. See Fig. 1. Now since

$$\mathbf{u} = \frac{H(z)}{1-H(z)F(z)}G(z)\mathbf{w} + \frac{1}{1-H(z)F(z)}K(z)\mathbf{v}$$

we see that $\mathbf{u}$ and $\mathbf{w}$ are **correlated** unless $H(z) = 0$.

Definition 1. (Granger). There is *no feedback* from $\mathbf{y}$ to $\mathbf{u}$ if the future of $\mathbf{u}$ is conditionally uncorrelated with the past of $\mathbf{y}$, given the past of $\mathbf{u}$. Equivalently, $\forall\ t \in \mathbb{Z}$,

$$\mathrm{E}\{\mathbf{y}(t) \mid \mathbf{u}(s),\ s \in \mathbb{Z}\} = \mathrm{E}\{\mathbf{y}(t) \mid \mathbf{u}(s),\ s \leq t\},$$

It is well-known that, generically (no cancellations in the loop), absence of feedback is equivalent to $H(z) \equiv 0$.

2.1 Basic Notations

Let $\mathcal{H}$ be a Hilbert space of real zero-mean random variables with inner product

$$\langle \mathbf{x},\,\mathbf{z}\rangle := \mathrm{E}\,\{\mathbf{xz}\}$$

and let $\mathcal{A}, \mathcal{B}$ be subspaces of $\mathcal{H}$ s.t. $\mathcal{A} \cap \mathcal{B} = \{0\}$, so that

$$\mathrm{E}\,[\,\mathbf{z} \mid \mathcal{A} + \mathcal{B}\,] = \mathbf{z}_{\mathcal{A}} + \mathbf{z}_{\mathcal{B}}$$

The two terms on the right are the **oblique projections** of $\mathbf{z}$ onto $\mathcal{A}$ along $\mathcal{B}$ and conversely, and are denoted

$$\mathrm{E}_{\,\|\mathcal{B}}\,[\mathbf{z} \mid \mathcal{A}] := \mathbf{z}_{\mathcal{A}} \quad , \quad \mathrm{E}_{\,\|\mathcal{A}}\,[\mathbf{z} \mid \mathcal{B}] := \mathbf{z}_{\mathcal{B}}.$$

• **Data subspaces:** for $-\infty < t < +\infty$ define the linear subspaces of second order random variables

$$\mathcal{U}_t^- := \overline{\text{span}}\,\{\mathbf{u}_k(s);\ k = 1,\ldots,p,\ s < t\}$$

$$\mathcal{Y}_t^- := \overline{\text{span}}\,\{\mathbf{y}_k(s);\ k = 1,\ldots,m,\ s < t\}$$

$$\mathcal{P}_t := \mathcal{U}_t^- \vee \mathcal{Y}_t^-$$

and let $\mathcal{P}_t$ be the *joint past space* of the input and output processes at time t. The *Present* of $\mathbf{u}$ at time t is denoted is denoted $\mathcal{U}_t :=$ span $\{\mathbf{u}_k(t);\ k = 1,\ldots,p,\ \}$ similarly we introduce $\mathcal{Y}_t$, etc. The spaces generated by the *whole time history* of $\mathbf{u}$ and $\mathbf{y}$ are instead denoted $\mathcal{U}$, $\mathcal{Y}$, etc..

The *shift operator* by t time units, induced by the (stationary) joint process $(\mathbf{y},\,\mathbf{u})$ is denoted σ^t. It maps $\mathbf{u}(k)$ and $\mathbf{y}(h)$ *isometrically* into $\mathbf{u}(t+k)$, $\mathbf{y}(t+h)$ respectively. Let $\mathcal{X} \subset \mathcal{H}$. The family of *stationary translates* of $\mathcal{X}$

$$\mathcal{X}_t := \sigma^t\mathcal{X}, \quad t \in \mathbb{Z}.$$

has a *past* and *future* spaces at time t defined as

$$\mathcal{X}_t^- := \vee_{s<t}\mathcal{X}_s, \quad \mathcal{X}_t^+ := \vee_{s \geq t}\mathcal{X}_s$$

At time $t = 0$ we shall drop the subscript, i.e. $\mathcal{X}_0 \equiv \mathcal{X}$.

• **The Generating white noise.** Let $\mathcal{S} = \mathcal{P}^- \vee \mathcal{X}^-$. This is an invariant subspace for the adjoint $\sigma^t)^* = \sigma^{-t}$. For short any such subspace will also be called an *incoming subspace.*

Definition 2. The incoming family $\{\mathcal{S}_t\}$ is (forward) *purely-non-deterministic* (p.n.d.) if

$$\bigcap_{t<0} \mathcal{S}_t = \{0\} \qquad (4)$$

In particular we shall assume in the following that both $\{\mathcal{Y}_t^-\}$ and $\{\mathcal{U}_t^-\}$ are p.n.d.. The *(conditional) wandering subspace* $\mathcal{W}_t$ of $\{\mathcal{S}_t\}$ *given* $\mathcal{U}_t$ is

$$\mathcal{W}_t := \mathcal{S}_{t+1} \ominus (\mathcal{S}_t + \mathcal{U}_t)\,. \qquad (5)$$

Lemma 1. If the incoming family $\{\mathcal{S}_t\}$ is p.n.d. the wandering subspaces $\mathcal{W}_t = \sigma^t\mathcal{W}$, $t \in \mathbb{Z}$, are pairwise orthogonal, i.e. $\mathcal{W}_t \perp \mathcal{W}_s, \forall t \neq s$ and we have the decomposition

$$\mathcal{S}_{t+1} = (\mathcal{S}_t + \mathcal{U}_t) \oplus \mathcal{W}_t$$

If $\mathcal{S}$ is of finite multiplicity m, then $\mathcal{W}$ is finite-dimensional of dimension m and hence admits an orthonormal basis, say $\mathbf{w}(0)$ so that

$$\mathcal{W}_t^- = \overline{\text{span}}\{\mathbf{w}_k(s);\ k = 1, \ldots, m,\ s < t\}$$

The stationary (normalized) white noise process $\mathbf{w}(t) = \sigma^t \mathbf{w}(0)$ is called the *generating process* of $\mathcal{S}$.

3. THE LOCAL (OBLIQUE) MARKOVIAN SPLITTING PROPERTY

Definition 3. The subspace $\mathcal{X}_t$ is *Locally Oblique Markovian Splitting*, if

$$E\left[\mathcal{X}_{t+1} \vee \mathcal{Y}_t \,|\, \mathcal{X}_{t+1}^- \vee \mathcal{U}_t \vee \mathcal{P}_t^-\right] = \\ = E[\mathcal{X}_{t+1} \vee \mathcal{Y}_t \,|\, \mathcal{X}_t + \mathcal{U}_t]. \quad (6)$$

which is equivalent to the *conditional orthogonality* property:

$$(\mathcal{X}_{t+1} \vee \mathcal{Y}_t) \perp (\mathcal{X}_t^- \vee \mathcal{P}_t^-) \mid (\mathcal{X}_t + \mathcal{U}_t) \quad (7)$$

If $\mathcal{U}_t \cap \mathcal{X}_t = \{0\}$ then (7) implies

$$E_{\|\mathcal{U}_t}[\,\mathcal{X}_{t+1} \,|\, \mathcal{X}_t^- \vee \mathcal{U}_t^-\,] = E_{\|\mathcal{U}_t}[\,\mathcal{X}_{t+1} \,|\, \mathcal{X}_t\,].$$

which is a Markov property, *conditional also on* $\mathcal{U}_t$. An oblique Markovian splitting subspace is *causal* if $\mathcal{X}_t \subseteq \mathcal{P}_t^-$ and *purely non deterministic* (p.n.d.) if the family $\{\mathcal{X}_t^-\}$ has the p.n.d. property (4).

3.1 Coordinate-Free State Space Equations

The local Oblique Markovian splitting property is precisely what is needed for the space $\mathcal{X}$ to qualify as a state space for a stochastic model described by the standard state equations (1).

Theorem 1. Let $\mathcal{X}_t$ be a p.n.d. Local oblique Markovian splitting subspace for $(\mathcal{Y}, \mathcal{U})$; then

$$\mathcal{X}_{t+1} \subseteq (\mathcal{X}_t + \mathcal{U}_t) \oplus \mathcal{W}_t \quad (8)$$
$$\mathcal{Y}_t \subseteq (\mathcal{X}_t + \mathcal{U}_t) \oplus \mathcal{W}_t \quad (9)$$

conversely, if these inclusions hold $\mathcal{X}_t$ is a locally oblique Markovian splitting subspace.

Assume $\mathcal{X}_t$ is finite dimensional. To get state equations just choose a basis $\mathbf{x}(t)$ in $\mathcal{X}_t$. Then there exist matrices A, B, C, D, G, J s.t.

$$\begin{cases} \mathbf{x}(t+1) = A\mathbf{x}(t) + B\mathbf{u}(t) + G\mathbf{w}(t) \\ \mathbf{y}(t) = C\mathbf{x}(t) + D\mathbf{u}(t) + J\mathbf{w}(t) \end{cases}$$

As we have seen, the local (oblique) Markovian splitting property is equivalent to existence of representation by state equations. The question then is how do we *construct* local Markovian splitting subspaces from the external measurable signals $(\mathbf{y},\ \mathbf{u})$?

To this purpose one needs to introduce stronger conditions: intuitively the subspace $\mathcal{X}_t$ should split (in an appropriate sense) the joint "full" past and the full future of the output process. "Appropriate sense" means that everything should hold *conditionally* given the input process $\mathbf{u}$. (This amounts to considering the future $\mathbf{u}$'s "known").

4. THE GLOBAL (OBLIQUE) MARKOVIAN SPLITTING PROPERTY

Define the *Extended Future Input Space* of an oblique Markovian splitting subspace $\mathcal{X}$,

$$\mathcal{F}_t^+ := \mathcal{U}_t^+ \vee \mathcal{W}_t^+.$$

which is the future space spanned by the observable and unobservable inputs in the model (1) corresponding to $\mathcal{X}$.

Definition 4. A subspace $\mathcal{X}$ is *globally oblique Markovian splitting* for $(\mathcal{Y}, \mathcal{U})$, if

$$E\left[\mathcal{X}_{t+1}^+ \vee \mathcal{Y}_t^+ \mid \mathcal{X}_{t+1}^- \vee \mathcal{P}_t^- \vee \mathcal{F}_t^+\right] = \\ E[\mathcal{X}_{t+1}^+ \vee \mathcal{Y}_t^+ \,|\, \mathcal{X}_t + \mathcal{F}_t^+] \quad . \quad (10)$$

which can also be written in form of conditional orthogonality as

$$(\mathcal{X}_{t+1}^+ \vee \mathcal{Y}_t^+) \perp (\mathcal{X}_t^- \vee \mathcal{P}_t^-) \mid (\mathcal{X}_t + \mathcal{F}_t^+) \quad (11)$$

Note that (11) implies the conditional splitting condition of past and future data given the extended future inputs

$$\mathcal{Y}_t^+ \perp \mathcal{P}_t^- \mid [\,\mathcal{X}_t \vee \mathcal{F}_t^+\,]$$

as well as the conditional Markovianness condition

$$\mathcal{X}_t^+ \perp \mathcal{X}_t^- \mid [\,\mathcal{X}_t \vee \mathcal{F}_t^+\,].$$

Note also that, provided the "global causality condition"

$$\mathcal{X}_t \cap \mathcal{F}_t^+ = \{0\} \quad (12)$$

holds, the global oblique splitting property can be expressed using oblique projections i.e.

$$E_{\|\mathcal{F}_t^+}\left[\mathcal{X}_{t+1}^+ \vee \mathcal{Y}_t^+ \,|\, \mathcal{X}_t \vee \mathcal{P}_t^-\right] = \\ E_{\|\mathcal{F}_t^+}\left[\mathcal{X}_{t+1}^+ \vee \mathcal{Y}_t^+ \,|\, \mathcal{X}_t\right]. \quad (13)$$

Unfortunately, there may be situations in which condition (12) does not hold. As we shall see, this is so in the presence of unstable poles of F and/or H.

4.1 Local Vs Global

In general the *global* oblique Markovian splitting is *weaker* than the "more natural" condition

$$E\left[\mathcal{X}_{t+1}^{+} \vee \mathcal{Y}_t^{+} \mid \mathcal{X}_{t+1}^{-} \vee \mathcal{P}_t^{-} \vee \mathcal{U}_t^{+}\right] = E[\mathcal{X}_{t+1}^{+} \vee \mathcal{Y}_t^{+} \mid \mathcal{X}_t \vee \mathcal{U}_t^{+}]. \quad (14)$$

which does not involve the future of the unmeasurable noise process. However, it can be shown that if there is feedback, this condition leads in general to a larger state space than (10).

Theorem 2. If condition (12) holds the concepts of Local and Global oblique Markovian splitting coincide, i.e. $\mathcal{X}$ is a Local oblique Markovian splitting subspace if and only if it is also global oblique Markovoan splitting.

5. THE OBLIQUE PREDICTOR SPACE

By assumption $\mathcal{P}_t^{-}$ is p.n.d.. Let $\mathcal{E}_t$ be the conditional wandering subspace generating $\mathcal{P}_t^{-}$, i.e.

$$\mathcal{E}_t := \mathcal{P}_{t+1}^{-} \ominus (\mathcal{P}_t^{-} + \mathcal{U}_t) \quad \Rightarrow$$
$$\mathcal{P}_{t+1}^{-} = (\mathcal{P}_t^{-} + \mathcal{U}_t) \oplus \mathcal{E}_t \quad (15)$$

Clearly the (forward) **innovation process**

$$\mathbf{e}(t) := \mathbf{y}(t) - E\left[\mathbf{y}(t) \mid \mathcal{P}_t^{-} \vee \mathcal{U}_t\right]$$

is a white noise process such that $\mathbf{e}(t)$ a basis for $\mathcal{E}_t$. Hence $\mathbf{e}$ is the generating process of the family $\{\mathcal{P}_t^{-}\}$. Consider the extended future space

$$\mathcal{G}_t^{+} := \mathcal{U}_t^{+} \vee \mathcal{E}_t^{+}$$

Theorem 3. Assume that

$$\mathcal{P}_t^{-} \cap \mathcal{G}_t^{+} = \{0\}. \quad (16)$$

Then the **oblique predictor space**

$$\mathcal{X}_t^{+/-} := E_{\|\mathcal{G}_t^{+}}\left[\mathcal{Y}_t^{+} \mid \mathcal{P}_t^{-}\right] \quad (17)$$

is a causal oblique Markovian splitting subspace (both in the local and in the global sense).

Note that $\mathcal{P}_t^{-} \cap \mathcal{G}_t^{+} = \{0\}$ implies $\mathcal{X}_t^{+/-} \cap \mathcal{G}_t^{+} = \{0\}$. Hence, whenever (16) holds, this theorem provides a recipe for constructing a state space for the process $\mathbf{y}$ which works in the presence of feedback.

5.1 Acausality and Feedback

Due to feedback $(\mathbf{y}, \mathbf{u})$ may well be stationary even if $F(z)$ and or $H(z)$ are not stable. Hence, in general, when there is feedback $\mathbf{y}$ *will not depend causally on the input signals* $\mathbf{u}$, $\mathbf{w}$, in the sense that inclusions of the type $\mathcal{Y}_t^{-} \subset \mathcal{U}_t^{-} \vee \mathcal{W}_t^{-}$ need not necessarily be true. By looking at the I/O model (2) one sees that causality holds as long as $F(z)$ and $G(z)$ are *stable*, or, equivalently, $|\lambda(A)| < 1$ in the corresponding (minimal) state-space model.

To form stationary processes, unstable modes have to be integrated *backwards*. Hence, when there is feedback, past outputs may be influenced by future inputs and future noise, which means that the model is *not causal*.

Consider the minimal state space model (1) and bring A to the block-diagonal form

$$A = \begin{pmatrix} A_{-} & 0 & 0 \\ 0 & A_0 & 0 \\ 0 & 0 & A_{+} \end{pmatrix} \quad (18)$$

where $|\lambda(A_{-})| < 1$, $|\lambda(A_0)| = 1$, $|\lambda(A_{+})| > 1$. Correspondingly let,

- $\mathcal{X}_{-}$: the "stable manifold"
- $\mathcal{X}_{+}$: the "unstable manifold"
- $\mathcal{X}_0$: the "central manifold" of the state space.

The symbols $\mathbf{x}_{-}$, $\mathbf{x}_{+}$ and $\mathbf{x}_0$ will denote bases in these invariant subspaces. Similar meaning have the symbols B_{-}, B_{+}, B_0, G_{-}, G_{+} and G_0. We have

$$\begin{cases} \mathbf{x}_{-}(t+1) = A_{-}\mathbf{x}_{-}(t) + B_{-}\mathbf{u}(t) + G_{-}\mathbf{w}(t) \\ \mathbf{x}_0(t+1) = A_0\mathbf{x}_0(t) + B_0\mathbf{u}(t) + G_0\mathbf{w}(t) \\ \mathbf{x}_{-}(t+1) = A_{+}\mathbf{x}_{+}(t) + B_{+}\mathbf{u}(t) + G_{+}\mathbf{w}(t) \end{cases}$$

Clearly the first subsystem is causal and $\mathbf{x}_{-}(t) \in \mathcal{U}_t^{-} \vee \mathcal{W}_t^{-}$.

The second subsystem is neither causal nor anticausal : $\mathbf{x}_0(t) \in \mathcal{U}_t^{-} \vee \mathcal{W}_t^{-}$ and also $\mathbf{x}_0(t) \in \mathcal{U}_t^{+} \vee \mathcal{W}_t^{+}$.

The third subsystem is anticausal: $\mathbf{x}_{+}(t) \in \mathcal{U}_t^{+} \vee \mathcal{W}_t^{+}$.

Note that in the case $|\lambda(A)| < 1$ we have a feedback interconnection with a stable forward loop transfer function $F(z)$.

Proposition 1. If the joint spectrum of $\mathbf{u}$ and $\mathbf{w}$ is coercive and the poles of $F(z)$ lie strictly inside the unit circle then the state space of the model (2) (a Markovian oblique splitting subspace) satisfies condition (12). In particular, if the same holds for the I/O innovations model, then the oblique predictor $\mathcal{X}_t^{+/-}$ is a bona-fide state space for the process $\mathbf{y}$.

In the rest of the paper we shall need condition (12) and therefore assume that the hypothesis of proposition 1 are satisfied. We postpone a discussion of the general case to future work.

6. OBSERVABILITY, CONSTRUCTIBILITY AND MINIMALITY

Definition 5. An oblique Markovian splitting subspace $\mathcal{X}$ is *minimal* if it does not contain properly other oblique Markovian splitting subspaces.

Define the (adjoint of the) *Observability operator*

$$\mathbb{O}^* : \mathcal{Y}^+ \to \mathcal{X} \quad , \quad \mathbb{O}^* \lambda := E_{\|\mathcal{F}^+}[\lambda \mid \mathcal{X}], \quad \lambda \in \mathcal{Y}^+$$

The *observable subspace of* $\mathcal{X}_t$ *(given* $\mathcal{F}_t^+$*)* is

$$\mathcal{X}_t^o := \overline{\text{Range } \mathbb{O}^*} = E_{\|\mathcal{F}_t^+}\left[\mathcal{Y}_t^+ \mid \mathcal{X}_t\right]$$

Then $\mathcal{X}_t$ is *observable* (given $\mathcal{F}_t^+$) if $\mathcal{X}_t^o = \mathcal{X}_t$, i.e. $\mathbb{O}^*$ has dense range.

The *Constructibility operator* is defined dually, as

$$\mathbb{K} : \mathcal{X} \to \mathcal{P}^- \quad , \quad \mathbb{K}\xi := E_{\|\mathcal{F}^+}\left[\xi \mid \mathcal{P}^-\right], \quad \xi \in \mathcal{X}$$

$\mathbb{K}$ measures the predictability of $\mathcal{X}$ based on the joint past $\mathcal{P}^-$, given the future inputs $\mathcal{F}^+$. The "constructible part" $\mathcal{X}_t^c$ is

$$\mathcal{X}_t^c = \mathcal{X}_t \ominus \text{Ker } \mathbb{K}$$

so that $\mathcal{X}_t^c$ is the closure of the range of the adjoint constructibility operator. We shall say that $\mathcal{X}_t$ is *constructible* if $\mathcal{X}_t^c = \mathcal{X}_t$.

Theorem 4. An oblique Markovian splitting subspace $\mathcal{X}$ is minimal if and only if it is both observable and constructible.

7. SCATTERING REPRESENTATION

Let us define $\bar{\mathcal{S}} = \mathcal{Y}^+ \vee \mathcal{X}^+$ and consider the pair $(\mathcal{S}, \bar{\mathcal{S}})$ associated to a Markovian splitting subspace $\mathcal{X}$. We shall say that $(\mathcal{S}, \bar{\mathcal{S}})$ are the *incoming* and *outgoing* subspaces associated to $\mathcal{X}$. The following theorem generalizes a well-known characterization in classical stochastic realization theory.

Theorem 5. Let $\mathcal{H}$ be a Hilbert space of random variables with shift operator σ and let $\mathcal{X}$ be a subspace of $\mathcal{H}$ such that

$$\mathcal{H} = \mathcal{Y} \vee \mathcal{U} \vee \left(\bigvee_t \mathcal{X}_t\right).$$

Then $\mathcal{X}$ is an oblique Markovian splitting subspace, if and only if

$$\mathcal{X} = \bar{\mathcal{S}} \cap \mathcal{S}$$

for some pair of subspaces $\mathcal{S}$, $\bar{\mathcal{S}}$ such that the following properties hold

(1) The extended past and future property

$$\begin{cases} \mathcal{Y}^+ \subseteq \bar{\mathcal{S}} \\ \mathcal{P}^- \subseteq \mathcal{S} \end{cases} , \quad \mathcal{S} \cap \mathcal{F}^+ = \{0\}$$

(2) Shift-invariance

$$\begin{cases} \sigma \bar{\mathcal{S}} \subseteq \bar{\mathcal{S}} \\ \sigma^* \mathcal{S} \subseteq \mathcal{S} \end{cases}$$

(3) Oblique intersection at $\mathcal{X}$

$$\bar{\mathcal{S}} \perp \mathcal{S} \mid ((\bar{\mathcal{S}} \cap \mathcal{S}) + \mathcal{F}^+)$$

Conversely, given an oblique Markovian splitting subspace $\mathcal{X}$, there is a pair of subspace satisfying conditions 1), 2), 3), which can be constructed so that

$$\mathcal{S} = \mathcal{P}^- \vee \mathcal{X}^- \quad , \quad \mathcal{Y}^+ \vee \mathcal{X}^+ \subseteq \bar{\mathcal{S}} \subseteq \mathcal{Y}^+ \vee \mathcal{X}^+ \vee \mathcal{U}^+.$$

The minimal subspace $\bar{\mathcal{S}}$ satisfying 1), 2), and 3) (i.e. contained in any other $\bar{\mathcal{S}}$ satisfying 1), 2), and 3)), is given by

$$\bar{\mathcal{S}} = \mathcal{Y}^+ \vee \mathcal{X}^+$$

7.1 Relations with feedback-free stochastic realization theory

Lemma 2. Let $[\mathbf{y}^\top \ \mathbf{u}^\top]^\top$ be a stationary process, and assume that there is no feedback form $\mathbf{y}$ to $\mathbf{u}$. Then there exist joint Markovian splitting subspaces $\mathcal{X}_J \equiv (\mathcal{S}_J, \bar{\mathcal{S}}_J)$ such that

$$\mathcal{S}_J \perp \mathcal{U}^+ \mid \mathcal{U}^- \tag{19}$$

A joint Markovian splitting subspace $\mathcal{X}_J$ which satisfies this condition is called *feedback-free.*

Theorem 6. Let $\mathcal{X}_J := (\mathcal{S}_J, \bar{\mathcal{S}}_J)$ be a feedback-free Markovian splitting subspace for the joint stationary process $[\mathbf{y}^\top \ \mathbf{u}^\top]^\top$. Then $\mathcal{X}_J$ is an oblique Markovian splitting subspace.

Note that in general $\mathcal{X}_J$ is *not minimal.* In particular the joint state space $\mathcal{X}_J$ will not be conditionally observable given $\mathbf{u}$ as it also models the dynamics of $\mathbf{u}$. Se we need to reduce it.

A constructive geometric procedure to reduce the subspace and to obtain a minimal oblique Markovian splitting subpace can be given see (Chiuso and Picci, 2002).

8. CONCLUSIONS

We have presented some basic ideas for a theory of stochastic realization with feedback. The concept of oblique Markovian splitting subspace plays a central role in the construction of state space models in the presence of feedback. We have provided a formula by which one can in certain situations, construct such an oblique Markovian splitting subspace starting from the available input/output data of the system. We have also discussed and generalized the notions of observability

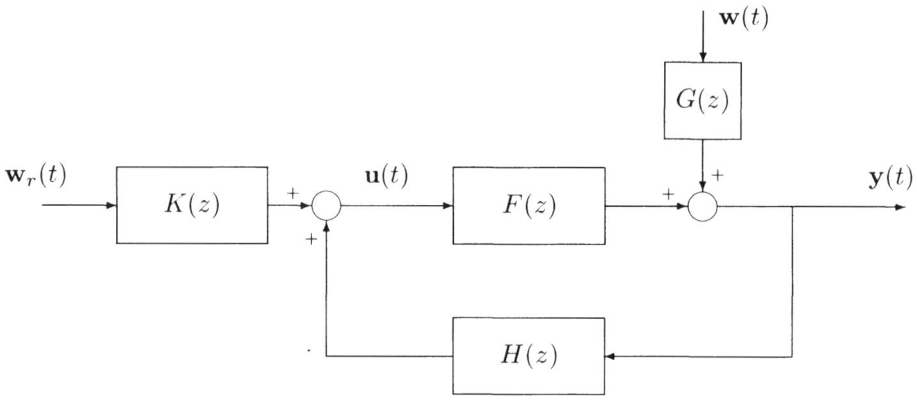

Fig. 1. Full feedback model of the processes $\mathbf{y}$ and $\mathbf{u}$.

and constructibility of classical stochastic realization theory and provided a criterion for minimality in terms of these two properties. Finally we have made contact with the classical state space construction for the joint input/output process. Many points remain open:

(1) How to deal with mixed causality structure (what happens when $F(z)$ is not strictly stable)
(2) How to deal with finite data: e.g. construction of a *transient* Kalman filter state with feedback

REFERENCES

A. C. Van der Klauw, M. Verhaegen and P. P. J. Van den Bosch (1997). State space identification of closed loop systems. In: *Proc. 30th IEEE Conference on Decision & Control.* Brighton, U.K.. pp. 1327–1332.

Chiuso, A. and G. Picci (2002). Geometry of oblique splitting, minimality and hankel operators. In: *Directions in Mathematical Systems Theory and Optimization.* pp. 85–124. Number 286. Springer Verlag. New York.

Chou, C. T. and M. Verhaegen (1999). Closed-loop identification using canonical correlation analysis. In: *Proceedings of the 1999 European Control Conference.* Karlsruhe, FRG. pp. paper F-162.

Gevers, M.R. and B.D.O. Anderson (1981). Representation of jointly stationary feedback free processes. *Intern. Journal of Control* **33**, 777–809.

Larimore, W.E. (1983). System identification, reduced-order filtering and modeling via canonical variate analysis. In: *Proc. American Control Conference.* pp. 445–451.

Lindquist, A. and G. Picci (1996). Canonical correlation analysis approximate covariance extension and identification of stationary time series. *Automatica* **32**, 709–733.

Overschee, P. Van and B. De Moor (1993). Subspace algorithms for the stochastic identification problem. *Automatica* **29**, 649–660.

Overschee, P. Van and B. De Moor (1994). N4SID: Subspace algorithms for the identification of combined deterministic– stochastic systems. *Automatica* **30**, 75–93.

Overschee, P. Van and B. De Moor (1997). Closed loop subspace systems identification. In: *Proc. 36th IEEE Conference on Decision & Control.* San Diego, Ca.. pp. 1848–1853.

Picci, G. (1996). Geometric methods in stochastic realization and system identification. **9**, 205–240.

Picci, G. (1997). Stochastic realization and system identification. In: *Statistical Methods in Control and Signal Processing* (T. Katayama and I. Sugimoto, Eds.). pp. 205–240. M. Dekker. N.Y.

Verhaegen, M. (1993). Application of a subspace model identification technique to identify lti systems operating in closed-loop. *Automatica* **29**, 1027–1040.

Verhaegen, M. (1994). Identification of the deterministic part of mimo state space models given in innovations form from input-output data. *Automatica* **30**, 61–74.

IFAC
Publications
www.elsevier.com/locate/ifac

CLOSED-LOOP SUBSPACE IDENTIFICATION WITH INNOVATION ESTIMATION

S. Joe Qin * and **Lennart Ljung** **

* *Department of Chemical Engineering*
The University of Texas at Austin
Austin, TX 78712, USA
e-mail: qin@che.utexas.edu.
** *Department of Electrical Engineering*
Linkoping University
Linkoping, Sweden

Abstract: Most subspace identification algorithms are not applicable to closed-loop identification because they require future input to be uncorrelated with past innovation. In this paper, we propose a new subspace identification method that remove this requirement by using a parsimonious model formulation with innovation estimation. A simulation example is included to show the effectiveness of the proposed method. *Copyright © 2003 IFAC*

Keywords: subspace identification; closed-loop identification; parsimonious models; innovation estimation

1. INTRODUCTION

Subspace identification methods (SIM) have gone through tremendous development over the last decade (Chou and Verhaegen, 1997; Moor *et al.*, 1988; Larimore, 1990; Moonen *et al.*, 1989; Overschee and Moor, 1993; Overschee and Moor, 1994; Verhaegen, 1991; Verhaegen, 1994; Viberg, 1994). Among these algorithms canonical variate analysis (CVA)(Larimore, 1983; Larimore, 1990), N4SID (Overschee and Moor, 1994), MOESP (Verhaegen and Dewilde, 1992), and IV-4SID (Viberg, 1995) are some of the representing algorithms. These SIM algorithms are now well explained in many papers and several books (Overschee and Moor, 1996), (Ljung, 1999). Typically, an SIM estimates the extended observability matrix with or without estimating the state sequence. Then the system matrices and the disturbance characteristics are estimated. The unifying theorem (Overschee and de Moor, 1995; Jansson and Wahlberg, 1998) formulates many of the SIM algorithms in a singular value decomposition framework, with differences in weighting matrices. The CVA and MOESP weighting matrices have been proven to be approximately optimal weighting (Gustafsson, 2002). In addition, statistical properties such as consistency have recently been explored (Bauer *et al.*, 1999; Deistler *et al.*, 1995; Heij and Scherrer, 1999; Jansson and Wahlberg, 1998).

Although the SIM algorithms are attractive because the state space form is very convenient for estimation, filtering, prediction of multivariable systems, severe drawbacks have been experienced. In general, the SIM estimates are not as accurate as the prediction error methods (PEM). Further, very few, if any, SIM methods are applicable to closed-loop identification, even though the data satisfy identifiability conditions for traditional methods such as PEMs.

In a companion paper (Qin and Ljung, 2003), we give the reasons why subspace identification approaches exhibit these drawbacks and propose parallel parsimonious SIM (PARSIM-P) for open-loop applications. With the analysis of existing subspace formulation using the linear regression formulation (Jansson and Wahlberg, 1996; Jansson and Wahlberg, 1998; Knudsen, 2001), we reveal that the typical SIM algorithms actually use non-parsimonious model formulation, with extra terms in the model that appear to be non-causal. These terms, although conveniently included for performing subspace projection, are the causes for inflated variance in the estimates and partially responsible for the loss of closed-loop identifiability.

Removing the non-causal terms in the SIM formulation makes the model parsimonious, but it does not make the SIM methods automatically applicable to closed-loop identification. As pointed out in (Qin and Ljung, 2003), the PARSIM-P algorithm still requires that there is no correlation between future input u_k and past innovation e_k, which is not the case for closed-loop data.

In this paper, we propose a new parsimonious method that removes this requirement, thus making it applicable to closed-loop identification. We propose to estimate the innovation process e_k using the intermediate results in SIMs. Then the estimated innovation sequence is used in the subsequent projections in the SIM procedure. This method will be referred to as PARSIM-E, which means that the innovation process e_k is estimated first. A simulation example is used to demonstrate the effectiveness of the proposed PARSIM-E method for closed-loop identification with comparison to the PARSIM-P and MOESP algorithms.

2. SUBSPACE MODEL

2.1 *Conventional Subspace Models*

We begin with an innovation model formulation,

$$x_{k+1} = Ax_k + Bu_k + Ke_k \tag{1a}$$

$$y_k = Cx_k + Du_k + e_k \tag{1b}$$

where $y_k \in R^{n_y}$, $x_k \in R^n$, $u_k \in R^{n_u}$, and $e_k \in R^{n_y}$ are the system output, state, input, and innovation, respectively. A,B,C,D and K are system matrices with appropriate dimensions.

An extended state space model can be formulated as

$$Y_f = \Gamma_f X_k + H_f U_f + G_f E_f \tag{2a}$$

$$Y_p = \Gamma_p X_{k-p} + H_p U_p + G_p E_p \tag{2b}$$

where the extended observability matrix

$$\Gamma_f = \begin{bmatrix} C \\ CA \\ \vdots \\ CA^{f-1} \end{bmatrix} \tag{3}$$

and the Toeplitz matrices are

$$H_f = \begin{bmatrix} D & 0 & \cdots & 0 \\ CB & D & \cdots & 0 \\ \vdots & \vdots & \ddots & \vdots \\ CA^{f-2}B & CA^{f-3}B & \cdots & D \end{bmatrix} \tag{4a}$$

$$G_f = \begin{bmatrix} I & 0 & \cdots & 0 \\ CK & I & \cdots & 0 \\ \vdots & \vdots & \ddots & \vdots \\ CA^{f-2}K & CA^{f-3}K & \cdots & I \end{bmatrix} \tag{4b}$$

The input and output data are arranged in the following Hankel form:

$$U_f = \begin{bmatrix} u_k & u_{k+1} & \cdots & u_{k+N-1} \\ u_{k+1} & u_{k+2} & \cdots & u_{k+N} \\ \vdots & \vdots & \ddots & \vdots \\ u_{k+f-1} & u_{k+f} & \cdots & u_{k+f+N-2} \end{bmatrix} \tag{5a}$$

$$\triangleq \left[u_f(k) \; u_f(k+1) \; \cdots \; u_f(k+N-1) \right] \tag{5b}$$

$$U_p = \begin{bmatrix} u_{k-p} & u_{k-p+1} & \cdots & u_{k-p+N-1} \\ u_{k-p+1} & u_{k-p+2} & \cdots & u_{k-p+N} \\ \vdots & \vdots & \ddots & \vdots \\ u_{k-1} & u_k & \cdots & u_{k+N-2} \end{bmatrix} \tag{5c}$$

$$\triangleq \left[u_p(k-p) \; u_p(k-p+1) \; \cdots \; u_p(k-p+N-1) \right] \tag{5d}$$

Denoting

$$L^1 = \begin{bmatrix} L^1_{11} & L^1_{12} & \cdots & L^1_{1p} \\ L^1_{21} & L^1_{22} & \cdots & L^1_{2p} \\ \vdots & & \ddots & \\ L^1_{f1} & L^1_{f1} & & L^1_{fp} \end{bmatrix} \triangleq \begin{bmatrix} L^1_1 \\ L^1_2 \\ \vdots \\ L^1_f \end{bmatrix} \tag{6a}$$

$$L^2 = \begin{bmatrix} L^2_{11} & L^2_{12} & \cdots & L^2_{1p} \\ L^2_{21} & L^2_{22} & \cdots & L^2_{2p} \\ \vdots & & \ddots & \\ L^2_{f1} & L^2_{f1} & & L^2_{fp} \end{bmatrix} \triangleq \begin{bmatrix} L^2_1 \\ L^2_2 \\ \vdots \\ L^2_f \end{bmatrix} \tag{6b}$$

$$L^3 = \begin{bmatrix} L^3_{11} & L^3_{12} & \cdots & L^3_{1f} \\ L^3_{21} & L^3_{22} & \cdots & L^3_{2f} \\ \vdots & & \ddots & \\ L^3_{f1} & L^3_{f1} & & L^3_{ff} \end{bmatrix} \triangleq \begin{bmatrix} L^3_1 \\ L^3_2 \\ \vdots \\ L^3_f \end{bmatrix} \tag{6c}$$

the above problem is equivalent to f separate subproblems:

$$\left[\hat{L}^1_i \; \hat{L}^2_i \; \hat{L}^3_i \right] = \arg\min\{J\} \tag{7}$$

where

$$J = \sum_{j=0}^{N-1} \left\| y(k+j+i-1) - \begin{bmatrix} L_i^1 & L_i^2 & L_i^3 \end{bmatrix} \begin{bmatrix} y_p(k-p+j) \\ u_p(k-p+j) \\ u_f(k+j) \end{bmatrix} \right\|^2$$
$$\text{for } i = 1, 2, \ldots, f \quad (8)$$

For the case of $i = 1$, for example, the problem implies that the following model is specified:

$$y(k) = \begin{bmatrix} L_1^1 & L_1^2 & L_1^3 \end{bmatrix} \begin{bmatrix} y_p(k-p) \\ u_p(k-p) \\ u_f(k) \end{bmatrix} + v(k)$$
$$= \begin{bmatrix} L_1^1 & L_1^2 \end{bmatrix} \begin{bmatrix} y_p(k-p) \\ u_p(k-p) \end{bmatrix} + L_{11}^3 u(k)$$
$$+ \sum_{j=2}^{f} L_{1j}^3 u(k+j-1) + v(k) \quad (9)$$

Note that the third term on the RHS of the above equation is non-causal and unnecessary. Therefore, the model format used in SIM during the projection step is non-causal. This would result in non-causal models in the projection step. Although the non-causal terms are ignored at the step to estimate B, D, all the model parameters estimate have inflated variance due to the fact that extra and unnecessary terms are included in the model, making the model non-parsimonious. For $i > 1$ the number of non-causal terms will reduce, but they are unnecessary as long as $i < f$.

To avoid these problems the SIM model must not include these non-causal terms. The PARSIM-P algorithms remove: these terms by enforcing triangular structure of the Toeplitz matrix H_f at every step of the SIM procedure. The approach are referred to as parsimonious subspace identification methods (PARSIM) as it uses parsimonious model formulation.

2.2 *Parsimonious Subspace Models*

The key idea in the proposed method is to exclude those non-causal terms of U_f. To accomplish this we partition the extended state space model row-wise as follows:

$$Y_f = \begin{bmatrix} Y_{f1} \\ Y_{f2} \\ \vdots \\ Y_{ff} \end{bmatrix}; \; Y_i \triangleq \begin{bmatrix} Y_{f1} \\ Y_{f2} \\ \vdots \\ Y_{fi} \end{bmatrix}; i = 1, 2, \ldots, f \quad (10)$$

Partition U_f and E_f in a similar way to define U_{fi}, U_i, E_{fi}, and E_i, respectively, for $i = 1, 2, \ldots, f$. Denote further

$$\Gamma_f = \begin{bmatrix} \Gamma_{f1} \\ \Gamma_{f2} \\ \vdots \\ \Gamma_{ff} \end{bmatrix} \quad (11a)$$
$$H_{fi} \triangleq \begin{bmatrix} CA^{i-2}B & \cdots & CB & D \end{bmatrix} \quad (11b)$$
$$\triangleq \begin{bmatrix} H_{i-1} & \cdots & H_1 & H_0 \end{bmatrix} \quad (11c)$$
$$G_{fi} \triangleq \begin{bmatrix} CA^{i-2}K & \cdots & CK & I \end{bmatrix} \quad (11d)$$
$$\triangleq \begin{bmatrix} G_{i-1} & \cdots & G_1 & G_0 \end{bmatrix} \quad (11e)$$
$$\forall i = 1, 2, \cdots, f$$

where H_i and G_i are the Markov parameters for the deterministic input and innovation sequence, respectively. We have the following partitioned equations:

$$Y_{fi} = \Gamma_{fi} X_k + H_{fi} U_i + G_{fi} E_i$$
$$\forall i = 1, 2, \cdots, f \quad (12)$$

Note that each of the above equation is guaranteed causal.

2.3 *Parallel Estimation of* Γ_{fi} *and* H_{fi}

By eliminating $e(k)$ in the innovation model through iteration, it is straightforward to derive the following relation (Knudsen, 2001),

$$X_k = L_z Z_p + A_K^p X_{k-p} \quad (13)$$

where

$$L_z \triangleq \begin{bmatrix} \Delta_p(A_K, K) & \Delta_p(A_K, B_K) \end{bmatrix} \quad (14a)$$
$$\Delta_p(A, B) \triangleq \begin{bmatrix} A^{p-1}B & \cdots & AB & B \end{bmatrix} \quad (14b)$$
$$A_K \triangleq A - KC \quad (14c)$$
$$B_K \triangleq B - KD \quad (14d)$$

Substituting this equation into Eq. 12, we obtain

$$Y_{fi} = \Gamma_{fi} L_z Z_p + \Gamma_{fi} A_K^p X_{k-p} + H_{fi} U_i + G_{fi} E_i$$
$$\forall i = 1, 2, \cdots, f \quad (15)$$

Since the second term in the RHS of Eq. 15 tends to zero as p tends to infinity, we have the following least squares estimates:

$$\begin{bmatrix} \hat{\Gamma}_{fi} L_z & \hat{H}_{fi} \end{bmatrix} = Y_{fi} \begin{bmatrix} Z_p \\ U_i \end{bmatrix}^+ \quad (16)$$
$$\forall i = 1, 2, \cdots, f$$

Qin and Ljung (Qin and Ljung, 2003) point out that the PARSIM-P algorithm requires that the input $u(k)$ and innovation sequence $e(k)$ are uncorrelated, i.e.,

$\frac{1}{N}E_iU_i^T \to 0$ as $N \to \infty$, to be unbiased. Because of this requirement, the PARSIM-P algorithm is biased for closed-loop identification. In the next section we propose a new PARSIM algorithm, PARSIM-E, that estimates the past innovation process first. The estimated innovation is treated as known data and the subsequent projections do not require future input to be uncorrelated with past innovation, hence the PARSIM-E method is applicable to closed-loop identification.

3. PARSIM WITH INNOVATION ESTIMATION

By ignoring the second term on the RHS of Eq. 15 and setting $i = 1$, we have

$$Y_{f1} = \begin{bmatrix} \Gamma_{f1}L_z \; H_{f1} \end{bmatrix} \begin{bmatrix} Z_p \\ U_1 \end{bmatrix} + E_1 \qquad (17)$$

Therefore, a least squares estimate of the innovation process is:

$$\hat{E}_1 = Y_{f1} - \begin{bmatrix} \hat{\Gamma}_{f1}L_z \; \hat{H}_{f1} \end{bmatrix} \begin{bmatrix} Z_p \\ U_1 \end{bmatrix} \qquad (18)$$

Now return to Eq. 15 for a general $i = 2, 3, \ldots, f$. Noticing that

$$E_i = \begin{bmatrix} E_{f1} \\ E_{f2} \\ \vdots \\ E_{fi} \end{bmatrix} = \begin{bmatrix} E_{i-1} \\ E_{fi} \end{bmatrix} \qquad (19)$$

and replacing E_{i-1} with $\hat{E}_{i-1}$, Eq. 15 becomes,

$$Y_{f1} = \begin{bmatrix} \Gamma_{fi}L_z \; H_{fi} \; G_{fi}^- \end{bmatrix} \begin{bmatrix} Z_p \\ U_i \\ \hat{E}_{i-1} \end{bmatrix} + E_{fi} \qquad (20)$$

where

$$G_{fi}^- = \begin{bmatrix} CA^{i-2}K \; CA^{i-3}K \; \ldots \; CK \end{bmatrix}. \qquad (21)$$

The least squares estimate

$$\begin{bmatrix} \hat{\Gamma}_{fi}L_z \; \hat{H}_{fi} \; \hat{G}_{fi}^- \end{bmatrix} = Y_{f1} \begin{bmatrix} Z_p \\ U_i \\ \hat{E}_{i-1} \end{bmatrix}^+ \qquad (22)$$

now does not require future input u_k to be uncorrelated with past innovation e_k. It only requires that future innovation to be independent of past input, which is always true for both open-loop and closed-loop data. The innovation data are calculated recursively using

$$\hat{E}_i = \begin{bmatrix} \hat{E}_{i-1} \\ \hat{E}_{fi} \end{bmatrix} \qquad (23)$$

With the least squares estimates from Eq. 22, the system matrices A, B, C, D, K can be estimated similarly to the procedures given in (Qin and Ljung, 2003).

4. SIMULATION RESULTS

We simulate the following process

$$y_k + ay_{k-1} = bu_{k-1} + e_k + ce_{k-1} \qquad (24)$$

with a feedback controller

$$u_k = -Ky_k + r_k \qquad (25)$$

where $a = -0.9$, $b = 1$, and $c = 0.9$. The standard deviation for e_k is one and that for r_k is two; both of the signals are Gaussian white noise. Open-loop experiments are simulated with $K = 0$ and closed-loop experiments with $K = 0.6$. In both cases 2000 data points are collected and 20 Monte-Carlo simulations are performed. Figure 1 shows the pole estimates from PARSIM-E, PARSIM-P and MOESP for open-loop and closed-loop data. There is no observed difference for open-loop identification, while the closed-loop identification results are very different. The PARSIM-E gives the best estimate without bias.

Figure 2 shows the box plots of the parameter estimates from 20 simulations. In the open-loop case, all methods estimate a equally well. PARSIM-E and PARSIM-P give much better estimates for b than MOESP, showing the benefit of parsimonious formulation. PARSIM-E and PARSIM-P give equally good estimates for c, while MOESP does not estimate the stochastic parameters. In the closed-loop case, the PARSIM-E algorithm gives unbiased estimates for a and b. Both PARSIM-P and MOESP fail on closed-loop identification, with MOESP giving the worst results.

To examine the frequency responses of the identified models, Figure 3 gives the Bode plots by averaging the 20 closed-loop experiments. It is clearly shown that MOESP and PARSIM-P method fail to identify the steady state gain, while the PARSIM-E method is unbiased in all frequencies.

5. CONCLUSIONS

The proposed new subspace identification method with parsimonious models and innovation estimation gives unbiased results for closed-loop identification. For open-loop data both PARSIM-E and PARSIM-P algorithms give superior results than the contentional subspace model formulation.

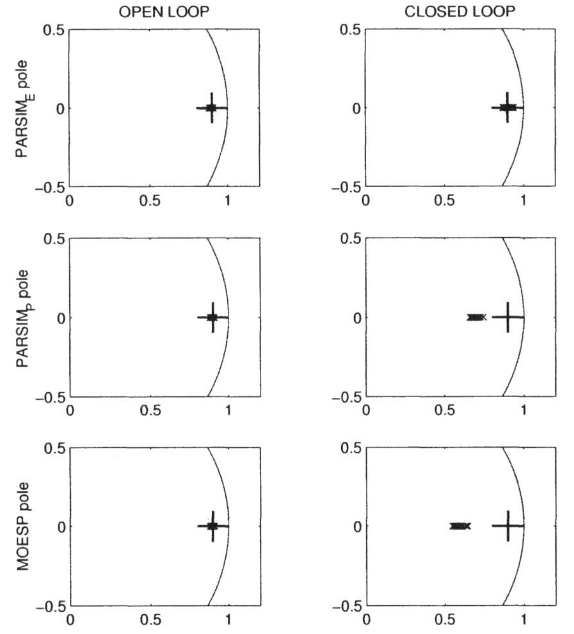

Fig. 1. Pole estimates for the simulation example.

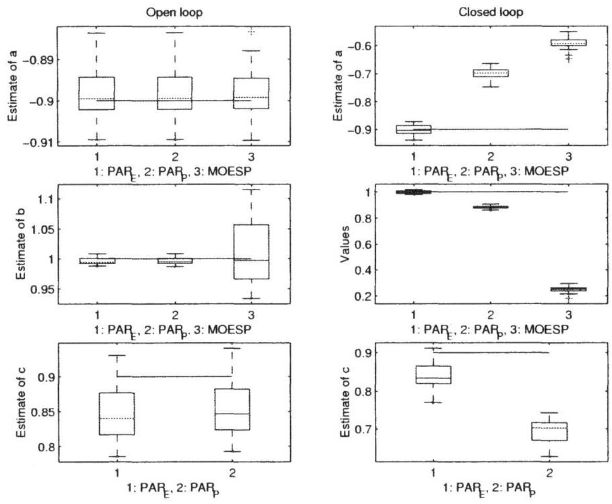

Fig. 2. Parameter estimates for the simulation example.

ACKNOWLEDGMENTS

Financial support from National Science Foundation under CTS-9985074 and a Faculty Research Assignment grant from University of Texas is gratefully acknowledged.

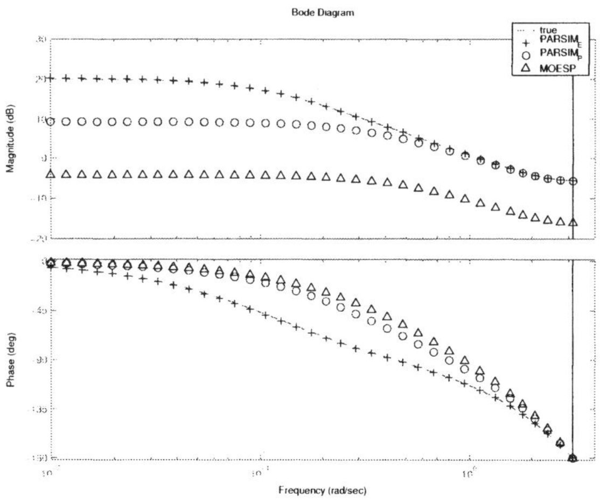

Fig. 3. Bode diagram for the closed-loop identification results

6. REFERENCES

Bauer, D., M. Deistler and W. Scherrer (1999). Consistency and asymptotic normality of some subspace algorithms for systems without observed inputs. Automatica **35**, 1243–1254.

Chou, C.T. and Michel Verhaegen (1997). Subspace algorithms for the identification of multivariable dynamic errors-in-variables models. *Automatica* **33**(10), 1857–1869.

Deistler, M., K. Peternell and W. Scherrer (1995). Consistency and relative efficiency of subspace methods. *Automatica* **31**, 1865–1875.

Gustafsson, Tony (2002). Subspace-based system identification: weighting and pre-filtering of instruments. *Automatica* **38**, 433–443.

Heij, C. and W. Scherrer (1999). Consistency of system identification by global total least squares. *Automatica* **35**, 993–1008.

Jansson, Magnus and Bo Wahlberg (1996). A linear regression approach to state-space subspace system. *Signal Processing* **52**, 103–129.

Jansson, Magnus and Bo Wahlberg (1998). On consistency of subspace methods for system identification. *Automatica* **34**(12), 1507–1519.

Knudsen, Torben (2001). Consistency analysis of subspace identification methods based on a linear regression approach. *Automatica* **37**, 81–89.

Larimore, Wallace. E. (1983). System identification, reduced-order filtering and modeling via canonical variate analysis. In: *Proceedings of the 1983 American Control Conference.* pp. 445–451.

Larimore, Wallace. E. (1990). Canonical variate analysis in identification, filtering and adaptive control. In: *Proceedings of the 29th Conference on Decision and Control.* pp. 596–604.

Ljung, L. (1999). *System Identification: Theory for the User.* Prentice-Hall, Inc.. Englewood Cliffs, New Jersey.

Moonen, M., B. DeMoor, L. Vandenberghe and J. Vandewalle (1989). On and off-line identification of linear state-space models. *International Journal of Control* **49**, 219–232.

Moor, B. De, J. Vandewalle, L. Vandenberghe and P. Van Mieghem (1988). A geometrical strategy for the identification of state space models of linear multivariable systems with singular value decomposition. In: *Proceedings of the 8th IFAC Symp. on Identification and System Parameter Estimation.* pp. 700–704.

Overschee, P. Van and B. De Moor (1993). Subspace algorithms for the stochastic identification problem. *Automatica* **29**, 649–660.

Overschee, P. Van and B. De Moor (1994). N4SID: Subspace algorithms for the identification of combined deterministic-stochastic systems. *Automatica* **30**(1), 75.

Overschee, Peter Van and Bart de Moor (1995). A unifying theorem for three subspace system identification algorithms. *Automatica* **31**(12), 1853–1864.

Overschee, Peter Van and Bart De Moor (1996). *Subspace Identification for Linear Systems.* Kluwer Academic Publishers.

Qin, S. J. and L. Ljung (2003). Parallel qr implementation of subspace identification with parsimonious models. In: *IFAC Symposium on System Identification.*

Verhaegen, M. (1991). A novel non-iterative mimo state space model identification techniques. In: *Proceedings of the 9th IFAC/IFORS Symp. on Identification and System Parameter Estimation.* pp. 1453–1458.

Verhaegen, M. and P. Dewilde (1992). Subspace model identification. part i: the output-error state-space model identification class of algorithms. *International Journal of Control* **56**, 1187–1210.

Verhaegen, Michel (1994). Identification of the deterministic part of MIMO state space models given in innovations form from input-output data. *Automatica* **30**(1), 61–74.

Viberg, M. (1994). Subspace methods in system identification. In: *Proceedings of the 10th IFAC Symp. on Identification and System Parameter Estimation.* pp. 1–12.

Viberg, Mats (1995). Subspace-based methods for the identification of linear time-invariant systems. *Automatica* **31**(12), 1835–1851.

IFAC
Publications
www.elsevier.com/locate/ifac

A FREQUENCY DOMAIN SUBSPACE ALGORITHM FOR MIXED CAUSAL, ANTI-CAUSAL LTI SYSTEMS

Rufus Fraanje, Michel Verhaegen, Vincent Verdult *,[1]
and Rik Pintelon **

* *Delft University of Technology, Faculty of Information Technology and Systems, Control Systems Laboratory, PO-Box 5031, 2600 GA Delft, The Netherlands, Phone: ++31 15 27 82087, Fax: ++31 15 27 86679, Email:* R.Fraanje@ITS.TUDelft.NL
** *Vrije Universiteit Brussel, dept. ELEC, Pleinlaan 2, 1050 Brussels, Belgium, Phone: ++32 2 629 2944, Fax: ++32 2 629 2850, Email:* Rik.Pintelon@vub.ac.be

Abstract: The paper extends the subspace identification method to estimate state-space models from frequency response function (FRF) samples, proposed by McKelvey *et al.* (1996) for mixed causal/anti-causal systems, and shows that other frequency domain subspace algorithms can be extended similarly. The method is demonstrated by simulation experiments.
Copyright © 2003 IFAC

Keywords: frequency domain identification, subspace method, descriptor system, Kronecker canonical form, state-space model

1. INTRODUCTION

Subspace identification methods are powerful methods in identifying linear multi variable systems. This is because these methods are based on numerically reliable algorithms as SVD and QR-decomposition and directly yield a state-space model. Advantages of a state-space model over a transfer function model are e.g. that a resonance mode of the system, which is observed at multiple outputs, is modeled only once and its numerical sensitivity to round-off errors is in general significantly smaller (see e.g. Gevers and Li (1993)). Further, most modern control methods are based on state-space models.

Originally, subspace identification methods were based on time-domain measured input/output data (Verhaegen, 1994; Overschee and Moor, 1996). Not much time later, subspace identification methods were proposed which are based on frequency-domain data: Fourier transformed input/output data (McKelvey, 1997) or FRF samples McKelvey *et al.* (1996); see Pintelon (2002) for new results on consistency and convergence. Using frequency domain data the number of samples may be significantly reduced, especially in case of systems with (many) widely separated resonance modes (e.g. stiff systems and systems with a high number of resonances like acoustical systems) and a non-uniform frequency grid can be exploited to accurately model the system at specific frequencies (see Pintelon and Schoukens (2001) for more details on system identification in the frequency domain).

Using frequency domain subspace identification methods systems with anti-causal/unstable modes can be identified too, which is an advantage over time-domain subspace algorithms. However, to study the causal and/or anti-causal behavior, additional post-processing has to be a applied to separate the causal and anti-causal modes. This separation is also necessary, when only a model of the causal part of the system has to be used for control/filter design, such as in the Causal Wiener filter, see e.g. (Fraanje *et al.*, 2001).

[1] The research of R. Fraanje is supported by the Knowledge Center 'Sound and Vibration UT-TNO', a joint initiative of TNO, Delft, The Netherlands and the University of Twente, Enschede, The Netherlands.

This paper proposes a frequency domain subspace method to identify a state-space model of the causal/stable part and the anti-causal/unstable part of a system *directly*. The method is obtained by adjusting *Algorithm 2* from McKelvey (1997) using techniques from Verhaegen (1996), who proposed a *time-domain* algorithm to identify mixed causal, anti-causal systems. We will base our method for mixed causal, anti-causal systems on FRF samples, however the same reasoning can be used to extend other subspace methods based on e.g. discrete Fourier transforms (DFT's) of input and output data as discussed by McKelvey (1997).

Problems with identification of anti-causal (or unstable) systems do arise e.g. in closed-loop identification of a stabilized unstable system, in discretized systems with fractional delay (Laakso *et al.*, 1996), in estimating the inverse of non-minimum phase systems (usually due to time delay) or direct estimation of a deconvolution filter. Furthermore, this problem arises in the frequency domain implementation of a method we proposed to estimate the Causal Wiener filter (Fraanje *et al.*, 2001).

The paper is organized as follows. Section 2 describes the problem of frequency domain subspace algorithm for mixed causal, anti-causal systems in more detail. Section 3 derives a solution of the problem based on FRF samples. Section 4 illustrates the method by simulation examples.

2. PROBLEM DESCRIPTION

Consider the following discrete time mixed causal, anti-causal state-space system

$$x^c(k+1) = A^c\, x^c(k) \;+ B^c\, u(k) \quad \text{(causal)} \tag{1}$$
$$x^{ac}(k-1) = A^{ac}x^{ac}(k) + B^{ac}u(k) \quad \text{(anti-causal)} \tag{2}$$
$$y(k) = C^c\, x^c(k) \;+ C^{ac}x^{ac}(k) + Du(k) \tag{3}$$

with $u(k) \in \mathbb{R}^m$, $y(k) \in \mathbb{R}^l$, $x^c(k) \in \mathbb{R}^{n_c}$ and $x^{ac}(k) \in \mathbb{R}^{n_{ac}}$ and A^c, B^c, C^c, A^{ac}, B^{ac}, C^{ac} and D of appropriate dimensions. Equation (1) models the causal and (2) the anti-causal dynamics. The order of the system is given by $n = n_c + n_{ac}$. Furthermore, the eigenvalues of A^c and A^{ac} are inside the unit-circle and we assume that the state-space description is minimal, i.e. there are no unobservable or uncontrollable modes. This class of systems is a special case of descriptor systems, described in the so-called Kronecker canonical form (Verhaegen, 1996). The transfer function of the system is given by

$$G(z) = \underbrace{\sum_{i=-\infty}^{1} C^{ac}A^{ac(1-i)}B^{ac}z^{-i}}_{anti-causal} + \underbrace{D + \sum_{i=1}^{\infty} C^c A^{c(i-1)}B^c z^{-i}}_{causal} \tag{4}$$

and clearly consists of a causal and an anti-causal part. The FRF of the system is given by

$$\begin{aligned} G(e^{j\omega}) = {} & D + C^c(e^{j\omega}I_{n_c} - A^c)^{-1}B^c + \\ & + C^{ac}(e^{-j\omega}I_{n_{ac}} - A^{ac})^{-1}B^{ac} \end{aligned} \tag{5}$$

The problem is to estimate the matrices A^c, B^c, C^c (up to a similarity transformation T_c), A^{ac}, B^{ac}, C^{ac} (up to a similarity transformation T_{ac}) and D using M noise corrupted estimates of the frequency response

$$G_k \;=\; G(e^{j\omega_k}) + n_k, \quad k = 1, 2, \cdots, M \tag{6}$$

at a given but arbitrary number of distinct frequencies ω_k (cf. McKelvey *et al.* (1996)).

Note, that if $\omega_k = \pi k/M$, $k = 1, \cdots, M$ (uniformly spaced frequencies) the impulse response coefficients g_i of $G(z) = \sum_{i=-\infty}^{\infty} g_i z^{-i}$ can be calculated by the two-sided inverse discrete Fourier transform

$$g_i = \frac{1}{2M} \sum_{k=1-M}^{M} G_k e^{j2\pi ik/2M}, \quad i = 1 - M, \cdots, M$$

with $G_{-k} = G_k^*$, $(k = 1, \cdots, M-1)$. Then, D can be set to $D = g_0$. A_T^c, B_T^c and C_T^c can be calculated from g_i $(i = 1, \cdots, M)$ by well known realization algorithms (e.g. Kung (1978)). Dually A_T^{ac}, B_T^{ac} and C_T^{ac} can be calculated from g_{-i} $(i = 1, \cdots, M)$ by these same realization algorithms. This is basically an extension of *Algorithm 1* of McKelvey *et al.* (1996) for mixed causal, anti-causal systems. The following Section derives an alternative algorithm, which can also be used for non-uniformly spaced frequencies, which is basically an extension of *Algorithm 2* of McKelvey *et al.* (1996).

3. DERIVATION OF THE ALGORITHM

First consider the DFT of (1)–(3), where we shifted equation (2) $i-1$ samples forward in time

$$e^{j\omega}X^c(\omega) = A^cX^c(\omega) + B^cU(\omega) \tag{7}$$
$$e^{-j\omega}X^{ac,i}(\omega) = A^{ac}X^{ac,i}(\omega) + B^{ac}e^{j(i-1)\omega}U(\omega) \tag{8}$$
$$\begin{aligned} Y(\omega) = {} & C^cX^c(\omega) + C^{ac}e^{-j(i-1)\omega}X^{ac,i}(\omega) + \\ & + DU(\omega) \end{aligned} \tag{9}$$

where $X^c(\omega)$, $X^{ac,i}(\omega)$, $U(\omega)$ and $Y(\omega)$ denoted the DFT of $x^c(k)$, $x^{ac}(k+i-1)$, $u(k)$ and $y(k)$ respectively and let $i > n$.

As in (McKelvey *et al.*, 1996), let $X_i^c(\omega)$ the resulting state transform when $U(\omega) = e_r$, with e_r the unit-vector with 1 on the rth position, $X_r^{ac,i}(\omega)$ is defined similarly. By defining the compound state matrix

$$X_C^c(\omega) = [X_1^c(\omega)\, X_2^c(\omega) \cdots X_m^c(\omega)]$$

and $X_C^{ac,i}$ similarly, the transfer function can be implicitly described as

$$G(e^{j\omega}) = C^c X_C^c(\omega) + C^{ac} e^{-j(i-1)\omega} X_C^{ac,i}(\omega) + D \tag{10}$$

with

$$e^{j\omega} X_C^c(\omega) = A^c X_C^c(\omega) + B^c \tag{11}$$

$$e^{-j\omega} X_C^{ac,i}(\omega) = A^{ac} X_C^{ac,i}(\omega) + B^{ac} e^{j(i-1)\omega} \tag{12}$$

By iterative substituting the state-equations we obtain the relation

$$\begin{bmatrix} G(e^{j\omega}) \\ e^{j\omega} G(e^{j\omega}) \\ \vdots \\ e^{j(i-2)\omega} G(e^{j\omega}) \\ e^{j(i-1)\omega} G(e^{j\omega}) \end{bmatrix} = \mathcal{O}_i \begin{bmatrix} X_C^c(\omega) \\ X_C^{ac,i}(\omega) \end{bmatrix} + \Gamma_i \begin{bmatrix} I_m \\ e^{j\omega} I_m \\ \vdots \\ e^{j(i-2)\omega} I_m \\ e^{j(i-1)\omega} I_m \end{bmatrix} \tag{13}$$

with

$$\mathcal{O}_i = \begin{bmatrix} C^c & C^{ac} A^{ac(i-1)} \\ C^c A^c & C^{ac} A^{ac(i-2)} \\ \vdots & \vdots \\ C^c A^{c(i-2)} & C^{ac} A^{ac} \\ C^c A^{c(i-1)} & C^{ac} \end{bmatrix} \tag{14}$$

$$= [\, \mathcal{O}_i^c \; \mathcal{O}_i^{ac} \,] \tag{15}$$

and the following Toeplitz matrix filled with impulse response coefficients

$$\Gamma_i = \begin{bmatrix} D & C^{ac} B^{ac} & \cdots & C^{ac} A^{ac(i-2)} B^{ac} \\ C^c B^c & D & \ddots & \vdots \\ \vdots & \ddots & \ddots & C^{ac} B^{ac} \\ C^c A^{c(i-2)} B^c & \cdots & C^c B^c & D \end{bmatrix}$$

By repeating (13) for all ω_k ($k = 1, \cdots, M$) and using the frequency response estimates G_k, the range space of $\mathcal{O}_i$ (and thus also the order n of the system) can be determined by a QR factorization and an SVD as explained by McKelvey *et al.* (1996). If the covariance function $E[n_k n_s^H] = R_k \delta_{ks}$ of the noise n_k is known, a weighting can be performed to compensate for n_k.

Let U_n be such that its columns span the range space of $\mathcal{O}_i$. Then, we look for an invertible $n \times n$ transformation matrix P such that

$$U_n P = [\, \mathcal{O}_i^c | \mathcal{O}_i^{ac} \,] \left[\begin{array}{c|c} T_c & 0 \\ \hline 0 & T_{ac} \end{array} \right] \tag{16}$$

This problem to separate U_n in a causal part fully determined by the pair (A^c, C^c) and an anti-causal part fully determined by the pair (A^{ac}, C^{ac}) is exactly the problem which also appears in mixed causal, anti-causal subspace identification using time-domain data, and is solved e.g. by Verhaegen (1996).

Verhaegen (1996) calculates the matrix P and n_c, n_{ac}. Then from the first l rows in $U_n' = U_n P$, C_T^c and C_T^{ac} can be picked up:

$$C_T^c = U_n'(1:l, 1:n_c), \tag{17}$$

$$C_T^{ac} = U_n'(1:l, n_c+1:n_c+n_{ac}) \tag{18}$$

and A_T^c, A_T^{ac} can be calculated by solving

$$U_n'(1:(i-1)l, 1:n_c) A_T^c = U_n'(l+1:il, 1:n_c) \tag{19}$$

and

$$\begin{aligned} U_n'(1:(i-1)l, n_c+1:n_c+n_{ac}) A_T^{ac} &= \\ = U_n'(l+1:il, n_c+1:n_c+n_{ac}) \end{aligned} \tag{20}$$

If A_T^c, C_T^c, A_T^{ac}, C_T^{ac} are calculated, B_T^c, B_T^{ac} and D can be calculated by solving a least squares problem using the samples (6), because B_T^c, B_T^{ac} and D appear linearly in $G(e^{j\omega})$, cf. (5). Again a weighting using the noise covariance matrices R_k can be used to compensate for n_k (McKelvey *et al.*, 1996).

Let us summarize the method in the following *Algorithm 2 (Causal/Anti-causal)*.

Algorithm 2 (C/AC):

(1) Given: Samples G_k of the frequency response, and the covariance R_k at frequency ω_k for $k = 1, \cdots, M$.
(2) Calculate the estimate U_n of the extended observability matrix $\mathcal{O}_i$ as in *Algorithm 2* in (McKelvey *et al.*, 1996).
(3) Calculate the matrix P, which separates the causal and anti-causal part in U_n, and the orders of the causal and anti-causal part n_c and n_{ac} respectively as in Section 3.3 and 3.5 in (Verhaegen, 1996).
(4) Calculate $U_n' = U_n P$ and select C_T^c and C_T^{ac} according to (17) and (18) respectively. Further solve A_T^c and A_T^{ac} from (19) and (20) respectively.
(5) Solve B_T^c, B_T^{ac} and D from:

$$(B_T^c, B_T^{ac}, D) = \arg\min \sum_{k=1}^{M} \|R_k^{-1/2} \cdot \Big(G_k - D - C_T^c (e^{j\omega_k} I_{n_c} - A_T^c)^{-1} B_T^c + -C_T^{ac} (e^{-j\omega_k} I_{n_{ac}} - A_T^{ac})^{-1} B_T^{ac} \Big) \|_F^2$$

We remark, that solving the least squares problem for B_T^c, B_T^{ac} and D might be ill-conditioned, especially if the system has poles close to the unit circle (McKelvey *et al.*, 1996). Regularization with a small $\epsilon > 0$ parameter can improve the conditioning of the least squares problem, at the expense of a small bias, see e.g. Golub and van Loan (1996).

4. SIMULATION EXAMPLES

We will consider two simulation examples: the identification of an academic system and of an acoustic system, both with a stable and an unstable resonance modes.

4.1 *Academic example*

The discrete time academic system has a stable mode at 100Hz and an unstable mode at 300Hz, the sampling rate is $F_s = 1000$Hz, and its transfer function is given by

$$G(z) = \frac{(z - 0.9e^{-j2\pi 0.15})(z - 0.9e^{+j2\pi 0.15})}{(z - 0.95e^{-j2\pi 0.1})(z - 0.95e^{+j2\pi 0.1})} \cdot \frac{(z - 0.5e^{-j2\pi 0.4})(z - 0.5e^{+j2\pi 0.4})}{(z - 1.1e^{-j2\pi 0.3})(z - 1.1e^{+j2\pi 0.3})}$$

The system was excited with a Schroeder multi sine, with frequencies ranging from .5Hz to 500Hz in steps of $\Delta F = .5Hz$. The number of samples in one block to estimate the frequency response was chosen to be 2000 ($= F_s/\Delta F$) such that leakage due to Fourier transforming a finite block of samples is prevented (Pintelon and Schoukens, 2001). The measured output was corrupted with unit variance Gaussian white noise $v(k)$ filtered by $H(z)$

$$y(k) = G(z)u(k) + H(z)v(k)$$

with $H(z)$ given by

$$H(z) = \frac{0.3z^2}{(z - 0.95e^{-j2\pi 0.1})(z - 0.95e^{+j2\pi 0.1})} \cdot \frac{z^2}{(z - 0.91e^{-j2\pi 0.3})(z - 0.91e^{+j2\pi 0.3})}$$

The output data was generated by splitting G in a stable/causal part and an unstable/anti-causal part. The latter was simulated by filtering anti-causally, to prevent the output from exploding.

Based on 100 blocks, three methods were used to calculate the state-space model of G:

PO-MOESP: Time-domain PO-MOESP (Verhaegen, 1996);
A2: *Algorithm 2 (C/AC)* without knowledge of R_k;
A2wi: *Algorithm 2(C/AC)* with estimated R_k.

The i parameter of (13) was chosen to be $i = 10$. To make a fair comparison, for PO-MOESP the 100 blocks each of 2000 samples were used to average the output to reduce. In A2 and A2wi, the 100 blocks we used to average the estimated frequency responses, and in A2wi also to calculated the variance R_k ($k = 1, \cdots, 2000$).

Each experiment of 100 blocks was repeated 1000 times. Figure 4.1 shows the real frequency response of G, the average of the frequency response error made by PO-MOESP, A2 and A2wi. From this Figure, we clearly see that the stable as well as the unstable mode

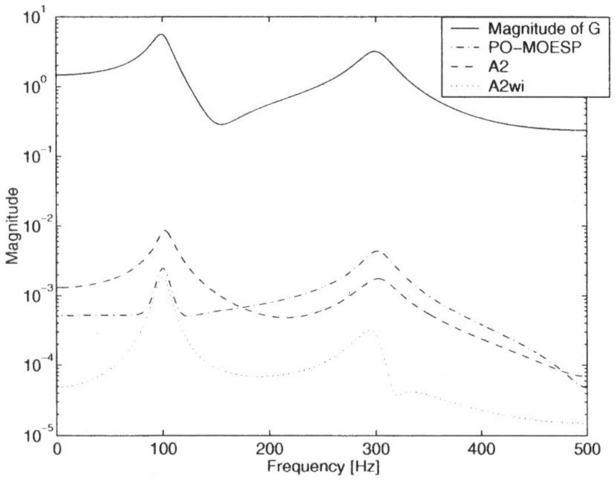

Fig. 1. Magnitude of G, and the frequency response estimation errors obtained by using 100 blocks of 2000 samples, which were averaged over 1000 experiments.

are accurately modeled by all three methods. Further, we infer that by taking the covariance information R_k of the noise into account, the model is more accurately estimated, as was also concluded in (McKelvey *et al.*, 1996) for the causal method. Finally, we infer that on the average using the causal/anti-causal frequency domain with covariance information, a more accurate model was estimated then by using the causal/anti-causal time domain PO-MOESP method.

4.2 *Acoustic system*

The acoustic system to be identified is a transfer function in an acoustical duct, which contains unstable modes due to the inversion of delays between actuators and sensors which are not collocated, for more details we refer to Fraanje *et al.* (2001). The sampling rate is again 1000Hz and the frequency response of the real system G and the noise coloring H is shown in Figure 2. The unstable poles of the system are given in Table 1.

Table 1. Frequency and magnitude of unstable poles.

Frequency	Magnitude	Frequency	Magnitude
0	1.64	± 171	1.31
0	3.65	± 278	1.41
0	8.21	± 388	1.46
± 32.1	1.01	500	1.59
± 33.5	1.08	500	2.44
± 76.0	1.04	500	12.6
± 91.7	1.00		

The excitation signal was chosen the same as in the previous example, a Schroeder multi sine with frequencies ranging from 0.5Hz to 500Hz with steps of 0.5Hz. Each of the 100 blocks consists of 2000 samples. The results of the three methods, PO-MOESP, A2 and A2wi, were averaged over 200 experiments and the i parameter of (13) was chosen to be $i = 100$. We note, that the least squares problem to solve B_T^c, B_T^{ac} and D in A2 and A2wi was ill conditioned, due to poles close to the unit circle.

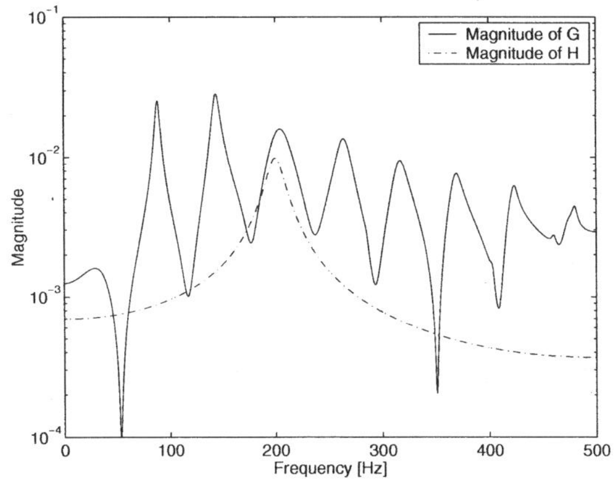

Fig. 2. Magnitude of the real system G and the noise model H.

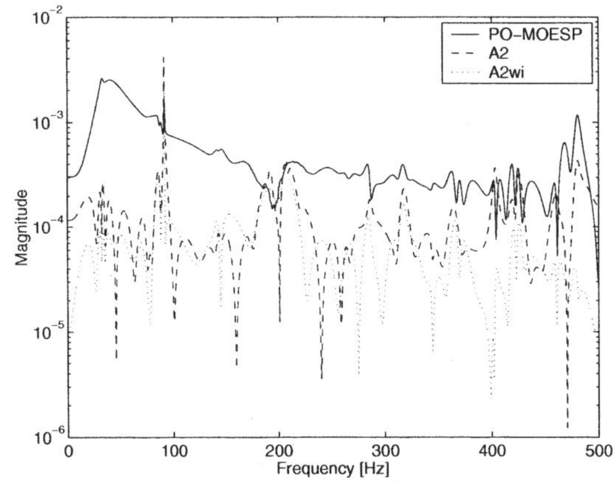

Fig. 3. Magnitude of the frequency response estimation errors obtained by the 44th order models identified by using 100 blocks of 2000 samples, which were averaged over 200 experiments.

Figure 3 shows the average frequency response estimation error when the order of the state-space model was chosen to be 44, which is the order of the real system G. We observe, that again A2wi gives the smallest estimation error, but the difference with A2 is not that large as in the previous academic example. The causal, anti-causal PO-MOESP method gives less accurate models, which is currently under study. Finally, Figure 4 gives the average estimation error, which is defined as (McKelvey *et al.*, 1996):

$$||\widehat{G} - G||_2 = \sqrt{\frac{1}{M}\sum_{k=1}^{M} |\widehat{G}(e^{j\omega_k}) - G(e^{j\omega_k})|^2}$$

for different model orders. From the Figure, we infer that for orders above 32 A2wi yields the best result, closely followed by A2. It is remarkable that for orders between 20 and 30, A2wi yields significantly less accurate results, whereas A2 gives reasonable good results for these orders.

Though, some questions remain on the precise interpretation of the simulation results, the simulation experiments show that the extension of frequency domain subspace identification methods for mixed causal, anti-causal systems was successful.

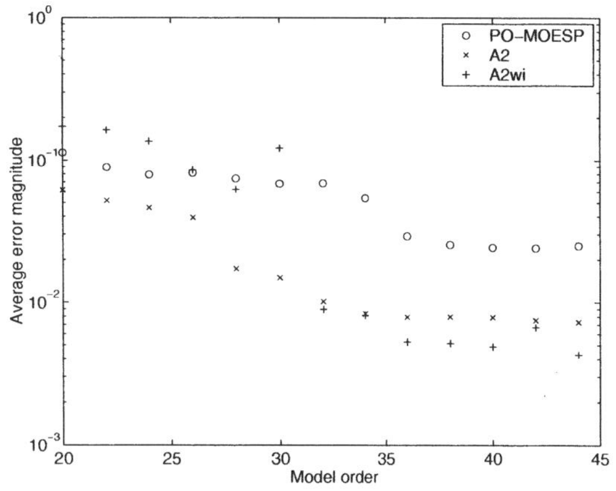

Fig. 4. Estimation error obtained by using 100 blocks of 2000 samples, averaged over 200 experiments.

5. CONCLUSIONS

It has been shown how subspace identification methods based on frequency domain data, can be adjusted to estimate a state-space model which models the causal and anti-causal part of the system separately. The crucial step in extending the frequency domain methods is to split up the extended observability matrix in a part due to causal modes and a part due to anti-causal modes. The two simulation experiments demonstrated that with the derived mixed causal, anti-causal subspace identification algorithm using FRF samples the causal and anti-causal modes of the systems could be accurately identified. We also observed, that including noise weighting to compensate for noise on the FRF samples, the obtained model error was better than using time-domain PO-MOESP for mixed causal, anti-causal systems.

6. REFERENCES

Fraanje, P. R., M. Verhaegen and N. J. Doelman (2001). Subspace identification for active noise control in a duct. In: *Proceedings of The 2001 International Congress and Exhibition on Noise Control Engineering*. The Hague, the Netherlands.

Gevers, M. and G. Li (1993). *Parametrizations in Control, Estimation and Filtering Problems : Accuracy Aspects*. Communication and Control Engineering Series. Springer Verlag. New York.

Golub, G. H. and C. F. van Loan (1996). *Matrix Computations*. The Johns Hopkins University Press.

Kung, S.K. (1978). A new low-order approximation algorithm via singular value decomposition. In: *Proceedings 12th Asilomar Conference on Circuits, Systems and Computers*. pp. 705–714.

Laakso, T., V. Välimäki, M. Karjalainen and U.K. Laine (1996). Splitting the unit delay – tools for fractional delay filter design. *IEEE Signal Processing Magazine* **13**(1), 30–60.

McKelvey, T. (1997). Frequency domain system identification with instrumental variable based subspace algorithm. In: *Proceedings of the 1997 ASME design engineering technical conferences*. Sacramento, CA.

McKelvey, T., H. Akçay and L. Ljung (1996). Subspace-based multivariable system identification from frequency response data. *IEEE Trans. on Automatic Control* **41**(7), 960–979.

Overschee, P. Van and B. De Moor (1996). *Subspace Identification for Linear Systems*. Kluwer.

Pintelon, R. (2002). Frequency-domain subspace system identification using non-parametric noise models. *Automatica* **38**, 1295–1311.

Pintelon, R. and J. Schoukens (2001). *System Identification – A Frequency Domain Approach*. IEEE Press.

Verhaegen, M. (1994). Identification of the deterministic part of MIMO state space models given in innovations form from input-output data. *Automatica* **30**(1), 61–74.

Verhaegen, M. (1996). A subspace model identification solution to the identification of mixed causal anti-causal lti systems. *SIAM Journal on Matrix Analysis* **17**(2), 332–347.

IFAC Publications
www.elsevier.com/locate/ifac

A STOCHASTIC REALIZATION IN A HILBERT SPACE BASED ON "LQ DECOMPOSITION" WITH APPLICATION TO SUBSPACE IDENTIFICATION

Hideyuki TANAKA * **Tohru KATAYAMA** *

* *Department of Applied Mathematics and Physics,*
Graduate School of Informatics, Kyoto University,
Yoshida-Hommachi, Sakyo, Kyoto, 606-8501, Japan.
{htanaka,katayama}@amp.i.kyoto-u.ac.jp

Abstract: In this paper, we develop a new stochastic realization algorithm using canonical correlation analysis, thereby deriving the forward innovation representation along the line of (Desai *et al.*, 1985) by means of "LQ decomposition" in a Hilbert space generated by a second-order stationary random process. As an application, we show that our abstract result is easily adapted to the case where a finite string of a time-series data is available to derive a stochastic subspace identification algorithm. *Copyright © 2003 IFAC*

Keywords: Stochastic realization, Subspace identification, Canonical Correlation Analysis, Balanced realization, Hilbert space, LQ decomposition

1. INTRODUCTION

Stochastic realization problem is to find a set of Markov models whose output covariance matrix matches a given covariance matrix of a stationary random process (Faurre, 1976). A novel method of stochastic realization has been developed by using the canonical correlation analysis (CCA) (Akaike, 1975), and a CCA-based balanced reduced order realization technique has been derived in (Desai *et al.*, 1985). Also, the subspace identification methods of state space models for time series have extensively been studied based on the CCA (Lindquist and Picci, 1996*a*), including nontrivial problems related to positivity of covariance matrices. The conditional CCA is employed to derive a stochastic realization algorithm in the presence of exogenous inputs in (Katayama and Picci, 1999).

It is well known that stochastic realization theory has played a basic role in deriving stochastic subspace identification algorithms (Overschee and Moor, 1993; Overschee and Moor, 1996), in which they have derived subspace algorithms to identify stochastic state space models from given output data without forming the covariance matrix.

The subspace identification methods have brought some problem, from the view point of the stochastic realization theory; because a nontrivial positivity issue arises when only a set of finite sequence of covariance matrices is given (Byrnes and Lindquist, 1997; Byrnes *et al.*, 2001).

Lindquist and Picci have analyzed state space identification algorithms in the light of geometric theory of stochastic realization (Lindquist and Picci, 1996*a*; Lindquist and Picci, 1996*b*). They discussed the state space modeling of time series, by separating three different cases: **(i)** an infinite complete covariance sequence is available, **(ii)** a finite complete covariance data is available, and **(iii)** a finite string of time-series data is available. Especially, they have defined a Hilbert space of observed infinite string of time sequence data for Cases **(i)** and **(ii)**.

In this paper, we re-visit the stochastic realization problem in a Hilbert space of a second-order stationary time-series, under the assumption that **(i)** the infinite data is available. We re-derive the forward innovation representation obtained in (Desai *et al.*, 1985) by means of the "LQ decomposition" in the Hilbert space. We briefly show that the present method can be adapted to **(iii)** the finite time-series data.

The paper is organized as follows. In section 2, we review the stochastic realization theory and state a problem setting. Section 3 gives an "LQ decomposition" in the Hilbert space generated by an infinite string of the time sequence data. In Section 4, we derive a stochastic realization algorithm based on the "LQ decomposition". A stochastic subspace identification method is suggested in Section 5. Section 6 concludes the paper.

2. PROBLEM STATEMENT

Consider a second-order stationary process $\{y(t), t = 0, \pm 1, \cdots\}$, where $y(t)$ is a p-dimensional nondeterministic process with mean zero and covariance matrix

$$\Lambda_k = \mathrm{E}\left(y(t+k)y(t)^T\right), \quad k = 0, \pm 1, \cdots \tag{1}$$

It is assumed that a set of covariance matrices $\{\Lambda_k, k = 0, \pm 1, \cdots\}$ is a positive real sequence in the sense that $\sum_{i,j} u_i^T \Lambda_{i-j} u_j > 0$, $u_i \in \mathbb{R}^p$ ($u_i \not\equiv 0$).

Suppose that there exists a finite dimensional realization for y, so that the covariance matrix has the decomposition

$$\Lambda_k = HF^{k-1}G, \quad k = 1, 2, \cdots \tag{2}$$

where we assume that (F, G, H) is a minimal realization with $F \in \mathbb{R}^{n\times n}$.

Define the stacked vectors

$$Y^-(t) = \begin{bmatrix} y(t-1) \\ y(t-2) \\ y(t-3) \\ \vdots \end{bmatrix}, \quad Y^+(t) = \begin{bmatrix} y(t) \\ y(t+1) \\ y(t+2) \\ \vdots \end{bmatrix}$$

where $Y^-(t)$ and $Y^+(t)$ are respectively called the past and the future of observed data. Furthermore, we define infinite covariance matrices [1]

$$\mathcal{H} := \begin{bmatrix} \Lambda_1 & \Lambda_2 & \Lambda_3 & \cdots \\ \Lambda_2 & \Lambda_3 & \Lambda_4 & \cdots \\ \Lambda_3 & \Lambda_4 & \Lambda_5 & \cdots \\ \vdots & \vdots & \vdots & \ddots \end{bmatrix} = \mathrm{E}\left(Y^+(t)Y^-(t)^T\right),$$

$$\Phi := \begin{bmatrix} \Lambda_0 & \Lambda_1 & \Lambda_2 & \cdots \\ \Lambda_1^T & \Lambda_0 & \Lambda_2 & \cdots \\ \Lambda_2^T & \Lambda_1^T & \Lambda_0 & \cdots \\ \vdots & \vdots & \vdots & \ddots \end{bmatrix} = \mathrm{E}\left(Y^-(t)Y^-(t)^T\right),$$

$$\Psi := \begin{bmatrix} \Lambda_0 & \Lambda_1^T & \Lambda_2^T & \cdots \\ \Lambda_1 & \Lambda_0 & \Lambda_1^T & \cdots \\ \Lambda_2 & \Lambda_1 & \Lambda_0 & \cdots \\ \vdots & \vdots & \vdots & \ddots \end{bmatrix} = \mathrm{E}\left(Y^+(t)Y^+(t)^T\right).$$

[1] We assume that an infinite matrix $\begin{bmatrix} \Phi & \mathcal{H}^T \\ \mathcal{H} & \Psi \end{bmatrix}$ has an inverse.

Compute the weighted singular value decomposition (SVD) of the Hankel matrix $\mathcal{H}$ as [2]

$$\begin{aligned} \Psi^{-\frac{1}{2}}\mathcal{H}\Phi^{-\frac{T}{2}} &= \begin{bmatrix} \bar{U} & \tilde{U} \end{bmatrix} \begin{bmatrix} \bar{\Sigma} & 0 \\ 0 & \tilde{\Sigma} \end{bmatrix} \begin{bmatrix} \bar{V}^T \\ \tilde{V}^T \end{bmatrix} \\ &= \bar{U}\bar{\Sigma}\bar{V}^T, \quad \bar{\Sigma} \in \mathbb{R}^{n\times n}. \end{aligned} \tag{3}$$

Define extended observability and controllability matrices respectively as

$$\mathcal{O} := \Psi^{\frac{1}{2}}\bar{U}\bar{\Sigma}^{\frac{1}{2}}, \quad \mathcal{C} := \bar{\Sigma}^{\frac{1}{2}}\bar{V}^T\Phi^{\frac{T}{2}}. \tag{4}$$

Then, $\mathcal{H}$ has a canonical decomposition

$$\mathcal{H} = \mathcal{O}\mathcal{C} \tag{5}$$

and $\mathcal{O}$ and $\mathcal{C}$ are described by certain matrices $A \in \mathbb{R}^{n\times n}$, $B \in \mathbb{R}^{n\times p}$ and $C \in \mathbb{R}^{p\times n}$ as follows

$$\mathcal{O} = \begin{bmatrix} C^T & (CA)^T & (CA^2)^T & \cdots \end{bmatrix}^T, \tag{6}$$

$$\mathcal{C} = \begin{bmatrix} B & AB & A^2B & A^3B & \cdots \end{bmatrix}. \tag{7}$$

Since the decomposition in (2) is minimal, we have

$$\operatorname{rank}\mathcal{O} = n, \quad \operatorname{rank}\mathcal{C} = n \tag{8}$$

and $\operatorname{rank}\bar{\Sigma} = n$. The matrix $\bar{\Sigma}$ is diagonal and satisfies both the forward Riccati equation

$$\begin{aligned} \bar{\Sigma} &= A\bar{\Sigma}A^T \\ &+ (B - A\bar{\Sigma}C^T)(\Lambda_0 - C\bar{\Sigma}C^T)^{-1}(B - A\bar{\Sigma}C^T)^T \end{aligned} \tag{9}$$

and the backward Riccati equation

$$\begin{aligned} \bar{\Sigma} &= A^T\bar{\Sigma}A \\ &+ (C - B^T\bar{\Sigma}A)^T(\Lambda_0 - B^T\bar{\Sigma}B)^{-1}(C - B^T\bar{\Sigma}A). \end{aligned} \tag{10}$$

It can be shown (Desai *et al.*, 1985), (Lindquist and Picci, 1996*a*) that y has stochastic balanced realizations called a forward innovation model

$$\begin{bmatrix} \hat{x}(t+1) \\ y(t) \end{bmatrix} = \begin{bmatrix} A & \hat{K} \\ C & I \end{bmatrix} \begin{bmatrix} \hat{x}(t) \\ \hat{v}(t) \end{bmatrix} \tag{11}$$

and a backward innovation model

$$\begin{bmatrix} \check{x}(t-1) \\ \eta(t) \end{bmatrix} = \begin{bmatrix} A^T & \check{K}^T \\ B^T & I \end{bmatrix} \begin{bmatrix} \check{x}(t) \\ \check{v}(t) \end{bmatrix}. \tag{12}$$

where $\eta(t)$ is defined as $\eta(t) := y(t-1)$, and $\hat{x}(t)$ and $\check{x}(t)$ are the state vectors with the covariance matrix $\bar{\Sigma}$, and $\hat{v}(t)$ and $\check{v}(t)$ are the forward and backward innovations with the covariance matrices

$$\hat{R} := \Lambda_0 - C\bar{\Sigma}C^T > 0, \tag{13}$$

$$\check{R} := \Lambda_0 - B^T\bar{\Sigma}B > 0, \tag{14}$$

respectively, and $\hat{K}$ and $\check{K}$ are the steady state Kalman gains given by

$$\hat{K} := (B - A\bar{\Sigma}C^T)(\Lambda_0 - C\bar{\Sigma}C^T)^{-1}, \tag{15}$$

$$\check{K} := (\Lambda_0 - B\bar{\Sigma}B^T)^{-1}(C^T - A^T\bar{\Sigma}B)^T. \tag{16}$$

[2] For an infinite matrix $A = (a_{ij})$ with $A = A^T$, $a_{ij} \in \mathbb{R}$, we define the quadratic form $x^TAx = \sum_{i,j} x_i a_{ij} x_j$, where $x = \begin{bmatrix} x_1 & x_2 & \cdots \end{bmatrix}^T$, $x_i \in \mathbb{R}$. If $x^TAx > 0$ for all $x \not\equiv 0$, then A is strictly positive real and is written as $A > 0$. A square root of A denotes a matrix X with $A = XX^T, X = X^T$. If A is invertible, so is $A^{\frac{1}{2}}$.

Note that $\hat{x}(t)$ and $\check{x}(t)$ are stochastically balanced states described by

$$\hat{x}(t) = \mathcal{C}\Phi^{-1}Y^{-}(t), \quad \check{x}(t) = \mathcal{O}^{T}\Psi^{-1}Y^{+}(t),$$

and have the same covariance matrix

$$\mathcal{C}\Phi^{-1}\mathcal{C}^{T} = \bar{\Sigma} = \mathcal{O}^{T}\Psi^{-1}\mathcal{O}. \tag{17}$$

Substituting (13), (14), (15) and (16) into Riccati equations (9) and (10), we have

$$\bar{\Sigma} = A\bar{\Sigma}A^{T} + \hat{K}\hat{R}^{-1}\hat{K}^{T}, \tag{18}$$

$$\bar{\Sigma} = A^{T}\bar{\Sigma}A + \check{K}^{T}\check{R}^{-1}\check{K}. \tag{19}$$

Since, from $\operatorname{rank}\bar{\Sigma} = n$, $(A, \hat{K}, \check{K})$ is minimal, it follows that

$$\operatorname{rank}\hat{\mathcal{F}} = n, \quad \operatorname{rank}\check{\mathcal{F}} = n, \tag{20}$$

where $\hat{\mathcal{F}}$ and $\check{\mathcal{F}}$ are defined by

$$\hat{\mathcal{F}} := \begin{bmatrix} \hat{K} & A\hat{K} & A^{2}\hat{K} & \cdots \end{bmatrix}, \tag{21}$$

$$\check{\mathcal{F}} := \begin{bmatrix} \check{K} & A\check{K} & A^{2}\check{K} & \cdots \end{bmatrix}. \tag{22}$$

In this paper, we develop a new method of deriving the innovation model of (11) by means of "LQ decomposition" in a Hilbert space generated by a second-order stationary random process.

3. "LQ DECOMPOSITION" IN A HILBERT SPACE

Given an infinite data $\{y_t, t = 0, \pm 1, \cdots\}$, we define a Hilbert space according to (Lindquist and Picci, 1996*a*). First, let the tail matrix be defined by

$$\boldsymbol{y}_t := \begin{bmatrix} y_t & y_{t+1} & y_{t+2} & \cdots \end{bmatrix} \in \mathbb{R}^{p\times\infty}.$$

We then define a vector space by

$$\mathcal{Y} := \left\{ \sum a_k^{T}\boldsymbol{y}_k \mid a_k \in \mathbb{R}^{p},\ k = 0, \pm 1, \cdots \right\}$$

which is a linear space spanned by all finite linear combinations of $\{\boldsymbol{y}_t\}$. Define a bilinear form (inner product) by

$$\langle a^{T}\boldsymbol{y}_i, b^{T}\boldsymbol{y}_j\rangle_{\frac{1}{\infty}} := \lim_{\nu\to\infty}\frac{1}{\nu}\sum_{k=t_0}^{t_0+\nu-1} a^{T}y_{k+i}y_{k+j}^{T}b$$

$$= a^{T}\Lambda_{i-j}b \tag{23}$$

where the right hand side is independent of t_0 by stationarity. By completing this vector space with the norm induced by the inner product (23), we get a Hilbert space, which is also written as $\mathcal{Y}$.

Let $\mathcal{U}$ be a Hilbert subspace of $\mathcal{Y}$. Let the orthogonal projection of $\boldsymbol{y} \in \mathbb{R}^{p\times\infty}$ onto the space $\mathcal{U}$ be denoted by $\hat{\mathrm{E}}_{\frac{1}{\infty}}(\boldsymbol{y} \mid \mathcal{U})$. The row space spanned by a matrix U be expressed as $\operatorname{span}(U)$ and the orthogonal projection is also written as $\hat{\mathrm{E}}_{\frac{1}{\infty}}(\boldsymbol{y} \mid U) := \hat{\mathrm{E}}_{\frac{1}{\infty}}(\boldsymbol{y} \mid \operatorname{span}(U))$. If $\langle U, U\rangle_{\frac{1}{\infty}}$ has an inverse, then

$$\hat{\mathrm{E}}_{\frac{1}{\infty}}(\boldsymbol{y} \mid U) = \langle \boldsymbol{y}, U\rangle_{\frac{1}{\infty}}\langle U, U\rangle_{\frac{1}{\infty}}^{-1}U. \tag{24}$$

We define the past and future data matrices as

$$Y_t^{-} := \begin{bmatrix} \boldsymbol{y}_{t-1} \\ \boldsymbol{y}_{t-2} \\ \boldsymbol{y}_{t-3} \\ \vdots \end{bmatrix}, \quad Y_t^{+} := \begin{bmatrix} \boldsymbol{y}_{t} \\ \boldsymbol{y}_{t+1} \\ \boldsymbol{y}_{t+2} \\ \vdots \end{bmatrix}. \tag{25}$$

By the definition (25), it clearly follows that

$$\mathcal{H} = \langle Y_t^{+}, Y_t^{-}\rangle_{\frac{1}{\infty}}, \tag{26}$$

$$\Phi = \langle Y_t^{-}, Y_t^{-}\rangle_{\frac{1}{\infty}}, \tag{27}$$

$$\Psi = \langle Y_t^{+}, Y_t^{+}\rangle_{\frac{1}{\infty}}. \tag{28}$$

Define the residuals as

$$\hat{\boldsymbol{v}}_t := \boldsymbol{y}_t - \hat{\mathrm{E}}_{\frac{1}{\infty}}(\boldsymbol{y}_t \mid Y_t^{-}). \tag{29}$$

Lemma 1: The process $\hat{\boldsymbol{v}}_t$ defined by (29) is an innovation process satisfying

$$\langle \boldsymbol{y}_i, \hat{\boldsymbol{v}}_j\rangle_{\frac{1}{\infty}} = \begin{cases} 0 & (i < j) \\ \hat{L}_{i-j}\hat{R} & (i \geq j) \end{cases} \tag{30}$$

$$\langle \hat{\boldsymbol{v}}_i, \hat{\boldsymbol{v}}_j\rangle_{\frac{1}{\infty}} = \hat{R}\delta_{ij} \tag{31}$$

where the matrices $\hat{L}_k$ are given by

$$\hat{L}_0 = I; \quad \hat{L}_k = CA^{k-1}\hat{K}, \quad k = 1, 2, \cdots \tag{32}$$

and where $\hat{K}$ is defined by (15).

Proof: See Appendix A.1. □

In terms of $\hat{L}_k$ of (32), we define infinite matrices $\hat{\mathcal{L}}^{-}$, $\hat{\mathcal{L}}^{+}$ and $\hat{\mathcal{S}}$ as

$$\hat{\mathcal{L}}^{-} := \begin{bmatrix} \hat{L}_0 & \hat{L}_1 & \hat{L}_2 & \hat{L}_3 & \cdots \\ 0 & \hat{L}_0 & \hat{L}_1 & \hat{L}_2 & \cdots \\ 0 & 0 & \hat{L}_0 & \hat{L}_1 & \cdots \\ 0 & 0 & 0 & \hat{L}_0 & \cdots \\ \vdots & \vdots & \vdots & \vdots & \ddots \end{bmatrix},$$

$$\hat{\mathcal{L}}^{+} := \begin{bmatrix} \hat{L}_0 & 0 & 0 & 0 & \cdots \\ \hat{L}_1 & \hat{L}_0 & 0 & 0 & \cdots \\ \hat{L}_2 & \hat{L}_1 & \hat{L}_0 & 0 & \cdots \\ \hat{L}_3 & \hat{L}_2 & \hat{L}_1 & \hat{L}_0 & \cdots \\ \vdots & \vdots & \vdots & \vdots & \ddots \end{bmatrix},$$

$$\hat{\mathcal{S}} := \begin{bmatrix} \hat{L}_1 & \hat{L}_2 & \hat{L}_3 & \hat{L}_4 & \cdots \\ \hat{L}_2 & \hat{L}_3 & \hat{L}_4 & \hat{L}_5 & \cdots \\ \hat{L}_3 & \hat{L}_4 & \hat{L}_5 & \hat{L}_6 & \cdots \\ \hat{L}_4 & \hat{L}_5 & \hat{L}_6 & \hat{L}_7 & \cdots \\ \vdots & \vdots & \vdots & \vdots & \ddots \end{bmatrix}.$$

Moreover, we define tail matrices $\hat{V}_t^{-}$ and $\hat{V}_t^{+}$ as

$$\hat{V}_t^{-} := \begin{bmatrix} \hat{\boldsymbol{v}}_{t-1} \\ \hat{\boldsymbol{v}}_{t-2} \\ \hat{\boldsymbol{v}}_{t-3} \\ \vdots \end{bmatrix}, \quad \hat{V}_t^{+} := \begin{bmatrix} \hat{\boldsymbol{v}}_{t} \\ \hat{\boldsymbol{v}}_{t+1} \\ \hat{\boldsymbol{v}}_{t+2} \\ \vdots \end{bmatrix},$$

and the covariance matrix $\hat{\mathcal{R}}$ as

$$\hat{\mathcal{R}} := \langle \hat{V}_t^{-}, \hat{V}_t^{-}\rangle_{\frac{1}{\infty}} = \langle \hat{V}_t^{+}, \hat{V}_t^{+}\rangle_{\frac{1}{\infty}}. \tag{33}$$

Theorem 1: In terms of the innovation processes $\hat{V}_t^{-}$ and $\hat{V}_t^{+}$, the past Y_t^{-} and the future Y_t^{+} are decomposed as

$$\begin{bmatrix} Y_t^- \\ Y_t^+ \end{bmatrix} = \begin{bmatrix} \hat{\mathcal{L}}^- & 0 \\ \hat{\mathbb{S}} & \hat{\mathcal{L}}^+ \end{bmatrix} \begin{bmatrix} \hat{V}_t^- \\ \hat{V}_t^+ \end{bmatrix} \tag{34}$$

where $\hat{V}_t^-$ and $\hat{V}_t^+$ satisfy

$$\left\langle \begin{bmatrix} \hat{V}_t^- \\ \hat{V}_t^+ \end{bmatrix}, \begin{bmatrix} \hat{V}_t^- \\ \hat{V}_t^+ \end{bmatrix} \right\rangle_{\frac{I}{\infty}} = \begin{bmatrix} \hat{\mathfrak{R}} & 0 \\ 0 & \hat{\mathfrak{R}} \end{bmatrix}. \tag{35}$$

Furthermore, the orthogonal projection of the future onto the past is written as

$$\hat{\mathrm{E}}_{\frac{I}{\infty}}(Y_t^+ \mid Y_t^-) = \hat{\mathbb{S}}\hat{V}_t^-. \tag{36}$$

Proof: See Appendix A.2. □

4. STOCHASTIC REALIZATION BY "LQ DECOMPOSITION"

We show that the system matrices in the innovation representation can be derived from the decomposition of the Hankel matrix $\hat{\mathbb{S}}$ in Theorem 1.

Lemma 2: The block Hankel matrix $\hat{\mathbb{S}}$ has rank n, and satisfies

$$\hat{\mathbb{S}} = \mathcal{O}\hat{\mathfrak{F}} \tag{37}$$

where $\mathcal{O}$ and $\hat{\mathfrak{F}}$ are given by and (6) and (21), respectively.

Proof: See Appendix B. □

It should be noted that A, C, $\hat{K}$ are the matrices for the stochastically balanced forward realization of y in (11).

Theorem 2: Given $\hat{\mathbb{S}}$, $\hat{\mathfrak{R}}$, and Ψ, we compute the weighted SVD

$$\Psi^{-\frac{1}{2}}\hat{\mathbb{S}}\hat{\mathfrak{R}}^{\frac{1}{2}} = \acute{U}\acute{\Sigma}\acute{V}^T, \quad \acute{\Sigma} \in \mathbb{R}^{n \times n}. \tag{38}$$

Then, the matrices $\mathcal{O}$ and $\hat{\mathfrak{F}}$ are given by

$$\mathcal{O} = \Psi^{\frac{1}{2}}\acute{U}\acute{\Sigma}^{\frac{1}{2}}, \quad \hat{\mathfrak{F}} = \acute{\Sigma}^{\frac{1}{2}}\acute{V}^T\hat{\mathfrak{R}}^{-\frac{1}{2}}. \tag{39}$$

Proof: See Appendix B. □

This theorem implies that the SVD of (38) yields a desired decomposition of $\hat{\mathbb{S}}$, in which $\mathcal{O}$ is the extended observability matrix of the innovation representation. We can therefore derive a stochastic realization algorithm, if we are able to calculate the decomposition in (34).

Stochastic Realization Algorithm

Step 1: Given Y_t^- and Y_t^+, we compute $\hat{V}_t^-$, $\hat{V}_t^-$ and $\hat{\mathbb{S}}$ by (34).

Step 2: Compute the weighted SVD of (38), and obtain $\mathcal{O}$ and $\hat{\mathfrak{F}}$ from (39).

Step 3: Compute A, C, $\hat{K}$, and $\hat{R}$ by

$$\mathcal{O}(1:\infty,:)A = \mathcal{O}(p+1:\infty,:),$$
$$C = \mathcal{O}(1:p,:),$$
$$\hat{K} = \hat{\mathfrak{F}}(:,1:p),$$
$$\hat{R} = \hat{\mathfrak{R}}(1:p,1:p).$$

The system (11) with matrices A, C, $\hat{K}$ given above is a forward innovation representation of y.

5. EXTENSION TO A SUBSPACE IDENTIFICATION METHOD

In this section, we extend the abstract stochastic realization algorithm to a numerical procedure of a subspace identification method.

Define the bilinear form for $\boldsymbol{x}, \boldsymbol{y} \in \mathbb{R}^{\bullet \times \nu}$ as

$$\langle \boldsymbol{x}, \boldsymbol{y} \rangle_{\frac{I}{\nu}} := \frac{1}{\nu}\boldsymbol{x}\boldsymbol{y}^T.$$

Suppose that we have finite observations

$$\{y_0, y_1, y_2, \cdots, y_{\nu+2\tau-2}\}$$

for a sufficiently large ν and τ. For $t = 0, 1, \cdots, 2\tau - 1$, define

$$\boldsymbol{y}_t := \begin{bmatrix} y_t & y_{t+1} & \cdots & y_{t+\nu-1} \end{bmatrix} \in \mathbb{R}^{p \times \nu}$$

and matrices

$$Y_\tau^- = \begin{bmatrix} \boldsymbol{y}_{\tau-1} \\ \boldsymbol{y}_{\tau-2} \\ \vdots \\ \boldsymbol{y}_0 \end{bmatrix}, \quad Y_\tau^+ = \begin{bmatrix} \boldsymbol{y}_\tau \\ \boldsymbol{y}_{\tau+1} \\ \vdots \\ \boldsymbol{y}_{2\tau-1} \end{bmatrix}$$

where we assume that the positivity condition

$$\left\langle \begin{bmatrix} Y_\tau^- \\ Y_\tau^+ \end{bmatrix}, \begin{bmatrix} Y_\tau^- \\ Y_\tau^+ \end{bmatrix} \right\rangle_{\frac{I}{\nu}} > 0$$

is satisfied.

Since Λ_{i-j} is approximated by $\Lambda_{i,j} := \langle \boldsymbol{y}_i, \boldsymbol{y}_j \rangle_{\frac{I}{\nu}}$, we can derive a subspace method of obtaining an innovation representation for the process y based on the finite data as follows.

Subspace Identification Method

Step 1: Compute the following decomposition

$$\begin{bmatrix} Y_\tau^- \\ Y_\tau^+ \end{bmatrix} = \begin{bmatrix} \hat{\mathcal{L}}_\tau^- & 0 \\ \hat{\mathbb{S}}_\tau & \hat{\mathcal{L}}_\tau^+ \end{bmatrix} \begin{bmatrix} \hat{V}_\tau^- \\ \hat{V}_\tau^+ \end{bmatrix} \tag{40}$$

where $\hat{\mathcal{L}}_\tau^-$, $\hat{\mathcal{L}}_\tau^+$, $\hat{\mathbb{S}}_\tau$ are described as

$$\hat{\mathcal{L}}_\tau^- = \begin{bmatrix} \hat{L}_{\tau-1,\tau-1} & \hat{L}_{\tau-1,\tau-2} & \cdots & \hat{L}_{\tau-1,0} \\ & \hat{L}_{\tau-2,\tau-2} & \cdots & \hat{L}_{\tau-2,0} \\ & & \ddots & \vdots \\ 0 & & & \hat{L}_{0,0} \end{bmatrix},$$

$$\hat{\mathcal{L}}_\tau^+ = \begin{bmatrix} \hat{L}_{\tau,\tau} & & & 0 \\ \hat{L}_{\tau+1,\tau} & \hat{L}_{\tau+1,\tau+1} & & \\ \vdots & \vdots & \ddots & \\ \hat{L}_{2\tau-1,\tau} & \hat{L}_{2\tau-1,\tau+1} & \cdots & \hat{L}_{2\tau-1,2\tau-1} \end{bmatrix},$$

$$\hat{\mathbb{S}}_\tau = \begin{bmatrix} \hat{L}_{\tau,\tau-1} & \hat{L}_{\tau,\tau-2} & \cdots & \hat{L}_{\tau,0} \\ \hat{L}_{\tau+1,\tau-1} & \hat{L}_{\tau+1,\tau-2} & \cdots & \hat{L}_{\tau+1,0} \\ \vdots & \vdots & \vdots & \vdots \\ \hat{L}_{2\tau-1,\tau-1} & \hat{L}_{2\tau-1,\tau-2} & \cdots & \hat{L}_{2\tau-1,0} \end{bmatrix},$$

$$\hat{L}_{i,i} = I,$$

and where $\hat{\mathfrak{R}}_\tau^-$ and $\hat{\mathfrak{R}}_\tau^+$ are described as

$$\hat{\mathfrak{R}}_\tau^- = \text{block-diag}\,(\hat{R}_{\tau-1}, \hat{R}_{\tau-2}, \cdots, \hat{R}_0),$$
$$\hat{\mathfrak{R}}_\tau^+ = \text{block-diag}\,(\hat{R}_\tau, \hat{R}_{\tau+1}, \cdots, \hat{R}_{2\tau-1}).$$

Step 2: Define $\Psi_{-\tau}$ as $\Psi_{-\tau} := \langle Y_\tau^+, Y_\tau^+ \rangle_{\frac{1}{v}}$ and compute the SVD of $\Psi_{-\tau}^{-\frac{1}{2}} \hat{S}_\tau (\hat{\mathcal{R}}_\tau^-)^{\frac{1}{2}}$ as

$$\Psi_{-\tau}^{-\frac{1}{2}} \hat{S}_\tau (\hat{\mathcal{R}}_\tau^-)^{\frac{1}{2}} = \begin{bmatrix} U_1 & U_2 \end{bmatrix} \begin{bmatrix} \Sigma_1 & 0 \\ 0 & \Sigma_2 \end{bmatrix} \begin{bmatrix} V_1 \\ V_2 \end{bmatrix}^T$$
$$= U_1 \Sigma_1 V_1^T.$$

Step 3: Define $\mathcal{O}_\tau$ and $\hat{\mathcal{F}}_\tau$ as

$$\mathcal{O}_\tau = \Psi_{-\tau}^{\frac{1}{2}} U_1 \Sigma_1^{\frac{1}{2}},$$
$$\hat{\mathcal{F}}_\tau = \Sigma_1^{\frac{1}{2}} V_1^T (\hat{R}_\tau^-)^{-\frac{1}{2}}.$$

Step 4: Compute $\hat{A}$, $\hat{C}$, $\hat{K}_{\tau-1}$ and $\hat{R}_{\tau-1}$ as

$$\mathcal{O}_\tau(1:(\tau-1)p,:)\hat{A} = \mathcal{O}_\tau(p+1:\tau p,:),$$
$$\hat{C} = \mathcal{O}_\tau(1:p,:),$$
$$\hat{K}_{\tau-1} = \hat{\mathcal{F}}_\tau(:,1:p),$$
$$\hat{R}_{\tau-1} = \hat{\mathcal{R}}_\tau^-(1:p,1:p).$$

We see that the system

$$\begin{bmatrix} \hat{x}(t+1) \\ y(t) \end{bmatrix} = \begin{bmatrix} \hat{A} & \hat{K}_{\tau-1} \\ \hat{C} & I \end{bmatrix} \begin{bmatrix} \hat{x}(t) \\ \hat{v}(t) \end{bmatrix} \tag{41}$$

$$\mathrm{E}\{\hat{v}(s)\hat{v}(t)^T\} = \hat{R}_{\tau-1}\delta_{st}$$

is an approximation for the balanced stochastic realization of a stationary process y.

6. CONCLUDING REMARKS

We have re-derived a stochastic realization algorithm of (Desai *et al.*, 1985) based on the LQ decomposition in a Hilbert space of stationary random processes, and briefly suggested how the abstract realization algorithm is useful for deriving a stochastic subspace identification method when a finite string of a time series data is given.

REFERENCES

Akaike, H. (1975). Markovian representation of stochastic processes by canonical variables. *SIAM J. Control* **13**(1), 162–173.

Byrnes, C. I. and A. Lindquist (1997). On the partial stochastic realization problem. *IEEE Trans. on Automat. Contrl.* **AC-42**(8), 1049–1070.

Byrnes, C. I., S. V. Guesev and A. Lindquist (2001). From finite covariance windows to modeling filters: A convex optimization approach. *SIAM REVIEW* **43**(4), 645–675.

Desai, U. B., D. Pal and R. D. Kirkpatrick (1985). A realization approach to stochastic model reduction. *Int. J. Control* **42**(4), 821–838.

Faurre, P. L. (1976). Stochastic realization algorithms. In: *System Identification: Advances and Case Studies* (R. Mehra and D. Lainiotis, Eds.). pp. 1–25. Academic Press.

Katayama, T. and G. Picci (1999). Realization of stochastic systems with exogenous inputs and subspace identification methods. *Automatica* **35**(10), 1635–1652.

Lindquist, A. and G. Picci (1996*a*). Canonical correlation analysis, approximate covariance extension, and identification of stationary time series. *Automatica* **32**(5), 709–733.

Lindquist, A. and G. Picci (1996*b*). Geometric methods for state space identification. In: *Identification, Adaptation, Learning* (S. Bittanti and G. Picci, Eds.). pp. 1–69. Springer-Verlag.

Overschee, P. Van and B. De Moor (1993). Subspace algorithms for the stochastic identification problem. *Automatica* **29**(3), 649–660.

Overschee, P. Van and B. De Moor (1996). *Subspace Identification for Linear Systems*. Kluwer Academic Pub.

Appendix A. PROOFS OF LEMMA 1 AND THEOREM 1

A.1 Proof of Lemma 1

First, we prove (31). It follows from the definition in (29) that

$$\langle \hat{\boldsymbol{v}}_i, \hat{\boldsymbol{v}}_j \rangle_{\frac{1}{\infty}} = 0 \quad (\text{for } i \neq j).$$

For $i \geq j$, we have

$$\begin{aligned} &\langle \boldsymbol{y}_i, Y_j^- \rangle_{\frac{1}{\infty}} \\ &= \begin{bmatrix} \langle \boldsymbol{y}_i, \boldsymbol{y}_{j-1} \rangle_{\frac{1}{\infty}} & \langle \boldsymbol{y}_i, \boldsymbol{y}_{j-2} \rangle_{\frac{1}{\infty}} & \langle \boldsymbol{y}_i, \boldsymbol{y}_{j-3} \rangle_{\frac{1}{\infty}} & \cdots \end{bmatrix} \\ &= \begin{bmatrix} \Lambda_{i-j+1} & \Lambda_{i-j+2} & \Lambda_{i-j+3} & \cdots \end{bmatrix} \\ &= \begin{bmatrix} CA^{i-j}B & CA^{i-j+1}B & CA^{i-j+2}B & \cdots \end{bmatrix} \\ &= CA^{i-j} \begin{bmatrix} B & AB & A^2B & \cdots \end{bmatrix} \\ &= CA^{i-j}\mathcal{C}. \end{aligned} \tag{A.1}$$

Define $\hat{x}_t$ as

$$\hat{\boldsymbol{x}}_t := \mathcal{C}\Phi^{-1}Y_t^-. \tag{A.2}$$

Using (17) and (27), we have

$$\langle \hat{\boldsymbol{x}}_t, \hat{\boldsymbol{x}}_t \rangle_{\frac{1}{\infty}} = \mathcal{C}\Phi^{-1}\mathcal{C}^T = \bar{\Sigma}, \tag{A.3}$$

Define $\hat{\boldsymbol{y}}_t := \hat{\mathrm{E}}_{\frac{1}{\infty}}(\boldsymbol{y}_t \mid Y_t^-)$. From (A.2) and (A.1) for $i = j = t$, we have

$$\begin{aligned} C\hat{\boldsymbol{x}}_t &= C\mathcal{C}\Phi^{-1}Y_t^- = \langle \boldsymbol{y}_t, Y_t^- \rangle_{\frac{1}{\infty}} \langle Y_t^-, Y_t^- \rangle_{\frac{1}{\infty}}^{-1} Y_t^- \\ &= \hat{\mathrm{E}}_{\frac{1}{\infty}}(\boldsymbol{y}_t \mid Y_t^-) = \hat{\boldsymbol{y}}_t. \end{aligned}$$

Since $\boldsymbol{y}_t = \hat{\boldsymbol{y}}_t + \hat{\boldsymbol{v}}_t$ and $\langle \hat{\boldsymbol{y}}_t, \hat{\boldsymbol{v}}_t \rangle_{\frac{1}{\infty}} = 0$ hold, we have

$$\begin{aligned} \langle \hat{\boldsymbol{v}}_t, \hat{\boldsymbol{v}}_t \rangle_{\frac{1}{\infty}} &= \langle \boldsymbol{y}_t, \boldsymbol{y}_t \rangle_{\frac{1}{\infty}} - \langle \hat{\boldsymbol{y}}_t, \hat{\boldsymbol{y}}_t \rangle_{\frac{1}{\infty}} \\ &= \Lambda_0 - C\langle \hat{\boldsymbol{x}}_t, \hat{\boldsymbol{x}}_t \rangle_{\frac{1}{\infty}} C^T = \hat{R}, \end{aligned} \tag{A.4}$$

where we have used (13) and (A.3) to obtain (A.4). We therefore have (31).

Next, we prove (30) for $i < j$. Since the row spaces spanned by Y_t^- and $\hat{V}_t^-$ must be the same by the definition in (29), we obtain

$$\mathrm{span}\,(Y_t^-) = \mathrm{span}\,(\hat{V}_t^-), \qquad \text{(A.5)}$$

for all t. Specifically, $\mathrm{span}\,(Y_{i+1}^-) = \mathrm{span}\,(\hat{V}_{i+1}^-)$ holds for $t = i+1$. From (31), we obtain $\langle \boldsymbol{y}_i, \hat{\boldsymbol{v}}_j \rangle_{\frac{I}{\infty}} = 0$ for $i < j$.

We prove (30) for $i \geq j$. We have the following equations:

$$\begin{aligned} \hat{\boldsymbol{y}}_j &= \hat{\mathrm{E}}_{\frac{I}{\infty}}(\boldsymbol{y}_j \mid Y_j^-) \\ &= \langle \boldsymbol{y}_j, Y_j^- \rangle_{\frac{I}{\infty}} \langle Y_j^-, Y_j^- \rangle_{\frac{I}{\infty}}^{-1} Y_j^- \\ &= C\mathcal{C}\boldsymbol{\Phi}^{-1} Y_j^-. \end{aligned}$$

Using (A.1) and (A.3), we obtain

$$\begin{aligned} \langle \boldsymbol{y}_i, \hat{\boldsymbol{y}}_j \rangle_{\frac{I}{\infty}} &= \langle \boldsymbol{y}_i, C\mathcal{C}\boldsymbol{\Phi}^{-1} Y_j^- \rangle_{\frac{I}{\infty}} \\ &= \langle \boldsymbol{y}_i, Y_j^- \rangle_{\frac{I}{\infty}} \boldsymbol{\Phi}^{-1} \mathcal{C}^T C^T \\ &= CA^{i-j} \mathcal{C} \boldsymbol{\Phi}^{-1} \mathcal{C}^T C^T \\ &= CA^{i-j} \bar{\Sigma} C^T. \end{aligned}$$

From (13), (15) and (32), it follows that

$$\begin{aligned} \langle \boldsymbol{y}_i, \hat{\boldsymbol{v}}_j \rangle_{\frac{I}{\infty}} &= \langle \boldsymbol{y}_i, \boldsymbol{y}_j \rangle_{\frac{I}{\infty}} - \langle \boldsymbol{y}_i, \hat{\boldsymbol{y}}_j \rangle_{\frac{I}{\infty}} \\ &= \Lambda_{i-j} - CA^{i-j} \bar{\Sigma} C^T \\ &= CA^{i-j-1}(B - A\bar{\Sigma}C^T) \\ &= CA^{i-j-1} \hat{K} \hat{R} = \hat{L}_{i-j} \hat{R}. \end{aligned}$$

Thus, we have proved (30). □

A.2 Proof of Theorem 1

By the definition in (33), $\hat{\mathfrak{R}}$ is expressed as

$$\hat{\mathfrak{R}} = \text{block-diag}(\hat{R}, \hat{R}, \hat{R}, \cdots).$$

Since $\langle \hat{V}^+(t), \hat{V}^-(t) \rangle_{\frac{I}{\infty}} = 0$, we have (35).

Using (30), we obtain

$$\left\langle \begin{bmatrix} Y_t^- \\ Y_t^+ \end{bmatrix}, \begin{bmatrix} \hat{V}_t^- \\ \hat{V}_t^+ \end{bmatrix} \right\rangle_{\frac{I}{\infty}} = \left\langle \begin{bmatrix} \boldsymbol{y}_{t-1} \\ \boldsymbol{y}_{t-2} \\ \boldsymbol{y}_{t-3} \\ \vdots \\ \hline \boldsymbol{y}_t \\ \boldsymbol{y}_{t+1} \\ \boldsymbol{y}_{t+2} \\ \vdots \end{bmatrix}, \begin{bmatrix} \hat{\boldsymbol{v}}_{t-1} \\ \hat{\boldsymbol{v}}_{t-2} \\ \hat{\boldsymbol{v}}_{t-3} \\ \vdots \\ \hline \hat{\boldsymbol{v}}_t \\ \hat{\boldsymbol{v}}_{t+1} \\ \hat{\boldsymbol{v}}_{t+2} \\ \vdots \end{bmatrix} \right\rangle_{\frac{I}{\infty}}$$

$$= \left[\begin{array}{cccc|cccc} \hat{L}_0\hat{R} & \hat{L}_1\hat{R} & \hat{L}_2\hat{R} & \cdots & 0 & 0 & 0 & \cdots \\ 0 & \hat{L}_0\hat{R} & \hat{L}_1\hat{R} & \cdots & 0 & 0 & 0 & \cdots \\ 0 & 0 & \hat{L}_0\hat{R} & \cdots & 0 & 0 & 0 & \cdots \\ \vdots & \vdots & \vdots & \ddots & \vdots & \vdots & \vdots & \ddots \\ \hline \hat{L}_1\hat{R} & \hat{L}_2\hat{R} & \hat{L}_3\hat{R} & \cdots & \hat{L}_0\hat{R} & 0 & 0 & \cdots \\ \hat{L}_2\hat{R} & \hat{L}_3\hat{R} & \hat{L}_4\hat{R} & \cdots & \hat{L}_1\hat{R} & \hat{L}_0\hat{R} & 0 & \cdots \\ \hat{L}_3\hat{R} & \hat{L}_4\hat{R} & \hat{L}_5\hat{R} & \cdots & \hat{L}_2\hat{R} & \hat{L}_1\hat{R} & \hat{L}_0\hat{R} & \cdots \\ \vdots & \vdots & \vdots & \ddots & \vdots & \vdots & \vdots & \ddots \end{array} \right]$$

$$= \begin{bmatrix} \hat{\mathfrak{L}}^-\hat{\mathfrak{R}} & 0 \\ \hat{\mathcal{S}}\hat{\mathfrak{R}} & \hat{\mathfrak{L}}^+\hat{\mathfrak{R}} \end{bmatrix} = \begin{bmatrix} \hat{\mathfrak{L}}^- & 0 \\ \hat{\mathcal{S}} & \hat{\mathfrak{L}}^+ \end{bmatrix} \begin{bmatrix} \hat{\mathfrak{R}} & 0 \\ 0 & \hat{\mathfrak{R}} \end{bmatrix}. \qquad \text{(A.6)}$$

Since the row spaces spanned by $\left[\begin{smallmatrix} Y_t^- \\ Y_t^+ \end{smallmatrix}\right]$ and $\left[\begin{smallmatrix} \hat{V}_t^- \\ \hat{V}_t^+ \end{smallmatrix}\right]$ must be the same by the definition in (29), we have

$$\mathrm{span}\left(\begin{bmatrix} Y_t^- \\ Y_t^+ \end{bmatrix}\right) = \mathrm{span}\left(\begin{bmatrix} \hat{V}_t^- \\ \hat{V}_t^+ \end{bmatrix}\right) \qquad \text{(A.7)}$$

for all integer t. From (A.6) and (A.7), we obtain

$$\begin{aligned} \begin{bmatrix} Y_t^- \\ Y_t^+ \end{bmatrix} &= \hat{\mathrm{E}}_{\frac{I}{\infty}}\left(\begin{bmatrix} Y_t^- \\ Y_t^+ \end{bmatrix} \middle| \begin{bmatrix} \hat{V}_t^- \\ \hat{V}_t^+ \end{bmatrix} \right) \\ &= \left\langle \begin{bmatrix} Y_t^- \\ Y_t^+ \end{bmatrix}, \begin{bmatrix} \hat{V}_t^- \\ \hat{V}_t^+ \end{bmatrix} \right\rangle_{\frac{I}{\infty}} \left\langle \begin{bmatrix} \hat{V}_t^- \\ \hat{V}_t^+ \end{bmatrix}, \begin{bmatrix} \hat{V}_t^- \\ \hat{V}_t^+ \end{bmatrix} \right\rangle_{\frac{I}{\infty}}^{-1} \begin{bmatrix} \hat{V}_t^- \\ \hat{V}_t^+ \end{bmatrix} \\ &= \begin{bmatrix} \hat{\mathfrak{L}}^- & 0 \\ \hat{\mathcal{S}} & \hat{\mathfrak{L}}^+ \end{bmatrix} \begin{bmatrix} \hat{V}_t^- \\ \hat{V}_t^+ \end{bmatrix}. \end{aligned} \qquad \text{(A.8)}$$

Thus, we have obtained (34). □

Appendix B. PROOFS OF LEMMA 2 AND THEOREM 2

First, we prove Lemma 2. From (32),

$$\hat{\mathcal{S}} = \begin{bmatrix} C \\ CA \\ CA^2 \\ \vdots \end{bmatrix} [\,\hat{K} \;\; A\hat{K} \;\; A^2\hat{K} \;\; \cdots\,] = \mathcal{O}\hat{\mathfrak{F}} \qquad \text{(B.1)}$$

holds. We therefore have (37). □

Next, we prove Theorem 2. From (34), we have

$$\begin{aligned} \begin{bmatrix} \boldsymbol{\Phi} & \mathcal{H}^T \\ \mathcal{H} & \boldsymbol{\Psi} \end{bmatrix} &= \left\langle \begin{bmatrix} Y_t^- \\ Y_t^+ \end{bmatrix}, \begin{bmatrix} Y_t^- \\ Y_t^+ \end{bmatrix} \right\rangle_{\frac{I}{\infty}} \\ &= \begin{bmatrix} \hat{\mathfrak{L}}^- & 0 \\ \hat{\mathcal{S}} & \hat{\mathfrak{L}}^+ \end{bmatrix} \begin{bmatrix} \hat{\mathfrak{R}} & 0 \\ 0 & \hat{\mathfrak{R}} \end{bmatrix} \begin{bmatrix} (\hat{\mathfrak{L}}^-)^T & \hat{\mathcal{S}}^T \\ 0 & (\hat{\mathfrak{L}}^+)^T \end{bmatrix} \\ &= \begin{bmatrix} \hat{\mathfrak{L}}^-\hat{\mathfrak{R}}(\hat{\mathfrak{L}}^-)^T & \hat{\mathfrak{L}}^-\hat{\mathfrak{R}}\hat{\mathcal{S}}^T \\ \hat{\mathcal{S}}\hat{\mathfrak{R}}(\hat{\mathfrak{L}}^-)^T & \hat{\mathcal{S}}\hat{\mathfrak{R}}\hat{\mathcal{S}}^T + \hat{\mathfrak{L}}^+\hat{\mathfrak{R}}(\hat{\mathfrak{L}}^+)^T \end{bmatrix}. \end{aligned} \qquad \text{(B.2)}$$

From the (1,1) block of the matrix in (B.2), we have

$$\boldsymbol{\Phi}^{\frac{1}{2}} = \hat{\mathfrak{L}}^-\hat{\mathfrak{R}}^{\frac{1}{2}}, \quad \boldsymbol{\Phi}^{\frac{T}{2}} = \hat{\mathfrak{R}}^{\frac{T}{2}}(\hat{\mathfrak{L}}^-)^T,$$

and from the (2,1) block, we obtain

$$\begin{aligned} \boldsymbol{\Psi}^{-\frac{1}{2}}\mathcal{H}\boldsymbol{\Phi}^{-\frac{T}{2}} &= \boldsymbol{\Psi}^{-\frac{1}{2}}(\hat{\mathcal{S}}\hat{\mathfrak{R}}(\hat{\mathfrak{L}}^-)^T)\boldsymbol{\Phi}^{-\frac{T}{2}} \\ &= \boldsymbol{\Psi}^{-\frac{1}{2}}\hat{\mathcal{S}}\hat{R}^{\frac{1}{2}}. \end{aligned}$$

Since (3) and (38) are the SVD of the same matrix, we can have

$$\bar{U} = \acute{U}, \quad \bar{\Sigma} = \acute{\Sigma}, \quad \bar{V} = \acute{V}.$$

From (4) we have $\mathcal{O}$ in (39); hence we obtain $\hat{\mathfrak{F}}$ from (37). □

IFAC Publications
www.elsevier.com/locate/ifac

SUBSPACE-BASED IDENTIFICATION METHODS USING SCHUR COMPLEMENT APPROACH

Yoshinori Takei * **Hidehito Nanto** *
Shunshoku Kanae ** **Zi-Jiang Yang** **
Kiyoshi Wada **

* *Kanazawa Inst. of Tech., Kanazawa, Japan*
** *Kyushu University, Fukuoka, Japan*

Abstract: This paper shows a new interpretation of the subspace-based identification methods by using Schur complement approach. MOESP (MIMO output error state space model identification) algorithms are considered. Instead of the data matrices, we start to consider a data product moment consisted of the Hankel matrices of input-output data. It is shown that the instrumental variables (IV) extensions of data matrices in the MOESP can be expressed as modifications of the data product moment. It enables us to treat the IV-MOESP algorithms under a unified framework, and we also show that the interpretation can be applicable to the errors-in-variables (EIV) problems and the framework still can be kept.

Keywords: Interpretation, Subspace methods, Identification, Stochastic systems

1. INTRODUCTION

Subspace-based identification methods have lately attracted much attention because of being essentially suitable for the identification of MIMO systems. Many applications and efficient implementations of the methods have been developed in recent years. MOESP (MIMO output error state space model identification) algorithms by Verhaegen et al. (Verhaegen *et al.*, 1992a,b) are well known as an elementary subspace method. The methods are characterized by the determination of the extended observability matrix from input-output data. The QR factorization and the singular value decomposition (SVD) are used as the principal computational tools in the methods. The MOESP, extended by instrumental variables (IV) (Verhaegen, 1993), have been proposed. It is known that the methods can be useful to solve stochastic identification problems. The IV-MOESP schemes extract the column space of the extended observability matrix of the system by using different procedures, the schemes have been treated separately.

In this paper, a new interpretation of the MOESP algorithms are shown by using the Schur complement approach. We start to consider a data product moment consisted of the Hankel matrices of input-output data instead of the data matrices. The Schur complement of the data product moment, involving the column space of the extended observability matrix of the unknown system, is extracted. The eigenvalue decomposition (EVD) of the Schur complement matrix is computed to estimate the extended observability matrix of the unknown system. It is shown that the extensions of the data matrix in the MOESP can be expressed as modifications of the data product moment and IV matrices form the weighting matrices of the data product moment. Then we show the Schur complement approach can be treated the IV-MOESP schemes under a unified framework.

The interpretation for errors-in-variables (EIV) models is also discussed. The identification problems of the noisy input and output systems are known as EIV problems. Many identification methods assume that the measured input is noise-free and the measured output is only corrupted by additive noise. This is not valid in realistic cases and the identification methods with such an assumption give biased estimates if both the measured input and output are disturbed. The subspace identification method for the EIV models have been proposed (Chou *et al.*, 1997) as the extension of the MOESP algorithms. The above interpretation considered the output error model as the system, that is, all the noise is added to the output. Here we also show that the Schur complement approach can be applicable to the noisy input-output systems.

Furthermore, considering the relationship between the Schur complement matrix and the least squares residual, the recursive computational methods of the Schur complement matrices can be derived, which are based on the recursive least squares (RLS) method (Takei *et al.*, 1999).

This paper is organized as follows. In the following section, we present preliminaries. Section 3 provides an overview of the MOESP algorithm and its variants for stochastic cases. Section 4 presents an interpretation of the subspace-based identification by using the Schur complement approach. Finally section 5 concludes this paper.

2. PRELIMINARIES

We consider the following n-th order discrete time linear time-invariant state-space model:

$$x_{k+1} = Ax_k + B\tilde{u}_k + f_k, \qquad (1)$$

$$\tilde{y}_k = Cx_k + D\tilde{u}_k, \qquad (2)$$

where x_k is the n-th order state vector and the noise-free m inputs and l outputs collected in $\tilde{u}_k$ and $\tilde{y}_k$, respectively. f_k is the process noise. The unknown system matrices A, B, C and D have appropriate dimensions. It is assumed that the model is minimal, that is, the system is completely reachable and observable. The observations from the system are related to the noise-free input and output as follows:

$$u_k = \tilde{u}_k + w_k, \qquad (3)$$

$$y_k = \tilde{y}_k + v_k. \qquad (4)$$

The noise processes are assumed to be zero-mean, white noise and they are assumed to be statistically independent of the past noise-free input $\tilde{u}_k$. We assume that these noise sequences are correlated, and their noise covariance is given as following matrix:

$$E\left\{\begin{bmatrix} f_k \\ w_k \\ v_k \end{bmatrix}\begin{bmatrix} f_j \\ w_j \\ v_j \end{bmatrix}^T\right\} = \begin{bmatrix} R_{ff} & R_{fw} & R_{fv} \\ R_{fw}^T & R_{ww} & R_{wv} \\ R_{fv}^T & R_{wv}^T & R_{vv} \end{bmatrix}\delta_{k,j} \qquad (5)$$

where $E\{\cdot\}$ denotes statistical expectation, and $\delta_{k,j}$ is the Kronecker delta.

Here we introduce a stacked input vector:

$$u_i(k+N) \triangleq \left[u_{k+N}^T, u_{k+N+1}^T, \cdots, u_{k+N+i-1}^T\right]^T, \qquad (6)$$

where $u_i(k+N) \in \mathbb{R}^{mi\times 1}$. Stacked vectors of the output and the noise are defined similarly. Then we define the Hankel matrix:

$$U_{k,i,N} \triangleq [u_i(k), u_i(k+1), \cdots, u_i(k+N-1)]. \qquad (7)$$

Corresponding definitions for the output and the noises are in a similar way. In relation to the order of the system n, the pair i and N satisfy $i > n$ and $N \gg n$. We define a state vector sequence:

$$X_{k,N} \triangleq [x_k \; x_{k+1} \; \cdots \; x_{k+N-1}]. \qquad (8)$$

Then we obtain a following relation:

$$Y_{k,i,N} = \Gamma_i X_{k,N} + H_i U_{k,i,N} - H_i W_{k,i,N} + G_i F_{k,i,N} + V_{k,i,N}. \qquad (9)$$

The extended observability matrix Γ_i is defined as

$$\Gamma_i = \begin{bmatrix} C \\ CA \\ \vdots \\ CA^{i-1} \end{bmatrix} \qquad (10)$$

and the block Toeplitz matrices H_i and G_i are defined as follows:

$$H_i = \begin{bmatrix} D & & & 0 \\ CB & D & & \\ \vdots & \ddots & \ddots & \\ CA^{k-2}B & \cdots & CB & D \end{bmatrix}, \qquad (11)$$

$$G_i = \begin{bmatrix} 0 & & & 0 \\ C & 0 & & \\ \vdots & \ddots & \ddots & \\ CA^{k-2} & \cdots & C & 0 \end{bmatrix}. \qquad (12)$$

The key problem of the subspace method is to estimate the column space of the extended observability matrix Γ_i.

We also assume the input u_k be such that the following condition is satisfied.

$$\text{rank}\begin{bmatrix} U_{k,i,N} \\ X_{k,N} \end{bmatrix} = mi + n \qquad (13)$$

3. REVIEW OF MOESP ALGORITHMS

To extract the column space of the extended observability matrix, the QR factorization and the singular value decomposition (SVD) as the principal computational tools are used in the MOESP algorithms. In this section, we focus on the QR factorization and review the MOESP algorithms.

3.1 Deterministic case

We assume that all noise sequences are equal to zero, that is, $w_k = 0$, $f_k = 0$ and $v_k \doteq 0$. In the elementary MOESP algorithm, the following QR factorization compresses the data matrix constructed from the Hankel matrices of input-output data (Verhaegen *et al.*, 1992a):

$$\begin{bmatrix} U_{k,i,N} \\ Y_{k,i,N} \end{bmatrix} = \begin{bmatrix} R_{11} & 0 \\ R_{21} & R_{22} \end{bmatrix} \begin{bmatrix} Q_1 \\ Q_2 \end{bmatrix}, \tag{14}$$

where R_{11} and R_{22} are lower triangular, and Q_1 and Q_2 are orthogonal. To apply the SVD to the R_{22} matrix, the estimate of the extended observability matrix is obtained.

3.2 Stochastic case

Assuming that $w_k = 0$, $f_k \neq 0$ and $v_k \neq 0$, we next consider the stochastic case. The following data matrix extended by the IV matrix Φ_N is used in place of the one shown in equation (14) (Takei *et al.*, 1999):

$$\begin{bmatrix} U_{k,i,N} \\ Y_{k,i,N} \\ \Phi_N \end{bmatrix} = \begin{bmatrix} R_{11} & 0 & 0 \\ R_{21} & R_{22} & 0 \\ R_{31} & R_{32} & R_{33} \end{bmatrix} \begin{bmatrix} Q_1 \\ Q_2 \\ Q_3 \end{bmatrix}, \tag{15}$$

where $\Phi_N \in \mathbb{R}^{p \times N}$ and the p depends on the selection of the IV matrix, which we don't define here. Then the $R_{22}R_{32}^T$ matrix is used to extract the column space of the extended observability matrix.

3.3 EIV case

3.3.1. White noise input case In the MOESP for EIV models, the QR factorization is used for the data matrix with the IV (Chou *et al.*, 1997):

$$\begin{bmatrix} U_{k,i,N} \\ U_{k-i,i,N} \\ Y_{k-i,i,N} \\ Y_{k,i,N} \end{bmatrix} = \begin{bmatrix} \tilde{R}_{11} & O & O & O \\ \tilde{R}_{21} & \tilde{R}_{22} & O & O \\ \tilde{R}_{31} & \tilde{R}_{32} & \tilde{R}_{33} & O \\ \tilde{R}_{41} & \tilde{R}_{42} & \tilde{R}_{43} & \tilde{R}_{44} \end{bmatrix} \begin{bmatrix} \tilde{Q}_1 \\ \tilde{Q}_2 \\ \tilde{Q}_3 \\ \tilde{Q}_4 \end{bmatrix} \tag{16}$$

The SVD is applied to the matrix $\begin{bmatrix} \tilde{R}_{42} & \tilde{R}_{43} \end{bmatrix}$ obtained by the factorization to estimate the extended observability matrix.

3.3.2. Colored noise input case Considering the following data matrix, the IV-MOESP scheme for this case is also discussed, which the noise-free input is not discrete time zero-mean white noise (Chou *et al.*, 1997):

$$\begin{bmatrix} U_{k,i,N}U_{k-i,i,N}^T & U_{k,i,N}Y_{k-i,i,N}^T \\ Y_{k,i,N}U_{k-i,i,N}^T & Y_{k,i,N}Y_{k-i,i,N}^T \end{bmatrix} = \begin{bmatrix} \bar{R}_{11} & 0 \\ \bar{R}_{21} & \bar{R}_{22} \end{bmatrix} \begin{bmatrix} \bar{Q}_1 \\ \bar{Q}_2 \end{bmatrix}. \tag{17}$$

The SVD is applied to the matrix $\bar{R}_{22}$ to estimate the extended observability matrix Γ_i.

4. SUBSPACE METHODS USING SCHUR COMPLEMENT APPROACH

In this section, we show an interpretation of the MOESP algorithms by using the Schur complement approach. The interpretation consists of constructing the data product moment instead of the data matrix in the MOESP schemes and extracting the Schur complement of the data product moment. The following subsections provide Schur complement approaches for deterministic and stochastic cases, and we also discuss the interpretation for EIV models. We show that IV-MOESP algorithms can be treated under the unified framework. The interpretation for EIV models is considered for two assumptions of the input. One is the case the noise-free input is discrete time zero-mean white noise, and the other is the case the true input is not discrete time zero-mean white noise.

4.1 Deterministic case

Alternatively, we consider the following data product moment instead of the data matrix (14):

$$\begin{bmatrix} U_{k,i,N} \\ Y_{k,i,N} \end{bmatrix} \begin{bmatrix} U_{k,i,N}^T & Y_{k,i,N}^T \end{bmatrix} = \begin{bmatrix} U_{k,i,N}U_{k,i,N}^T & U_{k,i,N}Y_{k,i,N}^T \\ Y_{k,i,N}U_{k,i,N}^T & Y_{k,i,N}Y_{k,i,N}^T \end{bmatrix}. \tag{18}$$

Then the Schur complement matrix of $U_{k,i,N}U_{k,i,N}^T$ in (18) can be represented by

$$\begin{aligned} S &= Y_{k,i,N}Y_{k,i,N}^T - Y_{k,i,N}U_{k,i,N}^T(U_{k,i,N}U_{k,i,N}^T)^{-1} \\ &\quad \times U_{k,i,N}Y_{k,i,N}^T \\ &= Y_{k,i,N}\Pi_{U_{k,i,N}}^{\perp}Y_{k,i,N}^T, \end{aligned} \tag{19}$$

where $\Pi_{U_{k,i,N}}^{\perp}$ is defined as

$$\Pi_{U_{k,i,N}}^{\perp} \triangleq I - \Pi_{U_{k,i,N}}, \tag{20}$$

$$\Pi_{U_{k,i,N}} \triangleq U_{k,i,N}^T(U_{k,i,N}U_{k,i,N}^T)^{-1}U_{k,i,N}. \tag{21}$$

Using the result of equation (14), the Schur complement matrix obtained by (19) can be rewritten as follows:

$$\begin{aligned} S &= R_{22}Q_2Q_2^TR_{22}^T \\ &= R_{22}R_{22}^T. \end{aligned} \tag{22}$$

Therefore we see that the column space of the Schur complement matrix is identical to the matrix R_{22} obtained by the QR factorization of the data matrix in the MOESP. Therefore we can also obtain the estimate of the extended observability matrix by using the eigenvalue decomposition (EVD) of the Schur complement matrix.

4.2 Stochastic case

Considering a data product moment corresponding to the data matrix shown in (15) by using the replacement of the output with the linear combination with the input and the IV.

To solve the minimization problems of the Frobenius norm $\|Y_{k,i,N} - \mathcal{L}_1U_{k,i,N} - \mathcal{L}_2\Phi_N\|_F^2$ in terms of coefficient matrices $\mathcal{L}$, the approximated output can be represented by

$$\begin{aligned} \hat{Y}_{k,i,N} &= Y_{k,i,N}\Omega_{k,i,N}^T(\Omega_{k,i,N}\Omega_{k,i,N}^T)^{-1}\Omega_{k,i,N} \\ &= Y_{k,i,N}\Pi_{\Omega_{k,i,N}}, \end{aligned} \tag{23}$$

where $\Omega_{k,i,N}$ and $\Pi_{\Omega_{k,i,N}}$ are defined as

$$\Omega_{k,i,N} \triangleq \left[U_{k,i,N}^T \; \Phi_N^T\right]^T, \tag{24}$$

$$\Pi_{\Omega_{k,i,N}} \triangleq \Omega_{k,i,N}^T(\Omega_{k,i,N}\Omega_{k,i,N}^T)^{-1}\Omega_{k,i,N}. \tag{25}$$

In the data product moment shown in (18), substituting the matrix $\hat{Y}_{k,i,N}$ obtained in (23) for the matrix $Y_{k,i,N}$, we have the data product moment for the stochastic case as follows:

$$\begin{aligned} &\begin{bmatrix} U_{k,i,N} \\ \hat{Y}_{k,i,N} \end{bmatrix} \begin{bmatrix} U_{k,i,N}^T & \hat{Y}_{k,i,N}^T \end{bmatrix} \\ &= \begin{bmatrix} U_{k,i,N} \\ Y_{k,i,N} \end{bmatrix} \Pi_{\Omega_{k,i,N}} \begin{bmatrix} U_{k,i,N}^T & Y_{k,i,N}^T \end{bmatrix} \end{aligned} \tag{26}$$

where

$$U_{k,i,N}\Pi_{\Omega_{k,i,N}} = U_{k,i,N}. \tag{27}$$

Note that the data product moment shown in (26) is formed by weighting the one shown in (18) with $\Pi_{\Omega_{k,i,N}}$. Then the Schur complement matrix of $U_{k,i,N}U_{k,i,N}^T$ in (26) can be represented by

$$\begin{aligned} S &= \hat{Y}_{k,i,N}\Pi_{U_{k,i,N}}^{\perp}\hat{Y}_{k,i,N}^T \\ &= Y_{k,i,N}\Pi_{U_{k,i,N}}^{\perp}\Phi_N^T(\Phi_N\Pi_{U_{k,i,N}}^{\perp}\Phi_N^T)^{-1} \\ &\quad \times\Phi_N\Pi_{U_{k,i,N}}^{\perp}Y_{k,i,N}^T. \end{aligned} \tag{28}$$

The above matrix can be rewritten by using the R matrices obtained in (15):

$$S = R_{22}R_{32}^T\Lambda^{-1}R_{32}R_{22}^T, \tag{29}$$

where

$$\Lambda = R_{32}R_{32}^T + R_{33}R_{33}^T. \tag{30}$$

Then it have been shown that the column space of the Schur complement matrix is also identical to the matrix obtained by the QR factorization shown in (15). Hence we see that the Schur complement approach can be applicable to the stochastic case as similar as the deterministic case.

4.3 EIV case

4.3.1. White noise input case It is assumed that white noise sequence is used as the input $\tilde{u}_k$. Based on the assumptions of the process noise and the measurement noise, we have:

$$\lim_{N\to\infty}\frac{1}{N}U_{k,i,N}U_{k,i,N}^T = \sigma_u^2 I. \tag{31}$$

The input measurement noise is not correlated, and thus the followings is held:

$$\lim_{N\to\infty}\frac{1}{N}U_{k,i,N}\Phi_N^T = 0. \tag{32}$$

The equation (9) is multiplied by the Φ_N^T and then we obtain:

$$\begin{aligned} Y_{k,i,N}\Phi_N^T &= \Gamma_iX_{k,N}\Phi_N^T + H_iU_{k,i,N}\Phi_N^T - H_iW_{k,i,N}\Phi_N^T \\ &\quad + G_iF_{k,i,N}\Phi_N^T + V_{k,i,N}\Phi_N^T. \end{aligned} \tag{33}$$

From the assumptions of the noise sequences, the following equations are obtained:

$$\lim_{N\to\infty}\frac{1}{N}W_{k,i,N}\Phi_N^T = 0, \tag{34}$$

$$\lim_{N\to\infty}\frac{1}{N}F_{k,i,N}\Phi_N^T = 0, \tag{35}$$

$$\lim_{N\to\infty}\frac{1}{N}V_{k,i,N}\Phi_N^T = 0. \tag{36}$$

Hence, the data equation shown in (33) converges as follows:

$$\begin{aligned} &\lim_{N\to\infty}\frac{1}{N}Y_{k,i,N}\Phi_N^T \\ &= \lim_{N\to\infty}\frac{1}{N}\Gamma_iX_{k,N}\Phi_N^T + \lim_{N\to\infty}\frac{1}{N}H_iU_{k,i,N}\Phi_N^T \\ &- \lim_{N\to\infty}\frac{1}{N}H_iW_{k,i,N}\Phi_N^T + \lim_{N\to\infty}\frac{1}{N}G_iF_{k,i,N}\Phi_N^T \\ &+ \lim_{N\to\infty}\frac{1}{N}V_{k,i,N}\Phi_N^T \\ &= \lim_{N\to\infty}\frac{1}{N}\Gamma_iX_{k,N}\Phi_N^T. \end{aligned} \tag{37}$$

Therefore the consistent estimate of the extended observability matrix can be obtained by using the SVD of the matrix $Y_{k,i,N}\Phi_N^T$.

Then we consider the data product moment for the stochastic case again:

$$\begin{bmatrix} U_{k,i,N} \\ \hat{Y}_{k,i,N} \end{bmatrix} \begin{bmatrix} U_{k,i,N}^T & \hat{Y}_{k,i,N}^T \end{bmatrix} = \begin{bmatrix} U_{k,i,N} \\ Y_{k,i,N} \end{bmatrix} \Pi_{\Omega_{k,i,N}} \begin{bmatrix} U_{k,i,N}^T & Y_{k,i,N}^T \end{bmatrix}, \quad (38)$$

where the IV matrix Φ_N for this case, included in the weighting matrix $\Pi_{\Omega_{k,i,N}}$, is selected as follows:

$$\Phi_N \triangleq \begin{bmatrix} U_{k-i,i,N} \\ Y_{k-i,i,N} \end{bmatrix}. \quad (39)$$

The above selection of the IV is identical to the PO-MOESP algorithm (Verhaegen, 1994).

Then the Schur complement S of $U_{k,i,N}U_{k,i,N}^T$ in (38) is represented by

$$\begin{aligned} S &= Y_{k,i,N}\left(\Pi_{\Omega_{k,i,N}} - \Pi_{U_{k,i,N}}\right) Y_{k,i,N}^T \\ &= Y_{k,i,N}\Pi_{U_{k,i,N}}^{\perp}\Phi_N^T(\Phi_N^T\Pi_{U_{k,i,N}}^{\perp}\Phi_N^T)^{-1} \\ &\quad \times\Phi_N\Pi_{U_{k,i,N}}^{\perp}Y_{k,i,N}^T. \end{aligned} \quad (40)$$

Then we have

$$\begin{aligned} &\lim_{N\to\infty}\frac{1}{N}Y_{k,i,N}\Pi_{U_{k,i,N}}^{\perp}\Phi_N^T \\ &= \lim_{N\to\infty}\frac{1}{N}Y_{k,i,N} \\ &\times\left[I - U_{k,i,N}^T(U_{k,i,N}U_{k,i,N}^T)^{-1}U_{k,i,N}\right]\Phi_N^T \\ &= \lim_{N\to\infty}\frac{1}{N}Y_{k,i,N}\Phi_N^T \end{aligned} \quad (41)$$

and

$$\begin{aligned} &\lim_{N\to\infty}\frac{1}{N}\Phi_N\Pi_{U_{k,i,N}}^{\perp}\Phi_N^T \\ &= \lim_{N\to\infty}\frac{1}{N}\Phi_N\left[I - U_{k,i,N}^T(U_{k,i,N}U_{k,i,N}^T)^{-1}U_{k,i,N}\right] \\ &\times\Phi_N^T \\ &= \lim_{N\to\infty}\frac{1}{N}\Phi_N\Phi_N^T. \end{aligned} \quad (42)$$

Therefore the Schur complement matrix obtained in (40) converges as follows:

$$\begin{aligned} &\lim_{N\to\infty}\frac{1}{N}S \\ &= \lim_{N\to\infty}\frac{1}{N}Y_{k,i,N}\Pi_{U_{k,i,N}}^{\perp}\Phi_N^T(\Phi_N\Pi_{U_{k,i,N}}^{\perp}\Phi_N^T)^{-1} \\ &\times\Phi_N\Pi_{U_{k,i,N}}^{\perp}Y_{k,i,N}^T \\ &= \lim_{N\to\infty}\frac{1}{N}Y_{k,i,N}\Phi_N^T(\lim_{N\to\infty}\frac{1}{N}\Phi_N\Phi_N^T)^{-1} \\ &\times\lim_{N\to\infty}\frac{1}{N}\Phi_N Y_{k,i,N}^T \\ &= \lim_{N\to\infty}\frac{1}{N}Y_{k,i,N}\Phi_N^T(\Phi_N\Phi_N^T)^{-1}\Phi_N Y_{k,i,N}^T. \end{aligned} \quad (43)$$

We see that the extended observability matrix can be estimated by the SVD or the EVD of the Schur complement matrix and we will show that the Schur complement approach can be applicable to the case that the noise-free input is not the white noise sequence.

The $\tilde{R}$ matrix obtained by the QR factorization shown in (16) can rewrite the Schur complement matrix in (40), and we have

$$\begin{aligned} S &= \begin{bmatrix} \tilde{R}_{41} & \tilde{R}_{42} & \tilde{R}_{43} \end{bmatrix} \begin{bmatrix} \tilde{R}_{41}^T \\ \tilde{R}_{42}^T \\ \tilde{R}_{43}^T \end{bmatrix} \\ &\quad -\tilde{R}_{41}\tilde{R}_{11}^T\left(\tilde{R}_{11}\tilde{R}_{11}^T\right)^{-1}\tilde{R}_{11}\tilde{R}_{41}^T \\ &= \begin{bmatrix} \tilde{R}_{42} & \tilde{R}_{43} \end{bmatrix} \begin{bmatrix} \tilde{R}_{42}^T \\ \tilde{R}_{43}^T \end{bmatrix}. \end{aligned} \quad (44)$$

Therefore the column space of the Schur complement matrix is also identical to the one of the matrix obtained by the QR factorization in the MOESP and it enable us to estimate the extended observability matrix.

4.3.2. Colored noise input case The input $\tilde{u}_k$, undisturbed by the noise, is not assumed to be white noise sequence. Then the equation (32) is not satisfied but the equations (34), (35) and (36) still hold. Therefore, from the equation (33), we have:

$$\begin{aligned} &\lim_{N\to\infty}\frac{1}{N}Y_{k,i,N}\Phi_N^T \\ &= \lim_{N\to\infty}\frac{1}{N}\Gamma_i X_{k,N}\Phi_N^T + \lim_{N\to\infty}\frac{1}{N}H_i U_{k,i,N}\Phi_N^T. \end{aligned} \quad (45)$$

Here we introduce the following replacements for the equation (45):

$$Y_{k,i,N}^* \triangleq Y_{k,i,N}\Phi_N^T, \quad (46)$$

$$X_{k,N}^* \triangleq X_{k,N}\Phi_N^T, \quad (47)$$

$$U_{k,i,N}^* \triangleq U_{k,i,N}\Phi_N^T. \quad (48)$$

Then the alternative data equation can be obtained as $N \to \infty$:

$$Y_{k,i,N}^* = \Gamma_i X_{k,N}^* + H_i U_{k,i,N}^*. \quad (49)$$

Multiplying the equation (49) by the $\Pi_{U_{k,i,N}^*}^{\perp}$, we have the following equation:

$$Y_{k,i,N}^*\Pi_{U_{k,i,N}^*}^{\perp} = \Gamma_i X_{k,N}^*\Pi_{U_{k,i,N}^*}^{\perp}, \quad (50)$$

where the $\Pi_{U_{k,i,N}^*}^{\perp}$ is defined as

$$\Pi_{U_{k,i,N}^*} \triangleq (U_{k,i,N}^*)^T(U_{k,i,N}^*(U_{k,i,N}^*)^T)^{-1}U_{k,i,N}^*, \quad (51)$$

$$\Pi_{U_{k,i,N}^*}^{\perp} \triangleq I - \Pi_{U_{k,i,N}^*}. \quad (52)$$

Hence the consistent estimate of the extended observability matrix can be obtained by applying the SVD to the matrix $Y^*_{k,i,N}\Pi^{\perp}_{U^*_{k,i,N}}$. This scheme refers to papers (Okada *et al.*, 1996) and (Gustafsson, 1997).

Then, substituting both $U^*_{k,i,N}$ and $Y^*_{k,i,N}$ for the input-output Hankel matrices in the equation (18), we have the following data product moment:

$$\begin{bmatrix} U^*_{k,i,N} \\ Y^*_{k,i,N} \end{bmatrix} \left[(U^*_{k,i,N})^T \ (Y^*_{k,i,N})^T\right] = \begin{bmatrix} U_{k,i,N} \\ Y_{k,i,N} \end{bmatrix} \Phi_N^T \Phi_N \left[U^T_{k,i,N} \ Y^T_{k,i,N}\right].. \quad (53)$$

The Schur complement of the $U^*_{k,i,N}(U^*_{k,i,N})^T$ in (53) can be obtained as

$$\begin{aligned} S &= Y^*_{k,i,N}(Y^*_{k,i,N})^T - Y^*_{k,i,N}(U^*_{k,i,N})^T \\ &\quad \times (U^*_{k,i,N}(U^*_{k,i,N})^T)^{-1} U^*_{k,i,N}(Y^*_{k,i,N})^T \\ &= Y^*_{k,i,N}\Pi^{\perp}_{U^*_{k,i,N}} {Y^*_{k,i,N}}^T. \end{aligned} \quad (54)$$

From the result in the equation (50), we see that the consistent estimate of the extended observability matrix Γ_i can be yielded by using the EVD of the Schur complement matrix in (54).

Using the result of the QR factorization in the equation (17), the Schur complement matrix in (54) can be rewritten as follows:

$$\begin{aligned} S &= \left[\bar{R}_{21} \ \bar{R}_{22}\right] \begin{bmatrix} \bar{R}^T_{21} \\ \bar{R}^T_{22} \end{bmatrix} \\ &\quad - \bar{R}_{21}\bar{R}^T_{11}(\bar{R}_{11}\bar{R}^T_{11})\bar{R}_{11}\bar{R}^T_{12} \\ &= \bar{R}_{22}\bar{R}^T_{22}. \end{aligned} \quad (55)$$

Therefore we see that the column space of the Schur complement matrix is also identical to the one of the matrix obtained by the QR factorization of the data matrix in the IV-MOESP. This result shows that the Schur complement matrix provides the estimate of the extended observability matrix.

5. CONCLUSION

In this paper, an interpretation of subspace-based identification methods using the Schur complement approach has been proposed. Considering the data product moment, we have shown that the IV-MOESP schemes can be interpreted as weighting of the data product moments.The IV matrices that extend the MOESP algorithms are reformulated as the weighting matrix in the framework. The weighting matrix $\mathcal{W}$ corresponding to each MOESP schemes can be listed as follows:

$$\begin{aligned} \text{Deterministic case}: \mathcal{W} &= I, \\ \text{Stochastic case}: \mathcal{W} &= \Pi_{\Omega_{k,i,N}}, \\ \text{EIV case}: \mathcal{W} &= \Phi_N^T\Phi_N, \end{aligned}$$

where the stochastic case includes the EIV case using a white noise input. The extended observability matrix can be estimated by applying the EVD to the Schur complement matrix of the data product moment.

Therefore we can treat IV-MOESP algorithms under a unified framework, which consists of constructing the data product moment and extracting the Schur complement of the matrix.

REFERENCES

Verhaegen, M. and P. Dewilde (1992a). Subspace Model Identification, Part1: the Output-Error State-Space Model Identification Class of Algorithms, *Int. J. Control*, **56**(5), pp. 1187-1210.

Verhaegen, M. and P. Dewilde (1992b). Subspace Model Identification, Part2: Analysis of the elementary output-error state space model identification algorithm,*Int. J. Control*,**56**(5), pp. 1211-1241.

Verhaegen, M. (1993). Subspace Model Identification, Part3: Analysis of the ordinary output-error state space model identification algorithm, *Int. J. Control*, **58**(3), pp. 555-586.

Verhaegen, M. (1994). Identification of the deterministic part of MIMO state space models given in innovation form from input-output data, *Automatica*, **30**(1), pp. 61-74.

Chou, C. T. and M. Verhaegen (1997). Subspace algorithms for the identification of multivariable dynamic errors-in-variables models, *Automatica*, **33**(10), pp. 1857-1869.

Takei, Y., J. Imai and K. Wada (1999). Recursive Computation for Error Covariance Matrix in Subspace Identification Methods, *Proc. 31st ISCIE Symp. Stochastic Systems Theory and Its Applications*, pp. 19-24.

Okada, M. and T. Sugie (1996). Subspace System Identification considering both Noise Attenuation and Use of Prior Knowledge, *Proc. 35th IEEE Conference on Decision and Control*, pp. 3662-3667.

Gustafsson, T. (1997). System Identification using Subspace-based Instrumental Variable Methods, *Proc. 11th IFAC Symposium on System Identification*, pp. 1119-1124.

IFAC Publications
www.elsevier.com/locate/ifac

RECURSIVE SUBSPACE IDENTIFICATION FOR CONTINUOUS-/DISCRETE-TIME STOCHASTIC SYSTEMS

Akira Ohsumi,* Yûji Matsuüra, and Kentaro Kameyama

Department of Mechanical & System Engineering,
Graduate School of Science and Technology, Kyoto Institute of Technology,
Matsugasaki, Sakyo, Kyoto 606-8585, Japan
E-mail: ohsumi@ipc.kit.ac.jp; Tel/Fax: +81-75-724-7352

Abstract: A recursive algorithm is derived for subspace system identification which is flexible and applicable to continuous/discrete-time, time-invariant/varying stochastic systems. Efficacy of the algorithm is demonstrated by simulations.

Keywords: Subspace identification, recursive algorithm, stochastic system, continuous-time system, time-varying system.

1. INTRODUCTION

Subspace-based identification of state space models has attracted a great deal attention in the last few years. The surge of interest has been mainly concentrated on the time-invariant discrete- or continuous-time linear systems. However, in practice, most existing systems show time-varying and/or nonlinear behaviour. For the identification of such systems not off-line but on-line computations are necessarily required even from the viewpoint of computational burden. To realize this, recursive algorithms for subspace identification is advisable.

For instance, a recursive algorithm by updating the LQ-factorization and making use of the classical method of Givens rotations and applied it to the discrete-time time-varying system is derived (Verhaegen and Deprettere, 1991). Under the assumption that the system order is a priori known a recursive algorithm by updating directly the extended observability matrix (see Gustafsson 1997), and algorithms based on the instrumental variables and on the use of subspace tracking for the update of the singular value decomposition (SVD) (Lovera, Gustafsson and Verhaegen 2000) is proposed. Oku and Kimura proposed an algorithm by updating the projection matrix called the compressed input-output data matrix, and they also gave a method using gradient type subspace tracking (e.g., see Oku and Kimura, 1999, 2002), Takei, Imai and Wada (2000) presented the algorithm based on the estimate of extended obsevability matrix by subspace extraction via Schur complement.

* Corresponding author.

On the other hand, actual phenomena show complex behaviours and their mathematical models are written by time-varying and/or non-linear equations. The recursive algorithm is also useful to identify such systems.

So, in this direction, we derive a recursive subspace identification algorithm which requires low computational cost, and is simple, flexible and applicable to continuous / discrete-time, time-invariant/varying stochastic systems.

2. PROBLEM STATEMENT

Consider the following input-output algebraic equation appearing in the subspace identification problems:

$$Y_i = \Gamma_i X_i + H_i U_i + \Sigma_i W_i + V_i, \qquad (1)$$

where $Y_i \in R^{i\ell \times N}$ and $U_i \in R^{im \times N}$ are output and input data matrices which are constructed by arranging properly N (column) vectors of the output $y \in R^{\ell}$ and N vectors of input $u \in R^m$; $X_i \in R^{n \times N}$ is the matrix constructed by the system states $x \in R^n$; $W_i \in R^{in \times N}$, $V_i \in R^{i\ell \times N}$ are matrices constructed similarly to Y_i or U_i from the system and observation noises $w \in R^n$, $v \in R^{\ell}$; $\Gamma_i \in R^{i\ell \times n}$ is the extended observability matrix, and $H_i \in R^{i\ell \times im}$, $\Sigma_i \in R^{i\ell \times in}$ are lower block triangular matrices consisting of system matrices (the detail structures are found in references).

The input-output algebraic matrix equation (1) can be obtained from the discrete- or continuous-time, time-invariant/varying linear system. For instance, when the system is described by the discrete-time, time-invariant stochastic systems subjected to white Gaussian noises w_k and v_k with zero-means,

$$\Sigma_D : \begin{cases} x_{k+1} = Ax_k + Bu_k + w_k \\ y_k = Cx_k + Du_k + v_k, \end{cases} \tag{2}$$

then, using $(N + i - 1)$ past and current data, i.e., $\{y_j\}_{j=k-N-i+2,\cdots,k-1,k}$, the matrix Y_i is constructed by the Hankel matrix,

$$Y_i = [\, y_i(k - N + 1), \cdots, y_i(k - 1), y_i(k) \,],$$

where $y_i(j) = [y_{j-i+1}^T, \cdots, y_{j-1}^T, y_j^T]^T$ $(j = k-N+1, \cdots, k-1, k)$ is the $i\ell$-vector; U_i, W_i and V_i have same structures;

$$X_i = [\, x_{k-N-i+2}, \cdots, x_{k-i}, x_{k-i+1} \,].$$

On the other hand, if we employ the distribution-based subspace identification (Ohsumi, Kameyama, and Yamaguchi, 2002), the continuous-time stochastic systems,

$$\Sigma_C : \begin{cases} \dot{x}(t) = Ax(t) + Bu(t) + w(t) \\ y(t) = Cx(t) + Du(t) + v(t), \end{cases} \tag{3}$$

where $w(t)$, $v(t)$ are zero-mean white Gaussian random processes, leads us to (1) in which the output matrix Y_i denoted by $Y_i(\varphi)$ indicating the used C^∞-class test function $\varphi(t;\cdot)$ is constructed by

$$Y_i(\varphi) = [\, y_i(\varphi)(t_{k-N+1}), \cdots, y_i(\varphi)(t_{k-1}), y_i(\varphi)(t_k) \,],$$

where $y_i(\varphi)(t_j) = [\, y^T(\varphi)(t_j), \cdots, (-1)^{i-1} y^T(\varphi^{(i-1)})(t_j) \,]^T$, $y(\varphi)(t_j) := \int_{-\infty}^{\infty} y(t)\varphi(t : t_j)dt$, $(\{t_j\}_{j=k-N+1,\cdots,k-1,k})$ are (not necessarily equidistant) different time instants;

$$X_i(\varphi) = [\, x(\varphi)(t_{k-N+1}), \cdots, x(\varphi)(t_{k-1}), x(\varphi)(t_k) \,];$$

and other matrices are defined similarly.

For the purpose of deriving recursive subspace identification algorithms, let t_k or k be the current time instant or step for Σ_C or Σ_D, respectively, and let the integer N be fixed as a (sufficiently) large number. Since no confusion will arise, we write Y_i or $Y_i(\varphi)$ as $Y_i(k|k{-}N{+}1)$ for both cases to indicate explicitly the current time t_k or k and the number of columns N of the data employed. And, furthermore, in order to cope with the time-varying systems, write the matrices Γ_i, H_i, Σ_i as $\Gamma_i(k)$, $H_i(k)$ and $\Sigma_i(k)$. Hence, instead of (1), the input-output algebraic equation is expressed as:

$$\begin{aligned} &Y_i(k|k{-}N{+}1) \\ &\quad = \Gamma_i(k)X_i(k|k{-}N{+}1) + H_i(k)U_i(k|k{-}N{+}1) \\ &\qquad + \Sigma_i(k)W_i(k|k{-}N{+}1) + V_i(k|k{-}N{+}1). \end{aligned} \tag{4}$$

Then, our problem is to derive the recursive algorithm for determining the quadruple system matrices (up to within a similarity transformation) and the system order n. The size of i is selected tacitly such that $i > n$.

3. DERIVATION OF RECURSIVE ALGORITHMS

3.1 Bona fide recursive algorithm

Let $U_h(k|k{-}N{+}1) \in R^{hm \times N}$ be a matrix constructed from the input data such that it is different from the matrix $U_i(k|k{-}N{+}1)$, and perform the LQ factorization with the following partitioning:

$$\begin{bmatrix} U_i(k|k{-}N{+}1) \\ U_h(k|k{-}N{+}1) \\ Y_i(k|k{-}N{+}1) \end{bmatrix} = \begin{bmatrix} L_{11}(k) & 0 & 0 \\ L_{21}(k) & L_{22}(k) & 0 \\ L_{31}(k) & L_{32}(k) & L_{33}(k) \end{bmatrix} \begin{bmatrix} Q_1^T(k) \\ Q_2^T(k) \\ Q_3^T(k) \end{bmatrix}. \tag{5}$$

As is well known this LQ factorization is critical to determine the column space of the extended observability matrix $\Gamma_i(k)$, since for sufficiently large N the relation

$$\frac{1}{\sqrt{N}} L_{32}(k) = \Gamma_i(k) \frac{1}{\sqrt{N}} X_i(k)Q_2(k) \tag{6}$$

holds. So, by means of the singular value decomposition (SVD) of the matrix $L_{32}(k)$ or eigenvalue decomposition of the matrix $L_{32}(k)L_{32}^T(k)$, the pair of matrices (A_T, C_T) and system order n can be determined (where the suffix T indicates the similarity transformation by a nonsingular matrix T). Furthermore, the equation

$$\{U_n^\perp(k)\}^T L_{31}(k) L_{11}^{-1}(k) = \{U_n^\perp(k)\}^T H_i(k), \tag{7}$$

where $U_n^\perp(k)$ is the orthonormal complement of the estimates of $\Gamma_i(k)$ is derived. From the matrices $L_{31}(k)$ and $L_{11}(k)$ the pair of (B_T, D) can be also determined (Verhaegen and Dewilde, 1992) (where $A_T = TAT^{-1}, B_T = TB, C_T = CT^{-1}$). In short, only $L_{32}(k)L_{32}^T(k)$ and $L_{31}(k)L_{32}^T(k)$ are needed to derive estimates of quadruplet recursively.

Given a set of new input and output data $\{u_{k+1}, y_{k+1}\}$ or $\{u(t_{k+1}), y(t_{k+1})\}$, construct vectors $u_i(k+1)$, $u_h(k+1)$ and $y_i(k+1)$ as updated data. Then, obviously we have the relation

$$\begin{bmatrix} U_i(k|k-N+1) \vdots u_i(k+1) \\ U_h(k|k-N+1) \vdots u_h(k+1) \\ Y_i(k|k-N+1) \vdots y_i(k+1) \end{bmatrix} = \begin{bmatrix} u_i(k-N+1) \vdots U_i(k+1|k-N+2) \\ u_h(k-N+1) \vdots U_h(k+1|k-N+2) \\ y_i(k-N+1) \vdots Y_i(k+1|k-N+2) \end{bmatrix} \quad (8)$$

and this is the key equation to derive the recursive algorithm. The LQ-factorization at the time step $k+1$

$$\begin{bmatrix} U_i(k+1|k-N+2) \\ U_h(k+1|k-N+2) \\ Y_i(k+1|k-N+2) \end{bmatrix} = \begin{bmatrix} L_{11}(k+1) & 0 & 0 \\ L_{21}(k+1) & L_{22}(k+1) & 0 \\ L_{31}(k+1) & L_{32}(k+1) & L_{33}(k+1) \end{bmatrix} \begin{bmatrix} Q_1^T(k+1) \\ Q_2^T(k+1) \\ Q_3^T(k+1) \end{bmatrix} \quad (9)$$

together with (5) yields from the relation (8)

$$\left[L_{11}(k)Q_1^T(k) \vdots u_i(k+1)\right] = \left[u_i(k-N+1) \vdots L_{11}(k+1)Q_1^T(k+1)\right] \quad (10)$$

$$\left[L_{21}(k)Q_1^T(k) + L_{22}(k)Q_2^T(k) \vdots u_h(k+1)\right] = \left[u_h(k-N+1) \vdots L_{21}(k+1)Q_1^T(k+1) + L_{22}(k+1)Q_2^T(k+1)\right] \quad (11)$$

$$\left[L_{31}(k)Q_1^T(k) + L_{32}(k)Q_2^T(k) + L_{33}(k)Q_3^T(k) \vdots y_i(k+1)\right] = \left[y_i(k-N+1) \vdots L_{31}(k+1)Q_1^T(k+1) + L_{32}(k+1)Q_2^T(k+1) + L_{33}(k+1)Q_3^T(k+1)\right]. \quad (12)$$

Post-multiplying (12) by the transpose of (11), we have

$$L_{32}(k+1) = \left[L_{31}(k)L_{21}^T(k) + L_{32}(k)L_{22}^T(k) - L_{31}(k+1)L_{21}^T(k+1) + \{y_i(k+1)u_h^T(k+1)\} - y_i(k-N+1)u_h^T(k-N+1)\right]\left[L_{22}^T(k+1)\right]^{-1}. \quad (13)$$

Furthermore, post-multiplying (13) by its transpose, we obtain $L_{32}(k+1)L_{32}^T(k+1)$. In the similar way, after computing $L_{11}(k+1)L_{11}^T(k+1)$, $L_{22}(k+1)L_{22}^T(k+1)$ and $L_{31}(k+1)L_{21}^T(k+1)$, we have the recursive form for $L_{32}(k+1)L_{32}^T(k+1)$, and hence, with the help of eigenvalue decomposition, the pair (A_T, C_T) and the system order n can be determined.

By the similar procedure, having obtained the recursive form for $L_{31}(k+1)L_{11}^{-1}(k+1)$, we can also determine the pair (B_T, D). The procedure is summarised as follows:

Recursive Subspace Identification Algorithm

Assume that at time step k we are given input and output data matrices $Y_i(k|k-N+1)$, $U_i(k|k-N+1)$ and $U_h(k|k-N+1)$ for assumed i and h and obtained the L-factors $\{L_{ij}(k); i,j = 1,2,3\}$ by the LQ-factorization (5).

Step 1: With newly constructed input and output vectors $u_i(k+1)$ and $y_i(k+1)$, compute the following matrices successively:

$$L_{11}(k+1)L_{11}^T(k+1) = L_{11}(k)L_{11}^T(k) - u_i(k-N+1)u_i^T(k-N+1) + u_i(k+1)u_i^T(k+1) \quad (14)$$

$$L_{21}(k+1)L_{21}^T(k+1) = \hat{L}_1(k+1)\{L_{11}(k+1)L_{11}^T(k+1)\}^{-1}\hat{L}_1^T(k+1) \quad (15)$$

$$L_{31}(k+1)L_{21}^T(k+1) = \hat{L}_2(k+1)\{L_{11}(k+1)L_{11}^T(k+1)\}^{-1}\hat{L}_1^T(k+1) \quad (16)$$

$$L_{22}(k+1)L_{22}^T(k+1) = \hat{L}_3(k+1) - L_{21}(k+1)L_{21}^T(k+1), \quad (17)$$

where $\hat{L}_1(k) := L_{21}(k)L_{11}^T(k)$, $\hat{L}_2(k) := L_{31}(k) \cdot L_{11}^T(k)$ and $\hat{L}_3(k) := L_{21}(k)L_{21}^T(k) + L_{22}(k)L_{22}^T(k)$, and these are obtained recursively as

$$\hat{L}_1(k+1) = \hat{L}_1(k) - u_h(k-N+1)u_i^T(k-N+1) + u_h(k+1)u_i^T(k+1) \quad (18)$$

$$\hat{L}_2(k+1) = \hat{L}_2(k) - y_i(k-N+1)u_i^T(k-N+1) + y_i(k+1)u_i^T(k+1) \quad (19)$$

$$\hat{L}_3(k+1) = \hat{L}_3(k) - u_h(k-N+1)u_h^T(k-N+1) + u_h(k+1)u_h^T(k+1). \quad (20)$$

Step 2: Compute $L_{32}(k+1)L_{32}^T(k+1)$ by

$$L_{32}(k+1)L_{32}^T(k+1) = \left[\hat{L}_4(k+1) - L_{31}(k+1)L_{21}^T(k+1)\right] \cdot \left[L_{22}(k+1)L_{22}^T(k+1)\right]^{-1} \cdot \left[\hat{L}_4(k+1) - L_{31}(k+1)L_{21}^T(k+1),\right]^T, \quad (21)$$

where $\hat{L}_4(k) := L_{31}(k)L_{21}^T(k) + L_{32}(k)L_{22}^T(k)$ and this is computed from

$$\hat{L}_4(k+1) = \hat{L}_4(k) - y_i(k-N+1)u_h^T(k-N+1)$$

$$+ y_i(k{+}1)u_h^T(k{+}1). \qquad (22)$$

Step 3: Perform the eigenvalue decomposition of $L_{32}(k{+}1)L_{32}^T(k{+}1)$ obtained in Step 4 as follows:

$$L_{32}(k{+}1)L_{32}^T(k{+}1) = \begin{bmatrix} U_n(k) & U_n^{\perp}(k) \end{bmatrix} \begin{bmatrix} S_1^2(k) & 0 \\ 0 & S_2^2(k) \end{bmatrix} \begin{bmatrix} U_n^T(k) \\ \{U_n^{\perp}(k)\}^T \end{bmatrix} \qquad (23)$$

which gives the information of the system order n.

Step 4: From the knowledge $U_n \in R^{li \times n}$ obtained in Step 3, compute the pair (A_T, C_T).

Step 5: Compute $L_{31}(k)L_{11}^{-1}(k)$ by

$$L_{31}(k{+}1)L_{11}^{-1}(k{+}1) = \hat{L}_2(k{+}1)\{L_{11}(k{+}1)L_{11}^T(k{+}1)\}^{-1}, \qquad (24)$$

and obtain the pair (B_T, D) from this and $U_n(k)$. Repeating these steps, estimates of the quadruple matrices for (A_T, B_T, C_T, D) and obtained recursively.

We call this algorithm the *bona fide* recursive algorithm in contrast to other approaches because our approach is based on the direct recursion of L-factors of LQ factorization.

3.2 Algorithm Using Cholesky Factorization

The most primitive and simplest approach to derive the recursive algorithm will be the use of Cholesky factorization in the data matrix. To fix the idea, write the relation (8) as (Sawada, 2003)

$$\begin{bmatrix} \mathcal{H}_0(k) & h_0(k{+}1) \end{bmatrix} = \begin{bmatrix} h_0(k{-}N{+}1) & \mathcal{H}_0(k{+}1) \end{bmatrix}. \qquad (25)$$

Post-multiplying the L.H.S. of (25) by its transpose, we have

$$\begin{aligned} \mathcal{H}_0(k{+}1)\mathcal{H}_0^T(k{+}1) &= \mathcal{H}_0(k)\mathcal{H}_0^T(k) + h_0(k{+}1)h_0^T(k+1) \\ &\quad - h_0(k{-}N{+}1)h_0^T(k{-}N{+}1). \qquad (26) \end{aligned}$$

This gives the recursion form for $\mathcal{H}_0(k)\mathcal{H}_0^T(k)$.

Performing the Cholesky factorization of $\mathcal{H}_0(k) \cdot \mathcal{H}_0^T(k)$ matrix, we have the relation

$$\mathcal{H}_0(k)\mathcal{H}_0^T(k) = L(k)L^T(k), \qquad (27)$$

where $L(k)$ is a lower triangular matrix.

This approach is simple, however the matrix $\mathcal{H}_0(k)\mathcal{H}_0^T(k)$ may become rank-deficient whenever the data are free from the observation noise or the noise intensity is so small, so that the Cholesky factorization may fail (Westwick, Kearney and Verhaegen, 1996).

4. ALTERNATIVE RECURSIVE ALGORITHM

In the previous section, two recursive algorithms have been proposed based on the key relation (8) which implies that the total amount of input and output data employed in the current identification is kept constant. Instead, alternative algorithm is still available if we allow the increase of the amount of input and output data employed in current identification. Such algorithm may be rather preferable for time-invariant systems because the increase in the amount data implies inevitably the higher accuracy in the identification results.

Let (for Σ_D),

$$Y_i(k|k_0) = [y_i(k_0), \cdots, y_i(k-1), y_i(k)], \qquad (28)$$

where k_0 is some fixed time step, and $k = k_0, k_0 + 1, \cdots$; and perform the LQ-factorization,

$$\begin{bmatrix} U_i(k{+}1|k_0) \\ U_h(k{+}1|k_0) \\ Y_i(k{+}1|k_0) \end{bmatrix} = \begin{bmatrix} L_{11}(k{+}1) & 0 & 0 \\ L_{21}(k{+}1) & L_{22}(k{+}1) & 0 \\ L_{31}(k{+}1) & L_{32}(k{+}1) & L_{33}(k{+}1) \end{bmatrix} \begin{bmatrix} Q_1^T(k{+}1) \\ Q_2^T(k{+}1) \\ Q_3^T(k{+}1) \end{bmatrix}. \qquad (29)$$

This time the relation

$$\begin{bmatrix} U_i(k{+}1|k_0) \\ U_h(k{+}1|k_0) \\ Y_i(k{+}1|k_0) \end{bmatrix} = \begin{bmatrix} U_i(k|k_0) & \vdots & u_i(k+1) \\ U_h(k|k_0) & \vdots & u_h(k+1) \\ Y_i(k|k_0) & \vdots & y_i(k+1) \end{bmatrix} \qquad (30)$$

is the key. Similar procedure taken in Subsection 3.1 is applicable. In this algorithm the number of columns of each data matrix increases according to $N = k - k_0 + 1$ as a function of k.

5. IDENTIFICATION OF TIME-VARYING SYSTEMS

It will be needless to say that the recursive algorithm is valuable, especially, for time-varying systems. As mentioned above, the input-output algebraic equation (4) can cope with the time-varying case. "Time-varying" means in this paper that all system matrices $(A(t), B(t), C(t), D(t))$ for Σ_C or $(A(k), B(k), C(k), D(k))$ for Σ_D change slowly with time. Qualitatively speaking, the instinctive word "slowly" implies that each matrix changes smoothly and continuously, and does never abruptly or randomly.

Indeed, consider an interval I_k around the time step k, and assume that the system matrix $A(t)$ or

$A(k)$ behaves to be approximately time-invariant during this interval, i.e., $A(t) = A(t_k) = A_k$, or $A(k) = A_k$. Under this local stationarity assumption, (4) is derived as the input-output algebraic relation in which matrices $\Gamma_i(k), H_i(k)$ and $\Sigma_i(k)$ are given in terms of A_k, B_k, C_k and D_k (Ohsumi and Kawano, 2002).

6. SIMULATION STUDIES

Two cases for the continuous-time system Σ_C are considered to illustrate the proposed recursive algorithm. Simulations for the discrete-time system Σ_D can be done similarly *mutatis mutandis*.

As test functions, the Gaussian function $\varphi(t; t_j) = \exp\{-(t-t_j)^2/2\sigma_0\}$ is employed for constructing U_i, Y_i and U_h ($\sigma_0 = 0.8$ for U_i and Y_i, and $\sigma_0 = 5.0$ for U_h). The matrix size parameters were set $i = 6$, $h = 10$. The system Σ_C was discretized with time-partition $\Delta t = 0.01$ sec, and the sampling time for identification was equi-partitioned as $t_j = t_{j-1} + \nu M \Delta t$ with $M = 15$, $\nu = 20$.

6.1 Identification of time-invariant system

The true system is the second-order linear system with

$$A = \begin{bmatrix} -0.2 & 1 \\ -0.4 & -1 \end{bmatrix}, \quad B = \begin{bmatrix} 2 & 1.5 \\ -1 & 2 \end{bmatrix}$$

$$C = \begin{bmatrix} 1 & 2 \\ -1 & 2 \end{bmatrix}, \quad D = \begin{bmatrix} 1.5 & 3 \\ 2 & 1 \end{bmatrix};$$

and $Q = R = 0.1^2 I_{2\times 2}$, $S = 0$ ($I_{2\times 2}$ denotes the 2×2-identity matrix).

Figures 1 and 2 depict typical sample paths of the identified real part of the pole using the *bona fide* recursive algorithm proposed in Subsection 3.1 for $N = 500$ and $N = 10000$, respectively. True trajectories are depicted by dashed lines and estimated ones by solid lines. Clearly, it is known from these figures that the more the amount of data is, the higher the estimation accuracy becomes.

On the other hand, Fig. 3 depicts the results by the algorithm stated in Section 4. The estimation accuracy becomes higher as the data amount, i.e. $N = N(k)$, increases. The data amount used in Figs. 1 and 2 correspond to that at $t = 252.5$ and $t = 5000.25$ sec in Fig. 3, respectively.

6.2 Identification of time-varying system

First, consider the case where the $(1, 1)$-component of the system matrix $A(t)$ changes slowly such that

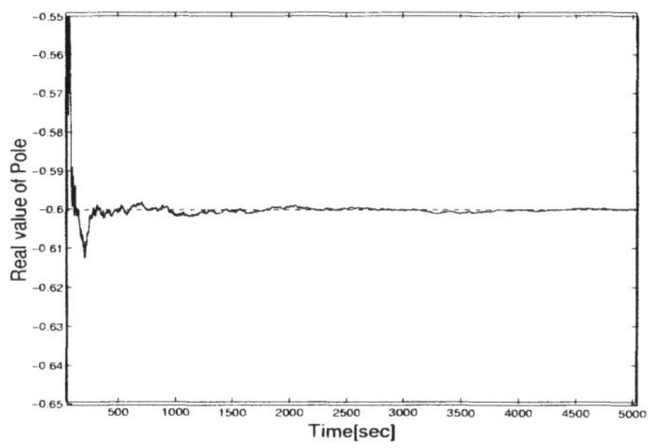

Fig.1. Estimated real part of the pole by the *bona fide* recursive algorithm (Time-invariant case; $N = 500$).

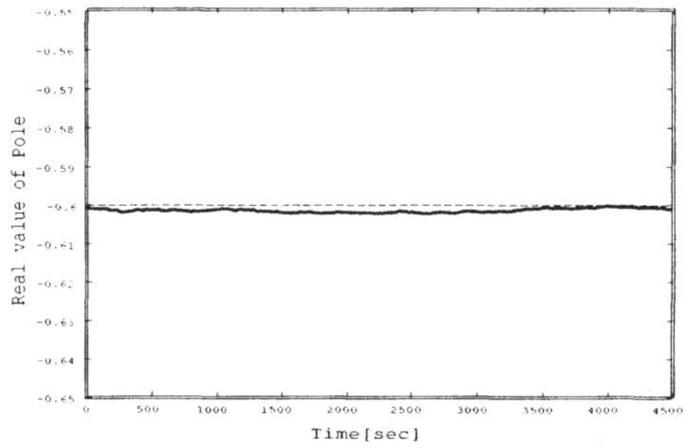

Fig.2. Estimated real part of the pole by the *bona fide* recursive algorithm (Time-invariant case; $N = 10000$).

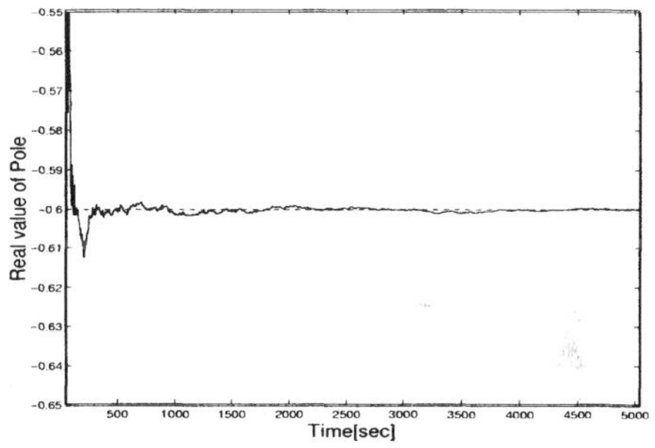

Fig.3. Estimated real part of the pole by the recursive algorithm stated in Section 4 (Time-invariant case).

$$a_{11}(t) = 0.1 - 0.3\exp(t/1500) \qquad (31)$$

(note that if we set as $t = 0$, this is equal to $a_{11} = -0.2$).

Figures 4 and 5 show the results of sample means of 100 experiments by the algorithms stated in Subsection 3.1 and Section 5, respectively. It can be seen in Fig. 5 that in the time-varying case the algorithm stated in Section 5 degrades its estimation accuracy as time goes on, in other words, as data amount increases.

It should be noted that the *bona fide* recursive algorithm took 23.6 seconds for simulation, while the off-line algorithm (Ohsumi and Kawano, 2002) did 184.3 seconds. This proves the efficacy of the proposed recursive algorithm.

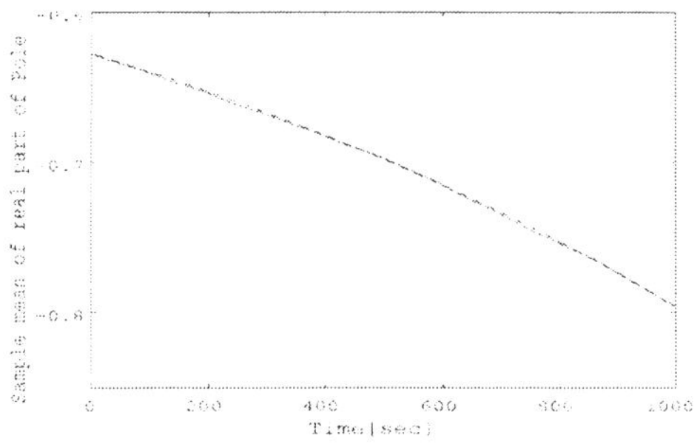

Fig.4. Estimated real part of the pole by the *bona fide* recursive algorithm (Time-varying case; $N = 500$).

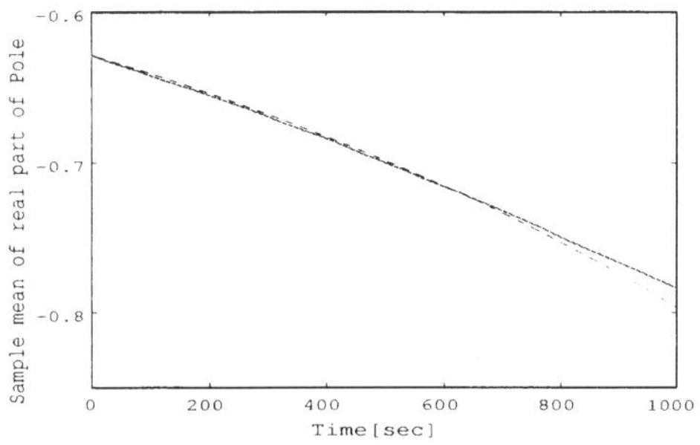

Fig.5. Estimated real part of the pole by the recursive algorithm stated in Section 4 (Time-varying case).

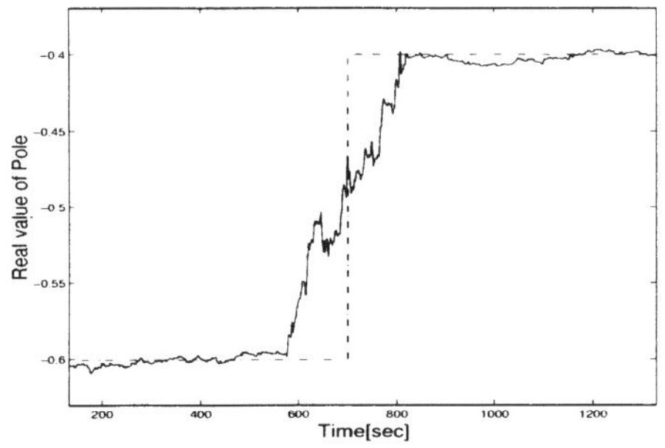

Fig.6. Estimated real part of the pole by the *bona fide* recursive algorithm (sudden change case; $N = 500$).

Figure 6 illustrates the simulation result obtained by the *bona fide* algorithm when the component $a_{11}(t)$ changes suddenly as

$$a_{11}(t) = \begin{cases} -0.2 & (0 < t \leq 700) \\ 0.2 & (t > 700). \end{cases} \quad (32)$$

Even in this case the proposed *bona fide* algorithm tracks the system fairly well.

7. CONCLUSION

An efficient recursive algorithm has been derived by updating the L-factors in the LQ-factorization based on the obvious relation (8). It should be noted that the amount of the data matrices is unchanged at every updating time step, while all other approaches mentioned in Introduction seem to increase the matrix size. The efficiency of the proposed method has been shown by simulation experiments.

8. REFERENS

Verhaegen, M. and E. Deprettere (1991). A Fast, Recursive MIMO State Space Model Identification Algorithm, *Proc. 30th Conference on Decision and Control*, Brighton, England, 1349/1354.

Gustafsson, T. (1997). Recursive System Identification Using Instrumental Variable Subspace-Tracking, *Proc. 11th IFAC Symp. System Identification* (SYSID'97), Fukuoka, 1683/1688.

Lovera, M., T. Gustafsson, and M. Verhaegen (2000). Subspace Identification of Linear and Nonlinear Wiener State-space Models, *Automatica*, vol. 36, 1639/1650.

Oku, H. and H. Kimura (1999). A Recursive 4SID from the Input-Output Point of View, *Asian J. Control*, vol. 1, 258/269.

Oku, H. and H. Kimura (2002). Recursive 4SID Algorithms Using Gradient Type Subspace Tracking, *Automatica*, vol. 38, 1035/1043.

Takei, Y., J. Imai, and K. Wada (2000). Recursive Subspace Identification Based on Subspace Extraction via Schur Complement, *Proc. 32nd ISCIE Int. Symp. Stochastic System Theory and Its Applications*, Tottori, 27/32.

Ohsumi, A., K. Kameyama, and K.-I. Yamaguchi (2002). Subspace Identification for Continuous-time Stochastic Systems via Distribution-based Approach, *Automatica*, vol. 38, 63/79.

Ohsumi, A. and T. Kawano (2002). Subspace Identification for a Class of Time- varying Continuous- time Stochastic Systems via Distribution-based Approach, *Preprint of 15th IFAC World Congress.*

Sawada, K. (2003). Recursive Subspace Identification Using Cholesky Factorization, Bachelor Thesis, Kyoto Inst. Tech. Kyoto, Feb.

Verhaegen, M. and P. Dewilde (1992). Subspace Model Identification, Part 1: The Output-Error State-Space Model Identification Class of Algorithms, *Int. J. Control*, vol. 56, no. 5, pp.1187-1210.

Westwick, D.T., R. E. Kearney and M. Verhaegen (1996). An Efficient Implementation of the PI Subspace State-Space System Identification Algorithm. In: *Proc. 18th Annual Int. Conf. IEEE Engineering and Biology Soc.*, Amsterdam, pp.1674-1675.

IFAC Publications
www.elsevier.com/locate/ifac

"PLANT-FRIENDLY" SYSTEM IDENTIFICATION: A CHALLENGE FOR THE PROCESS INDUSTRIES

Daniel E. Rivera, *,1 **Hyunjin Lee**, * **Martin W. Braun** *,2
and Hans D. Mittelmann **

* *Control Systems Engineering Laboratory*
Department of Chemical and Materials Engineering
** *Department of Mathematics and Statistics*
Arizona State University, Tempe, Arizona 85287

Abstract: The term "plant-friendly" system identification has been used within the chemical process control research community in reference to the broad-based goal of accomplishing informative identification testing while meeting the demands of industrial practice. While many different identification topics (such as control-relevant identification, closed-loop identification and optimal input design) can be said to contribute to plant-friendliness in identification, the problem has some unique character of its own. This paper describes some of the issues that motivate plant-friendly identification and presents an overview of some approaches that have been proposed in this topic. The problem of *identification test monitoring* is presented as a novel means for accomplishing plant-friendly identification.

Keywords: plant-friendly system identification, chemical process control, identification test monitoring

1. INTRODUCTION

The term "plant-friendly" identification is one that seems to be exclusive to the chemical process control community. The exact origin of the term is not clear, but it first appears mentioned in print in the paper by Pearson *et al.* (1993). However, articulation of issues related to plant-friendly considerations can be found in literature based on the experience of industrial practitioners (Rivera *et al.*, 1992) or influenced by the demands of practice (Godfrey, 1993).

The concept of plant-friendliness in system identification for the process industries stems from the fundamental need for informative experiments despite practical requirements to the contrary. Broadly speaking, a plant-friendly identification test will produce data leading to a suitable model within an acceptable time period, while keeping the variation in both input and output signals within user-defined constraints. Examples of plant-friendly constraints (and their impact on process operations) include:

- keeping output deviations low to minimize variability in product quality,
- implementing a signal of sufficiently short duration to minimize the amount of off-spec product and reduce engineering time associated with an identification test,
- keeping move sizes small to satisfy actuator constraints and minimize "wear and tear" on process equipment.

These practical considerations are often in conflict with theoretical requirements (e.g., asymptotic operation, persistence of excitation, etc.) that demand long identification tests under high signal-to-noise ratios. As a result, plant-friendliness often involves a compromise between the demands of theory (which are for the most part "plant-hostile") and the demands of

[1] to whom all correspondence should be addressed; phone:(480) 965-9476, email:daniel.rivera@asu.edu
[2] Currently with Texas Instruments Inc., 13570 N. Central Expressway, MS 3701, Dallas, TX, 75265

plant engineers (who would prefer no changes in the process as a result of identification testing).

The main objective of this paper is to create increased awareness of the plant-friendly identification problem within the larger system identification community. The paper is organized as follows: Section 2 discusses the motivation for plant-friendliness, using industrially-relevant examples to justify why the problem goes beyond classical input signal design approaches. Section 3 presents two plant-friendliness measures and surveys some existing approaches to the problem. In Section 4 the problem of *identification test monitoring* is proposed as a novel means for accomplishing plant-friendly identification; this discussion is supported with some analysis and an example. The paper concludes with a summary section. Ultimately, it is our desire that this paper will stimulate multidisciplinary research efforts in this important problem to the practicing process control community.

2. MOTIVATION FOR PLANT-FRIENDLY IDENTIFICATION

System identification in practice is an iterative procedure. The lack of *a priori* information regarding the plant model will require that initially each step be examined in a superficial manner. After each stage, the user must discern if the previous stages were properly accomplished; if this is not the case, the stage(s) must be redone until a "satisfactory" model is obtained. A satisfactory model is one that meets the requirements of the intended application (e.g., simulation, prediction, or control).

The quality of the data generated from the experimental design stage is critical to the success of the comprehensive system identification and subsequent control design procedures. In the chemical process industries, identification testing is by necessity conducted while the plant is in normal operation, and as such represents one of the most expensive and time consuming steps in the application of advanced control in the process industries. There are many industrial examples of the significant disruption and cost that identification testing represents to plant operations; here we just discuss a few. As a research engineer at Shell Development Company in the late 1980's, the first author was involved in implementing individual identification tests lasting five days or more. More recently, while consulting for a major chemical company, the first author was presented with the identification testing guidelines of a major control software vendor. To identify an ethylene furnace characterized by 17 independent variables and a six hour settling time, the vendor's guideline suggested nearly a month (25.5 days) of identification tests. Mathur and Conroy (2002) cite an example where the total costs of step testing (taking into account reductions in throughput, off-spec product, and engineering time) were estimated at $270,000. It comes as no surprise that model development has been reported to account for 75% of the costs associated with an advanced control project (Hussain, 1999).

In addition, "fanout" issues are a problem - there are a large percentage of loops in industrial practice for which significant performance improvements would be possible as a result of system identification (Ogawa and Douke, 2002). This task, however, has to be done in an acceptable manner to operations. This represents yet another significant incentive for accomplishing plant-friendly, informative identification testing.

One could expect that for a mature area such as system identification, existing literature would address these problems. For example, the identification literature abounds with optimal input design theorems which, one would expect, should be well suited to address the plant-friendly identification problem. However, the majority of these "optimal" solutions are not formulated in a manner meaningful to the chemical process engineer. For starters, most optimal input design results depend on knowledge of the true plant and noise models (Ljung, 1999), which is information that is largely unknown to the user (at least at the start of identification testing). Furthermore, process control engineers tend to think more in terms of maintaining high/low limits, move size constraints, and test duration during identification testing and less in terms of the norm criteria that are typically used in the classical optimal experimental design formulations.

Control-relevant system identification has been an important research topic in the community since the late 1980's. However, control-relevance in and of itself does not automatically imply plant-friendliness. Taking control performance requirements into consideration during identification testing will often lead to more focused input signal with a narrower bandwidth and emphasizing the important regimes of time and frequency needed for the intended application (Rivera *et al.*, 1992; Cooley *et al.*, 1998). However, control-relevant input designs require additional scrutiny (principally in the form of constraint enforcement) to insure that they will promote plant-friendliness. Similar statements can be made for closed-loop identification and its role in plant-friendly identification.

Ultimately, the main question in plant-friendly identification (and which distinguishes it from other problems in the field) is this: how does one build process knowledge relevant to system identification in a systematic and nearly automatic way, with little user intervention and without demanding significant computational time and effort? If adequate process knowledge is absent at the start of experimental testing, how can this knowledge be systematically acquired in the course of identification testing, for the purposes of improving the identification test?

3. SURVEY OF PLANT-FRIENDLY IDENTIFICATION APPROACHES

Current approaches to plant-friendly identification range from the use of simple measures to sophisticated constrained formulations. Doyle *et al.* (1999) first proposed a *friendliness index* f for an arbitrary input sequence $u(k)$, $k = 1 \cdots N$. The measure is defined as a percentage according to

$$f = 100 \times \left(1 - \frac{n_T}{N-1}\right) \tag{1}$$

where N is the sequence length and n_T constitutes the number of transitions (i.e., situations where $u(k) \neq u(k-1)$) in the input signal. This measure is also examined in Rengasamy *et al.* (2000) and Parker *et al.* (2001). According to this measure, a constant sequence is "100% plant-friendly", while any sequence that changes value at every instant is "0% plant friendly." A stochastic interpretation and its use in the design of input signals for identifying Volterra series models are presented in the aforementioned papers.

Another measure that has been proposed to determine plant-friendliness is the Crest Factor (CF) (Guillaume *et al.*, 1991). The crest factor is defined as the ratio of the ℓ_∞ norm and the ℓ_2-norm of a signal

$$CF(u) = \frac{\ell_\infty(u)}{\ell_2(u)} \tag{2}$$

and provides a measure of how well distributed the signal values are over the input span. A low crest factor indicates that most of the elements in the input sequence are distributed near the minimum and maximum values of the sequence. Reducing the crest factor of the input or output signals (or both) can significantly contribute to plant-friendliness during experimental testing. For example, if two signals with equivalent power spectral densities are to be evaluated for identification purposes, the one with lower crest factor is preferred because it will deliver the same power over a lower overall span. An alternative measure to crest factor is the Performance Index for Perturbation Signals (PIPS) (Godfrey *et al.*, 1999).

$$PIPS(\%) = 200\frac{\sqrt{u_{rms}^2 - u_{mean}^2}}{u_{max} - u_{min}} \tag{3}$$

The PIPS measure ranges between 0 and 100% (compared to 1 versus ∞ for crest factor), which gives it an intuitive, practical appeal.

In Braun *et al.* (2002), crest factor minimization is used as the basis for designing multisine inputs meaningful for plant-friendly identification of both linear and nonlinear systems. Constrained extensions have been presented for both the single input (Rivera *et al.*, 2002) and multivariable cases (Lee *et al.*, 2003); these formulations allow specifying both frequency domain signal requirements (meaningful from a system theoretic perspective) and time-domain constraints (meaningful from the user's standpoint).

We previously noted that control-relevant approaches to input signal design can be formulated to promote plant-friendliness; examples of this approach include the work of J.H. Lee and co-workers (Chikkula and Lee, 1997; Cooley *et al.*, 1998; Cooley and Lee, 2001). Actuator constraints are recognized in these problem formulations. Other control-relevant approaches taking into account constraints include the work of Li and Georgakis (2002) who focus on highly interactive systems. The recent work Narasimhan *et al.* (2003) considers a multi-objective framework in which the tradeoffs associated with maximizing input-output friendliness, constraints, and other criteria are explored in the context of a mixed-integer nonlinear programming problem.

4. IDENTIFICATION TEST MONITORING

Identification test monitoring is proposed as a novel approach to the plant-friendly identification problem. Factors that influence this approach include the following:

(1) *a priori* knowledge of the system dynamics available to the engineer is often sketchy at the start of identification testing. To maintain flexibility in the design the user is forced to make some generous assumptions. A framework is needed that allows the user to systematically acquire process knowledge which in turn is used to refine the experimental test, without resorting to a completely new experimental testing procedure.

(2) The use of periodic, deterministic inputs such as multisine or pseudo-random signals provides distinct advantages. The timespan defined by one period of a periodic signal provides a natural window or examination point for analysis of the data and signals. As a result, monitoring procedures can be established that work in "quasi-real" time, where the data resulting from prior cycles of a multisine signal are analyzed while the current cycle is being implemented.

The iterative evaluation and refinement of identification signals implied by the identification test monitoring problem contrasts much of current industrial practice. The tendency in practice is to collect and analyze data in one batch, and determine (after pursuing the complete identification cycle) if the data is adequate. This results in costly re-testing and requires significant user intervention and effort. In contrast, a systematic monitoring approach based on periodic inputs allows the opportunity for users to improve tests without having to perform a full comprehensive analysis.

4.1 *Towards Identification Test Monitoring: Integrating Identification and Robust Loopshaping*

In addition to system identification, it is possible to draw from the fields of signal processing, robust control, and optimization to synergistically create novel frameworks for identification test monitoring. Data generated from periodic inputs can be used to calculate an empirical transfer function estimate (ETFE) (Ljung, 1999). According to Bayard (1993), a confidence region in the Nyquist plane for the ETFE $p^*(\omega_i)$ is a perfect circle centered at $p^*(\omega_i)$ of radius ϵ_i where

$$\epsilon_i^2 = \frac{\hat{\sigma}^2|\overline{W}(e^{-j\omega_i T})|^2 F_{1-\kappa}(2,\nu)}{\Phi_u(\omega_i)m} \quad (4)$$

$\overline{W}(z)$ is the estimated noise model while $\hat{\sigma}^2$ is the estimated variance from the residual output spectrum. m corresponds to the number of cycles of the periodic input, while $F_{1-\kappa}(2,\nu)$ is the 2-way Fisher statistic computed for a specified statistical confidence of $(1-\kappa)\times 100\%$. Noting that $F_{1-\kappa}(2,\nu)$ is bounded as ν becomes large (e.g., $F_{1-\kappa}(2,\nu)\leq 3$ for $1-\kappa=.95$ and $\nu > 120$), it becomes clear that the uncertainty region increases with the noise-to-signal ratio $\hat{\sigma}^2|\overline{W}|/\Phi_u$ and decreases as the number of input signal cycles m increases. Thus (4) provides significant insight regarding the important practical issues of signal magnitude and test length in system identification. Noise in the data set can be overcome by either increasing signal power or lengthening the test duration. The decision to follow one approach over the other is dependent upon the circumstances being faced during identification testing, such as operational restrictions, and so forth. These are the types of tradeoffs that the identification test monitoring problem seeks to address.

The confidence regions defined by (4) can be expressed as norm-bounded multiplicative uncertainty

$$|(p(e^{j\omega_i T}) - p^*(\omega_i))p^{*-1}(\omega_i)| \leq \bar{\ell}_m(\omega_i) = \epsilon_i/|p^*(\omega_i)| \quad (5)$$

which in turn can be used to assess robust performance, such as the μ analysis measure

$$\mu^* = \sup_{\omega}|\eta^*(e^{j\omega T})\bar{\ell}_m(\omega)| + |\epsilon^*(e^{j\omega T})w_P(j\omega)| \quad (6)$$

w_P weights the sensitivity function $\epsilon = (1+pc)^{-1}$; $\eta = pc(1+pc)^{-1}$ is the complementary sensitivity function. η^* and ϵ^* are the frequency responses of the closed-loop transfer functions based on the estimated frequency response p^*. Whenever $\mu^* < 1$, the following condition is satisfied for the closed-loop system

$$|\epsilon| \leq 1/|w_P| \quad \forall p \in \bar{\ell}_m \quad 0 \leq \omega \leq \pi/T$$

The paper by Braatz *et al.* (1991) provides a procedure based on robust loopshaping for determining necessary and sufficient bounds on the nominal closed-loop transfer functions from knowledge of the process uncertainty and performance specifications on the closed-loop system. For a SISO system, sufficient bounds on η^* and ϵ^* are

$$|\eta^*| < \frac{1-|w_P|}{|w_P|+\bar{\ell}_m} \qquad |\epsilon^*| < \frac{1-\bar{\ell}_m}{|w_P|+\bar{\ell}_m} \quad (7)$$

Additional bounds (namely necessary upper and lower bounds on $|\eta^*|$ and $|\epsilon^*|$) can be found in Braatz *et al.* (1991). Using this analysis, one realizes that the robust loopshaping bounds on η^* and ϵ^* can be computed in real-time, *during identification testing*. The decision to halt or modify an identification test can be determined on the basis of how these bounds evolve with increasing number of cycles of the input m.

4.2 *An Identification Test Monitoring Example*

The "thought process" that may be involved in the case of identification test monitoring of a SISO system is presented in this section using a simulated representative scenario. This is depicted in Figure 1 with relevant statistics summarized in Table 1. The simulation considers a first-order system per the transfer function

$$p(s) = \frac{1}{10s+1}$$

the parameters for this system are assumed to be vaguely known *a priori*. We further assume that the plant is facing significant disturbances, but plant operating restrictions dictate that the test be carried out under low Signal-to-Noise Ratios (SNRs). Multisine input signals relying on the constrained minimum crest factor design procedure per Rivera *et al.* (2002) and Lee *et al.* (2003) are used to construct this identification test monitoring scheme. The initial signal (for Stage 1) is designed per the guidelines in (Rivera *et al.*, 2002; Lee *et al.*, 2003) with $\alpha_s = 2, \beta_s = 3, \tau_{dom}^L = 5$ min, $\tau_{dom}^H = 20$ min. The result is a signal of very wide bandwidth, implemented using a low amplitude span (± 0.75) to avoid undesirable levels of off-spec product. This signal is represented by Stage 1 in Table 1 and Figure 1. Analysis of the Stage 1 data results in a preliminary model that can then serve as a basis for the design of the signal in Stage 2. This preliminary model will most likely display high variance in the parameters, as a result of the low signal-to-noise ratio of the data. Nonetheless the information contained in this model is useful for determining the design parameters of Stage 2. In this simulated scenario the model time constant range is narrowed to between 8 and 12 minutes; the resulting use of the guidelines produces a signal of much shorter duration. Furthermore, the initial estimate of system gain grants the user confidence to increase the input span to ± 1.5, improving the signal to noise ratio. The model estimated from the the two cycles of data can be used to generate the third and final stage; in this simulated scenario an input design with a control-relevant power spectrum as defined per Rivera *et al.* (1992) and with higher span (± 2.5) is used. Given the improved model knowledge from analysis of the first two cycles, the increased input span does not result in unacceptable swings in the output. This final design takes advantage of improved process knowledge to incorporate control requirements which ultimately results in a high performance control system.

It was previously stated that robust loopshaping measures can be used to determine in real-time, during identification testing, limitations to achievable control performance from the data. Consider the signal design per Stage 1 in the previous example. Here a Type-I weight function per Skogestad and Postlethwaite (1996) is chosen:

$$w_P = \frac{s/M_p + \omega_B^*}{s + \omega_B^* A} \tag{8}$$

where M_p and ω_B^* represent the upper maximum peak and lower bandwidth bounds in ϵ, respectively. A (set to zero in this analysis) is a zero-frequency upper bound. Figure 2 presents the results of the various bounds in Braatz *et al.* (1991) for two cycles of the signal per Stage 1 with SNR = 1 and performance specs defined by $M_p = 2$ and $\omega_B^* = 0.1$. Figure 2 shows that a nominal control design with sensitivity and complementary sensitivity operators per

$$\epsilon(s) = \frac{10s}{10s+1}, \qquad \eta(s) = \frac{1}{10s+1}$$

satisfies robust performance conditions. Having confirmed via this analysis that the data is sufficiently informative to yield a model leading to acceptable performance requirements, the experimental test can be halted.

5. SUMMARY AND CONCLUSIONS

We have presented the plant-friendly identification problem as an important issue meaningful to the process industries. In spite of the significant advances in the system identification field, this problem remains open to further avenues of research. Identification test monitoring was presented as one approach to address this gap, and a basic identification test monitoring scheme relying on the integration of identification concepts with robust loopshaping was presented. These measures, while useful, are conservative, which is a consideration that should be addressed in future work. Furthermore, non-expert usage is critical in the implementation of this concept in practice.

6. ACKNOWLEDGEMENT

Support for this work from the American Chemical Society - Petroleum Research Fund, Grant No. ACS PRF# 37610-AC9 is gratefully acknowledged.

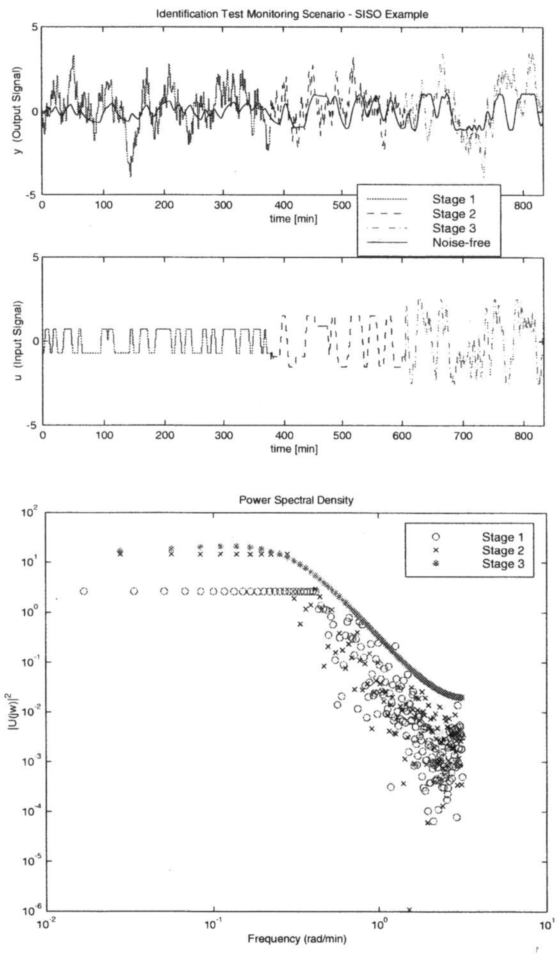

Fig. 1. Simulated identification test monitoring scenario for a single-input, single-output plant. a) time series b) power spectra. The "snow effect" is enabled only in the high frequency range of Stages 1 and 2.

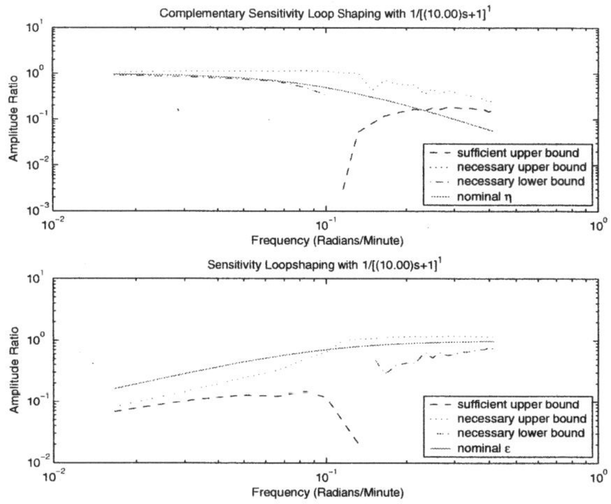

Fig. 2. Robust loopshaping example results.

7. REFERENCES

Bayard, D.S. (1993). Statistical plant set estimation using schroeder-phased multisinusoidal input design. *J. Applied Mathematics and Computation* **58**, 169.

Braatz, R.D., M. Morari and J.H. Lee (1991). Necessary/sufficient loopshaping bounds for robust

Type	Signal (x)	CF(x)	PIPS(%)	max $\|\Delta x\|$	max x	min x	SNR (db)
Stage 1: flat PSD (with snow);	u	1.0898	91.7598	0.4999	0.7042	-0.7042	
$\|\Delta u\| \leq 0.5$	y	2.0623	50.4701	0.2437	0.6174	-0.6705	-18.99
Stage 2: flat PSD (with snow);	u	1.2665	78.9565	0.4999	1.4999	-1.4999	
$\|\Delta u\| \leq 0.5$, $\|u\| \leq 1.5$	y	1.4816	67.4283	0.6497	1.0248	-1.0247	-13.67
Stage 3: control-relevant PSD;	u	1.6620	60.1671	0.6999	2.4999	-2.4999	
$\|\Delta u\| \leq 0.7$, $\|u\| \leq 2.5$	y	1.3619	75.3906	0.3079	1.0850	-1.1457	-13.46

Table 1. Summary of results for the simulated SISO identification test monitoring problem scenario. All statistics for y (except SNR) are calculated on the noise-free portion of the signal.

performance. In: *1991 AIChE Annual Meeting*. Los Angeles, CA.

Braun, M.W., R. Ortiz-Mojica and D.E. Rivera (2002). Application of minimum crest factor multisinusoidal signals for "plant-friendly" identification of nonlinear process systems. *Control Engineering Practice* **10**, 301.

Chikkula, Y. and J.H. Lee (1997). Input sequence design for parametric identification of nonlinear systems. In: *American Control Conference*. Albuquerque, New Mexico. pp. 3037–3041.

Cooley, B.L. and J.H. Lee (2001). Control-relevant experiment design for multivariable systems described by expansions in orthonormal bases. *Automatica* **37**, 273–281.

Cooley, B.L., J.H. Lee and S.P. Boyd (1998). Control-relevant experiment design: a plant-friendly, LMI-based approach. In: *American Control Conference*. Vol. 2. Philadelphia , PA. pp. 1240–1244.

Doyle, F.J., R.S. Parker, R.K. Pearson and B.A. Ogunnaike (1999). Plant-friendly identification of second-order volterra models. In: *European Control Conference*. Karlsruhe, Germany.

Godfrey, K., Ed.) (1993). *Perturbation Signals For System Identification*. Prentice Hall International (UK) Limited. Hertfordshire, UK.

Godfrey, K.R., H.A Barker and A.J. Tucker (1999). Comparison of perturbation signal for linear system identification in the frequency domain. *IEE. Proc. Control Theory Appl.* **146**, 535.

Guillaume, P., J. Schoukens, R. Pintelon and I. Kollár (1991). Crest-factor minimization using nonlinear chebyshev approximation methods. *IEEE Trans. on Inst. and Meas.* **40**(6), 982–989.

Hussain, M.A. (1999). Review of the applications of neural networks in chemical process control-simulation and on-line implementation. *Artificial Intelligence in Engineering* **13**(1), 55–68.

Lee, H., D.E. Rivera and H. Mittelmann (2003). Constrained minimum crest factor multisine signals for plant-friendly identification of highly interactive systems. In: *SYSID 2003*. Rotterdam, The Netherlands.

Li, T. and C. Georgakis (2002). Design of multivariable identification signals for constrained systems. In: *Annual AIChE 2002 Meeting*. Indianapolis, IN. paper 255g.

Ljung, L. (1999). *System Identification: Theory for the User*. 2nd ed.. Prentice-Hall. New Jersey.

Mathur, U. and R.J. Conroy (2002). Multivariable control without plant tests. In: *Annual AIChE 2002 Meeting*. Indianapolis, IN. paper 254g.

Narasimhan, S., S. Rengaswamy and R. Rengasamy (2003). Multiobjective input signal design for plant-friendly identification. In: *SYSID 2003*. Rotterdam, The Netherlands.

Ogawa, M. and H. Douke (2002). Process control at Mitsubishi. `http://www.che.utexas.edu / twmcc / Presentation0203/ Douke03spring.pdf` .

Parker, R.S., D. Heemstra, J.D. Doyle III, R.K. Pearson and B.A. Ogunnaike (2001). The identification of nonlinear models for process control using tailored "plant-friendly" input sequences. *J. of Process Control* **11**(2), 237–250.

Pearson, R.K., B.A. Ogunnaike and F.J. Doyle III (1993). Identification of nonlinear input/output models using non-gaussian input sequences. In: *American Control Conference*. San Francisco, CA. pp. 1465–1469.

Rengasamy, R., R.S. Parker and F.J. Doyle III (2000). Issues in design of input signals for the identification of nonlinear models of process systems. In: *ADCHEM 2000*. Pisa, Italy.

Rivera, D.E., J.F. Pollard and C.E. García (1992). Control-relevant prefiltering: A systematic design approach and case study: Special issue on system identification for robust control design. *IEEE Trans. Autom. Cntrl.* **37**, 964–974.

Rivera, D.E., M.W. Braun and H.D. Mittelmann (2002). Constrained multisine inputs for plant-friendly identification of chemical processes. *15th IFAC World Congress, Barcelona, Spain*.

Skogestad, S. and I. Postlethwaite (1996). *Multivariable feedback control: analysis and design*. Wiley. New York, NY.

IFAC
Publications
www.elsevier.com/locate/ifac

MULTI-OBJECTIVE INPUT SIGNAL DESIGN FOR PLANT-FRIENDLY IDENTIFICATION

Sridharakumar Narasimhan * **Ranganathan Srinivasan** *
Raghunathan Rengaswamy *,[1]

* *Dept of Chemical Engineering, Clarkson University, Potsdam, NY 13699-5705, USA*

Abstract: The choice of perturbation inputs is critical in the identification and model building exercise. One of the major objectives of system identification is accurate estimation of the system parameters. Accurate identification requires that the input be persistently exciting so as to excite all modes of interest. The practitioner would however prefer a 'plant friendly' input signal. Typically, in identification, an input is chosen based on some criteria. We propose a unified multi-objective formulation for the input design. The input can be evaluated as a solution to a constrained multi-objective optimization problem. *Copyright © 2003 IFAC*

Keywords: Input signal design, optimal, Multi-objective optimization, identification

1. INTRODUCTION

One of the primary goals of system identification is to ensure that the identified model has good predictive capabilities and the model parameters are accurately estimated. It is common practice to perturb the system with specially tailored inputs and the consequential input output data are used to build the system model. The quality of the model depends strongly on the experiment design and identification and hence the input used for perturbing the system should be carefully selected.

A number of excellent reviews on system identification and input signal design are available (Ljung, 1999; Norton, 1986; Godfrey, 1993; Pintelon and Schoukens, 2001). A considerable amount of literature exists on statistical experiment design. The input design problem for dynamic systems in system identification has received much less attention (Kalaba and Spingarn, 1982; Goodwin and Payne, 1977; Zarrop, 1979; Mehra, 1981).

Parker et al. (2001) proposed the use of 'plant friendly' input signals and appropriately defined the input friendliness. The use of multi-sinusoidal ,'plant friendly' input sequences for 'Model on Demand ' estimation and Model Predictive Control has been reported (Braun *et al.*, 2002). A multi-objective constrained optimization approach for identification of FIR models has been proposed (Johansen, 2000). This approach to system identification has not been extended to the problem of designing the input signal. We propose a multi-objective optimization formulation for input signal design which would enable the user to perform the experiment in an optimal manner so as to maximize benefit. It would also help the user uncover the mathematical relationships between the various objectives and constraints in a single unified manner.

2. OBJECTIVES

2.1 Parameter space criteria

Consider a Linear system described by the Finite Impulse Response Model of the following form

[1] Corresponding author. Email: raghu@clarkson.edu

$$y_r = \sum_{i=0}^{L} u_{r-i} h_i + n_r \tag{1}$$

where y_r, u_r, n_r are the output, input and noise respectively and h_i is the i th impulse response coefficient.

One of the primary objectives of a good system identification exercise is to accurately estimate the unknown parameters by reducing the bias and the covariance of the parameter estimates. For input design purposes, it is convenient to assume an unbiased efficient estimator, so that the covariance of the parameter estimates is given by the Cramer-Rao bound, viz., the inverse of the Fisher information matrix M. The problem of optimal input design is to choose an input or set of inputs that minimizes the variance of the parameter estimates, or equivalently, a suitable scalar norm of M^{-1}, like $tr(M^{-1})$ or $\det M^{-1}$ or $\lambda_{\max}(M^{-1})$ where λ is an eigenvalue of M^{-1}. It is well known, that subject to energy or amplitude constraints, an input with ideal autocorrelation (similar to white noise) is optimal with respect to the above criteria (Levin, 1960; Goodwin and Payne, 1977).

For a q input, p output MIMO system ,the input signal should have the following properties:

- The cross-correlation functions $R_{u_iu_j}$ between inputs i and j be ideally zero or negligible for times less than the settling time of the process
- The autocorrelation functions $R_{u_iu_i}$ of the inputs are of approximately impulsive form, similar to that of white noise

In addition, it has been shown that it is not possible to obtain independent estimates of the weighing functions of MIMO systems unless the common period is at least as long as sum of the settling times of the process (Briggs and Godfrey, 1976). A slightly different justification for the sequences to be uncorrelated with each other can be obtained by studying the discrete frequency spectra of the signals. The simplest way to ensure that they are uncorrelated is to prevent any coincidence of their line spectral components.

2.2 *Plant friendly Identification*

Multi-variable model based control strategies are commonly used in chemical process industries. Plant friendly identification has received the attention of researchers in recent times (Parker *et al.*, 2001; Braun *et al.*, 2002). Identification experiments in process industries are carried out on running plants. While a persistently rich excitation with high signal to noise ratio is theoretically preferred, operational, safety, environmental and economic considerations have to be taken into account during identification.

- An input requiring aggressive and frequent movement of valves and actuators is not desirable as this can lead to equipment wear and tear.
- Identification experimentation time has to be kept to a minimum so as to minimize off-spec products and consumption of utilities. Tests using the popular Pseudo-Random Binary Signals (PRBS) usually require days to conduct (Smith, 2003).
- Output deviations should be reduced to ensure that the product quality differs as little as possible from the set point

The extent of plant friendliness of an input signal can be quantified in terms of its crest factor, which is defined as the ratio of the Chebyshev norm ℓ_∞ and the ℓ_2 norm (Braun *et al.*, 2002). Techniques for designing multi-sine input signals having a low crest factor have been reported (Pintelon and Schoukens, 2001).

Multi-level pseudo-random sequences are easy to generate and because of their low auto-correlation properties, they are attractive from a system identification point of view. The use of crest factor to quantify the plant friendliness of a multi-level signal is not sufficient as can be seen from the following example. Consider two input signals u and y defined as follows

$$\begin{aligned} u_i &= -1 \;\; i = 1, \ldots n & (2)\\ &= 1 \;\; i = n+1, \ldots 2n & \\ y_{2k-1} &= 1 \;\; y_{2k} = -1 \;\; k = 1, \ldots n & (3) \end{aligned}$$

Clearly, both the sequences have the same crest factor, while the second sequence would be considered less plant friendly, as the it would require more frequent valve or actuator movement. So, for a discrete multi-level sequence, one may define the plant friendliness differently (Parker *et al.*, 2001) in terms of the number of changes from one level to another. The most plant friendly possible sequence is a constant sequence. For a deterministic sequence, the input plant friendliness can be defined as

$$\Phi_i = 100 \left(1 - \frac{n_t}{N-1}\right) \tag{4}$$

where N is the total length and n_t is the total number of switches. For a discrete level stochastic signal, the friendliness factor is simply the probability that the signal continues to be in the same state. Although the above definition is intuitively appealing, it might be difficult to represent this quantity in a closed form in an optimization formulation. A closed form definition which retains the same intuitive appeal could be

$$\begin{aligned} \Phi_i &= 1 - \frac{\sum_{k=2}^{N}(u_k - u_{k-1})^2}{(N-1)\max(u_k - u_{k-1})^2} & (5)\\ &= 1 \;\text{ for } u_k = u_{k-1} \;\; k = 2, \ldots n & \end{aligned}$$

A suitable output friendliness factor Φ_o that takes into account the output variability, the time spent in out of control region, the spectral energies or crest factor needs to be investigated. On intuitive grounds, one would expect a plant friendly input to result in a plant friendly output.

Typically, in identification, the input chosen for perturbing the system is based on some criteria. Multiple and possibly conflicting objectives and constraints associated with the input design problem suggest a multi-objective optimization approach for its solution. A multi-objective optimization framework that addresses these issues is presented subsequently.

3. MIXED INTEGER NONLINEAR PROGRAMMING FORMULATION

Consider a system represented by the linear discrete finite impulse response model as described above. The input is constrained to be an n level discrete sequence characterized by its logical state $0,1,2,\dots,n-1$. The signal can change its state at regularly spaced times, $0,\triangle t,2\triangle t\dots$, where $\triangle t$ is the switching time. The transformation from sequence logic levels to actual signal levels is given by

$$0 \rightarrow g(0),\ 1 \rightarrow g(1)\dots\ n-1 \rightarrow g(n-1) \quad (6)$$

We wish to determine the optimal switching sequence $s_1,s_2\dots s_N$ where s_i denotes the logical level (integer variable) of the sequence and the mapping $g(0),g(1),\dots g(n-1)$ (continuous variables). The input signal to the system u_i is then

$$u_i = \sum_{k=0}^{n-1} g(k) \prod_{j=0\ j\neq k}^{n-1} \frac{(s_i-j)}{(k-j)} \quad (7)$$

The input sequence is converted by a zero-order hold with clock interval $\triangle t$.

$$u(t) = u_i,\ (i-1)\triangle t \leq t \leq i\triangle t \quad (8)$$

Now we can pose the signal design problem as a Mixed Integer Nonlinear Programming (MINLP) problem, where the objective is to choose an input sequence or vector $U^t = [u_1,\dots,u_N]$ (which has s_i and $g(j)$ as the integer and continuous variables respectively) so as to attain the following objectives

- Minimize a scalar norm of M^{-1}, $f(M^{-1})$
- Maximize Input friendliness Φ_i
- Maximize output friendliness Φ_o

The formulation has input amplitude, energy and output amplitude constraints. Further, if the plant is nonlinear, one can identify a good linear approximation to the plant by suppressing certain harmonic multiples in the frequency spectrum. In particular, the use of a signal with inverse repeat property is desirable for elimination of even order nonlinearities (Srinvasan and Rengaswamy, 1999). Generally, we would fix the length of the sequence to determine the optimal switching sequence and mapping. It would be desirable to let this be floating so that the length is also minimized. This is a more difficult problem and formulations for incorporating this need to be investigated.

3.1 Problem formulation

The input can be evaluated as the solution to a constrained multi-objective Mixed Integer Nonlinear Programming problem using the following methods.

Weighted cost function
In this approach, a single weighted cost function $J = f(M^{-1}) - \mu_1\Phi_i - \mu_2\Phi_o$ is minimized. The choice of the weights is indicative of the relative importance that is assigned to each objective.

$$\text{Minimize}\{J\}\quad \text{s.t} \begin{cases} a_i \leq u_i \leq b_i\ \ \forall i \\ U^tU \leq b \\ c_i \leq y_i \leq d_i\ \ \forall i \\ \text{Suppression of} \\ \text{Nonlinearities} \end{cases}$$

Goal Programming
In goal programming, conflicting objectives are converted into constraints with an associated target value γ_i and x, a measure of the deviation is minimized. s_i is the relative slack in the constraint.

$$\text{Minimize}\{x\}\quad \text{s.t} \begin{cases} f(M^{-1}) - xs_1 \leq \gamma_1 \\ \Phi_i + xs_2 \geq \gamma_2 \\ \Phi_o + xs_3 \geq \gamma_3 \\ a_i \leq u_i \leq b_i\ \ \forall i \\ U^tU \leq b \\ c_i \leq y_i \leq d_i\ \ \forall i \\ x \geq 0 \\ \text{Suppression of} \\ \text{Nonlinearities} \end{cases}$$

Evolutionary algorithms for determining Pareto optimal sets
In single objective optimization problems, the aim is to determine the global optimal solution, if it exists. The above techniques essentially convert a multi-objective optimization problem into a single objective one by either weighting the costs or converting the costs into constraints and minimizing the deviations from a target or goal. Unlike single objective optimization, in optimization with conflicting objectives, there is no single optimal solution. The interaction among different objectives gives rise to a set of solutions, called the Pareto optimal solutions (Ref Fig 1). Solutions A, D, B form a Pareto optimal front and no one solution in this set can be said to be better than another in pure quantitative terms. However, solution C is dominated by solution D as solution D is better than C in both objectives. A set is called a global Pareto-optimal set, if no solution in the search space dominates any member in it. The optimization algorithm should attain two goals :- search for the

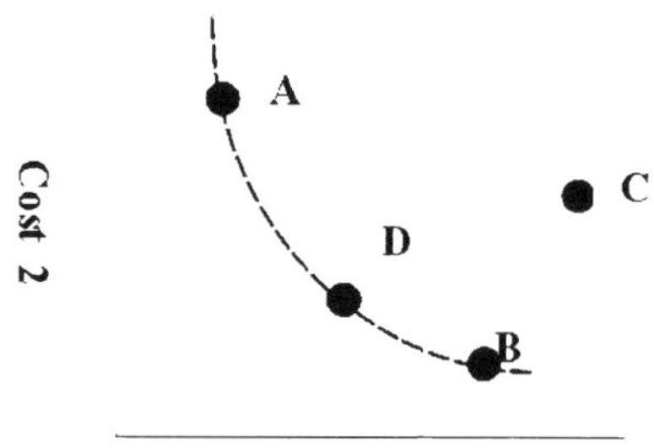

Fig. 1. Pareto optimal solutions

global Pareto-optimal front, and maintain population diversity in the optimal front so as to remove any bias towards any particular objective function. The final solution that is chosen for implementation is based on some other criteria. Evolutionary algorithms have been particularly useful in multi-objective optimization problems- especially for problems involving discrete variables (Deb, 1999). Hence, they are a strong candidate for solving multi-objective discrete multi-level input design problems.

4. CASE STUDY -FCC

In this section, we substantiate our arguments and demonstrate the inherent multi-objective nature of the input design problem by simulations of a Fluidised Catalytic Cracking Unit. Different input signals are chosen for identification and compared with respect to different objectives. The nonlinear, multi-variable state space model proposed in Balchen et al (1992) is used for simulations.

We consider a 5 input, 3 output system with the following inputs:

- Mass flow rate of air to regenerator, Fa
- Mass flow rate of spent catalyst, Fsc
- Temperature of feed gas oil, Toil
- Mass flow rate of gas oil feed, Foil
- Temperature of air to regenerator,Ta

The outputs of the system are

- Temperature at riser outlet, Tri
- Oxygen mole fraction in regenerator, Od
- Temperature of catalyst in regenerator, Trg

The base length of the input signal is chosen based on the maximum settling time, 15000 seconds. The clock time for the input signals and sampling time were chosen to be 300 seconds. The criterion for choice of input signals is that they have near ideal auto-correlation and minimal cross-correlation over the common period.The different input signals tried were

- Delayed pseudo-random sequences: For a q input system, the input to the lth channel is delayed by $N\Delta t/q$ relative to the input at the $l-1$ th input channel. Provided that T_{smax}, the maximum settling time of the system is less than $N\Delta t/q$, the measured cross-correlation for lags less than $N\Delta t/q$ is zero
- A Pseudo-Random Binary Sequence (PRBS) of period $N\Delta t$ and an Inverse Repeat Sequence (IRS) derived from it, by doubling the PRBS and inverting every other bit are uncorrelated over their common period $2N\Delta t$. A third uncorrelated sequence can be obtained by increasing the original sequence length four fold and inverting every $3, 4, 7, 8\ldots$ bit in the sequence. This idea can be used to generate further uncorrelated sequences. The drawback of the approach is that for a q input system, the lengths of the inputs increase by powers of 2 over the original sequence length to 2^q
- Hadamard sequences. A composite sequence c_i is defined as a sequence generated by the term-by-term multiplication of two component sequences,a_i and b_i of co-prime periods M and N. The autocorrelation $R_{cc}(i)$ of the composite sequence can be shown to be (Darnell, 1989)

$$R_{cc}(i) = R_{aa}(i)R_{bb}(i) \qquad (9)$$

  If the component sequences are selected to be a PRBS and a sequence derived from the rows or columns of a Hadamard matrix, the resulting sequence would be have an inverse repeat property and near-ideal autocorrelation. The composite sequences also would have minimal or zero cross-correlation
- Further sets of uncorrelated 4 level sequences can be generated using a similar modified matrix approach (Darnell, 1989).

4.1 Comparison of different signals

The range of input friendliness factors for the various input signals are as shown in Table 1. A segment of the delayed m sequence and 4 level sequence for dFa, the perturbed flow rate of air to regenerator (Fa) is shown in Fig 2.

Table 1. Input friendliness

Signal type	Minimum	Maximum
Hadamard Sequence	0.4930	0.5070
4-level Sequence	0.7180	0.8951
Binary IRS	0.4926	0.5074
Delayed Sequence	0.8997	0.9005

The variabilities in the output- Tri resulting from the different perturbation inputs are tabulated in Table 2.

Table 3 shows the experiment time and Mean Squared Errors (MSE) for the outputs corresponding to the models identified by the various input signals. Fig

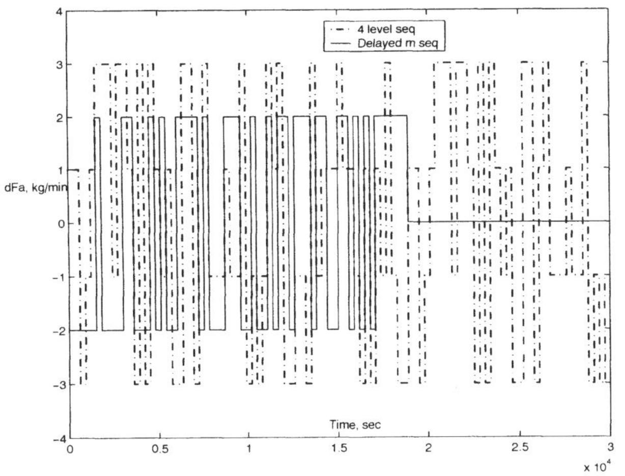

Fig. 2. Perturbed Input-dFa

Table 2. Output variability for Tri

Signal type	Min	Max	Variance
Hadamard Sequence	822.6	739.8	433.3
4-level Sequence	853	727.7	440.9
Binary IRS	820.5	735.16	420.5
Delayed Sequence	810.6	745.7	112.6

Table 3. Time and Mean Squared Errors

Signal Type	Time (sec)	Tri1	Trg	Od
Hadamard Sequence	150900	1.16	0.0006	1.05
4-level Sequence	302100	0.96	0.0006	1.2
Binary IRS	302100	1.06	0.0005	0.92
Delayed m Sequence	94200	17.79	0.008	14.03

3 shows the model plant comparison for the models identified by using delayed m sequences and 4 level sequences. As can be seen from the figure, agreement between the actual values and model prediction is much better for the model identified by using 4 level sequences.

The above observations are conveniently summarized in Fig 4. It is seen that the delayed sequences require the shortest time, while the 4 level/IRS sequences require the maximum time. However, the predictive capability of the model identified using the delayed

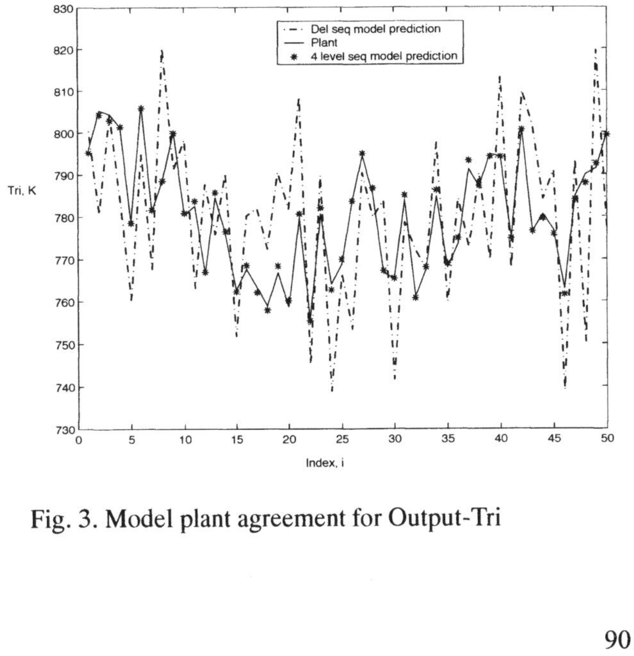

Fig. 3. Model plant agreement for Output-Tri

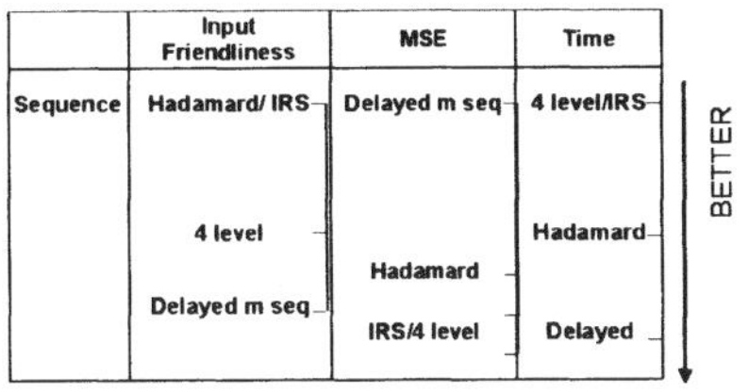

Input Friendliness
MSE
Time
Sequence
Hadamard/ IRS
Delayed m seq
4 level/IRS
4 level
Hadamard
Delayed m seq
IRS/4 level
Delayed
BETTER

Fig. 4. Overall comparison of different input signals

sequence is poor and this results in a high MSE. IRS/4 level sequences yield models with good predictive capabilities and low MSE. The delayed m sequences are most input friendly while the Hadamard/ IRS signals are the least. The Hadamard signals however, require much less time as compared to the IRS signals and the identified models are of acceptable quality.

It must be noted that, as in conventional input design, the main design requirement was that the input signals have near ideal auto-correlation and minimal cross-correlation. Plant friendliness was not accounted for at the design stage. The table ranks the suitability of the different input signals for model building and plant friendliness. The above table shows that among these signals, some are more plant friendly than the other and that some are more useful from a model building point of view. Thus, when multiple objectives and constraints are present, it is imperative to arrive at a solution in a transparent manner that explicitly addresses these issues. The standard methods of generation of these input sequences do not allow the user this flexibility.

It is clear from the table that the user would have to arrive at a trade-off between adequate identification and the multiple issues concerning plant friendly implementation. While this issue has been recognized by Braun *et al.* (2002), the authors have concentrated on designing single criterion input signals with minimum crest factors for the specified amplitude spectrum. Explicit handling of the trade-offs or objectives has not been considered.

The experiment time and hence, input signal length in system identification is governed primarily by the largest settling time of the system. The length of a multi-sine signal is decided by the number of harmonics and the fundamental frequency. The multi-level sequences used in the above simulations are typically available for certain lengths only, usually, $q^n - 1$ or a multiple. The restricted lengths are a consequence of a single criterion design requirement :- the amplitude spectrum or the autocorrelation properties. Often, the user is forced to opt for an input of length much larger than what may be theoretically or practically sufficient. Since the overall experiment time is crucial in identification, it is necessary to explicitly account for the time or signal length in the design formulation itself.

Design problems in engineering are really multi-objective in nature. In the mathematical treatment of single-objective optimization problems, the aim is to search for a unique global minimum, if it exists. In multi-objective optimization problems, there is no unique optimal solution and the goal is to search for the set of non-inferior solutions or the Pareto set. In many single objective optimization problems, conflicting objectives or trade-offs are accounted for implicitly. In some implementations, the multiple objectives are weighted relatively by using some kind of physical insight or higher level of knowledge. Multi-objective problems have addressed in control, for eg., (El-Kady *et al.*, 2003) and recently in identification (Johansen, 1996; Johansen, 2000; Weiland and Stoorvogel, 1997). However, the issue of multi-objective design of input signals for identification has not received attention in this regard. Considering the inherent multi-objective nature of designing plant friendly input signals, it is natural to pose the problem in a multi-objective optimization framework.

5. CONCLUSION

The multi-objective nature of the input signal design problem for plant friendly identification was motivated and substantiated with simulations on a FCC case study. A suitable multi-objective optimization problem was formulated and different candidate methods of solution outlined which would enable the user to identify the suitable input in a transparent procedure.

REFERENCES

Balchen, J.G., D. Ljungquiest and S. Strand (1992). State space predictive control. *Chemical Engineering Science* **47**, 787–807.

Braun, M.W, R. Ortiz-Mojica and D.E. Rivera (2002). Application of minimum crest factor multisinusoidal signals for "plant-friendly" identification of nonlinear process systems. *Control Engineering Practice* **10**, 301–31.

Briggs, P.A.N. and K.R. Godfrey (1976). Design of uncorrelated signals. *Electonics Letters* **12**, 555–556.

Darnell, M. (1989). The theory and generation of uncorrelated digital sequences. In: *Cryptography and Coding* (H. Beker and F.C. Piper, Eds.). Clarendon Press.

Deb, K. (1999). Evolutionary algorithms for multi-criterion optimization in engineering design. In: *Proceedings of Evolutionary Algorithms in Engineering and Computer Science* (Miettinen K., M.M. Mäkelä, P. Neittaanmäki and J. Periaux, Eds.). John Wiley and Sons.

El-Kady, M.M., M.S. Salim and A.M. El-Sagheer (2003). Numerical treatment of multiobjective opitmal control problems. *Automatica* **39**, 47–55.

Godfrey, K. (1993). *Perturbation Signals for System Identification*. Prentice Hall, USA.

Goodwin, G.C. and R.L. Payne (1977). *Dynamic System Identification: Experiment Design and Data Analysis*. Academic Press, USA.

Johansen, T.A. (1996). Identification of non-linear systems using empirical data and prior knowledge- an optimization approach. *Automatica* **32**, 337–356.

Johansen, T.A (2000). Multi-objective identification of fir models. In: *IFAC SYSID 2000*. Santa Barbara, USA.

Kalaba, R. and K. Spingarn (1982). *Control, Identification, and Input Optimization*. Plenum Press, New York.

Levin, M.J. (1960). Optimal estimation of impulse response in the presence of noise. *IRE Transactions on Circuit Theory* **7**, 50–56.

Ljung, L. (1999). *System Identification, Theory for the User, 2nd Edition*. Prentice Hall, USA.

Mehra, R.K. (1981). Choice of input signals. In: *Trends and Progress in Systems Identification* (Eykhoff, Ed.). Pergamon Press, New York.

Norton, J.P. (1986). *An Introduction to Identification*. Academic Press: UK.

Parker, R.S., D. Heemstra, F.J. Doyle III, Pearson R.K. and B.A. Ogunnaike (2001). The identification of nonlinear models for process control using tailored "plant-friendly" input sequences. *J. Process Control* **11**, 237–250.

Pintelon, R. and J. Schoukens (2001). *System Identification : A Frequency Domain Approach*. John Wiley & Sons, USA.

Smith, C.L. (2003). Intelligently tune pid controllers, part 2. *Chemical Engineering* **110**, 54–59.

Srinvasan, R. and R. Rengaswamy (1999). Use of inverse repeat sequence (irs) for identification in chemical process systems. *Industrial and Engineering Chemistry Research* **38**, 3420–3429.

Weiland, S. and A.A. Stoorvogel (1997). Optimal hankel norm identification of dyanamical systems. *Automatica* **33**, 1235–1246.

Zarrop, M.B. (1979). *Optimal Experiment for Dynamic System Identification*. Springer-Verlag, USA.

IFAC Publications
www.elsevier.com/locate/ifac

CONTROL-RELEVANT DESIGN OF PERIODIC TEST INPUT SIGNALS FOR ITERATIVE OPEN-LOOP IDENTIFICATION OF MULTIVARIABLE FIR SYSTEMS

Jay H. Lee [*,1]

[*] *School of Chemical Engineering, Georgia Institute of Technology, Atlanta, GA 30332, U.S.A.*

Abstract: In this paper, the problem of control-relevant test input design for iterative system identification is considered. Two approaches are presented for the L-optimal input test signal design by Cooley and Lee (2001), which minimizes the *a posteriori* closed-loop error based on *a priori* information given by the data already collected from previous iterations. To simplify the computation, we limit our design to a period signal of period n, which is the number of FIR coefficients. The first approach is direct time-domain search for optimal test input values, which can be formulated as a nonconvex optimization or a concave Quadratic Program. The second approach considers frequency domain solution of the L-optimal design problem to obtain the optimal discrete spectra of the test signals, which is subsequently realized as a periodic signal. *Copyright © 2003 IFAC*

Keywords: Control-relevant identification; test input design; periodic signals; frequency domain solution

1. INTRODUCTION

The design of test input signals has been an active research topic for a long time (Mehra, 1974; Zarrop, 1979; Goodwin and Payne, 1977; Gevers and Ljung, 1986; Yuan and Ljung, 1985; Schoukens *et al.*, n.d.; Forssell and Ljung, 2000; Antoulas and Anderson, 1999; Zhu and van Bosch, 2000) due to the importance of obtaining accurate and reliable models. Especially with the emerging popularity of model-based control in the process industries, system identification is increasingly being recognized as the critical bottleneck, with an estimated 80 to 90 % of the cost and time involved in implementing model-based controllers devoted to it (Andersen *et al.*, 1991). The choice of input test signals is critical as they have a large bearing on the nature and accuracy of the identified system characteristics (Mehra, 1974).

Most of the existing literature on optimal test input design focuses on minimizing some measure (*i.e.*, "size") of the parameter error covariance matrix (Goodwin and Payne, 1977; Zarrop, 1979). The premise behind this approach is that sufficiently accurate parameter estimates will lead to high performances in different applications. However, the distribution of model error in the parameter space or in the frequency domain is just as important as the total size, if not more (Li and Lee, 1996). Since the test input affects the distribution of both variance and bias, the test input signals used for identification should preferably be designed so that the shape of their power spectra is optimal with respect to the plant's open-loop characteristics and desired closed-loop characteristics.

[1] The author to whom correspondence should be addressed. Email: jay.lee@che.gatech.edu

The determination of model's quality should consider the end application of the model, which is the intended closed-loop control (Hjalmarsson *et al.*, 1996; Gevers and Ljung, 1986). Gevers and Ljung (1986) argued that the optimal test signal for minimum variance control is that generated by the very minimum variance controller in closed loop. Based on this finding, they proposed to perform iterative identification, in which controller design and closed-loop identification are alternated. Forssell and Ljung (2000) derived an explicit solution to the experiment design problem of minimizing the variance error in the dynamic model under power constraints on the inputs and outputs. Their results can be applied for open-loop and closed-loop experiments. Cooley and Lee (2001) formulated a control-relevant, *open-loop* test input design problem for general Finite Impulse Response (FIR) systems and showed that the solution leads to an iterative *L-optimal* experiment design scheme. The optimization, when formulated to solve for the test input sequence directly, is non-convex and is computationally prohibitive in many cases.

Motivated by the large computational burden associated with the latter approach, we propose to restrict the design at each iteration to periodic signals of period n, where n is the number of FIR coefficients needed to model the system. This restriction is reasonable since one needs a persistently exciting signal of order n or higher to obtain a consistent estimate of n parameters. Under the restriction, two different ways of solving the problem are proposed. First approach involves direct search of optimal test input values in time domain. This is a nonconvex optimization problem. We show that this problem can be reformulated as a concave quadratic program under a reasonable approximation. Even though concave QP is NP-hard, efficient algorithms are available and may be considerably easier to handle than the original nonconvex problem. Second approach (section 4) involves solving the L-optimal design problem in the frequency domain to obtain an explicit solution in the form of discrete spectra for the periodic test signals. Spectral factorization on the optimal solution can then be performed to realize the solution as actual input sequences.

Examples and case studies illustrating the efficacy of the methods, which are skipped here due to space limitation, will be presented in the conference presentation and also in the full version of this paper.

2. PROBLEM DESCRIPTION

In this section, we present the objective function adopted in the iterative design of control-relevant test input signal, as derived in (Cooley and Lee, 2001).

2.1 *General framework*

We assume that the underlying system has the following finite impulse response (FIR) representation:

$$y(t) = \Theta^T \phi(t) + \varepsilon(t) \tag{1}$$

Here, $y \in \Re^{n_y}$ is the vector of outputs and $u \in \Re^{n_u}$ is the vector of inputs. ε representing the noise/disturbance effects is modeled as a zero-mean i.i.d random sequence with an identity matrix as its covariance. Θ and ϕ are given by

$$\Theta^T = \begin{bmatrix} \theta_1 & \cdots & \theta_n \end{bmatrix} \tag{2}$$

$$\phi(t) = \begin{bmatrix} u(t-1) \\ \vdots \\ u(t-n) \end{bmatrix} \tag{3}$$

where $\theta_i \in \Re^{n_y \times n_u}$ is the (*real-valued*) expansion coefficient matrices.

Note: The assumption of zero-mean white noise is limiting, but in the case that the noise spectrum is fully known *a priori*, one can always formulate the problem in the above way through data filtering, as discussed in (Cooley and Lee, 2001). In the more general case of unknown noise spectrum, some iterations of the subsequently presented method may be needed to identify it.

2.1.1. *Least squares solution* The least squares estimates of the FIR parameters are given by

$$\begin{aligned} \hat{\Theta} &= P^{-1} \sum_{t=1}^{N} \phi(t) y^T(t) \\ P &\stackrel{\text{def}}{=} \sum_{t=1}^{N} \phi(t) \phi^T(t). \end{aligned} \tag{4}$$

2.1.2. *Iterative solution* In this work, an *iterative design* procedure will be adopted, in which the input signal for a future experiment (the kth experiment) is designed based on the parameter estimates and the covariance matrix computed with the data collected from prior experiments (*see* Fig. 1). Once the data for the kth experiment becomes available, the parameter estimates and the covariance matrix should be updated according to

$$\hat{\Theta}_k = P_k^{-1} \left(\sum_{t=1}^{\Delta N_k} \phi_k(t) y_k^T(t) + P_{k-1} \hat{\Theta}_{k-1} \right) \tag{5}$$

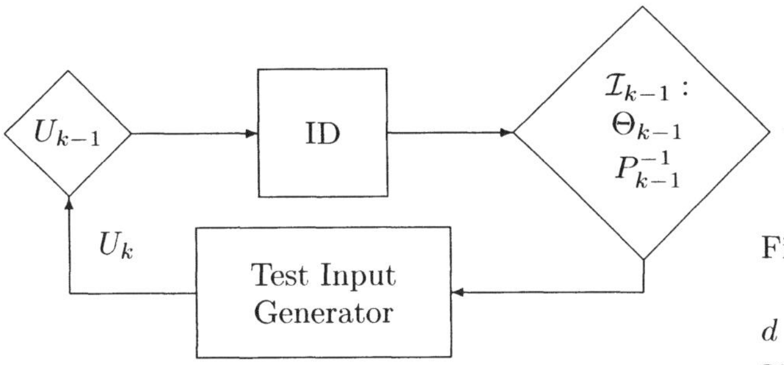

k: index for iteration in each experiment

Fig. 1. Iterative Design Scheme for Input Signal Generation

where the information matrix P_k is given by

$$P_k = \underbrace{\sum_{t=1}^{\Delta N_k} \phi_k(t)\phi_k^T(t)}_{\Delta P_k} + P_{k-1} \qquad (6)$$

The subscript k represents that the data is obtained from a kth experiment. Thus, ΔN_k is the number of data points collected during the kth experiment. ΔP_k denotes the incremental change in the information matrix due to the data from the kth experiment and P_{k-1} represents the information matrix based on the inputs implemented before the kth experiment.

2.1.3. *Model Error* Assuming that there is no structural mismatch between the model and the plant, which is reasonable with a high order FIR model, (4) relates the input signal to the model error because the covariances of parameter error and transfer function error are related to P in the following way:

$$E\left\{(\hat{\Theta}-\Theta)(\hat{\Theta}-\Theta)^T\right\} = n_y P^{-1} \qquad (7)$$

$$E\left\{\tilde{G}^*(\omega)\tilde{G}(\omega)\right\} = n_y \Lambda^*(\omega) P^{-1}\Lambda(\omega) \quad (8)$$

where $\tilde{G}(\omega) = \hat{G}(\omega) - G(\omega)$ and $\Lambda(\omega)$ is a Discrete Fourier Transform matrix defined as

$$\Lambda(\omega) = \left[e^{-j\omega} I_{n_u} \cdots e^{-j\omega n} I_{n_u}\right]^T. \qquad (9)$$

$\Lambda^*(\omega)$ denotes the conjugate transpose. This result gives an explicit relationship between the test signal and the model error. Note that this only valid if a true system is in the model set. From now on, assume that the true system is in the model set.

2.2 *Formulation of design objective*

For the feedback system in Fig. 2, one possible choice for closed-loop performance index, when

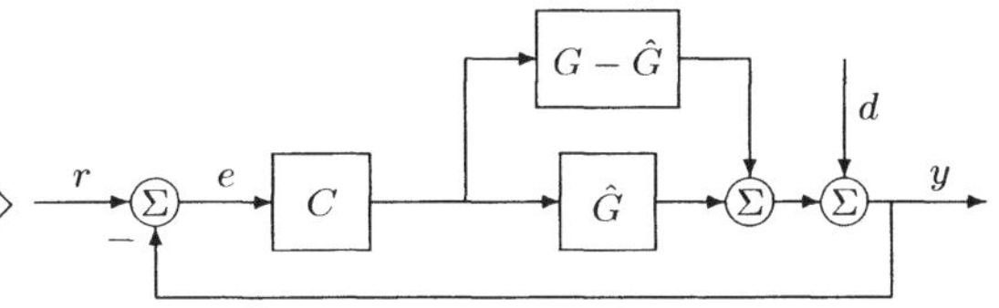

Fig. 2. Closed-loop System

d and r are quasi-stationary signals, is the mean square error

$$V = E\left\{e^T e\right\} = E\left\{trace\left(ee^T\right)\right\} \qquad (10)$$

The error e is given by

$$e = \hat{S}\left(I + \left(G-\hat{G}\right)\hat{G}^{-1}\hat{H}\right)^{-1}\underbrace{(r-d)}_{\eta}, \quad (11)$$

where $\hat{S} \overset{\text{def}}{=} (I+\hat{G}C)^{-1}$ is the sensitivity function and $\hat{H} \overset{\text{def}}{=} I - \hat{S}$ is the complementary sensitivity function. Our development will be done entirely in discrete time and using discrete spectrum.

Using the Parseval's relation, we can express the performance index in frequency domain too. We note that V being a function of the unknown model $\hat{G}$ and unknown controller C, we must specify some *controller design strategy.* Here we use the *"direct synthesis"* approach, where one fixes the sensitivity function $\hat{S}$ according to the desired closed loop performance.

Even though the design problem is intrinsically a multi-stage problem since one should consider all the remaining tests in designing a current test, the multi-stage nature complicates the formulation significantly since designs of test signals for future tests are to be based on future information (yet to be available) and hence stochastic. A reasonable simplification is to assume that the current test considered is the last one and the model obtained after the test will be implemented as a controller. This way we can formulate the optimal test input design as a deterministic optimization where the search variables are the values of the test inputs during a current test. Further simplification is possible by linearizing the performance index by taking second order Taylor series expansion at $\hat{G}_k = \hat{G}_{k-1}$ and $G = \hat{G}_{k-1}$. Details of this procedure can be found elsewhere (see pp. 275–278 in Cooley and Lee (2001)). The resulting approximation of the performance index is a L-optimal criterion and can be found there. The frequency domain counterpart is:

$$\hat{J} = E\left\{\text{tr}\left(\frac{1}{2\pi}\int_{-\pi}^{\pi} \hat{\Xi}_{k-1}\Phi_\eta\hat{\Xi}_{k-1}^* d\omega\right) \mid \mathcal{I}_{k-1}\right\} (12)$$

where $\hat{\Xi}_{k-1} = \hat{S} + \hat{S}(\hat{G}_k - G)\hat{G}_{k-1}^{-1}\hat{H}$ and $\mathcal{I}_{k-1} = \{\hat{\Theta}_{k-1}, P_{k-1}\}$. The integrand of (12) is a linear combination of the elements of the covariance matrix of the transfer function error and therefore P_k^{-1}, as indicated by (8). Recall that P_k^{-1} is related explicitly to u_k according to (6).

Different approaches can be pursued to obtain the optimal input sequence based on minimization of (12). Cooley and Lee (2001) have formulated the following experiment design:

Problem 1. Given the *a priori* estimate $\hat{\Theta}_{k-1}$, the information matrix P_{k-1}, the target sensitivity function $\hat{S}$, and the disturbance/reference signal spectrum Φ_η, the optimal input sequence is generated by solving the optimization,

$$\min_{u_k(1),u_k(2),\ldots,u_k(\Delta N_k)} \operatorname{tr}(W_k P_k{}^{-1}) \quad (13)$$

subject to the input constraints. P_k is given by (6), and the weighting matrix W_k is defined by

$$W_k = \frac{1}{2\pi}\int_{-\pi}^{\pi} \operatorname{tr}\left(\hat{S}^*\hat{S}\right) \Upsilon_k \Phi_\eta \Upsilon_k^* d\omega \quad (14)$$

where $\Upsilon_k = \Lambda \hat{G}_{k-1}^{-1}\hat{H}$, and Λ is defined as in expression (9).

3. DIRECT TIME-DOMAIN SEARCH OF PERIODIC TEST INPUTS

Problem 1 is a non-convex optimization problem. Generating the optimal input sequence by solving it presents a computational challenge. Motivated by the computational challenge in solving Problem 1, we propose to restrict the design to periodic signals of period n. This restriction is reasonable as one only needs persistently exciting signals of order n or higher in order to obtain consistent estimates of the FIR coefficients.

Problem 2. Given the *a priori* estimate $\hat{\Theta}_{k-1}$, the information matrix P_{k-1}, the target sensitivity function $\hat{S}$, and the disturbance/reference signal spectrum Φ_η, the optimal periodic input sequence is generated by solving the optimization,

$$\min_{u_k(1),u_k(2),\ldots,u_k(n)} \operatorname{tr}(W_k P_k{}^{-1}) \quad (15)$$

subject to the input constraints. For periodic signals of period n, the covariance matrix P will have the following structure, assuming that the signal length ΔN is an *integer multiple* of the period and the periodic signal is implemented starting at time $-n+1$, that is, $u(-n+1),\cdots,u(0)$ represents the first period at every iteration step:

$$P_k = \quad (16)$$

$$\begin{bmatrix} \sum_{i=1}^{k}\Delta N_i R_i(0) & \cdots & \sum_{i=1}^{k}\Delta N_i R_i(n-1) \\ \sum_{i=1}^{k}\Delta N_i R_i^T(1) & \cdots & \sum_{i=1}^{k}\Delta N_i R_i(n-2) \\ \vdots & \ddots & \vdots \\ \sum_{i=1}^{k}\Delta N_i R_i^T(n-1) & \cdots & \sum_{i=1}^{k}\Delta N_i R_i(0) \end{bmatrix}$$

where $R_i(\tau)$ is the τth covariance matrix for the periodic signal used at the ith iteration defined as:

$$R_i(\tau) = \frac{1}{n} U_i^T M_\tau U_i \quad (17)$$

$$U_i^T = \left[\, u_i(1)\ u_i(2)\ \cdots\ u_i(n) \,\right] \quad (18)$$

$$M_\tau U_i^T = \begin{bmatrix} u_i^T(\tau+1) \\ \vdots \\ u_i^T(n) \\ u_i^T(1) \\ \vdots \\ u_i^T(\tau) \end{bmatrix} \quad (19)$$

Here $u_i(j)$ is a n_u valued vector containing the values of the periodic inputs at the jth time point of each period at the ith iteration step.

Equation (13) is a highly nonlinear function of $u_k(t)$. One can reduce the computational complexity further by employing the approximation

$$\begin{aligned} P_k^{-1} &= [P_{k-1} + \Delta P_k]^{-1} \\ &= P_{k-1}^{-1}\left[I + \Delta P_k P_{k-1}^{-1}\right]^{-1} \\ &\approx P_{k-1}^{-1}\left[I - \Delta P_k P_{k-1}^{-1}\right] \end{aligned}$$

The above approximation should be valid, provided $\|\Delta P_k P_{k-1}^{-1}\| \ll I$. Thus, we can reformulate (13) as

Problem 3. $$\min_{u_k(1),u_k(2),\ldots,u_k(n)} \operatorname{tr}(-W_k P_{k-1}^{-1}\Delta P_k P_{k-1}^{-1}) \quad (20)$$

subject to (17) and input constraints.

The above is a concave QP, which is NP-hard but still easier to handle than the original nonconvex problem.

4. FREQUENCY DOMAIN DESIGN

A straightforward manipulation on (12) gives

$$\hat{J} = \frac{1}{2\pi}\int_{-\pi}^{\pi} \operatorname{tr}\left(\tilde{W}_k \Lambda P_k{}^{-1}\Lambda^*\right) d\omega, \quad (21)$$

where $\tilde{W}_k = \operatorname{tr}\left(\hat{S}^*\hat{S}\right)\hat{G}_{k-1}^{-1}\hat{H}\Phi_\eta\hat{H}^*\hat{G}_{k-1}^{-*}$.

For a P matrix of (17), a key result is that the following equality holds at $\omega = i\frac{2\pi}{n}, i = 0, \cdots, n-1$:

$$(\Lambda^* P_k \Lambda)^{-1} = \frac{1}{n^2}\Lambda^* P_k^{-1}\Lambda \quad (22)$$

Using (6)

$$\frac{1}{n^2}\Lambda^* P_k^{-1}\Lambda = (\Lambda^*\Delta P_k\Lambda + \Lambda^* P_{k-1}\Lambda)^{-1} \quad (23)$$

Multiplying both sides by n^2 and rearranging the resulting expression give

$$\Lambda^* P_k^{-1}\Lambda = \frac{n}{N_k}\left[\frac{\Delta N_k}{N_k}\Phi_k^u + \frac{N_{k-1}}{N_k}\hat{\Phi}_{k-1}^u\right]^{-1} \quad (24)$$

where N_k is the total number of data points collected during the entire k sets of experiments, and

$$\Phi_k^u \stackrel{\text{def}}{=} \frac{1}{n\Delta N_k}\Lambda^*\Delta P_k\Lambda \quad (25)$$

$$\hat{\Phi}_{k-1}^u \stackrel{\text{def}}{=} \frac{1}{nN_{k-1}}\Lambda^* P_{k-1}\Lambda \quad (26)$$

The above relationship (24) is valid for data of any finite size that is an integer multiple of the period n. Φ_k^u is the discrete spectra of the test signals U_k.

Since $\tilde{W}_k$ and Λ are periodic functions with period of 2π, we can approximate the integral in (21) by the following sum:

$$\hat{J} \approx \frac{1}{n}\sum_{i=0}^{n-1}\left\{\mathrm{tr}\left(\tilde{W}_k\Lambda P_k{}^{-1}\Lambda^*\right)\Big|_{\omega=\frac{2\pi}{n}\cdot i}\right\} \quad (27)$$

The approximation is reasonable if n, the number of impulse coefficients, is large relative to the rate at which $\tilde{W}_k$ with frequency. Based on the above approximation, we can restate the problem of optimal input test design as follows:

Problem 4. The approximate iterative input test design of Problem 1 corresponds to the following optimization problem:

$$\min_{\Phi_k^u(\omega_i)} \frac{1}{n}\sum_{i=0}^{n-1}\mathrm{trace} \quad (28)$$

$$\left(\tilde{W}_k\left[\frac{\Delta N_k}{N_k}\Phi_k^u + \frac{N_{k-1}}{N_k}\hat{\Phi}_{k-1}^u\quad\right]^{-1}\right)$$

for $\omega_i = i\frac{2\pi}{n}, i = 0, \ldots, n-1$, and subject to input constraints.

Once we obtain the solution, the problem reduces to finding the input sequence $u_k(t)$ that leads to ΔP_k matching the optimal solution for Φ_k^u according to (25).

Input constraint will typically be power constraints on the individual input signals:

$$\frac{1}{n}\sum_{i=0}^{n-1}\mathrm{diag}\left\{[\Phi_k^u(\omega)]|_{\omega=i\cdot\frac{2\pi}{n}}\right\} \leq \Gamma, \quad (29)$$

where Γ is an $n_u \times n_u$ diagonal matrix whose elements are the power limits on the individual inputs. The notation diag $[\cdot]$ represents a diagonal matrix formed by using the diagonal elements of the argument matrix.

Proposition 1. The explicit solution to Problem 4 with the constraint (29) is given by

$$\Phi_k^{u\,opt}(\omega_i) = \frac{N_k}{\Delta N_k}\left[\Psi_k\tilde{W}_k^{1/2}(\omega_i) - \frac{N_{k-1}}{N_k}\hat{\Phi}_{k-1}^u(\omega_i)\right] \quad (30)$$

for $\omega_i = i\cdot\frac{2\pi}{n}$ and Ψ_k is defined as

$$\Psi_k = \left[n\cdot\frac{\Delta N_k}{N_k}\Gamma + \frac{N_{k-1}}{N_k}\sum_{\ell=0}^{n-1}\mathrm{diag}\left(\hat{\Phi}_{k-1}^u(\omega_\ell)\right)\right]\times\left[\sum_{\ell=0}^{n-1}\mathrm{diag}\left(\tilde{W}_k^{1/2}(\omega_\ell)\right)\right]^{-1}$$

Proof is omitted for the sake of brevity. A full length paper will be made available at `http://dot.che.gatech.edu/information/research/issicl/files/sysid03.pdf` (note: there is no hyphen in the url).

Next, we describe one possible procedure for obtaining periodic input signals matching a given $\Phi_k^u(\omega)$. Justification and shortcomings of this procedure are omitted for brevity.

(1) Perform a spectral factorization of Φ_k^u at $\omega_i = \frac{i\cdot 2\pi}{n}, i = 0, \cdots, n-1$ to obtain $L(\omega_i)$ so that $\Phi_k^u(\omega_i) = L(\omega_i)L^*(\omega_i)$.
(2) Perform the following inverse DFT to obtain the coefficients of the FIR filter:

$$H_L(\ell) = \frac{1}{M}\sum_{i=0}^{n-1}L(\omega_i)e^{j\omega_i\ell}; \quad (31)$$
$$0 \leq \ell \leq (n-1)$$

(3) Generate a PRBS of period n, which is denoted by $\epsilon(t)$.
(4) Generate the test input sequence as the periodic signal that results from filtering the PRBS through the FIR filter according to

$$u_k(t) = \underbrace{\left(\sum_{\ell=0}^{n-1} H_L(\ell) q^{-\ell}\right)}_{L(q)} \epsilon(t) \quad (32)$$

For an FIR system of order n, the output should converge to a periodic signal after n time steps. $u_k(t)$ becomes just a linear combination of the PRBS shifted by different amounts.

5. TIME DOMAIN LMI / BMI DESIGN

Alternatively, a direct time-domain approach can be pursued by formulating it as LMI and BMI problems, for which rich sets of solution algorithms have been developed. Some re-formulations to the developed objective function are necessary for this.

First, we note that (13) can be re-formulated as

$$\min_{\Delta P_k, Z} \text{trace}(Z) \quad (33)$$

subject to the constraint,

$$\begin{bmatrix} Z & I \\ I & P_k W_k^{-1} \end{bmatrix} \geq 0 \quad (34)$$

Since $P_k = P_{k-1} + \Delta P_k$, the above is an LMI problem formulated in terms of ΔP_k. However, it is difficult to ensure that the solution will have the right structure, so that it can be translated into actual periodic signals.

When definition of P_k from (6) is used, we can transform it into the following BMI problem:

Problem 5. Given the prior estimate $\hat{\Theta}_{k-1}$, the information matrix P_{k-1}, the nominal sensitivity function $\hat{S}$, and the disturbance/reference signal spectrum Φ_η, the optimal input covariance sequence is obtained by performing the following optimization:

$$\min_{u_k(1),\cdots,u_k(n),Z} \text{trace}(Z) \quad (35)$$

subject to the constraint,

$$\begin{bmatrix} Z & I \\ I & P_k W_k^{-1} \end{bmatrix} \geq 0 \quad (36)$$

where $P_k = \Delta P_k + P_{k-1}$ is defined in (17) to (19)

Time domain constraints can be incorporated into the above BMI problem.

6. REFERENCES

Andersen, H. W., K. H. Rasmussen and S. B. Jorgensen (1991). Advances in process identification. In: *Proc. 4th Int'l Conf. on Chemical Process Control - CPC IV* (Y. Arkun and W. Ray, Eds.). South Padre Island, TX. pp. 157–162.

Antoulas, A. C. and B. D. O. Anderson (1999). On the choice of inputs in identification for robust control. *Automatica* **35**(6), 1009–1031.

Cooley, B. L. and J. H. Lee (2001). Control-relevant experiment design for multivariable systems decsribed by expansions in orthonormal bases. *Automatica* **37**(2), 273–281.

Forssell, U. and L. Ljung (2000). Some results on optimal experiment design. *Automatica* **36**(5), 749–756.

Gevers, M. and L. Ljung (1986). Optimal experiment designs with respect to the intended model application. *Automatica* **22**(5), 543–554.

Goodwin, G. C. and R. L. Payne (1977). *Dynamic System Identification : Experiment Design and Data Analysis.* Academic Press. NY.

Hjalmarsson, H., M. Gevers and F. De Bruyne (1996). For model-based control design, closed-loop identification gives better performance. *Automatica* **32**(12), 1659–1673.

Li, W. and J.H. Lee (1996). Frequency domain closed-loop identification of multivariable systems for feedback control. *AIChE Journal* **42**, 2813–2827.

Mehra, R. K. (1974). Optimal input signals for parameter estimation - survey and new results. *EEE Trans. on Automatic Control* **AC-19**(6), 753–768.

Schoukens, J., P. Guillaume and R. Pintelon (n.d.). *Design of broadband excitation signals.* Chap. 3, pp. 126–159.

Yuan, Z. D. and L. Ljung (1985). Unprejudiced optimal open loop input design for identification of transfer functions. *Automatica* **21**, 233–256.

Zarrop, M. B. (1979). *Optimal Experiment Design for Dynamic System Identification.* Springer-Verlag. NY.

Zhu, Y. and P. P. J. van Bosch (2000). Optimal closed-loop identification test design for internal model control. *Automatica* **36**(8), 1237–1241.

IFAC Publications
www.elsevier.com/locate/ifac

CONSTRAINED SIGNAL DESIGN USING APPROXIMATE PRIOR MODELS WITH APPLICATION TO THE TENNESSEE EASTMAN PROCESS

Tong Li * **Christos Georgakis** **

* *Department of Chemical Engineering, Lehigh University, Bethlehem, PA 18015, USA*
** *Department of Chemical Engineering, Chemistry, Material Science, Polytechnic University, Brooklyn, NY 11201, USA*

Abstract: Identification input design is the first and very important step towards a successful data-driven modeling task. In this paper, a previously communicated input design method, based on an approximate a-priori steady state model, is extended to the case that the known approximate model at hand is a dynamic one. Several additional issues are also discussed including the iterative signal design and identification process and the derivation of the model uncertainty. As an important case study, the methodology is applied to the challenging Tennessee Eastman problem, which is widely used in process control study. The identification results of the proposed and prior methods are compared.

Keywords: Dynamic models, Input signals, System identification, Uncertainty

1. INTRODUCTION

The first step in any identification task is the design of the input signal. Such a signal needs to be large enough to have the best signal to noise ratio and, at the same time, should not cause the process to violate its output constraints. These constraints are necessary because the product produced during the multi-week identification task needs to meet customer specifications.

Based on preliminary and approximate steady state model of a MIMO process to be identified, the simplest way of signal design is to use uncorrelated pseudo-random binary sequences (PRBS) to perturb each input. In order to improve the robust stability and performance, Koung and Macgregor (1994) proposed a design method based on the singular value decomposition (SVD) of the prior steady state gain matrix. This approach considers the gain directionality and designs MIMO PRBS signals that excite all the output directions equally well. Zhan and Georgakis (2000) summarized the existing techniques of signal design and proposed an optimal steady state design method, which aims to design the signal with the largest possible input area and accommodates both the input and output constraints at the same time. The input signal is designed by solving a constrained optimization problem that maximizes a D-optimal design objective.

The system to be identified usually does not reach its steady state if sufficiently high frequency contents are in the input variables. In the MIMO input design, this is especially the case because the clock time of the basic PRBS is usually designed based on the shortest input-output time constant (Rivera, 1992). Thus when the time constants of other input-output pairs are much larger than the smallest one, which is often the case in practise, some of the outputs are always quite far from their steady state conditions.

Since certain components of a multivariable system are far from their steady state conditions most of the time, signal design based on a known steady state model might not be accurate. For a system whose input-output components have overdamped dynamic characteristics, the signal size will be conservative. On the other hand, if an input-output component of a system has underdamped dynamic, the system response may violate the output constraints if the steady state design is used. The present paper assumes that an approximate dynamic model is available for an $n \times m$ system. Then a set of $n \times m$ "signatures" of the output profiles are first calculated to characterize the responses of the known dynamic model to a nominal input signal, such as a one-dimensional PRBS of unit magnitude. The largest possible multivariable input signal is then designed by solving a constrained optimization problem.

In order to test the identification results under different input designs, the Tennessee Eastman (TE) Process is selected in this paper as a case study. The Tennessee Eastman Process was proposed by Downs and Vogel (1993) as a challenging problem in the process control and related areas. For this process, Lyman and Georgakis (1995) gave a successful control structure with 21 PI controllers. In the present study, it is assumed that some of these SISO PI controllers need to be replaced by a MIMO model based constrained controller, e.g. MPC. Then the first step is to identify the open-loop MIMO model. In this paper, an iterative identification process will be carried out to show the effectiveness of the proposed input design method.

2. THE TENNESSEE EASTMAN PROCESS

According to Downs and Vogel (1993), there are eight components, including 4 reactants, 2 products, an inert and a byproduct, and four reactions in the TE process. The process has five major unit operations: a reactor, a product condenser, a vapor-liquid separator, a recycle compressor and a product stripper. In the paper, a number of possible process disturbances and set point changes are also given. The process is open-loop unstable, so a prerequisite for most studies is a stabilizing control strategy for operating the plant.

In Lyman and Georgakis (1995), 4 control structures are given for the problem, among which structure 2 can provide effective control under all circumstances, so it is chosen to be studied in this paper. In the control structure 2 (Figure1), 21 SISO PI controllers are designed as the result of the synthesis of the plant-wide control problem. Among all these, the controller on the condenser cooling water flow is chosen as the primary production rate manipulator. Under this control structure, the plant can be operated stably with all the process variables maintained at desired values and most of the requirements proposed by Downs and Vogel (1993) can be satisfied, such as recovering quickly and smoothly from disturbances, production rate changes or product mix changes, etc.

3. PRELIMINARY IDENTIFICATION OF THE STEADY STATE GAIN AND TIME CONSTANTS

Here we treat the TE problem as a real process from which we collect data by running the simulation program (Lyman and Georgakis, 1995). When we face such a plant without any knowledge about its dynamics, the first thing to do is to perturb the inputs with some conservative step changes to obtain the steady-state gain matrix, as well as the time constants of the input-output pairs for the further input design.

From the control structure above, we choose to replace 2 SISO controllers by one MIMO MPC. The two SISO controllers are the controller on the composition of component B in purge gas (XC19) and the controller on the composition of component E in the product (XC20). These two are chosen because they are composition controllers to provide set points to other regulators and an MPC often proves better than PI controllers in such a condition.

After these two controllers are removed from the whole system, we face a 2×2 open-loop system without any knowledge of its dynamic. The corresponding inputs and outputs are:

Input:

- Set point of FC6;
- Set point of TC16.

Output:

- Composition of B in the purge flow;
- Composition of E in the product flow.

Since we will use PRBS to to perturb the inputs up and down around its normal operating points, the constraints are the maximum deviations allowed from the operating point for both the inputs and outputs. The following input and output constraints are developed from the operating condition described in Downs and Vogel (1993),

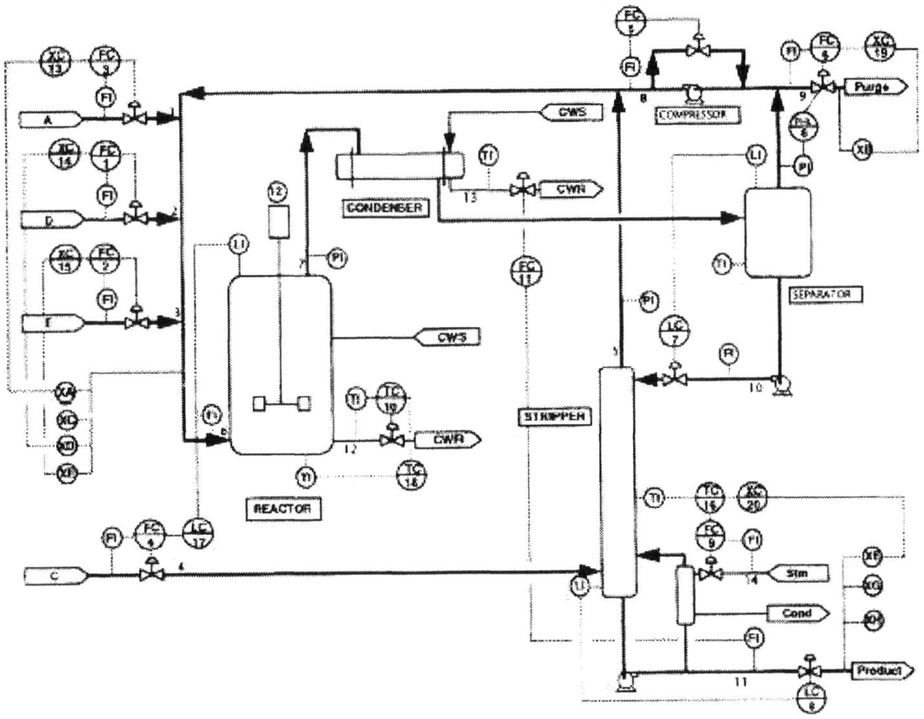

Fig. 1. Control structure 2 on the TE process proposed by Lyman and Georgakis

or designated by the author if they are not there: $\Delta u_1 - 0.3kscmh$, $\Delta u_2 - 5°C$, $\Delta y_1 - 5mol\%$, and $\Delta y_2 - 0.1mol\%$

The first task is to perform some up and down step input tests to estimate the steady state gain and time constants of the MIMO system. Since it is well known that the TE process is highly nonlinear and the nonlinearity is not the focus of this paper, we handle this problem as following: we perturb the two inputs not too far from their steady state values, assuming that we are within the linear region, and run the TE simulation (Lyman and Georgakis, 1995) program under the noise free condition. We thus can obtain a steady state gain matrix by taking the mean of gains obtained from both the up and down step change experiments and attribute the nonlinearity as the model uncertainty. The resulting steady state gain matrix together with model uncertainty can be represent as:

$$G_{ss} = \begin{bmatrix} -(0.95 \sim 1.05) \cdot 1.3 & -(0.84 \sim 1.16) \cdot 0.01 \\ -(0.90 \sim 1.10) \cdot 1.1 & -(0.96 \sim 1.04) \cdot 1.0 \end{bmatrix}$$

This result is already normalized by the input and output constraint values. Also the input and output constraints become

$$-1 \le \Delta u_1 \le 1 \quad -1 \le \Delta u_2 \le 1$$
$$-1 \le \Delta y_1 \le 1 \quad -1 \le \Delta y_2 \le 1$$

after the normalization.

The time constants of each input-output pair are also obtained from the simulation result.

4. STEADY STATE SIGNAL DESIGN

Given an prior steady state gain matrix G of a MIMO system, the input signal design aims to generate a signal that is large enough but does not violate either the input nor the output constraints. Since PRBS has a lot of advantages in system identification (Godfrey, 1993), an input design usually appears as an transformed MIMO PRBS. In the steady state design, a single unit PRBS is represented just by the two levels that it can take: $[1 \; -1]$, and a unit n-dimensional PRBS can then be represented by all the possible combinations of $[1 \; -1]$ taken by its n inputs as:

$$\begin{bmatrix} 1 & 1 & \cdots & -1 \\ 1 & 1 & \cdots & -1 \\ \vdots & \vdots & \ddots & \vdots \\ 1 & -1 & \cdots & -1 \end{bmatrix}$$

Zhan and Georgakis (2000) proposed a steady state method on input signal design. The method, which is called Constrained Transform Multi-input Signal (CTMIS), makes a generalization of all the previous designs and aims to maximize the input space volume under input and output constraints.CTMIS designs an MIMO input signal as $u = T \cdot PRBS$, where T is an $n \times n$ matrix and can be determined from the optimization:

$$\max_{t_{ij}} |det(\mathbf{T})|$$

with the constraints of

$$\mathbf{u}_{low} \leq \mathbf{T} \cdot \mathbf{PRBS} \leq \mathbf{u}_{high}$$

$$\mathbf{y}_{low} \leq \mathbf{G} \cdot \mathbf{T} \cdot \mathbf{PRBS} \leq \mathbf{u}_{high}$$

The simple MIMO design and the design proposed by Koung and Macgregor (1994) can be represented as special cases by taking special $\mathbf{T}$ matrices. In the simple MIMO design,

$$\mathbf{T} = \begin{bmatrix} a_1 & 0 & \cdots & 0 \\ 0 & a_2 & \cdots & 0 \\ \vdots & \vdots & \ddots & \vdots \\ 0 & 0 & \cdots & a_n \end{bmatrix}$$

and in Koung and Macgregor's design,

$$\mathbf{T} = k \cdot V\Sigma^{-1}$$

Furthermore, Koung and Macgregor's method can be generalized into the orthogonal rotated MIMO (ORMIS) design by taking different coefficients for the orthogonal directions, where

$$\mathbf{T} = V \cdot \begin{bmatrix} k_1/\sigma_1 & 0 & \cdots & 0 \\ 0 & k_2/\sigma_2 & \cdots & 0 \\ \vdots & \vdots & \ddots & \vdots \\ 0 & 0 & \cdots & k_n/\sigma_n \end{bmatrix}$$

Here $G = U \cdot \Sigma \cdot V'$ is the singular value decomposition of the steady state gain matrix and $[\sigma_1 \sigma_2 \cdots \sigma_n]$ are the sigular values.

In Zhan and Georgakis (2000), it is argued that a linear transformation by multiplying a matrix T expands or shrinks the volume of a region at rate of $|det(T)|$. So it is also argued that the maximization of the absolute value of the determinant of $\mathbf{T}$, which leads to the D-optimal design (Ljung, 1987), is the same as the maximization of the input space geometrically.

5. ITERATIVE IDENTIFICATION PROCESS AND DYNAMIC INPUT DESIGN FOR THE TE PROBLEM

5.1 Iterative Identification Process

After performing the preliminary step test, we just have the steady state gain matrix in hand. At this time, the input design should follow what is described above. However, after the designed signal is introduced into the system for a period of time and some input and output data are collected, a dynamic model can be calculated by an identification algorithm, such as subspace identification. The N4SID algorithm of the Matlab Identification Toolbox (Ljung, 2000) is used in our work. With this preliminary model in hand, we can now redesign the input signal. Although the first dynamic model may not be accurate enough, it is better than the steady state design. And then we can update the design after it is introduced into the system for some time, more data are collected and an updated model is identified. As a result, this idea of iterative identification process can be represented as in Fig 2.

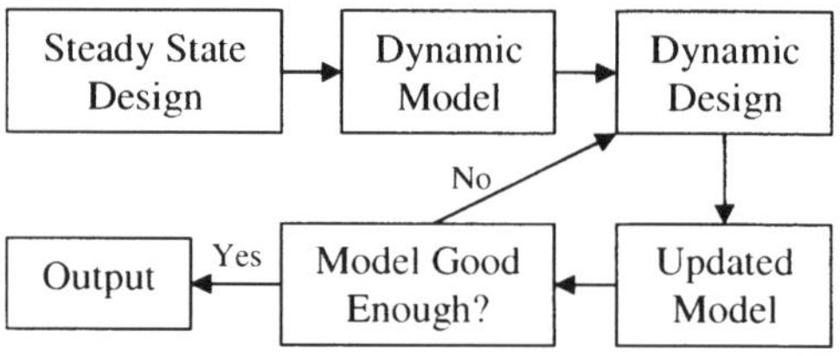

Fig. 2. Iterative Identification Process

5.2 Input Design Using Dynamic Models

When we move on from a steady state to a dynamic design, we assume that the input still takes the form of $\mathbf{u} = \mathbf{T} \times \mathbf{PRBS}$, where the **PRBS** now represents a real time sequence with two levels $[1 \; -1]$. As a result, the objective function and the constraints on the input signal will still take the same form as before. The output constraints must be changed because the outputs do not reach their steady state. For example, Fig 3 shows the dynamic responses of a typical first and second order system, whose steady state gains are 1 and -1, under the perturbation of 4th order PRBS of the unit amplitude.

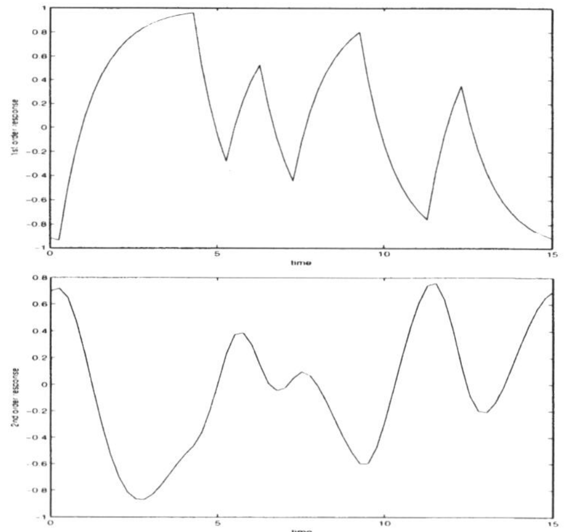

Fig. 3. Typical dynamical responses under a 4th order PRBS

Generally speaking, all of the outputs usually do not reach their steady state at the same time, making the steady state constraints too conservative for overdamped systems. At the same time, there are overshoots in the outputs of the underdamped systems. For these systems, steady state design may cause them to go beyond their limits.

In order to solve this problem, our method utilizes the available approximate dynamic model. For each input-output pair, we calculate the response to a unit-magnitude PRBS, whose order and clock time have been specified according to the approximate time constants of the known model components (Rivera, 1992). The discrete form of the responses will be denoted as "signature"s that represent the expected behavior of the system under the perturbation of the specific input. So the outputs constraints can be updated as:

$$\mathbf{y}_{low} \le \mathbf{G}\left(q^{-1}\right) \cdot \mathbf{T} \cdot \mathbf{PRBS}\left(t\right) \le \mathbf{y}_{high}$$

where $\mathbf{G}(q^{-1})$ represents the dynamic model of system which can be derived from the corresponding state space model, and $\mathbf{PRBS}(t)$ is the time sequence of a unit multiple PRBS. As a result, if we increase the number of signatures for one period, the design result will be more accurate, and the optimization task becomes more challenging at the same time since the number of constraints is increased.

5.3 Model Uncertainty in the Dynamic Condition

Model uncertainty is important in the input signal design because of the inaccuracy of the available prior model. We will assume that the noise terms that are calculated in the stochastic-deterministic version of the state space identification algorithm represent the model uncertainty. In the following analysis, we assume that the output is one dimensional. If the output is multi-dimensional, we can apply the same analysis on each of the components.

From the state space form of the model,

$$\begin{aligned} x_{k+1} &= A \cdot x_k + B \cdot u_k + K \cdot e_k \\ y_k &= C \cdot x_k + D \cdot u_k + e_k \end{aligned} \tag{1}$$

we can derive an input-output form:

$$\begin{aligned} y(k) &= G(q)u(k) + v_y(k) \\ G(q) &= C(qI - A)^{-1}B \\ v_y(k) &= [C(qI - A)^{-1}K + 1]e(k) \end{aligned} \tag{2}$$

Then if the statistical properties of $v_y(k)$ can be determined, the model uncertainty can be easily derived after that. Because $v_y(k)$ is the linear combination of $e(k)$'s, which are assumed to be Gaussian noises, v_y itself is also a Gaussian noise of zero mean. Then if the variances of the $v_y(k)$ can be calculated, it is easy to find a interval range for $v_y(k)$ to be located in with a given uncertain level, e.g. 95% or 99%.

However, it must be brought to notice that y is not a stationary process since $Ey(k) = G(q)u(k)$. As a result, the statistic properties for each $y(k)$ are different for different k's. In order to handle this problem, here we adopt the concept of "quasi-stationarity" (Ljung, 1987), whose key idea is to study the mean of the statistical properties of a non-stationary process.

By using Theorem 2.2 in Ljung (1987), the spectrum of v_y is:

$$\Phi(\omega) = \lambda |C(e^{i\omega} \cdot I - A)^{-1}K + 1|^2$$

where λ is the variance of e. Letting ω take the discrete values from 0 to $N-1$, where N is number of data points, we can evaluate the discrete power spectrum of v_y. Then after performing inverse discrete Fourier Transform and letting $\tau = 0$, the mean value of the variances of v_y is obtained. This completes the major part of the determination of the model uncertainty.

It should be noted that not all the design methods can incorporate model uncertainty as easily as the CTMIS method. As a result, when we make the comparison in the present paper, the model uncertainty is neglected in order to compare all the methods equally. However, in practical applications, model uncertainty should be included by the rule described above.

5.4 Comparison of Identification Result on the TE Process

As we have stated above, this process is highly nonlinear and nonlinearity is not our major concern for this paper. For this reason, the 2×2 subsystem in the TE process was perturbed slightly around the normal operating point by a MIMO PRBS under the noise-free condition for a certain period of time. Then a 5th order dynamic model together with the noise properties were identified from these initial experiment data. From this point on, it is assumed that this model is the true model of the linearized 2×2 subsystem in the TE process, and it is used in the comparison of the different signal designs.

Then the iterative identification process and dynamic input design method described before was tried on the above 2×2 linearized process and the whole simulation time was 930 hours with a sampling time of 3 minutes. The whole period was divided into three equal parts. Within the first part, the steady state design was introduced, and within the second and third parts the dynamic and updated dynamic designs were used to perturb the system. The noise used in the simulation is 20% of the amplitude of a white Gaussian noise. The experiment time and amplitude of the noise were selected in order for all the methods to generate stable identified models that is not too far from the original one.

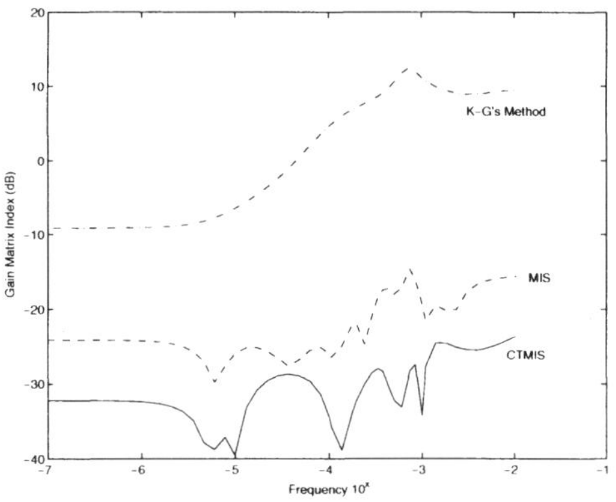

Fig. 4. Identification Result Comparison

As comparison, we also perturbed the system with simple MIMO signal and orthogonal rotated MIMO signal (Koung and Macgregor, 1994). These two signals were designed based only the steady state model and each of them were introduced to perturb the system for 930 hours too. The accuracy of the models from different input designs can be compared through the coefficient of determination R^2 (Juricek *et al.*, 2002). However, the more complete comparison among different models should be the frequency responses. In order to evaluate the accuracy of the whole MIMO model instead of comparing each input-output pair, we use the following form as the index:

$$\frac{\|G_0(\omega) - G_{id}(\omega)\|_2}{\|G_0(\omega)\|_2}$$

Here the matrix G is composed of the frequency response magnitudes of all the input-output pairs under different frequencies, which can be obtained from the Bode plot of the system. G_0 is the true system and G_{id} is the identified model. The result is plotted in Fig 4. In order to reduce the impact of the specific noise, we run the process 3 times and the mean values were used to generate Fig 4.

6. CONCLUSION

This paper presents a methodology for identification input signal design of constrained dynamic MIMO systems. It is tested in a linearized subsystem in the challenging TE process. From the above results plotted in Figure 4 , it can be seen that the proposed input design procedure is better than the existing ones in that it provides a more accurate identification result.

REFERENCES

Downs, J. J. and E. F. Vogel (1993). A plant-wide industrial process control problem. *Comp. Chem. Eng.* **17**(3), 245–255.

Godfrey, K. (1993). *Perturbation signals for system identification.* Prentice Hall. Hemel Hempstead, Hertfordshire, UK.

Juricek, B. C., D. E. Seborg and W. E. Larimore (2002). Identification of multivariable, linear, dynamice models: comparing regression and subspace techniques. *Ind. Eng. Chem. Res.* **41**(9), 2185–2203.

Koung, C. and J. F. Macgregor (1994). Identification for robust multivariable control: the design of experiments. *Automatica* **30**(10), 1541–1554.

Ljung, L. (1987). *System identification-theory for the user.* Prentice hall.

Ljung, L. (2000). *System Identificatin Toolbox. For Use with Matlab.* Natrick, MA.

Lyman, P. R. and C. Georgakis (1995). Plant-wide control of the tennessee eastman problem. *Computers Chem. Eng.* **19**(3), 321–331.

Rivera, D. E. (1992). Identification of distillation systems. In: *Practical Distillation Control* (W. L. Luyben, Ed.). Chap. 7. Van Nostrand Reinhold. New York.

Zhan, Q. and C. Georgakis (2000). Steady state optimal test signal design for multivariable model based control. In: *12th Symposium on System Identification.*

IFAC Publications
www.elsevier.com/locate/ifac

CONSTRAINED MINIMUM CREST FACTOR MULTISINE SIGNALS FOR "PLANT-FRIENDLY" IDENTIFICATION OF HIGHLY INTERACTIVE SYSTEMS

Hyunjin Lee *, **Daniel E. Rivera** *,[1], **Hans D. Mittelmann** **

* *Dept. of Chemical and Materials Engineering,*
Arizona State University, Tempe, Arizona 85287-6006
Email: Hyunjin.Lee@asu.edu; daniel.rivera@asu.edu

** *Dept. of Mathematics and Statistics,*
Arizona State University
Email: mittelmann@asu.edu

Abstract: Highly interactive systems are ill-conditioned and highly sensitive to model uncertainty, which imposes limitations to achievable closed-loop performance. In this paper, the goal is to develop an identification testing framework meaningful to highly interactive systems based on the application of constrained minimum crest factor multisine signals. "Plant-friendliness" in the design procedure is accomplished by imposing constraints on the overall span, move size, and variability of both input and output signals. A modified "zippered" power spectral density is proposed which contains both correlated and uncorrelated harmonics to simultaneously excite low- and high-gain directions in the data. This signal design procedure is applied to a simple 2-by-2 multivariable system that exhibits challenging interaction and gain directionality issues. Validation of these signals in a closed-loop setting is accomplished by evaluating the performance of decentralized PI with steady-state decoupler controller designs based on models estimated from noisy open-loop experiments. The modified "zippered" spectrum signals present a clear advantage over standard zippered designs for short data sets under noisy conditions. *Copyright © 2003 IFAC*

Keywords: multivariable systems, system identification, plant-friendliness, input signal design

1. INTRODUCTION

Multivariable systems may exhibit ill-conditioning and be highly interactive, which is manifested in models displaying large condition numbers and RGA values. Highly interactive processes have a natural tendency to respond in the high-gain direction, which makes it very difficult to obtain low-gain directionality information using conventional open-loop identification tests. As a result, these systems have been considered challenging cases for multivariable system identification and robust control design in the process industries (Andersen and Kümmel, 1992; Skogestad and Morari, 1988).

In this paper, the primary purpose is to develop an identification testing framework that is applicable to highly interactive multivariable systems. The framework relies on the use of constrained minimum crest factor multisine signals that have been previously proposed for SISO systems (Braun *et al.*, 2002). Signals conducive to "plant-friendly" tests (Rivera *et al.*, 2003) are generated through the use of powerful constrained optimization techniques that minimize crest factor while enforcing constraints on magnitude and move sizes of both input and output signals.

[1] To whom all correspondence should be addressed; phone: (480)965-9476, Email: daniel.rivera@asu.edu

Rivera *et al.* (1997) presented the use of a "zippered" power spectrum relying on orthogonal Fourier coefficients as a means to generate multisine signals for simultaneous multichannel testing in multivariable system identification. Recently, Stec and Zhu (2001) have proposed the use of sequential cycles of high-magnitude correlated and low-magnitude uncorrelated input signals in order to increase the low-gain information in the data, resulting in wider spread in the output state-space plot. In this paper, a similar philosophy is implemented in the frequency domain via the use of a modified "zippered" power spectrum. The modified "zippered" power spectrum contains both correlated and uncorrelated harmonics to simultaneously excite low- and high-gain directions. By enhancing the information content in the low-gain direction, the effectiveness of these signals for identification under noisy conditions is vastly improved. These modified "zippered" designs ultimately lead to shorter identification tests compared to standard signals, an important consideration in industrial practice.

This paper is organized as follows: Section 2 describes the design guidelines for standard and modified zippered multisine signals. Section 3 focuses on constrained optimization problem formulations for generating plant-friendly signals, while Section 4 presents a case study showing the application of the design procedure and closed-loop validation. Section 5 consists of a summary and conclusions.

2. MULTISINE SIGNALS WITH "ZIPPERED" POWER SPECTRA

2.1 *Multisine Signal Design Parameter Specification*

Multisine signals are deterministic, periodic signals whose power spectrum can be directly specified by the user. A multisine input $u_j(k)$ for the *j*-th channel of a multivariable system with m inputs can be defined as,

$$u_j(k) = \sum_{i=1}^{m\delta} \hat{\delta}_{ji} \cos(\omega_i kT + \phi_{ji}^{\delta}) + \sum_{i=m\delta+1}^{m(\delta+n_s)} \alpha_{ji} \cos(\omega_i kT + \phi_{ji}) + \sum_{i=m(\delta+n_s)+1}^{m(\delta+n_s+n_a)} \hat{a}_{ji} \cos(\omega_i kT + \phi_{ji}^{a}),\ j = 1, \cdots, m \quad (1)$$

where T is sampling time, N_s is the sequence length, m is the number of channels, δ, n_s, n_a are the number of sinusoids per channel ($m(\delta + n_s + n_a) = N_s/2$), $\phi_{ji}^{\delta}, \phi_{ji}, \phi_{ji}^{a}$ are the phase angles, α_{ji} represents the Fourier coefficients defined by the user, $\hat{\delta}_{ji}, \hat{a}_{ji}$ are the "snow effect" Fourier coefficients, and $\omega_i = 2\pi i/N_sT$ is the frequency grid. The "snow effect" (Guillaume *et al.*, 1991) refers to sinusoids where both Fourier coefficients and phases are selected by the optimizer.

In the design procedure presented in this paper, the primary frequency bound of interest for excitation is determined by the dominant time constants of the system to be identified and the desired closed-loop speed-of-response,

$$\omega_* = \frac{1}{\beta_s \tau_{dom}^H} \leq \omega \leq \omega^* = \frac{\alpha_s}{\tau_{dom}^L} \quad (2)$$

α_s and β_s are parameters that specify the high and low frequency ranges of interest in the signal, respectively for a given range of low and high dominant time constants (defined by τ_{dom}^L and τ_{dom}^H). The primary band of excitation is bounded by the following inequality based on the choice of design parameters,

$$\frac{2\pi m(1+\delta)}{N_sT} \leq \omega_* \leq \omega \leq \omega^* \leq \frac{2\pi m n_s(1+\delta)}{N_sT} \leq \frac{\pi}{T} \quad (3)$$

which in turn translates into the following inequalities for sampling time, number of sinusoids, and sequence length (T, n_s, and N_s, respectively):

$$(1+\delta)\frac{\omega^*}{\omega_*} \leq n_s \leq \frac{N_s}{2m} \quad (4)$$

$$T \leq \min\left(\frac{\pi}{\omega^*}, \frac{\pi}{\omega^* - \omega_*}(1 - \frac{1+\delta}{n_s})\right) \quad (5)$$

$$\max\left(2mn_s, \frac{2\pi m(1+\delta)}{\omega_* T}\right) \leq N_s \leq \frac{2\pi m n_s}{\omega^* T} \quad (6)$$

As shown in Figure 1, the shape of the power spectrum in a multisine input is specified by the choice of Fourier coefficients. In this design procedure, a "notch" spectrum design is used, with potentially variable number of Fourier coefficients in the low frequency area, primary frequency band, and high frequency area. Theoretical requirements such as persistence of excitation, harmonic suppression (a key consideration in the identification of nonlinear systems), and control-relevance can be satisfied without loss of generality through the specification of Fourier coefficients (Rivera *et al.*, 2002).

2.2 *Modified "Zippered" Power Spectrum Design*

In the standard "zippered" design procedure described in Rivera *et al.* (1997), Fourier coefficients for each channel are defined independently over the frequency grid of interest. The resulting orthogonal multifrequency signal can be simultaneously introduced to all channels of the plant to be identified.

In this paper, the philosophy presented by Stec and Zhu (2001) in the time domain is adapted to define a modified "zippered" power spectrum meaningful to highly interactive systems. In addition to orthogonal Fourier coefficients, the modified zippered spectrum includes correlated harmonics with high levels of power (Figure 1) which serve to emphasize low gain information in the data. Determining *a priori* the relative degree of power between the correlated and uncorrelated harmonics is a subject of current investigation; the case study in Section 4 provides proof-of-concept for the idea.

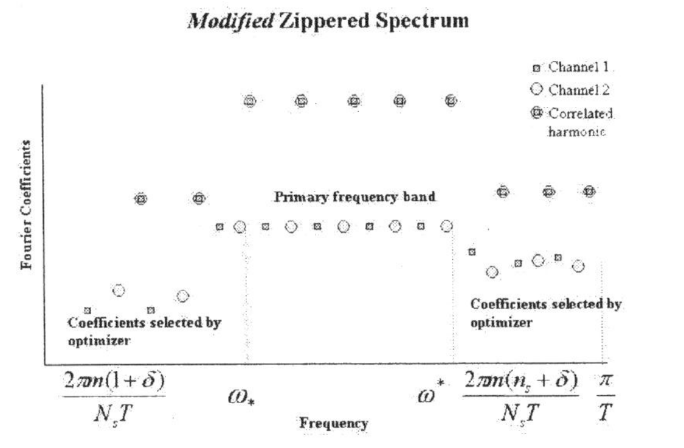

Fig. 1. Conceptual design of a modified "zippered" spectrum for a two-channel signal.

3. CONSTRAINED OPTIMIZATION FOR PLANT-FRIENDLY SIGNAL DESIGN

Optimization techniques are used in the design procedure to obtain multisine signals which satisfy desirable plant-friendly requirements in the time domain with user-specified power spectra. The optimization problems consist of minimizing the worst-case crest factor (of either the input channels, the output channels, or a combination of both) given time-domain constraints on move size, maximum and minimum signal limits, and so forth.

3.1 *Plant-Friendly Signal Evaluation*

The crest factor is a criterion which indicates how well a signal is distributed within its span; the crest factor of a signal x is defined as the ratio of the infinity-norm versus the 2-norm of a signal,

$$CF(x) = \frac{\ell_\infty(x)}{\ell_2(x)} \tag{7}$$

In particular, a low crest factor value (i.e., close to 1) indicates that most elements of the signal are distributed near its minimum and maximum values. Lowering crest factor in general contributes to "plant-friendliness" (Braun *et al.*, 2002), since identification tests with equivalent information content can be performed over lower input (or output) spans. Alternatively, the Performance Index for Perturbation Signals (PIPS) (Godfrey *et al.*, 1999) defined by

$$PIPS(\%) = 200\frac{\sqrt{u_{rms}^2 - u_{mean}^2}}{u_{max} - u_{min}} \tag{8}$$

is a convenient measure of signal distribution. The PIPS measurement ranges only between 0 and 100% (compared to 1 versus ∞ for crest factor), which gives it an intuitive, practical appeal.

3.2 *Constrained Minimum Crest Factor Optimization Problem Formulations*

Given a multisine signal structure per (1) for a channel j and a desired power spectral density (standard or modified, defined by the Fourier coefficients α_{ji} for n_s spectral lines), one optimization problem which can be solved is to minimize the maximum crest factor over all the input channels

$$\min_{\{\phi^a_{ji}\},\{\phi^\delta_{ji}\},\{\phi_{ji}\},\{\hat{a}_{ji}\},\{\hat{\delta}_{ji}\}} \max_j \text{CF}(u_j) \quad j = 1,\cdots,m \tag{9}$$

subject to maximum move size constraints on the input sequence $\{u_j(k)\}$

$$|\Delta u_j(k)| \le \Delta u_j^{max} \quad \forall\, k,j \tag{10}$$

and high/low limits on $\{u_j(k)\}$

$$u_j^{min} \le u_j(k) \le u_j^{max} \quad \forall\, k,j \tag{11}$$

Minimizing output crest factor is of great value in the process industries, since the nature of the output signal often has the most influence on product quality and profitability. If a dynamic model is available *a priori*, optimization problems minimizing the maximum crest factor over all output channels can be formulated as

$$\min_{\{\phi^a_{ji}\},\{\phi^b_{ji}\},\{\phi_{ji}\},\{\hat{a}_{ji}\},\{\hat{b}_{ji}\}} \max_z \text{CF}(y_z) \tag{12}$$
$$j = 1,\cdots,m \qquad z = 1,\cdots,N_{outs}$$

subject to (in addition to the input signal constraints (10) and (11)) constraints on both changes and upper/lower values of the output signal,

$$|\Delta y_z(k)| \le \Delta y_z^{max} \quad \forall\, k,z \tag{13}$$

$$y_z^{min} \le y_z(k) \le y_z^{max} \quad \forall\, k,z \tag{14}$$

Alternatively, it is possible to minimize the maximum crest factor of both input and output signals,

$$\min_{\{\phi^a_{ji}\},\{\phi^b_{ji}\},\{\phi_{ji}\},\{\hat{a}_{ji}\},\{\hat{b}_{ji}\}} \max_{j,\,z}\{\text{ CF}(u_j),\ \text{CF}(y_z)\} \tag{15}$$
$$j = 1,\cdots,m \qquad z = 1,\cdots,N_{outs}$$

subject to (10) - (11) and (13) - (14).

These optimization problems are formulated in the modeling language AMPL, which provides exact, automatic differentiation up to second derivatives. Initial approaches for minimizing crest factor (Rivera *et al.*, 2002) were patterned after the work of Guillaume *et al.* (1991) which is based on Pólya's algorithm. Recent approaches involve a direct min-max solution where the nonsmoothness in the problem is transferred to the constraints. In particular, the trust region, interior point method by (Byrd *et al.*, 1999) has been used for all examples in this paper. This method, as implemented in versions 2.1 and most recently 3.0 of KNITRO, performed in the overall most efficient and robust fashion compared to several other well-known NLP solvers with AMPL interface.

4. CASE STUDY

A simple, highly interactive problem based on the dynamics of a high-purity distillation column (Skogestad and Morari, 1988) is examined in this paper. The model corresponds to the transfer function,

$$P(s) = \frac{1}{75s+1}\begin{bmatrix} 87.8 & -86.4 \\ 108.2 & -109.6 \end{bmatrix} \quad (16)$$

In spite of its simplicity, the model represents an interesting case study for evaluating the input signal design procedure presented in the paper. In this case study, a series of problem formulations based on the system per (16) are considered, as noted below,

(1) Minimize the maximum crest factor of u using a standard zippered spectrum (Rivera *et al.*, 1997),
(2) Minimize the maximum crest factor of u using a modified zippered spectrum (per Figure 1),
(3) Minimize the maximum crest factor of y using a modified zippered spectrum subject to constraints on Δu, y, and Δy.
(4) Minimize the maximum crest factor of both u and y using a modified zippered spectrum subject to constraints on Δu, y, and Δy.

In all cases involving output crest factor, the model (16) is assumed to be known *a priori*. Model estimates obtained from noisy data are validated in a closed-loop setting using a decentralized PI with steady-state decoupler control system.

For the system per (16), we assume $\tau^L_{dom} = \tau^H_{dom} = 75$ min and select $\delta = 0$, $\alpha_s = 7.5$, and $\beta_s = 3.33$. Based on the guidelines of Section 2 we further specify $T = 15$ min, $n_s = n_a = 26$, and $N_s = 210$. The ratio of correlated to uncorrelated harmonic Fourier coefficients is 64.5. The "snow" effect is not examined in this signal, but all harmonics in the secondary band of excitation possess Fourier coefficient values that are 10% of those of the primary. Figure 2 shows the power spectral densities for both standard and modified zippered signals examined in the case study.

4.1 *Open-Loop Identification Results*

Statistics for all four signals are summarized in Table 1; output state-space plots are shown in Figure 3, while the time series for one cycle are shown in Figures 4 and 5. Evaluating the open-loop experiment data, the standard zippered spectrum {min CF(u)} signal has information primarily in the high-gain direction which is reflected via a very thinly spread concentration of points in the state-space plot and large output spans (0.6819 to -0.7003 for y_1 and 0.8553 to -0.8712 for y_2) despite low input magnitude changes (Figure 4 (left)). On the contrary, the modified zippered spectrum {min CF(u)} signal (Figure 4 (right)) produces a much better directional distribution of the data in the state-space and despite a much larger input span results in only modest increases in the output span (30% and 20% greater

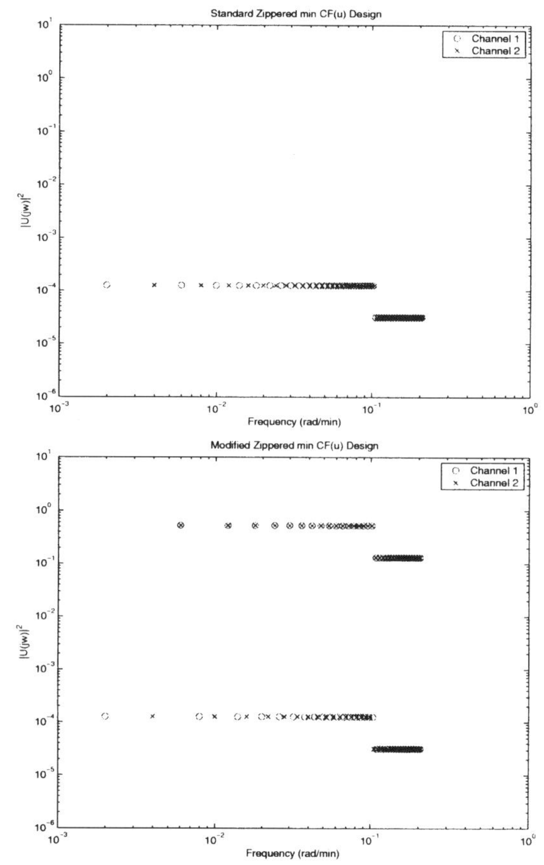

Fig. 2. Power spectrum for the standard zippered (top) and modified zippered (bottom) signal designs presented in the case study.

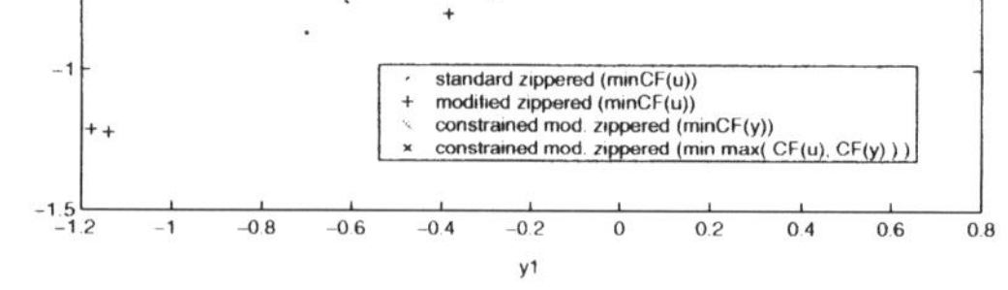

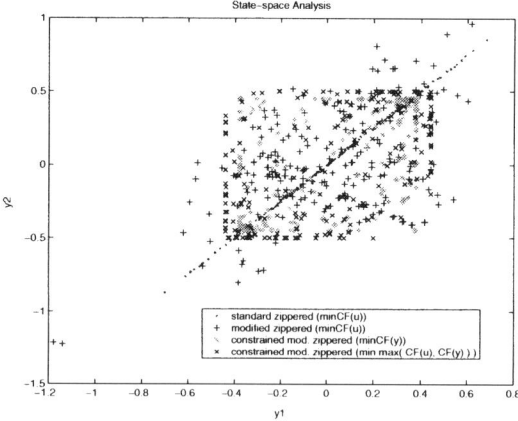

Fig. 3. State-space plots from simulations of standard "zippered" spectrum min CF(u) (•, red), modified "zippered" spectrum min CF(u) (+, blue), modified "zippered" spectrum min CF(y) ($*$, green), and modified "zippered" spectrum min max(CF(u), CF(y)) ($\times$, black)

for y_1 and y_2, respectively) compared to the standard zippered spectrum signal. Minimizing the output crest factor and enforcing input and output constraints on the optimization with a modified zippered spectrum {min CF(y)} (Figure 5, (left)), lowers the output spans to desirable levels (0.378893 to -0.378896 for y_1 and 0.448301 to -0.448300 for y_2) and generates

Type	Signal (x)	CF(x)	PIPS(%)	max Δx	max x	min x
min CF(u) design; standard zippered spectrum	u_1	1.132425	88.306039	0.014106	0.007079	-0.007079
	u_2	1.136376	87.999046	0.014201	0.007100	-0.007100
	y_1	2.383208	42.520481	0.318468	0.681899	-0.700354
	y_2	2.372958	42.530317	0.397652	0.855283	-0.871213
min CF(u) design; modified zippered spectrum	u_1	1.130450	88.460763	0.745004	0.372522	-0.372526
	u_2	1.130632	88.446404	0.745163	0.372585	-0.372583
	y_1	3.924442	33.358140	0.763222	0.619802	-1.174440
	y_2	3.459978	32.254284	0.863657	0.966114	-1.219642
min CF(y) design; modified zippered, $\|\Delta u, y, \Delta y\| \leq 0.5$	u_1	3.018388	36.782253	0.499983	0.789886	-0.985599
	u_2	3.004964	36.887273	0.499986	0.789231	-0.981237
	y_1	1.221784	81.847791	0.499897	0.378893	-0.378896
	y_2	1.221781	81.847838	0.499989	0.448301	-0.448300
min max (CF(u) CF(y)) design; modified zippered, $\|\Delta u, y, \Delta y\| \leq 0.5$	u_1	1.423676	70.240749	0.499997	0.464209	-0.464209
	u_2	1.423675	70.240777	0.499995	0.464217	-0.464217
	y_1	1.423637	70.242767	0.407867	0.441263	-0.441262
	y_2	1.363417	73.346052	0.499934	0.499980	-0.499967

Table 1. Summary of open-loop case study results

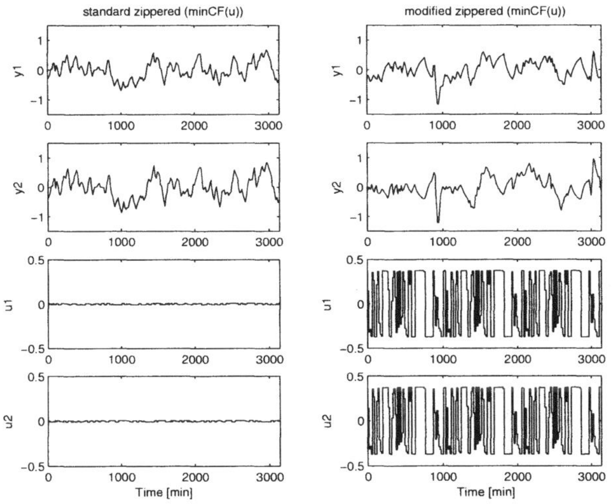

Fig. 4. min CF(u) multisine signals using standard (left) and modified (right) "zippered" spectra.

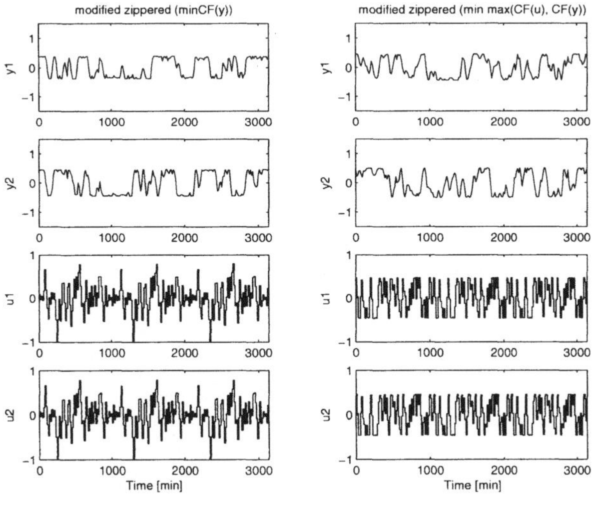

Fig. 5. Constrained multisine signals using modified zippered spectrum design minCF(y) (left) and min max(CF(u), CF(y)) (right)

a nice, bounded rectangular distribution in the state-space plot. Crest factors for both channels of u, however, have increased substantially. The signal corresponding to the {min max (CF(u), CF(y))} problem with constraints (Figure 5, (right)) provides a reasonable balance between input and output crest factors that appears to promote plant friendliness. Compared to the {min CF(y)} case, input spans (and correspondingly, crest factor) are reduced dramatically (by over 50%), while the output span increases by a small amount (approximately 12% higher for each output).

From these case study results we can summarize that while the correlated harmonics in the modified zippered spectrum significantly increase the input span, the output span does not increase proportionately, and in the case of the constrained min CF(y) and min max (CF(u), CF(y)) problems, the output span is actually lower than the standard spectrum signal, even though the modified spectrum has a much higher level of average power (Figure 2). Thus, we are able to show that the proposed design procedure represents a flexible approach to plant-friendly multisine input design in a manner appealing to the user.

4.2 *Closed-Loop Evaluation Controller Design*

The control-relevant quality of the signals is evaluated through model estimation and closed-loop validation. First, a model is estimated from one cycle of noisy data using a multivariable ARX $[na = 1, nb = 1, nk = 1]$ structure. This model is used to design a decentralized PI controller with steady-state decoupler which is then applied to the true system. The PI controllers are tuned using the Prett-García tuning rules (Prett and García, 1988),

$$C_{PI}(z) = D_{ss} \times \begin{bmatrix} Kc_{11}\dfrac{z - exp(-T/\tau_{11})}{z-1} & 0 \\ 0 & Kc_{22}\dfrac{z - exp(-T/\tau_{22})}{z-1} \end{bmatrix} \quad (17)$$

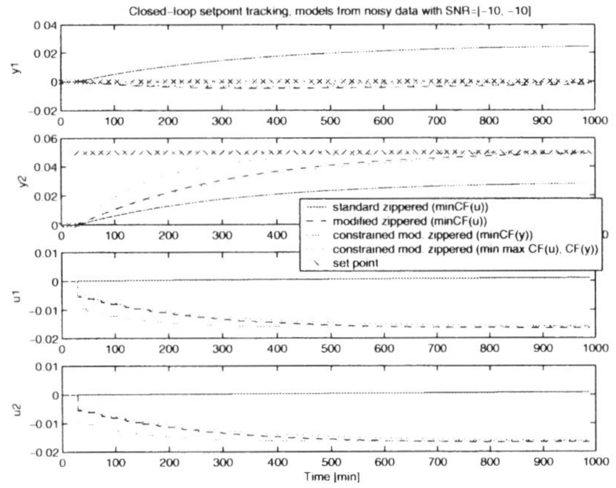

Fig. 6. Closed-loop simulation of controllers designed from models estimated from the standard zippered spectrum (min CF(u), — red), modified zippered spectrum (min CF(u, –, blue), constrained modified zippered spectrum (min CF(y), -·, green), and constrained modified zippered spectrum (min max (CF(u), CF(y)), $\cdots$, black) signals.

where

$$D_{ss} = K^{-1} \times diag(k_{11}, k_{22}) \quad K = P(0) = \begin{bmatrix} k_{11} & k_{12} \\ k_{21} & k_{22} \end{bmatrix}$$

$$Kc_{11} = \frac{1 - exp(-T/\lambda)}{k_{11}}, \quad Kc_{22} = \frac{1 - exp(-T/\lambda)}{k_{22}}$$

Under noise-free conditions, all the controllers based on the multisine signals result in an equivalent closed-loop performance. With the addition of significant noise during identification testing (SNR $= -10dB$ per channel), the difference in the closed-loop results becomes significant. As seen in Figure 6, the control system based on the model generated from the standard zippered spectrum data is not able to track the set points properly, while all the models arising from the modified zippered spectrum signals yield stable results with no offset in the controlled variables.

5. SUMMARY AND CONCLUSIONS

A modified "zippered" power spectrum in conjunction with constrained optimization has been presented as means to generate multi-channel multisine inputs that achieve plant-friendliness while addressing the requirements of highly interactive systems. The use of correlated harmonics in the modified zippered spectrum increases the output power in the low-gain direction to be comparable to that of the high-gain direction. Simulation results show that under noisy conditions the increased information content in the low-gain direction leads to useful models from shorter identification tests in comparison to standard designs. Current research includes a more systematic definition of the relationship between the correlated and uncorrelated harmonics, as well as the application to more sophisticated distillation column models, such as that presented in Weischedel and McAvoy (1980).

6. ACKNOWLEDGEMENT

This research has been supported by the American Chemical Society - Petroleum Research Fund, Grant No. ACS PRF# 37610-AC9.

7. REFERENCES

Andersen, H.W. and M. Kümmel (1992). Evaluating estimation of gain directionality Parts 1 and 2: A case study of binary distillation. *J. Proc. Cont.* **2**, 59–86.

Braun, M.W., R. Ortiz-Mojica and D.E. Rivera (2002). Application of minimum crest factor multisinusoidal signals for "plant-friendly" identification of nolinear process systems. *Control Engineering Practice* **10**, 301.

Byrd, R., M.E. Hribar and J. Nocedal (1999). An interior point method for large scale nonlinear programming. *SIAM J. Optim.* **9**, 877–900.

Godfrey, K.R., H.A Barker and A.J. Tucker (1999). Comparison of perturbation signal for linear system identification in the frequency domain. *IEE. Proc. Control Theory Appl.* **146**, 535.

Guillaume, P., J. Schouken, R. Pintelon and I. Kollar (1991). Crest-factor minimization using nonlinear chebyshev approximation methods. *IEEE Transactions on Instrument and Measurement* **40**, 982.

Prett, D.M. and C.E. García (1988). *Fundamental Process Control.* Butterworth. Stoneham, M.A.

Rivera, D.E., H. Lee, M.W. Braun and H.D. Mittelmann (2003). "plant-friendly" system identification: a challenge for the process industries. *IFAC Symposium on System Identification (SYSID 2003), Rotterdam, The Netherlands.*

Rivera, D.E., M.W. Braun and H.D. Mittelmann (2002). Constrained multisine inputs for plant-friendly identification of chemical process. *IFAC World Congress, Barcelona, Spain.*

Rivera, D.E., S. Zong and W.M. Ling (1997). A control-relevant multivariable system identification methodology based on orthogonal multi-frequency input perturbations. *IFAC Symposium on System Identification (SYSID'97), Fukuoka, Japan* pp. 595–600.

Skogestad, S. and M. Morari (1988). LV-control of a high-purity distillation column. *Chem. Eng. Sci.* **43**, 33.

Stec, P. and Y. Zhu (2001). Some study on identification of ill-conditioned processes for control. *Proc. of the ACC, Arlington, VA.*

Weischedel, K. and T.J. McAvoy (1980). Feasibility of decoupling in conventionally controlled distillation columns. *Ind. Eng. Chem. Fund.* **19**, 379.

IFAC Publications
www.elsevier.com/locate/ifac

ONLINE IDENTIFICATION OF A ROBOT USING BATCH ADAPTIVE CONTROL

Björn Bukkems,[1] Dragan Kostić,[2] Bram de Jager,[3] and Maarten Steinbuch[4]

*Technische Universiteit Eindhoven, Department of Mechanical Engineering,
Dynamics and Control Technology Group, P.O. Box 513, 5600 MB Eindhoven, The Netherlands
[1]B.H.M.Bukkems@tue.nl, [2]D.Kostic@tue.nl, [3]A.G.de.Jager@wfw.wtb.tue.nl, and [4]M.Steinbuch@tue.nl*

Abstract: A technique to identify parameters of a robot dynamic model is presented in this paper. It is based on a batch adaptive control algorithm that, using a model of the robot dynamics, realizes a repetitive robot trajectory. The tracking error decreases due to a feedforward control input generated from the dynamic model. This feedforward input is computed after adaptation of the model parameters at the end of each trial. As the algorithm is effective, even if the model parameters are all initially set to zero, it can be used to recover their physical values. For that purpose, an identification experiment is carried out during which the robot is excited persistently. The estimation technique admits an online implementation without a delay between trials and is quite appealing for use in practice. Its merits are experimentally demonstrated on a spatial direct-drive robotic manipulator with 3 rotational joints. *Copyright © 2003 IFAC*

Keywords: Robotics, Identification, Dynamics, Application, Adaptive, Model-based control

1. INTRODUCTION

Growing interest in model-based robot control is boosted by the computational power of modern digital processors and by advances in the theory on robot modelling and identification (Armstrong, 1989; Gautier and Khalil, 1992; Kozlowski, 1998; Slotine and Li, 1991; Swevers, *et al.*, 1997; Calafiore, *et al.*, 2001; Olsen and Petersen, 2001). A dynamic model simplifies analysis of the control problem at hand and facilitates design of a solution to that problem. The model may compensate for nonlinear dynamic couplings between robot axes and enables robust robot operation of high performance, even if linear feedback control designs are used (Kostić, *et al.*, 2002a). Adaptive, sliding-mode and other nonlinear control strategies also make use of a dynamic model.
To use a dynamic model for control of a robotic system, one must be sure that the model closely matches the real dynamics. To enable a close match, the structure of the model should be capable of describing the relevant aspects of the physics and accurate values for the model parameters are needed. So far, a number of theoretical and experimental studies on estimating model parameters have been reported. Each exploits the well-known property that a model of the robot dynamics can be represented linearly in a minimum set of identifiable parameters, called the base parameter set, BPS, (Mayeda, *et al.*, 1990). The elements of the BPS are nonlinear combinations of robot inertial parameters, such as mass and moments of inertia of the robot links, as well as the Cartesian coordinates of the center of mass. Friction effects should be taken into account as well if model-based control of high quality is desired. Friction may cause problems, e.g., steady state errors and limit cycles. To avoid these, the friction force is counteracted using model-based or non-model based techniques (Ray, 2001). If model-based techniques are used, then relevant values of the friction parameters are needed in addition to the BPS.

In general, we can distinguish between the least-squares-like (Kozlowski, 1998; Swevers, *et al.*, 1997; Calafiore, *et al.*, 2001; Olsen and Petersen, 2001) and adaptive techniques (Slotine and Li, 1991) to estimate elements of the BPS and the friction parameters.

In this paper we suggest to use a particular batch adaptive control technique for online estimation of both the BPS elements and the parameters of a friction model. This friction model must admit a linear representation in its parameters. The technique is proposed by (Hamamoto and Sugie, 2001) as a method to accurately realize a repetitive trajectory, even when correct parameters of the dynamic model with friction are unknown. It is an improved version of an already known algorithm suggested by (Arimoto, 1996), and a notable advantage of this technique is its relatively simple online implementation. Two other advantages are its robustness against measurement noise and faster convergence of the tracking error. When a trajectory is used that excites the robot dynamics persistently (Slotine and Li, 1987), the technique can deliver accurate estimates of the BPS elements and the friction parameters. Design rules for such a trajectory, that we will call the exciting trajectory, are well known in robotics (Armstrong, 1989; Gautier and Khalil, 1992; Swevers, *et al.*, 1997; Calafiore, *et al.*, 2001) and will be particularly addressed later in this paper.

To demonstrate the suggested technique in estimating parameters of robot dynamic and friction models, we perform experiments on a robotic arm with 3 rotational joints, implemented as a waist, shoulder and elbow. As the considered kinematic structure is often met in industry, results of this study are representative for industrial cases. Furthermore, due to direct-drive actuation of the robotic arm, nonlinear dynamic couplings between the robot axes are fully expressed. Online estimation is tested for different robot loads. We validate the estimated models in a demanding writing task (Potkonjak, *et al.*, 2000). The results obtained are affirmative.

The paper is organized as follows. In the next section we describe a batch adaptive technique suitable for online robot identification. How to use this technique in estimating the BPS elements and friction parameters is explained in Section 3. In Section 4 we demonstrate its effectiveness by experiments. Conclusions will come at the end.

2. BATCH ADAPTIVE CONTROL

Let the dynamics of a robot with n actuated joints be represented using a standard Euler-Lagrange formalism (Fu, *et al.*, 1987):

$$\mathbf{D}(\mathbf{q}(t))\ddot{\mathbf{q}}(t)+\mathbf{C}(\mathbf{q}(t),\dot{\mathbf{q}}(t))\dot{\mathbf{q}}(t)+ \\ +\mathbf{f}(\dot{\mathbf{q}}(t))+\mathbf{g}(\mathbf{q}(t))=\tau(t), \tag{1}$$

where $\mathbf{q},\dot{\mathbf{q}}$ and $\ddot{\mathbf{q}}$ are the $n\times 1$ vectors of joint motions, speeds, and accelerations, respectively, $\mathbf{D}$ is the $n\times n$ inertia matrix, $\mathbf{C}\dot{\mathbf{q}}$, $\mathbf{f}$, and $\mathbf{g}$ are $n\times 1$ vectors of Coriolis/centripetal, friction, and gravitational forces and τ is an $n\times 1$ vector of control inputs (joint forces/torques). The linear parameterisation of (1) has the following form:

$$\mathbf{R}(\mathbf{q}(t),\dot{\mathbf{q}}(t),\ddot{\mathbf{q}}(t))\,\mathbf{p}=\tau(t), \tag{2}$$

where $\mathbf{R}\in\mathbb{R}^{n\times m}$ is a regressor matrix, m is the number of parameters to identify, and $\mathbf{p}\in\mathbb{R}^{m}$ is the vector of BPS elements and friction parameters. If the tracking error is defined as the difference between reference and actual motions:

$$\mathbf{e}(t)=\mathbf{q}_{\mathbf{r}}(t)-\mathbf{q}(t), \tag{3}$$

then the robot can be stabilised using a PD-feedback controller:

$$\tau(t)=\mathbf{K}_{\mathbf{p}}\mathbf{e}(t)+\mathbf{K}_{\mathbf{d}}\dot{\mathbf{e}}(t)+\mathbf{v}(t), \tag{4}$$

where $\mathbf{K_p}$ and $\mathbf{K_d}$ are $n\times n$ positive definite diagonal matrices, $\dot{\mathbf{e}}$ is the time derivative of the tracking error, and $\mathbf{v}$ denotes a $n\times 1$ vector with new input signals. The closed-loop system (2),(4) takes the form:

$$\mathbf{R}(\mathbf{q}(t),\dot{\mathbf{q}}(t),\ddot{\mathbf{q}}(t))\mathbf{p}-\mathbf{K}_{\mathbf{p}}\mathbf{e}(t)-\mathbf{K}_{\mathbf{d}}\dot{\mathbf{e}}(t)=\mathbf{v}(t). \tag{5}$$

Let $\mathbf{v_r}$ be obtained from the closed-loop system (5), when all motions, speeds and accelerations are chosen as reference ones:

$$\mathbf{R}(\mathbf{q}_{\mathbf{r}}(t),\dot{\mathbf{q}}_{\mathbf{r}}(t),\ddot{\mathbf{q}}_{\mathbf{r}}(t))\,\mathbf{p}=\mathbf{v}_{\mathbf{r}}(t). \tag{6}$$

The residual error dynamics, with input $\delta\mathbf{v}$ defined as $\mathbf{v}_{\mathbf{r}}(t)-\mathbf{v}(t)=\delta\mathbf{v}(t)$, is given as follows:

$$\mathbf{R}(\mathbf{q}_{\mathbf{r}}(t),\dot{\mathbf{q}}_{\mathbf{r}}(t),\ddot{\mathbf{q}}_{\mathbf{r}}(t))\mathbf{p}-\mathbf{R}(\mathbf{q}(t),\dot{\mathbf{q}}(t),\ddot{\mathbf{q}}(t))\mathbf{p}+ \\ +\mathbf{K}_{\mathbf{p}}\mathbf{e}(t)+\mathbf{K}_{\mathbf{d}}\dot{\mathbf{e}}(t)=\delta\mathbf{v}(t). \tag{7}$$

With an appropriate choice of $\mathbf{K_p}$ and $\mathbf{K_d}$ (Arimoto, 1996), (7) can be made passive for the input $\delta\mathbf{v}$ and for the output

$$\delta\mathbf{y}=\dot{\mathbf{e}}+\beta\,\mathbf{s}(\mathbf{e}), \tag{8}$$

where β is a positive constant scalar and $\mathbf{s}(\mathbf{e})$ is the $n\times 1$ vector of partial derivatives of a suitable potential function. This potential function is semi-positive definite, twice continuously differentiable and has to satisfy additional conditions given in (Arimoto, 1996).

Before showing a batch adaptive update law suitable for estimation of the BPS elements and parameters of the friction model, several definitions and assumptions are needed. Define

$$\mathbf{R}_{\mathbf{r}}:=\mathbf{R}(\mathbf{q}_{\mathbf{r}}(t),\dot{\mathbf{q}}_{\mathbf{r}}(t),\ddot{\mathbf{q}}_{\mathbf{r}}(t)) \tag{9}$$

and $\mathbf{L}_{\mathbf{r}}\in\mathbb{R}^{m\times m}$ as:

$$\mathbf{L}_{\mathbf{r}}:=\int_{0}^{t_f}\mathbf{R}_{\mathbf{r}}^{\mathrm{T}}\mathbf{R}_{\mathbf{r}}\,\mathrm{d}t. \tag{10}$$

Here, t_f denotes the duration of one trial of the

repetitive trajectory. The matrix $\mathbf{L_r}$ is assumed to be non-singular, which is a valid assumption if $\mathbf{p}$ in (2) is a minimum set of dynamic and friction parameters and if the reference trajectory excites the robot dynamics persistently.

With these definitions and assumptions, the batch adaptive update law is given as follows (Hamamoto and Sugie, 2001):

$$\begin{aligned} \mathbf{p}^{k+1} &= \mathbf{p}^{k} + \mathbf{K_r}\left\langle \mathbf{R_r}, \delta\mathbf{y}^{k} \right\rangle \\ \mathbf{v}^{k+1} &= \mathbf{R_r}\mathbf{p}^{k+1}, \end{aligned} \qquad (11)$$

where k denotes the number of the trial (k=0,1,2,...), $\mathbf{K_r}$ is an $m \times m$ positive definite matrix called the 'adaptation gain' and $\left\langle \mathbf{R_r}, \delta\mathbf{y}^{k} \right\rangle$ represents the inner product between the matrix $\mathbf{R_r}$ and the vector $\delta\mathbf{y}^k$. For two matrices of equal number of rows, this inner product is defined as:

$$\left\langle \mathbf{U}, \mathbf{V} \right\rangle_{\mathbf{\Phi}} = \begin{bmatrix} \left\langle \mathbf{u}_1, \mathbf{v}_1 \right\rangle_{\mathbf{\Phi}} & \left\langle \mathbf{u}_1, \mathbf{v}_2 \right\rangle_{\mathbf{\Phi}} & \cdots \\ \left\langle \mathbf{u}_2, \mathbf{v}_1 \right\rangle_{\mathbf{\Phi}} & \left\langle \mathbf{u}_2, \mathbf{v}_2 \right\rangle_{\mathbf{\Phi}} & \cdots \\ \vdots & \vdots & \ddots \end{bmatrix}, \qquad (12)$$

where $\mathbf{u}_i$ and $\mathbf{v}_j$ are the i-th column of $\mathbf{U}$ and the j-th column of $\mathbf{V}$, respectively and $\mathbf{\Phi} \in \mathbb{R}^{n \times n}$ is a positive definite matrix, which is omitted if it is identity. The inner product of two columns is defined as follows:

$$\left\langle \mathbf{u}_i, \mathbf{v}_j \right\rangle_{\mathbf{\Phi}} = \int_0^{t_f} \mathbf{u}_i^{\mathrm{T}}(t)\mathbf{\Phi}\mathbf{v}_j(t)\mathrm{d}t. \qquad (13)$$

If the adaptation gain $\mathbf{K_r}$ satisfies

$$\mathbf{K_r} = \mathbf{L_r}^{-1}\left\langle \mathbf{R_r}, \mathbf{R_r} \right\rangle_{\mathbf{\Phi}} \mathbf{L_r}^{-1}, \qquad (14)$$

convergence of the tracking error (3),

$$\left\| \mathbf{e}^k(t) \right\| \to 0 \ (k \to \infty), \qquad (15)$$

can be proven, with $\|.\|$ denoting a signal norm, see (Hamamoto and Sugie, 2001). In this algorithm, convergence of the tracking error implies convergence of the elements in $\mathbf{p}^k$ to steady state values ($\mathbf{p}^k = \mathbf{p}^{k+1}$). Theoretically, there is no restriction on the initial value $\mathbf{p}^0$ required in (11). Practically, when control inputs τ are bounded, $\mathbf{p}^0$ cannot be chosen arbitrarily, but should be constrained to a certain feasible region corresponding to limits on the control inputs.

3. ROBOT IDENTIFICATION PROBLEM

With the algorithm presented in the previous section, all elements of the vector $\mathbf{p}^k$ converge after a sufficient repetition of trials. We make use of this property to obtain an estimate of $\mathbf{p}$ that provides a close match between the model (1) and the real robot dynamics. Obtaining such an estimate is the goal of the robot identification problem. A relevant solution to this problem requires the robot to realize an appropriate exciting trajectory.

To design such a trajectory, one should optimise some property of the information matrix. This matrix is formed by vertically stacking the regressor matrices that correspond to different points of the exciting trajectory. Some choices of the performance criteria to be optimised are the condition number of the information matrix, its extreme singular value, its Frobenius condition number, and the determinant of the weighted product between the information matrix and its transpose. No matter which performance criterion is adopted, the resulting optimisation problem is essentially non-convex, which means that the global optimum for the exciting trajectory cannot be obtained easily. Different initial conditions may lead to different exciting trajectories, all corresponding to local minima of the optimisation problem. As argued in (Armstrong, 1989), "engineering judgment, patience, and good luck are required to find a good trajectory."

In our opinion, "a good trajectory" is one that provides us with the estimates that are relevant for at least those motions that the robot is supposed to realise in practice, for example, for all motions with a frequency content up to 10 [Hz]. With such reasonning, a successful outcome of the robot identification problem would be a dynamic model closely reconstructing dynamics of the real robot for all covered frequencies.

An exciting trajectory that meets the given requirements can be determined by postulating it in the form of a finite Fourier series:

$$q_{r,i}(t) = q_{0,i} + \sum_{j=1}^{N} \frac{1}{j\omega}[a_{ij}\sin(j\omega t) - b_{ij}\cos(j\omega t)] \quad (i = 1,2,\ldots,n), \qquad (16)$$

with prescribed fundamental frequency ω and the parameters $q_{0,i}$, a_{ij}, and b_{ij} to be computed via optimisation. No matter what performance criterion is adopted, constraints on mechanical limits in the robot joints and permissible levels of joint speeds and accelerations must be taken into account. The fundamental frequency defines the resolution of the frequency content of the exciting trajectory. A resolution of 0.1 [Hz] is usually taken as sufficient (Swevers, *et al.*, 1997; Calafiore, *et al.*, 2001). Together with the number of harmonics N, which affects persistency of excitation, the fundamental frequency determines the bandwidth of excitation. This bandwidth should be high enough to cover all possible motions of the robot, but also well below the eigenfrequency of the first flexible mode of the robot structure.

4. CASE STUDY

The robotic arm, which is the subject of our case study, is shown in Fig. 1. The photo and kinematic parameterisation according to the well-known Denavits-Hartenberg's (DH) notation (Fu, et al., 1987) reveal three revolute degrees of freedom (d.o.f.), what makes such a kinematic structure referred to as RRR. Each d.o.f. is actuated by a brushless DC direct-drive motor and has an infinite range of motions, thanks to the use of slip-rings for the transfer of power and sensor signals (Van Beek and De Jager, 1997; Van Beek and De Jager, 1999). Joint motions are measured using incremental encoders with a resolution of approximately 10^{-5} [rad]. A PC-based platform is used for implementation of control algorithms using Matlab/Simulink software. Controlled current invertors amplify signals generated on the PC, before they are applied to the motors.

A detailed description of the robot kinematics and dynamics is available in (Kostić *et al.*, 2003). From Fig. 1, one can determine the DH parameters as presented in Table 1. The DH parameters are: twist angles α_i, link lengths a_i, joint displacements q_i, and link offsets d_i.

The exciting trajectories have been designed using (16). The performance criterion was the condition number of the information matrix and the constraints were:

$$|\dot{q}_1| \le 2\pi\,[\text{rad/s}],\ |\dot{q}_2| \le 3\pi\,[\text{rad/s}],$$
$$|\dot{q}_3| \le 3\pi\,[\text{rad/s}],\ |\ddot{q}_i| \le 30\,[\text{rad/s}^2]\ (i = 1, 2, 3). \qquad (17)$$

All three trajectories contain frequencies up to 10 [Hz] and therefore all possible trajectories for the RRR robotic arm are well covered. The motions of the exiting trajectories have a duration of 10 [s] and are depicted in Fig. 2.

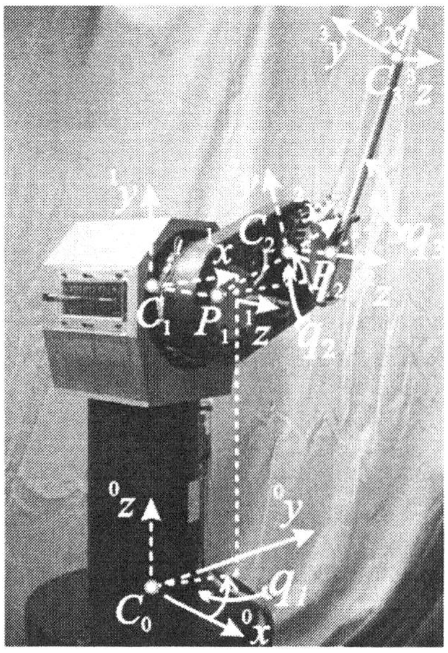

Fig. 1. The RRR robotic arm and the kinematic parameterisation according to the DH notation.

Table 1: DH parameters of the experimental RRR robotic arm

i	α_i [rad]	a_i [m]	q_i	d_i [m]
1	$\pi/2$	0	q_1	$C_0C_1 = 0.56$
2	0	$P_1C_2 = 0.2$	q_2	$C_1P_1 = 0.169$
3	0	$P_2C_3 = 0.415$	q_3	$C_2P_2 = 0.09$

The model of (Kostić *et al.*, 2003) has been used, together with a friction model consisting of Coulomb and viscous friction:

$$\mathbf{f}(\dot{\mathbf{q}}(t)) = \mathbf{f_c}\,sign(\dot{\mathbf{q}}(t)) + \mathbf{B}\dot{\mathbf{q}}(t), \qquad (18)$$

where $\mathbf{f}$ is the 3×1 vector of total friction torques, $\mathbf{f_c}$ and $\mathbf{B}$ are 3×3 positive definite diagonal matrices of Coulomb friction torques and viscous friction parameters, respectively. Since this friction model admits a linear representation, its parameters can be concatenated with the BPS elements of the dynamic model, resulting in the vector $\mathbf{p}^k$ to be estimated.

After online implementation of the batch adaptive update law (11), β and $\mathbf{s(e)}$ in (8) are tuned in order to achieve fast convergence of the parameters to steady state values, whereas the matrix $\mathbf{\Phi}$ in (14) was chosen to be equal to the matrix $\mathbf{K_d}$ in (4). In order to show the full potential of the algorithm, three different experiments have been conducted. In the first experiment, no load has been mounted onto the third joint, in the second experiment a steel rod of 1.22 [kg], shown in Fig. 1, has been attached, whereas in the third one an extra load of 0.45 [kg] has been attached onto the tip of the steel rod. In all three experiments, the initial value of the vector of parameters to be estimated, $\mathbf{p}^0$, is set to zero, so the extra input $\mathbf{v}$ in (4) during the first trial is zero as well.

In the model of (Kostić *et al.*, 2003), 15 BPS elements are distinguished. As in that paper, only 13 BPS elements are identified in this case study, since the physical value of two of the original BPS elements are already known to be zero. Fig. 3 shows the results of the three experiments with different loads. The parameters p_1, ..., p_{13} represent the BPS elements, whereas the parameters p_{14}, ..., p_{19} represent the friction parameters.

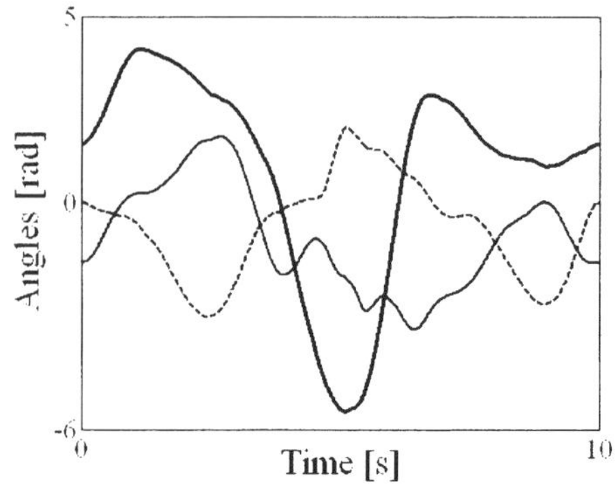

Fig. 2. Motions of the exciting trajectories: q_1 – thick solid, q_2 – thin solid, and q_3 – dashed.

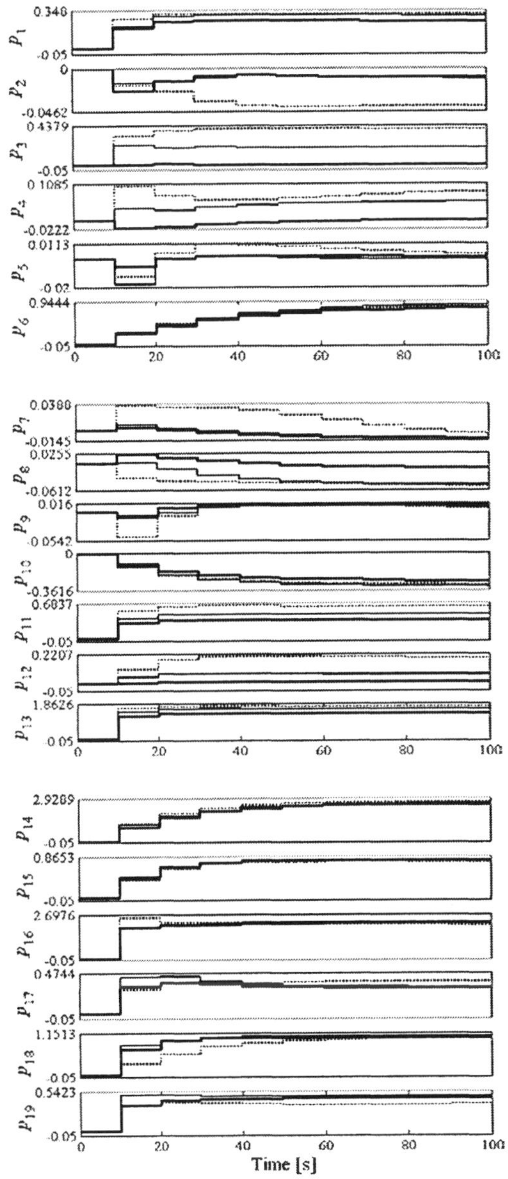

Fig. 3. Convergence of the estimates: without the last link (thick solid), with the last link (thin solid), and with the extra load attached to the last link (dotted)

From Fig. 3, it can be seen that almost all parameters have reached their steady state values after eight trials. When the analytical expressions of the parameters with large differences in steady state values at the end of the three experiments are evaluated, it appears that these expressions mainly contain terms of the third link. One may also notice differences between some friction parameters of the first two experiments and the third one.

To validate whether the obtained parameters provide a close match between the model (1) and the real robot dynamics for all three loads, validation experiments have been conducted. In these experiments, the robotic arm had to perform a writing task, which is often recognized as very demanding for the dynamics of a mechanical system due to its fast motions.

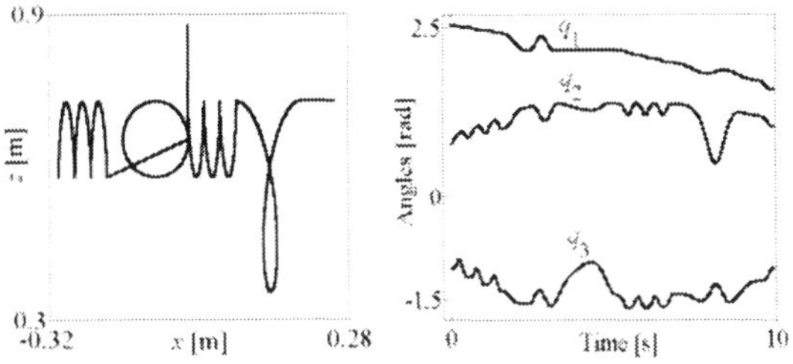

Fig. 4. Reference path of the robot tip (left) and corresponding joint motions (right).

In this writing task, the tip of the robotic arm has to track a sequence of letters in the vertical plane, shown on the left hand side of Fig. 4. The corresponding joint motions are depicted on the right hand side of this figure.

During the validation experiments, the robotic arm is stabilised using the following PD-feedback controller:

$$\tau(t) = \mathbf{K}_{\mathbf{p}}\mathbf{e}(t) + \mathbf{K}_{\mathbf{d}}\dot{\mathbf{e}}(t)\,, \tag{19}$$

where the $n \times n$ positive definite diagonal matrices $\mathbf{K}_\mathbf{p}$ and $\mathbf{K}_\mathbf{d}$ are tuned corresponding to the load attached to the third joint. The values of $\mathbf{K}_\mathbf{p}$ and $\mathbf{K}_\mathbf{d}$ in the validation experiments are not the same as their values in the identification experiments in order to assure stability of the closed-loop system in every experiment. To verify whether a close match between the model (1) and the actual robot dynamics is obtained, the output of the PD controller is compared with the reconstructed torques calculated using the model (1). To calculate these torques, the joint motions, speeds and accelerations are reconstructed online using the Kalman filter described in (Kostić *et al.*, 2002a). In Fig. 5, outputs of the PD controllers of the second joint have been depicted for the three different validation experiments, together with the reconstructed torques in that joint. Fig. 5a, 5b and 5c represent the validation results of the experiments without any load, with the steel rod and with steel rod and an extra load attached on the tip, respectively. From this figure, it can be seen that in all cases the torques reconstructed from the model closely match the outputs of the PD controllers, which shows that relevant estimates have been obtained. In Fig. 5c, it can be seen that the output of the feedback controller is clipped at its maximum permissible value. Under normal circumstances, i.e. when no extra load has been attached at the tip of the steel rod, this does not happen, which proves that the model predicts the output of the PD controller well, even under 'abnormal' circumstances.

5. CONCLUSIONS

An online batch adaptive algorithm for the estimation of robot parameters was presented. It requires a model linear in the dynamic and friction parameters. The estimation technique admits online implementation and in turn it is quite appealing for use in practice.

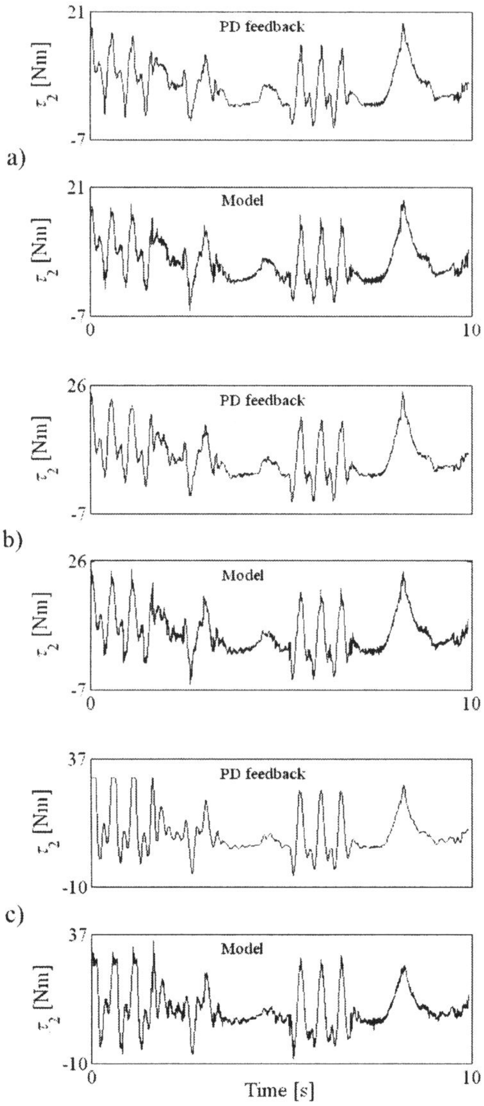

Fig. 5. Validation of the estimates in the writing task for three different cases: output of the feedback controller acting on the second joint (up) and the corresponding reconstruction from the model (bottom).

Its merits, i.e. its relatively simple online implementation, its robustness against measurement noise and the fast convergence of the parameters to steady-state values, are experimentally demonstrated on a spatial direct-drive robotic manipulator with three rotational joints. A set of identification experiments corresponding to different robot loads show equally fast estimation of all model parameters. The estimates are validated in a demanding writing task.

REFERENCES

Arimoto, S. (1996). *Control theory of non-linear mechanical systems: A passivity-based circuit-theoretic approach.* Oxford Univ. Press, Oxford.

Armstrong, B. (1989). On Finding Exciting Trajectories for Identification Experiments Involving Systems with Nonlinear Dynamics, *Int. Jour. Robot. Res.*, **Vol 8, No. 6**, 28-48.

Calafiore, G., M. Indri and B. Bona (2001). Robot Dynamic Calibration: Optimal Excitation Trajectories and Experimental Parameter Estimation, *Jour. of Rob. Systems*, **Vol. 18, No. 2,** 55-68.

Fu, K.S., R.C. Gonzales and C.S.G. Lee (1987). *Robotics: Control, sensing, vision and intelligence.* McGraw-Hill, London.

Gautier, M. and W. Khalil (1992). Exciting Trajectories for the Identification of Base Inertial Parameters of Robots, *Int. Jour. Robot. Res.*, **Vol. 11, No. 4,** 362-375.

Hamamoto, K and T. Sugie (2001). Iterative Learning Control for Robot Manipulators Using The Finite Dimensional Input Subspace, *IEEE Conf. Dec. Control*, Orlando, FL, USA, 4926-4931.

Kostić, D., B. de Jager and M. Steinbuch (2002a). Experimentally Supported Control Design for a Direct Drive Robot, *IEEE Int. Conf. Control Appl.*, Glasgow, Scotland, 186-191.

Kostić, D., B. de Jager and M. Steinbuch (2003). Modeling and Identification for Model-based Robot Control: An RRR-Robotic Arm Case Study, *Submitted to IEEE Trans. Control Sys. Tech.*

Kozlowski, K. (1998). *Modelling and identification in robotics*, Springer, London.

Mayeda, H., K. Yoshida and K. Osuka (1990). Base Parameters of Manipulator Dynamic Models, *IEEE Trans. Rob. Autom.*, **Vol. 6, No. 3,** pp. 312-321.

Olsen, M.M. and H.G. Petersen (2001). A New Method for Estimating Parameters of a Dynamic Robot Model, *IEEE Trans. Rob. Autom.*, **Vol. 17, No. 1,** 95-100.

Potkonjak, V., S. Tzafestas and D. Kostić (2000). Concerning the primary and secondary objectives in robot task definition – The "learn from humans" principle, *Math. and Comp. in Simul.*, **Vol. 54, No. 1-3,** 145-157.

Ray, L.R., A. Ramasubramanian and J. Townsend (2001). Adaptive Friction Compensation Using Extended Kalman-Bucy Filter Friction Compensation, *Control Eng. Pract.*, **Vol. 9, No. 2,** 169-179.

Slotine, J.J.E. and W. Li (1987). Theoretical issues in adaptive manipulator control. *Proc. 5th Yale Workshop on Apl. Adaptive Systems,* New Haven, USA.

Slotine, J.J.E. and W. Li (1991). *Applied nonlinear control*, Prentice Hall, London.

Swevers, J., C. Ganseman, D.B. Tukel, J. de Schutter and H. van Brussel (1997). Optimal Robot Excitation and Identification, *IEEE Trans. Rob. Autom.*, **Vol. 13, No. 5,** 730-740.

Van Beek, B. and B. de Jager (1997). RRR-Robot design: Basic outlines, servo sizing, and control, *Proc. IEEE Int. Conf. Control Appl.*, Hartford, CT, USA, 36-41.

Van Beek, B. and B. de Jager (1999). An experimental facility for nonlinear robot control, *Proc. IEEE Int. Conf. Control Appl.*, Hawaii, USA, 668-673.

IFAC Publications
www.elsevier.com/locate/ifac

DYNAMIC IDENTIFICATION OF A COMPACTOR USING SPLINES DATA PROCESSING

Charles-Eric Lemaire * **Pierre-Olivier Vandanjon** *
Maxime Gautier **

* *Laboratoire Central des Ponts et Chaussées (LCPC)*
Route de Bouaye, BP 4129, 44341 Bouguenais, France
charles-eric.lemaire/pierre-olivier.vandanjon@lcpc.fr
** *Institut de Recherche en Cybernétique et Communication de Nantes (IRCCyN)*
1 rue de la Noë, BP 92101, 44321 Nantes Cedex 3, France
maxime.gautier@irccyn.ec-nantes.fr

Abstract: This paper deals with the identification of a dynamic model of a compactor. This model is linear in relation to a set of dynamic parameters which can be identified using a Weighted Least Squares method. With this method, the measurement of the trajectory must be accurate enough. In order to respect this accuracy in an industrial environnement, a specific data processing, based on cubic spline interpolation, as been developed to compute the trajectory of the compactor from the device mounted on standard compactors.

Keywords: Dynamic model, Identification, Splines

1. INTRODUCTION

Up to now, different systems have been developed in order to manage compaction control (Floss and Kloubert, 2000). Compaction control is important to ensure the quality of asphalt compaction. In order to control the compaction process, the degree of compaction of the asphalt must be measured. In fact, a relationship between the rolling resistance and the degree of compaction of the asphalt had been highlighted (Delclos *et al.*, 2001). What is the rolling resistance ? It is the necessary force needed to make the compactor roll. Computation of the rolling resistance is based on a dynamic model of the compactor .

This article deals with an identification method of the dynamic model of a compactor. This identification is an important step in the development of a new system of compaction measurement. Dynamic identification of the compactor has been already done (Guillo, 2000). However, this identification needed high-cost and low-toughness sensors (incremental encoder for example). Such sensors are not compliant with industrial constraints, a method with tougher sensors is developed in this article.

Fig. 1. A typical compactor

2. COMPACTOR MODELS

According to classical robot manipulator description (Canudas de Wit *et al.*, 1997), the compactor is considered as a mechanical system Σ composed of a tree structure of n rigid bodies C_i where C_0 is the base body and where :

- C_0, is a virtual body attached to a Galilean reference frame,
- C_1, is the chassis,
- C_2, C_3 are the front and rear drum.

2.1 Geometric description

Classical tree structure description using the modified Denavit-Hartenberg notations (Khalil, 1999) applied to the system Σ defines the geometric parameters of the compactor (see Table 1) In that case the homogeneous transform (1) of the frame R_i with respect to $R_{a(i)}$ (where a_i is the antecedent of i) is expressed as a function of the 4 following parameters:

i	$a(i)$	σ	μ	α	d	θ	r
1	0	1	0	0	0	0	r_1
2	1	0	1	$\pi/2$	0	θ_2	0
3	1	0	1	$\pi/2$	0	θ_3	0

Table 1. Geometric parameters of the compactor

- α_i: angle between $\underline{z}_{a(i)}$ and $\underline{z}_i$, corresponding to a rotation about $\underline{x}_{a(i)}$,
- d_i: distance from $\underline{z}_{a(i)}$ to $\underline{z}_i$ along $\underline{x}_{a(i)}$,
- θ_i: angle between $\underline{x}_{a(i)}$ and $\underline{x}_i$, corresponding to a rotation about $\underline{z}_i$,
- r_i: distance from $\underline{x}_{a(i)}$ to $\underline{x}_i$ along $\underline{z}_i$.

$$ {}^{a_i}\underline{\underline{Z}}_i = \begin{bmatrix} {}^{a_i}\underline{\underline{A}}_i & {}^{a_i}\underline{P}_i \\ 0_{1\times 3} & 1 \end{bmatrix} \qquad (1) $$

where ${}^{a_i}\underline{\underline{A}}_i$ is the (3×3) rotation matrix which defines the orientation of the frame R_i with respect to the frame R_{a_i} and ${}^{a_i}\underline{P}_i$ is the origin of the frame R_i expressed in the frame R_{a_i}.

Remark 1. $\mu_i = 1$ means that the i-joint is actuated and $\mu_i = 0$ that it is not. σ_i specifies the type of the joint ($\sigma_i = 0$ if rotational, $\sigma_i = 1$ if translational, $\sigma_i = 2$ if fixed).

According to the previous description of the compactor, the vehicle motion is completely described by the following vector of three generalized coordinates

$$ q = \begin{bmatrix} r_1 & \theta_2 & \theta_3 \end{bmatrix}^T \qquad (2) $$

2.2 Dynamic model

When the Lagrange equations are calculated for the system Σ, they yield a dynamic equation which can be written in the form

$$ M(q)\ddot{q} + H(q,\dot{q}) = U + Q \qquad (3) $$

where :

- $M(q)$ is the (3×3) mass matrix of the system Σ,
- $H(q,\dot{q})$ is a (3×1) vector of centrifugal, Coriolis and gravity terms,
- U is a (3×1) vector depending on the motor torques on joints 2 and 3 (see Table 1)

$$ U = \begin{bmatrix} 0 & u_2 & u_3 \end{bmatrix}^T \qquad (4) $$

  where u_i is the motor torque on i-joint.
- Q is a (3×1) vector depending on the bonding strengths between the layer and the drums C_i.

Relation (3) gives the general expression of the dynamic model, but it is a very simple model with only one possible trajectory : a straight line. In these conditions, the dynamic model can be develop to the following expression

$$ \begin{cases} M\ddot{r}_1 = S_x^2 - S_r^2 + S_x^3 - S_r^3 + G_r \\ ZZ_2\ddot{\theta}_2 + F_{v2}\dot{\theta}_2 + F_{s2}\,\mathrm{sign}(\dot{\theta}_2) = u_2 - R_{c2}S_x^2 \\ ZZ_3\ddot{\theta}_3 + F_{v3}\dot{\theta}_3 + F_{s3}\,\mathrm{sign}(\dot{\theta}_3) = u_3 - R_{c3}S_x^3 \end{cases} \qquad (5) $$

where for $(i = 2,3)$:

- M is the total mass of the compactor,
- ZZ_i is the inertia of the drum C_i,
- S_r^i is the rolling resistance on the drum C_i,
- S_x^i is the propelling force of the drum C_i,
- F_{vi} and F_{si} are the viscous and striction friction parameters for the i-joint,
- G_r is the gravity term,
- R_{ci} is the radius of the drum C_i.

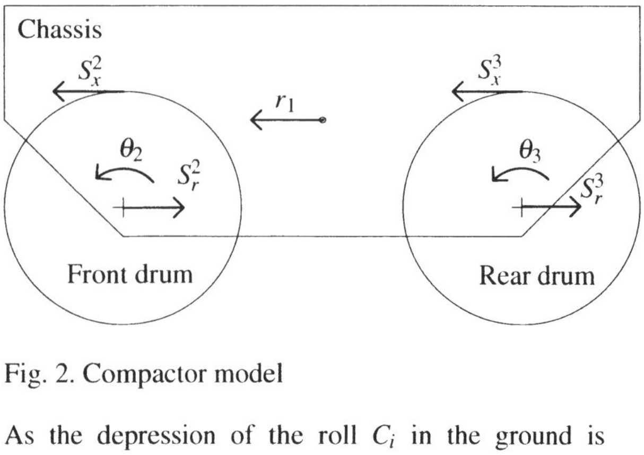

Fig. 2. Compactor model

As the depression of the roll C_i in the ground is constant and small compared to its radius R_{ci} then it is possible to consider that

$$ S_r^i = -K_r^i\,\mathrm{sign}(\dot{r}_1) \qquad (6) $$

2.3 *Identification model*

The same Least Squares (LS) techniques as those generally used to identify the dynamic parameters of robots (Khalil, 1999; Kozlowski, 1998) are applied to the compactor. Parameters are estimated as the Weighted Least Squares (WLS) solution of an overdetermined linear system of r equations in n_p unknowns, obtained from the sampling of the dynamic model along a known trajectory $(q, \dot{q}, \ddot{q})$

$$Y = W_s X_s + \rho \tag{7}$$

where :

- W_s is the $(r \times n_p)$ observation matrix,
- n_s is the number of samples,
- r is the number of equations $(r = n_e n_s > n_p)$.

The Least Squares (LS) solution $\hat{X}_s$ minimizes the 2-norm of the error vector ρ in Eq. (7)

$$\hat{X}_s = \min_{X_s} \|W_s X_s - Y\|_2 \tag{8}$$

Classical results from statistics are used to calculate of standard deviation $\hat{\sigma}_{X_s}$. The observation matrix W_s is considered as a deterministic one and ρ to be a vector of zero-mean additive independent noise with standard deviation σ_ρ, such that errors covariance matrix is

$$C_{\rho\rho} = \sigma_\rho^2 I_r \tag{9}$$

It is excessive to consider that standard deviation of errors are the same on each row of the identification model (equation 7). (Gautier, 1997) proposed to use a Weighted Least Squares (WLS) method to take into account different standard deviation for each equation of the model. Applied to (equation 7) it consists in weighting each row by $\hat{\sigma}_{\rho j}$ for $(j = 1, \ldots, n_e)$

$$\hat{\sigma}_{\rho j}^2 = \frac{\left\|W_s^{(j)} \hat{X}_s^{(j)} - Y^{(j)}\right\|_2}{n_s - n_{pj}} \tag{10}$$

where $\hat{X}_s^{(j)}$ is the least squares solution of the subsystem $Y^{(j)} = W_s^{(j)} \cdot X_s^{(j)} + \rho^{(j)}$ and n_{pj}, the number of dynamic parameters that take part in equation j. So, the errors covariance matrix is defined as following

$$C_{\rho\rho} = (G^T G)^{-1} \quad G = \text{diag}(S) \tag{11}$$

with :

$$S = \left[S_1 \cdots S_{n_e} \right] \quad S_j = \left[1/\hat{\sigma}_{\rho j} \cdots 1/\hat{\sigma}_{\rho j} \right]$$

The WLS solution $\hat{X}_{sw}$ minimizes the 2-norm of the vector of weighted errors ρ_w

$$\hat{X}_{sw} = \min_{X_s} \|W_{sw} X_s - Y_w\|_2 \tag{12}$$

with : $Y_w = GY$, $W_{sw} = GW_s$ and $\rho_w = G\rho$.

3. EXPERIMENTATION

The matter is to identify parameters of the dynamic model of the compactor. To carry out this identification, it is necessary to measure the velocity and the acceleration of rotation of the drums. In an experimental environment, it is possible to mount specific device on the compactor to carry out such a measurement. In our case, the objective is to compare the acceleration measurement given by a encoder or by a toothed wheel combined with an inductive proximity sensor which detects each tooth. Such a wheel is mounted on standard compactors.

Acceleration computation has been tested on a MC08 Poclain hydraulic motor from the CB544 Caterpillar compactor. Two kind of sensors were used to measure the rotation of the motor. From one part, a 100000 pulses/rev incremental encoder and from the other part, a 63 teeth wheel with an inductive proximity sensor. The objective is to evaluate the acceleration precision obtained with the toothed wheel with the encoder as a reference. Motor torque has been measured with two pressure sensors plug at the input and the output of the hydraulic motor because the torque is computed from the pressure difference across the motor (Guillo *et al.*, 1997).

$$\Gamma = D_m \Delta P \tag{13}$$

with :

- Γ is the motor torque,
- D_m is the volumetric displacement of the motor,
- ΔP is the pressure difference across the motor.

Data acquisition has been performed with a DS1103 dSPACE system and the sampling frequency was set to 10kHz.

4. COMPUTATION OF THE ACCELERATION

In the previous section, the dynamic identification model (equation 7) implies that the noise is mainly on the torque. From a theoric point of view, it should not have any noise on the observation matrix. From a practical point of view, this noise has to be estimated and has to be small compared to the noise on the torque. In this observation matrix, the most critical part is the computation of the acceleration. In this section, it is described firstly our classical method using encoders. Afterward, a method based on toothed wheel sensor more suited to civil engineering is described.

4.1 Encoder case

In the frame of identification of robot manipulator, accelerations of joints are often computed from encoders (Kozlowski, 1998). In this frame, we suppose that at each sampling time $t_k = k \times \mathrm{T}$, the position $n_k \times \Delta_q$ is given by the encoder. Δ_q is the accuracy of the encoder ($\Delta_q = \frac{2\times\pi}{N}$).Therefore, the following information on the real position q(t_k) is obtained :

$$(n_k - 1) \times \Delta_q < q(t_k) < (n_k + 1) \times \Delta_q$$

It is reasonable to model this information with a random model : q(t_k) has an uniform probability between $(n_k - 1) * \Delta_q$ and $(n_k + 1) * \Delta_q$. It implies that

$$E(q(t_k)) = n_k \times \Delta_q \mathrm{and} Var(q(t_k)) = {\Delta_q}^2/3 \quad (14)$$

In order to have an off line estimation of the acceleration, a centered finite difference is used :

$$\hat{\ddot{q}}(t_k) = \frac{q(t_{(k+2)}) + q(t_{(k-2)}) - 2q(t_k)}{4\mathrm{T}^2} \quad (15)$$

The error on the estimation of the acceleration can be analyzed in a stochastic way and in a deterministic way.Indeed,it comes from the last equation :

$$Var(\ddot{q}(t_k)) = \frac{{\Delta_q}^2}{(3\mathrm{T}^4)} \quad (16)$$

Moreover, it is clear that the equation 15 does not give the exact acceleration. By using a Taylor's development, it comes :

$$\|\hat{\ddot{q}}(t_k) - \ddot{q}(t_k)\| \leq \frac{2}{3}T \max_{t\in[t_{(k+2)} - t_{(k-2)}]} \|\dddot{q}(t)\| \quad (17)$$

The equation of the stochastic error (equation 16) implies that with a sampling time equal to 10 kHz, the standard deviation is multiplied by about 100 Million. It is the reason why, these signals are filtered. Usually, a forward and reverse low pass Butterworth filter is used.

The cut pulsation is chosen according to the maximum mechanical dynamic of the system : the identification process is focused in the frequency domaine where the model is valid. The acceleration is computed by filtering and differentiating twice the position signal. As linear filters are used, it is possible to obtain the standard deviation on the acceleration : it is simply the standard deviation on the position multiplied by the norm of the filters used.

In our case, the main part of dynamic is under 10 Hz according to a priori information on the parameters. Under this frequency and by taking into account a sampling frequency of 10 kHz , the standard deviation on the acceleration is only multiplied by about 70 (to be compared to one million without filtering)

4.2 Toothed wheel case

The direct application of the previous method in our case leads to get a bad estimation of the acceleration. As the wheel includes 63 teeth, the accuracy of the position is $\Delta_q = 2 \times \pi/63$rd. This implies a standard deviation of 4rd/s^2 for the estimation of the acceleration with a 10 kHz sampling frequency. In this case, the model of an uniform probability is too rough compared to the reality.

It is the reason why, we changed our analysis by focusing on the detection of a new tooth. In this frame, the encoder position is not computed at each sampling time but the sampling time is required at each detection of a new tooth. The previous way is a synchronous acquisition and the new way can be seen as an asynchronous mode. In this frame, the system of measurement for the wheel gives : $qw(t_k) = i$ (i : number of the tooth on the wheel) with $t_{(k+1)}$-$t_k$can be different of the sampling time. If we note q(t_k) the real measurement and $\tilde{qw} = qw(t_k) - q(t_k)$ the error on q(t_k). This error has mainly two sources. The first one is due do the sampling time :

$$\|q(t_k) - qw(t_k)\| < T \times \|\dot{q}(t)\|_{\max t\in[t_{(k-1)} - t_k]} \quad (18)$$

The second one comes from the shape of teeth, this error can be modelled as a non centered noise. q(t_k) has an uniform probability between $qw(t_k) + \delta_i - \delta$ and $qw(t_k) + \delta_i + \delta$. δ_i is the bias induced by less of accuracy in the shape of the teeth, δ is a random error due to the level of detection of teeth by the sensor according to the rotational speed.

In order to approximate q(t) from qw(t_k), natural splines are used (Nürnberger, 1989). The natural spline qs(t) has the following property qs(t_k)=qw(t_k), qs(t) is $C^2$and $\int \|qs(t)\|^2 dt$ is minimal. This last property is interesting to reject pertubations which induce too important acceleration.

As our signal is figured by a spline, the acceleration can be computed at any time by analytical derivation. Therefore, the acceleration is computed at the sampling frequency. The acceleration signal computed from the spline is filtered in order to focus the identification process in a frequency domain compliant with the real system (exactly, as it was done in the previous subsection).

The computation of the standard deviation of the results cannot be obtained analytically as in the case of the encoder. Moreover, this computation requires assumption on the approximation of the real signal by the natural splines. It is the reason why, simulation are used. We simulate the acceleration computed by the method described above applied on a signal given by a virtual non-perfect toothed wheel on a trajectory planned for the identification process. Our wheel is not perfect because noise are added according to two sources of error described above : sampling time and shape of teeth. As far shape of teeth is concerned, this information can be got either from the data sheet or by identification. This second way was done in the experimental case studied in this paper.

5. IDENTIFICATION

The objective is to carry out the identification of a one degree of freedom system, the compactor drum (the second or the third equation of (5)). In this case the drum is isolated from the ground, so the propelling force is equal to zero ($S_x^i = 0$). Residuals analysis leads to take into account a term in $\dot{\theta}^2$ in the friction model. The identification model is :

$$ZZ\ddot{\theta} + F_{v2}\dot{\theta}^2 + F_v\dot{\theta} + F_s\,\mathrm{sign}(\dot{\theta}) = u \qquad (19)$$

5.1 acceleration computation error analysis

In order to set up parameters of acceleration computation methods a reference trajectory has been simulated. This trajectory (equation 20) is close to a real trajectory (Fig. 3) used to the identification. The objective of the simulation was to evaluate the precision of the reference, i.e. the angle, the velocity and the acceleration given by the 100000 pulses/rev encoder.

$$\dot{\theta} = \dot{\theta}_{mean} + A_0\sin(\omega_0 t) + A_1\sin(\omega_1 t) \qquad (20)$$

with :

- $\dot{\theta}_{mean} = 4$ rad/s,
- $A_0 = 2$ rad/s, $\omega_0 = 2\pi$ rad/s,
- $A_1 = 0.4$ rad/s, $\omega_1 = 10\pi$ rad/s.

To simulate the encoder, a uniform noise with $\frac{2\pi}{100000\sqrt{3}}$ rd standard deviation has been added to the reference angle of rotation. Velocity and acceleration have been calculated by centered finite difference. Each signal has been filtered with a forward and reverse digital fourth order Butterworth filter. The table (2) gives, simulated and computed from the filter norm, standard deviation for the position, velocity and acceleration (in rd, rd/s and rd/s^2) according to different cutoff frequencies.

	$F_c = 5$ Hz		$F_c = 10$ Hz		$F_c = 20$ Hz	
	σ_{simu}	σ_{comp}	σ_{simu}	σ_{comp}	σ_{simu}	σ_{comp}
σ_θ	$1.9e^{-5}$	$1.1e^{-6}$	$2.3e^{-6}$	$1.5e^{-6}$	$2.2e^{-6}$	$2.2e^{-6}$
$\sigma_{\dot{\theta}}$	$1.8e^{-5}$	$1.9e^{-5}$	$5.4e^{-5}$	$5.3e^{-5}$	$1.5e^{-4}$	$1.5e^{-4}$
$\sigma_{\ddot{\theta}}$	$4.5e^{-4}$	$4.5e^{-4}$	$2.7e^{-3}$	$2.6e^{-3}$	$1.5e^{-2}$	$1.4e^{-2}$

Table 2. Standard deviation of the encoder signals

Taking into account these results and the bandwidth of the system, a cutoff frequency of 10Hz was chosen. At this frequency, the accuracy of the acceleration is excellent. This implies that the acceleration computed with the encoders can be considered as a reference. Therefore, we will compare the acceleration computed with the wheel to this reference.

As the encoder is the reference, it can be used to estimate the precision of position, velocity and acceleration computed from the toothed wheel. From one part they are computed with an encoder model (a nearest interpolation), and from the other part they are computed with natural splines. Table (3) gives the standard deviation of results for the two methods (with the 100000 pulses/rev encoder as the reference) on a real trajectory (Fig. 3). For each cutoff frequency, the standard deviation has been computed with an encoder model (left column) and with Splines (**right column**).

	$F_c = 5$ Hz		$F_c = 10$ Hz		$F_c = 20$ Hz	
σ_θ	$5.8e^{-4}$	**$3.4e^{-4}$**	$6.1e^{-4}$	**$3.5e^{-4}$**	$1.1e^{-3}$	**$3.8e^{-4}$**
$\sigma_{\dot{\theta}}$	$6.7e^{-3}$	**$2.7e^{-3}$**	$1.1e^{-2}$	**$5.4e^{-3}$**	$1.0e^{-1}$	**$1.3e^{-2}$**
$\sigma_{\ddot{\theta}}$	$1.3e^{-1}$	**$4.5e^{-2}$**	$4.8e^{-1}$	**$2.4e^{-1}$**	$1.3e^{1}$	**1.3**

Table 3. Error to the reference

Table (4) presents other interesting results, the expected precision with an encoder model (see 4.1) compared with precision obtained in a real case.

	63 teeth encoder model	Real toothed wheel	Splines
σ_θ	$1.2e^{-3}$	$6.1e^{-4}$	$3.5e^{-4}$
$\sigma_{\dot{\theta}}$	$4.2e^{-2}$	$1.1e^{-2}$	$5.4e^{-3}$
$\sigma_{\ddot{\theta}}$	2	$4.8e^{-1}$	$2.4e^{-1}$

Table 4. Error to the reference according to the cutoff frequency

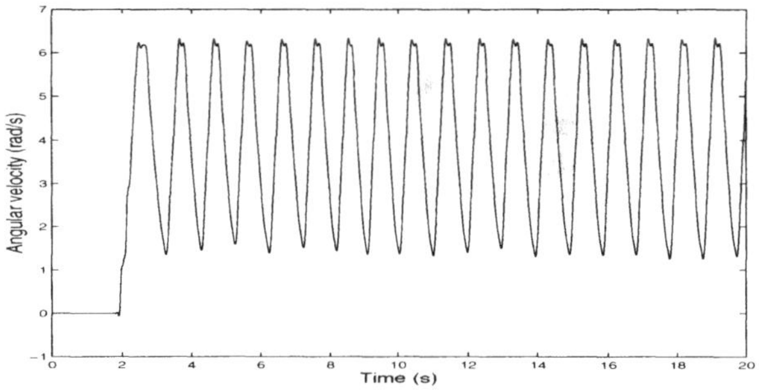

Fig. 3. Angular velocity of a real trajectory

Two conclusions are obtained from these results. The first one is that the model used in the case of a 100000 pulses/rev encoder doesn't match in the case of a 63 teeth wheel. Actually, best results than expected are obtained. It could be explained by the fact that each tooth of the wheel is located with a standard deviation on the position of about $3.7e-4$ rad. This is about the standard deviation of the position obtained with the spline interpolation, whatever is the cutoff frequency used (see table 3).

The second conclusion is that the standard deviation of velocity and acceleration is reduced with the spline interpolation. It could be explained with the fact that spline interpolation introduces regularity conditions on the motion which correspond to the real case. Actually, a spline interpolation on the position makes the hypothesis that acceleration is continuous.

5.2 Hydraulic motor identification

In the previous section, results on the precision of position, velocity and acceleration were presented. In order to carry out the identification of the hydraulic motor of the compactor, results about velocity and acceleration have been used.

Parameters of the equation (19) have been identified. Data processing parameters were these ones :

- Forward and reverse low pass digital fourth order Butterworth filter,
- 10 Hz cutoff frequency.

Y and W_s are calculated to get equation (7). Estimated value of dynamic parameters X_s are given in table (5) with their relative standard deviation $\hat{\sigma}_r = \frac{\hat{\sigma}_{X_s}}{\hat{X}_s}$.

Parameters	Units	$\hat{X}_s$	$\hat{\sigma}_{X_s}$	$\hat{\sigma}_r$ (%)
ZZ	$kg \cdot m^2$	1.77	0.031	1.71
F_{v^2}	$N \cdot m \cdot s^2$	9.97	0.17	1.63
F_v	$N \cdot m \cdot s$	45.4	1.3	2.87
F_s	$N \cdot m$	198	2.3	1.13
$cond(W_s) = 172$	$cond(\Phi) = 34$	$\Phi = W_s \cdot diag(\hat{X}_s)$		

Table 5. Identified dynamic parameters

According to the identification results (see table 5), all parameters are well estimated. A cross validation is carried out with an other trajectory than the one used to the identification. The figure (4) shows that the estimated torque matches closely the measured torque.

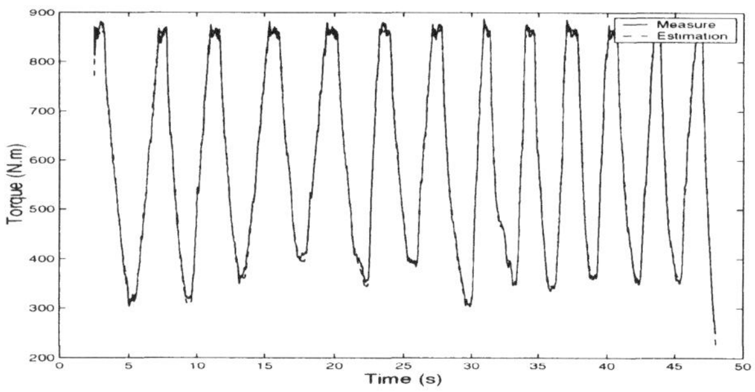

Fig. 4. Dynamic model of the hydraulic motor

To be sure that a least square identification method is valid, the standard deviation on $W_s X_s$ (which is supposed to be zero) is compared with the standard deviation of ρ (equation (7)). $\frac{\sigma_\rho}{\sigma_{W_s X_s}} = 8$ so least squares identification method is valid.

6. CONCLUSION

The dynamic modeling and identification of an earth-moving mobile engine has been presented and applied to the compactor. Its geometric description using Modified Denavit-Hartenberg notations can be extended to a more complex model of the compactor which take into account 3D motion. This formulation let a computer perform automatic symbolic calculation on models. The dynamic model linear in relation to parameters allows to identify them using Least Squares methods. In order to achieve these method in an industrial environnement, specific data processing has been develop using spline interpolation. In near future, other trials will be done to identify all parameters of the straight line model of the compactor. Then the model will be extended to a two dimension model to take into account the vertical force between the ground and the compactor which is as important as the rolling resistance in the compaction process.

REFERENCES

Canudas de Wit, C., B. Siciliano and G. Bastin (1997). *Theory of Robot Control.* Chap. 1, pp. 4–29. Springer-Verlag. London.

Delclos, A, P-O Vandanjon, F Peyret and M Gautier (2001). Estimating the degree of compaction of asphalt using proprioceptive sensor and dynamic model. In: *Proceedings of International Symposium on Automation in Road Construction.* Krakow, Poland.

Floss, R. and H-J. Kloubert (2000). Newest developments in compaction technology. In: *Le Compactage Des Sols et Des Matériaux Granulaires.* SISMG / Comité ETC 11. pp. 247–261.

Gautier, M. (1997). Dynamic identification of robots with power model. In: *Proceedings of IEEE International Conference on Robotics and Automation.* Albuquerque, USA.

Guillo, É., M. Gautier, F. Louveau and C. Bidard (1997). Dynamic modelling and identification of a hydraulic servoactuator. In: *Proceedings of the 5th SYmposium on RObot COntrol.* Vol. 1. Nantes, France. pp. 125–130.

Guillo, Eric (2000). Modélisation et identification dynamique des engins mobiles de construction de routes. Thèse de doctorat. Université de Nantes/École Centrale de Nantes.

Khalil, W. and Dombre, E. (1999). *Modélisation, identification et commande des robots.* second ed.. Hermès Science Publications. Paris.

Kozlowski (1998). *Modelling and Identification in Robotics.* Springer-Verlag.

Nürnberger, Günther (1989). *Approximation by spline function.* first ed.. Springer-Verlag. Berlin.

IFAC Publications
www.elsevier.com/locate/ifac

NON-STATIONARY MECHANICAL VIBRATION MODELING AND ANALYSIS VIA FUNCTIONAL SERIES TARMA MODELS ⋆

Aggelos G. Poulimenos *and* **Spilios D. Fassois**

Stochastic Mechanical Systems (SMS) Group
Department of Mechanical & Aeronautical Engineering
University of Patras, GR 265 00 Patras, Greece
E-mail: {poulimen,fassois}@mech.upatras.gr
Internet: http://www.mech.upatras.gr/~sms

Abstract: The problem of modeling and analysis of the non-stationary random vibration of a time-varying "bridge-like" laboratory structure is considered via a Functional Series Time-dependent ARMA (FS-TARMA) approach. The results of the study demonstrate high modeling accuracy, as well as superiority to adaptive Recursive Maximum Likelihood – Recursive ARMA (RML-RARMA) and non-parametric Short-Time Fourier Transform (STFT) analysis. The applicability, effectiveness, and high potential of the FS-TARMA approach, which is capable of modeling fast or slow variations in the dynamics while simultaneously achieving high accuracy and resolution, are also demonstrated. *Copyright © 2003 IFAC*

Keywords: Time-varying systems, vibration, mechanical systems, ARMA models, non-stationary signals.

1. INTRODUCTION

Non-stationary vibration is characterized by features that vary with time and require time-frequency methods for their analysis (Hammond and White 1996). It occurs in systems with time-dependent properties and/or non-linearities. Notable examples include traffic-excited bridge vibration, earthquake-excited vibration, vibration in surface vehicles, airborne structures and sea vessels, in robotic devices, rotating machinery, and so on.

The most widely used non-stationary analysis methods are *non-parametric*, such as the Short-Time Fourier Transform (STFT) and its ramifications (Hammond and White 1996), Priestley's evolutionary spectrum (Priestley 1988), Wigner-Ville distributions and their extensions (Hammond and White 1996), and wavelet-based methods (Newland 1994). Yet, *parametric* methods offer a number of advantages, such as representation parsimony, improved accuracy, resolution, and tracking.

A notable class of parametric methods is based upon *Functional Series Time-dependent AutoRegressive Moving Average (FS-TARMA) models* (Grenier 1989, Ben Mrad *et al.* 1998). Although resembling their conventional, stationary, ARMA counterparts, FS-TARMA models have a much richer structure, originating from the fact that their parameters are explicit functions of time, belonging to specific functional subspaces. FS-TARMA models are physically motivated for representing non-stationary mechanical vibration due to the fact that the underlying physical mechanisms responsible for the non-stationary behaviour change in a smooth, deterministic, way that is best captured by the functional parameter variation of FS-TARMA models.

Compared to potentially alternative parametric methods, such as *adaptive* (Ljung 1999) or *stochastic*

⋆ Research sponsored by the VolkswagenStiftung.

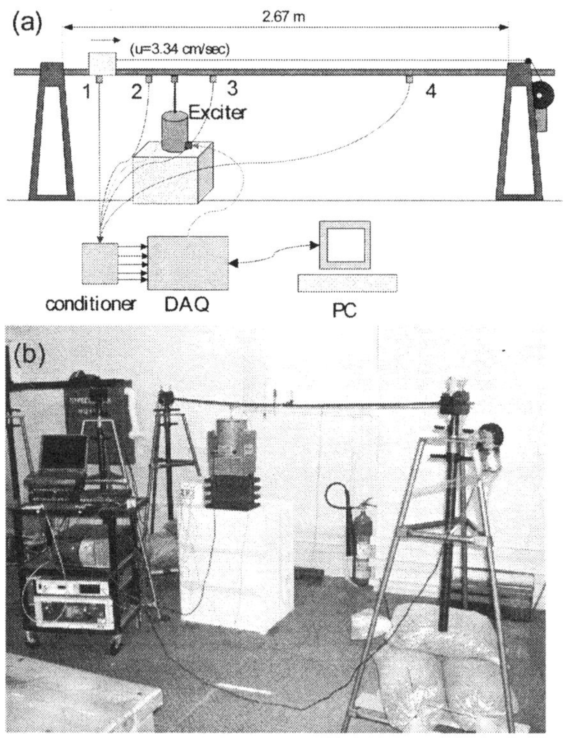

Fig. 1. (a) Schematic diagram, and, (b) photo of the laboratory set-up.

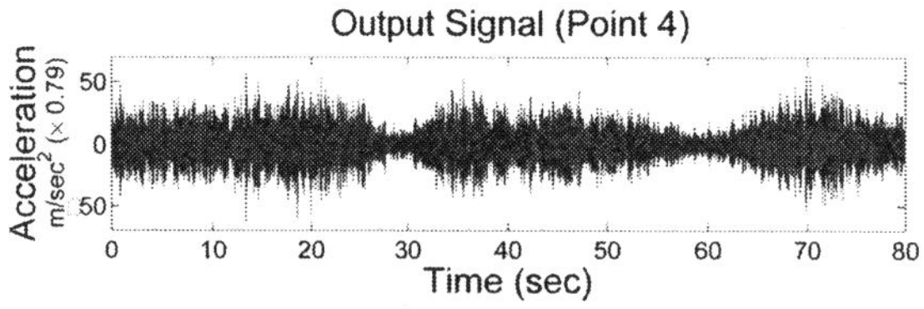

Fig. 2. The non-stationary random vibration signal.

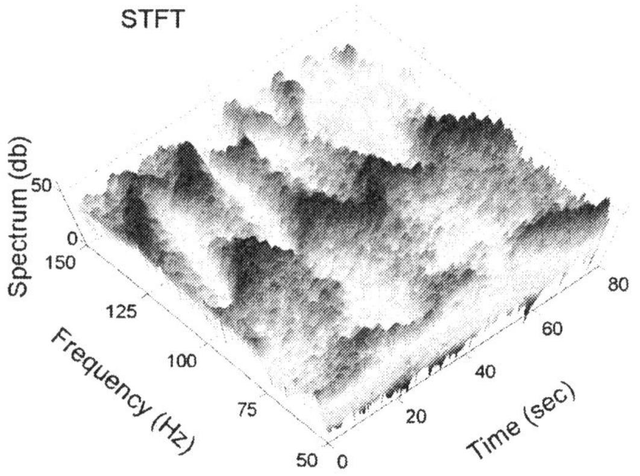

Fig. 3. Non-parametrically estimated time-dependent power spectral density function (STFT method).

parameter evolution methods (Kitagawa and Gersh 1996), FS-TARMA methods impose maximum "structure" on parameter evolution, achieving a high degree of parsimony and the capability of tracking fast and abrupt variations while maintaining high accuracy and resolution (Ben Mrad *et al.* 1998, Petsounis and Fassois 2000).

Yet, FS-TARMA models have been, thus far, used in a very limited number of studies: Fouskitakis and Fassois (2002) use FS-TARMA models for the modeling and analysis of earthquake ground motion. Conforto and D'Alessio (1999) use partial FS-TAR models for the modeling of mechanical vibration in rotating machinery, and Petsounis and Fassois (2000) use FS-TARMA models for the modeling and analysis of simulated robot vibration.

The goal of the present study is to demonstrate the applicability and explore the effectiveness of refined FS-TARMA methods, recently introduced by Fouskitakis and Fassois (2001) and Poulimenos and Fassois (2003), for the modeling and analysis of the non-stationary random vibration of a time-varying "bridge-like" laboratory structure. Interesting comparisons with the non-parametric Short-Time Fourier Transform (STFT) and the parametric adaptive Recursive Maximum Likelihood – Recursive ARMA (RML-RARMA) method, are also made.

2. THE NON-STATIONARY VIBRATION AND PRELIMINARY ANALYSIS

The set-up. The laboratory set-up is shown in Figure 1: It consists of a steel beam of dimensions $2670 \times 50 \times 70$ mm ($L \times W \times H$), which is clamped close to its both ends on two vertical stands. A steel cylindrical mass of dimensions 52.5×75.0 mm ($R \times H$), slides on it, being pulled by a DC motor at a selected speed u. As the ratio of the two masses is significant ($m/M = 0.4$), the system is clearly non-stationary (time-varying), with the rate of variation depending upon the selected speed u.

The beam is subject to random, zero-mean, and approximately white Gaussian force excitation which is vertically exerted via an electromechanical exciter equipped with a stinger (Figure 1). The vertical beam vibration is measured at four positions (positions $1-4$) via piezoelectric accelerometers. The measured vibration signals are conditioned and subsequently driven into the data acquisition (DAQ) system.

The experiment and signal pre-processing. In a single experiment the cylindrical mass traverses the beam (from left to right) once, at a constant speed of $u = 33.4$ mm/sec (total experiment time $\Delta t = 79.998$ $secs$). At this speed, the system may be characterized as relatively *slowly* time-varying. The vertical vibration (acceleration) signals are sampled at $f_s = 512$ Hz, each one being $N = 40,960$ $samples$ (79.998 $secs$) long. The study focuses on the $50 - 145$ Hz frequency range, hence the signals are therein digitally band-pass filtered.

The vibration signal and preliminary analysis. The modeling and analysis of the signal measured at the rightmost position (position 4; see Figure 1) is considered. The measured signal (Figure 2) is evidently variance non-stationary. Non-parametric analysis based upon the Short-Time Fourier Transform (STFT; moving window of 2048 $samples$, Hamming data window) yields the time-dependent power spectral density function of Figure 3, from which non-stationarity in the frequency content is also evident.

3. FUNCTIONAL SERIES TARMA MODELING

3.1 *FS-TARMA Models*

The Functional Series Time-dependent AutoRegressive Moving Average (FS-TARMA) models constitute conceptual extensions of their conventional (stationary) ARMA counterparts (?), in that their parameters are explicit functions of time, belonging to functional subspaces spanned by selected deterministic functions (*basis functions*). A TARMA$(n_a, n_c)_{[p_a, p_c]}$ model, with n_a, n_c indicating its AutoRegressive (AR) and Moving Average (MA) orders, respectively, and p_a, p_c the corresponding functional basis dimensionalities, is of the form:

$$\underbrace{\left[1+\sum_{i=1}^{n_a} a_i[t]\cdot B^i\right]}_{A[B,t]}\cdot x[t] = \underbrace{\left[1+\sum_{i=1}^{n_c} c_i[t]\cdot B^i\right]}_{C[B,t]}\cdot w[t] \quad (1)$$

with t designating discrete time, $x[t]$ the non-stationary signal modeled, and $w[t]$ an innovations (uncorrelated) sequence with zero mean and time-dependent variance $\sigma_w^2[t]$. B stands for the backshift operator ($B^i \cdot x[t] \stackrel{\Delta}{=} x[t-i]$), and $a_i[t]$ and $c_i[t]$ for the model's AR and MA parameters, respectively. $A[B,t]$, $C[B,t]$ designate the corresponding time-varying AR and MA polynomial operators.

The AR and MA parameters belong to functional subspaces with respective bases:

$$\mathcal{F}_{AR} \stackrel{\Delta}{=} \{ G_{b_a(1)}[t],\ G_{b_a(2)}[t],\ \ldots,\ G_{b_a(p_a)}[t]\} \quad (2)$$

$$\mathcal{F}_{MA} \stackrel{\Delta}{=} \{ G_{b_c(1)}[t],\ G_{b_c(2)}[t],\ \ldots,\ G_{b_c(p_c)}[t]\} \quad (3)$$

and similarly does the innovations variance $\sigma_w^2[t]$. In these expressions the indices $b_a(i)$ $(i = 1,\ldots,p_a)$ and $b_c(i)$ $(i = 1,\ldots,p_c)$ designate the functions (from a properly ordered set, such as Chebyshev, Legendre, Laguerre, and so on) that are included in each basis. The time-dependent AR and MA parameters may be thus expressed as:

$$a_i[t] \stackrel{\Delta}{=} \sum_{j=1}^{p_a} a_{i,j}\cdot G_{b_a(j)}[t],\ c_i[t] \stackrel{\Delta}{=} \sum_{j=1}^{p_c} c_{i,j}\cdot G_{b_c(j)}[t] \quad (4)$$

with $a_{i,j}$, $c_{i,j}$ designating the corresponding *coefficients of projection*.

3.2 *FS-TARMA Model Estimation*

The estimation of the projection coefficient vector[1] $\boldsymbol{\theta}$, consisting of the corresponding AR and MA vectors, $\boldsymbol{a}$ and $\boldsymbol{c}$, respectively:

$$\boldsymbol{\theta} \stackrel{\Delta}{=} [\boldsymbol{a}^T \vdots \boldsymbol{c}^T]^T \stackrel{\Delta}{=} [a_{1,1} \ldots a_{1,p_a} \vdots \ldots \vdots c_{n_c,1} \ldots c_{n_c,p_c}]^T$$

is considered based upon available signal samples $x[t]$ $(t = 1,2,\ldots,N)$.

[1] Bold face symbols designate column vector quantities.

The estimation approach utilized employs the Two Stage Least Squares (2SLS) (Grenier 1989) or the Polynomial-Algebraic (P-A) (Ben Mrad *et al.* 1998, Fouskitakis and Fassois 2001, Poulimenos and Fassois 2003) method for obtaining initial estimates, which are subsequently refined via the Prediction Error (PE) method. Brief descriptions of these methods follow.

<u>The Two-Stage Least Squares (2SLS) method.</u>

Step 1: Residual Series Estimation.
The TARMA model of Equation (1) may be re-written in terms of the inverse function operator $I[B,t]$ (Ben Mrad *et al.* 1998, Fouskitakis and Fassois 2001) as:

$$I[B,t]\cdot x[t] = \left[1 + \sum_{q=1}^{\infty} i_q[t]\cdot B^q\right]\cdot x[t] = w[t] \quad (5)$$

with $i_q[t]$ $(q = 1,2,\ldots)$ designating the model's time-varying inverse function. As illustrated in Poulimenos and Fassois (2003), the functional space of $i_q[t]$ is (in the polynomial case) related to the AR and MA functional subspaces through the expressions:

$$\mathcal{F}_{i_1[t]} = \mathcal{F}_{AR} \cup \mathcal{F}_{MA} \quad (6)$$

$$\mathcal{F}_{i_q[t]} = \{G_{b_{i_q}(1)}[t],\ldots,G_{b_{i_q}(p_{i_q})}[t]\} \quad (7)$$

where p_{i_q} designates the functional space dimensionality and:

$$b_{i_q}(1),\ \ldots,\ b_{i_q}(p_{i_q}) : \text{consecutive integers} \quad (8)$$

$$b_{i_q}(1) = 0 \quad (9)$$

$$b_{i_q}(p_{i_q}) = (q-1)\cdot\max_j\{b_c(j)\} + \max_{k,j}\{b_a(k), b_c(j)\} \quad (10)$$

A truncated (n_i-order) inverse function model of the form [compare with Equation (5)]:

$$I[B,t,\boldsymbol{i}]\cdot x[t] = e[t,\boldsymbol{i}] \quad (11)$$

is then estimated based upon minimization of the residual ($e[t,\boldsymbol{i}]$) sum of squares via linear regression. In this expression $\boldsymbol{i}$ designates the projection coefficient vector of the truncated inverse function.

Step 2: AR/MA Projection Coefficient Estimation.
Once the residual series $e[t,\hat{\boldsymbol{i}}]$ has been obtained[2], a FS-TARMA model corresponding to that of Equation (1) may be approximated as follows (replacing the *past*, but not the current, values of the prediction error $e[t,\boldsymbol{\theta}]$, corresponding to the proper FS-TARMA model structure, by the obtained $e[t,\hat{\boldsymbol{i}}]$):

$$A[B,t,\boldsymbol{\theta}]\cdot x[t] = (1 - C[B,t,\boldsymbol{\theta}])\cdot e[t,\hat{\boldsymbol{i}}] + e[t,\boldsymbol{\theta}]$$

$$\Longrightarrow x[t] + \sum_{i=1}^{n_a}\sum_{j=1}^{p_a} a_{i,j}\cdot G_{b_a(j)}\cdot x[t-i] =$$

$$= \sum_{i=1}^{n_c}\sum_{j=1}^{p_c} c_{i,j}\cdot G_{b_c(j)}\cdot e[t-i,\hat{\boldsymbol{i}}] + e[t,\boldsymbol{\theta}]$$

$$\Longrightarrow x[t] = \boldsymbol{\phi}^T[t]\cdot\boldsymbol{\theta} + e[t,\boldsymbol{\theta}] \quad (12)$$

where:

[2] The hat designates estimator/estimate.

$$\phi[t] \triangleq \Big[-G_{b_a(1)}\cdot x[t-1]\ldots - G_{b_a(p_a)}\cdot x[t-n_a] \vdots$$

$$\vdots\, G_{b_c(1)}\cdot e[t-1,\hat{\boldsymbol{i}}]\ldots G_{b_c(p_c)}\cdot e[t-n_c,\hat{\boldsymbol{i}}]\Big]^T \quad (13)$$

The AR/MA projection coefficient vector may be then estimated by minimizing the residual ($e[t,\boldsymbol{\theta}]$) sum of squares in Equation (12) via linear regression.

Step 3: Residual Variance Estimation.
The variance of the estimated [using a model expression corresponding to that of Equation (1) with the estimated parameters] FS-TARMA residual series is obtained via a moving average filter (sliding time-window), and subsequently fitted to a properly selected functional basis (of the type used for the AR/MA parameters).

The Polynomial-Algebraic (P-A) method.

Step 1: Inverse Function Estimation.
The estimation of a truncated (n_i-th order) inverse function operator $I[B,t,\boldsymbol{i}]$ is accomplished as in the 2SLS method (step 1).

Step 2: Initial AR/MA Projection Coefficient Estimation.
Initial estimates of the AR and MA projection coefficients (vectors $\boldsymbol{a}$ and $\boldsymbol{c}$) are obtained by replacing the theoretical inverse function by the estimated $I[B,t,\hat{\boldsymbol{i}}]$ in the relationship:

$$A[B,t,\boldsymbol{a}] = C[B,t,\boldsymbol{c}] \circ I[B,t,\boldsymbol{i}] \quad (14)$$

and performing the necessary deconvolution. Notice that $\circ$ designates the skew multiplication operator of the time-varying polynomial operator algebra (Ben Mrad *et al.* 1998, Fouskitakis and Fassois 2001).

Step 3: Signal Filtering.
Once initial estimates of the AR and MA polynomial operators are available, the filtering operations:

$$C[B,t,\hat{\boldsymbol{c}}]\cdot z[t] = A[B,t,\hat{\boldsymbol{a}}]\cdot x[t] \quad (15)$$

$$A[B,t,\hat{\boldsymbol{a}}]\cdot \bar{x}[t] = z[t] \quad (16)$$

are performed and the signal $\bar{x}[t]$ is obtained.

Step 4: Final AR/MA Projection Coefficient Estimation.
The filtered signal $\bar{x}[t]$ theoretically obeys the FS-TAR part of the original FS-TARMA model, that is:

$$A[B,t]\cdot \bar{x}[t] = w[t] \quad (17)$$

Hence the AR projection coefficient estimates are obtained by minimizing the residual sum of squares of the corresponding model:

$$A[B,t,\boldsymbol{a}]\cdot \bar{x}[t] = e[t,\boldsymbol{a}] \quad (18)$$

via linear regression. The final MA coefficient of projection estimates (vector $\boldsymbol{c}$) are subsequently obtained based upon Equation (14) using the updated AR estimates (deconvolution).

Step 5: Residual Variance Estimation.
The residual variance is estimated as in the 2SLS method (step 3).

The Prediction Error (PE) method.

The Prediction Error (PE) method is based upon minimization of the FS-TARMA model residual sum of squares ($e[t,\boldsymbol{\theta}]$) using non-linear optimization [the Levenberg-Marquardt scheme (Ljung 1999)].

4. NON-STATIONARY RANDOM VIBRATION MODELING AND ANALYSIS

4.1 *Random Vibration Modeling*

The FS-TARMA modeling of the non-stationary random vibration signal is now considered using Chebyshev type II functional subspaces (selected to reflect smooth variability in the dynamics). An "extended", "complete" (in the sense of including all consecutive polynomials from degree zero on), functional basis of dimensionality 25 is initially adopted for the AR and MA functional subspaces. The modeling procedure may be then described as follows:

(a) FS-TARMA($n, n-1$) models, featuring the initially selected functional subspace, are estimated for $n = 2, 3, \ldots$ ($n_i = 18$).
(b) Once a proper value of the AR order is selected, the possibility of MA order reduction is examined, and a final MA order is selected.
(c) Following AR/MA model order determination, dimensionality reduction for both the AR and MA functional subspaces is considered, and adequate final dimensionalities are selected.

For model order and functional subspace dimensionality selection, the condition number of the model regression (information) matrix (Ljung 1999), prediction error criteria, and evaluation of the changes in the model parameter trajectories, are used.

The finally obtained model is a FS-TARMA$(8,5)_{[15,6]}$, characterized by AR and MA orders $n_a = 8$ and $n_c = 5$, respectively, and corresponding functional subspaces consisting of the first $p_a = 15$ and $p_c = 6$ Chebyshev type II polynomials. Notice that the selected AR order ($n_a = 8$) is consistent with the STFT-based time-dependent power spectral density estimate in which four modes are observed (Figure 3).

In order to assess the FS-TARMA$(8,5)_{[15,6]}$ model accuracy, model-based one-step-ahead predictions are, for a segment of the signal, presented in Figure 4, and are, evidently, very accurate. For purposes of comparison, a Recursive ARMA (RARMA) model of the same AR/MA orders is also estimated by the Recursive Maximum Likelihood (RML) method (Ljung 1999). The FS-TARMA$(8,5)_{[15,6]}$ and RARMA$(8,5)$ residuals (one-step-ahead prediction errors) are presented in Figure 5, along with their estimated time-dependent variance (estimation window length 201; in the FS-TARMA case the variance is expanded on the Chebyshev type II functional basis of dimensionality $p_{\sigma_w^2} = 15$). It is evident that the FS-TARMA$(8,5)_{[15,6]}$ model surpasses its RARMA$(8,5)$ counterpart in terms of

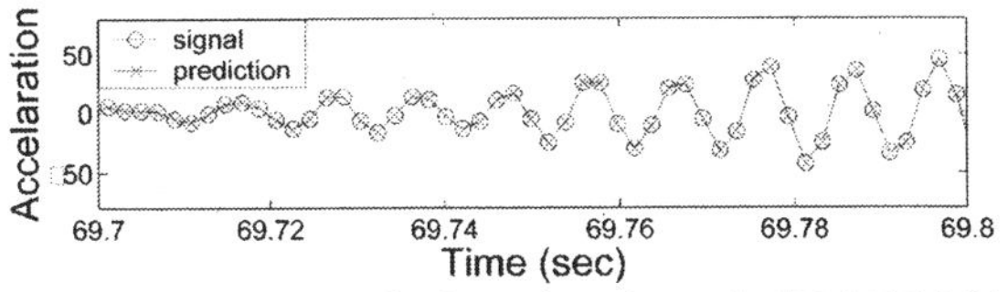

Fig. 4. Segment of the signal and FS-TARMA $(8,5)_{[15,6]}$ based one-step-ahead predictions.

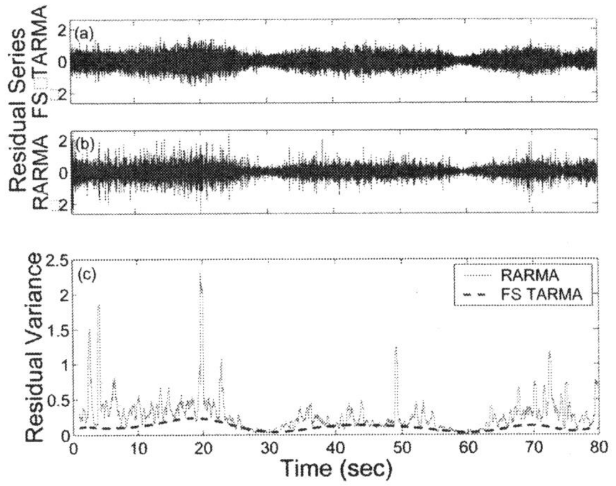

Fig. 5. (a) FS-TARMA$(8,5)_{[15,6]}$ residuals, (b) RARMA$(8,5)$ residuals, (c) residual variances.

Table 1. Aggregate error (RSS/SSS) for the FS-TARMA$(8,5)$ and RARMA$(8,5)$ models.

	FS-TARMA$(8,5)$	RARMA$(8,5)$
RSS/SSS (%)	0.0731	0.1653

achieved accuracy (smaller FS-TARMA residuals); a fact clearly reflected in Table 1 in which the Residual Sum of Squares (RSS) normalized by the Series Sum of Squares (SSS), is significantly smaller (by factor of 2) for the FS-TARMA$(8,5)_{[15,6]}$ model.

4.2 *Model-Based Analysis*

The signal's "frozen" time-dependent power spectral density may be obtained from the estimated FS-TARMA$(8,5)_{[15,6]}$ and RARMA$(8,5)$ models via the expression:

$$S(\omega,t) = \left| \frac{1+\sum_{i=1}^{n_c} c_i[t]\cdot e^{-j\omega T_s i}}{1+\sum_{i=1}^{n_a} a_i[t]\cdot e^{-j\omega T_s i}} \right|^2 \cdot \sigma_w^2[t] \quad (19)$$

in which the model parameters and innovations variance are replaced by their respective estimates, ω designates frequency in $rads/sec$, T_s the sampling period in $secs$, and j the imaginary unit.

The system's time-dependent global modal parameters (natural frequencies and damping ratios) may be also obtained as:

$$\omega_{ni}[t] = \frac{\left|\ln\left(\lambda_i[t]\right)\right|}{T_s} \quad (rad/sec) \quad (20)$$

$$\zeta_i[t] = -\cos\Big(\arg\Big(\ln\big(\lambda_i[t]\big)\Big)\Big) \quad (21)$$

with $\lambda_i[t]$ designating the model's i-th "frozen" pole.

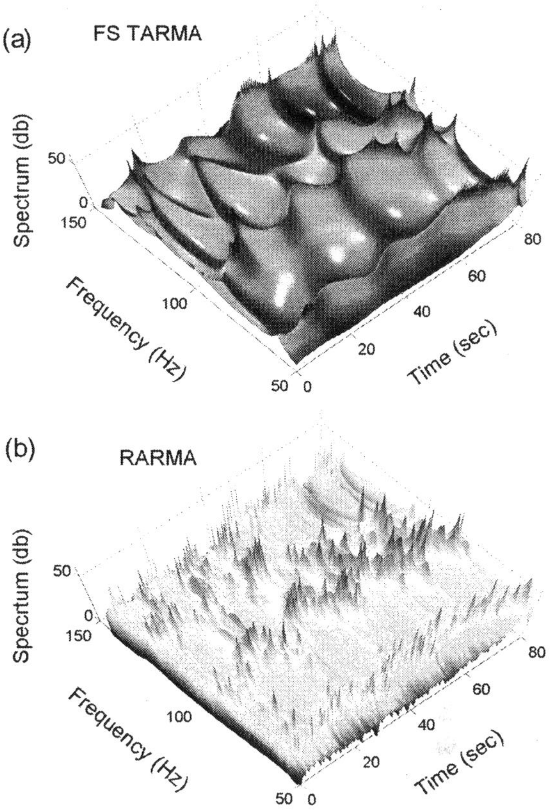

Fig. 6. Estimated time-dependent power spectral density functions: (a) FS-TARMA$(8,5)_{[15,6]}$, (b) RARMA$(8,5)$.

The FS-TARMA$(8,5)_{[15,6]}$ and RARMA$(8,5)$ time-dependent power spectral densities are depicted in Figure 6. The FS-TARMA$(8,5)_{[15,6]}$ based density is in rough overall agreement with the non-parametric STFT estimate (Figure 3), exhibiting resonances that are time-dependent but in the gross neighborhoods of 60, 90, 115 (downward trend), and 135 Hz. The FS-TARMA$(8,5)_{[15,6]}$ model based density is of high resolution, and yet much more clear, smooth and informative than its non-parametric counterpart or the parametric RARMA$(8,5)$ density – notice that smoothness makes physical sense as the inherent variation in the system dynamics is smooth and slow.

The system's time-dependent global modal parameters (natural frequencies and damping ratios) based upon the estimated FS-TARMA$(8,5)_{[15,6]}$ and RARMA$(8,5)$ models are depicted in Figure 7. The natural frequencies are within the previously mentioned ranges, and the damping ratios are (expectedly) low. The rough agreement of the trajectories, as obtained via the FS-TARMA and RARMA approaches, is quite remarkable and certainly encouraging.

5. CONCLUDING REMARKS

The problem of modeling and analysis of the non-stationary random vibration of a "bridge-like" laboratory structure was considered. The main results of the study may be summarized as follows:

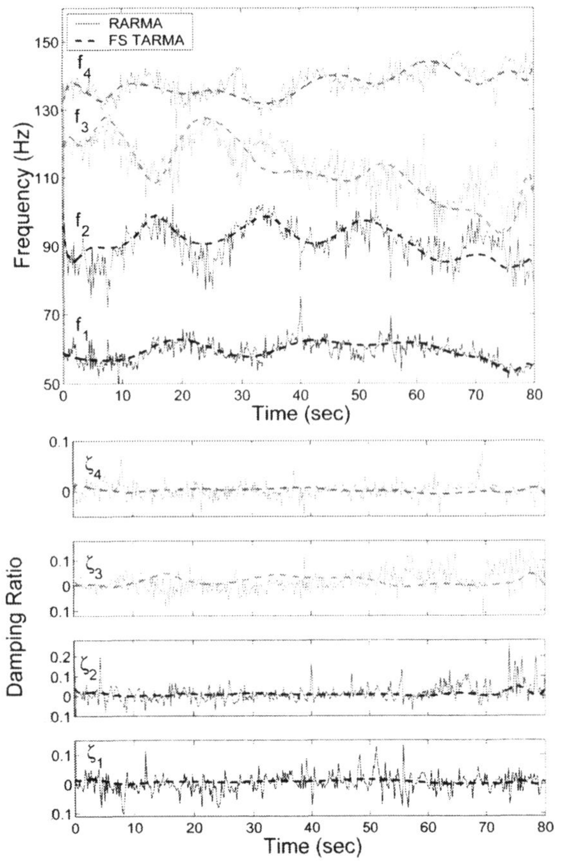

Fig. 7. FS-TARMA$(8,5)_{[15,6]}$ and RARMA$(8,5)$ natural frequency and damping ratio estimates.

1. The applicability, effectiveness, and overall benefits of the FS-TARMA approach for modeling the non-stationary vibration via a FS-TARMA$(8,5)$ model characterized by Chebyshev type II functional subspaces were demonstrated.
2. The FS-TARMA$(8,5)_{[15,6]}$ based time-dependent power spectral density was of high resolution, smooth (in agreement with the physics of the problem), and clearly depicting the modal information. The global modal parameter trajectories were also smooth and informative.
3. The FS-TARMA$(8,5)_{[15,6]}$ model was shown to be superior to its RARMA$(8,5)$ counterpart, achieving an aggregate prediction error (RSS) that is half of that of the latter. It was quite positive that the characteristics of the two models were in good overall agreement, although the time-dependent power spectral density and modal parameter trajectories of the FS-TARMA model were smoother, and yet detailed and informative.
4. Although in rough agreement with its non-parametric STFT counterpart, the FS-TARMA$(8,5)_{[15,6]}$ model based time-dependent power spectral density was judged to be superior in terms of achievable accuracy and resolution.
5. The study demonstrated the benefits and high potential of the FS-TARMA approach for the modeling and analysis of non-stationary random vibration. Three of its important advantages are: (a) Its ability to model fast or slow variations in the dynamics; (b) its ability to model the variations via smooth curves; and, (c) its simultaneous achievement of high accuracy and resolution. Additional research on both modeling (identification) and analysis is necessary for further development of the approach.

Acknowledgements

The authors wish to acknowledge the financial support of this study via a VolkswagenStiftung award, as well as the help of Mr. Demos Rizos of the University of Patras in preparing the experimental set-up.

6. REFERENCES

Ben Mrad, R., S.D. Fassois and J.A. Levitt (1998). A polynomial-algebraic method for non-stationary TARMA signal analysis. Part I: The method. *Signal Processing* **65**, 1–19.

Conforto, S. and T. D'Alessio (1999). Spectral analysis for non-stationary signals from mechanical measurements: A parametric approach. *Mechanical Systems and Signal Processing* **13**, 395–411.

Fouskitakis, G.N. and S.D. Fassois (2001). On the estimation of non-stationary functional series TARMA models: an isomorphic matrix algebra based method. *ASME Journal of Dynamic Systems, Measurement, and Control* **123**, 601–610.

Fouskitakis, G.N. and S.D. Fassois (2002). Functional series TARMA modelling and simulation of earthquake ground motion. *Earthquake Engineering and Structural Dynamics* **31**, 399–420.

Grenier, Y. (1989). Parametric time-frequency representations. In: *Time and Frequency Representations of Signals and Systems* (G. Longo and B. Picinbono, Eds.). Springer-Verlag.

Hammond, J.K. and P.R. White (1996). The analysis of non-stationary signals using time-frequency methods. *Journal of Sound and Vibration* **190**, 419–447.

Kitagawa, G. and W. Gersh (1996). *Smoothness Priors Analysis of Time Series*. Springer-Verlag. New York.

Ljung, L. (1999). *System Identification: Theory for the User (second edition)*. Prentice-Hall PTR.

Newland, D.E. (1994). Wavelet analysis of vibration: parts 1 and 2. *Journal of Vibration and Acoustics* **116**, 409–425.

Petsounis, K.A. and S.D. Fassois (2000). Non-stationary functional series TARMA vibration modelling and analysis in a planar manipulator. *Journal of Sound and Vibration* **231**, 1355–1376.

Poulimenos, A.P. and S.D. Fassois (2003). Estimation and identification of non-stationary functional series TARMA models. *These conference proceedings*.

Priestley, M.B. (1988). *Non-Linear and Non-Stationary Time Series Analysis*. Academic Press.

IFAC
Publications
www.elsevier.com/locate/ifac

GLOBALLY CONVERGENT ADAPTIVE TRACKING OF ANGULAR VELOCITY WITH INERTIA IDENTIFICATION AND ADAPTIVE LINEARIZATION

Amit K. Sanyal, Madhusudhan Chellappa, Jean Luc Valk, Jasim Ahmed, Jinglai Shen, Dennis S. Bernstein [1]

Aerospace Engineering, University of Michigan, Ann Arbor, MI-48105

Abstract: The problem of a rigid body tracking a desired angular velocity trajectory is addressed using adaptive feedback control. An adaptive controller is developed for a planar rotating body tracking a desired angular velocity command. Lyapunov analysis is used to show that tracking is achieved globally. A periodic angular velocity command is then used to identify the inertia parameter. The adaptive controller is implemented on a triaxial attitude control testbed with fan thrusters. A piecewise linear approximation of an observed input nonlinearity is inverted to obtain improved angular velocity tracking and inertia identification. To eliminate residual tracking error, an adaptive algorithm is used for improved feedback linearization. Lyapunov analysis is used to show boundedness of the angular velocity and inertia estimate errors. The approach is validated by numerical simulation.

Keywords: Adaptive Control, Identification, Nonlinear Systems, Tracking, Units

1. INTRODUCTION

Stabilization of a single free rigid body in three dimensions is a widely studied and fundamental problem in spacecraft dynamics. Although the problem is trivial in the presence of three control torques, significant research has been devoted to the cases of two torques and one torque. If minimum fuel or minimum time performance is required in addition to stabilization, then this problem is challenging even in the case of three torque inputs.

The above remarks are based on the assumption that the spacecraft mass distribution is known and constant. However, in practice there are limitations due to fuel usage, moving appendages, and complex geometry on the ability to determine the mass distribution. Hence it is of interest to determine stabilizing controllers that can operate with as little inertia information as possible.

In the present paper the inertia-uncertainty problem is addressed by deriving an adaptive controller that provides asymptotic tracking of arbitrary angular velocity commands without mass distribution information. For a rotating spacecraft modeled as a rigid body, this controller is effectively a PI control law involving an integrator. The integrator state converges to the actual spacecraft inertia under sufficiently persistent excitation.

It is important to point out that angular velocity tracking does not provide a guarantee of inertial spacecraft attitude control. The inclusion of attitude states within an inertia-independent adaptive controller is given in Ahmed *et al.* (1998). The tracking problem considered in the present paper can be viewed as an extension of Ahmed *et al.*

[1] Research supported in part by the Air Force Office of Scientific Research under Grant F49620-98-1-0037 and the National Science Foundation under grant ECF 0140053

(1998) to the case in which attitude measurements are not available.

2. ADAPTIVE CONTROL OF A PLANAR ROTATING BODY

Consider a rigid body constrained to rotate about a fixed axis passing through its center of mass. For $t \geq 0$, the equation of motion is given by

$$\dot{\omega} = \frac{1}{J}\tau, \tag{1}$$

where ω is the angular velocity of the body about its rotation axis, J is the unknown moment of inertia of the body about its rotation axis, and τ is the applied torque.

Let $\nu : [0, \infty) \to \mathcal{R}$ denote the desired angular velocity of the body. Assume that ν is C^1. Defining the angular velocity error $\tilde{\omega}$ by $\tilde{\omega} \triangleq \omega - \nu$, it follows from (1) that $\tilde{\omega}$ satisfies

$$\dot{\tilde{\omega}} = -\dot{\nu} + \frac{1}{J}\tau. \tag{2}$$

The control objective is to determine τ such that $\tilde{\omega} \to 0$ as $t \to \infty$ for all initial conditions $\omega(0)$ and without knowledge of J. The following result provides an adaptive controller for angular velocity tracking based on an estimate $\hat{J}(t)$ of J.

Theorem 1. Assume that $\dot{\nu}$ is bounded, let $k > 0$ and $q > 0$, and consider the closed-loop system consisting of (2) and the adaptation control law

$$\dot{\hat{J}} = -q\dot{\nu}\tilde{\omega}, \tag{3}$$
$$\tau = -k\tilde{\omega} + \dot{\nu}\hat{J}. \tag{4}$$

Then $\tilde{\omega} \to 0$ as $t \to \infty$ for all $\omega(0)$ and $\hat{J}(0)$. Furthermore, $\hat{J}$ is bounded for all $t \geq 0$, and $\dot{\hat{J}} \to 0$ as $t \to \infty$.

Note that the control law given by (3) does not require knowledge of the inertia J of the body. Although $\dot{\hat{J}}$ converges to zero, $\hat{J}$ does not necessarily converge, and, if it does converge, it does not necessarily converge to the actual inertia J. The following result gives a sufficient condition under which $\hat{J}$ converges to J.

Proposition 1. Assume that ν is not constant and periodic. Then, under the control and adaptation law given by (3) and (4), $\hat{J} \to J$ as $t \to \infty$.

Proposition 2. Assume that $\dot{\nu}$ is periodic, and assume there exist $\alpha > 0$ and $T > 0$ such that, for all $t > 0$,

$$\int_t^{t+T} \dot{\nu}^2(s)ds \geq \alpha. \tag{5}$$

Then, under the control and adaptation law (3) and (4), $\tilde{J} \to 0$ as $t \to \infty$ exponentially.

To illustrate Theorem 1 and Proposition 1, consider the sinusoidal angular velocity command

$$\nu(t) = \frac{a}{\omega_0}(1 - \cos(\omega_0 t)),$$

where $a = 1.2$ rad^2/sec^2 and $\omega_0 = 1.0$ rad/sec. The inertia of the planar rotating body is taken to be $J = 1.0$ kg-m^2/rad^2, and its initial estimate is $\hat{J}(0) = 0.7$ kg-m^2/rad^2. The gains are $k = 4.8$ kg-m^2/rad^2-sec and $q = 2.8$ kg-m^2-sec^2/rad^4. The initial angular velocity error is given by $\tilde{\omega}(0) = 0.35$ rad/s. The angular velocity tracking error, inertia estimate error, and applied input torque are shown in Figure 1. It can be seen that $\tilde{\omega}$ converges to zero and $\hat{J}$ converges to J within 15 seconds. The torque is seen to have an initial transient with $\tau(0) = -1.68$ N-m/rad.

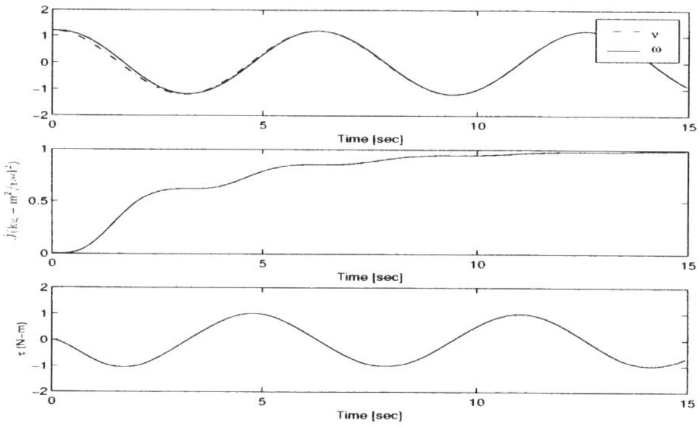

Fig. 1. Angular Velocities, Inertia Estimate and Control Torque for the Planar Rotating Body Tracking a Sinusoidal Angular Velocity Command

3. EXPERIMENTAL RESULTS

3.1 *Preliminary Analysis*

The adaptive control algorithm described in Section 2 was tested on the triaxial attitude control testbed. Experimental results are detailed and practical implementation issues touched upon in this section. As already mentioned we consider only yaw motion of the testbed with thrusters used for actuation.

First, we have

$$\omega = K_{\text{gyro}} V_{\text{gyro}}, \qquad \tau = K_{\text{fan}} V_{\text{fan}}, \tag{6}$$

where V_{gyro} is the voltage output of the gyro, K_{gyro} is the conversion coefficient from V_{gyro} in volts to ω in deg/sec, V_{fan} is the voltage input to the thruster amplifiers, and K_{fan} is the conversion coefficient from V_{fan} in volts to the control torque τ in N-m. From equation (1) we see that

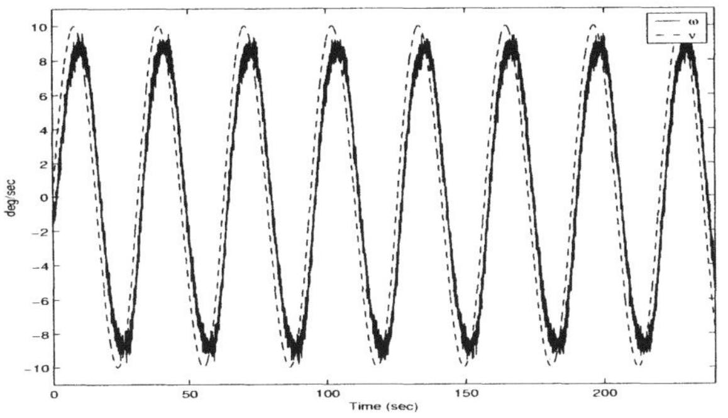

Fig. 2. Angular Velocities $\omega(t)$ and $\nu(t)$ Using the Proportional Controller

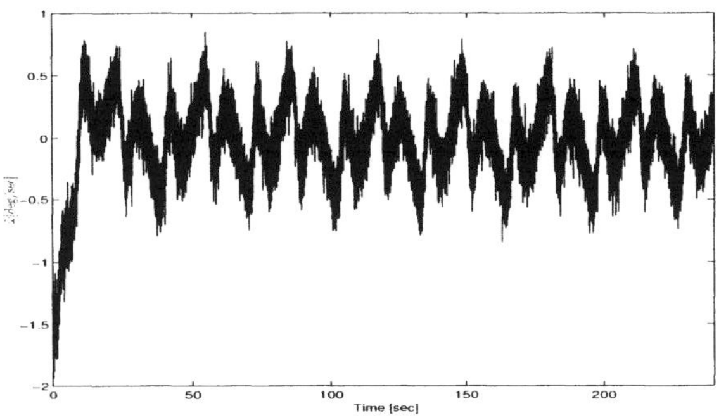

Fig. 3. Angular Velocity Tracking Error Using the Adaptive Controller

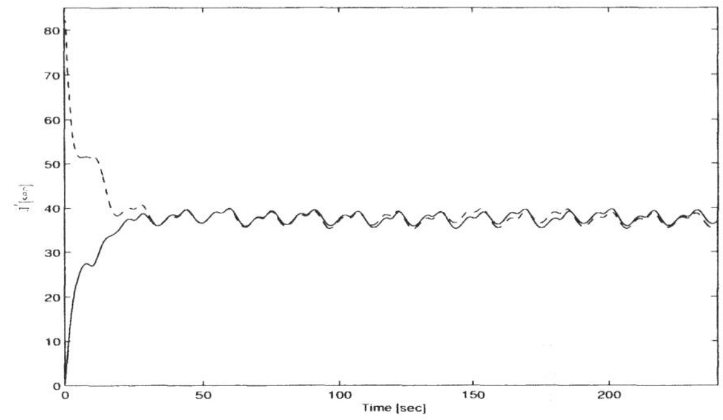

Fig. 4. Scaled Inertia Estimate Using the Adaptive Controller for $\hat{J}(0) = 0$ sec, and $\hat{J}(0) = 100$ sec

$\dot{V}_{\text{gyro}} = V_{\text{fan}}/J'$, where the scaled inertia $J' \triangleq JK_{\text{gyro}}/K_{\text{fan}}$. Note that the units of J' are sec. We define $V_{\text{ref}} \triangleq \nu/K_{\text{gyro}}$ and $\tilde{V} \triangleq V_{\text{gyro}} - V_{\text{ref}}$. We can thus rewrite (3) as

$$V_{\text{fan}} = -k'\tilde{V} + \dot{V}_{\text{ref}}\hat{J}', \tag{7}$$

where $k' \triangleq kK_{\text{gyro}}K_{\text{fan}}$ and $\hat{J}'$ is the estimate of J'. The adaptive law (4 can be written as

$$\dot{\hat{J}}' = -q'\dot{V}_{\text{ref}}\tilde{V}, \tag{8}$$

where $q' \triangleq qK^3_{\text{gyro}}/K_{\text{fan}}$. Comparing (4) and (5) with (7) and (8), it follows that the conversion coefficients are incorporated within the constants q' and k'. It can be seen that k' is dimensionless and q' has units of V/sec^2. Hence, we can apply the adaptive control algorithm of Section 2 without further calibration. However, to relate our results to physical motion, we calibrated the gyro voltage and found $K_{\text{gyro}} = 45.5$ deg/V-sec. For the remainder of this section we view V_{fan} as the control signal.

3.2 *Control Experiments*

When ν is constant, the adaptive controller specializes to the proportional controller

$$V_{\text{fan}} = -k'\tilde{V}. \tag{9}$$

Since the plant (1) is an integrator, the closed-loop system with the proportional controller (9) yields zero steady-state error for step commands.

The angular velocity $\omega(t)$ for the sinusoidal command $\nu(t) = 10\sin .2t$ and a proportional gain of $k' = 20$ sec/V^2 is shown in Figure 2. The angular velocity $\omega(t)$ converges to a periodic signal with rms value of about 2.6 deg/sec. Now we use the adaptive controller for the same command $\nu(t) = 10\sin .2t$ deg/sec. Figure 3 shows the angular velocity error for $k' = 3$, $q' = 1000$ sec/V^2, and initial scaled inertia estimate $\hat{J}'(0) = 0$ sec. The angular velocity error $\tilde{\omega}(t)$ shown in Figure 3 converges to a periodic signal with rms value of about .25 deg/sec and mean value of 0.0061 deg/sec. Figure 4 gives the scaled inertia estimates obtained with $\hat{J}'(0) = 100$ sec and $\hat{J}'(0) = 0$ sec. The scaled inertia estimate converges to a periodic signal with mean value about 37.9 sec and a peak-to-peak amplitude of about 4.5 sec. Simulations and experiments (not shown) indicate that the value of the inertia estimate varies with the frequency of the command signal.

3.3 *Actuator Nonlinearity*

The oscillation of $\hat{J}'$ suggests the possible presence of an actuator nonlinearity. We therefore plotted $\dot{V}_{\text{gyro}}$ V/sec (obtained by numerically differentiating the measured V_{gyro}) versus V_{fan} V as computed by the adaptive algorithm during an experiment. From this plot we found that $\dot{V}_{\text{gyro}}$ is a nonlinear function of the computed moment V_{fan}, that is,

$$\dot{V}_{\text{gyro}} = \mathcal{N}(V_{\text{fan}}). \tag{10}$$

The data was fit by the cubic polynomial

$$\mathcal{N}(x) = 0.1677x^3 + 0.0117x^2 + 0.6747x + 0.0078. \tag{11}$$

From (1), (10), (4) and (6), the moment input is given by

$$\tau = JK_{\text{gyro}}\mathcal{N}\left(\frac{k\tilde{\omega} + \dot{\nu}\hat{J}}{K_{\text{fan}}}\right). \tag{12}$$

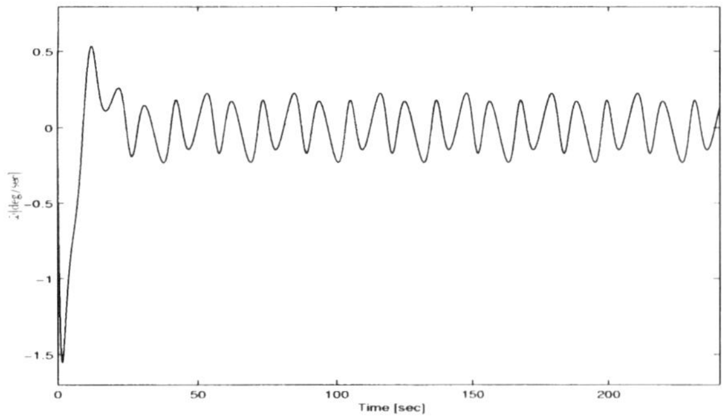

Fig. 5. Simulated Angular Velocity Error with Actuator Nonlinearity

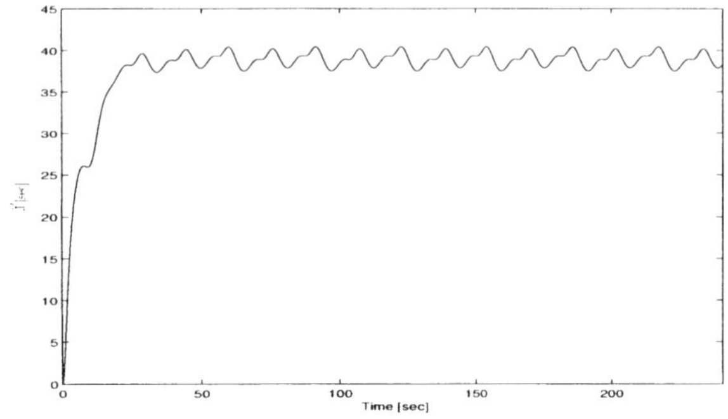

Fig. 6. Simulated Scaled Inertia Estimate with Actuator Nonlinearity

To check whether the nonlinearity (11) could cause the oscillations shown in Figure 4, the cubic nonlinearity (11) is included in a simulation of the adaptive closed-loop system for the command $\nu(t) = 10 \sin .2t$ deg/sec. From Figures 5 and 6 it is seen that this nonlinearity could indeed cause oscillations similar to those observed from the testbed. The angular velocity error $\tilde{\omega}$, shown in Figure 5, converges to a periodic signal with rms value of about .14 deg/sec and mean value of 0.0066 deg/sec. The scaled inertia estimate converges to a periodic signal with value of about 39 sec and a peak-to-peak amplitude of about 3 sec.

Noting the difficulty involved in inverting the cubic, we obtain a piecewise linear approximation of the cubic nonlinearity, and invert this piecewise linear function. The cubic nonlinearity and the inverse of the piecewise linear approximation are shown in Figure 7. The simulated response of the closed-loop system is shown in Figures 8 and 9. From these plots it can be seen that the inverse function approximately linearizes the nonlinearity and reduces oscillations in the angular velocity error and scaled inertia estimate. The rms value of $\tilde{\omega}$ is about .02 deg/sec and the mean value is 0.01 deg/sec, which is below the noise level of the gyro. The mean value of the scaled inertia estimate is about 40.5 sec with a peak-to-peak amplitude of about .9 sec. Furthermore, simulations show that the scaled inertia estimates converge to the same value for different values of the frequency of

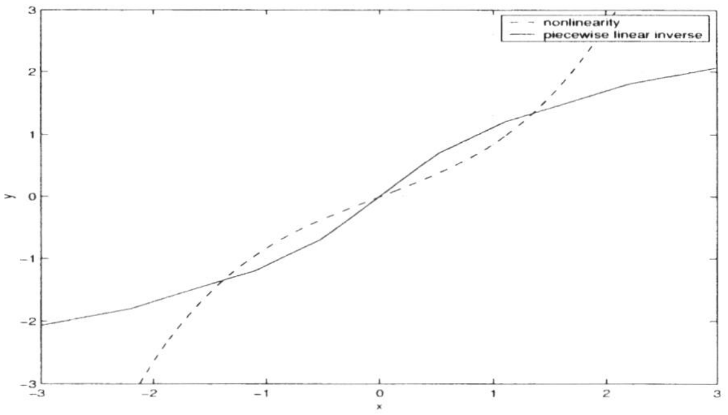

Fig. 7. Cubic Actuator Nonlinearity and an Approximate Piecewise Linear Inverse

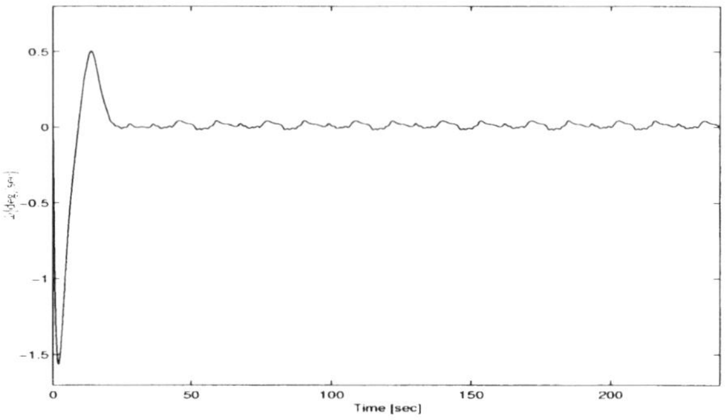

Fig. 8. Simulated Angular Velocity Error with Inverted Actuator Nonlinearity

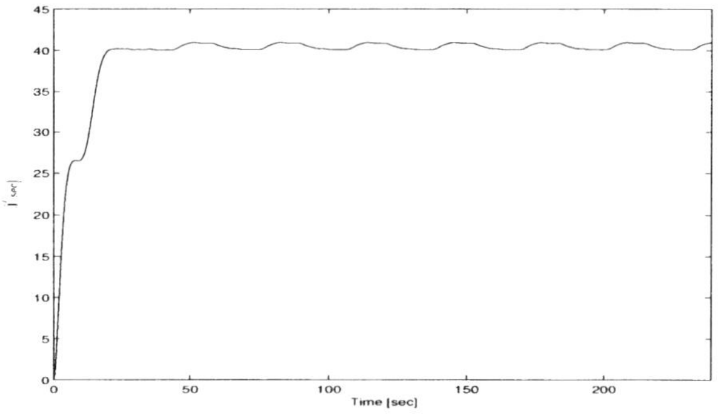

Fig. 9. Simulated Scaled Inertia Estimate with Inverted Actuator Nonlinearity

$\nu(t)$. The inverted actuator nonlinearity with the adaptive controller is implemented on the triaxial testbed. The results are shown in Figures 10 - 11. The rms value of $\tilde{\omega}$ is about .15 deg/sec and the mean value is 0.011 deg/sec. The mean value of the scaled inertia is about 39.6 sec and the peak-to-peak amplitude of oscillation is about 2.5 sec. Note that while oscillations in the angular velocity error and inertia estimates are reduced, they are not entirely eliminated. Sensor noise may account for some part of the oscillations seen in Figure 10.

3.4 *Real Inertia of the Triaxial Testbed*

To determine the actual inertia in kg-m^2/rad^2, test masses are added at known distances from the rotational axis. Let ΔJ, J' and J'' denote the change in inertia, the scaled inertia and the

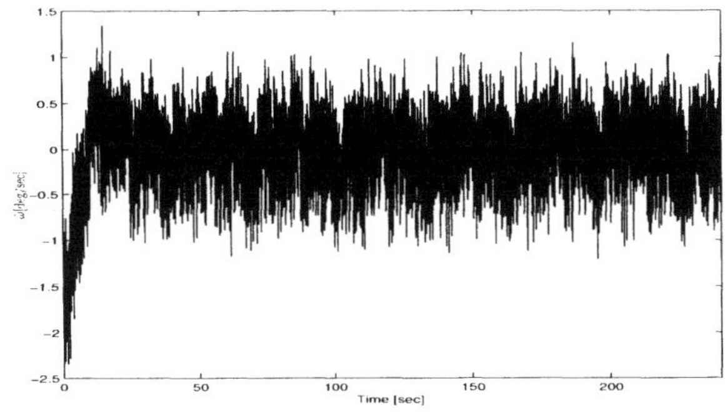

Fig. 10. Experimental Angular Velocity Error with Inverted Actuator Nonlinearity

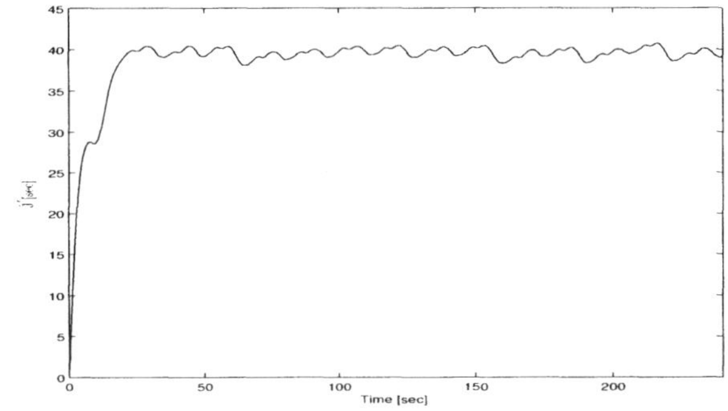

Fig. 11. Experimental Scaled Inertia Estimate with Inverted Actuator Nonlinearity

scaled inertia from an experiment with added test masses, respectively. Hence J'' can be written as

$$J'' = (J + \Delta J)\frac{K_{\text{gyro}}}{K_{\text{fan}}}, \qquad (13)$$

and the inertia in kg-m^2/rad^2 is given by

$$J = \frac{-J'\Delta J}{J' - J''}. \qquad (14)$$

A total mass of 5.11 kg is added to the two square mounting plates of the testbed, each at a distance of 0.75 m from the rotational axis. Hence, $\Delta J = 2.87$ kg-m^2/rad^2. Since the scaled inertia estimate $\hat{J}''$ is about 39.6 sec, the real moment of inertia is found to be $J = 66.6$ kg-m^2/rad^2 using equation (14).

4. APPROXIMATE FEEDBACK LINEARIZATION

4.1 *Preliminary Analysis*

The single degree of freedom spacecraft dynamics with nonlinear actuation is modeled by

$$\dot{\omega} = \frac{1}{J}h(u), \qquad (15)$$

where ω is the spacecraft angular velocity, u is the control signal, J is the moment of inertia, and h: $\mathcal{R} \to \mathcal{R}$ is an unknown one-to-one nonlinear input mapping. With these definitions, the torque applied to the spacecraft axis is $\tau = h(u)$.

Let $\hat{h} : \mathcal{R} \to \mathcal{R}$ be a one-to-one approximation of h with inverse $\hat{h}^{-1}$. For example, $\hat{h}$ may denote the inverted piecewise linear approximation of the cubic nonlinear fit of h as shown in Figure 7. An approximately linearizing feedback control law is then given by

$$u = \hat{h}^{-1}(v), \qquad \text{or} \qquad v = \hat{h}(u). \qquad (16)$$

Motivated by Hovakimyan *et al.* (2002) , we define $\Delta : \mathcal{R} \to \mathcal{R}$ by

$$\Delta(u) \triangleq h(u) - \hat{h}(u) \qquad (17)$$

so that $\Delta(\hat{h}^{-1}(v)) = h(\hat{h}^{-1}(v)) - v$ and $\tau = h(u) = \Delta(u) + v$. Then

$$\dot{\omega} = \frac{1}{J}h(\hat{h}^{-1}(v)) = \frac{1}{J}[v + \Delta(\hat{h}^{-1}(v))]. \qquad (18)$$

Now choose $v = v_a - v_c$ so that

$$\dot{\omega} = \frac{1}{J}[v_a + \Delta(\hat{h}^{-1}(v)) - v_c]. \qquad (19)$$

Here v_a is the torque specified by the adaptive algorithm (4),(5) for the system (1), and v_c is the torque used to cancel $\Delta(\hat{h}^{-1}(v))$.

4.2 *Adaptive Feedback Linearization Control*

Assumption 5.1 There exists a known function σ: $\mathcal{R} \to \mathcal{R}^l$ and an unknown vector $M \in \mathcal{R}^l$ such that, for all $v \in \mathcal{R}$,

$$\Delta(\hat{h}^{-1}(v)) = M^{\mathrm{T}}\sigma(v). \qquad (20)$$

Using (20), (18) can be written as

$$\dot{\omega} = \frac{1}{J}[v_a + M^{\mathrm{T}}\sigma(v) - v_c]. \qquad (21)$$

To approximately cancel $M^{\mathrm{T}}\sigma(v)$ in (19) we use an estimate of $M^{\mathrm{T}}\sigma(v)$ given by $v_c = \hat{M}^{\mathrm{T}}\sigma(v)$, where $\hat{M}$ is an estimate of M. Hence, from (21) we have

$$\dot{\omega} = \frac{1}{J}[v_a + (M - \hat{M})^{\mathrm{T}}\sigma(v)].$$

The vector $\hat{M}$ is updated according to the adaptation law

$$\dot{\hat{M}} = G\tilde{\omega}\sigma(v), \qquad (22)$$

where $G \in \mathcal{R}^{l\times l}$ is a positive-definite adaptation gain matrix.

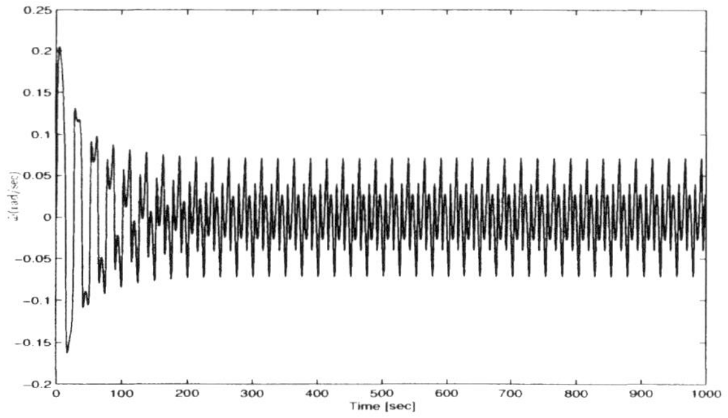

Fig. 12. Simulated Angular Velocity Error $\tilde{\omega}$ without Adaptive Feedback Linearization

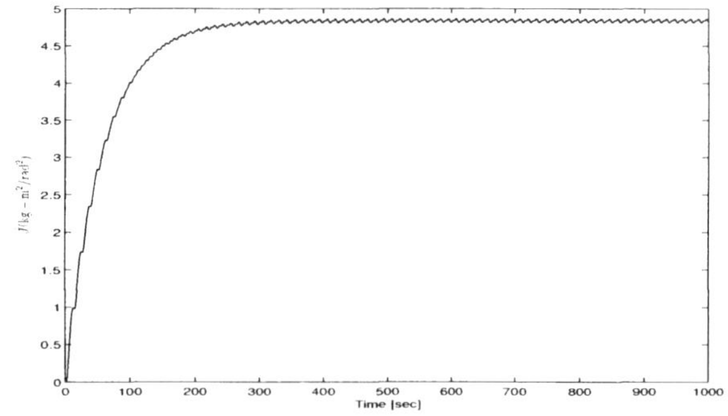

Fig. 13. Simulated Estimated Inertia $\hat{J}$ without Adaptive Feedback Linearization

4.3 *Stability Results*

Theorem 2. Assume that $\nu(t)$ is C^1, $\dot{\nu}$ is bounded, and $\sigma(v)$ is bounded for all bounded $v \in \mathcal{R}$. Let $k > 0$. Consider the system (15), the control (16) and the adaptation law (3), (4) and (22). Then $\tilde{J}$ and $\tilde{M}$ are bounded and $\tilde{\omega} \to 0$ as $t \to \infty$.

The following result presents a method for identifying the inertia J and the coefficients M.

Proposition 3. Consider the system (15). Assume that $k > 0$, $\nu(t)$ is C^1, $\nu(t)$ and $\dot{\nu}(t)$ are bounded, periodic with period T, but not constant, and $\sigma(v)$ is bounded for all $v \in \mathcal{R}$. Let $v(t)$ satisfy $v(t) + \bar{M}^T\sigma(v(t)) = \dot{\nu}(t)\bar{J}$, where $\bar{J} \in \mathcal{R}$ and $\bar{M} \in \mathcal{R}^l$. Furthermore, assume that, for all positive integers k, there exist $l+1$ time instants $t_1, t_2, \ldots, t_{l+1}$ satisfying $kT \le t_1 < t_2 < \ldots < t_{l+1} \le (k+1)T$ such that the $(l+1) \times (l+1)$ matrix

$$\begin{bmatrix} \dot{\nu}(t_1) & \dot{\nu}(t_2) & \cdots & \dot{\nu}(t_{l+1}) \\ -\sigma(v(t_1)) & -\sigma(v(t_2)) & \cdots & -\sigma(v(t_{l+1})) \end{bmatrix}$$

is nonsingular for all $\bar{J} \neq 0$ and all $\bar{M} \in \mathcal{R}^l$. Then, under the adaptive control law (3), (4) and (22), $\hat{J} \to J$ and $\hat{M} \to M$ as $t \to \infty$.

4.4 *Example and Simulation*

In this example, we assume $J = 1$ and $h(u) = 0.2u^3$, $\hat{h}(u) = u$ and we choose 11 spline func-

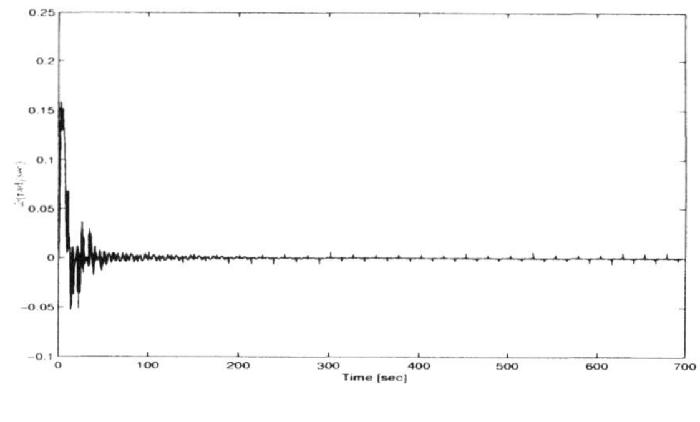

Fig. 14. Simulated Angular Velocity Error $\tilde{\omega}$ with Adaptive Feedback Linearization

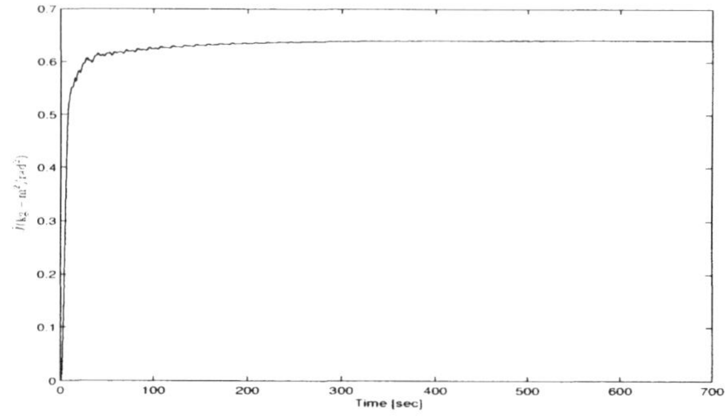

Fig. 15. Simulated Estimated Inertia $\hat{J}$ with Adaptive Feedback Linearization

tions for σ. The control signal is selected as $\nu(t) = \sin 0.25t$, and the control parameters in the adaptive control law are chosen as k = 1 kg-m^2/rad^2-sec, q = 4.8 kg-m^2-sec^2/rad^4, G = $80 I_{11\times 11}$ N^{-1}m^{-1}rad^{-1}.

The simulation results are shown in Figures 14-15. It is clear that $\tilde{\omega}$ converges to zero and $\tilde{J}$ is bounded as expected. For comparison, we show the angular velocity error $\tilde{\omega}$ and the estimated inertia $\hat{J}$ without the adaptive law in Figures 12 and 13. It can be seen that the adaptive law improves responses of the angular velocity tracking and inertia identification.

5. ACKNOWLEDGEMENTS

The last-named author wishes to thank Dr. Naira Hovakimyan for helpful discussions.

6. REFERENCES

Ahmed, J., V. T. Coppola and D. S. Bernstein (1998). Asymptotic tracking of spacecraft attitude motion with inertia matrix identification. *J. Guid. Contr. Dyn.* **21**, 684–691.

Hovakimyan, N., F. Nardi and A. Calise (2002). A novel error observer-based adaptive output feedback approach for control of uncertain systems. *IEEE Trans. Automatic Contr.* **47**, 1310–1314.

IFAC Publications
www.elsevier.com/locate/ifac

ON VISION-BASED KINEMATIC CALIBRATION OF n-LEG PARALLEL MECHANISMS

P. Renaud [1], N. Andreff [1], G. Gogu [1], P. Martinet [2]

[1] *Laboratoire de Recherches et Applications en Mécanique Avancée*
IFMA – Université Blaise Pascal, 63175 Aubière, France
[2] *LAboratoire des Sciences et Matériaux pour l'Electronique et d'Automatique*
Université Blaise Pascal – CNRS, 63175 Aubière, France

Abstract: A vision-based kinematic calibration algorithm is proposed for parallel mechanisms with end-effector connected to the base by *n* legs. The joint between corresponding leg ends can be a passive or actuated prismatic joint, which include constant-length legs. Information on the position and orientation of the mechanism legs is extracted from the observation of these elements with a standard camera. No workspace limitation nor installation of additional proprioceptive sensors are required. The algorithm is first detailed, then an evaluation of the method is achieved for a Stewart-Gough platform, with experimental measurement accuracy evaluation and simulation of the identification process. *Copyright © 2003 IFAC*

Keywords: robotics, parameter identification, physical parameters, nonlinear equations, computer vision.

1. INTRODUCTION

Compared to serial mechanisms, parallel structures exhibit a much better repeatability (Merlet, 1997), but not better accuracy (Wang and Masory, 1993). A kinematic calibration is thus also needed. Among the algorithms proposed to conduct calibration for these structures, methods based on the use of additional proprioceptive sensors on the passive joints are interesting, because they enable one to have a unique solution to the direct kinematic model (Tancredi, 1995), and then to use a criterion based on this model. An alternate way is to use the additional sensors on some legs to express a direct or inverse kinematic model as a function of the parameters of these legs and the redundant information. Calibration can then be achieved in a single process (Wampler and Arai, 1992, Zhuang, 1997) or in two steps (Daney, 2000). The main advantages of these methods are then the absence of workspace limitation and the analytical expression of the identification criterion. However, practically speaking, the design of the mechanism has to take into account the use of these sensors. Furthermore, for some mechanisms, the passive joints can not be equipped with additional sensors, for instance spherical joints. Consequently, the proposed method combines the advantage of information redundancy on the legs with non-contact measurements to perform the kinematic calibration.

The parallel mechanisms are designed with slim, often cylindrical, legs that link the end-effector and the base. The kinematic behavior of the mechanism is closely related to the movement of these legs. Hence the study of their geometry has already led to singularity analysis based on line geometry (Merlet, 1988). For such geometrical entities, the image obtained with a camera can be bound to their position and orientation with respect to the camera. By observing simultaneously several legs, it is then possible to get information on their relative position. Calibration can be achieved by deriving an identification algorithm adapted to this information. No workspace limitation is introduced, nor modification of the mechanism.

In this article, an algorithm is introduced for vision-based kinematic calibration of parallel mechanisms by observation of the mechanism legs. The method is developed in the context of mechanisms with n legs. The joints between corresponding leg ends can be passive or actuated prismatic joints, which include constant-length legs, and joints at leg ends may be revolute, spherical or universal joints. The method is composed of four steps: the first one consists of determining in the camera frame the parameters of the joints linked to the base. The second step enables one to estimate these parameters in the base frame, then in a third step the actuator encoder offsets are identified. Finally, in the fourth step, the parameters of the joints between the legs and the end-effector are identified.

The second section presents the mechanism modelling. The identification algorithm is then detailed in the third section, recalling first the relation between position and orientation of a cylindrical axis and its image projection. The four steps of the identification process are then detailed. In the fourth section, an evaluation of the proposed method is achieved by an experimental estimation of the measurement accuracy and a simulation of the identification of a Deltalab Stewart-Gough platform. Conclusions are then finally given on the performance and further developments of this method.

2. KINEMATIC MODELLING

The mechanism to identify is a parallel structure with n legs between the base and the end-effector (Fig. 1). The joint between corresponding leg ends can be a passive or actuated prismatic joint, which include constant-length legs. The desired end-effector pose is achieved by modifying the actuated leg lengths. The legs are considered connected to the base with revolute (R), spherical (S) or universal (U) joints. Many mechanisms have such a structure: 3-(RPR), 3-(UPU) (Fig. 1), 6-(SPU), ... According to the analysis achieved for the Stewart-Gough platform (Wang and Masory, 1993), these joints are supposed to be perfect. A revolute joint is then defined by its joint center and its axis direction. A universal joint is composed of two consecutive perpendicular revolute joints having a common intersection point. This joint is therefore defined by its joint center and the direction of the first rotation axis, linked either to the base or to the end-effector. Eventually, a spherical joint is defined by its joint center.
For manipulators, the controlled pose is the Euclidean rigid transformation between the world frame R_w and the tool frame R_t (Fig. 1). Noting R_b the frame defined by the joints between the legs and the base and R_e the frame defined by the joints between effector and legs, two transformations ${}^wT_b$ and ${}^eT_t$ can be defined between world and base frames and between end-effector and tool frames. These transformations are however dependent on the application and must be identified for each tool and relocation of the mechanism. Therefore the only considered kinematic parameters define relatively the joint parameters on the base and the end-effector, and the actuator encoder offsets. The transformations ${}^wT_b$ and ${}^eT_t$ can be identified by other techniques (Tsai and Lenz, 1989, Andreff *et al.*, 2001).

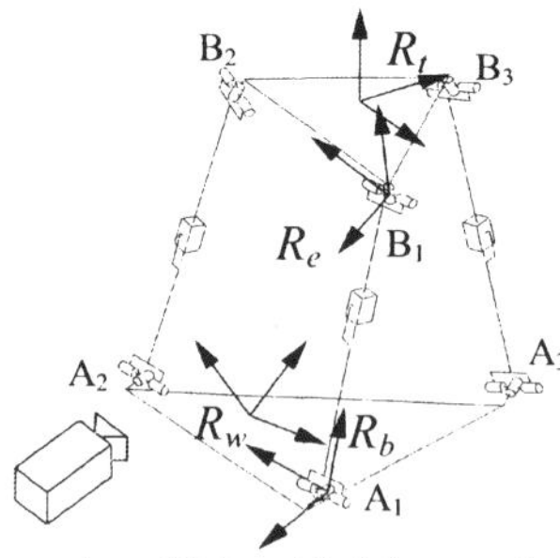

Fig. 1. Example of identifiable mechanism: 3-(UPU) parallel mechanism and the camera.

3. ALGORITHM

3.1 Vision-based Information Extraction

Projection of a cylinder In this paragraph, the relationship between the position and orientation of the legs of the mechanism, supposed to be cylindrical of known radius R, and their image is expressed. The image formation is represented by the pinhole model (Faugeras, 1993) and the camera is assumed to be calibrated. In such a context, a cylinder image is composed of two lines (Fig. 2), generally intersecting except if the cylinder axis is going through the center of projection. Each corresponding generating line D_i, $i \in [1,2]$ can be defined in the camera frame R_c $(\boldsymbol{C}, \boldsymbol{x_c}, \boldsymbol{y_c}, \boldsymbol{z_c})$ by its Plücker coordinates (Pottmann *et al.*, 1998) $(\underline{\boldsymbol{u}_i}, \boldsymbol{h}_i)$ with $\underline{\boldsymbol{u}_i}$ the unit axis direction vector and $\boldsymbol{h}_i$ defined by:

$$\boldsymbol{h}_i = \underline{\boldsymbol{u}_i} \times \boldsymbol{CP} \qquad (1)$$

where $\boldsymbol{P}$ is an arbitrary point of D_i, and $\times$ represents the vector cross product.

Each generating line image d_i can be defined by a triplet (a_i, b_i, c_i) such that this line is defined in the sensor frame R_s $(\boldsymbol{O}, \boldsymbol{x_s}, \boldsymbol{y_s})$ by the relationship:

$$\begin{cases} (a_i \quad b_i \quad c_i)(x \quad y \quad 1)^T = 0 \\ a_i^2 + b_i^2 + c_i^2 = 1 \end{cases} \qquad (2)$$

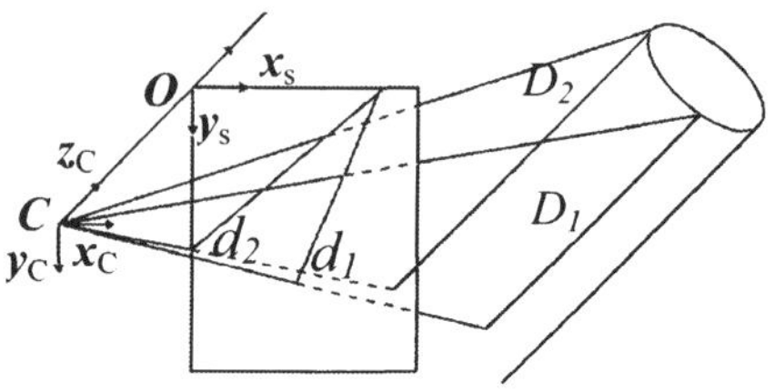

Fig. 2: Perspective projection of a cylinder and its outline in the sensor frame.

Due to perspective geometry, (a_i,b_i,c_i) and $\boldsymbol{h_i}$ are colinear. Provided that lines are oriented, one has:

$$(a_i,b_i,c_i)^T = \frac{\boldsymbol{h}_i}{\|\boldsymbol{h}_i\|} = \underline{\boldsymbol{h}}_i \qquad (3)$$

Determining the cylinder axis direction from the image Since the projection $(\underline{\boldsymbol{h}}_1, \underline{\boldsymbol{h}}_2)$ of the cylinder is now known, the cylinder axis direction $\underline{\boldsymbol{u}}$ can be computed by:

$$\underline{\boldsymbol{u}} = \frac{\underline{\boldsymbol{h}}_1 \times \underline{\boldsymbol{h}}_2}{\|\underline{\boldsymbol{h}}_1 \times \underline{\boldsymbol{h}}_2\|} \qquad (4)$$

Determining the cylinder axis position from the image Furthermore the distances between the cylinder axis and the generating lines are equal to the cylinder radius. Let $\boldsymbol{M}(x_M,y_M,z_M)$ be a point of the cylinder axis. As $\underline{\boldsymbol{h}}_i$ is computed as a unit vector, the belonging of $\boldsymbol{M}$ to the axis can be expressed by the two equations:

$$\underline{\boldsymbol{h}}_i^T \boldsymbol{M} = \varepsilon_i R \,, i \in [1,2] \qquad (5)$$

with $\varepsilon_1=\pm1$, $\varepsilon_2=-\varepsilon_1$. The determination of ε_1 is performed in the grayscale image by analyzing the position of the cylinder with respect to the generating line d_1. As the lines are chosen with the same orientation, ε_1 and ε_2 are of opposite sign.

It can be easily proved that the kernel dimension of $[\underline{\boldsymbol{h}}_1 \ \ \underline{\boldsymbol{h}}_2]^T$ is equal to one, by decomposing $\boldsymbol{M}$ on the orthogonal basis $(\underline{\boldsymbol{u}}, \underline{\boldsymbol{h}}_1, \underline{\boldsymbol{u}} \times \underline{\boldsymbol{h}}_1)$. The system (5) is therefore under-determined. The position of $\boldsymbol{M}$ can be computed in several ways, for instance by choosing a particular point as $\boldsymbol{M}_{LS}(x_{MLS}, 0, z_{MLS})$, under condition of its existence.

From the observation of one leg with a camera, it is thus possible to determine the position and orientation of its axis in the camera frame.

3.2 Joint Parameters Estimation in the Camera Frame

In this section, the relationships necessary to determine the joint parameters in the camera frame are derived. For each joint j, N_l images of the corresponding leg are stored for different end-effector poses, which enables one to compute in the camera frame the leg axis orientation $\underline{\boldsymbol{u}}_{j,k}$ and the leg axis point $\boldsymbol{M}_{j,k}$, $k \in [1, N_l]$.

Joint Center For spherical, universal and revolute joints, the position of the joint center $\boldsymbol{A}_j$ in R_c can be computed by expressing its belonging to the axis for the N_l poses:

$$\boldsymbol{A}_j\boldsymbol{M}_{j,k} \times \underline{\boldsymbol{u}}_{j,k} = \boldsymbol{0}, k \in [1, N_l] \qquad (6)$$

The joint center $\boldsymbol{A}_j$ is determined by solving the over-determined system obtained by concatenation of the $3T$ equations expressed in (6). As the three equations provided by the cross product are not independent, the solution is obtained by singular value decomposition. At least two different axis orientations are necessary to estimate the joint center.

Joint axis For a revolute joint, the joint axis $\underline{\boldsymbol{v}}_j$ is perpendicular to the leg axis orientation vectors $\underline{\boldsymbol{u}}_{j,k}$:

$$\underline{\boldsymbol{v}}_j \cdot \underline{\boldsymbol{u}}_{j,k} = 0 \,, k \in [1, N_l] \qquad (7)$$

The joint axis $\underline{\boldsymbol{v}}_j$ can be determined by solving the over-determined system obtained by concatenation of the N_l equations expressed in (7). At least two different leg orientations are necessary to estimate the parameters.

For a universal joint, if the first joint axis direction has an influence on the mechanism kinematics, its orientation will be computed in the fourth step (3.5).

3.3 Joint Parameters Estimation in the Base Frame

Identification criterion Because of leg visibility conditions, it may be necessary to move the camera around the mechanism. N_C different camera positions are therefore considered. The end-effector poses are not supposed to be identical for each camera position.

The base frame is defined using N_d joint centers (N_d=2 for a planar mechanism, N_d=3 for a spatial mechanism).

For a camera position defined by the camera frame $R_{c\alpha}$, n_α legs can be observed for any end-effector pose, which include r_α legs with revolute joints on the base. Let N_α the whole observable leg set, and R_α the set of observable legs connected to the base with revolute joints. The joint parameters have been computed in $R_{c\alpha}$ using (6-7). From the n_α joint centers on the base, $(n_\alpha - 1)$ independent vectors $\boldsymbol{A}_j\boldsymbol{A}_g$, $(j,g) \in N_\alpha$ can be computed.

Let $\boldsymbol{V}$ be the union of this vector set and the revolute joint axes:

$$\boldsymbol{V} = \{\boldsymbol{A}_j\boldsymbol{A}_g, (j,g) \in N_\alpha\} \cup \{\boldsymbol{v}_j, j \in R_\alpha\} \qquad (8)$$

With its elements, $C_{r_\alpha}^2 + C_{n_\alpha}^2 + r_\alpha(n_\alpha - 1)$ independent scalar products can be computed. Using the scalar product invariance with frame transformation, the joint parameters in the base frame are computed by non-linear minimization of the criterion C_1 :

$$C_1 = \sum_{\alpha=1}^{N_C} \sum_{p,q} \left[\boldsymbol{V}_p \cdot \boldsymbol{V}_q \big|_{R_{c\alpha}} - \boldsymbol{V}_p \cdot \boldsymbol{V}_q \big|_{R_b} \right]^2 \qquad (9)$$

with N_C the number of camera positions, $\boldsymbol{V}_j$ the j-th element of V and $\cdot|_R$ denoting the reference frame (R) in which the vectors V_j are expressed.

Identifiability conditions To perform joint parameters determination in the base frame, two conditions have to be fulfilled: Firstly, all the legs have to observed at least once. Secondly, the number of equations has to be greater or equal to the number of parameters to identify. For a planar mechanism, two joint centers define the base frame, and one parameter defines the plane perpendicular. The number N_P of parameters to identify is hence equal to:

$$N_P = 2 + 2(n - N_d) \quad (10)$$

In the same way, for a spatial mechanism:

$$N_P = 3 + 3(n - N_d) + 2r \quad (11)$$

with r the total number of revolute joints. The second identifiability condition is then:

$$\sum_{\alpha=1}^{N_C} \left(C_{r_\alpha}^2 + C_{n_\alpha}^2 + r_\alpha (n_\alpha - 1) \right) \geq N_P \quad (12)$$

At the end of this second step, the joint parameters in the base frame are determined, without any other assumption on the kinematics than the absence of joint clearance.

If the previously outlined identifiability conditions can not be fulfilled, the use of an additive calibration board linked to the base enables one to compute for each leg its position and orientation *w.r.t* the camera frame and, simultaneously, the pose of the camera *w.r.t* the calibration board (Dhome *et al.*, 1989). The gathering of the data for the different camera positions is then possible.

3.4 *Actuator Encoder Offsets Estimation*

Identification criterion In this third step, the actuator encoder offsets and the constant leg lengths are identified. For each successive camera frame $R_{c\alpha}$, the joint center positions A_j on the base and the axis orientations $\underline{u}_{j,k}$ are known. The position of the leg end $B_{j,k}$ can therefore be computed for the N_l poses as a function of only the offsets q_{0j} in the camera frame:

$$\boldsymbol{B}_{j,k\,R_{c\alpha}} = \boldsymbol{A}_{j\,R_{c\alpha}} + (q_{j,k} + q_{0j})\underline{\boldsymbol{u}}_{j,k\,R_{c\alpha}}, k \in [1, N_l] \quad (13)$$

Let n_{AC} the number of legs with actuated prismatic joints or constant-length legs, and AC_α the set of n_{AC_α} such legs observed for the camera position α. A number $C_{n_{AC\alpha}}^2$ of distances $\|\boldsymbol{B}_j\boldsymbol{B}_g\|$ can be expressed and, by comparing the value of these distances between two consecutive positions, an error function C_2 can then be expressed as a function of the n_{AC} offsets $q_{0j}, j \in [1, n_{AC}]$:

$$C_2 = \sum_{\alpha=1}^{N_C} \sum_{k=1}^{N_l - 1} \sum_{\substack{(j,g) \in AC_\alpha \\ g > j}} \left[\left\| \boldsymbol{B}_{j,k+1} \boldsymbol{B}_{g,k+1} \right\|_{R_{c\alpha}} - \left\| \boldsymbol{B}_{j,k} \boldsymbol{B}_{g,k} \right\|_{R_{c\alpha}} \right]^2 \quad (14)$$

with $\boldsymbol{B}_{j,k}$ the position of $\boldsymbol{B}_j$ for the k-th end-effector pose. The offsets are obtained by nonlinear optimization of C_2.

Notice that this includes the case of constant-length legs, where the actuator encoder value is equal to zero in the criterion C_2 and the joint offset is equal to the leg length.

Identifiability conditions The offsets identification can only be achieved if each leg can be observed with the camera for at least one camera position, which is already necessary in the previous step. The number of relationships has also to be greater or equal to the number n_{AC} of joint offsets:

$$\sum_{\alpha=1}^{N_C} \left(C_{n_{AC_\alpha}}^2 \right) \geq n_{AC} \quad (15)$$

3.5 *Joint Parameters Estimation in the End-Effector Frame*

Joint centers The determination of the joint offsets enables one to compute the average value of $\|\boldsymbol{B}_j\boldsymbol{B}_g\|$ for n_{AC} legs and therefore the relative position of the joints on the end-effector. Using the distance invariance with frame transformation the joint center positions in the end-effector frame can be identified by non-linear minimization of the criterion C_3:

$$C_3 = \sum_{\alpha=1}^{N_C} \sum_{(j,g) \in AC_\alpha, g > j} \left[\left\| \boldsymbol{B}_j \boldsymbol{B}_g \right\|_{R_{c\alpha}} - \left\| \boldsymbol{B}_j \boldsymbol{B}_g \right\|_{R_e} \right]^2 \quad (16)$$

To perform this joint center determination, two conditions have to be fulfilled. Each leg has to observed for at least one camera position. Furthermore, the number of equations has to be greater or equal to the number E of parameters, with $E=1+2(n-N_d)$ for a planar mechanism, and $E=3+3(n-N_d)$ for a spatial mechanism:

$$\sum_{\alpha=1}^{N_C} C_{n_{AC\alpha}}^2 \geq E \quad (17)$$

Revolute joint axes The transformation between the camera frame and the end-effector frame ${}^{R_C}T_{R_e} = ({}^{R_C}\boldsymbol{R}_{R_e}, {}^{R_C}\boldsymbol{t}_{R_e})$ can be computed from the positions $\boldsymbol{B}_{j,k}$ in these frames ($j \in N_{R_\alpha}, k \in [1, N_l]$):

$$\left({}^{R_C}\boldsymbol{R}_{R_e} \boldsymbol{B}_{j,k}\big|_{R_e} + {}^{R_C}\boldsymbol{t}_{R_e} \right) \times \boldsymbol{B}_{j,k}\big|_{R_C} = \boldsymbol{0} \quad (18)$$

The computation is achieved by solving the nonlinear system obtained by concatenation of the equation (18) for the observable leg set N_{R_α}. Three legs, including the one with a revolute joint on the end-effector, need to be observed for one camera position: $N_{R_\alpha} \geq 3$.

This enables one to express the leg axis orientation $\underline{\boldsymbol{u}}_{j,k}$ and the axis point $\boldsymbol{M}_k$ in the end-effector frame for each position k. The determination of the joint axis is then similar to the determination achieved in the second step (3.2). For a planar mechanism, the joint axis directions are already identified on the base.

Universal joint axes The computation of the transformation between camera and base frame ${}^{R_C}T_{R_b}$ is similar to the estimation of ${}^{R_C}T_{R_e}$ in the previous paragraph. Three legs have to be observed simultaneously. If so, the transformation between base and end-effector frames ${}^{R_b}T_{R_e}$ can then be estimated, and the use of the inverse kinematic model enables one to identify the universal joint axes.

Passive legs From the knowledge of ${}^{R_C}T_{R_e}$ solving equation (18), the position of a passive leg end on the end-effector can be expressed in the camera frame as a function of this transformation and its position in the end-effector frame. Three actuated legs need to be observed simultaneously. The belonging of the passive leg end to the axis can then be expressed by:

$$\underline{\boldsymbol{u}}_{j,k}\Big|_{R_C} \times \boldsymbol{A}_j\boldsymbol{B}_j\Big|_{R_C} = \boldsymbol{0}, k \in [1, N_I] \qquad (19)$$

The joint center is computed by solving the over-determined linear system obtained by concatenation of the equation (19). At least two different axis orientations are necessary.

4. METHOD EVALUATION

The proposed method is evaluated for the Deltalab Stewart-Gough platform (Fig. 3) as follows: First the calibration conditions are detailed. Then the measurement accuracy is experimentally evaluated, and the simulation of the identification process with the formerly evaluated measurement noise is achieved. To estimate the calibration method performance, analysis of the identified parameters and accuracy improvement is eventually conducted.

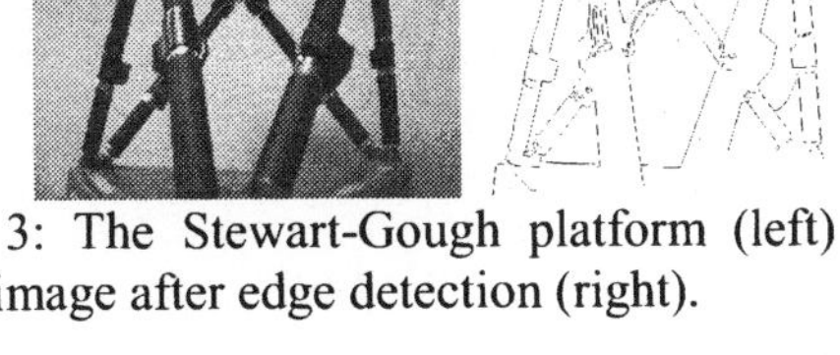

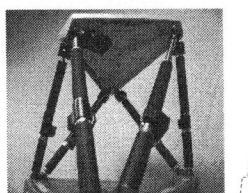

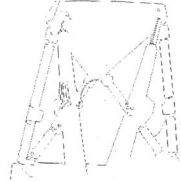

Fig. 3: The Stewart-Gough platform (left) and its image after edge detection (right).

4.1 Calibration Conditions

The structure is a 6-(SPU) parallel mechanism. The kinematic model is however not sensitive to the joint axis direction of the U-joints. Consequently, the only identified parameters are the joint locations on the base and the end-effector and the six actuator offsets: n=6, N_d=3. Because of the symmetry of the mechanism (Fig. 3), three different camera positions are considered (*i.e.* N_C=3). From (12), the simultaneous observation of four legs is then sufficient: n_α=4.

4.2 Measurement Accuracy

Six equally-spaced leg orientations are considered within the extremal values. The measurement accuracy is evaluated from a set of consecutive measurements, for each leg position.

A *1024×768* camera with a *6mm* lens is used to acquire the images, connected to a PC via an IEEE1394 bus. Cylinder outline detection is achieved by means of a Canny filter (Canny, 1986) (Fig. 3). Lines are then computed from the detected points by a least-squares method.

In Table 1, the upper-bound of the estimated standard deviations of the cylinder position and orientation are listed. The orientation is described with the Euler angles (ψ,θ). The position is obtained by estimating the point $\boldsymbol{M}_{LS}(x_{MLS}, 0, z_{MLS})$.

Table 1 Upper-bound of the standard deviations

Parameter	ψ	θ	x_{MLS}	z_{MLS}
Est. st. dev.	0.05rad	0.06rad	0.05mm	0.1mm

It must be stated that the image processing could be improved by the use of a subpixel detection filter (Steger, 1997) and now available higher CCD resolution sensor, since the accuracy is intrinsically bound to this resolution.

4.3 Simulation

Performance Evaluation Simulation allows one to evaluate directly the knowledge improvement of the kinematic parameter values. Let ξ_{gt_i} be the ground-truth value of the i-th kinematic parameter ($i \in [1,30]$), and ξ_{id_i} its identified value. The calibration gain can then be computed by the estimation error $\left|\xi_{id_i} - \xi_{gt_i}\right|$.

In order to evaluate the influence of a parameter estimation error, the displacement error ΔX and the orientation error ΔE are computed for ten randomly chosen poses:

$$\begin{cases} \Delta X = \left\| \boldsymbol{A_1 B_1}_{\xi_{gt}} - \boldsymbol{A_1 B_1}_{\xi_{id}} \right\| \\ \Delta E = \left\| \Delta\psi \quad \Delta\theta \quad \Delta\varphi \right\| \end{cases} \qquad (20)$$

where ($\Delta\psi$, $\Delta\theta$, $\Delta\varphi$) are the Euler angles defining the difference between the end-effector orientation computed with the kinematic parameter sets ξ_{gt} and the one computed with ξ_{id}.

Simulation Process Fifteen end-effector poses are generated by randomly selecting configurations with extreme leg lengths. These leg lengths are corrupted with noise to simulate proprioceptive sensor measurements (uniformly distributed noise, variance equal to *3μm*). The leg orientation and axis points $\boldsymbol{M}_i$ are modified by addition of white noise with standard deviation equal to those previously estimated in Table 1. For each end-effector pose, three images are acquired with the camera to reduce the measurement noise. Initial kinematic parameter values are obtained by addition to the model values of a uniform noise with variance equal to 2mm. The base and end-effector frames are defined using joint centers 1, 3 and 5.

Results Figure 4 represents the ground-truth parameter values and the mean estimation errors $Mean\left|\xi_{id_i} - \xi_{gt_i}\right|$, computed by 100 simulations of the calibration. A sharp improvement of the knowledge of the kinematic parameters is observed, except for the *z* component of the joint locations. The parameter estimation errors are however low with an average error between *0.04mm* and *1mm*.

It must also be underlined that the accuracy improvement is significant with an average displacement error reduced from 1mm for the initial kinematic parameters to 0.08mm, and an orientation error reduced from 0.12rad to 0.018rad.

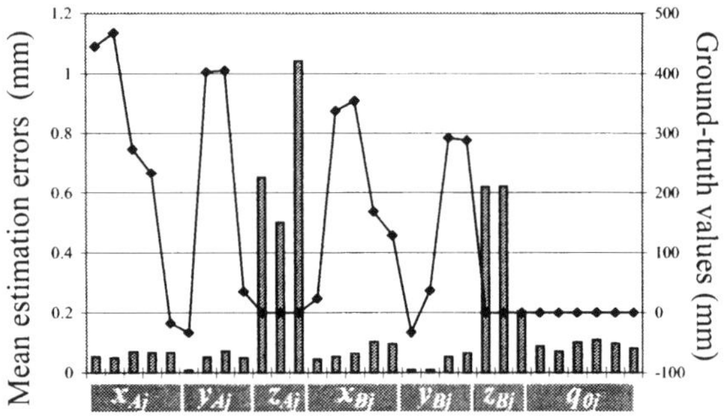

Fig. 4: Mean estimation errors (bars) and ground-truth values (line) of the thirty parameters.

5. CONCLUSION

In this article, a vision-based calibration method for mechanisms with *n* legs between the base and the end-effector has been proposed. Using an exteroceptive sensor, the kinematic parameters of the structure are identified. No mechanical constraint nor additional proprioceptive sensor are required. The method is low-cost as standard off-the-shelf cameras are used. The identification criteria and identifiability conditions have been derived. The experimental evaluation of the measurement accuracy and the simulation results show a significant accuracy improvement for a Deltalab Stewart-Gough platform. The algorithm performance can be improved by using more accurate detection algorithms, and a better selection of the end-effector poses for calibration, which will soon be implemented. The method will be also validated for other mechanisms.

ACKNOWLEDGEMENT

This study was jointly funded by CPER Auvergne 2001-2003 and by the CNRS-ROBEA program through the MAX project.

REFERENCES

Andreff N., Horaud R. and Espiau B. (2001). Robot hand-eye calibration using structure-from-motion, *Int. J. Rob. Research*, **20**(3),pp. 228-248.

Canny J.F. (1986). A computational approach to edge detection, *IEEE Trans. Pattern Analysis and Machine Intelligence*, **8**(6), pp.679-698.

Daney D. (2000). Etalonnage géométrique des robots parallèles. PhD Thesis, Université de Nice.

Dhome M., Richetin M., Lapreste J.T. and Rives G. (1989). Determination of the Attitude of 3-D Objects From A Single Perspective View, *IEEE Trans on Pattern Analysis and Machine Intelligence*, **11**(12), pp.1265-1278.

Faugeras O. (1993). *Three-dimensional Computer Vision: A Geometric Viewpoint*, The MIT Press.

Merlet J.P. (1988). Parallel manipulators, Part 2, Singular Configurations and Grassmann geometry, *Res. Report RR-0791T*, INRIA.

Merlet J.P. (1997). *Les Robots Parallèles*, Hermès.

Pottmann H., Peternell M. and Ravani B. (1998). Approximation in line space – applications in robot kinematics and surface reconstruction, In: *Advances in Robot Kinematics: Analysis and Control,* pp. 403-412, Strobl.

Steger C. (1997). Removing the Bias from Line Detection, In: *Computer Vision and Pattern Recognition '97*, pp. 116-122, Puerto Rico.

Tancredi L. (1995). De la simplification et la résolution du modèle géométrique direct des robots parallèles, PhD Thesis, Ecoles des Mines de Paris.

Tsai R.Y. and Lenz R.K. (1989). A new technique for fully autonomous and efficient 3D robotics hand/eye calibration. *IEEE Trans. On Rob. and Automation*, **5**(3), pp. 345-358.

Wampler C. and Arai T. (1992). Calibration of Robots Having Kinematic Closed Loops Using Non-Linear Least-Squares Estimation, In: *Proc. IFToMM-jc Int. Symp. On Theory of Machine and Mechanisms*, pp.153-158, Nagoya.

Wang J. and Masory O. (1993). On the Accuracy of a Stewart Platform – Part I: The Effect of Manufacturing Tolerances, In: *Proc. of ICRA*, pp.114-120, Atlanta.

Zhuang H. (1997). Self Calibration of Parallel Mechanisms With a Case Study on Stewart Platforms, *IEEE Trans. On Robotics and Automation*, **13**(3):387-397.

IFAC Publications
www.elsevier.com/locate/ifac

A GEOMETRIC APPROACH TO MOTION TRACKING IN MANIFOLDS[1]

J. G. Silva*, J. S. Marques†, J. M. Lemos‡

*ISEL, R. Conselheiro Emídio Navarro, 1949-019 Lisboa, Portugal
jgs@isel.ipl.pt

†IST/ISR, Av. Rovisco Pais 1949-001, Lisboa, Portugal
jsm@isr.ist.utl.pt

‡INESC-ID/IST, R. Alves Redol, 9, 1000-029 Lisboa, Portugal
jlml@inesc.pt

Abstract: In many multi-dimensional tracking problems, the quantities of interest are restricted to a manifold in observation space. Learning the manifold shape is a necessary step for dimensionality reduction, which in turn allows faster and more robust tracking performance. For manifolds with arbitrary topology, learning the shape from noisy scattered data is not trivial. This paper presents a geometric approach that is valid for arbitrary manifold dimension and topology. An approximation of the tangent bundle is computed by region growing, making it possible to estimate a set of manifold charts. A tracking algorithm which takes advantage of the geometric information thus found is also presented. *Copyright © 2003 IFAC*

Keywords: Multi-dimensional, Modeling, Target tracking.

1. INTRODUCTION

Many tracking problems in control and computer vision, involving observations in high-dimensional spaces, can have their complexity reduced by incorporating geometric information in the tracking approach.

Assuming the observed trajectory to lie in a manifold, embedded in a high dimensional space, and to be corrupted by noise, the dimensionality reduction provided by using inherent geometric constraints is expected to result in higher robustness and smaller computational load (Marques *et al.*, 1999). This paper presents a tracking algorithm which makes use of such geometric information. It is an extension of previous work on 1-D manifolds (Silva *et al.*, 2000).

The algorithm described here is exemplified for a 2-torus in 3-D space, but it is valid for other topologies and higher dimensions, both of the manifold and the embedding space. The manifold is assumed to be orientable and compact, that is, to have a unit normal vector field and to allow covering by a finite number of charts. It may or not have a boundary.

The tangent bundle of the manifold is the set of tangent hyperplanes at all manifold points (O'Neill 1997). There are infinitely many such points and tangent hyperplanes.

The main, novel idea contained in the present work is to use a region growing approach in the non-trivial task of approximating the tangent bundle of the manifold by a finite number of tangent hyperplanes, when the manifold has arbitrary topology. This simplifies estimation of the charts.

An overview of the algorithm is given in the next section. The following sections describe the major steps in detail and present experimental results, as well as conclusions.

2. OVERVIEW

Beginning with a set of scattered, noisy points in observation space, used as a training set, it is

[1] This work was partially supported by FCT POCTI, under project 37844.

intended to track trajectories that evolve in the same manifold as the training set.

An example training set is shown in Figure 1, where several noisy observations lie on a torus.

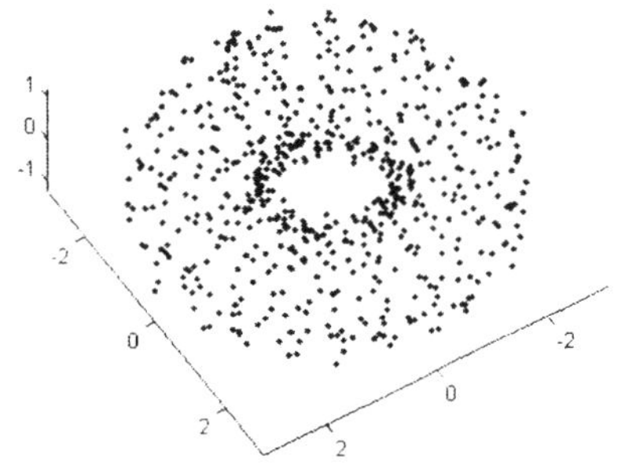

Fig. 1. Example training set: noisy observations **y** on a torus.

The steps involved in the algorithm are the following:

- Approximating the tangent bundle of the manifold;
- Finding local parametrizations and charts;
- Performing tracking in the lower dimensional parametric domains.

Each of these steps is addressed in detail in the following sections.

The manifold M is modelled through a set of diffeomorphisms $g_i : U_i \subset R^n \rightarrow R^m$, where n is the manifold dimension, m is the dimension of the embedding space and the U_i are open sets. These functions are *charts* of the manifold. Collectively, their overlapping images $g_i(U_i)$, also called *patches*, cover the manifold. Being diffeomorphisms, the charts also admit inverses, g_i^{-1}, which are called *parametrizations*.

First, the unit normal vector field of the whole manifold is estimated by computing, for each data point, the smallest eigenvectors of the local covariance matrix.

Next, the set of overlapping patches is found. This is done by region growing, until all data points belong to at least one patch. Each patch grows by appending all neighbouring points where the normal vector field does not deviate, in angle, more than a set threshold from the normal at the initial seed. In general, a given data point may belong to more than one patch. This yields local coordinate systems, by using using Principal Component Analysis (PCA) to find the best fitting hyperplane for each patch.

Charts are then estimated by thin-plate spline approximation, using the coordinate systems found above. Since the normal was only allowed to change direction up to a specified angular limit, it is guaranteed that the charts are bijective. They are also differentiable, because of thin-plate spline properties, so they are diffeomorphisms, as intended.

Tracking is then performed in the chart domains. The estimated trajectories can be lifted back to the embedding observation space by using the charts.

3. TANGENT BUNDLE APPROXIMATION

In this step, hyperplanes are found from scattered data points $\mathbf{y}_i=(y^1,...,y^m)$, allowing locally flat descriptions of the manifold.

When flattening a manifold with arbitrary topology, it is, in general, necessary to partition the manifold into more than one patch to avoid the so-called cartographer's dillema, which is due to metric distortion. Note that in some situations, such as the torus example, which has zero Gaussian curvature, one single patch would be theoretically enough. However, the present algorithm is intended for a broader class of problems.

Each patch can be associated to an hyperplane, and the collection of hyperplanes will be an approximation to the tangent bundle. The hyperplanes provide local coordinate systems valid in different regions. The best fitting hyperplane for each patch, in a least squares sense, is spanned by the the n largest eigenvectors returned by the PCA procedure, performed on all patch member points.

In order to make chart estimation easier, at a later stage, it is required for simple projection to give a one-to-one mapping between the hyperplane and the corresponding manifold region. This can be ensured by not allowing the hypersurface normal to vary more than a set threshold τ, in angle. It is thus necessary to compute the normals.

Normals are computed by visiting all data points and, for each one, finding the m-n smallest eigenvectors (smaller than a threshold) of the covariance in a neighbourhood of radius ε, as shown in Figure 2. It is assumed that the data are sufficiently dense to leave no gaps greater than δ and observation noise has standard deviation nowhere greater than σ. The radius ε is chosen to account for both δ and σ.

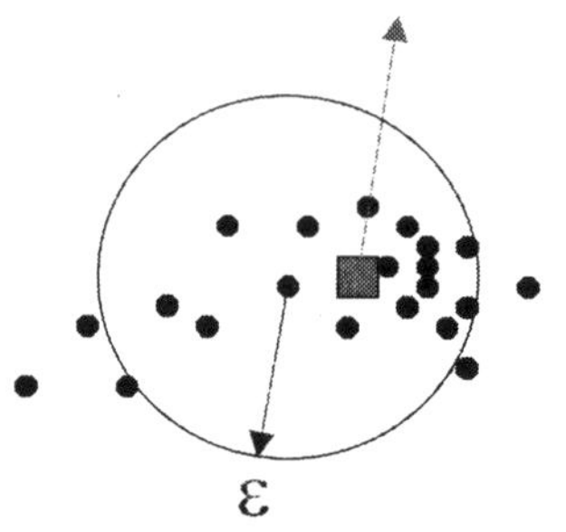

Fig. 2. Surface normal from PCA in a neighbourhood.

Since the data points are noisy, the point of application of the normal is set to the mean of the neighbourhood, which is a form of low-pass filtering.

The eigenvectors thus returned have arbitrary orientations, so it would be necessary ro revisit all data points to maintain consistency between nearby normals, which is a NP-complete problem (Hoppe 1994). However, since it is just necessary to compute angles, direction is all that matters and orientation can be discarded. After normals are found, region growing takes place as following:

```
while M not covered
    P = new patch
    y0 = choose a new seed from data points not in
          any patch
    n0 = normal at y0
    while NOT all points visited
       y1 = choose nearest neighbor
       n1 = normal at y1
       if angle(n0,n1) < τ AND
                              distance(y0,y1) < ε
          append y1 to P
       end if
    end while
end while
```

The end result is a covering of M by a finite number, p, of overlapping patches. Within each patch, the normal doesn't deviate more than τ, and the distance test ensures that each patch is a connected set.

4. CHARTS AND PARAMETRIZATION

It is desirable that the manifold charts be diffeomorphisms, so that smoothness is ensured. There are many alternatives for non-linear function approximation that meet these requirements. The results presented here were obtained using thin-plate splines, which are described in (Duchamp 2002).

In short, a thin-plate spline g is the function that minimizes the weighted sum

$$\rho E(\tilde{g}) + (1-\rho) R(\tilde{g}) \qquad (1)$$

where ρ is a smoothing parameter and

$$E(\tilde{g}) = \sum_j (y_j - \tilde{g}(x_j))^2 \qquad (2)$$

$$R(\tilde{g}) = \int_{R^n} tr(H_g^2)\, dx_1 \ldots dx_n \qquad (3)$$

denote an error measure and a roughness measure, respectively. H_g is the Hessian of $\tilde{g}$. The l-degree solution comes in the form

$$\tilde{g}(x) = p(x) + \sum_j^{l-k} a_j \Psi(x_j - c_j) \qquad (4)$$

where $p(x)$ is a polynomial term involving k of the a_j coefficients, while the remaining l-k coefficients multiply the radial basis functions ψ, which are given by

$$\Psi(x_j - c_j) = \left\|x_j - c_j\right\|^2 \log\left(\left\|x_j - c_j\right\|^2\right), \qquad (5)$$

with centres c_j.

Having previously partitioned the manifold M in p patches, it is possible to find p charts $g_i(x)$ in the form given by (4), followed by a translation and a rotation.

Figure 3 shows some of the charts estimated in this fashion.

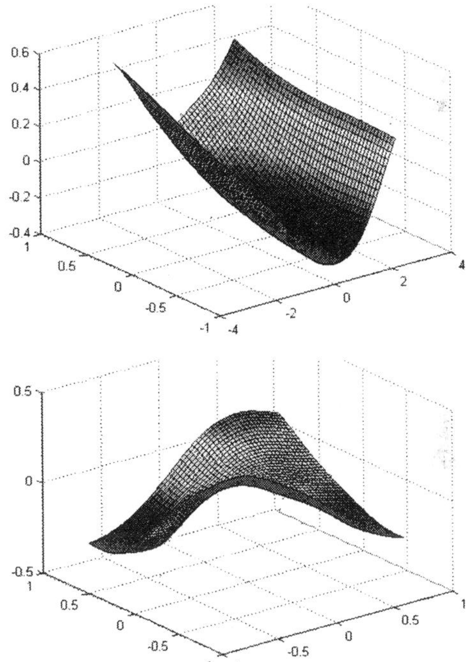

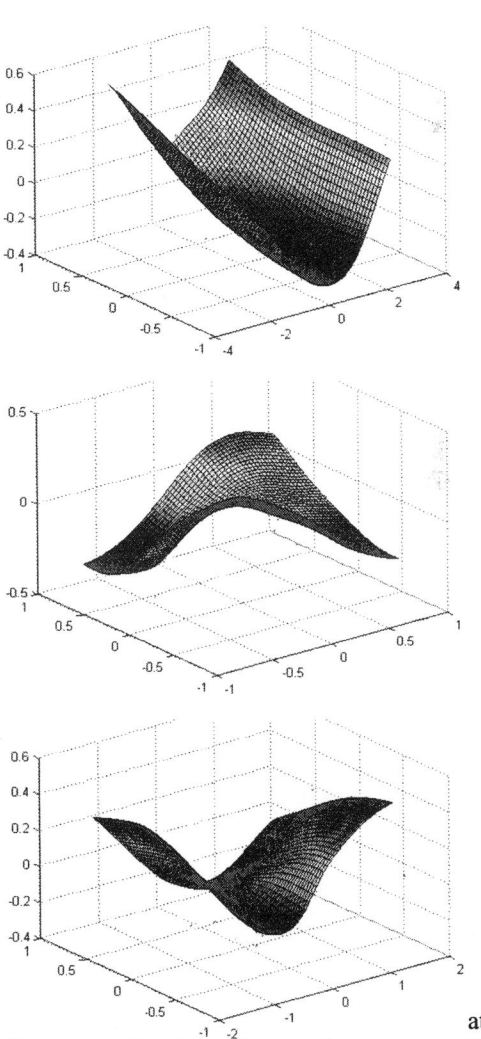

Fig. ate spline approximation. The graphs represent the functions, $\tilde{g}(x)$, which is this case are scalar, since they were obtained for a 2-torus embedded in 3-D space.

To summarize, with points $y=(y_1,\ldots,y_m)$ belonging to a given patch i, and having previously performed

PCA, a matrix V_i of eigenvectors and a mean vector μ_i are available. Projecting y on the hyperplane associated with patch i is a matter of computing

$$\tilde{x} = V_i^T (y - \mu_i) \qquad (6)$$

$$x = \begin{bmatrix} \tilde{x}_1 \\ \vdots \\ \tilde{x}_n \end{bmatrix}. \qquad (7)$$

Equation (6) is an isometry, in this case, a translation followed by a rotation, while (7) describes simple projection.

As for the the chart g_i, which allows the inverse mapping of (7), it follows the expression

$$g_i(x) = V_i \begin{bmatrix} x_1 \\ \vdots \\ x_n \\ \tilde{g}(x) \end{bmatrix} + \mu_i \qquad (8)$$

The remaining m-n components of $\tilde{x}$, instead of being set to zero, which would yield a rough, piecewise linear approximation of M, are thus preserved. Locally, the manifold parametrization is

$$x \rightarrow \begin{bmatrix} x & \tilde{g}(x) \end{bmatrix}^T \qquad (9)$$

with $\tilde{g}$ given by (4).

5. TRACKING

At this stage, an observed trajectory can be projected onto the previously found hyperplanes, and all tracking can be done in n-D instead of m-D.

In order to track the projected trajectories, dynamic models are needed. A linear, discrete state model is used that includes dynamic and observation noise, both assumed Gaussian for simplicity.

It must be stressed that, while the original, m-D observations y_k are non-linear functions of x_k, it is assumed that x itself is directly observed, by projection of y, so it is possible to use a linear observation model.

A time-varying Kalman filter is used to estimate the trajectory from the noisy observations, and lifting back to the manifold with charts g allows recovery of the m-D trajectory.

Motion is descibed by the following stochastic difference equation:

$$s_{k+1} = A s_k + B w_k \qquad (10)$$

where k denotes time, s is the state vector and the scalar w is dynamic noise. The observation equation is

$$o_k = C s_k + D n_k \qquad (11)$$

where o is the observation vector (made equal to x_k) and n is sensor noise. A white-noise acceleration model (Kalata, 1984) is followed, and the same dynamics are assumed for all patches. The A, B, C and D matrices are, therefore, constant.

Finally, since the $\mathbf{y}_k$ are projected onto p different hyperplanes, there are p different observations at time k. The problem of how to select the best hyperplane is solved by nearest neighbour classification. The training data point nearest to y_k is found (a computationally expensive procedure) and, among the patches it may belong to, the one that yields the least squared reconstruction error, that is

$$e_k = \| y_k - g_i(x_k) \|^2, \qquad (12)$$

is selected.

6. RESULTS

For the experiment described in this section, the purpose is two-fold: it is intended to approximate a synthetically generated manifold, and also to recover an experimental trajectory, generated on the same manifold and deliberately corrupted with additive Gaussian noise.

The synthetic manifold is, in this case, a 2-torus. The learning dataset is a collection of points that lie on the surface as seen above, in Figure 1.

The results from the approximation step can also be seen above, in Figure 3. At this stage, the charts become available.

The next step in the experiment is motion recovery. A synthetic trajectory was generated, independently from the training set, according to a simple dynamic behaviour:

$$\begin{bmatrix} \theta_{k+1} \\ \varphi_{k+1} \end{bmatrix} = \begin{bmatrix} \theta_k \\ \varphi_k \end{bmatrix} + u_k + w_k, \qquad (13)$$

including a constant velocity u_k term and Gaussian dynamic noise w_k. The trajectory, as illustrated in Figure 4, is a slightly noisy straight line in (θ, φ) space, which wraps around at angles $-\pi$ and π.

Through use of toroidal coordinates

$$
\begin{aligned}
r &= 2 + \cos(\varphi) \\
x &= r\cos(\theta) \\
y &= r\sin(\theta) \\
z &= \sin(\varphi)
\end{aligned}
\quad , \qquad (14)
$$

the synthetic trajectory was transported to the manifold, and Gaussian observation noise was finally also added. The time sequence of resulting data points, **y**=(x, y, z), constitutes the { $\boldsymbol{y}_1, \ldots, \boldsymbol{y}_k, \ldots$} experimental trajectory.

Tracking was performed after projection of the observations on the appropriate planes, learned during tangent bundle approximation. The resulting projected motion is shown in Figure 5.

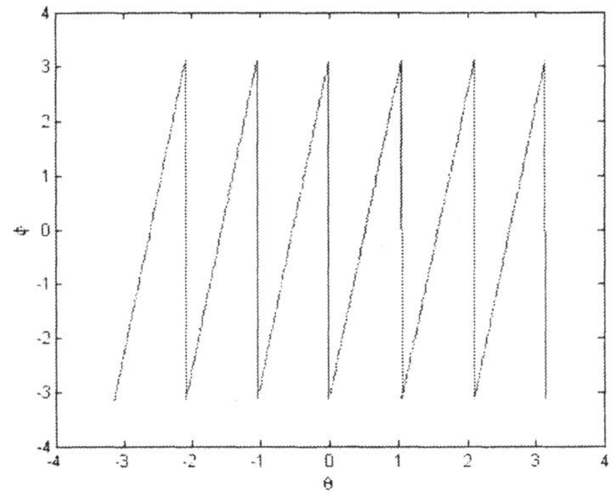

Fig. 4. Synthetic trajectory in (θ, φ), - π < θ, φ < π. The φ period is six times shorter than the θ period.

After lifting to observation space (using the charts), the accurately estimated trajectory can be seen in Figures 6 and 7, overlayed on the experimental one.

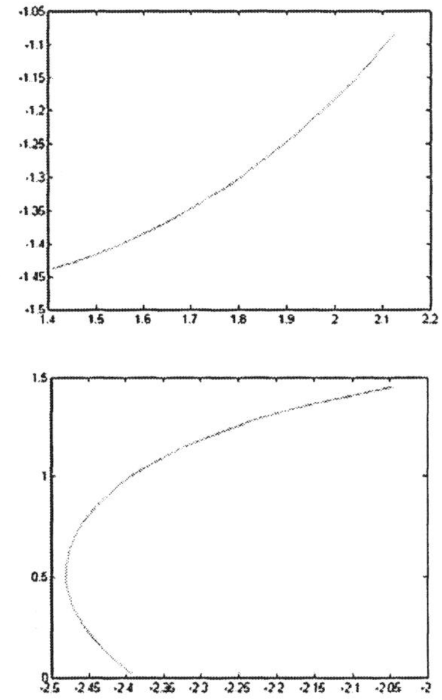

Fig. 5. Parts of the experimental trajectory, projected into some of the estimated planes.

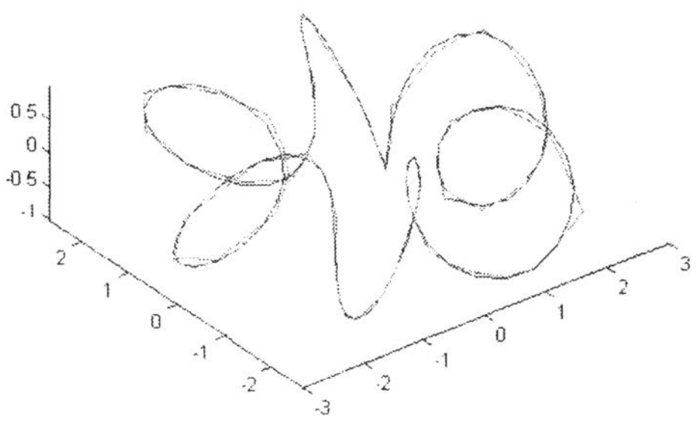

Fig. 6. Tracking results in the observation space. The experimental and estimated trajectories are blue and red, respectively.

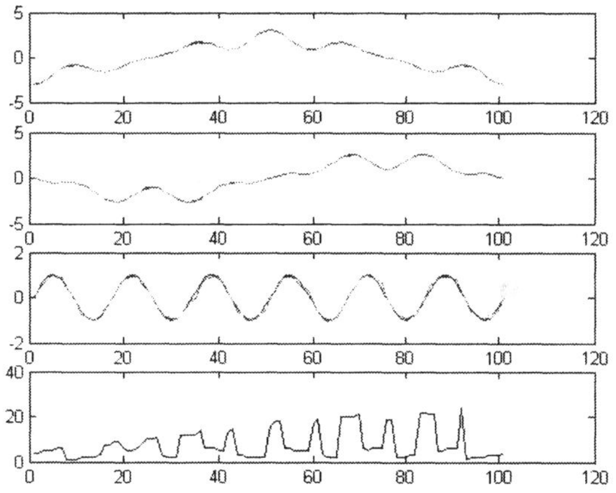

Fig. 7. Time plots of the (*x,y,z*) experimental trajectory coordinates (blue) and the estimated trajectory coordinates (red). Below, the index of the nearest hyperplane at each instant.

The dynamic model parameters for expressions (10) and (11), with 3-D data points being in this case projected onto 2-D planes, are

$$
A = \begin{bmatrix} 1 & 1 & 1 & 1 \\ 0 & 1 & 1 & 1 \\ 0 & 0 & 1 & 1 \\ 0 & 0 & 0 & 1 \end{bmatrix}, \quad B = \begin{bmatrix} 0 \\ 0 \\ 1 \\ 1 \end{bmatrix} \qquad (15)
$$

and the state vector is $s = [x_1 \quad x_2 \quad dx_1 \quad dx_2]^T$, with dx_1 and dx_2 denoting the increments in the x_1 and x_2 directions respectively, while the observation model parameters are

$$
C = [1 \quad 1 \quad 0 \quad 0], \quad D = 1. \qquad (16)
$$

7. CONCLUSIONS

This paper presents a method for manifold learning, applicable to trajectory tracking in *n*-D manifolds embedded in m-D space, that relies on geometric information to reduce complexity. The manifold learning results are promising and make accurate trajectory tracking simpler to achieve.

An important direction of future work is modelling non-linear dynamics, with possibly distinct behaviours in different manifold regions, instead of the present method of using the white-noise acceleration model in all patches. A non-linear observer approach is being considered, rather than any form of Extended Kalman Filtering.

Also, a way to combine a probabilistic model with the current geometric model is being studied. This is needed in order to detect outliers and to select the best hyperplane for projection in a principled way, without having to use the computationally expensive nearest neighbour approach. One possible solution is to use mixtures of Gaussians on the hyperplanes, using Expectation-Maximization (EM) to estimate the parameters. This is simpler, both computationally and in terms of convergence, if done in the lower dimensional space, rather than directly in the observation space.

REFERENCES

Duchamp, T. and W. Stuetzle (2002). Spline Smoothing on Surfaces (preprint), To appear in: *Proceedings of ACM Siggraph.*

Hoppe, H. (1994). *Surface reconstruction from unorganized points* (PhD dissertation). University of Washington.

Kalata, P. (1984). The tracking index: A generalized parameter for α-β and α-β-γ trackers, *IEEE Transactions on Aerospace and Electronic Systems,* pp. 174-182.

Marques, J., J. M. Lemos and A. Abrantes (1999). Estimation of random trajectories on manifolds: Application to object tracking, In: *Proceedings of the IEEE International Conference on Image Processing,* pp. 103-107.

O'Neill, B. (1997). *Elementary Differential Geometry,* Academic Press.

Silva, J., J. Marques and J. M. Lemos, (2000). Robust motion tracking on one-dimensional manifolds, In: *Proceedings of the 4th Portuguese Conference on Automatic Control,* pp. 279-283.

IFAC Publications
www.elsevier.com/locate/ifac

VERSION 6 OF THE SYSTEM IDENTIFICATION TOOLBOX

Lennart Ljung *

* *Division of Automatic Control, Linköping University, SE-58183, Linköping, Sweden, email:* `ljung@isy.liu.se`

Abstract: This paper describes the new developments in Mathwork's SYSTEM IDENTIFICATION TOOLBOX to be run with MATLAB. There are three main additions to the new version: (1) Frequency domain input/output data as well as frequency response data can be directly used to fit models. (2) Simple continuous-time process models of the type Static Gain, Time Constant, Dead Time can be directly estimated as a new model object class. (3) A function `advice` can be applied both to data sets and to estimated models, in order to provide guidance and advice through the sometime complex identification process. *Copyright © 2003 IFAC*

1. INTRODUCTION

THE SYSTEM IDENTIFICATION TOOLBOX (SITB) for use with MATLAB was first released in 1987. It has gone through several updates, some of which have been reported at earlier IFAC Symposia on System Identification; see (Ljung 1994), (Ljung 1997), and (Ljung 2000).

Version 6 of the SYSTEM IDENTIFICATION TOOLBOX, (Ljung 2003) has the following three major additions

- A new model object, `idproc`, is introduced. It covers simple *Process Models* of the kind Static Gain + Time Constant + Time Delay.
- Input-Output data in the frequency domain can be used for identification as well as frequency response data, (`frd` and `idfrd` objects).
- A new feature, the function `advice` has been introduced to guide the inexperienced user through the many options and choices of system identification. It is also intended to give succinct summaries also for the experienced user. It can be applied to any `iddata` and `idmodel` object.

These three items will be described in somewhat more detail in this contribution.

2. SIMPLE PROCESS MODELS

2.1 *General Motivation*

Simple continuous time process models of the kind Static Gain + Time Constant + Time Delay are dominating for control design in process industry. No doubt, the most common identification method in practice is to perform a step response experiment and adjust two or three parameters to the measured response. Nevertheless, such models have not been treated very much in the System Identification literature. The situation resembles that in control design, where the practice is dominated by PID-regulator tuning, which, with some exceptions, e.g. (Åström and Hägglund 1995), is not widely treated in the literature.

The recent research area of *Identification for control*, e.g. (Gevers 1993), (van den Hof and Schrama 1995), (Kosut *et al.* 1992), (Zang *et al.* 1995), (Ljung 1998) has produced many interesting results around the interplay between (reduced complexity) model estimation and control design. In industrial practice *identification for control* really is construction of a simple two- or three-parameter model, most often from a transient or possibly relay experiment, followed by tuning of a PI(D)-regulator, based on these two-three parameters.

2.2 *Typical Models and Methods*

Perhaps the most commonly used process model is

$$G(s) = \frac{K}{1 + sT_{p1}} e^{-sT_d} \qquad (1)$$

Among variants of this model, we can have a model without delay ($T_d = 0$):

$$G(s) = \frac{K}{1 + sT_{p1}} \qquad (2)$$

and/or introduce an enforced integration (self-regulating process)

$$G(s) = \frac{K}{s(1 + sT_{p1})} e^{-sT_d} \qquad (3)$$

Moreover, on can postulate two real poles with or without a zero

$$G(s) = \frac{K(1 + sT_z)}{(1 + sT_{p1})(1 + sT_{p2})} e^{-sT_d} \qquad (4)$$

A further possibility is to allow resonant poles ("under-damped models"):

$$G(s) = \frac{K(1 + sT_z)}{1 + 2\zeta sT_r + (sT_r)^2} \qquad (5)$$

Clearly a variety of models can be defined based on these components.

Several papers and books discuss how to estimate models like (1) from transient response data (e.g. (Åström and Hägglund 1995), (Rake 1980), (Ziegler *et al.* 1943)). Most of the classical methods are graphical or semi-graphical, like finding the steepest tangent to the step response and calculate its intersection with the time axis, etc, or computing areas below the response curve and so on. See e.g. (Åström and Hägglund 1995) for a recent overview of such approaches.

2.3 *A Prediction Error Identification Perspective*

In a standard system identification framework, e.g. (Ljung 1999*b*), estimation of process models like (1) - (5) is of course no different from estimating any other parameterized linear model. Any of the models of the previous section can be written as

$$G(s, \theta) \qquad (6)$$

where θ comprises the model parameters K, T_{p1}, T_d etc. To estimate the parameters we have collected a data set $Z^N = \{u(1), y(1), \ldots, u(N), y(N)\}$ of sampled inputs and outputs. Suppose the sampling interval is constant and equal to T. The model (6) is sampled with this sampling interval, according to the input inter-sample behavior (zero-order-hold, first-order-hold, band-limited) giving the discrete time model

$$G_T(q, \theta) \qquad (7)$$

(q is the shift operator) and the model outputs

$$\hat{y}(t|\theta) = G_T(q, \theta)u(t), \quad t = 1, \ldots N. \qquad (8)$$

The parameters are then estimated by solving the non-linear least squares problem

$$\hat{\theta}_N = \arg\min_\theta \sum_{t=1}^{N} (y(t) - \hat{y}(t|\theta))^2 \qquad (9)$$

It is also straightforward to include a model of additive noise

$$y(t) = G(p, \theta)u(t) + H(p, \theta)e(t) \qquad (10)$$

where p denotes the differentiation operator (replacing s). Determining the proper sampled predictor from (10) and letting $\hat{y}(t|\theta)$ denote the corresponding predicted outputs, gives a method (9) that also estimates the noise model.

Moreover, an estimation focus can be defined as discussed in (Ljung 1999*a*). For a model with $H = 1$, the estimation focus filter L simply means that the inputs and outputs are first filtered through L.

The asymptotic properties of the estimated model are well known: Suppose the true frequency function for the sampled system is $G_0^{(T)}(e^{i\omega})$ (or, more generally, the frequency function of the linear time invariant second order equivalent of the true system, see (Ljung 2001)). Then for $H = 1$ we have, (Ljung 1999*b*)

$$\hat{\theta}_N \to \arg\min_\theta \int_{-\pi}^{\pi} |G_T(e^{i\omega}, \theta) - G_0^{(T)}(e^{i\omega})|^2 \times \Phi_u(\omega)|L(e^{i\omega})|^2 d\omega \qquad (11)$$

Here L is the "focus filter", and Φ_u is the input spectrum. The expression describes exactly in what way the simple process model like (1) approximates the true system. We also see how the focus filter may steer the fit to important frequency ranges.

In a sense, (11) also explains the success of simple process models. Even a three-parameter model like (1) has substantial "local flexibility". The delay term may pick up the true system's phase, even if there is no dead-time in the system. For successful control design it is often sufficient to have a rough picture of the Nyquist curve in a limited, but important frequency region, and (11) illustrates how this can be achieved.

2.4 *Basic Syntax*

The way to refer to the different types of process models is through acronyms built up from the basic symbols:

- A leading 'P' (for "Process Model")

- An integer denoting the number of poles (not counting a possible integration)
- 'D' for time Delay
- 'I' for Integration
- 'Z' for a zero (numerator term)
- 'U' for under-damped (complex) poles.

For example, the models (1) to (5) are denoted by `P1D`, `P1`, `P1ID`, `P2ZD` and `P2ZU`, respectively. The basic syntax to create such a model is

```
m0 = idproc('P1D')
```

and for estimating it from data

```
m = pem(data,'P1D')
```

The standard property/value pairs for various options can be added to the list of arguments. This includes the possibility to estimate initial conditions, add noise models, etc. Also, fixing certain parameters is possible, as well as defining upper and lower bounds for them.

For multi-input systems, different type of models can be applied to each input. This is handled by letting the model descriptions be cell arrays as in

```
m = pem(data,{'P1D','P2U'})
```

2.5 *GUI support*

The estimation of process models is also included in the standard Graphical User Interface (GUI)-setup of the SYSTEM IDENTIFICATION TOOLBOX. Estimated process models are included in the "Model Board" and can be subjected to any of the examination and visualization tools.

The definition of the process models is also supported by a GUI. It has pop-up menus and check-boxes for defining the structure and also allows fixing and bounding the parameter values. See Figure 1 for an illustration.

3. FREQUENCY DOMAIN DATA

3.1 *Input-Output Fourier Data*

If the input-output data are given in the frequency domain as Fourier transforms, the prediction error approach to estimating process models still can be applied. See, e.g. Section 7.7 in (Ljung 1999*b*) or chapters 5–8 in (Pintelon and Schoukens 2001).

The `iddata` object now allows definition of input/output data in the frequency domain over arbitrary frequencies as in

```
dat = iddata(Y,U,'Frequencies',W,...
        'Domain','Freq','Ts',0);
```

Note that the sampling interval, T, (`'Ts'`) is still relevant, since it has information of how the signal

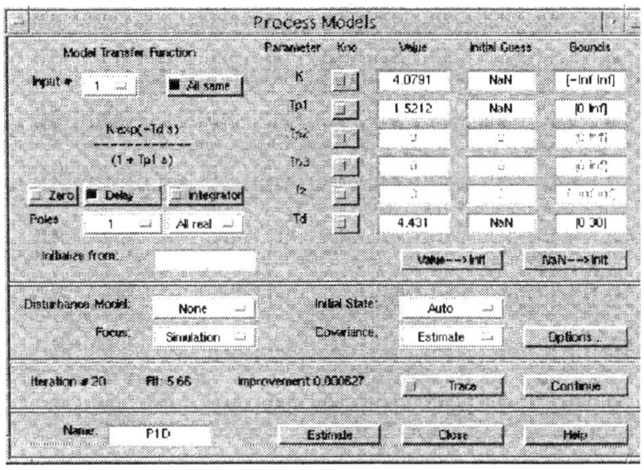

Fig. 1. The GUI for defining process models. For multi-input the pop-up menu at the top left selects the input, or you declare that all inputs have the same structure. The check boxes below determine whether Integration, Time Delay and/or a Zero should be included. The text above writes out the currently chosen structure. Up to the right, initial or fixed parameter values are selected, and also possible upper and lower bounds on the parameters. The zone below this concerns a number of options about disturbance models, initial states etc. Below that is a zone for iteration information and control.

Fourier transforms `Y` and `U` have been computed from time domain data. Discrete time Fourier Transforms conceptually have the frequency argument $e^{i\omega T}$. Note however, that frequency domain data, unlike time domain data allow continuous time signals ($T = 0$).

With frequency domain data objects, several MATLAB commands are naturally overloaded:

```
DF = fft(dat)
da = abs(dat)
df = phase(dat)
```

etc.

Models can now be estimated directly from `iddata` objects in the frequency domain. The syntax for this is entirely transparent. If the frequency domain data is denoted as continuous time, a continuous time model is estimated directly (without `d2c` transformations). All the toolbox commands for simulation and validation handle time/frequency domain data in a transparent manner. The only restriction is that, for the moment, estimation of noise models and time-series models is not supported for frequency domain data. The reason is the more complex criteria of fit that must be used for continuous-time and non-equidistantly spaced frequency data. See, e.g. page 230 in (Ljung 1999*b*).

3.2 *Frequency Response Data*

The estimation code also accepts frequency response data objects, like `frd` in the CONTROL SYSTEM TOOLBOX or `idfrd` in the SITB. It is often the case that experimental frequency domain response data are stored as frequency functions rather that as inputs and outputs, separately. This also could be more economical, e.g. by using logarithmically spaced frequencies. There is also data acquisition equipment, *frequency analyzers*, that collect and deliver the result of the experiment as frequency response data.

A new function that estimates `idfrd` objects (frequency functions and disturbance spectra) from time or frequency `iddata` objects is

```
g = spafdr(data)
```

which allows Frequency Dependent Resolution, with a logarithmic frequency grid as default along with a resolution that as adopted to the grid. This could be an efficient way of compressing measured data. It is often the case that a courser resolution (in rad/s) can be used at higher frequencies, and that a constant *relative resolution* is to be preferred. Figure 2 illustrates the effect of the frequency dependent resolution.

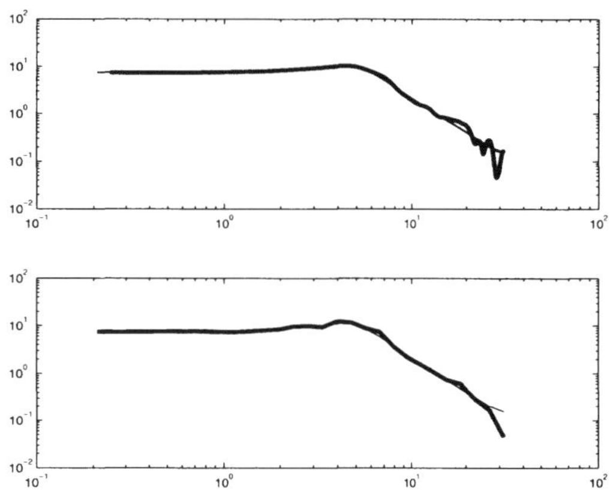

Fig. 2. Estimate of the frequency function for the data set IDDATA1, using SPA (above) and SPAFDR (below) with default arguments for frequency vector and resolution. Thin line: the true frequency response.

Frequency domain data offer useful potentials also for other problems:

- The focus filter can be implemented as specific frequencies for which the fit should be made. For example,
  `m = oe(dat,[2 2 1],'focus',[0.2 1])`
  will concentrate the fit to the pass band from 0.2 to 1 rad/s. The desired frequency bands may not necessarily be known a priori, but could be selected from a preliminary model, like using frequencies that correspond to the Nyquist curve being in the third quadrant, or being close to the critical point -1. Example:
  ```
  m = n4sid(data,5);
  f = idfrd(m);
  ph = phase(squeeze(f.resp));
  fs = fselect(f,find(ph>-pi & ph<-pi/2));
  mp = pem(fs,'p1d');
  ```
- If the inter-sample input behavior is band-limited, moving to the frequency domain will be the easiest way to handle the sampling. The FFT (discrete Fourier transform) of the input will then be equal to the Fourier transform of the underlying continuous time input signal. The FFT of the output will similarly correctly describe the continuous time Fourier transform of that part of the output that originates from the input, and we can directly fit a continuous time model:
  ```
  z = iddata(y,u,0.5);
  zf = fft(z);
  zf.Ts = 0;
  mp = oe(zf,[2 2 1])
  ```

3.3 *GUI support*

Frequency domain `iddata` and frequency response data as `frd` or `idfrd` objects can be imported into the GUI in the same way as time domain data. The icons for the different types of data sets are marked by different background colors. The `data preprocessing` menus allow the transformation between the various representations. Also the use of data objects of different types for estimation and validation is entirely transparent. For example, if an IDFRD object is chosen as validation data, the `Model output view` shows the frequency responses of the models, together with the data.

4. THE `ADVICE` FEATURE

The function `advice` in (Ljung 2003) takes any data object or any model object as an argument and delivers a text that comments the objects and gives some hints how to proceed.

This is accomplished by a hidden property (called `utility`) of these objects, that is a structure with several fields. The MATLAB functions `assignin` and `inputname` allows you to update a workspace variable (in this case its `utility`-property) even if it is not the output of a function called. This makes it possible to adjoin to a model or a data set a "diary" of what has been done, and thus give relevant advice of further things that can be tested.

For data objects, `advice` tests things like

- excitation

- feedback in data
- if a step or impulse occurs "too early" in the set. There is a technical side that is more treacherous: It is quite common to see applications where the data is a single step or impulse response, where the step/impulse occurs in the first few data points. Such data could very well be used for reasonable modeling if the signal–to–noise ratio is good. However, many estimation techniques (like estimating ARX-model and subspace methods) effectively shift the data (up to a number of samples that is about the model order) in order to avoid estimation data points prior to the start of the data collection. This makes the data look like a response to a constant input, and no reasonable models can be built. It is my experience that this problem is the single most common reason for complaints about the identification process from first-time users.
- possible detrending
- points to the possibility to assign input intersample behavior
- etc

For model objects `advice`

- points out that `compare` and `resid` should be run if this has not been done. It also checks whether this was done on a separate validation set.
- comments on what has been seen in the residual analysis test `resid`. Based on χ^2-tests it gives advice whether higher order dynamics and noise models are required.
- checks if the model order can be reduced.
- allows comparisons of several models, `advice(m1,m2,m3)` and points out which one should be preferred.
- etc

The function is best illustrated by an example of how it works:

```
>>load iddata1
>>advice(z1)

All your inputs have been denoted as
'zero order hold' ('zoh'), i.e.
they are assumed to be piecewise
constant over the sampling interval.
If the input is a sampled continuous
signal and you plan to build or
convert to time continuous models,
it is recommended to mark the
Intersample property as 'First
order hold': Data.int = 'foh' or
Data.int = {'foh','foh', ...}
for multiinput signals.
You may mix 'zoh' with 'foh' for
the different inputs in the latter
case.

All inputs and outputs are not zero
mean. It it generally recommended to
remove the means by DAT = DETREND(DAT),
except in the following two cases:

1. The signals are measured relative
   to a level that corresponds to a
   physical equlibrium. This could e.g.
   be the case if step responses
   are recorded from an equlibrium point.
2. There is an integrator in the
   system, and the input and output
   levels are essential to describe
   the effect of the integration.

There is no significant indication of
feedback in the data.

The input is persistantly exciting of
order more than 50. This means that you
will not encounter identifiability
problems if estimating models of order
lower than so.

>>m=arx(z1,[1 1 1]);
>>advice(m)

You should run a comparison test

  COMPARE(VDATA,m)

where VDATA preferably is a different
data set from the estimation data z1.
You can then run ADVICE(m) again.

You should run a residual test

  RESID(VDATA,m)

where VDATA preferably is a different
data set from the estimation data z1.
You can then run ADVICE(m) again.

>>compare(z1,m);
>>resid(z1,m);
>>advice(m)

There is a very strong indication that
the dynamics of the model is
not adequately described.
A first general advice is to run
RESID(VDATA,m,'FR') to check in which
frequency ranges the model error is
present. If the model error is unacceptable,
you will have to increase the model order.

In particular you should pay attention to
lags 2  3  6 from input 1.
```

```
Modify KU and the orders NB so that
these lags are included in the model.


There is a very strong indication that
the residuls are not white. To get a
good noise model you need to increase
the orders associated with the noise
parameters, or just increase
the order of a state-space model.
```

5. CONCLUSIONS

The three new features that have been added to the SYSTEM IDENTIFICATION TOOLBOX, viz. estimating and validation with frequency domain data, simple process models, and the `advice` function correspond to the most common requests from users. There is a risk that adding new features to a software package will make the use of it more complicated. To handle this risk, it has been important to make the syntax for frequency and time domain data entirely transparent, and also to make sure that the estimation and examination of process models follow the same syntax as for other models.

6. REFERENCES

Gevers, M. (1993). Towards a joint design of identification and control?. In: *Essays on control: Perspectives in the theory and its applications* (H L Trentelman and J C Willems, Eds.). ECC '93 Groningen.

Kosut, R.L., G. C. Goodwin and M. P. Polis (Eds) (1992). *Special Issue on System Identification for Robust Control Design, IEEE Trans. Automatic Control, Vol 37.*

Ljung, L. (1998). Identification for control – what is there to learn?. In: *Learning, Control and Hybrid Systems* (Y. Yamamoto and S. Hara, Eds.). Vol. 241 of *Springer Lecture Notes in Control and Information Sciences.* Springer Verlag. Berlin. pp. 207–221.

Ljung, L. (1999*a*). Estimation focus in system identification: Prefiltering, noise models, and prediction. In: *IEEE Conference on Decision and Control.*

Ljung, L. (1999*b*). *System Identification - Theory for the User.* 2nd ed.. Prentice-Hall. Upper Saddle River, N.J.

Ljung, L. (2001). Estimating linear time invariant models of non-linear time-varying systems. *European Journal of Control* **7**(2-3), 203–219. Semi-plenary presentation at the European Control Conference, Sept 2001.

Ljung, L. (2003). *System Identification Toolbox for use with* MATLAB. *Version 6.*. 6th ed.. The MathWorks, Inc. Natick, MA.

Ljung, Lennart (1994). A graphical user interface (gui) to the system identification toolbox. In: *Proc. 10th IFAC Symposium on System Identification (SYSID'94).* Vol. 4. Copenhagen, Denmark. pp. 29–33.

Ljung, Lennart (1997). Developments for the system identification toolbox for matlab. In: *Proc. of the 11th IFAC Symposium on System Identification.* Vol. 1. Kitakyushi, Japan. pp. 969–975. SYSID.

Ljung, Lennart (2000). Version 5 of the system identification toolbox for use with matlab - with object orientation. In: *Proc IFAC Symposium SYSID 2000.* Santa Barbara. pp. ThPM5–5.

Pintelon, R. and J. Schoukens (2001). *System Identification – A Frequency Domain Approach.* IEEE Press. New York.

Rake, H. (1980). Step response and frequency response methods. *Automatica* **16**, 519–526.

Åström, K. J. and T. Hägglund (1995). *PID Controllers: Theory, Design, and Tuning.* 2nd ed.. Instrument Society of America. Triangle Research Park, N.C.

van den Hof, P. M. J. and R.J.P. Schrama (1995). Identification and control–closed loop issues. *Automatica* **31**(12), 1751–1770.

Zang, Z., R. R. Bitmead and M. Gevers (1995). Iterative Weighted Least-squares Identification and Weighted LQG Control Design. *Automatica* **31**(11), 1577–1594.

Ziegler, J. G., N. B. Nichols and N.Y. Rochester (1943). Process lags in automatic-control circuits. *Trans. ASME* **65**, 443–444.

IFAC
Publications
www.elsevier.com/locate/ifac

PROCESS IDENTIFICATION, CONTROLLER TUNING AND CONTROL CIRCUIT SIMULATION USING MS EXCEL

H. M. Schaedel

University of Applied Sciences Koeln, Faculty of Information, Media and Electrical Engineering
Betzdorfer Str. 2, 50769 Koeln, e-mail: HMSchaedel@t-online.de

Abstract: The paper presents a tool for process identification, controller tuning and control circuit simulation based on spreadsheets like MS-Excel. Process identification and modelling can be carried out for the open and closed-loop control circuit. Controller tuning is done according to the criterion of cascaded damping ratios. The design is based on a direct relation between the parameters of the process and the controller. Tuning for optimal set-point control as well disturbance rejection is provided. Single-input/single output (SISO) and dual-input/dual-output (DIDO) systems can been simulated for proportional and integral plants with dead-time. *Copyright © 2003 IFAC*

Keywords: linear, data-based tuning, process control systems, computer/software

1. INTRODUCTION

Office programs are installed on nearly every PC or laptop. Spreadsheets provide numerous mathematical and graphical tools that can be used for solving problems in engineering sciences. Based on Microsoft Excel a tool has been developed for process identification, controller tuning and control circuit simulation.

2. PROCESSI DENTIFICATION

Process identification and modelling can be carried out for the open and closed-loop control circuit. The values of the input and output signals of the plant can be represented in ASCII-format.

2.1 Least square approximation using the solver function of EXCEL

For the open loop the step response of the plant is approximated by a second order system plus dead-time

$$G_P(s)=\frac{K_P\ e^{-sT_t}}{(1+s\tau_1)(1+s\tau_2)} \quad (1)$$

with least square fitting between the step response of the plant and the model using the solver function of Excel. From this the characteristic frequency parameters of the plant follow as

Sum of time constants:

$$T_1=\tau_1+\tau_2+T_t \quad (2)$$

Product sum of time constants:

$$T_2^2=\tau_1\tau_2+(\tau_1+\tau_2)T_t+0.5\ T_t^2 \quad (3)$$

Figure 1 shows the step response of a flow process that is provided by the Online Lab of J. Henry at the University of Chattanouga (Henry). The results of the process identification are given below as

$K_p = 0.514$ lb/min/%
$\tau_1 = 0.233$s $T_1 = 0.866$ s
$\tau_2 = 0.233$s $T_2^2 = 0.321$ s²
$T_t = 0.400$s.

These parameters are used for controller tuning according the criterion of cascaded damping ratios.

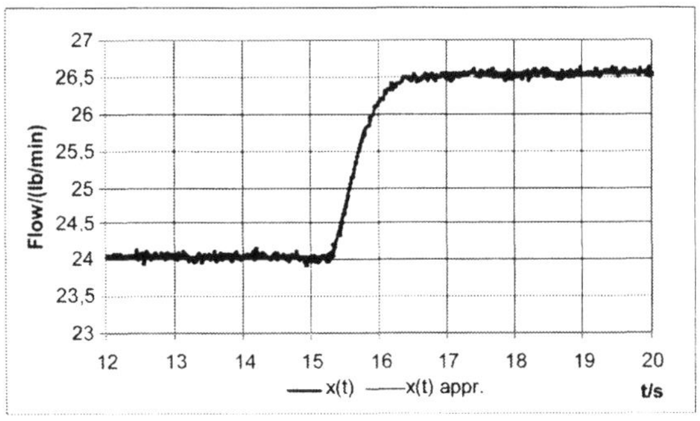

Fig. 1. Step response of a flow process and approximation by a SOPDT-model

2.2 Method of inflection tangent

Another method of parameter estimation is provided on the basis of the inflection tangent applied to the step response (Schaedel 1999).

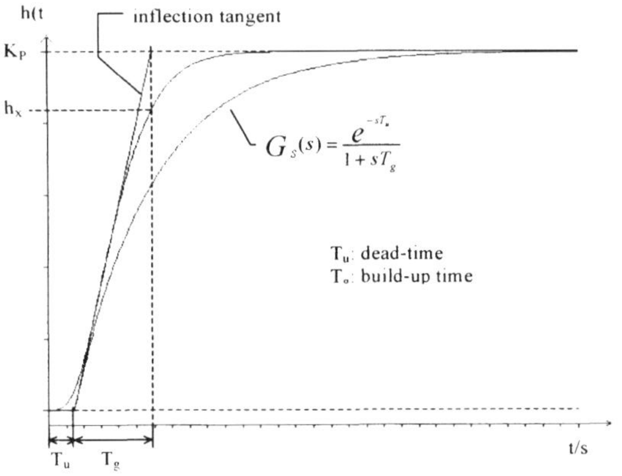

Fig. 2. Step response of a lag process and first order plus dead-time approximation (FOPDT) with identical T_u and T_g

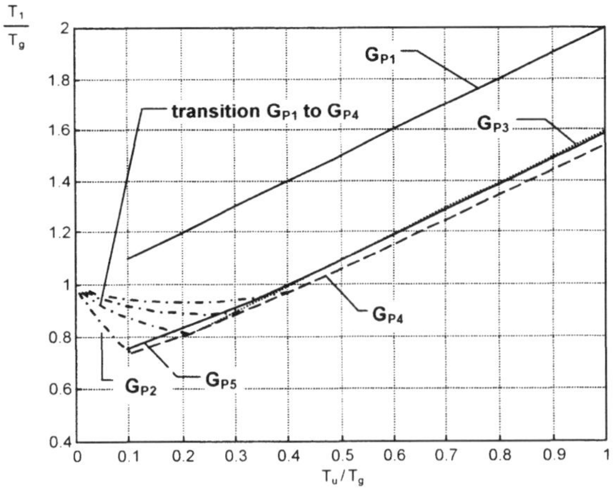

Fig. 3. Sum of time constants T_1 as a function of $\mu=T_u/T_g$

Ziegler and Nichols (1942) have demonstrated that information obtained from the inflection tangent can be used for a controller tuning. Their investigations were based on plants with first order lag plus dead-time (FOPDT). It is obvious that for the same values of T_u and T_g one can find numerous configurations of different lag orders with and without dead-time (Fig. 2). The FOPDT-approximation with identical values of dead-time T_u and build-up time T_g proves to be insufficient In order to locate the specific response function within the family of curves possible an additional parameter is needed.

The problem has to be looked at from the point of process identification. A controller design is only possible if a process identification can be carried out on the basis of parameters obtained by the inflection tangent. With this background a controller design then can be accomplished. Process identification has not to be highly accurate because PID- controllers are rather robust against deviations of the plant parameters.

Figures 3 shows the normalised characteristic time constant T_1/T_g as a function of the normalised lag-time $\mu = T_u/T_g$ for typical processes in table 1.

Table 1 Frequency transfer functions of typical processes

$$G_{P1}(s)=\frac{K}{1+s\tau}e^{-sT_L} \qquad G_{P2}(s)=\frac{K}{(1+s\tau_1)(1+s\alpha\tau_1)}$$

$$G_{P3}(s)=\frac{Ke^{-sT_L}}{(1+s\tau_1)(1+s\alpha\tau_1)^2} \quad \text{with } \alpha = 0...1$$

$$G_{P4}(s)=\frac{K}{(1+s\tau)^n} \qquad G_{P5}(s)=\frac{K}{\prod_{i=1}^{n}\left(1+s\frac{\tau}{i}\right)^i}$$

The dead-time term is approximated using Taylor series expansion. A set of curves results that is confined at the upper border by the first order lag with time delay (G_{P1}) and at the lower border by the nth order lag with equal time constants (G_{P4}). The left border for low ratios T_u/T_g is confined by the transition from the first order lag to the second order lag with varying ratios of the two time constants (G_{P2}). In order to determine the sum of time constants T_1 for a given ratio $\mu = T_u/T_g$ an additional information is needed. For practical application this characteristic information has to be taken in a simple way from the step response. It is rather obvious to take the value of the unit step response h_x for the time $t_x = T_u + T_g$. This build-up value h_x can be taken with rather good accuracy.

The simplest way of approximating the step response is a polygonal approach using the inflection tangent

$$G_P(s) \approx \frac{K_I}{s} e^{-sT_u}\left(1-e^{-sT_g}\right) \quad \text{with} \quad K_I = \frac{K_P}{T_g}. \quad (4)$$

This a superposition of two integrating terms with different delays T_u and T_t = Tu+T_g. This approach is modified by introducing a time lag into the integral term.

$$G_S(s) \approx \frac{K_I}{s} e^{-sT_u} - \frac{K_I}{s(1+s\tau)} e^{-sT_t}; \quad \tau + T_t = T_u + T_g \quad (5)$$

The parameters T_t and τ have to be adjusted in such a manner that the unit step response function passes through the build-up value $h_x = h(T_u+T_g)$. From this one finds the relation for time constant τ as a function of the build-up value h_x Using apparent dead-time T_u, build-up time T_g and build-up value $h(T_u+T_g)$ the characteristic frequency parameters T_1 and T_2^2 according equations (2) and (3) may be estimated

$\mu \le 0.104$	$\mu > 0.104$	
$T_1 = (\mu + A)T_g$	$T_1 = (\mu + A)T_g$	(6)
$T_2^2 = \frac{\mu}{A^2} T_1^2$	$T_2^2 = \frac{1}{2} \frac{\mu}{\mu + 0{,}2A} T_1^2$	

where

$$H = \frac{\tau}{T_g} = e(1 - h_x / K_P) \qquad A = 0.5(1 + H^2) \quad (7)$$

and

T_u lag time (apparent dead-time)
T_g build-up time
μ T_u/T_g
h_x $h(T_u+T_g)$ build-up value.

The results of identification through the inflection tangent for the flow process under test in section 2.1 differ only slightly from those according to least square approximation using the solver function of Excel.

2.3 Closed loop identification

For the closed loop the characteristic frequency parameters T_1 and T_2^2 of the process can be estimated from the input signals u(t) and output signals x(t) applying least square approximation of the frequency response $G_P(s)$ (Golubev and Horowitz, 1982).

$$\int_0^\infty \left[\frac{x(s)}{u(s)} - \frac{K_S}{1 + s \cdot T_1 + s^2 \cdot T_2^2}\right]^2 ds = \min \quad (8)$$

This leads to

$$\int_0^\infty [x(s) \cdot \{1 + s \cdot T_1 + s^2 \cdot T_2^2\} - u(s) \cdot K_S]^2 ds = \min \quad (9)$$

Using integration instead of differentiation the methods proves to be very insensitive against noise.

$$\int_0^\infty \left[\frac{x(s)}{s^2} + T_1 \cdot \frac{x(s)}{s} + T_2^2 \cdot x(s) - K_S \cdot \frac{u(s)}{s^2}\right]^2 ds = \min \quad (10)$$

Applying Parseval theorem

$$\frac{1}{2\pi} \int_{-\infty}^{+\infty} |X(\omega)|^2 d\omega = \int_{-\infty}^{+\infty} x^2(t) dt \quad (11)$$

the problem is transferred into the time domain

$$\int_0^\infty [I^2(x) + T_1 \cdot I(x) + T_2^2 \cdot x(t) - K_S \cdot I^2(u)]^2 dt = \min \quad (12)$$

where

$$I(x) = \int_0^\infty x(t) dt \qquad I^2(x) = \int_0^\infty \int_0^t x(t) dt' dt$$
$$I(u) = \int_0^\infty u(t) dt \qquad I^2(u) = \int_0^\infty \int_0^t u(t) dt' dt \quad (13)$$

are the integrals of the input and output data of the plant. The discrete time version follows as

$$\sum_{k=0}^{N} [I^2(x_k) + T_1 \cdot I(x_k) + T_2^2 \cdot x_k - K_S \cdot I^2(u_k)]^2 = \sum_{k=0}^{N} e^2(k)$$
$$\text{with} \quad I(x_k) = I(x_{k-1}) + (x_k + x_{k-1}) \cdot \frac{T_0}{2} \quad (14)$$

using trapezium rule for integration. Looking for least square error by taking the differentials to the characteristic values K_P, T_1 and T_2^2 leads to a matrix relation between input and output data and frequency parameters (15) from which the characteristic frequency parameters of the plant can be determined by matrix inversion (16).

$$\frac{\partial}{\partial K_S} \sum_{k=0}^{N} e_k^2 = -2 \cdot \sum_{k=0}^{N} [I^2(x_k) - K_S \cdot I^2(u_k) + T_1 \cdot I(x_k) + T_2^2 \cdot x_k] \cdot I^2(u_k) = 0$$
$$\frac{\partial}{\partial T_1} \sum_{k=0}^{N} e_k^2 = -2 \cdot \sum_{k=0}^{N} [I^2(x_k) - K_S \cdot I^2(u_k) + T_1 \cdot I(x_k) + T_2^2 \cdot x_k] \cdot I(x_k) = 0 \quad (15)$$
$$\frac{\partial}{\partial T_2^2} \sum_{k=0}^{N} e_k^2 = -2 \cdot \sum_{k=0}^{N} [I^2(x_k) - K_S \cdot I^2(u_k) + T_1 \cdot I(x_k) + T_2^2 \cdot x_k] \cdot x_k = 0$$

$$\underbrace{\begin{bmatrix} \sum_{k=0}^{N}[I^2(u_k)]^2 & -\sum_{k=0}^{N} I(x_k)\cdot I^2(u_k) & -\sum_{k=0}^{N} x_k\cdot I^2(u_k) \\ \sum_{k=0}^{N} I^2(u_k)\cdot I(x_k) & -\sum_{k=0}^{N}[I(x_k)]^2 & -\sum_{k=0}^{N} x_k\cdot I(x_k) \\ \sum_{k=0}^{N} I^2(u_k)\cdot x_k & -\sum_{k=0}^{N} I(x_k)\cdot x_k & -\sum_{k=0}^{N}[x_k]^2 \end{bmatrix}}_{\text{Matrix A}} \times \begin{bmatrix} K_S \\ T_1 \\ T_2^2 \end{bmatrix} = \underbrace{\begin{bmatrix} \sum_{k=0}^{N} I^2(x_k)\cdot I^2(u_k) \\ \sum_{k=0}^{N} I^2(x_k)\cdot I(x_k) \\ \sum_{k=0}^{N} I^2(x_k)\cdot x_k \end{bmatrix}}_{\text{Matrix C}} \tag{16}$$

Figure 4 shows the results for a PI-controlled flow process with poor tuning.

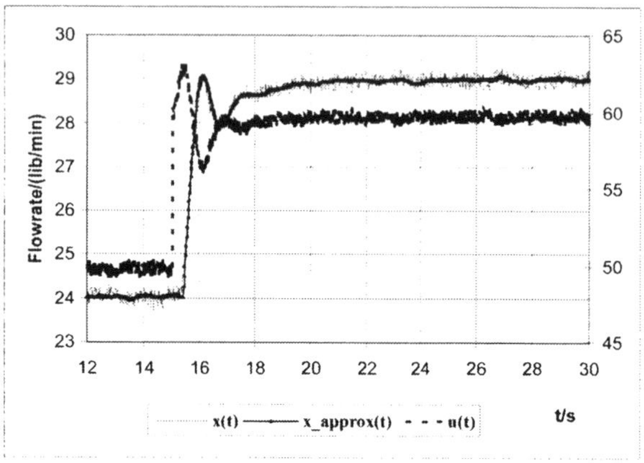

Fig. 4. Step response of a PI-controlled flow process with poor parameter tuning

Parameter estimation based on input and output data gives the characteristics of the plant

$K_P = 0.5$ lb/min/% $\quad T_1 = 0.826$ s $\quad T_2^2 = 0.246$ s

From these data a first-order plus dead-time model for the plant has be calculated in order to compare the response of the real process with the model identified.

$K_P = 0.5$ lb/min/% $\quad T_t = 0.38$s $\quad \tau = 0.446$ s

The referred dead-time $d = T_t/T_o$ is varied for optimal fitting manually. Figure 4 shows excellent agreement between approximating model and real plant.

3. CONTROLLER TUNING

Controller tuning is done according to the criterion of cascaded damping ratios (Schaedel, 1997a, 1997b, 1998). The design is based on a direct relation between the parameters of the process and the controller and enables the use of classical filter design (e.g. Butterworth, Tschebyscheff) and standard forms (e.g. ITAE, IAE). A design for optimal set-point control as well as optimal disturbance rejection provided. Table 2 and 3 give tuning rules for the parameters of the PI-controller proportional gain K_C and reset time T_r.

Table 2 PI-controller for optimal set-point control

Butterworth (normal design)	
$K_C = \frac{0.5}{K_P}\frac{T_r}{T_1 - T_r}$	$T_r = \sqrt{T_1^2 - 2T_2^2}$
Tschebyscheff 0,5 db (sharp design)	
$K_C = \frac{0.375}{K_P}\frac{T_r}{T_1 - T_r}$	$T_r = T_1 - \frac{T_2^2}{T_1}$
ITAE	
$K_C = \frac{0.375}{K_P}\frac{T_r}{T_1 - T_r}$	$T_r = -0.64T_1 + 1.64T_1\sqrt{1 - 1.2\frac{T_2^2}{T_1^2}}$

Table 3 PI-contoller for optimal disturbance rejection

Butterworth (normal design)	
$K_C = \frac{1}{K_P}\left(\frac{1}{2}\frac{T_1^2}{T_2^2} - 1\right)$	$T_r = 4\frac{T_2^2}{T_1}\left(1 - 2\frac{T_2^2}{T_1^2}\right)$
Tschebyscheff 0.1 db	
$K_C = \frac{1}{K_S}\left(0.7\frac{T_1^2}{T_2^2} - 1\right)$	$T_r = 3.11\frac{T_2^2}{T_1}\left(1 - 1.43\frac{T_2^2}{T_1^2}\right)$
ITAE	
$K_C = \frac{1}{K_P}\left(0.69\frac{T_1^2}{T_2^2} - 1\right)$	$T_r = 3.86\frac{T_2^2}{T_1}\left(1 - 1.46\frac{T_2^2}{T_1^2}\right)$

From parameter estimation in section 2.1 the tuning of the PI-controller of the sharp design is obtained for optimal reference control of the flow process as

$K_C = 0.973$ % /(lb/min) $T_r = 0.495$ s.

Figure 5 and 6 show the response of the control circuit to a set-point change at 15s and a change in disturbance at 20 s.

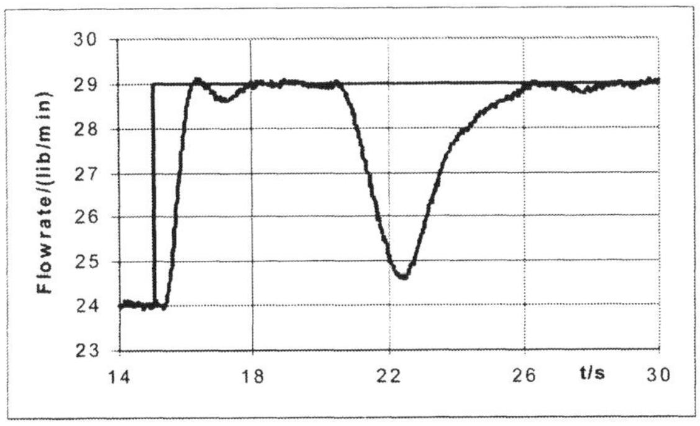

Fig. 5. Response of the PI-controlled process to a set-point change and a disturbance change

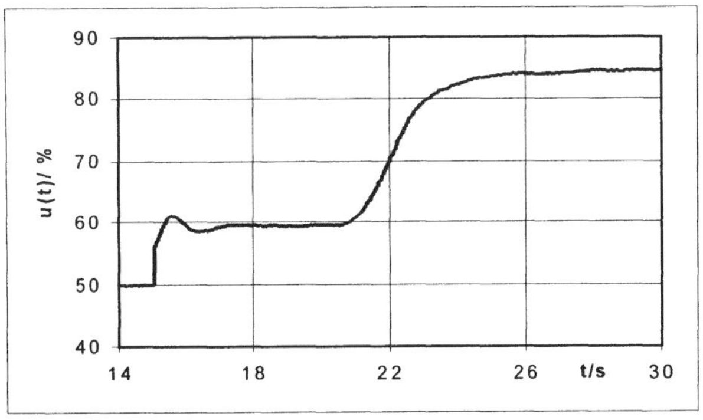

Fig. 6. Input signal (manipulated variable) of the PI-controlled flow process

4. CONTROL CIRCUIT SIMULATION

There are excellent possibilities in a spreadsheet to simulate the time response of a control circuit using difference equations. Single-input/single-output and dual-input/dual-output systems have been realised so far for proportional and integral plants with dead-time. Controllers are of PID type. Signal generators provide a step and a ramp signal. If a model of the plant has been obtained from measured data a controller design can be made and a simulation of the closed control circuit carried out before the real experiment. The closed loop data then again can be used to check and modify the controller tuning.

4.1 SISO control circuit simulation

Figure 7 and 8 demonstrates the simulation of a PI-controlled plant with a second order delay plus dead-time. The PI-controller is designed for optimal set-point control (sharp design, Tschebyscheff 0.5 db) and optimal disturbance rejection (ITAE). The set-point tuning shows fast settling of the controlled variable for changes in set-point. The tuning for optimal disturbance rejection responds with a significant overshoot for changes in set-point but has a much faster settling for disturbance rejection.

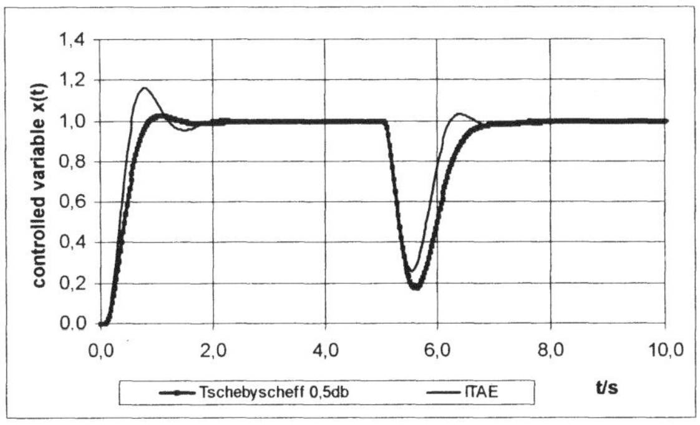

Fig. 7. Simulation of the response a PI-controlled process to set-point and disturbance changes

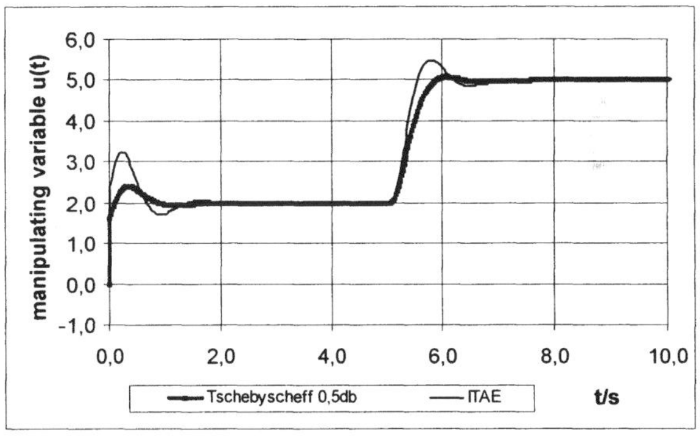

Fig. 8. Simulation of the input signal (manipulated variable) of a PI-controlled process

4.2 DIDO control circuit simulation

For the DIDO control circuit structures with main controllers of the PI-type and with additional de-coupling blocks are realised. Figure 9 shows the simple case, where only two PI-controllers are provided for controlling the process. The click-boxes in the cross-coupling branches can be used for activating and deactivating these branches. Figure 10 gives an example where the PI-controllers are designed for optimal set-point control of the upper circuit and for optimal disturbance rejection of the lower circuit when there is no cross-coupling. The transfer functions of the main and coupling branches are given in Table 4.

Table 4 Transfer functions of the DIDO control process

$$G_1(s)=\frac{1.5}{(1+s)(1+2s)} \qquad G_2(s)=\frac{2.0}{(1+2s)(1+4s)}$$

$$G_{12}(s)=\frac{0.5}{(1+s)(1+2s)} \qquad G_{21}(s)=\frac{1}{(1+2s)(1+3s)}$$

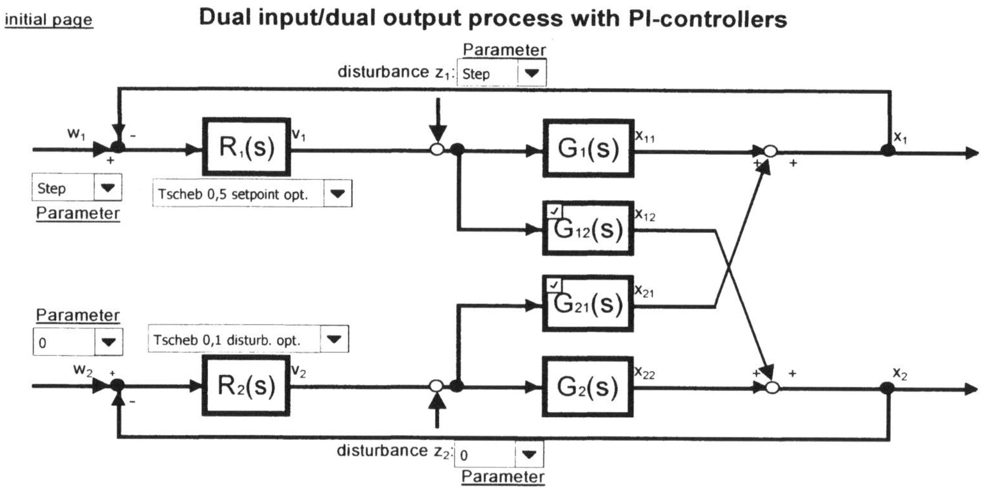

Fig. 9. Dual input/dual output control circuit

Activating the cross-coupling branches through the click-boxes results in a significant interference between the two circuits (Figure 11). The whole control circuit remains stable.

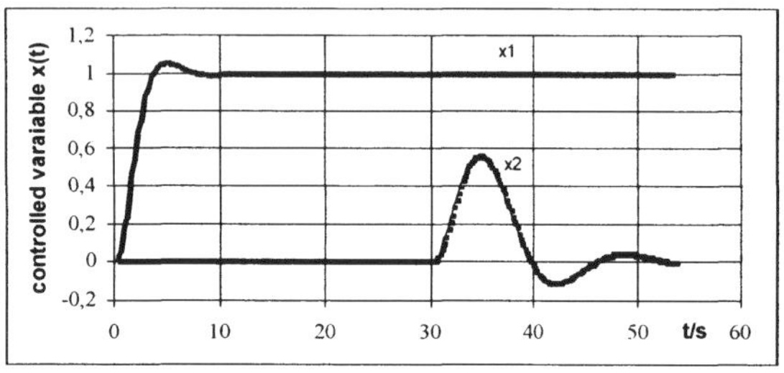

Fig. 10. Response of the DIDO-control circuit without cross coupling

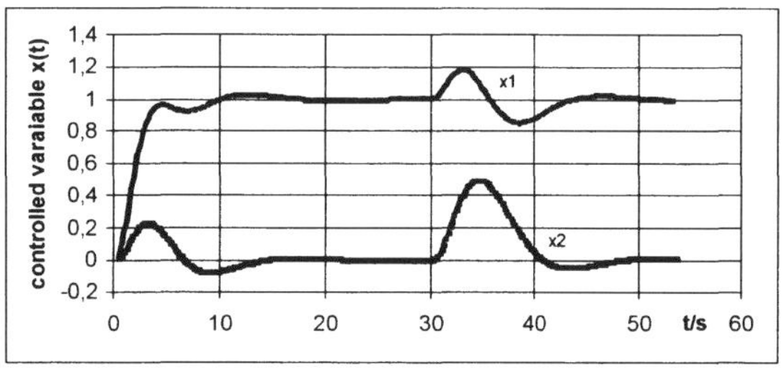

Fig. 11. Response of the DIDO-control circuit without cross coupling

CONCLUDING REMARKS

The Spreadsheet Control Tool is designed for use in teaching as well as industrial application. An extension of the program may be done quite easily by adding sheets and introducing hyperlinks. Cells may be protected to avoid unintentional changes. The program will be continuously extended and examples will included for educational purposes in order to illustrate modern methods of control engineering for application in industry.

REFERENCES

Golubec, B. and Horowitz, I. (1982). *Plant rational transfer approximation from input-output data.* Int. J. Control, **Vol. 36**, No. 4, pp 711-723

Henry, J. *Online Lab*, University of Tennesse at Chattanouga, http://chem.engr.utc.edu

Schaedel, H.M. (1997a). *Parameterschätzung über die Wendetangente und direkter Reglerentwurf in den CAE-Werkzeugen SimTool und SIMID.* 2. VDI/VDE Aussprachetag "Rechnergestützter Entwurf von Regelsystemen", 16./17.Sept., Kassel, GMA-Bericht 32, pp. 9-18.

Schaedel, H.M. (1997b). *A new method of direct PID controller design based on the principle of cascaded damping ratios.* Proc. 4th European Control Conference, Brussels, 1.-4. July 1997,, Paper WE-A H4, BELWARE Information Technology, Waterloo (B).

Schaedel, H.M. (1998). *Neue Prinzipien des direkten Entwurfs parameteroptimierter Regler für stabile, schwingungsfähige und instabile Strecken mit dem CAE-Werkzeug SimTool.* GMA-Kongress '98 Mess- und Automatisierungstechnik, Ludwigsburg, VDI-Berichte 1397, pp. 103-110.

Schaedel, H.M. (1999), *Prozessidentification und Regleroptimierung auf der Basis minimaler Prozessinformation aus der Übergangsfunktion des offenen und geschlossenen Regelkreises.* Internal Report, FH Koeln.

Ziegler, J.G. and G.A. Nichols, (1942). *Optimum settings for automatic controllers*, Trans. ASME, 64, pp. 759-768.

IFAC
Publications
www.elsevier.com/locate/ifac

DEVELOPMENTS FOR THE MATLAB CONTSID TOOLBOX

Hugues Garnier, Marion Gilson and Eric Huselstein

Centre de Recherche en Automatique de Nancy (CRAN),
CNRS UMR 7039
Université Henri Poincaré, Nancy 1, BP 239, F-54506
Vandœuvre-lès-Nancy Cedex, France,
email: `firstname.surname@cran.uhp-nancy.fr`

Abstract: The CONtinuous-Time System IDentification (CONTSID) toolbox is a successful implementation of the methods developed over the last twenty years for estimating continuous-time transfer function or state-space models directly from sampled data. This paper gives a short overview of the toolbox, describes the latest developments and illustrates them on a few examples. Finally, the future plans are briefly summarized. *Copyright © 2003 IFAC*

Keywords: black-box model, continuous-time model, linear system, Matlab toolbox, system identification, software tools.

1. INTRODUCTION

The CONTSID toolbox for Matlab was first presented at ECC'99 (Garnier and Mensler, 1999) and SYSID'2000 (Garnier and Mensler, 2000). The toolbox was also used to compare most of the implemented methods from simulated data (Mensler *et al.*, 1999) and from pilot plant data (Mensler *et al.*, 2000). Recently, the efficiency of the CONTSID identification algorithms has been illustrated on three sets of laboratory process data but also on an industrial binary distillation column (Garnier, 2002). The relevance of direct continuous-time (CT) model identification methods available in the toolbox has been recently illustrated with extensive numerical simulation (Rao and Garnier, 2002).

In this paper, some of the new features of the CONTSID toolbox are described. The purpose is to summarize the novelties and to illustrate them. The main additions first concern some new demonstration programs illustrating the interest of the implemented methods. Secondly the optimal instrumental variable method (*srivc*) developed by P.C. Young along with an associated method for model order selection are now implemented. Thirdly, most of the functions have been adapted to handle the case of non-uniformly sampled data. A last but important improvement of the updated version of the CONTSID toolbox concerns its rewriting to make use of the Matlab's objects provided by the System IDentification Toolbox (SITB).

This paper is organized as follows. A brief overview of the CONTSID toolbox is given in section 2. Section 3 is devoted to the presentation of the recent developments. These latter are illustrated with the help of examples in section 4. In section 5, some of the development plans are briefly discussed.

2. A SHORT OVERVIEW

The CONTSID Matlab toolbox contains time-domain identification methods of CT parametric models for linear-time-invariant SISO and MIMO systems operating in open-loop from sampled data. Version 3.0 is compatible with Matlab 6.x and operates exclusively in the Matlab command window. This updated version has been entirely rewritten, making use of the IDDATA, IDPOLY and IDSS objects used in the SITB. Note that the old syntax does not work anymore. This updated version of the CONTSID toolbox is freely available for academic researchers and can be downloaded from:

http://www.cran.uhp-nancy.fr/

The general scheme for direct CT model identification can be divided into two distinct stages:

- the first stage is specific to CT model identification. It consists in applying to the input/output data a linear transformation (LT) in order to avoid the differential issue.
- the second stage concerns the parameter estimation where most algorithms developed for DT model identification can be used.

There are a multitude of choice for the LT required in the primary stage (Garnier *et al.*, 2004). The toolbox contains indeed most of the LT methods developed over the last twenty years. Parameter estimation techniques required in the second stage and available in the toolbox can be subdivided into the following two families:

(1) transfer function model estimation schemes to identify SISO or MISO systems,
(2) state-space model estimation schemes to identify MIMO systems.

2.1 *Schemes for transfer function identification*

The implemented schemes consider both Equation Error (EE) and Output Error (OE) model structure-based methods.

Several parameter estimation algorithms using all implemented LT are available for the EE approach. First, conventional least-squares (*ls*) based LT algorithms have been completed. In order to overcome the bias problem associated with *ls*-based estimation in presence of noisy data, a bootstrap estimator of instrumental variable (*iv*) type where the *iv* are constructed using an auxiliary model, has also been coupled with all available LT. Some LT approaches have also been associated with a bias compensating least-squares (*bcls*) method.

In the case of OE model structure-based approach, a function to estimate the parameters of MISO models has been implemented using the Levenberg-Marquardt or Gauss-Newton algorithm via sensitivity functions. Table 1 summarizes the methods available in the CONTSID toolbox for SISO and MISO CT transfer function model identification. Acronym meaning and main references can be found in (Garnier and Mensler, 1999).

Methods	**Canonical state-space**			*4sid*
	ls	***iv***	***bcls***	
gpmf	✓	✓	✓	✓
fmf				✓
hmf				✓
rpm				✓
lif				✓

Table 2. Available methods for state-space model identification

2.2 *Schemes for state-space model identification*

A first methodology is based on the *a priori* knowledge of structural indices, and considers the estimation of CT canonical state-space models. It consists first in transforming the canonical state-space model into an equivalent input/output polynomial description which is linear-in-its-parameters and therefore more suitable for the parameter estimation problem. A LT method may then be used to convert the differential equation into a set of linear algebraic equations. The unknown model parameters can finally be estimated by *ls*, *bcls* or *iv*-based algorithms (Garnier *et al.*, 1995).

A second class of multivariable system identification schemes is based on subspace estimation methods. The most commonly known subspace methods were developed for DT model identification. The association of the more efficient LT methods with subspace methods for CT model identification has been recently developed (Johansson *et al.*, 1999; Bastogne *et al.*, 2001)

Table 2 summarizes the methods available in the CONTSID toolbox for CT state-space model identification.

2.3 *Utility programs*

In addition to the parameter estimation functions, the toolbox provides several functions in order to make easier the use of CT models (see table 3). Many routines have been recently added for model validation purpose (*comparec*, *residc*, *srivcstruc*, *selcstruc*). Note however that the old *thetac* format has been replaced by new model objects and is therefore not available anymore. Associated old functions as *presentc* and *thc2poly* have also been removed.

Classification		Methods	Equation error (EE)						Output error (OE)
			ls		*iv*		*bcls*	*srivc*	*coe*
			SISO	MISO	SISO	MISO	SISO	MISO	MISO
linear filters		other						✓	
		gpmf	✓	✓	✓	✓	✓		
		svf	✓	✓	✓	✓			
modulating functions		fmf	✓	✓	✓	✓	✓		
		hmf	✓	✓	✓	✓			
integral methods	specific	rpm	✓	✓	✓	✓			
		lif	✓	✓	✓	✓			
	numerical integration	bpf	✓		✓				
		tpf	✓		✓				
		simpson	✓		✓				
	orthogonal functions	fourier	✓		✓				
		walsh	✓		✓				
	orthogonal polynomials	chebychev 1	✓		✓				
		chebychev 2	✓		✓				
		hermite	✓		✓				
		laguerre	✓		✓				
		legendre	✓		✓				
OE method		Gauss-Newton							✓
		Lev.-Marquardt							✓

Table 1. Available methods for transfer function model identification

Program	Description
idcdemo	is the main program for the CONTSID toolbox demonstrations
idsimc	simulates a system under its continuous-time idpoly or idss form
simc	simulates a system under its transfer function form
comparec	compares measured and model outputs
residc	computes and plots the residuals (simulation error) of a continuous-time model
srivcstruc	computes fit between simulated and measured outputs for a group of model structure of CT ARX type estimated using the srivc method
selcstruc	helps to choose a model structure from the information obtained as the output from srivcstruc

Table 3. Utility programs

3. RECENT DEVELOPMENTS

3.1 *Demonstration programs illustrating the relevance of the CONTSID toolbox approaches*

The relevance of continuous-time model identification methods has been recently illustrated with extensive numerical simulation (Rao and Garnier, 2002). CT model identification approaches available in the CONTSID toolbox are particularly well suited in the case of:

- dominant system modes with different widely natural frequencies,
- fast sampled data,
- non-uniformly sampled data,
- or when the input does not respect the zero-order hold assumption.

The main demonstration program called *idcdemo.m* has thus been modified and provides now several examples illustrating the avantages of the CONTSID toolbox approaches.

3.2 *The srivc method and a model order selection procedure*

The *iv* approach to continuous-time linear model order estimation and identification (*srivc*) suggested by Young and Jakeman (1980) has been recently revisited (Young, 2002). It involves a method of adaptive prefiltering based on the optimal *iv* solution to the problem. It has been used successfully for many years in practical applications and was first available in the CAPTAIN Matlab toolbox[1]. The *srivc* method has been recently added to the CONTSID toolbox. Model order selection is one of the difficult task of the system identification procedure. A natural way to find the most appropriate model order is to compare the results obtained from model structures with different order and delays. A *srivcstruc* order estimation algorithm associated to the *srivc* routine allows the user to automatically search over a whole range of different model orders. Two statistical measures are used for the analysis: the R_T^2 and YIC criteria which are defined as (Young, 2002):

[1] http://www.es.lancs.ac.uk/cres/captain/

$$R_T^2 = 1 - \frac{\hat{\sigma}_\epsilon^2}{\hat{\sigma}_y^2} \quad (1)$$

$$YIC = ln\left\{\frac{\hat{\sigma}_\epsilon^2}{\hat{\sigma}_y^2}\right\} + ln\frac{1}{n_p}\sum_{j=1}^{n_p}\frac{\hat{\sigma}_\epsilon^2 p_{jj}}{\hat{\theta}_j^2} \quad (2)$$

where $\hat{\sigma}_\epsilon^2$ and $\hat{\sigma}_y^2$ denote the variance of the model residuals and of the output signal respectively; $\hat{\theta}_j^2$ is the squared value of the j-th estimated parameter, p_{jj} is the j-th diagonal element of the *srivc* estimated parametric error covariance matrix and n_p is the number of parameters to be estimated. The R_T^2 will be recognized as the coefficient of determination based on the simulated model error. The YIC coefficient is more complex and provides a measure of how well the parameters are defined statistically.

The application result of this model order selection procedure to a SISO thermal process is described in section 4.

3.3 *Identification from non-uniformly sampled data*

The problem of system identification from non-uniformly sampled data is of importance as this case occurs in several applications. It has received, in the last decade, considerable attention in case of discrete-time model identification (see e.g. Sanchis and Albertos (2002) and the references herein). However, the corresponding problem for CT model identification has received less attention. We showed recently how the case of non-uniformly sampled data can be easily handled by some of the most efficient CONTSID toolbox methods (Huselstein and Garnier, 2002). This is illustrated in section 4.

4. EXAMPLES OF NEW FEATURES

The new feature related to the model order selection is first illustrated on a pilot-scale system. The second part of this section is then devoted to the identification of a simulated system from non-uniformly sampled data. Both are parts of the demonstration program *idcdemo.m*.

4.1 *Dryer identification*

4.1.1. *Pilot description* This SISO laboratory set-up is a bench-scale hot air-flow device. It has been used many times to illustrate the performances of other identification methods. Air is pulled by a fan into a 30 cm tube through a valve and heated by a mesh of resistor wires at the inlet. The output is the voltage delivered by a thermocouple proportional to the air temperature at the outlet of the tube. The input is the voltage over the heating device.

4.1.2. *Experiment design* The input signal is chosen to be a Pseudo Random Binary Signal (PRBS) of maximum length. The sampling period is set to 100 ms. Two data sets of 1905 measurements collected in the same conditions is used to perform the model estimation and validation. Mean and linear trend of the signals is removed.

4.1.3. *Model order selection* For continuous-time models estimated by the *srivc* function, various orders and delays can be efficiently studied with the help of the *srivcstruc* routine. The in-line help specifies the required input parameters for the *srivcstruc* function:

```
ze=iddata(y,u,Ts);
V=srivcstruc(ze,[],nn,lambda0);
```

The routine collects in a matrix `nn` all the CT ARX model you want to investigate so that each row of `nn` is of the type `[nb nf nk]`, where `nb` and `nf` are the number of parameters for the numerator and denominator respectively and `nk` represents the number of samples for the delay [2]. Then, a continuous-time model is fitted to the iddata set `ze` for each of the structures in `nn`. For each of these estimated models, two different loss functions YIC and R_T^2 (see equations (1) and (2)) are computed from this estimation data set. `lambda0` is the value of the gpmf filter cut-off frequency required to get an initial estimate. The 10 best model structures that minimize the YIC coefficient are obtained with:

```
selcstruc(V,'YIC',nu,10);
```

where `nu`= 1 here, to indicate the number of input. The 10 best model structures according to YIC are given in table 4. The R_T^2 values are given in complement.

The best fit is clearly obtained for the model of order [1, 1, 5] which has the most negative YIC (-12.08) with an associated R_T^2 equal to 0.945.

4.1.4. *Identification results* The process identification is performed with the *srivc* algorithm *srivc.m*. The model can then be estimated with:

```
m = srivc(ze,[1 1 5],lambda0);
```

where m is the estimated idpoly object.

4.1.5. *Cross-validation results* The simulated output from the estimated models can be compared with the actual output by invoking:

---

[2] Note here an important change in the updated version of the CONTSID toolbox: `nb` and `nf` now represent the *number of parameters* for the numerator and denominator respectively and not *the order* of the numerator and denominator as in the previous version of the toolbox

nb	nf	nk	YIC	R_T^2
1	1	5	-12.08	0.945
1	1	4	-11.89	0.941
1	1	6	-11.58	0.937
1	1	3	-11.26	0.919
1	2	3	-10.39	0.951
2	1	6	-10.09	0.948
2	1	3	-9.88	0.935
1	2	4	-9.30	0.950
2	1	4	-8.70	0.945
2	1	5	-7.27	0.949

Table 4. 10 best model structures according to YIC

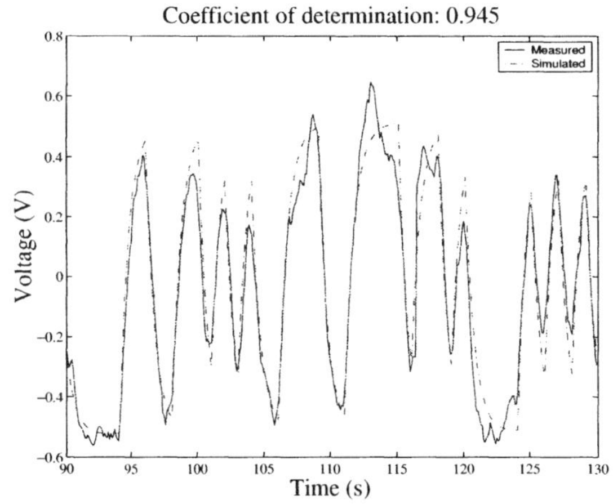

Fig. 1. Cross-validation results for the dryer

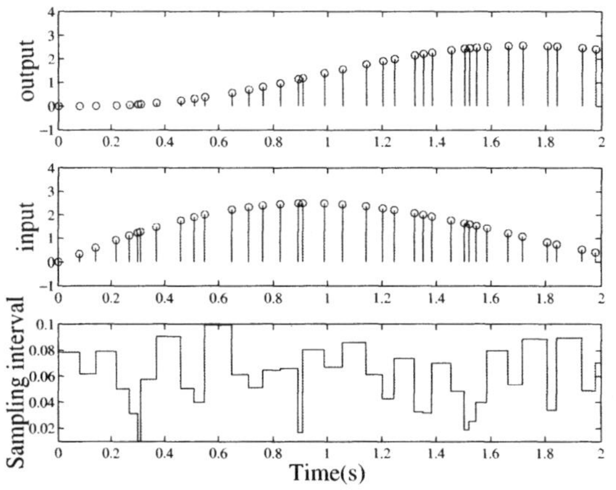

Fig. 2. Zoom on the noise-free output, input, and sampling interval for the simulated system

```
comparec(zv,m);
```

It may be noticed from figure 1 that the simulated output matches quite well to the measured one.

4.2 *Identification from non-uniformly sampled data*

Let us consider a CT second order SISO system without delay described by the following transfer function $G(s)$, where s represents the Laplace variable,

$$G(s) = \frac{5}{s^2 + 2.8s + 4}.$$

The input signal is chosen as the following sum of three sinusoidal signals,

$$u(t) = sin(0.714t) + sin(1.428t) + sin(2.142t).$$

A non-uniform sampling set-up similar as the one used in (Larsson and Söderström, 2002) is chosen here. The distance between two sampling instants is denoted by

$$h_k = t_{k+1} - t_k.$$

We assume that

$$\underline{h} \leq h_k \leq \overline{h},$$

where $\underline{h} > 0$ and $\overline{h}$ are the finite lower and upper bounds respectively. A uniform probability density function $U(\underline{h}, \overline{h})$ is used to describe the variations of the sampling interval, i.e.

$$h_k \sim U(0.01s, 0.1s).$$

3000 data points are used for the identification. Analytic expressions are chosen to compute the noise-free output in order to avoid errors due to numerical simulations. A zoomed plot of the sampled records (figure 2) reveals the non-uniform sampling intervals. A zero-mean white noise is then added to the system output in order to get a signal-to-noise ratio of 10 dB. The simulated output is stored in `y`.

Let us now identify a CT model for this system from the non-uniformly sampled data `z=iddata(y,u,[],'SamplingInstants',t)`, with the optimal IV algorithm *srivc*. The extra information needed are

- the estimated orders and delay of the model stored in the variable `nn=[nb nf nk]=[1 2 0]`,
- the sampling-instant vector `t` containing the time-instants at which the signals are acquired (in s),
- `lambda0` which is the cut-off frequency (in rad/s) of the filter used to get an initial parameter estimate (`lambda0=2`).

This algorithm can now be used as follows:

```
m=srivc(z,nn,lambda0);
```

The estimated parameters can then be displayed with:

```
m;
```

which leads to:

```
Continuous-time IDPOLY model:
y(t) = [B(s)/F(s)]u(t)
B(s) = 5.145
F(s) = s^2 + 2.88 s + 4.004
Loss function 0.1705 and FPE 0.1708
Estimated using SRIVC
```

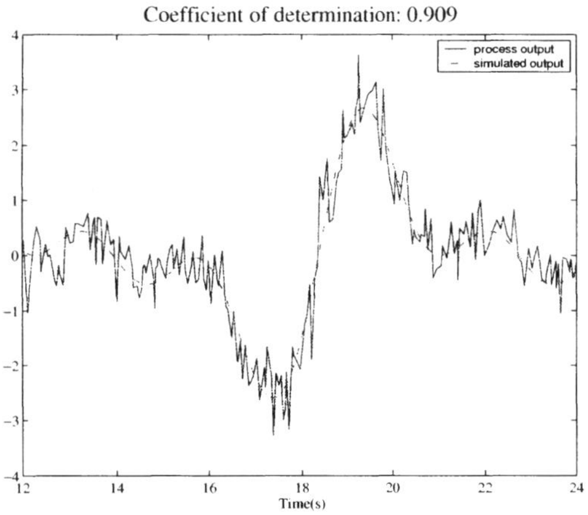

Fig. 3. Validation results for the simulated system

The estimated parameters are very close to the true ones.
For model validation purpose, it is also interesting to compare the simulated and system outputs. This can be done directly with:
`comparec(z,m);`
Figure 3 confirms the good fit of the estimated model.

5. DEVELOPMENT PLANS

The above new possibilities and in particular the rewriting of the CONTSID toolbox functions as object-oriented to ensure the exchange with the Matlab System Identification toolbox (SITB) already significantly extend the usability of the toolbox. Therefore, we plan in the near future to concentrate on easy user-friendly use. We foresee one main solution for that:

- implementation of a graphical user interface. We are working on the development of an easy-to-use interface which allows the user to perform identification, data and model analysis, as well as model validation by "click and mouse" operations.

6. REFERENCES

Bastogne, T., H. Garnier and P. Sibille (2001). A PMF-based subspace method for continuous-time model identification. Application to a multivariable winding process. *International Journal of Control* **74**(2), 118–132.

Garnier, H. (2002). Continuous-time model identification of real-life processes with the CONTSID toolbox. In: *15th Triennial IFAC World Congress on Automatic Control.* Barcelona (Spain).

Garnier, H. and M. Mensler (1999). CONTSID: a continuous-time system identification toolbox for Matlab. In: *5th European Control Conference (ECC'99).* Karlsruhe (Germany).

Garnier, H. and M. Mensler (2000). The CONTSID toolbox: a Matlab toolbox for CONtinuous-Time System IDentification. In: *12th IFAC Symposium on System Identification (SYSID'2000).* Santa Barbara (USA).

Garnier, H., M. Mensler and A. Richard (2004). Continuous-time model identification from sampled data. Implementation issues and performance evaluation. *International Journal of Control. To appear.*

Garnier, H., P. Sibille and A. Richard (1995). Continuous-time canonical state-space model identification via Poisson moment functionals. In: *34th IEEE Conference on Decision and Control (CDC'95).* Vol. 2. New Orleans (USA). pp. 3004–3009.

Huselstein, E. and H. Garnier (2002). An approach to continuous-time model identification from non-uniformly sampled data. In: *41st IEEE Conference on Decision and Control (CDC'02).* Las-Vegas, Nevada (USA).

Johansson, R., M. Verhaegen and C.T. Chou (1999). Stochastic theory of continuous-time state-space identification. *IEEE Transaction on Signal Processing* **47**(1), 41–50.

Larsson, E.K. and T. Söderström (2002). Identification of continuous-time AR processes from unevenly sampled data. *Automatica* **38**(4), 709–718.

Mensler, M., H. Garnier, A. Richard and P. Sibille (1999). Comparison of sixteen continuous-time system identification methods with the CONTSID toolbox. In: *5th European Control Conference (ECC'99).* Karlsruhe (Germany).

Mensler, M., H. Garnier and E. Huselstein (2000). Experimental comparison of continuous-time model identification methods on a thermal process. In: *12th IFAC Symposium on System Identification (SYSID'2000).* Santa Barbara (USA).

Rao, G.P. and H. Garnier (2002). Numerical illustrations of the relevance of direct continuous-time model identification. In: *15th Triennial IFAC World Congress on Automatic Control.* Barcelona (Spain).

Sanchis, R. and P. Albertos (2002). Recursive identification under scarce measurements - convergence analysis. *Automatica* **38**(3), 535–544.

Young, P.C. (2002). Optimal IV identification and estimation of continuous-time TF models. In: *15th Triennial IFAC World Congress on Automatic Control.* Barcelona (Spain).

Young, P.C. and A.J. Jakeman (1980). Refined instrumental variable methods of time-series analysis: Part III, extensions. *Int. J. Control* **31**, 741–764.

IFAC
Publications
www.elsevier.com/locate/ifac

detectNARMAX: A GRAPHICAL USER INTERFACE FOR STRUCTURE DETECTION OF NARMAX MODELS USING THE BOOTSTRAP METHOD

Esfandiar Shafai, Mikael Bianchi and Hans Peter Geering

Measurement and Control Laboratory,
Swiss Federal Institute of Technology (ETH) Zurich, Switzerland

Abstract: This paper describes a MATLAB-based program called detectNARMAX that utilizes the bootstrap method, a numerical procedure for estimating the distribution of statistical parameters, to find the best NARMAX model structure representing the nonlinear behavior of a system using its noisy input-output data. The performance of the bootstrap-based structure detection techniques is demonstrated by using synthetic data generated by simulation as well as real data measured to determine a model structure for the thermal impact on a sensor system. *Copyright © 2003 IFAC*

Keywords: nonlinear, identification, computer/software

1. INTRODUCTION

Many systems that have nonlinear behaviour may be modelled by a NARMAX model structure (Nonlinear AutoRegressive, Moving Average with eXogenous input signal). This structure describes both the stochastic and deterministic behaviour of the system. It models the input-output relationship of the system as the nonlinear, but linear-in-the-parameters (LIP), difference equation of the form

$$\begin{aligned} y(t) = F^p[&y(t-1), \ldots, y(t-n_y), u_1(t), \\ &u_1(t-1), \ldots, u_1(t-n_u), \\ &\vdots \\ &u_m(t), u_m(t-1), \ldots, u_m(t-n_u), \\ &e(t-1), \ldots, e(t-n_e)] + e(t), \end{aligned} \tag{1}$$

where y is the output, $u_1(t)\ldots u_m(t)$ are the controlled inputs (i.e., exogenous variables), $e(t)$ is the noise input, and F is a nonlinear mapping. This nonlinear mapping may include a variety of nonlinear terms, such as terms raised to an integer power ($u_3^2(t-3)$), products of past inputs ($u_1(t)u_1(t-1)$), past outputs ($y(t-1)y(t-2)$), or cross-terms ($u_2^2(t)y(t-1)$).

As introduced in equation (2), the maximum number of terms in a NARMAX model depends on the maximum lag on the inputs (n_u), the maximum lag on the output (n_y), the maximum lag on the error (n_e), the number of the inputs (m_u), and the maximum nonlinearity order (p):

$$\begin{aligned} \nu &= \sum_{i=1}^{p} \nu_i \\ \nu_i &= \frac{\nu_{i-1}[n_y + m_u(n_u+1) + n_e + i - 1]}{i}, \quad \nu_0 = 1\,. \end{aligned} \tag{2}$$

Based on this formula, it can be stated that even for moderately complex models the number of candidate terms becomes very large. In contrast, many systems are described by NARMAX models having only a few terms. Therefore, in the process of structure detection, a subset of the total possible candidate terms has to be selected. For this highly overparametrized model identification that consists of structure detection as well as parameter estimation, a bootstrap-based algorithm has been proposed in (Kukreja *et al.*, 1999).

Utilizing this algorithm in the MATLAB-based program detectNARMAX, we developed a graphical user interface that allows to keep track of both the structure detection process and the parameter estimation. This is particularly helpful in the case of a higher model order with a large number of candidate terms. The program also allows the user to make a preselection of candidate terms that may be known in advance from the physical phenomena of the system.

The paper is organized as follows: After a short introduction to the bootstrap sampling method in Section 2, the Bootstrap structure detection algorithm implemented in detectNARMAX is presented in Section 3. Section 4 introduces the graphical user interface. We demonstrate in Section 5 the performance of the bootstrap-based structure detection technique by using synthetic data generated by simulation as well as by using real data obtained from experiment to determine a model structure for the thermal impact on a sensor system.

2. BOOTSTRAP SAMPLING METHOD

The bootstrap method, introduced first by Efron (Efron, 1979), is a numerical procedure for estimating the distribution of statistical parameters. Introductions to bootstrap can be found in (Efron and Tibshirani, 1993) and (Davison and Hinkley, 1997).

The main idea behind the bootstrap method can be described as follows. Let $x = [\xi_1, \xi_2, \ldots, \xi_N]$ be a sample of N independent, identically distributed (i.i.d.) random variables with the unknown distribution function F. This sample is called an i.i.d. sample. We are interested in estimating some parameter $\theta = \theta(F)$ associated with F. As an estimator for θ, the statistic $s(F)$ is used to get the estimated value $\hat{\theta} = s(x)$ from the observed data x. The same statistic will also be used to find the accuracy of the estimated value $\hat{\theta}$. Since the distribution function F is unknown, the empirical distribution $\hat{F}(x)$ is utilized to generate B bootstrap resamples, $(x_1^*, x_2^*, \ldots, x_B^*)$ and estimate B parameters $(\hat{\theta}_1^*, \hat{\theta}_2^*, \ldots, \hat{\theta}_B^*)$, where asterisks indicate that we are working with bootstrap samples. From the empirical distribution $\hat{F}_\theta(x)$ of the parameter, the confidence intervals for $\theta(F)$ can be computed.

3. BOOTSTRAP STRUCTURE DETECTION ALGORITHM

1. The extended least squares (ELS) method is used to compute an initial estimate of the unknown parameters of the initially assumed NARMAX model structure and the corresponding residuals:

$$\hat{\theta}_{ELS} = [\Phi^T\Phi]^{-1}\Phi^T Y, \tag{3}$$

$$\hat{\varepsilon} = Y - \hat{Y} = Y - \Phi\hat{\theta}_{ELS}. \tag{4}$$

2. Since it is assumed that the residuals $\hat{\varepsilon}$ are independent and identically distributed (i.i.d.), they are resampled B times with replacement

$$\hat{F} \rightarrow (\varepsilon_1^*, \varepsilon_2^*, \ldots, \varepsilon_B^*) = \varepsilon^* \tag{5}$$

and added to the predicted output to generate B bootstrap replications of the output

$$Y_i^* = \Phi_i\hat{\theta}_{ELS} + \varepsilon_i^*, \quad \text{for } i = 1, 2, \ldots, B. \tag{6}$$

3. Utilizing the ELS method the bootstrap data sets generated in step 2 are used for the estimation of bootstrap parameter replications:

$$\hat{\theta}^*_{ELS,\,i} = [\Phi_i^T\Phi_i]^{-1}\Phi_i^T Y_i^*, \quad \text{for } i = 1, 2, \ldots, B. \tag{7}$$

4. The significance of the parameters is determined by forming percentile intervals for each parameter (see Figure 1). The estimates from the B parameter replications $\hat{\theta}^*_{ELS}$ are ranked in increasing order and the $B \cdot \alpha$th and $B \cdot (1-\alpha)$th values in the ordered list of the B replications are used as upper and lower bounds for the parameter deviation with an αth and a $(1-\alpha)$th level of significance, respectively.

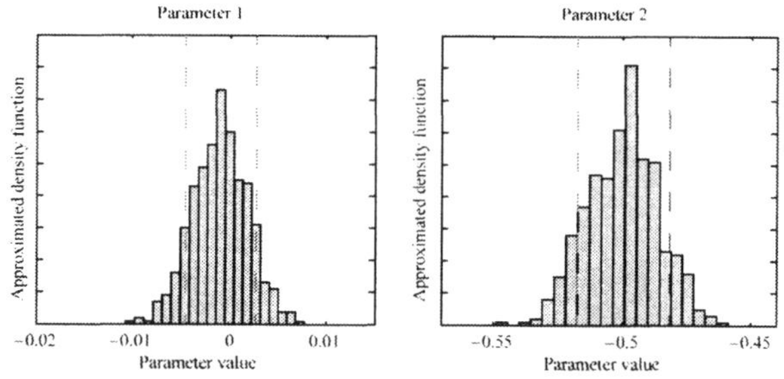

Fig 1. Percentile intervals for two parameters

5. The significance of each parameter is determined by checking if zero lies in its interval. This is the case for the Parameter 1 in Figure 1.

6. If zero lies in the interval of a parameter, the parameter is removed from the regression.

7. The ELS method is used to compute a new estimate of the unknown parameters of the new NARMAX model structure generated in the previous step and their residuals:

$$\hat{\theta}_{ELS,\,new} = [\Phi_{new}^T\Phi_{new}]^{-1}\Phi_{new}^T Y, \tag{8}$$

$$\hat{\varepsilon} = Y - \hat{Y}_{new} = Y - \Phi_{new}\hat{\theta}_{ELS,\,new}. \tag{9}$$

8. Repeat from step 2 until convergence.

4. DESCRIPTION OF THE PROGRAM

At program start, the detectNARMAX main information window opens (Figure 1).

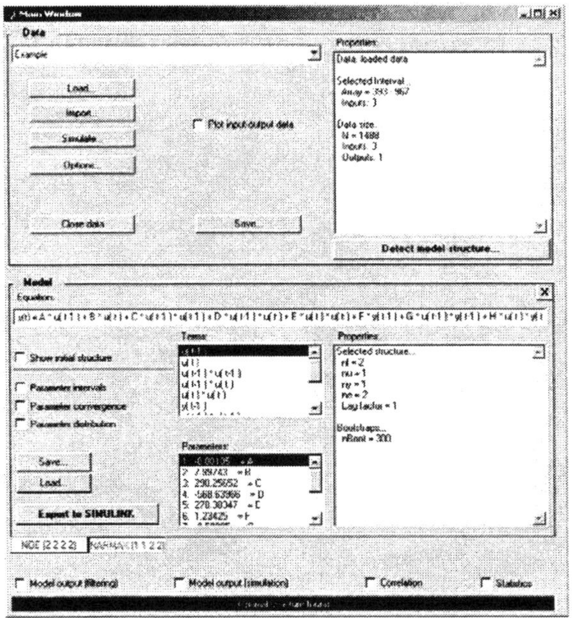

Fig 2. Main information window of detectNARMAX

This window is divided in two sections. In the upper section the input-output data can be loaded from a previously saved mat-file (load...), imported from the MATLAB workspace (import...), or generated by a simulation (simulate...). In the simulation mode, the model equation is entered directly in the corresponding field (Figure 3). The data simulated may be saved for future usage as a mat-file (save...).

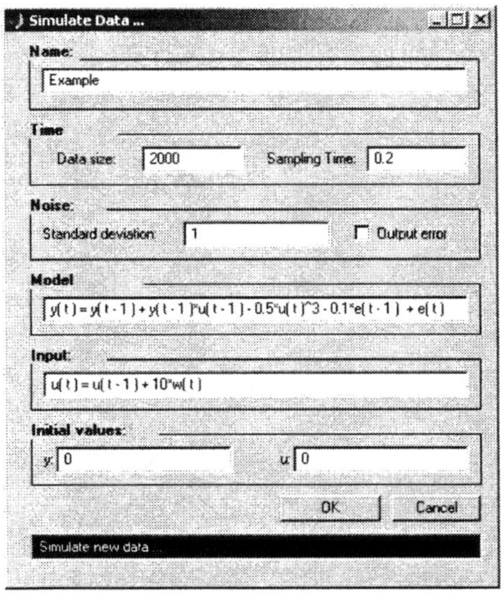

Fig 3. Generating data by simulation

By pressing the button "options..." the user may select a subset of data to be used for the structure detection. This can be done either by defining the values of the starting and the stopping time numerically (Figure 4) or selecting the data range graphically (Figure 5) by pressing the button "select..." shown in Figure 4.

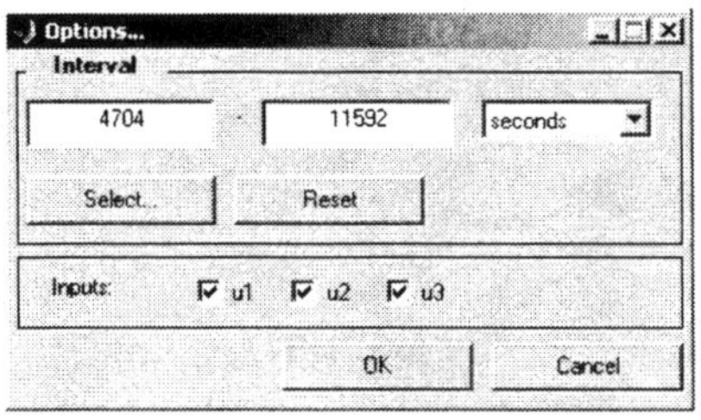

Fig 4. Numerical selection of the data range

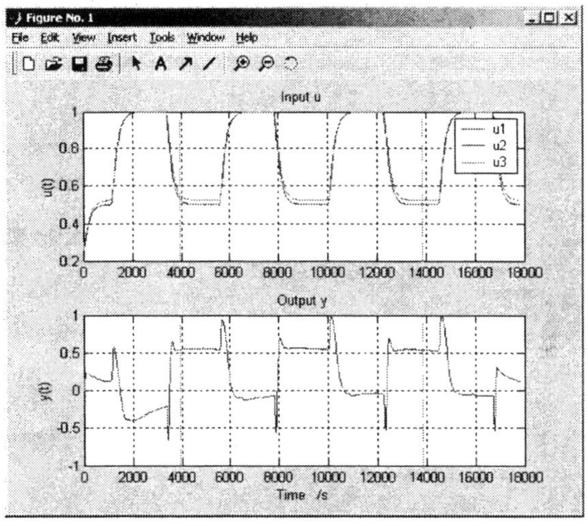

Fig 5. Graphical selection of the data range

Once the input-output data is defined, the model structure detection can be started with the button "detect model structure...". A corresponding window is opened where the parameters required for the bootstrap-based structure detection are defined (Figure 6).

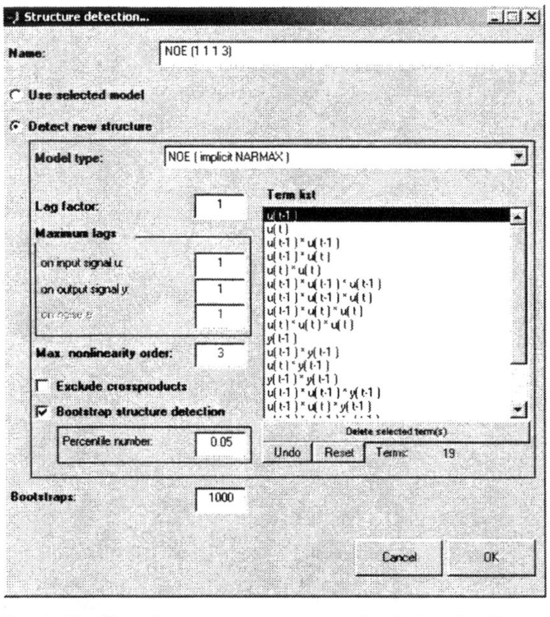

Fig 6. Define the parameters required for the bootstrap-based structure detection

In the main frame of this window the NARMAX model order can be defined by inserting desired values for the maximum lags n_u, n_y, n_e and the maximum nonlinearity order p. The corresponding candidate terms for this model order are then listed in the listbox "Term list". This list may be modified by the user. Candidate terms that are known in advance from the physical phenomena of the system can be preselected while the invalid terms are removed from the list. Both the parameters defining the percentile intervals and the number of bootstrap replications are also defined in this window. A new structure is only detected if the radio button "Detect new structure" is selected before "OK" is pressed. If the radio button "Using selected model" is chosen, the program only estimates the parameters of a preselected model structure that is loaded from an earlier session (without structure detection).

The model structures generated can be viewed and analyzed in the lower section of the main information window shown in Figure 2. More than one model may be available in a session. Each model can be activated by pressing the corresponding button (NARMAX 1122 or NOE 2222 in Figure 2). The model equation is presented in a field at the top of this section. The terms selected for the model structure as well as the corresponding parameter values are listed in two listboxes ("terms" and "parameters") that are interconnected. In order to view the distribution and the percentile intervals of the parameters estimated, the user first makes a selection of the terms or parameters from the corresponding listbox and then chooses the appropriate checkboxes "parameter distribution" and "parameter intervals" (Figure 7 and Figure 8). In addition, the resulting model can be exported to the SIMULINK program for further simulations ("Export to SIMULINK...").

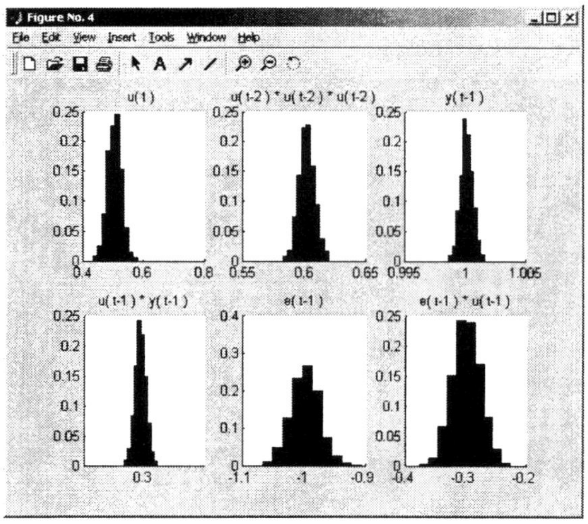

Fig 7. Parameter distributions

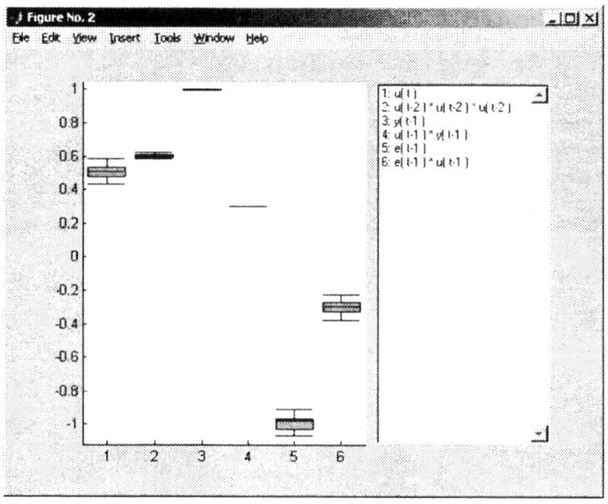

Fig 8. Parameter intervals

5. EXAMPLES

Two examples are selected to show the performance of the bootstrap-based structure detection technique. For a comparison with other techniques, such as t-test and stepwise regression, please refer to (Kukreja *et al.*, 1999).

5.1 Synthetic data generated by simulation

For generating input-output data by simulation we choose in this example a system with nonlinear output-error structure (NOE):

$$\begin{aligned} z(t) &= a_1 z(t-1)u(t-1) + a_2 z(t-2) \\ &\quad + b_0 u(t) + b_1 u(t-1)^2 \\ y(t) &= z(t) + e(t), \end{aligned} \tag{10}$$

which may also be written in the general NARMAX model structure:

$$\begin{aligned} y(t) &= a_1 y(t-1)u(t-1) + a_2 y(t-2) \\ &\quad + b_0 u(t) + b_1 u(t-1)^2 \\ &\quad + c_1 e(t-1)u(t-1) + c_2 e(t-2) + e(t)\,. \end{aligned} \tag{11}$$

In the simulation, zero-mean Gaussian white-noise sequences with a variance of 1 are used for the input $u(t)$ as well as for the error signal $e(t)$. The parameters for the simulation are chosen to be $a_1 = 0.5$, $a_2 = -0.8$, $b_0 = 1$, $b_1 = 0.05$, $c_1 = -a_1 = -0.5$, and $c_2 = -a_2 = 0.8$.

For the bootstrap structure detection and identification of the parameters a data length of $N = 5000$, a number of $B = 300$ bootstrap replications, and an initial NARMAX structure with $n_u = 2$, $n_y = 2$, $n_e = 2$, $m_u = 1$, and $p = 2$ are used.

According to equation (2), a model of this order has $\nu = 35$ candidate terms. However, as indicated in equation (11) only six "true parameters" suffice to describe this system. Every candidate parameter is tested for significance at the 90% confidence level.

After two iteration steps, where each step consists of 30 ELS iterations, the correct structure is detected and the parameters are identified. After the first iteration step all of the 29 "false parameters" are removed from the candidate list. For each parameter Figure 9 shows the maximum range, the confidence interval, and the parameter value estimated ($\hat{a}_1 = 0.495$, $\hat{a}_2 = -0.800$, $\hat{b}_0 = 1.005$, $\hat{b}_1 = 0.064$, $\hat{c}_1 = -0.503$, and $\hat{c}_2 = 0.800$). The 29 "false parameters" containing zero in their confidence interval and the correponding terms in the list at the top of the Figure are quite evident.

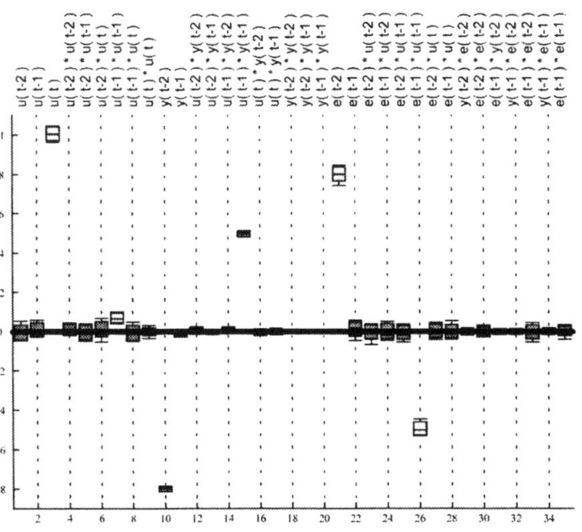

Fig 9. Intervals for all of the candidate parameters

Figure 10 shows the estimated distribution of the six "true parameters" as well as the estimated distribution of three parameters out of the 29 "false parameters".

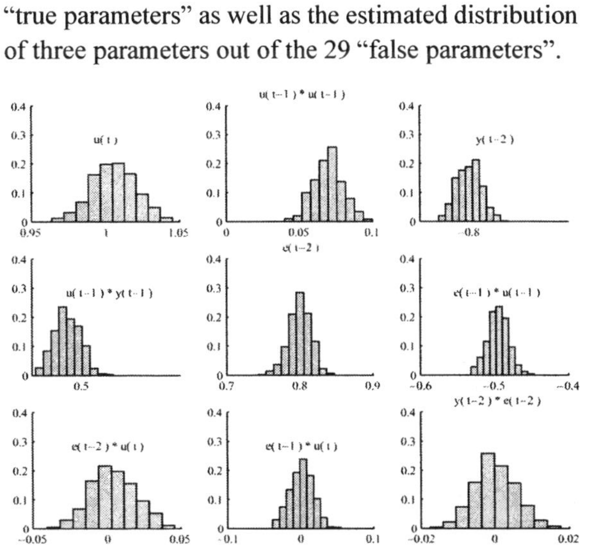

Fig 10. Estimated distribution for all six of the "true parameters", and estimated distribution of three parameters out of the 29 "false parameters"

5.1 Thermal impact on a sensor system

In this example a model structure for the thermal impact on a sensor system has to be detected. For this purpose the input-output data are obtained from an experiment on a test bench. It is assumed that only the temperatures at two different locations of the sensor system determine the thermal impact. Figure 11 shows this dependency of the sensor signal $y(t)$ on the temperature signals (T_1 and T_2) measured on the test bench.

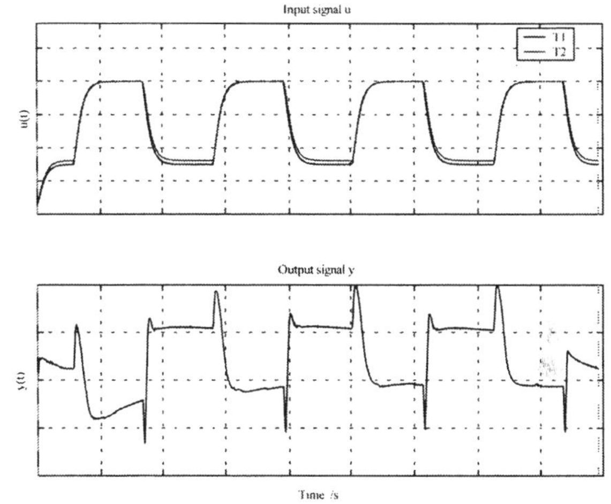

Fig 11. Thermal impact of a sensor system illustrated by the dependency of the sensor signal $y(t)$ on the temperature signals T_1 and T_2

In order to fulfill the condition of i.i.d. for the residual ε, an initial NARMAX structure with $n_u = 2$, $n_y = 2$, $n_e = 2$, $m_u = 2$, and $p = 3$ is used, where all nonlinear terms on y and e are manually removed from the structure in advance. This model structure is referred to as '$\mathcal{M}_i$' which consists of 87 terms. Figure 12 shows the autocorrelation function of the residuals resulting from '$\mathcal{M}_i$'.

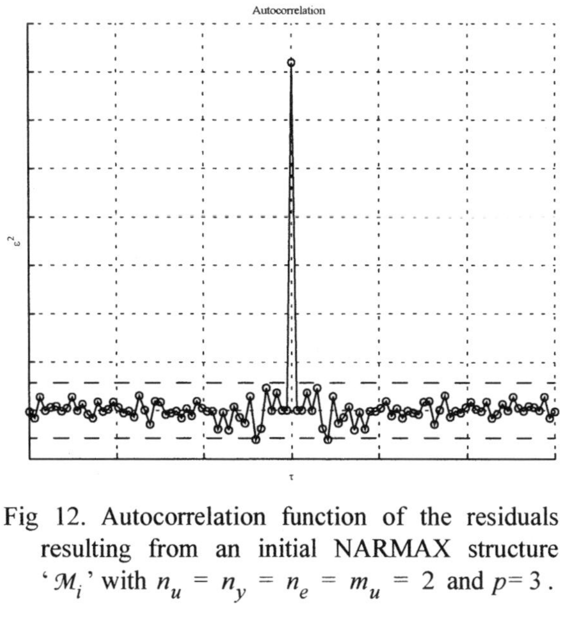

Fig 12. Autocorrelation function of the residuals resulting from an initial NARMAX structure '$\mathcal{M}_i$' with $n_u = n_y = n_e = m_u = 2$ and $p = 3$.

Starting with '$\mathcal{M}_i$', the bootstrap structure detection algorithm detects a final NARMAX model structure '$\mathcal{M}_f$' with 23 terms using a data length of $N = 1488$ and a number of $B = 300$ bootstrap replications. The confidence intervals of the parameters indentified and the correponding terms are shown in Figure 13.

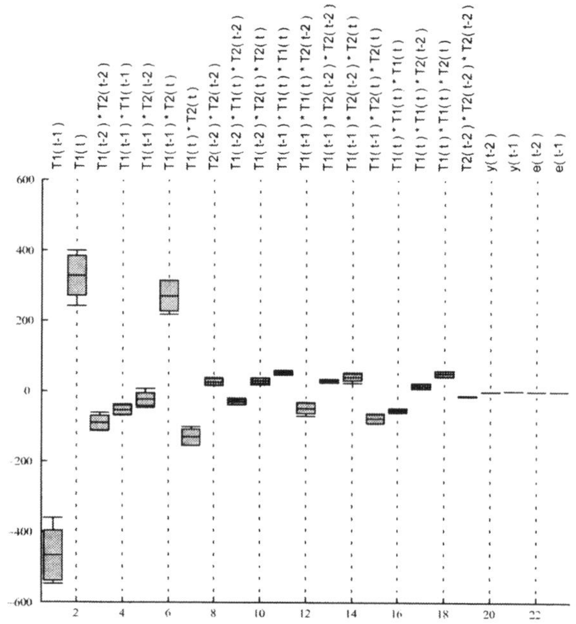

Fig 13. Confidence intervals of the parameters identified and the correponding terms for the final model structure '$\mathcal{M}_f$' with 23 terms.

To check the qualities of the estimated model '$\mathcal{M}_f$', the whiteness of the corresponding residuals as well as the correlation between the residuals and the two input signals (T_1 and T_2) are calculated and depicted in Figure 14.

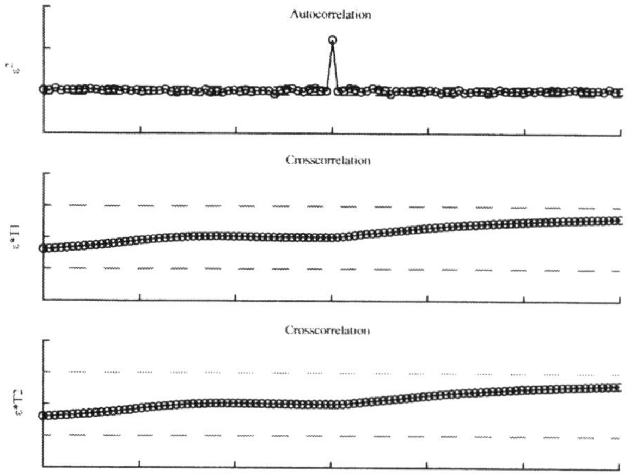

Fig 14. Autocorrelation of the residuals and cross-correlations between the residuals and the two input signals T_1 and T_2 for the final model structure '$\mathcal{M}_f$' with 23 terms.

REFERENCES

Kukreja, S.L., H.L. Galiana and R.E. Kearney (1999). Structure Detection of NARMAX Models using Bootstrap Method. In: *Proceedings of the 38th Conference on Decision & Control*, 1071-1076. IEEE, Phoenix, AZ, USA.

Tjärnström, F. (1999). *The Use of Bootstrap in System Identification*. Report number LiTH-ISY-R-2113, Department of Electrical Engineering, Linköping University, Sweden.

Efron, B. (1979). Bootstrap methods: another look at the jackknife. *The Annals of Statistics*, 7, 1-26.

Efron, B. and R.J. Tibshirani (1993). *An Introduction to the Bootstrap*. Chapman&Hall, New York.

Davision, A.C. and D.V. Hinkley (1997). *Bootstrap Methods and their Applications*. Cambridge University Press.

IFAC
Publications
www.elsevier.com/locate/ifac

DATA-BASED METHODS FOR PROCESS ANALYSIS, MONITORING AND CONTROL

John F. MacGregor

McMaster Advanced Control Consortium
Department of Chemical Engineering
McMaster University, Hamilton, ON, Canada L8S 4L7

Abstract: This paper gives an overview of methods for utilizing the massive amounts of highly correlated data available in most process databases. These data matrices are almost always of less than full statistical rank, and therefore latent variable methods are shown to be well suited to obtaining useful subspace models for treating a variety of important industrial problems. The following problems are discussed and illustrated with industrial examples: (i) the analysis of historical databases and trouble-shooting process problems; (ii) process monitoring and FDI; (iii) building soft sensors or inferential models; (iv) using of multivariate information from novel sensors; (v) subspace identification; and (vi) process control in reduced dimensional subspaces. In each of these problems latent variable models provide the framework on which solutions are based.

Keywords: Latent variables, PCA, PLS, subspace models, soft sensors, monitoring, control, identification

1. INTRODUCTION

System identification involves the building of empirical dynamic models of processes from data collected from designed experiments on a set of manipulated variables (MV's). If the design satisfies certain identifiability conditions (Soderstrom and Stoica, 1989), then one can obtain causal models which estimate the effect of changes in the manipulated variables on the controlled variables. As important as system identification is, it only represents one area of empirical model building that is important to systems engineering.

This paper looks at another equally important empirical modeling problem, namely that of utilizing the massive amounts of process operating data collected routinely by process computers. These data sets are usually much larger, often involving hundreds to thousands of variables collected over long periods of time. The data do not result from designed experiments, but from the result of routine process operation. As a result, the data are not causal in nature (i.e. models built from them cannot be used to infer the causal effects of any variable on any other). The measured variables are usually very highly correlated with one another. This correlation among the variables is usually of such a high degree that the resulting data matrices are typically of very low statistical rank (i.e., the signal or predictable component of them is non-full rank). This reduced rank nature of the data arises from the fact that even though hundreds of variables may be measured, their variation is the result of independent variations in only a small number of underlying latent variables. Consider the example of a single distillation column in the petrochemical industry. One might measure up to a hundred variables on it such as the temperatures on many of the trays, pressures, compositions and flowrates of the feeds, tops, bottoms and side draws, etc. However, if under normal closed-loop operation, the only things that are changing are feed composition, flow and an environmental factor (e.g., temperature), then it might be expected that the statistical rank of the data matrix (**X**) would be on only 3 or 4 rather than 100. In fact, this type of reduced dimension is the norm with such process data. Furthermore, in most process databases, there is usually a large amount of missing data. Data may be missing in certain multivariate observations due to sensors that are off-line because of faults or routine maintenance, due to sensors having different sampling rates, etc. As a rule-of-thumb, we are usually dealing with datasets where 10% to 20% of the data is missing, and at some times or with some variables, the percent missing can be much higher. If the methods used to address such datasets cannot handle missing data in a very easy and efficient method, then they will be of very little use. A final characteristic of these process data is that the signal to noise ratio is often quite low in any one variable. Consider the case of monitoring a process that is operating around some steady-state condition. If the control system and the process engineers are doing their job, there will be little variation in many of the variables. It is only by utilizing efficient multivariate statistical methods to extract some information from all the variables that the signal to noise ratio becomes larger. Even if one can model only 40% to 50% of the variation in such cases, that is usually sufficient to obtain very useful information on the process.

The objective of this paper is to present an overview of multivariate latent variable models and their use in exploiting such data. In doing so, several classes of problems will be treated, including: (i) the analysis and troubleshooting of process problems; (ii) process monitoring, fault detection and isolation (FDI); (iii) building soft sensors or inferential models; (iv) extracting information from new multivariate sensors and using it for process monitoring and control; (v) process identification; and (v) process control in reduced dimensional spaces.

2. LATENT VARIABLE MODELS

2.1 Concept of Latent Variables

Consider a dataset $\mathbf{x} = [x_1, x_2, ..., x_k]$ where k variables are measured. The concept behind a latent variable model for the data is that the process under observation is actually driven by a set of $a < k$ latent variables $\mathbf{z} = [z_1, z_2, \ldots, z_a]$. These variables are not observable but their influence can be seen in the measured variables, $\mathbf{x}$. Their relationship is modeled by:

$$\mathbf{x} = \mathbf{z}\mathbf{P}^T + \mathbf{e} \tag{1}$$

where $\mathbf{z}$ is $1 \times a$, $\mathbf{P}$ is $k \times a$, and $\mathbf{x}$ is $1 \times k$. The last term in the model, $\mathbf{e}$, is considered to be random error. This would be made up of unstructured sources of variability, such as measurement error, sampling error, and unobservable process disturbances. Since $\mathbf{z}$ is unmeasured and $\mathbf{P}$ is unknown, $\mathbf{z}$ is not identifiable in equation (1). In fact, the same values for $\mathbf{x}$ would arise if $\mathbf{z}$ and $\mathbf{P}$ are, respectively, replaced with $\mathbf{z}^* = \mathbf{zC}$ and $(\mathbf{P}^*)^T = \mathbf{C}^{-1}\mathbf{P}^T$, where $\mathbf{C}$ is any non-singular $A \times A$ matrix. Thus, the model is more commonly given as:

$$\mathbf{x} = \mathbf{t}\mathbf{P}^T + \mathbf{e} \tag{2}$$

where $\mathbf{t}$ is understood to be some transform $\mathbf{zC}$ of the actual latent variables $\mathbf{z}$. The transformation of $\mathbf{z}$ to $\mathbf{t}$ is simply a change of basis so that the points in $\mathbf{t}$ will lie in the same vector space as those in $\mathbf{z}$ but expressed in a different basis. In general, the actual latent variables are not as important as the overall space they generate. Therefore, any basis, $\mathbf{t}$, will be sufficient to define this space. For a given set of n observations following equation (2), the model can be written:

$$\mathbf{X} = \mathbf{T}\mathbf{P}^T + \mathbf{E} \tag{3}$$

where $\mathbf{X}$ is $n \times k$, $\mathbf{T}$ is $n \times a$, and $\mathbf{P}$ is $k \times a$.

2.2 The LVMR Model

The LVMR model is an extension of model (3) obtained by considering two spaces $\mathbf{X}$ ($n \times k$) and $\mathbf{Y}$ ($n \times m$) with a common underlying latent structure as follows:

$$\mathbf{X} = \mathbf{T}\mathbf{P}^T + \mathbf{E} \tag{4}$$

$$\mathbf{Y} = \mathbf{T}\mathbf{Q}^T + \mathbf{F} \tag{5}$$

where $\mathbf{E}$ and $\mathbf{F}$ are assumed to random error and the latent variables are modeled as linear combinations of the x's:

$$\mathbf{T} = \mathbf{XW} \tag{6}$$

The latent variable space generated by the columns of $\mathbf{T}$ is usually of much smaller dimension than $\mathbf{X}$ (and possibly $\mathbf{Y}$).

In the LVMR model, there is no intrinsic difference between the $\mathbf{X}$ and $\mathbf{Y}$ spaces. The LV model (4) and (5) is symmetric in $\mathbf{X}$ and $\mathbf{Y}$, They are both functions of the latent variables, they are both measured with error, and there is no assumption of a causality direction. The division of the data into $\mathbf{X}$ and $\mathbf{Y}$ matrices is arbitrary, and arises mainly from the intended use of the model rather than in the features of the data modeled. Often, the $\mathbf{Y}$ data are available only for the building of the model, and will not be available when the model is to be used. In such cases, the $\mathbf{Y}$ data may be used for model building simply because it is expected that they will help one obtain better estimates of the latent variable space $\mathbf{T}$. In other cases the y variables may be important quality and productivity variables (that are only measured off-line in a quality control lab), but it is desired to obtain latent variables that are highly predictive of $\mathbf{Y}$. The model can always be rearranged to express the prediction of $\mathbf{Y}$ in terms of $\mathbf{X}$ as:

$$\hat{\mathbf{Y}} = \mathbf{T}\mathbf{Q}^T = \mathbf{XW}\mathbf{Q}^T = \mathbf{X}\hat{\mathbf{B}} \tag{7}$$

but it is important to remember that this is not the model, rather just a rearrangement of the prediction equation for $\mathbf{Y}$.

There are many nonlinear versions of latent variable models. The simplest just use transformations of the

x and y's or just expand the **X** matrix with quadratic or other nonlinear functions of the x variables. Augmenting X with such additional terms does not cause problems with the LV models since only a small number (a) of latent variables will be used to describe the operational space. Other nonlinear approaches (Wold et. al. 1989) use linear LV models for the **X** and **Y** spaces, but relate the latent variables of the two spaces by nonlinear models. Dynamic ARX models are easily obtained by simply including lagged values of past x and y variables in the **X** matrix (e.g. MacGregor et al. 1991).

3. ESTIMATION OF LV MODELS

In this section, we consider estimation methods for the score matrix (**T**) and loading matrices (**P**, **W**, **Q**) in the latent variable model. A more complete discussion and some general objective function frameworks for these latent variable multivariate regression models are given in Burnham et al. (1996).

Principal Component Analysis (PCA)

Consider first the case where one has only a single ($N \times K$) data matrix **X**. In this case Principal Component Analysis (PCA) or Singular Value Decomposition (SVD) can be used to decompose the **X** matrix into the latent variable form of equation (3). The principal components are the largest variance directions in the column space of **X**. They are defined as linear combinations of the original variables (usually scaled and mean centered) and the loading matrix **P** which relates the **X** matrix to the score matrix **T** (**W**=**P** in this case) are the orthonormal eigenvectors of $\mathbf{X}^T\mathbf{X}$ ordered according to the size of the eigenvalues (variances of the scores). For small data matrices all the principal components (min(N, K)) can be estimated via SVD, but for large data matrices, the NIPALS algorithm [Geladi and Kowalski, 1986] is more commonly used to estimate them one at a time in a recursive manner.

The number of principal components ($a << \min(n, k)$) required to adequately model the covariance structure of **X** is often decided by a cross-validation procedure [Wold, 1978] which yields statistical tests for their significance based on a prediction criterion. The first a principal components

$$t_l = \mathbf{x}\mathbf{w}_l \quad (l = 1, 2, \ldots a)$$

define a set of latent variables that summarizes the movement of original variables in a low dimensional hyper-plane. Their observed values (scores) defined by $\mathbf{T} = \mathbf{XW}$ give the location of the objects on the plane, and the squared perpendicular distances of the observations from the plane (residuals) are given by

$$\text{SPE}_i = \sum_{j=1}^{K} (x_{ij} - \hat{x}_{ij})^2 \tag{8}$$

where the predicted values are given by

$$\hat{\mathbf{X}} = \mathbf{TP}^T \tag{9}$$

Therefore, the most significant variation in the **X**-space can be summarized by the variation in the much lower dimensional **T**-space, and the consistency of each multivariate observation with the PCA model can be assessed by the magnitude of its residual (SPE_i). This is illustrated in Figure 1 for a three-dimensional **X**-space and a two-dimensional **T**-space.

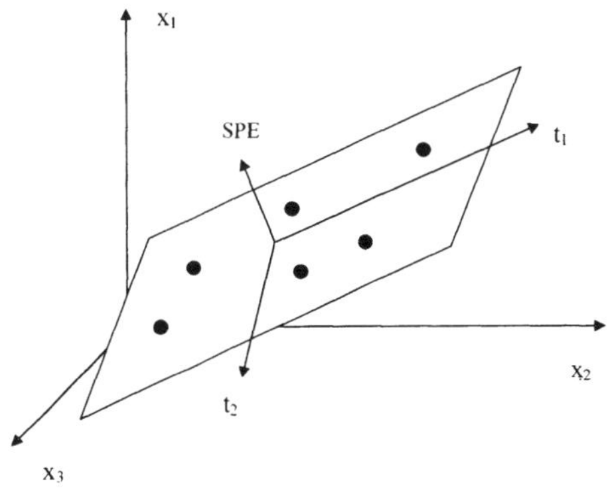

Figure 1 Illustration of reduced dimensional latent variable space

If one then wishes to relate **X** to a response matrix **Y**, the columns of **Y** can be regressed on the score matrix **T**, yielding the Principal Component Regression (PCR) model:

$$\hat{\mathbf{Y}} = \mathbf{T}\hat{\boldsymbol{\theta}} = \mathbf{XW}\hat{\boldsymbol{\theta}} = \mathbf{X}\hat{\mathbf{B}} \tag{10}$$

where $\hat{\boldsymbol{\theta}}$ and $\hat{\mathbf{B}}$ are parameter matrices.

Projection to Latent Structures or Partial Least Squares (PLS)

Unlike PCA which considers only the **X**-space variation when computing the latent variables, PLS considers both the **X** and **Y** spaces to obtain the **T**-space. Its objective function is less clear (Burnham et al., 1996), but it can be shown to maximize the covariance between the latent vectors of the **X**-space defined as $\mathbf{T} = \mathbf{XW}$ and those of the **Y** space. In this sense it models the high variance directions in the **X**-space which are most highly correlated with the latent variable space of **Y**. In effect it estimates the matrices **T**, **P**, **Q** and **W** simultaneously such that the latent vector space (**T**) best approximates both the **X** and **Y** spaces. An iterative NIPALS algorithm is again most commonly used to estimate the latent variables sequentially until a cross-validation test no longer deems the subsequent components to be statistically significant. However, the latent vectors can also be obtained as the eigenvectors of certain covariance matrices (Burnham et al., 1996).

For situations in which one of the matrix dimensions (n or k) is very large then various kernel algorithms exist that make the computation of the loadings and scores much simpler (Lingren et. al. 1993, Rannar et. al. 1994, Dayal and MacGregor, 1997).

Reduced Rank Regression

Reduced rank regression can be derived from the concept of multivariate regression between the **X** and **Y** matrices subject to the constraint that the least squares solution is of rank at most a. This can also be shown to be equivalent to regressing **Y** on the first A redundancy variates $t_l = \mathbf{xh}_l$ (l = 1, 2, ...,a) which maximize the fraction of the variation explained in the **y**-variables [Tso, 1981]. The solution again corresponds to the eigenvectors of certain matrix [Burnham et. al., 1996].

Canonical Correlation Analysis (CCA)

Canonical correlation analysis [Mardia et al., 1979] finds pairs of latent vectors in the **X** and **Y** space which are most highly correlated with one another. The first pair has the highest correlation and each subsequent pair has the next highest correlation, subject to being orthogonal to the previous canonical variates. From a regression viewpoint it is a generalization of multiple linear regression to the case of a multivariate **Y**-space.

Discussion of Latent Variable Methods

Although all of the above estimation methods provide a set of orthogonal latent variables, they do so by optimizing quite different objective functions (Burnham et al., 1996), and these differences play a vital role in the appropriateness of the methods for use in various problems to be discussed throughout the remainder of the paper. In particular, PLS and PCA are the only latent variable models that provide a good model for the **X**-space, since this was part of their objective function. This is an extremely important point since, in many problems, the model for the **X**-space will be the most important part of the LV model (equations (3)-(6)). This may seem strange to the statistical and identification communities because they are used to working only with datasets in which the **X**-space is full rank, by virtue of the way in which the data were collected (e.g. using designed experiments). In such situations, a model for the **X**-space is unnecessary since it is fully explained by the **X**-data directly. However, with the large data sets that we are dealing with in this paper, that arise from routine operation of the plant, the **X**-space typically has a statistical rank $a <<$ Min (n, k), and any models built from such data are only valid in this reduced dimensional space of the latent variables, and even then only in the restricted region of this space spanned by the training data.

For example, the model for the **X**-space (4) provides the basis for processing monitoring and FDI, in which abnormal behavior in the process is identified by comparing new process measurement against the LV model for the **X**-space. It also provides the basis for treating missing data in the **X**-space, and for testing for non-representative data (outliers) in soft sensors or inferential models. This heavy reliance on the model for the **X**-space will be emphasized in subsequent sections of the paper. It is this modeling of the reduced dimensional spaces of both **X** and **Y** that make the latent variable methods of PLS and PCR so different from standard regression approaches such as multiple linear regression (MLR), neural networks (NN), etc., and makes them much more powerful for treating the problems being addressed here.

In the following sections, some of the major areas in which these LV models have been applied in industry will be presented.

4. ANALYSIS OF PROCESS DATABASES

Perhaps the main area of industrial application for these LV models has been, and continues to be, the analysis of historical databases collected routinely by process computers. One of the justifications for the installation of almost every process computer has been that, by collecting all the measured process data and storing them in a database, process engineers would be able to analyze them and use the information gained from that analysis to improve the process. However, until recently, almost no analysis was ever performed on these data. The reasons always come down to the difficulties with handling such a large number of variables, dealing with large amounts of missing data and outliers, with the extreme correlation (non-full rank) nature of the data, and with interpreting the results of any analysis on data that are non-causal in nature. These points coupled with the fact that poorly configured data compression algorithms often destroyed what was initially good data, has turned data historians into data graveyards. Only with the recent application of latent variable methods, which are capable of overcoming many of the difficulties, has progress been made by industry in analyzing their historical data.

There are two common ways in which industrial engineers currently use these LV methods to analyze such data. The first is where nobody has every looked seriously at the historical data to see if things can be learned that might lead to process improvements. This retrospective analysis is often performed over several very different time scales. Daily averages on all variables enable one to investigate what has happened over a period of years, hourly averages over a period of weeks and months, and using filtered data on observations every minute or second allows one to study short term dynamic effects. A second usage is for short term trouble-shooting. Immediately after a process upset, a local LV model can be built over the recent period leading up to and covering the upset, and the local model used to analyze for possible causes.

The tools for analysis consist of looking at score plots of the latest variables (e.g. t_1 vs. t_2) to study how the process has moved in the reduced dimensional space, and at the corresponding loading plots (p_1 vs. p_2) to interpret the groups of variables that are related to movements in certain directions. Contribution plots are particularly useful for highlighting which group of variables is highly related to a movement in the score space or the residual space [MacGregor et al., 1996; Kourti and MacGregor, 1996; Miller et al., 1998].

The contribution of variable x_j to the residual SPE_i is

$$\text{Contribution } (x_j) \textit{ to } \text{SPE}_i = x_{ij} - \hat{x}_{ij} \qquad (11)$$

and the contribution of x_j to a movement in any latent variable Δt_l over any period is

$$\text{Contribution } (x_j) \textit{ to } \Delta t_l = \Delta x_j p_{jl} \qquad (12)$$

where p_{jl} is the loading of the variable x_j in the loading vector p_l of the LV, and Δx_j is the change in x_j over that period. These contribution plots do not show causal relationships. They only reveal which group of variables, in which part of the plant, are related to the movement or event that occurred. However, by narrowing down the possible variables and location in the plant it is usually much easier for the engineer or operator to diagnose some possible reasons for the event.

There are numerous applications of this use of LV models to analyze industrial databases in both continuous processes [e.g. Yacoub & MacGregor, 2002] and batch processes [Nomikos, 1996; Garcia et al. 2003]. In this section, a batch fermentation process will be used to illustrate the basic concepts. Typically data from batch processes are of the form illustrated in Fig. 2. For each batch one has data on initial conditions, prior processing history, etc. (**Z** matrix), histories on the time varying trajectories of process variables during each batch (**X** array), and data on the final product quality and productivity (**Y** matrix). Although the **Z**-matrix is important in this problem we ignore it here and illustrate the analysis using only **X** and **Y** data. The three dimensional **X** array can be unfolded to give an **X** matrix with each row containing the time histories of all the variables for a given batch. Figure 3 illustrates a plot of the scaled values of 6 of the variables trajectories for one of the batches over approximately 350 time periods.

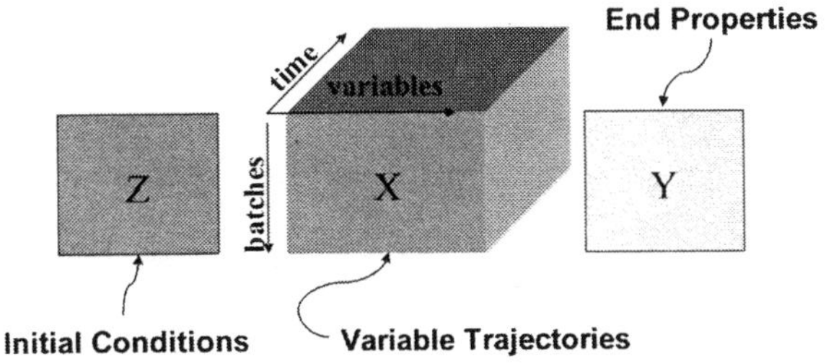

Figure 2 Nature of batch data

Although there are over 350,000 observations in this data set, a PLS model (using scaled and mean centered data) explains the statistically significant information with only a=2 latent variables. In other words, for each batch, the variations among all the variables, over the time history of the batch can be summarized by the score values of two latent variables (t_1, t_2). A score plot of these two LV's is shown in Fig. 4. It reveals a clear separation in the t_1 direction between batches with good **Y** results (high t_1 values – indicated by the region of the dark ellipse) and those with poor Y results (low t_1 values). A plot of the p_1 loading vector shown in Fig. 5 can be used to help interpret why these two groups of batches are different. For each process variable there are 350 loading values corresponding to the 350 time intervals during the batch. From this plot it is clear that good performance (a high value of t_1) is associated with trajectories of variables x_1 and x_3 lying above their mean trajectories throughout the last two thirds of the batch and with the trajectory of variable x_4 lying below its mean over the same period.

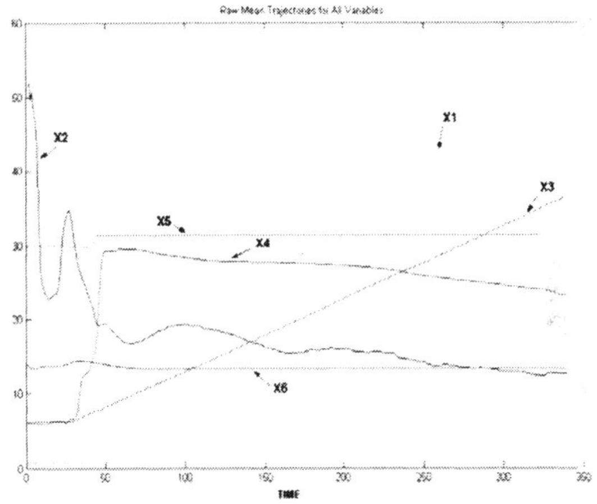

Figure 3 Measured trajectories on 6 variables over the duration of one batch

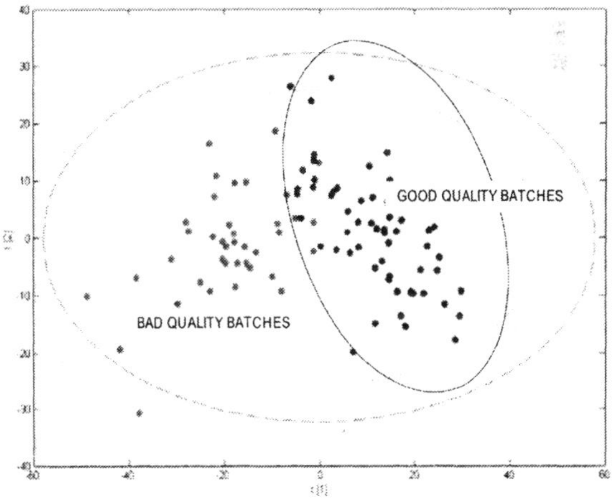

Figure 4 Score plot for all the batches (good batches in solid elliptical region)

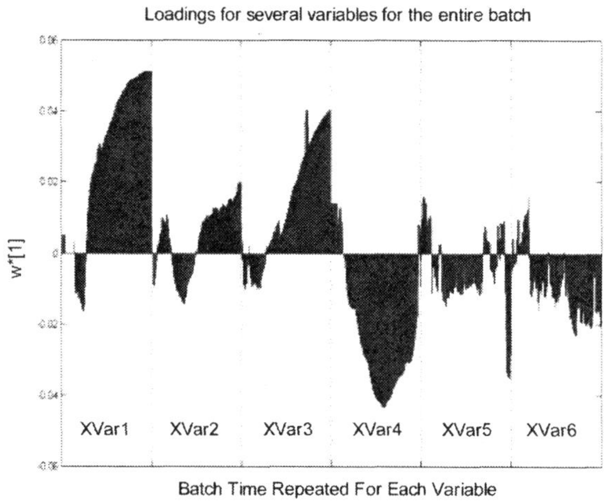

Figure 5 Plot of the elements of the first loading vector p_1 consisting of a loading value for each variable at each time period over the time history of the batch

5. PROCESS MONITORING AND FDI

Once the analysis of the historical process data is complete and improvements made to the process, one is interested in monitoring the process to ensure that any gains are maintained, and that any new problems are detected and identified as early as possible. The scope of process monitoring is wider than detecting simple hardware and sensor faults. Its purpose is to detect any type of complex atypical behavior such as might be associated with the effects of changes in impurities, or surface chemistry on the performance of a chemical process.

There is a large literature on FDI methods such as those based on analytical redundancy [Gertler, 1998] that involve the use of causal models from theory or identification experiments. However, multivariate statistical approaches based on LV models use non-causal models built from normal operating data. They are multivariate extensions of statistical process control (SPC) methods. They compare the behaviour of future operation of the plant against a LV model built from past behaviour where only "common cause variation" was present (i.e. from data collected where the performance was acceptable) [Kresta et.al., 1991; MacGregor and Kourti, 1995]. Any abnormal behaviour can be detected in the residual SPE plot or in the score plots or their Hotelling's T^2 equivalent

$$T^2 = \sum_{l=1}^{a} t_l^2 / s_l^2 \qquad (13)$$

where s_l^2 is the variance of t_l from the training data. Control limits on these plots [MacGregor and Kourti, 1996; Nomikos and MacGregor, 1995] can be obtained using the F-distribution or using the empirical reference distribution of the training data. An abnormal situation is detected when any of the control limits on these charts is violated, and then contribution plots can be used to identify that group of variables associated with the fault.

To illustrate the concepts we again use the batch fermentation process of the last section. To develop a PLS model for monitoring, we now only use the cluster of batch data with good performance shown in Fig. 4. Control charts for the SPE and Hotelling's T^2 statistics for a new (bad) batch are shown in Fig. 6.

An abnormal event is clearly detected by time 277 when the 99% control limit on both the T^2 and SPE charts are violated. The SPE contribution plot for the process variables at that time period is given in Fig. 7. High values of variables x_6 are clearly related to the fault.

6. SOFT SENSORS / INFERENTIAL MODELS

Because of the lack of on-line sensors for critical responses, "soft sensors" or inferential models are often built to predict these responses using available process measurements. Models based on theory are usually best for these soft sensors, but empirical models can be built from process data when such theory is lacking.

A very important point in the development of empirical inferential models, that is often overlooked, is the nature of the data required to build them. The suitability of the model as a soft sensor, in any particular application, is highly dependent upon the nature of the data used in building it. A general rule is that the data for building the model should be collected from the process while it is operating in a manner that resembles, as closely as possible, the way in which the soft sensor will also be using the data [Kresta et al., 1994]. For example, using data collected under open-loop conditions with designed experiments, will not lead to inferential models with good performance in a closed-loop environment.

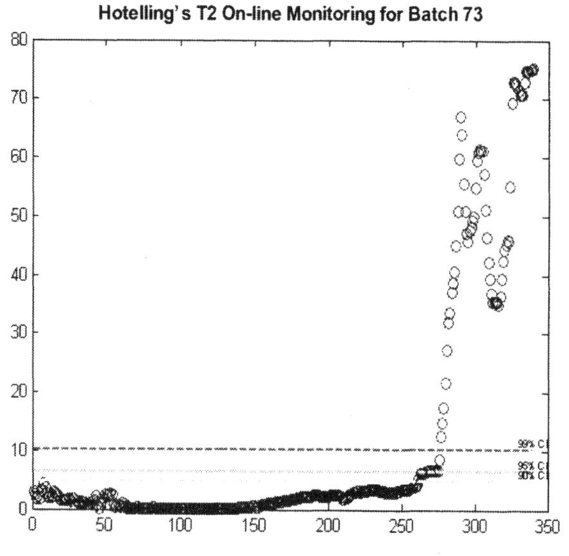

Figure 6 Hotelling's T2 plot for batch #73 with control limits shown

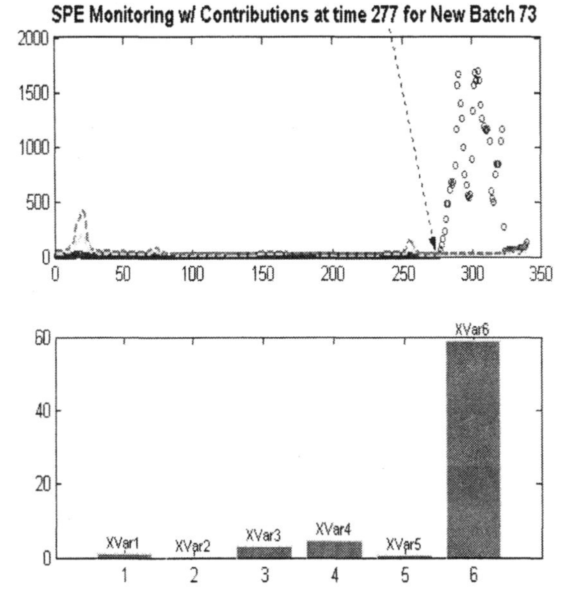

Figure 7 SPE plot for batch 73, and contribution plot at time 277 where fault is first detected

Linear and nonlinear regression and neural networks are often used to build the empirical models. However, linear or nonlinear PLS models have advantages when the number of process variables used as regressors is large, and when they are highly correlated. However, the greatest advantage of a PLS model for this problem is its ability to continue to perform well in the presence of missing data, and its ability to check for outliers in each set of new process variable measurements. These latter two advantages come from the fact that PLS not only models the **Y**-space but also the **X**-space. This allows for easy imputation of missing data [Nelson et al., 1996] and for straightforward checks for the validity of new incoming *x* data (SPE test).

The following example is presented to illustrate the power of latent variable models for developing soft sensors when there are a large number of highly correlated measurements. (Their advantages in the presence of missing data and outliers is discussed in Kresta et al. 1994, and Burnham et al., 1999). The monitoring of the off-gas pollutants (NO_x, SO_2, etc.) from combustion processes in boiler systems is an important environmental problem. Inferential models based on neural networks are well established for predicting these off-gas pollutants using process measurements taken around the boiler system. In this example we consider the use of colour (RGB) digital images of the turbulent flame in the boiler as a predictor variable [Yu and MacGregor, 2003a,c]. Figure 8 shows two flame images taken one second apart during a time when no changes were occurring. The difficulty in extracting information from these highly time varying flame images is obvious. The digitalized images are large three dimensional arrays of highly correlated data. However, by performing a PCA on these images and projecting them into the LV score space, the score plots of successive images (such as those in Fig. 8) are very stable at any given operating condition, but they do change significantly as the process conditions and waste fuel feeds change. Using masking methods, features can be extracted from the PCA score plots that summarize the changes in the flame [Yu and MacGregor, 2003a,c], and these can be used as predictors in a PLS model for the NO_x and SO_2 off-gas concentrations. The predicted versus observed plot for NO_x from a feasibility study is shown in Fig. 9. The predictions obtained from the flame data alone were as good as using both the flame and the process data and much better than just using the latter.

7. USING INFORMATION FROM NOVEL SENSORS

There is currently a major revolution occurring in new sensor technologies that will have major impacts on process control. In the chemical industry a whole new generation of sensors referred to as micro-sensors or molecular sensors are being developed by process analytical chemists. These sensors generate large amounts of data with information on the detailed molecular properties of the streams or products being sampled, and they do so at a greatly reduced cost per sample over traditional laboratory analysis. In the solids processing industries the lack of on-line sensors has greatly limited the ability to implement control systems. However, digital imaging systems, based on the availability of inexpensive cameras and computers, are starting to have an impact.

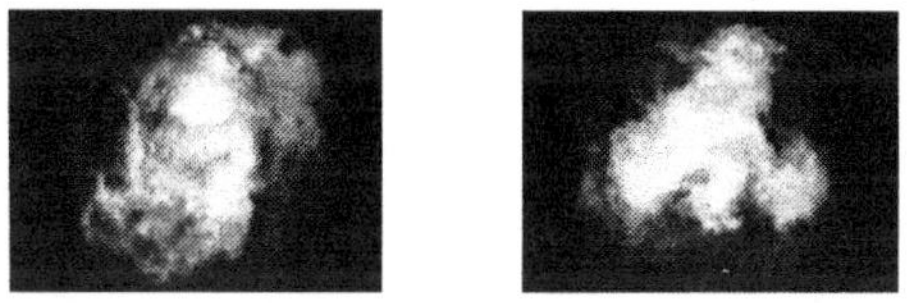

Figure 8 Flame images taken 1 second apart in an industrial boiler

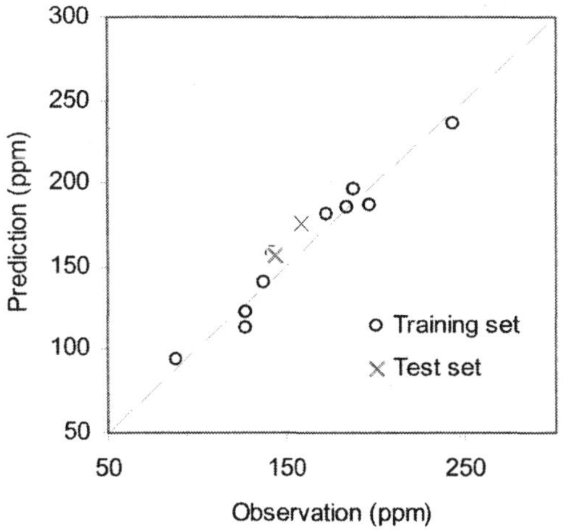

Figure 9 vs. observed NO_x concentration in the off-gas. Prediction done using only RGB flame images

In both of the above examples, the new sensors are not specific in nature, i.e. they do not simply provide a measure of one property of interest. Rather, they are multivariate sensors that can provide megabytes of data at each sampling interval. The information content of the data is very high, but it must be extracted using data analytic methods such as multivariate LV methods. The greatly expanded nature and quality of the information that can be obtained from them will have a major impact on process control over the coming years.

As an illustration consider the use of on-line digital colour imaging for the monitoring and feedback control of product quality in the snack food industry [Yu et al., 2003b,c]. A schematic of the system is shown in Fig. 10, and an image of one of the products, collected by the online camera, are shown in Fig. 11. Multi-way PLA and PLS methods are used to extract information from the images that is related to both the concentration of coating materials on the base product and to the distribution of the coatings over the product. This image information is then used for the on-line feedforward/feedback control of the process. Feedback control of the coating concentration based on the imaging systems is currently in use on several industrial lines. Table 1 compares the results of the image-based control on one of the lines against the results from the previous control system.

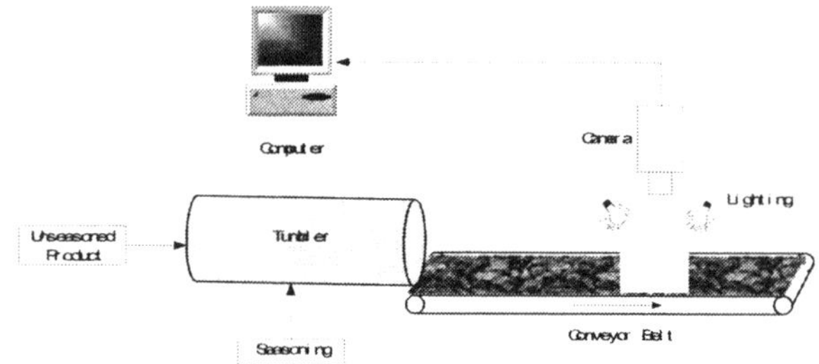

Figure 10 Imaging system for feedback control of snack food quality

Figure 11 On-line RGB image of snack food product

8. SYSTEM IDENTIFICATION

Latent variable methods have also found considerable application in the area of system identification. They have been used to overcome the ill-conditioning problems inherent in fitting FIR and ARX models [e.g. MacGregor et al. 1991], but most recently have found extensive use in subspace identification methods. In these subspace methods the objective is to extract a low dimensional state

Table 1: Mean Absolute Error (MAE) for the three cases

	Prior Control	Image-based Feedback Control (regulation)	Image-based Feedback Control (set point tracking)
MAE	0.8523	0.4481	0.4769

space representation from high dimensional matrices of lagged input/output data. In all these methods, the basis vectors for the low order (n) state space are simply taken to be the latent variable score vectors of the (n=a) statistically significant latent variable dimensions of one of the LV methods. For example Larimore's (1990) CVA method uses the first n canonical covariates of CCA as the state basis, and it can be shown (Shi and MacGregor, 2001) that different N4SID algorithms all correspond to using the dominant latent variables of RRR as the state basis. One can also use PLS, but unlike CCA and RRR, it will not lead to a minimal order state space model for the simple reason that it tries to model the input space of past data as well as the predictable subspace of the outputs. For this reason LV methods such as CCA and RRR are better suited for the identification problem since only the correlation between the two spaces is of interest [Shi and MacGregor, 2001].

9. PROCESS CONTROL IN REDUCED DIMENSIONAL LV SPACES

Latent variable methods are also useful in process control situations, particularly where the controlled (CV) and manipulated (MV) variable spaces are high dimensional, but of less than full rank. The simplest situation, where the CV space is high dimensional, but of low rank, include problems such as the cross-directional control of properties on paper machines or polymer films [Rigopoulos et. al., 1997], control of polymer molecular weight distributions [Clarke-Pringle and MacGregor, 1998] and particle size distributions [Flores-Cerrillo and MacGregor, 2003a], and control of polymer end properties in continuous reactors [Roffel et al., 1989]. In these cases some variation of PCA has usually been employed on the collected data to reduce the dimension of the CV space to a full rank space (e.g. use the latent variables of PCA, or a subset of the real variables which best define the PCA space, as new CV's), and the MV's are used to control this space.

A more complex situation occurs where the MV space is also of less than full rank. This situation will arise when there are operational constraints preventing all the MVs from acting independently. A classic example occurs in the control of batch reactors where the control at any time (θ_t) during the batch calls for the adjustment of the trajectories of all the MV's over the entire remainder of the batch (for example see the trajectories in Fig. 3). The MV vector therefore consists of an extremely high dimensional vector, but this vector of MV trajectories must respect many operational constraints both with respect to the shape of the trajectories over time, and with respect to the covariance among the MV's at all time points (e.g. Fig. 12).

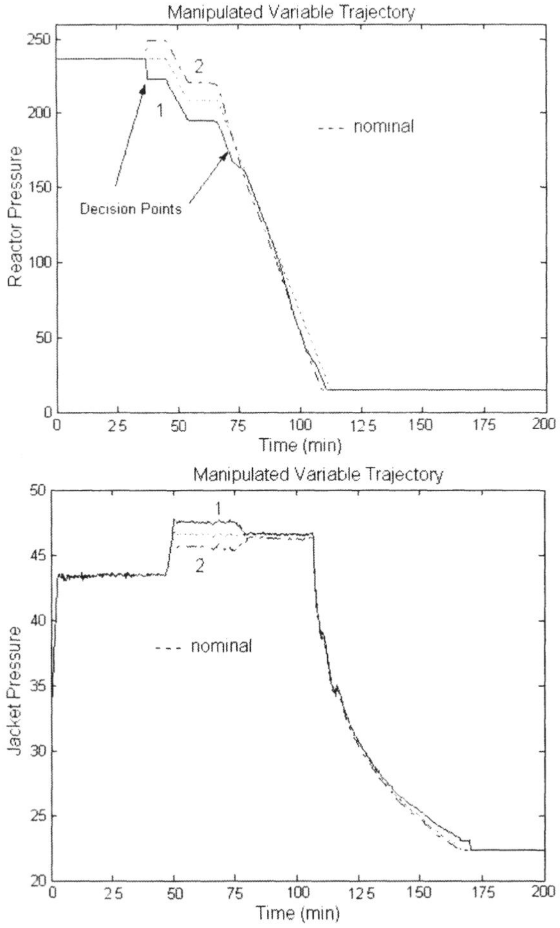

Figure 12 Control of polymer quality using LV models to adjust the MV trajectories at two decision times

This problem can easily be formulated and solved in the latent variable space of PLS models. An example from Flores-Cerrillo and MacGregor [2003b] on the control of product quality in batch nylon polymerization reactors is used here to illustrate the approach. A PLS model is built to allow for prediction of the vector of final product quality variables ($\hat{\mathbf{y}}$) using the measured trajectories on all the process and manipulated variables (e.g. Fig. 3) up to any decision time (θ_i) during the batch run. The missing data imputation feature of PLS models is used to impute the as yet unknown process variables for the uncompleted portion of the batch. A control action is then computed in the LV space (t_1, t_2, t_3) of the PLS model that will optimize the quadratic objective

$$\min_{\Delta \mathbf{t}(\theta_i)} [(\hat{\mathbf{y}}(\theta_i) - \mathbf{y}_{sp})\ R_1 (\hat{\mathbf{y}}(\theta_i) - \mathbf{y}_{sp})^T + \Delta \mathbf{t}\ R_2 \Delta \mathbf{t}^T]$$

where $\mathbf{y}$ is a 1x m vector and $\mathbf{t}$ a 1xa vector. The optimization is subject to the constraint that the new computed score values for the batch ($t_l = t_l(\theta_i) + \Delta t_l$; $l = 1, 2, \ldots, a$) fall within the range of historical values given by

$$T^2 = \sum_{l=1}^{a} \frac{(t_l + \Delta t_l)}{s_l^2} < \varepsilon$$

and

$$\mathbf{y}(\theta_i) = (\mathbf{t}(\theta_i) + \Delta \mathbf{t})\mathbf{Q}^T$$

Once the new score values have been computed from the control algorithm, the adjustments $\Delta \mathbf{u}$ to the remainder of all the actual MV's can then be computed using the PLS model for the **X**-space [Flores-Cerrillo and MacGregor, 2003b]. The plots in Fig. 12 show the resulting trajectories of two MV's resulting from solving the control problem at two decision times (θ_1 = 35 min and θ_2 = 75 min) for the nominal conditions (0) and two disturbances (1 and 2). In both situations the final controlled variables were returned to their target values at the end of the batch (Fig. 13).

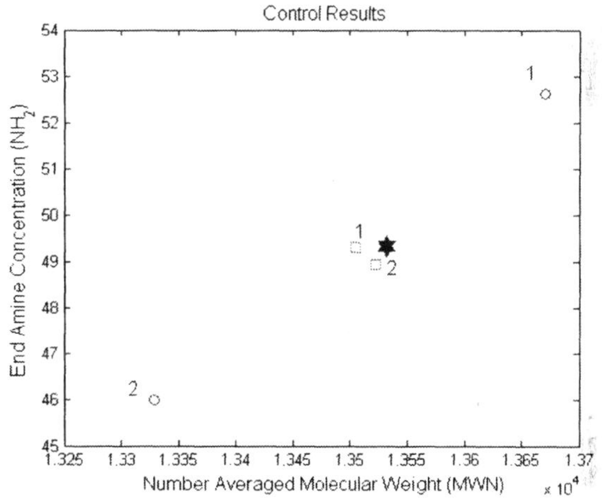

Figure 13 Control results on final polymer quality: 1 and 2 are results from two disturbances: o is when no control is applied; □ is when LV control is applied; ✱ denotes the setpoint

10. CONCLUSIONS

An overview of data based methods for the analysis, monitoring and control of processes has been presented. In particular, emphasis has been placed on the use of latent variable models, based on PLS and PCA, to treat situations where the data matrices are less than full statistical rank. In essence all these problems are subspace problems where the analysis, control etc. must be confined to the low dimensional subspaces defined by the latent vectors. Such situations dominate most industrial data base problems.

The latent variable model is shown to be the natural approach to treating these problems. It provides a model for both the **X** space (necessary because of its non-full rank nature) and the **Y** space. The LV models are able to easily handle missing data problems, which arise most of the time, and they allow one to test for the validity of incoming data.

The model for the **X** space plays a key role in all these problems except for the identification problem where, because of the design of experiments, the **X** matrix is of full statistical rank. Although linear LV models have been discussed here, this is not really an issue, since nonlinear LV models can just as easily be used. However, it is our experience that the nonlinear models do not offer much advantage in most of the industrial applications.

Several industrial problems are discussed in this paper that illustrate the power of these approaches. These include the problems of analyzing large historical databases and trouble-shooting process problems; process monitoring and FDI; building soft sensors or inferential models; process identification; and control in reduced dimensional subspaces. Many other important problems utilizing industrial databases have also been addressed in the literature. All of these problems involve building empirical latent variable models, and many of them are as important to systems engineering as identification. Hopefully this presentation will encourage some researchers in systems identification to look into these other interesting empirical modeling problems.

11. ACKNOWLEDGEMENTS

In keeping with the well known fact that, behind all good professors there lies many very good graduate students and researchers, I wish to express my thanks to the many good students and researchers who have contributed to this research over the years. Particular thanks in preparing this manuscript go to Honglu Yu, Salvador Garcia and Jesus Flores-Cerrillo.

REFERENCES

Burnham, A.J., R. Viveros-Aquilera and J.F. MacGregor, (1996) "Frameworks for Latent Variable Multivariate Regression", J. Chemometrics, 10, 31-46.

Burnham, A.; Viveros, R.; MacGregor, J.F. (1999) "Latent Variable Multivariate Regression Modeling. Chemometrics & Intell. Lab. Sys.", 48, 167.

Clarke-Pringle, T.L. and J.F. MacGregor, "Product Quality Control in Reduced Dimensional Spaces", Ind. Eng. Chem. Res., 37, 3992-4002,1998

Dayal, B.S. and J.F. MacGregor, 1997, "Improved PLS Algorithms", J. Chemometrics, 11, 73-85.

Dayal, B.S. and J.F. MacGregor, 1997, "Recursive Exponentially Weighted PLS and its Applications to Adaptive Control and Prediction", J. Process Control, 7, 169-179.

Garcia-Munoz, S., T. Kourti and J.F. MacGregor, A.G. Mateos, and G. Murphy, (2003) "Trouble-shooting of an industrial batch process using multivariate methods", Ind. & Eng. Chem. Res., In press.

Flores-Cerrillo, J. and J.F. MacGregor (2003a) "Within-batch and batch-to-batch inferential adaptive control of semi-batch reactors", Ind. & Eng. Chem. Res., In Press.

Flores-Cerrillo, J. and J.F. MacGregor (2003b) "Semi-batch trajectory control in reduced dimensional spaces", Proc. IFAC ADCHEM'2003, Hong Kong

Geladi, P. and B.R. Kowalski (1986) "Partial Least-Squares Regression: A Tutorial", Anal. Chim. Acta, 185 (1), 1.

Geladi, P. and Grahn, H. (1996) Multivariate Image Analysis. John Wiley & Sons, Chichester, England

Gertler, J. (1998), Fault detection and diagnosis in Engineering Systems, Marcel Dekker, New York

Kresta, J., J.F. MacGregor and T.E. Marlin, (1994) "Building Inferential Models Using PLS Regression, Computers Chem. Eng., 18, 597-611.

Kourti, T.; MacGregor, J.F., (1996) "Recent Developments in Multivariate SPC Methods for Monitoring and Diagnosing Process and Product Performance", Journal Of Quality Tech, 28, 409.

Lingren F., P. Geladi and S. Wold (1993) "The kernel algorithm for PLS", J. Chemometrics, 7, 45.

MacGregor, J. F., T. Kourti and J. V. Kresta, (1991), "Multivariate Identification: A Study of several Methods,"Proc. of IFAC Symposium ADCHEM-91, Toulouse, France.

MacGregor, J. F., C. Jaeckle, C. Kiparissides and M. Koutoudi, (1994) "Process monitoring and diagnosis by multiblock PLS methods," Amer. Inst.Chem. Eng. J., 40, 826-838.

MacGregor, J.F. and T. Kourti (1995), "Statistical process control of multivariable processes," Control Eng. Practice, 3, 403-414

MacGregor, J.F. (2003) "Multivariate statistical approaches to fault detection and identification", Proc. IFAC Safeprocess'2003, Washington, DC.

Mardia, K.V., J.T. Kent and J.M. Bibby (1979) Multivariate Analysis, Academic Press, London.

Miller, P., R.E. Swanson, and C.E. Heckler (1998), "Contribution Plots: The missing link in multivariate quality control," Appl. Math. & Comp. Sci, 8, 775-782

Nelson, P.A. Taylor and J.F. MacGregor, 1996. "Missing Data Methods in PCA and PLS: Score Calculations with Incomplete Observations", J. Chemometrics & Intell. Lab. Syst., 35, 45-65

Nomikos, P. and J.F. MacGregor (1994), "Monitoring batch processes using multiway principal component analysis," Amer. Inst.Chem. Eng. J., 40, 1361-1375

Nomikos, P., and J. F. MacGregor, (1995) "Multivariate SPC charts for monitoring batch processes," Technometrics, 37(1), 41-59.

Nomikos, P. (1996). Detection and Diagnosis of Abnormal Batch Operations Based on Multiway Principal Component Analysis. ISA Transactions., 35, 259-267

Rannar, S., F. Lingren, P. Geladi and S. Wold (1994), "A PLS kernel algorithm for data sets with many variables and fewer objects", J. Chemometrics, 8, 111.

Rigopoulos, A., A. Arkun and F. Kayihan (1997) "Identification of full profile disturbance models for sheet forming processes", AIChE J., 43, 727.
Roffel, J., J.F. MacGregor and T.W. Hoffman (1989) Design and implementation of a multivariable internal model controller for a continuous polybutadiene polymerization train", Proc. IFAC DYCOD'89, Eds. Rijnsdorp, MacGregor, Tyreus and Takamatsu, Pergamon Press
Shi, R. and J.F. MacGregor, (2001) "A Unifying Framework for Subspace Identification Methods", Proc. American Control Conference, Maryland, USA.
Soderstrom, T. and P. Stoica (1989) System Identification, Prentice Hall
Tso, M.K. (1981) "Reduced rank regression and canonical analysis", J. Roy. Statis. Soc., B 43, 183-189.
Wold, S., (1978) "Cross-Validatory Estimation of the number of components in factor and principal component models", Technometrics , 20, 397-405.
Wold, S., N., Kettaneh-Wold, and B. Skagerberg, (1989) "Nonlinear PLS modeling" Chemometrics and Intell. Lab. Syst., 7, 53.
Yacoub, F. and J.F. MacGregor, (2002) "Analysis and optimization of a polyurethane reaction injection molding (RIM) process using multivariate projection methods", Chem. & Intell. Lab. Syst., 65, 17-33.
Yoon, S. and J.F. MacGregor, (2000) "Relationships between Statistical and Causal Model-Based Approaches to Fault Detection and Isolation", Amer. Inst. Chem. Eng. J., 46, 1813-1824.
Yu, H. and J.F.MacGregor, (2003a) "Monitoring Turbulent Nonpremixed Flames in an Industrial Boiler Using Multivariate Image Analysis (MIA)", Proc. Safeprocess'2003, Washington, DC.
Yu, H. and J.F.MacGregor, (2003b) "Digital imaging for on-line monitoring and control of industrial snack food processes", Ind. & Eng. Chem. Res., In press.
Yu, H. and J.F.MacGregor, (2003c) "Digital imaging for process monitoring and control with industrial applications", Proc. IFAC ADCHEM'2003, Hong Kong

IFAC
Publications
www.elsevier.com/locate/ifac

SUBSPACE ALGORITHMS

Dietmar Bauer *,[1]

* *Institute for Econometrics, Operations Research
and System Theory,
TU Wien, Argentinierstr. 8,
A-1040 Vienna, Austria*

Abstract: Subspace algorithms have been established in the last decades as an alternative to prediction error methods for the estimation of linear dynamical systems. Conceptual simplicity and numerical feasability have been the main arguments in favor of the approach. This article gives a presentation of the mainstream approach and tries to convince the reader, that this class of algorithms has its virtues. Strengths and weaknesses of the approach are discussed. *Copyright © 2003 IFAC*

Keywords: subspace algorithms, estimation, linear dynamical systems

1. INTRODUCTION

'Subspace algorithms' is a technical term, which is both, too broad and misleading. Too broad, since the term is used in many different contexts in totally different meanings. Misleading, because even if one adds the context the term is not connected to a particular algorithm or class of algorithms, but rather to a general idea. Subspace algorithms have their origins in the algorithms of Zeiger and McEwen (1974) and Ho and Kalman (1966). As such, they bear elements of realization algorithms. However, the main idea centers around the concept of the state, as being an interface – in a sense to be made more clear below – between the past and the future, stated loosely. These early ideas have been developed further leading to the three most well known algorithms:

- `N4SID` (numerical algorithms for subspace state space system identification) proposed by Van Overschee and DeMoor (1994)
- `MOESP` (multivariable output error state space) system identification procedure proposed by Verhaegen (1994)
- `CCA` (canonical correlation analysis) proposed as `CVA` (canonical variate analysis) by Larimore (1983).

All three of them are used in the context of linear dynamical systems operating in open loop. Following the suggestion of the algorithms in parallel the analysis of the properties of these algorithms and the adaptation to different model classes occured. The general idea of the method has been adapted to lead to algorithms for the closed loop case (Chou and Verhaegen, 1999; Verhaegen, 1993; Ljung and McKelvey, 1996*b*), frequency domain data (McKelvey, 1995), bilinear models (Favoreel, 1999; Chen and Macicjowksi, 2000; Chou, 1994), piecewise linear models (Babuska *et al.*, 1997), time-varying parameters (Gustafsson, 1999; Verdult and Verhaegen, 2002; Oku and Kimura, 2002), Hammerstein models (Gomez and Baeyens, 2002), continuous-time models (Haverkamp *et al.*, 1997; Ohsumi and Kawano, 2002), errors-in-variables problems (Chou and Verhaegen, 1997), integrated processes (Bauer and Wagner, 2002), hidden markov chains (Andersson, 2002). Here we will only discuss the case of stationary, linear, discrete time, time invariant systems.

The aim of this paper is to present the concept of subspace algorithms in a unified way in order to highlight

[1] Support by the Austrian FWF under the project number P14438-INF is gratefully acknowledged.

the similarities between the various algorithms. The discussion will present the algorithms in much detail, while trying to keep the exposition self contained in order to allow also readers from related areas to follow. At many places the comparison to prediction error methods will be considered, since the subspace methods are an alternative to these methods. It is the purpose of this paper to point out situations, where there are advantages of the subspace approach over the prediction error approach.

2. STATE SPACE MODELS

In this paper the model class considered will always be the class of linear, discrete time, finite dimensional, time invariant state space models, given by

$$\begin{aligned} x_{t+1} &= Ax_t + Bu_t + K\varepsilon_t \\ y_t &= Cx_t + Du_t + \varepsilon_t \end{aligned} \tag{1}$$

Here $(y_t)_{t\in\mathbb{Z}}$ denotes the s-dimensional output process, observed for $t = 1,\ldots,T$, $(\varepsilon_t)_{t\in\mathbb{Z}}$ denotes the s-dimensional innovations, which for simplicity are assumed to be i.i.d. Gaussian random variables with zero mean and variance matrix $\Omega > 0$. As usual the noise is assumed to be unobserved. Furthermore $(u_t)_{t\in\mathbb{Z}}$ denotes the m-dimensional input process, observed for $t = 1,\ldots,T$. The n dimensional state process $(x_t)_{t\in\mathbb{Z}}$ is also not observed. The system matrices $A \in \mathbb{R}^{n\times n}, B \in \mathbb{R}^{n\times m}, C \in \mathbb{R}^{s\times n}, D \in \mathbb{R}^{s\times m}$ and $K \in \mathbb{R}^{n\times s}$ are to be estimated.

In the following some important properties of state space systems are discussed. Since it is assumed, that these concepts are known to all readers, the discussion is very brief. For a more detailed discussion we refer to (Hannan and Deistler, 1988, Chapter 1). Central to the definition of the model is the concept of a state: x_t is introduced in order to describe all the dynamics present in the model, as the observation equation is a static one. For so called *white box* models derived on the basis of physical principles, the state possesses a specific interpretation. In this talk only the so called *black box* modelling approach will be considered. Here only the input/output map is of interest and the state does not have any physical meaning, but is only a mathematical object to conveniently decsribe the dynamics of the system. As such, the state of a system is not unique: Any change from x_t to $z_t = Tx_t$ using a nonsingular matrix $T \in \mathbb{R}^{n\times n}$ results in a different model $(TAT^{-1}, TB, CT^{-1}, D, TK)$, which represents the same input/output map. In this case, (A,B,C,D,K) and $(TAT^{-1}, TB, CT^{-1}, D, TK)$ are called *observationally equivalent.* A state space representation of an input/output map is called *minimal*, if there exists no state space representation of the same input/output map with lower state dimension. In that case the integer n is called the *order* of the system.

The significance of the state in the state space models comes from the fact, that it summarizes all the dynamics in the model: Given the state trajectory, the ouput is obtained from the static observation equation. As a note we remark, that state space systems hence can be seen as very special hidden markov models. The state is not observed, but given the model and trajectories of the input u_t and the output y_t for $t = 1,\ldots,T$, the state x_{T+1} can be estimated. Assume, that the system is stable and strictly minimum-phase, i.e. that $|\lambda_{max}(A)| < 1$ and $|\lambda_{max}(A-KC)| < 1$ hold. Here $\lambda_{max}(.)$ denotes an eigenvalue of maximal modulus of a matrix. The best estimate of the state in the mean square sense (assuming the input to be a covariance stationary stochastic process) is calculated by the Kalman filter. If the input/output data is available for $t = T, T-1, T-2, \ldots$, then the steady state Kalman filter estimate of the state coincides with the state, since

$$x_{T+1} = \sum_{j=0}^{\infty} (A-KC)^j \left[Ky_{T-j} + (B-KD)u_{T-j}\right]$$

considering time T to be the 'present'. Therefore, the state can be recovered from the knowledge of the history of the input/output data. On the other hand, the prediction of the output is one of the main goals for identification. Consider (for $f \geq 0$)

$$\begin{aligned} y_{T+f} &= Cx_{T+f} + Du_{T+f} + \varepsilon_{T+f} \\ &= C(Ax_{T+f-1} + Bu_{T+f-1} + K\varepsilon_{T+f-1}) \\ &\quad + Du_{T+f} + \varepsilon_{T+f} \\ &= \cdots \\ &= CA^f x_T + Du_{T+f} + \varepsilon_{T+f} \\ &\quad + \sum_{j=0}^{f-1} (CA^j Bu_{T+f-j-1} + CA^j K\varepsilon_{T+f-j-1}) \\ &= CA^f x_T + \sum_{j=0}^{f} L_j u_{T+f-j} + \sum_{j=0}^{f} K_j \varepsilon_{T+f-j} \end{aligned} \tag{2}$$

where the last equation defines the impulse response sequences $L_j \in \mathbb{R}^{s\times m}, K_j \in \mathbb{R}^{s\times s}, j \geq 0$, i.e. $L_0 = D, K_0 = I, L_j = CA^{j-1}B, K_j = CA^{j-1}K, j > 0$. This equation decomposes the output y_{T+f} into three components: $CA^f x_T$ gives the contribution of the state at initial time T, $\sum_{j=0}^{f} L_j u_{T+f-j}$ the contribution of the future and the present of the input and $\sum_{j=0}^{f} K_j \varepsilon_{T+f-j}$ the contribution of the future and present of the noise. If one assumes open loop operation, then the input is uncorrelated with the noise. The state, being a function of the past input/output data also is uncorrelated with the noise. Therefore the best linear mean square prediction of y_{T+f} based on the whole input sequence $u_t, t \in \mathbb{Z}$ and the past of the output $y_s, s < T$, say $y(T+f|T)$, equals

$$y(T+f|T) = CA^f x_T + \sum_{j=0}^{f} L_j u_{T+f-j}$$

The following two facts constitute the role of the state in state space models:

(1) The state is a function of the past input/output data.
(2) The state summarizes all information contained in the past input/output measurements that is relevant for the prediction of the future output.

In this sense, the state is the interface between the past and the future. This basic fact lies at the heart of all subspace algorithms.

Choosing two integers f and p, the following vectors can be defined for arbitrary time instant t:

$$Y_{t,f}^{+} = \begin{bmatrix} y_t \\ y_{t+1} \\ \vdots \\ y_{t+f-1} \end{bmatrix} \in \mathbb{R}^{fs}, Z_{t,p}^{-} = \begin{bmatrix} y_{t-1} \\ u_{t-1} \\ \vdots \\ y_{t-p} \\ u_{t-p} \end{bmatrix} \in \mathbb{R}^{p(m+s)}$$

Additionally $U_{t,f}^{+}$ is defined using u_t analogously to $Y_{t,f}^{+}$ and $E_{t,f}^{+}$ is defined using the innovations ε_t. Let $\mathscr{K}_p$ denote the matrix corresponding to the finite Kalman filter, such that

$$n_t = x_t - \mathscr{K}_p Z_{t,p}^{-}$$

is orthogonal to $Z_{t,p}^{-}$, i.e. uncorrelated. From projection arguments in combination with ($\bar{A} = A - KC$)

$$x_t = \bar{A}^p x_{t-p} + \sum_{j=0}^{p-1} \bar{A}^j (K y_{t-j-1} + (B - KD) u_{t-j-1})$$

it follows that $\|n_t\| \leq \|(A-KC)^p\| \|x_{t-p}\|$ and therefore for p large the strict minimum-phase assumption implies that the error term n_t is small. Combining the equations (2) for $y_{t+j}, j = 0, \ldots, f-1$ one obtains the following central equation:

$$\begin{aligned} Y_{t,f}^{+} &= \mathscr{O}_f x_t + \mathscr{U}_f U_{t,f}^{+} + \mathscr{E}_f E_{t,f}^{+} \\ &= \mathscr{O}_f \mathscr{K}_p Z_{t,p}^{-} + \mathscr{U}_f U_{t,f}^{+} + \mathscr{E}_f E_{t,f}^{+} + \mathscr{O}_f n_t \quad (3) \\ &= \mathscr{O}_f \mathscr{K}_p Z_{t,p}^{-} + (\mathscr{U}_f + \mathscr{O}_f \mathscr{N}_{f,p}) U_{t,f}^{+} + N_t^{\perp} \end{aligned}$$

where $N_t^{\perp} = \mathscr{E}_f E_{t,f}^{+} + \mathscr{O}_f (n_t - \mathscr{N}_{f,p} U_{t,f}^{+})$, denoting the projection in mean square sense of n_t onto $U_{t,f}^{+}$ by $\mathscr{N}_{f,p} U_{t,f}^{+}$. Here $\mathscr{O}_f = [C', A'C', \cdots, (A^{f-1})'C']'$ denotes the truncated observability matrix, $\mathscr{U}_f = [L_{i-j}]_{i,j=1,\ldots,f}$ the Toeplitz matrix of the impulse responses L_j, where $L_j = 0, j < 0$ is used. $\mathscr{E}_f = [K_{i-j}]_{i,j=1,\ldots,f}, K_j = 0, j < 0$. This equation is a vector equation for $t \in \mathbb{Z}$. Often this equation is written as a matrix equation having the above equation (3) for $t = p+1, p+2, \ldots, T-f$ as its columns. The structure of the matrices containing the data caused the term 'data Hankel matrices'. We will put forward a different view of the equation.

The central equation (3) decomposes the vector $Y_{t,f}^{+}$ into three components: $\mathscr{O}_f \mathscr{K}_p Z_{t,p}^{-}, (\mathscr{U}_f + \mathscr{O}_f \mathscr{N}_{f,p}) U_{t,f}^{+}$ and $N_t^{\perp}$. For $t = p+1, \ldots, T-f$ the vectors $Y_{t,f}^{+}, U_{t,f}^{+}$ and $Z_{t,p}^{-}$ can be built using input/output data $y_t, u_t, t = 1, \ldots, T$. Hence the equation

$$Y_{t,f}^{+} = \beta_z Z_{t,p}^{-} + \beta_u U_{t,f}^{+} + N_t^{\perp}, t = p+1, \ldots, T-f$$

has the following interesting features:

- Under the assumption of open loop operation, the vector $N_t^{\perp}$ is uncorrelated with the remaining terms on the right hand side of the equation. Under the closed loop assumption, $U_{t,f}^{+}$ and $N_t^{\perp}$ are correlated, but $Z_{t,p}^{-}$ and $N_t^{\perp}$ remain uncorrelated.
- The matrix β_z has rank n, the system order.
- $\beta_u = \mathscr{U}_f + \mathscr{O}_f \mathscr{N}_{f,p} \to \mathscr{U}_f$ for $p \to \infty$.
- $N_t^{\perp} \to \mathscr{E}_f E_{t,f}^{+}$ for $p \to \infty$.
- $\mathscr{E}_f E_{t,f}^{+}$ is an MA(f) process.

These observations build the basis for the subspace algorithms.

3. DESCRIPTION OF THE ALGORITHMS

Most subspace algorithms share a common outline. They can be decomposed into three main steps [2]:

(1) Use the central equation to estimate β_z, β_u by regressing $Y_{t,f}^{+}$ onto $Z_{t,p}^{-}$ and $U_{t,f}^{+}$ for the open loop case. In the closed loop case, given an estimate $\hat{\beta}_u$ of $\mathscr{U}_f$, an estimate of β_z is obtained using regression of $Y_{t,f}^{+} - \hat{\beta}_u U_{t,f}^{+}$ onto $Z_{t,p}^{-}$. This leads to estimates $[\hat{\beta}_z, \hat{\beta}_u]$.
(2) The estimate $\hat{\beta}_z$ will typically be of full rank, whereas $\mathscr{O}_f \mathscr{K}_p$ is of rank n. Hence a rank n approximation $\hat{\mathscr{O}}_f \hat{\mathscr{K}}_p$ of $\hat{\beta}_z$ is obtained.
(3) Based on the estimates $\hat{\mathscr{O}}_f, \hat{\mathscr{K}}_p$ and $\hat{\beta}_u$, estimates of the system matrices are obtained.

This outline is shared by most of the commonly used subspace procedures. In particular MOESP and CCA fit into this framework, whereas N4SID uses a slightly different third step while using the same first two steps. Note, that the description was given for the open loop case and the closed loop case, whereas most of the literature only considers the open loop case.

In the following, we will describe the various approaches to the three steps in more detail, where the emphasis will be on a discussion with respect to applicability to real world data sets, numerical aspects and also asymptotic properties, above all consistency issues and asymptotic variance considerations.

3.1 Step 1: Regression

The first step in the procedure is a regression. Least squares regression is maybe the best understood statistical method. Efficient numerical procedures exist. The pitfalls are understood to a large extent. Using recursive regression methods, one immediately obtains recursive subspace methods (cf. e.g. Oku and Kimura, 2002). In particular, the regression faced in subspace methods has

- lagged output variables as regressors
- residuals, which are not white.
- reduced rank coefficient matrices
- for $p = \infty$ the matrix $\mathscr{U}_f$ has a rich structure, i.e. it is block Toeplitz.

[2] This decomposition has been given in (Peternell *et al.*, 1996). A similar view of the algorithms and in particular the use of the regression interpretation is independently given in (Shi, 2002).

- potential problems with illconditioning due to multicollinearity.

The consequences of these facts are discussed next.

3.1.1. Lagged Output Variables Different possible choices with respect to the initial and end conditions are possible. The regression equation was written for $t = p+1,\ldots,T-f$. As for ARX systems, setting the initial and end conditions to be equal to zero, the regressions can be calculated using the estimates of the covariance sequence, since in $Y^+_{t,f}, U^+_{t,f}$ and $Z^-_{t,p}$ lagged versions of the two processes y_t and u_t appear. This is equivalent to extending the regression equation to $t = 1,\ldots,p$ and $t = T-f+1,\ldots,T$, while replacing missing values with zeros. It will be clear from the following, that throughout the algorithms not the observed processes themselves are needed, but the estimates of the first $f+p-1$ covariances are sufficient. A different approach is to discard the time instants, where some variables are not observed and to use the regression equation only for $t = p+1,\ldots,T-f$. Arguments paralleling the autoregressive case could be made keeping in mind, that the obtained estimate is $\hat{\beta}_z$ rather than the system matrices themselves: Setting initial and end conditions to zero definitely leads to serious distortions for f and p relatively large in comparison to T. Additionally a different argument has been put forward in favor of not using covariance estimates: For linear equations it is known, that solving the least squares problem using the QR decomposition is numerically favorable to solving the normal equations. This is the reason for using the QR decomposition in the original versions of `MOESP` and `N4SID`. It is the belief of the author that the contribution of the numerical errors to the total error is minor. Therefore the decision, how to choose the initial and end conditions in the regression should be based on statistical grounds rather than for numerical reasons.
There are cases, where the idea of viewing subspace algorithms as being a nonlinear function of the estimated sample covariances is beneficial. First of all, this view is very convenient for the derivation of asymptotic properties of the estimators obtained from using the subspace approach. Basically this idea underlies all results proving consistency and asymptotic normality (with the exception of the case that there is an integrator present in the data generating process). But secondly, and more important for the practitioner, basing the estimation on estimated covariances brings many convenient features:

- Huge sample sizes can be dealt with: Calculating the sample covariances can be done even for very large data sets in a few seconds. On the contrary, the regression matrix can become huge even for moderate sample sizes: Choosing $f = p = 30$ e.g. for a three dimensional input and three dimensional output observed for 5000 time instants results in $Z^-_{t,p}$ being of dimension 180 and with t varying between 31 and 4970 the corresponding regression matrix would be of dimension 180 times 4940 containing approx. 890.000 entries. The QR decomposition of the original `MOESP` algorithm, which calculates in effect the regression, would thus have to be performed on a matrix of size 360 times 4940 having more than 1.6 million entries.
- Missing values: Due to the dynamic structure of the regression single irregularily missing values might reduce the effective sample size substantially. The estimated covariances, however, do not suffer such a loss in accuracy.
- Outliers: Covariance estimators exist, which are robust with respect to outlying data points. These might be easier to apply than robustifications for the regression itself.
- Time varying parameters: Recursive covariance estimators can be used in order to cope with time varying parameters, although this might not be preferable.

3.1.2. Nonwhite residuals Recall that the residuals are equal to $\mathscr{E}_f E^+_{t,f} + \mathscr{O}_f(n_t - \mathscr{N}_{f,p}U^+_{t,f})$. These are nonwhite, since $E^+_{t,f}$ and $E^+_{t-1,f}$ have a considerable overlap and because $n_t - \mathscr{N}_{f,p}U^+_{t,f}$ is nonwhite. Choosing p large, the second problem can be made negligible, whereas the first problem remains. The usual solution to the problem of correlated errors is to use the GLS estimator rather than the OLS estimator with an estimate of the covariance matrix of the residuals. However, there exist cross restrictions between the regression parameters and the noise covariance matrix, which is a problem for GLS estimation. Furthermore it is noted, that $\hat{\beta}_z$ is only an intermediate estimate, therefore it might not prove essential to use an optimal estimator in this stage.

3.1.3. Reduced rank regression It has been noted, that $\mathscr{O}_f\mathscr{K}_p$ has rank equal to the system order n, whereas the estimate $\hat{\beta}_z$ typically has full rank (for f and p sufficiently large). A natural idea would be to incorporate the reduced rank property already in the regression. Consider a regression problem

$$y_t = \beta_z z_t + \beta_u u_t + n_t$$

where y_t, u_t and z_t are observed for $t = 1,\ldots,T$ and the rank of $\beta_z = \alpha\beta'$ is restricted to be equal to n. In order to simplify notation, let $\langle a_t, b_t\rangle = \sum_{t=1}^T a_t b_t'$, where a_t and b_t here stand for any of the processes y_t, z_t or u_t. First consider the criterion function ($\hat{n}_t(\alpha,\beta,\beta_u) = y_t - \alpha\beta' z_t - \beta_u u_t$)

$$L_T(\alpha,\beta,\beta_u) = \mathrm{tr}[W_f\langle\hat{n}_t(\alpha,\beta,\beta_u),\hat{n}_t(\alpha,\beta,\beta_u)\rangle]$$

for some positive definite weighting matrix $W_f = W_f'$. The solution to this problem can be found e.g. in (Reinsel, 1998) and is given using the SVD of [3]

[3] Here the symmetric square root of a matrix is used.

$$W_f^{1/2}\langle y_t^{\perp}, z_t^{\perp}\rangle \hat{W_p^-} = \hat{U}\hat{\Sigma}\hat{V}' = \hat{U}_n\hat{\Sigma}_n\hat{V}_n' + \hat{R}_n \quad (4)$$

where $\hat{U} \in \mathbb{R}^{fs\times fs}$ denotes the matrix of left singular vectors and $\hat{U}_n$ is the principal submatrix constituted of the first n columns, $\hat{V}$ and $\hat{V}_n$ are the corresponding quantities corresponding the right singular vectors and $\hat{\Sigma}_n = \mathrm{diag}(\hat{\sigma}_1, \hat{\sigma}_2, \ldots, \hat{\sigma}_n)$, where $\hat{\sigma}_1 \geq \ldots \geq \hat{\sigma}_n > \hat{\sigma}_{n+1} \geq 0$ are the estimated singular values ordered decreasing in size. The residuals from a regression of y_t onto u_t are denoted by $y_t^{\perp} = y_t - \langle y_t, u_t\rangle\langle u_t, u_t\rangle^{-1}u_t$. Similarily $z_t^{\perp}$ is defined. Further $\hat{W_p^-} = \langle z_t^{\perp}, z_t^{\perp}\rangle^{-1/2}$. This leads to a minimum of $\hat{\beta} = (\hat{W_p^-})^{-1}\hat{V}_n$ and $\hat{\alpha} = (W_f)^{-1/2}\hat{U}_n\hat{\Sigma}_n$. Clearly the minimum is not unique.
Alternatively, pseudo maximum likelihood estimation, i.e. estimation based on the Gaussian likelihood for i.i.d. white noise n_t as the criterion function, can be used. Note, that this criterion function leads to reasonable estimators, even if n_t is not Gaussian white noise. It follows from similar arguments to the ones given above, that the solution in this case is identical to the one given above for $W_f = \langle y_t^{\perp}, y_t^{\perp}\rangle^{-1}$.
This procedure applied to the regression used in the subspace approach leads to the SVD

$$W_f^{1/2}\langle Y_{t,f}^{+,\perp}, Z_{t,p}^{-,\perp}\rangle \hat{W_p^-}$$

where $\hat{W_p^-} = \langle Z_{t,p}^{-,\perp}, Z_{t,p}^{-,\perp}\rangle^{-1/2}$ and $Z_{t,f}^{+,\perp}, Z_{t,p}^{-,\perp}$ denote the residuals from regression onto $U_{t,f}^{+}$. It will be seen below, that these choices are used in some procedures.

3.1.4. Structure in $\mathscr{U}_f$ It has been suggested to use the block Toeplitz structure in $\mathscr{U}_f$ in the regression in order to obtain better estimates of β_z in (Peternell *et al.*, 1996). In that paper $p \to \infty$ has been used as a justification for neglecting $\mathscr{O}_f(n_t - \mathscr{N}_{f,p}U_{t,p}^{+})$. In some simulation examples also advantages in the accuracy have been shown. There is no general result backing the intuition of better estimates obtained by using the structure. In the case of white noise inputs, moreover, the restricted regression approach does not lead to more accurate estimates, as is shown in (Bauer, 1998). A disadvantage of the restricted regression method is the significant increase in the computational complexity.

3.1.5. Multicollinearity There are two different kinds of multicollinearity problems, and each has to be dealt with differently. The first kind is the obvious problem of perfectly correlated regressors. This occurs e.g. if certain deterministic terms such as the constant are included as inputs. The solution in this case is simply to omit the corresponding variables and this multicollinearity does not introduce any serious problems. The second problem is concerned with almost perfect collinearities. Again, the viewpoint of a regression analysis is helpful in this respect: Ridge regression techniqes can be used in this case and in fact have been proposed (Shi, 2002; Gustafsson, 1999).

3.2 Step 2: Rank n approximation

In the second step of the subspace algorithms the estimate $\hat{\beta}_z$ is approximated by a rank n matrix. This is usually accomplished using a weighted singular value decomposition: Let $\hat{W_f^+} \in \mathbb{R}^{fs\times fs}$ and $\hat{W_p^-} \in \mathbb{R}^{p(s+m)\times p(s+m)}$ be two symmetric positive definite matrices. Then consider the SVD

$$\hat{W_f^+}\hat{\beta}_z\hat{W_p^-} = \hat{U}\hat{\Sigma}\hat{V}' = \hat{U}_n\hat{\Sigma}_n\hat{V}_n' + \hat{R}_n$$

Note, that this SVD is totally analogous to the SVD in (4), which uses a special choice of $\hat{W_p^-}$. Therefore the reduced rank regression approach leads to the same result as the unrestricted regression approach combined with a weighted rank n approximation. `N4SID` does not have this interpretation, since it uses a different weighting $\hat{W_p^-}$, but both `MOESP` and `CCA` fall into this category.
In this step, (almost) all the user choices to be taken in subspace algorithms are of crucial importance. The integers f and p define the dimensions of the matrix on which the SVD is performed. A lower bound on these integers has to be imposed, in order to make sure, that the essential dynamics of the system can be estimated. The weighting matrices $\hat{W_f^+}$ and $\hat{W_p^-}$ have to be chosen, which influence the approximation quality. And finally the order of the estimated system, n say, has to be prescribed in this step.
Order estimation is a - in my opinion - neglected topic. There have been two different approaches proposed: Estimating the order using criterion minimization and alternatively statistical rank testing (for a discussion see e.g. Camba-Mendez and Kapetanios, 2001). Estimating the order has been based mostly on criterion functions comparing the norm of the neglected part of the SVD, i.e. $\hat{R}_n$, to a penalty function:

$$IC(n) = \|\hat{R}_n\|^2 + \frac{C(T)d(n)}{T}$$

where $d(n) = ns + n(s+m) + sm$ denotes the number of parameters needed to parametrize the state space systems of the form (1). $C(T) > 0, C(T)/T \to 0$ is a term penalizing large models. The choice of $C(T)$ determines the properties of the estimates of the order, obtained as the minimizing argument of the criterion. With respect to the norm, the two norm ($\|\hat{R}_n\|^2 = \hat{\sigma}_{n+1}^2$, `SVC`, Bauer (2001)) and the Frobenius norm ($\|\hat{R}_n\|^2 = \sum_{j=n+1}^{M}\hat{\sigma}_j^2, M = \min\{fs, p(s+m)\}$, `NIC`, Peternell (1995)) have been proposed. Based on different grounds also $\|\hat{R}_n\|^2 = 1 - \sum_{j=n+1}^{M}\log(1 - \hat{\sigma}_j^2)$ has been proposed (Camba-Mendez and Kapetanios, 2001).
The statistical testing approach is based on a series of tests on the rank of a matrix according to the ideas of Gragg and Donald (1997): The series is started at the hypothesis of the order of the system being equal to null. If the null hypothesis is rejected, the null hypothesis is adapted, now saying that the order is equal to one. This procedure is continued as long as the null is rejected.

For both methods the asymptotical properties have been derived, mainly proving consistency. Small simulation studies compare the various approaches, but to the best of my knowledge no procedure has been found to be superior. Also the motivation for the estimation methods is relatively weak, basically only hinging on consistency. But of course, many consistent procedures can be defined.

3.3 Step 3: Estimation of system matrices

From the previous steps, the estimates

$$\hat{\mathcal{O}}_f = (\hat{W}_f^+)^{-1/2}\hat{U}_n\hat{\Sigma}_n, \hat{\mathcal{K}}_p = ((\hat{W}_p^-)^{-1}\hat{V}_n)'$$

and $\hat{\beta}_u$ have been obtained. Up to now, the discussion did not distinguish between the various different approaches, except for pointing to different choices of weighting matrices, which however only apply for the default algorithms. There is no difficulty in applying, say, CCA using the weighting scheme put forward in MOESP. The estimation of the system matrices, however, is where the differences in the algorithms show up. Hence this section is divided into two subsections, the first one dealing with the MOESP type of methods, whereas the second one deals with state based approaches.

3.3.1. MOESP *type of methods* The distinctive feature of this type of algorithms is the usage of the matrix $\hat{\beta}_u$ in the estimation. The estimation hinges on the estimates $\hat{\mathcal{O}}_f$ and $\hat{\beta}_u$ and most algorithms in this class only estimate (A,B,C,D), the subsystem describing the effects of the input on the output. The noise model is included in the estimation, since it is hoped that by doing this the estimation accuracy is increased. It is debatable, whether this is really true. Chiuso and Picci (2002) find examples, where the joint modelling approach leads to worse estimates, than seperately modelling the systems (A,B,C,D) and (A,K,C).
In the first part of step 3, the MOESP type of approach uses the structure of the matrix $\mathcal{O}_f$: Define $\overline{\mathcal{O}_f}$ as the submatrix of $\mathcal{O}_f$, which is obtained by omitting the first block row. Then obviously

$$\overline{\mathcal{O}_f} = \mathcal{O}_{f-1}A$$

Letting $\underline{\hat{\mathcal{O}}_f}$ be defined as the first $f-1$ block rows of $\hat{\mathcal{O}}_f$ this equation can be used in order to obtain an estimate $\hat{A}$ of A as the least squares solution to

$$\overline{\hat{\mathcal{O}}_f} = \underline{\hat{\mathcal{O}}_f}A + r$$

The estimate $\hat{C}$ is defined as the first block row of $\hat{\mathcal{O}}_f$. This procedure is usually termed 'shift invariance approach'.
Given these two estimates, a number of different procedures for the estimation of B and D have been proposed and it does not seem to be clear, which approach is to be favored. (Ljung and McKelvey, 1996*a*) propose to use the representation $y_t = \sum_{j=0}^{\infty} L_j u_{t-j} + v_t$ as the basis for the estimation of B and D, as $L_0 = D, L_j = CA^{j-1}B, j > 0$ are linear in B and D and hence given estimates of A and C the estimates of B and D are obtained as

$$(\hat{B},\hat{D}) = \arg\min \sum_{t=1}^{T} \|y_t - L(u_t,B,D)\|^2$$

where $L(u_t,B,D) = \sum_{j=0}^{t-1} L_j u_{t-j}$ is linear in B and D. Closed form expressions for the solution exist.
Alternatively the structure of $\beta_u = \mathcal{U}_f + \mathcal{O}_f\mathcal{N}_{f,p}$ can be used to construct estimates of B and D. This is in fact done in the original MOESP procedure. Note, that $\mathcal{U}_f$, being a matrix whose entries are L_j, is linear in B and D. Let $\mathcal{O}_f^{\perp} \in \mathbb{R}^{fs\times(fs-n)}$ denote a matrix, such that $\mathcal{O}_f'\mathcal{O}_f^{\perp} = 0$, while $\mathcal{O}_f^{\perp}$ is of full column rank. Further let $\mathcal{U}_f = L(A,B,C,D)$. Then

$$(\mathcal{O}_f^{\perp})'\beta_u = (\mathcal{O}_f^{\perp})'\mathcal{U}_f = (\mathcal{O}_f^{\perp})'L(A,B,C,D)$$

If estimated quantitites replace true quantities, the equation only holds approximately and B and D can be determined using least squares fitting on the vectorized equations:

$$\text{vec}(\widehat{(\mathcal{O}_f^{\perp})}'\hat{\beta}_u) = \text{vec}(\widehat{(\mathcal{O}_f^{\perp})}'L(\hat{A},B,\hat{C},D)) + r$$

There are two issues related to this equation: The first issue is the choice of the estimate $\widehat{(\mathcal{O}_f^{\perp})}$, which is based on an estimate of $\mathcal{O}_f$. Two possible choices are $\hat{\mathcal{O}}_f$ and $[\hat{C}',\hat{A}'\hat{C}',\ldots,(\hat{A}^{f-1})'\hat{C}]'$. The second issue is the distribution of r. Given (A,C), Chiuso and Picci (2002) give the variance of r and find an estimate of the variance in order to obtain GLS estimates.
For all these procedures it is unclear, which is the preferable one. Except for a few simulation examples, no evidence exists. Moreover, the estimation of the system matrices corresponding to the noise characteristics usually is neglected. There exist procedures to estimate the matrix K, however, we will not present them. For the case of no observed inputs present only realization methods can be seen to fall into the MOESP type of algorithms.

3.3.2. The state approach Contrary to the MOESP type of approach, the state approach uses the estimate $\hat{\mathcal{K}}_p$ and neglects $\hat{\mathcal{O}}_f$ and $\hat{\beta}_u$. Recalling that the state is equal to

$$x_t = \mathcal{K}_p Z_{t,p}^- + n_t$$

an estimate of the state can be given as $\hat{x}_t = \hat{\mathcal{K}}_p Z_{t,p}^-, t = p+1,\ldots,T$. This estimate can be used in the observation equation in place of the true state x_t in order to obtain an estimate of C and D from

$$y_t = \hat{C}\hat{x}_t + \hat{D}u_t + \hat{\varepsilon}_t$$

This also defines an estimate $\hat{\varepsilon}_t$ of the innovations. Secondly, if an estimate $\tilde{x}_{t+1}, t = p+1,\ldots,T$ is available, the state equation could be used in order to obtain estimates of A, B and K using the regression equation

$$\tilde{x}_{t+1} = \hat{A}\hat{x}_t + \hat{B}u_t + \hat{K}\hat{\varepsilon}_t + r_t$$

One obvious estimate is $\tilde{x}_{t+1} = \hat{x}_{t+1}, \tilde{x}_{T+1} = 0$ using the shifted estimated state sequence. Alternative estimates for $\tilde{x}_{t+1}$ have been proposed in the original N4ISD procedure and recently by Chiuso and Picci (2002). In the case of no observed inputs, the formulae are valid without a change, setting $m = 0$.

4. SOME ASYMPTOTIC RESULTS

After having described the algorithm in detail, in this section the main theoretical results are cited, which are important in order to obtain an understanding of the possible applications of the algorithms. We will not state the results in full technical detail, but rather refer to the original sources for the interested reader.

4.1 MOESP *type of procedures*

Corresponding to this class of procedures, there only exists a limited set of results on the asymptotic properties. The effects of the user choices (f, p, the weighting matrices) on the asymptotic properties are not well understood. For all procedures, consistency has been fairly well investigated and cases, where the algorithms are not consistent, have been singled out (Jansson and Wahlberg, 1998). Also asymptotic normality has been proved (Bauer and Jansson, 2000) and the calculations of the asymptotic variance described in detail (Jansson, 2000). It is known, that the choice of $\hat{W}_f^+$ does not influence the accuracy of the estimated poles of the system, i.e. the eigenvalues of A (Jansson, 1997).
Corresponding to the effects of the weighting matrices on the asymptotic bias in the case of underspecification of the order, there exist a number of examples, which show that in some cases, the bias can be affected to the favor of the modeller. However, there do not exist any results making this observation more concrete than rules of thumb based on a couple of observations. Also the effects of the choices of the weightings on the asymptotic variance are not well understood. The expressions derived so far do not provide good insights. With respect to the effects of the choice of f and p almost no advice exists to the best of the knowledge of the author. An invited session at this conferences is dedicated to these topics, which are an area of ongoing research, which also needs impact from applications.

4.2 State approach

For the state approach, there is much more knowledge present on the effects of the various user choices. Two different cases are distinguished: For the case of no observed inputs or white noise inputs, the asymptotic properties of subspace algorithms are well understood. For the case of coloured input, the situation ressembles much the situation for the MOESP type of procedures.

4.2.1. No observed inputs or white noise inputs

In this case, the procedure based on $\tilde{x}_{t+1} = \hat{x}_{t+1}$ is understood quite completely with respect to the asymptotic properties. A necessary condition in order to achieve consistency in this setting is to let p tend to infinity as a function of the sample size (cf. Deistler *et al.*, 1995). If additionally asymptotic normality of the estimators is to be ensured, then $p \geq -d \log T / (2 \log |\lambda_{max}(A-KC)|)$ for some arbitrary $d > 1$ is assumed in the proofs (cf. Bauer *et al.*, 1999). This bound depends on unknown system quantities. However, it can be shown (cf. e.g. Hannan and Deistler, 1988, Theorem 6.6.3) that $2\hat{p}_{AIC}$ fulfills this bound almost surely, where $\hat{p}_{AIC}$ is the order estimated using AIC in an autoregressive approximation of the output process y_t.
In the case of no observed inputs there exist expressions for the asymptotic bias term for underspecification of the order, which also include some results on the dependence of the bias distribution over frequency on the choice of the weighting matrix (cf. Bauer, 1998, Chapter 2). However, these results are not very sharp and particularly not of much use in practice. For the case of correctly specified order, (Bauer and Ljung, 2002) provide very transparent expressions for the asymptotic variance, which reveal the influence of the weighting $\hat{W}_f^+$ and the choice of f on the asymptotic accuracy. $\hat{W}_p^-$ has been shown to be of no concern in this case in (Bauer *et al.*, 2000). The bottom line of these results is that the CCA choice of the weighting matrices is optimal for each fixed f. Furthermore, the asymptotic accuracy for the CCA estimates increases monotonically with f. This implies, that an optimal procedure has to use $f \to \infty$. (Bauer, 2000) finally shows, that in the case of no observed inputs, CCA together with $f = p = 2\hat{p}_{AIC}$ leads to a procedure, which asymptotically is equivalent to prediction error methods and hence in the case of Gaussian innovations achieves the optimal accuracy given by the Cramer Rao lower bound.
(Dahlen and Scherrer, 2001) show, that CCA is asymptotically equivalent to a procedure, which performs balanced model reduction on a preliminary AR estimate in the sense, that the difference of the obtained estimates tends to zero faster than $1/\sqrt{T}$. This provides an alternative interpretation of the procedure, giving also some motivation to the choice of p as suggested above.
In the case of white noise observed inputs, the variance expressions given in (Bauer and Ljung, 2002) are still valid. Asymptotic equivalence to prediction error methods also has been shown. Therefore in these cases, CCA can be seen as an equivalent of (pseudo) maximum likelihood methods *under the assumption of correctly specified order.*

4.2.2. Coloured inputs In the case of coloured observed inputs some examples have been given, which show, that in this case CCA does not achieve optimal accuracy. The knowledge about the asymptotic properties is limited to the basic results of consistency (Peternell *et al.*, 1996) and asymptotic normality (Bauer, 1998). Expressions for the asymptotic variance exist, but are computationally demanding.

5. APPLICATION TO STOCK RETURN DATA

In order to illustrate the advantages of subspace based state space modeling we use a data set of so called 'high frequency' stock returns, provided by Tim Bollerslev. The data set is further described in Bollerslev and Zhang (2003). The data set consists of five-minute returns on a value weighted market portfolio consisting of more than 6000 of the largest issues traded on the NYSE, NASDAQ and AMEX stock exchanges. The sample consists of 1761 trading days from January 2, 1993 through December 1, 1999. For each trading day, 79 five minute returns are given, resulting in a total of 139119 observations. The data set hence covers an extended period in time and is also quite demanding with respect to its size.
One commonly used hypothesis in financial econometrics is the so called 'efficient market hypothesis', which basically means that the knowledge of past return data cannot be used to obtain a forecast of future returns, which beats the no change prediction systematically. For daily return data this hypothesis seems to be rather accurate and modelling the mean return is not a rewarding task. For five minute returns by contrast, even a simple AR(1) model already beats the no change prediction on average. Hence, fitting an ARMA model to the five minute return series seems to make sense.
Financial data sets share a number of commonly found characteristics:

- Heteroskedasticity: Conditional on the past, the variance of the innovations varies.
- 'Fat tails': The amount of 'large' innovations is higher than would be expected from a normal density, i.e. the distribution of the innovations is leptokurtic.

Subspace algorithms are known to be robust with respect to certain forms of heteroskedasticity including the commonly used GARCH models (Bauer, 2002). The leptokurtosis can be dealt with using outlier robust covariance estimators. Usually the analysis of financial data is done in two steps: First a model for the conditional mean is derived in order to obtain an estimate of the innovations. Secondly a model for the conditional variance (or respectively a model for the squared innovations) is derived based on the estimated innovations. For efficiency of estimation the final model is estimated jointly.
Therefore, consider a state space model estimated for

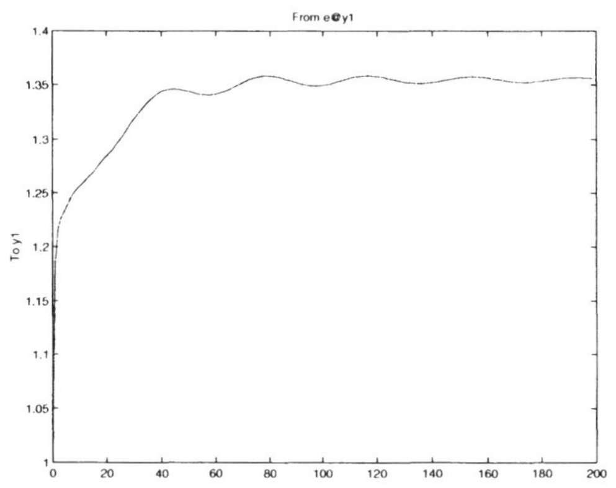

Fig. 1. Step function for the estimated 11-th order transfer function.

the whole data set: As the outlier robust estimator of the covariance sequence we used the trimmed mean, where we neglect 5% percent of the data. These covariance estimates are then inserted into the subspace algorithm and a model of order 11 (estimated from the data) is estimated for the full data set. In this step the integers f and p have been specified externally. A plot of the estimated step function can be seen in picture 1. The plot shows that the immediate effects dominate, while there is some influence on the first halfday. The step response levels out at a value of approximately 1.35 after half a day, although there is some fluctuation at the period length one half day.
Due to the extended time span (7 years), time constancy of the system is highly questionable. As an alternative to the constant parameter model, the data set has been partitioned into blocks of 10 consecutive trading days, resulting in 176 data sets of 790 data points each. For each data block, a seperate model has been specified and estimated. A number of different techniques for the specification have been tried out, including the estimation of robust estimators for the covariances, fully automated model selection procedures, and robust estimation for fixed model structure (eleventh order model). As a criterion to compare the time varying models to the constant parameter model, the one step ahead prediction error on the following ten trading days for the time varying models is compared to the prediction error on the same data set for the constant parameter model. This comparison is friendly to the constant model, since there all the data was used for estimation. Nevertheless, the time varying models perform better, as can be seen in figure 2, showing the quotient of the standard deviation of the one step ahead prediction error on the 10 consecutive trading days to the standard deviation of the one step ahead prediction using the constant parameter model. The plot shows, that the time varying models outperform the constant parameter model in the first half of the observation period, while it is somewhat worse on the second half. Overall the difference is negligible. Thus on these grounds the constant parameter hypothesis is rejected and the model for the mean is given by the time varying parameter models. These models

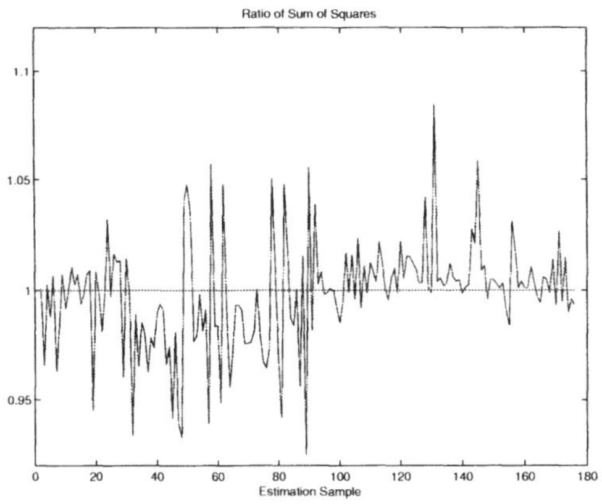

Fig. 2. Ratio of standard deviation of prediction error corresponding to the time varying parameter models against the standard deviation of the prediction error corresponding to the constant parameter model.

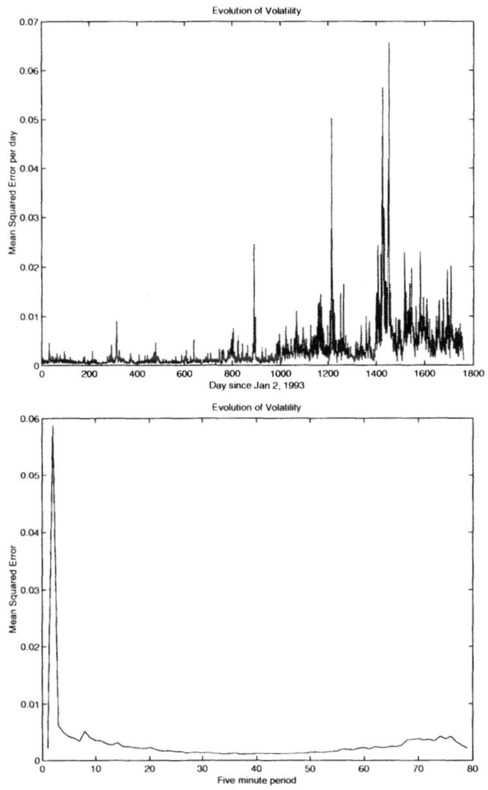

Fig. 3. Mean of squared residuals for the various days (upper plot) and for different five minute intervals (lower plot).

result in an overall R^2 of 0.11.

Figure 3 shows, that the variability as measured by the daily means of the squared errors is not constant over the various days. Also the variability is not constant for the various five minute intervals, as documented by the lower plot. It can be clearly seen, that the most variability in the stock returns occurs five minutes after the opening of the market. One commonly used model for this sort of data is the so called GARCH model (Bollerslev, 1986). Squared GARCH

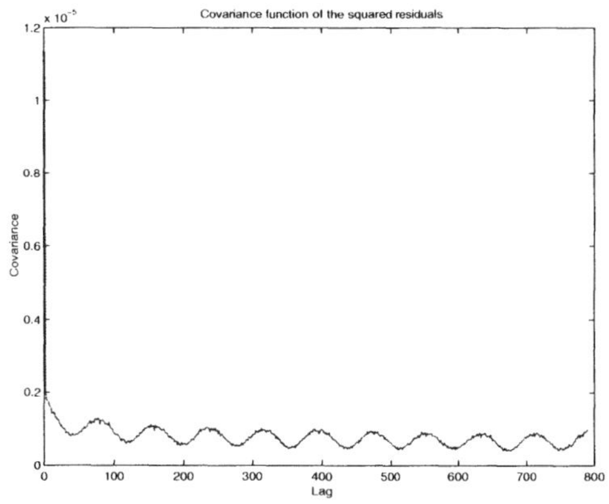

Fig. 4. Estimated covariance sequence of the squared innovations adjusted for intra day means.

processes have been interpreted as a linearily filtered heteroskedastic white noise including a nonzero intercept. Therefore the usual modeling techniques can be used. Clearly in our case, the deterministic term depends on the particular five minute interval and cannot be chosen as constant. Two alternatives are the introduction of one dummy for each interval or a parametric model for the distribution over time. We will here only deal with the dummy approach, implemented by subtracting the trimmed mean (10%) of the squared innovations for each five minute time period. The (trimmed, 10 %) estimated covariance sequence can be seen in figure 4. The plot shows some of the features, that are often found in these data sets: The covariance at lag one is already rather small compared to the covariance at lag 0 (correlation of about 0.18). The remaining covariances show a slow decay and a cyclical behaviour at the daily frequency.

In principle there are a number of different strategies to model the conditional variability: These include one model for all instances, a different model for each five minute interval or a multivariate model for the vector of all 79 five minute returns jointly. Each of these has its drawbacks: Building one univariate model is the most restrictive model. On the other hand, one model for each five minute interval leads to a large amount of work due to the necessity to specify 79 models. The multivariate model also has a number of drawbacks: First of all, the information set is different: Whereas in the univariate models, the prediction is performed on the basis of all returns up to time t, the multivariate model predicts on the basis of all data up to the last day and hence does not take into account the returns of the current day. Secondly, the properties of multivariate GARCH models are largely unknown. Especially the positivity constraint is problematic. Hence logarithms are taken, which also leviates the problems due to the leptokurtosis.

Using subspace methods, multivariate models can be estimated and specified for output dimension 79 without big problems. The specification step is numerically feasible, as the main computational load in this case lies in the calculation of the covariance sequence, which can be done prior to model estimation. The

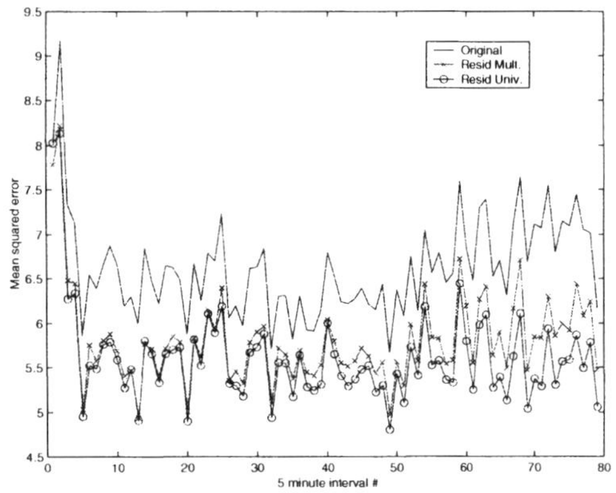

Fig. 5. Mean squared errors for the squared adjusted residuals, the residuals from the multivariate model and the residuals from the univariate model.

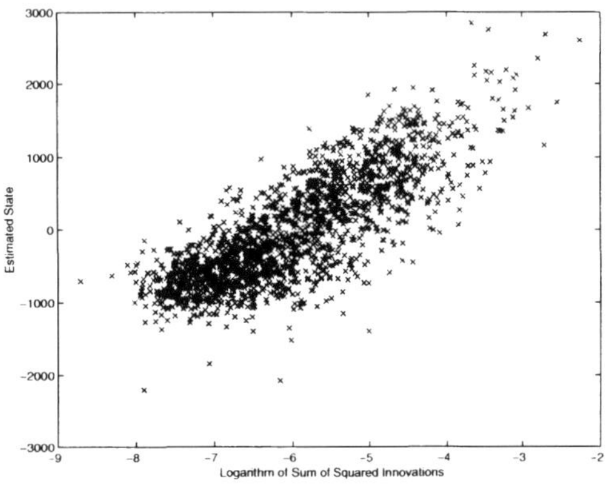

Fig. 6. Logarithm of the sum of the squared innovations plotted against the estimates of the state in the multivariate model.

automatic procedure estimates the order to equal 1. Plot 5 shows the mean squared errors of the logarithm of the squared residuals, which have been corrected for the mean for given five minute interval, the mean squared errors for the multivariate model and the mean squared error for the univariate model. It is observed, that both models substantially reduce the mean squared error over the naive model of constant volatility. The univariate model performs a bit worse for the early returns, but better for the late returns. The explanation for this lies in the different amount of information on which the prediction is based. Nevertheless, if one considers the estimated logarithm of the determinant of the residual variance, the multivariate model results in 133.01, whereas the univariate model only achieves 133.48. The number of parameters is $2 * 79 = 158$ for the multivariate model and 166 for the univariate model. The one dimensional state can be interpreted as an estimate of the logarithm of the mean squared errors of the various days, as can also be seen in a scatter plot (cf. Figure 6).

6. COMPARISON TO PREDICTION ERROR METHODS

Subspace methods are an alternative to prediction error methods in the sense, that they can be used to fit linear dynamical models to input/output data. It has been cited, that for the case of no observed inputs one particular method, namely `CCA` achieves the same asymptotic accuracy as the prediction error method, while being much more computationally efficient (especially for large sample sizes). For the case of coloured observed inputs, no subspace method has been proven up to now to provide asymptotically efficient estimates. All the results given above only correspond to asymptotic reasoning. The question remains, what place subspace methods should take in the toolbox of the model builder? My personal views, which are definitely not shared by all people in the community, are the following: At the very least, subspace methods for properly chosen user parameters lead to good initial estimates, which can then be used in gradient based optimization methods. This view is implementated in the fully automatic `pem` procedure in the system identification toolbox of `MATLAB`. Based on theoretical arguments, however, for the case of no observed inputs, there is no reason for rejecting the estimates obtained using `CCA` in favor of estimates obtained from numerical optimization of criterion functions.
Additionally there are a number of situations, where subspace methods can be a useful alternative:

- Model specification: Subspace algorithms provide an additional possibility for estimation of the order.
- Systems with moderate output dimension. For output dimension say up to $s = 5$ prediction error methods are probably still computationally feasible, while subspace algorithms provide a quick second look at the data.
- Automatic modelling: Providing rules of thumb for the choice of f and p and the weighting matrices, combined with order estimation procedures, immediately renders the subspace methods into an automatic modelling method: data in, estimated system out (cf. Bauer and de Waele, 2003).

In a number of cases, subspace algorithms seem to be the only choice, since prediction error methods for state space models are not feasible:

- Very large data sets. In the example the number of sampling instants was equal to 139119. This data set can still be dealt with using standard prediction error methods. Order estimation on this data set using prediction error methods nevertheless becomes infeasible. Using the subspace approach, much larger data sets can be dealt with.
- Many outputs. In the application example a system for a 79 dimensional output has been es-

timated. Using prediction error methods this would be numerically infeasible. Even autoregressive modelling would not be feasible, since for an AR(1) system, $79^2 = 6241$ parameters would need to be estimated. Hence, reduced rank regression in an autoregressive setting represents the only alternative in this case.

- Many models at a time. Again, the example considered used a model for two consecutive weeks, resulting in a total of 176 models to be estimated. Using prediction error methods, the only choice due to time restrictions would be to use the same model structure for all models. Subspace methods on the contrary allow for specification of each model at a time.
- Input selection. In econometrics (but also in other disciplines) it is common, that there exists only a set of regressors, which are seen as influental, but not necessary each in fact is. Hence the first step usually is input selection, i.e. the specification, which of the potential inputs contributes to the output. Typically in this step a huge number of models has to be estimated and hence only a computationally feasible method is of use.

It should be stressed again, that this list is only my personal belief. It definitely is not complete and some of the points are debatable.

Finally I would like to correct a possible misunderstanding: I do not argue, that subspace methods are superior to prediction error methods. In many cases, prediction error methods are still the better tool. In particular simulations indicate, that the small sample properties of subspace methods are quite poor for very small samples, such as the ones typically found in macroeconometrical applications. In this case, the estimates are of use only as initial estimates. Additionally, it is not possible to include prior information into algorithms easily. Therefore, they are only of interest for black box modelling. And last, but not least, there are still many topics, which are not completely clarified: (Bauer and de Waele, 2003) show, that although nice on a theoretical ground, the automatic modelling approach based on the recommendations of this paper works surprisingly poor in certain test cases. This is an indication, that especially the choice of the user supplied quantities $f, p, \hat{W}_f^+, \hat{W}_p^-$ and the order of the system has to be analyzed in more depth, also from the perspective of finite sample properties, which are relevant for applications.

ACKNOWLEDGEMENT

Even though my name is on the front page of this article, it really is the work of the group in Vienna, that is the root of this contribution. Almost all of my work in the area has been done in cooperation with and under the guidance of Manfred Deistler and Wolfgang Scherrer from the TU Wien. I am very grateful for their support. I am also indebted to Thomas Ribarits for careful proof reading. The data set has been provided by Tim Bollerslev, which is gratefully acknowledged.

REFERENCES

Andersson, S. (2002). Hidden Markov Models - Traffic Modeling and Subspace Methods. PhD thesis. Lund University, Sweden.

Babuska, R., J. Keizer and M. Verhaegen (1997). Identification of nonlinear dynamic systems as a composition of local linear parametric or state space models. In: *Proceedings of the SYSID'97 Conference, Fukuoka, Japan*. pp. 703–708.

Bauer, D. (1998). Some Asymptotic Theory for the Estimation of Linear Systems Using Maximum Likelihood Methods or Subspace Algorithms. PhD thesis. TU Wien, Austria.

Bauer, D. (2000). Asymptotic efficiency of the CCA subspace method in the case of no exogenous inputs. Technical report. Department of Automatic Control, Linköping Universitetet.

Bauer, D. (2001). Order estimation for subspace methods. *Automatica* **37**, 1561–1573.

Bauer, D. (2002). Identification of state space systems with conditionally heteroskedastic innovations. In: *Proceedings of the 15th IFAC World Congress*. pp. T–Mo–M02.

Bauer, D. and L. Ljung (2002). Some facts about the choice of the weighting matrices in Larimore type of subspace algorithms. *Automatica* **38**, 763–773.

Bauer, D. and M . Wagner (2002). Estimating cointegrated systems using subspace algorithms. *Journal of Econometrics* **111**, 47–84.

Bauer, D. and M. Jansson (2000). Analysis of the asymptotic properties of the MOESP type of subspace algorithms. *Automatica* **36**(4), 497–509.

Bauer, D. and St. de Waele (2003). A finite sample comparison of automatic model selection methods. In: *Proceedings of the SYSID'03 conference, August 2003, Rotterdam, The Netherlands.*

Bauer, D., M. Deistler and W. Scherrer (1999). Consistency and asymptotic normality of some subspace algorithms for systems without observed inputs. *Automatica* **35**, 1243–1254.

Bauer, D., M. Deistler and W. Scherrer (2000). On the impact of weighting matrices in subspace algorithms. In: *Proceedings of the SYSID'2000 conference*, Santa Barbara, California.

Bollerslev, T. (1986). Generalized autoregressive conditional heteroskedasticity. *Journal of Econometrics* **31**, 307–327.

Bollerslev, T. and B. Zhang (2003). Measuring and Modeling Systematic Risk in Factor Pricing Models using High-Frequency Data. *Journal of Empirical Finance*, No. 10, forthcoming.

Camba-Mendez, G. and G. Kapetanios (2001). Testing the rank of the hankel covariance matrix: A statistical approach. *IEEE Transactions on Automatic Control* **46**, 331–336.

Chen, H. and J. Maciejowksi (2000). Subspace identification methods for combined deterministic-stochastic bilinear systems. In: *Proceedings of the SYSID'2000 conference,* Stanta Narbara California.

Chiuso, A. and G. Picci (2002). Asymptotic variances of subspace identification by data orthogonalization and model decoupling. Technical report. University of Padua, Italy.

Chou, C. and M. Verhaegen (1997). Subspace algorithms for the identification of multivariable dynamic errors-in-variables models. *Automatica* **33**, 1857–1869.

Chou, C. T. (1994). Geometry of Linear Systems and Identification. PhD thesis. University of Cambridge.

Chou, C. T. and M. Verhaegen (1999). Closed-loop identification using canonical correlation analysis. In: *Proceedings of the ECC'99 Conference, Karlsruhe, Germany.*

Dahlen, A. and W. Scherrer (2001). The relation of the CCA subspace method to a balanced reduction of an autoregressive model. Submitted to Journal of Econometrics.

Deistler, M., K. Peternell and W. Scherrer (1995). Consistency and Relative Efficiency of Subspace Methods. *Automatica* **31**(12), 1865–1875.

Favoreel, W. (1999). Subspace Methods for Identification and Control of Linear and Bilinear Systems. PhD thesis. Katholieke Universiteit Leuven.

Gomez, J. and E. Baeyens (2002). Subspace identification of multivariable hammerstein and wiener models. In: *Proceedings of the 15th IFAC World Congress*. pp. T–Th–M01.

Gragg, J. and S. Donald (1997). Inferring the rank of a matrix. *Journal of Econometrics* **76**, 223–250.

Gustafsson, T. (1999). Subspace Methods for System Identification and Signal Processing. PhD thesis. Chalmers University, Gothenburg, Sweden.

Hannan, E. J. and M. Deistler (1988). *The Statistical Theory of Linear Systems*. John Wiley. New York.

Haverkamp, B., M. Verhaegen, C. Chou and R. Johansson (1997). Continuous-time subspace model identification method using laguerre filtering. In: *Proceedings of the SYSID'97 Conference, Fukuoka, Japan*. pp. 1143–1148.

Ho, B. and R. E. Kalman (1966). Efficient construction of linear state variable models from input/output functions. *Regelungstechnik* **14**, 545–548.

Jansson, M. (1997). On Subspace Methods in System Identification and Sensor Array Signal Processing. PhD thesis. KTH, Stockholm.

Jansson, M. (2000). Asymptotic variance analysis of subspace identification methods. In: *Proceedings of the SYSID'2000 Conference*. Santa Barbara, California.

Jansson, M. and B. Wahlberg (1998). On consistency of subspace methods for system identification. *Automatica* **34**(12), 1507–1519.

Larimore, W. E. (1983). System identification, reduced order filters and modeling via canonical variate analysis. In: *Proc. 1983 Amer. Control Conference 2*. (H. S. Rao and P. Dorato, Eds.). Piscataway, NJ. pp. 445–451.

Ljung, L. and T. McKelvey (1996*a*). A least squares interpretation of sub-space methods for system identification. In: *Proccedings of the CDC96 Conference*. Kobe, Japan.

Ljung, L. and T. McKelvey (1996*b*). Subspace identification from closed loop data. *Signal Processing, Special Issue on Subspace Methods, Part II: System Identification* **52**(2), 209–216.

McKelvey, T. (1995). Identification of State-Space Models from Time and Frequency Data. PhD thesis. Dept. of Electr. Eng., Linköping.

Ohsumi, A. and T. Kawano (2002). Subspace identification for a class of time-varying continuous-time stochastic systems via distribution-based approach. In: *Proceedings of the 15th IFAC World Congress*. pp. T–Mo–M02.

Oku, H. and H. Kimura (2002). Recursive 4SID algorithms using gradient type subspace tracking. *Automatica* **38**, 1035–1043.

Peternell, K. (1995). Identification of Linear Dynamic Systems by Subspace and Realization-Based Algorithms. PhD thesis. TU Wien.

Peternell, K., W. Scherrer and M. Deistler (1996). Statistical analysis of novel subspace identification methods. *Signal Processing* **52**, 161–177.

Reinsel, G. (1998). *Multivariate Reduced-Rank Regression*. Springer, New York.

Shi, R. (2002). Subspace Identification Methods for Process Dynamic Modeling. PhD thesis. McMaster University, Canada.

Van Overschee, P. and B. DeMoor (1994). N4sid: Subspace algorithms for the identification of combined deterministic-stochastic systems. *Automatica* **30**, 75–93.

Verdult, V. and M. Verhaegen (2002). Subspace identification of multivariable linear parameter-varying systems. *Automatica* **38**, 805–814.

Verhaegen, M. (1993). Application of a subspace model identification technique to identify lti systems operating in closed loop. *Automatica* **29**(4), 1027–1040.

Verhaegen, M. (1994). Identification of the deterministic part of mimo state space models given in innovations form from input-output data. *Automatica* **30**(1), 61–74.

Zeiger, H. P. and A. J. McEwen (1974). Approximate linear realizations of given dimension via Ho's algorithm. *IEEE Transaction on Automatic Control* **19**, 153.

IFAC Publications
www.elsevier.com/locate/ifac

OPTIMAL FILTERING FOR LINEAR SYSTEMS WITH MULTIPLE DELAYS IN OBSERVATIONS

Michael Basin * **Rodolfo Martinez-Zuniga** **

* *Autonomous University of Nuevo Leon, Mexico*
** *Autonomous University of Coahuila, Mexico*

Abstract: In this paper, the optimal filtering problem for a linear system over observations with multiple delays is treated proceeding from the general expression for the stochastic Ito differential of the optimal estimate and its variance. As a result, the optimal filtering equations similar to the traditional Kalman-Bucy ones are obtained in the form dual to the Smith predictor, commonly used for robust control design in time delay systems. In the example, the obtained optimal filter over observations with multiple delays is verified for a sample system and compared with the best Kalman-Bucy filter available for delayed measurements. *Copyright © 2003 IFAC*

Keywords: Linear system, Time delay, Filtering

1. INTRODUCTION

The optimal filtering and control problems for linear systems with measurement delays and its dual optimal control problem remain theoretically unsolved in their most general formulation with multiple and time-varying delays, although the importance of the optimal filtering problem for linear dynamic systems with observation delays was recognized a long time ago. The duality of the control and filtering problems in linear systems implies that the optimal state estimation for the system with measurement delays is closely related to the optimal quadratic regulator problem with delays in inputs, which was extensively studied (see (Eller *et al.*, 1969; Delfour, 1986; Alford and Lee, 1986; Uchida *et al.*, 1988) and references therein). A significantly smaller number of publications consider the problem of optimal filtering (the state and observation equations are corrupted with stochastic noises) for systems with measurement delays, mostly with a single delayed measurement. Comprehensive reviews of theory and algorithms for time delay systems are given in (Kolmanovskii and Shaikhet, 1996; Kolmanovskii and Myshkis, 1999; Malek-Zavarei and Jashmidi, 1987; Dion *et al.*, 1999; Boukas and Liu, 2002).

In this paper, the optimal filtering problem for a linear system over observations with multiple delays is treated proceeding from the general expression for the stochastic Ito differential of the optimal estimate and its variance (Pugachev and Sinitsyn, 1987; Pugachev and Sinitsyn, 2001). As a result, the optimal filtering equations similar to the traditional Kalman-Bucy ones are derived.

The obtained equations contain specific adjustments in the filter gain matrix and the quadratic term of the variance equation, which are calculated in view of linear functional dependence between the system states taken at different time moments, i.e., using linearity of the state equation. That form of the filtering equations is dual to the Smith predictor (Smith, 1958), commonly used for robust control design in time delay systems. Note that although the absolute majority of paper presented at the last 3rd IFAC Workshop on Time Delay Systems (Abdallah and Gu, 2001) have had a deal with the optimal and robust control design for systems with delays, the dual filtering problems have almost not been considered.

The paper is organized as follows. Section 2 and 3 present the filtering problem statement for a linear system state over observations with multiple delays and its solution, respectively. In Section 4, performance of the obtained optimal filter over observations with delay is verified for a sample system with delayed observations and compared to the performance of the best Kalman-Bucy filter available for delayed measurements. The simulation results show asymptotic convergence of the estimate calculated using the obtained optimal filter over observations with multiple delays to the reference variable as time tends to infinity, whereas the Kalman-Bucy estimates calculated without delay adjustment do not converge.

2. FILTERING PROBLEM OVER OBSERVATIONS WITH MULTIPLE DELAYS

Let (Ω, F, P) be a complete probability space with an increasing right-continuous family of σ-algebras $F_t, t \geq 0$, and let $(W_1(t), F_t, t \geq 0)$ and $(W_2(t), F_t, t \geq 0)$ be independent Wiener processes. The partially observed F_t-measurable random process $(x(t), y(t))$ is described by an ordinary differential equation for the dynamic system state

$$dx(t) = (a_0(t) + a(t)x(t))dt + b(t)dW_1(t), \quad (1)$$

$$x(t_0) = x_0,$$

and a differential equation with multiple delays for the observation process:

$$dy(t) = (A_0(t) + A(t)x(t) + \sum_{i=1}^{p} A_i(t)x(t - h_i))dt + \quad (2)$$

$$B(t)dW_2(t),$$

where $x(t) \in R^n$ is the state vector, $y(t) \in R^m$ is the observation process, the initial condition $x_0 \in R^n$ is a Gaussian vector such that x_0, $W_1(t)$, $W_2(t)$ are independent. The observation process $y(t)$ depends on delayed states $x(t - h_i), i = 1, \ldots, p$, where h_i are delay shifts, as well as non-delayed state $x(t)$, which assumes that collection of information on the system state for the observation purposes is made not only at the current time but also after certain time lags $h_i, i = 1, \ldots, p$. The vector-valued function $a_0(s)$ describes the effect of system inputs (controls and disturbances). It is assumed that $A(t)$ is a nonzero matrix and $B(t)B^T(t)$ is a positive definite matrix. All coefficients in (1)-(2) are deterministic functions of appropriate dimensions.

The estimation problem is to find the estimate of the system state $x(t)$ based on the observation process $Y(t) = \{y(s), 0 \leq s \leq t\}$, which minimizes the Euclidean 2-norm

$$J = E[(x(t) - \hat{x}(t))^T(x(t) - \hat{x}(t))]$$

at each time moment t. In other words, our objective is to find the conditional expectation

$$m(t) = \hat{x}(t) = E(x(t) \mid F_t^Y).$$

As usual, the matrix function

$$P(t) = E[(x(t) - m(t))(x(t) - m(t))^T \mid F_t^Y]$$

is the estimate variance.

The proposed solution to this optimal filtering problem is based on the formulas for the Ito differential of the conditional expectation $E(x(t) \mid F_t^Y)$ and its variance $P(t)$ (cited after (Pugachev and Sinitsyn, 1987; Pugachev and Sinitsyn, 2001)) and given in the following section.

3. OPTIMAL FILTER OVER OBSERVATIONS WITH MULTIPLE DELAYS

In the situation of multiple delays, the optimal filtering equations could be obtained directly from the formula for the Ito differential of the conditional expectation $m(t) = E(x(t) \mid F_t^Y)$ (see (Pugachev and Sinitsyn, 1987; Pugachev and Sinitsyn, 2001))

$$dm(t) = E(\varphi(x) \mid F_t^Y)dt +$$

$$E(x[\varphi_1 - E(\varphi_1(x) \mid F_t^Y)]^T \mid F_t^Y) \times$$

$$(B(t)B^T(t))^{-1}(dy(t) - E(\varphi_1(x) \mid F_t^Y)dt),$$

where $\varphi(x)$ is the drift term in the state equation equal to $\varphi(x) = a_0(t) + a(t)x(t)$ and $\varphi_1(x)$ is the drift term in the observation equation equal to $\varphi_1(x) = A_0(t) + A(t)x(t) + \sum_{i=1}^{p} A_i(t)x(t - h_i)$. Upon performing substitution and noticing that $E(x(t - h_i) \mid F_t^Y) = E(x(t - h_i) \mid F_{t-h_i}^Y) = m(t - h_i)$ for any $h_i > 0$, the estimate equation takes the form

$$dm(t) = (a_0(t) + a(t)m(t))dt + E(x(t)[A(t)(x(t) - m(t)) +$$

$$\sum_{i=1}^{p} A_i(t)(x(t - h_i) - m(t - h_i))]^T \mid F_t^Y)(B(t)B^T(t))^{-1} \times$$

$$(dy(t) - (A_0(t) + A(t)m(t) + \sum_{i=1}^{p} A_i(t)m(t - h_i))dt) =$$

$$(a_0(t) + a(t)m(t))dt + [E(x(t)(x(t) - m(t))^T \mid F_t^Y)A^T(t) +$$

$$\sum_{i=1}^{p} E(x(t)(x(t - h_i) - m(t - h_i))^T \mid F_t^Y)A_i^T(t)] \times$$

$$(B(t)B^T(t))^{-1}(dy(t) - (A_0(t) + A(t)m(t) +$$

$$\sum_{i=1}^{p} A_i(t)m(t - h_i))dt).$$

The obtained form of the optimal estimate equation is similar to the Kalman filter one, except the term $E(x(t)(x(t) - m(t))^T \mid F_t^Y)A^T(t) + \sum_{i=1}^{p} E(x(t)(x(t - h_i) - m(t - h_i))^T \mid F_t^Y)A_i^T(t) = P(t)A^T(t) + \sum_{i=1}^{p} E(x(t)(x(t - h_i) - m(t - h_i))^T \mid F_t^Y)A_i^T(t)$ standing instead of $P(t)A^T(t) = E((x(t) - m(t))(x(t) - m(t))^T \mid F_t^Y)A^T(t)$. However, the former term can be expressed as a function of the variance, using the Cauchy formula for $x(t)$ as the solution of the linear equation (1) and $m(t)$ as its conditional expectation. Indeed,

$$x(t) = \Phi(t, t-h)x(t-h) + \int_{t-h}^{t} \Phi(t,\tau)a_0(\tau)d\tau + \quad (3)$$

$$\int_{t-h}^{t} \Phi(t,\tau)b(\tau)dW_1(\tau),$$

where $\Phi(t,\tau)$ is the matrix of fundamental solutions of the homogeneous equation (1), that is solution of the matrix equation

$$\frac{d\Phi(t,\tau)}{dt} = a(t)\Phi(t,\tau), \quad \Phi(t,t) = I,$$

where I is the identity matrix. In other words, $\Phi(t, t-h) = \exp \int_{t-h}^{t} a(s)ds$. Thus, the delayed term in the estimate equation is equal to

$$\sum_{i=1}^{p} E(x(t)(x(t-h_i) - m(t-h_i))^T \mid F_t^Y)A_i^T(t) =$$

$$\sum_{i=1}^{p} E(x(t)(x(t)-m(t))^T \mid F_t^Y)\exp(-\int_{t-h_i}^{t} a^T(s)ds)A_i^T(t)$$

$$= \sum_{i=1}^{p} P(t)\exp(-\int_{t-h_i}^{t} a^T(s)ds)A_i^T(t),$$

and the entire equation takes the form

$$dm(t) = (a_0(t) + a(t)m(t))dt + P(t)[A^T(t) + \quad (4)$$

$$\sum_{i=1}^{p} \exp(-\int_{t-h_i}^{t} a^T(s)ds)A_i^T(t)]\times$$

$$\left(B(t)B^T(t)\right)^{-1}(dy(t) - (A_0(t) + A(t)m(t) +$$

$$\sum_{i=1}^{p} A_i(t)m(t-h_i))dt).$$

So far, the optimal estimate equation, similarly to the classic Kalman-Bucy case, includes the gain matrix $P(t)[A^T(t) + \sum_{i=1}^{p} \exp(-\int_{t-h_i}^{t} a^T(s)ds)A_i^T(t)]\left(B(t)B^T(t)\right)^{-1}$ depending on the estimate variance, but now also depending on the delay adjustment $[A^T(t) + \sum_{i=1}^{p} \exp(-\int_{t-h_i}^{t} a^T(s)ds)A_i^T(t)]$. The problem now is to find the equation for $P(t)$ in a closed form.

For this purpose, the formula for the Ito differential of the conditional expectation variance $P(t) = E((x(t) - m(t))(x(t)-m(t))^T \mid F_t^Y)$ could be used (cited again after (Pugachev and Sinitsyn, 1987; Pugachev and Sinitsyn, 2001)):

$$dP(t) = (E((x(t)-m(t))\varphi^T(x) \mid F_t^Y) +$$

$$E(\varphi(x)(x(t)-m(t))^T) \mid F_t^Y) + b(t)b^T(t) -$$

$$E(x(t)[\varphi_1 - E(\varphi_1(x) \mid F_t^Y)]^T \mid F_t^Y)(B(t)B^T(t))^{-1}\times$$

$$E([\varphi_1 - E(\varphi_1(x) \mid F_t^Y)]x^T(t) \mid F_t^Y))dt +$$

$$E((x(t)-m(t))(x(t)-m(t))[\varphi_1 -$$

$$E(\varphi_1(x) \mid F_t^Y)]^T \mid F_t^Y)(B(t)B^T(t))^{-1}\times$$

$$(dy(t) - E(\varphi_1(x) \mid F_t^Y)dt),$$

where the last term should be understood as a 3D tensor (under the expectation sign) convoluted with a vector, which yields a matrix. Upon substituting the expressions for φ and φ_1, the last formula takes the form

$$dP(t) = (E((x(t)-m(t))x^T(t)a^T(t) \mid F_t^Y) +$$

$$E(a(t)x(t)(x(t)-m(t))^T) \mid F_t^Y) + b(t)b^T(t) -$$

$$[E(x(t)(x(t)-m(t))^T \mid F_t^Y)A^T(t) +$$

$$\sum_{i=1}^{p} E(x(t)(x(t-h_i) - m(t-h_i))^T \mid F_t^Y)A_i^T(t)]\times$$

$$\left(B(t)B^T(t)\right)^{-1}[A(t)E((x(t)-m(t))x^T(t)) \mid F_t^Y) +$$

$$\sum_{i=1}^{p} A_i(t)E((x(t-h_i) - m(t-h_i))x^T(t)) \mid F_t^Y)])dt +$$

$$E((x(t)-m(t))(x(t)-m(t))\times$$

$$([A(t)(x(t)-m(t)) + \sum_{i=1}^{p} A_i(t)(x(t-h_i) -$$

$$m(t-h_i))]^T) \mid F_t^Y)(B(t)B^T(t))^{-1}\times$$

$$(dy(t) - E(\varphi_1(x) \mid F_t^Y)dt).$$

Using again the formula (3) for delayed values of the state and considering that

$$E(x(t)(x(t)-m(t))^T \mid F_t^Y)A^T(t) +$$

$$\sum_{i=1}^{p} E(x(t)(x(t-h_i) - m(t-h_i))^T \mid F_t^Y)A_i^T(t) =$$

$$P(t)A^T(t) + \sum_{i=1}^{p} E(x(t)(x(t)-m(t))^T \mid F_t^Y)\times$$

$$\exp(-\int_{t-h_i}^{t} a^T(s)ds)A_i^T(t) =$$

$$P(t)A^T(t) + \sum_{i=1}^{p} P(t)\exp(-\int_{t-h_i}^{t} a^T(s)ds)A_i^T(t),$$

the equation for $P(t)$ is reduced to

$$dP(t) = (P(t)a^T(t) + a(t)P(t) + b(t)b^T(t) -$$

$$[P(t)A^T(t) + \sum_{i=1}^{p} P(t)\exp(-\int_{t-h_i}^{t} a^T(s)ds)A_i^T(t)]\times$$

$$\left(B(t)B^T(t)\right)^{-1}[A(t)P(t) +$$

$$\sum_{i=1}^{p} A_i(t)\exp(-\int_{t-h_i}^{t} a(s)ds)P(t)])dt +$$

$$E((x(t)-m(t))(x(t)-m(t))(x(t)-m(t) \mid F_t^Y)\times$$

$$[A^T(t) + \sum_{i=1}^{p} \exp(-\int_{t-h_i}^{t} a^T(s)ds)] \times$$

$$A_i^T(t)(B(t)B^T(t))^{-1}(dy(t) - E(\varphi_1(x) \mid F_t^Y)dt).$$

The last term in this formula contains the conditional third central moment $E((x(t)-m(t))(x(t)-m(t))(x(t)-m(t)) \mid F_t^Y)$ of $x(t)$ with respect to observations, which is equal to zero, because $x(t)$ is conditionally Gaussian, in view of Gaussianity of the noises and the initial condition and linearity of the state and observation equations. Thus, the entire last term is vanished and the following variance equation is obtained

$$dP(t) = (P(t)a^T(t) + a(t)P(t) + b(t)b^T(t) - P(t)\times \quad (5)$$

$$[A^T(t)+\sum_{i=1}^{p} \exp(-\int_{t-h_i}^{t} a^T(s)ds)A_i^T(t)](B(t)B^T(t))^{-1}\times$$

$$[A(t) + \sum_{i=1}^{p} A_i(t)\exp(-\int_{t-h_i}^{t} a(s)ds)]P(t))dt.$$

The obtained system of filtering equations (4) and (5) should be complemented with the initial conditions $m(t_0) = E[x(t_0) \mid F_{t_0}^Y]$ and $P(t_0) = E[(x(t_0) - m(t_0)(x(t_0) - m(t_0)^T \mid F_{t_0}^Y]$. As noted, this system is similar to the conventional Kalman-Bucy filter, except the adjustments for delays in the estimate and variance equations, calculated due to the Cauchy formula for the state equation. It closely resembles the Smith predictor (Smith, 1958) commonly used for robust control design in time delay systems. Nevertheless, the obtained filter is optimal with respect to the introduced form of the observation process, since it is obtained from the exact Ito differential for the conditional expectation and variance.

In the case of a constant matrix a in the state equation, the optimal filter takes the especially simple form ($\exp(-\int_{t-h}^{t} a^T ds) = \exp(-a^T h)$)

$$dm(t) = (a_0(t) + am(t))dt + P(t)\times \quad (6)$$

$$[A^T(t) + \sum_{i=1}^{p} \exp(-a^T h_i)A_i^T(t)] \times (B(t)B^T(t))^{-1}$$

$$(dy(t) - (A_0(t) + A(t)m(t) + \sum_{i=1}^{p} A_i(t)m(t-h_i))dt),$$

$$dP(t) = (P(t)a^T + aP(t) + b(t)b^T(t) - \quad (7)$$

$$P(t)[A^T(t) + \sum_{i=1}^{p} \exp(-a^T h_i)A_i^T(t)]\times$$

$$(B(t)B^T(t))^{-1}[A(t) + \sum_{i=1}^{p} A_i(t)\exp(-ah_i)]P(t))dt.$$

4. EXAMPLE

This section presents an example of applying the obtained filter over linear observations with multiple delays to estimation of the state variable, provided that there are two observation channels with different delay shifts. This situation is very common and can be frequently encountered in various applications (see, for instance, a mixing tank example in Chapter 6 of (Kwakernaak and Sivan, 1972)).

Let the unobservable state $x(t)$ be given by

$$\dot{x}(t) = 0.1x(t), \quad x(0) = x_0, \quad (8)$$

and there are two observation devices measuring the system state with different delay shifts

$$y_1(t) = x(t-2) + \psi_1(t), \quad (9)$$

$$y_2(t) = x(t-20) + \psi_2(t).$$

where $\psi_1(t)$ and $\psi_2(t)$ are white Gaussian noises which are weak mean square derivatives of standard Wiener processes. Thus, the equations (8) and (9) present the conventional form for the equations (1) and (2), which is actually used in practice.

The estimation problem is to find the optimal estimate for the variable $x(t)$ using direct linear observations with delays (9) confused with independent and identically distributed disturbances modeled as white Gaussian noises. As noted, the delay shifts are different in each observation channel, so the filtering equations (6)–(7) should be employed.

Since $a = 0.1$ in (8) and $A_1^T = [1\ 0]$ and $A_2^T = [0\ 1]$ in (9), the equations (6)–(7) take the following particular form

$$\dot{m}(t) = 0.1m(t) + P(t)[\exp(-0.2)(y_1(t) - \quad (10)$$

$$m(t-2)) + \exp(-2)(y_2(t) - m(t-20))],$$

$$m(0) = m_0,$$

$$\dot{P}(t) = 0.2P(t) - [\exp^2(-0.2) + \exp^2(-2)]P^2(t),$$

$$P(0) = P_0.$$

The estimates obtained upon solving the equations (10) are compared to the conventional Kalman-Bucy estimates satisfying the following filtering equations for the unobservable state (8) over linear observations with multiple delays (9), provided that the filter gain matrix is considered equal to $P(t)A^T(t)(B(t)B^T(t))^{-1}$, as in the standard Kalman-Bucy filter, without the delay adjustment term $[\sum_{i=1}^{p} \exp(-a^T h_i)A_i^T(t)]$:

$$\dot{m}_1(t) = 0.1m_1(t) + P_1(t)[(y_1(t) - m_1(t-2)) + \quad (11)$$

$$(y_2(t) - m_1(t-20))], \quad m_1(0) = m_{10},$$

$$\dot{P}_1(t) = 0.2P_1(t) - P_1^2(t), \quad P_1(0) = P_{10}.$$

Numerical simulation results are obtained solving the systems of filtering equations (10) and (11). The obtained values of the estimates $m(t)$ and $m_1(t)$ satisfying (10) and (11) are compared to the real values of the state variable $x(t)$ in (8).

For each of the two filters (10) and (11) and the reference system (8) involved in simulation, the following initial values are assigned: $x_0 = 1$, $m_0 = m_{10} = 10$, $P_0 = P_{10} = 100$. Gaussian disturbances $\psi_1(t)$ and $\psi_2(t)$ in (9) are realized as sinusoidal signals: $\psi_1(t) = \psi_2(t) = \sin t$, although the standard MatLab function for white noise can also be used.

The following graphs are obtained: graphs of the reference state variable $x(t)$ for the system (8); graphs of the Kalman-Bucy filter estimate $m_1(t)$ satisfying the equations (11); and graphs of the optimal linear filter with multiple delays estimate $m(t)$ satisfying the equations (10). The graphs of all those variables are shown on the entire simulation interval from $T = 0$ to $T = 100$ (Fig. 1), and around the reference time points: $T = 60$ (Fig. 2), $T = 80$ (Fig. 3), and $T = 100$ (Fig. 4).

The following values of the reference state variable $x(t)$ and the estimates $m(t)$ and $m_1(t)$ are obtained at the reference time points: for $T = 60$, $x(60) = 405.30$, $m(60) = 405.83$, $m_1(60) = 408.01$; for $T = 80$, $x(80) = 2994.80$, $m(80) = 2994.82$, $m_1(80) = 2985.88$; for $T = 100$, $x(80) = 22026$, $m(80) = 22026$, $m_1(80) = 22053$.

Thus, it can be concluded that the optimal filter for linear systems over observations with multiple delays (10) yield definitely better estimates than the conventional Kalman-Bucy filter. The simulations have also been made for the Kalman filter with the gain matrix $P_2(t)$ satisfying the equation

$$\dot{P}_2(t) = 0.2P_2(t) - 2P_2^2(t), \quad P_2(0) = 100,$$

considering that $A^T A = [1\ 1][1\ 1]^T = 2$, and the estimate $m_2(t)$ given by the first equation in (11)

$$\dot{m}_2(t) = 0.1m_2(t) + P_2(t)[(y_1(t) - m_2(t-2))+$$
$$(y_2(t) - m_2(t-20))], \quad m_1(0) = 10.$$

However, the results obtained in this case occur to be even worse than for the Kalman filter (11). The values of the estimate $m_2(t)$ in the reference time points are: $m_2(60) = 461.80$, $m_2(80) = 3070.66$, $m_2(100) = 22140$.

Subsequent discussion of the obtained simulation results can be found in Conclusions.

5. CONCLUSIONS

The simulation results show that the values of the estimate calculated by using the obtained optimal filter over observations with multiple delays are noticeably closer to the real values of the reference variable than the values of the Kalman-Bucy estimates. Moreover, it can be seen that the estimate produced by the optimal filter over observations with multiple delays asymptotically converges to the real values of the reference variable as time tends to infinity, although the reference system (8) itself is unstable. On the contrary, the conventionally designed (non-optimal) Kalman-Bucy estimates without delay adjustment diverge from the real values. This significant improvement in the estimate behavior is obtained due to the more careful selection of the filter gain matrix using the delay adjustment term, which compensates for unstable dynamics of the reference system, as it should be in the optimal filter. Although this conclusion follows from the developed theory, the numerical simulation serves as a convincing illustration.

6. REFERENCES

Abdallah, C. T. and K. Gu (2001). *Proc. 3rd IFAC Workshop on Time Delay Systems.* OMNIPRESS. Madison.

Alford, R. L. and E. B. Lee (1986). Sampled data hereditary systems: linear quadratic theory. *IEEE Trans. Automat. Contr.* **31**, 60–65.

Boukas, E.-K. and Z.K. Liu (2002). *Deterministic and Stochastic Time-Delay Systems.* Birkhauser.

Delfour, M. C. (1986). The linear quadratic control problem with delays in state and control variables: a state space approach. *SIAM J. Control and Optim.* **24**, 835–883.

Dion, J.M., J.L.Dugard and M. Fliess (1999). *Linear Time-Delay Systems.* Pergamon. London.

Eller, D.H., J.K. Aggarwal and H.T. Banks (1969). Optimal control of linear time-delay systems. *IEEE Trans. Automat. Contr.* **14**, 678–687.

Kolmanovskii, V. B. and A. D. Myshkis (1999). *Introduction to the Theory and Applications of Functional Differential Equations.* Kluwer. New York.

Kolmanovskii, V. B. and L. E. Shaikhet (1996). *Control of Systems with Aftereffe.* American Mathematical Society. Providence.

Kwakernaak, H. and R. Sivan (1972). *Linear Optimal Control Systems.* Wiley–Interscience. New York.

Malek-Zavarei, M. and M. Jashmidi (1987). *Time-Delay Systems: Analysis, Optimization and Applications.* North-Holland. Amsterdam.

Pugachev, V. S. and I. N. Sinitsyn (1987). *Stochastic Differential Systems: Analysis and Filtering.* John Wiley & Sons. Chichester.

Pugachev, V. S. and I. N. Sinitsyn (2001). *Stochastic Systems: Theory and Applications.* World Scientific. Singapore.

Smith, O. J. M. (1958). *Feedback Control Systems.* McGraw Hill. New York.

Uchida, K., E. Shimemura, T. Kubo and N. Abe (1988). The linear-quadratic optimal control approach to feedback control design for systems with delay. *Automatica* **24**, 773–780.

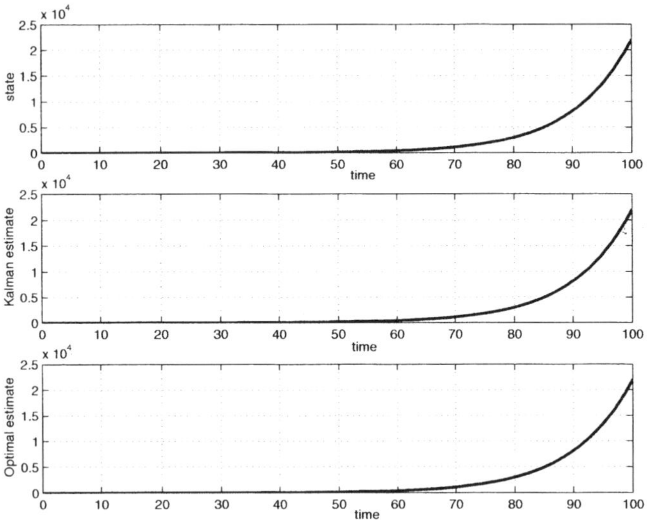

Fig. 1. Graphs of the reference state variable $x(t)$ and the estimates $m_1(t)$ and $m(t)$ on the entire simulation interval $[0, 100]$.

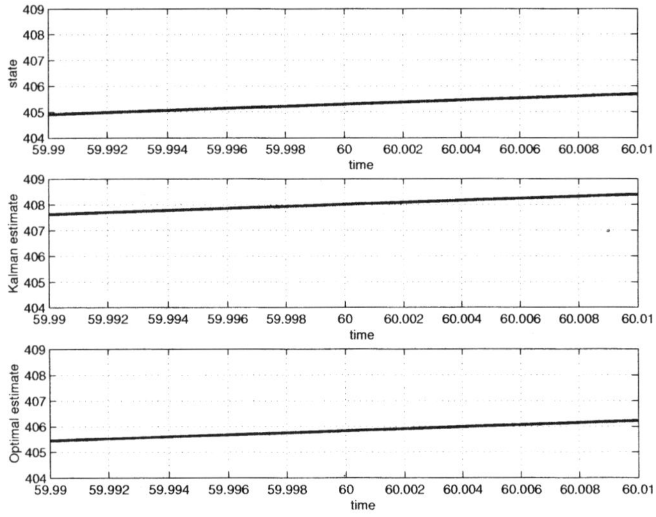

Fig. 2. Graphs of the reference state variable $x(t)$ and the estimates $m_1(t)$ and $m(t)$ around the reference time point $T = 60$.

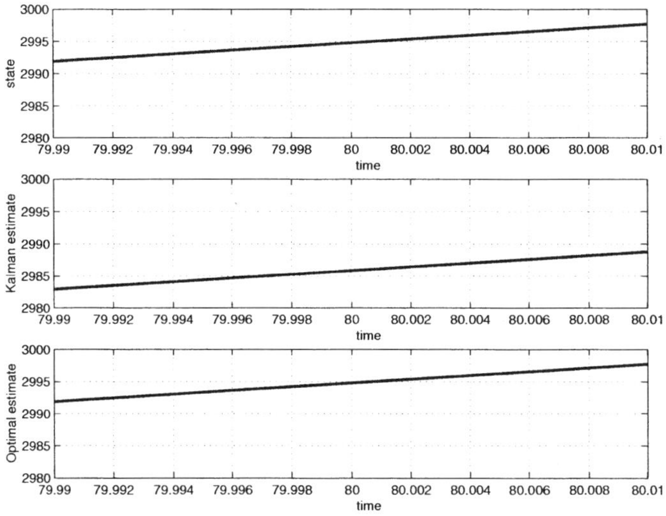

Fig. 3. Graphs of the reference state variable $x(t)$ and the estimates $m_1(t)$ and $m(t)$ around the reference time point $T = 80$.

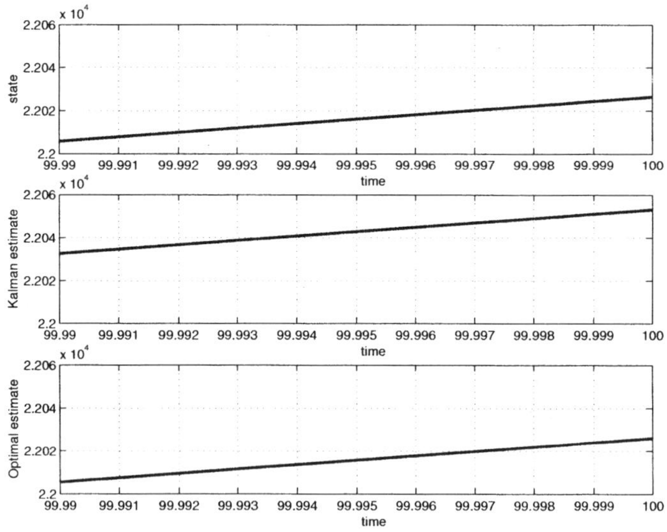

Fig. 4. Graphs of the reference state variable $x(t)$ and the estimates $m_1(t)$ and $m(t)$ around the reference time point $T = 100$.

IFAC Publications
www.elsevier.com/locate/ifac

THE INFORMATION ANALYSIS IN JOINT PROBLEM OF CONTINUOUS-DISCRETE FILTERING AND GENERALIZED EXTRAPOLATION

N.S. Dyomin[*], I.E. Safronova[*], S.V. Rozhkova[**]

[*] *Department of Applied Mathematics and Cybernetics,*
Tomsk State University, 36 Lenin ave., 634050 Tomsk, Russia
e-mail: sev@vmm.tsu.ru
[**] *Department of Higher Mathematics,*
Tomsk Polytechnic University, 30 Lenin ave., 634034 Tomsk, Russia
e-mail: onm@cam.tpu.ru

Abstract: This paper considers the information aspect of the problem of the joint filtering and extrapolation, when the output of observation channels is the set of realizations of the processes with continuous and discrete time, which depend on both the current and the past values of unobservable process. The expressions, defined time evolution of Shannon information measures, are obtained and the solution structure are researched. The problem of optimal transmission of Gaussian Markov signal on a continuous-discrete time memory channel and the example of information efficiency of observations with memory are considered. *Copyright © 2003 IFAC*

Keywords: signal processing, communication, stochastic systems, filtering, extrapolation, information amount.

1. INTRODUCTION

In the Kalman systems (Kalman, 1960) the pair of vector processes $\{x_t ; y_t\}$ with continuous or discrete time, where x_t is an unobservable process (state vector, useful signal) and y_t is an observable process (observation vector, output transmission signal), is the basic object. The situation is generalized when x_t is process with continuous time, and $y_t = y(t,t_m) = \{z_t, \eta(t_m); m = 0,1,\cdots\}$, i.e. one can observe set of processes with continuous z_t and discrete $\eta(t_m)$ time, which possess the memory relatively unobservable process and depend on the current and the past values of process x_t. For similar class of processes this paper considers an information aspect of the joint filtering (Abakumova *et. al.*, 1995) and generalized extrapolation (Dyomin *et. al.*, 1997) problem.

Used notations: $P\{\cdot\}$ is event probability; $E\{\cdot\}$ is expectation operator; $\mathrm{N}\{\cdot\}$ denotes Gaussian probability density function with given parameters a and B; $|\cdot|$ is a determinant of the matrix; $tr[A]$ is a trace matrix of A; D^{-1} is the inversion matrix of D; D^T denotes transpose of a matrix or a vector; $\tilde{\tau}_N = (\tau_1, \tau_2, \cdots, \tau_N)$, $\tilde{s}_L = (s_1, s_2, \cdots, s_L)$;

$$\tilde{x}_\tau^N = \left[x_{\tau_k}\right],\ \tilde{x}_{t,\tau}^{N+1} = \begin{bmatrix} x_t \\ \tilde{x}_\tau^N \end{bmatrix},$$

$$\tilde{x}_s^L = \left[x_{s_l}\right],\ \tilde{x}_{t,s}^{L+1} = \begin{bmatrix} x_t \\ \tilde{x}_s^L \end{bmatrix}, \tag{1}$$

$$\tilde{x}_{t,\tau,s}^{N+L+1} = \begin{bmatrix} \tilde{x}_{t,\tau}^{N+1} \\ \tilde{x}_s^L \end{bmatrix},\ k = \overline{1;N},\ l = \overline{1;L}.$$

2. STATEMENT OF THE PROBLEM

The unobservable n-dimensional process x_t (useful signal) and the observable l - dimensional process z_t with continuous time (an output signal of a continuous transmission channel) are described by the stochastic differential equations in the Ito sense (Liptser and Shiryayev, 1977)

$$dx_t = f(t,x_t)dt + \Phi_1(t)dw_t,\ t \ge 0, \tag{2}$$

$$dz_t = h\left(t,x_t,\tilde{x}_\tau^N,z\right)dt + \Phi_2(t,z)dv_t, \tag{3}$$

and the observable q-dimensional process with discrete time $\eta(t_m)$ (an output signal of a discrete transmission channel) $(m = 0,1,\cdots)$ has the form

$$\eta(t_m) = g\left(t_m,x_{t_m},\tilde{x}_\tau^N,z\right) + \Phi_3(t_m,z)\xi(t_m). \tag{4}$$

In (2)-(4) $0 < \tau_N < \cdots < \tau_1 < t_m \le t$, $t_k = const$, $k = \overline{1;N}$, i.e. memory is fixed (Dyomin, *et al.*, 1997), w_t and v_t are r_1 – dimensional and r_2 – dimensional standard Wiener processes, $\xi(t_m)$ is a standard white Gaussian r_3 – dimensional sequence,

$$f(\cdot) = f(t) + F(t)x_t,\quad p_0(x) = \mathrm{N}\{x;\mu_0,\Gamma_0\}, \tag{5}$$

$$h(\cdot) = h(t,z) + H_{0,N}(t,z)\tilde{x}_{t,\tau}^{N+1},$$

$$g(\cdot) = g(t_m,z) + G_{0,N}(t_m,z)\tilde{x}_{t_m,\tau}^{N+1}, \tag{6}$$

$$H_{0,N}(\cdot) = \left[H_0(t,z) \mid H_1(t,z) \mid \cdots \mid H_N(t,z)\right] = \left[H_0(\cdot) \mid H_{1,N}(\cdot)\right],$$

$$G_{0,N}(\cdot) = \left[G_0(t_m,z) \mid G_1(t_m,z) \mid \cdots \mid G_N(t_m,z)\right] = \left[G_0(\cdot) \mid G_{1,N}(\cdot)\right].$$

The other statement conditions are the same as in (Dyomin, *et al.*, 1997).

Problem: the relations for joint information amount (Gallager, 1968)

$$I_s^t\left[x_t,\tilde{x}_s^L;z_0^t,\eta_0^m\right] = E\left\{\ln\left[p_s^t\left(x_t;\tilde{x}_s^L\right)/p\left(t,x_t;\tilde{s}_L,\tilde{x}_s^L\right)\right]\right\}, \tag{7}$$

$$p_s^t\left(x;\tilde{x}^L\right) = \partial^{L+1}P\left\{x_t \le x;\tilde{x}_s^L \le \tilde{x}^L|z_0^t,\eta_0^m\right\}/\partial x\partial\tilde{x}^L,$$

$$p\left(t,x;\tilde{s}_L,\tilde{x}^L\right) = \partial P^{L+1}\left\{x_t \le x;\tilde{x}_s^L \le \tilde{x}^L\right\}/\partial x\partial\tilde{x}^L,$$

about current x_t and future $\tilde{x}_s^L = \left[x_{s_1},x_{s_2},\cdots,x_{s_L}\right]$, $t < s_1 < \cdots < s_L$, values of the unobservable process, which is contained in the realizations' set $z_0^t = \{z_\sigma; 0 \le \sigma \le t\}$ and $\eta_0^m = \{\eta(t_0),\eta(t_1),\cdots,\eta(t_m)\}$ of observable processes are to be found. In this case $s_l = const$, $l = \overline{1;L}$, i.e. the extrapolation is inverse (Dyomin, *et al.*, 1997).

3. MAIN RESULT

Proposition 1. For the posterior density $p_s^t\left(x;\tilde{x}_N;\tilde{x}^L\right) = \partial^{N+L+1}P\left\{x_t \le x;\tilde{x}_\tau^N \le \tilde{x}_N;\tilde{x}_s^L \le \tilde{x}^L|z_0^t,\eta_0^m\right\}/\partial x\partial\tilde{x}_N\partial\tilde{x}^L$ and the prior density $p\left(t,x;\tilde{\tau}_N,\tilde{x}_N;\tilde{s}_L,\tilde{x}^L\right) = \partial^{N+L+1}P\left\{x_t \le x;\tilde{x}_\tau^N \le \tilde{x}_N;\tilde{x}_s^L \le \tilde{x}^L\right\}/\partial x\partial\tilde{x}_N\partial\tilde{x}^L$ the following properties are valid (see (1))

$$p_s^t\left(x;\tilde{x}_N;\tilde{x}^L\right) = \mathrm{N}\left\{\tilde{x}_{N+L+1};\tilde{\mu}_{N+L+1}\left(\tilde{\tau}_N,t,\tilde{s}_L\right),\ \tilde{\Gamma}_{N+L+1}\left(\tilde{\tau}_N,t,\tilde{s}_L\right)\right\} = \mathrm{N}\left\{\begin{bmatrix} x \\ \tilde{x}_N \\ \tilde{x}^L \end{bmatrix};\begin{bmatrix} \mu(t) \\ \tilde{\mu}_N\left(\tilde{\tau}_N,t\right) \\ \tilde{\mu}_L\left(t,\tilde{s}_L\right) \end{bmatrix},\right. \tag{8}$$

$$\left.\begin{bmatrix} \Gamma(t) & \tilde{\Gamma}_{0N}\left(\tilde{\tau}_N,t\right) & \tilde{\Gamma}_{0,N+1}^L\left(t,\tilde{s}_L\right) \\ \tilde{\Gamma}_{0N}^T(\cdot) & \tilde{\Gamma}_N\left(\tilde{\tau}_N,t\right) & \tilde{\Gamma}_{N,N+1}^L\left(\tilde{\tau}_N,t,\tilde{s}_L\right) \\ \left(\tilde{\Gamma}_{0,N+1}^L(\cdot)\right)^T & \left(\tilde{\Gamma}_{N,N+1}^L(\cdot)\right)^T & \tilde{\Gamma}^L\left(t,\tilde{s}_L\right) \end{bmatrix}\right\},$$

$$p\left(t,x;\tilde{\tau}_N,\tilde{x}_N;\tilde{s}_L,\tilde{x}^L\right) = \mathrm{N}\left\{\tilde{x}_{N+L+1};\tilde{a}_{N+L+1}\left(\tilde{\tau}_N,t,\tilde{s}_L\right),\tilde{D}_{N+L+1}\left(\tilde{\tau}_N,t,\tilde{s}_L\right)\right\}. \tag{9}$$

The block components of the parameters of distribution (8) are defined by the system of differential-recurrence equations of Theorems 1, 2 in (Abakumova *et. al.*, 1995), Theorem 3 and Corollary 2 in (Dyomin *et. al.*, 1997). The structure of $\tilde{a}_{N+L+1}(\cdot)$ and $\tilde{D}_{N+L+1}(\cdot)$ similar to the structure of $\tilde{\mu}_{N+L+1}(\cdot)$ and $\tilde{\Gamma}_{N+L+1}(\cdot)$ with replacement μ and Γ by a and D, and the parameters of (9) are found obviously (Meditch, 1969).

Theorem 1. The information amount (7) on the time intervals $t_m \le t < t_{m+1}$ is defined by the equation

$$\begin{aligned} dI_s^t\left[x_t,\tilde{x}_s^L;z_0^t,\eta_0^m\right]/dt &= (1/2)tr\left[E\left\{R^{-1}(t,z)\right.\right. \\ &\quad \left.\left.\times\tilde{H}_{L+1}(t,z)\left(\tilde{\Gamma}^{L+1}\left(t,\tilde{s}_L\right)\right)^{-1}\tilde{H}_{L+1}^T(t,z)\right\}\right] \\ &\quad -(1/2)tr\left[Q(t)E\left\{\Gamma^{-1}\left(t|\tilde{s}_L\right)\right\} - D^{-1}\left(t|\tilde{s}_L\right)\right] \end{aligned} \tag{10}$$

subject to the initial condition

$$I_s^{t_m}\left[x_{t_m},\tilde{x}_s^L;z_0^{t_m},\eta_0^m\right] = I_s^{t_m-0}\left[x_{t_m},\tilde{x}_s^L;z_0^{t_m},\eta_0^{m-1}\right] + \Delta I_s^{t_m}, \tag{11}$$

where $Q(t) = \Phi_1(t)\Phi_1^T(t)$, $R(t,z) = \Phi_2(t,z)\Phi_2^T(t,z)$,

$$\widetilde{\Gamma}^{L+1}(t,\widetilde{s}_L)=\begin{bmatrix}\Gamma(t) & \widetilde{\Gamma}^L_{0,N+1}(t,\widetilde{s}_L)\\ \left(\widetilde{\Gamma}^L_{0,N+1}(\cdot)\right)^T & \widetilde{\Gamma}^L(t,\widetilde{s}_L)\end{bmatrix},$$

$$\Gamma(t|\widetilde{s}_L)=\Gamma(t) - \widetilde{\Gamma}^L_{0,N+1}(t,\widetilde{s}_L)\left(\widetilde{\Gamma}^L(t,\widetilde{s}_L)\right)^{-1}\left(\widetilde{\Gamma}^L_{0,N+1}(t,\widetilde{s}_L)\right)^T,$$

$$D(t|\widetilde{s}_L)=D(t) - \widetilde{D}^L_{0,N+1}(t,\widetilde{s}_L)\left(\widetilde{D}^L(t,\widetilde{s}_L)\right)^{-1}\left(\widetilde{D}^L_{0,N+1}(t,\widetilde{s}_L)\right)^T,$$

$$\widetilde{H}_{L+1}(t,z)=\left[\widetilde{H}_0(\cdot)\,\vdots\,\widetilde{H}_L(\cdot)\right] = \left[\widetilde{H}_0(\cdot)\,\vdots\,\widetilde{H}_{N+1}(\cdot)\,\vdots\,\cdots\,\vdots\,\widetilde{H}_{N+L}(\cdot)\right],$$

$$\widetilde{H}_0(t,z)=H_0(\cdot)\Gamma(t)+H_{1,N}(\cdot)\widetilde{\Gamma}^T_{0N}(\widetilde{\tau}_N,t),$$

$$\widetilde{H}_{N+l}(t,z)=H_0(\cdot)\Gamma^l_{0,N+1}(t,s_l)+H_{1,N}(\cdot)\widetilde{\Gamma}^l_{N,N+1}(\widetilde{\tau}_N,t,s_l),$$

$\Gamma^l_{0,N+1}(\cdot)$ is the $l-$th matrix element of the matrix $\widetilde{\Gamma}^L_{0,N+1}(\cdot)$, $\widetilde{\Gamma}^l_{N,N+1}(\cdot)$ is the $l-$th matrix column of the matrix $\widetilde{\Gamma}^L_{N,N+1}(\cdot)$, $l=\overline{1;L}$,

$$\Delta I_s^{t_m}=(1/2)\times E\left\{\ln\left[\left|\widetilde{\Gamma}^{L+1}(t_m-0,\widetilde{s}_L)\right|/\left|\widetilde{\Gamma}^{L+1}(t_m,\widetilde{s}_L)\right|\right]\right\}. \quad (12)$$

Proof. Corollary 1 in (Dyomin, *et al.*, 1997) implies $p_s^t(x;\widetilde{x}^L)$ on the time intervals $t_m \le t < t_{m+1}$ is defined by the equation

$$d_t p_s^t(x;\widetilde{x}^L)=\mathbf{L}_{t,x}\left[p_s^t(x;\widetilde{x}^L);p_t(x)\right]dt + p_s^t(x;\widetilde{x}^L)\left[\overline{h(\widetilde{\tau}_N,z|x,\widetilde{x}^L)}-\overline{h(t,z)}\right]^T R^{-1}(t,z)d\widetilde{z}_t \quad (13)$$

subject to the initial condition

$$p_s^{t_m}(x;\widetilde{x}^L) = \left[c(\eta(t_m),z|x,\widetilde{x}^L)/c(\eta(t_m),z)\right]p_s^{t_m-0}(x;\widetilde{x}^L), \quad (14)$$

where

$p_t(x)=\partial P\{x_t\le x|z_0^t,\eta_0^m\}/\partial x$, $R(t,z)=\Phi_2(\cdot)\Phi_2^T(\cdot)$, $d\widetilde{z}_t=dz_t-\overline{h(t,z)}dt$, $\overline{h(t,z)}=E\{h(t,x_t,\widetilde{x}_\tau^N,z)\,|z_0^t,\eta_0^m\}$,

$\overline{h(\widetilde{\tau}_N,z|x,\widetilde{x}^L)}=E\{h(\cdot)|x_t=x,\widetilde{x}_s^L=\widetilde{x}^L,\,z_0^t,\eta_0^m\}$,

$c(\eta(t_m),z)=E\{c(x_{t_m},\widetilde{x}_{\tau}^N,\eta(t_m),z)|z_0^{t_m},\eta_0^{m-1}\}$,

$c(\eta(t_m),z|x,\widetilde{x}_s^L)=E\{c(\cdot)|x_{t_m}=x,\ \widetilde{x}_s^L=\widetilde{x}^L,z_0^{t_m},\eta_0^{m-1}\}$,

$c(x,\widetilde{x}_N,\eta(t_m),z)=\exp\{-(1/2)[\eta(t_m)-g(t_m,x,\widetilde{x}_N,z)]^T \times V^{-1}(t_m,z)[\cdot]\}$, $V(t_m,z)=\Phi_3(\cdot)\Phi_3^T(\cdot)$,

$\mathbf{L}_{\tau,y}[\varphi_1(\tau,y,\cdot);\varphi_2(\tau,y,\cdot)]=[\varphi_1(\cdot)/\varphi_2(\cdot)]L_{\tau,y}[\varphi_2(\cdot)] - \varphi_2(\cdot)L^*_{\tau,y}[\varphi_1(\cdot)/\varphi_2(\cdot)]$, $L_{\tau,y}[\cdot]$ and $L^*_{\tau,y}[\cdot]$ are the direct and inverse Kolmogorov operators corresponding to process x_t. Since innovation process $\widetilde{z}_t$ is such that $\widetilde{Z}_t=(\widetilde{z}_t,\boldsymbol{F}_t^z)$ is Wiener process with $E\{\widetilde{z}_t\widetilde{z}_t^T|\boldsymbol{F}_t^z\}=\int_0^t R(\tau,z)d\tau$ (Liptser and Shiryayev, 1977) and $p(t,x;\widetilde{s}_L,\widetilde{x}^L)$ is defined by the equation $d_t p(t,x;\widetilde{s}_L,\widetilde{x}^L)=\mathbf{L}_{t,x}[p(t,x;\widetilde{s}_L,\widetilde{x}^L);p(t,x)]dt$, $p(t,x)=\partial P\{x_t\le x\}/\partial x$, then differentiating according to Ito formula (Liptser and Shiryayev, 1977) of process $\chi_t=\ln[p_s^t(x;\widetilde{x}^L)/p(t,x;\widetilde{s}_L,\widetilde{x}^L)]$ and, then using Ito-Ventzel formula (Ocone and Pardoux, 1989) similar to (Dyomin and Korotkevich, 1983) yields for $t_m\le t<t_{m+1}$

$$dI_s^t[\cdot]/dt = (1/2)tr\left[E\left\{R^{-1}(t,z)\left[\overline{h(\widetilde{\tau}_N,z|x_t,\widetilde{x}_s^L)}-\overline{h(t,z)}\right][\cdot]^T\right\}\right]$$
$$-\frac{1}{2}tr\left[Q(t)E\left\{\frac{\partial\ln p_s^t(x_t;\widetilde{x}_s^L)}{\partial x_t}(\cdot)^T-\frac{\partial\ln p(t,x_t;\widetilde{s}_L,\widetilde{x}_s^L)}{\partial x_t}(\cdot)^T\right\}\right]$$
$$+tr\left[Q(t)E\left\{\left[\frac{\partial\ln p_s^t(x_t;\widetilde{x}_s^L)}{\partial x_t}-\frac{\partial\ln p_t(x_t)}{\partial x_t}\right]\left(\frac{\partial\ln p_t(x_t)}{\partial x_t}\right)^T\right.\right.$$
$$\left.\left.-\left[\frac{\partial\ln p(t,x_t;\widetilde{s}_L,\widetilde{x}_s^L)}{\partial x_t}-\frac{\partial\ln p(t,x_t)}{\partial x_t}\right]\left(\frac{\partial\ln p(t,x_t)}{\partial x_t}\right)^T\right\}\right]. \quad (15)$$

The use of the properties (8), (9) and the representation $h(t,x_t,\widetilde{x}_\tau^N,z)$ in the form (6) in (15) taking into account (1), (8) yields (10). From (8) and (13), we have

$\left[c(\eta(t_m),z|x,\widetilde{x}^L)/c(\eta(t_m),z)\right]=\left[\mathrm{N}\{\widetilde{x}^{L+1};\widetilde{\mu}^{L+1}(t_m,\widetilde{s}_L),\widetilde{\Gamma}^{L+1}(t_m,\widetilde{s}_L)\}/\mathrm{N}\{\widetilde{x}^{L+1};\widetilde{\mu}^{L+1}(t_m-0,\widetilde{s}_L),\widetilde{\Gamma}^{L+1}(t_m-0,\widetilde{s}_L)\}\right.$. Then (12) follows from (7) and (13).

Corollary 1. The information amounts $I_t[x_t;z_0^t,\eta_0^m]$ and $I_s^t[\widetilde{x}_s^L;z_0^t,\eta_0^m]$ for separate problems of filtering and generalized extrapolation on the time intervals $t_m\le t<t_{m+1}$ are defined by the equations

$$dI_t[x_t;z_0^t,\eta_0^m]/dt = (1/2)tr\left[E\{R^{-1}(t,z)\widetilde{H}_0(t,z)\Gamma^{-1}(t)\widetilde{H}_0^T(t,z)\}\right] - (1/2)tr\left[Q(t)\left[E\{\Gamma^{-1}(t)\}-D^{-1}(t)\right]\right], \quad (16)$$

$$dI_s^t[\widetilde{x}_s^L;z_0^t,\eta_0^m]/dt = (1/2)tr\left[E\left\{R^{-1}(t,z)\widetilde{H}_L(t,z)\left(\widetilde{\Gamma}^L(t,\widetilde{s}_L)\right)^{-1}\widetilde{H}_L^T(t,z)\right\}\right], \quad (17)$$

$Q(t)=\Phi_1(\cdot)\Phi_1^T(\cdot)$, subject to the initial conditions

$$I_{t_m}\left[x_{t_m};z_0^{t_m},\eta_0^m\right] = I_{t_m-0}\left[x_{t_m};z_0^{t_m},\eta_0^{m-1}\right] + \Delta I_{t_m}, \quad (18)$$

$$I_s^{t_m}\left[\tilde{x}_s^L;z_0^{t_m},\eta_0^m\right] = I_s^{t_m-0}\left[\tilde{x}_s^L;z_0^{t_m},\eta_0^{m-1}\right] + \Delta I_s^{t_m}, \quad (19)$$

$$\Delta I_{t_m} = (1/2)E\left\{\ln\left[\left|\Gamma(t_m-0)\right|/\left|\Gamma(t_m)\right|\right]\right\}, \quad (20)$$

$$\Delta I_s^{t_m} = (1/2)E\left\{\ln\left[\left|\tilde{\Gamma}^L(t_m-0,\tilde{s}_L)\right|/\left|\tilde{\Gamma}^L(t_m,\tilde{s}_L)\right|\right]\right\}. \quad (21)$$

The formulated statement obviously follows from Theorem 1. Division in the structure $I_s^t\left[x_t;\tilde{x}_s^L;z_0^t,\eta_0^m\right]$ of components $I_t\left[x_t;z_0^t,\eta_0^m\right]$ and $I_s^t\left[\tilde{x}_s^L;z_0^t,\eta_0^m\right]$ is of interest. The answer to this question provides the following statement.

Theorem 2. The joint information amount (7) has the forms

$$I_s^t\left[x_t;\tilde{x}_s^L;z_0^t,\eta_0^m\right] = I_t\left[x_t;z_0^t,\eta_0^m\right] + I_{s|t}^t\left[\tilde{x}_s^L;z_0^t,\eta_0^m|x_t\right], \quad (22)$$

$$I_s^t\left[x_t,\tilde{x}_s^L;z_0^t,\eta_0^m\right] = I_s^t\left[\tilde{x}_s^L;z_0^t,\eta_0^m\right] + I_{t|s}^t\left[x_t;z_0^t,\eta_0^m|\tilde{x}_s^L\right]. \quad (23)$$

The conditional information amounts on the time intervals $t_m \le t < t_{m+1}$ are defined by the equations

$$\begin{aligned} dI_{s|t}^L\left[\tilde{x}_s^t;z_0^t,\eta_0^m|x_t\right]/dt &= \frac{1}{2}tr\Big[E\Big\{R^{-1}(t,z) \\ &\times \tilde{H}_{L+1}(t,z)\left(\tilde{\Gamma}^{L+1}(\tilde{s}_L,t)\right)^{-1}\tilde{H}_{L+1}^T(t,z) - \tilde{H}_0(t,z) \\ &\times \Gamma^{-1}(t)\tilde{H}_0^T(t,z)\Big\}\Big] - \frac{1}{2}tr\Big[Q(t)\Big[E\left\{\Gamma^{-1}(t|s_L)\right\} \\ &- E\left\{\Gamma^{-1}(t)\right\} - D^{-1}(t|\tilde{s}_L) + D^{-1}(t)\Big]\Big], \end{aligned} \quad (24)$$

$$\begin{aligned} &dI_{t|s}^t\left[x_t;z_0^t,\eta_0^m|\tilde{x}_s^L\right]/dt \\ &= \frac{1}{2}tr\Big[E\Big\{R^{-1}(t,z)\tilde{H}_{L+1}(t,z)\left(\tilde{\Gamma}^{L+1}(\tilde{s}_L,t)\right)^{-1} \\ &\times \tilde{H}_{L+1}^T(t,z) - \tilde{H}_L(t,z)\left(\tilde{\Gamma}^L(\tilde{s}_L,t)\right)^{-1}\tilde{H}_L^T(t,z)\Big\}\Big] \\ &- \frac{1}{2}tr\Big[Q(t)\Big[E\left\{\Gamma^{-1}(t|s_L)\right\} - D^{-1}(t|\tilde{s}_L)\Big]\Big] \end{aligned} \quad (25)$$

subject to the initial conditions

$$\begin{aligned} &I_{s|t_m}^{t_m}\left[\tilde{x}_s^L;z_0^{t_m},\eta_0^m|x_{t_m}\right] \\ &= I_{s|t_m}^{t_m-0}\left[\tilde{x}_s^L;z_0^{t_m},\eta_0^{m-1}|x_{t_m}\right] + \Delta I_{s|t_m}^{t_m}, \end{aligned} \quad (26)$$

$$\begin{aligned} &I_{t_m|s}^{t_m}\left[x_{t_m};z_0^{t_m},\eta_0^m|\tilde{x}_s^L\right] \\ &= I_{t_m|s}^{t_m-0}\left[x_{t_m};z_0^{t_m},\eta_0^{m-1}\right] + \Delta I_{t_m|s}^{t_m}, \end{aligned} \quad (27)$$

$$\Delta I_{s|t_m}^{t_m} = (1/2)E\left\{\ln\left[\left|\tilde{\Gamma}^L(\tilde{s}_L|t_m-0)\right|/\left|\tilde{\Gamma}^L(\tilde{s}_L|t_m)\right|\right]\right\}, \quad (28)$$

$$\Delta I_{t_m|s}^{t_m} = (1/2)E\left\{\ln\left[\left|\Gamma(t_m-0|\tilde{s}_L)\right|/\left|\Gamma(t_m|\tilde{s}_L)\right|\right]\right\}, \quad (29)$$

where $\tilde{\Gamma}^L(\tilde{s}_L|t) = \tilde{\Gamma}^L(t,\tilde{s}_L) - \left(\tilde{\Gamma}_{0,N+1}^L(t,\tilde{s}_L)\right)^T\Gamma^{-1}(t) \times \tilde{\Gamma}_{0,N+1}^L(t,\tilde{s}_L)$.

Remark 1. Theorem 2 of the considered problem illustrates the general property, regarding the relations between unconditional and conditional measures of Shannon information amount (Gallager, 1968). The proof is similar to Theorem 1 proof, taking into account $p_s^t(x;\tilde{x}^L) = p_{s|t}^t(\tilde{x}^L|x)p_t(x) = p_{t|s}^t(x|\tilde{x}^L) \times p_s^t(\tilde{x}^L)$ and $p(t,x;\tilde{s}_L,\tilde{x}^L) = p(\tilde{s}_L,\tilde{x}^L|t,x)\,p(t,x) = p(t,x|\tilde{s}_L,\tilde{x}^L)p(\tilde{s}_L,\tilde{x}^L)$.

Remark 2. If in $h(\cdot)$, $g(\cdot)$, $\Phi_2(\cdot)$, $\Phi_3(\cdot)$ (see (3), (4), (6)) be absent dependence on z, then everywhere in the formulated propositions dependences on z and operator $E\{\cdot\}$ are excluded. Thus an exact calculation of information amounts in the considered problem is possible only in the case of absence of the feedback with respect to the process z_t.

Example. It is assumed (Dyomin *et. al.*, 1997): 1) x_t, z_t, $\eta(t_m)$ are the scalar stationary processes; 2) x_t is a Gaussian Markov process with the exponential correlation function and correlation time $\alpha_k = 1/a$; 3) z_t is defines continuous observations without memory $\left(H_k = 0, k = \overline{1;N}\right)$; 4) $N = 2$; 5) $L = 1$.

As a measure of information efficiency of memory observations $\eta(t_m)$ in relation to without memory observations $\tilde{\eta}(t_m)$, when $G_1 = 0$, $G_2 = 0$, can be taken the value $\Delta = \Delta I_s^{t_m}\left[x_s;z_0^{t_m},\eta(t_m)\right] - \Delta\tilde{I}_s^{t_m}\left[x_s;z_0^{t_m},\eta(t_m)\right]$, where $\Delta I_s^{t_m}[\cdot]$ and $\Delta\tilde{I}_s^{t_m}[\cdot]$ are the increment of an information amount $I_s^t\left[\tilde{x}_s^L;z_0^t,\eta_0^m\right]$ at $L = 1$ in an instant t_m incoming from $\eta(t_m)$ and $\tilde{\eta}(t_m)$, respectively.

We carry out the further study for the case of scarce discrete-time observations. In this case (see (21) and (2.28) in (Dyomin, *et al.*, 1997))

$$\Delta = (1/2)\ln\left[\tilde{\gamma}^{11}(s,t_m)/\gamma^{11}(s,t_m)\right], \quad (30)$$

$$\begin{aligned} &\gamma^{11}(s,t_m) = \gamma^{11}(T) - [\alpha/\beta], \\ &\alpha = \left[G_0\gamma_0^1(T) + G_1\gamma_1^1(t_1^*,T) + G_2\gamma_2^1(t_2^*,T)\right]^2, \\ &\beta = V + G_0^2\gamma + G_1^2\gamma_{11}(t_1^*) + G_2^2\gamma_{22}(t_{22}^*) \\ &+ 2G_0G_1\gamma_{01}(t_1^*) + 2G_0G_2\gamma_{02}(t_2^*) + 2G_1G_2\gamma_{12}(t_1^*,t_1^*), \end{aligned} \quad (31)$$

$$\tilde{\gamma}^{11}(s,t_m) = \gamma^{11}(T) - \left[G_0^2(\gamma_0^1(T))^2\right] / \left[V + G_0^2\gamma\right], \quad (32)$$

where γ, $\gamma_{01}(t_1^*)$, $\gamma_{02}(t_1^*)$, $\gamma_{11}(t_1^*)$, $\gamma_{22}(t_2^*)$, $\gamma_{12}(t_1^*,t_2^*)$, $\gamma^{11}(T)$, $\gamma_0^1(T)$, $\gamma_1^1(t_1^*,T)$, $\gamma_2^1(t_2^*,T)$ are determined by the formulae (3.2) of (Dyomin, *et al.*, 1997), $t_1^* = t_m - \tau_1$ and $t_2^* = t_m - \tau_2$ are memory depths, $T = s - t_m$ is an interval of extrapolation.

There are two extreme situations concerning memory depth: a case of a small memory depth, when $t_1^* \to 0$, $t_2^* \to 0$, and a case of great memory depth, when $t_1^* \to \infty$, $t_2^* \to \infty$. Let $\Delta_0 = \lim\Delta$ at $t_1^* \to 0$, $t_2^* \to 0$, and $\Delta_\infty = \lim\Delta$ at $t_1^* \to \infty$, $t_2^* \to \infty$. From (30)-(32) and (3.2) of (Dyomin, *et al.*, 1997) it follows that

$$\Delta_0 = \frac{1}{2}\ln\left[\frac{1}{1-\delta_0}\right], \quad \Delta_\infty = \frac{1}{2}\ln\left[\frac{1}{1+\delta_\infty}\right], \quad (33)$$

$$\delta_0 = \frac{2aV\gamma^2\left((G_1+G_2)^2 + 2G_0(G_1+G_2)\right)\exp\{-2aT\}}{\left[V+\gamma(G_0+G_1+G_2)^2\right]} \times \frac{1}{\left[Q(V+\gamma G_0^2)(1-\exp\{-2aT\}) + 2aV\gamma\exp\{-2aT\}\right]}, \quad (34)$$

$$\delta_\infty = \frac{\kappa\gamma^3 G_0^2\left(G_1^2+G_2^2\right)\exp\{-2aT\}}{\left[V+\gamma\left(G_0^2+\kappa G_1^2+\kappa G_2^2\right)\right]} \times \frac{1}{\left[Q(V+\gamma G_0^2)(1-\exp\{-2aT\}) + 2aV\gamma\exp\{-2aT\}\right]}, \quad (35)$$

where $\gamma = (1/\delta)(\lambda - a)$, $\delta = H_0^2/R$, $\lambda = \sqrt{a^2+\delta Q}$, $\kappa = (\lambda+a)/2\lambda$.

Investigation of behaviour Δ for close t_1^* and t_2^* $(t_1^* = t_2^* = t^*)$ as functions of memory depth t^* on the basis (28)-(30) gives following result using (3.2) from (Dyomin, *et al.*, 1997).

Proposition 2. Let

$$M = \left\{(G_0,G_1,G_2): (G_1+G_2)^2 + 2G_0(G_1+G_2) \le 0\right\}. \quad (36)$$

If $(G_0,G_1,G_2) \notin M$, then $\Delta(t^*)$ is monotonically decreasing function of memory depth from a value $\Delta_0 > 0$ up to a value $\Delta_\infty < 0$, vanishing at the point

$$t_{eff}^* = \frac{1}{\lambda}\ln\frac{(G_1+G_2)(V+\kappa\gamma G_0^2)}{G_0\left(\sqrt{V^2+\kappa\gamma(G_1+G_2)^2(V+\kappa\gamma G_0^2)} - V\right)} \quad (37)$$

and which can be defined as effective memory depth. If $(G_0,G_1,G_2) \in M$, that $\Delta(t^*) \le 0$ for all $t^* \ge 0$.

The influencing of continuous observations on information efficiency of memory discrete observations is carried out through the parameter $\delta = H_0^2/R$, which is proportional to a signal-noise ratio on intensity in the continuous observation channel. At $\delta \to \infty$ $\Delta I_s^{t_m}[\cdot] \to 0$ and $\Delta\tilde{I}_s^{t_m}[\cdot] \to 0$, therefore $\Delta \to 0$. Thus at reaching absolute precise measurements in the continuous channel, discrete observations with memory and without memory do not introduce new information on values x_s at any T. At $\delta \to 0$, that corresponds to a case of lack of continuous observations, the formulas (30)-(33) hold, in which $\gamma = Q/2a$, $\lambda = a$, $\kappa = 1$, i.e. in this case a explicit dependence t_{eff}^* of correlation time $\alpha_k = 1/a$ of the process x_t takes place.

4. OPTIMAL TRANSMISSION OF GAUSSIAN MARKOV SIGNAL

Let x_t, z_t, $\eta(t_m)$ are the scalar processes, and $f(\cdot) = F(t)x_t$, $\Phi_2(\cdot) = \Phi_2(t)$, $\Phi_3(\cdot) = \Phi_3(t_m)$, $N = 1$.

Theorem 3. In the class $\boldsymbol{K} = \{\boldsymbol{H};\boldsymbol{G}\}$ of coding functionals

$$\begin{aligned}\boldsymbol{H} &= \{h(\cdot): h(\cdot) = h(t,z) + H_0(t,z)x_t + H_1(t,z)x_\tau\},\\ \boldsymbol{G} &= \{g(\cdot): g(\cdot) = g(t_m,z) + G_0(t_m,z)x_{t_m} + G_0(t_m,z)x_{t_m} + G_1(t_m,z)x_\tau\}\end{aligned} \quad (38)$$

with noiseless feedback relative to the process z_t under limitations

$$E\{h^2(\cdot)\} \le \tilde{h}(t), \quad E\{g^2(\cdot)\} \le \tilde{g}(t_m) \quad (39)$$

the functionals $h^0(t,x_t,x_\tau,z^0)$ and $g^0(t_m,x_{t_m},x_\tau,z^0)$ with the corresponding information amount $I_t^0\left[x_t;(z^0)_0^t,(\eta^0)_0^m\right]$ in the filtering problem, and which are in the following representations

$$\begin{aligned}h^0(t,z^0) &= -H_0^0(t,z^0)\mu^0(t),\\ H_0^0(t,z^0) &= \left[\tilde{h}(t)/\Delta^0(t)\right]^{1/2},\ H_1^0(t,z^0) = 0,\\ g^0(t_m,z^0) &= -G_0^0(t_m,z^0)\mu^0(t_m-0),\\ G_0^0(t_m,z^0) &= \left[\tilde{g}(t_m)/\Delta^0(t_m-0)\right]^{1/2},\ G_1^0(t_m,z^0) = 0,\end{aligned} \quad (40)$$

have the property

$$I_t^0\left[x_t;(z^0)_0^t,(\eta^0)_0^m\right] = \sup I_t\left[x_t;z_0^t,\eta_0^m\right], \quad (41)$$

where supremum is taken for all $\{h(\cdot);g(\cdot)\} \in \boldsymbol{K}$ and

$$I_t^0\left[x_t;(z^0)_0^t,(\eta^0)_0^m\right]=(1/2)\sum_{t_i\le t}\ln\left[1+\left(\tilde{g}(t_i)/V(t_i)\right)\right]$$

$$+\frac{1}{2}\left[\int_0^t\left(R^{-1}(\sigma)\tilde{h}(\sigma)-Q(\sigma)\left[\left(\Delta^0(\sigma)\right)^{-1}-D^{-1}(\sigma)\right]\right)d\sigma\right]. \quad (42)$$

The optimal message $\{z_t^0;\eta^0(t_m)\}$ is defined by the equation

$$dz_t^0=\left[\tilde{h}(t)/\Delta^0(t)\right]^{1/2}\left[x_t-\mu^0(t)\right]dt+\Phi_2(t)dv_t,$$

$$\eta^0(t_m)=\left[\tilde{g}(t_m)/\Delta^0(t_m-0)\right]^{1/2}\left[x_{t_m}-\mu^0(t_m-0)\right] \quad (43)$$

$$+\Phi_3(t_m)\xi(t_m),$$

and the optimal decoding $\mu^0(t)$ and the minimal decoding error $\Delta^0(t)$ on the time intervals $t_m\le t<t_{m+1}$ are defined by the equations

$$d\mu^0(t)=F(t)\mu^0(t)dt+R^{-1}(t)\left[\tilde{h}(t)\Delta^0(t)\right]^{1/2}dz_t^0,$$

$$d\Delta^0(t)/dt=\left[2F(t)-R^{-1}(t)\tilde{h}(t)\right]\Delta^0(t)+Q(t) \quad (44)$$

subject to the initial conditions

$$\mu^0(t_m)=\mu^0(t_m-0)+\left[\tilde{g}(t_m)\Delta^0(t_m-0)\right]^{1/2}$$

$$\times\left[V(t_m)+\tilde{g}(t_m)\right]^{-1}\eta^0(t_m), \quad (45)$$

$$\Delta^0(t_m)=V(t_m)\left[V(t_m)+\tilde{g}(t_m)\right]^{-1}\Delta^0(t_m-0).$$

Besides the coding (40) provides the minimal decoding error $\Delta^0(t)=\inf\Delta(t)$, where $\Delta(t)=E\left\{\left[x_t-\mu(t)\right]^2\right\}=E\{\gamma(t)\}$ is the decoding error in the filtering problem corresponding to the arbitrary coding $\{h(\cdot);g(\cdot)\}\in \boldsymbol{K}$.

The proof of this statement is similar to Lemma 16.7 and Theorem 16.6 proofs in (Liptser and Shiryayev, 1978), taking into account inequalities of Jensen, Fisher, Cauhy-Bunyakovskii, and is made on the basis of Corollary 1 (see (16), (18), (20)).

Remark 3. If in Corollary 1 the expressions (17), (19), (21) for $L=1$ are used then one can analogously prove that coding functionals (40) provide analogous properties in the extrapolation problem as well, i.e.

$$I_s^t\left[x_s;(z^0)_0^t,(\eta^0)_0^m\right]=\sup I_s^t\left[x_s;z_0^t,\eta_0^m\right],$$

$$\Delta^0(t,s)=\inf\Delta(t,s), \quad (46)$$

where $\Delta(t,s)=E\left\{\left[x_s-\mu(t,s)\right]^2\right\}=E\{\gamma_{22}(t,s)\}$.

5. CONCLUSION

The obtained result can be applied for research of the information efficiency of stochastic process filtering and extrapolation (Dyomin *et. al.*, 1997), an optimal signal transmission (Liptser and Shiryayev, 1978), maximization of transmission channel traffic capacity (Gallager, 1968), information foundation of estimation problems (Tomita *et. al.*, 1976) in the case of memory continuous-discrete channels of observation and transmission. Considered example demonstrated that presence of memory may both increase and decrease information efficiency.

REFERENCES

Abakumova, O.L., N.S. Dyomin and T.V. Sushko (1995). Filtering of stochastic processes for continuous and discrete observations with memory. *Autom. and Remote Control,* **56**(10), 1383-1393.

Dyomin, N.S. and V.I. Korotkevich (1983). Amount of information in problems of filtering the components of Markov processes. *Autom. and Remote Control,* **44**(7), 899-907.

Dyomin, N.S., T.V. Sushko and A.V. Yakovleva (1997). Generalized inverse extrapolation of stochastic processes by an aggregate of continuous-discrete observations with memory. *Comput. Syst. Sci. Intern.* **36**(4), 543-554.

Gallager, R.G. (1980) *Information theory and reliable communication*. Wiley, New-York.

Kalman R.E. (1960). A new approach to linear filtering and prediction problems. *Trans. ASME. J.Basic Eng., Ser.D.,* **82**(March), 35-45.

Liptser, R.Sh. and A.N. Shiryayev (1977, 1978). *Statistics of Random Processes*. Springer-Verlag, New-York.

Meditch, J.S. (1969). *Stochastic optimal linear estimation and control.* McGraw-Hill, New York.

Ocone, D. And E. Pardoux (1989). A generelized Ito-Ventzel formula. *Ann. Inst. Henri Poincare,* **25**(1), 39-71.

Tomila, Y., S. Ohmatsu and T. Soeda (1976). An application of the information theory to estimation problems. *Inform. and Control,* **32**, 101-111.

IFAC
Publications
www.elsevier.com/locate/ifac

GUARANTEED ELLIPSOIDAL STATE ESTIMATION FOR UNCERTAIN MIMO MODELS

Boris T. Polyak *, **Sergey A. Nazin** *,
Cécile Durieu **, **Éric Walter** ***

* *Institute of Control Sciences, Russian Academy of Sciences, Moscow, Russia*
** *SATIE, CNRS – ENS Cachan, France*
*** *Laboratoire des Signaux et Systèmes, CNRS – Supélec – UPS, Gif-sur-Yvette, France*

Abstract: This paper extends to uncertain models the classical ellipsoidal outer-bounding of the set of all feasible state vectors with unknown but bounded state perturbations and measurement noise. The technique applies to linear discrete-time dynamic systems and could be extended to weakly non-linear systems. Combined quadratic constraints on model uncertainty and additive disturbances are considered in order to simplify analysis and to allow an analytical solution of the basic problems involved in parameter or state estimation. The results obtained for combined quadratic constraints are extended to other types of model uncertainty. *Copyright ©2003 IFAC*

Keywords: bounded noise, discrete time, ellipsoidal bounding, parameter estimation, set-membership uncertainty, state estimation, uncertain dynamic systems.

1. INTRODUCTION

Set-membership estimation, first considered more than 30 years ago (Schweppe, 1968; Witsenhausen, 1968; Bertsekas and Rhodes, 1971; Schweppe, 1973), is an alternative to stochastic estimation when the perturbations are unknown but bounded. It aims at characterizing the set of all values of the parameter or state vector that are consistent with the data, the model structure and the hypotheses on the process perturbations and measurement noise. Ellipsoidal outer-bounding, which computes ellipsoids guaranteed to contain this set, is now well established when the plant model (its structure in the case of parameter estimation) is assumed precisely known and all the uncertainty relates to process perturbations and measurement noise (Kurzhanskii, 1977; Fogel and Huang, 1982; Chernousko, 1994; Kurzhanskii and Valyi, 1996; Walter, 1990; Norton, 1994, 1995; Milanese *et al.*, 1996; Walter and Pronzato, 1997; Durieu *et al.*, 2001). This assumption, however, is unrealistic for real-life problems. The lack of precise information is a fundamental paradigm of modern control theory, where the concept of robustness plays a key role. Model uncertainty has been taken into account with ellipsoidal techniques for particular cases in (Clement and Gentil, 1990; Cerone, 1993; Kurzhanskii and Valyi, 1996; Chernousko, 1996; Rokityanskii, 1997; Norton, 1999; Chernousko and Rokityanskii, 2000; Polyak *et al.*, 2002).

The aim of this paper is to present additional new results along the lines of (Polyak *et al.*, 2002). It was assumed there that the initial ellipsoids used during prediction were centered at the origin, a suggestion that is natural in the context of guaranteed estimation of attainability sets, but rather artificial when there are measurements leading to alternate prediction and correction. A first contribution of this paper is to remove this restriction. As in (Polyak *et al.*, 2002) the model uncertainties, process perturbations and measurement noise are combined in quadratic constraints. This again may sometimes seem too restrictive, as one may only have at one's disposal separate inequalities on model uncertainty, process perturbations and measurement noise. So a second contribution of this paper is an extension of this approach to the case of separate uncertainties. A final contribution is the explicit description of the state estimation algorithm based on the resulting building blocks. The paper is organized as follows. Section 2 states the problem. Section 3 is devoted to prediction in the general case where the initial ellipsoid is not centered at the origin. Section 4 extends the results to the case of separate uncertainties and Section 5 presents the resulting global algorithm for guaranteed state estimation.

2. PROBLEM STATEMENT

The systems considered here are assumed to be described by the linear discrete-time uncertain state-space model

$$\begin{cases} \mathbf{x}_{t+1} = (\mathbf{A}_t + \mathbf{\Delta A}_t)\,\mathbf{x}_t + \mathbf{B}_t\mathbf{u}_t + \mathbf{v}_t, \\ \mathbf{y}_t = (\mathbf{C}_t + \mathbf{\Delta C}_t)\,\mathbf{x}_t + \mathbf{D}_t\mathbf{u}_t + \mathbf{w}_t, \end{cases} \quad (1)$$

where $\mathbf{x}_t$ is the unknown n-vector of the system state at time t, $\mathbf{y}_t$ is a known p-vector of measured outputs, $\mathbf{u}_t$ is a known input vector, and $\mathbf{A}_t$, $\mathbf{B}_t$, $\mathbf{C}_t$ and $\mathbf{D}_t$ are known matrices while $\mathbf{v}_t$ is an unknown process perturbation vector, $\mathbf{w}_t$ is an unknown measurement noise vector, and $\mathbf{\Delta A}_t$ and $\mathbf{\Delta C}_t$ are unknown matrices due to model uncertainty. The only information available regarding unknown quantities is under the form of bounds. The initial state vector is assumed to belong to some known compact set. Important particular cases are: the estimation of attainability sets (without measurements), parameter estimation (without dynamics: $\mathbf{A}_t \equiv \mathbf{I}$, the identity matrix, $\mathbf{\Delta A}_t \equiv \mathbf{0}$, $\mathbf{B}_t \equiv \mathbf{0}$, $\mathbf{v}_t \equiv \mathbf{0}$, $\mathbf{x}_t \equiv \boldsymbol{\theta}_t$) and parameter tracking ($\mathbf{A}_t \equiv \mathbf{I}$, $\mathbf{\Delta A}_t \equiv \mathbf{0}$, $\mathbf{B}_t \equiv \mathbf{0}$, $\mathbf{x}_t \equiv \boldsymbol{\theta}_t$).

Kalman filtering cannot be applied to take into account information about bounds, which is more conveniently considered in the context of set estimation. The approach to be followed here is recursively to compute the smallest possible ellipsoid guaranteed to contain $\mathbf{x}_t$ given bounds on the unknown quantities. More precisely it is assumed that the model uncertainties are combined with those due to process perturbations and measurement noise in the following ellipsoidal constraints

$$\begin{cases} \dfrac{\|\mathbf{\Delta A}_t\|^2}{\delta_{\mathbf{A}}^2} + \dfrac{\|\mathbf{v}_t\|^2}{\delta_{\mathbf{v}}^2} \le 1, \\ \dfrac{\|\mathbf{\Delta C}_t\|^2}{\delta_{\mathbf{C}}^2} + \dfrac{\|\mathbf{w}_t\|^2}{\delta_{\mathbf{w}}^2} \le 1, \end{cases} \quad (2)$$

where $\delta_{\mathbf{A}}$, $\delta_{\mathbf{v}}$, $\delta_{\mathbf{C}}$ and $\delta_{\mathbf{w}}$ are prespecified weights. In (2) and below, the vector norm is understood as Euclidean: $\|\mathbf{x}\|^2 = \sum x_i^2$, while the spectral norm is used for matrices: for $\mathbf{M} \in \mathbb{R}^{m\times n}, \|\mathbf{M}\| = \max_{\|\mathbf{x}\|\le 1} \|\mathbf{Mx}\| = \sqrt{\max(\mathrm{eig}(\mathbf{M}^{\mathrm{T}}\mathbf{M}))}$. In (2) $\delta_{\mathbf{A}} = 0$ is understood as $\mathbf{\Delta A}_t = \mathbf{0}$ and $\|\mathbf{v}_t\|^2 \le \delta_{\mathbf{v}}^2$, while $\delta_{\mathbf{v}} = 0$ means that $\mathbf{v}_t = \mathbf{0}$ and $\|\mathbf{\Delta A}_t\|^2 \le \delta_{\mathbf{A}}^2$. To simplify notation, and without loss of generality, the time dependency of the weights is not indicated but can easily be introduced. Similar models arise in other problems related to systems under uncertainty, such as total least squares (Golub and Van Loan, 1980; El Ghaoui and Lebret, 1997) and robust optimization (Ben-Tal *et al.*, 2000). A more general Linear-Fractional Representation of uncertainty has been considered in (El Ghaoui and Calafiore, 1999), where the search of the optimal outer-bounding ellipsoid was reduced to Semi-Definite Programming.

The sets of all values of $\mathbf{x}_{t+1}$ consistent with the information available at time $t+1$ are approximated recursively by the smallest prior and posterior ellipsoids, respectively denoted by $\widehat{\mathcal{E}}_{t+1/t} = \mathcal{E}(\widehat{\mathbf{c}}_{t+1/t}; \widehat{\mathbf{M}}_{t+1/t})$ and $\widehat{\mathcal{E}}_{t+1/t+1} = \mathcal{E}(\widehat{\mathbf{c}}_{t+1/t+1}; \widehat{\mathbf{M}}_{t+1/t+1})$, where $\mathcal{E}\,(\mathbf{c}; \mathbf{M})$ is a non-degenerate ellipsoid defined by

$$\mathcal{E}\,(\mathbf{c};\mathbf{M}) = \left\{x : (\mathbf{x}-\mathbf{c})^{\mathrm{T}}\mathbf{M}\,(\mathbf{x}-\mathbf{c}) \le 1\right\}, \quad (3)$$

with $\mathbf{M} > 0$(*i.e.* real, symmetric and positive definitive). So the objective is to find the smallest ellipsoid $\widehat{\mathcal{E}}_{t+1/t}$ containing

$$\mathcal{S}_{t+1} = \left\{ \begin{array}{l} \mathbf{x}_{t+1} : \mathbf{x}_{t+1} = \mathbf{A}_t\mathbf{x}_t + \mathbf{\Delta A}_t\mathbf{x}_t \\ \quad +\mathbf{B}_t\mathbf{u}_t + \mathbf{v}_t, \quad \mathbf{x}_t \in \widehat{\mathcal{E}}_{t/t}, \\ \dfrac{\|\mathbf{\Delta A}_t\|^2}{\delta_{\mathbf{A}}^2} + \dfrac{\|\mathbf{v}_t\|^2}{\delta_{\mathbf{v}}^2} \le 1 \end{array} \right\}, \quad (4)$$

and the smallest ellipsoid $\widehat{\mathcal{E}}_{t+1/t+1}$ containing

$$\mathcal{I}_{t+1} = \left\{ \begin{array}{l} \mathbf{x}_{t+1} : \mathbf{y}_{t+1} = \mathbf{C}_{t+1}\mathbf{x}_{t+1} + \\ \quad \mathbf{\Delta C}_{t+1}\mathbf{x}_{t+1} + \mathbf{D}_{t+1}\mathbf{u}_{t+1} + \\ \quad \mathbf{w}_{t+1}, \quad \mathbf{x}_{t+1} \in \widehat{\mathcal{E}}_{t+1/t}, \\ \dfrac{\|\mathbf{\Delta C}_{t+1}\|^2}{\delta_{\mathbf{C}}^2} + \dfrac{\|\mathbf{w}_{t+1}\|^2}{\delta_{\mathbf{w}}^2} \le 1 \end{array} \right\}. \quad (5)$$

The sets $\mathcal{S}_{t+1}$ and $\mathcal{I}_{t+1}$ are not ellipsoids; they are often not even convex (see Example 1). The most natural measures of size of $\mathcal{E}\,(\mathbf{c};\mathbf{M})$ are

$$f_t\,(\mathbf{M}) = \mathrm{tr}\,\mathbf{M}^{-1} \quad (6)$$

and

$$f_d\,(\mathbf{M}) = -\ln\det\mathbf{M}. \quad (7)$$

$f_t(\mathbf{M})$ is the sum of the squares of the lengths of the ellipsoid semi-axe (trace criterion) and $f_d(\mathbf{M})$ relates to the volume of the ellipsoid (determinant criterion). The algorithm for the correction step of ellipsoidal state estimation was established in (Polyak *et al.*, 2002). It will be recalled in Section 5 and we shall now concentrate on the prediction step.

3. PREDICTION

Starting from $\widehat{\mathcal{E}}_{t/t}$ guaranteed to contain $\mathbf{x}_t$, the prediction step of the state estimator must compute $\widehat{\mathcal{E}}_{t+1/t}$ by minimizing (6) or (7) with $\mathbf{M} = \widehat{\mathbf{M}}_{t+1/t}$, subject to the constraint $\widehat{\mathcal{E}}_{t+1/t} \supset \mathcal{S}_{t+1}$ as defined in (4). The result below reduces this problem to one-dimensional optimization.

Theorem 1. Each ellipsoid in the family $\mathcal{E}(\mathbf{c}_\alpha; \mathbf{M}_\alpha)$ with

$$\begin{cases} \mathbf{c}_\alpha = (1-\alpha)\,\mathbf{A}_t\mathbf{Q}_\alpha^{-1}\widehat{\mathbf{M}}_{t/t}\widehat{\mathbf{c}}_{t/t} + \mathbf{B}_t\mathbf{u}_t, \\ \mathbf{M}_\alpha = \dfrac{\left(\mathbf{A}_t\mathbf{Q}_\alpha^{-1}\mathbf{A}_t^{\mathrm{T}} + \alpha^{-1}\mathbf{I}\right)^{-1}}{1-\delta_\alpha}, \end{cases} \tag{8}$$

where

$$\begin{cases} \mathbf{Q}_\alpha = (1-\alpha)\,\widehat{\mathbf{M}}_{t/t} - \alpha\delta_{\mathbf{A}}^2\mathbf{I}, \\ \delta_\alpha = \alpha - \alpha\delta_{\mathbf{v}}^2 + (1-\alpha)\,\widehat{\mathbf{c}}_{t/t}^{\mathrm{T}}\widehat{\mathbf{M}}_{t/t}\widehat{\mathbf{c}}_{t/t} \\ \quad - (1-\alpha)^2\,\widehat{\mathbf{c}}_{t/t}^{\mathrm{T}}\widehat{\mathbf{M}}_{t/t}\,\mathbf{Q}_\alpha^{-1}\widehat{\mathbf{M}}_{t/t}\widehat{\mathbf{c}}_{t/t}, \end{cases} \tag{9}$$

contains $\mathcal{S}_{t+1}$ for all α such that

$$0 < \alpha < \alpha_{\max} = \frac{\lambda_{\min}(\widehat{\mathbf{M}}_{t/t})}{\lambda_{\min}(\widehat{\mathbf{M}}_{t/t}) + \delta_{\mathbf{A}}^2}, \tag{10}$$

where $\lambda_{\min}(\widehat{\mathbf{M}}_{t/t})$ is the minimal eigenvalue of $\widehat{\mathbf{M}}_{t/t}$.

Proof. Any $\mathbf{x}_{t+1} \in \mathcal{S}_{t+1}$ satisfies $\mathbf{x}_{t+1} = \mathbf{x}^+ + \mathbf{B}_t\mathbf{u}_t$ with $\mathbf{x}^+ = \mathbf{A}_t\mathbf{x}_t + \mathbf{z}$, where $\mathbf{x}_t \in \widehat{\mathcal{E}}_{t/t}$ and $\mathbf{z} = \mathbf{\Delta A}_t\mathbf{x}_t + \mathbf{v}_t$, with $\mathbf{\Delta A}_t$ and $\mathbf{v}_t$ satisfying the first ellipsoidal condition in (2). Then $\mathbf{A}_t\mathbf{x}_t \in \mathcal{E}(\mathbf{A}_t\widehat{\mathbf{c}}_{t/t}; \mathbf{A}_t^{-\mathrm{T}}\widehat{\mathbf{M}}_{t/t}\mathbf{A}_t^{-1})$ (where $\mathbf{A}_t^{-\mathrm{T}} = (\mathbf{A}^{-1})^{\mathrm{T}}$) and $\mathbf{z}$ satisfies the quadratic inequality $\|\mathbf{z}\|^2 \le \delta_{\mathbf{A}}^2\|\mathbf{x}_t\|^2 + \delta_{\mathbf{v}}^2$, see Lemma 1 in (Polyak *et al.*, 2002). (For simplicity, it is assumed in this proof that $\mathbf{A}_t$ is invertible; otherwise one may consider the prediction step in a subset of $\mathbb{R}^{n\times n}$ where $\mathbf{A}_t$ is not degenerate.) Moreover the condition $\mathbf{x}_{t+1} \in \mathcal{E}(\mathbf{c}; \mathbf{M})$ is equivalent to $\mathbf{x}^+ \in \mathcal{E}(\mathbf{c} - \mathbf{B}_t\mathbf{u}_t; \mathbf{M})$.

Thus the problem is to check when the two quadratic inequalities $(\mathbf{x}_t - \widehat{\mathbf{c}}_{t/t})^{\mathrm{T}}\widehat{\mathbf{M}}_{t/t}(\mathbf{x}_t - \widehat{\mathbf{c}}_{t/t}) \le 1$ and $\|\mathbf{z}\|^2 \le \delta_{\mathbf{A}}^2\|\mathbf{x}_t\|^2 + \delta_{\mathbf{v}}^2$ imply

$$\mathbf{a}^{\mathrm{T}}\mathbf{M}_\alpha\mathbf{a} \le 1, \tag{11}$$

where

$$\mathbf{a} = \mathbf{A}_t\mathbf{x}_t + \mathbf{z} - (\mathbf{c}_\alpha - \mathbf{B}_t\mathbf{u}_t). \tag{12}$$

To write this in standard form, introduce $\mathbf{s} \in \mathbb{R}^{2n}$, $\mathbf{s}_i \in \mathbb{R}^{2n}$ and $\mathbf{M}_i \in \mathbb{R}^{2n\times 2n}$, $i = 0, 1, 2$, defined by

$$\begin{cases} \mathbf{s} = \begin{pmatrix} \mathbf{A}_t\mathbf{x}_t \\ \mathbf{z} \end{pmatrix}, \\ \mathbf{s}_0 = \begin{pmatrix} \mathbf{c}_\alpha - \mathbf{B}_t\mathbf{u}_t \\ \mathbf{0} \end{pmatrix}, \\ \mathbf{M}_0 = \begin{pmatrix} \mathbf{M}_\alpha & \mathbf{M}_\alpha \\ \mathbf{M}_\alpha & \mathbf{M}_\alpha \end{pmatrix}, \\ \mathbf{s}_1 = \begin{pmatrix} \mathbf{A}_t\widehat{\mathbf{c}}_{t/t} \\ \mathbf{0} \end{pmatrix}, \\ \mathbf{M}_1 = \begin{pmatrix} \mathbf{A}_t^{-\mathrm{T}}\widehat{\mathbf{M}}_{t/t}\mathbf{A}_t^{-1} & \mathbf{0} \\ \mathbf{0} & \mathbf{0} \end{pmatrix}, \\ \mathbf{s}_2 = \mathbf{0}, \\ \mathbf{M}_2 = \begin{pmatrix} -\delta_{\mathbf{A}}^2\mathbf{A}_t^{-\mathrm{T}}\mathbf{A}_t^{-1} & \mathbf{0} \\ \mathbf{0} & \mathbf{I} \end{pmatrix}, \end{cases} \tag{13}$$

and the functions $f_i(\mathbf{s}) = (\mathbf{s} - \mathbf{s}_i)^{\mathrm{T}}\mathbf{M}_i(\mathbf{s} - \mathbf{s}_i)$, $i = 0, 1, 2$. Now, the condition $\mathbf{x}^+ \in \mathcal{E}(\mathbf{c}_\alpha - \mathbf{B}_t\mathbf{u}_t; \mathbf{M}_\alpha)$ can be rewritten as

$$\begin{cases} f_1(\mathbf{s}) \le 1 \text{ and } f_2(\mathbf{s}) \le \delta_{\mathbf{v}}^2 \\ \Rightarrow\; f_0(\mathbf{s}) \le 1. \end{cases} \tag{14}$$

Consider the linear combination of the quadratic functions $f_1(.)$ and $f_2(.)$ defined by $\tilde{f}_\alpha(\mathbf{s}) = (1-\alpha)\,f_1(\mathbf{s}) + \alpha f_2(\mathbf{s})$. Obviously for all $\alpha \in [0, 1]$

$$\begin{cases} f_1(\mathbf{s}) \le 1 \text{ and } f_2(\mathbf{s}) \le \delta_{\mathbf{v}}^2 \\ \Rightarrow\; \tilde{f}_\alpha(\mathbf{s}) \le 1 - \alpha + \alpha\delta_{\mathbf{v}}^2. \end{cases} \tag{15}$$

Let $\mathcal{E}_\alpha = \{\mathbf{s} : \tilde{f}_\alpha(\mathbf{s}) \le 1 - \alpha + \alpha\delta_{\mathbf{v}}^2\}$. By definition $f_1(\mathbf{s}_1) \le 1$ and $f_2(\mathbf{s}_1) \le \delta_{\mathbf{v}}^2$, so $\mathbf{s}_1$ belongs to $\mathcal{E}_\alpha$. As $\widehat{\mathbf{M}}_{t/t} > 0$, there exists $\mathbf{\Delta x} \in \mathbb{R}^n$ such that $f_1(\mathbf{s}) \le 1$ with $\mathbf{s} = \mathbf{s}_1 + (\mathbf{\Delta x}^{\mathrm{T}}\ \mathbf{0})^{\mathrm{T}}$. Moreover $f_2(\mathbf{s}) \le \delta_{\mathbf{v}}^2$, so $\mathbf{s}$ also belongs to $\mathcal{E}_\alpha$ and $\mathcal{E}_\alpha$ is non-empty for all $\alpha \in [0, 1]$. After simple transformations, $\mathcal{E}_\alpha$ can be written as an ellipsoid $\mathcal{E}(\widehat{\mathbf{s}}_\alpha; \widehat{\mathbf{M}}_\alpha)$ in $\mathbb{R}^{2n}$ with

$$\begin{cases} \widehat{\mathbf{s}}_\alpha = (1-\alpha)\,\widehat{\mathbf{Q}}_\alpha^{-1}\mathbf{M}_1\mathbf{s}_1, \\ \widehat{\mathbf{M}}_\alpha = (1-\delta_\alpha)^{-1}\,\widehat{\mathbf{Q}}_\alpha, \end{cases} \tag{16}$$

where

$$\begin{cases} \widehat{\mathbf{Q}}_\alpha = (1-\alpha)\,\mathbf{M}_1 + \alpha\mathbf{M}_2, \\ \delta_\alpha = \alpha - \alpha\delta_{\mathbf{v}}^2 + \\ \quad (1-\alpha)\,\mathbf{s}_1^{\mathrm{T}}\mathbf{M}_1\mathbf{s}_1 - \widehat{\mathbf{s}}_\alpha^{\mathrm{T}}\widehat{\mathbf{Q}}_\alpha\widehat{\mathbf{s}}_\alpha, \end{cases} \tag{17}$$

for all $\alpha \in [0, 1]$ such that $\widehat{\mathbf{Q}}_\alpha > 0$ and $1 - \delta_\alpha > 0$. Since $\mathcal{E}_\alpha$ is non-empty, $\widehat{\mathbf{Q}}_\alpha > 0$ implies that $1 - \delta_\alpha > 0$ and thus that $\widehat{\mathbf{M}}_\alpha > 0$. The condition $\widehat{\mathbf{Q}}_\alpha > 0$ is equivalent to $\alpha > 0$ and $(1-\alpha)\,\widehat{\mathbf{M}}_{t/t} - \alpha\delta_{\mathbf{A}}^2\mathbf{I} > 0$, and this last condition is equivalent to that $\alpha < \alpha_{\max}$ with $\alpha_{\max}$ given by (10). The implication (14) is satisfied, *i.e.* $f_0(\mathbf{s}) \le 1$, if $\mathbf{s}_0 = \widehat{\mathbf{s}}_\alpha$ and $\mathbf{M}_0 \le \widehat{\mathbf{M}}_\alpha$, where $\widehat{\mathbf{s}}_\alpha$ and $\widehat{\mathbf{M}}_\alpha$ are given by (16). $\mathbf{M}_0 \le \widehat{\mathbf{M}}_\alpha$ is equivalent to

$$\begin{pmatrix} \mathbf{M}_\alpha - \mathbf{M} & \mathbf{M}_\alpha \\ \mathbf{M}_\alpha & \mathbf{M}_\alpha - \dfrac{\alpha\mathbf{I}}{(1-\delta_\alpha)} \end{pmatrix} \le 0, \tag{18}$$

where

$$\begin{cases} \mathbf{M} = (1-\delta_\alpha)^{-1} \mathbf{A}_t^{-\mathrm{T}} \mathbf{Q}_\alpha \mathbf{A}_t^{-1}, \\ \mathbf{Q}_\alpha = (1-\alpha) \widehat{\mathbf{M}}_{t/t} - \alpha \delta_{\mathbf{A}}^2 \mathbf{I}. \end{cases} \tag{19}$$

Lemma 2 in (Polyak *et al.*, 2002) implies that

$$\mathbf{M}_\alpha^{-1} \geq (1-\delta_\alpha) \left(\mathbf{A}_t \mathbf{Q}_\alpha^{-1} \mathbf{A}_t^{\mathrm{T}} + \alpha^{-1} \mathbf{I} \right) \tag{20}$$

provided that $\alpha (1-\delta_\alpha)^{-1} > 0$ and $\mathbf{M} > 0$, which implies $0 < \alpha < \alpha_{\max}$. One seeks the minimal ellipsoid in the family $\mathcal{E}(\widehat{\mathbf{s}}_\alpha; \widehat{\mathbf{M}}_\alpha)$ according to (6) or (7) with $\mathbf{M} = \widehat{\mathbf{M}}_\alpha$. Therefore (20) is converted into the second equation in (8), where $\mathbf{Q}_\alpha$ and δ_α are given by (9). The equation giving $\mathbf{s}_0$ in (13), where $\mathbf{s}_0$ corresponds to $\widehat{\mathbf{s}}_\alpha$ in (16), leads to the first equation of (8), which completes the proof. □

The search for the suboptimal outer-bounding ellipsoid containing $\mathcal{S}_{t+1}$ will now be performed over the parametrized family $\mathcal{E}(\mathbf{c}_\alpha; \mathbf{M}_\alpha)$. This yields a one-dimensional optimization problem, with the computation of either $\alpha^* = \arg \min_{0<\alpha<\alpha_{\max}} \operatorname{tr} \mathbf{M}_\alpha^{-1}$ or $\alpha^* = \arg \min_{0<\alpha<\alpha_{\max}} \ln \det \mathbf{M}_\alpha^{-1}$. Finally $\widehat{\mathcal{E}}_{t+1/t} = \mathcal{E}(\widehat{\mathbf{c}}_{t+1/t}; \widehat{\mathbf{M}}_{t+1/t})$ is defined by

$$\widehat{\mathbf{c}}_{t+1/t} = \mathbf{c}_{\alpha^*}, \; \widehat{\mathbf{M}}_{t+1/t} = \mathbf{M}_{\alpha^*}. \tag{21}$$

This is a complete analog of the situation with no model uncertainty, where one-parametric optimization is required to construct the best ellipsoid for the prediction step, compare (Chernousko, 1994; Kurzhanskii and Valyi, 1996; Polyak, 1998; Durieu *et al.*, 2001). So this theorem provides a useful solution for the prediction step of ellipsoidal state estimation, which can be incorporated directly into a state estimator for systems with measurements. In general the solution is only suboptimal because the optimal ellipsoid does not necessarily belong to the family $\mathcal{E}(\mathbf{c}_\alpha; \mathbf{M}_\alpha)$.

A simpler method of outer approximation of $\mathcal{S}_{t+1}$ was proposed by Chernousko and Rokityanskii (2000). Their algorithm computes the ellipsoid $\mathcal{E}(\mathbf{c}; \mathbf{M}) \supset \mathcal{S}_{t+1}$ as an approximation of the sum of $\mathcal{E}(\mathbf{A}_t \widehat{\mathbf{c}}_{t/t} + \mathbf{B}_t \mathbf{u}_t; \mathbf{A}_t^{-\mathrm{T}} \widehat{\mathbf{M}}_{t/t} \mathbf{A}_t^{-1})$ and a ball of radius r given by $\mathcal{E}(\mathbf{0}; r^{-2}\mathbf{I})$ with

$$\begin{aligned} r^2 &= \delta_{\mathbf{A}}^2 \max_{\mathbf{x} \in \widehat{\mathcal{E}}_{t/t}} \|\mathbf{x}\|^2 + \delta_{\mathbf{v}}^2 \\ &= \delta_{\mathbf{A}}^2 \left(\lambda_{\min}^{-1} \left(\widehat{\mathbf{M}}_{t/t} \right) + \right. \\ & \left. 2\sqrt{\widehat{\mathbf{c}}_{t/t}^{\mathrm{T}} \widehat{\mathbf{M}}_{t/t} \widehat{\mathbf{c}}_{t/t}} + \|\widehat{\mathbf{c}}_{t/t}\|^2 \right) + \delta_{\mathbf{v}}^2. \end{aligned} \tag{22}$$

The sum of these two ellipsoids is already convex and contains $\mathcal{S}_{t+1}$, and it is easy to construct the smallest ellipsoid that contains this sum. The simplicity of this approach is obviously attractive. But, as illustrated by the next example, such an estimate may be much more conservative than the one obtained by our method.

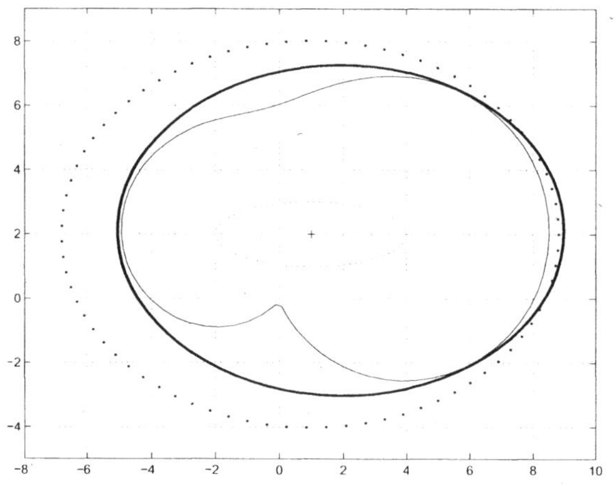

Fig. 1. Outer approximations of $\mathcal{S}_{t+1}$.

Example 1. For $\widehat{\mathbf{M}}_{t/t} = \operatorname{diag}\{1/9, 1\}$, $\widehat{\mathbf{c}}_{t/t} = (1,2)^{\mathrm{T}}$, $\mathbf{A}_t = \operatorname{diag}\{1,1\}$, $\mathbf{u}_t = \mathbf{0}$, $\delta_{\mathbf{A}} = 1$ and $\delta_{\mathbf{v}} = 0.5$, the set $\mathcal{S}_{t+1}$ is non-convex (see Fig. 1, thin solid line) and the ellipsoid $\mathcal{E}(\mathbf{c}_\alpha; \mathbf{M}_\alpha)$ with $\mathbf{c}_\alpha$ and $\mathbf{M}_\alpha$ from (8) contains $\mathcal{S}_{t+1}$ for all α such that $0 < \alpha < \alpha_{\max} = 1/10$. For brevity, the exact expressions for $\mathbf{c}_{\alpha^*}$ and $\mathbf{M}_{\alpha^*}$ are omitted. The minimal-trace ellipsoid $\mathcal{E}(\mathbf{c}_{\alpha^*}; \mathbf{M}_{\alpha^*})$ in the family of Theorem 1 (thick solid line on Fig. 1) gives an outer approximation of $\mathcal{S}_{t+1}$. Compare with the significantly more pessimistic outer approximation obtained as in (Chernousko and Rokityanskii, 2000) and represented on Fig. 1 by a dotted line. The difference in pessimism between the two algorithms is even more pronounced if they are used recursively for long-range prediction. □

Theorem 1 simplifies into Theorem 1 of (Polyak *et al.*, 2002) if the ellipsoid $\widehat{\mathcal{E}}_{t/t}$ is centered at the origin. $\widehat{\mathcal{E}}_{t+1/t}$ is then optimal. Moreover for deterministic models ($\delta_{\mathbf{A}} = 0$) Theorem 1 simplifies into Theorem 4.2 of (Durieu *et al.*, 2001) with $K = 2$.

4. EXTENSIONS

4.1 *Frobenius norm for uncertainties*

Theorem 1 still holds true if the spectral matrix norm is replaced by the Frobenius norm $\|.\|_F$ in the model uncertainty constraints. The proof remains the same because $\|\mathbf{Hx}\| \leq \|\mathbf{H}\|_F \|\mathbf{x}\|$ and $\|\mathbf{zx}^{\mathrm{T}}\|_F = \|\mathbf{z}\|\|\mathbf{x}\|$, where $\mathbf{z}$ and $\mathbf{x}$ are vectors. Thus the prediction step derived from Theorem 1 extends trivially to models of uncertainties where the Frobenius matrix norm is used in (2). Theorem 2 in (Polyak *et al.*, 2002) concerning the correction step also remains true.

4.2 *Prediction with separate uncertainties*

The first combined quadratic constraint in (2) may sometimes seem too restrictive and inflexible. Often, only separate inequalities on model uncertainty and additive disturbances are available. The approach followed in this paper could also help to construct reliable ellipsoidal estimates in this case.

Consider a linear state-space model

$$\mathbf{x}_{t+1} = (\mathbf{A}_t + \Delta\mathbf{A}_t)\,\mathbf{x}_t + \mathbf{B}_t\mathbf{u}_t + \mathbf{v}_t, \quad (23)$$

with $\mathbf{x}_t \in \widehat{\mathcal{E}}_{t/t}$, $\|\Delta\mathbf{A}_t\| \leq \delta_\mathbf{A}$ and $\|\mathbf{v}_t\| \leq \delta_\mathbf{v}$. The spectral norm is again used for matrices. The prediction step for such a process must find an ellipsoid $\widehat{\mathcal{E}}_{t+1/t}$ as small as possible containing

$$\mathcal{S}_{t+1} = \left\{ \begin{array}{l} \mathbf{x}_{t+1} : \mathbf{x}_{t+1} = \mathbf{A}_t\mathbf{x}_t + \Delta\mathbf{A}_t\mathbf{x}_t \\ \quad +\mathbf{B}_t\mathbf{u}_t + \mathbf{v}_t, \quad \mathbf{x}_t \in \widehat{\mathcal{E}}_{t/t}, \\ \|\Delta\mathbf{A}_t\| \leq \delta_\mathbf{A}, \|\mathbf{v}_t\| \leq \delta_\mathbf{v}. \end{array} \right\} (24)$$

A suboptimal solution consits of performing the next three operations.

(1) Find an ellipsoid $\mathcal{E}^+ = \mathcal{E}(\mathbf{c}^+, \mathbf{M}^+)$ containing the vector sum $\mathbf{x}^+ = \mathbf{A}_t\mathbf{x}_t + \Delta\mathbf{A}_t\mathbf{x}_t$ directly from Theorem 1 with $\delta_\mathbf{v} = 0$.
(2) Based on classical algorithms compute an ellipsoidal outer approximation $\mathcal{E}^{++}$ of the sum of $\mathcal{E}^+$ and $\mathcal{E}_\mathbf{v} = \{\mathbf{v}_t : \|\mathbf{v}_t\| \leq \delta_\mathbf{v}\}$ (see, for instance, (Durieu *et al.*, 2001) or Theorem 1 with $\delta_\mathbf{A} = 0$).
(3) Translate $\mathcal{E}^{++}$ by $\mathbf{B}_t\mathbf{u}_t$ to obtain $\widehat{\mathcal{E}}_{t+1/t} = \mathcal{E}(\widehat{\mathbf{c}}_{t+1/t}; \widehat{\mathbf{M}}_{t+1/t})$.

Note that the order of the operations in the procedure can be modified, leading to a different suboptimal solution.

Hence, the dynamic system (23) with general separate constraints on model uncertainty and additive process perturbation vector is also tractable.

4.3 *Correction with separate uncertainties*

The second combined quadratic constraint in (2) leads to a quadratic inequality (see Section 3.2 in (Polyak *et al.*, 2002)). Unfortunately, this is no longer the case when the constraints are separate. Indeed, the set of all state vectors $\mathbf{x}_t$ consistent with the measurement equation

$$\mathbf{y}_t = (\mathbf{C}_t + \Delta\mathbf{C}_t)\,\mathbf{x}_t + \mathbf{D}_t\mathbf{u}_t + \mathbf{w}_t, \quad (25)$$

with $\mathbf{x}_t \in \widehat{\mathcal{E}}_{t/t-1}$, $\|\Delta\mathbf{C}_t\| \leq \delta_\mathbf{C}$ and $\|\mathbf{w}_t\| \leq \delta_\mathbf{w}$, is equivalently defined by

$$\|\mathbf{y}_t - \mathbf{C}_t\mathbf{x}_t - \mathbf{D}_t\mathbf{u}_t\| \leq \delta_\mathbf{C}\,\|\mathbf{x}_t\| + \delta_\mathbf{w}, \quad (26)$$

which is not a quadratic constraint. For this reason, Theorem 2 in (Polyak *et al.*, 2002) does not apply directly. Nevertheless, (26) implies that

$$\|\mathbf{y}_t - \mathbf{C}_t\mathbf{x}_t - \mathbf{D}_t\mathbf{u}_t\|^2 \leq \delta_\mathbf{C}^2\,\|\mathbf{x}_t\|^2 + \tilde{\delta}^2, \quad (27)$$

where

$$\tilde{\delta}^2 = \delta_\mathbf{w}^2 + 2\delta_\mathbf{C}\delta_\mathbf{w} \max_{\mathbf{x}\in\widehat{\mathcal{E}}_{t/t-1}} \|\mathbf{x}\|. \quad (28)$$

Theorem 2 in (Polyak *et al.*, 2002) can then be applied with δ replaced by $\tilde{\delta}$, which can be easily computed.

An ellipsoidal approximation of intersection at the correction step for a dynamic system with separate constraints on model uncertainty and additive measurement noise vector can thus be obtained. Of course, the resulting approximation is suboptimal.

5. STATE ESTIMATOR

This section summarizes an iteration of the recursive estimator for the case where uncertainty is described by (2). Similarly to what is usual in Kalman filtering, the algorithm can be initialized using a very large ellipsoid $\widehat{\mathcal{E}}_{0/0}$ centered at the origin. Starting from $\widehat{\mathcal{E}}_{t/t} = \mathcal{E}(\widehat{\mathbf{c}}_{t/t}; \widehat{\mathbf{M}}_{t/t})$ the prediction step is a follows.

- Compute

$$\begin{aligned}
\alpha_{\max} &= \left(\lambda_{\min}(\widehat{\mathbf{M}}_{t/t}) + \delta_\mathbf{A}^2\right)^{-1} \lambda_{\min}(\widehat{\mathbf{M}}_{t/t}),\\
\mathbf{Q}_\alpha &= (1-\alpha)\,\widehat{\mathbf{M}}_{t/t} - \alpha\delta_\mathbf{A}^2\mathbf{I},\\
\delta_\alpha &= \alpha - \alpha\delta_\mathbf{v}^2 + (1-\alpha)\,\widehat{\mathbf{c}}_{t/t}^\mathrm{T}\widehat{\mathbf{M}}_{t/t}\widehat{\mathbf{c}}_{t/t}\\
&\quad - (1-\alpha)^2\,\widehat{\mathbf{c}}_{t/t}^\mathrm{T}\widehat{\mathbf{M}}_{t/t}\mathbf{Q}_\alpha^{-1}\widehat{\mathbf{M}}_{t/t}\widehat{\mathbf{c}}_{t/t},\\
\mathbf{c}_\alpha &= (1-\alpha)\,\mathbf{A}_t\mathbf{Q}_\alpha^{-1}\widehat{\mathbf{M}}_{t/t}\widehat{\mathbf{c}}_{t/t} + \mathbf{B}_t\mathbf{u}_t,\\
\mathbf{M}_\alpha &= \left((1-\delta_\alpha)\left(\mathbf{A}_t\mathbf{Q}_\alpha^{-1}\mathbf{A}_t^\mathrm{T} + \alpha^{-1}\mathbf{I}\right)\right)^{-1}.
\end{aligned}$$

- Determine $\alpha^* = \arg\min_{0<\alpha<\alpha_{\max}} \operatorname{tr}\mathbf{M}_\alpha^{-1}$ or $\alpha^* = \arg\min_{0<\alpha<\alpha_{\max}} \ln\det\mathbf{M}_\alpha^{-1}$.
- Compute $\widehat{\mathcal{E}}_{t+1/t} = \mathcal{E}(\widehat{\mathbf{c}}_{t+1/t}; \widehat{\mathbf{M}}_{t+1/t})$, where $\widehat{\mathbf{c}}_{t+1/t} = \mathbf{c}_{\alpha^*}$ and $\widehat{\mathbf{M}}_{t+1/t} = \mathbf{M}_{\alpha^*}$.

Starting from $\widehat{\mathcal{E}}_{t+1/t}$, the correction step is as follows (Polyak *et al.*, 2002).

- Compute

$$\begin{aligned}
\mathbf{y} &= \mathbf{y}_{t+1} - \mathbf{D}_{t+1}\mathbf{u}_{t+1},\\
\mathbf{R} &= \mathbf{C}_{t+1}^\mathrm{T}\mathbf{C}_{t+1} - \delta_\mathbf{C}^2\mathbf{I},\\
\mathbf{S} &= \mathbf{R}\left(\mathbf{y}^\mathrm{T}\mathbf{C}_{t+1}\mathbf{R}^{-1}\mathbf{C}_{t+1}^\mathrm{T}\mathbf{y} - \mathbf{y}^\mathrm{T}\mathbf{y} + \delta_\mathbf{w}^2\right)^{-1},\\
\mathbf{d} &= \mathbf{R}^{-1}\mathbf{C}_{t+1}^\mathrm{T}\mathbf{y},\\
\mathbf{Q}_\alpha &= (1-\alpha)\,\widehat{\mathbf{M}}_{t+1/t} + \alpha\mathbf{S},\\
\mathbf{c}_\alpha &= \mathbf{Q}_\alpha^{-1}\left((1-\alpha)\,\widehat{\mathbf{M}}_{t+1/t}\widehat{\mathbf{c}}_{t+1/t} + \alpha\mathbf{S}\mathbf{d}\right),\\
\delta_\alpha &= (1-\alpha)\,\widehat{\mathbf{c}}_{t+1/t}^\mathrm{T}\widehat{\mathbf{M}}_{t+1/t}\widehat{\mathbf{c}}_{t+1/t} +\\
&\quad \alpha\mathbf{d}^\mathrm{T}\mathbf{S}\mathbf{d} - \mathbf{c}_\alpha^\mathrm{T}\mathbf{Q}_\alpha\mathbf{c}_\alpha,\\
\mathbf{M}_\alpha &= (1-\delta_\alpha)^{-1}\,\mathbf{Q}_\alpha.
\end{aligned}$$

Let $\lambda_{\min}$ be the minimum generalized eigenvalue of the matrix pair $(\mathbf{S}, \widehat{\mathbf{M}}_{t+1/t})$ (the generalized eigenvalues λ_i and eigenvectors $\mathbf{v}_i$ of the matrix pair $(\mathbf{S}, \mathbf{M})$ satisfy $\mathbf{S}\mathbf{v}_i = \lambda_i \mathbf{M}\mathbf{v}_i$). Compute

$$\alpha_{\max} = \min\left(1, (1 - \lambda_{\min})^{-1}\right).$$

- Determine $\alpha^* = \arg \min_{0<\alpha<\alpha_{\max}} \operatorname{tr} \mathbf{M}_\alpha^{-1}$ or $\alpha^* = \arg \min_{0<\alpha<\alpha_{\max}} \ln\det \mathbf{M}_\alpha^{-1}$.
- Compute $\widehat{\mathcal{E}}_{t+1/t+1} = \mathcal{E}(\widehat{\mathbf{c}}_{t+1/t+1}; \widehat{\mathbf{M}}_{t+1/t+1})$, where $\widehat{\mathbf{c}}_{t+1/t+1} = \mathbf{c}_{\alpha^*}$ and $\widehat{\mathbf{M}}_{t+1/t+1} = \mathbf{M}_{\alpha^*}$.

6. CONCLUSIONS

An outer-bounding ellipsoidal technique for the estimation of the state of a linear discrete-time dynamic system under model uncertainty has been proposed. This paper generalizes the results presented in (Polyak *et al.*, 2002) for the prediction step in order to construct a recursive state estimator for systems with measurements, while the correction step remains the same. The algorithm readily extends to other models of system uncertainty.

7. ACKNOWLEDGMENTS

This work was supported by Grant INTAS-97-10782. The work of S.A. Nazin was supported in part by INTAS grant YSF 2002-181.

8. REFERENCES

Ben-Tal, A., L. El Ghaoui and A.S. Nemirovskii (2000). Robust semidefinite programming. *Handbook of Semidefinite Programming*, R. Saigal, L. Vanderberghe, H. Wolkowicz (Eds), Kluwer, Waterloo, 139–162.

Bertsekas, D.P. and I.B. Rhodes (1971). Recursive state estimation for a set-membership description of uncertainty. *IEEE Trans. on Automatic Control*, **16**, 117–128.

Cerone, V. (1993). Feasible parameter set for linear models with bounded errors in all variables. *Automatica*, **29**, 1551–1555.

Chernousko, F.L. (1994). *State Estimation for Dynamic Systems*. CRC Press, Boca Raton.

Chernousko, F.L. (1996). Ellipsoidal approximation of attainability sets for linear system with uncertain matrix (in Russian). *Prikl. Matem. i Mekh.*, **80**, 940–950.

Chernousko, F.L. and D.Ya. Rokityanskii (2000). Ellipsoidal bounds on reachable sets of dynamical systems with matrices subjected to uncertain perturbations. *J. Optim. Theory and Appl.*, **104**, 1–19.

Clement, T. and S. Gentil (1990). Recursive membership set estimation for output-errors models. *Math. and Comput. in Simul.*, **32**, 505–513.

Durieu, C., E. Walter and B.T. Polyak (2001). Multi-input multi-output ellipsoidal state bounding. *J. Optim. Theory and Appl.*, **111**, 273–303.

El Ghaoui, L. and H. Lebret (1997). Robust solutions to least-squares problems with uncertain data. *SIAM J. Matrix Anal. and Appl.*, **18**, 1035–1064.

El Ghaoui, L. and G. Calafiore (1999). Worst-case simulation of uncertain systems. *Robustness in Identification and Control*, A. Garulli, A. Tesi, A. Vicino (Eds), Springer, London, 134–146.

Fogel, E. and Y.F. Huang (1982). On the value of information in system identification — bounded noise case. *Automatica*, **18**, 229–238.

Golub, G.H. and C.F. Van Loan (1980). An analysis of the total least squares problems. *SIAM J. Numer. Anal.*, **17**, 883–893.

Kurzhanskii, A.B. (1977). *Control and Observation under Uncertainty* (in Russian). Nauka, Moscow.

Kurzhanskii, A.B. and I. Valyi (1996). *Ellipsoidal Calculus for Estimation and Control*. Birkhauser, Basel.

Milanese, M., J. Norton, H. Piet-Lahanier and E. Walter (Eds), (1996). *Bounding Approaches to System Identification*. Plenum, New York.

Norton, J. (Ed.) (1994, 1995). Special issues on bounded-error estimation, 1, 2. *Intern. J. of Adaptive Control and Signal Proc.*, **8**, No. 1, **9**, No. 2.

Norton, J. (1999). Modal robust state estimation with deterministic specification of uncertainty. *Robustness in Identification and Control*, A. Garulli, A. Tesi, A. Vicino (Eds), Springer, London, 62–71.

Polyak, B.T. (1998). Convexity of quadratic transformations and its use in control and optimization. *J. Optim. Theory and Appl.* **99**, 553–583.

Polyak, B.T., S. Nazin, C. Durieu and E. Walter (2002). Ellipsoidal estimation under model uncertainty. *15th IFAC Triennal World Congress, Barcelona, Spain*, 1090–1095.

Rokityanskii, D.Ya. (1997). Optimal ellipsoidal estimates of attainability sets for linear systems with uncertain matrix (in Russian). *Izvestiya RAN. Theor. i Syst. Upr.*, No. 4, 17–20.

Schweppe, F. C. (1968). Recursive state estimation: unknown but bounded errors and system inputs. *IEEE Trans. on Automatic Control*, **13**, No. 1, 22-28.

Schweppe, F.C. (1973). *Uncertain Dynamic Systems*. Prentice Hall, Englewood Cliffs.

Walter, E. (Ed.) (1990). Special issue on parameter identification with error bound. *Math. and Comput. in Simul.*, **32**, No. 5–6.

Walter, E. and L. Pronzato (1997). *Identification of Parametric Models from Experimental Data*. Springer, London.

Witsenhausen, H. S. (1968). Sets of possible states of linear systems given perturbed observations. *IEEE Trans. on Automatic Control*, **13**, No. 5, 556-558.

IFAC Publications
www.elsevier.com/locate/ifac

Regularized Robust Estimators for Time Varying Uncertain Discrete-Time Systems

Ananth Subramanian and Ali H. Sayed

Abstract—**This paper addresses the issue of robust filtering for time varying uncertain discrete time systems. The proposed robust filters are based on a regularized least-squares formulation and guarantee minimum state error variances. Simulation results indicate their superior performance over other robust filter designs.**

keywords: regularization, least-squares, robust filter, regularization parameter, parametric uncertainty.

I. Introduction

The Kalman filter is the optimal linear least-mean-squares estimator for systems that are described by linear state-space Markovian models [1]. However, when the model is not accurately known, the performance of the filter can deteriorate appreciably. This filter sensitivity to modeling errors has led to several works in the literature on the development of robust state-space filters; robust in the sense that they attempt to limit, in certain ways, the effect of model uncertainties on the overall filter performance. Some of the well known approaches to state-space estimation in this regard are $\mathcal{H}_\infty$ filtering, mixed $\mathcal{H}_2/\mathcal{H}_\infty$ filtering, set-valued estimation, guaranteed-cost designs and minimum variance filtering (see [2]-[11]). In [11], a robust filter design framework was proposed that performs regularization as opposed to de-regularization. The design in [11] involved choosing certain Ricatti variables so as to enforce a local optimality and robustness property. In this paper, we pursue the design of such regularized robust filters and consider two general classes of uncertain state-space models. We consider uncertain model descriptions that involve norm bounded uncertainties for the output matrices, and stochastic and polytopic uncertainties for the state matrices; both descriptions are common in applications. For each class, we shall design robust filters that bound the state error covariance matrix *globally*. The robustness criterion used is different from prior robust designs (e.g., $\mathcal{H}_\infty$, guaranteed-cost or set-valued estimation) in that, it is based on robust *regularization*. In this way, the resulting filters are well suited for online/real-time filtering applications involving both time-invariant and time-variant models. Simulation results are included to illustrate the superior performance of the proposed robust filters over other robust designs.

II. Least-Squares with Uncertainties

Let $J(x) = x^T Qx + R(x)$ denote a cost function with

$$R(x) = \Big((A+\delta A)x - (b+\delta b)\Big)^T W \Big((A+\delta A)x - (b+\delta b)\Big) \quad (1)$$

where δA denotes an $N \times n$ perturbation to A, δb denotes an $N \times 1$ perturbation to b, and $\{\delta A, \delta b\}$ are assumed to satisfy a model of the form

$$\begin{bmatrix} \delta A & \delta b \end{bmatrix} = H\Delta \begin{bmatrix} E_a & E_b \end{bmatrix} \quad (2)$$

where Δ is an arbitrary contraction, $\|\Delta\| \leq 1$, and $\{H, E_a, E_b\}$ are known quantities of appropriate dimensions. Consider then the constrained two player game problem

$$\hat{x} = \arg\min_x \max_{\{\delta A, \delta b\}} J(x) \quad (3)$$

The authors are with the Department of Electrical Engineering, University of California at Los Angeles, Box 951594, Los Angeles, CA 90095-1594.
E-mail: {msananth, sayed}@ee.ucla.edu.

Supported by NSF grant ECS-9820765 and CCR-0208573.

subject to (2). The following result is proven in [12].

Theorem 1: The problem (2)–(3) has a unique solution $\hat{x}$ that is given by

$$\hat{x} = \left[\widehat{Q} + A^T \widehat{W} A\right]^{-1} \left[A^T \widehat{W} b + \widehat{\beta} E_a^T E_b\right] \quad (4)$$

where

$$\widehat{Q} = Q + \widehat{\beta} E_a^T E_a \quad (5)$$

$$\widehat{W} = W + WH(\widehat{\beta} I - H^T W H)^\dagger H^T W \quad (6)$$

and the scalar $\widehat{\beta}$ is determined from the optimization

$$\widehat{\beta} = \arg \min_{\beta \geq \|H^T W H\|} G(\beta) \quad (7)$$

where the function $G(\beta)$ is defined as follows:

$$\begin{aligned} G(\beta) = & x^T(\beta) Q x(\beta) + \beta \|E_a x(\beta) - E_b\|^2 \\ & + [Ax(\beta) - b]^T W(\beta) [Ax(\beta) - b] \end{aligned} \quad (8)$$

with

$$W(\beta) = W + WH(\beta I - H^T W H)^\dagger H^T W$$

$$Q(\beta) = Q + \beta E_a^T E_a$$

and

$$x(\beta) = \left[Q(\beta) + A^T W(\beta) A\right]^{-1} \left[A^T W(\beta) b + \beta E_a^T E_b\right] \quad (9)$$

[The notation $X^\dagger$ denotes the pseudo-inverse of X.]

$\diamond$

It was shown in [12], [16] that the function $G(\beta)$ has a unique global minimum (and no local minima) over the interval $\beta \geq \|H^T W H\|$, which means that the determination of $\widehat{\beta}$ can be pursued by employing standard search procedures without worrying about convergence to undesired local minima. It was further argued in [11] that a good approximation for $\widehat{\beta}$ is to choose it as $\widehat{\beta} = (1+\alpha)\beta_l$ for some $\alpha > 0$ and where $\beta_l = \|H^T W H\|$.

III. The State Space Models

We now use Thm. 1 to design two robust filters. Each filter will be applicable to a particular class of model uncertainties. Thus consider an $n-$dimensional state-space model of the form:

$$x_{k+1} = F_k x_k + G_k w_k \quad (10)$$

$$y_k = (H_k + \Delta H_k) x_k + v_k, \quad k \geq 0 \quad (11)$$

where $\{w_k, v_k\}$ are uncorrelated white zero-mean random processes with variances

$$E w_k w_k^* = W_k, \qquad E v_k v_k^* = V_k$$

and x_0 is a zero-mean random variable that is uncorrelated with $\{w_k, v_k\}$ for all k. The uncertainties ΔH_k are modelled as

$$\Delta H_k = M_k \Delta_k E_k \quad (12)$$

where M_k and E_k are known matrices, while Δ_k is an arbitrary contraction, $\Delta_k^T \Delta_k \leq I$.

We shall consider two types of uncertainty descriptions for the state matrices F_k: one is in terms of polytopic uncertainties and the other is in terms of norm bounded stochastic uncertainties. In the first case, we assume that F_k lies inside a convex bounded polyhedral domain $\mathcal{K}_k$ described by m vertices as follows:

$$\mathcal{K}_k = \left\{ F_k = \sum_{i=1}^{i=m} \alpha_{i,k} F_{i,k}, \ \alpha_{i,k} \geq 0, \ \sum_{i=1}^{i=m} \alpha_{i,k} = 1 \right\} \tag{13}$$

Observe that $\mathcal{K}_k$ is allowed to vary with k. In the second case, we assume that F_k is described by

$$F_k = F_{k,c} + \Delta F_k, \quad \Delta F_k = N_k \bar{\Delta}_k J_k \tag{14}$$

for some known $F_{k,c}$ and where $\bar{\Delta}_k$ is a random matrix whose entries are zero mean and uncorrelated with each other, and such that

$$E \bar{\Delta}_k \bar{\Delta}_k^* \leq \rho_{\bar{\Delta}} I \tag{15}$$

for some known $\rho_{\bar{\Delta}}$.

IV. Robust State Space Filtering

When uncertainties are not present in the model (10)–(11), it is known that the optimal linear estimator for the state is the Kalman filter [18]. This filter admits a deterministic interpretation as the solution to a regularized least-squares problem as follows. Let [1]

$$\hat{x}_{k|k-1} \stackrel{\Delta}{=} \text{an estimate of } x_k \text{ given } \{y_0, y_1, \ldots, y_{k-1}\}$$

$$\hat{x}_{k|k} \stackrel{\Delta}{=} \text{an estimate of } x_k \text{ given } \{y_0, y_1, \ldots, y_{k-1}, y_k\}$$

Given the predicted estimate $\hat{x}_{k|k-1}$ and an observation y_k, the filtered estimate $\hat{x}_{k|k}$ that is computed by the Kalman filter is the solution of

$$\min_x \left[\|x - \hat{x}_{k|k-1}\|^2_{P_k^{-1}} + \|y_k - H_k x\|^2_{R_k^{-1}} \right] \tag{16}$$

where P_k and R_k are the state error covariance and the measurement noise covariance matrices, respectively. When uncertainties are present in $\{H_k, F_k\}$, we formulate a robust version of (16) by solving instead the min-max problem :

$$\min_x \max_{\delta H_k, \delta F_k} \left(\|x - \hat{x}_{k|k-1}\|^2_{P_k^{-1}} + \|y_k - (H_k + \delta H_k) x\|^2_{R_k^{-1}} \right) \tag{17}$$

This formulation was proposed in [11]. Compared with other robust designs, it has the advantage of performing regularization as opposed to de-regularization, a property that is useful for on-line/real-time operation since the resulting filter will not require existence conditions. In [11], the weighing matrices P_k in (17) were propagated through a Ricatti recursion that enforces a local optimality criterion. In our first filter below, we shall instead determine P_k so as to minimize the state error covariance matrix *globally*. We do so by showing how to re-parametrize P_k and R_k in terms of a single parameter Q_k, over which the global minimization of the error covariance matrix reduces to a linear convex problem. In our second filter, we shall derive a more efficient procedure for updating P_k. The procedure does not require solving a linear convex problem at each iteration and has the same computational complexity as the Kalman filter.

[1] When uncertainties are not present, the qualification "estimate" refers to the linear-least-mean-squares estimate.

A. Polytopic Uncertainties

We consider first the case of polytopic uncertainties in F_k as in (13). Our objective is to design a robust linear estimator for the state variable x_k of the form

$$\hat{x}_{k|k} = F_{p,k} \hat{x}_{k|k-1} + K_{p,k} y_k, \quad k \geq 0 \tag{18}$$

$$\hat{x}_{k+1|k} = F_{k,c} \hat{x}_{k|k} \tag{19}$$

for some matrices $F_{p,k}$ and $K_{p,k}$ to be determined in order to minimize the worst case error variance of the state for all uncertainties, and where $F_{k,c}$ denotes the centroid of the polytope $\mathcal{K}_k$:

$$F_{k,c} = \frac{1}{m} \sum_{i=1}^{i=m} F_{i,k} \tag{20}$$

Assume first that the F_k are fixed; we will incorporate the uncertainties in F_k soon. With uncertainties in the output matrices H_k alone, problem (17) becomes

$$\min_x \max_{\delta H_k} \left(\|x - \hat{x}_{k|k-1}\|^2_{P_k^{-1}} + \|y_k - (H_k + \delta H_k) x\|^2_{R_k^{-1}} \right) \tag{21}$$

which can be written more compactly in the form (1)–(3) with the identifications:

$$\begin{array}{lcl} x & \longleftarrow & \{x_k - \hat{x}_{k|k-1}\}, \quad b \longleftarrow y_k - H_k \hat{x}_{k|k-1} \\ \delta A & \longleftarrow & M_k \Delta_k E_k \\ \delta b & \longleftarrow & -M_k \Delta_k E_k \hat{x}_{k|k-1}, \quad Q \longleftarrow P_k^{-1} \\ W & \longleftarrow & R_k^{-1}, \ H \longleftarrow M_k, \ E_a \longleftarrow E_k \\ E_b & \longleftarrow & -E_k \hat{x}_{k|k-1}, \ \Delta \longleftarrow \Delta_k, \ A \longleftarrow H_k \end{array}$$

From Thm. 1, the solution $\hat{x}_{k|k}$ of (21) is given by

$$\begin{aligned} \hat{x}_{k|k} = \hat{x}_{k|k-1} + &(P_k^{-1} + \hat{\beta} E_k^T E_k + H_k^T \hat{R}_k^{-1} H_k)^{-1} \\ &\{H_k^T \hat{R}_k^{-1} (y_k - H_k \hat{x}_{k|k-1}) - \hat{\beta} E_k^T E_k \hat{x}_{k|k-1}\} \end{aligned} \tag{22}$$

where

$$\hat{R}_k^{-1} = (R_k - \hat{\beta}^{-1} M_k M_k^T)^{-1} \tag{23}$$

If we now introduce the matrix

$$Q_k \stackrel{\Delta}{=} (P_k^{-1} + \hat{\beta} E_k^T E_k + H_k^T \hat{R}_k^{-1} H_k)^{-1} \tag{24}$$

then the expression for $\hat{x}_{k|k}$ becomes

$$\begin{aligned} \hat{x}_{k|k} = &(I - \hat{\beta} Q_k E_k^T E_k - Q_k H_k^T \hat{R}_k^{-1} H_k) \hat{x}_{k|k-1} \\ &+ Q_k H_k^T \hat{R}_k^{-1} y_k \end{aligned} \tag{25}$$

in terms of the parameter Q_k. Noting that w_k is a zero-mean white random process, we let the following be an estimate for x_{k+1} given the measurement y_k:

$$\hat{x}_{k+1|k} \stackrel{\Delta}{=} F_{k,c} \hat{x}_{k|k} \tag{26}$$

We then get

$$\hat{x}_{k+1|k} = F_{p,k} \hat{x}_{k|k-1} + K_{p,k} y_k \tag{27}$$

where $F_{p,k}$ and $K_{p,k}$ are defined in terms of Q_k as

$$F_{p,k} = F_{k,c}(I - \hat{\beta} Q_k E_k^T E_k - Q_k H_k^T \hat{R}_k^{-1} H_k) \tag{28}$$

$$K_{p,k} = F_{k,c} Q_k H_k^T \hat{R}_k^{-1} \tag{29}$$

Denoting $\tilde{x}_k = x_k - \hat{x}_{k|k-1}$, we define the extended weight vector

$$\eta_k \triangleq \begin{pmatrix} x_k \\ \tilde{x}_k \end{pmatrix} \tag{30}$$

Then η_k satisfies

$$\eta_{k+1} = \bar{F}_k \eta_k + \bar{G}_k u_k \tag{31}$$

where

$$u_k = \begin{pmatrix} w_k \\ v_k \end{pmatrix}, \quad \bar{G} = \begin{pmatrix} G & 0 \\ G & -K_{p,k} \end{pmatrix} \tag{32}$$

$$\bar{F}_k = \begin{pmatrix} F_k & 0 \\ F_k - F_{p,k} - K_{p,k}H_k & F_{p,k} \end{pmatrix} \tag{33}$$

and the covariance matrix of η_k satisfies

$$\Pi_{k+1} = \bar{F}_k \Pi_k \bar{F}_k^T + \bar{G}_k S_k \bar{G}_k^T \tag{34}$$

where

$$S_k = \begin{pmatrix} W_k & 0 \\ 0 & V_k \end{pmatrix} \tag{35}$$

and Π_0 is the covariance matrix of η_0. Now observe that the expressions for $\{F_{p,k}, K_{p,k}\}$ are parametrized in terms of the single parameter Q_k. We shall then choose Q_k so as to minimize the covariance of η_k. In this way, the resulting filter will satisfy the robustness condition (21), in addition to minimizing the state error covariance. This is achieved as follows. First note that Q_k in (24) is to problem (21) as the matrix $\widehat{Q} + A^T\widehat{W}A$ in (4) is to problem (1)–(3). Therefore, Q_k must be positive definite so that the $\hat{x}_{k|k}$ is guaranteed to be the minimum of (21). Then we shall choose $Q_k > 0$ so as to minimize Π_{k+1} of (34). This can be obtained by solving

$$\min_{Q_k > 0} \quad \mathrm{Trace}(\Pi_{k+1}) \tag{36}$$

subject to the inequality

$$\Pi_{k+1} \geq \bar{F}_k \Pi_k \bar{F}_k^T + \bar{G}_k S_k \bar{G}_k^T \tag{37}$$

or, equivalently,

$$\begin{pmatrix} -\Pi_{k+1} & \bar{F}_k\Pi_k & \bar{G}_k S_k^{1/2} \\ \Pi_k \bar{F}_k^T & -\Pi_k & 0 \\ S_k^{T/2}\bar{G}_k^T & 0 & -I \end{pmatrix} \leq 0 \tag{38}$$

In order to incorporate the polytopic uncertainties in the F_k, as defined by the sets $\mathcal{K}_k$ in (13), we need to solve the above optimization problem with F_k taking values at the m vertices of the convex polytope $\mathcal{K}_k$, i.e., from the set $\{F_{1,k}, F_{2,k}, \ldots\ldots, F_{m,k}\}$. Since the inequality (38) is affine in F_k, the Q_k thus found will ensure minimum error covariance Π_k over all possible F_k in $\mathcal{K}_k$. Therefore, the time varying robust filter is given by (65)–(67), where Q_k is the positive definite solution of (51)–(38) with F_k taking values on the vertices of the convex polytope $\mathcal{K}_k$, and initializing $\Pi_0 = \mathrm{diag}\{P_o, \epsilon I\}$ for some positive definite P_o. Note that there always exists a solution to (51)–(38). This is because, at every time instant k, $Q_k = \epsilon I$ for $\epsilon > 0$ is a feasible solution. The filter is summarized in Table 1.

Infinite horizon case : In this paper, by the notion of stability in the infinite horizon case, we mean that the variables associated with the filter are bounded for all k. Assume $\|F_{k,c}\| < 1$ and choose Q_k to satisfy (51)–(38) as well as $\|\bar{F}_k\| < 1$. This additional constraint is easily represented in terms of a linear matrix inequality in the variable Q_k as

$$\begin{pmatrix} I & \bar{F}_k^T \\ \bar{F}_k & I \end{pmatrix} > 0 \tag{39}$$

This condition guarantees that Π_k will remain bounded for all k.

Assumed uncertain model. Eqs. (10)–(13).
Initial conditions: $\hat{x}_0 = 0$, $\Pi_0 = \mathrm{diag}\{P_o, \epsilon I\}$. Step 1. If $M_k = 0$, then set $\widehat{\beta}_k = 0$. Otherwise, set instead $\hat{\beta} = (1+\alpha)\beta_{l,k}$ where $\beta_{l,k} = \|M_k^T R_k^{-1} M_k\|$.
Step 2. Using Π_k, compute $\{Q_k, \Pi_{k+1}\}$ by solving $$\min_{Q_k > 0} \quad \mathrm{Trace}(\Pi_{k+1})$$ subject to the inequality $$\begin{pmatrix} -\Pi_{k+1} & \bar{F}_k\Pi_k & \bar{G}_k S_k^{1/2} \\ \Pi_k \bar{F}_k^T & -\Pi_k & 0 \\ S_k^{T/2}\bar{G}_k^T & 0 & -I \end{pmatrix} \leq 0$$ where $\{\bar{F}_k, \bar{G}_k, S_k\}$ are defined by (32),(33) and (35). Step 3. Update $\hat{x}_k$ to $\hat{x}_{k+1}$ as $$\hat{x}_{k+1} = F_{p,k}\hat{x}_k + K_{p,k}y_k$$ where $$F_{p,k} = F_{k,c}(I - \hat{\beta}Q_k E_k^T E_k - Q_k H_k^T \hat{R}_k^{-1} H_k)$$ $$K_{p,k} = F_{k,c}Q_k H_k^T \hat{R}_k^{-1}$$ $$\hat{R}_k^{-1} = (R_k - \hat{\beta}^{-1}M_k M_k^T)^{-1}$$ with $F_{k,c}$ from (20).

Table 1: Regularized robust filter for polytopic uncertainties.

B. Stochastic Uncertainties

We now consider the case of norm bounded stochastic uncertainties in F_k as in (14). We derive two filters; one through convex optimization method and the other through a suboptimal ricatti equation method. We first look at a filter that is derived in terms of convex optimization.

B.1 A robust filter

Here again, our objective is to design a robust linear estimator for the state variable x_k of the form

$$\hat{x}_{k|k} = F_{p,k}\hat{x}_{k|k-1} + K_{p,k}y_k, \quad k \geq 0 \tag{40}$$

$$\hat{x}_{k+1|k} = F_{k,c}\hat{x}_{k|k} \tag{41}$$

for some matrices $F_{p,k}$ and $K_{p,k}$ to be determined in order to minimize the worst case error variance of the state for all uncertainties, and where $F_{k,c}$ denotes the nominal state matrix. Proceeding in the same manner as in the previous section from the robustness condition (21), we know that the expressions for $\{F_{p,k}, K_{p,k}\}$ will be parametrized in terms of the parameter Q_k. We shall then choose Q_k so as to minimize the covariance of η_k. Here η_k satisfies :

$$\eta_{k+1} = (\bar{F}_{k,c} + \bar{N}_k\bar{\Delta}\bar{J}_k)\eta_k + \bar{G}_k u_k \tag{42}$$

where

$$u_k = \begin{pmatrix} w_k \\ v_k \end{pmatrix} \quad (43)$$

$$\bar{F}_{k,c} = \begin{pmatrix} F_{k,c} & 0 \\ F_{k,c} - F_{p,k} - K_{p,k}H_k & F_{p,k} \end{pmatrix} \quad (44)$$

$$\bar{N}_k = \begin{pmatrix} N_k & 0 \\ N_k & 0 \end{pmatrix} \quad (45)$$

$$\bar{J}_k = \begin{pmatrix} J_k & 0 \\ 0 & 0 \end{pmatrix} \quad (46)$$

and the covariance matrix of η_k satisfies

$$\Pi_{k+1} = E\{(\bar{F}_{k,c}+\bar{N}_k\bar{\Delta}\bar{J}_k)\Pi_k(\bar{F}_{k,c}+\bar{N}_k\bar{\Delta}\bar{J}_k)^T\}+\bar{G}_kS_k\bar{G}_k^T \quad (47)$$

where E denotes the expectation operator and

$$S_k = \begin{pmatrix} W_k & 0 \\ 0 & V_k \end{pmatrix} \quad (48)$$

Let $\check{\Pi}_k$ satisfy

$$\check{\Pi}_{k+1} = \bar{F}_{k,c}\check{\Pi}_k\bar{F}_{k,c}^T + \bar{G}_kS_k\bar{G}_k^T + \rho_{\bar{\Delta}}\hat{\alpha}_k\bar{N}_k\bar{N}_k^T \quad (49)$$

and choose $\hat{\alpha}_k$ such that $\hat{\alpha}_k - \bar{J}_k\check{\Pi}_k\bar{J}_k^T > 0$. Then

$$\hat{\alpha}_kI - \bar{J}_k\Pi_k\bar{J}_k^T > 0 \quad (50)$$

and

$$\Pi_{k+1} \leq \bar{F}_{k,c}\Pi_k\bar{F}_{k,c}^T + \bar{G}_kS_k\bar{G}_k^T + \rho_{\bar{\Delta}}\hat{\alpha}_k\bar{N}_k\bar{N}_k^T$$

This suggests that we can choose Q_k by constructing a sequence of matrices $\hat{\Pi}_k$ so as to solve:

$$\min_{Q_k>0} \quad \text{Trace}(\hat{\Pi}_{k+1}) \quad (51)$$

subject to the inequality

$$\hat{\Pi}_{k+1} \geq \bar{F}_k\hat{\Pi}_k\bar{F}_k^T + \bar{G}_kS_k\bar{G}_k^T + \rho_{\bar{\Delta}}\hat{\alpha}_k\bar{N}_k\bar{N}_k^T \quad (52)$$

or, equivalently,

$$\begin{pmatrix} -\hat{\Pi}_{k+1} + \rho_{\bar{\Delta}}\hat{\alpha}_k\bar{N}_k\bar{N}_k^T & \bar{F}_k\hat{\Pi}_k & \bar{G}_kS_k^{1/2} \\ \hat{\Pi}_k\bar{F}_k^T & -\hat{\Pi}_k & 0 \\ S_k^{T/2}\bar{G}_k^T & 0 & -I \end{pmatrix} \leq 0 \quad (53)$$

Note that $\Pi_k \leq \hat{\Pi}_k$ for every k. The filter hence derived is summarized in Table 2.

Infinite horizon case : We now describe some sufficient conditions for infinite horizon stability (i.e., for the matrices $\{\hat{\Pi}_k, \check{\Pi}_k\}$ to remain bounded). Assume that $\|F_{k,c}\| < 1$ and choose Q_k to satisfy (51)–(53) as well as $\|\bar{F}_{k,c}\| < 1$. This additional constraint is easily represented in terms of a linear matrix inequality in the variable Q_k as

$$\begin{pmatrix} I & \bar{F}_{k,c}^T \\ \bar{F}_{k,c} & I \end{pmatrix} > 0 \quad (54)$$

Let a fixed $\hat{\alpha}$ be chosen in the following way. Consider the recursion

$$\check{\Pi}_{k+1} = \bar{F}_{k,c}\check{\Pi}_k\bar{F}_{k,c}^T + \bar{G}_kS\bar{G}_k^T + \rho_{\bar{\Delta}}\hat{\alpha}_k\bar{N}_k\bar{N}_k^T$$

Then

$$\|\check{\Pi}_{k+1}\| \leq \|\bar{F}_{k,c}\|^2\|\check{\Pi}_k\| + \|\bar{G}_kS\bar{G}_k^T\| + \rho_{\bar{\Delta}}\hat{\alpha}_k\|\bar{N}_k\bar{N}_k^T\|$$

Let $\beta = \|\bar{F}_{k,c}\|^2 < 1$, $\|\bar{N}_k\bar{N}_k^T\| < \gamma$ and since Q_k lies inside a bounded set $\mathcal{C}$, assume $\|\bar{G}_kS_k\bar{G}_k^T\| < \zeta$. Then

$$\|\check{\Pi}_{k+1}\| \leq \beta^k\|\check{\Pi}_0\| + \sum_{i=0}^{i=k-1}\beta^{k-1-i}\zeta + \sum_{i=0}^{i=k-1}\beta^{k-1-i}\rho_{\bar{\Delta}}\hat{\alpha}_k\gamma$$

or

$$\|\check{\Pi}_{k+1}\| \leq \|\check{\Pi}_0\| + \frac{1}{1-\beta}(\zeta + \rho_{\bar{\Delta}}\hat{\alpha}_k\gamma)$$

We then have

$$\begin{aligned} \|\bar{J}_{k+1}\check{\Pi}_{k+1}\bar{J}_{k+1}^T\| &\leq \|\check{\Pi}_{k+1}\|\|\bar{J}_{k+1}\|^2 \\ &\leq \chi(\|\check{\Pi}_0\| + \frac{1}{1-\beta}(\zeta + \rho_{\bar{\Delta}}\hat{\alpha}_k\gamma)) \end{aligned}$$

where $\|\bar{J}_{k+1}\|^2 \leq \chi$. If $\hat{\alpha}_k = \hat{\alpha}$ can be chosen for each k such that

$$\hat{\alpha} > \chi\left\{\|\check{\Pi}_0\| + \frac{1}{1-\beta}(\zeta + \rho_{\bar{\Delta}}\hat{\alpha}\gamma)\right\} \quad (55)$$

we see that $\|\bar{J}_{k+1}\check{\Pi}_{k+1}\bar{J}_{k+1}^T\|$ is bounded for all k. A fixed $\hat{\alpha}$ chosen according to condition (55) along with (54) guarantee that $\hat{\Pi}_k$ and $\check{\Pi}_k$ will remain bounded for all k.

Assumed uncertain model. Eqs. (14).

Initial conditions: $\hat{x}_0 = 0$, $\hat{\Pi}_0 = \check{\Pi}_0 = \text{diag}\{P_o, \epsilon I\}$.

Step 1. If $M_k = 0$, then set $\hat{\beta}_k = 0$. Otherwise, set instead $\hat{\beta} = (1+\alpha)\beta_{l,k}$ where $\beta_{l,k} = \|M_k^TR_k^{-1}M_k\|$. Choose $\hat{\alpha}_k$ such that $\hat{\alpha}_k - \bar{J}_k\check{\Pi}_k\bar{J}_k^T > 0$ for $\check{\Pi}_k$ satisfying

$$\check{\Pi}_{k+1} = \bar{F}_{k,c}\check{\Pi}_k\bar{F}_{k,c}^T + \bar{G}_kS_k\bar{G}_k^T + \rho_{\bar{\Delta}}\hat{\alpha}_k\bar{N}_k\bar{N}_k^T$$

Step 2. Using $\hat{\Pi}_k$ and $\hat{\alpha}_k$ compute, $\{Q_k, \hat{\Pi}_{k+1}\}$ by solving

$$\min_{Q_k>0} \quad \text{Trace}(\hat{\Pi}_{k+1})$$

subject to the inequality

$$\begin{pmatrix} -\hat{\Pi}_{k+1} + \rho_{\bar{\Delta}}\hat{\alpha}_k\bar{N}_k\bar{N}_k^T & \bar{F}_k\hat{\Pi}_k & \bar{G}_kS_k^{1/2} \\ \hat{\Pi}_k\bar{F}_k^T & -\hat{\Pi}_k & 0 \\ S_k^{T/2}\bar{G}_k^T & 0 & -I \end{pmatrix} \leq 0$$

where $\{\bar{F}_k, \bar{G}_k, S_k\}$ are defined by (32),(33) and (35).

Step 3. Update $\hat{x}_k$ to $\hat{x}_{k+1}$ as

$$\hat{x}_{k+1} = F_{p,k}\hat{x}_k + K_{p,k}y_k$$

where

$$\begin{aligned} F_{p,k} &= F_{k,c}(I - \hat{\beta}Q_kE_k^TE_k - Q_kH_k^T\hat{R}_k^{-1}H_k) \\ K_{p,k} &= F_{k,c}Q_kH_k^T\hat{R}_k^{-1} \\ \hat{R}_k^{-1} &= (R_k - \hat{\beta}^{-1}M_kM_k^T)^{-1} \end{aligned}$$

Table 2: A regularized robust filter for stochastic uncertainties.

B.2 Another Robust Filter

Consider again equations (47)–(51). We will now show how to generate a *new* sequence of matrices $\tilde{\Pi}_k$ and $\hat{\Pi}_k$ such that $\Pi_k \leq \tilde{\Pi}_k \leq \hat{\Pi}_k$. This construction will enable us to avoid the solution of the optimization problem (51)-(53) at each iteration thus reducing the computational complexity. At every iteration we would find a suboptimal Q_k that minimizes the bound $\tilde{\Pi}_k$ of the error covariance matrix Π_k. At time instant $k+1$, assuming we have $\Pi_k \leq \tilde{\Pi}_k \leq \hat{\Pi}_k$, then the state error covariance is bounded by the $(2,2)$ block element of the matrix $\tilde{\Pi}_{k+1}$ defined by

$$\tilde{\Pi}_{k+1} = F_{k,c}\hat{\Pi}_k \bar{F}_{k,c}^T + \bar{G}_k S_k \bar{G}_k^T + \rho_{\bar{\Delta}}\hat{\alpha}_k \bar{N}_k \bar{N}_k^T$$

The sequence of matrices $\hat{\Pi}_k$ are restricted to have a structure of the form:

$$\hat{\Pi}_k = \begin{pmatrix} Y_k & Y_k - Z_k \\ Y_k - Z_k & Y_k - Z_k \end{pmatrix} \tag{56}$$

and are generated as will be explained in the sequel. Moreover, with at time instant k, with $\hat{\Pi}_k$ given as in (56) we have that

$$\tilde{\Pi}_{k+1} = \begin{pmatrix} \tilde{Y}_{k+1} & \tilde{X}_{k+1} \\ \tilde{X}_{k+1}^T & \tilde{Y}_{k+1} - \tilde{Z}_{k+1} \end{pmatrix} \tag{57}$$

where

$$\begin{aligned}
\tilde{Y}_{k+1} &= F_{k,c}Y_kF_{k,c}^T + \rho_{\bar{\Delta}}\hat{\alpha}_kN_kN_k^T + G_kW_kG_k^T \\
\tilde{Z}_{k+1} &= F_{k,c}Z_kF_{k,c}^T - F_{k,c}Q_kH_k^TV_k^{-1}H_kQ_kF_{k,c}^T \\
&\quad +F_{k,c}(Y_k - Z_k)H_k^TV_k^{-1}H_kQ_kF_{k,c}^T \\
&\quad +F_{k,c}Q_kH_k^TV_k^{-1}H_k(Y_k - Z_k)F_{k,c}^T \\
&\quad -\beta^2F_{k,c}Q_kE_k^TE_kZ_kE_k^TE_kQ_kF_{k,c} \\
&\quad -F_{k,c}Q_kH_k^TV_k^{-1}H_k(Y_k - Z_k)H_k^TV_k^{-1}H_kQ_kF_{k,c}^T \\
\tilde{X}_{k+1} &= \beta F_{k,c}Y_kE_k^TE_kQ_kF_{k,c}^T + F_{k,c}(Y_k - Z_k)F_{p,k}^T \\
&\quad +\rho_{\bar{\Delta}}\hat{\alpha}_kN_kN_k^T + G_kW_kG_k^T
\end{aligned}$$

Also, the $(2,2)$ block element of $\tilde{\Pi}_{k+1}$ is given by

$$\tilde{\Pi}_{k+1}^{2,2} = C_{1,k} + C_{2,k} \tag{58}$$

where

$$\begin{aligned}
C_{1,k} &= F_{k,c}(Y_k - Z_k)F_{k,c}^T + \rho_{\bar{\Delta}}\hat{\alpha}_kN_kN_k^T \\
&\quad +G_kW_kG_k^T + F_{k,c}Q_kH_k^TV_k^{-1}H_kQ_kF_{k,c}^T \\
&\quad -F_{k,c}(Y_k - Z_k)H_k^TV_k^{-1}H_kQ_kF_{k,c}^T \\
&\quad -F_{k,c}Q_kH_k^TV_k^{-1}H_k(Y_k - Z_k)F_{k,c}^T \\
&\quad +F_{k,c}Q_kH_k^TV_k^{-1}H_k(Y_k - Z_k)H_k^TV_k^{-1}H_kQ_kF_{k,c}^T \\
C_{2,k} &= \beta^2F_{k,c}Q_kE_k^TE_kZ_kE_kE_k^TQ_kF_{k,c}
\end{aligned}$$

In deriving the above expressions, without loss of generality, we have chosen the weighing matrices R_k such that $\hat{R}_k = V_k$. It is usually hard to find a positive-definite Q_k that minimizes $\tilde{\Pi}_{k+1}^{2,2}$. Hence, we will find a suboptimal solution as follows. We shall bound $C_{2,k}$ by λI for some $\lambda > 0$. The choice of λI will become clear in the sequel. Then we will find a lower bound for $C_{1,k} + \lambda I$, which occurs at the lower bound of $C_{1,k}$. After some considerable algebra, we can show that for

$$Q_{k,opt} = (Y_k - Z_k) - (Y_k - Z_k)H_k^T\bar{R}_{e,k}^{-1}H_k(Y_k - Z_k)$$

and $\bar{R}_{e,k} = V_k + H_k(Y_k - Z_k)H_k^T$, we have

$$\frac{\partial C_{1,k}}{Q_k} = 0 \quad \text{and} \quad \frac{\partial^2 C_{1,k}}{Q_k} > 0 \tag{59}$$

That is, $Q_{k,opt}$ minimizes $C_{1,k}$. It can be seen that $Q_{k,opt}$ is positive definite and hence guarantees a unique solution to the problem (21). Now note that $\tilde{\Pi}_{k+1}^{2,2}$ is quadratic in the variable Q_k and hence its value, for any arbitrary Q_k, is proportional to $\|Q_k - Q_{k,opt}\|_F$, where $\|.\|_F$ denotes the Frobenius norm of the argument. Also note that the set of all Q_k that guarantee $C_{2,k} < \lambda I$ is a convex bounded set $\mathcal{C}$ about the origin in the normed vector space of all matrices of dimension $n \times n$. If $Q_{k,opt}$ lies inside $\mathcal{C}$, we choose Q_k as $Q_{k,opt}$. Otherwise, Q_k is chosen as a matrix that is closest to $Q_{k,opt}$ in the Frobenius norm and simultaneously lying inside the set $\mathcal{C}$. Now we will derive an upper bound for $\tilde{\Pi}_{k+1}$ in the form (which is compatible with the form we started with in (56)):

$$\hat{\Pi}_{k+1} = \begin{pmatrix} Y_{k+1} & Y_{k+1} - Z_{k+1} \\ Y_{k+1} - Z_{k+1} & Y_{k+1} - \hat{Z}_{k+1} \end{pmatrix} \tag{60}$$

for some matrices Y_{k+1} and Z_{k+1}. Choose ψ_k as the maximum singular value of $I + B$ where

$$\begin{aligned}
B &= F_{k,c}Q_kH_k^TV_k^{-1}H_k(Y_k - Z_k)H_k^TV_k^{-1}H_kQ_kF_{k,c}^T \\
&\quad +\beta^2F_{k,c}Q_kE_k^TE_kZ_kE_k^TE_kQ_kF_{k,c} \\
&\quad +F_{k,c}Q_kH_k^TV_k^{-1}H_kQ_kF_{k,c}^T \\
&\quad -F_{k,c}Q_kH_k^TV_k^{-1}H_k(Y_k - Z_k)F_{k,c}^T \\
&\quad -\beta F_{k,c}Z_kQ_kE_k^TE_kF_{k,c}^T
\end{aligned}$$

<u>Assumed uncertain model</u>. Eqs. (10)–(13) and (14)–(15).
<u>Initial conditions</u>: $\hat{x}_0 = 0, Y_0 = I, Z_0 = \mu I$,

$$\Pi_0 = \begin{pmatrix} Y_0 & Y_0 - Z_0 \\ Y_0 - Z_0 & Y_0 - Z_0 \end{pmatrix}$$

<u>Step 1a</u>. Using $\{V_k, H_k, Y_k, Z_k\}$ compute $\{R_{e,k}, Q_k\}$:

$$\begin{aligned}
R_{e,k} &= V_k + H_k(Y_k - Z_k)H_k^T \\
Q_{k,opt} &= (Y_k - Z_k) - (Y_k - Z_k)H_k^T\bar{R}_{e,k}^{-1}H_k(Y_k - Z_k)
\end{aligned}$$

<u>Step 1b</u>. If $M_k = 0$, then set $\widehat{\beta}_k = 0$. Otherwise, set $\widehat{\beta}_k = (1+\alpha)\beta_{l,k}, \alpha > 0$. Determine the largest $Q_k = \xi Q_{k,opt}$ for some positive ξ such that $C_{2,k} < \lambda I$ for some small $\lambda > 0$.
<u>Step 2</u>. Compute the parameters:

$$\begin{aligned}
F_{p,k} &= F_{k,c}(I - Q_k\beta E_k^TE_k - Q_kH_k^T\hat{R}_k^{-1}H_k) \\
K_{p,k} &= F_{k,c}Q_kH_k^T\hat{R}_k^{-1} \\
B &= F_{k,c}Q_kH_k^TV_k^{-1}H_k(Y_k - Z_k)H_k^TV_k^{-1}H_kQ_kF_{k,c}^T \\
&\quad +\beta^2F_{k,c}Q_kE_k^TE_kZ_kE_k^TE_kQ_kF_{k,c} \\
&\quad +F_{k,c}Q_kH_k^TV_k^{-1}H_kQ_kF_{k,c}^T \\
&\quad -F_{k,c}Q_kH_k^TV_k^{-1}H_k(Y_k - Z_k)F_{k,c}^T \\
&\quad -\beta F_{k,c}Z_kQ_kE_k^TE_kF_{k,c}^T \\
\psi_k &= \|I + B\| \\
\hat{\alpha}_k &= \|\bar{J}_k\check{\Pi}_k\bar{J}_k^T\|
\end{aligned}$$

<u>Step 3</u>. Now update $\{Y_k, Z_k, \Pi_k, \hat{x}_k\}$ to $\{Y_{k+1}, Z_{k+1}, \Pi_{k+1}, \hat{x}_{k+1}\}$ as follows:

$$\begin{aligned}
Y_{k+1} &= F_{k,c}Y_kF_{k,c}^T + \rho_{\bar{\Delta}}\hat{\alpha}_kN_kN_k^T + G_kW_kG_k^T + \psi_k^2I \\
Z_{k+1} &= F_{k,c}Z_kF_{k,c}^T - F_{k,c}Q_kH_k^TV_k^{-1}H_kQ_kF_{k,c}^T \\
&\quad +F_{k,c}(Y_k - Z_k)H_k^TV_k^{-1}H_kQ_kF_{k,c}^T \\
&\quad +F_{k,c}Q_kH_k^TV_k^{-1}H_k(Y_k - Z_k)F_{k,c}^T \\
&\quad -F_{k,c}Q_kH_k^TV_k^{-1}H_k(Y_k - Z_k)H_k^TV_k^{-1}H_kQ_kF_{k,c}^T \\
&\quad +\psi_k^2I - I \\
\check{\Pi}_{k+1} &= \bar{F}_{k,c}\check{\Pi}_k\bar{F}_{k,c}^T + \bar{G}_kS_k\bar{G}_k^T + \rho_{\bar{\Delta}}\hat{\alpha}_k\bar{N}_k\bar{N}_k^T \\
\hat{x}_{k+1} &= F_{p,k}\hat{x}_k + K_{p,k}y_k
\end{aligned}$$

Table 3: A second regularized robust filter for stochastic uncertainties.

TABLE IV

ERROR VARIANCE WITH POLYTOPIC UNCERTAINTIES IN F_k.

Filters	error variance
Proposed filter	22.8dB
Regularized robust filter of [11]	26.9dB
Guaranteed-cost filter [2]	30dB
Set-valued filter [3]	34.47dB
Kalman filter with nominal model	31.18dB

TABLE V

ERROR VARIANCE WITH STOCHASTIC UNCERTAINTIES IN F_k.

Filters	error variance
Proposed filter from table 2	21.85dB
Proposed filter from table 3	21.20dB
Regularized robust filter of [11]	22.68dB
Guaranteed-cost filter [2]	25.3dB
Set-valued filter [3]	25.9dB
Kalman filter with nominal model	39.5dB

Now with

$$Y_{k+1} = \psi_k^2 I + \tilde{Y}_{k+1} \quad (61)$$

$$Z_{k+1} = \tilde{Z}_{k+1} + \psi_k^2 I - I \quad (62)$$

$\hat{\Pi}_{k+1}$ is an upper bound of $\tilde{\Pi}_{k+1}$. This is because

$$\hat{\Pi}_{k+1} - \tilde{\Pi}_{k+1} = \begin{pmatrix} \psi_k^2 I & I+B \\ (I+B)^T & I \end{pmatrix} > 0 \quad (63)$$

We will now in fact choose $\hat{\alpha}_k$ such that $\hat{\alpha}_k - \bar{J}_k T_k \bar{J}_k^T > 0$ for T_k satisfying

$$T_k = \bar{F}_{k-1,c} T_{k-1} \bar{F}_{k-1,c}^T + \bar{G}_{k-1} S_{k-1} \bar{G}_{k-1}^T + \rho_\Delta \hat{\alpha}_{k-1} \bar{N}_{k-1} \bar{N}_{k-1}^T \quad (64)$$

Note that $\hat{\alpha}_k$ chosen as explained above implies that $\hat{\alpha}_k$ satisfies (50). We now state the filter :

$$\hat{x}_{k+1|k} = F_{p,k} \hat{x}_{k|k-1} + K_{p,k} y_k \quad (65)$$

where $F_{p,k}$ and $K_{p,k}$ are defined in terms of Q_k as

$$F_{p,k} = F_{k,c}(I - \hat{\beta} Q_k E_k^T E_k - Q_k H_k^T \hat{R}_k^{-1} H_k) \quad (66)$$

$$K_{p,k} = F_{k,c} Q_k H_k^T \hat{R}_k^{-1} \quad (67)$$

and Q_k is determined in terms of Y_k and Z_k at every k as explained before. The designed filter is shown in Table 3.

V. SIMULATIONS

To illustrate the operation of the filter developed for deterministic uncertainties, we choose an implementation of order 2 with $E_k = [.12 \quad .12]$, and $M_k = 1$ for all k. The uncertain state matrices F_k are assumed to lie inside the convex polytope

$$F_k = \begin{pmatrix} .9802 & .0196 + \delta \\ 0 & .5802 + \delta \end{pmatrix} \quad (68)$$

with $|\delta| \leq 0.4982$ The uncertainties in the output matrices H_k are determined by $M_k = 1$, $E_k = [.4 \quad .4]$ and $G_k = [-6 \quad 1]$. Table 2 shows the average squared state-error values (averaged over 50 experiments) for the Kalman filter, the proposed filter, the set-valued estimation filter [3], the guaranteed cost filter [2] and the filter of [11]. To illustrate the filter developed for stochastic uncertainties, we choose an implementation of order 2 with $E_k = [.8 \quad .8]$, $M_k = 1$ for all k. The uncertain state matrices F_k are assumed to be

$$F_k = \begin{pmatrix} .9802 + \bar{\Delta} & .4196 + \bar{\Delta} \\ \bar{\Delta} & .8802 + \bar{\Delta} \end{pmatrix} \quad (69)$$

with $|\bar{\Delta}| \leq 0.4982$. Table 3 shows the average squared state-error values in this case.

VI. CONCLUSION

In this paper we developed two regularized robust filters for state-space estimation. The design procedure is through the solution of a regularized weighted recursive least squares problem and it enforces minimum state error variance. The proposed filters outperform earlier robust designs and are suitable for on-line/real-time filtering applications since they do not require existence conditions.

REFERENCES

[1] T. Kailath, A. H. Sayed, and B. Hassibi. *Linear Estimation*. Prentice-Hall, NJ, 2000.

[2] I. R. Petersen and A. V. Savkin. *Robust Kalman Filtering for Signals and Systems with Large Uncertainties*. Birkhauser, Boston, 1999.

[3] A. V. Savkin and I. R. Petersen. Robust state estimation and model validation for discrete time uncertain systems with a deterministic description of noise and uncertainties. *Automatica*, vol. 34, no. 2, pp 271–274, 1998.

[4] L. Xie, Y. C. Soh, and C. E. de Souza. Robust Kalman filtering for uncertain discrete-time systems. *IEEE Trans. Automat. Contr.*, vol. 39, no. 6, pp. 1310–1314, 1994.

[5] P. Bolzern, P. Colaneri, and G. De Nicolao. Optimal design of robust predictors for linear discrete-time systems. *Systems & Control Letters*, vol. 26, pp. 25–31, 1995.

[6] M. Fu, C. E. de Souza and Z. Luo. Finite horizon robust kalman filter design. *Proc. IEEE Conf. Decision Control*, pp 4555–4560, Phoenix, AZ, 1999.

[7] Y. Theodor and U. Shaked. Robust discrete time minimum variance filtering. *IEEE Trans. Signal Processing*, vol. 44, pp 181–189, Feb. 1996.

[8] F. Yang, Z. Wang and Y.S. Hung. Robust Kalman filtering for discrete time-varying uncertain systems with multiplicative noises. *IEEE Trans. Automat. Contr.*, vol. 47, no. 7, pp. 1179-1184, 2002.

[9] D. P. Bertsekas and I. B. Rhodes. Recursive state estimation for a set-membership description of uncertainty. *IEEE Trans. Automat. Contr.*, vol. 16, no. 2, pp. 117–128, 1971.

[10] P. P. Khargonekar and K. M. Nagpal. Filtering and smoothing in an $\mathcal{H}_\infty$ setting. *IEEE Transactions on Automatic Control*, vol 36, pp. 151-166, 1991.

[11] A. H. Sayed. A framework for state space estimation with uncertain models. *IEEE Trans. Automat. Contr.*, vol. 46, no. 7, pp. 998–1013, July 2001.

[12] A. H. Sayed, V. H. Nascimento, and F. A. M. Cipparrone. A regularized robust design criterion for uncertain data. *SIAM J. Matrix Anal. Appl.*, vol. 23, no. 4 pp 1120–1142, 2002.

[13] A. Garulli, A. Vicino, and G. Zappa. Conditional central algorithms for worst case set-membership identification and filtering. *IEEE Trans. Automat. Contr.*, vol. 45, no. 1, pp. 14–23, Jan. 2000.

[14] J.C. Geromel. Optimal linear filtering under parameter uncertainty. *IEEE Trans. Signal Processing*, vol. 47, no. 1, Jan. 1999.

[15] A. H. Sayed, V. H. Nascimento, and S. Chandrasekaran. Estimation and control with bounded data uncertainties. *Linear Algebra and Its Applications*, vol. 284, pp. 259–306, Nov. 1998.

[16] A. H. Sayed and H. Chen. A uniqueness result concerning a robust regularized least-squares solution. *Systems and Control Letters*, vol. 46, no. 5, pp 361–369, Aug. 2002.

[17] S. Boyd, L. El Ghaoui, E. Feron and V. Balakrishnan. *Linear Matrix Inequalities in System and Control Theory*. SIAM Studies in Applied Mathematics, 1994.

[18] A. E. Bryson and Y.-C. Ho. *Applied Optimal Control: Optimization, Estimation, and Control*. Taylor & Francis, revised printing, 1975.

IFAC Publications
www.elsevier.com/locate/ifac

MINIMAX L_2-E_2 FIR FILTERS FOR DETERMINISTIC CONTINUOUS-TIME STATE SPACE SIGNAL MODELS

Soo Hee Han * Wook Hyun Kwon *,[1]

* *Engr. Research Center for Advanced Contr. and Instru.,and School of Electrical Engr. & Computer Science, Seoul Nat'l Univ., Seoul, 151-742, Korea*

Abstract: In this paper, a new minimax L_2-E_2 performance criterion is introduced, which is represented as a worst case gain between the disturbance during the recent time interval and the current estimation error. Minimax L_2-E_2 FIR filter (MLEFF) is proposed for deterministic continuous-time state space signal models. The MLEFF is designed to minimize the maximum value of the L_2-E_2 performance criterion together with linearity, unbiased property in the deterministic sense, FIR structure, and independence of the initial state information, simultaneously. The MLEFF is shown to be robust with respect to model uncertainties due to FIR structure and disturbances due to the worst case design. For efficient implementation, it is shown that the MLEFF can be expressed recursively. It is also shown to be optimal for the performance criterion for each component of the state. *Copyright © 2003 IFAC*

Keywords: Minimax L_2-E_2 performance criterion, Minimax L_2-E_2 FIR filter (MLEFF), FIR structure.

1. INTRODUCTION

Unbiasedness from the real value and insensitivity or robustness to signal model uncertainties and disturbances have been considered to be desirable properties.

For a filer to be unbiased using the stochastic information, we mean that, no matter what the true value, the filter will yield it on the average. Similarly to the stochastic case, filters based on deterministic systems can adopt the unbiased property in the deterministic sense. In this paper, disturbances of deterministic systems are assumed not to be biased from zero. This means that the systems with disturbances behave around nominal systems without disturbances. Thus, it is desirable that filters are required to be exact to the states of nominal systems without disturbances. This will be called an unbiased property in the deterministic sense. For short, "the unbiased property" will be used through this paper.

Among filters with the unbiased property, one can be chosen according to the performance criterion. For insensitivity to disturbances, the worst case estimation error due to some worst case elements of possible disturbances can be considered as a performance criterion. For the deterministic systems that are dealt in this paper, H_∞ filters have been used for the worst case (Li and Fu, 1997) (Nagpal and Khargonekar, 1991) (Fu and de Souza, 1992). The H_∞ filter is designed such that the H_∞ norm, which reflects the worst case estimation error, is minimized. The H_∞ filter as the optimal worst case filter may be insensitive

[1] To whom all correspondence should be addressed: phone: +82-2-880-7307, fax: +82-2-871-7010, e-mail: whkwon@cisl.snu.ac.kr
This research work was supported by Brain Korea 21

to exterior disturbance. However, the estimation error may be unnecessarily large for large class of systems including the nominal system since it is designed only for the worst case disturbances. Additionally, an optimal solution of H_∞ filters is difficult to implement so that suboptimal solutions are often obtained. We may need a new measure of the worst case estimation error differently from the H_∞ performance criterion, if necessary.

Filters can be of the infinite impulse response (IIR) type or the finite impulse response (FIR) type. FIR filters make use of finite measurements and inputs on the most recent time interval $[t-T,t]$, called the receding horizon, or the window. It has been generally accepted that the FIR structure in filters is more robust to temporary modeling uncertaint parameters and numerical errors than the IIR structure.

For most existing state estimation filters, initial state information is often assumed to be known even if the initial state is also a state to be estimated. This is not reasonable. Therefore, in this paper the initial state information is assumed to be completely unknown. The suggested filters will be obtained independently of the initial state information.

In this paper, the linear FIR filter which is independent of the initial state information can be represented by

$$\begin{aligned}\hat{x}(t) = & \int_{t-T}^{t} H(t-\sigma)y(\sigma)d\sigma \\ & + \int_{t-T}^{t} L(t-\sigma)u(\sigma)d\sigma \end{aligned} \quad (1)$$

at the present time t for some gains $H(\cdot)$ and $L(\cdot)$.

The suggested filter (1) will not have a state term and the filter gains $H(\cdot)$ and $L(\cdot)$ will be independent of the initial state information. It is noted that standard Kalman filters have an initial state term and the gain $H(\cdot)$ also depends on the initial state information.

In this paper, among linear FIR filters with the unbiased property, the optimal filter will be chosen according to a following new performance criterion for exterior disturbances:

$$\min_{H(\cdot)L(\cdot)} \max_{w(\cdot)\neq 0} \frac{[x(t)-\hat{x}(t)]^T[x(t)-\hat{x}(t)]}{\int_{t-T}^{t} w^T(\tau)w(\tau)d\tau}. \quad (2)$$

The performance criterion (2) and the corresponding optimal filter will be called the minimax L_2-E_2 performance criterion and the minimax L_2-E_2 FIR filter (MLEFF), respectively. In (Han *et al.*, 2002), the deterministic discrete system is considered based on a similar performance criterion to (2), which only requires algebraic manipulation. However, there is no result for the corresponding problem in case of continuous-time systems since an entailed functional optimization problem is more difficult to solve. This paper proposes a new minimax performance criterion (2) assuming an unknown initial state and the requirement of the unbiased property for continuous-time systems.

It is noted that the criterion (2) differs from the existing criterion for H_∞ problems for IIR filters as

$$\inf_{H(\cdot)L(\cdot)} \sup_{w(\cdot)\neq 0} \frac{\int_{t_0}^{t}[x(\tau)-\hat{x}(\tau)]^T[x(\tau)-\hat{x}(\tau)]d\tau}{\int_{t_0}^{t} w^T(\tau)w(\tau)d\tau} \quad (3)$$

where t_0 is the initial time. Note that the numerator of (2) considers only the current estimation error compared with (3) that uses the distributed terms in the numerator. Since the current estimation error is an issue, it is reasonable to take only the current estimation error in the numerator of the cost function instead of the accumulated estimation error including the previous estimation errors. As pointed out, H_∞ problems are difficult to solve and thus, solved only by considering an upper bound on the H_∞ norm, which yields solutions of a differential game. However, the proposed MLEFF will be shown to provide an optimal solution explicitly, even with the unbiased property.

The MLEFF is both unbiased and optimal *by design* for the proposed cost. The '*by design*' means that the unbiased property and optimality are built into the MLEFF during its design. The proposed MLEFF will be represented in both a standard batch form and an iterative form. It will be shown that the MLEFF for deterministic systems is similar in form to the existing receding horizon unbiased FIR filter (RHUFF) for stochastic systems (Han *et al.*, 2001)(Kwon *et al.*, 2002).

This paper is organized as follows. In Section 2, the MLEFF for continuous-time state space models is proposed in a standard batch form. In Section 3, iterative forms will be obtained and it is shown that the MLEFF is also optimal for the performance criterion considering one component of the state. Finally, conclusions are presented in Section 4.

2. MINIMAX L_2-E_2 FIR FILTERS

Consider a linear continuous-time state space model with control input:

$$\dot{x}(t) = Ax(t) + Bu(t) + Gw(t), \quad (4)$$
$$y(t) = Cx(t) + Dw(t) \quad (5)$$

where $x(t) \in \Re^n$, $u(t) \in \Re^l$, $y(t) \in \Re^q$, and $w(t) \in \Re^p$ are the state, the input, the measurement, and the disturbance, respectively. $GD^T = O$ and $DD^T = I$ are satisfied to decouple the system disturbance and the measurement disturbance.

The systems (4) and (5) will be represented on the most recent time interval $[t-T, t]$, called the horizon. The current state $x(t)$ is given by a solution of (4) as follows:

$$x(t) = e^{A(t-\sigma)}x(\sigma) + \int_\sigma^t e^{A(t-\tau)}Bu(\tau)d\tau$$
$$+ \int_\sigma^t e^{A(t-\tau)}Gw(\tau)d\tau, \quad t-T \le \sigma \le t. \quad (6)$$

Then, $x(\sigma)$ in (6) is written as

$$x(\sigma) = e^{A(\sigma-t)}x(t) - \int_\sigma^t e^{A(\sigma-\tau)}Bu(\tau)d\tau$$
$$- \int_\sigma^t e^{A(\sigma-\tau)}Gw(\tau)d\tau.$$

Therefore, on the horizon $[t-T, t]$, the finite measurements and inputs can be expressed in terms of the state $x(t)$ at the current time t as follows:

$$y(\sigma) = Cx(\sigma) + Dw(\sigma)$$
$$= C\left[e^{A(\sigma-t)}x(t) - \int_\sigma^t e^{A(\sigma-\tau)}Bu(\tau)d\tau\right.$$
$$\left.- \int_\sigma^t e^{A(\sigma-\tau)}Gw(\tau)d\tau\right] + Dw(\sigma). \quad (7)$$

Now, we require a constraint that $\hat{x}(t)$ in (1) becomes an unbiased filter in the deterministic sense as

$$\hat{x}(t) = x(t) \quad \text{for all states.} \quad (8)$$

Using (8), the left side of (8) are given as

$$\hat{x}(t) = \int_{t-T}^t H(t-\sigma)Ce^{A(\sigma-t)}d\sigma x(t)$$
$$- \int_{t-T}^t \int_\sigma^t H(t-\sigma)Ce^{A(\sigma-\mu)}Bu(\mu)d\mu d\sigma$$
$$+ \int_{t-T}^t L(t-\sigma)u(\sigma)d\sigma.$$

Since the constraint (8) should be satisfied for all states and all controls, the following relations are obtained:

$$L(t-\sigma) = \int_{t-T}^\sigma H(t-\mu)Ce^{A(\mu-\sigma)}Bd\mu, \quad (9)$$
$$I = \int_{t-T}^t H(t-\sigma)Ce^{A(\sigma-t)}d\sigma \quad (10)$$

which are called unbiased constraints. Using the unbiased constraints, the MLEFF for the current state $x(t)$ can be expressed as a linear functional of the finite measurements and inputs on the horizon $[t-T, t]$ as follows:

$$\hat{x}(t) = \int_{t-T}^t H(t-\sigma)\left[y(\sigma) + C\int_\sigma^t e^{A(\sigma-\tau)}Bu(\tau)d\tau\right]d\sigma.$$

Note that constraints (9) and (10) must hold regardless of the information on the horizon initial state $x(t-T)$ on the horizon $[t-T, t]$. This constraint may be too strict but, surprisingly, we were able to obtain the solution.

The objective now is to obtain the best gain matrix $H^*(t-\sigma)$, subject to the unbiased constraint (10), based on the following criterion:

$$H^*(t-\sigma)$$
$$= \arg \min_{H(t-\sigma)} \max_{w(\cdot)\neq 0} \left\{\frac{[x(t)-\hat{x}(t|t)]^T[x(t)-\hat{x}(t|t)]}{\int_{t-T}^t w^T(\tau)Fw(\tau)d\tau}\right\}. \quad (11)$$

To solve the above state estimation problem with the unbiased constraint, an optimization problem with constraints will be introduced.

Replacing $y(\sigma)$ with the right hand side of (7), we obtain the estimate:

$$\hat{x}(t|t)$$
$$= \int_{t-T}^t H(t-\sigma)\left[y(\sigma) + C\int_\sigma^t e^{A(\sigma-\tau)}Bu(\tau)d\tau\right]d\sigma$$
$$= \int_{t-T}^t H(t-\sigma)\left[Cx(\sigma) + Dw(\sigma) + C\int_\sigma^t e^{A(\sigma-\tau)}Bu(\tau)d\tau\right]d\sigma$$
$$= \int_{t-T}^t H(t-\sigma)\left[C\{e^{A(\sigma-t)}x(t) - \int_\sigma^t e^{A(\sigma-\tau)}Bu(\tau)d\tau\right.$$
$$\left.- \int_\sigma^t e^{A(\sigma-\tau)}Gw(\tau)d\tau\} + Dw(\sigma) + C\int_\sigma^t e^{A(\sigma-\tau)}Bu(\tau)d\tau\right]d\sigma. \quad (12)$$

Using the unbiased constraints (9)- (10) and rearranging the terms, the error between the real current state and the estimate can be expressed as

$$e(t) \triangleq x(t) - \hat{x}(t)$$
$$= \int_{t-T}^t H(t-\sigma)\{C\int_\sigma^t e^{A(\sigma-\tau)}Gw(\tau)d\tau - Dw(\sigma)\}d\sigma$$
$$= \int_{t-T}^t \{\int_{t-T}^\sigma H(t-\tau)Ce^{A(\tau-\sigma)}Gd\tau - H(t-\sigma)D\}w(\sigma)d\sigma$$
$$= \int_{t-T}^t \{F(\sigma)G - H(t-\sigma)D\}w(\sigma)d\sigma \quad (13)$$

where

$$F(\sigma) = \int_{t-T}^\sigma H(t-\tau)Ce^{A(\tau-\sigma)}d\tau. \quad (14)$$

The integral form (14) can be changed to the differential form with boundary conditions as

$$\dot{F}(\sigma) = -F(\sigma)A + H(t-\sigma)C \qquad (15)$$

with $F(t-T) = 0$ and $F(t) = I$.

Before proceeding to optimization of the performance criterion (2), define the following matrix variable:

$$\zeta(\sigma) = \left[G^T F^T(\sigma) - D^T H^T(t-\sigma)\right]\Xi_T^{-1} \left[\int_{t-T}^{t} \{F(\sigma)G - H(t-\sigma)D\}w(\sigma)d\sigma\right]$$

where

$$\Xi_T = \left[\int_{t-T}^{t} \left[F(\sigma)GG^T F^T(\sigma) + H(t-\sigma)H^T(t-\sigma)\right]d\sigma\right].$$

It can easily be seen that the variable $\zeta(\sigma)$ always satisfies the following equation:

$$\int_{t-T}^{t} \left[F(\sigma)G - H(t-\sigma)D\right]\left[w(\sigma) - \zeta(\sigma)\right]d\sigma = 0. \qquad (16)$$

Using $\int_{t-T}^{t} \zeta^T(\sigma)(w(\sigma) - \zeta(\sigma))d\sigma = 0$ derived easily from (16), the inequality for the L_2-E_2 performance criterion can be obtained as

$$\begin{aligned} &\frac{\left[x(t) - \hat{x}(t)\right]^T \left[x(t) - \hat{x}(t)\right]}{\int_{t-T}^{t} w^T(\sigma)w(\sigma)d\sigma} \\ &= \frac{\Phi^T\Phi}{\int_{t-T}^{t} \zeta^T(\sigma)\zeta(\sigma)d\sigma + \int_{t-T}^{t} \pi(\sigma)^T\pi(\sigma)d\sigma} \\ &\le \frac{\Phi^T\Phi}{\int_{t-T}^{t} \zeta^T(\sigma)\zeta(\sigma)d\sigma} \\ &= \frac{\Phi^T\Phi}{\Phi^T\Xi_T^{-1}\Phi} \le \lambda_{max}(\Xi_T) \end{aligned}$$

where an equality is satisfied for $w(\sigma) = \{F(\sigma)G - H(t-\sigma)D\}\alpha$ for some constant vector α and

$$\Phi = \int_{t-T}^{t} \{F(\sigma)G - H(t-\sigma)D\}w(\sigma)d\sigma$$
$$\pi(\sigma) = w(\sigma) - \zeta(\sigma).$$

All we have to do is to minimize the maximum eigenvalue of Ξ_T. Maximum eigenvalue can be characterized as

$$\lambda_{max}(\Xi_T) = \sup_{\beta} \; \{ \; \beta^T\Xi_T\beta \mid \beta^T\beta = 1 \; \}. \qquad (17)$$

For any constant vector β, the following equation should be minimized:

$$\int_{t-T}^{t} \beta^T \left[F(\sigma)GG^T F^T(\sigma) + H(t-\sigma)H^T(t-\sigma)\right]\beta d\sigma$$

subject to the constraint (15) with $F(T) = 0$ and $F(0) = I$.Using the basic results for optimality and the definition as

$$\begin{aligned} T(\sigma) &\triangleq F_\beta^T(\sigma)GG^T F_\beta(\sigma) + H_\beta^T(t-\sigma)H_\beta(t-\sigma) \\ &+ \lambda(\sigma)^T(\dot{F}_\beta(\sigma) - C^T H_\beta(t-\sigma) + A^T F_\beta(\sigma)) \end{aligned}$$

with $F_\beta(t-\sigma) = F^T(t-\sigma)\beta$ and $H_\beta(t-\sigma) = H^T(t-\sigma)\beta$, we need to calculate the following value:

$$\frac{T}{\partial F_\beta} = 2GG^T F_\beta(\sigma) + A\lambda(\sigma) \qquad (18)$$
$$\frac{T}{\partial \dot{F}_\beta} = \lambda(\sigma) \qquad (19)$$
$$\frac{T}{\partial H_\beta} = 2H_\beta(t-\sigma) - C\lambda(\sigma). \qquad (20)$$

We obtain a Hamiltonian matrix form

$$\begin{aligned} \frac{d}{d\sigma}\begin{bmatrix} F_\beta(\sigma) \\ \lambda(\sigma) \end{bmatrix} &= \begin{bmatrix} -A^T & \frac{1}{2}C^T C \\ 2GG^T & A \end{bmatrix} \begin{bmatrix} F_\beta(t-T) \\ \lambda(t-T) \end{bmatrix} \\ &\triangleq \mathbf{H} \begin{bmatrix} F_\beta(t-T) \\ \lambda(t-T) \end{bmatrix}. \end{aligned}$$

From (20), $H_\beta(t-\sigma)$ is of the form

$$\begin{aligned} H_\beta(t-\sigma) &= \frac{1}{2}C\lambda(\sigma) \\ &= \frac{1}{2}C\begin{bmatrix} 0 & I \end{bmatrix} e^{H(\sigma-t+T)} \begin{bmatrix} F_\beta(t-T) \\ \lambda(t-T) \end{bmatrix}. \end{aligned}$$

Using $F_\beta(t-T) = 0$ and $e^{\mathbf{H}(\sigma-t+T)}$ defined by

$$e^{\mathbf{H}(\sigma-t+T)} \triangleq \begin{bmatrix} X(\sigma-t+T) & Y(\sigma-t+T) \\ Z(\sigma-t+T) & W(\sigma-t+T) \end{bmatrix}, \qquad (21)$$

$H_\beta(t-\sigma)$ can be expressed as

$$H_\beta(t-\sigma) = \frac{1}{2}CW(\sigma-t+T)\lambda(t-T).$$

Using the unbiased condition

$$\int_{t-T}^{t} \frac{1}{2}\lambda^T(t-T)W^T(\sigma-t+T)C^T Ce^{A(\sigma-t)}d\sigma = \beta,$$

we have

$$\lambda^T(t-T) = \beta^T\left[\int_{T}^{t} \frac{1}{2}W^T(\sigma-t+T)C^T Ce^{A(\sigma-t)}d\sigma\right]^{-1}$$

where it can be shown that the inverse exists (Han *et al.*, 2001).

Theorem 1. The MLEFF for the observable system (4) and (5) can be expressed as

$$\hat{x}(t|t) = \int_{t-T}^{t} H(t-\sigma)y(\sigma)d\sigma + \int_{t-T}^{t} L(t-\sigma)u(\sigma)d\sigma \quad (22)$$

where $H(t-\sigma)$ and $L(t-\sigma)$ are as follows:

$$H(t-\sigma) = \left[\int_{t-T}^{t} W^T(\tau - t + T)C^T C e^{A(\tau - t)} d\tau\right]^{-1} W^T(\sigma - t + T)C^T \quad (23)$$

and

$$L(t-\sigma) = \int_{t-T}^{\sigma} H(t-\tau)Ce^{A(\tau-\sigma)}Bd\tau \quad (24)$$

where $W(\tau - t + T)$ is given by (21). ∎

It is surprising that there exists a closed form solution (23) and (24) even under the strong condition (10) and that the gain $H(\cdot)$ is independent of the initial state information.

Remark 2. If G is replaced by γG where γ reflects the degree of disturbances, the value of γ can be used as a design parameter. The smaller the disturbances are , the smaller the value of γ should be. For special case $\gamma = 0$, the filter gain of the MLEFF is reduced to the following form:

$$H(t-\sigma) = \left[\int_{t-T}^{t} e^{A^T(\tau - t + T)} C^T C e^{A(\tau - t)} d\tau\right]^{-1} e^{A^T(\sigma - t + T)} C^T. \quad (25)$$

Remark (2) shows that the MLEFF is simplified under zero system disturbance and the inverse is guaranteed for the observability of $(A.C)$.

In the next section, an iterative form for a batch form (23) and (24) is developed.

3. ITERATIVE FORMS AND PROPERTIES

In the previous section, the MLEFF is represented as the batch form. For efficient implementation, it is shown that the MLEFF can be expressed recursively.

Consider $S(\sigma - t + T)$ and $\eta(\sigma - t + T)$ defined by

$$S(\sigma - t + T) \triangleq W^{-T}(\sigma - t + T) \left[\int_{t-T}^{\sigma} W^T(\tau - t + T)C^T C e^{A(\tau - \sigma)} d\tau\right]$$

$$\hat{\eta}(\sigma|t) \triangleq \int_{t-T}^{\sigma} W^{-T}(\sigma - t + T)W^T(s - t + T)C^T \left[y(s) + C\int_{s}^{\sigma} e^{A(s-\tau)}Bu(\tau)d\tau\right] ds$$

for $t - T \le \sigma \le t$. From (Han *et al.*, 2001), the recursive form of $(S(\sigma - t + T)$ and $\eta(\sigma - t + T)$ can be easily obtained so that the proof will be left out. The following theorem shows the recursive form of the MLEFF.

Theorem 3. Assume that $\{A, C\}$ is observable and $T > 0$. The MLEFF $\hat{x}(t|t)$ for continuous-time state space models (4), (5) is given on the horizon $[t - T, t]$ as follows:

$$\hat{x}(t|t) = S^{-1}(T)\hat{\eta}(t|t) \quad (26)$$

where $\hat{\eta}(t|t)$ and $S(T)$ are obtained as follows:

$$\frac{dS(\tau)}{d\tau} = -S(\tau)A - A^T S(\tau) - S(\tau)GG^T S(\tau) + C^T C$$

$$\frac{\partial \hat{\eta}(\tau + t - T|t)}{\partial \tau} = -[A^T + S(\tau)GG^T]\hat{\eta}(\tau + t - T|t) + C^T y(\tau + t - T) + S(\tau)Bu(\tau + t - T)$$

where $0 \le \tau \le T$, $S(0) = 0$, and $\hat{\eta}(t - T|t) = 0$. ∎

Remark 4. In case of the H_∞ filter, the filter exists only if initial state information is available. In other words, filters without respect to initial state information cannot exist for H_∞ performance criterion.

Now, a new performance criterion is introduced and checked for the proposed MLEFF. The performance criterion (2) reflects the worst case between all components of the state and the disturbance. Additionally, it is meaningful to consider the worst case for only one components of the state as

$$\min_{H(\cdot)L(\cdot)} \max_{w(\cdot)\neq 0} \frac{\{x_i(t) - \hat{x}_i(t|t)\}^2}{\int_{t-T}^{t} w^T(\tau)w(\tau)d\tau}. \quad (27)$$

In solving the above performance criterion for $H(t-\sigma)$, it is convenient to define $H(t-\sigma)$ consisting of the row vector $h_i^T(t-\sigma)$ for $1 \le i \le n$ as

$$H(t-\sigma) \triangleq \begin{bmatrix} h_1^T(t-\sigma) \\ h_2^T(t-\sigma) \\ h_3^T(t-\sigma) \\ \vdots \\ h_n^T(t-\sigma) \end{bmatrix}. \quad (28)$$

The error of the i-th state $x_i(t)$ can now be expressed in terms of the vector components of $H(t-\sigma)$ as

$$\{x_i(t) - \hat{x}_i(t|t)\}^2 = \left[\int_{t-T}^{t} h_i^T(t-\sigma)\{C\int_{\sigma}^{t} e^{A(\sigma-\tau)}Gw(\tau)d\tau - Dw(\sigma)\}d\sigma\right]^2 \tag{29}$$

Although, as previously stated, the MLEFF is required to be exact to the states of nominal systems, the exact estimation is also possible in the disturbance corrupted systems if the integral kernel and the $w(\cdot)$ in (29) are orthogonal to each other. This shows that the proposed MLEFF can be a exact estimator for all $w(\cdot)$ belonging to the orthogonal space with respect to the integral kernel.

By the Cauchy-Schwartz inequality, the following inequality is obtained from (29):

$$\{x_i(t) - \hat{x}_i(t|t)\}^2 \le \left[\int_{t-T}^{t} \{\int_{t-T}^{\sigma} h_i^T(t-\tau)Ce^{A(\tau-\sigma)}Gd\tau - h_i^T(t-\sigma)D\}^2 d\sigma\right]\int_{t-T}^{t} w^T(\tau)w(\tau)d\tau.$$

Thus

$$\frac{\{x_i(t) - \hat{x}_i(t|t)\}^2}{\int_{t-T}^{t} w^T(\tau)w(\tau)d\tau} \le \int_{t-T}^{t} \{\int_{t-T}^{\sigma} h_i^T(t-\tau)Ce^{A(\tau-\sigma)}Gd\tau - h_i^T(t-\sigma)D\}^2 d\sigma.$$

Note that an equality is satisfied for some $w(\tau)$ which is linearly dependent on an error. Therefore,

$$\max_{w(\cdot)\neq 0} \frac{\{x_i(t) - \hat{x}_i(t|t)\}^2}{\int_{t-T}^{t} w^T(\tau)w(\tau)d\tau} = \int_{t-T}^{t} \{\int_{t-T}^{\sigma} h_i^T(t-\tau)Ce^{A(\tau-\sigma)}Gd\tau - h_i^T(t-\sigma)D\}^2 d\sigma. \tag{30}$$

In (Han *et al.*, 2001), minimization of the right side of the equation (30) was tried, which provides the same solution as the proposed MLEFF. It is shown that the MLEFF proposed in this paper is also optimal for the performance criterion (27).

4. CONCLUSIONS

In this paper, a new minimax L_2-E_2 performance criterion and an unbiased property are introduced and a minimax L_2-E_2 FIR filter (MLEFF) is proposed for deterministic continuous-time state space signal models. The proposed MLEFF is linear with the most recent finite measurements and inputs, does not require *a priori* information about the horizon initial state, and has the unbiased property. It is surprising in that a closed form solution exists for MLEFF even with the unbiased condition. Due to the worst case design, MLEFF is robust to exterior disturbances design and, due to FIR structure, is believed to be robust against temporary modeling uncertainties or numerical errors, while other minimax filters and H_∞ filters with an IIR structure may show poor robustness in these cases. It is shown that the MLEFF for deterministic systems is similar in form to the existing RHUFF for stochastic systems with unit covariance matrices of both the system noise and the measurement noise. Although the performance criterion is similar to the H_∞ problem, the solution is quite different from that of the H_∞ problem. The proposed MLEFF is first represented in a standard batch form and then an iterative form that has computational advantages. To consider the worst case with respect to each component of the state, another performance criterion is proposed. It is shown that the proposed MLEFF is also optimal for this performance criterion.

The proposed MLEFF with FIR structure can be a substitute for the commonly used H_∞ and the conventional deadbeat filters.

5. REFERENCES

Fu, M. and C. E. de Souza (1992). H_∞ estimation for uncertain systems. *Int. J. Robust and Nonlinear Contr.* **2**, 87–105.

Han, Soo Hee, Wook Hyun Kwon and Pyung Soo Kim (2001). Receding Horizon Unbiased FIR Filters for Continuous-Time State Space Models without A Priori Initial State Information. *IEEE Trans. Automat. Contr.* **46**(5), 766–770.

Han, Soo Hee, Wook Hyun Kwon and Pyung Soo Kim (2002). Quasi-Deadbeat Minimax Filters for Deterministic State Space Models. To appear in *IEEE Trans. Automat. Contr.*

Kwon, Wook Hyun, Pyung Soo Kim and Soo Hee Han (2002). A Receding Horizon Unbiased FIR Filter for Discrete-Time State Space Models. *Automatica* **38**(3), 545–551.

Li, Huaizhong and Minyue Fu (1997). A Linear Matrix Inequality Approach to Robust H_∞ Filtering. *IEEE Trans. on Signal Processing* **45**(9), 2338–2350.

Nagpal, K. M. and P.P. Khargonekar (1991). Filtering and smoothing in an H_∞ setting. *IEEE Trans. Automat. Contr.* **36**, 152–166.

IFAC Publications
www.elsevier.com/locate/ifac

NUMERICALLY RELIABLE H_∞– SYNTHESIS OF ESTIMATORS BASED ON J– LOSSLESS FACTORISATIONS

Piotr Suchomski *

* *Department of Automatic Control, WETI, Gdansk University of Technology, Poland*

Abstract: The paper deals with a problem of numerically reliable synthesis of H_∞–optimal discrete-time observers based on dual J–lossless factorisations of delta-domain models of estimated processes. Both the regular problem concerning models with no zeros on the stability boundary and the extended problem of models with such zeros are discussed. Solutions are obtained via solving a delta-domain algebraic Riccati equation and a relative condition number of this equation is used as a measure of its numerical conditioning. An example is given to show that solutions obtained for the delta operator are much better conditioned than its counterpart versions based on the common forward shift operator. *Copyright © 2003 IFAC*

Keywords: discrete time, state estimation, process control systems

1. INTRODUCTION

The use of the so-called delta (δ) operator in formulation and solving many discrete-time problems (control, estimation, modelling, signal processing) has a number of advantages as opposed to using the conventional forward shift operator (q) (Middleton and Goodwin, 1990).

First, the δ–operator formulation has better numerical conditioning at higher sampling rate and is less sensitive to arithmetic round-off errors. Second, the δ–operator formulation allows for describing the asymptotic behaviour of discrete-time models of continuous-time systems as the sampling period converges to zero.

In this paper, an approach to the H_∞ suboptimal synthesis of state estimators for δ–operator representations of continuous-time processes is presented. The approach is based on a dual chain scattering description of a discrete-time model of the process and results in a filter with a cost function strictly less than a prespecified bound. Both the regular problem concerning models having no zeros on the stability circle in δ–domain and the extended problem of models with such zeros are examined. For the first case, an approach based on a dual J–lossless factorisation is proposed while in the second case an extended dual J–lossless factorisation of the process model is required. Suitable factorisations are obtained by solving algebraic Riccati equations given in the δ–domain formulation. A relative condition number of this equation is used as a measure of its numerical conditioning. An example is given to show that solutions obtained for the delta operator are much better conditioned than its counterpart versions based on the common forward shift operator.

1.1 Basic properties of the delta operator

The delta operator $\delta : l_2 \to l_2$ is defined as the first-order divided difference

$$\delta = (q-1)/\Delta \quad (1)$$

where q is the forward shift operator and $\Delta > 0$ is the sampling interval (Middleton and Goodwin, 1990). Let (q, z) and (δ, ζ) denote the pairs of discrete-time operators q and δ, and the corresponding complex variables z and ζ. Let $D_\Delta = \{\zeta : |\zeta + 1/\Delta| < 1\}$ be the open shifted circle. The closed circle is denoted as $\bar{D}_\Delta$ with the boundary $\partial\bar{D}_\Delta$. The set of all eigenvalues of a square matrix A is denoted by $\lambda(A)$. A matrix A is said to be stable if $\lambda(A) \subset D_\Delta$ while a transfer matrix $G(\zeta)$ is stable if all its poles belong to D_Δ. The conjugate of $G(\zeta)$ is defined as $G^\sim(\zeta) = G^T(-\zeta/(1+\Delta\zeta))$. The Hermitian conjugate is $G^*(\zeta) = G^T(\zeta^*)$.

1.2 Delta-domain modelling

Let a linear continuous-time ($\rho = d/dt$) system be described be the following state-space model

$$\begin{cases} \rho x(t) = A_\rho x(t) + B_\rho u(t) \\ y(t) = C_\rho x(t) + D_\rho u(t) \end{cases} \quad (2)$$

here $x(t)$ denotes the state vector, $u(t)$ is the input, and $y(t)$ is the output. If $u(t)$ is piece-wise constant and right-continuous the following state-space model can be derived

$$\begin{cases} \delta x_k = A x_k + B u_k \\ y_k = C x_k + D u_k \end{cases} \quad (3)$$

where $A = \Delta^{-1}\Gamma_\Delta A_\rho$, $B = \Delta^{-1}\Gamma_\Delta B_\rho$, $C = C_\rho$, $D = D_\rho$, and $\Gamma_\Delta = \int_0^\Delta e^{\tau A_\rho} d\tau$. The $q-$domain model takes the form of (A_q, B_q, C_q, D_q) with $A_q = I_n + \Delta A_\delta$, $B_q = \Delta B_\delta$, $C_q = C_\delta$, and $D_q = D_\delta$.

2. DELTA-DOMAIN RICCATI EQUATIONS

Consider the discrete-time Riccati equation (ARE)

$$\begin{aligned} & P_q^T X_q P_q - X_q - (P_q^T X_q Q_q + S_q) \\ & (T_q + Q_q^T X_q Q_q)^{-1} \\ & (P_q^T X_q Q_q + S_q)^T + R_q = 0_{n\times n} \end{aligned} \quad (4)$$

where P_q, $R_q = R_q^T \in \mathrm{R}^{n\times n}$, Q_q, $S_q \in \mathrm{R}^{n\times m}$ and $T_q = T_q^T \in \mathrm{R}^{m\times m}$. Assuming that ($P_q = I_n + \Delta P$, $Q_q = Q$, $R_q = \Delta^2 R$, $S_q = \Delta S$, $T_q = T$) where $P \in \mathrm{R}^{n\times n}$, $R = R^T \in \mathrm{R}^{n\times n}$, $Q, S \in \mathrm{R}^{n\times m}$ and $T = T^T \in \mathrm{R}^{m\times m}$ leads to the corresponding $\delta-$domain Riccati equation (δARE) in $X = X_q/\Delta$

$$\begin{aligned} & P^T X + XP + \Delta P^T X P - \\ & ((I_n + \Delta P^T)XQ + S)(T + \Delta Q^T X Q)^{-1} \\ & ((I_n + \Delta P^T)XQ + S)^T + R = 0_{n\times n} \end{aligned} \quad (5)$$

Let (U, W) denote a pair of real matrices

$$(U, W) = \left(\begin{bmatrix} P & 0_{n\times n} & Q \\ -R & -P^T & -S \\ S^T & Q^T & T \end{bmatrix}, \begin{bmatrix} I_n & 0_{n\times n} & 0_{n\times m} \\ 0_{n\times n} & I_n + \Delta P^T & 0_{n\times m} \\ 0_{m\times n} & -\Delta Q^T & 0_{m\times m} \end{bmatrix} \right) \quad (6)$$

Eigenvalues $\lambda(U, W)$ of the $(2n+m)\times(2n+m)$ extended pencil associated with (U, W) are defined by $\lambda(U, W) = \{z \in \mathrm{C} : \det(U - zW) = 0\}$. Let $X_-(U, W)$ of dimension $n_- = \dim(X_-(U, W)) \le n$ denote the invariant subspace corresponding to stable eigenvalues of $U - \lambda W$.

Let $[\,X_1^T \; X_2^T \; X_3^T\,]^T \in \mathrm{R}^{(n+n+m)\times n_-}$ be a matrix of full column rank whose columns form a basis for $X_-(U, W)$. The domain of δRic, denoted by dom(δRic), consists of all pairs (U, W) such that $n_- = n$ and $X_1 \in \mathrm{R}^{n\times n}$ is non-singular.

Let $\kappa(P, Q, R, S, T)$ denote the relative condition number of the δARE of (5), which measures the sensitivity of X with respect to perturbations in (P, Q, R, S, T) and $\kappa_q(P_q, Q_q, R_q, S_q, T_q)$ denote the relative condition number of the $q-$domain ARE of (4). The following lemma that completely explains a superiority of $\delta-$domain solutions to their counterparts based on the forward-shift operator (Suchomski, 2001).

Lemma 1: *For a sufficiently small sampling period Δ it holds*

$$\kappa_q(P_q, Q_q, R_q, S_q, T_q) \propto \kappa(P, Q, R, S, T)/\Delta \quad (7)$$

Hence, the $q-$domain ARE of the assumed type of parameterisation is ill conditioned for $\Delta \to 0$.

3. H_∞ OPTIMISATION IN DELTA-DOMAIN

Let $RL_\infty^{p\times r}$ denote the space of proper real-rational $p\times r$-matrix-valued functions of $\zeta \in \mathrm{C}$ that are analytical in $\partial\bar{D}_\Delta$. $RH_\infty^{p\times r}$ is the subspace of $RL_\infty^{p\times r}$ consisting of all stable matrices. The set of all unitary bounded matrices in $RH_\infty^{p\times r}$ is defined by $BH_\infty^{p\times r} = \{\Phi \in RH_\infty^{p\times r}: ||\Phi||_\infty < 1\}$, where $||\cdot||_\infty$ is the $RH_\infty^{p\times r}$ infinity norm, while the units of $RH_\infty^{p\times p}$ are denoted by GH_∞^p. Moreover, let $J_{mn} \in \mathrm{R}^{(m+n)\times(m+n)}$ be a signature matrix defined as $J_{mn} = I_m \oplus (-I_n)$.

3.1 Standard problem

Consider a linear discrete-time generalised plant described by its *scattering matrix*

$$P : \begin{bmatrix} w \\ u \end{bmatrix} \to \begin{bmatrix} z \\ y \end{bmatrix}, \; P(\zeta) = \begin{bmatrix} P_{zw}(\zeta) & P_{zu}(\zeta) \\ P_{yw}(\zeta) & P_{yu}(\zeta) \end{bmatrix} \quad (8)$$

with four input/output signals: w is the exogenous input of dimension r, u of dimension p is the controlling input, z of dimension m is the controlled output (objective) and y is the measured output of dimension q (Kimura, 1997; Zhou *et al.*, 1996). Assume that

$$P(\zeta) = \left[\begin{array}{c|c} A & B \\ \hline C & D \end{array}\right] = \left[\begin{array}{c|cc} A & B_w & B_u \\ \hline C_z & D_{zw} & D_{zu} \\ C_y & D_{yw} & D_{yu} \end{array}\right] \quad (9)$$

where $A \in R^{n \times n}$. The closed-loop system given in Fig. 1 can be described by a *linear fractional transformation* $LF(P, K)$: $w \to z$ of a state feedback filter K : $y \to u$ with respect to P. The standard problem of optimisation in H_∞ is to find a causal linear K, being a $m \times q$ transfer matrix, which internally stabilises the closed-loop system and enforces the norm bound $||LF(P, K)||_\infty < \gamma$ for a prespecified parameter $\gamma > 0$ (Kimura, 1997; Zhou *et al.*, 1996).

Consider the following common conditions for plant regularity (Stoorvogel, 1992):

(*C1*) (A, B_u, C_y) is stabilisable and detectable,

(*C2*) D_{zu} is injective and D_{yw} surjective,

(*C3*) rank $\begin{bmatrix} \bar{A}(\omega) & B_u \\ C_z & D_{zu} \end{bmatrix} = n + p$, $\forall \omega \geq 0$,

(*C4*) rank $\begin{bmatrix} \bar{A}(\omega) & B_w \\ C_y & D_{yw} \end{bmatrix} = n + q$, $\forall \omega \geq 0$, where

$$\bar{A}(\omega) = A - \frac{e^{j\omega\Delta} - 1}{\Delta} I_n$$

(*C5*) $D_{yu} = 0$. $\diamondsuit$

In the case of dual J−lossless factorisation approach it is assumed that all above conditions (*C1-C5*) are satisfied while in the approach based on dual extended J−lossless factorisation the fourth condition (*C4*) is not valid.

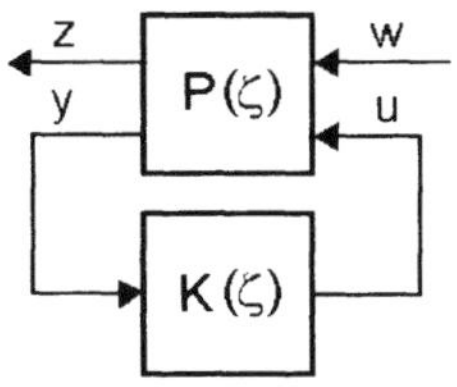

Fig. 1. System with generalised plant.

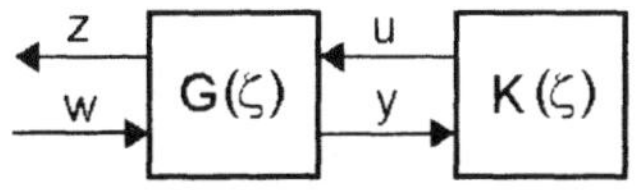

Fig. 2. Dual chain scattering modelling.

3.2 Dual chain-scattering models for H_∞

The plant P of (8) with $m = p$ and an invertible $P_{zu}(\zeta)$ can be characterised via its *dual chain-scattering* representation (Kimura, 1997)

$$G : \begin{bmatrix} z \\ w \end{bmatrix} \to \begin{bmatrix} u \\ y \end{bmatrix} \quad (10)$$

The closed-loop system given in Fig. 2 can thus be described by a *dual homographic transformation* $DHM(G, K)$: $w \to z$ of a filter K : $y \to u$ with respect to G. The standard H_∞ optimisation problem is to find a causal K that stabilises the closed-loop system and enforces (Kimura, 1997)

$$||DHM(G, K)||_\infty < \gamma. \quad (11)$$

4. J-LOSSLESS FACTORISATION SOLUTIONS

The key role in the theory of H_∞ optimisation is played by the so-called J−lossless factorisations of transfer matrices of a plant (Kimura, 1997). The considered approach based on dual J−lossless factorisations is basically analogous to those for continuous-time and discrete-time (q) cases (Kimura, 1997; Kongprawechnon and Kimura, 1998; Tsai *et al.*, 1993).

Definition 1:

(*i*) A $G(\zeta) \in RL_\infty^{(m+q)\times(m+r)}$ is said to be *dual* (J_{mq}, J_{mr})*−unitary*, if $G(\zeta) J_{mr} G^{\sim}(\zeta) = J_{mq}$, $\forall \zeta$.

(*ii*) A dual (J_{mq}, J_{mr})−unitary $G(\zeta)$ is said to be *dual* (J_{mq}, J_{mr})*−lossless*, if $G(\zeta) J_{mr} G^*(\zeta) \geq J_{mq}$, $\forall \zeta \notin D_\Delta$. $\diamondsuit$

4.1 Dual J− lossless approach

Definition 2: If $G(\zeta) \in RL_\infty^{(m+q)\times(m+r)}$ can be represented as $G(\zeta) = \Omega(\zeta)\Psi(\zeta)$ where $\Psi(\zeta) \in RL_\infty^{(m+q)\times(m+r)}$ is dual (J_{mq}, J_{mr})−lossless and $\Omega(\zeta) \in GH_\infty^{m+q}$, then $G(\zeta)$ is said to have a *dual* (J_{mq}, J_{mr})*−lossless factorisation.* $\diamondsuit$

The following theorem can be proved in a similar way to q−domain (Kongprawechnon and Kimura, 1998) and δ−domain (Suchomski, 2002) discrete-time cases.

Theorem 1: *Let* $G(\zeta) = \left[\begin{array}{c|c} A & B \\ \hline C & D \end{array}\right]$ *be a minimal realisation of* $G(\zeta) \in RL_\infty^{(m+q)\times(m+r)}$ *with no zeros on* $\partial\bar{D}_\Delta$*. It has a dual* (J_{mq}, J_{mr})*−lossless factorisation* $G(\zeta) = \Omega(\zeta)\Psi(\zeta)$ *iff the following conditions hold:*

(i) $(U_x, W_x) \in \mathrm{dom}(\delta\mathrm{Ric})$ *and* $X = \delta\mathrm{Ric}(U_x, W_x) \geq 0$*, where*

$$\begin{aligned} P_x = A^T,\ Q_x = C^T,\ R_x = -BJ_{mr}B^T, \\ S_x = -BJ_{mr}D^T,\ T_x = -DJ_{mr}D^T; \end{aligned} \quad (12)$$

(ii) $(U_{\bar{x}}, W_{\bar{x}}) \in \mathrm{dom}(\delta\mathrm{Ric})$ *and* $\bar{X} = \delta\mathrm{Ric}(U_{\bar{x}}, W_{\bar{x}}) \geq 0$*, where*

$$\begin{aligned} P_{\bar{x}} = A,\quad Q_{\bar{x}} = B,\quad R_{\bar{x}} = 0_{n\times n}, \\ S_{\bar{x}} = 0_{n\times(m+q)},\quad T_{\bar{x}} = J_{mr}; \end{aligned} \quad (13)$$

(iii) $||X\bar{X}||_s < 1$*;*

(iv) there exists a non-singular M_x *such that*

$$M_x(T_x + \Delta Q_x^T X Q_x)M_x^T = -J_{mq}. \quad (14)$$

Let $G_\gamma(\zeta)$ denote the plant model scaled with γ and assume that $G_\gamma(\zeta)$ has a dual $(J_{mq}, J_{mr})-$lossless factorisation $G_\gamma(\zeta) = \Omega(\zeta)\Psi(\zeta)$. The set of filters $K(\zeta)$, for which $||DHM(G_\gamma, K)||_\infty < 1$ holds, is parameterised with an arbitrary $\Phi(\zeta) \in BH_\infty^{m\times q}$

$$K = DHM(\Omega^{-1}, \Phi) \quad (15)$$

4.2 Extended dual J−lossless approach

A necessary condition for the existence of the stabilising solution X of Theorem 1 is that $G(\zeta)$ has no zeros on $\partial\bar{D}_\Delta$. We will discuss the case in which this assumption does not hold. The following definitions of the so-called *extended dual J−lossless* factorisation are basically analogous to this for the continuous-time (Hara *et al.*, 1992).

Definition 3: If $G(\zeta) \in RL_\infty^{(m+q)\times(m+r)}$ is represented as a product $G(\zeta) = \Omega(\zeta)\Psi(\zeta)$ where $\Psi(\zeta) \in RL_\infty^{(m+q)\times(m+r)}$ is dual $(J_{mq}, J_{mr})-$lossless and $\Omega(\zeta) \in RH_\infty^{(m+q)\times(m+q)}$ does not have any zeros outside $\bar{D}_\Delta$, then $G(\zeta)$ is said to have an *extended dual* $(J_{mq}, J_{mr})-$*lossless factorisation.* ◊

Assume that $G(\zeta) \in RL_\infty^{(m+q)\times(m+r)}$ has n_z invariant zeros on $\partial\bar{D}_\Delta$.

An extended dual $(J_{mq}, J_{mr})-$ lossless factorisation of $G(\zeta)$ (if exists) can be obtained by using a technique similar to that called 'a zero compensation' (Copeland and Safonov, 1995).

Suppose that a left zero compensator $U(\zeta)$ of a minimal realisation of dimension n_z exists, for which $U(\zeta)^{-1} \in RH_\infty^{(m+q)\times(m+q)}$ and $\tilde{G}(\zeta) = U(\zeta)G(\zeta) \in RL_\infty^{(m+q)\times(m+r)}$ with no zeros on $\partial\bar{D}_\Delta$ has a dual $(J_{mq}, J_{mr})-$lossless factorisation $\tilde{G}(\zeta) = \tilde{\Omega}(\zeta)\Psi(\zeta)$ where $\tilde{\Omega}(\zeta) \in GH_\infty^{m+q}$. It follows that $G(\zeta) = U(\zeta)^{-1}\tilde{G}(\zeta) = \Omega(\zeta)\Psi(\zeta)$ with a stable $\Omega(\zeta) = U(\zeta)^{-1}\tilde{\Omega}(\zeta) \in RH_\infty^{(m+q)\times(m+q)}$ can stand for an extended dual $(J_{mq}, J_{mr})-$lossless factorisation of $G(\zeta)$.

On account of the above, we can see that all poles of $U(\zeta)$ are on $\partial\bar{D}_\Delta$ and all zeros are in D_Δ. Moreover, $\Omega(\zeta)$ can be represented by a realisation of dimension of $n + n_z$. Seeking for a minimal realisation of dimension of n, we can observe that the unimodularity of $\tilde{\Omega}(\zeta)$ implies that the only way that allows for such a simplification of $\Omega(\zeta)$ is a stable pole-zero cancellation between poles of $U(\zeta)^{-1}$ and zeros of $\tilde{\Omega}(\zeta)$. The set of all filters $K(\zeta)$ satisfying $||DHM(G_\gamma, K)||_\infty < 1$ is given by (15), where $\Phi(\zeta) \in BH_\infty^{m\times q}$ is such that $K(\zeta) \in RH_\infty^{m\times q}$. We expect that in a rational method for synthesis of $K(\zeta)$ a minimal realisation of $\Omega(\zeta)^{-1} = \tilde{\Omega}(\zeta)^{-1}U(\zeta)$ should be derived without obtaining a left zero compensator.

Let $S_{G^T}(\zeta)$ denote the system matrix of $G(\zeta)^T$. Using a QZ transformation with unitary matrices Q_z and Z_z gives

$$Q_z^T S_{G^T}(\zeta) Z_z = \begin{bmatrix} S_z - \zeta T_z & * \\ 0_{(n+m+r-n_z)\times n_z} & * \end{bmatrix} \quad (16)$$

where $S_z - \zeta T_z$ with $S_z, T_z \in \mathrm{R}^{n_z\times n_z}$ is a regular pencil containing all the elementary divisors associated with the $\partial\bar{D}_\Delta$ zeros of $G(\zeta)$. Therefore, $\lambda(S_z, T_z) = \lambda(T_z^{-1}S_z) \subset \partial\bar{D}_\Delta$. Let Z_z be partitioned in conformity with $S_{G^T}(\zeta)$

$$Z_z = \begin{bmatrix} Z_{11} & Z_{12} \\ Z_{21} & Z_{22} \end{bmatrix} \quad (17)$$

The following theorem holds which we give without the proof.

Theorem 2: *Let* $G(\zeta) = \left[\begin{array}{c|c} A & B \\ \hline C & D \end{array}\right]$ *be a minimal realisation of* $G(\zeta) \in RL_\infty^{(m+q)\times(m+r)}$ *having* n_z *zeros on* $\partial\bar{D}_\Delta$*. Let* $[\,S_1^T\ S_2^T\ S_3^T\,]^T \in \mathrm{R}^{(n+n+(m+q))\times(n-n_z)}$ *denote a basis of a stable invariant* $(n - n_z)-$ *dimensional subspace of the extended pencil* $U_x - \zeta W_x$ *of (12).* $G(\zeta)$ *has an extended dual* $(J_{mq}, J_{mr})-$*lossless factorisation* $G(\zeta) = \Omega(\zeta)\Psi(\zeta)$ *iff the following conditions hold:*

(i) $[\,S_1\ Z_{11}\,] \in \mathrm{R}^{n\times n}$ *is non-singular and* $X = [\,S_2\ 0_{n\times n_z}\,][\,S_1\ Z_{11}\,]^{-1} \geq 0$, $X \in \mathrm{R}^{n\times n}$*;*

(ii) $(U_{\bar{x}}, W_{\bar{x}}) \in \mathrm{dom}(\delta\mathrm{Ric})$ *and* $\bar{X} = \delta\mathrm{Ric}(U_{\bar{x}}, W_{\bar{x}}) \geq 0$ *for* $(U_{\bar{x}}, W_{\bar{x}})$ *defined by (13);*

(iii) $||X\bar{X}||_s < 1$*;*

(iv) there exists a non-singular M_x *satisfying (14).*

5. H_∞ ESTIMATION IN DELTA-DOMAIN

Consider a linear discrete-time plant (Fig. 3) with three vector-valued input/output signals: w_1 and

w_2 are the exogenous inputs (disturbances) of dimension r_1 and r_2, respectively, and y is the measured output of dimension q. Let x denote the observed state vector of dimension n_1 and $v = Lx$ be a weighted state vector of dimension m, where $L \in R^{m \times n_1}$ stands for a weighting matrix. A measurement noise channel is represented by $C_2(\zeta I_{n_2} - A_2)^{-1}B_2 + D_2$. An approximate weighted state $\hat{v}$ is generated via employing the filter (estimator) $K(\zeta)$. Defining a residual $z = v - \hat{v}$ as the objective we obtain the generalised plant (Fig. 2) with $u = \hat{v}$, $w = [\, w_1^T \; w_2^T \,]^T \in R^r$, and $r = r_1 + r_2$. A disturbance d acting directly on the output is also given in Fig. 3. The corresponding dual chain-scattering model is

$$G_\gamma(\zeta) = \left[\begin{array}{c|c} A & B \\ \hline C & D \end{array}\right] \tag{18}$$

$$= \left[\begin{array}{cc|cc} A_1 & 0_{n_1 \times n_2} & 0_{n \times m} & \left[\begin{array}{cc} B_1 & 0_{n_1 \times r_2} \\ 0_{n_2 \times r_1} & B_2 \end{array}\right] \\ 0_{n_2 \times n_1} & A_2 & & \\ \hline L & 0_{m \times n_2} & -\gamma I_m & 0_{m \times r} \\ C_1 & C_2 & 0_{q \times m} & [\, 0_{q \times r_1} \; D_2 \,] \end{array}\right]$$

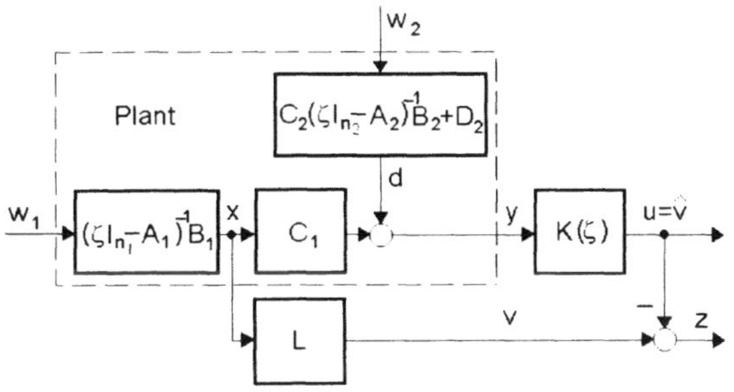

Fig. 3. Estimation problem formulation.

Remarks:

1. From (18) it follows that condition (*C1*) holds if and only if both A_1 and A_2 are stable.

2. For a stable $G_\gamma(\zeta)$ the zero solution $\bar{X} = 0_{n \times n}$ satisfies the second Riccati equation of *Theorem 1* and *Theorem 2*. Moreover, condition (*iii*) of these theorems is also satisfied with $\bar{X} = 0_{n \times n}$.

3. For a stable $G_\gamma(\zeta)$ we have

$$\Omega(\zeta) = \left[\begin{array}{c|c} A + B & H_x \\ \hline -C & I_{m+q} \end{array}\right] M_x^{-1} \tag{19}$$

where

$$H_x = -((I_n + \Delta P_x^T)XQ_x + S_x)$$
$$(T_x + \Delta Q_x^T X Q_x)^{-1} \tag{20}$$

This implies

$$\Omega(\zeta)^{-1} = M_x \left[\begin{array}{c|c} A + H_x C & H_x \\ \hline C & I_{m+q} \end{array}\right] \tag{21}$$

Note that for $G_\gamma(\zeta)$ having no zeros on $\partial \bar{D}_\Delta$ matrix $A + H_x C$ is stable while in the case of $G_\gamma(\zeta)$ with such zeros this matrix is unstable ($\lambda(T_z^{-1}S_z) \subset \lambda(A + H_x C)$) and X of *Theorem 2* is not a stabilising solution to the $\delta-$domain Riccati equation corresponding to (U_x, W_x).

4. For a sufficiently large γ we can find M_x of the following convenient block triangular form

$$M_x = \begin{bmatrix} M_{11} & 0_{m \times q} \\ M_{21} & M_{22} \end{bmatrix} \tag{22}$$

$$M_{11} = L_{11}^{-T}, \; M_{22} = L_{22}^{-T} \tag{23}$$

$$M_{21} = -M_{22}E_{12}^T M_{11}^T M_{11} \tag{24}$$

$$E = \begin{bmatrix} \gamma^2 I_m - \Delta \bar{L} X \bar{L}^T & -\Delta \bar{L} X \bar{C}^T \\ -\Delta \bar{C} X \bar{L}^T & -D_2 D_2^T - \Delta \bar{C} X \bar{C}^T \end{bmatrix}$$
$$= \begin{bmatrix} E_{11} & E_{12} \\ E_{12}^T & E_{22} \end{bmatrix} \tag{25}$$

$$\bar{L} = [\, L \; 0_{m \times n_2} \,], \; \bar{C} = [\, C_1 \; C_1 \,] \tag{26}$$

and $L_{11} \in \mathrm{R}^{m \times m}$ is the Cholesky factor of E_{11}, $L_{22} \in \mathrm{R}^{q \times q}$ is the Cholesky factor of $-\nabla E_{11}$, while $\nabla E_{11} = E_{22} - E_{12}^T E_{11}^{-1} E_{12}$ denotes the Schur complement of E_{11}. Hence, M_x described by (22) exists if and only if $E_{11} > 0$ and $\nabla E_{11} < 0$.

5. Consider $G_\gamma(\zeta)$ with no zeros on $\partial \bar{D}_\Delta$. If M_x of (22) exists, zeroing the termination in (15), i.e. taking $\Phi(\zeta) = 0_{m \times q}$, gives the so called central solution of the simplest strictly proper form

$$K(\zeta) = \left[\begin{array}{c|c} A + H_q \bar{C} & -H_q \\ \hline \bar{L} & 0_{m \times q} \end{array}\right] \tag{27}$$

where $H_q \in \mathrm{R}^{n \times q}$ is taken from $H_x = [\, * \; H_q \,]$.

6. NUMERICAL EXAMPLE & CONCLUSION

Consider a continuous-time stable plant

$$P_c(s) = \left[\begin{array}{ccc|c} -2.3 & -0.4 & -1.3 & 12 \\ -1 & -2 & -1.15 & 12 \\ -1.7 & 0.4 & -2.7 & 4 \\ \hline 1 & -1 & 0 & 0 \\ 0 & 0 & 1 & 0 \end{array}\right] \tag{28}$$

with two different models of measurement noises:

(*i*) no zeros on $\partial \bar{D}_\Delta$

$$D_c(s) = \left[\begin{array}{cc|cc} -0.04 & 0 & 1 & 0 \\ 0 & -0.02 & 0 & 1 \\ \hline 0.04 & 0 & 0.1 & 0 \\ 0 & 0.005 & 0 & -0.1 \end{array}\right] \tag{29}$$

(*ii*) one zero on $\partial \bar{D}_\Delta$

$$D_c(s) = \left[\begin{array}{cc|cc} -0.5 & 0 & 1 & 0 \\ 0 & -0.03 & 0 & 1 \\ \hline -0.05 & 0 & 0.1 & 0 \\ 0 & -0.018 & 0 & 0.6 \end{array}\right] \tag{30}$$

Assume $L = [\,1.25\ -1\ 0\,]$ and $\Delta = 0.02$ s. Moreover, let z_d denote the residual signal obtained for the H_∞ estimator derived for the case we have no knowledge about the measurement noises ($d = w_2$). For a feasible $\gamma = 1.4$ we obtain two sets of results:

i) no zeros on $\partial\bar{D}_\Delta$ (Fig. 4a, b)

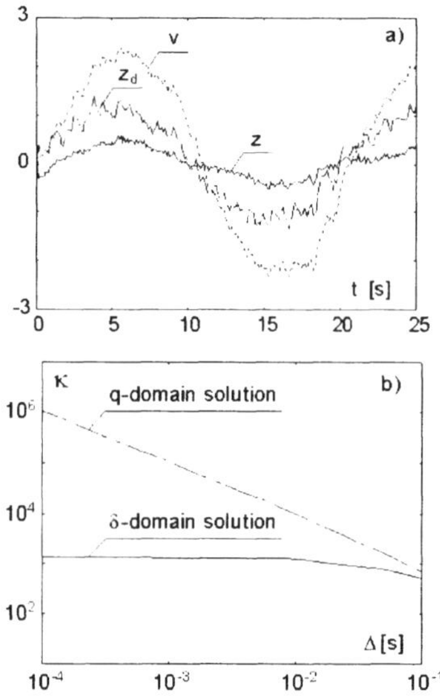

Fig. 4. a) Reference (v) and residuals (z, z_d), b) comparison of condition numbers.

ii) one zero on $\partial\bar{D}_\Delta$ (Fig. 5a-d)

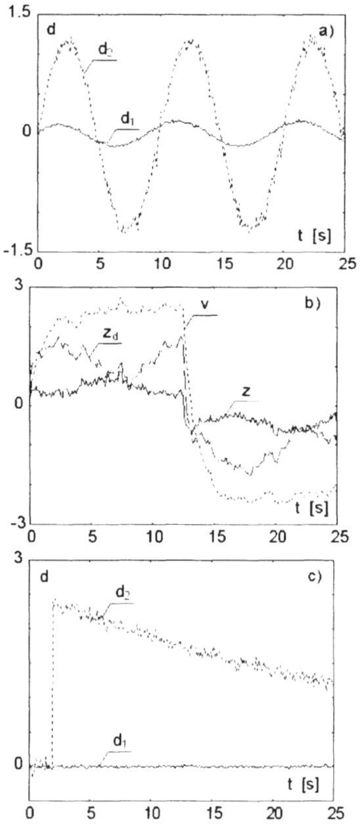

Fig. 5. a,c) Measurement noise (d), b) reference (v) and residuals (z, z_d).

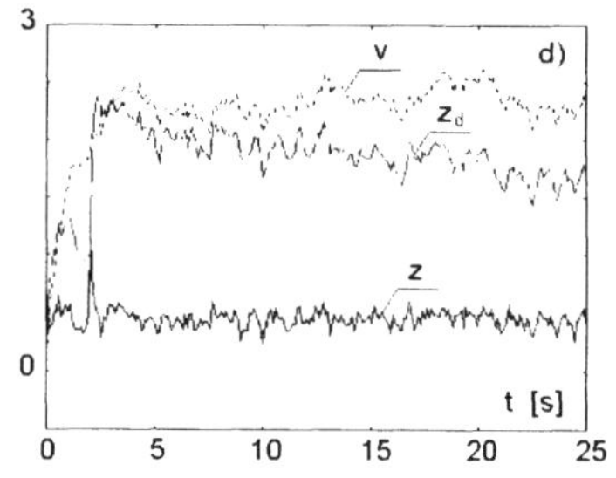

Fig. 5. d) reference (v) and residuals (z, z_d),

The (extended) dual J–lossless factorisation approach to suboptimal H_∞ estimation has been presented. For both the regular and the extended cases the fundamental solvability conditions have been derived in terms of the δ–domain algebraic Riccati equations. Based on their relative condition numbers we have demonstrated (*Lemma 1*) the superiority of the proposed delta-domain formulation as opposed to the traditional q–operator approach.

REFERENCES

Copeland, R. B., and M. G. Safonov (1995). A zero compensation approach to singular H_2 and H_∞ problems. *Int. J. Robust and Nonlinear Control*, vol. 6, pp. 71-106.

Hara, S., T. Sugie, and R. Kondo (1992). H_∞ control problem with $j\omega$–axis zeros. *Automatica*, vol. 28, pp. 55-70.

Kimura, H. (1997) *Chain-scattering approach to H_∞ control.* Birkhauser, Boston, Basel.

Kongprawechnon, W., and H. Kimura (1998). J–lossless factorization and H_∞ control for discrete-time systems. *Int. J. Control*, vol. 70, pp. 423-446.

Middleton, R. H., and G. C. Goodwin (1990). *Digital control and estimation.* Prentice Hall, Englewood Cliffs, N.J.

Stoorvogel, A. (1992). *The H_∞ control problem. A state space approach.* Prentice Hall, New York.

Suchomski, P. (2001). Numerical conditioning of delta-domain Lyapunov and Riccati equations. *IEE Proc. Control Theory and Applic.*, vol. 148, pp. 497-501.

Suchomski, P. (2002). A J–lossless factorization approach to H_∞ control in delta domain. *Automatica*, vol. 38, pp. 1807-1814.

Tsai, M. C., C. S. Tsai, and Y. Y. Sun (1993). On discrete-time H_∞ control: A J-lossless coprime factorization approach. *IEEE Trans. Autom. Control*, vol. 38, pp. 1143-1147.

Zhou, K., J.C. Doyle, and K. Glover (1996). *Robust and optimal control.* Prentice Hall, Upper Saddle River, N.J.

IFAC
Publications
www.elsevier.com/locate/ifac

STATISTICAL ANALYSIS OF SUBSPACE-BASED METHOD FOR DIRECTION ESTIMATION WITHOUT EIGENDECOMPOSITION

Jingmin Xin † **and Akira Sano** §

† *Mobile Communications Development Labs., Fujitsu Laboratories Ltd.*
5-5 Hikari-no-oka, Yokosuka 239-0847, Japan
§ *Department of System Design Engineering, Keio University*
3-14-1 Hiyoshi, Kohoku-ku, Yokohama 223-8522, Japan

Abstract: A computationally simple direction-of-arrival (DOA) estimation method with good statistical performance is attractive in many practical applications of array processing. This paper studies the statistical properties of a new computationally efficient subspace-based method without eigendecomposition (SUMWE) for estimating the directions of narrowband signals impinging on a uniform linear array (ULA), where he asymptotic mean-squared-error (MSE) expression of the estimation error is derived. The theoretical analysis is substantiated through numerical examples, and it is shown that the SUMWE can resolve closely spaced coherent signals better with a smaller number of snapshots and at lower signal-to-noise ratio (SNR). *Copyright © 2003 IFAC*

Keywords: array processors, discrete time, signal processing, statistical analysis, subspace methods, target tracking

1. INTRODUCTION

The directions-of-arrival (DOAs) estimation of signals impinging on an array of sensors is a fundamental problem in array processing used in many fields such as radar, sonar, communications, seismic data processing, and a computationally simple direction estimation method with good statistical performance is much attractive in most practical applications. Although subspace-based methods have received widely attention because of their relatively high resolution and computational simplicity, most of these methods require the eigendecomposition to estimate the signal or noise (null) subspace (Krim and Viberg, 1996). However, the eigendecomposition process is computationally intensive and time-consuming, which means that subspace-based methods are usually limited in many situations particularly when the number of array elements (sensors) is large and/or the directions of incident signals should be tracked in an on-line manner.

Recently some computationally simpler subspace-based methods without eigendecomposition such as the BEWE (Yeh, 1987), propagator method (Marcos *et al.*, 1995), and SWEDE (Eriksson *et al.*, 1994) were developed, where the signal or null subspace is obtained from the array data by a linear operation. Unfortunately, these methods (except a variant of BEWE) suffer serious degradation when the incident signals are mutually coherent, which is often encountered in many practical scenarios. Although a simpler direction estimation method without eigendecomposition was earlier presented for the coherent cyclostationary signals (Xin and Sano, 2000), but it is based on the inherently temporal property of impinging signals and its performance is poor when the number of snapshots is small.

Therefore, a new computationally efficient subspace-based method without eigendecomposition (SUMWE) is proposed for the DOA estimation of coherent narrowband signals impinging on a uniform linear array (ULA) (Xin and Sano, 2002). The coherency of incident signals is decorrelated through subarray averaging, and the null subspace is obtained through a linear operation of a matrix formed from the correlations between some sensor data, where the effect of additive noise is eliminated. Hence the directions can be estimated without performing eigendecomposition, and there is no need to evaluate all correlations of the array data. Furthermore, the SUMWE is also suitable for the case of partly coherent or incoherent signals, and it can be extended to the spatially correlated noise by choosing appropriate subarrays. This paper concentrates on the statistical analysis of the SUMWE and derives the explicit expression of asymptotic mean-squared-error (MSE) (or variance) of the estimation error. The estimation performance of the SUMWE is demonstrated, and the theoretical analysis is substantiated through numerical examples.

2. SUBSPACE-BASED METHOD WITHOUT EIGENDECOMPOSITION — SUMWE

2.1 Data Model and Basic Assumptions

Consider a ULA of M identical and omnidirectional sensors with spacing d, and suppose that p narrowband signals $\{s_k(n)\}$ with the centre frequency f_0 are in the field far from the array and impinge on the array from distinct directions $\{\theta_k\}$. Under the narrowband assumption, the received noisy signal $y_i(n)$ at the ith sensor can be expressed as

$$y_i(n) = \sum_{k=1}^{p} s_k(n) e^{j\omega_0 (i-1)\tau(\theta_k)} + w_i(n) \qquad (1)$$

where $w_i(n)$ is the additive noise, $\omega_0 = 2\pi f_0$, $\tau(\theta_k) = (d/c)\sin\theta_k$, c is the propagation speed, and $\{\theta_k\}$ are measured relative to the normal of array. The received signals can be reexpressed compactly as

$$\boldsymbol{y}(n) = \boldsymbol{A}(\theta)\boldsymbol{s}(n) + \boldsymbol{w}(n) \qquad (2)$$

where $\boldsymbol{y}(n)$, $\boldsymbol{s}(n)$, and $\boldsymbol{w}(n)$ are the vectors of the received signals, incident signals, and additive noise, $\boldsymbol{A}(\theta)$ is the array response matrix given by $\boldsymbol{A}(\theta) = [\boldsymbol{a}(\theta_1), \boldsymbol{a}(\theta_2), \cdots, \boldsymbol{a}(\theta_p)]$ with $\boldsymbol{a}(\theta_k) = [1, e^{j\omega_0\tau(\theta_k)}, \cdots, e^{j\omega_0(M-1)\tau(\theta_k)}]^T$, and $(\cdot)^T$ denotes the transpose.

In this paper, the array is calibrated, and the array response matrix $\boldsymbol{A}(\theta)$ is unambiguous (equivalently $\boldsymbol{A}(\theta)$ has full rank). Without loss of generality, the signals $\{s_k(n)\}$ are all coherent, and then under the flat fading multipath propagation, they can be expressed as (Shan *et al.*, 1985; Xin and Sano, 2001)

$$s_k(n) = \beta_k s_1(n), \quad \text{for } k = 1, 2, \cdots, p \qquad (3)$$

where β_k is the complex attenuation coefficient with $\beta_k \neq 0$ and $\beta_1 = 1$. The incident signal $s_1(n)$ and additive noise $\{w_i(n)\}$ are independent and are complex circularly white Gaussian random processes with zero-mean and the variances given by $E\{s_1(n)s_1^*(t)\} = r_s \delta_{n,t}$, $E\{s_1(n)s_1(t)\} = 0$, $E\{\boldsymbol{w}(n)\boldsymbol{w}^H(t)\} = \sigma^2 \boldsymbol{I}_M \delta_{n,t}$, and $E\{\boldsymbol{w}(n)\boldsymbol{w}^T(t)\} = \boldsymbol{O}_{M\times M}$, $\forall n,t$, where $E\{\cdot\}$, $(\cdot)^*$, $\delta_{n,t}$, $\boldsymbol{I}_m$, $\boldsymbol{O}_{m\times q}$, and $(\cdot)^H$ denote the expectation, the complex conjugate, Kronecker delta, the $m\times m$ identity matrix, the $m \times q$ null matrix, and Hermitian transpose. Additionally the number of incident signals p is known or estimated by some techniques (e.g., Xin and Sano, 2001), and it satisfies the inequality that $p < M/2$.

2.2 SUMWE for Direction Estimation

Here the correlation r_{ik} of signals $y_i(n)$ and $y_k(n)$ is defined as $r_{ik} = E\{y_i(n)y_k^*(n)\}$, where $r_{ik} = r_{ki}^*$, and the array covariance matrix $\boldsymbol{R}$ is given by

$$\boldsymbol{R} = E\{\boldsymbol{y}(n)\boldsymbol{y}^H(n)\} = \boldsymbol{A}(\theta)\boldsymbol{R}_s\boldsymbol{A}^H(\theta) + \sigma^2\boldsymbol{I}_M \qquad (4)$$

where $\boldsymbol{R}_s = E\{\boldsymbol{s}(n)\boldsymbol{s}^H(n)\}$. By dividing the full array into L overlapping subarrays with p sensors in the forward and backward directions, where $L = M - p + 1$, the signal vectors $\boldsymbol{y}_{fl}(n)$ and $\boldsymbol{y}_{bl}(n)$ of the received signals in the lth forward and backward subarrays can be expressed as (Shan *et al.*, 1985; Xin and Sano, 2001)

$$\boldsymbol{y}_{fl}(n) = \boldsymbol{A}_1(\theta)\boldsymbol{D}^{l-1}\boldsymbol{s}(n) + \boldsymbol{w}_{fl}(n) \qquad (5)$$

$$\boldsymbol{y}_{bl}(n) = \boldsymbol{A}_1(\theta)\boldsymbol{D}^{-(M-l)}\boldsymbol{s}^*(n) + \boldsymbol{w}_{bl}(n) \qquad (6)$$

where $\boldsymbol{y}_{fl}(n) = [y_l(n), y_{l+1}(n), \cdots, y_{l+p-1}(n)]^T$, $\boldsymbol{y}_{bl}(n) = [y_{M-l+1}(n), y_{M-l}(n), \cdots, y_{L-l+1}(n)]^H$, $\boldsymbol{w}_{fl}(n) = [w_l(n), w_{l+1}(n), \cdots, w_{l+p-1}(n)]^T$, $\boldsymbol{w}_{bl}(n) = [w_{M-l+1}(n), w_{M-l}(n), \cdots, w_{L-l+1}(n)]^H$, $\boldsymbol{D} = \mathrm{diag}(e^{j\omega_0\tau(\theta_1)}, e^{j\omega_0\tau(\theta_2)}, \cdots, e^{j\omega_0\tau(\theta_p)})$, and $\boldsymbol{A}_1(\theta)$ is the submatrix of $\boldsymbol{A}(\theta)$ consisting of the first p rows with the column $\boldsymbol{a}_1(\theta_k) = [1, e^{j\omega_0\tau(\theta_k)}, \cdots, e^{j\omega_0(p-1)\tau(\theta_k)}]^T$.

Under the basic assumptions, the correlation φ_{fl} between the signal vector $\boldsymbol{y}_{fl}(n)$ of the lth forward subarray and the signal $y_M^*(n)$ is obtained

$$\varphi_{fl} = E\{\boldsymbol{y}_{fl}(n)y_M^*(n)\} = \rho_M r_s \boldsymbol{A}_1(\theta)\boldsymbol{D}^{l-1}\boldsymbol{\beta} \qquad (7)$$

for $l = 1, 2, \cdots, L-1$, where $\rho_i = \boldsymbol{\beta}^H \boldsymbol{b}_i^*(\theta)$, $\boldsymbol{b}_i(\theta) = [e^{j\omega_0(i-1)\tau(\theta_1)}, e^{j\omega_0(i-1)\tau(\theta_2)}, \cdots, e^{j\omega_0(i-1)\tau(\theta_p)}]^T$, and $\boldsymbol{\beta} = [\beta_1, \beta_2, \cdots, \beta_p]^T$. Hence by concatenating the correlations $\{\varphi_{fl}\}$ in (7) and by performing some algebraic manipulations, a Hankel matrix $\boldsymbol{\Phi}_f$ can be formed as (Xin and Sano, 2002)

$$\boldsymbol{\Phi}_f = [\varphi_{f1}, \varphi_{f2}, \cdots, \varphi_{f,L-1}]^T = \rho_M r_s \bar{\boldsymbol{A}}(\theta)\boldsymbol{B}\boldsymbol{A}_1^T(\theta) \qquad (8)$$

where $\boldsymbol{B} = \mathrm{diag}(\beta_1, \beta_2, \cdots, \beta_p)$, and $\bar{\boldsymbol{A}}(\theta)$ is the submatrix of $\boldsymbol{A}(\theta)$ consisting of the first $L-1$ rows with the column $\bar{\boldsymbol{a}}(\theta_k) = [1, e^{j\omega_0\tau(\theta_k)}, \cdots, e^{j\omega_0(L-2)\tau(\theta_k)}]^T$. Obviously the matrix $\boldsymbol{\Phi}_f$ is composed of the correlations $\{r_{1M}, r_{2M}, \cdots, r_{M-1,M}\}$. In a similar way, the correlation $\bar{\varphi}_{fl}$ between the signal vector $\boldsymbol{y}_{fl}(n)$ and the signal $y_1^*(n)$ can be got

$$\bar{\varphi}_{fl} = E\{\boldsymbol{y}_{fl}(n)y_1^*(n)\} = \rho_1 r_s \boldsymbol{A}_1(\theta)\boldsymbol{D}^{l-1}\boldsymbol{\beta} \qquad (9)$$

for $l = 2, 3, \cdots, L$. Then a Hankel matrix $\bar{\boldsymbol{\Phi}}_f$ can be obtained from the correlations $\{\bar{\varphi}_{fl}\}$ (Xin and Sano, 2002)

$$\bar{\boldsymbol{\Phi}}_f = [\bar{\varphi}_{f2}, \bar{\varphi}_{f3}, \cdots, \bar{\varphi}_{fL}]^T = \rho_1 r_s \bar{\boldsymbol{A}}(\theta)\boldsymbol{B}\boldsymbol{D}\boldsymbol{A}_1^T(\theta) \qquad (10)$$

Obviously the matrix $\bar{\boldsymbol{\Phi}}_f$ consists of $\{r_{21}, r_{31}, \cdots, r_{M1}\}$.

By evaluating the correlation $\varphi_{bl} = E\{y_1(n)\boldsymbol{y}_{bl}(n)\}$ between $y_1(n)$ and $\{\boldsymbol{y}_{bl}(n)\}$ and the correlation $\bar{\varphi}_{bl} = E\{y_M(n)\boldsymbol{y}_{bl}(n)\}$ between $y_M(n)$ and $\{\boldsymbol{y}_{bl}(n)\}$, the correlation matrices Φ_b and $\bar{\Phi}_b$ for the backward subarrays can be obtained (Xin and Sano, 2002)

$$\begin{aligned}\Phi_b &= [\varphi_{b1}, \varphi_{b2}, \cdots, \varphi_{b,L-1}]^T \\ &= \rho_1^* r_s \bar{A}(\theta) B^* D^{-(M-1)} A_1^T(\theta) \qquad (11)\\ \bar{\Phi}_b &= [\bar{\varphi}_{b2}, \bar{\varphi}_{b3}, \cdots, \bar{\varphi}_{bL}]^T \\ &= \rho_M^* r_s \bar{A}(\theta) B^* D^{-(M-2)} A_1^T(\theta) \qquad (12)\end{aligned}$$

It is apparent that the Hankel matrices Φ_b and $\bar{\Phi}_b$ consist of $\{r_{1M}, r_{1,M-1}, \cdots, r_{12}\}$ and $\{r_{M,M-1}, r_{M,M-2}, \cdots, r_{M1}\}$, respectively. Furthermore it is easily found that $\Phi_b = J_{L-1}\bar{\Phi}_f^* J_p$ and $\bar{\Phi}_b = J_{L-1}\Phi_f^* J_p$, where J_m is an $m \times m$ reversal matrix.

Clearly the correlation matrices Φ_f, $\bar{\Phi}_f$, Φ_b, and $\bar{\Phi}_b$ are not affected by the additive noise $\{w_i(n)\}$. Because $M > 2p$ (i.e. $L-1 > p$), $\beta_k \neq 0$, and $A(\theta)$ is a Vandermonde matrix with full rank, it can be found that the ranks of these $(L-1)\times p$ correlation matrices equal p, i.e. the dimension of their signal subspace equals the number of coherent signals. Hence the DOAs $\{\theta_k\}$ of coherent signals can be estimated from the subspaces of these matrices.

Since it is assumed that $M > 2p$, the $(L-1)\times p$ matrix $\bar{A}(\theta)$ can be divided into two parts as

$$\bar{A}(\theta) = \begin{bmatrix} A_1(\theta) \\ A_2(\theta) \end{bmatrix} \begin{matrix} \}p \\ \}L-p-1 \end{matrix} \qquad (13)$$

where $A_2(\theta)$ has the column $a_2(\theta_k) = [e^{j\omega_0 p\tau(\theta_k)}, e^{j\omega_0 (p+1)\tau(\theta_k)}, \cdots, e^{j\omega_0 (L-2)\tau(\theta_k)}]^T$. As $\bar{A}(\theta)$ and $A_1(\theta)$ are two submatrices of the Vandemonde matrix $A(\theta)$ with full rank, the rows of $A_2(\theta)$ can be expressed as a linear combination of linearly independent rows of $A_1(\theta)$; equivalently there is a $p\times(L-p-1)$ linear operator P between $A_1(\theta)$ and $A_2(\theta)$ (Marcos *et al.*, 1995; Eriksson *et al.*, 1994)

$$P^H A_1(\theta) = A_2(\theta) \qquad (14)$$

Hence it follows from (14) that

$$Q^H \bar{A}(\theta) = O_{(L-p-1)\times p} \qquad (15)$$

where $Q = [P^T, -I_{L-p-1}]^T$. Because the matrix Q has a full rank of $L-p-1$, the columns of Q in fact form the basis for the null space of $\bar{A}^H(\theta)$, and clearly the orthogonal projector onto this subspace is given by $\Pi_Q = Q(Q^H Q)^{-1} Q^H$ which implies that

$$\Pi_Q \bar{a}(\theta) = \mathbf{0}_{(L-1)\times 1}, \quad \text{for } \theta = \theta_k \qquad (16)$$

where $\bar{a}(\theta) = [1, e^{j\omega_0\tau(\theta)}, \cdots, e^{j\omega_0(L-2)\tau(\theta)}]^T$, and $\mathbf{0}_{m\times 1}$ is an $m\times 1$ null vector.

The next problem is how to find the null space of $\bar{A}^H(\theta)$ (i.e. P) from the available array data. Based on the partition of $\bar{A}(\theta)$, the $(L-1)\times p$ correlation matrices Φ_f, $\bar{\Phi}_f$, Φ_b, and $\bar{\Phi}_b$ can be also divided as

$$\Phi_f = \begin{bmatrix} \Phi_{f1} \\ \Phi_{f2} \end{bmatrix} \begin{matrix} \}p \\ \}L-p-1 \end{matrix}, \quad \bar{\Phi}_f = \begin{bmatrix} \bar{\Phi}_{f1} \\ \bar{\Phi}_{f2} \end{bmatrix} \begin{matrix} \}p \\ \}L-p-1 \end{matrix} \qquad (17)$$

$$\Phi_b = \begin{bmatrix} \Phi_{b1} \\ \Phi_{b2} \end{bmatrix} \begin{matrix} \}p \\ \}L-p-1 \end{matrix}, \quad \bar{\Phi}_b = \begin{bmatrix} \bar{\Phi}_{b1} \\ \bar{\Phi}_{b2} \end{bmatrix} \begin{matrix} \}p \\ \}L-p-1 \end{matrix} \qquad (18)$$

From (8) and (10)-(14), the relation between the submatrices of Φ_f, $\bar{\Phi}_f$, Φ_b, and $\bar{\Phi}_b$ is obtained

$$P^H \Phi_1 = \Phi_2 \qquad (19)$$

where $\Phi_1 = [\Phi_{f1}, \bar{\Phi}_{f1}, \Phi_{b1}, \bar{\Phi}_{b1}]$, and $\Phi_2 = [\Phi_{f2}, \bar{\Phi}_{f2}, \Phi_{b2}, \bar{\Phi}_{b2}]$. Thus the matrix P can be found from Φ_1 and Φ_2 as

$$P = A_1^{-H}(\theta) A_2^H(\theta) = (\Phi_1\Phi_1^H)^{-1}\Phi_1\Phi_2^H \qquad (20)$$

Therefore when the finite array data are available, the directions $\{\theta_k\}$ can be estimated by minimizing the following cost function

$$f(\theta) = \bar{a}^H(\theta)\Pi_{\hat{Q}}\bar{a}(\theta) \qquad (21)$$

where $\Pi_{\hat{Q}} = \hat{Q}(\hat{Q}^H\hat{Q})^{-1}\hat{Q}^H$, and $\hat{P} = (\hat{\Phi}_1\hat{\Phi}_1^H)^{-1}\hat{\Phi}_1\hat{\Phi}_2^H$.

Remarks: Clearly number of needed corrlations $\{r_{ik}\}$ is $4M-6$, which is rather less than that $M+(m-1)\cdot(2M-m)$ needed by the SS-based MUSIC (Shan *et al.*, 1985), where m is the number of sensors in the subarray with $m > p$, and the implementation of the proposed method is given by Xin and Sano (2002). The proposed SUMWE is suitable for the case of partly coherent or incoherent signals, and it can be extended to the spatially correlated noise by choosing appropriate correlations of array data (see (Xin and Sano, 2002) for details). □

3. STATISTICAL ANALYSIS

In this section, the statistical properties of the proposed SUMWE algorithm are studied. First the following Lemma on the consistency of the SUMWE estimates can be obtained easily.

Lemma: As the number of snapshots N tends to infinity, the estimates $\{\hat{\theta}_k\}$ obtained by minimizing the cost function $f(\theta)$ in (21) approach the true parameters $\{\theta_k\}$ with probability one (w.p.1).

Proof: This lemma can be readily established by adopting the proof of Lemma 1 in (Stoica and Söderström, 1992). Clearly $f(\theta)$ converges to the true cost function $f_o(\theta) = \bar{a}^H(\theta)\Pi_Q\bar{a}(\theta)$ w.p.1 and uniformly in θ when $N\to\infty$, and the minima of $f_o(\theta)$ are achieved if and only if $\theta = \theta_k$. Thus the estimates $\{\hat{\theta}_k\}$ approach the true parameters $\{\theta_k\}$ w.p.1 as $N\to\infty$. ■

From this Lemma, it is found that the estimates $\{\hat{\theta}_k\}$ of the SUMWE are consistent. Then for a sufficiently large number of snapshots N, the expression of the asymptotic MSE (or variance) of the SUMWE estimates can be obtained by the following Theorem.

Theorem: For the estimate $\hat{\theta}_k$ obtained by minimizing the function $f(\theta)$ in (21), the large-sample MSE (or variance) of the estimation error $\hat{\theta}_k - \theta_k$ is given by

$$\mathrm{MSE}(\theta_k) = \frac{1}{2N\bar{H}_{kk}}\mathrm{Re}\{2g_k^H\left(r_{1M}(I_{L-1}\otimes h_{k1}^T)M_{fb}\right.$$

$$\cdot(\boldsymbol{I}_{L-1}\otimes\boldsymbol{h}_{k3}) + r_{MM}(\boldsymbol{I}_{L-1}\otimes\boldsymbol{h}_{k1}^T)\bar{\boldsymbol{M}}_{fb}(\boldsymbol{I}_{L-1}\otimes\boldsymbol{h}_{k4})$$
$$+r_{11}(\boldsymbol{I}_{L-1}\otimes\boldsymbol{h}_{k2}^T)\tilde{\boldsymbol{M}}_{fb}(\boldsymbol{I}_{L-1}\otimes\boldsymbol{h}_{k3}) + r_{M1}(\boldsymbol{I}_{L-1}\otimes\boldsymbol{h}_{k2}^T)$$
$$\cdot\breve{\boldsymbol{M}}_{fb}(\boldsymbol{I}_{L-1}\otimes\boldsymbol{h}_{k4})\Big)\boldsymbol{g}_k^* + \boldsymbol{g}_k^H\Big(r_{MM}(\boldsymbol{I}_{L-1}\otimes\boldsymbol{h}_{k1}^T)\boldsymbol{M}_{ff}$$
$$\cdot(\boldsymbol{I}_{L-1}\otimes\boldsymbol{h}_{k1}^*) + r_{11}(\boldsymbol{I}_{L-1}\otimes\boldsymbol{h}_{k2}^T)\bar{\boldsymbol{M}}_{ff}(\boldsymbol{I}_{L-1}\otimes\boldsymbol{h}_{k2}^*)$$
$$+r_{11}(\boldsymbol{I}_{L-1}\otimes\boldsymbol{h}_{k3}^T)\boldsymbol{M}_{bb}(\boldsymbol{I}_{L-1}\otimes\boldsymbol{h}_{k3}^*) + r_{MM}(\boldsymbol{I}_{L-1}\otimes\boldsymbol{h}_{k4}^T)$$
$$\cdot\bar{\boldsymbol{M}}_{bb}(\boldsymbol{I}_{L-1}\otimes\boldsymbol{h}_{k4}^*)\Big)\boldsymbol{g}_k + 2\boldsymbol{g}_k^H\Big(r_{1M}(\boldsymbol{I}_{L-1}\otimes\boldsymbol{h}_{k1}^T)\tilde{\boldsymbol{M}}_{ff}$$
$$\cdot(\boldsymbol{I}_{L-1}\otimes\boldsymbol{h}_{k2}^*) + (\boldsymbol{I}_{L-1}\otimes\boldsymbol{h}_{k1}^T)\boldsymbol{M}_w(\boldsymbol{I}_{L-1}\otimes\boldsymbol{h}_{k3}^*)$$
$$+(\boldsymbol{I}_{L-1}\otimes\boldsymbol{h}_{k2}^T)\bar{\boldsymbol{M}}_w(\boldsymbol{I}_{L-1}\otimes\boldsymbol{h}_{k4}^*) + r_{1M}(\boldsymbol{I}_{L-1}\otimes\boldsymbol{h}_{k3}^T)\tilde{\boldsymbol{M}}_{bb}$$
$$\cdot(\boldsymbol{I}_{L-1}\otimes\boldsymbol{h}_{k4}^*)\Big)\boldsymbol{g}_k\Big\} \qquad (22)$$

where $\bar{\boldsymbol{H}} = \bar{\boldsymbol{D}}^H(\theta)\boldsymbol{\Pi}_Q\bar{\boldsymbol{D}}(\theta)$, $\bar{\boldsymbol{D}}(\theta) = [\bar{\boldsymbol{d}}(\theta_1), \bar{\boldsymbol{d}}(\theta_2), \cdots, \bar{\boldsymbol{d}}(\theta_p)]$ with $\bar{\boldsymbol{d}}(\theta) = d\bar{\boldsymbol{a}}(\theta)/d\theta = j\omega_0(d/c)\cos\theta_k[0, e^{j\omega_0\tau(\theta)}, \cdots, (L-2)e^{j\omega_0(L-2)\tau(\theta)}]^T$, while $\boldsymbol{g}_k = \boldsymbol{\Pi}_Q\bar{\boldsymbol{d}}(\theta_k)$, $\boldsymbol{h}_{k1} = \boldsymbol{\Phi}_{f1}^H \cdot\boldsymbol{\Psi}_1^{-1}\boldsymbol{a}_1(\theta_k)$, $\boldsymbol{h}_{k2} = \bar{\boldsymbol{\Phi}}_{f1}^H\boldsymbol{\Psi}_1^{-1}\boldsymbol{a}_1(\theta_k)$, $\boldsymbol{h}_{k3} = \boldsymbol{\Phi}_{b1}^H\boldsymbol{\Psi}_1^{-1}\boldsymbol{a}_1(\theta_k)$, $\boldsymbol{h}_{k4} = \bar{\boldsymbol{\Phi}}_{b1}^H\boldsymbol{\Psi}_1^{-1}\boldsymbol{a}_1(\theta_k)$, and $\boldsymbol{\Psi}_1 = \boldsymbol{\Phi}_1\boldsymbol{\Phi}_1^H$. Furthermore, the matrices $\boldsymbol{M}_{fb}$, $\bar{\boldsymbol{M}}_{fb}$, $\tilde{\boldsymbol{M}}_{fb}$, $\breve{\boldsymbol{M}}_{fb}$, $\boldsymbol{M}_{ff}$, $\bar{\boldsymbol{M}}_{ff}$, $\tilde{\boldsymbol{M}}_{ff}$, $\boldsymbol{M}_{bb}$, $\bar{\boldsymbol{M}}_{bb}$, and $\tilde{\boldsymbol{M}}_{bb}$ denote the correlations between the forward/backward subarrays given by

$$\boldsymbol{M}_{fb} = E\{z_f(n)z_b^T(n)\},\ \bar{\boldsymbol{M}}_{fb} = E\{z_f(n)\bar{z}_b^T(n)\},$$
$$\tilde{\boldsymbol{M}}_{fb} = E\{\bar{z}_f(n)z_b^T(n)\},\ \breve{\boldsymbol{M}}_{fb} = E\{\bar{z}_f(n)\bar{z}_b^T(n)\},$$
$$\boldsymbol{M}_{ff} = E\{z_f(n)z_f^H(n)\},\ \bar{\boldsymbol{M}}_{ff} = E\{\bar{z}_f(n)\bar{z}_f^H(n)\},$$
$$\tilde{\boldsymbol{M}}_{ff} = E\{z_f(n)\bar{z}_f^H(n)\},\ \boldsymbol{M}_{bb} = E\{z_b(n)z_b^H(n)\},$$
$$\bar{\boldsymbol{M}}_{bb} = E\{\bar{z}_b(n)\bar{z}_b^H(n)\},\ \tilde{\boldsymbol{M}}_{bb} = E\{z_b(n)\bar{z}_b^H(n)\},$$

where $z_f(n) = [\boldsymbol{y}_{f1}^T(n), \boldsymbol{y}_{f2}^T(n), \cdots, \boldsymbol{y}_{f,L-1}^T(n)]^T$, $\bar{z}_f(n) = [\boldsymbol{y}_{f2}^T(n), \boldsymbol{y}_{f3}^T(n), \cdots, \boldsymbol{y}_{fL}^T(n)]^T$, $z_b(n) = [\boldsymbol{y}_{b1}^T(n), \boldsymbol{y}_{b2}^T(n), \cdots, \boldsymbol{y}_{b,L-1}^T(n)]^T$, and $\bar{z}_b(n) = [\boldsymbol{y}_{b2}^T(n), \boldsymbol{y}_{b3}^T(n), \cdots, \boldsymbol{y}_{bL}^T(n)]^T$, the ikth elements of the $(L-1)p\times(L-1)p$ matrices $\boldsymbol{M}_w$ and $\bar{\boldsymbol{M}}_w$ are given by respectively

$$(\boldsymbol{M}_w)_{ik} = \begin{cases}\sigma^4, & \text{for } i=k=1\\ 0, & \text{others}\end{cases} \qquad (23)$$

$$(\bar{\boldsymbol{M}}_w)_{ik} = \begin{cases}\sigma^4, & \text{for } i=k=(L-1)p\\ 0, & \text{others}\end{cases} \qquad (24)$$

and $\otimes$ denotes the Kronecker operation.

Proof: As the estimate $\hat{\theta}_k$ is obtained by minimizing $f(\theta)$ and it is a consistent estimate, for a sufficiently large number of snapshots N, the derivative of $f(\theta)$ can be approximated using two terms in its Taylor series expansion about the true value θ_k as (Stoica and Nehorai, 1989, 1990)

$$0 = f'(\hat{\theta}_k) \approx f'(\theta_k) + f''(\theta_k)(\hat{\theta}_k - \theta_k) \qquad (25)$$

where the second- and higher order terms in (25) can be neglected, and the first- and second-order derivatives of $f(\theta)$ with respect to θ are given by

$$f'(\theta) = df(\theta)/d\theta = 2\mathrm{Re}\{\bar{\boldsymbol{d}}^H(\theta)\boldsymbol{\Pi}_{\hat{Q}}\bar{\boldsymbol{a}}(\theta)\} \qquad (26)$$
$$f''(\theta) = df'(\theta)/d\theta$$
$$= 2\mathrm{Re}\{\dot{\bar{\boldsymbol{d}}}^H(\theta)\boldsymbol{\Pi}_{\hat{Q}}\bar{\boldsymbol{a}}(\theta) + \bar{\boldsymbol{d}}^H(\theta)\boldsymbol{\Pi}_{\hat{Q}}\bar{\boldsymbol{d}}(\theta)\} \qquad (27)$$

in which $\dot{\bar{\boldsymbol{d}}}(\theta) = d\bar{\boldsymbol{d}}(\theta)/d\theta$, and $\mathrm{Re}\{\cdot\}$ denotes the real part of the bracketed quantity. From (25)-(27), the first-order expression for the estimation error $\Delta\theta_k = \hat{\theta}_k - \theta_k$ can be obtained as

$$\Delta\theta_k \approx -\frac{f'(\theta_k)}{f''(\theta_k)} \approx -\frac{\mathrm{Re}\{\bar{\boldsymbol{d}}^H(\theta_k)\boldsymbol{\Pi}_{\hat{Q}}\bar{\boldsymbol{a}}(\theta_k)\}}{\bar{\boldsymbol{d}}^H(\theta_k)\boldsymbol{\Pi}_Q\bar{\boldsymbol{d}}(\theta_k)} \qquad (28)$$

where the estimated orthogonal projector $\boldsymbol{\Pi}_{\hat{Q}}$ in the denominator of (28) can be replaced with the true one $\boldsymbol{\Pi}_Q$ without affecting the asymptotic property of estimate $\hat{\theta}_k$ (Stoica and Nehorai, 1989).

Following the idea in (Stoica and Söderström, 1992), the first-order approximate of the orthogonal projector $\boldsymbol{\Pi}_{\hat{Q}}$ can be got

$$\boldsymbol{\Pi}_{\hat{Q}} \approx (\boldsymbol{I}_{L-1} - \boldsymbol{\Pi}_Q)\hat{\boldsymbol{Q}}(\boldsymbol{Q}^H\boldsymbol{Q})^{-1}\boldsymbol{Q}^H$$
$$+\boldsymbol{Q}(\boldsymbol{Q}^H\boldsymbol{Q})^{-1}\hat{\boldsymbol{Q}}^H(\boldsymbol{I}_{L-1} - \boldsymbol{\Pi}_Q) + \boldsymbol{\Pi}_Q \qquad (29)$$

Then by using the fact that $\boldsymbol{Q}^H\bar{\boldsymbol{a}}(\theta) = \boldsymbol{0}_{(L-p-1)\times 1}$ and substituting (29) into (28), the estimation error $\Delta\theta_k$ can be approximately given by

$$\Delta\theta_k \approx -\frac{\mathrm{Re}\{\bar{\boldsymbol{d}}^H(\theta_k)\boldsymbol{Q}(\boldsymbol{Q}^H\boldsymbol{Q})^{-1}\hat{\boldsymbol{Q}}^H\bar{\boldsymbol{a}}(\theta_k)\}}{\bar{\boldsymbol{d}}^H(\theta_k)\boldsymbol{\Pi}_Q\bar{\boldsymbol{d}}(\theta_k)}$$
$$= -\frac{\mathrm{Re}\{\mu_k\}}{\bar{\boldsymbol{H}}_{kk}} \qquad (30)$$

where

$$\mu_k = \bar{\boldsymbol{d}}^H(\theta_k)\boldsymbol{Q}(\boldsymbol{Q}^H\boldsymbol{Q})^{-1}\hat{\boldsymbol{Q}}^H\bar{\boldsymbol{a}}(\theta_k) \qquad (31)$$

Consequently because the estimate $\hat{\theta}_k$ is consistent, from (30), its MSE (or variance) is given by

$$\mathrm{MSE}(\theta_k) = E\{(\Delta\theta_k)^2\} = \mathrm{var}(\theta_k)$$
$$\approx \frac{1}{2\bar{\boldsymbol{H}}_{kk}^2}\mathrm{Re}\{E\{\mu_k^2\} + E\{|\mu_k|^2\}\} \qquad (32)$$

where the fact that $\mathrm{Re}\{\mu_i\}\mathrm{Re}\{\mu_k\} = 0.5(\mathrm{Re}\{\mu_i\mu_k\} + \mathrm{Re}\{\mu_i\mu_k^*\})$ is used implicitly.

By letting $\boldsymbol{\Psi}_2 = \boldsymbol{\Phi}_2\boldsymbol{\Phi}_1^H$ and $\boldsymbol{\Psi} = [\boldsymbol{\Psi}_1^T, \boldsymbol{\Psi}_2^T]^T$, an approximation of $\hat{\boldsymbol{\Psi}}$ can be obtained (Eriksson *et al.*, 1994)

$$\hat{\boldsymbol{\Psi}} = \hat{\boldsymbol{\Phi}}\hat{\boldsymbol{\Phi}}_1^H \approx \boldsymbol{\Phi}(\hat{\boldsymbol{\Phi}}_1^H - \boldsymbol{\Phi}_1^H) + \hat{\boldsymbol{\Phi}}\boldsymbol{\Phi}_1^H \qquad (33)$$

where $\boldsymbol{\Phi} = [\boldsymbol{\Phi}_1^T, \boldsymbol{\Phi}_2^T]^T = [\boldsymbol{\Phi}_f, \bar{\boldsymbol{\Phi}}_f, \boldsymbol{\Phi}_b, \bar{\boldsymbol{\Phi}}_b]$. And from (20), the following approximation is got as well

$$\hat{\boldsymbol{P}}^H - \boldsymbol{P}^H = \hat{\boldsymbol{\Psi}}_2\hat{\boldsymbol{\Psi}}_1^{-1} - \boldsymbol{P}^H = (\hat{\boldsymbol{\Psi}}_2 - \boldsymbol{P}^H\hat{\boldsymbol{\Psi}}_1)\hat{\boldsymbol{\Psi}}_1^{-1}$$
$$\approx (\hat{\boldsymbol{\Psi}}_2 - \boldsymbol{P}^H\hat{\boldsymbol{\Psi}}_1)\boldsymbol{\Psi}_1^{-1} = -\boldsymbol{Q}^H\hat{\boldsymbol{\Psi}}\boldsymbol{\Psi}_1^{-1} \qquad (34)$$

where the term of order $O(1/N)$ is neglected. In addition, the estimated matrices $\hat{\boldsymbol{\Phi}}_f$, $\hat{\bar{\boldsymbol{\Phi}}}_f$, $\hat{\boldsymbol{\Phi}}_b$, and $\hat{\bar{\boldsymbol{\Phi}}}_b$ can be expressed by

$$\hat{\boldsymbol{\Phi}}_f = \frac{1}{N}\sum_{n=1}^{N}\boldsymbol{Y}_f(n)y_M^*(n),\ \hat{\bar{\boldsymbol{\Phi}}}_f = \frac{1}{N}\sum_{n=1}^{N}\bar{\boldsymbol{Y}}_f(n)y_1^*(n) \qquad (35)$$

$$\hat{\boldsymbol{\Phi}}_b = \frac{1}{N}\sum_{n=1}^{N}\boldsymbol{Y}_b(n)y_1(n),\ \hat{\bar{\boldsymbol{\Phi}}}_b = \frac{1}{N}\sum_{n=1}^{N}\bar{\boldsymbol{Y}}_b(n)y_M(n) \qquad (36)$$

where $\boldsymbol{Y}_f(n) = [\boldsymbol{y}_{f1}(n), \boldsymbol{y}_{f2}(n), \cdots, \boldsymbol{y}_{f,L-1}(n)]^T$, $\bar{\boldsymbol{Y}}_f(n) = [\boldsymbol{y}_{f2}(n), \boldsymbol{y}_{f3}(n), \cdots, \boldsymbol{y}_{fL}(n)]^T$, $\boldsymbol{Y}_b(n) = [\boldsymbol{y}_{b1}(n), \boldsymbol{y}_{b2}(n), \cdots, \boldsymbol{y}_{b,L-1}(n)]^T$, and $\bar{\boldsymbol{Y}}_b(n) = [\boldsymbol{y}_{b2}(n), \boldsymbol{y}_{b3}(n), \cdots, \boldsymbol{y}_{bL}(n)]^T$.

Then by using the formula $\boldsymbol{X}\boldsymbol{c} = (\boldsymbol{I}\otimes\boldsymbol{c}^T)\mathrm{vec}(\boldsymbol{X}^T)$ for matrix $\boldsymbol{X}$ and vector $\boldsymbol{c}$ with compatible dimensions, and by noting that $\bar{\boldsymbol{a}}(\theta)$ can be partitioned as $\bar{\boldsymbol{a}}(\theta) = [\boldsymbol{a}_1^T(\theta), \boldsymbol{a}_2^T(\theta)]^T$, where $\mathrm{vec}(\boldsymbol{X}^T)$ is a matrix operation stacking the columns of matrix $\boldsymbol{X}^T$ one under the other to form a single column, μ_k in (31) can be approximated for a sufficiently large N as

$$\begin{aligned}\mu_k &= \bar{\boldsymbol{d}}^H(\theta_k)\boldsymbol{Q}(\boldsymbol{Q}^H\boldsymbol{Q})^{-1}(\hat{\boldsymbol{Q}}^H - \boldsymbol{Q}^H)\bar{\boldsymbol{a}}(\theta_k)\\ &= \bar{\boldsymbol{d}}^H(\theta_k)\boldsymbol{Q}(\boldsymbol{Q}^H\boldsymbol{Q})^{-1}(\hat{\boldsymbol{P}}^H - \boldsymbol{P}^H)\boldsymbol{a}_1(\theta_k)\\ &\approx -\bar{\boldsymbol{d}}^H(\theta_k)\boldsymbol{\Pi}_Q(\boldsymbol{\Phi}(\hat{\boldsymbol{\Phi}}_1^H - \boldsymbol{\Phi}_1^H) + \hat{\boldsymbol{\Phi}}\boldsymbol{\Phi}_1^H)\boldsymbol{\Psi}_1^{-1}\boldsymbol{a}_1(\theta_k)\\ &= -\boldsymbol{g}_k^H\hat{\boldsymbol{\Phi}}\boldsymbol{\Phi}_1^H\boldsymbol{\Psi}_1^{-1}\boldsymbol{a}_1(\theta_k)\\ &= \mu_{k1} + \mu_{k2} + \mu_{k3} + \mu_{k4} \end{aligned} \qquad (37)$$

where

$$\mu_{k1} = -\boldsymbol{g}_k^H\hat{\boldsymbol{\Phi}}_f\boldsymbol{h}_{k1} = \frac{-\boldsymbol{v}_{k1}^H}{N}\left(\sum_{n=1}^{N}\mathrm{vec}(\boldsymbol{Y}_f^T(n))y_M^*(n)\right) \qquad (38)$$

$$\mu_{k2} = -\boldsymbol{g}_k^H\hat{\bar{\boldsymbol{\Phi}}}_f\boldsymbol{h}_{k2} = \frac{-\boldsymbol{v}_{k2}^H}{N}\left(\sum_{n=1}^{N}\mathrm{vec}(\bar{\boldsymbol{Y}}_f^T(n))y_1^*(n)\right) \qquad (39)$$

$$\mu_{k3} = -\boldsymbol{g}_k^H\hat{\boldsymbol{\Phi}}_b\boldsymbol{h}_{k3} = \frac{-\boldsymbol{v}_{k3}^H}{N}\left(\sum_{n=1}^{N}\mathrm{vec}(\boldsymbol{Y}_b^T(n))y_1(n)\right) \qquad (40)$$

$$\mu_{k4} = -\boldsymbol{g}_k^H\hat{\bar{\boldsymbol{\Phi}}}_b\boldsymbol{h}_{k4} = \frac{-\boldsymbol{v}_{k4}^H}{N}\left(\sum_{n=1}^{N}\mathrm{vec}(\bar{\boldsymbol{Y}}_b^T(n))y_M(n)\right) \qquad (41)$$

and $\boldsymbol{v}_{ki} = (\boldsymbol{I}_{L-1}\otimes\boldsymbol{h}_{ki}^*)\boldsymbol{g}_k$. Hence from (37), the terms $E\{\mu_k^2\}$ and $E\{|\mu_k|^2\}$ in (32) are obtained as

$$\begin{aligned}E\{\mu_k^2\} &= E\{\mu_{k1}^2 + \mu_{k2}^2 + \mu_{k3}^2 + \mu_{k4}^2\\ &\quad +2(\mu_{k1}\mu_{k2} + \mu_{k1}\mu_{k3} + \mu_{k1}\mu_{k4}\\ &\quad +\mu_{k2}\mu_{k3} + \mu_{k2}\mu_{k4} + \mu_{k3}\mu_{k4})\}\end{aligned} \qquad (42)$$

$$\begin{aligned}E\{|\mu_k|^2\} &= E\{|\mu_{k1}|^2 + |\mu_{k2}|^2 + |\mu_{k3}|^2 + |\mu_{k4}|^2\\ &\quad +2(\mu_{k1}\mu_{k2}^* + \mu_{k1}\mu_{k3}^* + \mu_{k1}\mu_{k4}^*\\ &\quad +\mu_{k2}\mu_{k3}^* + \mu_{k2}\mu_{k4}^* + \mu_{k3}\mu_{k4}^*)\}\end{aligned} \qquad (43)$$

Under the basic assumptions on data model, by using the formula for the expectation of the product of four complex Gaussian random matrices with zero-mean and compatible dimensions that $E\{\boldsymbol{ABCD}\} = E\{\boldsymbol{AB}\}\cdot E\{\boldsymbol{CD}\} + E\{\boldsymbol{C}\otimes\boldsymbol{A}\}E\{\boldsymbol{D}\otimes\boldsymbol{B}\} + E\{\boldsymbol{A}E\{\boldsymbol{BC}\}\boldsymbol{D}\}$ (Janssen and Stoica, 1988), and by using the fact in (15), the expectations $E\{\mu_{ki}\mu_{kq}\}$ and $E\{\mu_{ki}\mu_{kq}^*\}$ can be obtained after some straightforward manipulations

$$E\{\mu_{k1}^2\} = E\{\mu_{k2}^2\} = E\{\mu_{k3}^2\} = E\{\mu_{k4}^2\} = 0 \qquad (44)$$

$$E\{\mu_{k1}\mu_{k3}\} = (1/N)r_{1M}\boldsymbol{v}_{k1}^H\boldsymbol{M}_{fb}\boldsymbol{v}_{k3}^* \qquad (45)$$

$$E\{\mu_{k1}\mu_{k4}\} = (1/N)r_{MM}\boldsymbol{v}_{k1}^H\bar{\boldsymbol{M}}_{fb}\boldsymbol{v}_{k4}^* \qquad (46)$$

$$E\{\mu_{k2}\mu_{k3}\} = (1/N)r_{11}\boldsymbol{v}_{k2}^H\tilde{\boldsymbol{M}}_{fb}\boldsymbol{v}_{k3}^* \qquad (47)$$

$$E\{\mu_{k2}\mu_{k4}\} = (1/N)r_{M1}\boldsymbol{v}_{k2}^H\breve{\boldsymbol{M}}_{fb}\boldsymbol{v}_{k4}^* \qquad (48)$$

$$E\{\mu_{k1}\mu_{k2}\} = E\{\mu_{k3}\mu_{k4}\} = 0 \qquad (49)$$

$$E\{|\mu_{k1}|^2\} = (1/N)r_{MM}\boldsymbol{v}_{k1}^H\boldsymbol{M}_{ff}\boldsymbol{v}_{k1} \qquad (50)$$

$$E\{|\mu_{k2}|^2\} = (1/N)r_{11}\boldsymbol{v}_{k2}^H\bar{\boldsymbol{M}}_{ff}\boldsymbol{v}_{k2} \qquad (51)$$

$$E\{|\mu_{k3}|^2\} = (1/N)r_{11}\boldsymbol{v}_{k3}^H\boldsymbol{M}_{bb}\boldsymbol{v}_{k3} \qquad (52)$$

$$E\{|\mu_{k4}|^2\} = (1/N)r_{MM}\boldsymbol{v}_{k4}^H\bar{\boldsymbol{M}}_{bb}\boldsymbol{v}_{k4} \qquad (53)$$

$$E\{\mu_{k1}\mu_{k2}^*\} = (1/N)r_{1M}\boldsymbol{v}_{k1}^H\tilde{\boldsymbol{M}}_{ff}\boldsymbol{v}_{k2} \qquad (54)$$

$$E\{\mu_{k3}\mu_{k4}^*\} = (1/N)r_{1M}\boldsymbol{v}_{k3}^H\tilde{\boldsymbol{M}}_{bb}\boldsymbol{v}_{k4} \qquad (55)$$

$$E\{\mu_{k1}\mu_{k3}^*\} = (1/N)r_{1M}\boldsymbol{v}_{k1}^H\boldsymbol{M}_{w}\boldsymbol{v}_{k3} \qquad (56)$$

$$E\{\mu_{k2}\mu_{k4}^*\} = (1/N)\boldsymbol{v}_{k2}^H\bar{\boldsymbol{M}}_{w}\boldsymbol{v}_{k4} \qquad (57)$$

$$E\{\mu_{k1}\mu_{k4}^*\} = E\{\mu_{k2}\mu_{k3}^*\} = 0 \qquad (58)$$

By substituting (42)-(58) into (32), the asymptotic variance $\mathrm{MSE}(\theta_k)$ (or $\mathrm{var}(\theta_k)$) of the estimation error $\Delta\theta_k$ in (22) can be obtained immediately. ■

4. NUMERICAL EXAMPLES

The ULA with M sensors is separated by a half-wavelength, and two coherent signals with equal power come from angles θ_1 and θ_2. The signal-to-noise ratio (SNR) is defined as the ratio of the power of the source signals to that of the additive noise at each sensor. For comparing the estimation performance of the SUMWE, the SS-based root-MUSIC (Shan *et al.*, 1985), and BEWE (variant for coherent signals) (Yeh, 1987) are carried out, and the stochastic Cramér-Rao lower bound (CRB) (Stoica and Nehorai, 1990) is calculated. The shown results are based on 1000 independent trials.

Example 1 — Performance versus SNR

The incident directions of two coherent signals are $\theta_1 = 5°$ and $\theta_2 = 12°$, and their SNR is varied from -10 to 25 dB. The number of sensors is $M = 10$, and the number of snapshots is $N = 128$. Additionally the subarray size is set at $m = 7$ for the SS-based root-MUSIC. The empirical root-MSEs (RMSEs) of the estimates $\hat{\theta}_1$ and $\hat{\theta}_2$ are shown in Fig. 1. It is found that the proposed SUMWE outperforms the more common SS-based root-MUSIC with EVD and the BEWE without eigendecomposition, and the empirical RMSEs of the SUMWE are very close to the theoretical ones (except at lower SNR). The theoretical and empirical RMSEs of the SUMWE decrease monotonically with the increasing SNR, and it is shown that the difference between the theoretical RMSEs of the SUMWE and the CRBs is small, and the empirical RMSEs of the SUMWE approach the theoretical ones at medium and larger SNRs.

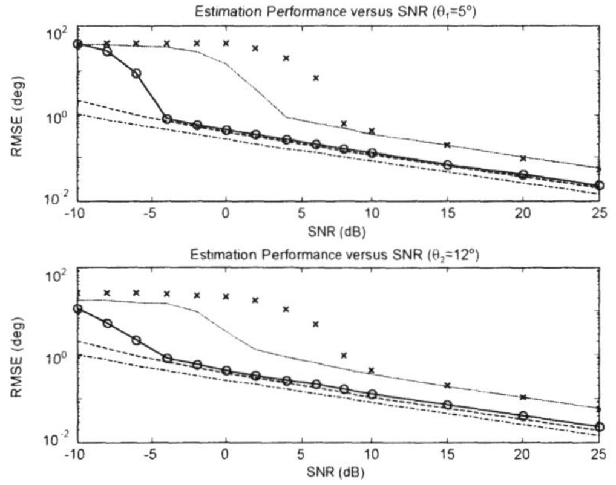

Fig. 1. RMSEs of the estimates versus the SNR (dotted line: SS-based root-MUSIC; "x": BEWE; solid line with "o": SUMWE; dashed line: theoretical RMSE of SUMWE; and dash-dot line: CRB) for Example 1.

Example 2 — Performance versus Snapshot Number

The simulation conditions are similar to those in Example 1, except that the SNR is set at 10 dB, and the number of snapshots is varied from $N = 10$ to $N = 1000$. From Fig. 2, it can be seen that the SUMWE is superior to the other methods even for a small number of snapshots, and that its performance agrees well with the derived theoretical analysis, where the empirical and theoretical RMSEs of the SUMWE decrease monotonically with the number of snapshots.

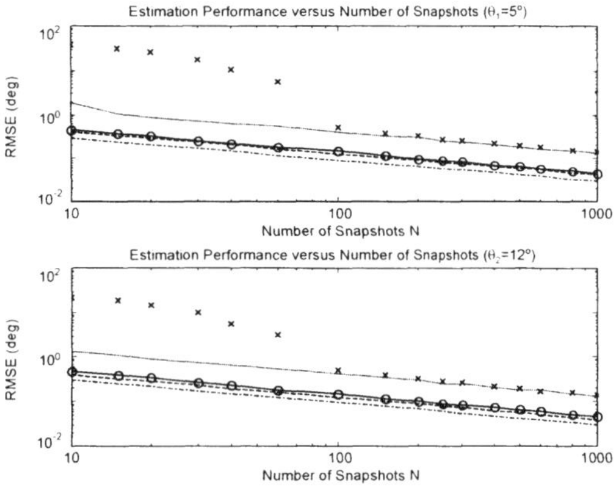

Fig. 2. RMSEs of the estimates versus the number of snapshots (dotted line: SS-based root-MUSIC; "x": BEWE; solid line with "o": SUMWE; dashed line: theoretical RMSE of SUMWE; and dash-dot line: CRB) for Example 2.

Example 3 — Performance versus Angle Separation

Two coherent signals impinge on the array along $\theta_1 = 5^{\circ}$ and $\theta_2 = \theta_1 + \Delta\theta$, where $\Delta\theta$ is varied from $\Delta\theta = 1^{\circ}$ to $\Delta\theta = 14^{\circ}$, and the other simulation parameters are the same as those in Example 1 except that the SNR is fixed at 10 dB. As shown in Fig. 3, the SUMWE generally estimates the directions of closely spaced signals more accurately with a much smaller RMSE than the other methods, and the empirical RMSEs of the SUMWE are much near to the theoretical ones for larger angular separation.

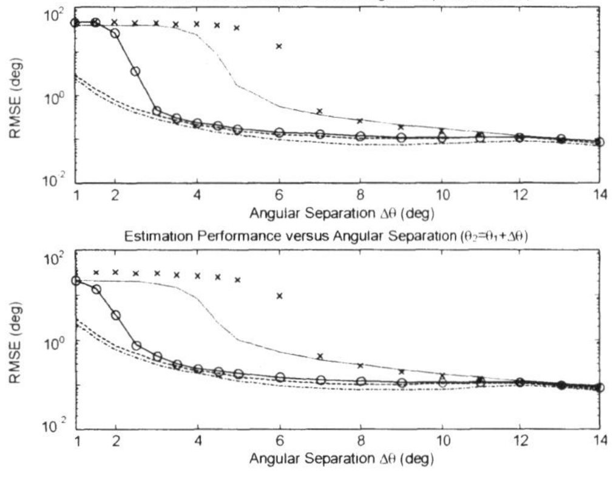

Fig. 3. RMSEs of the estimates versus the angular separation (dotted line: SS-based root-MUSIC; "x": BEWE; solid line with "o": SUMWE; dashed line: theoretical RMSE of SUMWE; and dash-dot line: CRB) for Example 3.

5. CONCLUSION

In this paper, the statistical analysis of a new computationally efficient subspace-based method called SUMWE was studied, and the explicit expression of asymptotic MSE (or variance) of the estimation error was derived. The estimation performance of the SUMWE was demonstrated and the theoretical analysis was substantiated through numerical examples. The simulation results showed that the SUMWE algorithm has the advantages of reduced computational load and superior estimation performance in resolving closely spaced coherent signals with short length of data and at low SNR.

REFERENCES

Eriksson, A., P. Stoica and T. Söderström (1994). On-line subspace algorithms for tracking moving sources. *IEEE Trans. Signal Processing*, **42**, 2319-2330.

Janssen, P.H.M. and P. Stoica (1988). On the expectation of the product of four matrix-valued Gaussian random variables. *IEEE Trans. Automatic Control*, **33**, 867-870.

Krim, H. and M. Viberg (1996). Two decades of array signal processing research: The parametric approach. *IEEE Signal Processing Mag.*, **13/4**, 67-94.

Marcos, S., A. Marsal and M. Benidir (1995). The propagator method for source bearing estimation. *Signal Processing*, **42**, 121-138.

Shan, T.-J., M. Wax and T. Kailath (1985). On spatial smoothing for direction-of-arrival estimation of coherent signals. *IEEE Trans. Acoust., Speech, Signal Processing*, **33**, 806-811.

Stoica, P. and A. Nehorai (1989). MUSIC, maximum likelihood, and Cramer-Rao bound. *IEEE Trans. Acoust., Speech, Signal Processing*, **37**, 720-741.

Stoica, P. and A. Nehorai (1990). Performance study of conditional and unconditional direction-of-arrival estimation. *IEEE Trans. Acoust., Speech, Signal Processing*, **38**, 1783-1795.

Stoica, P. and T. Söderström (1992). Statistical analysis of a subspace method for bearing estimation without eigendecomposition. *IEE Proc.*, Pt. F, **139**, 301-305.

Xin, J. and A. Sano (2000). Direction estimation of coherent signals without eigendecomposition and spatial smoothing. *Proc. IFAC 12th Symp. System Identification*, **1**, 349-354, Santa Barbara.

Xin, J. and A. Sano (2001). MSE-based regularization approach to direction estimation of coherent narrowband signals using linear prediction. *IEEE Trans. Signal Processing*, **49**, 2481-2497.

Xin, J. and A. Sano (2002). Computationally efficient subspace-based method for direction-of-arrival estimation without eigendecomposition. Submitted to *IEEE Trans. Signal Processing*, 2002.

Yeh, C.-C. (1987). Simple computation of projection matrix for bearing estimations. *IEE Proc.*, Pt. F, **134**, 146-150.

IFAC Publications
www.elsevier.com/locate/ifac

FAULT DETECTION OF NON-LINEAR SYSTEMS BASED ON MULTI-FORM QUASI-ARMAX MODELING AND ITS APPLICATION TO THE SHIP BENCHMARK

K.Kumamaru * K.Inoue * Y.Hosoyamada * and T.Söderström **

** Faculty of Computer Science and Systems Engineering, Kyushu Institute of Technology Kawazu 680-4, Iizuka 820-8502, Fukuoka, Japan e-mail:kumamaru@ces.kyutech.ac.jp*
*** Department of Automatic Control and System Analysis, Institute of Technology, Uppsala University, P.O.Box 27, Uppsala, Sweden*

Abstract: This paper is concerned with an application study of model-based fault detection method to a ship propulsion system. When modeling the object system, Quasi-ARMAX model with multi-model form is used. In this model, the system non-linearity is incorporated into model parameters by using non-linear non-parametric models (NNMs). Kullback discrimination Information (KDI) is introduced as fault detection index to evaluate the distortion in identified model, which is caused by a fault. The effectiveness of the method is verified through simulation studies on the ship propulsion system. *Copyright © 2003 IFAC*

Keywords: Fault detection, Kullback discrimination information, non-linear system, parameter estimation, quasi-ARMAX model

1. INTRODUCTION

There have been proposed many kinds of model-based fault detection method for dynamic systems (Izermann and Balle(1996)). Most of faults in dynamic systems can be represented as unexpected variations in system operating modes which are caused by changes in system configuration parameters. Based on this assumption, Kumamaru and Söderström (1986) proposed a fault detection method for linear systems. In the method, fault detection is carried out by evaluating model distortion due to a fault via Kullback discrimination information (KDI). The KDI is defined as a distortion measure for two probability density functions. It can be used for model discrimination by introducing likelihood functions corresponding to identified models and explicitly analyzed under the assumption of gaussian distribution.

When dealing with the fault detection for black-box type non-linear systems, it is not possible to express the characteristics precisely by the linear model, e.g. ARAMX model. In order to describe black-box type non-linear systems as precisely as possible, a Quasi-ARMAX model has been proposed (Hu, Kumamaru and Hirasawa(2001)), in which the system non-linearity is incorporated into ARMAX model parameters by using non-linear non-parametric models (NNMs). Thus the model has a linear structure and can effectively be used for control design and system analysis of general non-linear systems in the framework of linear system theory. However, the KDI can no longer be directly applied to evaluate the distortion of identified Quasi-ARMAX model, since the model is essentially non-linear, and the corresponding likelihood function becomes non-gaussian.

To solve this problem, a fault detection method based on multi-model has been proposed for general non-linear systems (Kumamaru *et al*(1997), (1999)). The basic idea of the method is as follows: when unified NNMs are used for expression of all parameters in Quasi-ARMAX model, it can be transformed to a multi-model form that consists of several local linear ARMAX models. Let us perform identification based on the Quasi-ARMAX model with the multi-model form. Then, the model distortion due to a fault will be reflected in the local ARMAX model. Therefore, the model distortion in the local model can be evaluated by the KDI and distortion of the identified model is estimated by considering an appropriate weight for each KDI.

So far, simulation studies have been carried out on mathematically described non-linear systems in order to verify the effectiveness of the method. In this paper the fault detection method is applied to the ship propulsion system as more realistic plant model, which was constructed for benchmark test (Izadi-Zamanabadi and Blanke (1999)). Thus the paper covers some extensions of our recent work (Kumamaru, Abe and Inoue (2001)). Although the plant model can be described by state-space form, it is treated as a black-box system in the modeling and identification. In this case, deciding an optimal model structure becomes an important issue. The KDI is able to be applied not only to fault detection but also to model validation (see Kumamaru and Söderström (1986)).

The paper is organized as follows: in section 2, the outline of the Quasi-ARMAX model, and its multi-ARMAX form are briefly introduced. In section 3, the KDI analysis based on multi-model is developed in detail, together with model validation and fault detection schemes. The outline of the ship propulsion system is introduced in section 4. In section 5, simulation studies on the ship propulsion system have been carried out under both of deterministic and stochastic situations. Finally, section 6 is devoted to conclusions.

2. QUASI-ARMAX MODEL

Consider a discrete-time non-linear SISO system described by

$$\mathcal{S} : y(t) = g(\varphi(t)) + v(t) \tag{1}$$

$$\varphi(t) = [-y(t-1) \cdots -y(t-n) \quad u(t-1) \cdots u(t-m)]^T$$

where $y(t)$ is the output at time $t(t = 1,2...)$, $u(t)$ the input, $v(t)$ the system disturbance, $\varphi(t)$ the regression vector and $g(\cdot)$ the unknown nonlinear function. When the system is modeled, the Quasi-ARMAX model is introduced as follows,

$$\mathcal{M} : A(q^{-1},\phi(t))y(t) = B(q^{-1},\phi(t))u(t) + C(q^{-1})e(t) \tag{2}$$

$$e(t) \in \mathcal{N}(0,\sigma^2)$$

$$\phi(t) = [y(t-1) \cdots y(t-n') \quad u(t-1) \cdots u(t-m')]$$

$$A(q^{-1},\phi(t)) = 1 + a_1(\phi(t))q^{-1} + \cdots + a_n(\phi(t))q^{-n}$$
$$B(q^{-1},\phi(t)) = b_1(\phi(t))q^{-1} + \cdots + b_m(\phi(t))q^{-m}$$
$$C(q^{-1}) = 1 + c_1 q^{-1} + \cdots + c_l q^{-l}$$

where q^{-1} is the backward shift operator. The model (2) is the same structure as a linear ARMAX model. However, $a_i(\phi(t))$ and $b_i(\phi(t))$ are the non-linear functions of $\phi(t)$ which consists of past input-output data up to m' and n', respectively. In order to describe the non-linear terms, non-linear non-parametric models (NNMs) are used as follows,

$$a_i(\phi(t)) = f_i(\phi(t)) \qquad (i = 1,\cdots,n)$$
$$b_j(\phi(t)) = f_{n+j}(\phi(t)) \qquad (j = 1,\cdots,m)$$

$$f_i(\phi(t)) = \sum_{j=1}^{M} \omega_{ij} N_f(p_j,\phi(t)) \tag{3}$$
$$(i = 1,2,\cdots n+m)$$

where $N_f(\cdot)$ is the basis functions, ω_{ij}'s are the coordinate parameters to be estimated, p_j's are the scale and position parameters specifying the basis functions that are to be pre-assigned based on available information about the system dynamics. The model (NNMs) is equipped with flexibility for representing system non-linearity. As an example of NNMs, Adaptive Fuzzy System (AFS) is given by

$$f_i(\phi(t)) = \frac{\sum_{i=1}^{M} \omega_{ij}(\wedge_{k=1}^{r} \mu_{A_k^j} x_k(t))}{\sum_{i=1}^{M}(\wedge_{k=1}^{r} \mu_{A_k^j} x_k(t))} \tag{4}$$

where $x_k(t)$ is the k-th element of $\phi(t)$. M and r are the number of fuzzy rules and input to AFS, respectively. $\mu_{A_k^j}$ is the membership function corresponding to a j-th fuzzy rule.

Using (3), (4) in (2), the Quasi-ARMAX model (2) can be transformed into the following forms,

$$\mathcal{M}(\Omega_e) :$$
$$y(t) = \sum_{j=1}^{M} \varphi(t)^T \Omega_j N_f(p_j,\phi(t)) + C(q^{-1})e(t)$$
$$= \sum_{j=1}^{M} (\varphi_e(t)^T \Omega_{ej} + e(t)) N_f(p_j,\phi(t)) \tag{5}$$

$$\begin{aligned} \varphi_e(t) &= [\varphi(t)^T \; e(t-1) \; \cdots \; e(t-l)]^T \\ \Omega_j &= [\omega_{1j} \; \cdots \; \omega_{(n+m)j}]^T \\ \Omega_{ej} &= [\Omega_j^T c_1 \; \cdots \; c_l]^T \\ \Omega_e &= [\Omega_{e1}^T \; \cdots \; \Omega_{eM}^T]^T \qquad (j = 1,\cdots,M) \\ \sum_{j=1}^{M} N_f(p_j,\phi(t)) &= 1 \end{aligned}$$

The parameter estimation for (5) can be performed by existing method of recursive identification, e.g., the prediction error method (Ljung and Söderström (1983)). When the basis function have compact supports, the identified model has a multi-ARMAX model structure interpolated by the basis functions, which consist of M local identified models

$$\mathcal{M}(\hat{\Omega}_{ej}) : A(q^{-1} : \hat{\Omega}_{ej})z_j(t) = B(q^{-1} : \hat{\Omega}_{ej})u(t) + C(q^{-1} : \hat{\Omega}_{ej})\varepsilon(t) \quad (6)$$

$$\varepsilon(t) \in \mathcal{N}(0, \hat{\sigma}^2)$$
$$A(q^{-1} : \hat{\Omega}_{ej}) = 1 + \hat{\omega}_{1j}q^{-1} + \cdots + \hat{\omega}_{nj}q^{-n}$$
$$B(q^{-1} : \hat{\Omega}_{ej}) = \hat{\omega}_{(n+1)j}q^{-1} + \cdots + \hat{\omega}_{(n+m)j}q^{-m}$$
$$C(q^{-1} : \hat{\Omega}_{ej}) = 1 + \hat{c}_1 + \cdots + \hat{c}_l q^{-l}$$
$$(j = 1, \cdots M)$$

where the symbol $\hat{}$ means the estimated value and $z_j(t)$ stands for the output variable of the j-th local identified model.

3. KULLBACK DISCRIMINATION INFORMATION (KDI)

When a fault has occurred in the system, the effect will appear as a change in identified model. We can thus detect the fault by discriminating the difference between two models which are identified in disjunctive intervals. Such a difference can be evaluated by using KDI. This is basic idea of our fault detection scheme based on the KDI. The KDI, a distortion measure of two probability density functions, was developed as an effective index to detect the fault in linear dynamics system with gaussian stochastic properties (Kumamaru and Söderström(1986)). As is mentioned in section 2, when the objective system is a black-box type non-linear system, the Quasi-ARMAX model with multi-model form (5) is used to identify the system. And the fault effect will appear as a change in estimated parameter $\hat{\Omega}_e$. However the KDI can not be applied to discriminate the identified multi-model, because the model is really a non-linear one. In order to solve this problem, the KDI is used to evaluate the difference in each identified local linear ARMAX model described by (6), then the overall difference in the identified multi-ARMAX model $\mathcal{M}(\hat{\Omega}_e)$ can be estimated from these local model differences.

3.1 *KDI analysis*

Assume that data from the system are available from two disjunctive time intervals I_1 and I_2 with number of data set N_1, N_2, respectively. Perform identification for the multi-ARMAX model (5) using data obtained from I_1 and I_2, and denote by $\hat{\Omega}_{eji}$ $(j = 1, \cdots, M)$, $(i = 1, 2)$ the estimates of j-th local model parameters.
We thus have identified local linear ARMAX models $\mathcal{M}(\hat{\Omega}_{ej1})$, $\mathcal{M}(\hat{\Omega}_{ej2})$ $(j = 1, \cdots, M)$. Introduce the KDI defined for the likelihood functions corresponding to these two models

$$I_{N_1}[1,2](j) = \int p(Z_{N_1}^{(j)} | \hat{\Omega}_{ej1}, U_{N_1 - 1}) \cdot \ln \frac{p(Z_{N_1}^{(j)} | \hat{\Omega}_{ej1}, U_{N_1-1})}{p(Z_{N_1}^{(j)} | \hat{\Omega}_{e2j}, U_{N_1-1})} dZ_{N_1}^{(j)} \quad (7)$$

where $U_{N_1-1} = [u(1), \cdots, u(N_1 - 1)]^T$ are input data sets in interval I_1, and $Z_{N_1}^{(j)} = [z_j(1), \cdots, z_j(N_1)]^T$ are augmented output variables of j-th local identified model. The index in (7) hence indicates how well the model using $\hat{\Omega}_{ej2}$ describes the data in the interval I_1. Due to the gaussian assumption on $\varepsilon(t)$, the likelihood functions corresponding to identified local linear ARMAX models are also gaussian distributed. Therefore, using Bayesian rule, (7) can be analyzed into an explicit form, by which the KDI's for each identified local linear ARMAX model can be evaluated in a feasible way for finite but fairly large data set. The result of such an analysis is:

$$I_{N_1}[1,2](j) = I_{N_1}^{(1)}[1,2](j) + I_{N_1}^{(2)}[1,2](j) + I_{N_1}^{(3)}[1,2](j) \quad (8)$$

where each component of $I_{N_1}[1,2](j)$ is given by

$$I_{N_1}^{(1)}[1,2](j) = \frac{N_1}{2}[(\hat{\sigma}_1/\hat{\sigma}_2 - 1) - ln(\hat{\sigma}_1/\hat{\sigma}_2)] \quad (9)$$

$$I_{N_1}^{(2)}[1,2](j) = \frac{1}{2}\sum_{k=0}^{N_1-1} \left\| (H_2^{(j)})^{-1} \cdot (G_1^{(j)} - G_2^{(j)})u(k+1) \right\|_{\hat{\sigma}_2^{-2}}^2 \quad (10)$$

$$I_{N_1}^{(3)}[1,2](j) = \frac{N_1}{2}\left[\frac{\hat{\sigma}_2^{-2}}{2\pi i}\oint \left\{(H_2^{(j)}(z))^{-1}H_1^{(j)}(z) - 1\right\} \cdot \hat{\sigma}_1^2 \left\{H_1^{(j)}(z^{-1})(H_2^{(j)}(z^{-1}))^{-1} - 1\right\}\frac{dz}{z}\right] \quad (11)$$

$$(j = 1, \cdots M)$$

and each simplified variable is defined by

$$G_i^{(j)} = B(q^{-1} : \hat{\Omega}_{eji})/A(q^{-1} : \hat{\Omega}_{eji})$$
$$H_i^{(j)} = C(q^{-1} : \hat{\Omega}_{eji})/A(q^{-1} : \hat{\Omega}_{eji})$$
$$H_i^{(j)}(z) = H(q^{-1} : \hat{\Omega}_{eji})/_{q^{-1}=z}$$
$$(i = 1, 2), (j = 1, \cdots M)$$

The difference measure for the identified multi-ARMAX models are estimated from $I_{N_1}[1,2](j)$ in (8), by weighting average of them over the number of multi-models M.

$$I_{FD} = \frac{1}{M}\sum_{j=1}^{M} W(j) I_{N_1}[1,2](j) \quad (12)$$

where $W(j)$'s are weighting factors. It is expected that an appropriate choice of weighting factors in (12) will make the index more sensitive to the change of identified multi-ARMAX model. Considering the structure of the multi-model (5) interpolated by the basis functions, we will set the weights by

$$W(j) = \frac{1}{N_1} \sum_{t=1}^{N_1} N_f(p_j, \phi(t)) \tag{13}$$

In this way, the index I_{FD} can be used to evaluate the difference between two identified Quasi-ARMAX models with multi-model form. Such a model discrimination based on the KDI can be applied to two problems, model validation and fault detection problems.

3.2 *Model Validation*

In the model-based approach to fault detection, it is important task to test the adequateness of the model, this is so called "model validation". Let us assume that the object system is time-invariant under the normal operation. Then, when the model structure is appropriately chosen, the difference between two identified models obtained from disjunctive intervals I_1 and I_2 under the normal mode may be comparatively small. On the other hand, if the model structure is not good enough to describe the system, the model parameter estimates will fluctuate strongly depending on the identification conditions. Therefore, the difference measure I_{FD} in (12) for two identified models can also be used as the index for model validation.

3.3 *Fault Detection Scheme*

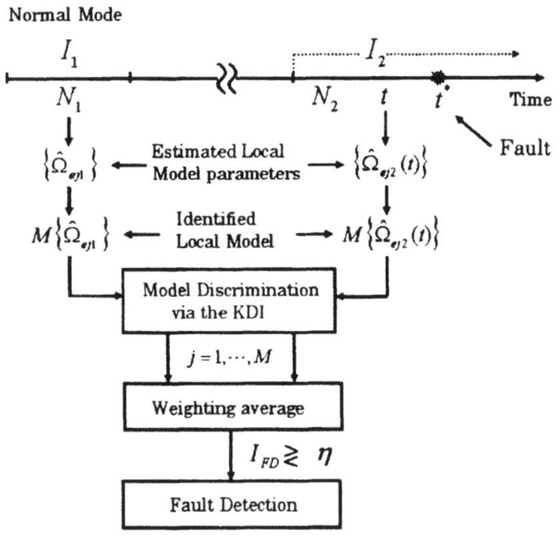

Fig. 1. Fault detection scheme

Fig.1 explains the procedure of KDI-based fault detection. Assume that the data are available from two disjunctive intervals I_1 and I_2 with number of data sets N_1 and N_2, respectively, where the system is under normal mode in the interval I_1 and is monitored for fault detection in the interval I_2. Perform the identification based on the Quasi-ARMAX model with multi-model form. Denote the identified model obtained in I_1 by $\mathcal{M}\{\hat{\Omega}_{ej1}\}$, and the on-line identified model obtained in I_2 by $\mathcal{M}\{\hat{\Omega}_{ej2}(t)\}$, respectively. The differences between two identified local ARMAX models are evaluated by the KDI, $I_{N_1}[1,2](j)$ in (8) and the differences measure I_{FD} for the two identified Quasi-ARMAX models can be obtained from weighting average of these KDI's. Then the fault detection is executed online based on the threshold decision approach by using the detection index I_{FD}.

4. SHIP-PROPULSION SYSTEM

As the plant to be diagnosed, a ship-propulsion system with one engine and one propeller is considered. This system-model has been constructed as the benchmark test of fault diagnosis (Izadi-Zamanabadi and Blanke(1999)). This system has following several features.

- The model of a marine vehicle. (length:147.2m,displacement:12,840m^3)
- Various non-linearities are faithfully modeled.
- 6 ordered non-linear system.
- Various stochastic factors are modeled. (measurement noise, external force etc...)

The schematic diagram of this system is shown in Fig.2.

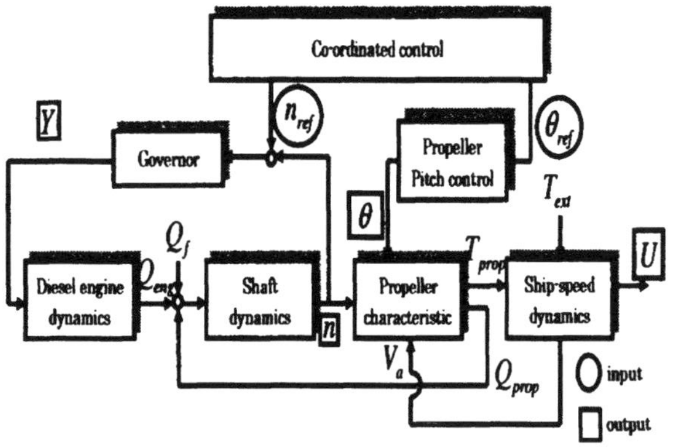

Fig. 2. Schematic diagram of the ship-propulsion system

Main components/subsystems of the system are:

- Propeller pitch control system.
- Governor.
- Diesel engine.
- Shaft dynamics.
- Propeller characteristics.
- Ship speed dynamics.

This system has two inputs and four measurable outputs (See Fig.2). Inputs and outputs are following

- inputs
  - the shaft speed reference: n_{ref}
  - the propeller pitch reference: θ_{ref}

- outputs
  - the diesel engine shaft speed: n
  - propeller pitch position: θ
  - the fuel index: Y
  - the ship speed: U

In the figure, Q_f and T_{ext} denote the friction torque and the external thrust force, respectively. Disturbance factors such as v_n, v_θ, v_Y and v_U are considered in n, θ, Y and U, respectively as the gaussian distributed variables with zero mean. The friction torque is random valued so it is treated as a disturbance. The ship plant model can realize many kinds of fault mode including both of dynamics and sensor faults.

5. SIMULATION STUDIES

For convenience's sake, the part of "Governor - Diesel Engine - Propeller Shaft" is selected as the objective system for fault detection. Where the input variable u_1 is the shaft speed reference n_{ref} and the torque Q_{prop} developed from propeller dynamics stands for another input variable u_2, while the output variable y is the measurement of the diesel engine shaft speed n. Note that the nonlinear elements included in the system are saturation and delay type ones.

5.1 *Quasi-ARMAX Modeling*

To describe the objective system, let us consider the following Quasi-ARMAX model;

$$A(q^{-1},\phi(t))y(t) = B_1(q^{-1},\phi(t))u_1(t) + B_2(q^{-1})u_2(t) + C(q^{-1})e(t) \qquad (14)$$

There are several factors specifying the model structure of the Quasi-ARMAX model. They are the model order n, m_1, m_2, l, the components of the vector ϕ, i.e. fuzzy variables, and the number of fuzzy rule M. These factors are determined appropriately according to the model validation procedure as is mentioned in Section 3.2. Such the model validation test has been executed during the experimental identification procedure under the normal mode in the interval I_1, where the number of the data set N_1 is set to be $N_1 = 40,000$. As to the data acquisition for the identification, the object system is driven by a given input sequence $u_1(t)$ with rectangular form, and resulting output sequence $y(t)$ is observed, while another input $u_2(t)$ is treated as known valued internal variable of the ship propulsion system. See Fig.3 for the sample of these input-output sequences.

Note here that the object system has the saturation or the limiter type nonlinear characteristics, which appear depending on the system operating condition.

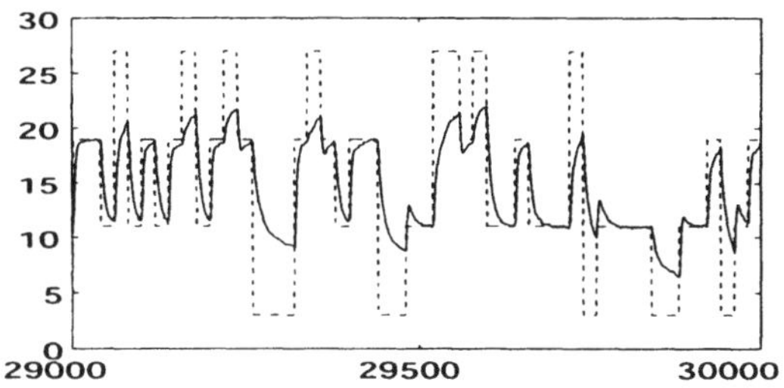

Fig. 3. Realized input-output sequences

Therefore, in order to treat the system as the nonlinear system, it should be excited by input $u_1(t)$ with relatively large variations. From Fig. 3, it can be seen that the nonlinear characteristics of the object system are clearly appearing in the output behavior. Thus the system can be considered as nonlinear system and the Quasi-ARMAX model is effectively used for the modeling and identification. Under these experimental conditions, the identification tests for model validation have been executed. As the result, we have had the following model structure:

- model order : $n = 3, m_1 = 4, m_2 = 2, l = 1$
- number of fuzzy rule : $M = 16$
- fuzzy variable : $\phi(t) = [y(t-1) \;\; u_1(t-1)]^T$
  (n' = 1 , m' = 1)

5.2 *Fault Mode*

Simulations on fault detection have been carried out for many kinds of dynamics fault under both of deterministic and stochastic situations. Among them, the result for a fault due to abrupt change in the engine gain k_y is introduced here. Where the fault occurrence time t^* is taken to be $1000th$ step in the interval I_2 with number of data set $N_2 = 2000$ and the degree of the parameter change is specified to $+20\%$ of its rating value. On the other hand, the KDI in (8) is calculated using input data sets U_{N_1-1} in the interval I_1 with $N_1 = 100$.

5.3 *Simulations on Fault Detection*

The following two cases are considered as simulation studies on fault detection;

Case 1 : Fault detection under a deterministic situation.

In this case, all the disturbance factors v_n, v_θ, v_Y and v_U are deleted from the model together with another random factor Q_f, and the object system is driven by a input sequence $u_1(t)$ with rectangular form as is shown in Fig. 3. Using the data, perform the fault detection based on the scheme explained in section 3.3. The result is shown in Fig.4 as the evaluated KDI sequences.

Case 2 : Fault Detection under a stochastic situation.

In this case, the plant dynamics suffer from disturbances v_n, v_θ, v_Y, v_U and Q_f. These disturbances are realized by samples from zero mean gaussian distributions with rating variance values. Thus the object system is operated under the stochastic situation. The simulation results (a), (b) and (c) are shown in Fig.5. Each of them corresponds to variance value of v_n, i.e., (a) is for the case $v_n \in \mathcal{N}(0,0.025)$ with rather small variance, (b) is for $v_n \in \mathcal{N}(0,0.050)$ and (c) is for $v_n \in \mathcal{N}(0,0.075)$ with rather large variance ,respectively. While the variance values of other disturbances are kept to be constant rating values.

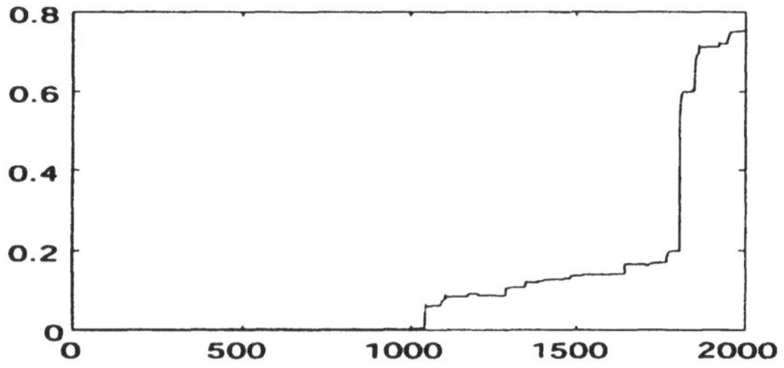

Fig. 4. Caluculated KDI (I_{FD}) under deterministic situation

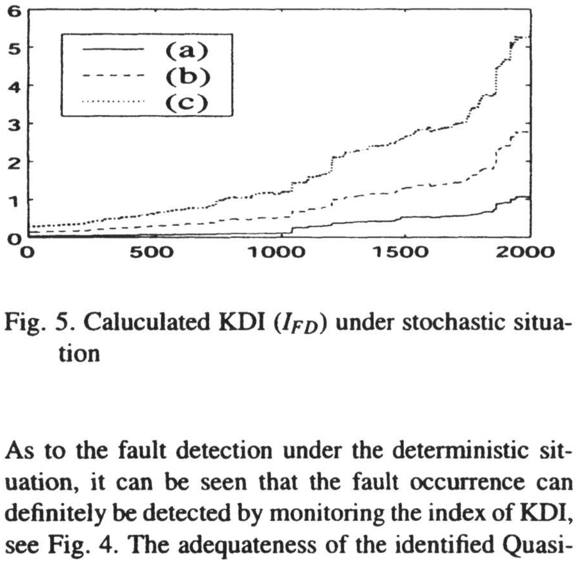

Fig. 5. Caluculated KDI (I_{FD}) under stochastic situation

As to the fault detection under the deterministic situation, it can be seen that the fault occurrence can definitely be detected by monitoring the index of KDI, see Fig. 4. The adequateness of the identified Quasi-ARMAX model with structure specified in the simulation can also be confirmed from the result shown in Fig. 4, since the model distortion measure KDI evaluated under the normal mode is kept to be small value. This means that the system parameters do not change, i.e. there is no fault.

On the other hand, in the case 2, the fault detection under the stochastic situation is still fairly possible in the case of weak disturbance, see the result of (a) in Fig. 5. However, the evaluated KDI becomes more larger than the deterministic case, and the fault detection ability may degrade as disturbances become larger and larger, see the results of (b) and (c) in Fig. 5. This is because that the identification performance is strongly affected by the disturbances.

6. CONCLUSIONS

In this paper, a fault detection method using Quasi-ARMAX model with multi-model form was proposed, and simulation studies on the ship propulsion system were carried out for various dynamics fault modes under both of deterministic and stochastic situations.

As the result, the effectiveness of the method and its applicability to realistic plant have been confirmed. However, the object system treated in the simulation is restricted to rather small-scaled subsystem of the plant. Simulation studies to examine the applicability of our method to more larger-scaled plant are under investigation.

REFERENCES

Hu,J.,K.Kumamaru and K.Hirasawa(2001). A Quasi-ARMAX approach to modelling of non-linear systems. *Interna. J. Control,* 74-18,1754-1766.

Izadi-Zamanabadi, R. and M.Blanke(1999). Ship propulsion system as a benchmark for fault tolerant control. *Control Eng. Practice*,7-3, 227-239.

Izermann,R. and P.Balle(1996). Trends in the application of model based fault detection and diagnosis of technical processes. In: *Proc of 13th World Congress of IFAC*, Vol.N,1-12.

Kumamaru,K. and T.Söderström(1986). Fault detection and model validation using index of Kullback discrimination information. *Trans. of the Society of Instrument and Control Engineers*, 22-10,1135-1140.

Kumamaru,K., J.Hu, K.Inoue and T.Söderström (1997). Fault detection of nonlinear systems by using hybrid quasi-ARMAX models. In: *Proc. of the IFAC Symposium-Safeprocess*,1126-1131.

Kumamaru,K., J.Hu, S.Furukawa and K.Inoue (1999). A fault detection method for nonlinear black-box systems based on multi-ARMAX modeling. In: *Proc. of the 31st ISCIE Interna. Symp. on Stochastic Systems Theory and Its Applications, Yokohama*, 67-72.

Kumamaru,K. K.Abe and K.Inoue(2001). KDI-based fault detection of non-linear systems by using quasi-ARMAX model and its application to ship propulsion system. In: *Proc. of the 33rd ISCIE Interna. Symp. on Stochastic Systems Theory and Its Applications, Tochigi*, 75-80.

Ljung,L. and T.Söderström(1983). *Theory and Practice of Recursive Identification.* The MIT Press, Cambridge,Mass..

IFAC
Publications
www.elsevier.com/locate/ifac

A COMPARISON OF TWO METHODS FOR STOCHASTIC FAULT DETECTION: THE PARITY SPACE APPROACH AND PRINCIPAL COMPONENTS ANALYSIS

Anna Hagenblad, Fredrik Gustafsson, Inger Klein

Department of Electrical Engineering, Linköpings universitet,
SE-581 83 Linköping, Sweden
Email: {annah, fredrik, inger,}@isy.liu.se

Abstract: This paper reviews and compares two methods for fault detection and isolation in a stochastic setting, assuming additive faults on input and output signals and stochastic unmeasurable disturbances. The first method is the parity space approach, analyzed in a stochastic setting. This leads to Kalman filter like residual generators, but with a FIR filter rather than an IIR filter as for the Kalman filter. The second method is to use principal component analysis (PCA). The advantage is that no model or structural information about the dynamic system is needed, in contrast to the parity space approach. We explain how PCA works in terms of parity space relations. The methods are illustrated on a simulation model of an F-16 aircraft, where six different faults are considered. The result is that PCA has similar fault detection and isolation capabilities as the stochastic parity space approach. *Copyright © 2003 IFAC*

Keywords: Fault detection, fault isolation, diagnosis, Kalman filtering, adaptive filters, linear systems, parity space, principal components analysis, PCA

1. INTRODUCTION

This paper concerns the detection and identification of additive faults on input and output signals, for a system that is described by a state space model. The parity space approach, (Basseville and Nikiforov, 1993; Chow and Willsky, 1984; Ding *et al.*, 1999; Gertler, 1997; Gertler, 1998) is a well-known method for this kind of problem, which is based on simple algebraic projections and geometry. The method computes a residual vector that is zero when no fault is present, and non-zero otherwise, to detect that a fault has occurred. The residual will also be different for different faults, to enable diagnosing which fault has occurred.

The parity space approach often shows very good results in simulations, but it can be highly sensitive to measurement noise and process noise, since these are not taken into consideration in the design of the parity space. We will briefly review the results in (Gustafsson, 2002), where a state space model which inludes both deterministic and stochastic unmeasurable disturbances is used and a statistical fault detection and isolation algorithm is derived. The probability for incorrect diagnosis can be computed explicitly for this method, given that only a single fault has occurred.

A singular value decomposition (SVD) is used in computing the parity space, and this is instrumental in many approaches to fault detection, see (Lou *et al.*, 1986) for another example. SVD is also the basic step in PCA.

If no model is available a priori, an alternative to estimating a state space model from data is to use principal components analysis, PCA. By a SVD of the covariance matrix for input output data, we can split the data into two parts, model and residual. The residual part can be used for fault detection similarly to the parity space residual. The covariance matrix can be estimated from normal operation data. To be able to isolate different faults, we also need data from

[1] This work was supported by VINNOVA's center of excellence, ISIS, Information Systems for Industrial Control and Supervision

typical fault cases, to estimate how the faults affects the residuals.

In this paper, we describe the two methods, and apply them to a model of a F-16 aircraft. The aircraft is described by a fifth order state-space model, with three inputs and three outputs.

2. THE PARITY SPACE APPROACH

2.1 *Stochastic Parity Spaces*

The linear system is defined as the state space model

$$\begin{aligned} x_{t+1} &= A_t x_t + B_{u,t} u_t + B_{f,t} f_t + B_{v,t} v_t \\ y_t &= C_t x_t + D_{u,t} u_t + D_{f,t} f_t + e_t \end{aligned} \tag{1}$$

We separate the following types of input:

- Deterministic known input u_t. This is common in control applications.
- Deterministic unknown fault input f_t, which is used in the fault detection literature. f_t is here assumed to be zero, or proportional to the unit vector, $f_t = m_t f^i$. The vector f^i corresponds to fault number i, and is zero except for element i which is one. m_t corresponds to the size of the fault. The matrices $B_{f,t}$ and $D_{f,t}$ determines which part of the system will be affected by the different faults.
- Stochastic unknown disturbances v_t and e_t, process noise and measurement noise, respectively, which are used in the Kalman filter setting. Both will here be assumed to be independent and Gaussian, with zero mean and covariance matrices Q_t and R_t, respectively.

Furthermore, the initial state x_0 is treated as an unknown variable. In the Kalman filter literature it is assumed to be Gaussian.

The diagnosis task can be formulated as a recursive problem applied to a sliding window. Stack L signal values to define the signal vectors $Y_t = (y_{t-L+1}^T, \ldots, y_t^T)^T$, etc for all signals $s \in \{u, f, v\}$. Also define the Hankel matrices

$$H_s = \begin{pmatrix} D_s & 0 & \ldots & 0 \\ CB_s & D_s & \ldots & 0 \\ \vdots & & \ddots & \vdots \\ CA^{L-2}B_s & \ldots & CB_s & D_s \end{pmatrix} \tag{2}$$

for all signals s and the observability matrix

$$\mathcal{O} = \begin{pmatrix} C \\ CA \\ \vdots \\ CA^{L-1} \end{pmatrix}. \tag{3}$$

Equation (1) can then be written as

$$Y_t - H_u U_t = \mathcal{O} x_{t-L+1} + H_f F_t + H_v V_t + E_t. \tag{4}$$

Next, a residual to be used for detection and diagnosis can be defined as

$$\begin{aligned} r_t &= W^T(Y_t - H_u U_t) & (5) \\ &= W^T(\mathcal{O} x_{t-L+1} + H_f F_t + H_v V_t + E_t) \\ &= W^T(H_f F_t + H_v V_t + E_t) & (6) \end{aligned}$$

where the last equality is obtained by construction of W. W is selected as a basis for the nullspace of $\mathcal{O}$, i.e., $W^T\mathcal{O} = 0$.

The residual is thus designed to be insensitive to the initial state. We have $r_t = 0$ for any initial state x_{t-L+1}, provided that we have no stochastic disturbance and no fault. If the residual is different from zero, this is due either to the noise v_t and e_t or to a fault f_t (or both). The diagnosis task aims to distinguish these causes.

Note that the residual can be regarded as the output of an FIR filter. This is in contrast to a traditional Kalman filter, which is IIR. Since the residual in Equation (5) is FIR, an input will only affect the residual a finite number of time steps. This means the residual will be able to faster react on new faults etc. (Gustafsson, 2002)

A parity space of non-zero dimension will always exist if L is chosen large enough. If the size of a signal s_t is denoted $n_s = \dim(s_t)$, the maximal dimension of the residual vector is given by

$$\max n_r = L n_y - n_x \tag{7}$$

2.2 *Diagnosis Algorithm*

The residual defined in Equation (5) is assumed to be Gaussian. Assume to facilitate notation that the measurement noise and process noises are time invariant, so the involved covariance matrices can be written $\mathbf{Cov}(E_t) = I_L \otimes R$ and $\mathbf{Cov}(V_t) = I_L \otimes Q$, respectively, where $\otimes$ denotes the Kronecker product. For a unity fault with constant magnitude m, the fault vector F_t in Equation (1) will be $F_t = mF^i$. We then get

$$\begin{aligned} (r_t | mf^i) &= W^T(H_v V_t + E_t + mH_f F^i) \\ &\in \mathbf{N}(mW^T H_f F^i, W^T S W) \end{aligned} \tag{8}$$

The Gaussian distribution requires that both V_t and E_t are Gaussian, which will be used for computing the probabilities for incorrect diagnosis, but is not required for derivation of the algorithm.

We let μ^i denote the vector $W^T H_f F^i$. The matrix S in the expression for the covariance is

$$S = H_v(I_L \otimes Q)H_v^T + I_L \otimes R. \tag{9}$$

The Equations (8) and (9) show that each fault f^i is mapped onto a vector μ^i with a covariance matrix $W^T S W$. To normalize the uncertainty in the residual, and to get a minimum variance residual, we define the new residuals

$$\bar{r}_t = (W^T S W)^{-1/2} r_t \quad (10a)$$
$$= \underbrace{(W^T S W)^{-1/2} W^T}_{\bar{W}^T} (Y_t - H_u U_t) \quad (10b)$$

This can be interpreted as selecting one particular null space for $\mathcal{O}$. The normalized residuals $\bar{r}_t$ will be

$$(\bar{r}_t | mf^i) = \bar{W}^T (H_v V_t + E_t + m H_f F^i) \in \mathbf{N}(m\bar{\mu}^i, I) \quad (11)$$

where $\bar{\mu}^i = \bar{W}^T H_f F^i$. The residuals are now whitened spatially by this normalization. The residuals are, however, correlated over the time window L by the FIR construction. We have

$$(\bar{r}_t | f = 0) \in \mathbf{N}(0, I) \quad \text{and} \quad (12)$$
$$(\bar{r}_t^T \bar{r}_t | f = 0) \in \chi^2(n_r) \quad (13)$$

We can use a χ^2-test with threshold h for detection of faults, and isolate the faults by finding the fault vector closest to the residual. This gives the following (well-known) algorithm:

Algorithm 1. **On-line diagnosis**

(1) Compute a normalized parity space $\bar{W}$, see Equation (10).
(2) Compute the normalized fault vectors $\bar{\mu}^i$ in the parity space. See Equation (11).
(3) Compute recursively:

$$\text{Residual:} \quad \bar{r}_t = \bar{W}^T (Y_t - H_u U_t)$$
$$\text{Detection:} \quad \bar{r}_t^T \bar{r}_t > h$$
$$\text{Isolation:} \quad \hat{i} = \arg\min_i \left\| \frac{\bar{r}_t}{\|\bar{r}_t\|} - \frac{\bar{\mu}^i}{\|\bar{\mu}^i\|} \right\|^2 = \arg\min_i \text{angle}(\bar{r}_t, \bar{\mu}^i)$$

Here, $\text{angle}(\bar{r}_t, \bar{\mu}^i)$ denotes the angle between the two vectors $\bar{r}_t$ and $\bar{\mu}^i$. It is possible to improve the false alarm rate by rejecting a detection if the angle is too large, i.e., no suitable isolation is found.

Using the Gaussian noise assumption, it is possible to compute the risk of incorrect diagnosis, in the case of only two faults. The expression is approximate if there are more than two possible faults, but in general the approximation is good, at least if the residuals are far from parallel. See (Gustafsson, 2002) for details and motivation for the following algorithm:

Algorithm 2. **Off-line diagnosis analysis**

(1) Compute a normalized parity space $\bar{W}$, see Equation (10).
(2) Compute the normalized fault vectors $\bar{\mu}^i$ in the parity space. See Equation (11).
(3) The probability of incorrect diagnosis is approximately

$$\text{prob}(\text{diagnosis } i | \text{ fault } mf^j) = \frac{1}{2}\text{erfc}\left(m \left\| \bar{\mu}^j - \frac{(\bar{\mu}^j, \bar{\mu}^j + \bar{\mu}^i)}{(\bar{\mu}^j + \bar{\mu}^i, \bar{\mu}^j + \bar{\mu}^i)} (\bar{\mu}^j + \bar{\mu}^i) \right\| \right) \quad (14)$$

erfc denotes the Gaussian error function,

$$\text{erfc}(x) = 2 \int_x^\infty \frac{1}{\sqrt{2\pi}} e^{-x^2/2} dx$$

$(\bar{\mu}^j, \bar{\mu}^i)$ denotes the scalar product of the vectors $\bar{\mu}^j$ and $\bar{\mu}^i$ and m denotes the magnitude of the fault. A similar expression is obtained if the magnitude is not constant, see (Gustafsson, 2002).

3. PRINCIPAL COMPONENTS ANALYSIS

If no model is available for the diagnosis, it may be possible to identify a state space model from data, and then apply the parity space methods described in the previous section. An alternative to this is to use principal components analysis, PCA (Dunia *et al.*, 1996). Stack the inputs and outputs into a data vector $Z_t = \begin{pmatrix} Y_t^T & U_t^T \end{pmatrix}^T$. PCA splits the data into two parts, model and residual:

$$Z_t = \begin{pmatrix} Y_t \\ U_t \end{pmatrix} = \hat{Z}_t + \tilde{Z}_t = \mathcal{O} x_t + W r_t \quad (15)$$

The notation has been chosen to show the resemblence with the model-based approach, though the relation is rather informal. We first describe how to compute this representation, and then comment on properties, relations and applications.

A singular value decomposition (SVD) is applied to the estimated covariance matrix of Z_t as follows:

$$\hat{R}_Z = \frac{1}{N-L} \sum_{t=L+1}^{N} Z_t Z_t^T = U D U^T \quad (16)$$

Here U is a square unitary matrix, that is $U^T U = U U^T = I$, and D is a diagonal matrix containing the singular values of $\hat{R}_Z$. We will split the SVD into two parts as

$$U = \begin{pmatrix} \mathcal{O} & W \end{pmatrix}, \quad D = \begin{pmatrix} D_x & 0 \\ 0 & D_r \end{pmatrix} \quad (17)$$

The split assigns the n_x largest singular values to the model, and the other n_r singular values are assumed to belong to the residual space. By construction, we have $\mathcal{O}^T \mathcal{O} = I_{n_x}$, $\mathcal{O}^T W = 0$, $W^T \mathcal{O} = 0$, $W^T W = I_{n_r}$ and $W W^T + \mathcal{O}\mathcal{O}^T = I_{n_x + n_r}$.

The split in (15) is computed by

$$\hat{Z}_t = \mathcal{O}\mathcal{O}^T Z_t \quad (18a)$$
$$\tilde{Z}_t = W W^T Z_t. \quad (18b)$$

The first term $\mathcal{O} x_t$ in (15) is the 'model', where the data belong to an observability space $\mathcal{O}$, where the 'state' x_t denotes the coordinates of the data at time t. The usual notion of observability applies, so a state observer is given by $\mathcal{O}^T Z_t = x_t$.

The second term $W r_t$ in (15) is the residual space spanned by W, which as before is a basis for the null space of $\mathcal{O}$, and r_t denotes the coordinates for the residual at time t.

For fault identification, we take the residuals

$$r_t = W^T Z_t \tag{19}$$

$$\bar{r}_t = D_r^{-1/2} W^T Z_t, \tag{20}$$

where the transformation implies $\mathbf{Cov}(r_t) = I$ in the limit $N \to \infty$. Note that the data projection matrix W^T here, corresponds to $W^T[I, -H_u]$ in (5).

PCA does not use any a priori fault model, which makes isolation of the faults more difficult. The analytic fault vectors (c.f. Algorithm 2 and Figure 2) cannot be computed. If data from a particular fault is available, it can however be estimated, by calculating the corresponding residual and estimating its mean and covariance. If the system is linear and the faults are additive (as assumed for the parity space approach described previously), the covariance matrix does not change. That is, take

$$\mu_i = \mathbf{E}(r_t^i) = \mathbf{E}(W^T Z_t^i), \tag{21}$$

$$\bar{\mu}_i = \mathbf{E}(\bar{r}_t^i) = \mathbf{E}(D_r^{-1/2} W^T Z_t^i) \tag{22}$$

for data Z_t^i known to suffer from fault i.

4. EXAMPLE

Signal	Not.	Meaning	Size
Inputs	u_1	spoiler angle (0.1 deg)	1
	u_2	forward accelerations (m/s^2)	1
	u_3	elevator angle (deg)	1
Outputs	y_1	relative altitude (m)	10^{-4}
	y_2	forward speed (m/s)	10^{-6}
	y_3	pitch angle (deg)	10^{-6}
Disturb.	d_1	speed disturbance	–
States	x_1	altitude (m)	10^{-4}
	x_2	forward speed (m/s)	10^{-4}
	x_3	pitch angle (deg)	10^{-4}
	x_4	pitch rate (deg/s)	10^{-4}
	x_5	vertical speed (deg/s)	10^{-4}
Faults	f_1	spoiler angle actuator	0.5
	f_2	forward acceleration actuator	0.1
	f_3	elevator angle actuator	1
	f_4	relative altitude sensor	1
	f_5	forward speed sensor	1
	f_6	pitch angle sensor	1

Table 1. Signals in the F16 simulation study. Size means the variance for the inputs, measurement noise variance for the outputs, state noise variance for the states and constant magnitude for the faults, respectively.

The fault detection algorithm is applied to a model of the vertical dynamics of an F-16 aircraft. The model is taken from (Gustafsson, 2000), which is a sampled version of a model in (Maciejowski, 1989). The involved signals and their generation in the simulations are summarized in Table 1. Input, state and measurement noises are all simulated as independent Gaussian variables, whose variance is given in the same table.

We have the following numerical values for the matrices in Equation (1):

$$A = \begin{pmatrix} 1 & 0.0014 & 0.1133 & 0.0004 & -0.0997 \\ 0 & 0.9945 & -0.0171 & -0.0005 & 0.0070 \\ 0 & 0.0003 & 1.0000 & 0.0957 & -0.0049 \\ 0 & 0.0061 & -0.0000 & 0.9130 & -0.0966 \\ 0 & -0.0286 & 0.0002 & 0.1004 & 0.9879 \end{pmatrix} \tag{23}$$

$$B_u = \begin{pmatrix} -0.0078 & 0.0000 & 0.0003 \\ -0.0115 & 0.0997 & 0.0000 \\ 0.0212 & 0.0000 & -0.0081 \\ 0.4150 & 0.0003 & -0.1589 \\ 0.1794 & -0.0014 & -0.0158 \end{pmatrix} \tag{24}$$

$$B_d = \begin{pmatrix} 0 & 1 & 0 & 0 & 0 \end{pmatrix}^T \tag{25}$$

$$B_f = \begin{pmatrix} -0.0078 & 0.0000 & 0.0003 & 0 & 0 & 0 \\ -0.0115 & 0.0997 & 0.0000 & 0 & 0 & 0 \\ 0.0212 & 0.0000 & -0.0081 & 0 & 0 & 0 \\ 0.4150 & 0.0003 & -0.1589 & 0 & 0 & 0 \\ 0.1794 & -0.0014 & -0.0158 & 0 & 0 & 0 \end{pmatrix} \tag{26}$$

$$C = \begin{pmatrix} 1 & 0 & 0 & 0 & 0 \\ 0 & 1 & 0 & 0 & 0 \\ 0 & 0 & 1 & 0 & 0 \end{pmatrix}, D_f = \begin{pmatrix} 0 & 0 & 0 & 1 & 0 & 0 \\ 0 & 0 & 0 & 0 & 1 & 0 \\ 0 & 0 & 0 & 0 & 0 & 1 \end{pmatrix} \tag{27}$$

D_u and D_d are zero matrices of appropriate dimensions.

Residuals were computed for the fault-free case, and for the six different single faults described above, according to Algorithm 1, the stochastic parity space approach. The time window L was selected to 3. This gives a four-dimensional ($n_r = Ln_y - n_x = 3 \cdot 3 - 5 = 4$) residual, which is illustrated in Figure 1.

It is clear from the figure that some of the faults are easy to detect and isolate, while some (where the residuals are closer to the origin) are harder. Fault f_4, fault in the relative altitude sensor, gives a zero residual, so it cannot be detected. The threshold is chosen to $h = 9.3$ to get a false alarm rate of 0.05. The probability of correct isolation is in this simulation and for this threshold 1, 1, 0.96, 0.05, 0.72, 1, respectively. That is, fault 4 is not possible to isolate or detect. Note that the fault size, as well as the noise level, will affect the detectability and isolability of the faults. This can be analyzed using Algorithm 2.

Algorithm 2 gives the mean fault vector. For the normalized residuals, a unit circle corresponds to one standard deviation. This is illustrated in Figure 2. The arrows indicate the directions of the residuals for the different faults. A larger fault will give a residual with the same direction, but a longer vector, and vice versa for a smaller fault. To be able to isolate different faults, the angle between the fault vectors is thus important, something that is also seen in Algorithm 2, Equation (14), where the scalar product can be interpreted as this angle.

The probability of incorrect diagnosis, Equation (14), can be calculated analytically. The matrix below contains these probabilities, where $P^{(i,j)}$ denotes prob(diagnosis$i$|fault$j$). The residual for fault f_4 is zero, the relative altitude fault cannot be detected simply because we do not measure absolute height.

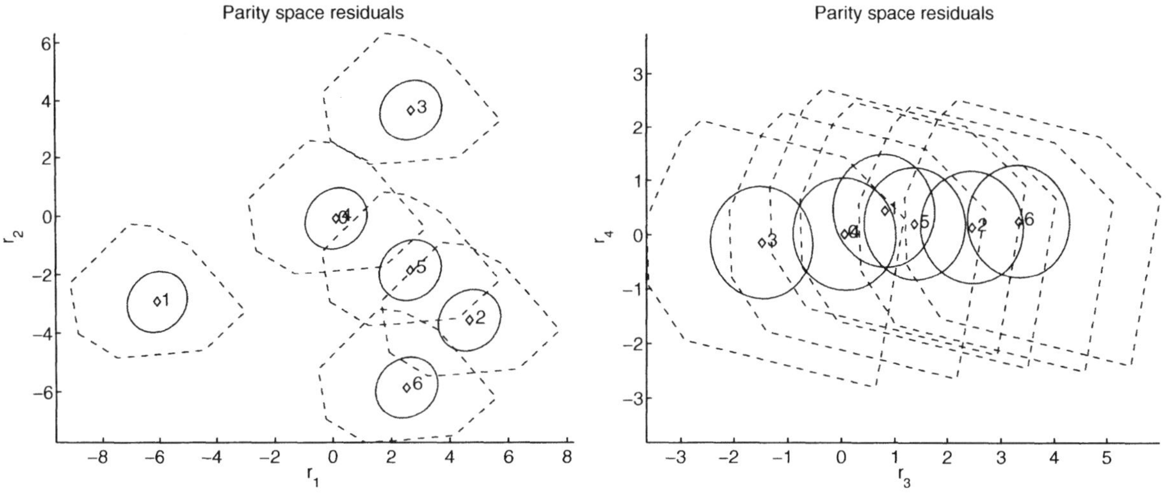

Fig. 1. Illustration of the residuals from parity space for no fault (0) and fault 1–6, respectively. The mean value, estimated covariance matrix and convex hull of each group of residuals are illustrated. Fault 4 is obviously not diagnosable, and residual r_4 contains almost no information.

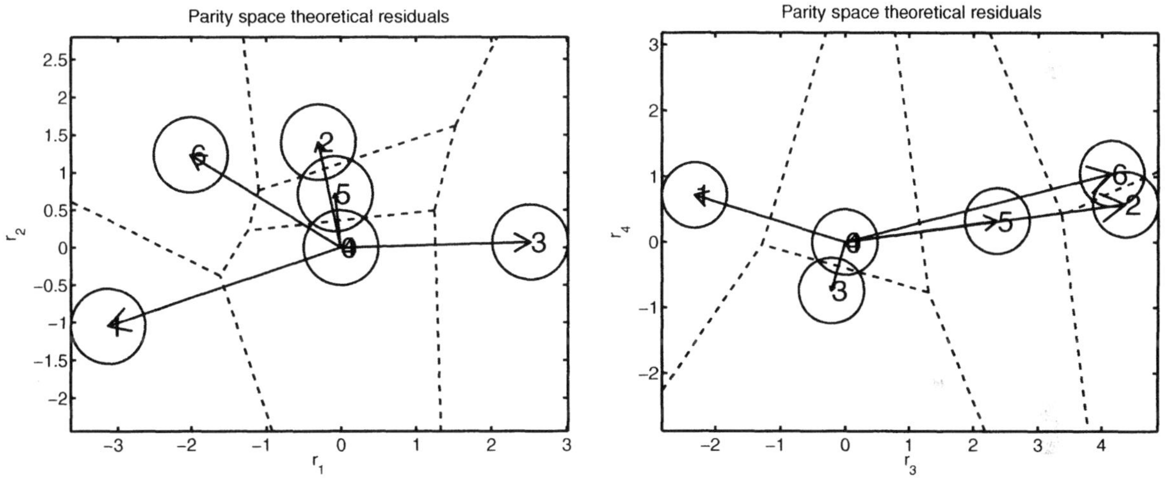

Fig. 2. Illustration of the residuals from parity space for no fault (0) and fault 1–6, respectively, but here in another basis. This confirms that fault 4 is not diagnosable. The decision lines for fault isolation are indicated.

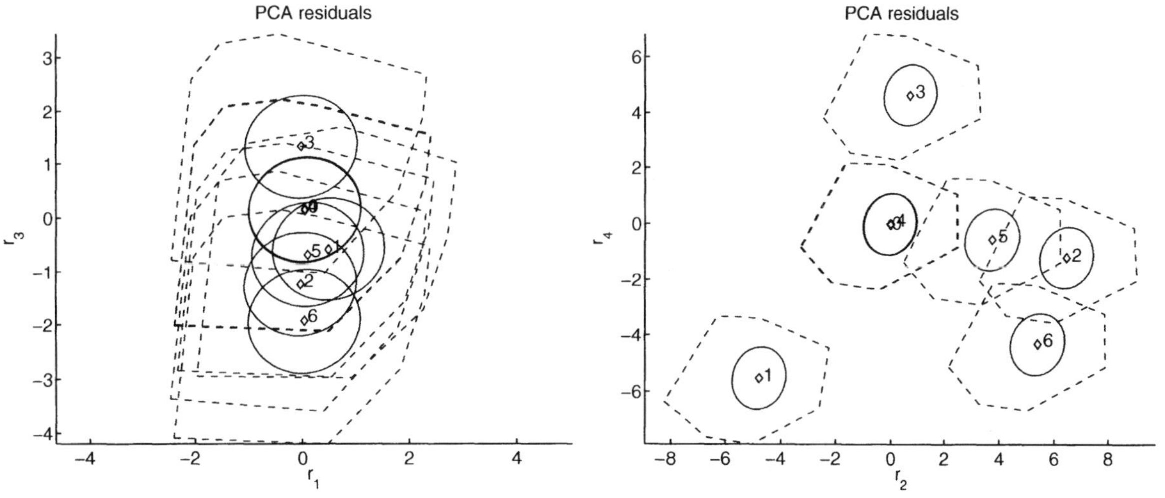

Fig. 3. Illustration of the residuals from PCA for no fault (0) and fault 1–6, respectively. The mean value, estimated covariance matrix and convex hull of each group of residuals are illustrated. These can however not directly be compared to the residual components in Figures 1 and 2 due to that the bases are different. Again, fault 4 is not diagnosable, and here residual r_1 contains little information.

This means that probability of incorrect as well as correct diagnosis all can be considered zero ($P^{(i,4)}$ and $P^{(4,i)}$).

$$P = \begin{pmatrix} 1.0000 & 0.0000 & 0.0000 & 0 & 0.0000 & 0.0000 \\ 0.0000 & 0.5980 & 0.0000 & 0 & 0.4020 & 0.0001 \\ 0.0000 & 0.0000 & 0.9999 & 0 & 0.0001 & 0.0000 \\ 0 & 0 & 0 & 0 & 0 & 0 \\ 0.0000 & 0.4020 & 0.0001 & 0 & 0.5415 & 0.0564 \\ 0.0000 & 0.0001 & 0.0000 & 0 & 0.0564 & 0.9436 \end{pmatrix} \tag{28}$$

The probability for incorrect diagnosis is very small in most cases. The case that poses the most problems is to distinguish faults f_2 and f_5. These two faults are also very close in Figure 2, in the sense that they are almost parallel.

Simulations of PCA are shown in Figure 3. The dimension of the residuals (the dimension of $\tilde{P}$ in Equation (16)) is selected to 4, to facilitate a comparison with the parity space approach. Figure 3 shows the residuals. Note that the residual components are not the same as in the parity space approach in Figure 1, since we have another basis for the residual space. The threshold is chosen to $h = 9.7$ to get a false alarm rate of 0.05. The probability of correct isolation is in this simulation and this threshold $1,\ 10.96,\ 0.05,\ 0.67,\ 1$, respectively. That is, compared to the parity space approach almost the same, and only a slightly worse performance for isolating fault 5.

The residual component r_1 from the PCA method is very small for all faults. This suggests that it does not contain information about the faults, and that the residual space is indeed only three-dimensional. From the simulations and analysis of the stochastic parity space approach, it appears that the residual component r_4 plays a similar role, and contain very little information for fault isolation.

5. CONCLUSIONS

In this paper, two approaches to fault detection and isolation are compared, the parity space approach and PCA, principle components analysis. The assumptions, advantages and drawbacks of these approaches are summarized below:

- The parity space approach starts with a state space model of the system. The use of prior model knowledge improves the performance compared to PCA. With a partially known model, system identification techniques can be applied. Generally, the more prior structural knowledge, the better performance. Another advantage is that *a priori* probabilities of incorrect diagnosis can be calculated.
- PCA requires absolutely no prior knowledge, not even causality (which ones of the known signals in z_t are inputs u_t and outputs y_t, respectively). The performance has been demonstrated to be only slightly worse compared to the case of perfect model knowledge. Determination of the state dimension is one critical step in PCA, and it is based on the singular values of the data correlation matrix. Over-estimating the state dimension gives too few residuals which decreases performance. Under-estimating state dimension can give very good performance, in that new residuals almost belonging to the parity space are used for detection and diagnosis. One major risk here, is that when the system enters a new operating point which was never reached in the training data, this residual might increase in magnitude.

A recently proposed analysis of the parity space approach in a stochastic setting was surveyed. One contribution is the detailed interpretation of PCA analysis in terms of parity space notation.

6. REFERENCES

Basseville, M. and I. V. Nikiforov (1993). *Detection of Abrupt Changes, Theory and Application*. Information and system sciences series. Prentice-Hall. Enlewood Cliffs, NJ.

Chow, A. Y. and A. S. Willsky (1984). Analytical redundancy and the design of robust failure detection systems. *IEEE Transactions on Automatic Control* **29**(7), 603–614.

Ding, X., L. Guo and T. Jeinsch (1999). A characterization of parity space and its application to robust fault detection. *IEEE Transactions on Automatic Control* **44**(2), 337–343.

Dunia, R., S. J Qin, T. F. Edgar and T. J. McAvoy (1996). Use of principal components analysis for sensor fault analysis. *Computers and Chemical Engineering* **20**(971), S713–S718.

Gertler, J. (1997). Fault detection and isolation using parity relations. *Control Engineering Practice* **5**(5), 653–661.

Gertler, J. J. (1998). *Fault Detection and Diagnosis in Engineering Systems*. Marcel Dekker, Inc.

Gustafsson, Fredrik (2000). *Adaptive Filtering and Change Detection*. Wiley. Baffins Lane, Chichester, West Sussex, PO 19 1UD, England.

Gustafsson, Fredrik (2002). Stochastic fault diagnosability in parity spaces. In: *Proceedings of the 15th IFAC World Congress*. Barcelona, Spain.

Lou, X-C., A.S. Willsky and G. Verghese (1986). Optimally robust redundancy relations for failure detection in uncertain systems. *Automatica* **22**(3), 333–344.

Maciejowski, J. M (1989). *Multivariable Feedback Design*. Addison Wesley.

IFAC Publications
www.elsevier.com/locate/ifac

IDENTIFICATION OF OBJECT'S MOVEMENT MODELS IN A RADAR TRACKING FILTER

Mirosław Sankowski * and Zdzisław Kowalczuk **

** Telecommunications Research Institute, Gdańsk Division, ul. Hallera 13, 80-401 Gdańsk, Poland, msankowski@pit.gda.pl*
*** Gdansk Univ. of Technology, Dept. of Automatic Control, ul. Narutowicza 11/12, 80-952 Gdańsk, Poland, kova@pg.gda.pl*

Abstract: The basic problem of tracking moving objects lies in unpredictability of object manoeuvres with respect to the type of trajectory and the time of occurrence. A method for identification of manoeuvre trajectory type is described, which is based on the analysis of the innovation process of the Kalman filter involved. *Copyright © 2003 IFAC*

Keywords: Nonlinear models, Identification, Kalman filters, Target tracking.

1. INTRODUCTION

Tracking filters (TFs) make a principal part in radar data processing systems. In fact, a TF is a state estimator of an object being tracked. Its task is to process radar measurements in order to reduce the measurement errors by means of time averaging, to estimate the object's velocity and acceleration, and to predict its future positions.

The basic problem of tracking manoeuvring moving objects lies in unpredictability of object manoeuvres, with respect to the type of their trajectory and the time of their occurrence, and is similar to the problem of failure detection and isolation (Kerr, 1989). A number of different approaches to tackle manoeuvring targets can be found in the vast literature on the subject (Bar-Shalom and Fortmann, 1988; Bar-Shalom and Rong Li, 1995; Blackman and Popoli, 1999; Farina and Studer, 1985). However, the most successful results have been obtained by using adaptive state-estimation techniques.

It is assumed in this work that a 'regular' object's movement fits the model of straight-line uniform motion, while manoeuvres appear rarely. This assumption means that the considered objects are supposed not to manoeuvre permanently. Moreover, the state of manoeuvring is interpreted as a result of driving the object by an unknown deterministic input process (Chan *et al.*, 1979; Kerr, 1989). By constructing an appropriate set of classes of analytical models of manoeuvres, e.g. a model of uniform speed change and a model of standard turn, it is possible to describe how different manoeuvres influence the state estimation process performed by a Kalman filter (KF), based on a non-manoeuvre model of motion. This influence can be directly observed in a bias of its innovation process.

The above methodology leads to a diagnostic algorithm (Kowalczuk and Sankowski, 2003), which provides for *detection* of the event of manoeuvre, estimation of its onset time instant, *isolation* of its trajectory within the considered classes of manoeuvre paths, as well as for *estimation* of trajectory parameters. This method allows for implementation of a multiple-model adaptive state-estimation algorithm, based on switched KFs designed for certain partial models of a movement trajectory (Kowalczuk and Sankowski, 2002). In this paper the method proposed by Kowalczuk and Sankowski (2003) is reformulated for a set of dynamic models with state vectors containing Cartesian position coordinates and polar velocity (CPPV) coordinates (Gustafsson and Isaksson, 1996; Rong Li and Jilkov, 2000). The resulting algorithms for isolation (structural identification) and estimation (parametric identification) of the manoeuvre model will be tested via simulation and compared to those obtained from the tracking filter based on a state vector consisting of Cartesian position and Cartesian velocity (CPCV) (Kowalczuk and Sankowski, 2003).

2. MODEL OF MEASUREMENT PROCESS

It is assumed that a radar provides unbiased measurements of the object positions (plots) in polar coordinates (slant range $\tilde{\varrho}[n]$ and azimuth $\tilde{\alpha}[n]$, where $\tilde{}$ designates a measured variable) in the discrete-time instants n. The polar *Range-Azimuth* (RA) coordinate system originates in the radar location with the line of sight pointing to the North for $\alpha = 0$. The measurements are assumed to be disturbed by additive stationary independent Gaussian-distributed random processes (noises) r_ϱ and r_α with zero expectations and variances σ_ϱ^2 and σ_α^2, respectively. Such measurements constitute the following 2-dimensional (2D) measurement vector

$$\boldsymbol{z}[n] = \begin{bmatrix} \tilde{\varrho}[n] \\ \tilde{\alpha}[n] \end{bmatrix} = \begin{bmatrix} \varrho[n] \\ \alpha[n] \end{bmatrix} + \begin{bmatrix} r_\varrho[n] \\ r_\alpha[n] \end{bmatrix} \tag{1}$$

characterised by a diagonal covariance matrix of the measurement noise

$$\boldsymbol{R} = \begin{bmatrix} \sigma_\varrho^2 & 0 \\ 0 & \sigma_\alpha^2 \end{bmatrix}. \tag{2}$$

The common method in modelling of aircraft trajectories for radar tracking systems is to use both a local Cartesian *East-North* (EN) coordinate system for describing the system dynamics and a first-order approximation of measurement nonlinearities. The EN coordinate system has its origin in the radar location and is defined by two axes: an x axis pointing to the East (E) and a y axis pointing to the North (N). This principle can be utilised in a *converted measurement* method (Blackman and Popoli, 1999), which is based on a transformation of the measured coordinates from the RA coordinate system to the EN frame:

$$\boldsymbol{y}[n] = \boldsymbol{g}(\boldsymbol{z}[n]) = \begin{bmatrix} \tilde{\varrho}[n]\sin(\tilde{\alpha}[n]) \\ \tilde{\varrho}[n]\cos(\tilde{\alpha}[n]) \end{bmatrix}. \tag{3}$$

The resulting pseudo-measurement vector

$$\boldsymbol{y}[n] = \boldsymbol{p}[n] + \boldsymbol{e}[n] \tag{4}$$

with

$$\boldsymbol{p}[n] = \begin{bmatrix} x[n] \\ y[n] \end{bmatrix}, \boldsymbol{e}[n] = \begin{bmatrix} e_x[n] \\ e_y[n] \end{bmatrix}. \tag{5}$$

of dimension $\mathrm{d}_y = 2$ will henceforth be referred to as the measurement vector.

Quantities $x[n]$ and $y[n]$, constituting the true position vector $\boldsymbol{p}[n]$, describe a real object position, while $e_x[n]$ and $e_y[n]$ can be interpreted as projections of the measurement errors $r_\varrho[n]$ and $r_\alpha[n]$ into the Cartesian frame. Since the transformation (3) is nonlinear, the random variables $e_x[n]$ and $e_y[n]$ are not Gaussian in general. Nevertheless, it is convenient to assume the Gaussian distribution of the measurement errors in the Cartesian coordinate system, as well. With this assumption the covariance matrix $\boldsymbol{E}[n]$ of the measurement errors in the Cartesian frame can be approximated by linearization of (3):

$$\boldsymbol{E}[n] = \boldsymbol{G}[n]\boldsymbol{R}\boldsymbol{G}^{\mathrm{T}}[n] \tag{6}$$

where $\boldsymbol{R}$ is defined in (2), while $\boldsymbol{G}[n]$ is the Jacobian matrix of the transformation $\boldsymbol{g}(\cdot)$ from the polar RA coordinates to the Cartesian EN ones, evaluated for the measurement $\boldsymbol{z}[n]$.

3. MODELS OF OBJECT MOTION

Based on the characteristics of civil aircrafts we assume that the moving-object trajectory consists of straight-line motions (travel paths) and certain manoeuvres. Precisely, we presume that the object moves according to the model of straight-line uniform motion and then, at a time instant t_0, it starts to manoeuvre according to a suitable physically-based model. This manoeuvre lasts until a time moment t_0'.

A continuous-time dynamic model of planar curvilinear motion can be shown to have the form

$$\boldsymbol{a}(t) = a_1(t)\boldsymbol{u}_1(t) + a_2(t)\boldsymbol{u}_2(t) \tag{7}$$

where $\boldsymbol{a}(t)$ is an acceleration vector, while

$$\boldsymbol{u}_1(t) = \begin{bmatrix} \sin(\psi(t)) \\ \cos(\psi(t)) \end{bmatrix}, \boldsymbol{u}_2(t) = \begin{bmatrix} \cos(\psi(t)) \\ -\sin(\psi(t)) \end{bmatrix} \tag{8}$$

denote unit vectors, which are respectively tangential and normal to the trajectory, $a_1(t)$ and $a_2(t)$ describe corresponding tangential and normal accelerations, and $\psi(t)$ is a course of the object. Fig. 1 shows a geometrical model of such curvilinear motion.

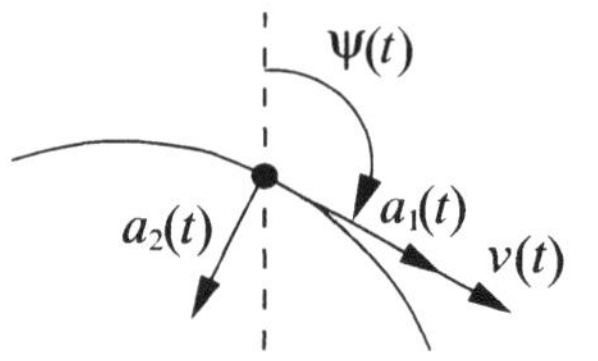

Figure 1. Geometry of curvilinear motion

In order to classify the models of the object's motion, three types of trajectories are associated with the following set of hypotheses $\{\mathrm{H}j\}$, $j \in \{0,1,2\}$ (Kowalczuk and Sankowski, 2003), including:

- hypothesis H0 (uniform motion at constant velocity CV) meaning that the object travels along a straight line ($a_1(t) = 0$ and $a_2(t) = 0$),

- hypothesis H1 (uniform speed change with constant acceleration CA) meaning that the object moves along a straight line with a constant tangential acceleration $a_1(t)=a_1$ and $a_2(t)=0$, for $t_0 \leq t < t_0'$,
- hypothesis H2 (standard or coordinated turn CT) that the object manoeuvres on a circular path with a constant normal acceleration $a_2(t) = a_2$ and $a_1(t)=0$, for $t_0 \leq t < t_0'$.

Fig. 2 explains the relationship between the defined object's trajectories within the assumed set of motion models.

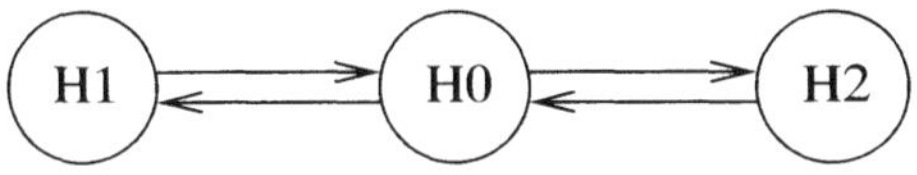

Figure 2. Graph of possible transitions between the object's trajectories (hypotheses).

Additionally, we shall distinguish a useful subset of analytical manoeuvre models within the motion-model set given above. Namely, a subclass consisting of hypotheses H1 and H2 can be associated with an aggregate hypothesis of manoeuvre HM={H1,H2}.

3.1 Constant-velocity model CV

Discretization of the model (7) with $a_1(t) = 0$ and $a_2(t)=0$ results in a discrete-time second-order linear state-space model, which can be extended by an additional forcing input $\boldsymbol{w}[n]$ and completed with an observation equation:

$$\boldsymbol{x}[n+1] = \boldsymbol{f}(n+1,n;\boldsymbol{x}[n]) + \boldsymbol{w}[n] \quad (9)$$

$$\boldsymbol{y}[n] = \boldsymbol{C}\boldsymbol{x}[n] + \boldsymbol{e}[n] \quad (10)$$

where

$$\boldsymbol{x}[n] = \begin{bmatrix} \boldsymbol{p}[n] \\ \boldsymbol{h}[n] \end{bmatrix}, \boldsymbol{h}[n] = \begin{bmatrix} v[n] \\ \psi[n] \end{bmatrix}, \boldsymbol{C} = \begin{bmatrix} \mathbf{I}_{2\times 2} \\ \mathbf{0}_{2\times 2} \end{bmatrix}^{\mathrm{T}} \quad (11)$$

$$\boldsymbol{f}(n+1,n;\boldsymbol{x}[n]) = \begin{bmatrix} \boldsymbol{p}[n] + T_n v[n] \boldsymbol{u}_1[n] \\ \boldsymbol{h}[n] \end{bmatrix} \quad (12)$$

In the above, $\boldsymbol{x}[n]$ is a state vector, $\boldsymbol{h}[n]$ is a polar velocity vector, $\{\boldsymbol{w}[n]\}$ is a stationary Gaussian white-noise vector sequence with a covariance matrix $\boldsymbol{W}$, $\boldsymbol{C}$ is an output matrix, while $T_n = t_{n+1} - t_n$ denotes a sampling period. We also assume that sequences $\{\boldsymbol{w}[n]\}$, $\{\boldsymbol{e}[n]\}$ and an initial Gaussian-distributed state $\boldsymbol{x}[t_1]$ are mutually uncorrelated. On a common basis, $\mathbf{0}_{r\times c}$ and $\mathbf{I}_{r\times c}$ denote null and identity matrices, respectively. The subscript $r\times c$ declares the size of the matrix having r rows and c columns.

3.2 Controlled constant-acceleration model CCA

Discretization of the model (7) with $a_1(t) = a_1$ and $a_2(t) = 0$ results in the following discrete-time state-space model with the exogenous input a_1:

$$\boldsymbol{x}_1^*[n+1] = \boldsymbol{f}(n+1,n;\boldsymbol{x}_1^*[n]) + \\ + \boldsymbol{b}_1[n+1,n;\cdot]a_1 + \boldsymbol{w}[n] \quad (13)$$

with

$$\boldsymbol{b}_1[n+1,n;\boldsymbol{x}_1^*[n_0]] = \begin{bmatrix} \frac{1}{2}T_n^2 \cdot \boldsymbol{u}_1[n_0] \\ a_1^{-1} \cdot \Delta\boldsymbol{h}_1[n+1,n] \end{bmatrix} \quad (14)$$

where

$$\Delta\boldsymbol{h}_1[n+1,n] = \begin{bmatrix} a_1 T_n \\ 0 \end{bmatrix} \quad (15)$$

is a change of polar velocity due to the control signal a_1.

3.3 Controlled coordinated turn model CCT

Discretization of the model (7) with $a_1(t) = 0$ and $a_2(t) = a_2$ results in the following discrete-time state-space model with the exogenous input a_2:

$$\boldsymbol{x}_2^*[n+1] = \boldsymbol{f}(n+1,n;\boldsymbol{x}_2^*[n]) + \\ + \boldsymbol{b}_2[n+1,n;\cdot]a_2 + \boldsymbol{w}[n] \quad (16)$$

for $n_0 \leq n < n_0'$ with

$$\boldsymbol{b}_2[n+1,n;\boldsymbol{x}_2^*[n_0],a_2] = \\ = \omega^{-1} \begin{bmatrix} -T_n \boldsymbol{u}_1[n] - \omega^{-1}\Delta\boldsymbol{u}_2[n] \\ v^{-1} \cdot \Delta\boldsymbol{h}_2[n+1,n] \end{bmatrix} \quad (17)$$

where

$$\Delta\boldsymbol{u}_2[n] = \boldsymbol{u}_2[n+1] - \boldsymbol{u}_2[n] \quad (18)$$

with

$$\Delta\boldsymbol{h}_2[n+1,n] = \begin{bmatrix} 0 \\ \omega T_n \end{bmatrix} \quad (19)$$

denoting a change of polar velocity due to the control a_2.

4. BASE KALMAN FILTER

The system (9) constitutes a design basis for tracking of non-manoeuvring vessels by the extended Kalman filter (Anderson and Moore, 1979):

$$
\begin{aligned}
\hat{\boldsymbol{x}}[n+1|n+1] &= \hat{\boldsymbol{x}}[n+1|n]+\boldsymbol{L}[n+1]\boldsymbol{\nu}[n+1] \\
\hat{\boldsymbol{x}}[n+1|n] &= \boldsymbol{f}\left(n+1,n;\hat{\boldsymbol{x}}[n|n]\right) \\
\boldsymbol{\nu}[n+1] &= \boldsymbol{y}[n+1]-\boldsymbol{C}\hat{\boldsymbol{x}}[n+1|n] \\
\boldsymbol{L}[n+1] &= \boldsymbol{P}[n+1|n]\boldsymbol{C}^{\mathrm{T}}\boldsymbol{\omega}^{-1}[n+1] \qquad (20) \\
\boldsymbol{\omega}[n+1] &= \boldsymbol{C}\boldsymbol{P}[n+1|n]\boldsymbol{C}^{\mathrm{T}}+\boldsymbol{E}[n+1] \\
\boldsymbol{P}[n+1|n+1] &= \boldsymbol{P}[n+1|n]-\boldsymbol{L}[n+1]\boldsymbol{C}\boldsymbol{P}[n+1|n] \\
\boldsymbol{P}[n+1|n] &= \boldsymbol{F}[n+1,n]\boldsymbol{P}[n|n]\boldsymbol{F}^{\mathrm{T}}[n+1,n]+\boldsymbol{W}
\end{aligned}
$$

The quantities $\hat{x}[n|n]$ and $\boldsymbol{P}[n|n]$ given above denote a filtered state estimate and a covariance matrix of its estimation errors, respectively, while $\hat{\boldsymbol{x}}[n|n-1]$ and $\boldsymbol{P}[n|n-1]$ denote a predicted state estimate and a covariance matrix of its prediction errors. The remaining parameters of this filter are: a gain matrix $\boldsymbol{L}[n]$, an innovation process $\boldsymbol{\nu}[n]$, its covariance $\boldsymbol{\omega}[n]$, and $\boldsymbol{F}[n+1,n]$ is the Jacobian matrix of the transformation $\boldsymbol{f}(n+1,n;\cdot)$.

It is clear that the above filter used for tracking manoeuvring objects will yield biased estimates. On the other hand, the innovation process can be used for detection of non-uniform movements (manoeuvres). An algorithm that tests the innovation process for detection of manoeuvres will be referred to as a *manoeuvre detector*. If such a manoeuvre detector accepts the hypothesis HM, meaning that the object is manoeuvring, a procedure of isolating the manoeuvre type is required in order to adapt the estimation process properly.

5. ISOLATION OF MANOEUVRES

A procedure, proposed for the qualification of the detected manoeuvre to one of the considered classes, utilises the IE method (described in the following subsection) for estimation of an unknown value of control signal and is based on the following reasoning. If the IE algorithm, based on the model CCT (H2), is used for estimating the normal acceleration of the object, whose trajectory corresponds to the hypothesis H2, the resulting estimate will, certainly, significantly differ from zero. If the same algorithm is used for the estimation of the normal acceleration of the object, whose trajectory corresponds to the hypothesis H1, the obtained estimate is supposed to be close to zero. It is clear that an analogous judgement concerning the IE algorithm based on the model CCA associated with the hypothesis H1 can be described in a similar way.

We, therefore, propose to construct two statistics, based on the estimates of the tangential and normal accelerations, that allow to test a statistical significance of the estimates. An expertise based on these statistics makes it possible to accept either H1 or H2. This means that the manoeuvre corresponds to either the CCA model of (13) or the CCT model of (16).

5.1 Input estimation (IE) method

A principal address to estimation of an unknown control signal (IE) can be found in (Chan *et al.*, 1979). In the derivation of the algorithm, two types of Kalman filters are employed:

- a *base* (H0) filter of (20) for the model (9–10), which assumes that the control $a_j[n]\equiv 0$,
- a *hypothetical* filter for the following general model of manoeuvre

$$
\begin{aligned}
\boldsymbol{x}_j^*[n+1] = \boldsymbol{F}[n+1,n]\boldsymbol{x}_j^*[n]+ \\
+\boldsymbol{b}_j[n+1,n]a_j+\boldsymbol{w}[n] \qquad (21)
\end{aligned}
$$

which, in particular, takes the form of (13) or (16), corresponding to the hypotheses Hj, $j\in\{1,2\}$, with the presumption that the control signal a_j is non zero and its value is known for a certain period of time.

It is thus assumed that a nonzero acceleration starts exerting its influence on the object at the time instant $n_0^N=n-N$. This influence can be observed within a sliding window of length N in moments $n_i^N=n-N+i$, $i=1,\ldots,N$; namely, from the time instant n_1^N till $n_N^N=n$.

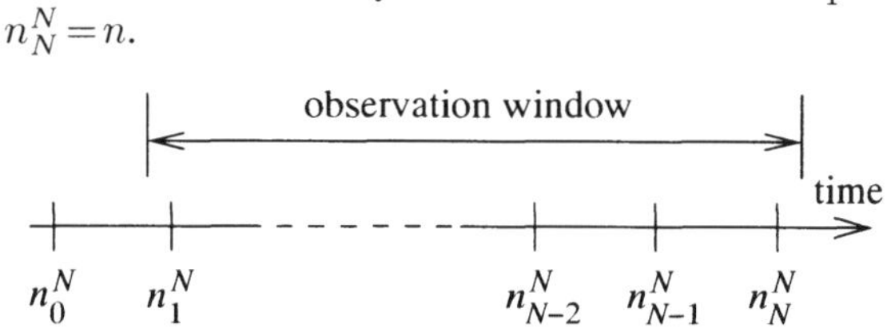

Figure 3. Sliding N-length window of observation

Let $\hat{\boldsymbol{x}}[n_{i+1}|n_i]$ be an estimate yielded by the *base* (H0) filter in the interval during which hypothesis Hj, $j\in\{1,2\}$, is true

$$
\begin{aligned}
\hat{\boldsymbol{x}}[n_{i+1}^N|n_i^N] = \boldsymbol{\Phi}[n_{i+1}^N,n_i^N]\hat{\boldsymbol{x}}[n_i^N|n_{i-1}^N]+ \\
+\boldsymbol{F}[n_{i+1}^N,n_i^N]\boldsymbol{L}[n_i^N]\boldsymbol{y}[n_i^N] \qquad (22)
\end{aligned}
$$

with

$$
\boldsymbol{\Phi}[n_{i+1}^N,n_i^N]=\boldsymbol{F}[n_{i+1}^N,n_i^N]\left(\mathbf{I}_{4\times 4}-\boldsymbol{L}[n_i^N]\boldsymbol{C}\right) \qquad (23)
$$

for $i=0,\ldots,N-1$. Moreover, let $\hat{\boldsymbol{x}}_j^*[n_{i+1}^N|n_i^N]$ be an estimate obtained from the *hypothetical* filter within the same circumstances.

By assuming $a_j[n_i^N]=0$, with $n_i^N\leq n_{-1}^N$, the initial condition of the recursion equation (22) is:

$$
\hat{\boldsymbol{x}}_j^*[n_0^N|n_{-1}^N]=\hat{\boldsymbol{x}}[n_0^N|n_{-1}^N] \qquad (24)
$$

It can be efficiently shown that if the acceleration does not change its value during the manoeuvre, the innovation process of the *base* filter is

$$
\boldsymbol{\nu}[n_i^N]=\boldsymbol{\psi}_j[n_i^N,n_0^N]a_j+\boldsymbol{\nu}_j^*[n_i^N] \qquad (25)
$$

where

$$\psi_j[n_i^N,n_0^N]=\boldsymbol{C}\boldsymbol{m}_j[n_i^N,n_0^N] \quad (26)$$

$$\boldsymbol{m}_j[n_i^N,n_0^N]= =\sum_{l=0}^{i-1}\left[\prod_{m=0}^{i-l-2}\boldsymbol{\Phi}[n_{i-m}^N,n_{i-m-1}^N]\right]\boldsymbol{b}_j[n_{l+1}^N,n_l^N] \quad (27)$$

for $i=1,\ldots,N$. The vector $\psi_j[n_i^N,n_0^N]$ describes the result of a mismatch between the *base* filter and the *hypothetical* one based on the manoeuvre model that is represented by a bias in the innovation process observed within the time interval from n_1^N to n_N^N. The vector $\boldsymbol{\nu}_j^*[n_i^N]$ is a random variable, Gaussian-distributed with the zero mean and the covariance matrix $\boldsymbol{\omega}[n_i^N]$.

As the acceleration exercises influence on the object for samples taken beginning with the time n_0^N, its effect can be observed in the innovation process (25) within the sliding window, n_1^N through n_N^N, where n_N^N is the actual instant. All such N (successive) equations can be shown in the form of an aggregate matrix equation:

$$\boldsymbol{V}_N[n]=\boldsymbol{\Psi}_{j,N}[n]a_j+\boldsymbol{V}_{j,N}^*[n] \quad (28)$$

where

$$\boldsymbol{\Psi}_{j,N}[n]=\begin{bmatrix}\psi_j[n_1^N,n_0^N]\\ \vdots\\ \psi_j[n_N^N,n_0^N]\end{bmatrix}_{Nd_y\times 1} \quad (29)$$

$$\boldsymbol{V}_N[n]=\begin{bmatrix}\boldsymbol{\nu}[n_1^N]\\ \vdots\\ \boldsymbol{\nu}[n_N^N]\end{bmatrix},\ \boldsymbol{V}_{j,N}^*[n]=\begin{bmatrix}\boldsymbol{\nu}_j^*[n_1^N]\\ \vdots\\ \boldsymbol{\nu}_j^*[n_N^N]\end{bmatrix} \quad (30)$$

and $\boldsymbol{V}_{j,N}^*[n]$ is a vector of independent random variables $\boldsymbol{\nu}_j^*[n_i^N]$, for $i=1,\ldots,N$. Thus the covariance matrix of the vector $\boldsymbol{V}_{j,N}^*[n]$ is

$$\boldsymbol{\Omega}_N[n]=\begin{bmatrix}\boldsymbol{\omega}[n_1^N] & \cdots & \boldsymbol{0}_{2\times 2}\\ \vdots & \ddots & \vdots\\ \boldsymbol{0}_{2\times 2} & \cdots & \boldsymbol{\omega}[n_N^N]\end{bmatrix}_{Nd_y\times Nd_y} \quad (31)$$

The estimate $\hat{a}_{j,N}[n]$ of the acceleration a_j can be computed by means of the generalised least squares (GLS) approach:

$$\hat{a}_{j,N}[n]=\frac{h_{j,N}[n]}{g_{j,N}[n]} \quad (32)$$

where $j\in\{1,2\}$, and

$$g_{j,N}[n]=\boldsymbol{\Psi}_{j,N}^{\mathrm{T}}[n]\boldsymbol{\Omega}_N^{-1}[n]\boldsymbol{\Psi}_{j,N}[n] \quad (33)$$
$$h_{j,N}[n]=\boldsymbol{\Psi}_{j,N}^{\mathrm{T}}[n]\boldsymbol{\Omega}_N^{-1}[n]\boldsymbol{V}_N[n]. \quad (34)$$

5.2 Nonlinear input estimation (NIE)

The IE method constitutes a convenient tool for identification of unknown values of the control signal in the model (21). However, it is not possible to obtain an accurate estimates of normal acceleration using the estimator (32) in the considered application, because the model (16) is nonlinear.

In order to cope with this nonlinearity, we propose to estimate the acceleration a_2 by using the following iterations, performed $\mathrm{K_m}$-times:

$$\hat{a}_{2,N}[n;k]=f_{2,N}(\hat{a}_{2,N}[n;k-1]) \quad (35)$$

for $N=1,\ldots,\mathrm{N_M}$, $k=1,\ldots,\mathrm{K_m}$ and $\hat{a}_{2,N}[\eta;0]=\hat{a}_{2,N}^0[\eta]$. Equation (35) describes the set of N estimators (32) resulting from the IE method and the model described in (13) (Kowalczuk and Sankowski, 2003).

This iterative procedure is repeated until the changes of the estimates $\hat{a}_{2,N}[\eta;k]$ obtained in the successive k iterations become sufficiently small.

5.3 Identification of a manoeuvre model

If the detector (at the time instant η) accepts the hypothesis HM meaning that the object is manoeuvring, the classification of the object manoeuvre is performed on the basis of the two sets of statistics

$$\mu_{1,N}[t_\eta]=\hat{a}_{1,N}^2[t_\eta;\mathrm{K_m}]g_{1,N}[t_\eta;\mathrm{K_m}] \quad (36)$$
$$\mu_{2,N}[t_\eta]=\hat{a}_{2,N}^2[t_\eta;\mathrm{K_m}]g_{2,N}[t_\eta;\mathrm{K_m}] \quad (37)$$

each for $N=1,\ldots,\mathrm{N_M}$, which can be approximated by random variables (χ^2-distributed with one degree of freedom). Using the denotations

$$\mu_{1,N}^{\mathrm{M}}=\max_{N=1,\ldots,\mathrm{N_M}}\{\mu_{1,N}[t_\eta]\} \quad (38)$$
$$\hat{N}_1=\arg\max_{N=1,\ldots,\mathrm{N_M}}\{\mu_{1,N}[t_\eta]\} \quad (39)$$
$$\mu_{2,N}^{\mathrm{M}}=\max_{N=1,\ldots,\mathrm{N_M}}\{\mu_{2,N}[t_\eta]\} \quad (40)$$
$$\hat{N}_2=\arg\max_{N=1,\ldots,\mathrm{N_M}}\{\mu_{2,N}[t_\eta]\}, \quad (41)$$

the decision procedure takes the following form:

if $\mu_{1,N}^{\mathrm{M}}>\mu_{2,N}^{\mathrm{M}}$
  hypothesis H1 is accepted
  $\hat{n}_0[\eta]=\eta-\hat{N}_1,\quad \hat{a}_1[\eta]=\hat{a}_{1,\hat{N}_1}[\eta;\mathrm{K_m}]$
else (42)
  hypothesis H2 is accepted
  $\hat{n}_0[\eta]=\eta-\hat{N}_2,\quad \hat{a}_2[\eta]=\hat{a}_{2,\hat{N}_2}[\eta;\mathrm{K_m}]$.

Formulas (36–41) with the decision logic (42) allows both structural and parametric identification of a target's manoeuvre model.

6. SIMULATION TEST

Efficiency of the proposed methods has been tested by using simulation. Two variants of state vector structure has been examined: the first with the Cartesian position and polar velocity (CPPV) coordinates, the second with the Cartesian position and velocity (CPCV).

In the following simulation, two target trajectories characterised below have been used. The target moves along a straight line trajectory at a constant speed $v = 154$ [m/s] (H0) up to the time instant $t_0 = 160$ [s], when it starts to accelerate with a: (1) tangential acceleration $a_t = 1$ [m/s^2] or (2) normal acceleration $a_n = 4$ [m/s^2].

6.1 Structural identification

Tab. 1 shows estimates of the efficiency of the isolation algorithm. A percentage of correct isolations has been chosen as a quality index.

Table 1. Efficiency of isolation of manoeuvres, in [%] of correct classifications

state vector coordinates	estimator memory length N						
	2	3	4	5	6	7	8
trajectory 1							
CPCV	45	56	63	68	84	97	100
CPPV	44	54	61	70	84	97	100
trajectory 2							
CPCV	52	67	87	99	100	100	100
CPPV	54	66	85	99	100	100	100

6.2 Parametric identification

Tabs. 2 and 3 present biases and RMS errors of the estimates of accelerations for the two trajectories and different estimator memory lengths.

Table 2. Mean values and standard deviations of errors $\frac{\hat{a}_1 - a_t}{a_t}$ [%]

error [%]	estimator memory length N					
	5	6	7	8	9	10
CPCV						
mean	16.0	9.0	2.0	3.0	3.0	3.0
std.	58.0	41.0	28.0	18.0	13.0	11.0
CPPV						
mean	15.0	8.0	2.0	3.0	3.0	3.0
std.	58.0	41.0	28.0	18.0	13.0	11.0

7. CONCLUSION

The presented simulation experiments have shown suitability of the proposed algorithms. It has been observed that the iterative NIE estimators based on the considered models converge quickly when used with data matched to the CCT model. With this, both the IE and NIE estimators are consistent. Moreover,

Table 3. Mean values and standard deviations of errors $\frac{\hat{a}_2 - a_n}{a_n}$ [%]

error [%]	estimator memory length N					
	3	4	5	6	7	8
CPCV						
mean	1.25	0.75	1.00	0.25	0.25	0.5
std.	49.3	27.5	16.3	10.3	7.75	6.0
CPPV						
mean	3.00	0.25	0.00	0.75	0.25	1.25
std.	50.0	28.3	16.5	10.8	8.0	6.25

the analysis of the accuracy of the algorithms, based on different state vectors (CPCV/CPPV), has shown that the effect of the choice of coordinates on the overall performance of the considered estimator is rather insignificant.

REFERENCES

Anderson, B.D.O. and J.B. Moore (1979). *Optimal Filtering*. Prentice Hall. Englewood Cliffs, USA.

Bar-Shalom, Y. and T.E. Fortmann (1988). *Tracking and Data Association*. Academic Press Inc.. Orlando, FL.

Bar-Shalom, Y. and X. Rong Li (1995). *Multitarget-Multisensor Tracking: Principles and Techniques*. YBS Publishing. Stoors, CT.

Blackman, S.S. and R. Popoli (1999). *Design and Analysis of Modern Tracking Systems*. Artech House. Norwood, MA.

Chan, Y.T., A.G.C. Hu and J.B. Plant (1979). A Kalman filter based tracking scheme with input estimation. *IEEE Trans. Aerosp. Electron. Syst.* **AES-15**(2), 237–242.

Farina, A. and F.A. Studer (1985). *Radar Data Processing*. Vol. 1/2. Research Studies Press Ltd. Letchworth, England.

Gustafsson, F. and A.J. Isaksson (1996). Best choice of coordinate system for tracking coordinated turns. In: *Proc. 35th IEEE Conf. on Decision and Control*. Vol. 3. Kobe, Japan. pp. 3145–3150.

Kerr, T.H. (1989). Duality between failure detection and radar/optical maneuver detection. *IEEE Trans. Aerosp. Electron. Syst.* **AES-25**(4), 581–584.

Kowalczuk, Z. and M. Sankowski (2002). Exclusive and non-exclusive multi-model tracking filters for air traffic control. In: *Automatic Control in Aerospace 2001* (G. Bertoni, Ed.). IFAC Proc. Volumes. Pergamon Press.

Kowalczuk, Z. and M. Sankowski (2003). Detection and isolation of maneuvers in adaptive tracking filtering based on multiple model switching. In: *Fault Diagnosis – Models, Artificial Intelligence, Applications* (J. Korbicz et al., Eds.). Chap. 20, pp. 701–739. Springer. To appear in.

Rong Li, X. and V. Jilkov (2000). Dynamic models for maneuvering taret tracking: A survey. In: *3rd ONR/GTRI Workshop on Target Tracking and Sensor Fusion*. Atlanta, GE.

IFAC Publications
www.elsevier.com/locate/ifac

ESTIMATION AND TRACKING OF QUASI-PERIODICALLY VARYING PROCESSES

Maciej Niedźwiecki * **Piotr Kaczmarek** *

* *Faculty of Electronics, Telecommunications and Computer Science, Department of Automatic Control, Technical University of Gdańsk*
ul. Narutowicza 11/12, Gdańsk , Poland
e-mail: maciekn@eti.pg.gda.pl

Abstract: The problem of identification/tracking of quasi-periodically varying processes is considered. Two solutions to this problem are presented. First, the global search algorithm is derived. Then its decoupled version, with a highly parallel computational structure, is proposed. *Copyright © 2003 IFAC*

Keywords: system identification, time-varying processes, frequency estimation

1. PROBLEM STATEMENT

Consider the problem of identification/tracking of coefficients of a complex time varying system governed by

$$y(t) = \sum_{l=1}^{n} \theta_l(t)u(t-l+1) + v(t)$$
$$= \boldsymbol{\varphi}^{\mathrm{T}}(t)\boldsymbol{\theta}(t) + v(t) \qquad (1)$$

where $t = 1, 2, \ldots$ denotes the normalized discrete time, $y(t)$ denotes the system output, $\boldsymbol{\varphi}(t) = [u(t), \ldots, u(t-n+1)]^{\mathrm{T}}$ is the regression vector made up of the past input samples, $v(t)$ is an additive (white) noise, uncorrelated with $u(t)$, and $\boldsymbol{\theta}(t) = [\theta_1(t), \ldots, \theta_n(t)]^{\mathrm{T}}$ denotes the vector of time varying impulse response coefficients, modeled as weighted sums of complex exponentials

$$\theta_l(t) = \sum_{i=1}^{k} a_{li}(t) e^{j\left(\sum_{s=1}^{t} \omega_i(s)\right)}, \quad l = 1, \ldots, n \quad (2)$$

Since the amplitudes and frequencies in (2) are time-varying, system parameters change over time in a periodic-like but not exactly periodic manner. We will assume that for every frequency component i, $i = 1, \ldots, k$ the quantities $a_{li}(t), l = 1, \ldots, n$ and $\omega_i(t)$ are slowly time-varying, i.e. they can be regarded approximately constant over a few 'cycles' of parameter variation. The system, governed by (1) - (2), which obeys the above-mentioned limitation, will be further referred to as *quasi-periodically* time-varying. It is important to note that even in the case where all amplitudes and frequencies in (2) are constant, system parameters are *not*, in general, periodic functions of time - they are periodic if all frequency ratios ω_i/ω_j, $i, j = 1, \ldots, k$ are rational numbers, otherwise they fall into the category of *almost periodic* functions (Wiener, 1930).

One of the challenging potential applications, which under certain conditions admits formulation presented above, is adaptive equalization of rapidly fading communication channels. In modern wireless systems, distortion introduced into the transmitted signals is caused mostly by the multipath effect - the signal reaches the receiver along different paths, i.e. with different time delays. When the multipath effects are dominated by few strong reflectors (scatterers) and when the transmitter and/or receiver moves with a constant speed, the impulse response of the channel (along with the transmitter and receiver filters) can be modeled in the form

$$\theta_l(t) = \sum_{i=1}^{k} a_{li} e^{j\omega_i t}, \quad l = 1, \dots, n \qquad (3)$$

In this particular case $y(t)$ denotes the sampled baseband signal, received by the mobile radio system, $u(t), t = 1, 2, \dots$ denotes the sequence of transmitted (complex) symbols, k stands for the number of different signal paths and $\omega_i, i = 1, \dots, k$ are the corresponding Doppler shifts. When the speed of the vehicle changes over time, Doppler shifts are also time-varying, which in a straightforward way leads to (2).

For mobile radio channels the sinusoidal model described above has a long history, which goes back to Aiken (Aiken, 1967). Quite recently, a number of papers explored the possibility of using it for equalization purposes - see e.g. (Tsatsanis and Giannakis, 1996), (Giannakis and Tepedelenlioğlu, 1998) and (Bakkoury et. al., 2000).

For $n = 1$ and $u(t) = 1, \forall t$ the model (1) - (2) becomes a description of a noisy nonstationary multifrequency signal

$$y(t) = \sum_{i=1}^{k} a_i(t) e^{j\left(\sum_{s=1}^{t} \omega_i(s)\right)} + v(t). \qquad (4)$$

The problem of either elimination or extraction of nonstationary sinusoidal signals buried in noise has attracted a great deal of attention of researchers in the field of signal processing (Nehorai, 1985), (Ng, 1987), (Dragošević and Stanković, 1995). Most of the solutions proposed so far are based on adaptive notch filtering. The conventional notch filter is designed to cancel a sinusoidal interference with known constant frequency ω_0. When the instantaneous frequency $\omega(t)$ of the signal is subject to small changes around ω_0, efficient cancellation is still possible, provided that the bandwidth of the notch filter is widened to accommodate for possible changes in $\omega(t)$. When the frequency changes are not local, e.g. there is a frequency drift, the frequency-adaptive versions of the notch filter must be designed. When restricted to the special case discussed above, the results developed in the paper offer a new solution to the problem of frequency tracking and adaptive notch filtering.

2. KNOWN FREQUENCIES

Suppose, for the time being, that both the amplitudes and angular frequencies in (2) are constant, i.e. that the changes in system parameters are governed by (3). Let

$$\boldsymbol{\alpha}_i = [a_{1i}, \dots, a_{ni}]^{\mathrm{T}}, \quad \boldsymbol{\psi}_i(t) = \boldsymbol{\varphi}(t) e^{j\omega_i t}$$
$$i = 1, \dots, k$$

Using the short-hand notation introduced above (1) can be rewritten in the form

$$y(t) = \sum_{i=1}^{k} \boldsymbol{\psi}_i^{\mathrm{T}}(t)\boldsymbol{\alpha}_i + v(t) = \boldsymbol{\psi}^{\mathrm{T}}(t)\boldsymbol{\alpha} + v(t) \quad (5)$$

where

$$\boldsymbol{\alpha} = [\boldsymbol{\alpha}_1^{\mathrm{T}}, \dots, \boldsymbol{\alpha}_k^{\mathrm{T}}]^{\mathrm{T}}$$
$$\boldsymbol{\psi}(t) = [\boldsymbol{\psi}_1^{\mathrm{T}}(t), \dots, \boldsymbol{\psi}_k^{\mathrm{T}}(t)]^{\mathrm{T}} = \mathbf{f}(t) \otimes \boldsymbol{\varphi}(t)$$
$$\mathbf{f}(t) = [e^{j\omega_1 t}, \dots, e^{j\omega_k t}]^{\mathrm{T}}$$

and $\otimes$ denotes the Kronecker product. Note that $\boldsymbol{\alpha}_i$ is the vector of coefficients associated with a particular frequency ω_i and *not* with a particular impulse response parameter $\boldsymbol{\theta}_i(t)$. Similarly, $\boldsymbol{\psi}_i(t)$ is the generalized regression vector associated with the ith frequency component. This nonstandard parameterization was adopted deliberately. Later on it will allow us to easily derive the decoupled version of the estimation algorithm.

Suppose now that the vector $\boldsymbol{\alpha}$ is slowly varying with time, slowly compared to the 'period' of parameter variation. It is known that, in the case considered, one can track $\boldsymbol{\alpha}(t)$ using the method of exponentially weighted least squares (EWLS). The EWLS estimate of $\boldsymbol{\alpha}(t)$ can be obtained from

$$\widehat{\boldsymbol{\alpha}}(t) = \arg\min_{\boldsymbol{\alpha}} \sum_{s=1}^{t} \gamma_{\alpha}^{t-s} \left| y(s) - \boldsymbol{\psi}^{\mathrm{T}}(s)\boldsymbol{\alpha} \right|^2 \quad (6)$$

where γ_α $(0 < \gamma_\alpha < 1, 1 - \gamma_\alpha \ll 1)$ denotes the so-called forgetting constant - the design parameter which controls the memory of the estimator, and hence allows one to trade off between its tracking speed and tracking accuracy. The recursive algorithm for evaluation of $\widehat{\boldsymbol{\alpha}}(t)$ is given by

$$\widehat{\boldsymbol{\alpha}}(t) = \widehat{\boldsymbol{\alpha}}(t-1) + \left(\mathbf{R}_\alpha^*(t)\right)^{-1} \boldsymbol{\psi}^*(t)\varepsilon(t)$$
$$\varepsilon(t) = y(t) - \boldsymbol{\psi}^{\mathrm{T}}(t)\widehat{\boldsymbol{\alpha}}(t-1)$$
$$\mathbf{R}_\alpha(t) = \gamma_\alpha \mathbf{R}_\alpha(t-1) + \boldsymbol{\psi}(t)\boldsymbol{\psi}^{\mathrm{H}}(t) \qquad (7)$$

Based on (7) one can estimate system parameters using

$$\widehat{\boldsymbol{\theta}}(t) = \mathbf{D}(t)\widehat{\boldsymbol{\alpha}}(t) \qquad (8)$$

where

$$\mathbf{D}(t) = \mathbf{f}^{\mathrm{T}}(t) \otimes \mathbf{I}_n$$

and $\mathbf{I}_n$ denotes the $n \times n$ identity matrix.

Since the algorithm (7) - (8) combines the basis function parameterization (it is assumed that system parameters can be expressed as linear combinations of known functions of time, called basis functions) with exponentially weighted least squares estimation, it will be further referred to as the exponentially weighted basis function (EWBF) algorithm (Niedźwiecki, 2000).

Another, equivalent form of the EWBF estimator, which will be very useful for our purposes, can be obtained by rewriting (7) - (8) in a different system of coordinates. Using the linear time-varying transformation

$$\widehat{\boldsymbol{\beta}}(t) = \mathbf{A}_n^{t+1}\widehat{\boldsymbol{\alpha}}(t)$$
$$\mathbf{R}_\beta(t) = \mathbf{A}_n^{-(t+1)}\mathbf{R}_\alpha(t)\mathbf{A}_n^{t+1}$$

where $\mathbf{A}_n = \mathbf{A}\otimes\mathbf{I}_n$ and $\mathbf{A} = \mathrm{diag}\{e^{j\omega_1}, \ldots, e^{j\omega_k}\}$, one can easily convert (7) - (8) into

$$\begin{aligned}
\widehat{\boldsymbol{\theta}}(t) &= \mathbf{D}_0\widehat{\boldsymbol{\beta}}(t)\\
\widehat{\boldsymbol{\beta}}(t) &= \mathbf{A}_n\widehat{\boldsymbol{\beta}}(t-1) + \left(\mathbf{R}_\beta^*(t)\right)^{-1}\mathbf{A}_n\boldsymbol{\varphi}_n^*(t)\varepsilon(t)\\
\varepsilon(t) &= y(t) - \boldsymbol{\varphi}_n^{\mathrm{T}}(t)\widehat{\boldsymbol{\beta}}(t-1)\\
\mathbf{R}_\beta(t) &= \mathbf{A}_n^*\left[\gamma_\beta\mathbf{R}_\beta(t-1) + \boldsymbol{\varphi}_n(t)\boldsymbol{\varphi}_n^{\mathrm{H}}(t)\right]\mathbf{A}_n
\end{aligned} \tag{9}$$

where $\boldsymbol{\varphi}_n(t) = \mathbf{A}_n^{-t}\boldsymbol{\psi}(t) = \mathbf{f}(0)\otimes\boldsymbol{\varphi}(t) = [\boldsymbol{\varphi}^{\mathrm{T}}(t), \ldots, \boldsymbol{\varphi}^{\mathrm{T}}(t)]^{\mathrm{T}}$, $\gamma_\beta = \gamma_\alpha$ and

$$\begin{aligned}
\mathbf{D}_0 = \mathbf{D}(t)\mathbf{A}^{-(t+1)} &= (\mathbf{f}^{\mathrm{T}}(t)\otimes\mathbf{I}_n)(\mathbf{A}^{-(t+1)}\otimes\mathbf{I}_n)\\
&= \mathbf{f}^{\mathrm{H}}(1)\otimes\mathbf{I}_n
\end{aligned}$$

The last transformation follows from the identity $(\mathbf{X}\otimes\mathbf{Y})(\mathbf{P}\otimes\mathbf{Q}) = \mathbf{XP}\otimes\mathbf{YQ}$ which holds for Kronecker products.

It is interesting to note that the reparameterized algorithm (9) corresponds to the following *backward-time* description of parameter changes

$$\theta_l(t-s+1) = \sum_{i=1}^{k} b_{li}(t)e^{-j\omega_i s}, \quad s\in T_t \tag{10}$$

where $T_t = [1, \ldots, t]$ and $b_{li}(t) = e^{j\omega_i(t+1)}a_{li}$, $l = 1, \ldots, n$, $i = 1, \ldots, k$. Since, for every $t \geq k$, the subspace spanned by the basis set $\{e^{j\omega_1 s}, \ldots, e^{j\omega_k s}, s\in T_t\}$ is identical with the subspace spanned by the time-reversed conjugate basis set $\{e^{-j\omega_1(t-s+1)}, \ldots, e^{-j\omega_k(t-s+1)}, s\in T_t\}$, the backward-time description (10) is equivalent to the forward-time description (3). Note that in the second case the basis set is always 'fixed' at $s = t$, i.e. at the end of the analysis interval. Following (Niedźwiecki, 1990) and (Niedźwiecki, 2000), the EWBF algorithm (9), based on the backward-time model, will be referred to as the fixed basis (FB) algorithm.

3. UNKNOWN FREQUENCIES

Note that when the constant coefficient a_{il} is replaced with the periodically varying coefficient $a_{il} = a_{il}^o e^{j\delta t}$ the corresponding term in (3) becomes $a_{il}e^{j\omega_l t} = a_{il}^o e^{j(\omega_l+\delta)t}$, i.e. the net effect is identical with that caused by changing the frequency from ω_l to $\omega_l + \delta$. Therefore, keeping the forgetting constant γ_β away from 1 guarantees that the EWBF algorithm will retain its tracking capabilities even if the corresponding frequencies $\omega_1, \ldots, \omega_k$ are subject to small changes around their nominal values. However, even though the EWBF filter is robust to small local changes in frequencies, it will fail to identify the system correctly in the presence of a frequency drift. For this reason in this section we will derive two frequency-adaptive EWBF algorithms, capable of tracking the time-varying frequencies $\omega_i(t), i = 1, \ldots, k$.

3.1 Global search

Denote by $\boldsymbol{\omega} = [\omega_1, \ldots, \omega_k]^{\mathrm{T}}$ the vector of unknown and/or time-varying frequencies and let $V(t, \boldsymbol{\omega})$ be the exponentially weighted measure of fit

$$V(t, \boldsymbol{\omega}) = \frac{1}{2}\sum_{s=1}^{t}\gamma_\omega^{t-s}\left|\varepsilon(s, \boldsymbol{\omega})\right|^2 \tag{11}$$

where γ_ω, $0 < \gamma_\omega < 1$, is the forgetting constant, which will be used to control the speed of the frequency adaptation.

To evaluate the estimate

$$\widehat{\boldsymbol{\omega}}(t) = \arg\min_{\boldsymbol{\omega}} V(t, \boldsymbol{\omega})$$

we will use the recursive prediction error (RPE) approach. According to Söderström and Stoica (Söderström and Stoica, 1988), the RPE algorithm can be expressed in the form

$$\widehat{\boldsymbol{\omega}}(t) = \widehat{\boldsymbol{\omega}}(t-1) +$$
$$- \left[V''(t, \widehat{\boldsymbol{\omega}}(t-1))\right]^{-1} V'(t, \widehat{\boldsymbol{\omega}}(t-1))$$

where

$$V'(t, \widehat{\boldsymbol{\omega}}(t-1)) \cong$$
$$\cong \mathrm{Re}\left[\varepsilon(t, \widehat{\boldsymbol{\omega}}(t-1))\frac{\partial\varepsilon^*(t, \widehat{\boldsymbol{\omega}}(t-1))}{\partial\boldsymbol{\omega}}\right],$$
$$V''(t, \widehat{\boldsymbol{\omega}}(t-1)) \cong \gamma_\beta V''(t-1, \widehat{\boldsymbol{\omega}}(t-2)) +$$
$$+ \mathrm{Re}\left[\frac{\partial\varepsilon(t, \widehat{\boldsymbol{\omega}}(t-1))}{\partial\boldsymbol{\omega}}\frac{\partial\varepsilon(t, \widehat{\boldsymbol{\omega}}(t-1))}{\partial\boldsymbol{\omega}^{\mathrm{H}}}\right]$$

and all derivatives are taken with respect to $\boldsymbol{\omega}$.

Let $\varepsilon(t) = \varepsilon(t, \widehat{\boldsymbol{\omega}}(t))$, $\boldsymbol{\eta}(t) = \partial\varepsilon(t, \widehat{\boldsymbol{\omega}}(t-1))/\partial\boldsymbol{\omega}$, $\boldsymbol{\Xi}(t) = \partial\boldsymbol{\beta}^{\mathrm{T}}(t, \widehat{\boldsymbol{\omega}}(t))/\partial\boldsymbol{\omega}$, $\mathbf{G}(t) = V''(t, \widehat{\boldsymbol{\omega}}(t-1))$ and

$$\widehat{\mathbf{A}}(t) = \mathrm{diag}\{e^{j\widehat{\omega}_1(t)}, \ldots, e^{j\widehat{\omega}_k(t)}\}$$

$$\widehat{\mathbf{A}}_n(t) = \widehat{\mathbf{A}}(t)\otimes\mathbf{I}_n$$

$$\widehat{\mathbf{D}}_0(t) = [\, e^{-j\widehat{\omega}_1(t)}, \ldots, e^{-j\widehat{\omega}_k(t)}\,]\otimes\mathbf{I}_n$$

Using the notation introduced above, the frequency-adaptive fixed basis EWBF algorithm can be written down in the form

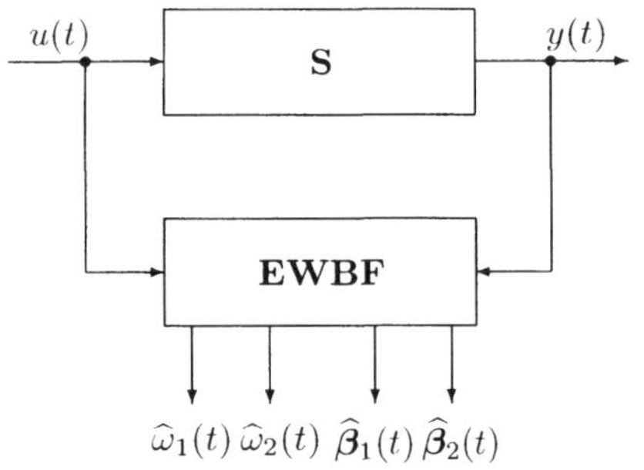

Fig. 1. Block diagram of the global search algorithm for two frequencies ($k = 2$).

$$
\begin{aligned}
\varepsilon(t) &= y(t) - \boldsymbol{\varphi}_n^{\mathrm{T}}(t)\widehat{\boldsymbol{\beta}}(t-1)\\
\boldsymbol{\eta}(t) &= -\boldsymbol{\Xi}(t-1)\boldsymbol{\varphi}_n(t)\\
\mathbf{G}(t) &= \gamma_\omega \mathbf{G}(t-1) + \mathrm{Re}[\boldsymbol{\eta}(t)\boldsymbol{\eta}^{\mathrm{H}}(t)]\\
\widehat{\boldsymbol{\omega}}(t) &= \widehat{\boldsymbol{\omega}}(t-1) - \mathbf{G}^{-1}(t)\mathrm{Re}[\varepsilon(t)\boldsymbol{\eta}^*(t)]\\
\mathbf{R}_\beta(t) &= \widehat{\mathbf{A}}_n^*(t)\left[\gamma_\beta \mathbf{R}_\beta(t-1) + \boldsymbol{\varphi}_n(t)\boldsymbol{\varphi}_n^{\mathrm{H}}(t)\right]\widehat{\mathbf{A}}_n(t)\\
\widehat{\boldsymbol{\beta}}(t) &= \widehat{\mathbf{A}}_n(t)\widehat{\boldsymbol{\beta}}(t-1) + (\mathbf{R}_\beta^*(t))^{-1}\widehat{\mathbf{A}}_n(t)\boldsymbol{\varphi}_n^*(t)\varepsilon(t)\\
\boldsymbol{\Xi}(t) &= \boldsymbol{\Xi}(t-1)\widehat{\mathbf{A}}_n(t) + \boldsymbol{\eta}(t)\boldsymbol{\varphi}_n^{\mathrm{H}}(t)\widehat{\mathbf{A}}_n(t)\mathbf{R}_\beta^{-1}(t)\\
&\quad +j\ \mathrm{diag}\{\widehat{\boldsymbol{\beta}}_1^{\mathrm{T}}(t-1),\ldots,\widehat{\boldsymbol{\beta}}_k^{\mathrm{T}}(t-1)\}\widehat{\mathbf{A}}_n(t)\\
&\quad +j\ \underset{k}{\mathrm{diag}}\{\boldsymbol{\varphi}^{\mathrm{H}}(t),\ldots,\boldsymbol{\varphi}^{\mathrm{H}}(t)\}\widehat{\mathbf{A}}_n(t)\mathbf{R}_\beta^{-1}(t)\varepsilon(t)\\
\widehat{\boldsymbol{\theta}}(t) &= \widehat{\mathbf{D}}_0(t)\widehat{\boldsymbol{\beta}}(t) \qquad (12)
\end{aligned}
$$

Using the well-known matrix inversion lemma (Söderström and Stoica, 1988) one can easily derive a recursive algorithm for computation of $\mathbf{P}_\beta(t) = \mathbf{R}_\beta^{-1}(t)$

$$
\mathbf{P}_\beta(t) = \frac{1}{\gamma_\beta}\widehat{\mathbf{A}}_n^*(t)\Bigg[\mathbf{P}_\beta(t-1) - \frac{\mathbf{P}_\beta(t-1)\boldsymbol{\varphi}_n(t)\boldsymbol{\varphi}_n^{\mathrm{H}}(t)\mathbf{P}_\beta(t-1)}{\gamma_\beta + \boldsymbol{\varphi}_n^{\mathrm{H}}(t)\mathbf{P}_\beta(t-1)\boldsymbol{\varphi}_n(t)}\Bigg]\widehat{\mathbf{A}}_n(t)
$$

The gradient search algorithms, which bear some resemblance to (12), were proposed in (Tsatsanis and Giannakis, 1996) and (Bakkoury et. al., 2000).

3.2 Local search

Denote by

$$y_i(t) = \boldsymbol{\psi}_i^{\mathrm{T}}(t)\boldsymbol{\alpha}_i + v(t)$$

the output of the ith subsystem of (5), i.e. subsystem associated with the frequency ω_i. Even though the signal $y_i(t)$ is not available, one can easily estimate it using the formula

$$\widehat{y}_i(t) = y(t) - \sum_{\substack{l=1\\ l\neq i}}^{k}\widehat{y}_l(t|t-1) \qquad (13)$$

where

$$\widehat{y}_i(t|t-1) = \boldsymbol{\psi}_i^{\mathrm{T}}(t)\widehat{\boldsymbol{\alpha}}_i(t-1) = \boldsymbol{\varphi}^{\mathrm{T}}(t)\widehat{\boldsymbol{\beta}}_i(t-1)$$

is the predicted value of $y_i(t)$ yielded by the estimation algorithm designed to track parameters of the ith subsystem.

Estimation of $y_i(t)$, in the way described above, allows one to decompose the tracking algorithm, i.e.

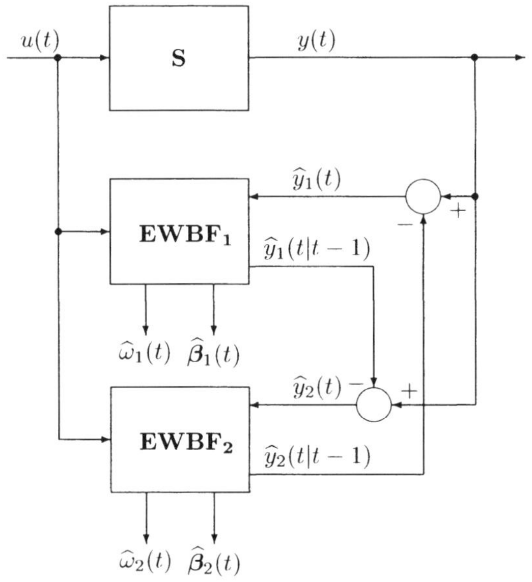

Fig. 2. Block diagram of the decoupled algorithm for two frequencies ($k = 2$).

to replace one 'global' adaptive search procedure (12) with k mutually coupled 'local' algorithms, each of which takes care of a particular subsystem - see Fig.1 and Fig.2. The ith component algorithm can be easily derived from (12) by setting $\widehat{\mathbf{A}}_n(t) = \widehat{\rho}_i(t)\mathbf{I}_n$, $\widehat{\rho}_i(t) = e^{j\widehat{\omega}_i(t)}$, $\boldsymbol{\Xi}(t) = \boldsymbol{\xi}^{\mathrm{T}}(t)$ and $\boldsymbol{\varphi}_n(t) = \boldsymbol{\varphi}(t)$. The resulting decoupled algorithm can be written down in the form

$$
\begin{aligned}
\varepsilon_i(t) &= \widehat{y}_i(t) - \boldsymbol{\varphi}^{\mathrm{T}}(t)\widehat{\boldsymbol{\beta}}_i(t-1)\\
\eta_i(t) &= -\boldsymbol{\varphi}^{\mathrm{T}}(t)\boldsymbol{\xi}_i(t-1)\\
g_i(t) &= \gamma_\omega g_i(t-1) + |\eta_i(t)|^2\\
\widehat{\omega}_i(t) &= \widehat{\omega}_i(t-1) - g_i^{-1}(t)\mathrm{Re}[\varepsilon_i(t)\eta_i^*(t)]\\
\widehat{\rho}_i(t) &= e^{j\widehat{\omega}_i(t)}\\
\widehat{\boldsymbol{\beta}}_i(t) &= \widehat{\rho}_i(t)\left[\ \widehat{\boldsymbol{\beta}}_i(t-1) + (\mathbf{R}^*(t))^{-1}\boldsymbol{\varphi}^*(t)\varepsilon_i(t)\ \right]\\
\boldsymbol{\xi}_i(t) &= \widehat{\rho}_i(t)\left[\ \boldsymbol{\xi}_i(t-1) + (\mathbf{R}^*(t))^{-1}\boldsymbol{\varphi}^*(t)\eta_i(t)\ \right]\\
&\quad + j\ \widehat{\boldsymbol{\beta}}_i(t) \qquad (14)
\end{aligned}
$$

$$i = 1, \ldots, k$$

$$\mathbf{R}(t) = \gamma_\beta \mathbf{R}(t-1) + \boldsymbol{\varphi}(t)\boldsymbol{\varphi}^{\mathrm{H}}(t)$$

$$\widehat{\boldsymbol{\theta}}(t) = \sum_{i=1}^{k}\widehat{\rho}_i^*(t)\widehat{\boldsymbol{\beta}}_i(t) \qquad (15)$$

The recursive algorithm for computation of $\mathbf{P}(t) = \mathbf{R}^{-1}(t)$ has the form

$$\mathbf{P}(t) =$$

$$= \frac{1}{\gamma_\beta}\left[\mathbf{P}(t-1) - \frac{\mathbf{P}(t-1)\boldsymbol{\varphi}(t)\boldsymbol{\varphi}^{\mathrm{H}}(t)\mathbf{P}(t-1)}{\gamma_\beta + \boldsymbol{\varphi}^{\mathrm{H}}(t)\mathbf{P}(t-1)\boldsymbol{\varphi}(t)}\right]$$

The decoupled algorithm (14) - (15) has two

advantages over the global search algorithm (12). First, it has a simpler computational structure. Note, for example, that all component algorithms use the same regression matrix $\mathbf{R}(t)$ (for this reason the subscript i was dropped) and that the matrix update for $\mathbf{G}(t)$ was replaced with simpler scalar updates for $g_i(t)$, $i = 1, \ldots, k$. Second, owing to its highly parallel structure, the modified scheme offers some extra design flexibility as it allows one to assign different forgetting factors $\gamma_i = \gamma_{\omega_i}$ to different frequencies.

In order to shed more light on the algorithm described above consider the problem of extraction of a complex sinusoidal signal $\theta(t) = ae^{j\omega_i t}$ from noisy measurements $y(t) = \theta(t) + v(t)$. When the frequency ω_i is known, the EWBF recursions (14) reduce down to ($k = 1$, $n = 1$, $u(t) = 1$)

$$\widehat{\theta}(t) = e^{-j\omega_i}\widehat{\beta}(t)$$
$$\widehat{\beta}(t) = e^{j\omega_i}\left[\widehat{\beta}(t-1) + \frac{\epsilon(t)}{r(t)}\right]$$
$$\varepsilon(t) = y(t) - \widehat{\beta}(t-1)$$
$$r(t) = \gamma_\beta r(t-1) + 1$$

Note that the gain $r(t)$ stabilizes at the value $1/(1-\gamma_\beta)$ as time goes to infinity, resulting in the following steady-state relationship between the prediction error $\varepsilon(t)$ and the input signal $y(t)$

$$\varepsilon(t) = N(q^{-1})y(t)$$

where the filter

$$N(q^{-1}) = \frac{1 - e^{j\omega_i}q^{-1}}{1 - \gamma_\beta e^{j\omega_i}q^{-1}}$$

can be easily recognized as the classical notch filter centered at the notch frequency ω_i, with bandwidth dependent on the forgetting factor γ_β. Hence, when used for extraction or cancellation of nonstationary multifrequency signals buried in noise, the EWBF algorithm (14) can be regarded as a bank of interconnected adaptive notch filters.

4. COMPUTER SIMULATIONS

One of the most important aspects of the problem of frequency estimation/tracking is correct frequency matching. The difficulty is caused by the fact that the minimized cost function has one global minimum, corresponding to the correct frequency assignment, and a large number of local minima, corresponding to incorrect or partially correct assignments. By partially correct assignments we mean all situations where two or more estimates $\widehat{\omega}_i(t)$ lock onto the same frequency, while some other frequencies are not matched at all. Basically, the incorrect frequency matching can occur in three situations: i) in the initial convergence phase, if the starting values of frequency estimates are far from the true frequency values;

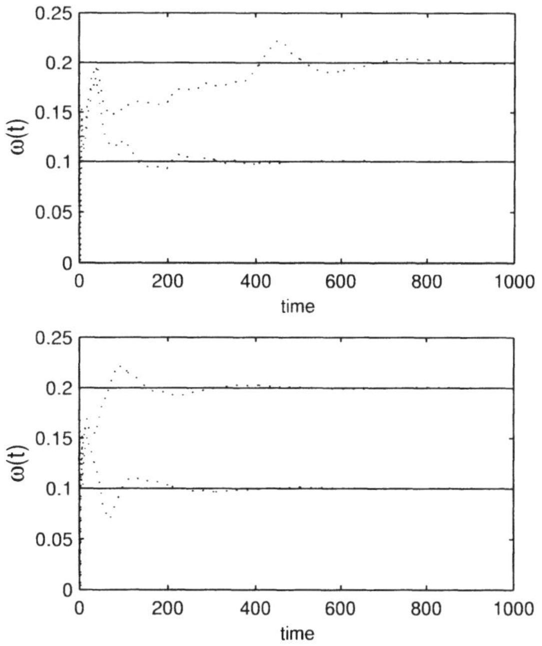

Fig. 3. True frequencies (solid lines) and their estimates (dotted lines) obtained in the initial convergence period using the global search algorithm (upper plot) and the decoupled algorithm (lower plot).

ii) in the 'steady state' tracking phase, when some of the frequencies become too closely spaced; iii) after a sudden frequency change - which, from the qualitative viewpoint, is similar to i) above.

A large number of simulation experiments were performed to check the frequency tracking/matching capabilities of both algorithms proposed in Section 3. The presented results were obtained for a hypothetical time-varying communication channel with two impulse response coefficients $\theta_1(t)$ and $\theta_2(t)$ (n=2), each of which was modeled as a linear combination of two complex exponentials (k=2). The weighting coefficients in (2) had constant values $\boldsymbol{\alpha}= [a_{11}, a_{12}, a_{21}, a_{22}]^{\mathrm{T}}= [2, 0.5j, -j, 1.5]^{\mathrm{T}}$. The input signal was the white 4-QAM sequence ($u(t) = \pm 1 \pm j$, $\sigma_u^2 = 2$) and the noise was complex Gaussian with variance $\sigma_v^2 = 0.2$. For both algorithms the same values of the forgetting factors were used, namely $\gamma_\beta = 0.99$ and $\gamma_\omega = 0.98$.

In all initial convergence tests the frequencies ω_1 and ω_2 were kept constant. To avoid the initial convergence "deadlock", which occurs when identical initial conditions are adopted for all blocks in the parallel estimation scheme, the starting values of the frequency estimates $\widehat{\omega}_1(0)$ and $\widehat{\omega}_2(0)$ were set to two *different* numbers close to zero. The typical results are shown in Fig. 3. For a majority of simulation runs (90%), corresponding to different realizations of the noise sequence, the frequency estimates, yielded by both algorithms, locked onto the true values. The global search

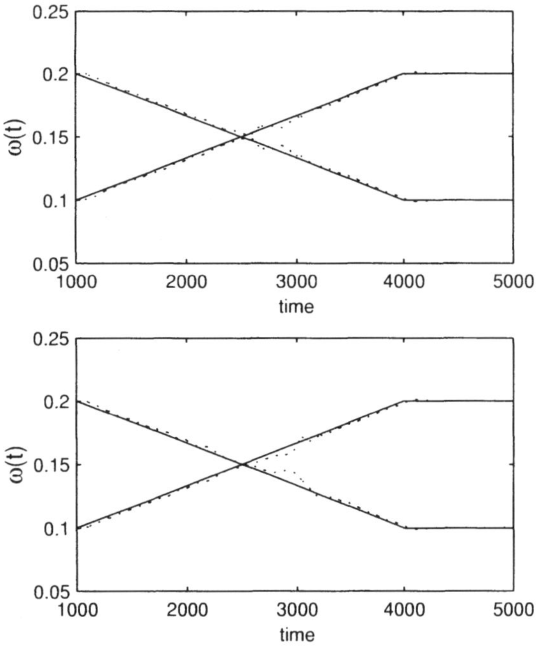

Fig. 4. Instantaneous frequencies (solid lines) and their estimates (dotted lines) obtained using the global search algorithm (upper plot) and the decoupled algorithm (lower plot).

algorithm usually showed a slower initial convergence than the decoupled algorithm.

To check the steady state frequency tracking/matching capabilities of both algorithms, linear changes in frequencies were enforced after the initial convergence period was over (to guarantee correct frequency matching in the startup phase, the initial frequency estimates were set to the true values). The trajectories of $\omega_1(t)$ and $\omega_2(t)$ intersected in the middle of the analysis interval. Our main concern was the behavior of the tracking algorithms after reaching the crossover point. The obtained results were on most occasions satisfactory, both in terms of frequency matching (see Fig. 4) and in terms of parameter tracking (see Fig. 5) - after passing the crossover point both frequency estimates $\widehat{\omega}_1(t)$ and $\widehat{\omega}_2(t)$ usually followed the correct values $\omega_1(t)$ and $\omega_2(t)$, respectively. The resulting parameter tracking performance was good. As expected, the tracking behavior of the decoupled algorithm was usually slightly worse than the tracking behavior of the global search algorithm. It should be noted, however, that the differences were not significant.

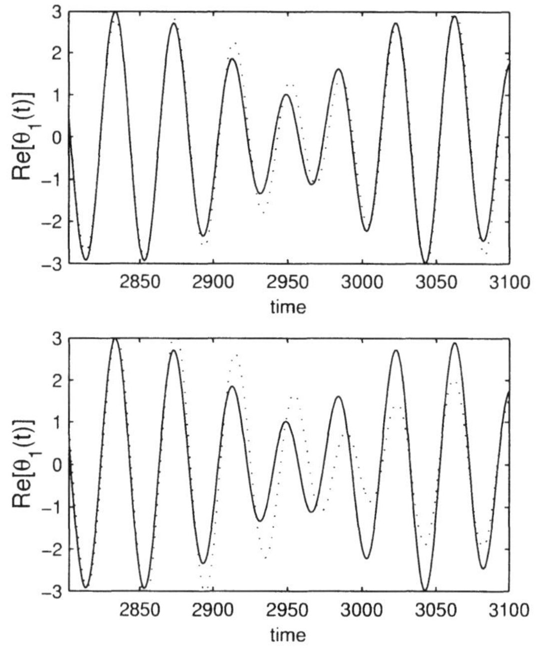

Fig. 5. Real part of the true system parameter $\theta_1(t)$ (solid lines) and its estimates (dotted lines) obtained using the global search algorithm (upper plot) and the decoupled algorithm (lower plot).

REFERENCES

Aiken, R.T. (1967). Communication over the discrete-path fading channel. *IEEE Trans. Inform. Theory* **IT-13**, 346–347.

Dragošević, M. V. and S. S. Stanković (1995). An adaptive notch filter with improved tracking properties. *IEEE Trans. on Signal Processing* **vol. 43**, 2068–2077.

Bakkoury, J., D. Roviras, M. Ghogho and F. Castanie (2000). Adaptive MLSE receiver over rapidly fading channels. *Signal Processing* **vol. 80**, 1347–1360.

Giannakis, G.B. and C. Tepedelenlioğlu (1998). Basis expansion models and diversity techniques for blind identification and equalization of time-varying channels. *Proc. IEEE* **vol. 86**, 1969–1986.

Nehorai, A. (1985). A minimal parameter adaptive notch filter with constrained poles and zeros. *IEEE Trans. on Acoustics, Speech and Signal Processing* **ASSP-33**, 158–161.

Ng, T. S. (1987). Some aspects of an adaptive digital notch filter with constrained poles and zeros. *IEEE Trans. on Acoustics, Speech and Signal Processing* **ASSP-35**, 158–161.

Niedźwiecki, M. (1990). Recursive functional series modeling approach to identification of time-varying plants - more bad news than good?. *IEEE Trans. Automat. Contr.* **AC-35**, 610–616.

Niedźwiecki, M. (2000). *Identification of Time-varying Processes.* Wiley. New York.

Söderström, T. and P. Stoica (1988). *System Identification.* Prentice Hall. Englewood Cliffs NJ.

Tsatsanis, M.K. and G.B. Giannakis (1996). Modeling and equalization of rapidly fading channels. *Int. J. Adaptive Contr. Signal Process.* **vol. 10**, 159–176.

Wiener, N. (1930). Generalised harmonic analysis. *Acta Math.* pp. 117–258.